Storm and Cloud Dynamics

This is Volume 44 in the
INTERNATIONAL GEOPHYSICS SERIES
A series of monographs and textbooks
Edited by RENATA DMOWSKA and JAMES R. HOLTON

A complete list of the books in this series appears at the end of this volume.

Storm and Cloud Dynamics

William R. Cotton
Department of Atmospheric Science
Colorado State University
Fort Collins, Colorado

Richard A. Anthes
National Center for Atmospheric Research
Boulder, Colorado

ACADEMIC PRESS, INC.
Harcourt Brace Jovanovich, Publishers
San Diego New York Boston
London Sydney Tokyo Toronto

This book is printed on acid-free paper. ♾

Academic Press, Inc.
1250 Sixth Avenue, San Diego, California 92101-4311

United Kingdom Edition published by
Academic Press Limited
24–28 Oval Road, London NW1 7DX

Library of Congress Cataloging-in-Publication Data

Cotton, William R.,
Storm and cloud dynamics.
(International geophysics series ; 44)
1. Clouds–dynamics. 2. Cumulus.
3. Precipitation (Meteorology) 4. Mesometeorology
I. Anthes, Richard A. II. Title.
III. Series.
QC921.6D95C67 1989 551.57'6 88-8049
ISBN 0-12-192530-7 (Hardcover)
ISBN 0-12-192531-5 (Paperback)

PRINTED IN THE UNITED STATES OF AMERICA
92 93 94 95 96 97 EB 9 8 7 6 5 4 3 2 1

Contents

Chapter 4 The Parameterization or Modeling of Microphysical Processes in Clouds

Chapter 5 Radiative Transfer in a Cloudy Atmosphere and Its Parameterization

Chapter 6 Cumulus Parameterization and Diagnostic Studies of Convective Systems

Part II The Dynamics of Clouds

Chapter 7 Fogs and Stratocumulus Clouds

Chapter 12 The Influence of Mountains on Airflow, Clouds, and Precipitation

Preface

The focus of this book is on the dynamics of clouds and of precipitating mesoscale meteorological systems. Mesoscale meteorology is concerned with weather systems that have spatial and temporal scales between the domains of macro- and micrometeorology. Generally, macrometeorology is concerned with weather systems having spatial scales greater than 1000 km and temporal scales on the order of several days or longer. Micrometeorology is the science dealing with atmospheric dynamics having spatial scales of tens to hundreds of meters and time scales on the order of minutes. Mesoscale meteorology can therefore be thought of as the science dealing with any weather system lying between these two extreme temporal and spatial scales. Orlanski (1975) subdivided the classification of mesoscale systems into three scales: meso-α, meso-β, and meso-γ (see the accompanying figure). He suggested that the term meso-α should be applied to weather systems such as frontal systems and hurricanes having horizontal scales of 200–2000 km and temporal scales of 1 day to 1 week. The term meso-β should be applied to such systems as the nocturnal low-level jet, squall lines, inertial waves, cloud clusters, and mountain and lake/coastal circulations. These systems have horizontal scales on the order of 20–200 km and temporal scales on the order of several hours to 1 day. Finally, he suggested that the meso-γ regime should include thunderstorms, internal gravity waves, clear air turbulence, and urban effects with horizontal scales of 2–20 km and temporal scales on the order of one-half hour to several hours. In this book, we shall generally adhere to this terminology.

Clouds and precipitating mesoscale systems represent some of the most important and scientifically exciting weather systems in the world. These are the systems that produce torrential rains, severe winds including downbursts and tornadoes, hail, thunder and lightning, and major snow storms. Forecasting such storms represents a major challenge since they are too small to

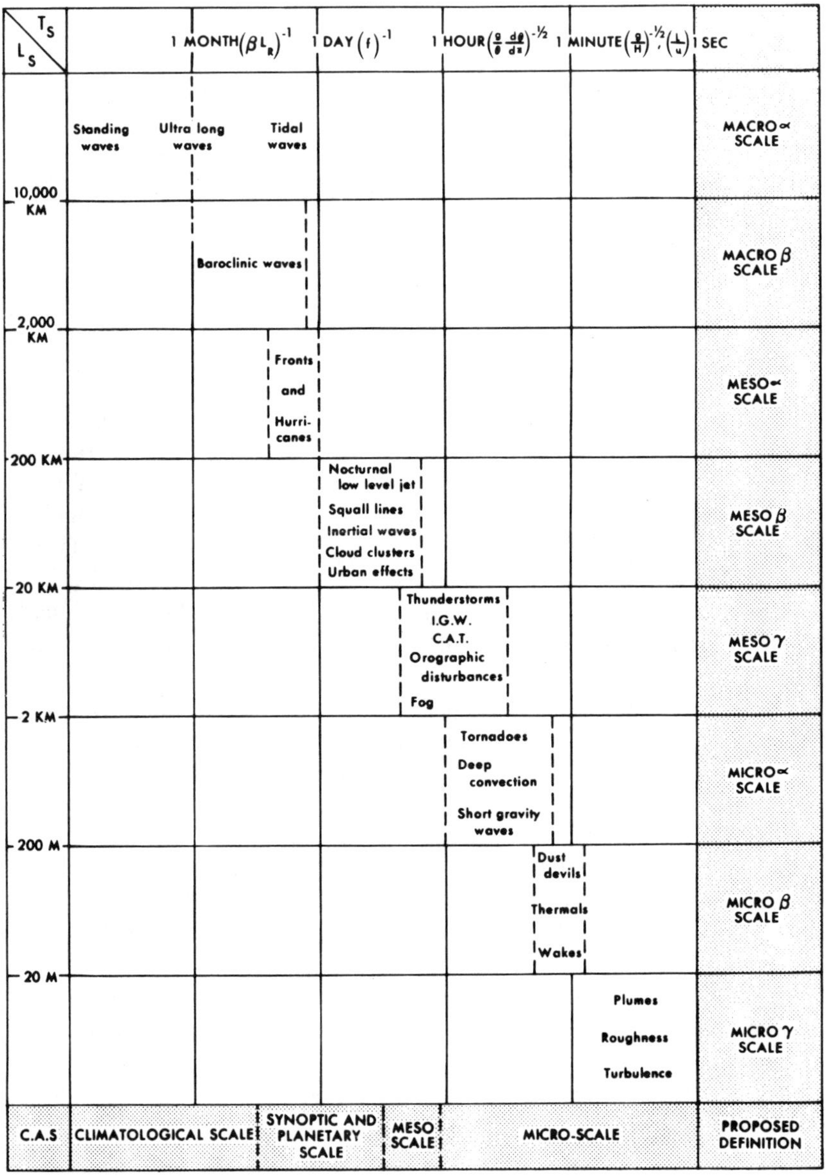

T_S
L_S
1 MONTH $(\beta L_R)^{-1}$
1 DAY $(f)^{-1}$
1 HOUR
1 MINUTE
1 SEC
Standing waves
Ultra long waves
Tidal waves
MACRO ∝ SCALE
10,000 KM
Baroclinic waves
MACRO β SCALE
2,000 KM
Fronts and Hurri-canes
MESO ∝ SCALE
200 KM
Nocturnal low level jet
Squall lines
Inertial waves
Cloud clusters
Urban effects
MESO β SCALE
20 KM
Thunderstorms
I.G.W.
C.A.T.
Orographic disturbances
Fog
MESO γ SCALE
2 KM
Tornadoes
Deep convection
Short gravity waves
MICRO ∝ SCALE
200 M
Dust devils
Thermals
Wakes
MICRO β SCALE
20 M
Plumes
Roughness
Turbulence
MICRO γ SCALE
C.A.S
CLIMATOLOGICAL SCALE
SYNOPTIC AND PLANETARY SCALE
MESO SCALE
MICRO-SCALE
PROPOSED DEFINITION

Scale definitions and different processes with characteristic time and horizontal scales. [Adapted from Orlanski (1975).]

be adequately resolved by conventional observing networks and numerical prediction models.

Less well known is the role that clouds and storms play in atmospheric chemistry on regional and global scales. Not only do they sweep large quantities of pollutants residing in the boundary layer into the middle and upper troposphere and even the lower stratosphere, they also mix the pollutants with cloud particles that form in the rising air parcels, thus serving as wet chemical reactors. It is now generally recognized, for example, that the annual budget of acid precipitation at a particular location is dominated by a few major storm events that strip the boundary layer of large quantities of sulfates and oxides of nitrogen, where they become incorporated in precipitation elements, undergo chemical reactions, and then settle to the ground in raindrops or snowflakes as highly concentrated acidic precipitation elements. Even the lightning produced in convective storms contributes significantly to global production of nitrates, NO_x, and other chemical species. Here again, one major mesoscale convective system can produce as much as 25% of the annual number of cloud-to-ground lightning strokes at a given location.

Clouds and precipitating mesoscale systems play an important role in the earth's general circulation, climate, and climatic variability. The latent heats released in clouds serve as "engines" which drive the global atmospheric circulations. Moreover, they are dominant components in the earth's hydrologic cycle, transporting water vertically and horizontally, changing the phase of water from vapor to liquid and solid phases, removing water from the atmosphere through precipitation, and shielding the earth's surface from direct solar radiation, thus altering surface evapotranspiration rates. Clouds are also a major factor in determining the earth's radiation budget by reflecting incoming solar radiation and absorbing upwelling terrestrial radiation. Variations in the coverage of clouds, the heights of clouds, and even the number and sizes of individual cloud particles all have large effects on the radiation budget of the earth.

Thus, clouds and precipitating mesoscale systems are not only fascinating subjects of study for their own sake, but the knowledge gained from studying them is essential to improving short- and long-range forecasting and gaining a quantitative understanding of atmospheric chemistry, the earth's general circulation, and climatic variability.

The book comprises two parts. Part I is a survey of the general theory that is appropriate to describing clouds and precipitating mesoscale systems. The theory may be described as small-scale dynamics and is, for the most part, concerned with nonhydrostatic dynamics. In addition to a review of the general theory of cloud dynamics, a review of the various physical processes relevant to the modeling of clouds and mesoscale systems is given. These processes include precipitation processes, radiative transfer, turbulent transport, and the effects of cumulus clouds on their larger scale environment. They are described not so much from the point of view of a detailed examination of the physics of the process as from the perspective of what essential elements of the processes most strongly influence the modeling of clouds and mesoscale systems. In other words, we discuss the parameterization of the detailed physical processes.

In Part II, we describe the physics and dynamics of various cloud systems, ranging from the least dynamic, fogs and stratocumulus clouds, to the most dynamic, severe convective clouds. A discussion follows of the physics and dynamics of the dominant precipitating mesoscale systems, including squall lines, cloud clusters, mesoscale convective systems, and the subsynoptic structure of tropical and extratropical cyclones. We conclude with discussions of the physics and dynamics of middle and high clouds and stable orographic clouds. In our discussion of each cloud type, we attempt to describe the current state of knowledge as derived from the marriage of theoretical modeling and observational analysis. The detailed examination of the theory in Part I is not essential to the interpretation of Part II. However, we recommend that the reader become familiar with the notations in Part I in order to follow more readily the discussions in Part II.

We appreciate the essential help of Brenda Thompson in typing the manuscript, requesting figure authorizations, editing, and keeping figures and the many drafts of this text organized. Lillian Vigil, Nancy Duprey, and Ann Modahl also assisted in typing. The editorial assistance of Annette Claycomb, Hope Hamilton, and Ann Modahl is acknowledged. Lucy McCall drafted the original figures in the text.

The manuscript was greatly improved by the comments of Mitch Moncrieff, who reviewed the entire text. Reviewers of selected portions of the text include Charlie Chappell, Paul Ciesielski, Terry Clark, Steve Cox, Leo Donner, Dale Durran, Piotr Flatau, Bob Houze, Dick Johnson, Joe Klemp, Kevin Knupp, Hung-Chi Kuo, Bob Maddox, Ray McAnelly, Mike Moran, Roger Pielke, Bob Rauber, Dave Reynolds, Wayne Schubert, Joanne Simpson, Ron Smith, Graeme Stephens, Duane Stevens, Sean Twomey, Melanie Wetzel, and Michio Yanai.

Original research results reported in the text were supported by the National Science Foundation (NSF) under grants ATM-8512480, ATM-8312077, and ATM-8113082; the National Center for Atmospheric Research, which is sponsored by NSF; Air Force Geophysics Lab Contract #F19628-84-C-0005; and Office of Naval Research Contract #N00014-83-K-0322.

Orlanski, I. (1975). A rational subdivision of scale for atmospheric processes. *Bull. Am. Meteorol. Soc.* **56**, 527–530.

Chapter 1 Clouds

Clouds are pictures in the sky
They stir the soul, they please the eye
They bless the thirsty earth with rain,
which nurtures life from cell to brain—
But no! They're demons, dark and dire,
hurling hail, wind, flood, and fire
Killing, scarring, cruel masters
Of destruction and disasters
Clouds have such diversity—
Now blessed, now cursed,
the best, the worst
But where would life without them be?

Vollie Cotton

1.1 Introduction

Since the late 1940s, when the experiments by Langmuir (1948) and Schaefer (1948) suggested that seeding of certain types of clouds could release additional precipitation, there has been intensive investigation into the physics of clouds. The major focus of these studies has been on the microphysical processes involved in cloud formation and the production of precipitation. As the studies have unraveled much about the detailed microphysics of clouds, it has become increasingly apparent that these processes are affected greatly by macroscale dynamics and thermodynamics of the cloud systems. We have also learned to appreciate that the microphysical processes can alter the macroscale dynamic and thermodynamic structure of clouds. Thus, while the focus of this book is on the dynamics of clouds, we cannot neglect cloud microphysical phenomena. The title of this book implies a perspective from which we view the cloud or cloud system as a whole. From this perspective, cloud microphysical processes can be seen as a swarm or ensemble of particles that contribute collectively and in an integrated way to the macroscale dynamics and thermodynamics of the cloud.

We take a similar perspective of small-scale air motions in clouds. Again, we will not use our highest power magnifying lens to view the smallest scale motions or turbulent eddies in clouds. We will instead examine the collective behavior or statistical contributions of the smallest cloud eddies (i.e., those less than a few hundred meters or so) to the energetics of clouds and to transport processes in clouds.

Following the same analogy, we view the meso-β-scale and meso-α-scale with a wide-angle lens, thus encompassing their contributions to the energetics and transport processes of a particular cloud as well as to neighboring clouds or cloud systems. The meso-β-scale and meso-α-scale can, for the most part, be considered the environment of the cloud scale that is generally the meso-γ-scale.

1.2 The Classification of Clouds

Since this text emphasizes the dynamics of clouds, it would seem appropriate that we adopt a classification of clouds that is based on the "dynamic" characteristics of clouds rather than on the physical appearance of clouds from the perspective of a ground observer. In fact, several scientists (Scorer, 1963; Howell, 1951; Scorer and Wexler, 1967) have attempted to design such a classification scheme based on cloud motions. However, since we wish to label the various cloud forms for later discussion, we shall generally adhere to the classifications given in the "International Cloud Atlas" (World Meteorological Society, 1956). This classification is based on 10 main groups called *genera*, and most of the genera are subdivided into *species*. Each subdivision is based on the shape of the clouds or their internal structure. The species is sometimes further divided into *varieties*, which define special characteristics of the clouds related to their transparency and the arrangements of the macroscopic cloud elements.

The definitions of the 10 genera are as follows:

Cirrus—Detached clouds in the form of white, delicate filaments or white or mostly white patches or narrow bands. These clouds have a fibrous (hairlike) appearance, or a silky sheen, or both.

Cirrocumulus—Thin, white patches, sheets, or layers of cloud without shading, composed of very small elements in the form of grains or ripples, merged or separate, and more or less regularly arranged; most of the elements have an apparent width of less than 1°.

Cirrostratus—Transparent, whitish cloud veil of fibrous or smooth appearance, totally or partially covering the sky, and generally producing halo phenomena.

Altocumulus—White or grey, or both white and grey, patches, sheets, or layers of cloud, generally with shading, composed of laminae, rounded masses, or rolls, which are sometimes partially fibrous or diffuse and which may or may not be merged; most of the regularly arranged small elements usually have an apparent width of 1–5°.

Altostratus—Greyish or bluish cloud sheet or layer of striated, fibrous, or uniform appearance, totally or partially covering the sky, and having parts thin enough to reveal the sun at least dimly, as through ground glass. Altostratus does not produce halos.

Nimbostratus—Grey cloud layer, often dark, the appearance of which is rendered diffuse by more or less continuously falling rain or snow, which in most cases reaches the ground. It is thick enough to completely obscure the sun. Low, ragged clouds frequently occur below the nimbostratus layer.

Stratocumulus—Grey or whitish, or both grey and whitish, patches, sheets, or layers of cloud which almost always have dark parts, composed of crenellations, rounded masses, or rolls, which are nonfibrous (except when virga—inclined trails of precipitation—are present) and which may or may not be merged; most of the regularly arranged small elements have an apparent width of more than 5°.

Stratus—Generally grey clouds with a fairly uniform base, which may produce drizzle, ice prisms, or snow grains. If the sun is visible through the cloud, its outline is clearly discernible. Stratus clouds do not produce halo phenomena except, possibly, at very low temperatures. Sometimes stratus clouds appear in the form of ragged patches.

Cumulus—Detached clouds, generally dense and with sharp outlines developing vertically in the form of rising mounds, domes, or towers, of which the bulging upper part often resembles a cauliflower. The sunlit parts of these clouds are mostly brilliant white; their base is relatively dark and nearly horizontal. Sometimes cumulus clouds are ragged.

Cumulonimbus—Heavy, dense clouds, with a considerable vertical extent, in the form of a mountain or huge tower. At least part of their upper portion is usually smooth, fibrous, or striated and is nearly always flattened; this part often spreads out in the shape of an anvil or vast plume. Under the base of these clouds, which is generally very dark, there are frequently low ragged clouds and precipitation, sometimes in the form of virga.

In general we will not have to refer to the definitions of the clouds species or varieties used in the "International Cloud Atlas." The exceptions mainly concern cumulus clouds, which we refer to as follows:

Cumulus humilis—Cumulus clouds of only a slight vertical extent; they generally appear flattened.

Cumulus mediocris—Cumulus clouds of moderate vertical extent, the tops of which show fairly small protuberances.

Cumulus congestus—Cumulus clouds which exhibit markedly vertical development and are often of great vertical extent; their bulging upper part frequently resembles a cauliflower.

We also may have occasion to refer to the following supplementary features and accessories of clouds:

Mamma—Hanging protuberances, like udders, on the under surface of a cloud.

Virga—Vertical or inclined trails of precipitation (fall streaks) falling from the base but reaching the earth's surface.

Pileus—An accessory cloud of small horizontal extent, in the form of a cap or hood above the top or attached to the upper part of a cumuliform cloud which often penetrates it.

Fog is not treated as a separate cloud genus in the "International Cloud Atlas." Instead it is defined in terms of its microstructure, visibility, and proximity to the earth's surface as follows:

Fog—Composed of very small water droplets (sometimes ice crystals) in suspension in the atmosphere; it reduces the visibility at the earth's surface generally to less than 1000 m. The vertical extent of fog ranges between a few meters and several hundred meters.

We include the discussion of fog in the chapter on stratocumulus clouds, since we shall see there is not always a clear distinction between the formative mechanisms of a marine stratocumulus cloud whose base is elevated from the surface and a fog which reaches the surface.

Another cloud form discussed in this text that is not treated in the "International Cloud Atlas" as a separate cloud genus is the *orographic* cloud. According to the "Glossary of Meteorology" (Huschke, 1959), an orographic cloud is a cloud whose form and extent is determined by the disturbing effects of orography upon the passing flow of air. Since orography can also initiate convective clouds, we shall often refer to a stable, orographic cloud as the cloud form typically encountered in the wintertime during periods when the atmosphere is stably stratified. The *cap* cloud is the least complicated form of the orographic cloud and refers to a nearly stationary cloud that hovers over an isolated peak. The *crest* cloud is like the cap cloud with the exception that it hovers over a mountain ridge. The *chinook arch* or *foehn* wall cloud refers to a bank or wall of clouds associated with a chinook or foehn wind storm. Finally, the *lenticular* cloud, or *lenticularis*, is a lens-shaped cloud that forms over or to the lee of orographic barriers as a result of mountain waves. As the name implies, lenticular clouds generally have a smooth shape with sharp outlines, sometimes vertically stacked with clear air separating each lenslike element.

1.3 Cloud Time Scales, Vertical Velocities, and Liquid-Water Contents

In this section we examine certain cloud characteristics that have a major controlling influence upon whether or not precipitation processes are important and whether diabatic processes such as condensational heating and radiative transfer dominate the cloud energetics. Because these physical processes affect the dynamics of the cloud, it is important to recognize under what conditions and in which cloud types these processes are important.

The macroscopic parameters of clouds that characterize precipitation and diabatic process are (i) cloud time scales, (ii) cloud vertical velocities, (iii) cloud liquid-water contents, (iv) cloud temperature, and (v) cloud turbulence. Time scales are important because precipitation processes are time dependent. Therefore, if the cloud lifetime is too short for the time it takes to form precipitation, the cloud will not precipitate even though other properties such as liquid-water content are sufficient to support precipitation. Two time scales are critical. One is the cloud lifetime, which we shall label T_c. The other represents the time it takes a parcel to enter the cloud and exit its top or sides. We shall label this time scale as T_P.

Cloud vertical velocities are important because the updrafts control the time scale T_P and determine the cloud's ability to suspend precipitation particles. The magnitude of vertical velocity also provides an estimate of the wet (saturated) adiabatic cooling rate. For example, in the middle troposphere the wet adiabatic lapse rate γ_m is approximately 0.5°C/100 m. Thus the wet adiabatic cooling rate CR_γ is

$$CR_\gamma \simeq (0.5°C/100\ m) \times W,$$

where W is the cloud vertical velocity in meters per second.

Both the potential for precipitation formation and the cooling rates of clouds depend on the liquid-water content (LWC) in a cloud for two reasons. First, the LWC determines the ultimate potential for a cloud to produce precipitation. Generally speaking, unless a cloud generates a liquid-water content in excess of 0.5 g m^{-3}, it is unlikely to precipitate. Of course, other factors such as whether the air mass is continental (see Chapter 4) and whether the cloud is supercooled also control the critical LWC for initiating precipitation. Second, the LWC is important because it determines the rates of shortwave or longwave radiational heating and cooling.

Cloud temperature also represents an important parameter in precipitation potential. The cloud-base temperature indicates the liquid-water-producing potential of the cloud. For example, a cloud with a base temperature of +20°C has a cloud-base saturation mixing ratio of ~15 g kg^{-1}, while a cloud with a base temperature of +4°C has a cloud-base saturation mixing ratio of only 5 g kg^{-1}. Thus, if these two clouds have equal depths, the one with the warmer cloud base has a much greater potential for producing rainfall. Cloud-top temperature is important for similar reasons, because the greater the difference between the cloud-base temperature and cloud-top temperature, the greater the potential for rainfall. Furthermore, if the cloud-top temperature is below 0°C, then ice is possible, which greatly affects precipitation and radiation processes.

Turbulence, the last consideration in our discussion of macroscopic cloud parameters, is important because it mixes properties of the cloud and interacts closely with the other parameters. When we speak of "characteristic" time scales, vertical velocities, liquid-water contents, and temperatures, the level of turbulence determines how representative these "characteristic" scales really are. For example, in some convective clouds the average updraft velocity may be 1 m s^{-1}, while the standard deviation of the vertical velocity may be as large as 3 m s^{-1}. The level of turbulence also affects the precipitation processes, due to the formation of higher peak supersaturations and to increased interactions among cloud particles of different types and sizes. Turbulence is also likely to affect the radiative properties of a cloud. In a turbulent cloud the cloud top is likely to be very lumpy, and large fluctuations in liquid water will exist. As a consequence, the cloud-top radiative emittance and absorptance will differ significantly from that found in a more homogeneous cloud.

Let us next consider how the characteristics just introduced differ in several cloud forms that we will study in this book.

1.3.1 Fog

Fog may be considered the least dynamic of clouds. Fogs typically have lifetimes (T_c) of 2 to 6 h. The mean vertical velocity in fog is usually quite small. If we assume a mean updraft of 0.01 m s^{-1} for a 100-m-deep fog, the time scale for a parcel entering cloud base and exiting cloud top would be

$$T_P = 100\ \text{m}/0.01\ \text{m s}^{-1} = 10^4\ \text{s},$$

that is, T_P is on the order of 3 h. This represents the time scale in which cloud microphysical processes must operate in order to generate precipitation.

However, the liquid-water content in fog typically ranges from 0.05 to 0.2 g m^{-3}. Thus, precipitation is unlikely in all but the deepest, wettest, and most maritime fogs, even though the mean vertical velocity might indicate a potential for precipitation. If we use our estimate of W of 0.01 m s^{-1}, we determine that the cooling rate due to wet adiabatic cooling is on the order of

$$\text{CR}_\gamma = (0.5°\text{C}/100\ \text{m})0.01\ \text{m s}^{-1} = 5\times10^{-5}\ °\text{C s}^{-1},$$

which is approximately 0.2°C h^{-1}. By comparison, the rate of cooling by longwave radiation flux divergence at the top of the fog can easily range from 1 to 4°C h^{-1}. Thus, we see that fogs can be dominated by radiative cooling.

The absolute magnitude of turbulence in fogs is usually small, although there have been reports of vertical velocity fluctuations in some valley fogs as large as 1 m s^{-1}. However, if we consider turbulence in terms of fluctuations from the mean motions, it appears that because both horizontal and vertical mean velocities are typically small in fogs, a fog is dominated by turbulence. Thus, turbulence affects transport and nearly all physical processes in fogs, even though its absolute magnitude is generally small.

1.3.2 Stratus and Stratocumulus Clouds

Stratus clouds and stratocumulus clouds do not differ markedly from fogs in terms of time scales, liquid-water contents, or turbulence levels. The lifetimes of stratus and stratocumulus clouds are longer, ranging from 6 to 12 h. As in fog, the time scale for a parcel to enter a stratus having a mean vertical velocity of 0.1 m s^{-1} and rising through a depth of, say, 1000 m may be 3 h. Typical liquid-water contents in stratus clouds range from 0.05 to 0.25 g m^{-3}, with some maxima of over 0.6 g m^{-3} reported. This combination of time scales and liquid-water contents results in precipitation in the deepest, wettest stratus and stratocumulus clouds in the form of drizzle.

Again, assuming vertical velocities of 0.1 m s^{-1}, the wet adiabatic cooling rates are on the order of 2°C h^{-1}. Thus, radiation and wet adiabatic cooling are approximately equal contributors to the destabilization of stratus and stratocumulus clouds.

The turbulence level in stratus clouds is low in absolute magnitude, just as it is in fog. However, since mean vertical velocities are also small, turbulence is a significant contributor to vertical transport processes, energetics, and the physics of stratus clouds.

1.3.3 Cumulus (Humilis and Mediocris) Clouds

Cumulus clouds whose vertical extent may be 1500 m have a lifetime (T_c) of 10–30 min, which is shorter than that for the preceding two types of clouds. If we consider an average vertical velocity of 3 m s^{-1}, the time scale for a parcel to enter cloud base and exit cloud top is on the order of

$$T_P = 1500\ \mathrm{m}/3\ \mathrm{m\ s^{-1}} = 500\ \mathrm{s} \simeq 10\ \mathrm{min}.$$

The liquid-water content of small cumuli rarely exceeds 1.0 g m^{-3} and is typically approximately 0.3 g m^{-3}. Thus, for such short time scales and low liquid-water contents, precipitation is unlikely in all but the most maritime, wettest cumuli.

Comparing wet adiabatic cooling rates to cloud-top radiation cooling, we estimate

$$CR_\gamma \simeq (0.5°C/100\ m) \times 3\ m\ s^{-1} = 1.5 \times 10^{-2}\ °C\ s^{-1} \simeq 50°C\ h^{-1},$$

which is considerably greater than the cloud-top radiation cooling rates for clouds of such liquid-water contents ($CR_{IR} \sim 4°C\ h^{-1}$). Thus, wet adiabatic cooling dominates radiative effects in such clouds.

The turbulence levels in small cumuli is relatively moderate, with root-mean-square (RMS) velocities ranging from 1 to 3 m s^{-1}. Thus, turbulence plays an important role in such clouds.

1.3.4 Cumulus Congestus Clouds

The lifetime of cumulus congestus clouds exceeds that of cumuli, from 20 to 45 min. However, the transit time T_P for a parcel entering cloud base, rising at 10 m s^{-1}, and exiting cloud top is similar to small cumuli, since

$$T_P = 5000\ m/10\ m\ s^{-1} = 500\ s \simeq 10\ min.$$

That is, higher updraft velocities in cumulus congestus clouds offset their greater depth in determining T_P. Because of the small T_P, precipitation would be unlikely if it were not for the higher liquid-water content of cumulus congestus clouds, which ranges from 0.5 to 2.5 g m^{-3}. Because the turbulence level in such clouds is often quite strong, it is possible for air parcels to spend a considerably longer residence time in the cloud than would be implied by T_P.

As in the smaller cumuli, radiative effects are secondary to wet adiabatic processes in the energetics of cumulus congestus clouds.

1.3.5 Cumulonimbus Clouds

Cumulonimbi are the longest living convective clouds. They have lifetimes from 45 min to several hours. However, the time scale for a parcel of air to enter cloud base and commence forming precipitation before exiting the top remains relatively short. Let us take, as an example, a cumulonimbus cloud that is 12,000 m deep and has an average updraft velocity of 30 m s^{-1}. The Lagrangian time scale is only

$$T_P = 12{,}000\ m/30\ m\ s^{-1} = 400\ s,$$

which is actually less than in the smaller cumuli. Because of the enormous cooling of air parcels rising through the great depths of the cloud, typical liquid-water contents in cumulonimbi range from 1.5 to 4.5 g m^3 and often greater. These high liquid-water contents compensate, to some extent, for the short time scale. The short time scale sometimes limits the formation of precipitation, which accounts for the weak echo regions (WERs) that are often observed by radars. It should be noted that such intense updrafts as those present in WERs are not characteristic of the entire convective storm. Because turbulence levels can be so intense, there is considerable opportunity for air parcels to experience much longer lifetimes than are encountered rising in the main updraft.

With the exception of the anvil outflow region of cumulonimbi, wet adiabatic processes dominate over radiative cooling. However, radiative cooling may contribute significantly to the destabilization and maintenance of the weak updraft regions of cumulonimbus anvils.

1.3.6 Stable Orographic Clouds

Let us consider now a wintertime stable orographic cloud that is above a 1400-m-high mountain with a half-width of 18 km (Fig. 1.1).

For this type of cloud, the cloud lifetime could be many hours or even days. However, the time scale for precipitation processes to operate if the winds are about 15 m s^{-1} is only

$$T_P = 18{,}000\ \mathrm{m}/15\ \mathrm{m\ s^{-1}} = 1200\ \mathrm{s} = 20\ \mathrm{min}$$

Thus, the time scale T_p is longer than that for cumuli but considerably shorter than that for stratus clouds. The liquid-water contents of wintertime stable orographic clouds do not differ substantially from those of stratocumuli; they are typically less than 0.2 g m^{-3}. It is only in highly efficient maritime clouds or the colder ice-phase-dominated clouds that precipitate occurs occasionally.

If we consider typical updraft speeds near the mountain barrier to be about 1 m s^{-1}, we have an estimate of wet adiabatic cooling rates of

$$\mathrm{CR}_\gamma = 18°\mathrm{C\ h^{-1}},$$

which is greater by an order of magnitude than radiative cooling rates.

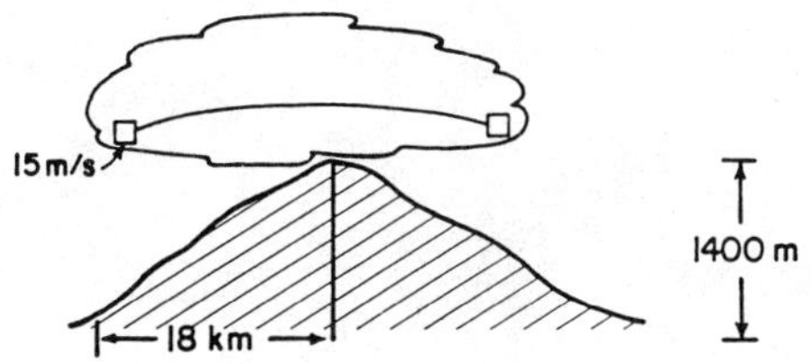

Fig. 1.1. Schematic diagram of a stable orographic cloud.

Thus, near the barrier crest, wet adiabatic processes remain dominant. At distances removed from the barrier crest, however, where a blanket cloud may reside, or in weaker wind situations, one can anticipate that radiative processes become more significant in such clouds.

There has been very little characterization of the levels of turbulence in wintertime stable orographic clouds. At cloud levels near the barrier crest, surface-generated turbulence could be quite significant. At higher cloud levels, however, turbulence levels can be expected to be relatively weaker under the typically stable conditions.

We can see from these simple comparisons and contrasts that clouds form in a broad range of conditions that control the ultimate destiny of the cloud. Depending on the vertical velocity, liquid-water content, and cloud time scale, precipitation processes may or may not affect significantly the dynamics of the cloud. Similarly, radiation processes may or may not be an important destabilizing influence on the cloud. It should be remembered that these are only rough estimates and that one can expect considerable variability within a given cloud category. To account for such variability we must construct sophisticated models of each of the cloud types. In the following chapters, we present the foundation for constructing such models of the dynamics and physics of various cloud systems.

References

Howell, W. E. (1951). The classification of cloud forms. Clouds, fogs and aircraft icing. *In* "Compendium of Meteorology" (T. F. Malone, ed.), pp. 1162–1166. Am. Meteorol. Soc., Boston, Massachusetts.

Huschke, R. E., ed. (1959). "Glossary of Meteorology." Am. Meteorol. Soc., Boston, Massachusetts.

Langmuir, I. (1948). The growth of particles in smokes and clouds and the production of snow from supercooled clouds. *Proc. Am. Philos. Soc.* **92**, 167.

Schaefer, V. J. (1948). The production of clouds containing supercooled water droplets or ice crystals under laboratory conditions. *Bull. Am. Meteorol. Soc.* **29**, 175.

Scorer, R. S. (1963). Cloud nomenclature. *Q. J. R. Meteorol. Soc.* **89**, 248–253.

Scorer, R. S., and H. Wexler (1967). "Cloud Studies in Colour." Pergamon, Oxford.

World Meteorological Society (1956). "International Cloud Atlas," Vol. 1. Geneva.

Part I Fundamental Concepts and Parameterizations

Chapter 2 Fundamental Equations Governing Cloud Processes

If your work is to withstand
tests of time and season
Your foundation must be planned
with patience, insight, reason.

Vollie Cotton

2.1 Introduction

In this chapter we develop a set of equations, parameterizations, and methodology necessary for establishing a theoretical understanding of clouds or cloud systems. The equations could be used to formulate a numerical model of a cloud, but it is not our intent to discuss numerical modeling techniques. Instead, our goal is to lay out a theoretical foundation that will aid the design of a model and involve some form of solution technique, be it pure analytic integration or finite-difference, spectral or finite-element techniques. We will avoid discussing the problem of defining boundary conditions for limited-area models, although we recognize that boundary conditions play an important part in the solution algorithm for any limited-area model, and even though the selection of boundary conditions must fit both the solution algorithms and the physics of the model.

2.2 General Equations

2.2.1 Equation of State

Often in meteorology we ignore the presence of gases other than dry air in describing the equation of state of the atmosphere. In clouds, however, water vapor plays a key role in determining the pressure and temperature of the system and can be a meaningful source of energy.

Let us consider a system composed of dry air with density ρ_a, water vapor density ρ_v, and total water condensate ρ_w (i.e., cloud water, rain water, and ice water) such that

$$\rho_t = \rho_v + \rho_w, \tag{2.1}$$

where ρ_t is the total water density.

Assuming that the system behaves as an ideal gas, the equation of state can be written:

$$p = R_a \rho_a T + R_v \rho_v T = p_d + e, \tag{2.2}$$

where R_a and R_v are the gas constants for dry air and water vapor (287.04 and 461.50 J kg^{-1} K^{-1}, respectively), and p_d and e are the partial pressure of dry air and water vapor.

Equation (2.2) can be rewritten as

$$p = \rho_a R_a T(1 + R_v \rho_v / R_a \rho_a). \tag{2.3}$$

The quantity ρ_v/ρ_a is defined as the mixing ratio of water vapor r_v, and $R_v/R_a \simeq 1.61$, thus,

$$p = \rho_a R_a T(1 + 1.61 r_v). \tag{2.4}$$

It should be noted that Eq. (2.4) is written in terms of the dry air density ρ_a and the dry gas constant R_a and that no approximations have been made.

Another more familiar form of the equation of state can be found by rewriting Eq. (2.2) as

$$p = \rho_m \frac{m_a R_a + m_v R_v}{m_a + m_v} T, \tag{2.5}$$

where m_a and m_v are the masses of dry air and water vapor, respectively, and

$$\rho_m = \frac{m_a + m_v}{v} = \rho_a + \rho_v, \tag{2.6}$$

and v is the volume of gas. Equation (2.5) can be rewritten as

$$p = \rho_m R_a \frac{1 + (m_v/m_a)(R_v/R_a)}{1 + m_v/m_a} T. \tag{2.7}$$

The mixing ratio of water vapor r_v can be written as

$$r_v = m_v/m_a, \tag{2.8}$$

so that Eq. (2.7) becomes

$$p = \rho_m R_a \frac{1 + 1.61 r_v}{1 + r_v} T, \tag{2.9}$$

where we have again let $R_v/R_a = 1.61$. Comparing Eqs. (2.4) and (2.9), one should note that the equation of state, Eq. (2.9), contains the density of the mixture ρ_m, whereas Eq. (2.4) contains the density of dry air ρ_a. Equation (2.9) can be rewritten as

$$p = \rho_m R_a \left(1 + \frac{0.61 r_v}{1 + r_v}\right) T. \tag{2.10}$$

We define the quantity specific humidity q_v as

$$q_v = \frac{m_v}{m_v + m_a} = \frac{r_v}{1 + r_v}. \tag{2.11}$$

Thus Eq. (2.10) can be written

$$p = \rho_m R_a (1 + 0.61 q_v) T. \tag{2.12}$$

Since vapor mixing ratios rarely exceed 22 g kg^{-1} even in the tropics, $q_v \ll 1$; therefore $q_v \sim r_v$ and Eq. (2.12) may be approximated as

$$p \sim \rho_m R_a (1 + 0.61 r_v) T. \tag{2.13}$$

Comparing Eq. (2.4) to Eq. (2.13), we note that the coefficient 0.61 multiplies r_v when the density of the mixture is used in the equation of state, and the coefficient 1.61 multiplies r_v when ρ_a is used. We emphasize that this is often a source of confusion for a moist system.

It is common for meteorologists to define a virtual temperature T_v as

$$T_v = (1 + 0.61 q_v) T \simeq (1 + 0.61 r_v) T, \tag{2.14}$$

thus Eq. (2.12) or (2.13) becomes

$$p = \rho_m R_a T_v. \tag{2.15}$$

It should be noted that in a cloud system containing hydrometeors, the equation of state—Eq. (2.10), (2.12), or (2.15)—only describes the relationship among p, T, and ρ_m, or the density of the gaseous components of the system. The total density of the system ρ is equal to

$$\rho = \rho_m + \rho_w = \rho_a + \rho_t. \tag{2.16}$$

2.2.2 *Equations of Motion*

The equations of motion of a parcel of moist air having density ρ_m can be written [see basic texts such as Dutton (1976) or Holton (1972) for derivation]

$$\rho_m (du_i/dt) = -\partial p/\partial x_i - \rho_m g \delta_{i3} - \rho_w g \delta_{i3} + \rho_m \varepsilon_{ijk} u_j f_k + \rho_m f_r, \tag{2.17}$$

where g is the acceleration due to gravity, δ is the delta function, ε_{ijk} is the permutation symbol (or Levi-Civaita density), and f_k is twice the earth's angular velocity Ω_k. (Throughout the text the Lagrangian operator is defined as d/d_t.) The last term on the RHS of Eq. (2.17) is the molecular viscosity, defined as

$$f_r = \nu \left[\frac{\partial^2 u_i}{\partial x_j \, \partial x_j} + \frac{1}{3} \left(\frac{\partial}{\partial x_i} \right) \left(\frac{\partial u_j}{\partial x_j} \right) \right],$$

where ν is the kinematic viscosity $\nu = \mu / \rho_m$ and μ is the dynamic viscosity.

The only additional force considered in our system is the gravitational force—the third term on the RHS of Eq. (2.17)—due to the mass density of suspended condensate.

Throughout this section basic Cartesian tensor notation is used for convenient, abbreviated notation. The indices $i = 1, 2, 3$ correspond to Cartesian vector coordinates x, y, and z, respectively. The following simple rules apply:

- Repeated indices are summed (e.g., $a_{ii} = a_{11} + a_{22} + a_{33}$ in three-dimensional space).
- Single indices in a term are called free indices and refer to the order of a tensor. The maximum value a free index can attain in a three-dimensional system is 3.
- Only tensors of the same order can be added.
- Multiplication of tensors can be performed as for scalars. They are commutative with respect to addition and multiplication.
- The delta function parameter is defined to simplify writing the gravitational acceleration term; thus

$$\delta_{ij} = \begin{cases} 1 & \text{for} \quad i = j. \\ 0 & \text{for} \quad i \neq j. \end{cases}$$

Similarly the permutation symbol ε_{ijk} is defined to express vector cross products. Thus,

$$\varepsilon_{ijk} = \begin{cases} 0 & \text{if} \quad i = j, \quad \text{or} \quad j = k, \quad \text{or} \quad i = k. \\ 1 & \text{if} \quad i, j, k \text{ are an even permutation of } 1, 2, 3. \\ -1 & \text{if} \quad i, j, k \text{ are an odd permutation of } 1, 2, 3. \end{cases}$$

Hence,

$$\varepsilon_{1,2,3} = 1, \qquad \varepsilon_{1,3,2} = -1, \qquad \varepsilon_{3,1,2} = 1, \qquad \varepsilon_{3,2,1} = -1, \qquad \varepsilon_{3,3,1} = \varepsilon_{2,1,2} = 0.$$

Cartesian tensor notation yields the following equivalents with Cartesian vector notation

$$\text{Divergence:} \qquad \nabla \cdot \mathbf{V} \rightarrow \partial u_j / \partial x_j.$$

Advection: $-\mathbf{V}\cdot\nabla\phi\rightarrow -u_j(\partial\phi/\partial x_j)$.

Coriolis: $-\mathbf{f}\times\mathbf{V}\rightarrow -\varepsilon_{ijk}f_j u_k$.

2.2.3 *The Equation for Conservation of Dry Air Mass*

The derivation of the equation of mass continuity is given in numerous basic dynamics texts (e.g., Holton, 1972; Dutton, 1976). For dry air it takes the form

$$\partial\rho_a/\partial t+\partial(\rho_a u_j)/\partial x_j=0. \tag{2.18}$$

We choose not to apply Eq. (2.18) to the total air density ρ_m, since water vapor ρ_v has numerous sources and sinks.

2.2.4 *Conservation of Water Substance*

To illustrate the formulation of conservation equations for a particular cloud species or contaminant in a mesoscale system, let us consider a parcel containing dry air, water vapor, and cloud droplets, which we assume move with the air. Ignoring molecular diffusion of water substance and precipitation, the mixing ratio of total water substance r_T is conserved, hence

$$dr_T/dt=0. \tag{2.19}$$

Expanding the substantial derivative gives us the conservation equation in Eulerian form

$$dr_T/dt=\partial r_T/\partial t+u_j(\partial r_T/\partial x_j)=0. \tag{2.20}$$

The total mixing ratio is the sum of the water vapor mixing ratio r_v, the liquid-water mixing ratio r_c, and the ice mixing ratio r_i,

$$r_T=r_v+r_c+r_i. \tag{2.21}$$

If we ignore the usually small degree of supersaturation in clouds, the water vapor in excess of the saturation vapor mixing ratio r_s condenses immediately to form cloud water, thus

$$r_c=(r_T-r_s)H(r_T-r_s), \tag{2.22}$$

where $H(x)$ is the Heaviside step function defined as

$$H(x)=\begin{cases}1, & x>0.\\ 0, & x\leq 0.\end{cases}$$

It should be noted that for this simple system, individual continuity equations do not have to be written or integrated for r_v and r_c since Eqs. (2.21) and (2.22) completely define their relative contributions to r_T, provided r_s is evaluated from the thermodynamic energy equation. A continuity equation for r_i must be written, however. The addition of precipitation variables greatly complicates the picture, since r_T can no longer be conserved and, therefore, continuity equations for precipitation elements must be formulated. Some examples of the formulation of precipitation continuity equations will be given in Chapter 4.

2.2.5 The Thermodynamic Energy Equation for Moist Convection

Let us consider the thermodynamic energy equation for a moist system containing water vapor, liquid water, and ice particles having mixing ratios r_v, r_l, and r_i, respectively. The liquid particles may be composed of raindrops and small cloud droplets. According to Dutton (1976, p. 284), the thermodynamic energy equation for an open thermodynamic system is

$$d\left(c_{pa}\ln T - R_a \ln p_a + \frac{r_v L_{lv}}{T}\right) - d_i \frac{r_i L_{il}}{T} + r_v d\frac{A_{lv}}{T} - r_i d\frac{A_{il}}{T} + (r_v + r_c + r_i)\frac{c_l\, dT}{T} = Q(\mathrm{R}) + Q(\mathrm{D}), \tag{2.23}$$

where c_{pa} and c_l are the heat capacities of dry air and liquid water, T is the temperature, R_a is the gas constant for dry air, p_a is the partial pressure of dry air, L_{lv} and L_{il} are the latent heat of liquid-to-vapor and ice-to-liquid phase changes, A_{lv} and A_{il} are the affinity of vaporization and melting, $Q(\mathrm{R})$ represents the heating rate due to radiative flux divergence, and $Q(\mathrm{D})$ represents the heat loss due to molecular dissipation. Following Dutton (1976), the subscript i following a derivative operator refers to changes in a quantity occurring internally within the parcel. The subscript e (for external) following the derivative operator refers to changes in a quantity due to a net flux of that quantity into or out of the parcel. No subscript following the derivative operator refers to the total change, which is the sum of the external and internal changes. Extensive variables can have external and internal derivatives if a flux of that quantity relative to the air parcel occurs. Intensive variables such as temperatures, however, have external derivatives identically zero. If vapor is assumed to move with the parcel, $d_e r_v = 0$. The affinity terms may be expressed as

$$A_{lv} = \mu_l - \mu_v, \tag{2.24}$$

$$A_{il} = \mu_i - \mu_l, \tag{2.25}$$

where μ_v, μ_l, and μ_i are the chemical potentials of water vapor, liquid water, and ice, respectively.

The chemical potentials may then be defined as

$$\mu_v = \mu_0 + R_v T \ln e_v, \tag{2.26}$$

$$\mu_l = \mu_0 + R_v T \ln e_s, \tag{2.27}$$

$$\mu_i = \mu_0 + R_v T \ln e_{si}, \tag{2.28}$$

where e_v is the partial pressure of water vapor, e_s and e_{si} are the saturation vapor pressures with respect to water and ice, respectively, and μ_0 is the base state chemical potential of water. Strictly speaking, e_s and e_{si} represent the saturated vapor pressures with respect to liquid and ice at the surface temperatures, which differ from the ambient temperature T. In the following discussion, we will ignore surface curvature and solution effects, however, and evaluate e_s and e_{si} with respect to *plane pure, water,* or *ice surfaces* at the ambient temperature T of the environment. From the equation of state, the following identity may be formulated:

$$r_v R_v \, d \ln e_v = (R_a + R_v r_v) \, d \ln p - R_a \, d \ln p_a. \tag{2.29}$$

Combining Eqs. (2.23)-(2.29) and using the relationships

$$\left.\frac{dL_{lv}}{dT}\right]_p = c_{pv} - c_l, \tag{2.30}$$

$$\left.\frac{dL_{il}}{dT}\right]_p = c_l - c_i, \tag{2.31}$$

$$L_{il} = L_{iv} - L_{lv}, \tag{2.32}$$

the first law may be rewritten

$$c_{pm} \, d \ln T - R_m \, d \ln p + \frac{L_{lv}}{T} \, dr_v - \frac{L_{il}}{T} \, d_i r_i + (r_v + r_i)\left[L_{lv} \, d\left(\frac{1}{T}\right) + R_v \, d \ln e_s \right]$$

$$- r_i \left[L_{iv} \, \mathbf{d}\left(\frac{1}{T}\right) + R_v \, d \ln e_{si} \right] = Q(\mathrm{R}) + Q(\mathrm{D}), \tag{2.33}$$

where $c_{pm} = c_{pa} + r_v c_{vp} + r_i c_i + r_l c_l$ and $R_m = R_a + r_v R_v$, and L_{iv} is the latent heat of ice-to-vapor phase changes.

Defining the variation in saturation vapor pressures from the Clausius-Clapeyron relation reduces the bracketed terms to zero. The final or rigorous form of the first law is then

$$c_{pm} \, d \ln T - R_m \, d \ln p + \frac{L_{lv}}{T} \, dr_v - \frac{L_{il}}{T} \, d_i r_i = Q(\mathrm{R}) + Q(\mathrm{D}). \tag{2.34}$$

It should be noted that Eq. (2.34) has been derived without assuming that the system is in vapor equilibrium. As a result, this formulation should be equally valid for subsaturated evaporation of liquid or ice water, supersaturated growth of ice, and other phase changes which may or may not be equilibrium processes.

From Poisson's equation, potential temperature may be defined as

$$d \ln \theta = d \ln T - (R_{\mathrm{a}}/c_{\mathrm{pa}})\, d \ln p. \tag{2.35}$$

Combining Eqs. (2.34) and (2.35) we can write

$$d \ln \theta = \left(\frac{R_{\mathrm{m}}}{c_{\mathrm{pm}}} - \frac{R_{\mathrm{a}}}{c_{\mathrm{pa}}}\right) d \ln p - \frac{L_{\mathrm{lv}}}{c_{\mathrm{pm}} T}\, dr_{\mathrm{v}} + \frac{L_{\mathrm{il}}}{c_{\mathrm{pm}} T}\, d_{\mathrm{i}} r_{\mathrm{i}} + \frac{1}{c_{\mathrm{pm}}} [Q(\mathrm{R}) + Q(\mathrm{D})]. \tag{2.36}$$

It should be noted that in deriving Eq. (2.36) it is assumed that the gas constants and heat capacities do not vary with temperature.

It is often desirable to formulate approximate forms of the thermodynamic energy equation. Commonly used assumptions are to neglect the heat stored in a condensed water substance and to neglect radiative heating $Q(\mathrm{R})$ and molecular dissipation $Q(\mathrm{D})$. Under these assumptions, Eq. (2.36) reduces to

$$d \ln \theta = -\frac{L_{\mathrm{lv}}}{c_{\mathrm{pa}} T}\, dr_{\mathrm{v}} + \frac{L_{\mathrm{il}}}{c_{\mathrm{pa}} T}\, d_{\mathrm{i}} r_{\mathrm{i}}. \tag{2.37}$$

If equilibrium is assumed, then $dr_{\mathrm{v}} = dr_{\mathrm{s}}$, and if we neglect the effects of the ice phase (i.e., $d_{\mathrm{i}} r_{\mathrm{i}} = 0$), then Eq. (2.37) becomes the classical wet (also called saturated) pseudoadiabatic relationship.

Assuming conservation of total water given by

$$dr_{\mathrm{v}} + d_{\mathrm{i}} r_{\mathrm{l}} + d_{\mathrm{i}} r_{\mathrm{i}} = 0, \tag{2.38}$$

and applying Eq. (2.32), we may rewrite Eq. (2.37) as

$$d \ln \theta = \frac{L_{\mathrm{lv}}}{c_{\mathrm{pa}} T}\, d_{\mathrm{i}} r_{\mathrm{l}} + \frac{L_{\mathrm{iv}}}{c_{\mathrm{pa}} T}\, d_{\mathrm{i}} r_{\mathrm{i}}. \tag{2.39}$$

Variations in θ given by Eqs. (2.36), (2.37), and (2.39) occur only due to internal changes of r_{i} or r_{l}.

It is often desirable to obtain a thermodynamic variable that is conservative under adiabatic liquid and ice transformations. Therefore, using Eq. (2.39), one can define the conservative variable ice–liquid-water potential temperature θ_{il}

$$d_{\mathrm{i}} \ln \theta_{\mathrm{il}} = d \ln \theta - \frac{L_{\mathrm{lv}}}{c_{\mathrm{pa}} T}\, d_{\mathrm{i}} r_{\mathrm{l}} - \frac{L_{\mathrm{iv}}}{c_{\mathrm{pa}} T}\, d_{\mathrm{i}} r_{\mathrm{i}} = 0. \tag{2.40}$$

The variable θ_{il} may be considered a generalization of the variable defined by Betts (1973), which is conservative for wet adiabatic motions only (i.e., $d_i r_i = 0$). It should be noted that because θ_{il} is defined both by θ, which is an intensive property, and by r_l and r_i, which are extensive properties, θ_{il} too is extensive. Therefore, external fluxes of θ_{il} may occur into and out of the parcel due to relative precipitation movement. From Eq. (2.40), θ_{il} will be conserved provided all condensate remains within the original parcel. If precipitation does occur, external fluxes of r_l and r_i and therefore θ_{il} may be taken into account when calculating the total change in θ_{il}. Variations in θ will also occur if radiative heating and molecular dissipation are also considered.

The virtue of θ_{il} or its non-ice equivalent θ_l is that, aside from being conservative, this variable reduces to θ in the absence of cloud or precipitation. Thus, this variable varies continuously from clear air to cloudy air.

Another thermodynamic variable frequently used for thermodynamic diagnostic studies of convective clouds is the equivalent potential temperature θ_e. Generalization of θ_e to an ice-phase system may be accomplished by defining the conservative variable under pseudoadiabatic motion in Eq. (2.37) as

$$d_i \ln \theta_{eiv} = d \ln \theta + \frac{L_{lv}}{c_{pa} T}\, dr_v - \frac{L_{il}}{c_{pa} T}\, d_i r_i = 0. \tag{2.41}$$

Equation (2.41) defines θ_e for the case when $d_i r_i = 0$. The variable θ_{eiv}, or its non-ice equivalent θ_e, is useful in diagnostic studies as a tracer of air parcel motions. Like θ_{il}, θ_{eiv} is conservative over phase changes but not if precipitation fluxes exist.

Further approximations to Eqs. (2.40) and (2.41) must be made in order to allow analytic integration. Thus, Eqs. (2.40) and (2.41) can be rewritten in the form

$$d_i \ln(\theta_{il}/\theta) = -d_i \left(\frac{L_{lv} r_l}{c_{pa} T} + \frac{L_{iv} r_i}{c_{pa} T} \right) + \varepsilon_1 + \varepsilon_2, \tag{2.42}$$

and

$$d_i \ln(\theta_{eiv}/\theta) = d_i \left(\frac{L_{lv} r_v}{c_{pa} T} - \frac{L_{il} r_i}{c_{pa} T} \right) + \varepsilon_3 + \varepsilon_4, \tag{2.43}$$

respectively, where

$$\varepsilon_1 = \frac{r_l}{c_{pa}}\, d\left(\frac{L_{lv}}{T} \right), \tag{2.44}$$

$$\varepsilon_2 = \frac{r_i}{c_{pa}}\, d\left(\frac{L_{iv}}{T} \right), \tag{2.45}$$

$$\varepsilon_3 = \frac{-r_v}{c_{pa}} d\left(\frac{L_{lv}}{T}\right), \tag{2.46}$$

$$\varepsilon_4 = \frac{r_i}{c_{pa}} d\left(\frac{L_{il}}{T}\right). \tag{2.47}$$

Neglect of ε_1, ε_2, ε_3, and ε_4, allows us to integrate Eqs. (2.42) and (2.43) into the forms

$$\theta_{il} = \theta \exp\left| -\left[\frac{L_{lv}(T_0) r_l}{c_{pa} T} + \frac{L_{iv}(T_0) r_i}{c_{pa} T}\right]\right| \tag{2.48}$$

and

$$\theta_{eiv} = \theta \exp\left| -\left[-\frac{L_{lv}(T_0) r_v}{c_{pa} T} + \frac{L_{il}(T_0) r_i}{c_{pa} T}\right]\right|, \tag{2.49}$$

where we have taken L_{il}, L_{lv}, and L_{iv} to be constant and equal to their value at $T_0 = 273.16$ K. Neglect of terms ε_1, ε_2, ε_3, and ε_4 introduces errors which are proportional to $-dT/T^2$ and grow nonlinearly as temperature decreases. Therefore ε_1, for example, increases nonlinearly with decreasing temperature in the presence of liquid water. The error ε_3, however, is weighted by vapor rather than liquid water. Thus, as temperature falls so does the vapor mixing ratio, and ε_3 will first increase with height and decrease as r_v diminishes exponentially.

If we use direct numerical solutions to Eq. (2.36) as our "norm" or "truth," we find that the approximations made in deriving Eq. (2.48) or (2.49) result in substantial errors in estimating θ or T in the upper troposphere. This was demonstrated by Wilhelmson (1977) and Tripoli and Cotton (1981). Figure 2.1 illustrates the magnitude of potential errors in estimating θ or T that these approximations introduce in the upper troposphere. There are several ways of avoiding these errors. First, one could attempt to obtain direct solutions to Eq. (2.36) or restrict the use of Eq. (2.48) or (2.49) to shallow clouds. However, direct solutions to Eq. (2.36) are often too computationally demanding for routine use in three-dimensional cloud or mesoscale models. Alternately, one can follow procedures outlined by Tripoli and Cotton (1981) in which series approximations to Eq. (2.48) are given

$$\theta_{il} = \theta\left[1 - \frac{L_{lv}(T_0) r_l}{c_{pa} T} - \frac{L_{iv}(T_0) r_i}{c_{pa} T}\right], \tag{2.50}$$

$$\theta = \theta_{il}\left[1 + \frac{L_{lv}(T_0) r_l}{c_{pa} T} + \frac{L_{iv}(T_0) r_i}{c_{pa} T}\right]. \tag{2.51}$$

Because ε_1 grows very large at low temperatures, Tripoli and Cotton improved the accuracy in diagnosing θ by holding the temperature constant

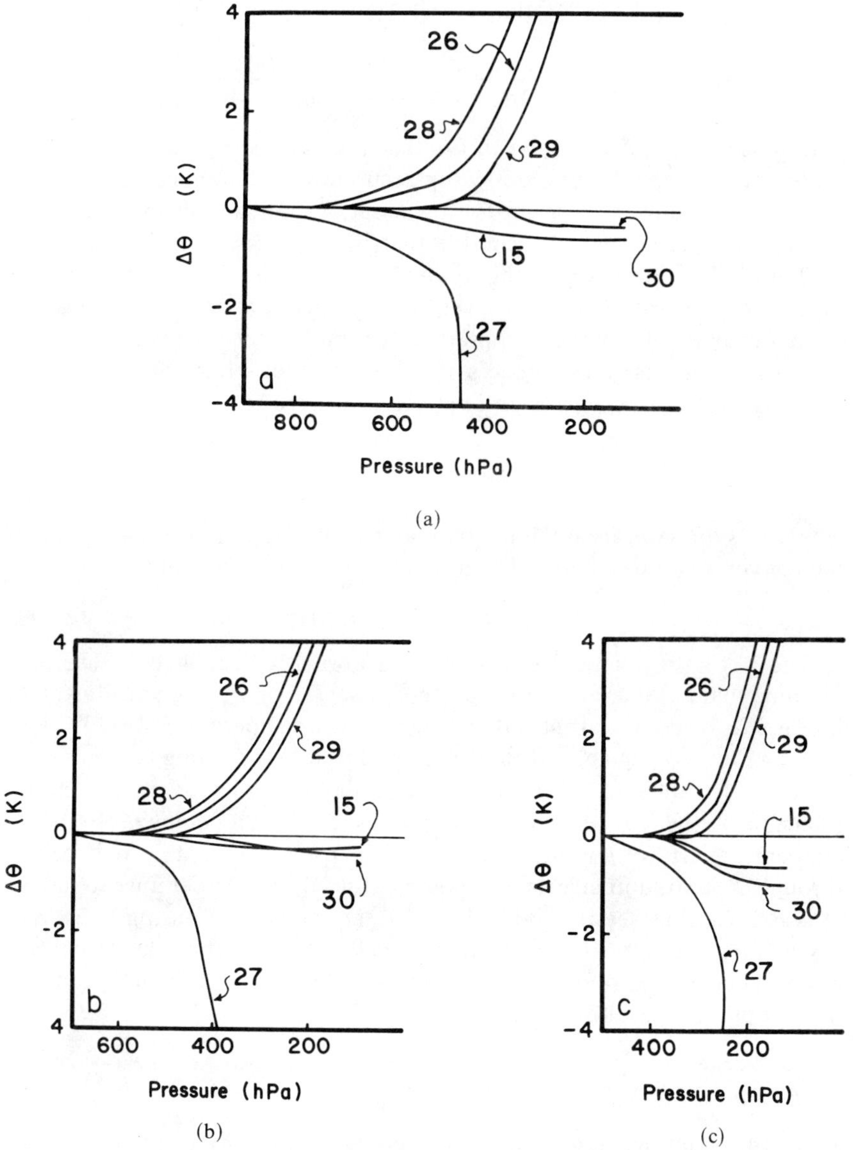

Fig. 2.1. Differences between θ diagnosed by Eq. (2.36), assuming $Q(\mathrm{R})$ and $Q(\mathrm{D})$ are zero, and θ diagnosed using Eq. (2.48), here labeled 26; Eq. (2.49), here labeled 27; Eq. (2.50), here labeled 28; Eq. (2.51), here labeled 29; Eq. (2.52), here labeled 30; and Eq. (2.39), here labeled 15. Calculations are performed for a parcel initially saturated with liquid water at (a) 90 kPa, (b) 70 kPa, and (c) 50 kPa pressure rising to 11 kPa. Initial potential temperature was 300 K in each case. [From Tripoli and Cotton (1981).]

when it falls below 253 K. Thus Eq. (2.51) becomes

$$\theta = \theta_{\mathrm{il}}\left[1+\frac{L_{\mathrm{lv}}(T_0)r_{\mathrm{l}}}{c_{\mathrm{pa}}\,\mathrm{Max}(T,253)}+\frac{L_{\mathrm{iv}}(T_0)r_{\mathrm{i}}}{c_{\mathrm{pa}}\,\mathrm{Max}(T,253)}\right]. \tag{2.52}$$

Solutions to Eq. (2.52) relative to Eq. (2.36) are also shown in Fig. 2.1. This represents one approach to obtaining accurate yet economical solutions to the thermodynamic energy equation when it is applied to deep convection.

Another thermodynamic variable that is frequently used in diagnostic studies of cloud processes is the wet-bulb temperature T_{w} and the wet-bulb potential temperature θ_{w}. The wet-bulb temperature results when air is cooled isobarically by the evaporation of liquid water (or ice).

Thus, for an isobaric process in which the ice phase is absent, and ignoring radiative processes and molecular dissipation, Eq. (2.33) gives

$$\int_{T}^{T_{\mathrm{w}}} c_{\mathrm{pm}}\,dT = -\int_{r_{\mathrm{v}}}^{r_{\mathrm{s}}} L_{\mathrm{lv}}\,dr_{\mathrm{v}}, \tag{2.53}$$

where r_{s} is the saturation mixing ratio at T_{w}. Integration of Eq. (2.53) gives the conventional definition of thermodynamic wet-bulb temperature

$$(c_{\mathrm{pa}}+r_{\mathrm{v}}c_{\mathrm{pv}})(T-T_{\mathrm{w}}) = L_{\mathrm{lv}}(r_{\mathrm{s}}-r_{\mathrm{v}}). \tag{2.54}$$

The wet-bulb potential temperature is normally evaluated by use of a thermodynamic diagram. Consider the skew T, Log P diagram illustrated in Fig. 2.2, where the air at, say, 800 mbar has a temperature of +10°C and a mixing ratio of 7 g kg^{-1}. If the air is first lifted to the lifting condensation level (LCL) and we follow the pseudoadiabat to its original pressure (800 mbar), the temperature arrived at is called the *adiabatic wet-bulb* temperature. If we continue following the pseudoadiabat, which passes through T_{w} until it intersects the 1000-mbar level, the temperature we arrive at is θ_{w}. We thus see that both θ_{w} and θ_{e} are conserved during moist and dry adiabatic processes. In practice, therefore, use of these two variables is interchangeable, and the investigators' selection criteria are generally based upon personal preference.

2.2.6 *Parcel Static Stability*

Using the equations derived in the preceding sections, we can now consider the concept of parcel static stability. Let us consider an infinitely small parcel of air that is thermally insulated from its environment. The parcel is allowed to ascend or descend through an environment having a constant temperature lapse rate

$$\gamma = -dT/dx_3. \tag{2.55}$$

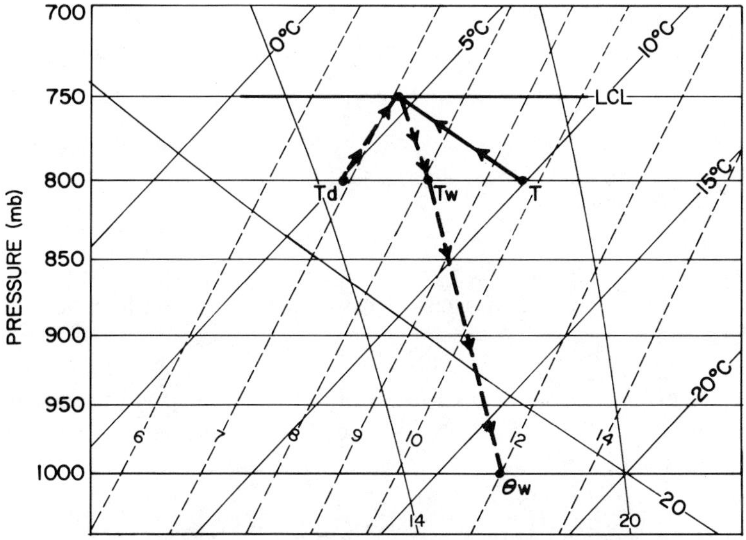

Fig. 2.2. The wet-bulb potential temperature is normally defined by use of a thermodynamic diagram. Consider this skew *T*, Log *P* diagram, where the air at, say, 800 mbar has a temperature of +10°C and a mixing ratio of 6.9 g/kg. If the air is first lifted to the lifting condensation level (LCL) and we follow the pseudoadiabat to its original pressure (800 mbar), then the temperature obtained is called the *adiabatic wet-bulb* temperature. If we continue following the pseudoadiabat that passes through T_w until it intersects the 1000-mbar level, the temperature we obtain is θ_w. We thus see that both θ_w and θ_e are conserved during moist and dry adiabatic processes. In practice, use of these two variables is interchangeable. An investigator's selection criteria are generally based upon personal preference.

As a parcel ascends or descends, it experiences adiabatic expansion or compression, thus cooling or warming accordingly. The rate of cooling of the rising parcel can be evaluated from Eq. (2.34) by assuming that radiative heating $Q(\mathrm{R})$ and molecular dissipation $Q(\mathrm{D})$ of the parcel are negligible. Furthermore, if we neglect the heat stored in a condensed water substance (i.e., $C_{\mathrm{pm}} = C_{\mathrm{pa}}$ and $R_{\mathrm{m}} = R_{\mathrm{a}}$) and consider only a liquid-phase system, then Eq. (2.34) becomes

$$C_{\mathrm{pa}}\, d \ln T - R_{\mathrm{a}}\, d \ln p = -L_{\mathrm{lv}}/T\, dr_{\mathrm{v}}. \tag{2.56}$$

Differentiating Eq. (2.56) with respect to height yields

$$C_{\mathrm{pa}}\, d \ln T/dx_3 - R_{\mathrm{a}}\, d \ln p/dx_3 = -(L_{\mathrm{lv}}/T)(dr_{\mathrm{v}}/dx_3). \tag{2.57}$$

A common assumption in parcel theory is that the parcel experiences only hydrostatic changes in pressure as it rises or ascends. Thus,

$$d \ln p/dx_3 \simeq -\rho g/p. \tag{2.58a}$$

Substitution of Eq. (2.58a) into Eq. (2.57) along with the equation of state for a dry atmosphere ($p = \rho_a R_a T$) yields

$$C_{pa}\, d \ln T/dx_3 + g/T = -(L_{lv}/T)(dr_v/dx_3). \tag{2.58b}$$

For an unsaturated parcel of air, the RHS of Eq. (2.58b) is zero and Eq. (2.58b) gives the well-known dry adiabatic lapse rate

$$\gamma_d = -dT/dx_3 = g/C_{pa} = 9.8 \quad \mathrm{K\,km^{-1}}. \tag{2.59}$$

Likewise, for an unsaturated parcel of air, Eq. (2.37) shows that the change in potential temperature of a parcel rising or ascending adiabatically is

$$d\theta/dx_3 = 0. \tag{2.60}$$

Table 2.1 gives the conditions for absolute parcel stability in a dry atmosphere, in terms of γ and $d\theta/dx_3$, where $d\theta/dx_3$ represents the stability of the environment.

For a saturated cloud, $dr_v = dr_s$, and Eq. (2.58b) becomes

$$dT/dx_3 = -g/C_{pa} - (L_{lv}/C_{pa})(dr_s/dx_3). \tag{2.61}$$

Using the chain rule, we can rewrite Eq. (2.61) as

$$dT/dx_3 = -g/C_{pa} - (L_{lv}/C_{pa})(dT/dz)(dr_s/dT). \tag{2.62}$$

Differentiating $r_s = \varepsilon e_s/p$ logarithmically, and introducing the Clausius–Clapeyron equation

$$d \ln e_s = (L/R_v)(dT/T^2), \tag{2.63a}$$

where R_v is the gas constant for moist air, Eq. (2.62) becomes, after some algebraic manipulation,

$$\frac{dT}{dx_3} = -\frac{g}{C_p}\left(1 + \frac{r_s L}{R_a T}\right) \Big/ \left(1 + \frac{r_s L^2}{C_{pa} R_v T^2}\right). \tag{2.63b}$$

The wet adiabatic lapse rate γ_w, can then be defined as

$$\gamma_w = -\frac{dT}{dx_3} = \gamma_d\left(1 + \frac{r_s L}{R_a T}\right) \Big/ \left(1 + \frac{r_s L^2}{C_{pa} R_v T}\right). \tag{2.64}$$

Based on Table 2.1 and Eq. (2.64), a parcel of air which ascends first dry adiabatically until it reaches the lifting condensation level where it becomes saturated, and then ascends wet adiabatically, will become *conditionally unstable* if

$$\gamma_d > \gamma > \gamma_w.$$

Conditional instability, as defined above, refers to the lifting of a parcel of air through its environment. *Convective instability*, refers to the lifting of

Table 2.1
Absolute Stability of a Parcel

Condition	Terms
Absolutely unstable	$\gamma > \gamma_d,\ d\theta/dx_3 < 0$
Neutral	$\gamma = \gamma_d,\ d\theta/dx_3 = 0$
Absolutely stable	$\gamma < \gamma_d,\ d\theta/dx_3 > 0$

an entire layer of air. If the temperature and moisture profiles are such that a lifted layer of air reaches saturation first at the bottom of the layer, subsequent lifting produces a very rapid destabilization because the upper part of the layer cools at the dry adiabatic rate while the lower part cools at the slower wet adiabatic rate. A layer of air which has this temperature and moisture structure is convectively unstable, and the condition for convective instability is that the equivalent potential temperature, Eq. (2.41) (with ice-phase heating neglected), decrease throughout the layer,

$$\partial\theta_e/\partial x_3 < 0. \tag{2.65}$$

Extension to an Entraining Finite Volume of Cloudy Air

Equation (2.63b) was derived under the assumption that an infinitesimal parcel of air does not mix with any neighboring parcels of air. Let us now consider a cloud volume of mass M. As the cloud volume ascends it isobarically (at constant pressure) entrains an amount dM of environment air having a temperature T_E and vapor mixing ratio r_E. The fractional change in the mass of the cloud is defined as the entrainment rate μ:

$$\mu = (1/M)(dM/dz). \tag{2.66}$$

Accordingly, Eq. (2.61) can be modified to obtain the lapse in temperature following a saturated finite volume of cloud entraining environmental air at the rate μ, as

$$\frac{dT}{dx_3} = -\frac{g}{C_{pa}} - \frac{L_{lv}}{C_{pa}}\left(\frac{dT}{dz}\right)\left(\frac{dr_s}{dT}\right) - \frac{\mu}{C_{pa}}[(T - T_E) + L_{lv}(r_s - r_E)]. \tag{2.67}$$

Following procedures similar to those used in deriving Eq. (2.64), the wet adiabatic lapse rate for an entraining cloud volume

$$\gamma_\mu = -\frac{dT}{dx_3} = \gamma_w + \mu\left[(T - T_E) + \frac{L_{lv}}{C_{pa}}(r_s - r_E)\right] \Big/ \left(1 + \frac{r_s L^2}{C_{pa} R_v T}\right). \tag{2.68}$$

Thus an ascending cloud volume entraining environmental air that is cooler and drier than the cloud volume will always be cooler than a corresponding parcel of air ascending wet adiabatically.

2.3 Scale Analysis and Approximate Equations

We often find that the derivation of more approximate equations than those given in Section 2.2 greatly simplifies cloud models or models of mesoscale systems. Let us reconsider the equation of state.

2.3.1 Equation of State

Consider the equation of state for a moist atmosphere in the form of Eq. (2.13),

$$p = \rho_m R_a(1+0.61 r_v) T. \tag{2.69}$$

First, we define an arbitrary, horizontally uniform dry reference state denoted by the subscript 0. The reference state might represent an atmospheric sounding taken at some distance remote from a cloud yet close enough to represent the environment of the cloud. Alternately, the reference state could represent an "average" sounding over a mesoscale region in which the cloud disturbance resides. Ogura and Phillips (1962) selected an isentropic atmosphere as a reference state. Following Dutton and Fichtl (1969), we have avoided this definition of a reference state, since departures from an isentropic atmosphere [i.e., $\theta'(z) = \theta(z) - \theta_0$] can be very large, whereas departures from an arbitrary reference state [i.e., $\theta'(z) = \theta(z) - \theta_0(z)$] will remain relatively small.

If we restrict our basic state to be dry, our analysis is greatly simplified. However, when the resultant system is applied to a moist atmosphere, the addition of moisture to the system immediately introduces density anomalies that, in turn, introduce pressure anomalies in an otherwise quiescent atmosphere. For such a dry reference state,

$$r_{v_0} = r_{T_0} = 0.$$

We can now express the thermodynamic variables as

$$\alpha_m = \alpha_0 + \alpha'_m; \qquad p = p_0 + p'; \qquad T = T_0 + T'; \qquad \text{and} \qquad r_v = r'_v, \tag{2.70}$$

where the prime quantities represent a deviation from the reference state, and α_m, the specific volume, is

$$\alpha_m = 1/\rho_m. \tag{2.71}$$

The reference state then obeys the equation of state

$$p_0\alpha_0 = R_a T_0. \tag{2.72}$$

Using Eq. (2.70) we can now rewrite Eq. (2.69) as

$$p_0(1+p'/p_0)\alpha_0(1+\alpha'_m/\alpha_0) = R_a(1+0.61r'_v)T_0(1+T'/T_0). \tag{2.73}$$

Taking the natural logarithm of both sides of Eq. (2.73) gives us

$$\ln(1+p'/p_0)+\ln(1+\alpha'_m/\alpha_0) = \ln(1+0.61r'_v)+\ln[1+T'/T_0]. \tag{2.74}$$

If we now expand Eq. (2.74) in a Taylor series $[\ln(1+x) = x - x^2/2 + x^3/3 - x^4/4 + \cdots$ for $|x|<1]$ and ignore higher order terms, we find

$$p'/p_0 + \alpha'_m/\alpha_0 = T'/T_0 + 0.61r'_v, \tag{2.75}$$

or

$$\alpha'_m/\alpha_0 = T'/T_0 + 0.61r'_v - p'/p_0, \tag{2.76}$$

which represents an equation of state for a moist system that is fully linearized. It is often desirable to express Eq. (2.76) in terms of potential temperature. From Poisson's equation we have

$$\theta = T(p_r/p)^{R_a/c_{pa}}, \tag{2.77}$$

where p_r is 1000 mbar. Taking logarithms of both sides of Eq. (2.77), expanding in a Taylor series, and dropping higher order terms gives us

$$\theta'/\theta_0 = T'/T_0 - (R_a/c_{pa})(p'/p_0). \tag{2.78}$$

Substitution of Eq. (2.78) into Eq. (2.76) gives an alternate form of the equation of state in terms of θ

$$\alpha'_m/\alpha_0 = \theta'/\theta_0 + 0.61r'_v - (c_{va}/c_{pa})(p'/p_0). \tag{2.79}$$

It should be noted that we have used the relation $c_{pa} = c_{va} + R_a$.

We now wish to develop an approximate form of the equations of motion.

2.3.2 *Equations of Motion*

We begin by multiplying both sides of Eq. (2.17) by α_m, giving us

$$\frac{du_i}{dt} = \left(-\alpha_m\frac{\partial p}{\partial x_i} - g\delta_{i3}\right) - \alpha_m\rho_w g\delta_{i3} + \varepsilon_{ijk}u_j f_k + f_r. \tag{2.80}$$

Consider now the terms in parentheses in the vertical component of Eq. (2.80). Expanding about the reference state gives

$$-\alpha_m\frac{\partial p}{\partial x_3} - g = -\alpha_0\left(1+\frac{\alpha'_m}{\alpha_0}\right)\frac{\partial p_0}{\partial x_3} - \alpha_0\left(1+\frac{\alpha'_m}{\alpha_0}\right)\frac{\partial p'}{\partial x_3} - g. \tag{2.81}$$

We require that

$$|\alpha'_{\mathrm{m}}/\alpha_0| \ll 1. \tag{2.82}$$

Also, the reference state is in hydrostatic equilibrium with no horizontal pressure gradient such that

$$\alpha_0 \frac{\partial p_0}{\partial x_3} = -g \qquad \text{and} \qquad \frac{\partial p_0}{\partial x_1} = \frac{\partial p_0}{\partial x_2} = 0. \tag{2.83}$$

Substitution of Eqs. (2.83) and (2.82) into Eq. (2.81) gives

$$-\alpha_{\mathrm{m}} \frac{\partial p}{\partial x_3} - g = -\alpha_0 \frac{\partial p'}{\partial x_3} + g \frac{\alpha'_{\mathrm{m}}}{\alpha_0}. \tag{2.84}$$

The third term on the RHS of Eq. (2.80) can be written as

$$-\alpha_{\mathrm{m}} \rho_{\mathrm{w}} g \delta_{i3} = -\alpha_0 \left(1 + \frac{\alpha'_{\mathrm{m}}}{\alpha_0}\right) \rho'_{\mathrm{w}} g \delta_{i3} \simeq -r'_{\mathrm{w}} g \delta_{i3}, \tag{2.85}$$

where we note that $r_{\mathrm{w}} = r'_{\mathrm{w}}$ and $r'_{\mathrm{w}} = \rho'_{\mathrm{w}}/\rho_0$.

Substituting Eqs. (2.84) and (2.85) into Eq. (2.80), we find

$$\frac{du_i}{dt} = -\alpha_0 \frac{\partial p'}{\partial x_i} + \left(\frac{\alpha'_{\mathrm{m}}}{\alpha_0} - r'_{\mathrm{w}}\right) g \delta_{i3} + \varepsilon_{ijk} u_j f_k + f_{\mathrm{r}}. \tag{2.86}$$

Equation (2.86) is now linear in the thermodynamic variables. This form of the equations of motion expresses the essence of the Boussinesq approximation; that is, variations in density or specific volume are ignored except when multiplied by gravity. Substitution of Eq. (2.79) into Eq. (2.86) and expansion of the substantial derivative gives

$$\frac{\partial u_i}{\partial t} + u_j \frac{\partial u_i}{\partial x_j} = -\alpha_0 \frac{\partial p'}{\partial x_i} + \left(\frac{\theta'}{\theta_0} + 0.61 r'_{\mathrm{v}} - \frac{c_{\mathrm{va}}}{c_{\mathrm{pa}}} \frac{p'}{p_0} - r'_{\mathrm{w}}\right) g \delta_{i3}$$
$$+ \varepsilon_{ijk} u_j f_k + f_{\mathrm{r}}. \tag{2.87}$$

The second term on the RHS of Eq. (2.87) is the so-called buoyancy term which is affected by perturbations of θ, r_{v}, p and by the presence of condensate r_{w}.

The equations of motion are often expressed in terms of the so-called Exner function defined as

$$\pi = c_{\mathrm{pa}}(p/p_{00})^k, \tag{2.88}$$

where p_{00} is an arbitrary reference pressure often taken to be 1000 mbar and $k = R_{\mathrm{a}}/c_{\mathrm{pa}}$.

Consider now the terms in parentheses in Eq. (2.80). From Eq. (2.88) we note that

$$\frac{\partial \pi}{\partial x_i} = \frac{c_{pa}}{p_{00}} k \left(\frac{p}{p_{00}}\right)^{k-1} \frac{\partial p}{\partial x_i} = c_{pa} k \left(\frac{p}{p_{00}}\right)^{k} \frac{1}{p} \frac{\partial p}{\partial x_i}. \tag{2.89}$$

Since $k = R_a/c_{pa}$, Eq. (2.89) becomes

$$\partial \pi/\partial x_i = R_a(p/p_{00})^k (1/p)(\partial p/\partial x_i). \tag{2.90}$$

Poisson's equation gives us

$$(p/p_{00})^k = T/\theta, \tag{2.91}$$

thus

$$\partial \pi/\partial x_i = (R_a T/\theta)(1/p)(\partial p/\partial x_i). \tag{2.92}$$

Multiplication of both sides of Eq. (2.92) by $\theta(1+0.61 r_v)$ gives us

$$\theta_v(\partial \pi/\partial x_i) = (R_a T_v/p)(\partial p/\partial x_i), \tag{2.93}$$

where T_v is defined by Eq. (2.14) and

$$\theta_v = \theta(1+0.61 r_v). \tag{2.94}$$

Substitution of Eq. (2.15) in the RHS of Eq. (2.93) gives us

$$\theta_v(\partial \pi/\partial x_i) = (1/\rho_m)(\partial p/\partial x_i). \tag{2.95}$$

Thus the terms in parentheses in Eq. (2.80) can be written as

$$[-\alpha_m(\partial p/\partial x_i) - g\delta_{i3}] = [-\theta_v(\partial \pi/\partial x_i) - g\delta_{i3}]. \tag{2.96}$$

Following a procedure similar to the above, the terms in parentheses can be linearized. For example, assuming a hydrostatic reference state, we have

$$du_3/dt = 0 = -\theta_0(\partial \pi_0/\partial x_3) - g,$$

or the hydrostatic relation in the π system is

$$\partial \pi_0/\partial x_3 = -g/\theta_0. \tag{2.97}$$

Furthermore, we may expand θ_v and π as follows

$$\theta_v = \theta_0 + \theta_v'; \qquad \pi = \pi_0 + \pi'. \tag{2.98}$$

Substitution of Eq. (2.98) into Eq. (2.80) gives us the equations of motion in terms of the Exner function, linearized in the thermodynamic variables

$$\frac{\partial u_i}{\partial t} + u_j \frac{\partial u_i}{\partial x_j} = -\theta_0 \frac{\partial \pi'}{\partial x_i} + \left[\frac{\theta_v'}{\theta_0} - r_w'\right] g\delta_{i3} + \varepsilon_{ijk} u_j f_k + f_r. \tag{2.99}$$

Comparison of Eqs. (2.99) and (2.87) shows that the main advantage of the π' system is that pressure anomalies do not appear in the buoyancy term.

2.3.3 *The Continuity Equation for Dry Air*

A consistent method of estimating the magnitude of the various terms in the relevant equations is given by Dutton and Fichtl (1969). All variables are represented by the Fourier components.

$$\phi(x, y, z, t) = \hat{\phi}(\omega, k_1, k_2, z)\, e^{i(\omega t + k_1 x_1 + k_2 x_2)}, \qquad (2.100)$$

where ω is the frequency of modulation of a wave and k_1 and k_2 are its wave numbers in the x_1 and x_2 directions, respectively. The quantity $\hat{\phi}$ represents the Fourier coefficient of a wave vector of "length" $L = 2\pi/|k|$ and frequency ω, where $k = (k_1^2 + k_2^2)^{1/2}$. The order of magnitude of $\hat{\phi}$ will be denoted by $|\hat{\phi}|_\mathrm{M}$. We can express the magnitude of the derivative of ϕ by employing the scale L_ϕ defined as

$$L_\phi \left|\frac{\partial \hat{\phi}}{\partial z}\right|_\mathrm{M} = |\hat{\phi}|_\mathrm{M} \qquad \text{or} \qquad \frac{1}{L_\phi} = \frac{1}{|\hat{\phi}|_\mathrm{M}} \left|\frac{\partial \hat{\phi}}{\partial z}\right|_\mathrm{M}. \qquad (2.101)$$

Consider the magnitude of the local component of the vertical velocity acceleration. We employ the linearized form of Eq. (2.86) with viscosity, water drag, and Coriolis terms ignored.

Substitute

$$\begin{aligned} u_3 &= \hat{u}_3\, e^{i(\omega t + k_1 x_1 + k_2 x_2)}, \\ p' &= \hat{p}'\, e^{i(\omega t + k_1 x_1 + k_2 x_2)}, \\ \alpha'_\mathrm{m} &= \hat{\alpha}'_\mathrm{m}\, e^{i(\omega t + k_1 x_1 + k_2 x_2)}, \end{aligned} \qquad (2.102)$$

into

$$\partial u_3/\partial t = -\alpha_0(\partial p'/\partial x_3) + (\alpha'_\mathrm{m}/\alpha_0)g, \qquad (2.103)$$

giving

$$\partial u_3/\partial t = u_3 i\omega = -\alpha_0(\partial p'/\partial x_3) + (\alpha'_\mathrm{m}/\alpha_0)g, \qquad (2.104)$$

and thus

$$i\omega\hat{u}_3 = -\alpha_0(\partial\hat{p}'/\partial x_3) + (\hat{\alpha}'_\mathrm{m}/\alpha_0)g. \qquad (2.105)$$

For convection, buoyancy is the principal driving force, thus

$$|g(\hat{\alpha}'_\mathrm{m}/\alpha_0)|_\mathrm{M} \simeq |\omega\hat{u}_3|_\mathrm{M}. \qquad (2.106)$$

Pressure-gradient forces are also known to play an important role in cloud-scale accelerations (Schlesinger, 1975; Orville and Kopp, 1977; Klemp and Wilhelmson, 1978; Cotton and Tripoli, 1978; Tripoli and Cotton, 1980).

Therefore,

$$|g(\hat{\alpha}'_{\text{m}}/\alpha_0)|_{\text{M}} \simeq |\alpha_0(\partial\hat{p}'/\partial x_3)|_{\text{M}}. \tag{2.107}$$

Vertical accelerations are controlled by both buoyancy and pressure-gradient terms, and Dutton and Fichtl (1969) have argued that they tend to be out of phase. For instance, at the top of a cloud the buoyancy tends toward zero and $|\alpha_0(\partial\hat{p}'/\partial x_3)|_{\text{M}}$ can be large and opposing buoyancy.

Equations (2.104) and (2.105) form the basis for further analysis. Let us consider the equation of continuity, Eq. (2.18), rewritten in the form

$$d\alpha_{\text{a}}/dt = \alpha_{\text{a}}(\partial u_i/\partial x_i). \tag{2.108}$$

Let $\alpha_{\text{a}} = \alpha_0 + a'_{\text{a}}$, then

$$\frac{d\alpha_{\text{a}}}{dt} = \frac{\partial\alpha_{\text{a}}}{\partial t} + u_i\frac{\partial\alpha_{\text{a}}}{\partial x_{i(i=1,2)}} + u_3\frac{\partial\alpha_{\text{a}}}{\partial x_3} = \alpha_{\text{a}}\frac{\partial u_i}{\partial x_{i(i=1,2)}} + \alpha_{\text{a}}\frac{\partial u_3}{\partial x_3}. \tag{2.109}$$

It is assumed that the basic state is horizontally homogeneous and steady; therefore

$$\frac{\partial\alpha_0}{\partial t} = 0; \qquad u_i\frac{\partial\alpha_0}{\partial x_{i(i=1,2)}} = 0. \tag{2.110}$$

It is also assumed that $|\alpha'_{\text{a}}/\alpha_0| \ll 1$. Equation (2.109) then becomes

$$\underset{\text{(a)}}{\frac{\partial\alpha'_{\text{a}}}{\partial t}} + \underset{\text{(b)}}{u_i\frac{\partial\alpha'_{\text{a}}}{\partial x_{i(i=1,2)}}} + \underset{\text{(c)}}{u_3\frac{\partial\alpha'_{\text{a}}}{\partial x_3}} + \underset{\text{(d)}}{u_3\frac{\partial\alpha_0}{\partial x_3}}$$

$$= \underset{\text{(e)}}{\alpha_0\left(1+\frac{\alpha'_{\text{a}}}{\alpha_0}\right)\frac{\partial u_i}{\partial x_{i(i=1,2)}}} + \underset{\text{(f)}}{\alpha_0\left(1+\frac{\alpha'_{\text{a}}}{\alpha_0}\right)\frac{\partial u_3}{\partial x_3}}. \tag{2.111}$$

We now define the density scale height H_α as

$$1/H_\alpha = (1/\alpha_0)(\partial\alpha_0/\partial x_3). \tag{2.112}$$

Applying the Fourier decomposition to Eq. (2.111) and dividing each term by the order of magnitude of term (d) or $|\hat{u}_3(\partial\alpha_0/\partial x_3)|_{\text{M}}$ gives us the following relations.

By dividing term (a) by (d) and recalling Eq. (2.106), we have

$$\frac{|\partial\hat{\alpha}'_{\text{a}}/\partial t|_{\text{M}}}{|\hat{u}_3(\partial\alpha_0/\partial x_3)|_{\text{M}}} = \frac{|\omega\hat{\alpha}'_{\text{a}}|}{|\hat{u}_3\alpha_0/H_\alpha|} = \frac{H_\alpha\omega^2}{\omega\hat{u}_3}\left|\frac{\hat{\alpha}'_{\text{a}}}{\alpha_0}\right| - \frac{H_\alpha\omega^2}{g}. \tag{2.113}$$

Dividing term (b) by (d) gives

$$\frac{|\hat{u}_i(\partial\hat{\alpha}'_{\text{a}}/\partial x_{i(i=1,2)})|_{\text{M}}}{|\hat{u}_3(\partial\alpha_0/\partial x_3)|_{\text{M}}} = \frac{|\hat{v}_{\text{H}}k_i\hat{\alpha}'_{\text{a}}|}{|\hat{u}_3\alpha_0/H_\alpha|} = \left|\frac{\hat{v}_{\text{H}}}{\hat{u}_3}\right|\left[\frac{2\pi H_\alpha}{L_i}\right]\left|\frac{\hat{\alpha}'_{\text{a}}}{\alpha_0}\right|, \tag{2.114}$$

where we have substituted for any horizontal velocity v_{H}, $v_{\mathrm{H}} = \hat{v}_{\mathrm{H}}\, e^{i(\omega t + k_i x_i)}$. Dividing term (c) by (d) gives

$$\left|\frac{u_3(\partial\alpha'_{\mathrm{a}}/\partial x_3)|_{\mathrm{M}}}{\hat{u}_3(\partial\alpha_0/\partial x_3)|_{\mathrm{M}}}\right| = \frac{|\hat{u}_3\hat{\alpha}'_{\mathrm{a}}/H'_{\alpha_{\mathrm{a}}}|}{|\hat{u}_3\alpha_0/H_\alpha|} = \frac{H'_\alpha}{H_{\alpha_{\mathrm{a}}}}\left|\frac{\hat{\alpha}'_{\mathrm{a}}}{\alpha_0}\right|. \tag{2.115}$$

Dividing term (e) by (d) gives

$$\frac{|\alpha_0(\partial u_i/\partial x_{i(i=1,2)})|_{\mathrm{M}}}{|\hat{u}_3(\partial\alpha_0/\partial x_3)|_{\mathrm{M}}} = \frac{|\alpha_0\hat{v}_{\mathrm{H}}k_i|}{|\hat{u}_3\alpha_0/H_\alpha|} = \frac{2\pi}{L_i}H_\alpha\left|\frac{\hat{v}_{\mathrm{H}}}{\hat{u}_3}\right|. \tag{2.116}$$

Dividing term (f) by (d) gives

$$\frac{|\alpha_0(\partial u_3/\partial x_3)|_{\mathrm{M}}}{|\hat{u}_3(\partial\alpha_0/\partial x_3)|_{\mathrm{M}}} = \frac{|\alpha_0\hat{u}_3/L_{u_3}|}{|\hat{u}_3\alpha_0/H_\alpha|} = \frac{H_\alpha}{L_{u_3}}. \tag{2.117}$$

We are now in a position to evaluate the relative importance of the various terms in Eq. (2.111). First of all, we will restrict the system to low frequencies, thus excluding sound waves. This leads to $|\omega^2| \ll g/H_\alpha$.

From Eq. (2.113) we see that term (a) is negligible. In general, we find that $|\hat{v}_{\mathrm{H}}/\hat{u}_3| \leq L_i/2\pi H_\alpha$ and since $|\hat{\alpha}'_{\mathrm{a}}/\alpha_0| \ll 1$, by Eq. (2.114) we find that term (b) is negligible.

For deep atmospheric motions (i.e., deep convection), $H_\alpha \sim H_{\alpha_{\mathrm{a}}}$, and $|\hat{\alpha}'_{\mathrm{a}}/\alpha_0| \ll 1$; thus from Eq. (2.115) we find that term (c) is small. Also for deep convection, $L_{u_3} \sim H_\alpha$ and $L_{u_3} \sim L_i$, showing us from Eqs. (2.116) and (2.117) that terms (d) and (e) are important.

The analysis shows us that an appropriate continuity equation for *deep convection* is

$$u_3(\partial\alpha_0/\partial x_3) = \alpha_0(\partial u_i/\partial x_i), \tag{2.118}$$

or

$$(\partial/\partial x_i)[(1/\alpha_0)u_i] = (\partial/\partial x_i)(\rho_0 u_i) = 0. \tag{2.119}$$

Equation (2.119) is also known as the *anelastic* continuity equation, because while it allows for variation in density of the basic state, it does not contain modulations in α'_{a}. The system Eq. (2.119), along with Eq. (2.88) and a pressure diagnostic equation, is soundproof.

Dutton and Fichtl (1969) extended the analysis to shallow atmospheric motions wherein $L_{u_3} \ll H_\alpha$. In this case an appropriate approximate continuity equation is

$$\partial u_i/\partial x_i = 0. \tag{2.120}$$

Equation (2.120) is known as the *incompressible* or *shallow form* of the continuity equation. Both Eqs. (2.120) and (2.119) make the system sound filtered. It should also be noted that turbulent motions are, in general, shallow (i.e., $L_{u_3} \ll H_\alpha$); therefore Eq. (2.120) is appropriate for turbulence modeling.

2.3.4 Approximations to the Buoyancy Term

We have earlier argued that, in moist convection, pressure-gradient and buoyancy forces are of the same order of magnitude

$$|g(\alpha'_m/\alpha_0)|_M \simeq |\alpha_0(\partial p'/\partial x_3)|_M.$$

This implies that

$$|\hat{\alpha}'_m/\alpha_0| \simeq (\alpha_0/g)|\hat{p}'/H_{p'}| \tag{2.121}$$

or

$$H_{p'}|\alpha'_m/\alpha_0|_M \sim (\alpha_0 p_0/g)|p'/p_0|_M. \tag{2.122}$$

The hydrostatic relationship gives us $|\partial p_0/\partial x_3|_M = |\rho_0 g|_M$ or $p_0/H_i = \rho_0 g$, thus $\alpha_0 p_0/g = H_i$. We note that $H_i \sim H_\alpha$ and $H_{p'} \sim L_{u_3}$. Thus Eq. (2.122) becomes

$$|\hat{\alpha}'_m/\alpha_0| \sim (H_\alpha/L_{u_3})|\hat{p}'/p_0| \tag{2.123}$$

or

$$|\hat{p}'/p_0| \sim |\hat{\alpha}'_m/\alpha_0|(L_{u_3}/H_\alpha). \tag{2.124}$$

For shallow convection, $L_{u_3} \ll H_\alpha$; thus we may neglect (p'/p_0) in the linearized equations of state [Eq. (2.76) or (2.77)] and in the buoyancy term of Eq. (2.87). For deep convection, $L_{u_3} \sim H_\alpha$ and the pressure perturbation term must be retained in these equations.

2.3.5 Summary

Based on the results of this analysis, we can now write a set of equations suitable for modeling deep or shallow moist convection or convective systems. They are as follows:

2.3.5.a Equation of State

$$\alpha'_m/\alpha_0 = \theta'/\theta_0 + 0.61 r'_v - (c_{va}/c_{pa})(p'/p_0)\delta_s, \tag{2.125}$$

where $\delta_s = 1$ for deep convection and $\delta_s = 0$ otherwise.

2.3.5.b Equations of Motion

$$\frac{du_i}{dt} = -\alpha_0 \frac{\partial p'}{\partial x_i} + \left(\frac{\alpha'_m}{\alpha_0} - r'_w\right) g\delta_{i3} + \varepsilon_{ijk} u_j f_k + f_r. \tag{2.126}$$

(Note that to be strictly consistent, the molecular viscosity term, $\mathbf{f}_r$, defined in Eq. (2.17), should include the reference state kinematic viscosity $\gamma_0 = \mu/\rho_0$.)

2.3.5.c Thermodynamic Energy Equation

From Eq. (2.40) we can write

$$d\theta_{il}/dt = p(\theta_{il}), \tag{2.127}$$

where $p(\theta_{il})$ represents the influence of precipitation fallout on θ_{il} and must be modeled or parameterized.

2.3.5.d The Mass Continuity Equation for Dry Air

Equations (2.119) and (2.120) give us

$$\partial u_i/\partial x_i + (u_i/\rho_0)(\partial \rho_0/\partial x_i)\delta_s = 0. \tag{2.128}$$

2.3.5.e The Continuity Equation for Total Water

Equation (2.20) gives us

$$dr_T/dt = p(r_T), \tag{2.129}$$

where $p(r_T)$ represents the effects of precipitation fallout on r_T; and

$$r_T = r_v + r_c + r_r + r_i + r_g + r_a, \tag{2.130}$$

where r_r, r_g, and r_a represent the mixing ratios of raindrops, graupel, and aggregates of snowflakes, respectively. Thus, assuming a cloud does not become supersaturated with respect to water,

$$r_c = (r_T - r_s - r_r - r_i - r_g - r_a)H(r_T - r_s - r_r - r_i - r_g - r_a). \tag{2.131}$$

Similar continuity equations for r_r, r_i, r_g, and r_a can also be written. These continuity equations must include models of the rates of transfer of water mass from r_c to r_r, r_r to r_g, and r_v or r_c to r_i, and r_v or r_c or r_i to r_a. We shall examine models of the transfer rates from one water category to another more fully in Chapter 4.

The system of equations, Eqs. (2.125)-(2.131), along with continuity equations for r_r, r_i, r_g, and r_a, form the basis for a model of a cloud or a mesoscale convective system. The system of equations is not closed, however, since we have not defined a means of predicting p'.

2.3.6 Evaluation of Pressure

Before attempting to integrate the set of equations [Eqs. (2.125)-(2.131)], one must evaluate the pressure field. In synoptic-scale meteorological systems, one typically evaluates the pressure field hydrostatically (i.e., by vertically integrating the mass field). In cumulus clouds and thunderstorm-scale systems, the vertical accelerations are so great that the pressure field must be evaluated nonhydrostatically. There are a number of mesoscale systems, however, such as mountain–valley circulations, mountain-wave clouds, and squall-line mesoscale systems, where it is not obvious whether pressure in such systems can be evaluated hydrostatically.

To examine the consequences of making the hydrostatic assumption, consider the dispersion diagram for waves in the atmosphere shown in Fig. 2.3. Plotted in the diagrams is the frequency of wave ω against horizontal wave number k. Tapp and White (1976) schematically showed the dispersion diagrams for a nonhydrostatic atmosphere when the stratification is (a) neutral, (b) uniform with a constant Brunt–Väisälä frequency $N = (g\theta^{-1}\,\partial\theta/\partial z)^{1/2}$, and (c) a uniformly stratified hydrostatic atmosphere. As shown, sound waves or acoustic waves exhibit a low frequency cutoff μ in a neutrally stratified atmosphere and $N_1 = (\mu^2 + N^2)^{1/2}$ in a stratified atmosphere. The hydrostatic assumption eliminates vertically propagating acoustic waves, but not Lamb waves. Lamb waves are horizontally propagating acoustic waves trapped near the earth's surface by unstable stratification and travel at the speed of sound.

Gravity waves are absent in a neutrally stratified atmosphere. In b and c, Fig. 2.3, it is shown that gravity waves propagate differently in hydrostatic and nonhydrostatic atmospheres. In a nonhydrostatic atmosphere there exists an upper bound N to the frequency of gravity waves, which does not exist in a hydrostatic atmosphere. As a consequence, the phase velocity (ω/k) and the group velocity $(\partial\omega/\partial k)$ decrease to zero as the wave number increases in a nonhydrostatic atmosphere, but they are dependent on wave number in the hydrostatic case. The differing dispersion properties of gravity

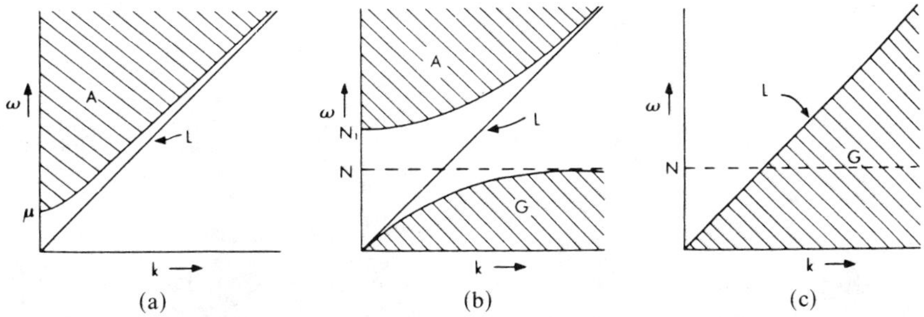

Fig. 2.3. Schematic dispersion diagrams for (a) a nonhydrostatic neutrally stratified atmosphere, (b) a nonhydrostatic stratified atmosphere, and (c) a hydrostatic stratified atmosphere. Shaded regions show where internal waves may occur: A, acoustic waves; G, gravity waves; L, Lamb waves. The effect of the earth's rotation is ignored. [From Tapp and White (1976).]

waves in the two atmospheres are particularly important when the lapse rate in the atmosphere varies with height in a nonuniform way. In a nonhydrostatic atmosphere (but not the hydrostatic), wave energy may become trapped in layers over certain ranges of wave number. This is responsible for the formation of "resonant" lee waves, which we shall describe more fully in Chapter 12. As discussed in Chapter 10, trapping of gravity-wave energy may also be responsible for shaping the particular character of some mesoscale convective systems and possibly the genesis of mesoscale convective complexes.

We shall now employ scale analysis to assist us in estimating the limits of the hydrostatic approximation.

Consider the vertical equation of motion, linearized in the thermodynamic variables, with the Coriolis, water load, and viscous terms ignored:

$$du_3/dt = -\alpha_0(\partial p'/\partial x_3) + (\alpha'_{\mathrm{m}}/\alpha_0)g. \tag{2.132}$$

Employing the Fourier decomposition analysis to the vertical equation of motion, we found previously [Eq. (2.105)] that

$$i\omega\hat{u}_3 = -\alpha_0(\partial\hat{p}'/\partial x_3) + g(\hat{\alpha}'_{\mathrm{m}}/\alpha_0). \tag{2.133}$$

We also noted in Eq. (2.107) that in convective disturbances,

$$|-\alpha_0(\partial\hat{p}'/\partial x_3)|_{\mathrm{M}} \simeq |g(\hat{\alpha}'_{\mathrm{m}}/\alpha_0)|_{\mathrm{M}}, \tag{2.134}$$

where

$$|-\alpha_0(\partial\hat{p}'/\partial x_3)|_{\mathrm{M}} \sim \alpha_0\hat{p}'/H_{p'}. \tag{2.135}$$

For a system to be in hydrostatic equilibrium, we require

$$|\omega\hat{u}_3| \ll \alpha_0\hat{p}'/H_{p'} \tag{2.136}$$

or

$$\frac{\omega\hat{u}_3}{\alpha_0\hat{p}'/H_{p'}} \ll 1. \tag{2.137}$$

We can now rewrite a horizontal component of the linearized equations of motion as

$$\partial u_1/\partial t = -\alpha_0(\partial p'/\partial x_1) + fu_2, \tag{2.138}$$

or, after applying the Fourier decomposition analysis, we have

$$|\omega\hat{u}_1|_{\mathrm{M}} = |-\alpha_0\hat{p}'/L_{p'}|_{\mathrm{M}} + |f\hat{u}_2|_{\mathrm{M}}. \tag{2.139}$$

We thus find from Eq. (2.139) that

$$|\hat{p}'| \sim |f\hat{u}_2|\rho_0 L_{p'}. \tag{2.140}$$

Substitution of Eq. (2.140) into Eq. (2.137) gives us

$$\frac{\omega \hat{u}_3}{\alpha_0 f \hat{u}_2 \rho_0 L_{p'}/H_{p'}} \ll 1, \tag{2.141}$$

and, after rearranging,

$$(\hat{u}_3/H_{p'})/(\hat{u}_2/L_{p'})(H_{p'}/L_{p'})^2(\omega/f) \ll 1. \tag{2.142}$$

If we concern ourselves with frequencies on the order of the Coriolis frequency (i.e., $\omega \sim f$), then sufficient conditions for justifying the hydrostatic approximation are

$$(\hat{u}_3/H_{p'})/(\hat{u}_2/L_{p'}) \leq 1 \qquad \text{and} \qquad (H_{p'}/L_{p'}) \ll 1. \tag{2.143}$$

We have noted previously that $H_{p'} \sim L_{u_3}$ and $L_{p'} \sim L_2 \sim L_1$. Thus the left-hand inequality is equivalent to

$$(\partial u_3/\partial x_3)/(\partial u_2/\partial x_2) \leq 1, \tag{2.144}$$

which is generally satisfied for approximately incompressible systems.

For synoptic-scale systems,

$$H_{p'}/L_{p'} \sim L_{u_3}/L_2 = 10/10^3 \text{ km} \ll 1. \tag{2.145}$$

Thus, the hydrostatic approximation is *quite valid.*

For cumulus clouds, however, $L_{u_3}/L_2 \sim 1$; thus, the hydrostatic approximation is *not valid.*

For shallow mesoscale systems such as sea-breeze disturbances

$$L_{u_3}/L_2 \sim 2/20 \text{ km} = 10^{-1}.$$

Strict justification for the hydrostatic approximation is not given, yet use of the hydrostatic approximation may not lead to serious errors.

It should be remembered that the hydrostatic approximation restricts the system to frequencies $\omega \sim f$, which removes vertically propagating acoustic waves from the system.

If we cannot justify the hydrostatic approximation for the system of interest, we must evaluate p' nonhydrostatically. One approach is to take the divergence of mass of Eq. (2.126), or of Eq. (2.99), where

$$\frac{\partial}{\partial x_i}\left[\frac{\partial(\rho_0 u_i)}{\partial t}\right] = \frac{\partial}{\partial t}\left[\frac{\partial(\rho_0 u_i)}{\partial x_i}\right] = 0, \tag{2.146}$$

and solve for $\partial^2 p'/\partial x_i^2$ or $\partial^2 \pi'/2x_i^2$, respectively. Pressure can then be evaluated by one of several techniques for inverting an elliptic equation (e.g., successive overrelaxation, matrix inversion, fast Fourier transform).

Another approach often used in two-dimensional cloud and mesoscale models is to eliminate pressure as an explicit variable in Eq. (2.126) by forming a horizontal vorticity equation. Consider, for example, the two-dimensional form of Eq. (2.126) in the (x_1, x_3) plane with Coriolis and viscous terms ignored.

Employing the continuity equation

$$\partial(\rho_0 u_1)/\partial x_1 + \partial(\rho_0 u_3)/\partial x_3 = 0, \tag{2.147}$$

we define a streamfunction ϕ such that

$$\rho_0 u_1 = -\partial\phi/\partial x_3; \qquad \rho_0 u_3 = \partial\phi/\partial x_1. \tag{2.148}$$

The component of density-weighted vorticity normal to the (x_1, x_3) plane is

$$\eta = -\nabla^2\phi = -(\partial^2\phi/\partial x_3^2 + \partial^2\phi/\partial x_1^2). \tag{2.149}$$

By ignoring p' in Eq. (2.125), one can derive a vorticity equation $\partial\eta/\partial t$ in which an explicit dependence on pressure is eliminated. An advantage of using the Exner form of the equations of motion [Eq. (2.99)] is that such an assumption is not necessary.

Several three-dimensional cloud models (Klemp and Wilhelmson, 1978; Cotton and Tripoli, 1978) have employed so-called time-splitting schemes originally developed by Marchuk (1974). The approach is simply to rewrite Eq. (2.87) or (2.99) in the form

$$\frac{\partial u_i}{\partial t} + \alpha_0 \frac{\partial p'}{\partial x_i} = -u_j \frac{\partial u_i}{\partial x_j} + \left(\frac{\alpha'_{\mathrm{m}}}{\alpha_0} - r_{\mathrm{w}}\right) g\delta_{i3} + \varepsilon_{ijk}\, u_j f_k + f_{\mathrm{r}} = Ru_i, \tag{2.150}$$

where the pressure gradient is on the LHS of Eq. (2.150) and the lower frequency advective time-scale terms are on the RHS. The approach is then either to form a pressure tendency equation using the continuity equation and a linearized form of the equation of state (see Klemp and Wilhelmson, 1978; or Cotton and Tripoli, 1978) or to integrate an approximate form of the continuity equation (Tripoli and Cotton, 1980). For example, consider the latter approach, in which we retain high-frequency fluctuations in specific volume α'_{a} in term (a) of Eq. (2.111). Based on our scaling analysis, terms (b) and (c) are negligible; therefore, an appropriate approximate continuity equation in which sound waves are not filtered out is

$$\partial\rho'_{\mathrm{a}}/\partial t - \partial(\rho_0 u_j)/\partial x_j = 0. \tag{2.151}$$

Linearization of the equation of state in the form of Eq. (2.4) and substitution of Poisson's equation gives us

$$(c_{\mathrm{va}}/c_{\mathrm{p}})(p'/p_0) = \rho'_{\mathrm{a}}/\rho_0 + \theta'/\theta_0 + 1.61 r_{\mathrm{v}}. \tag{2.152}$$

In this system the LHS of Eq. (2.150) can be integrated using a small time step in which acoustic waves are resolved. Equation (2.152) is then used to diagnose p'. The RHS of Eq. (2.150), along with the thermodynamic energy equation [Eq. (2.127)] and other scalar equations such as Eqs. (2.129)–(2.131), can be integrated on longer time steps that are controlled by advective time scales. Care must be taken in numerically integrating the system, Eqs. (2.150), (2.151), and (2.152), since a decoupling between the mass field, Eq. (2.151), and the thermodynamic field can occur. Tripoli and Cotton (1980) have assured such a coupling through their numerical procedure. Klemp and Wilhelmson (1978) and Cotton and Tripoli (1978) assured such a coupling by forming a pressure-tendency equation.

Yet another approach to simulating a nonhydrostatic system is to use an implicit numerical scheme on the terms which excite acoustic waves and use explicit numerical procedures on the remaining terms (Tapp and White, 1976). The use of an implicit scheme on the sound wave terms allows one to employ a time step that is considerably longer than one required in an explicit numerical scheme. One consequence of using an implicit numerical scheme with a longer time step on sound-wave-producing terms is that the phase speed of sound is artificially slowed down. The amount of "slow down" of the phase speed of sound is proportional to the amount that the time step is extended beyond that required for stability in an explicit scheme. Such an artificial retardation of the phase speed of sound, however, may not be very detrimental to the quality of the numerical solutions. Using the time-split compressible system discussed above, Droegemeier and Wilhelmson (1987) found that they could artificially reduce the phase speed of sound to 50 m s^{-1} without making any detectable difference in the dynamics of simulated thunderstorm outflows.

Cotton's group has likewise found that artificially reducing the phase speed of sound in a time-split system does not appreciably change the quality of solutions for either simulated large eddies in the atmospheric boundary layer or deep convective systems, but it does greatly increase the economy of the numerical calculations. The main advantage of artificially changing the phase speed of sound in a time-split system is that the user has direct control over the amount of retardation of the acoustic waves.

2.4 The Vertical Coordinate

Thus far we have discussed the equation set suitable for modeling clouds and mesoscale systems only in terms of the Cartesian vertical coordinate x_3, or height. It may be desirable, however, for certain applications to select a vertical coordinate that is either more consistent with the physics of the model or that allows a more accurate representation of terrain effects.

Miller (1974), for example, selected pressure as a "natural" vertical coordinate for a cumulonimbus model. He argued that a model with grid points at constant pressure levels allows accurate evaluation of T and θ without introducing computational problems. Several authors have noted that in Cartesian coordinates (Ogura and Phillips, 1962; List and Lozowski, 1970; Takeda, 1971; Wilhelmson and Ogura, 1972) an implicit relationship exists among the variables T, θ, p, and $r_s(T)$ that requires either some form of iterative solution or approximations in order to close the system. Miller pointed out that since p is known in the p-coordinate system, an explicit solution to T, θ, and $r_s(T)$ can be obtained without simplifying assumptions. However, as noted by Miller, the magnitudes of pressure anomalies predicted by cloud models are not sufficiently large to warrant serious concern about the effects of such simplifying assumptions.

While pressure as a vertical coordinate may have certain advantages in treating the thermodynamics of deep cloud systems, it nevertheless suffers from the disadvantages that terrain effects cannot be easily included in such a model. The standard Cartesian vertical coordinate suffers a similar disadvantage in mountainous terrain. Orville (1964, 1965) adapted a Cartesian coordinate model to a simple, two-dimensional linear ridge having a 45° slope. By selecting a ridge of simple geometry, he was able to avoid problems associated with boundary points which are not grid points in a rectangular finite-difference mesh. There have been attempts to apply a Cartesian coordinate to three-dimensional simulations over arbitrary terrain, but the techniques involved have either required severe constraints on terrain shape (Hirt and Cook, 1972) or involved complicated programming procedures having numerous decision processes that are not efficient on modern vector-based computers (Viecelli, 1971).

An alternate approach is to select a vertical coordinate transformation in which the lowest level maps the desired terrain features (the so-called σ coordinate). Phillips (1957) designed such a terrain-following coordinate system for use in hydrostatic, numerical prediction models. He defined a vertical coordinate σ as

$$\sigma = p/p_s, \tag{2.153}$$

where $p_s(x_1, x_2, t)$ represents the surface pressure. Thus, σ varies from a value $\sigma = 1$ at the ground to $\sigma = 0$ at the top of the atmosphere. Also, the lower boundary condition is

$$\dot{\sigma} = 0 \qquad \text{at} \quad \sigma = 1.$$

Utilizing the hydrostatic approximation, one can apply the chain rule separately in the vertical and horizontal dimensions to transform the system of equations from a Cartesian framework to the new σ system (Kasahara,

1974). Thus for any variable ϕ, the horizontal gradient along a σ surface is defined as

$$(\partial\phi/\partial x)_\sigma = (\partial\phi/\partial x)_p + (\partial\sigma/\partial p_s)(\partial p_s/\partial x)(\partial\phi/\partial\sigma). \qquad (2.154)$$

The above equation shows that in order to complete the transformation, the surface pressure must be calculated. The surface pressure can be obtained by integrating a surface pressure-tendency equation that involves the vertical integration of the continuity equation in σ coordinates (Holton, 1972, p. 153). The σ system derived by Phillips is valid only for hydrostatic systems. However, it is possible to employ approximations similar to those used by Miller (1974) to formulate a σ system that is valid for nonhydrostatic models (Miller and Moncrieff, 1983; Miller and White, 1984). In recent years, however, it has become more popular to use a so-called σ–z coordinate system in nonhydrostatic models.

Gal-Chen and Somerville (1975) formulated the equations of motion in a generalized non-Cartesian, nonorthogonal coordinate system. From this generalized system they derived a particular coordinate transformation that has the following properties:

(i) The transformed domain is a rectangular domain.

(ii) The transformation is reversible allowing a one-to-one relationship between the old and transformed coordinates.

(iii) The transformation, if the terrain is flat, reduces to the original Cartesian coordinates.

(iv) The top vertical level of the transformed coordinate becomes identical with the upper boundary at $z = H$.

(v) The transformation is continuous up to second derivatives.

The fifth property implies that the magnitude of the second derivative to topographic height must be no greater than that of the first derivative. The particular coordinate transformation they selected, which has now been used extensively by Clark (1977), Tripoli and Cotton (1982), Durran (1981) and Mahrer and Pielke (1975), is

$$x^* = x,$$

$$y^* = y,$$

and

$$z^* = [(z - z_s)/(H - z_s)]H, \qquad (2.155)$$

where quantities with an asterisk represent the transformed coordinates and those without an asterisk are Cartesian coordinates. The surface height above some reference level, usually taken to be sea level, is given by z_s, and the height of the model top at which the z^* coordinate becomes horizontal is given by H. This particular transformation has the virtue that the horizontal coordinates are identical to Cartesian coordinates and that the vertical coordinate ranges from $z^* = 0$ along the topographic surface to $z^* = H$ at the top of the model domain.

Gal-Chen and Somerville (1975) showed that the Jacobian of the transformation is defined as

$$a(x^*, y^*) = \partial z/\partial z^* = 1 - z_s(x^*, y^*)/H, \tag{2.156}$$

which leads to the following transformations of the spatial derivatives of some quantity A

$$\partial A/\partial x_i = (1/a)(\partial a b^{ij} A/\partial x_{j^*}), \tag{2.157}$$

where the tensor b^{ij} is defined as

$$b^{ij} = \begin{bmatrix} 1 & 0 & (1/a)(\partial z_s/\partial x)(z^*/H - 1) \\ 0 & 1 & (1/a)(\partial z_s/\partial y)(z^*/H - 1) \\ 0 & 0 & 1/a \end{bmatrix}. \tag{2.158}$$

The transformed velocity components then become

$$\begin{aligned} u^* &= u, \\ v^* &= v, \\ w^* &= (uab^{13} + vab^{23} + w)1/a. \end{aligned} \tag{2.159}$$

If we ignore viscous terms for the sake of simplicity, the transformed equations of motion given in Eq. (2.12) become

$$\frac{\partial u_i}{\partial t} = \mathrm{ADV}(u_i) - \frac{\alpha_0}{a}\frac{\partial a b^{ij} p'}{\partial x_{j^*}} + \left(\frac{\alpha'_{\mathrm{m}}}{\alpha_0} - r'_{\mathrm{w}}\right) g\delta_{i3} + \varepsilon_{ijk} f_k u_j, \tag{2.160}$$

where the advective operator ADV is defined

$$\mathrm{ADV}(u_i) = \frac{1}{a\rho_0}\frac{\partial(a b^{jk} \rho_0 u_j u_i)}{\partial x_{k^*}}. \tag{2.161}$$

Following Clark (1977), the advective operator has been cast in flux form. This differs from the approach taken by Gal-Chen and Sommerville (1975), in which the advective operator contains the transformed velocity w^*. Clark pointed out that this results in Christoffel symbols of the second kind in the w^* equation, which makes it very difficult to conserve momentum or kinetic energy in a numerical model.

One must be careful in adopting such a coordinate transformation, especially if the model is hydrostatic. Pielke and Martin (1981) have pointed out that if one applies the hydrostatic approximation to the equations of motion before performing a generalized vertical coordinate transformation, one obtains a different form of the equations than by applying the hydrostatic approximation following a transformation. In the latter case, additional terms appear in the equations that can only be neglected if the terrain slope is much less than 45°. They also noted that applying the hydrostatic approximation before transformation can lead to errors in the evaluation of kinetic energy.

The system of equations that we have developed thus far is nonlinear and time dependent. In addition, the system applies to infinitely small parcels of air rather than properties averaged over several hundred meters, which is a typical grid size in a numerical model of convection. Thus, further modification to the system of equations must be accomplished before some form of integrated solution can be obtained. In particular, we find that it is necessary to average the equation set.

References

Betts, A. K. (1973). Nonprecipitating cumulus convection and its parameterization. *Q. J. R. Meteorol. Soc.* **99**, 178–196.

Clark, Terry L. (1977). A small-scale dynamic model using a terrain-following coordinate transformation. *J. Comput. Phys.* **24**, 186–215.

Cotton, W. R., and G. J. Tripoli (1978). Cumulus convection in shear flow three-dimensional numerical experiments. *J. Atmos. Sci.* **35**, 1503–1521.

Droegemeier, K. K., and R. B. Wilhelmson (1987). Numerical simulation of thunderstorm outflow dynamics. Part I: Outflow sensitivity experiments and turbulence dynamics. *J. Atmos. Sci.* **44**, 1182–1210.

Durran, Dale Richard (1981). The effects of moisture on mountain lee waves. Coop. Thesis No. 65. Mass. Inst. Technol. and Nat. Cent. Atmos. Res. (NCAR-CT-65).

Dutton, J. A. (1976). "The Ceaseless Wind, An Introduction to the Theory of Atmospheric Motion." McGraw-Hill, New York.

Dutton, J. A., and G. H. Fichtl (1969). Approximate equations of motion for gases and liquids. *J. Atmos. Sci.* **26**, 241–254.

Gal-Chen, Tzvi, and R. C. J. Somerville (1975). On the use of a coordinate transformation for the solution of the Navier–Stokes equations. *J. Comput. Phys.* **17**, 209–228.

Hirt, C. W., and J. L. Cook (1972). Calculating three-dimensional flows around structure and over rough terrain. *J. Comput. Phys.* **10**, 324.

Holton, J. R. (1972). "An Introduction to Meteorology." Academic Press, New York.

Kasahara, A. (1974). Various vertical coordinate systems used for numerical weather prediction. *Mon. Weather Rev.* **102**, 509–522.

Klemp, J. B., and R. B. Wilhelmson (1978). Simulations of right- and left-moving storms produced through storm splitting. *J. Atmos. Sci.* **35**, 1097–1110.

List, R., and E. P. Lozowski (1970). Pressure perturbations and buoyancy in convective clouds. *J. Atmos. Sci.* **27**, 168–170.

Mahrer, Y., and R. A. Pielke (1975). A numerical study of the air flow over mountains using the two-dimensional version of the University of Virginia mesoscale model. *J. Atmos. Sci.* **32**, 2144–2155.

Marchuk, G. I. (1974). "Numerical Methods in Weather Prediction." Academic Press, New York.

Miller, M. J. (1974). On the use of pressure as vertical coordinate in modelling convection. *Q. J. R. Meteorol. Soc.* **100**, 155–162.

Miller, M. J., and M. W. Moncrieff (1983). Dynamics and simulation of organized deep convection. *Proc. NATO Adv. Study Inst. Mesoscale Meteorol.—Theories, Observations Models, Bonas, Fr., 1982* pp. 451–496. Dordrecht, Netherlands.

Miller, M. J., and A. A. White (1984). On the non-hydrostatic equations in pressure and sigma coordinates. *Quart. J. R. Meteorol. Soc.* **110**, 515–533.

Ogura, Y., and N. A. Phillips (1962). A scale analysis of deep and shallow convection in the atmosphere. *J. Atmos. Sci.* **19**, 173–179.

Orville, H. D. (1964). On mountain upslope winds. *J. Atmos. Sci.* **21**, 622–633.

Orville, H. D. (1965). A numerical study of the initiation of cumulus clouds over mountainous terrain. *J. Atmos. Sci.* **22**, 684–699.

Orville, H. D., and F. J. Kopp (1977). Numerical simulation of the life history of a hailstorm. *J. Atmos. Sci.* **34**, 1596–1618.

Phillips, N. A. (1957). A coordinate system having some special advantages for numerical forecasting. *J. Meteorol.* **14**, 184–185.

Pielke, Roger A., and Charles L. Martin (1981). The derivation of a terrain-following coordinate system for use in a hydrostatic model. *J. Atmos. Sci.* **38**, 1707–1713.

Schlesinger, R. E. (1975). A three-dimensional numerical model of an isolated deep convective cloud: Preliminary results. *J. Atmos. Sci.* **32**, 934.

Takeda, T. (1971). Numerical simulation of a precipitating convective cloud: The formation of a long-lasting cloud. *J. Atmos. Sci.* **28**, 350–376.

Tapp, M. C., and P. W. White (1976). A nonhydrostatic mesoscale model. *Q. J. R. Meteorol. Soc.* **102**, 277–296.

Tripoli, G. J., and W. R. Cotton (1980). A numerical investigation of several factors contributing to the observed variable intensity of deep convection over South Florida. *J. Appl. Meteorol.* **19**, 1037–1063.

Tripoli, G. J., and W. R. Cotton (1981). The use of ice-liquid water potential temperature as a thermodynamic variable in deep atmospheric models. *Mon. Weather Rev.* **109**, 1094–1102.

Tripoli, G. J., and W. R. Cotton (1982). The Colorado State University three-dimensional cloud/mesoscale model—1982. Part I: General theoretical framework and sensitivity experiments. *J. Rech. Atmos.* **16**, 185–220.

Viecelli, J. A. (1971). A computing method for incompressible flows bounded by moving walls. *J. Comput. Phys.* **8**, 119–143.

Wilhelmson, R. B. (1977). On the thermodynamic equation for deep convection. *Mon. Weather Rev.* **105**, 545–549.

Wilhelmson, R., and Y. Ogura (1972). The pressure perturbation and the numerical modelling of a cloud. *J. Atmos. Sci.* **29**, 1295–1307.

Chapter 3 On Averaging

3.1 Introduction

The nonlinear equations that have been developed in Chapter 2 contain information concerning atmospheric motion and transport over a broad range of spatial scales, ranging from that of the largest eddies on the globe to the smallest eddies contributing to molecular dissipation. At the present time there is no known mathematical technique for exactly integrating this set of equations. Moreover, feasible observational systems are incapable of resolving or defining all scales of motion in the atmosphere. Thus, meteorologists have been forced to distinguish between those eddies that are in a sense "resolvable" either by our observation systems or by some form of finite-difference representation of the atmosphere (here we include finite-element and truncated spectral representations of the atmosphere), and the remaining eddies that are not fully resolved either observationally or computationally. Such unresolved eddies we define as "turbulence."

Unfortunately, we must be content with describing or predicting only the statistical properties of the turbulent eddies. Otherwise a complete, unique description or prediction of their characteristics brings them back into the perspective of "resolvable" eddies. Whenever we deal with the statistical properties of a system, we must formally introduce an averaging operator. As it turns out, meteorologists have not agreed upon a single averaging operator that can be generally applied to all forms of meteorological modeling and observation.

Before we survey the various averaging operators used in meteorology, we list a set of criteria that any averaging operator should satisfy:

(i) The operator should provide a formal mechanism for distinguishing between "resolvable" and "unresolvable" eddies.

(ii) The operator should produce a set of equations that is more amenable to integration (either analytically or numerically) than the unaveraged system of equations.

(iii) The averaged set of atmospheric variables should be capable of being measured by current or anticipated atmospheric sensing systems.

In the following sections, we shall examine the proposed averaging operators and evaluate their potential for satisfying the above criteria.

Anthes (1977) has discussed several methods of averaging a set of equations describing a large-scale flow field with imbedded small-scale clouds. The first method utilizes the classical Reynolds averaging method by considering the convective clouds as eddies superimposed on a large-scale flow. The second method divides an area into a mean cloud and environment region and obtains the effect of cumulus clouds on this area by considering the equations for the two regions separately. The latter method is most suitable for large-scale models or diagnostic studies, but as the averaging domain is decreased in size such that clouds comprise 50 to 100% coverage over the area, the Reynolds averaging method becomes more suitable. When employing the Reynolds averaging method, several different averaging operators are possible.

3.2 Ensemble Average

If the fluctuating field or turbulence field is relatively *stationary* in time, a suitable averaging operator for a variable ϕ measured by a sensor located at a fixed coordinate point (x_1, y_1, z_1) is the time average

$$\bar{\phi}_{\mathrm{t}} = \lim_{T \to \infty} \frac{1}{2T} \int_{-T}^{+T} \phi(x_1, y_1, z_1, t)\, dt. \tag{3.1}$$

Alternatively, if the turbulence is relatively *homogeneous* in space, one could sample a population of eddies at a given time t_1, perhaps by aircraft, giving us a spatial average

$$\bar{\phi}_{\mathrm{s}} = \lim_{X \to \infty} \frac{1}{2X} \int_{-X}^{+X} \phi(x, t_1)\, dx. \tag{3.2}$$

If we only sample the atmosphere discretely, rather than continuously, we can define an ensemble-averaging operator in which a set of discrete observations, k (realizations of a variable ϕ), is made under "superficially look-alike" conditions at a specific location and time, giving us the ensemble-averaging operator

$$\bar{\phi}_e = \lim_{N \to \infty} \frac{1}{N} \sum_{k=1}^{N} \phi_k(x_1, y_1, z_1, t_1). \tag{3.3}$$

In surface-layer applications superficially look-alike conditions are defined with respect to certain nondimensional similarity parameters such as z/L, where L represents the Monin–Obukhov length scale (Monin and Obukhov, 1954).

According to the ergodic hypothesis, if the turbulence is stationary and homogeneous, the three averaging processes will be equivalent, i.e.,

$$\bar{\phi}_t = \bar{\phi}_s = \bar{\phi}_e.$$

For our purposes, ϕ could represent the three wind components (u, v, w), temperature (T), pressure (P), total water mixing ratios (r_T), or other variables.

In practice, the flow is neither stationary nor homogeneous, so that we are forced to seek alternate averaging operators that are suitable for typical cloud and mesoscale fields. In the case of numerical weather prediction (NWP) models, it has become common to define an averaging operator that is related to the grid scale.

3.3 Grid-Volume Average

In large mesoscale models or regional models for which the mean flow is largely two-dimensional, the grid-volume average can be defined as simply a horizontal average. Anthes (1977) and others have defined it as

$$\bar{\phi}_A(x, y) = \frac{1}{\Delta x \Delta y} \int_{x-\Delta x/2}^{x+\Delta x/2} \int_{y-\Delta y/2}^{y+\Delta y/2} \phi(x', y')\, dy'\, dx', \tag{3.4}$$

where Δx and Δy are the averaging intervals in the x and y directions and are generally taken to be the mesh size in the model. Deardorff (1970) and others have generalized the operator to a three-dimensional grid volume, and have applied it to modeling a small-scale planetary boundary layer (PBL) flow. Deardorff's version of the operator is

$$\bar{\phi}_v(x, y, z) = \frac{1}{\Delta x\, \Delta y\, \Delta z} \int_{z-\Delta z/2}^{z+\Delta z/2} \int_{y-\Delta y/2}^{y+\Delta y/2} \int_{x-\Delta x/2}^{x+\Delta x/2} \phi(x', y', z')\, dx'\, dy'\, dz', \tag{3.5}$$

where the average $\bar{\phi}_v$ represents a "running mean" in space and varies continuously from point to point.

As long as the meteorological system and the spatial averaging scales, Δ, are judiciously selected such that the energy-containing eddies of the system have wavelengths considerably greater than Δ, and energy flows through Δ in a spectrally continuous (not sporadic) manner, the grid-volume average is a well-behaved function. However, if the meteorological systems of interest have significant energy in scales close to Δ (e.g., hurricanes, convective mesoscale systems with $\Delta \sim 50$ km), the grid scale is in the midst of the energy-containing scales. Cumulus clouds and cumulonimbi may be converting moist, static energy into kinetic energy on scales close to Δ. In such circumstances, $\bar{\phi}$ exhibits strong spatial variability on scales of Δ. In other words, $\bar{\phi}$ behaves as a turbulent fluctuating field with only the smaller scale turbulence fluctuations filtered by the averaging operator. This is true not only of the variables $\bar{\theta}$, $\bar{u}$, $\bar{v}$, $\bar{w}$, and $\bar{q}$, but also of their covariances $\overline{w'\theta'}$, $\overline{u'\theta'}$, and so on. In some instances the variability of the grid-volume-averaged fields is large enough to make it impractical to model such systems using finite-difference techniques, unless the grid size is reduced. The technique generally used in large-eddy simulation (LES) models is to choose a grid scale that lies in the inertial subrange, so that kinetic energy is produced on scales larger than Δ and is dissipated on scales smaller than Δ. In a spectral sense, kinetic energy cascades through the truncation scale. The most energetic of the unresolved eddies will then have scales close to Δ; however, the energy in these eddies will be small. [For a description of atmospheric power spectra and spectral analysis, see Panofsky and Dutton (1984).]

Because the grid-volume average is defined over a finite volume at an instant in time, much like a snapshot of a region of the atmosphere, it is not measurable in any practical sense. Since there are no known techniques for simultaneously measuring u, v, w, θ, r, etc., over a finite volume, models based on the grid-volume average cannot be conveniently tested.

3.4 The Generalized Ensemble Average

A compromise for the averaging operator is thus the *generalized ensemble average*, defined by

$$\bar{\phi}_E = \frac{1}{\tau L_x L_y L_z} \lim_{N \to \infty} \frac{1}{N} \sum_{k=1}^{N} \times \int_{z-L_z/2}^{z+L_z/2} \int_{y-L_y/2}^{y+L_y/2} \int_{x-L_x/2}^{x+L_x/2} \int_{t-\tau/2}^{\tau+\tau/2} \phi_k(x', y', z', t')\, dx'\, dy'\, dz'\, dt', \quad (3.6)$$

where L_x, L_y, L_z, and τ represent the length and time intervals of the running mean, or some other appropriate length/time scale, and need not be associated with the model-defined grid intervals Δt, Δx, etc. In fact, in the case where the averaged equations are integrated by finite-difference techniques, L_x, L_y, L_z, and τ should be selected such that $\bar{\phi}_{\mathrm{E}}$ varies smoothly in time and in space. Most numerical schemes better represent $\bar{\phi}_{\mathrm{E}}$ if the grid scales are defined as some fraction of L, (i.e., $\Delta x = 0.2L_x$, $\Delta t = 0.1\tau$, etc.).

As in Eq. (3.3), the index k represents a realization of a random process. Thus, in addition to being a space/time average, this operator represents the average of an ensemble of observations of "superficially identical" conditions occurring within the domain defined by L_x, L_y, L_z, and τ. The primary difficulty is to identify suitable "superficially identical" states of the atmosphere, such as similar Richardson numbers, lapse rates, convergence fields, and so forth.

The primary advantage of the generalized ensemble-averaging operator is that it filters out all turbulence. As we shall see in Section 3.6, this has definite implications for the Reynolds averaging process. This operator has several other advantages. Consider first the application to general circulation-scale systems. We define L_x, L_y in accordance with a latitudinal and longitudinal belt of interest. Perhaps we are interested in making a simulation, forecast, or analysis with time resolution $\tau \sim 1$ h. If we wish to determine the mean vertical moisture flux through a layer L_z (e.g., subcloud layer, cloud layer, upper troposphere) under conditions where a horizontally homogeneous cloud field exists over the region, only a few realizations (e.g., aircraft cross sections or acoustic and microwave radar observations) may be needed to obtain a meaningful estimate of $\overline{w'q'}$. Perhaps even $N = 1$ is sufficient, in which case the average is similar to the ensemble average over a cloud field, such as that used by Arakawa and Schubert (1974) in their cumulus parameterization theory. This is particularly true if τ is quite small. For both small τ and a relatively homogeneous mean field, we may take $L_x = \Delta_x$, $L_y = \Delta_y$. Then the operator converges to a grid-volume average.

Consider, however, a diurnally forced mesoscale system containing active cumulonimbi. Suppose we wish to forecast precipitation, for example, with a time resolution $\tau \sim 30$ min and over a specific location having dimensions $L_x = L_y \simeq 10$ km and vertical resolution $L_z \sim 1.0$ km. It is likely that during a given time period (i.e., 1430 to 1500 LST) only one active cumulonimbus exists in our forecast domain. Clearly one cannot establish a statistically significant average from a single realization, an ensemble of cumulonimbus eddies does not exist in the region at any one time. Furthermore, we can expect $\bar{\phi}$ and $\overline{\phi'^2}$ to vary significantly in the horizontal or vertical from a single realization. Thus by increasing N, the number of observations of look-alike cumulonimbi, the mean will become a better behaved function.

However, the meaning of the predicted average must be carefully understood! The eddy fluxes $\overline{w''q''}$, or rainfall rates, etc., may be considered as forecasts of climatology over the region for a given period and under a given large-scale forcing. Using this operator, a forecast at a certain time will contain the uncertainty or variance of the climatological sample. For example, in a situation in which superficially look-alike conditions occur over a given domain (L_x, L_y, L_z, τ), the forecast mean rainfall $\bar{R}$ will also be accompanied by a variance in expected rainfall $\overline{R''^2}$. The variance is due not only to the variability of rainfall within the defined domain (L_x, L_y, L_z) at a given instant, but is also due to the variability resulting from the fact that we cannot precisely define superficially look-alike conditions. Use of this operator is thus a recognition of the fact that we cannot observe or predict initial conditions or boundary conditions over a finite domain (L_x, L_y, L_z, τ) with sufficient precision to predict uniquely rainfall at any given point (x, y, z, t) or even over a finite domain (L_x, L_y, L_z, τ). Thus, this operator introduces the concept of "inherent uncertainty" in any model forecast.

3.5 Average Equations by the "Top-Hat" Method

In the "top-hat" scheme, clouds are assumed to be characterized by constant values of the dynamic and thermodynamic variables, while the environment takes on other constant values. The fraction of a given area covered by cumulus clouds, for example, is designated by σ, while the cloud-free environment occupies a larger fraction $(1 - \sigma)$. Here σ may vary as a function of pressure, as illustrated in Fig. 3.1.

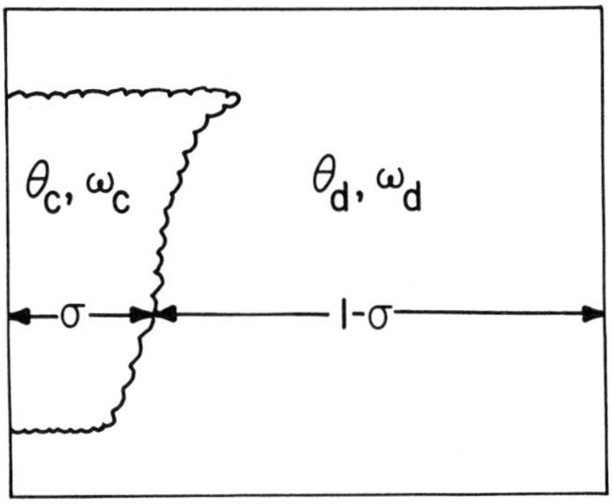

Fig. 3.1. Schematic diagram of a deep cumulus cloud and the percentage of the area covered by the cumulus cloud. [From Anthes (1977).]

Derivation of the equations governing the time rate of change of ϕ can take several forms (Anthes, 1977). In one method ϕ is assumed constant within the cloudy and environmental regions. Consequently, $\bar{\phi}$ does not vary continuously in space as do the Reynolds averages, but is obtained as

$$\bar{\phi}_{\mathrm{T}}(x)=\frac{1}{\Delta x}\int_{x_0-\Delta x/2}^{x_0+\Delta x/2}\phi(x')\,dx'=\sigma\phi_{\mathrm{c}}+(1-\sigma)\phi_{\mathrm{d}}, \tag{3.7}$$

where x lies in the interval $x_0-\Delta x/2$ to $x_0+\Delta x/2$.

For this definition, $\bar{\phi}_{\mathrm{T}}$ and σ are constant over the entire interval and $\bar{\phi}'\equiv 0$. Since $\bar{\phi}_{\mathrm{T}}$ is not continuous over the domain, this definition is not useful for deriving equations for the large-scale flow.

A second variation of the top-hat scheme assumes that ϕ_{c} and ϕ_{d} are constant but the horizontal average is a running average, similar to the Reynolds averaging technique.

$$\overline{\phi_{\mathrm{T}}(x)}=\frac{1}{\Delta x}\int_{x-\Delta x/2}^{x+\Delta x/2}\phi(x')\,dx'=\sigma\phi_{\mathrm{c}}+(1-\sigma)\phi_{\mathrm{d}}, \tag{3.8}$$

where σ represents the fractional area covered by clouds over the region centered about x. Note that the only difference between Eqs. (3.7) and (3.8) is the limits of integration. In Eq. (3.7) the limits correspond to fixed values x_0, where x varies continuously in Eq. (3.8). Since $\bar{\phi}_{\mathrm{T}}$ varies continuously, this definition is suitable for deriving equations for a large-scale model. We may write

$$d\phi_{\mathrm{c}}/dt=\partial\phi_{\mathrm{c}}/\partial t+\omega_{\mathrm{c}}(\partial\phi_{\mathrm{c}}/\partial p)=S_{\mathrm{c}} \tag{3.9}$$

and

$$d\phi_{\mathrm{d}}/dt=\partial\phi_{\mathrm{d}}/\partial t+\omega_{\mathrm{d}}(\partial\phi_{\mathrm{d}}/\partial p)=S_{\mathrm{d}}, \tag{3.10}$$

where S_{c} and S_{d} represent cloud and environmental sources and sinks of ϕ, respectively. The average vertical motion in pressure coordinates in the cloudy region and cloud-free environment are designated ω_{c} and ω_{d}, respectively.

Strict adherence to the top-hat assumption eliminates horizontal advection. Multiplying $\partial\phi_{\mathrm{c}}/\partial t$ by σ and $\partial\phi_{\mathrm{d}}/\partial t$ by $1-\sigma$, and adding the resultant equations, gives

$$\partial\bar{\phi}_{\mathrm{T}}/\partial t+\omega(\partial\bar{\phi}/\partial p)=\bar{S}+(\phi_{\mathrm{c}}-\phi_{\mathrm{d}})(\partial\sigma/\partial t). \tag{3.11}$$

This equation is useful for horizontally homogeneous conditions in the environment, but is not very useful for the general case where horizontal gradients exist in the environment.

A third variation on the top-hat scheme is obtained by allowing ϕ_c and ϕ_d to vary continuously in space. Thus

$$\bar{\phi}_T(x) = \frac{1}{\Delta x}\int_{x-\Delta x/2}^{x+\Delta x/2} \phi \, dx' = \sigma\bar{\phi}_0^{c} + (1-\sigma)\bar{\phi}_d^{d}, \tag{3.12}$$

where $\overline{()}^c$ and $\overline{()}^d$ denote horizontal averages over the cloud and environmental regions. With this definition all variables are continuous across cloud boundaries. Differentiating Eq. (3.12) we have

$$\frac{\partial\bar{\phi}}{\partial t} = \sigma\frac{\overline{\partial\phi_c}^{c}}{\partial t} + (1-\sigma)\frac{\overline{\partial\phi_d}^{d}}{\partial t} + (\bar{\phi}_c^{c} - \overline{\phi_d}^{d})\frac{\partial\sigma}{\partial t}. \tag{3.13}$$

Expressions for $\overline{\partial\phi_c}^c/\partial t$ and $\overline{\partial\phi_d}^d/\partial t$ will include horizontal advection terms. This expression for $\bar{\phi}_T$ is equivalent to the Reynolds averaged equation for $\bar{\phi}$.

Equations governing the interactions between cumulus clouds and the large scale have been derived by Ogura and Cho (1973), Yanai *et al.* (1973), Betts (1975), and Arakawa and Schubert (1974). Solution of these equations is basically the problem of parameterization. It requires the determination of the following parameters:

(i) Total condensation rate in an averaging volume.
(ii) Vertical distribution of cloud-scale heating or cooling.
(iii) Mean properties of temperature, moisture, and momentum in clouds.
(iv) Fractional area covered by clouds.

Simple cloud models have been employed to determine parameters (i)–(iv). Convective parameterization schemes such as that of Arakawa and Schubert (1974) compute the contribution of each cloud size in a cloud ensemble to the condensation and eddy flux terms. (See Chapter 6 for further discussion of convective parameterization schemes.)

It should be emphasized again that, strictly speaking, the top-hat approach is useful only if there is a distinct scale separation between the cloud and the environment, and/or if the large scale is free of clouds. For convective systems (e.g., tropical disturbances, tropical cyclones, midlatitude mesoscale convective systems) in which the averaging scale is on the order of tens of kilometers, $\sigma \to 1.0$, the use of the top-hat average does not simplify the system of equations.

3.6 An Example of the Reynolds Averaging Procedure

In Sections 3.1, 3.2, and 3.3 we discussed the philosophical concepts of averaging. In this section we shall apply the Reynolds averaging technique to the anelastic equations of motion in order to illustrate the basic procedure and the impact of the averaging operator. We begin with the equations of motion in the form of Eq. (2.86) or

$$\frac{du_i}{dt} = -\alpha_0 \frac{\partial p'}{\partial x_i} + \left(\frac{\alpha'_\mathrm{m}}{\alpha_0} - r'_\mathrm{w}\right) g\delta_{i3} + \varepsilon_{ijk}\, u_j\, f_k + f. \tag{3.14}$$

Multiplying Eq. (3.14) by ρ_0 and expanding the substantial derivative into local and advective changes gives

$$\rho_0 \frac{du_i}{dt} = \rho_0 \frac{\partial u_i}{\partial t} + \rho_0 u_j \frac{\partial u_i}{\partial x_j} = \frac{\partial}{\partial t}(\rho_0 u_i) + \frac{\partial}{\partial x_j}(\rho_0 u_j u_i)$$

$$- u_i \frac{\partial [\cancel{\rho_0 u_j}]^0}{\partial x_j} = \mathrm{RHS}, \tag{3.15}$$

where the third term on the LHS is zero by virtue of the anelastic continuity equation. Equation (3.15) then becomes

$$\frac{\partial}{\partial t}(\rho_0 u_i) = -\frac{\partial}{\partial x_j}(\rho_0 u_j u_i) - \frac{\partial p'}{\partial x_i} + \rho_0 \left(\frac{\alpha'_\mathrm{m}}{\alpha_0} - r'_\mathrm{w}\right) g\delta_{i3}$$

$$+ \rho_0 \varepsilon_{ijk} f_k u_j + \rho_0 \nu_0 \frac{\partial^2 u_i}{\partial x_j^2}, \tag{3.16}$$

where we have assumed that viscous forces behave incompressibly (i.e., $\partial u_j / \partial x_j = 0$). We now decompose each variable into a mean and a turbulent fluctuating component

$$\begin{aligned} u_i &= \bar{u}_i + u''_i, \\ p &= \bar{p} + p'' = p_0 + p' + p'', \\ \alpha_\mathrm{m} &= \bar{\alpha}_\mathrm{m} + \alpha''_\mathrm{m} = \alpha_0 + \alpha'_\mathrm{m} + \alpha''_\mathrm{m}, \end{aligned} \tag{3.17}$$

where p' and α'_m represent a mean departure from the horizontally homogeneous reference state values p_0 and α_0, respectively. In accordance with our scaling analysis, we assume that the turbulent fluctuations behave incompressibly. Substitution of Eq. (3.17) into Eq. (3.16) gives

$$\frac{\partial}{\partial t} \rho_0 (\bar{u}_i + u''_i) = -\frac{\partial}{\partial x_j} [\rho_0 (\bar{u}_j + u''_j)(\bar{u}_i + u''_i)]$$

$$- \frac{\partial}{\partial x_i}(p' + p'') + \rho_0 \left(\frac{\alpha'_\mathrm{m}}{\alpha_0} + \frac{\alpha''_\mathrm{m}}{\alpha_0} - r'_\mathrm{w} - r''_\mathrm{w}\right) g\delta_{i3}$$

$$+ \rho_0 \varepsilon_{ijk} f_k (\bar{u}_j + u''_j) + \rho_0 \nu_0 \frac{\partial^2}{\partial x_j^2}(\bar{u}_i + u''_i), \tag{3.18}$$

which can be written after averaging as

$$\overline{\frac{\partial}{\partial t}(\rho_0 \bar{u}_i)} + \overline{\frac{\partial}{\partial t}(\rho_0 u_i'')} = -\overline{\frac{\partial}{\partial x_j}(\rho_0 \bar{u}_j \bar{u}_i)} - \overline{\frac{\partial}{\partial x_j}(\rho_0 u_j'' \bar{u}_i)} - \overline{\frac{\partial}{\partial x_i}(\rho_0 \bar{u}_j\, u_i'')}$$

$$-\overline{\frac{\partial}{\partial x_j}(\rho_0 u_j'' u_i'')} - \overline{\frac{\partial p'}{\partial x_i}} - \overline{\frac{\partial p''}{\partial x_i}} + \rho_0 \overline{\left[\frac{\alpha_m'}{\alpha_0} - r_w'\right]} g\delta_{i3}$$

$$+\rho_0 \overline{\left(\frac{\alpha_m''}{\alpha_0} - r_w''\right)} g\delta_{i3} + \overline{\rho_0 \varepsilon_{ijk} f_k \bar{u}_j}$$

$$+\overline{\rho_0 \varepsilon_{ijk} u_j'' f_k} + \rho_0 \nu_0 \overline{\frac{\partial^2 u_i}{\partial x_j^2}} + \rho_0 \nu_0 \overline{\frac{\partial^2 u_i''}{\partial x_j^2}}, \qquad (3.19)$$

where the averaging operator is one of the operators, Eqs. (3.1) to (3.5). The usual procedure is to assume that for any variable ϕ

$$\bar{\bar{\phi}} = \bar{\phi},$$
$$\overline{\phi''} = 0, \qquad (3.20)$$
$$\overline{\phi'' \bar{u}_j} = 0 = \overline{u_j'' \bar{\phi}}.$$

However, in the case of the grid-volume average, where the mean field may contain turbulent fluctuations, it is not necessarily true that

$$\partial \overline{\phi''}/\partial t = 0,$$
$$(\partial/\partial x_j)\overline{u_j'' \bar{\phi}} = 0, \qquad (3.21)$$
$$(\partial/\partial x_j)\overline{\phi'' \bar{u}_j} = 0$$

or even

$$\overline{(\partial/\partial x_j)\bar{u}_i \bar{u}_j} = (\partial/\partial x_j)(\bar{u}_i \bar{u}_j).$$

That is, correlations can exist between the shorter wavelength fluctuating variables ϕ'' and the longer wavelength mean variables $\bar{u}_j$, which still behave as turbulent fluctuating variables. Leonard (1974) discussed the inequality of the previous equation in LES models and suggested procedures for more accurately estimating $\overline{(\partial/\partial x_j)\bar{u}_i \bar{u}_j}$.

In the case of the generalized ensemble-averaging operator, Eq. (3.6), all turbulence in the mean field is removed as long as the number of realizations, N, is large enough. Thus, mean and fluctuating variables are uncorrelated, and Eqs. (3.20) and (3.21) are exactly satisfied. Typically one assumes that Eqs. (3.20) and (3.21) are satisfied even for the grid-volume averaging operation. The magnitude of error that is introduced by this assumption, however, is not known. Thus, in practice, use of either operator results in Eq. (3.19) becoming

$$\frac{\partial \bar{u}_i}{\partial t} = -\frac{1}{\rho_0}\frac{\partial}{\partial x_j}(\rho_0 \bar{u}_j \bar{u}_i) - \frac{1}{\rho_0}\frac{\partial}{\partial x_j}(\rho_0 \overline{u_j'' u_i''}) - \frac{1}{\rho_0}\frac{\partial \bar{p}'}{\partial x_i} + \left(\frac{\overline{\alpha_m'}}{\alpha_0} - \bar{r}_w'\right) g\delta_{i3} + \varepsilon_{ijk}\bar{u}_j f_k, \tag{3.22}$$

where for large Reynolds number flow we neglect the effects of molecular diffusion on the mean dependent variables. The second term on the RHS of Eq. (3.22) is called the Reynolds stress term. It represents the horizontal and vertical transport of momentum by turbulent velocity fluctuations. It should be noted that had we not commenced our averaging with the equations of motion linearized with the thermodynamic variables, triple correlations among density and velocity fluctuations would have resulted.

Following a similar procedure, averaging Eqs. (2.127), (2.128), and (2.129) leads to

$$\frac{\partial \bar{\theta}_{il}}{\partial t} = -\frac{1}{\rho_0}\frac{\partial}{\partial x_j}(\rho_0 \bar{u}_j \bar{\theta}_i) - \frac{1}{\rho_0}\frac{\partial}{\partial x_j}(\rho_0 \overline{u_j'' \theta_{il}''}) + \overline{p(\theta_{il})}, \tag{3.23}$$

$$\frac{\partial}{\partial x_i}(\rho_0 \bar{u}_i) = 0, \tag{3.24}$$

$$\frac{\partial \bar{r}_T}{\partial t} = -\frac{1}{\rho_0}\frac{\partial}{\partial x_j}(\rho_0 \bar{u}_j \bar{r}_T) - \frac{1}{\rho_0}\frac{\partial}{\partial x_j}(\rho_0 \overline{u_j'' r_T''}) + \overline{p[r_T]}, \tag{3.25}$$

and so on for any scalar variable. The terms $p(\theta_{il})$ and $p(r_T)$ represent sources and sinks to θ_{il} and r_T, respectively, due to precipitation or, in the case of θ_{il}, radiation. Equations (3.22), (3.23), and (3.25) are not, however, closed. Some means of defining the correlations among velocity fluctuations and among velocity and scalar fluctuations must be found.

3.7 First-Order Closure Theory

The simplest approach to closing the averaged equations is to assume that the Reynolds stresses and eddy transport terms can be expressed in terms of gradients of the mean variables. Using the concept of eddy viscosity as introduced by Boussinesq (1877), it is often assumed that the Reynolds stress may be approximated by

$$-\overline{u_i'' u_j''} = K_m D_{ij}, \tag{3.26}$$

where K_m is the eddy viscosity or eddy exchange coefficient for momentum and D_{ij} is the mean rate of deformation tensor

$$D_{ij} = \tfrac{1}{2}[\partial \bar{u}_i/\partial x_j + \partial \bar{u}_j/\partial x_i]. \tag{3.27}$$

Hinze (1959) argued that Eq. (3.26) is physically inconsistent because, when the term is contracted for an incompressible fluid, the right-hand side is identically zero, whereas the left-hand side is zero only when there is no turbulence. A more consistent approximation would be

$$-\overline{u_i'' u_j''} \simeq -\tfrac{1}{3}\delta_{ij}\overline{u_l'' u_l''} + K_{\text{m}} D_{ij}, \tag{3.28}$$

where the first term on the right side of Eq. (3.28) is proportional to the average turbulent energy. Based on the work of Lilly (1967), Deardorff (1970) assumed that the turbulent energy is given by

$$\tfrac{1}{2}\overline{u_l'' u_l''} = K_{\text{m}}^2/(c_1\Delta)^2, \tag{3.29}$$

where $c_1 = 0.094$ and Δ represents a numerical model grid length. Following a similar line of reasoning, it is often assumed that the turbulent transport of any scalar property A (i.e., θ_{il}, r_{T}) can be expressed in terms of the scalar eddy viscosity (K_A) as

$$-\overline{u_j'' A''} \simeq K_A(\partial \bar{A}/\partial x_j). \tag{3.30}$$

Early cloud and mesoscale models contained values of K_{m} and K_A which were invariant in both time and space. Because the eddy viscosity must depend on the flow variables, and is not a property of the fluid itself, as is the case with molecular viscosity, such an assumption is unjustified.

Smagorinsky (1963) suggested that K_{m} be formulated as

$$K_{\text{m}} = [(c\Delta)^2/\sqrt{2}]|D|, \tag{3.31}$$

where $|D|$ is the magnitude of the mean deformation tensor. Lilly (1967) generalized the Smagorinsky concept by including the effects of static stability

$$K_{\text{m}} = [(c\Delta)^2/\sqrt{2}]|D_{ij}|[1 - (K_{\text{H}}/K_{\text{m}})R_{\text{i}}]^{1/2}, \tag{3.32}$$

where R_{i} is the gradient Richardson number and $K_{\text{H}}/K_{\text{m}}$ is the ratio of eddy diffusion for heat to momentum.

Following a somewhat different line of reasoning, Hill (1974) expressed K_{m} as

$$K_{\text{m}} = K_{\text{H}} = k\Delta^2[t_{\text{s}}^{-1} + t_{\text{B}}^{-1}], \tag{3.33}$$

where t_{s} represents the velocity-deformation time scale given by

$$t_{\text{s}} = |D_{ij}^2|^{-1/2}, \tag{3.34}$$

and t_{B} represents the buoyant time scale

$$t_{\text{B}} = -|(g/\theta)(\partial\theta/\partial z)|^{-1/2}. \tag{3.35}$$

Hill concluded that the effect of the buoyancy term will be much stronger than the shear term in regions where $\partial\theta/\partial z \lesssim 1°\text{C km}^{-1}$.

Klemp and Wilhelmson (1978), hereafter referred to as KW, expressed the eddy exchange coefficient for momentum in terms of the intensity of turbulence as

$$K_{\mathrm{m}} = c_{\mathrm{m}} \bar{i} L, \tag{3.36}$$

where

$$\bar{i} = \sqrt{(u_L'')^2},$$

and c_{m} is a coefficient selected to be $c_{\mathrm{m}} = 0.2$ and L is a turbulent length scale defined as a function of model-grid dimensions

$$L = (\Delta x\, \Delta y\, \Delta z)^{1/3}. \tag{3.37}$$

KW then formulated a prognostic equation for the turbulent kinetic energy. A derivation of this turbulent kinetic energy equation will be given in the next section. An advantage in their approach is that the Reynolds eddy stress terms are not only functions of mean variables ($\bar{u}_i$, $\bar{\theta}_{\mathrm{il}}$, $\bar{r}_{\mathrm{T}}$, etc.), but they are also a function of the locally predicted turbulent kinetic energy. Thus turbulent diffusion will occur only in those regions where active turbulence exists.

Alternatively, one can express an eddy exchange coefficient in terms of a turbulent time scale τ, i.e.,

$$K_{\mathrm{m}} = c' \bar{i}^2 \tau. \tag{3.38}$$

The time scale must be diagnosed by empirical or theoretical models. We shall discuss some models of turbulent time scales or length scales in subsequent chapters. The form of Eq. (3.36) or (3.38) is obtained from dimensional arguments.

One disadvantage of the eddy viscosity closure approach in a cloud system is that eddy exchange coefficients K_A for all mean prognostic variables must be specified or predicted. Generally, the specification of any K_A is done with little basis from observation or theory as far as most cloud systems are concerned. KW assumed that the ratio of eddy exchange coefficients for heat and momentum is a constant, namely

$$K_{\mathrm{H}}/K_{\mathrm{m}} = 3. \tag{3.39}$$

The factor of 3 is based on the results of Deardorff's (1972) numerical simulations of unstable and neutral planetary boundary layers. Exchange coefficients for a water substance were assumed identical to K_{H}. There is little evidence to support or refute such an assumption when applied to moist deep convection.

A fundamental weakness of eddy exchange theory is its underlying assumption that turbulence always acts to diffuse a property down the mean gradient of the property. There are many examples of counter gradient transport in planetary boundary layer studies and planetary-scale eddy transport analyses. While there exist few, if any, documented cases of countergradient turbulent transport associated with deep convection, most researchers would not be surprised to find not only that it exists, but that it prevails at times.

As a consequence, a number of investigators are seeking alternate means of closing the averaged equations. The approach which we shall discuss next is called *higher order closure theory.*

3.8 Higher Order Closure Theory

Rather than make a simple first-order approximation to the correlations $\overline{u_i''u_j''}$ or $\overline{u_i''A''}$ such as the down-gradient approximations, Eqs. (3.26) and (3.30), higher order closure theory involves the formulation of prognostic equations for these covariances. If a simple diagnostic model is used to evaluate the resulting triple-correlation terms $\overline{u_i''u_j''u_k''}$ or $\overline{u_i''u_j''A''}$, the model is referred to as a second-order closure model. If a predictive equation is formed on the triple-correlation terms and a diagnostic equation is formed on the resulting quadruple-correlation terms, the model is referred to as a third-order closure model, and so forth. To illustrate the procedure we return to Eq. (3.18),

$$\begin{aligned}\frac{\partial}{\partial t}(\rho_0\bar{u}_i)+\frac{\partial}{\partial t}(\rho_0 u_i'')={}&-\frac{\partial}{\partial x_j}(\rho_0\bar{u}_j\bar{u}_i)-\frac{\partial}{\partial x_j}(\rho_0 u_j''\bar{u}_i)-\frac{\partial}{\partial x_j}(\rho_0 u_i''\bar{u}_j)-\frac{\partial}{\partial x_j}(\rho_0 u_j''u_i'')\\&-\frac{\partial p'}{\partial x_i}-\frac{\partial p''}{\partial x_i}+\rho_0\left(\frac{\alpha_{\mathrm{m}}'}{\alpha_0}-r_{\mathrm{w}}'\right)g\delta_{i3}+\rho_0\left(\frac{\alpha_{\mathrm{m}}''}{\alpha_0}-r_{\mathrm{w}}''\right)g\delta_{i3}\\&+\rho_0\varepsilon_{ijk}\bar{u}_j f_k+\rho_0\varepsilon_{ijk}u_j''f_k+\cancelto{0}{\rho_0\nu_0\frac{\partial^2\bar{u}_i}{\partial x_j^2}}+\rho_0\nu_0\frac{\partial^2 u_i''}{\partial x_j^2},\end{aligned}\tag{3.40}$$

where we again ignore molecular diffusion of mean variables. If we now subtract Eq. (3.22) from Eq. (3.40) we obtain

$$\begin{aligned}\frac{\partial}{\partial t}(\rho_0 u_i'')={}&-\frac{\partial}{\partial x_j}(\rho_0 u_j''\bar{u}_i)-\frac{\partial}{\partial x_j}(\rho_0 u_i''\bar{u}_j)-\frac{\partial}{\partial x_j}(\rho_0 u_j''u_i'')\\&-\frac{\partial p''}{\partial x_i}+\rho_0\left(\frac{\alpha_{\mathrm{m}}''}{\alpha_0}-r_{\mathrm{w}}''\right)g\delta_{i3}+\rho_0\varepsilon_{ijk}u_j''f_k\\&+\rho_0\nu_0\frac{\partial^2 u_i''}{\partial x_j^2}+\frac{\partial}{\partial x_j}(\overline{\rho_0 u_j''u_i''}),\end{aligned}\tag{3.41}$$

Multiplication of Eq. (3.41) by u_k'' gives us

$$u_k'' \frac{\partial}{\partial t}(\rho_0 u_i'') = -u_k'' \frac{\partial}{\partial x_j}(\rho_0 u_j'' \bar{u}_i) - u_k'' \frac{\partial}{\partial x_j}(\rho_0 u_i'' \bar{u}_j) - u_k'' \frac{\partial}{\partial x_j}(\rho_0 u_j'' u_i'') - u_k'' \frac{\partial p''}{\partial x_i}$$

$$+ \rho_0 \left(\frac{\alpha_m''}{\alpha_0} u_k'' - u_k'' r_w'' \right) g\delta_{i3} + \rho_0 \varepsilon_{ijn} f_n u_j'' u_k''$$

$$+ \rho_0 \nu_0 u_k'' \frac{\partial^2 u_i''}{\partial x_j^2} + u_k'' \frac{\partial}{\partial x_j}(\rho_0 \overline{u_j'' u_i''}), \qquad (3.42)$$

where we have substituted $n = k$ in the Coriolis term without changing the results. Following a similar procedure, an equation on u_k'' can be obtained by interchanging indexes k for i in Eq. (3.14) in all terms in Eq. (3.14) except the Coriolis term, where the index n is substituted for k. After averaging and subtracting the mean form of the equation, an equation for u_k'' becomes

$$\frac{\partial}{\partial t}(u_k'') = -u_j'' \frac{\partial \overline{u_k}}{\partial x_j} - \overline{u_j} \frac{\partial u_k''}{\partial x_j} - u_j'' \frac{\partial u_k''}{\partial x_j} + \varepsilon_{kjn} f_n u_j''$$

$$- \frac{1}{\rho_0} \frac{\partial p''}{\partial x_k} + \left(\frac{\alpha_m''}{\alpha_0} - r_w'' \right) g\delta_{k3} + \nu_0 \frac{\partial^2 u_k''}{\partial x_j^2} + \overline{u_j'' \frac{\partial u_k''}{\partial x_j}}. \qquad (3.43)$$

If we now multiply Eq. (3.43) by $\rho_0 u_i''$ and add to Eq. (3.42) we obtain

$$\frac{\partial}{\partial t}(\rho_0 u_k'' u_i'') = u_k'' \frac{\partial}{\partial t}(\rho_0 u_i'') + \rho_0 u_i'' \frac{\partial u_k''}{\partial t}$$

$$= -\rho_0 u_i'' u_j'' \frac{\partial \bar{u}_k}{\partial x_j} - \overline{u_j} \rho_0 u_i'' \frac{\partial u_k''}{\partial x_j} - \rho_0 u_i'' u_j'' \frac{\partial u_k''}{\partial x_j} + \rho_0 \varepsilon_{kjn} f_n u_i'' u_j''$$

$$- u_i'' \frac{\partial p''}{\partial x_k} + \rho_0 \left(\frac{u_i'' \alpha_m''}{\alpha_0} - r_w'' u_i'' \right) g\delta_{k3} + \nu_0 \rho_0 u_i'' \frac{\partial^2 u_k''}{\partial x_j^2}$$

$$+ \rho_0 u_i'' \overline{u_j'' \frac{\partial u_k''}{\partial x_j}} - u_k'' \frac{\partial}{\partial x_j}(\rho_0 u_j'' \overline{u_i}) - u_k'' \frac{\partial}{\partial x_j}(\rho_0 u_i'' \overline{u_j})$$

$$- u_k'' \frac{\partial}{\partial x_j}(\rho_0 u_j'' u_i'') - u_k'' \frac{\partial p''}{\partial x_i} + \rho_0 \left(\frac{u_k'' \alpha_m''}{\alpha_0} - r_w'' u_k'' \right) g\delta_{i3}$$

$$+ \rho_0 \varepsilon_{ijn} f_n u_j'' u_k'' + \rho_0 \nu_0 u_k'' \frac{\partial^2 u_i''}{\partial x_j^2} + u_k'' \frac{\partial}{\partial x_j}(\rho_0 \overline{u_j'' u_i''}). \qquad (3.44)$$

We assume that the turbulent fluctuations behave incompressibly such that

$$\partial u_j'' / \partial x_j = (\partial / \partial x_j)(\rho_0 u_j'') = 0, \qquad (3.45)$$

and we rearrange the pressure–velocity as follows

$$-u_i''\frac{\partial p''}{\partial x_k}-u_k''\frac{\partial p''}{\partial x_i}=-\left[\frac{\partial}{\partial x_k}(u_i''p'')+\frac{\partial}{\partial x_i}(u_k''p'')\right]+p''\left(\frac{\partial u_i''}{\partial x_k}+\frac{\partial u_k''}{\partial x_i}\right), \tag{3.46}$$

Also, we rearrange the viscous terms as follows

$$\rho_0\nu_0\left(u_i''\frac{\partial^2 u_k''}{\partial x_j x_j}+u_k''\frac{\partial^2 u_i''}{\partial x_j x_j}\right)=\rho_0\nu_0\left[\frac{\partial^2(u_i''u_k'')}{\partial x_j^2}-2\left(\frac{\partial u_i''}{\partial x_j}\right)\left(\frac{\partial u_k''}{\partial x_j}\right)\right]. \tag{3.47}$$

We neglect the first term on the RHS of Eq. (3.47), the molecular diffusion term. After averaging and some rearranging, Eq. (3.44) becomes:

$$\frac{d}{dt}(\rho_0\overline{u_k''u_i''})=\frac{\partial}{\partial t}(\rho_0\overline{u_k''u_i''})+\overline{u_j}\frac{\partial}{\partial x_j}(\rho_0\overline{u_k''u_i''})$$

$$\begin{aligned}
&\overset{(a)}{=-\rho_0\overline{u_i''u_j''}\frac{\partial\overline{u_k}}{\partial x_j}}\ \overset{(b)}{-\rho_0\overline{u_k''u_j''}\frac{\partial\overline{u_i}}{\partial x_j}}\ \overset{(c)}{+g\left(\frac{\overline{u_i''\alpha_m''}}{\alpha_0}-\overline{u_i''r_w''}\right)\delta_{k3}\rho_0}\\
&\overset{(d)}{+\rho_0 g\left(\frac{\overline{u_k''\alpha_m''}}{\alpha_0}-\overline{u_k''r_w''}\right)\delta_{i3}}\ \overset{(e)}{-\rho_0\frac{\partial}{\partial x_j}(\overline{u_i''u_j''u_k''})}\\
&\overset{(f)}{-\left(\frac{\partial}{\partial x_k}\overline{u_i''p''}+\frac{\partial}{\partial x_i}\overline{u_k''p''}\right)}\ \overset{(g)}{+\overline{p''\left(\frac{\partial u_i''}{\partial x_k}+\frac{\partial u_k''}{\partial x_i}\right)}}\\
&\overset{(h)}{-\rho_0(\varepsilon_{ijn}f_n\overline{u_j''u_k''}+\varepsilon_{kjn}f_n\overline{u_i''u_j''})}\ \overset{(i)}{-2\rho_0\nu_0\overline{\left(\frac{\partial u_i''}{\partial x_j}\right)\left(\frac{\partial u_k''}{\partial x_j}\right)}}
\end{aligned} \tag{3.48}$$

Equation (3.48) is called the *Reynolds stress equation.* Thus instead of making an assumption regarding the behavior of the Reynolds stress $\rho_0\overline{u_k''u_i''}$, a prognostic equation for it is formed. The meaning of the various terms contributing to the time variation of the Reynolds stress is given below.

The LHS of Eq. (3.48) represents the substantive derivative of $\overline{\rho_0 u_k''u_i''}$ or the rate of change along a trajectory of mean velocity $\overline{u_j}$ space.

Terms (a) and (b) on the RHS of Eq. (3.48) represent *mechanical production* of turbulence. They represent the production of new velocity correlations by the interaction of the turbulence with the variation of the mean velocity field as well as the modification of existing correlations by the variation of the mean field. Terms (c) and (d), the so-called *buoyancy production terms*, represent the production or reduction of the velocity correlation due to interactions of velocity fluctuations with density fluctuations or water content fluctuations. In a cloud system, buoyancy production of turbulence can arise from the mixing of saturated and subsaturated air parcels. The resultant mixed parcels contain evaporating and/or condensing volumes of air which, in turn, can lead to further buoyant production of turbulence by the latent heats exchanged during the phase changes.

Term (e) represents the turbulent transport of the covariance $\overline{u_k'' u_i''}$ by the turbulent fluctuations and is often called the *velocity diffusion term*. It is the action of this term which causes the movement of air parcels from a turbulent region into a quiescent environment. This overshooting of turbulent air parcels causes the entrainment of quiescent air with differing properties into the turbulent field.

Term (f) is called the *pressure diffusion* term. The term is a redistribution term that often acts to destroy the existing stress.

Term (g) is called the *tendency toward isotropy* term. Its name arises from the fact that, in the total turbulent energy equation, this term vanishes because the divergence is zero in incompressible flow. Thus the term is thought to behave in such a way that the turbulent energy is rearranged among the various components.

Term (h) represents the production or reduction of $\rho_0 \overline{u_k'' u_i''}$ due to the earth's rotation.

Term (i) is the *molecular dissipation term* and represents the conversion of the energies $\overline{u_1''^2}$, $\overline{u_2''^2}$, and $\overline{u_3''^2}$ into heat by the action of molecular viscosity.

It should be noted that since $\alpha_m''/\alpha_0 = \theta''/\theta_0 + 0.61 r_v''$, the buoyancy production terms $\rho_0 g(\overline{u_i'' \alpha_m''}/\alpha_0 - \overline{u_i'' r_w''})\delta_{i3}$ introduce correlations among all the thermodynamic components of the system. Thus in a cloud system fluctuations in total condensate water r_w'' and phase changes causing anomalies in θ'', r_v'', r_c'' and r_i'' can make substantial contributions to buoyancy production of Reynolds stress. To model such processes requires the formulation of equations for $\overline{u_i'' r_w''}$, $\overline{u_i'' \theta_{il}''}$, and $\overline{u_i'' r_v''}$.

One of the problems of second-order turbulence theory is that Eq. (3.48) is not closed as it stands, even when equations forecasting correlations among velocity fluctuations and thermodynamic variable fluctuations are formulated. The triple-correlation production term (e) either requires the formulation of a prognostic equation for $(\partial/\partial t)\overline{u_i'' u_j'' u_k''}$ or the modeling of this term in analogy to the Reynolds stress terms [i.e., Eq. (3.26)]. In the former case, this introduces the formulation of an infinite set of moment equations, while, in the latter case, the artificiality of eddy diffusion is again introduced, but this time on the third-order terms. Model equations must also be devised for terms (f), (g), and (i).

Before we discuss models for closing the Reynolds stress equation, Eq. (3.48), let us first form an equation predicting the variation of turbulent kinetic energy of the flow. This may be accomplished by contracting Eq. (3.48) (i.e., replace the index k with i) before we introduce the averaging operator, giving

$$\begin{aligned}\frac{d}{dt}(\rho_0 u_i''^2) &= \frac{\partial}{\partial t}(\rho_0 u_i''^2) + \bar{u}_j \frac{\partial}{\partial x_j}(\rho_0 u_i''^2) \\ &= -2\rho_0 u_i'' u_j'' \frac{\partial \bar{u}_i}{\partial x_j} + 2g\left(\frac{u_i''\alpha_m''}{\alpha_0} - u_i'' r_w''\right)\delta_{i3} - \frac{\partial}{\partial x_j}(\rho_0 u_j'' u_i''^2) \\ &\quad - 2\frac{\partial}{\partial x_i}(u_i'' p'') + p''\left(\frac{\partial u_i''}{\partial x_i} + \frac{\partial u_i''}{\partial x_i}\right)^{0} - 2\rho_0 \nu_0 \left(\frac{\partial u_i''}{\partial x_j}\right)^2, \end{aligned} \tag{3.49}$$

where the Coriolis terms also go to zero. If we define $e = \frac{1}{2}\rho_0 u_i''^2$, then, after averaging, Eq. (3.49) becomes

$$\begin{aligned}\frac{d\bar{e}}{dt} = \frac{\partial \bar{e}}{\partial t} + \bar{u}_j \frac{\partial \bar{e}}{\partial x_j} &= -\rho_0 \overline{u_i'' u_j''}\frac{\partial \bar{u}_i}{\partial x_j} + \left(\frac{\overline{u_i''\alpha_m''}}{\alpha_0} - \overline{u_i'' r_w''}\right) g\delta_{i3} \\ &\quad - \frac{\partial}{\partial x_j}(\overline{e u_j''}) - \frac{\partial}{\partial x_i}(\overline{u_i'' p''}) - \rho_0 \nu_0 \overline{\left(\frac{\partial u_i''}{\partial x_j}\right)^2}. \end{aligned} \tag{3.50}$$

Equation (3.50) illustrates that turbulent kinetic energy is a species that can be advected by the mean flow in space. There are two sources of $\bar{e}$, namely, (a) mechanical production and (b) buoyant production. Term (e) represents dissipation of $\bar{e}$ by viscous forces. Turbulence can also act to redistribute itself in space by eddy transport (velocity diffusion), term (c), and through the interaction of velocity fluctuations and pressure fluctuations (pressure diffusion), or term (d).

As with the Reynolds stress equation, models must be devised to close terms such as (c), (d), and (e) in the turbulent kinetic energy equation.

Let us first consider term (e) in Eq. (3.48), or its equivalent, term (c), in Eq. (3.50). As mentioned previously, this term represents the transport of Reynolds stress or $\bar{e}$ by turbulent fluctuations. The earliest higher order closure models approximated the velocity diffusion term in analogy to modeling the Reynolds stress by an eddy viscosity [see Eq. (3.26)]. Thus we might approximate

$$-\overline{u_i'' u_j'' u_k''} \simeq K(\partial/\partial x_j)\overline{u_i'' u_k''}, \tag{3.51}$$

where K could be modeled in a manner similar to Eq. (3.36) or (3.38). However, several authors (André *et al.*, 1976a, b; Zeman and Lumley, 1976) have noted that such gradient-type diffusion models cannot represent vertical convective transport of turbulence. According to Zeman and Lumley (hereafter referred to as ZL), the gradient model often represents the wrong shape of the $\overline{u_i''u_j''u_k''}$ profiles and the wrong direction of eddy transport of turbulence. This can lead to serious misrepresentations of the entrainment process.

Thus ZL and André *et al.* (1976a, b; hereafter referred to as AN) have formulated a rate equation for the third-order moments. The derivation of the third-moment equation follows the same procedure used in deriving Eq. (3.48). The form of the equation is as follows:

$$
\begin{aligned}
\frac{\partial}{\partial t}\overline{u_i''u_j''u_k''} = & & :\frac{\partial F_i}{\partial t} \\
& \overline{u_i''u_j''}\frac{\partial}{\partial x_l}\overline{u_k''u_l''}+\overline{u_i''u_k''}\frac{\partial}{\partial x_l}\overline{u_j''u_l''} & \\
& +\overline{u_j''u_k''}\frac{\partial}{\partial x_l}\overline{u_i''u_l''}-\frac{\partial}{\partial x_l}(\overline{u_i''u_j''u_k''u_l''}) & :S_{\mathrm{G}} \\
& +(\overline{u_j''u_k''\alpha_{\mathrm{m}}''}-\overline{u_j''u_k''r_{\mathrm{w}}''})g\delta_{i3}+(\overline{u_i''u_k''\alpha_{\mathrm{m}}''}-\overline{u_i''u_k''r_{\mathrm{w}}''})g\delta_{i3} & \\
& +(\overline{u_i''u_j''\alpha_{\mathrm{m}}''}-\overline{u_i''u_j''r_{\mathrm{w}}''})g\delta_{k3} & :S_{\mathrm{B}} \\
& -\frac{1}{\rho_0}\left(\overline{\frac{\partial p''}{\partial x_i}u_k''u_j''}+\overline{\frac{\partial p''}{\partial x_j}u_i''u_k''}+\overline{\frac{\partial p''}{\partial x_k}u_i''u_j''}\right) & :P \\
& -2\nu_0\left(\overline{u_i''\frac{\partial u_j''}{\partial x_l}\frac{\partial u_k''}{\partial x_l}}+\overline{u_j''\frac{\partial u_k''}{\partial x_l}\frac{\partial u_k''}{\partial x_l}}+\overline{u_k''\frac{\partial u_i''}{\partial x_l}\frac{\partial u_j''}{\partial x_l}}\right) & :D. \qquad (3.52)
\end{aligned}
$$

AN and later Bougeault (1981b) integrated Eq. (3.52) in their models of the atmospheric boundary layer and trade-wind cloud layer. Whereas ZL formed a diagnostic equation from Eq. (3.52) by assuming that

$$\partial F_i/\partial t = S_{\mathrm{G}i}+S_{\mathrm{B}i}-F_i/T_3-D_i \simeq 0, \qquad (3.53)$$

where F_i represents a third moment, $S_{\mathrm{G}i}$ is the neutral source term (i.e., excludes buoyancy effects), $S_{\mathrm{B}i}$ is the source term accounting for buoyancy effects, F_i/T_3 is the pressure term or return-to-isotropy term, and D_i represents the molecular destruction of F_i. In order to close the system, Eq. (3.52), both AN and ZL had to develop models or parameterizations for terms P and D along with forming parallel equations for triple correlations with water and thermodynamic fields in the buoyancy terms S_{G}. Generally, the modeling of these terms requires the introduction of arbitrary time scales and/or length scales. Or, of course, one could form fourth-moment equations to Eq. (3.52). It appears from the experiences of AN and ZL that at least a diagnostic equation of Eq. (3.52) is needed to obtain a satisfactory simulation of the evolution of both daytime and nighttime boundary layers and turbulent cloud layers. For discussions on closure approximations to Eq. (3.52), we refer the reader to AN, ZL, or Chen and Cotton (1983a).

Returning now to Eq. (3.48), we find that models for terms (f) and (g) as well as (i) must be formulated. Term (g) is normally rather simply formulated to drive the system toward a state of isotropy over some turbulent time scale (see ZL or AN). Term (f), however, is thought to play a much more complicated role and reflects the generation of velocity-pressure correlations wherever turbulent nonhydrostatic pressure gradients can be expected to be generated. It is thought to play a particularly important role near the earth's surface. ZL, in fact, derived a model for (f) which involved a simplified solution to a Poisson equation.

The modeling of the dissipation term (i) in Eq. (3.48) can also involve varying degrees of sophistication. ZL, for example, formed a rate equation (i.e., $\partial\varepsilon/\partial t$) for the rate of energy dissipation (ε). Most researchers form simple models for dissipation terms in terms of the turbulence kinetic energy or turbulent intensity. Thus, AN modeled dissipation as

$$\bar{\varepsilon} = c_1(l)\,\bar{i}^3/l, \tag{3.54}$$

and Chen and Cotton (1983a) modeled it as

$$\bar{\varepsilon} = \mu\bar{i}^2/\tau. \tag{3.55}$$

These models are based on dimensional arguments and again introduce a turbulence time scale (τ) or length scale (l) for closure of the model.

Thus far we have only considered the problem of higher order closure of the momentum equations. In a complete cloud system, however, closure models must be formulated for mean prognostic equations for heat and water substance. Moreover, closure of the Reynolds stress equation, Eq. (3.48), required models for covariances with velocity fluctuations in order to model the buoyancy terms.

Following a procedure similar to that used in deriving Eq. (3.48), one can form tendencies on scalar covariances and fluxes on scalar quantities. Thus, for a cloud system composed of only cloud water (no rain) and a thermodynamic energy equation using θ_{il}, a second-order turbulence model can be expressed as follows.

For the scalar covariances, $\overline{\theta_{\mathrm{il}}''^2}$, $\overline{r_{\mathrm{T}}''^2}$, $\overline{\theta_{\mathrm{il}}'' r_{\mathrm{T}}''}$, etc., the tendencies written in terms of free variables a'', b'' are

$$\frac{\partial}{\partial t}(\overline{a''b''}) = -\frac{\partial}{\partial x_j}(\overline{u_j a''b''}) - \left|\overline{a''u_j''}\frac{\partial\bar{b}}{\partial x_j} + \overline{b''u_j''}\frac{\partial\bar{a}}{\partial x_j}\right|$$

$$-\frac{\partial}{\partial x_j}(\overline{u_j''a''b''}) - 2\bar{\varepsilon}_{ab}, \tag{3.56}$$

where $\bar{\varepsilon}_{ab}$ represents the rate of molecular dissipation of the scalar correlation $\overline{a''b''}$.

Similarly, the tendencies for the fluxes of the scalar quantities $\overline{u_i'''\theta_{il}''}$, $\overline{u_i''r_T''}$ are

$$\frac{\partial}{\partial t}(\overline{u_i''a''}) = -\frac{\partial}{\partial x_j}(\bar{u}_j\overline{u_i''a''}) + \varepsilon_{ijk}f_k\overline{u_j''a''} - \overline{a''u_j''}\frac{\partial \bar{u}_i}{\partial x_j}$$
$$-\overline{u_i''u_j''}\frac{\partial \bar{a}}{\partial x_j} - \frac{\partial}{\partial x_j}(\overline{u_j''u_i''a''}) - \alpha_0\frac{\partial}{\partial x_i}(\overline{p''a''}) + \alpha_0\overline{p''\frac{\partial a''}{\partial x_i}}$$
$$+\left|\frac{1}{\alpha_0}\overline{\alpha_m''a''} - \overline{r_w''a''}\right| g\delta_{i3} - 2\nu_{ua}\left(\overline{\frac{\partial u_i''}{\partial x_j}\frac{\partial a''}{\partial x_j}}\right). \quad (3.57)$$

As the physics of the cloud system is expanded to include rainwater, mixing ratio, and ice phase, the number of equations for scalar correlations and fluxes increases enormously.

It should also be noted that physical processes such as precipitation and radiation processes become intimately coupled in Eqs. (3.56) and (3.57). For example, the variance and transport of water substance quantities will vary depending on the models of cloud physical processes. Furthermore, as will be discussed in Chapter 4, the rates of production of precipitation will change depending on the turbulent structure of the cloud systems. Thus turbulence affects the precipitation structure of a cloud system, and precipitation affects its turbulent structure.

A number of approximations to the complete system of equations can be made. An example is ZL's equilibrium approximation to the third-order moments in Eq. (3.53). A similar philosophy can be applied to all or portions of the set of second-order moment equations.

For the purpose of illustration, consider the Reynolds stress equation, Eq. (3.48), for a neutrally stratified atmosphere. In a neutral atmosphere, buoyant production, or term (c), is negligible, and Eq. (3.48) reduces to the rate of change of Reynolds stress along a streamline of mean velocity $\bar{u}_j$ for steady flow. If mechanical production is balanced by dissipation and we ignore turbulence transport and pressure–velocity correlations, then

$$(d/dt)(\rho_0\overline{u_k''u_i''}) = 0. \quad (3.58)$$

Equation (3.48) thus reduces to the eddy viscosity model, Eq. (3.26). Extension of the equilibrium approximation to include a balance between mechanical and buoyant production of turbulence and dissipation gives rise to the generalized eddy viscosity model, Eq. (3.32).

It is obvious that these approximations may be overly restrictive when applied to various cloud systems. However, when dealing with the complete set of Eqs. (3.48), (3.56), and (3.57), one can experiment with various combinations of prognostic equations and diagnostic equations derived from equilibrium approximations such as Eq. (3.58). Mellor and Yamada (1974) have discussed the impact of forming various levels of prognostic/diagnostic equations in modeling the evolution of the planetary boundary layer. Similar analyses must be done for combinations of prognostic and diagnostic equations for models of stratocumuli, trade-wind cumuli, or other cloud systems based on the higher ordered equation set.

3.9 Partial Condensation over an Averaging Volume or Averaging Domain

Traditionally, various cloud models have been formulated such that the amount of cloud water is daignosed by an "all or nothing" scheme (see, e.g., Orville and Kopp, 1977; Klemp and Wilhelmson, 1978; Cotton and Tripoli, 1978). According to this scheme, condensation occurs only when the mixing ratio of the air, averaged over a horizontal grid volume, reaches the saturation mixing ratio determined from the average temperature over the grid volume. In a turbulent cloudy environment, however, one can expect to find local regions of condensate bordering subsaturated regions, all within a given averaging volume. An obvious example is the case where the averaging volume is large enough to encompass a field of fair-weather cumuli. Thus, if one were to make a horizontal traverse across the domain, one would expect to encounter local saturated cloudy regions followed by subsaturated regions. The "all or nothing" scheme would diagnose $\bar{r}_{\mathrm{T}} - r_{\mathrm{s}}(\bar{T})$ as being subsaturated and therefore free of clouds. In actual fact there would exist an average cloud liquid-water content $\overline{r_{\mathrm{c}}}$ in the domain which would contribute to the thermodynamics and dynamics of the cloud field even though $\bar{r}_{\mathrm{T}} - r_{\mathrm{s}}(\bar{T}) < 0$.

The application of such an "all or nothing" procedure in cloud or mesoscale models can lead to problems, since this treatment of condensation delays the release of latent heat during the early stages of cloud formation until a grid point is driven to saturation. Selected grid points then release latent heat explosively while neighboring ones may not. The result is that latent heat is released on the smallest horizontal scales resolved by the model, with a resultant generation of computational noise.

Several authors have derived expressions for average cloud water and fractional cloud coverage for a fluctuating water and thermal field (Sommeria and Deardorff, 1977; Manton and Cotton, 1977; Mellor, 1977; Oliver *et al.*, 1978; Bougeault, 1981b). Let us consider first the approach taken by Manton and Cotton (MC). For simplicity, we assume the system is nonprecipitating and that water substance is distributed into $r_{\mathrm{T}} = r_{\mathrm{v}} + r_{\mathrm{c}}$. MC assumed that the zero-mean random variable $r_{\mathrm{T}}'' - r_{\mathrm{s}}''$, where r_{s} is the saturation mixing ratio, is normally distributed with variance

$$\sigma_c^2 = \overline{(r_T'' - r_s'')^2}. \tag{3.59}$$

The corresponding average liquid-water content is

$$\overline{r_c} = (\overline{r_T} - \overline{r_s})h + (2\pi)^{1/2} \exp|-(\overline{r_T} - \overline{r_s})^2/2\sigma_c^2|, \tag{3.60}$$

where

$$h = \tfrac{1}{2}|a + \operatorname{erf}(\overline{r_T} - \overline{r_s})/2^{1/2}\sigma_c|. \tag{3.61}$$

The function h behaves as a Heaviside unit step function as σ_c approaches zero; i.e., $h = H(\overline{r_T} - \overline{r_s})$ as $\sigma_c \to 0$. Equation (3.60) allows the formation of cloud water contributing to a nonzero $\overline{r_c}$ even though $\overline{r_T} - \overline{r_s} < 0$. To define σ_c^2, we must determine $\overline{r_T''^2}$, $\overline{r_s''^2}$, and $\overline{r_T'' r_s''}$, since $\sigma_c^2 = \overline{r_T''^2} + \overline{r_s''^2} - 2\overline{r_T'' r_s''}$. The usefulness of this approach hinges on the validity of the assumption that $r_T'' - r_s''$ is normally distributed.

Banta (1979) examined this hypothesis by computing the distribution of $r_T'' - r_s''$ from aircraft data collected in the subcloud layer. Cloud-layer data were not used because of the difficulty of simultaneously measuring $r_T = r_v + r_c$ and $r_s(T)$ in cloudy air and clean air with instrument responses of comparable sensitivities. Figures 3.2 and 3.3 illustrate that for 1000-m averaging legs near the surface and well into the interior of the atmospheric boundary layer (ABL), all $r_T'' - r_s''$ are well represented by a normal distribution. While Banta's analysis lends confidence to MC's hypothesis, it does not confirm that $r_T'' - r_s''$ is normally distributed in the cloud layer or in more complex cloud regimes.

Banta also tested the validity of Sommeria and Deardorff's (SD) hypothesis that θ_l and r_T have joint normal probability distributions within a given grid volume. Under this assumption SD showed that

$$\overline{r_c} = \frac{1}{1 + \beta_1 \bar{r}_s} \left| R(\overline{r_T} - \overline{r_s}) + \frac{\sigma_c}{\sqrt{2\pi}} \exp\left[-\frac{(\overline{r_T} - \overline{r_s})^2}{2\sigma_c^2} \right] \right|, \tag{3.62}$$

where $\beta_1 = 0.622(L_{ev}/R_a T)(L_{ev}/c_{pa} T_l)$,

$$\sigma_c = (\sigma_T^2 + \alpha_1 \sigma_{\theta_l}^2 - 2\phi \alpha_1 \sigma_T \sigma_{\theta_l})^{1/2} = (\overline{r_T''^2} + \overline{r_s''^2} - 2\overline{r_T'' r_s''})^{1/2},$$

$$\phi = \overline{r_T'' \theta_l''}/(\sigma_{r_T} \sigma_{\theta_l}) \qquad \text{and} \qquad \alpha_1 = (\bar{p}/p_0)^{0.286} \left(\frac{\partial r_s}{\partial T} \right)_{T = T_l}.$$

The cloud fraction R is

$$R = \frac{1}{2\sqrt{2\pi}\sigma_{\theta_l}} \int_{\infty}^{\infty} \exp\left[-\frac{(\theta_l - \bar{\theta}_l)^2}{2\sigma_{\theta_l}^2} \right] \times \left| 1 + \operatorname{erf}\left[\frac{\overline{r_T} - \overline{r_s} - \alpha_1(\theta_l - \overline{\theta_l})}{\sqrt{2}\sigma_T} \right] \right| d\theta_l. \tag{3.63}$$

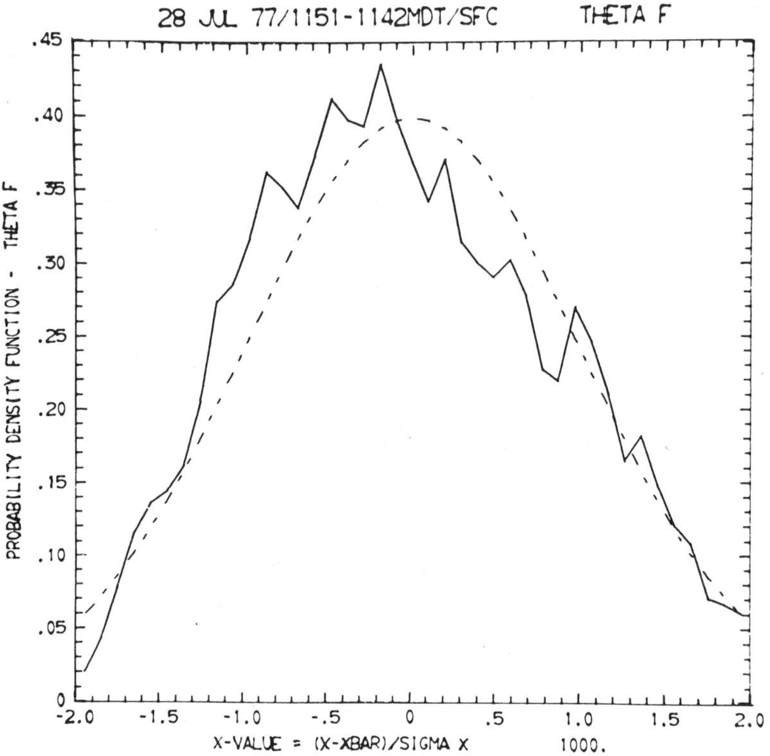

Fig. 3.2. Estimated probability density functions for θ and r_v, superimposed on a dashed curve of the standard normal density. Data are from 28 July for a flight leg over South Park, Colorado, at a height of 50 m off the surface, starting at 1152 MDT. Averaging interval was 1000 m. [From Banta (1979).]

Figure 3.4 illustrates Banta's evaluation of the joint distribution of θ_l, r_T in the subcloud layer. Departures from bivariant normality are quite pronounced. This does not refute SD's hypothesis, but it does introduce some doubt that the hypothesis can be substantiated in general. In both cases, MC's and SD's formulations are quite complex.

Using the assumption that θ_l and r_T can be represented by a bivariate Gaussian distribution, Mellor (1977) showed that the liquid-water content r_c depends only on a linear combination of θ_l'' and r_T. Thus,

$$r_c = a\Delta\overline{r_T} + 2s \qquad \text{if} \qquad s > -a\Delta\overline{r_T}/2 \tag{3.64}$$

and $r_c = 0$ otherwise;

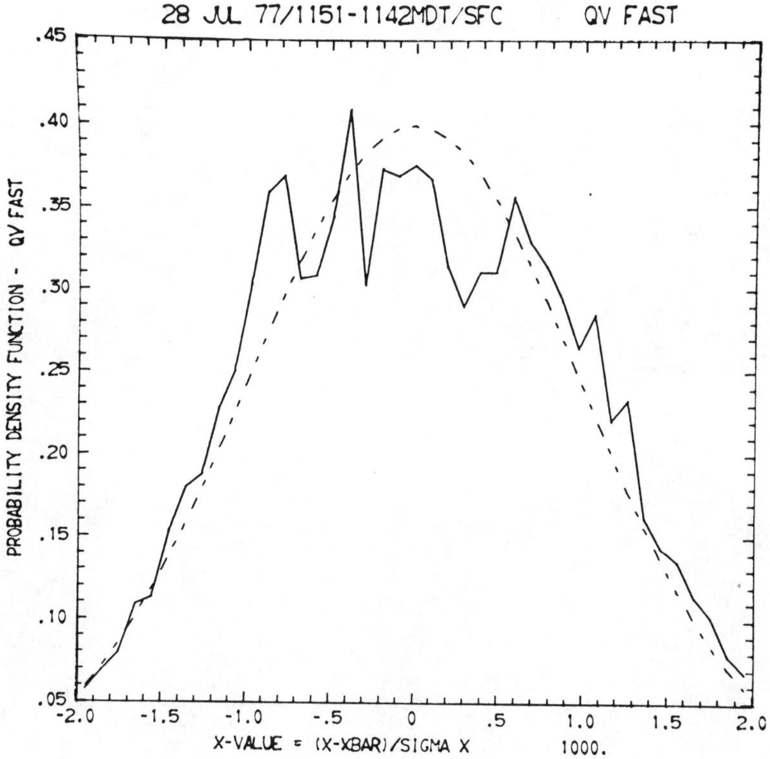

Fig. 3.2 *(continued)*

where

$$s = a(r''_{\mathrm{T}} - \alpha_1 \theta''_1)/2$$

and

$$a = (1 + L^2 \overline{r_{\mathrm{s}}} / R_{\mathrm{v}} c_{\mathrm{pa}} T_1^2)^{-1}.$$

Recognizing the absence of field observational data to evaluate thermodynamic and water substance parameters controlling cloud coverage and average liquid-water content, Bougeault (1981b) analyzed the statistics of data simulated with Sommeria's (1976) three-dimensional (3-D) model of the trade-wind cumulus layer.

Bougeault (1981a; hereafter referred to as B81) used the 3-D model data set obtained in a trade-wind cumuli simulation reported by Sommeria and

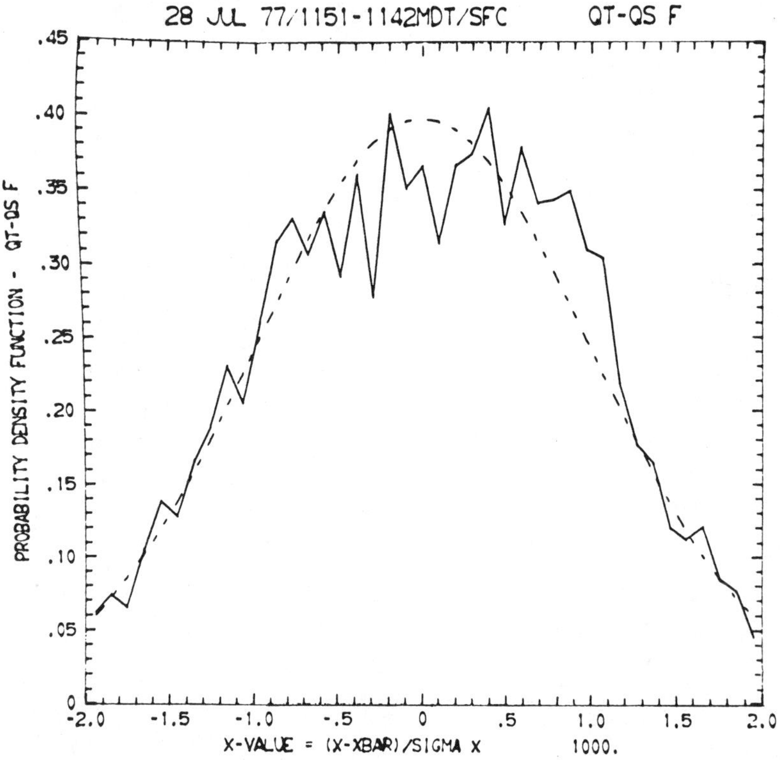

Fig. 3.3. Estimated probability density function for $r_T - r_s$ superimposed on standard normal curve (dashed line). Data are from the same flight leg as in Fig. 3.2. [From Banta (1979).]

LeMone (1978). Cloud cover in the 3-D simulations did not exceed 10%. To develop and test the cloud fraction model, B81 averaged the 3-D model-generated data over a given horizontal level and over a time interval sufficiently long to smooth out most of the variability associated with the largest eddies (or clouds) simulated by the model. In terms of the generalized ensemble-averaging operator, Eq. (3.6), this corresponds to the use of an infinitesimally small L_z, with L_x and L_y corresponding to the scale of the 3-D model domain and τ corresponding to his averaging time. The number of realizations k is one for a given set of boundary and initial conditions. This is true if the time interval is always selected at the same absolute initial time relative to the start of the simulation.

B81 used the 3-D data to examine three proposed models of the distribution of temperature and moisture fluctuations in a cloud layer. If we let t

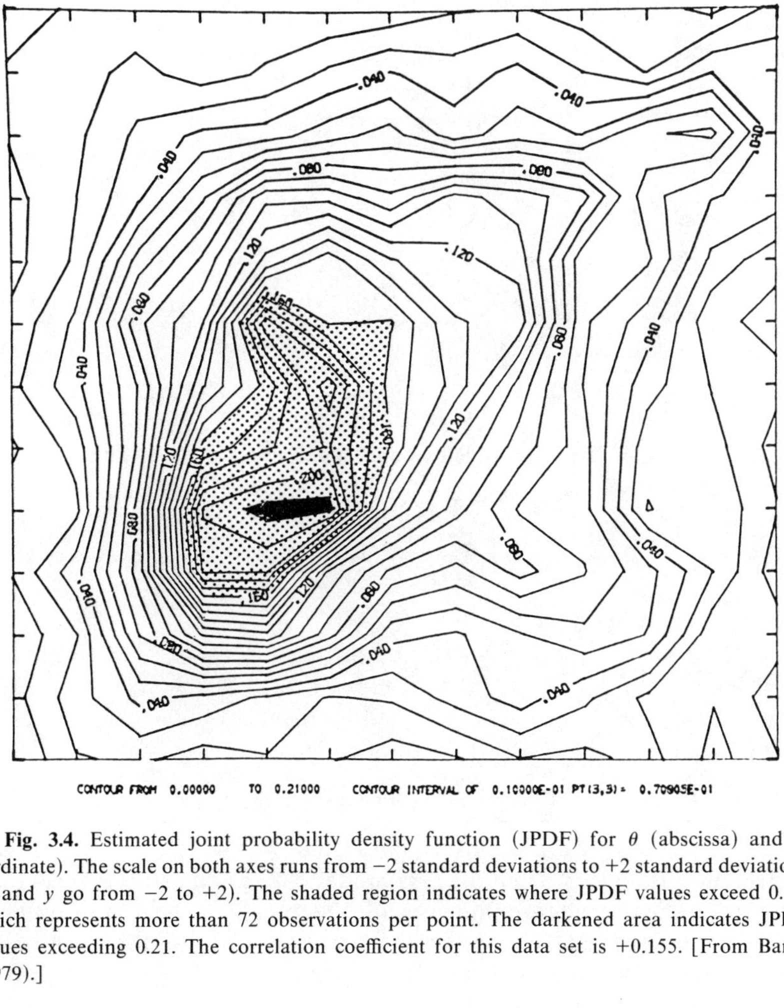

Fig. 3.4. Estimated joint probability density function (JPDF) for θ (abscissa) and r_v (ordinate). The scale on both axes runs from −2 standard deviations to +2 standard deviations (x and y go from −2 to +2). The shaded region indicates where JPDF values exceed 0.16, which represents more than 72 observations per point. The darkened area indicates JPDF values exceeding 0.21. The correlation coefficient for this data set is +0.155. [From Banta (1979).]

represent the normalized variable s/σ_s where

$$\sigma_s = (a/2)(\overline{r_T''^2} + \alpha_1^2\overline{\theta_l''^2} - 2\alpha_1\overline{\theta_l'' r_T''})^{1/2}, \tag{3.65}$$

$G(t)$ is the probability density and Q_1 is a measure of the departure of the mean state from saturation ($Q_1 = a\Delta\overline{r_T}/2\sigma_s$), then Eq. (3.64) can be written as

$$\begin{aligned} r_c/2\sigma_s &= Q_1 + t \qquad \text{if} \qquad t > -Q_1, \\ &= 0 \text{ otherwise.} \end{aligned} \tag{3.66}$$

The cloud fraction R and correlations between fluctuating liquid water content and other variables can then be expressed in terms of $G(t)$ as follows

$$R = \int_{-Q_1}^{\infty} G(t)\, dt, \tag{3.67}$$

$$r_c/2\sigma_s = \int_{-Q_1}^{\infty} (Q_1 + t)\, G(t)\, dt, \tag{3.68}$$

$$\overline{sr_c''}/2\sigma_s^2 = \int_{-Q_1}^{\infty} t(Q_1 + t)\, G(t)\, dt, \tag{3.69}$$

$$\overline{s^2 r_c''}/2\sigma_s^3 = \int_{-Q_1}^{\infty} t^2(Q_z + t)\, G(t)\, dt - \overline{r_c}/2\sigma_s. \tag{3.70}$$

B81 postulated three models to be tested against the 3-D data. They are the Gaussian model

$$G_1(t) = \frac{1}{\sqrt{2\pi}} \exp(-t^2/2), \tag{3.71}$$

the exponential model

$$G_2(t) = \frac{1}{\sqrt{2}} \exp(-|t|\sqrt{2}), \tag{3.72}$$

and a positively skewed distribution of the form

$$G_3(t) = H(t+1) \exp[-(t+1)]. \tag{3.73}$$

Each of these distributions is illustrated in Fig. 3.5. The computed distribution functions using the 3-D model data are illustrated in Fig. 3.6 along with the distribution functions for the three models at three different levels in the cloud layer. At the lowest level (500 m), the Gaussian model fits remarkably well. Higher in the cloud (1000 m), the computed distribution exhibits a pronounced tail. Thus, the skewed distribution function, Eq. (3.73), fits best. At 1250 m the distribution is not only skewed but a secondary mode in the distribution is evident. Bougeault concluded that a positively skewed distribution was more appropriate for a trade-wind cumulus layer. He suggested that the distribution of fluctuations of conservative variables should be expected to be skewed in a cumulus layer, since strong updrafts in clouds are compensated by slow descending motions outside of clouds.

Bougeault's results are actually consistent with Banta's finding that a Gaussian distribution fits observed distributions obtained below the cloud base. It appears from B81's analysis that the greatest departure from normality occurs well within the cloud layer. It is obvious, however, that further

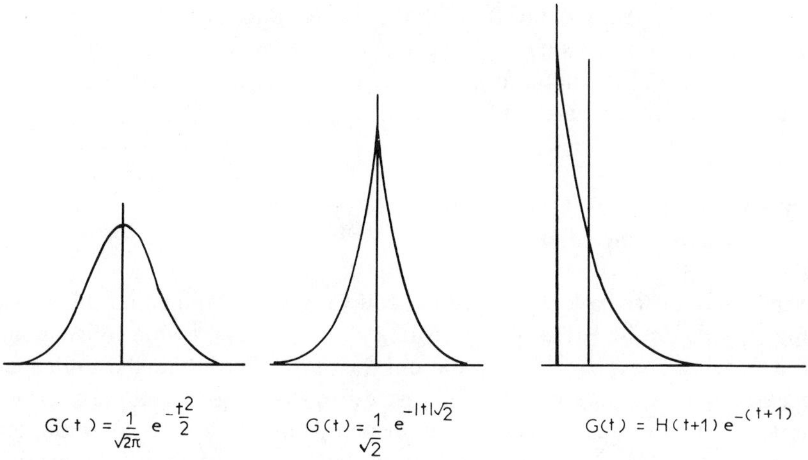

Fig. 3.5. Schematic representation of the three proposed models of distribution. [From Bougeault (1981a).]

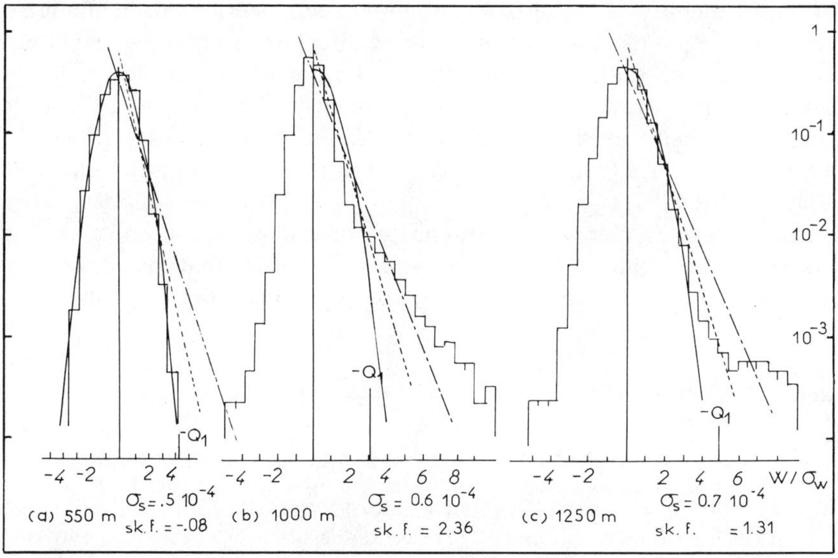

Fig. 3.6. Histograms of t from 3-D data at three levels inside the cloud layer. The three theoretical models have also been plotted: (a) 550 m, (b) 1000 m, and (c) 1250 m. [From Bougeault (1981a).]

advances in modeling cloud fractional coverage and mean cloud properties for a field of clouds using higher order closure models will require direct observations of the fluctuating thermal and water substance quantities at various levels in a cloud field.

3.10 Implications of Averaging to the Interpretation of Model-Predicted Data

In this chapter we have examined the concepts of averaging and the impact of averaging on the equations governing cloud processes. One consequence of averaging the governing equations is that inherent uncertainties are revealed in model forecasts. In the case of top-hat or grid-volume-averaged numerical models, uncertainties are present primarily due to the initial and boundary conditions supplied to the model and to variability across the model grid volume. Use of the generalized ensemble-averaging operator, moreover, introduces departures or variances from the mean quantities not only due to initial/boundary conditions, but also from variability introduced by our selection criteria for "look-alike" conditions. Thus the variances are also a measure of our inherent uncertainty in defining the existence criteria for a given cloud or cloud distribution. With time we may be able to learn to better identify "look-alike" conditions such as Richardson numbers, convergence parameters, and similarity scaling laws, so that the magnitude of this fundamental inherent uncertainty is gradually diminished. Until we can develop operational models with sufficient resolution to be able to simulate explicitly the finest details of cloud motions and physical processes, we must recognize that such inherent uncertainty exists in any forecast of a cloud system.

In the next chapter we will examine cloud microphysical processes and the parameterization of those processes. We shall see that the concepts of averaging also affect our perspective of how to model these processes.

References

André, J. C., G. DeMoor, P. Lacarrere, and R. DuVachat (1976a). Turbulence approximation for inhomogeneous flows. Part I: The clipping approximation. *J. Atmos. Sci.* **33**, 476–481.

André, J. C., G. DeMoor, P. Lacarrere, and R. DuVachat (1976b). Turbulence approximation for inhomogeneous flows. Part II: The numerical simulation of a penetration convection experiment. *J. Atmos. Sci.* **33**, 482–491.

Anthes, R. A. (1977). A cumulus parameterization scheme utilizing a one-dimensional cloud model. *Mon. Weather Rev.* **105**, 270–286.

Arakawa, A., and W. H. Schubert (1974). Interaction of a cumulus cloud ensemble with the large-scale environment. Part I. *J. Atmos. Sci.* **31**, 674–701.

Banta, R. M. (1979). Subgrid condensation in a cumulus cloud model. *Conf. Probab. Stat. Atmos. Sci., 6th, Banff, Alta*, pp. 197–202.

Betts, A. K. (1975). Parametric interpretation of trade-wind cumulus budget studies. *J. Atmos. Sci.* **32**, 1934.

Bougeault, Ph. (1981a). Modeling the trade-wind cumulus boundary layer. Part I: Testing the ensemble cloud relations against numerical data. *J. Atmos. Sci.* **38**, 2414–2428.

Bougeault, Ph. (1981b). Modeling the trade-wind cumulus boundary layer. Part II: A high-order one-dimensional model. *J. Atmos. Sci.* **38**, 2429–2439.

Boussinesq, J. (1877). Essai sûr la theorie des eaux courantes. *Mem. Acad. Sci.* **23**, 24–46.

Chen, C., and W. R. Cotton (1983a). A one-dimensional simulation of the stratocumulus-capped mixed layer. *Boundary-Layer Meteorol.* **25**, 289–321.

Cotton, W. R., and G. J. Tripoli (1978). Cumulus convection in shear flow three-dimensional numerical experiments. *J. Atmos. Sci.* **35**, 1503–1521.

Deardorff, J. W. (1970). A three-dimensional numerical investigation of the idealized planetary boundary layer. *Geophys. Fluid Dyn.* **1**, 377.

Deardorff, J. W. (1972). Numerical investigation of neutral and unstable planetary boundary layers. *J. Atmos. Sci.* **29**, 91–115.

Hill, G. E. (1974). Factors controlling the size and spacing of cumulus clouds as revealed by numerical experiments. *J. Atmos. Sci.* **31**, 646–673.

Hinze, J. O. (1959). "Turbulence," pp. 522–523. McGraw-Hill, New York.

Klemp, J. B., and R. B. Wilhelmson (1978). Simulations of right- and left-moving storms produced through storm splitting. *J. Atmos. Sci.* **35**, 1097–1110.

Leonard, A. (1974). Energy cascade in large-eddy simulations of turbulent fluid flows. *Adv. Geophys.* **18A**, 237–248.

Lilly, D. K. (1967). The representation of small-scale turbulence in numerical simulation experiments. *Proc. IBM Sci. Comput. Symp. Environ. Sci.*

Manton, M. J., and W. R. Cotton (1977). Parameterization of the atmospheric surface layer. *J. Atmos. Sci.* **34**, 331–334.

Mellor, G. L. (1977). The Gaussian cloud model relations. *J. Atmos. Sci.* **34**, 356–358; 1483–1484.

Mellor, G. L., and T. Yamada (1974). A hierarchy of turbulence closure models for planetary boundary layers. *J. Atmos. Sci.* **31**, 1791–1806.

Monin, A. S., and Obukhov, A. M. (1954). Basic laws of turbulent mixing in the atmosphere near the ground. *Akad. Nauk SSSR* **24**, 163–187.

Ogura, Y., and H. Cho (1973). Diagnostic determination of cumulus cloud populations from observed large-scale variables. *J. Atmos. Sci.* **30**, 1276–1286.

Oliver, D. A., W. S. Lewellen, and G. G. Williamson (1978). The interaction between turbulent and radiative transport in the development of fog and low-level stratus. *J. Atmos. Sci.* **35**, 310–316.

Orville, H. D., and F. J. Kopp (1977). Numerical simulation of the life history of a hailstorm. *J. Atmos. Sci.* **34**, 1596–1618.

Panofsky, H. A., and J. A. Dutton (1984). "Atmospheric Turbulence." Wiley, New York.

Smagorinsky, J. (1963). General circulation experiments with the primitive equations. 1: The basic experiment. *Mon. Weather Rev.* **91**, 99–164.

Sommeria, G. (1976). Three-dimensional simulation of turbulent processes in an undisturbed trade-wind boundary layer. *J. Atmos. Sci.* **33**, 216–241.

Sommeria, G., and J. W. Deardorff (1977). Subgrid-scale condensation in models of nonprecipitating clouds. *J. Atmos. Sci.* **34**, 344–355.

Sommeria, G., and M. A. LeMone (1978). Direct testing of a three-dimensional model of the planetary boundary layer against experimental data. *J. Atmos. Sci.* **35**, 25–39.

Yanai, M., S. Esbensen, and J. Chu (1973). Determination of bulk properties of tropical cloud clusters from large-scale heat and moisture budget. *J. Atmos. Sci.* **30**, 611–627.

Zeman, O., and J. L. Lumley (1976). Modeling buoyancy driven mixed layers. *J. Atmos. Sci.* **33**, 1974–1988.

Chapter 4 The Parameterization or Modeling of Microphysical Processes in Clouds

4.1 Introduction

As we have seen, the formulation of a model of a cloud or field of clouds requires a number of value judgments or compromises. The need for compromise, however, becomes most obvious when one is faced with the task of formulating models of the microstructure of clouds. If a modeler with access to the most advanced levels of computer power is developing a 3-D cloud model or even a 1-D model using higher order closure theory, he is likely to come to the conclusion that a sophisticated, explicit prediction of the evolution of cloud microstructure is either impossible or impractical.

The alternative is to develop a simple parameterization of cloud microphysical processes. As a general approach, detailed theoretical models and/or experimental data are used to formulate parameterizations of the physics. Ideally, the parameterizations should capture the essence of the known microphysics in simple formulations of the processes. The problem is that, in some cases, the physics is not sufficiently well enough known, or is too complex, to fully capture its essence in simple formulations.

In this chapter we review briefly the concepts and general theory of the microphysics of clouds. We then summarize approaches to parameterizing cloud microphysical processes. We conclude by discussing the interaction of cloud microphysical processes and the dynamics of clouds.

4.2 General Theory of the Microphysics of "Warm" Clouds

By "warm" clouds we refer to clouds in which the ice phase does not play a significant role in either the thermodynamics or precipitation processes. In general, the term refers to clouds whose tops are no colder than 0°C. However, the physical processes that are prevalent at temperatures warmer than 0°C can also operate quite effectively at colder temperatures or in supercooled clouds. Thus, we will not limit our discussion to clouds that are entirely warmer than 0°C.

Noting the differences in droplet concentration in cumuli formed in maritime and continental air masses, Squires (1956, 1958) introduced the concept of colloidal stability of warm clouds. He pointed out that similar clouds forming in a maritime air mass are more likely to produce rain than clouds forming in a continental air mass. Thus, maritime clouds are less colloidally stable than are their continental counterparts. The relationship between the cloud droplet concentration and the cloud nucleus population was demonstrated by Twomey and Squires (1959) and Twomey and Warner (1967). Hence, in a nucleus-rich continental air mass, a given liquid-water content must be distributed over numerous small droplets having small collection kernels or collection cross sections (i.e., low terminal velocities, collection efficiencies, and cross-sectional areas). Thus, in general, the collision and coalescence process is inhibited in nucleus-rich continental air masses. These concepts form the basis of many of the parameterizations of warm-cloud processes that we shall discuss shortly. The fundamental premise of many of the parameterizations is that the cloud droplet concentration or activated cloud condensation nuclei (CCN) concentration, at cloud base, determine whether or not a cloud will precipitate.

Recent studies (Johnson, 1980) have pointed out that the cloud-base temperature also influences the activation of cloud droplets. Other things being the same (i.e., aerosol distribution, cloud-base updraft velocity), clouds with colder cloud bases will activate more cloud droplets than those having warmer bases. This is a consequence of the nonlinear variation of saturation vapor pressure with temperature, which results in *higher peak supersaturations in cold-base clouds* than in warm-base clouds that are otherwise the same. It is the direction of this effect that is most interesting, because the temperature effect will accentuate the tendency for colloidal stability in continental clouds if those clouds also have cold bases (a common occurrence in midlatitude, continental regions). The aerosol distribution and updraft velocity at cloud base, however, remain as the most important factors controlling the concentration of activated droplets and, hence, the colloidal stability of a cloud.

Cloud-base temperature also figures into a cloud's colloidal stability because a cloud with a warm cloud-base temperature has a larger saturation mixing ratio at cloud base. Other things being the same (i.e., cloud depth, activated CCN concentration, etc.), a warm-based cloud will have a greater potential for producing a significant amount of condensed liquid water and, therefore, a greater chance of generating a few big droplets.

For some time it was thought that collision and coalescence could not proceed until the radius of droplets exceeded 19 μm (Hocking, 1959). More recent calculations of collision efficiency by Klett and Davis (1973), Hocking and Jonas (1970), and Jonas and Goldsmith (1972) suggest that droplets smaller than 19 μm in radius do exhibit finite collection efficiencies. Due to their small fall velocities and cross-sectional areas, however, the rate of collision among droplets of such a small size is very low. Thus, it is still thought that a few larger droplets ($r > 20$ μm) must form in a cloud in order to initiate significant growth rates through the relatively random collisions among small, comparably sized droplets. This initial phase of collision and coalescence has been modeled as a stochastic process (Telford, 1955; Gillespie, 1972) or what is now referred to as quasistochastic process (Berry, 1967).

Since condensation theory for a smooth, unmixed updraft predicts a narrowing of the droplet spectrum with time (Howell, 1949; Mordy, 1959; Neiburger and Chien, 1960; Fitzgerald, 1974), the search continues to explain sufficient broadening of the initial distribution to sustain vigorous collision and coalescence growth of precipitation. One school of thought pursued by a number of researchers (Beliaev, 1961; Mazin, 1965; Levin and Sedunov, 1966; Jaw, 1966; Wen, 1966; Stepanov, 1975, 1976) suggests that small-scale turbulence and associated fluctuations in supersaturation and turbulent mixing may initiate the broadening of the drop-size distribution. Simpler models proposed by Warner (1969), Bartlett and Jonas (1972), and Mason and Jonas (1974) predict only slight broadening of the droplet distribution or production of a droplet distribution of unrealistic shape.

Recently, Manton (1979) developed a more general theory of the interaction of the cloud droplet distribution with a turbulent cloud. The theories proposed by Mazin (1968) and Levin and Sedunov (1966), as well as Warner (1969) and Bartlett and Jonas (1972), represent special cases of Manton's theory. Manton showed that the theories of Mazin (1968) and Levin and Sedunov (1966), which represent supersaturation fluctuations by an eddy diffusivity, initially broaden a narrow droplet distribution in a homogeneous cloud, but the narrowing of the distribution by condensation eventually overwhelms this tendency. Similar to Warner (1969) and Bartlett and Jonas (1972), Manton also showed that vertical velocity fluctuations cannot of themselves induce broadening of the droplet distribution. This is because variations in time for an air parcel to reach a given level above cloud base are compensated by variations in supersaturation. By assuming that the convective cloud turbulence has a high vertical coherence and that fluctuations in the integral radius, or mean radius, of the droplet distribution are negatively correlated with vertical velocity, Manton showed that an initial unimodal distribution is transformed into a bimodal distribution at heights above cloud base. This theory predicts that the net dispersion increases with increasing turbulence intensity as a consequence of the increasing separation of the peaks when the turbulence intensity is large. Consistent with observations, the theory further predicts that the dominant mode of the distribution varies slowly as a function of height. Thus the mean droplet radius increases slowly with height, while the dispersion increases monotonically. The theory also predicts that more monomodal droplet distributions will prevail in clouds having low values of the large-scale turbulence intensity. Manton refers to large-scale turbulence as being generated by convective eddies which have high vertical coherence. In contrast, small-scale, vertically incoherent turbulence tends to attenuate and broaden the distinct modes and leads to a broad unimodal distribution whenever the large-scale turbulence intensity is low. Recently, Austin *et al.* (1985) showed that the fundamental premise of Manton's theory that fluctuations of droplet integral radius are negatively correlated with vertical velocity fluctuations is not observed in the continental cumuli they studied.

Based on the laboratory experiments of Latham and Reed (1977) and Baker and Latham (1979), Baker *et al.* (1980) have also calculated the effects of turbulence on the initial broadening of a droplet spectrum growing by condensation. In their model, entrainment and mixing are viewed as taking place inhomogeneously. Thus, finite blobs or streams of unsaturated air mix with nearly saturated blobs, resulting in the complete evaporation of some droplets of all sizes. Other droplets do not change in size. For better understanding, one can view this as a two-stage process in which a tongue or stream of dry air penetrates into the interior of a cloud either from the top or the sides. During the first stage, the dry tongue and cloudy air remain distinctly separate. At the interface between the two streams, complete evaporation of all droplets occurs. At some distance, for example, greater than one meter, all cloud droplets will remain unaffected by the presence of the dry tongue. During the second stage, the interface breaks down and the two air masses intermingle. The final concentration of the new mixture will be reduced in magnitude from the unmixed cloudy air by an amount which is in proportion to the weighted mean of the values of cloud-free and cloudy air. As illustrated in their laboratory experiments (see Fig. 4.1), the shape of the droplet-size distribution will change little; the main con-

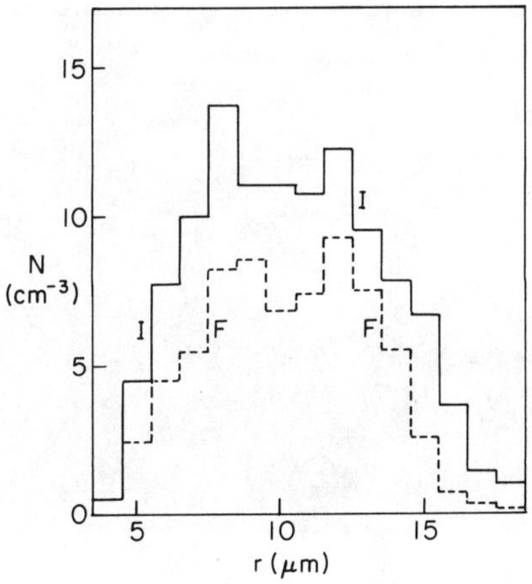

Fig. 4.1. Droplet-size distribution for experiment, with the large nozzle; I, before mixing; F, after mixing. [From Baker *et al.* (1980).]

sequence of the inhomogeneous mixing process is that the overall concentration is reduced. Similar to Manton's large-scale turbulence assumptions for which the turbulence correlation time scale is large compared to the droplet growth/evaporation time scale, Baker *et al.* assume that the time constant for turbulent mixing is large relative to the time scale for evaporation. Calculations with the model demonstrate that the theory predicts a bimodal droplet spectrum similar to observed spectra and that the dispersion of the spectra increases with height above cloud base. An interesting feature of the theory is that the largest droplets should grow much faster than predicted with either the adiabatic or homogeneous mixing theories. The authors attributed the more rapid growth of the largest droplets to local values of supersaturation much greater than either of the two latter cases. In the inhomogeneous mixing model, more droplets are completely evaporated, and the newly activated droplets that replace them cannot compete as effectively for the available water vapor. This leads to a local rise in the supersaturation, with the droplets least affected by the blob of dry air growing faster.

In Baker *et al.* (1980), the description of the characteristics of the turbulence contributing to inhomogeneous mixing was quite vague. Recently, Baker *et al.* (1984) adapted the Broadwell and Breidenthal (1982) model of turbulent mixing in a shear layer to inhomogeneous mixing in clouds. In this model, mixing occurs within discrete vortices in the shear layer (see Fig. 4.2) having an appearance similar to the Kelvin–Helmholtz billows occasionally observed on the top of stratus clouds. Extension of

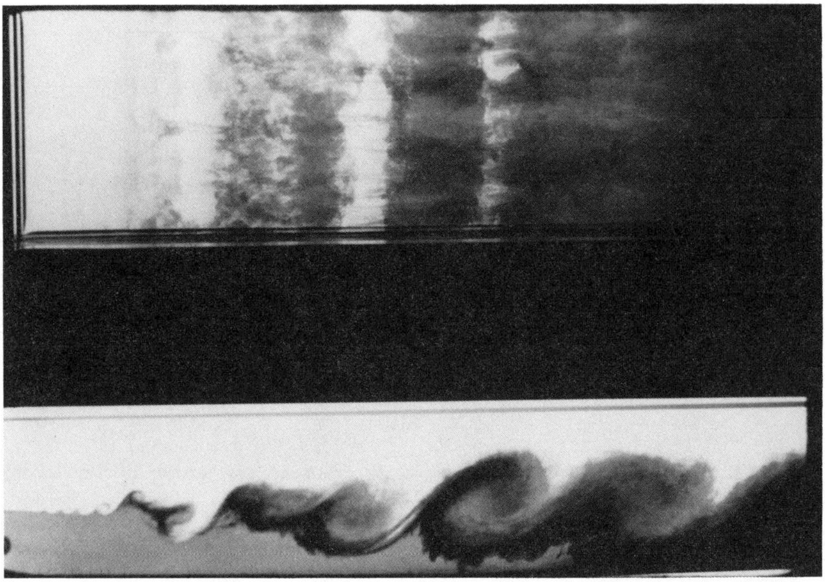

Fig. 4.2. Simultaneous plan (top) and side (bottom) views of a chemically reacting, plane shear layer. $\Delta UL/\nu \sim 10^4$. [From Baker *et al.* (1984).]

the Broadwell–Breidenthal model to a buoyant cloud provides a mathematical and pictorial framework for describing the concept of inhomogeneous mixing more clearly. We shall examine the concepts of entrainment and mixing more fully in Chapter 8.

Another view of the role of turbulence in droplet spectral broadening has been advanced by Telford and Chai (1980) and Telford *et al.* (1984). As in Manton's theory, Telford views a cloud as being composed of an ensemble of vertically coherent updraft and downdraft entities. The downdraft entities are created when saturated air near cloud top mixes with dry environmental air. The resultant evaporation of drops in the mixed parcel causes it to become denser than its surroundings, thereby initiating its descent. During descent, adiabatic heating causes further evaporation of drops. Some drops, partially evaporated during mixing at cloud top, are totally evaporated during descent; others will simply diminish in size. Telford hypothesizes that new drops are continually mixed into the descending entity from adjacent undiluted parcels. The droplets mixed from surrounding undiluted parcels thus constitute a continuing source of large drops in the descending entity. When the downdraft transforms into an updraft, the drops most recently mixed from the undilute surroundings will be larger than the other drops, and so grow bigger as the parcel approaches cloud top again. The largest of the recycled droplets will be bigger than the largest droplets formed during the initial undiluted ascent of the updraft. Thus, if there is sufficient vertical cycling and entrainment of dry air at cloud top, Telford and Chai (1980) predict that a broad droplet distribution will result.

Recent in-cloud measurements obtained during the 1984 Cooperative Convective Precipitation Experiment (CCOPE) have further fueled the controversy regarding the processes involved in the initial broadening of cloud droplet spectra (see, e.g., Telford, 1987; Paluch and Knight, 1987). Both Paluch and Knight (1986) and Hill and Choularton (1985) find that droplet spectral broadening does not occur in the most dilute or oldest portions of the cloud. Hill and Choularton (1985) find that the largest drops are produced in the wettest, least-mixed, rapidly ascending cloud turrents which reside on the upshear side of a cloud. This asymmetric structure of cumulus clouds is described more fully in Chapter 8. They calculate that the largest droplets are larger than would be produced, however, by condensation in adiabatic updrafts. There is no evidence, they also note, that the ascending turrents are mixing with the residues of older clouds as required by the Mason–Jonas (Mason and Jonas, 1974; Jonas and Mason, 1982) multiple-thermal cloud model.

These observations clearly do not support the Telford recycling process, since the largest drops should be found in downdrafts or in the more strongly mixed regions. As noted by Telford (1987), this does not mean that the recycling process does not operate in more maritime clouds. Telford suggests that the dryness of the environment of the continental clouds observed in CCOPE rapidly evaporates the cloud without allowing sufficient time for several cycles to occur with a particular cloud parcel. Although not supported by Telford, it is also likely that the narrow droplet spectrum initially produced in the unmixed rising parcels in continental air masses does not produce large enough droplets to survive the evaporation stages in the descending plumes or survive mixing between the wet rising plumes and drier descending plumes.

In spite of the fact that the broadest droplet spectra occurred in the least dilute regions of the cloud where liquid-water contents were as much as 70% of the adiabatic values, Hill and Choularton conclude that the observed small-scale variations in droplet spectra and droplet concentrations are evidence that inhomogeneous mixing is occurring. They suggest that the process may be more complicated than envisioned by Baker *et al.*, however. Paluch and Knight go one step further and speculate that there may exist a spectral broadening process not related to mixing. There is also the possibility that the spectral broadening exceeding adiabatic condensational growth calculations observed in the little-mixed portions of the CCOPE clouds may be an artifact of the FSSP sensor rather than real drop broadening. While one may question the details of the turbulence/drop-broadening models of Manton, Baker *et al.*, or Telford, they have clearly strengthened the case for turbulent mixing as a major factor in the formation of embryonic precipitation droplets.

Another hypothesis is that the aerosol distribution contains a number of giant or ultragiant aerosol particles which can act as the embryos for further coalescence growth. Observations reported by Nelson and Gokhale (1968), Hindman (1975), Johnson (1976, 1982) and Hobbs *et al.* (1977, 1978) have shown the presence of potentially significant concentrations of aerosol particles of sizes as large as 100 μm. Johnson (1979, 1982) calculated that these particles are sufficiently numerous to account for rapid development of precipitation-sized particles, even in colloidally stable, continental clouds. The calculated time for precipitation formation in the Miles City, Montana, region where CCOPE was performed was on the order of 45 min. Observations suggest, however, that the drop broadening occurs on time scales of the order of 10 min in clouds in that region. Furthermore, Woodcock *et al.* (1971) and Takahashi (1976) have concluded that giant salt nuclei do not contribute substantially to warm rain initiation in maritime clouds. It is, thus, still uncertain whether ultragiant aerosol particles are sufficiently ubiquitous or affect the droplet-broadening process sufficiently to be important to the observed rates of droplet broadening.

At present, this question is still unresolved: does the aerosol distribution, cloud turbulence, or some other process ultimately control the formation of embryonic precipitation droplets in cloud?

Once sufficiently large embryonic precipitation droplets are formed, turbulence, electric fields, and drop charges are all thought to influence the rates of drop collision and coalescence. An analysis of the effects of turbulence on droplet collision and coalescence has been performed by de Almeida (1975, 1976). His calculations suggest that a turbulent cloud with dissipation rates of 1 to 10 $cm^2\ s^{-3}$ can significantly enhance the collision efficiencies of 15-μm-radius collector droplets. Pruppacher and Klett (1978) argue that the effects of turbulence on drop collision is overestimated by de Almeida's model. Moreover, owing to the very small collection kernels of droplets with a radius of less than 20 μm, the influence of turbulence on collision and coalescence must be great in order to account for the initiation of precipitation embryos. The effects of drop charge and electric fields on collision and coalescence have been examined both theoretically and experimentally by numerous investigators over the years (Davis, 1964, 1965; Sartor, 1960; Krasnogorskaya, 1965; Lindblad and Semonin, 1963; Plumlee and Semonin, 1965; Semonin and Plumlee, 1966). The results of these studies suggest that the collision efficiencies for small, strongly charged drops can be considerably enhanced in field strengths that are characteristic of thunderstorms. However, Pruppacher and Klett (1978) conclude that the weak charges and weak fields present in developing warm cumulus clouds are probably not sufficient to promote the initiation of precipitation embryos.

Whatever the nature of the process of initial broadening of the droplet spectrum, the portions of a given cloud most favorable for the initiation of precipitation in warm clouds are the regions of highest liquid-water contents. Twomey (1976) showed that if locally enhanced regions of LWC comprise only 1% of the cloud volume, and exist for periods of a few minutes, such regions can produce significant concentrations of large drops averaged over the entire volume of the cloud. Thus, the presence of protected updrafts having nearly wet adiabatic liquid-water contents (Heymsfield *et al.*, 1978) can have significant bearing upon the initiation of precipitation in warm clouds. Of course, the ultimate amount of rainfall from a given cloud is controlled by the overall time–space character of its updrafts and its liquid-water content.

Langmuir (1948) suggested that once raindrops grow to a critical size of approximately 6 mm in diameter, they will break up due to hydrodynamic instability. He hypothesized that each breakup fragment will act as a new precipitation embryo which can grow to breakup size and create more raindrop embryos. He referred to this process as the "chain reaction" theory of warm-rain formation. Other observations (Blanchard, 1948; Magarvey and Geldhart, 1962; Cotton and Gokhale, 1967; Brazier-Smith *et al.*, 1972; McTaggart-Cowan and List, 1975) have suggested that collisions among droplets on the order of 2–3 mm in diameter and smaller can initiate breakup. Computations of the evolution of raindrop spectra reported by Brazier-Smith *et al.* (1973), Young (1975), and Gillespie and List (1976) have indicated the greater importance of collision-induced breakup over spontaneous breakup to the evolution of raindrop spectra. Srivastava (1978) calculated that for rainwater contents ($M > 1\ \mathrm{g\ m^{-3}}$), collision breakup results in raindrop size distributions which are approximately constant in slope and have an intercept of the distribution function which is proportional to M. Using a numerical cloud model, Farley and Chen (1975) have concluded that a necessary condition for the development of a Langmuir chain reaction requires that a cloud must develop sustained updrafts in excess of $10\ \mathrm{m\ s^{-1}}$.

It is a formidable task to model explicitly and realistically the evolution of the droplet spectra from nucleation, to initial broadening, to mature rain. Breakup processes must also be considered. One approach to the task is to assume that the droplet spectrum is continuous on the scale of the averaging domain of a cloud model. We then can predict the time variation of the spectral density $f(x)$ of cloud droplets of mass x to $x \pm \delta x/2$ at a given geometric position and at a given instant. An integral differential equation describing the evolution of the droplet spectrum takes the form

$$\frac{\partial f(x)}{\partial t} = N(x) - \frac{\partial[\dot{x} f(x)]}{\partial x} + G(x)|_{\text{gain}} + G(x)|_{\text{loss}} + B(x)|_{\text{gain}} + B(x)|_{\text{loss}} + \tau(x), \tag{4.1}$$

where N represents nucleation, G represents collection, B represents breakup, and τ represents the sum of both mean and turbulent transport processes.

The first term on the right-hand side of Eq. (4.1) is the production of droplets of mass x by the nucleation of such droplets on activated CCN. This term appears in Eq. (4.1) only if the droplet spectrum $f(x)$ is truncated at some small droplet mass.

The second term on the RHS of Eq. (4.1) is the divergence of $f(x)$ due to continuous vapor mass deposition on droplets growing at the rate $\dot{x}$, where $\dot{x}$ is a function of the droplet mass, its solubility in water, and the local cloud supersaturation as well as other factors (Byers, 1965; Mason, 1971; Pruppacher and Klett, 1978). If one chooses to extend the droplet spectrum to include soluble particles of mass x_s, the nucleation term (N) would be included explicitly in the second term. However, since $\dot{x}$ is a function of the solubility of a droplet, one should use a two-dimensional density function $f(x, x_s)$, as was done by Clark (1973).

The third and fourth terms on the RHS of Eq. (4.1) represent, respectively, the gain and loss integrals of $f(x)$ due to the collision and coalescence of cloud droplets. The gain and loss terms were formulated by Berry (1967) as follows

$$G(x)|_{\text{gain}} = \tfrac{1}{2} \int_0^x K(x_c, x') f(x_c) f(x')\, dx', \tag{4.2}$$

where $x_c = x - x'$ and

$$G(x)|_{\text{loss}} = -\int_0^\infty K(x, x') f(x) f(x')\, dx'. \tag{4.3}$$

The term $K(x, x')$ is the collision cross section, or collection kernel, often taken to be

$$K(x, x') = \pi(R_x + R_{x'})^2[v(x) - v(x')]E(x, x'), \tag{4.4}$$

where $E(x, x')$ is the collection efficiency, R_x and $R_{x'}$ are the radii of droplets of mass x and x', and $v(x)$ and $v(x')$ are the average terminal velocities of droplets of mass x and x'.

Integration of the droplet spectra evolution using Eqs. (4.2), (4.3), and (4.4) has been referred to by Gillespie (1975) as the quasistochastic model because it predicts a unique spectrum at a given time and point in space, whereas the pure stochastic model predicts fluctuations in the droplet spectrum at a given time and point in space.

The fifth and sixth terms on the RHS of Eq. (4.1) represent, respectively, the gain and loss of spectral density $f(x)$ due to breakup of droplets. Thus, the breakup of larger droplets whose fragments are of mass $x \pm \delta x/2$ is formulated as

$$B(x)|_{\text{gain}} = \int_x^\infty p(x')\, f(x')\, g(x|x')\, dx', \tag{4.5}$$

where $p(x')$ is the probability per unit/time that a droplet of mass x' to $x' \pm \delta x/2$ will break up due to internal hydrodynamic instability. The function $g(x|x')$ represents the number of fragments of mass $x \pm \delta x/2$ formed by the breakup of a droplet of mass $x' \pm \delta x/2$.

The loss of $f(x)$ due to the breakup of droplets of mass $x \pm \delta x/2$ is formulated as

$$B(x)|_{\text{loss}} = -p(x)\, f(x). \tag{4.6}$$

It was mentioned earlier that breakup due to collision of raindrops appears to be the dominant contributor to drop breakup, at least if the rainwater contents exceed 1 g m^{-3}. Brazier-Smith *et al.* (1973) formulated collision breakup by combining the collision and breakup terms in Eq. (4.1) to formulate a general stochastic interaction equation of the form

$$\begin{aligned} &G(x)|_{\text{gain}} + G(x)|_{\text{loss}} + B(x)|_{\text{gain}} + B(x)|_{\text{loss}} \\ &\quad = \tfrac{1}{2}\int_0^\infty \int_0^\infty K(x|x_\alpha, x_\beta)\, f(x_\beta)\, f(x_\alpha)\, dx_\alpha\, dx_\beta \\ &\qquad - f(x) \int_0^\infty \left| \int_0^\infty K(x_\alpha|x, x_\beta) x_\alpha\, dx_\alpha \right| \frac{f(x_\beta)}{x + x_\beta}\, dx_\beta, \end{aligned} \tag{4.7}$$

where $K(x|x_\alpha, x_\beta)$ may be considered a generalized interaction kernel related to the probability $[h(x|x_\alpha, x_\beta)]$ of forming a droplet of mass x because of the interaction of droplets of mass x_α and x_β. In this instance, the interaction could represent a pure coalescence problem in which case Eq. (4.7) degenerates to Eqs. (4.2) and (4.3). Alternatively, it could represent a collision event which promotes breakup.

A common approach to integrating Eq. (4.1) is to discretize $f(x)$ into 40 to 70 elements and then integrate the equations by finite-difference methods (Twomey, 1964, 1966; Bartlett, 1966, 1970; Warshaw, 1967; Berry, 1967;

Kovetz and Olund, 1969; Bleck, 1970; Chien and Neiburger, 1972). In several cases this technique has been applied to one-dimensional cloud models (Danielsen *et al.*, 1972; Takahashi, 1973), two-dimensional cloud models (Takahashi, 1974), and even three-dimensional cloud models (Takahashi, 1981). To do so, however, requires large amounts of computer time and storage. Fortunately, there are several other technqiues for integrating Eq. (4.1). These include moment-conserving techniques (Ochs and Yao, 1978; Ochs, 1978) and the use of a series of γ and double log-normal distribution functions as basis functions (Clark, 1976; Clark and Hall, 1983). The latter approach offers some promise of providing accurate and computationally efficient solutions to Eq. (4.1). It has not, however, been fully developed to include collision-induced breakup such as formulated in Eq. (4.7).

As computer power increases it may become more practical to obtain explicit solutions to the evolution of droplet spectrum by using Eq. (4.1). However, for complicated cloud or mesoscale models, there still remains a strong desire to develop simplified techniques for predicting the evolution of the droplet spectrum to form rain along with its sedimentation through the cloud.

4.3 Parameterizations of Warm-Cloud Physics

The primary goal in developing a simple parameterization of warm-cloud physics is to capture in a few simple formulas the essential physics embodied in more general theoretical models such as Eq. (4.1). We would like the parameterization to include the following physics:

(i) The nucleation of aerosol spectra into droplet spectra having features which characterize the properties of the environment.

(ii) The initial broadening of the droplet spectra into "embryonic" precipitation elements.

(iii) The development of a mature rain spectrum, including the processes of breakup in the spectrum evolution.

(iv) The differential sedimentation of cloud-sized droplets and raindrops.

(v) The evaporation of cloud droplets and raindrops.

The most popular approach to parameterizing these processes is to separate the partitioning of condensed liquid water into two domains. An effective way to illustrate such a partitioning is the use of Berry's (1965) liquid-water mass spectral density function shown in Fig. 4.3. The convenient feature of this spectral density function is that the area under the curve

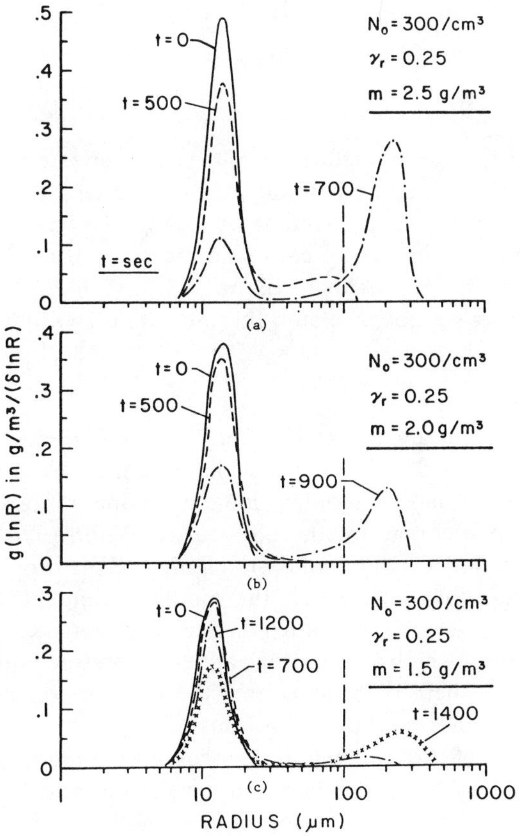

Fig. 4.3. Computed variation in the distribution of water mass density for an initial concentration of $100\ \text{cm}^{-3}$, a radius dispersion of 0.25, and LWCs of (a) 2.0, (b) 1.5, and (c) $1.0\ \text{g m}^{-3}$. [From Cotton (1972a).]

illustrated in Fig. 4.3 is equivalent to the total liquid-water content. Generally, we distinguish between cloud droplets, assumed to have sufficiently small terminal velocities to generally move with air parcels, and raindrops, which have significant settling speeds $v(R)$, where R is the drop radius. Kessler's (1969) identification of an arbitrary threshold separating cloud droplets falling at small terminal velocities from higher terminal velocity raindrops is illustrated as a vertical dashed line in Fig. 4.3. To the left of the vertical dashed line, water vapor is condensed on low-terminal-velocity cloud droplets, where turbulence-condensational broadening and collision and coalescence form raindrops which are on the right side of the line.

Berry and Reinhardt (1974) demonstrated that a natural break between cloud and raindrops occurs at a radius of 50 μm.

4.3.1 Conversion Parameterizations

Kessler (1969) first developed a simple parameterization of the rate of autoconversion of condensed liquid water from cloud droplets, having liquid-water content m (mass/volume), to raindrops, having water content M. He assumed that the rate of conversion was a linear function of m for liquid-water contents greater than 1.0 g m^{-3}. Transformed into mixing ratio quantities, Kessler's autoconversion formula can be written

$$\mathrm{CN}_{\mathrm{cr}} = -\rho_0^{-1}(dm/dt) = K_1(r_{\mathrm{c}} - a), \tag{4.8}$$

where $K_1 > 0$ if $r_{\mathrm{c}} > a$ and $K_1 = 0$ if $r_{\mathrm{c}} \leq a$. In our notation we have adopted the subscript cr ($\mathrm{CN}_{\mathrm{cr}}$), which indicates that raindrops having mixing ratios r_{r} are supplied by cloud droplets having mixing ratios r_{c}. Thus, this parameterization accounts for the physical condition (ii) set forth at the beginning of Section 4.3, the initial broadening of the droplet spectra into "embryonic" raindrops. However, the model is not able to distinguish between air masses which are continental and are therefore slow in condition (ii), and air/masses which are maritime and are therefore rapid in condition (ii). Furthermore, there is evidence that the rates of conversion of cloud water to rainwater are nonlinear functions of r_{c}.

Berry (1967, 1968) formulated an autoconversion model which allowed for condition (i) by being dependent upon the initial cloud droplet concentration N_{c} as well as the dispersion ν_{r} of the initial cloud droplet distribution. According to Berry, autoconversion should be formulated as

$$\mathrm{CN}_{\mathrm{cr}} = \frac{\rho_0 r_{\mathrm{c}}^2}{60}\left[2 + \left(\frac{0.0266}{\gamma_{\mathrm{r}}}\right)\frac{N_{\mathrm{c}}}{\rho_0 r_{\mathrm{c}}}\right]^{-1}. \tag{4.9}$$

Equation (4.9) was deduced from the average time for the sixth-moment radius r_6 of a theoretically calculated spectrum (Berry, 1967) to reach a 40-μm radius. It was then assumed that at this time all of r_{c} was converted to a spectrum centered about $r_{\mathrm{g}} = 40$ μm. In contrast to Eq. (4.8), Berry's formulation exhibits a roughly cubic dependence on r_{c} and furthermore demonstrates a direct dependence on the initial cloud droplet dispersion and an inverse dependence on the initial cloud droplet concentration.

Cotton (1972a) noted that Berry's formula consistently resulted in the formation of rain too low in simulated clouds. He then derived a conversion formula, also from detailed collision and coalescence calculations, that introduced a dependence upon the "age" of a rising parcel of cloud droplets.

While the concept of introducing the age of a population of droplets in estimating condition (ii) was desirable, the need to estimate an age of a droplet population made the scheme untenable for application in 3-D models.

Subsequently, Manton and Cotton (1977; see also Tripoli and Cotton, 1980) formulated another autoconversion formula which was able to respond to differing air masses as well as to retain the simplicity of Eq. (4.8). The Manton and Cotton (1977; hereafter referred to as MC) formulation of autoconversion was developed by simplifying the known theory involving droplet spectral broadening. The autoconversion rate was formulated as

$$\mathrm{CN}_{\mathrm{cr}} = f_{\mathrm{c}} r_{\mathrm{c}} H(r_{\mathrm{c}} - r_{\mathrm{cm}}), \tag{4.10}$$

where f_{c} represents the mean frequency of collision among cloud droplets which become raindrops. $H(x)$ is the Heaviside step function, so r_{cm} is a threshold cloud water content below which there is no conversion. It behaves in a manner similar to a in Kessler's formula, Eq. (4.8). However, in contrast to Kessler, both f_{c} and r_{cm} vary depending upon the microphysical properties of the cloud.

It is assumed, for example, that collision and coalescence are largely responsible for the conversion process. The role of condensation only broadens the droplet distribution beyond a critical size so that collision and coalescence can proceed at a significant rate. It is assumed that this occurs when the mean droplet radius exceeds R_{cm}. This provides an estimate of the critical liquid-water content needed to promote conversion as

$$r_{\mathrm{c_m}}\mathrm{m}_{\mathrm{cr}} = (4\pi/3)\rho_{\mathrm{l}} R_{\mathrm{cm}}^3 N_{\mathrm{c}}/\rho_0, \tag{4.11}$$

where ρ_{l} is the density of water and N_{c} represents the initial cloud droplet concentration near cloud base. MC estimated R_{cm} to be approximately 10 μm. We see that Eq. (4.10) satisfies condition (i) by introducing a dependence upon the concentration of active CCN which give rise to N_{c}.

Once $(\rho_0 \overline{r_{\mathrm{c}}})$ exceeds $(\rho_0 M_{\mathrm{cr}})$, the rate of conversion is controlled by $\overline{r_{\mathrm{c}}}$ and by the mean frequency of collision among embryonic raindrops. The latter process was formulated by MC to be

$$f_{\mathrm{c}} = \pi R_{\mathrm{a}}^2 E_{\mathrm{c}} V_{\mathrm{c}} N_{\mathrm{c}}, \tag{4.12}$$

where R_{a} represents the mean cloud droplet radius

$$R_{\mathrm{a}} = (3 r_{\mathrm{c}} \rho_0 / 4\pi\rho_{\mathrm{l}} N_{\mathrm{c}})^{1/3}, \tag{4.13}$$

V_{c} represents the terminal velocity of a droplet of radius R_{a}, and E_{c} represents an average collection efficiency associated with conversion. MC estimated E_{c} to be 0.55. Once the liquid water exceeds M_{cr}, Eq. (4.11) produces an autoconversion rate which is roughly

$$\mathrm{CN_{cr}} = \alpha N_c^{1/3} \bar{r}_c^{-4/3} r_c, \tag{4.14}$$

where we distinguish between the local random variable r_c and the mean liquid-water mixing ratio $\bar{r}_c$. The dependence of the conversion rate on $\bar{r}_c$ arises from assumptions used in estimating f_c in Eq. (4.10). Since Eq. (4.10) contains the local random variable r_c, one must be certain that Eq. (4.10) is averaged as we have done with other scalar variables in Chapter 3.

Equations (4.8), (4.9), and (4.10) represent extreme simplifications of the physical processes involved in the broadening of cloud droplet distributions to form raindrops. However, as noted previously, even in the most sophisticated models of warm-rain processes, there still remain major uncertainties concerning the processes which contribute to initial broadening of the cloud droplet distributions into raindrops. As discussed above, the relative roles of turbulence and the characteristics of the activated aerosol spectra in controlling the initial broadening of the droplet distribution remain unresolved.

A simple numerical evaluation of the conversion rate models given by Eqs. (4.8), (4.9), and (4.10) reveals that for a given liquid-water content they differ by several orders of magnitude. This suggests there is little skill in our ability to parameterize the conversion rates in warm clouds. Application of the conversion rate formulas to deep convective clouds reveals that these formulas serve mainly as triggers for warm-rain initiation. Once precipitation-size droplets form, the rain formation process is dominated by raindrops accreting smaller cloud droplets. As such, the absolute magnitude of the conversion rate may not be very important. Only in shorter lived, low-liquid-water cumuli will the magnitude of the conversion rate be essential to determining if such a cloud will produce precipitation. Recently, Chen and Cotton (1987) have shown that considerable skill in formulating conversion rates is required in longer lived, low-liquid-water content stratocumulus clouds. In this application the magnitude of the computed conversion rates is similar to computed accretion rates throughout the life cycle of the cloud system. Thus an unavoidable burden is placed on our ability to predict conversion rates in those clouds.

4.3.2 Parameterization of Accretion

Once embryonic precipitation particles are formed, Kessler (1969) hypothesized that the water content converted to rain is distributed in the inverse exponential distribution function formulated by Marshall and Palmer (1948) as

$$N(D) = N_0 e^{-\lambda D}, \tag{4.15}$$

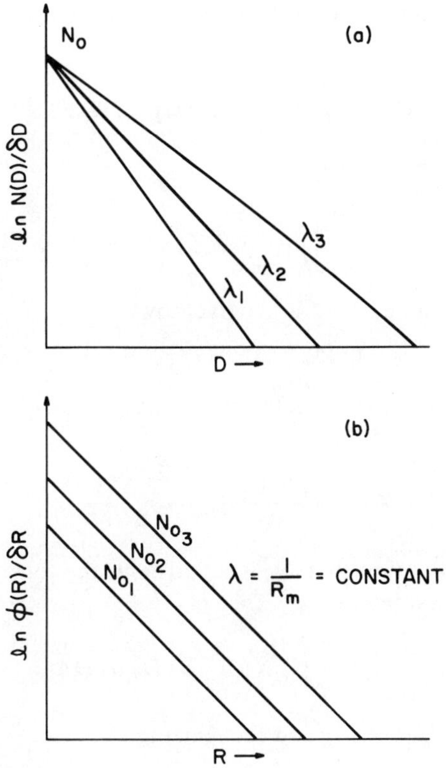

Fig. 4.4. Schematic illustration of the inverse exponential drop-size distribution function of Marshall–Palmer drop-size distribution. (a) Kessler's model in which N_0 is assumed constant and λ varies with rainwater content; (b) Manton–Cotton's model in which the slope $\lambda = 1/R_m$ is a constant and $N_0 = N_R/R_m$ varies with rainwater content.

where $N(D)$ represents the number of raindrops per unit/volume of diameter $D \pm \delta D/2$. Figure 4.4 illustrates the Marshall–Palmer distribution function.

Kessler assumed that the rate of mass growth of raindrops is primarily by accretion of cloud droplets of the form

$$(d/dt)[x(D)] = (\pi D^2/4)\bar{E} V_D \rho_0 r_c, \tag{4.16}$$

where $x(D)$ is the mass of a raindrop of diameter D, V_D is the terminal velocity of a raindrop of diameter D, and $\bar{E}$ represents an average collection efficiency between raindrops and cloud droplets.

The rate of change of rainwater mixing ratio by accretion or collection of cloud droplets is then

$$\mathrm{CL}_{\mathrm{cr}} = \frac{1}{\rho_0} \int_0^\infty \frac{d}{dt}[x(D)]\, N(D)\, dD, \tag{4.17}$$

or

$$\mathrm{CL}_{\mathrm{cr}} = \frac{1}{\rho_0} \int_0^\infty \frac{\pi D^2}{4} \bar{E} V_{\mathrm{D}} \rho_0 r_{\mathrm{c}}\, N(D)\, dD. \tag{4.18}$$

Kessler further assumed that V_{D} is given by

$$V(D) = 130.0 D^{1/2} \quad \mathrm{m\,s^{-1}}, \tag{4.19}$$

from Spilhaus (1948). Then, after substituting Eq. (4.19) into Eq. (4.18) and integrating, we find that

$$\mathrm{CL}_{\mathrm{cr}} = \frac{130.0\pi}{4} N_0 \bar{E} \rho_0 r_{\mathrm{c}} \frac{\Gamma(3.5)}{\lambda^{3.5}}. \tag{4.20}$$

Under the assumption that N_0 is a constant, the total rainwater mixing ratio r_{r} may be obtained as

$$r_{\mathrm{r}} = \frac{1}{\rho_0} \int_0^\infty X(D)\, N(D)\, d(D). \tag{4.21}$$

After substituting Eq. (4.15) and recognizing that

$$X(D) = \pi D^3 \rho_{\mathrm{l}}/6, \tag{4.22}$$

Eq. (4.21) becomes, after integrating,

$$r_{\mathrm{r}} = \pi \rho_{\mathrm{l}} N_0 \Gamma(4)/\rho_0 6 \lambda^4. \tag{4.23}$$

Solving Eq. (4.23) for λ and substituting into Eq. (4.20) gives us

$$\mathrm{CL}_{\mathrm{cr}} = \left\{ \frac{130.0\pi^{0.125}}{4} \left[\frac{6}{\Gamma(4)} \right]^{0.875} \Gamma(3.5) \right\} N_0^{0.125} \rho_0^{1.875} \bar{E} r_{\mathrm{c}} r_{\mathrm{r}}^{0.875}. \tag{4.24}$$

Equation (4.24) shows that raindrops accreting cloud droplets produce a rainwater mixing ratio at a rate proprotional to $(r_{\mathrm{c}} r_{\mathrm{r}}^{0.875})$.

MC (see also Tripoli and Cotton, 1980) formulated a slightly different model of raindrops accreting cloud droplets. They reexpressed Eq. (4.15) as

$$\phi(R) = (N_{\mathrm{R}}/R_{\mathrm{m}}) \exp(-R/R_{\mathrm{m}}), \tag{4.25}$$

where R_{m} is a characteristic radius of the distribution. Based on the experiments by Blanchard and Spencer (1970) and modeling experiments by Srivastava (1978), they assumed that the slope of the distribution $(1/R_{\mathrm{m}})$

is a constant, whereas the intercept parameter $N_0 = N_R/R_m$ varies with rainwater content (Fig. 4.4b). The mechanism for maintaining a constant slope has been suggested by Srivastava (1978), who found that once the rainwater content exceeded $1.0\,\mathrm{g\,m^{-3}}$, collisional breakup becomes so dominant that the slope of the distribution tends toward constant values. Figure 4.4 illustrates the difference in behavior of the constant slope and constant N_0 models with variations in raindrop liquid-water content.

Under the assumption of constant slope, and terminal velocity $V(R)$ given by

$$V(R) = 2.13(\rho_l/\rho_0)^{0.5}(gR)^{-0.5} \quad \mathrm{m\,s^{-1}}, \tag{4.26}$$

the equation expressing the rate of raindrop collection of cloud droplets becomes

$$\mathrm{CL_{cr}} = 0.884\bar{E}(g\rho_0/\rho_l R_m)^{-0.5} r_c r_r. \tag{4.27}$$

The main difference between Eqs. (4.27) and (4.24) is that Eq. (4.27) is a linear function of r_r, whereas Eq. (4.20) is a function of $r_r^{0.875}$. The fact that Eq. (4.27) is a product of r_c and r_r has a significant impact on the form of the accretion equation in an averaged system. For example, suppose we apply the Reynolds averaging procedures outlined in Chapter 3 to the collection equation. Equation (4.27) then becomes

$$\mathrm{CL_{cr}} = 0.884\bar{E}(g\rho_0/\rho_l R_m)^{-0.5}(\bar{r}_c\bar{r}_r + \overline{r_c'' r_r''}). \tag{4.28}$$

Thus, the mean rate of transfer of water mixing ratio from cloud droplets to raindrops is not only a function of the product of the mean mixing ratios, but also a function of the covariances between them. The covariances $\overline{r_c'' r_r''}$ may be obtained from a second-order turbulence model. Averaging Eq. (4.24), however, requires additional assumptions about the behavior of the variables r_c and r_r before a closure model can be derived.

The formulation of the raindrop distribution parameters affects other processes in the model system as well. For instance, by assuming that the distribution slope is a constant, the average terminal velocity of the raindrop distribution does not vary with r_r. By comparison, Kessler's theory predicts that the average fall velocity varies slowly with r_r, proportional to $r_r^{0.125}$.

4.3.3 Evaporation of a Population of Raindrops

Similarly, one can estimate the rate of change in raindrop water content due to the evaporation of raindrops. The rate of a single raindrop is given by

$$(d/dt)[x(R)] = 4\pi R\, G(T, P)(S-1)(1 + F\,\mathrm{Re}^{-0.5}), \tag{4.29}$$

where S is the saturation ratio (e/e_s), F is the ventilation coefficient normally taken to be 0.21, and Re is the Reynolds number. The function $G(T, P)$ is defined in Byers (1965) as

$$G(T, P) = 1 \Big/ \left(\frac{m_w L_{vl}^2}{K_T R_* T^2} + \frac{R_* T}{e_s(T) D_v m_w} \right),$$

where m_w is the molecular weight of water, K_T is the molecular diffusivity of heat, R_* is the universal gas constant, D_v is the molecular diffusivity for water vapor, and $e_s(T)$ is the saturation vapor pressure over a plane pure water surface at temperature T. Also, Re is defined as $\mathrm{Re} = 2R\, V(R)\, \rho_0/\mu$, where $V(R)$ can be estimated with Eq. (4.26) and μ is the kinematic viscosity.

The rate of change of raindrop mixing ratio due to evaporation of rain is given by

$$\mathrm{VD}_{rv} = \frac{1}{\rho_0} \int_0^\infty \frac{dx(R)}{dt} \phi(R)\, dR, \tag{4.30}$$

if the distribution function, Eq. (4.25), is used, or

$$\mathrm{VD}_{rv} = \frac{1}{\rho_0} \int_0^\infty \frac{dX(D)}{dt} N(D)\, dD, \tag{4.31}$$

if Eq. (4.15) is used. $X(D)$ is simply the reformulation of Eq. (4.26) in terms of drop diameter $D = 2R$.

Substituting Eqs. (4.26) and (4.25) into Eq. (4.31) and integrating yields the evaporation rate

$$\mathrm{VD}_{rv} = \left| 0.5 + 0.349 \left[\frac{(\rho_l g \rho_0 R_m^3)^{1/4}}{\mu^{1/2}} \right] \right| \frac{G(T, P)}{R_m^2 \rho_l} (S-1) r_r \tag{4.32}$$

for a raindrop distribution with constant slope. An alternative expression for VD_{rv} is obtained by substituting Eqs. (4.29) and (4.15) into Eq. (4.31),

$$\mathrm{VD}_{rv} = \left| \frac{2\pi\Gamma(2) N_0^{1/2}}{\left[\dfrac{\pi \rho_l \Gamma(4) \rho_0}{6} \right]^{1/2}} + \frac{(0.21)(g/2)^{1/4}\Gamma(11/4) N_0^{5/16} r_r^{3/16}}{\rho_0^{9/16} \rho_l^{7/16} \left[\dfrac{\pi\Gamma(4)}{6} \right]^{11/16}} \right| \times G(T, P)(S-1) r_r^{0.5}, \tag{4.33}$$

which is a function of $r_r^{0.5}$. Equation (4.33) is more complicated than Kessler's, since an explicit Reynolds number dependence has been included in the derivation. Kessler obtained a raindrop evaporation rate which was proportional to $r_r^{0.65}$.

Even if one is comfortable with the assumption that the raindrop distribution is of the form of Eq. (4.15) or (4.25), it is not likely that the assumption that the slope of the distribution is constant or that N_0 is constant will hold for all cloud types or throughout the lifetime of a given cloud. This is especially true for mesoscale convective systems which are composed of (1) a distinct convective-scale component exhibiting one characteristic precipitation structure and (2) a stratiform component exhibiting quite a different microphysical structure and history. For example, if the rainwater contents are less than 1.0 g m^{-3}, it becomes much less likely that collision-induced breakup will maintain a constant slope of the raindrop distribution, and a Marshall–Palmer-type raindrop distribution may not be found at all in such clouds.

Also, if the rain has its origins in the supercooled portions of "mixed-phase" clouds, collision-induced breakup may not have sufficient time to evolve the raindrop distribution into a stable distribution with a constant slope. Since the distribution is in transition, it may not have a constant N_0 either. In this case, it may be necessary to formulate equations in which both the shape and N_0 are predicted.

Srivastava (1978) formulated such a procedure for use in cloud models. Under the constraint that raindrops are distributed according to Eq. (4.15), he formulated equations which predict variations in N_0 and λ due to (i) coalescence between raindrops, (ii) collisional breakup of raindrops, (iii) spontaneous breakup of raindrops, (iv) the capture of cloud droplets, and (v) condensation or evaporation of raindrops. He also suggested that autoconversion could be introduced in the parameterization. As long as the raindrop distributions do not depart substantially from the inverse exponential form, this approach offers an opportunity for including greater generality in the formulation of warm-rain physics than is possible with the simpler parameterizations discussed previously.

Recently, Nickerson *et al.* (1986) formulated a mesoscale model in which the raindrop-size distribution was assumed to be log-normal,

$$N(D) = \frac{N_r}{(2\pi^{0.5}\sigma_0 D)} \exp\left[-\frac{1}{2\sigma_0^2}\ln^2(D/D_0)\right]. \tag{4.34}$$

This drop-size distribution function contains two parameters: the mean diameter D_0 and the standard deviation σ_0. The inverse exponential functions described previously, however, contain only one free parameter. As a result, the log-normal distribution function should be better able to approximate observed drop-size distributions than the more limited inverse exponential function. This was demonstrated recently by Feingold and Levin (1986). It does, however, result in more complicated expressions for collection, evaporation, and so on. Nickerson *et al.* also formulated separate prognostic equations for rainwater mixing ratio and total raindrop number concentration. Important to the simulation was the inclusion of self-collection among raindrops and collision-induced breakup. As noted by Berry and Reinhardt (1974), self-collection among raindrops is important in determining the ultimate concentration and shape of the raindrop spectrum.

Still, there remains the concern that the constraint of raindrop distributions to a single drop-size distribution will lead to errors. One such error could arise in estimating the settling rate of raindrops by using a single characteristic fall velocity of the raindrop distribution. Ultimately, we can expect that warm-cloud microphysical models constrained to single Marshall–Palmer or log-normal raindrop distribution functions will be abandoned in favor of more general approaches, such as the use of a series of log-normal distribution basis functions, under development by Clark (1976) and Clark and Hall (1983). Instead of using the autoconversion formulas discussed previously, Clark and Hall solve the quasistochastic collection equations explicitly in the transformed drop-size-coordinate space. This approach should yield more accurate predictions of the evolution of drop-size-spectra, yet greatly reduce the computational and storage requirements of simulating droplet spectral evolution relative to using finite-difference techniques.

4.4 Fundamental Principles of Ice-Phase Microphysics

The parameterization of ice-phase microphysical processes in a cloud model is greatly complicated by the variety of forms of the ice phase, as well as by the numerous physical processes that determine the crystal forms. Moreover, in contrast to the physics of warm clouds, our understanding of ice-phase physics is far less complete. This means that in many cases the formulation of simple parameterized models of the ice phase cannot be done using information derived from detailed theoretical/numerical models or from observations.

The physical processes that should be considered in formulating a model or parameterization of the ice phase are as follows:

(i) Primary and secondary nucleation of ice crystals.
(ii) Vapor deposition growth of ice crystals.
(iii) Riming growth of ice crystals.
(iv) Graupel or hail particle initiation from heavily rimed crystals.
(v) Graupel or hail particle initiation by the freezing of supercooled raindrops.
(vi) Graupel or hail particle riming and vapor deposition growth.
(vii) Graupel or hail particle–particle collision with supercooled raindrops.
(viii) Shedding of water drops from hailstones growing by wet growth or from partially wetting ice particles.

(ix) The initiation of aggregates of ice crystals by collision among ice crystals.

(x) Aggregate collection of ice crystals.

(xi) Aggregate riming of cloud droplets.

(xii) Melting of all forms of ice particles.

4.4.1 *Primary and Secondary Nucleation of Ice Crystals*

It is widely accepted that at temperatures warmer than −40°C the nucleation of the ice phase in the troposphere requires the presence of some form of nucleus or mote, in a process called *heterogeneous nucleation.* One possible exception occurs or is found in the neighborhood of a thunderstorm lightning discharge, where the breakup of drops in strong electric fields could initiate the ice phase (Peña and Hosler, 1971; Vonnegut and Moore, 1965). In less disturbed regions of the troposphere, it is believed that the principle ice nucleation mechanisms are (i) vapor-deposition (or sublimation) nucleation, (ii) sorption (or condensation-freezing) nucleation, (iii) immersion-freezing nucleation, and (iv) contact-freezing nucleation.

Vapor-deposition nucleation refers to the direct transfer of water vapor to a nucleus that results in the formation of an ice crystal. *Sorption nucleation* refers to the condensation of water vapor on a nucleus to form an embryonic droplet, followed by freezing. This is viewed as a two-step process, so the name "condensation freezing" is often used. In practice it is not easy to distinguish between these two modes of nucleation. Some researchers (see Koenig, 1962) argue that sorption nucleation predominates in the atmosphere, since the vapor–solid–liquid transition is energetically favored over the vapor–solid transition. Fletcher (1962) has argued that condensation or sorption is preferred at smaller supercooling and large supersaturation, while vapor deposition is preferred at large supercooling and small supersaturation.

Immersion freezing refers to the nucleation of a cloud droplet or raindrop on an ice nucleus which is immersed within the drop. Two theories of immersion freezing have emerged. One theory views the freezing process as stochastic, such that at a given degree of supercooling, not all drops of a population of drops will freeze at the same time (Bigg, 1953a, b, 1955; Carte, 1956; Dufuor and Defay, 1963). The second theory, called the singular theory, holds that at a given degree of supercooling, the probability that a drop of a given size will freeze depends solely on the likelihood that the drop contains an active freezing nucleus. This process is independent of time. Laboratory experiments reported by Vali and Stanbury (1966) have indicated that the freezing process is time dependent, though the amount of time dependence is small. Both theories predict that the probability of freezing increases exponentially with the degree of supercooling and with the volume (or size) of the drop. Thus, for small cloud droplets at small degrees of supercooling, this mechanism of nucleation is not very effective.

Contact nucleation refers to the nucleation of a supercooled drop by a nucleus that makes contact with the surface of the drop in the supercooled state. Observations have shown that dry particles, such as clays, sand, CuS, and organic compounds which make contact with a supercooled drop, are much more effective as contact-freezing nuclei than when immersed within the drop (Rau, 1950; Fletcher, 1962; Levkov, 1971; Gokhale and Spengler, 1972; Pitter and Pruppacher, 1973; Fukuta, 1972a, b). In the atmosphere, contact nucleation can produce considerable time dependence in the nucleation process, since it depends both on the probability that an aerosol particle makes contact with a supercooled drop and on the probability that the aerosol particle acts as an active freezing nucleus. Young (1974a) modeled contact nucleation by naturally and artificially generated aerosols. He considered cloud droplet scavenging of nuclei particles by *Brownian diffusion* and by the combined effects of *thermophoresis* and *diffusiophoresis.* Brownian diffusion refers to the chance encounter between a supercooled cloud droplet and an active nucleus (aerosol particle) due to the random motion of both species as a consequence of the thermal bombardment with gas molecules. The rate of collision is a function of the kinetic energy of the air molecules (or temperature) and the mobilities of both species. Thermophoresis refers to a net transport of particles in a thermogradient from warm toward colder regions. Diffusiophoresis refers to the net transport of aerosol particles in the direction of a vapor flux. In the case of a droplet growing by vapor deposition, diffusiophoresis is directed toward the droplet, whereas thermophoresis acts away from the droplet. The reverse is true of an evaporating droplet. Young (1974a) has noted that since thermophoresis dominates over diffusiophoresis, contact nucleation by nuclei in the size range $0.15 < 1.0$ μm would be suppressed in a growing cumulus cloud. He cites observations reported by Mee and Takeuchi (1968) and Koenig (1962) which indicate that ice is most prevalent in downdrafts and at the edges of clouds. This gives evidence, he suggests, that phoretic-contact nucleation is an effective mechanism in natural clouds.

Various laboratory cloud chambers such as expansion chambers or settling cloud chambers have been designed and constructed to examine the concentration of ice nuclei (IN) capable of initiating the ice phase. In general, these devices cannot distinguish between the various nucleation modes, although due to limited observation time they tend to discriminate against the contact freezing mode. The results of such measurements have been synthesized by Fletcher (1962) in a single empirical formula

$$N_{\mathrm{IN}} = A \exp(\beta T_{\mathrm{s}}), \tag{4.35}$$

where N_{IN} is the concentration of IN active at T_s, $\beta = 0.6\,(°C)^{-1}$, $A = 10^{-5}$ liter^{-1}, and T_s is the degree of supercooling ($T_s = T_0 - T$) where T_0 is 0°C and T is the cloud temperature. This equation states that the concentration of active IN in the atmosphere increases exponentially with the amount of supercooling. Roughly, the concentration of IN increases by one order of magnitude for each 4°C of supercooling.

There is some evidence that the concentration of IN also varies with the degree of supersaturation with respect to ice ($S_i - 1$). Thus, at a given temperature, Gagin (1972), Huffman (1973), and Huffman and Vali (1973) found that the concentration of IN varies according to the relation

$$N_{IN} = C(S_i - 1)^k, \tag{4.36}$$

where C and k are constants. As long as the atmosphere is supersaturated with respect to ice, ice nucleation can occur even if it is subsaturated with respect to water.

There is some concern that the ice-nucleating ability of atmospheric aerosols will differ, depending upon the particular mode of ice nucleation. As mentioned previously, the routinely used devices for counting IN do not discriminate among the various modes of nucleation. However, these devices do favor the deposition and sorption modes. The contact-freezing and inversion-freezing modes have little chance to operate effectively in those devices. There have been various attempts to examine the immersion-freezing ability of natural materials such as plant leaf litters (Schnell and Vali, 1976), which are ground up and dispersed in bulk water samples and then distributed in drops on a freezing stage. These experiments, however, do not tell us how such materials are actually distributed as aerosols in the atmosphere. There have also been several studies of the nucleating ability of artificial nuclei, such as AgI in the contact-freezing mode, but there have been few attempts to examine the ability of natural aerosols to act in the contact mode. Generally, one expects that the activity spectra of contact nuclei will be something like Eq. (4.35), since artificial contact nuclei exhibit such a behavior. However, the actual magnitude of the coefficients in Eq. (4.35) that are appropriate for contact nuclei remains unknown.

In recent years it has become increasingly evident that concentrations of ice crystals in "real" clouds are not always represented by the concentrations of IN measured or expected to be activated in such environments. In particular, it has been found that at temperatures warmer than −10°C, the concentration of ice crystals can exceed the concentration of IN activated at cloud-top temperature by as much as three or four orders of magnitude (Braham, 1964; Koenig, 1963, 1965; Mossop and Ono, 1969; Mossop *et al.*, 1967, 1968, 1970, 1972; Magono and Lee, 1973; Ono, 1972; Isono, 1965; Auer and Marwitz, 1969; Hobbs, 1969; Hobbs *et al.*, 1974). The effect is greatest in clouds with broad drop-size distributions (Koenig, 1963; Mossop *et al.*, 1968, 1972; Hobbs, 1974). However, in winter cumuli over Israel, Gagin (1971) observed that the expected activated IN concentrations were within a factor of 10 of observed ice crystal concentrations, even at warmer temperatures. Heymsfield (1972) and Heymsfield and Knollenberg (1972) observed crystal concentrations in cirrus clouds of 50 liter^{-1}, where activated IN concentration would be expected to be 10^3 liter^{-1}. Clearly, IN are not reliable predictors of ice crystal concentrations in real clouds.

Various hypotheses have been advocated to explain such discrepancies. In particular, the hypotheses attempt to account for the unexpectedly high ice crystal concentrations at warm temperatures. Some of the leading hypotheses are as follows:

(i) Fragmentation of large drops during freezing.
(ii) Mechanical fracture of fragile ice crystals (i.e., dendrites and needles) caused by collision of these crystals with graupel or other ice particles.
(iii) Splinter formation during riming of ice crystals.
(iv) Enhanced ice nucleation in regions of exceptionally high supersaturations in the presence of large quantities of supercooled raindrops or in a turbulent cloud environment.

The hypothesis that freezing supercooled drops frequently shatter and produce splinters is based on laboratory experiments. Mason and Maybank (1960) suggested that large quantities of splinters were produced during the freezing of suspended supercooled drops. However, Dye and Hobbs (1966, 1968) showed that the vigor of the splinter production process was influenced by the presence of CO_2 gas. Subsequent studies indicate that each drop-freezing event produces, on the average, less than two splinters (Brownscombe and Thorndyke, 1968; Takahashi and Yamashita, 1969, 1970; Pruppacher and Schlamp, 1975). Scott and Hobbs (1977) examined the drop-freezing-induced splintering mechanism in a one-dimensional, time-dependent cloud model. They found that if each freezing drop greater than 50 μm ejects four splinters in a maritime cloud, the graupel concentration increases by only a factor of 2. Even four splinters per freezing drop in a continental cloud would not promote any significant ice multiplication process. Therefore, this process does not appear to be capable of explaining crystal concentration 10^4 times greater than activated IN concentrations.

Circumstantial evidence that mechanical fracturing of delicate ice crystals may significantly enhance ice-particle concentration has been provided by field observations of fragmented crystals in the Colorado Rockies (Grant, 1968; Vardiman and Grant, 1972a, b). The rate of collision between ice particles (i.e., ice crystal–ice crystal, graupel–ice crystal, aggregate–ice crystals, and ice crystals with large drops) is proportional to the product of the concentration of ice crystals and the concentration of any hydrometeor element they may collide with. This is a self-perpetuating progress whose rate grows quadratically with the ice-particle concentration. However, if the initial concentrations are determined by the concentration of activated IN at warm temperatures, this process will have a very slow "spin-up" rate and therefore is unlikely to account for the very high crystal concentrations frequently observed at warm temperatures.

The hypothesis that secondary ice particles are produced during the riming growth of ice particles has been under laboratory investigation for a number of years (Macklin, 1960; Latham and Mason, 1961; Hobbs and Burrows, 1966; Aufdermauer and Johnson, 1972; Brownscombe and Goldsmith, 1972). However, these studies have produced conflicting results in regard to the quantity of secondary ice particles produced during riming. Recent laboratory studies by Hallett and Mossop (1974) and Mossop and Hallett (1974), confirmed by Goldsmith *et al.* (1976), have indicated that copious quantities of splinters are produced during ice-particle riming under highly selective conditions. These conditions are (i) temperature in the range −3°C to −8°C, (ii) a substantial concentration of large cloud droplets ($r > 12\ \mu m$), and (iii) large droplets coexisting with small cloud droplets. An optimum average splinter production rate of 1 secondary ice particle for 250 large droplet collisions occurred at a temperature of −5°C.

Several investigators have applied this mechanism to cloud models of varying degrees of sophistication in order to examine whether this process can account for the observed concentrations of ice crystals and the rapidity of glaciation of convective clouds (Mason, 1975; Chisnell and Latham, 1976a, b; Koenig, 1977). Chisnell and Latham (1976b) and Koenig (1977) computed that the process can account for significant production of secondary ice particles and complete glaciation of a cloud in less than 45 min, provided supercooled rain is present. In the absence of supercooled rain, Mason (1975) computed that crystal concentrations on the order of 10 liter^{-1} can be produced in about 1.5 h. The large supercooled drops capture small ice crystals formed on IN to become frozen drops, and then the large frozen drops rime cloud droplets at a high rate, thus producing many secondary splinters. These splinters are, in turn, captured by more of the larger supercooled drops, causing the freezing of those drops, which further accelerates the process. It appears that the Hallett–Mossop secondary-particle production process can account for many of the observations of high ice-particle concentrations at warm temperatures. It is not certain if the process can account for all such observations of high ratios of ice crystal concentrations to IN. It is clear, however, that this process depends on the details of the dynamics of a cloud and on the evolution of precipitation by warm-cloud processes. According to Koenig (1977), low updraft speeds are required to produce high concentrations of ice crystals and the rapid and complete glaciation of the cloud. This result was supported by Scott and Hobbs (1977), who found that ice multiplication did not occur until the simulated updrafts weakened enough to allow graupel to settle into the −4 to −8°C temperature zone. It is also important that the cloud circulations allow the downward transport of ice crystals formed in colder regions of the cloud into the critical −4 to −8°C temperature zone.

Finally, we have mentioned the possibility that ice nucleation could be enhanced in regions of spuriously high supersaturations in the presence of large quantities of supercooled raindrops or in a turbulent cloud environment. It is generally agreed that when precipitation is absent, peak supersaturations in convective clouds remain below 1%; this may not be the case, however, when precipitation is present. A number of researchers (Clark, 1973; Subbarao and Das, 1975; Young, 1975) have predicted that when precipitation drops accrete a substantial quantity of cloud droplets, the cloud drop population may no longer be able to condense enough water vapor to balance the production of supersaturation in rapidly rising convective towers. In this situation, the supersaturation may rise in excess of 5 to 10% with respect to water. If this occurs in the supercooled regions of a convective cloud, we would expect very high supersaturations with respect to ice, which would enhance ice nucleation by sorption or deposition. Moreover, if the cloud is quite turbulent, we would expect substantial transient supersaturation fluctuations as discussed in Section 4.1. It is not likely that such transient, locally high supersaturations would be large enough to account for ratios of ice crystal concentrations to IN on the order of 10^4 at warm temperatures. However, this could be a mechanism to initiate sufficiently high ice-particle concentrations to sustain a vigorous ice multiplication process. Unfortunately, this hypothesis has not been examined quantitatively.

4.4.2 Ice-Particle Growth by Vapor Deposition

Once ice crystals are nucleated by some mechanism of primary or secondary nucleation, and if the environment is supersaturated with respect to ice, the crystals can then grow by vapor deposition. Because the saturation vapor pressure with respect to ice is less than the saturation vapor pressure with respect to water, a cloud which is saturated with respect to water will be supersaturated with respect to ice. Figure 4.5 shows the variation in supersaturation with respect to ice as a function of temperature for a water-saturated cloud. Note that ice crystals in a water-saturated cloud can experience supersaturation in excess of 10%. This leads to a process com-

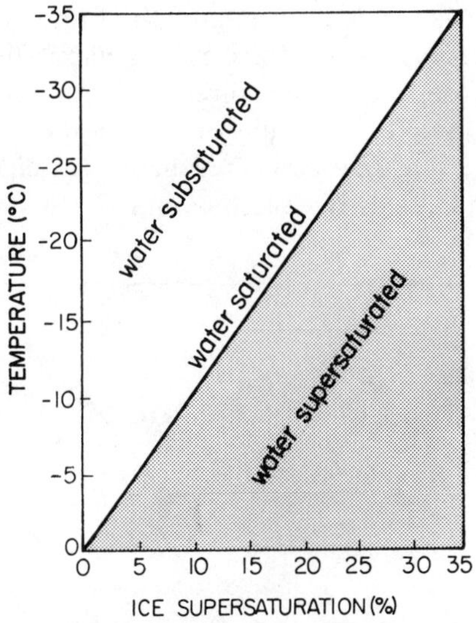

Fig. 4.5. Supersaturation with respect to ice as a function of temperature for a water-saturated cloud. The shaded area represents a water-supersaturated cloud.

monly known as the *Bergeron–Findeisen mechanism.* If a cloud is at water saturation at −10°C, for example, the supersaturation with respect to ice will be 10%. As the ice crystals grow by vapor deposition, they deplete the vapor content, thereby driving the environment below water saturation. The cloud droplets will then evaporate, which helps sustain a vapor pressure difference between ice and water. By this process ice crystals are then said to grow at the expense of cloud droplets. This process has been described by Wegener (1911), Bergeron (1935), and Findeisen (1938).

It should be noted that because the Bergeron–Findeisen process causes the evaporation of cloud droplets, it can also affect the contact nucleation process. Young (1974a, b) and Cotton *et al.* (1986) have pointed out that if the Bergeron–Findeisen process results in the partial evaporation of larger cloud droplets, phoretic scavenging of potential contact nuclei will be enhanced. Thus the Bergeron–Findeisen process and the contact nucleation process can form a positive-feedback loop. Vapor deposition growth of ice crystals lowers the saturation ratio below water saturation, causing droplet evaporation. This evaporation favors the phoretic-contact nucleation of the supercooled droplets that have not fully evaporated, that, in turn, grow as ice crystals, further lowering the saturation ratio below water saturation, and so on. This process is favored whenever the droplet spectrum is broad enough to allow the partial evaporation of the largest droplets.

Prediction of the vapor deposition growth of ice crystals is complicated by the fact that ice crystals exhibit differing habits or shapes depending upon the temperature and supersaturation (with respect to ice) of the environment. For example, the results of the laboratory experiments shown in Fig. 4.6 illustrate that the ice crystal habit can change from plates to needles or prisms to plates over less than 1°C (for the definition of the

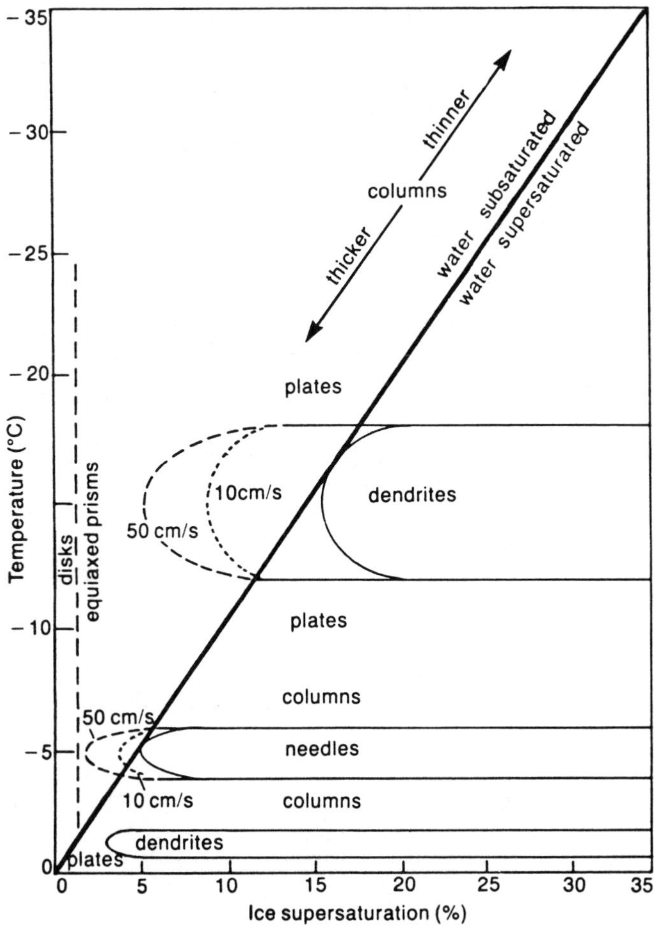

Fig. 4.6. The shape of a crystal is related to environmental conditions in a complicated manner: temperature, supersaturation of the atmosphere with water vapor, and speed of falling all have an effect. Most crystals grow under natural conditions not far removed from the diagonal line representing water saturation. The dotted lines to the left of "dendrites" and "needles" show how the speed of falling extends the zones in which those elongated forms grow. [After Keller and Hallett (1982); figure from Hallett (1984).]

various ice crystal habits, see Pruppacher and Klett, 1978; Mason, 1971). This figure also shows that the ambient temperature and supersaturation are not the only properties that determine the crystal habit. The fall velocity of the crystals and associated ventilation of the crystals can also extend the regimes of needle and dendritic forms of crystals into regions subsaturated with respect to water. The habit of growth of an ice crystal can have a pronounced influence on the rates of vapor-deposition growth. This is especially true for dendritic and needle growth habits, which greatly accelerate the deposition growth rate over that of an equivalent spherical particle.

Thus, the vapor-deposition (evaporation) equation, Eq. (4.29), for spherical particles must be modified to include the role of ice crystal habit. The conventional approach is to assume that the diffusion of heat and water vapor in the vicinity of a complex-shaped ice crystal behaves in a manner analogous to the rate of electrical charge dissipation from an electrically charged capacitor of similar shape (Jeffreys, 1916). Under this assumption, the rate of mass vapor deposition (sublimation) on an ice crystal can be formulated as

$$\left.\frac{dx_i}{dt}\right]_{\mathrm{VD}} = 4\pi C\, G_{\mathrm{i}}(T, P)(S_{\mathrm{i}} - 1)\, f(\mathrm{Re}) - \frac{M_{\mathrm{w}} L_{\mathrm{s}} L_{\mathrm{f}}\, G(T, P)}{K_{\mathrm{i}} R_{\mathrm{a}} T^2} \left.\frac{dx_{\mathrm{i}}}{dt}\right]_{\mathrm{RM}}, \quad (4.37)$$

where x_{i} is the crystal mass, C is the "capacitance" of an ice crystal, S_{i} is the saturation ratio with respect to ice, $f(\mathrm{Re})$ represents a ventilation function of an ice crystal, and $G(T, P)$ is a thermodynamic function similar to that in Eq. (4.29), but modified to include the saturation vapor pressure and the latent heat between vapor and ice. Both the ventilation function and the capacitance vary depending on the particular habit of the ice crystal. The second term on the RHS of Eq. (4.37) represents the contribution to the crystal heat balance by the latent heat released during riming. Cotton (1970) found that for clouds with liquid-water contents greater than 0.3 g m^{-3}, ice crystal riming $(dx_{\mathrm{i}}/dt)_{\mathrm{RM}}$ can warm the crystal sufficiently to drive the surface of the crystal below ice saturation, causing the crystal to sublimate mass. At this point, however, crystal riming growth dominates the growth of the crystal.

The crystal capacitance is generally computed from theoretical electrostatic capacitance models for simplified shapes such as spheres, disks, and prolate or oblate spheres of revolution. Thus, if we consider a to be the length of the basal plane, and c to be the length of the prism plane of an ice crystal as shown in Fig. 4.7, the capacitance can be approximated as follows:

for needles,

$$C = c/\ln(4c^2/a^2); \quad (4.38)$$

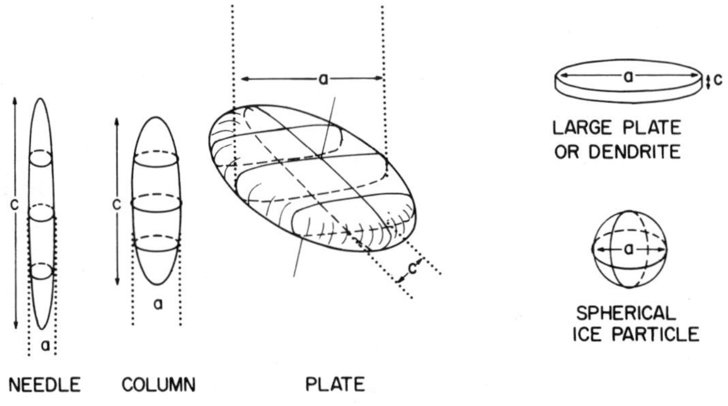

Fig. 4.7. Illustration of approximations to crystal shapes by spheroids of revolution.

for prismatic columns,

$$C = ce/[\ln|(1+e)/(1-e)|] \qquad \text{where} \quad e = \sqrt{1 - a^2/c^2}; \tag{4.39}$$

for hexagonal plates,

$$C = ae/2 \sin^{-1} e, \qquad \text{where} \quad e = \sqrt{1 - c^2/a^2}; \tag{4.40}$$

for thin hexagonal plates or dendrites, the disk approximation

$$C = a/\pi \tag{4.41}$$

may be used, while for spheres,

$$C = a/2. \tag{4.42}$$

In order to evaluate the capacitance of each crystal, the bulk geometry of the crystals must be diagnosed or predicted. This is generally done by using laboratory and/or field data to determine aspect ratios c/a of the crystals or crystal mass versus major dimension (Koenig, 1971; Cotton, 1972a; Young, 1974a; Scott and Hobbs, 1977; Cotton *et al.*, 1982; Jayaweera, 1971).

4.4.3 Riming Growth of Ice Particles

Once ice crystals become large enough, they can settle through a population of supercooled cloud droplets, colliding and coalescing with them. When they impinge upon the ice surface, the droplets immediately freeze, since ice is an "ideal" nucleator. A deposit of frozen droplets called rime accumulates on the surface of the ice crystal. The riming growth process is a collision–coalescence process, analogous to the collision–coalescence growth of liquid cloud droplets. The rate of growth of a single ice crystal of mass x_i can be thus described as

$$\left.\frac{dx_i}{dt}\right]_{RM} = \int_0^\infty A_i'(V_i - V_c)E(x_i/x)xf(x)\,dx, \tag{4.43}$$

where A_i' represents the geometric cross-sectional area that an ice crystal sweeps out relative to a cloud droplet of mass x, V_i is the terminal velocity of an ice crystal of mass x_i, V_c is the terminal velocity of a cloud droplet of mass x, $E(x_i/x)$ is the collection efficiency between the ice crystal and the cloud droplet x, and $f(x)$ is the spectral density of cloud droplets of mass $x \pm \delta x$.

Because ice crystals typically fall with their major dimension in the horizontal plane, the geometric cross section can be approximated as

$$A_i' = (a + r)(c + r)$$

for needles and columns of dimensions a and c falling through cloud droplets of radius r, and

$$A_i' = \pi(a/2 + r)^2$$

for plates, dendrites, and spherical ice particles.

By selecting a suitable spectral density function $f(x_i)$ to keep track of the mass and geometry of ice particles, one can compute the evolution of the ice spectrum by solving quasistochastic integral-differential growth equations similar to Eq. (4.1) used for cloud droplets. In this manner, Young (1974b) and Scott and Hobbs (1977) have simulated the evolution of the ice-particle spectrum by storing information about particle mass and geometry in a series of continuous "bins."

If the ice particles are considerably larger than the cloud droplets, then Eq. (4.43) may be simplified by ignoring the cloud droplet radius in evaluating A_i', by assuming $V_i \gg V_c$, and that an average $\bar{E}$ can be used between all cloud droplets and ice particles of mass x_i. Under these assumptions Eq. (4.43) becomes

$$\left.\frac{dx_i}{dt}\right]_{RM} = A_i' V_i \overline{E(i/c)}\rho_0 r_c, \tag{4.44}$$

where r_c is the cloud droplet mixing ratio.

It should be noted that the ice crystal vapor-deposition growth habits affect A_i', V_i, and E. Details on how to evaluate the terminal velocity and collection efficiency of ice crystals collecting cloud droplets can be found in texts by Pruppacher and Klett (1978) and Mason (1971). As an ice crystal grows by riming, the geometry of the ice crystal is modified, which, in turn, alters A_i', V_i, and E.

In convective clouds having high liquid-water contents (generally greater than 1.0 g m^{-3}) the riming growth of ice crystals can proceed to the point that the rime deposit nearly obscures the original crystal habit. Often such a particle tumbles and rimes to become nearly symmetric or conical in shape. We refer to such heavily rimed particles as *graupel.* The density of small graupel particles may be as low as 0.13 g cm^{-3} (Magono, 1953; Nakaya and Terada, 1935). However, as the graupel particle grows larger and falls faster, the density of the rime deposit may become as high as 0.9 g cm^{-3}. It is often thought that graupel forms from those ice crystals which have grown for a relatively long time by vapor deposition and riming. These "large" crystals are candidates for forming graupel. Reinking (1975) has observed that the embryos of graupel particles are predominantly a select few of relatively smaller crystals that collect droplets at comparably rapid rates. Perhaps these select few crystals randomly collect one or more larger than average cloud droplets. The large cloud droplets could upset the hydrodynamic stability of the crystals, causing them to tumble and form graupel particles. In contrast, crystals which have undergone longer growth times by vapor deposition would have large aspect ratios and be less susceptible to hydrodynamic instability by the chance collision with a droplet slightly larger in size than the average. Furthermore laboratory experiments performed by Fukuta *et al.* (1982, 1984) suggest that in the temperature range −6 to −10°C, where more isometrically shaped, columnar forms of vapor-grown crystals prevail, ice particles switch over more readily to the graupel mode of riming growth. It is interesting that the more rapid switchover to the graupel mode of growth compensates to some degree for the otherwise suppressed precipitation growth by vapor deposition in that temperature range.

There is also some indication (Holroyd, 1964) that aggregates of ice crystals can serve as embryos for graupel particles. We will discuss aggregation more fully in the next section. Graupel can also originate as frozen large cloud droplets or raindrops. The drops may freeze by contact or immersion freezing or by collecting a small ice crystal. These large frozen drops have a high density (perhaps greater than 0.9 g cm^{-3}) and therefore rapidly fall through the population of cloud droplets, collecting them to become high-density graupel particles.

If the cloud contains a high liquid-water content and vigorous updrafts as in cumulonimbi, such graupel particles can serve as embryos for hailstone growth. The rate of mass growth of a hailstone may be estimated from the accretion equation, Eq. (4.44), where $A_i' = \pi R_h^2$ and R_h is the radius of the hailstone. However, as the hailstone grows by collecting cloud droplets, the latent heat of freezing is liberated. This latent heat can warm the hailstone to such a degree that not all the accreted water can freeze. If this occurs, some of the unfrozen water can be shed from the hailstone, thereby limiting its growth (Ludlam, 1951). If the hailstone sheds all unfrozen water, it is said to be growing in the "wet regime," and its mass accumulation can be estimated from the thermodynamic budget of the hailstone (Pruppacher and Klett, 1978; Mason, 1971). If all the accreted water can freeze, then Eq. (4.44) represents the hailstone growth rate and the hailstone is said to be growing in the "dry regime." There remains, however, some uncertainty as to what fraction of the unfrozen water is actually "shed" from a hailstone growing in the wet regime. Macklin (1961) carried out a set of laboratory experiments simulating the growth of hail during the wet regime. He found that much of the unfrozen water was not shed in the wake of the accreting object, but instead was incorporated in the ice structure as a spongy or mushy ice deposit containing a substantial amount of liquid water. This process is apparently enhanced by the collection of ice crystals which, Macklin postulated, can form an interwoven mesh of dendritic crystal structures which trap liquid water. Macklin noted that only at temperatures of −1 to −2°C was any of the unfrozen accreted liquid shed in the wake of the simulated hailstones. At colder temperatures, all the unfrozen water was retained in the "spongy" ice deposit.

4.4.4 *Aggregation Growth of Ice Particles*

Snowflakes, a common form of precipitation in the wintertime in midlatitudes, are made up of clusters or aggregates of "pristine" ice crystals. These aggregates have formed by the collision and coalescence among ice crystals. The collection process can be described by a quasistochastic collection model similar to Eqs. (4.1), (4.2), (4.3) and (4.4). The problem, however, is complicated by the complex geometries and orientations of the falling pristine crystals, which affect the formulation of the collection kernel. An additional complication arises from the fact that once they have made physical contact, ice crystals do not always coalesce. In the case of cloud droplets, there is considerable evidence that the coalescence efficiency is reasonably high, approaching unity. In this case, the major problem in evaluating the collection kernel, Eq. (4.4), is in estimating the hydrodynamic collision efficiency between cloud droplets. This is a rather straightforward procedure but by no means a simple problem (Pruppacher and Klett, 1978). In the case of ice crystal aggregation, in addition to the challenging problem of estimating the hydrodynamic efficiency among ice crystals, we must also estimate their probability of sticking. Laboratory experiments and inferences from field studies suggest that the coalescence efficiency among ice crystals is higher at warm temperatures (Hallgren and Hosler, 1960; Hosler and Hallgren, 1960) and is a function of the habit of the ice crystals, with delicately branched dendrites having the highest efficiencies (Rogers, 1974). However, Latham and Saunders (1971) found no such temperature dependence in their laboratory experiments. Rauber (1985) has also suggested from field observational evidence that a mixture of plane dendrites and spatial dendrites, which have different fall-velocity spectra, favor the formation of aggregates. The exact magnitude of coalescence efficiencies among ice crystals and their variation with crystal habit and temperature remains unknown at the present time.

Another complicating aspect of snowflake aggregation theory is that ice crystals and aggregates of ice crystals exhibit both horizontal velocity fluctuations and fluctuations in terminal velocity. These velocity fluctuations are caused by the complex geometry of the crystals, which can produce aerodynamic lifting forces similar to that of an aircraft wing. Further causes of velocity fluctuations include changes in aerodynamic drag forces due to variations in orientation of the ice crystals and to the shedding of turbulent eddies in the wake of the particle. All of these factors influence the rate of collision among ice crystals. For example, when we discussed the coalescence among cloud droplets, we defined a collection kernel, Eq. (4.4), which was a function of the area swept out by the "parent" droplet and the differences in mean terminal velocities. However, in the case of ice crystals, the collection kernel is influenced by horizontal and vertical velocity fluctuations. Sasyo (1971) examined this effect by laboratory and field experiments as well as numerical calculations. He applied the kinetic theory of gases to this problem and formulated a collection kernel as

$$V(r_{\mathrm{i}}, r_{\mathrm{j}}) = 2\sigma\sqrt{\pi}(r_{\mathrm{i}} + r_{\mathrm{j}})^2, \tag{4.45}$$

where r_{i} and r_{j} represent the effective radii of ice crystals or aggregates, and σ is the standard deviation of horizontal and vertical velocity fluctuations. Sasyo estimated a standard deviation of 5 cm s^{-1} from field experiments. Based on these results he suggested that collisions of this type would be important during the first stages of aggregate formation when ice crystals are of similar size and shape and before a broad spectrum of crystals has formed.

Passarelli and Srivastava (1979; hereafter referred to as PS) have also noted that due to the particular structure of snowflakes, snowflakes of a given mass can have a spectrum of sizes, fall speeds, and shapes. In contrast to the standard collection kernel, Eq. (4.4), where particles of the same mass have zero probability of colliding, such a spectrum leads to a finite probability of collision of particles of the same mass. PS considered two models of aggregation. In the first model snowflakes of a given mass are considered to be spherical and have a unique diameter but a spectrum of fall speeds. This model bears some resemblance to Sasyo's model except that horizontal motions of snowflakes are not considered. In the second model snowflakes are also assumed to be spherical, but snowflakes of a given mass are assumed to have a spectrum of bulk densities, which results in a spectrum of diameters and fall speeds. Numerical experiments with the first model suggested that its contribution to snowflake aggregation was only 10% of that due to the standard kernel. However, experiments with the second model, which includes a spectrum of particle densities, indicated that the magnitude of the modified kernel is always greater than the standard kernel. Depending on the spectrum width, the modified kernel results in substantially more rapid aggregation.

It is clear from this discussion that there remains a great deal to be learned about snowflake aggregation processes. In addition to the uncertainties mentioned above, it should also be noted that aggregation commences from the collision among pristine crystals whose concentration cannot be consistently predicted even within several orders of magnitude of observed values. This is especially true at warmer temperatures, where aggregation seems to be the most prevalent. As can be seen from the collection equations [Eqs. (4.2) and (4.3)], the aggregation rate is proportional to the product of the concentration of ice crystals. Thus a threefold order of magnitude uncertainty in ice crystal concentration results in roughly a sixfold order of magnitude uncertainty in our estimates of the initial rate of aggregation. However, there is considerable motivation to continue to improve models of the aggregation process, since aggregates of dendrites are the most common type of snowfall in midlatitudes.

4.4.5 Melting of Ice Particles

There has recently been a resurgence of interest in the melting process because several researchers have indicated that the melting of ice particles (and associated cooling due to the latent heat of fusion) contributes substantially to the formation of mesoscale downdrafts in tropical squall lines (Houze, 1977) and in wintertime orographic cloud systems (Marwitz, 1983).

The melting of an ice particle is basically a thermodynamic process. Consider a graupel particle of mass X_g which has fallen through the 0°C isotherm into warmer temperatures. Suppose further that the layer that the graupel has fallen into contains cloud droplets at the ambient temperature T. Assuming a steady state and that the graupel maintains a surface temperature T_f of 0°C, the rate of latent heat release due to melting must be balanced by the rate of heat transfer through the layer of water on the graupel surface. Thus,

$$L_{\text{li}}\left[\frac{dX_{\text{g}}}{dt}\right]_{\text{melt}} = -2\pi D_{\text{g}} K_{\text{T}} f(\text{Re})(T - T_{\text{f}}) - 2\pi L_{\text{lv}} D_{\text{g}} D_{\text{v}} f(\text{Re})(\rho_{\text{v}} - \rho_{\text{vsfc}})$$
$$- \left[\frac{dX_{\text{g}}}{dt}\right]_{\text{RM}} c_{\text{w}}(T - T_{\text{f}}), \qquad (4.46)$$

where D_{g} is the graupel diameter, K_{T} and D_{v} are the diffusivities for heat and water vapor, respectively, and c_{w} is the thermal conductivity of liquid water. The first term on the RHS of Eq. (4.46) represents the diffusion of heat to the surface of the melting graupel particle at temperature T_{f}. The second term on the RHS of Eq. (4.46) represents the diffusion of water vapor and the corresponding transfer of latent heat from the graupel surface. A ventilation term $f(\text{Re})$ is included in both diffusion terms. The third term on the RHS of Eq. (4.46) represents the transfer of sensible heat to the graupel as it accretes cloud droplets at the rate $(dX_{\text{g}}/dt)_{\text{RM}}$. Mason (1956) considered the effects of the first two terms in his study of the melting process. Wisner *et al.* (1972) and Cotton *et al.* (1982) have included the last term in their cloud models. The transfer of sensible heat during collection can greatly accelerate the melting process, especially in clouds with low cloud bases and corresponding deep layers of cloud warmer than 0°C.

4.4.6 *Summary of Cloud Microphysical Processes*

In preceding sections we have seen that the evolution of precipitation in clouds can take on a variety of forms and involve numerous physical processes. The evolution of ice-phase precipitation processes is greatly dependent upon the prior or concurrent evolution of the liquid-phase spectrum of water. These processes, in turn, are dependent upon the characteristics of the air mass, the liquid-water production of the cloud, the vertical motion of the cloud, the turbulent structure, and the time scales of the cloud. Figure 4.8 is a flow diagram showing the various paths that precipitation processes may follow. Illustrated are the different precipitation paths that may occur depending upon whether the cloud is a cold-based continental cloud versus a warm-based maritime cloud. There is obviously a continuum of cloud types between these two extreme states. The heaviest precipitating clouds, and generally most efficient, are those that are warm-based and maritime in character.

An important goal in modeling precipitation processes is to develop the capability of predicting both the amounts of precipitation and the particular type of precipitation. It is quite important, for example, to be able to distinguish between the occurrence of graupel and freezing rain or the occurrence of numerous graupel and a few large hailstones. This places a

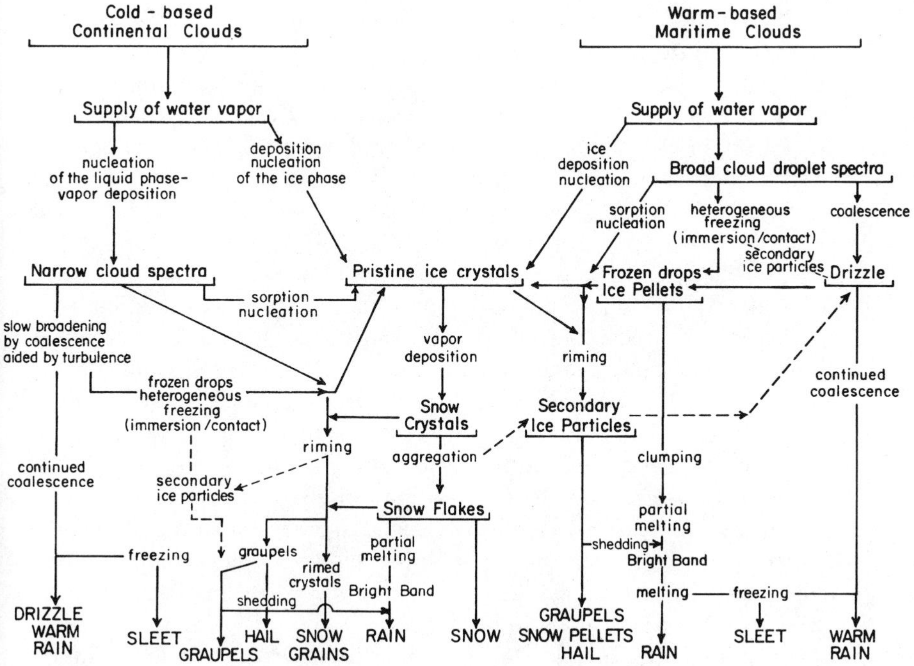

Fig. 4.8. Flow diagram describing microphysical processes, including different paths for precipitation formation. [Adapted from Braham (1968).]

major burden on our ability to formulate models and parameterizations of cloud microphysical processes.

4.5 Parameterization of Ice-Phase Microphysics

In contrast to warm-cloud physics, there is not a clear distinction between parameterization of ice microphysics and detailed explicit treatments. Even in the most sophisticated models a certain amount of parameterization is required. For example, in their simulation of ice crystal growth by vapor deposition, both Cotton (1972b) and Young (1974b) had to parameterize the aspect ratio of the modeled crystal dimensions. As with warm-cloud microphysics, the clearest distinction between a detailed theoretical model of the ice phase and a parameterization arises in the treatment of the distribution of particles. In detailed theoretical models the distribution of ice particles is not prespecified; rather, it is predicted either in a series of discrete classes of ice particles (Cotton, 1972b) or in finite-difference representations of spectral density functions of the ice phase such as the use of continuous storage bins (Young, 1974b; Scott and Hobbs, 1977). Several modelers have attempted to predict the spectral evolution of the ice phase by ignoring the particular habits of the ice crystals and assuming that all ice particles are spherical (Danielsen *et al.*, 1972; Takahashi, 1976; Ryan, 1973; Nelson, 1979). Others have followed a more conventional parameterization approach and defined a distribution function for graupel particles, hail particles, or aggregates (Weinstein and Davis, 1968; Simpson and Wiggert, 1969; Weinstein, 1970; Wisner *et al.*, 1972; Orville and Kopp, 1977; Koenig, 1977; Bennetts and Rawlins, 1981; Cotton *et al.*, 1982).

Generally, the ice-phase models which contain an explicit prediction of the habits of ice crystals and their size distribution have been limited to use in simple one-dimensional, steady-state parcel models (Cotton, 1972b; Young, 1974b, 1975) or, at the most in one-dimensional, time-dependent models (Scott and Hobbs, 1977). This is mainly due to the storage requirements and computing time required to compute such detailed microphysics. Young (1975), for example, stored information about ice crystal mass and habits (i.e., *a* axis and *c* axis dimensions) in some 400 storage bins and also stored the size distribution of spherical cloud droplets, rain, graupel, or hail in an additional 150 bins. Such an approach is clearly prohibitive for use in three-dimensional cloud models.

By assuming that all ice crystals are spherical or are of a common shape such as disks, Takahashi (1976) reduced the amount of storage, but he still required some 240 storage bins to keep track of all spherical ice particles and disklike ice crystals at every grid point in a two-dimensional, time-dependent cloud model.

The motivation, therefore, remains for developing simplified treatments or parameterizations of the ice phase. In order to organize our discussion of the parameterization of cloud microphysical processes, we use the notation employed by Cotton *et al.* (1986). We consider a mixed-phase cloud composed of the following species, having mixing ratios r_t for total water substance, r_r for raindrops, r_v for water vapor, r_c for cloud droplets, r_i for ice crystals, r_g for graupel or hail, and r_a for aggregates. The continuity equations for the time-dependent water variables can be written as follows:

Total water

$$\partial \bar{r}_T/\partial t = \mathrm{ADV}(\overline{r_T}) + \mathrm{TURB}(\overline{r_T}) + \mathrm{PR_r} + \mathrm{PR_g} + \mathrm{PR_i} + \mathrm{PR_a}. \qquad (4.47)$$

Raindrops

$$\begin{aligned}\partial \bar{r}_r/\partial t = {} & \mathrm{ADV}(\bar{r}_r) + \mathrm{TURB}(\bar{r}_r) + \mathrm{PR_r} \\ & - \mathrm{VD_{rv}} - \mathrm{CL_{ri}} + \mathrm{CL_{cr}} + \mathrm{CN_{cr}} + \mathrm{ML_{ir}} - \mathrm{CL_{rg}} - \mathrm{FR_{rg}} + \mathrm{ML_{gr}} \\ & - \mathrm{CL_{ra}} + \mathrm{ML_{ar}} + \mathrm{SH_{gr}}. \end{aligned} \qquad (4.48)$$

Ice crystals

$$\begin{aligned}\partial \bar{r}_i/\partial t = {} & \mathrm{ADV}(\bar{r}_i) + \mathrm{TURB}(\bar{r}_i) + \mathrm{PR}_i \\ & + \mathrm{NUA}_{vi} + \mathrm{NUB}_{vi} + \mathrm{NUC}_{vi} + \mathrm{NUD}_{vi} + \mathrm{SP}_{vi} + \mathrm{VD}_{vi} + \mathrm{CL}_{ci} \\ & + \mathrm{CL}_{ri} - \mathrm{ML}_{ir} - \mathrm{CL}_{ig} - \mathrm{CN}_{ig} - \mathrm{CL}_{ia} - \mathrm{CN}_{ia}. \end{aligned} \tag{4.49}$$

Graupel/hail

$$\begin{aligned}\partial \bar{r}_g/\partial t = {} & \mathrm{ADV}(\bar{r}_g) + \mathrm{TURB}(\bar{r}_g) + \mathrm{PR}_g \\ & - \mathrm{VD}_{gv} + \mathrm{CL}_{cg} + \mathrm{CL}_{rg} + \mathrm{FR}_{rg} - \mathrm{ML}_{gr} \\ & - \mathrm{SH}_{gr} + \mathrm{CL}_{ig} + \mathrm{CN}_{ig} + \mathrm{CL}_{ag} + \mathrm{CN}_{ag}. \end{aligned} \tag{4.50}$$

Aggregates

$$\begin{aligned}\partial \bar{r}_a/\partial t = {} & \mathrm{ADV}(\bar{r}_a) + \mathrm{TURB}(\bar{r}_a) + \mathrm{PR}_a \\ & - \mathrm{VD}_{va} + \mathrm{CL}_{ca} + \mathrm{CL}_{ra} - \mathrm{ML}_{ar} - \mathrm{SH}_{ar} \\ & + \mathrm{CL}_{ia} + \mathrm{CN}_{ia} - \mathrm{CL}_{ag} - \mathrm{CN}_{ag}. \end{aligned} \tag{4.51}$$

With the assumption that a cloud does not become supersaturated with respect to water, r_v and r_c can be obtained diagnostically.

In addition to the prognostic equations for the mixing ratios of water species, a prognostic equation for the concentration of ice crystals N_i can also be formulated. It is of the form

$$\begin{aligned}\partial \bar{N}_i/\rho_0/\partial t = {} & \mathrm{ADV}(\bar{N}_i/\rho_0) + \mathrm{TURB}(\bar{N}_i/\rho_0) + \mathrm{PR}_{N_i} \\ & + (1/m_{i0})(\mathrm{NUA}_{vi} + \mathrm{NUB}_{ci} + \mathrm{NUC}_{ci} + \mathrm{NUD}_{ci} + \mathrm{SP}_{ci}) \\ & - (N_i/\rho_0 \bar{r}_i)(\mathrm{ML}_{iv} + \mathrm{CL}_{ig} + \mathrm{CN}_{ig} + \mathrm{CL}_{ia} + \mathrm{CN}_{ia}), \end{aligned} \tag{4.52}$$

where ρ_0 represents a base state density which varies only with height. Similar predictive equations for the concentrations of other precipitation species can also be formulated in a manner similar to Nickerson *et al.* (1986) for an all-liquid cloud.

The source and sink terms used in Eqs. (4.47)–(4.52) are defined as CN for conversion; NUA for depositional nucleation; NUB, NUC, and NUD for Brownian, thermophoretic-contact, and diffusiophoretic-contact nucleation; SP for secondary production such as the Hallett and Mossop (1974) rime splinterings; ML for melting; FR for freezing; CL for collection; VD for vapor deposition or evaporation; SH for liquid-water shedding; and PR for precipitation. Each term includes a double subscript, where the first subscript is the water phase being depleted and the second subscript is the water phase which is growing. The subscripts, v, c, r, i, g, and a refer to vapor, cloud, rain, ice crystal, graupel, and aggregate water species, respectively. The ADV term represents the advection of these properties by the mean flow field while the term TURB represents turbulence diffusion.

4.5.1 Parameterization of Primary and Secondary Nucleation

The simplest approach to parameterizing the nucleation of the ice phase is to assume that a certain fraction of the available cloud liquid water freezes as a function of temperature. Simpson *et al.* (1965), for example, assumed that freezing did not occur naturally until −40°C, but "seeding" resulted in freezing of supercooled water linearly between −4 and −8°C.

Others such as Cotton (1972b), Koenig and Murray (1976), Bennetts and Rawlins (1981), and Cotton *et al.* (1982) have used Fletcher's (1962) formula, Eq. (4.35), to estimate the concentration of deposition or sorption nuclei activated as a function of temperature.

Young (1974c, 1975), however, has explicitly modeled nucleation of the ice phase by sorption, using a modified form of Fletcher's formulation and contact nucleation along with immersion freezing in a simple model of a wintertime orographic cloud (Young, 1974b) and a parcel model of a cumulonimbus (Young, 1975). Cotton *et al.* (1986) adapted Young's (1974c) simple form of contact nucleation along with a simple parameterization of deposition nucleation to the simulation of orographic clouds. Like Young's model, Cotton *et al.*'s model predicts enhanced contact nucleation by phoretic processes in conjunction with droplet evaporation caused by the Bergeron–Findeison process.

Hsie *et al.* (1980) have also considered contact nucleation in a two-dimensional cloud model. However, they neglected phoretic effects and considered only inertia impaction and Brownian collection of the aerosol. Their study was aimed mainly at the simulation of cloud seeding where, they argued, deposition nucleation was quite effective.

Both the singular and stochastic models of immersion freezing have been incorporated in cloud models. Cotton (1972b) included the singular model of raindrop freezing in a one-dimensional Lagrangian model, whereas Orville and Kopp (1977) applied the stochastic freezing theory to a two-dimensional cloud model. Subsequently, Cotton *et al.* (1982) neglected the heterogeneous model for raindrop freezing in a 2-D/3-D cloud model. They made this decision based on the work of Cotton (1972b) and Scott and Hobbs (1977), who noted that the simulated freezing of supercooled raindrops by collision with ice crystals dominated the heterogeneous freezing process. In fact, Scott and Hobbs (1977) found that only 1% of the predicted total number of frozen raindrops in their model originated by heterogeneous freezing.

As mentioned previously, a number of researchers have examined several theories of ice multiplication in cloud models of varying complexities. Cotton (1972b) examined the sensitivity of a one-dimensional, steady-state cloud model to enhanced ice crystal concentrations produced by some mechanism of ice multiplication. He simply enhanced the crystal concentrations by multiplying the crystal concentrations estimated with Fletcher's formula, Eq. (4.35), by the ratio of the observed concentration of ice particles to ice nuclei as a function of temperature as reported by Hobbs (1969). Scott and Hobbs (1977) examined both the drop-freezing-induced splintering mechanism and the Hallett–Mossop rime-splintering mechanism in a one-dimensional, time-dependent cloud model. The Hallett–Mossop mechanism was also studied in a one-dimensional, multiple-parcel model (Mason, 1975) and even in two-dimensional, time-dependent cloud models (Koenig, 1977; Cotton *et al.*, 1986). Thus, the theories of secondary ice multiplication are simple enough that they can be implemented in multidimensional cloud models. Unfortunately, the more complex models of ice crystal production require the formulation of complete continuity equations for ice crystal concentration such as given in Eq. (4.52). This requires the addition of another prognostic variable. Moreover, because ice crystal concentration can vary over many orders of magnitude over small distances, predicting ice crystal concentrations is quite a computational challenge!

4.5.2 *Parameterization of Ice Crystal Growth by Vapor Deposition and Riming*

The challenge of implementing an ice crystal vapor-deposition and riming growth model in a multidimensional cloud model is how to represent all the complex habits of ice crystals without using large amounts of computer storage. Koenig (1972) derived a parameterization of these processes from detailed calculations of ice crystal growth by vapor deposition and riming. For a water-saturated cloud, he divided the crystal growth process into three regimes in which (i) diffusion growth occurs only, and the growth rate is solely a function of temperature; (ii) both diffusion and riming occur; and (iii) only riming occurs.

Suppose that in a cloud model the mixing ratio of ice crystals r_i is predicted, and the concentration of ice crystals N_i is either predicted or specified. Then the average mass of an ice crystal $\overline{X_i}$ is

$$\overline{X_i} = \rho_0 r_i / N_i. \tag{4.53}$$

The rate of change in mixing ratio due to vapor deposition is then

$$\mathrm{VD}_{vi} = \rho_0 N_i \dot{x}]_{\mathrm{VD}}, \tag{4.54}$$

where $\dot{x}_i]_{VD}$ represents the rate of growth of an individual ice crystal by vapor deposition. Similarly, the change in ice crystal mixing ratio due to riming is

$$RM_{ci} = \rho_0 N_i \dot{x}_i]_{RM}, \tag{4.55}$$

where $\dot{x}_i]_{RM}$ represents the rate of riming growth of an ice crystal. In the vapor-deposition-only regime, Koenig represented vapor deposition growth by

$$\dot{x}_i]_{VD} = a_1 (X_i)^{a_2}, \tag{4.56}$$

where a_1 and a_2 are temperature-dependent coefficients derived from his detailed model calculations. Figure 4.9 is a schematic depiction of the various growth regimes.

In the second regime, illustrated by the B and C lines in Fig. 4.9, vapor deposition and riming both make a contribution to the mass growth of an

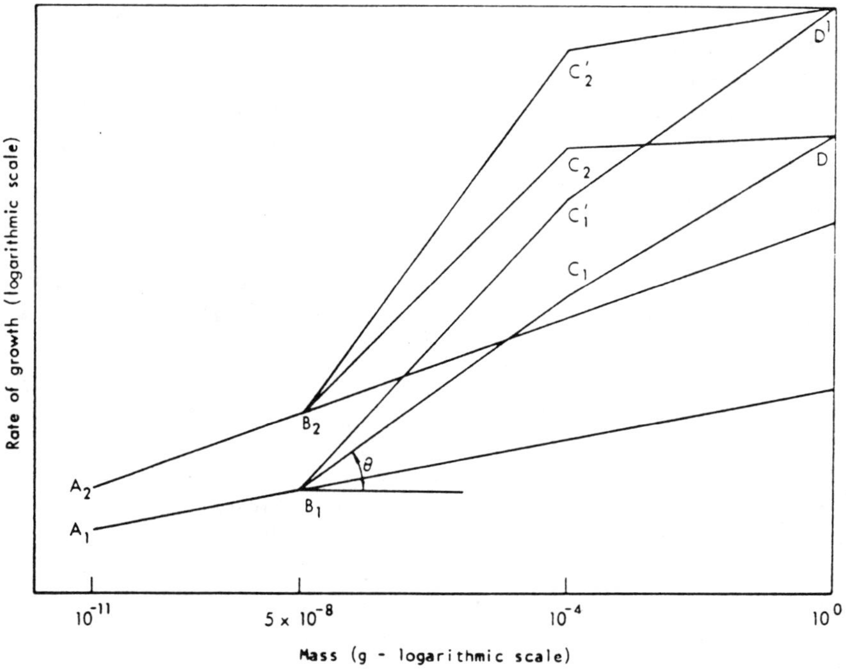

Fig. 4.9. Illustration of the concept of parameterization. Growth at constant temperature and liquid-water content follows path ABCD. Subscripts indicate growth properties at different temperatures; primed and unprimed points indicate growth properties at different liquid-water contents. [From Koenig (1972).]

ice crystal. Koenig assumed that if the average crystal mass lay in the range $5\times10^{-8}<\bar{X}_i\leq10^{-4}$ g, both riming and vapor deposition could occur. He determined the tangent of the angle θ between B and C (see Fig. 4.9) and the horizontal from B to be a function of the liquid-water mixing ratio as

$$\tan\theta=1+\ln(\rho_0 r_c\times10^3)/10. \tag{4.57}$$

The growth rate of an individual crystal by vapor deposition and riming is then

$$\dot{x}_i]_{VD}+\dot{x}_i]_{RM}=a_1(5\times10^{-8})^{a_2}(\dot{x}_0/5\times10^{-8})^{\tan\theta}\quad \mathrm{g\,s^{-1}}. \tag{4.58}$$

If the average ice crystal mass $\bar{X}_i>10^{-4}$ g, then the ice crystal was assumed to be in regime C–D, where the crystal growth is only a function of liquid-water content. In this regime

$$\dot{x}_i]_{RM}=b_1(\bar{X}_i/10^{-4})^{b_2}\quad \mathrm{g\,s^{-1}}, \tag{4.59}$$

where $b_1=a_1(5\times10^{-8})^{a_2}(2000^{\tan\theta})$ and

$$b_2=\ln(10^{-3}\rho_0 r_i/b_1)/\ln(10^4). \tag{4.60}$$

Other researchers have chosen to parameterize vapor deposition and riming growth of ice crystals following a more conventional approach. Cotton *et al.* (1982), for example, used Eq. (4.53) to diagnose an average ice crystal mass. They then modeled crystal growth in three forms:

(i) Unrimed hexagonal plate.
(ii) Lightly rimed hexagonal plates.
(iii) Graupel-like snow of hexagonal type.

Ice crystal growth by vapor deposition and riming were modeled by using Eqs. (4.37) and (4.43), where the ice crystal dimensions, fall velocity, and capacitances were determined from observational data. For example, if D_i and V_i represent the ice crystal diameter and fall velocity, then

$$D_i=K_1X_i^{1/2}, \tag{4.61}$$

$$V_i=k_2D_i(p_{00}/p)^{1/2}, \tag{4.62}$$

where p_{00} represents sea level pressure and the coefficients k_1 and k_2 are as follows (from Hobbs *et al.*, 1972):

$$k_1=0.515\quad \mathrm{m\,g^{-1/2}};\qquad k_2=304\quad \mathrm{s^{-1}}\qquad \text{for}\quad \bar{X}_i<1.7\times10^{-7}\ \mathrm{g}$$

$$k_1=0.192\quad \mathrm{m\,g^{-1/2}};\qquad k_2=1250\quad \mathrm{s^{-1}}\qquad \text{for}\quad 1.7\times10^{-7}\ \mathrm{g}\leq\bar{X}_i<10^{-5}\ \mathrm{g}.$$

For masses $\bar{X}_i>10^{-5}$ g, the equations pertaining to graupel-like snow of hexagonal type are (Locatelli and Hobbs, 1974)

$$D_i = k_1 \bar{X}_i^{0.417}, \tag{4.63}$$

$$V_i = k_2 D_i^{0.25} (p_{00}/p)^{k_1}, \tag{4.64}$$

where $k_1 = 8.89 \times 10^{-2}$ m g$^{-0.417}$ and $K_2 = 4.84$ m$^{3/4}$ s^{-1}. Such a procedure could also be extended to other crystal forms such as dendrites, needles, and columns.

Koenig and Murray (1976), Hsie *et al.* (1980), and Cotton *et al.* (1982) all assumed that ice crystal growth processes could be represented by the growth of an average ice particle. Thus, no distribution of pristine ice particles was considered. Bennetts and Rawlins (1981), however, apportioned ice crystals in three different size classes of radii, 0–100, 100–200, and 200–300 μm in a 3-D cloud model. In a sense these classes represent unrimed, lightly rimed, and heavily rimed ice particles. Ice was initiated in each class depending upon the heterogeneous freezing probability of an assumed droplet distribution.

Nickerson *et al.* (1978) employed three categories of ice in a 3-D mesoscale model. The categories were unrimed crystals, partially rimed crystals, and graupel. They assumed that each category of ice was distributed in a Marshall–Palmer distribution [Eq. (4.15)]. As far as unrimed and lightly rimed ice crystals are concerned, little research has been done to identify their size distributions. Recent observational studies in cirrus clouds (P. Flatau, personal communication) suggest that the size distribution of ice crystals may be bimodal in character. That is, a primary mode exists for very small crystals 10–15 μm in diameter, while a secondary mode exists for crystals 80 μm in diameter or greater. The primary mode of small crystals can play an important role in the radiative properties of a cloud, while the larger crystals play an active role in the precipitation process. If a bimodal size distribution of rimed and lightly rimed crystals is a common occurrence, then it may be desirable to model the ice crystal size spectra with two log-normal distribution functions similar to Clark and Hall's (1983) approach to modeling cloud droplet-size spectra.

4.5.3 Parameterization of Graupel- and Hail-Particle Initiation

Graupel particles are ice particles that have become so heavily rimed that their original vapor deposition growth habit is obscured. Graupel may have its origins in the heavy riming of pristine ice crystals, the freezing of large supercooled raindrops, and the riming of aggregates of snow crystals. In a model in which the distribution of ice particles is predicted in a finite spectral density function or an array of storage bins, the formation and evolution of the graupel/hail spectra are a direct product of the model.

However, as with warm-cloud physics, it is often desirable to define the distribution of graupel/hail and then parameterize their initiation.

Since graupel and hail often originate in supercooled raindrops, an obvious approach is to model graupel initiation by some mechanism of raindrop freezing. Both Cotton (1972b) and Wisner *et al.* (1972) modeled raindrop freezing by heterogeneous freezing. Cotton (1972b), Koenig and Murray (1976), and Bennetts and Rawlins (1981) modeled graupel initiation by collision between supercooled raindrops and ice crystals. As noted previously, Cotton *et al.* (1982) found this process so effective that the heterogeneous freezing model was not necessary. To illustrate the process, consider a supercooled raindrop distribution of the form of Eq. (4.25), having a mixing ratio r_r. Also consider a population of ice crystals having mixing ratio r_i and concentration N_i. Because the process is one of simple collection, we can estimate the change in the concentration of raindrops of radius $R+dR$ by collecting ice crystals and freezing as

$$\frac{d\phi(R)dR}{dt} = -\int_0^\infty \pi(R+D_i/2)^2 |V(R)-V_i| E(R/i) N_i \phi(R)\, dR\, dD_i, \tag{4.65}$$

where D_i is the diameter of an ice crystal, V_i is its fall velocity, and $E(R/i)$ is the collection efficiency between ice crystals and raindrops, normally assumed to be unity. Assuming that the collected ice crystal makes an insignificant contribution to the mass of a graupel particle, the rate of change of graupel mixing ratio by freezing of raindrops is

$$\mathrm{FR}_{rg} = -\frac{1}{\rho_0}\int_0^\infty X(R)\frac{d\phi(R)}{dt}\, dR, \tag{4.66}$$

where $x(R)$ is the mass of the supercooled drop.

Assuming further that $V(R) \gg V_i$ and substituting Eq. (4.65) into Eq. (4.66), Eq. (4.26) for $V(R)$, and integrating gives us

$$\mathrm{FR}_{rg} = \left[\frac{2.13\pi}{6}(g\rho_w)^{1/2}\Gamma(6.5)\right]\rho_0^{1/2} R_m^{2.5} N_i r_r. \tag{4.67}$$

This shows that the rate of production of graupel by collision between raindrops and ice crystals is proportional to the mixing ratio of supercooled raindrops and the concentration of ice crystals. This process is also a self-perpetuating one since graupel particles may also collide with supercooled raindrops, further enhancing the mixing ratio of graupel.

As mentioned earlier, graupel particles may also be initiated by the heavy riming of ice crystals or aggregates of ice crystals. Since graupel may form from ice crystals in steps of (i) lightly rimed crystals, (ii) to heavily rimed crystals, (iii) to conical or spherical graupel, the decision to categorize ice particles as graupel is somewhat arbitrary. Using Koenig's (1971) parameterization of vapor deposition, Hsie *et al.* (1980) formulated a conversion rate from cloud ice to graupel to be a function of the time it takes an ice crystal to grow from 40 to 50 μm. Cotton *et al.* (1982) estimated the conversion rate of ice crystals to graupel based on the riming growth rate of a single crystal. If the growth rate exceeds an arbitrary amount C_m, then such a particle is considered a graupel particle. Thus, if the rate of change of ice crystal mixing ratio by riming (RM_{ci}) exceeds a threshold amount,

$$RM_{ci} > (N_i/\rho_0)C_m, \tag{4.68}$$

then that excess is converted to graupel. The conversion rate is expressed as

$$CN_{ig} = MAX\left[\left(RM_{ci} - \frac{N_i}{\rho_0} C_m\right), 0\right], \tag{4.69}$$

where the coefficient C_m was arbitrarily selected as 10^{-6} g s^{-1}.

Nickerson *et al.* (1978) estimated the conversion from unrimed ice crystals to moderately rimed crystals to graupel particles based on the ratio

$$f = \dot{x}_i]_{RM}/\dot{x}_i]_{VD}, \tag{4.70}$$

which is the ratio of predicted riming rates to vapor-deposition rates. Whenever f exceeded a certain value (the value was not specified in their report), mass was transferred into the next largest category of ice. Unfortunately, the predicted composition of ice particles in a cloud is quite dependent upon the conversion rates. Too low a conversion rate may result in a cloud dominated by pristine ice crystals and aggregates, when in fact graupel may dominate. Too high a conversion rate may remove aggregates in cases where they may be prevalent.

4.5.4 *The Distribution of Graupel and Hail Particles*

Fundamental to the formulation of a simple parameterization of graupel or hail growth processes is the determination of an appropriate particle-size distribution function. Most observational studies of graupel- and hailstone-size distributions indicate that a Marshall and Palmer (1948) distribution of the form of Eq. (4.15) is appropriate. However, there is considerable disagreement concerning the behavior of the parameters N_0 and λ of the distribution. Marshall and Palmer concluded that N_0 was constant with the value 0.08 cm^{-4}. As a consequence, many cloud modelers have chosen to formulate graupel/hail models in which the distribution is of the form Eq. (4.15) with N_0 constant (Simpson and Wiggert, 1971; Wisner *et al.*, 1972; Koenig and Murray, 1976; Bennetts and Rawlins, 1981; Orville and Kopp,

1977). However, there are indications that the slope of graupel/hail distributions is constant. Douglas (1964) found that the slope of hail distributions was a constant with $\lambda = 2.93\ \text{cm}^{-1}$. Federer and Waldvogel (1975) analyzed the time variations of N_0, N_T (the total concentration), and λ during the passage of a hail swath. Of these parameters, the slope exhibited the least variation with a mean value of $0.42\ \text{cm}^{-1}$ and ranged from $0.33 \leq \lambda \leq 0.64\ \text{cm}^{-1}$. On the other hand, the intercept N_0 exhibited a mean value of $12.1\ \text{cm}^{-1}$ and ranged from $1.5 \leq N_0 \leq 52\ \text{cm}^{-1}$.

Jones (1960) reported on the results of analyses of graupel spectra obtained from aircraft flights in midlatitudes and in the tropics. He found his data fitted a truncated spectrum of the form

$$N(D) = N_0\, e^{-2(D - D_{\min})}, \tag{4.71}$$

where $D_{\min}$ represents the minimum-sized particle he observed with a foil impactor (250 μm). Jones demonstrated that Eq. (4.71) could fit his data provided

$$\lambda = 2.67 \times 10^3\ M_g^{-1/3}, \tag{4.72}$$

where λ has dimensions m^{-1} and M_g is the graupel water content in g m^{-3}. Cotton (1970) showed by dimensional analysis that Eq. (4.72) is equivalent to

$$\lambda = k\delta_i^{1/3} N_0^{1/4} M_g^{-1/3}, \tag{4.73}$$

where $k = 2.52 \times 10^{-6}$ if δ_i is assumed to be $0.6\ \text{g cm}^{-3}$, and $N_0 = 10^{7.1}\ \text{m}^{-4}$.

From the analysis of 41 hailstone samples in seven storms, Cheng and English (1983) concluded that neither a constant slope N nor a constant N_0 was a characteristic of the hail distributions. The confusion in the literature regarding the behavior of the parameters in graupel and hail distributions is probably a reflection of the fact that a physical process, such as drop breakup, does not exist for hail, thereby leading to an asymptotic distribution. Thus, depending on the origin of their distributions, graupel/hail may exhibit more or less stability of the distribution parameters. For example, if the graupel distribution has its origins in the freezing of raindrops, then the breakup process in the raindrop distribution may be reflected in the graupel/hail distribution as a constant slope. If, however, the graupel/hail distribution has its origins in the riming of snow crystals, then the graupel/hail distribution may exhibit a constant N_0. An alternative suggested by Cheng and English (1983) is that neither N_0 nor λ may be constant. It is, therefore, doubtful that graupel/hail-size spectra can be properly modeled with a single size-distribution function. Here again, either a family of log-normal basis functions or an explicit finite-difference representation of the hail spectrum such as used by Farley and Orville (1986) may be required.

4.5.5 Parameterization of the Change in Graupel Mixing Ratio by Riming, Vapor Deposition, and Melting

Once a suitable graupel/hail-size distribution function is selected, it is a relatively straightforward procedure to predict the change in mixing ratio of graupel particles by riming (CL_{cg}) or collection. For spherical particles, Eq. (4.44) can be written as

$$(dX_g/dt)_{RM} = (\pi D^2/4)\, V_g(D)\overline{E(g/c)}\rho_0 r_c, \tag{4.74}$$

where D is the diameter of the graupel particle.

Equation (4.74) represents an accretional growth process in which collision between graupel particles and larger droplets is ignored. Also, as graupel particles approach hailstones in size, incomplete freezing and shedding of the rain deposit may occur.

For small graupel particles one can estimate the terminal velocity from data tabulated by Locatelli and Hobbs (1974), in which

$$V_g(D) = 594 D^{0.46}(p_{00}/p)^{1/2}. \tag{4.75}$$

However, for larger graupel particles, the terminal velocity can be obtained from a balance between the drag force and the force due to gravity giving

$$V_g(D) = (4\rho_g g/3\rho_0 C_D)^{1/2} D^{1/2}, \tag{4.76}$$

where C_D is a drag coefficient. Macklin and Ludlam (1961) found that C_D for frozen particles varies over the range $0.45 \leq C_D \leq 0.8$. Magono's (1954) suggestion of ρ_g for graupel varies considerably, ranging from $\rho_g = 0.12\ \mathrm{g\ cm^{-3}}$ for small graupel to $0.92\ \mathrm{g\ cm^{-3}}$ for hail. The modeler must select representative values of these parameters in order to evaluate Eq. (4.76) for a spectrum of graupel particles. It is also convenient to estimate an average collection efficiency $E(g/c)$ between graupel particles and cloud droplets. This is often done by computing theoretical collection efficiencies (Pruppacher and Klett, 1978) between an average raindrop of diameter $\bar{D}$ and an average cloud droplet.

We are now in a position to estimate the change in graupel mixing ratio by riming. Suppose, for example, we assume that the graupel distribution has a constant slope $\lambda = 1/\bar{D}$, and a corresponding distribution function

$$N(D) = (N_g/\bar{D})\, e^{-D/\bar{D}}; \tag{4.77}$$

then

$$CL_{cg} = \frac{1}{\rho_0}\int_0^\infty \left.\frac{Dx_g}{dt}\right]_{RM} \frac{N_g}{\bar{D}}\, e^{-D/\bar{D}}\, dD. \tag{4.78}$$

The total concentration is related to the mixing ratio r_g and $\bar{D}$ as follows:

$$r_g = \frac{1}{\rho_0} \int_0^\infty X_g(D)\, N(D)\, dD. \tag{4.79}$$

Substituting Eq. (4.77) into Eq. (4.79) and defining the graupel mass in terms of its diameter and density (ρ_g) gives

$$r_g = \pi \rho_g N_g \bar{D}^3 \Gamma(4)/6\rho_0 \tag{4.80}$$

or

$$N_g = 6\rho_0 r_g / \pi \rho_g \bar{D}^3 \Gamma(4). \tag{4.81}$$

Substituting Eqs. (4.74), (4.76), and (4.81) into Eq. (4.78) and integrating over the entire droplet spectrum results in an equation of the form

$$\mathrm{CL}_{cg} = \left[6 \left(\frac{4g\rho_0}{3C_D\rho_g} \right)^{1/2} \frac{\Gamma(5/2)}{\Gamma(4)} \right] \frac{\bar{E}}{\bar{D}^{3/2}} r_g r_c. \tag{4.82}$$

That is, the rate of production of graupel by riming is a function of the product of the mixing ratios of cloud droplets and graupel. Equation (4.82) is similar in form to Eq. (4.27) for raindrops collecting cloud droplets. Had we assumed that N_0 was a constant, then the resultant integral would have contained noninteger powers of r_g similar to Eq. (4.24). Equation (4.82) represents the rate of production of graupel/hail mixing ratio by dry riming growth. If "wet" particle growth were considered, the collection rate would have to be reduced by the amount of water shed from the surface of the graupel/hail particles. Based on our earlier discussion, however, "spongy" hail formation would result in a net collection rate, which would be closer to Eq. (4.82).

Following a similar procedure, we could then apply Eq. (4.37) to the distribution of graupel particles to obtain the change in mixing ratio due to vapor deposition or evaporation of graupel (VD_{vg}) and apply Eq. (4.46) to estimate the rate of melting loss of graupel mixing ratio (ML_{gr}). We will leave these derivations to the reader.

One process requires further discussion. Let us consider the interaction between graupel particles and supercooled raindrops to form more graupel particles. This process is analogous to the production of graupel by collision between supercooled raindrops and ice crystals. However, in this case both species are assumed to be distributed in a Marshall–Palmer type of distribution function. The gain in mixing ratio of graupel is simply CL_{rg}, which is the loss of raindrop mixing ratio due to freezing upon contact with graupel particles. Although graupel particles act as the nucleating particle, they do not contribute to the graupel mixing ratio. Thus,

$$
\begin{aligned}
\mathrm{CL}_{\mathrm{rg}} &= \frac{1}{\rho_0} \int_0^\infty X_{\mathrm{r}}(R) \frac{d\phi(R)}{dt} \, dR, \\
&= \frac{1}{\rho_0} \int_0^\infty X_{\mathrm{r}}(R) \int_0^\infty \pi (R + D/2)^2 |V(R) \\
&\qquad - V_{\mathrm{g}}(D)| E(R/D) \phi(R) \, N(D) \, dR \, dD,
\end{aligned}
\tag{4.83}
$$

where $X_{\mathrm{r}}(R)$ is the mass of a raindrop of radius R_{r}. In order to make Eq. (4.83) easier to integrate, we assume that $E(R/D)$ can be replaced by $\bar{E} = 1$. As Wisner *et al.* (1972) did in the formulation of their hail model, one can approximate $|V(R) - V_{\mathrm{g}}(D)|$ in Eq. (4.83) by

$$
|V_{\mathrm{r}} - V_{\mathrm{g}}(D)| \simeq \overline{V_{\mathrm{r}}} - \overline{V_{\mathrm{g}}}
\tag{4.84}
$$

or the difference in the water-content-weighted terminal velocities. The water-content-weighted terminal velocities were defined by Srivastava (1967).

Lin *et al.* (1983) expressed concern about the impact of Eq. (4.84) on the integrated collection rate. P. Flatau (personal communication) has demonstrated that Eq. (4.83) can be analytically integrated without making the dubious assumption, Eq. (4.84). A quantitative evaluation of the impact of the approximation, Eq. (4.84), on $\mathrm{CL}_{\mathrm{rg}}$ has not, however, been assessed yet.

4.5.6 Parameterization of Ice Crystal Aggregation

Several researchers have included the ice crystal aggregation process in quasistochastic cloud models in which information regarding the spectrum of ice crystals and aggregates is stored in finite-difference spectral density functions or in storage bins. However, the use of such aggregation models has been limited to simple parcel models (Young, 1974b, c) or one-dimensional, time-dependent cloud models (Scott and Hobbs, 1977). Takahashi (1976) has included aggregation in a 2-D cloud model in which the spectrum of ice crystals and aggregates has been predicted. This effort was, however, at considerable computational expense.

Only a few attempts have been made to formulate simple parameterized models of aggregation which can be economically incorporated in 2-D and 3-D cloud or mesoscale models. This work has been mainly confined to two modeling groups: (1) the 2-D cloud model under the direction of H. Orville and (2) the 2-D/3-D Regional Atmospheric Modeling System (RAMS) (the cloud-scale model is nonhydrostatic while the mesoscale model is hydrostatic) under the direction of W. Cotton.

Orville's model, as reported by Hsie *et al.* (1980) and Lin *et al.* (1980), does not treat aggregates as a distinctly different class of ice particles. Instead, they classify a category of ice particles as "snow," which has the characteristics of graupel-like snow of hexagonal type and aggregates of ice crystals. As such, snow can have its origins in the self-collection of small ice crystals, which they define as cloud ice, the collision between cloud ice particles and small raindrops, and the deposition growth and riming of cloud ice particles. In Orville's model, cloud ice particles are assumed to take on a role analogous to cloud liquid water in which cloud ice particles do not have a precise size spectrum or concentration. They parameterize the self-collision of cloud ice particles to form snow crystals as an autoconversion process in which

$$\mathrm{CN}_{\mathrm{ia}} = \alpha_1(r_{\mathrm{i}} - r_{\mathrm{i0}}), \tag{4.85}$$

where α_1 is a temperature-dependent rate coefficient and r_{i0} is a threshold for aggregation to occur. Lin *et al.* set r_{i0} to be $1\ \mathrm{g\ kg^{-1}}$ and $\alpha_1 = 10^{-3}\exp[0.025(T - T_0)]$, where T_0 is 0°C. The form of α_1 is selected to reflect the fact that aggregation efficiencies vary for different temperature-dependent crystal structures.

In Orville's model, snow crystals can also form by the collision between cloud ice particles and supercooled raindrops, a mechanism we have previously discussed with regard to the formation of graupel/hail particles. However, if $r_{\mathrm{r}} < 10^{-1}\ \mathrm{g\ kg^{-1}}$, then they assume all raindrops will be small enough to become low-density particles which they refer to as snow. If $r_{\mathrm{r}} > 10^{-1}\ \mathrm{g\ kg^{-1}}$, then this mechanism results in the formation of spherical graupel/hail particles.

In a recent version of Orville's model (Lin *et al.*, 1980), the production of snow crystals by vapor deposition and riming of cloud ice particles is simulated. In the earlier version of the model described by Hsie *et al.* (1980) this same mechanism was used to generate spherical graupel/hail particles.

Once formed, snow crystals in Orville's model are assumed to be distributed in a Marshall–Palmer distribution, Eq. (4.15), with a constant N_0 whose value was taken from the measurements of Gunn and Marshall (1958) to be $3 \times 10^{-2}\ \mathrm{cm^{-4}}$. Following procedures similar to those outlined in the derivation of growth equations for the mixing ratios of raindrops and graupel, rate equations for the change of mixing ratio of snow were formulated for:

(i) Snow particles collecting cloud ice.
(ii) Snow particles accreting or riming cloud water.
(iii) Vapor deposition (sublimation) on (off) snow particles.
(iv) Snow-particle collection by raindrops.

(v) Snow-particle collection by graupel particles.
(vi) Melting of snow particles to form rain.
(vii) At temperatures warmer than 0°C, snow-particle collection of cloud water to form rain.
(viii) Sedimentation of snow.

Taking a somewhat different approach, aggregates of ice crystals are assumed to be a distinct snow species in the Cotton model (Cotton *et al.*, 1986). Thus, aggregates, having a mixing ratio r_a are first initiated by the self-collision among ice crystals having a mixing ratio r_i and concentration N_i. The conversion rate of ice crystals to aggregates is then given by

$$\mathrm{CN}_{ia} = -\frac{\bar{X}_i}{\rho_0} \frac{dN_i}{dt}\bigg]_a, \tag{4.86}$$

where $\bar{X}_i$ is the average ice crystal mass given by Eq. (4.53) and $(dN_i/dt)]_a$ represents the rate of aggregation or self-collection among pristine ice crystals to form aggregates. Because the ice crystal population in the Cotton model is not defined to have a particular size spectrum, the rate of aggregation is given by

$$(dN_i/dt)]_a = -K_i N_i^2, \tag{4.87}$$

where K_i represents the collision kernel between ice crystals. As noted previously, Eq. (4.87) illustrates the need for accurate prediction of ice crystal concentrations in order to estimate the initiation of aggregates of crystals, since $(dN_i/dt)]_a \alpha N_i^2$.

Cotton *et al.* estimated the collection kernel among the ice crystal population based on Passarelli and Srivastava's (1979) collection kernel for equal-sized ice crystals, in which a distribution of fall velocities is estimated by using particle density as a random variable. This model may be formulated as

$$K_i = (\pi D_i^2/6)\, V_i E_a(T_i) X, \tag{4.88}$$

where V_i is the average terminal velocity of an ice crystal of diameter D_i and X is a measure of the variance of the spectrum of ice-particle densities. Passarelli and Srivastava estimate X to be 0.25.

The spectrum of aggregate particles is also assumed to be in the Marshall–Palmer form in Cotton's model. However, in contrast to Orville's model, the slope of the distribution $\lambda = 1/D_m$, where D_m is the average diameter, is assumed to be constant. Thus, the distribution function is of the form

$$N(D) = (N_a/D_m)\, e^{-D/D_m}. \tag{4.89}$$

Snow aggregate distributions tabulated by Rogers (1974) suggested that D_m

is relatively constant. The parameter D_m was estimated to be 0.33 cm from Rogers' data. Discussions with R. E. Passarelli (personal communication) indicate that airborne observations of aggregate snowfall suggest that breakup of aggregates is prevalent enough to result in an asymptotic, constant-slope distribution. Recent observations in the stratiform region of mesoscale convective systems suggest that the slope of the aggregate size distribution function is not *constant* (Yeh *et al.*, 1986). It seems clear, therefore, that more research is needed in order to better determine the behavior of populations of aggregates.

One can derive rate equations similar to those used in Orville's model for aggregate mixing ratio for the following processes:

(i) Aggregates collecting ice crystals.
(ii) Aggregates collecting cloud droplets.
(iii) Vapor deposition (sublimation) of aggregates.
(iv) Raindrop collection of aggregates to form graupel.
(v) Graupel-particle collection of aggregates to form more graupel mixing ratio.
(vi) Conversion of aggregates to graupel by heavy riming.
(vii) Melting of aggregates to form rain.
(viii) Sedimentation of aggregates.

Derivation of the rate equations describing the change in mixing ratio of aggregates for the processes identified in (i) through (viii) follows procedures similar to those described earlier with respect to ice crystals and graupel particles interacting with other water species. We shall give only a few examples here. Consider the change in mixing ratio of aggregates due to collection of ice crystals ($\mathrm{CL_{ia}}$). Cotton *et al.* (1986) estimated the collection rate as

$$\mathrm{CL_{ia}} = -\frac{\bar{X}_i}{\rho_0} \frac{dN_i}{dt}\bigg]_{\mathrm{CL}}, \tag{4.90}$$

where $\bar{X}_i$ is the average mass of an ice crystal, and

$$\frac{dN_i}{dt}\bigg]_{\mathrm{CL}} = -\int_0^\infty K_i N_i\, N(D_a)\, dD_a. \tag{4.91}$$

For large aggregates collecting pristine ice crystals, the standard gravitational collection kernel was used,

$$K_i = (\pi/4)(D_a + D_i^2)|V_a - V_i| E(a/i). \tag{4.92}$$

With the approximation $|V_a - V_i| \simeq |\overline{V}_a - \overline{V}_i|$, and a substitution of Eqs. (4.90) and (4.87) into Eq. (4.89) and integrating, the change of mixing ratio due to collision between pristine ice crystals and aggregates may be written

$$\mathrm{CL}_{\mathrm{ia}} = \frac{0.5}{\beta_1} \rho_0 \overline{E_{\mathrm{a}}}(T_{\mathrm{a}}) |\overline{V_{\mathrm{a}}} - \overline{V_{\mathrm{i}}}| D_{\mathrm{m}}^{-2.4} (2D_{\mathrm{m}}^2 + 2D_{\mathrm{i}} D_{\mathrm{m}} + D_{\mathrm{i}}^2) r_{\mathrm{a}} r_{\mathrm{i}}. \quad (4.93)$$

The coefficient β_1 arises from the use by Cotton *et al.* of Passarelli and Srivastava's relationship between the density of aggregates and the diameter of aggregates. Equation (4.93) shows that the production of aggregate mixing ratio is proportional to the product of the mixing ratios of pristine ice crystals and aggregates.

Equations for the other physical processes listed above can be derived by similar procedures. Some of the processes require rather arbitrary decision making. For example, it must be decided what the "three-component process"—in which raindrops collide with aggregates (Lin *et al.*, 1983)—produce. Cotton *et al.* assumed that the resultant product of the process is graupel particles. One can imagine that fragmented aggregates and high-density frozen raindrops is a more likely consequence of such interactions. Also, when aggregates rime cloud droplets at a large rate, one must decide at what point the aggregates must be converted to graupel particles and at what rate. Unfortunately, sensitivity experiments with Cotton's model suggests that the contribution of aggregates to total precipitation is strongly dependent on this rather arbitrary conversion rate.

One can see at this point that even simple parameterized models of cloud microphysics become rather complicated when all the different physical processes are considered.

4.6 Impact of Cloud Microphysical Processes on Cloud Dynamics

What level of complication in the formulation of cloud microphysical processes is needed to simulate a particular cloud system? This question arises naturally, and the answer is influenced, in part, by the scientific objectives. If the goal is to simulate the average precipitation over a mesoscale region, it may not be necessary to simulate microphysics in great detail. If, however, it is important to distinguish among the various forms of precipitation—such as freezing rain versus graupel or aggregates versus pristine or lightly rimed ice crystals—then greater sophistication in the formulation of cloud microphysics is desirable. The answer to the above question is also driven by the impact of cloud microphysical processes on the dynamics of the cloud system. If cloud microphysical processes impact strongly on the dynamics of a cloud system, then an incorrect simulation of cloud microphysical process could lead, for example, to the simulation of short-lived multicellular thunderstorms in cases where severe, steady supercell storms are actually observed, or weak storm downdrafts where severe downbursts are observed. Frequently, numerical models of clouds experience a bifurcation in dynamic behavior, depending on the cloud microphysical structure. That is, local changes in cloud microphysical processes can eventually lead to a cloud system having completely different dynamical characteristics. Let us, therefore, summarize some of the interactions between cloud microphysical processes and cloud dynamics. More detailed examinations of such interactions will be made in Part II of this volume when we study the dynamics of various cloud systems.

4.6.1 *Water Loading*

In Chapter 2 we noted that the vertical equation of motion for a cloud system contains an additional term in the buoyancy term which accounts for the weight of suspended condensate having a total mixing ratio r_w. This term arises from the fact that if cloud droplets and precipitation particles are falling on the average at nearly their terminal velocities, then the sum of the drag forces on a parcel of air due to settling hydrometeors is equal to the weight of the condensate. As a rule of thumb, roughly $3\ \mathrm{g\,kg^{-1}}$ of condensate is equivalent to 1 K of negative thermal buoyancy. One consequence of a precipitation process, therefore, is that it unloads the cloud updraft at higher levels of its condensate and redistributes it to low levels. The additional condensate at lower levels may turn a thermally buoyant updraft into a downdraft.

4.6.2 *Redistribution of Condensed Water into Subsaturated Regions*

A possibly greater consequence of a precipitation process is the redistribution of condensation and evaporation processes associated with the condensate redistribution. Numerical experiments by Liu and Orville (1969) and Murray and Koenig (1972) suggest that the thermodynamic consequences of the precipitation process are far more important than water loading effects. Precipitation may settle into unsaturated air beneath the cloud base where evaporation commences. The evaporatively chilled air, in turn, can stimulate and intensify downdrafts in the subcloud layer, which can lead to the decay of the cloud or contribute to the propagation of the storm as air is lifted along the gust front. Recently Tripoli (1987) has proposed that evaporation of precipitation from an advancing anvil ahead of a mesoscale convective system can weaken or destroy inversions such as boundary layer capping inversions. The weakened inversions allow the convective system continued access to moist, unstable air beneath the capping inversion.

4.6.3 Cloud Supersaturation

In our derivation of the thermodynamics energy equation for a wet system in Chapter 2, we assumed that a cloud does not become supersaturated with respect to water. This is in accordance with theoretical studies (Howell, 1949; Squires, 1952; Mordy, 1959; Neiburger and Chien, 1960) and observations (Warner, 1969) in nonprecipitating clouds that peak supersaturations are generally less than 1% and more typical supersaturations are between 0.1 and 0.2%. Numerical experiments in a precipitating system suggest that there is a much greater tendency for the supersaturation to exceed nominal values (Clark, 1973; Young, 1974c). This is a result of the fact that the magnitude of supersaturation is a result of two competing processes: (1) supersaturation production by adiabatic cooling and (2) supersaturation reduction by condensation on cloud droplets. In a nonprecipitating cloud, numerous small cloud droplets are able to readily deplete supersaturation as condensation occurs on them. Because precipitation particles form at the expense of cloud droplets, the net surface area over which condensation takes place is reduced substantially when water is converted from numerous cloud droplets to fewer, large precipitation elements. As a result, supersaturations may exceed 5%, causing a delay in latent heat release compared to situations in which the cloud's humidity remains close to 100%.

4.6.4 Electrical Effects

The influence of cloud electrification processes on the dynamics of clouds is discussed more fully in Chapter 9. Here we will simply enumerate some of the hypothesized ways in which cloud electrification can affect cloud dynamics:

(i) Localized heating arising from lightning discharges.

(ii) Levitation of cloud particles, which alters the terminal velocity of particles, causing a redistribution of condensate.

(iii) Enhancement of droplet and ice-particle coalescence, thus enhancing precipitation formation in a redistribution of condensate.

4.6.5 Latent Heat Released during Freezing and Sublimation

At levels in the atmosphere colder than 0°C, the potential exists for ice crystals to be nucleated and grow by vapor deposition and collection of cloud drops. Also, supercooled cloud droplets and raindrops may freeze. The freezing of cloud droplets and raindrops results in the release of the latent heat of fusion. The amount of heat liberated is proportional to the amount of supercooled liquid water frozen. Furthermore, as ice crystals grow by vapor deposition, they release the latent heat of sublimation. If the cloud is water saturated and contains a substantial amount of supercooled cloud droplets, however, the full latent heat of sublimation is not absorbed by the cloudy air. This is because ice crystals grow by vapor deposition at the expense of cloud droplets, which must evaporate as the saturation vapor pressure is lowered locally below water saturation. As a result, the evaporating droplets absorb the latent heat of condensation. The net result of ice crystals growing by vapor deposition and of cloud droplets evaporating is that the cloud experiences heating only in proportion to the latent heat of fusion (i.e., $L_f = L_s - L_c$, where L_f is latent heat of fusion, L_s is latent heat of sublimation, and L_c is latent heat of condensation) and the water mass deposited on the ice crystals. In a glaciated cloud where liquid-water droplets are absent, a cloud will experience the full latent heat of sublimation as vapor is deposited on ice crystals.

What this means, of course, is any cloud will experience an additional source of buoyancy as freezing and ice vapor deposition takes place. The latent heating can be rather smoothly released if ice crystals grow by vapor deposition in an air mass that is cooling adiabatically by large-scale lifting. In contrast, it can take place as a burst of energy release in convective towers if large quantities of supercooled water suddenly freeze.

It is important to recognize that ice-phase-related latent heating begins to become important at levels in the atmosphere where the latent heat of condensation is greatly reduced. This is due to the fact that at colder temperatures the vertical variation of saturation vapor pressure with respect to water diminishes substantially as a parcel of air is cooled by adiabatic expansion. Ice-phase latent heating is also important in disturbed environments such as tropical cyclones (Lord *et al.*) and mesoscale convective systems (Chen and Cotton, 1988), where the environmental sounding is nearly wet adiabatic. In such an environment the cloud realizes little buoyancy gain from the latent heat of condensation, while ice-phase latent heating can contribute to substantial convective instability.

We will discuss the importance of ice-phase latent heating more fully in Chapters 8, 9, and 10.

4.6.6 *Cooling by Melting*

Melting is distinctly different from the freezing process, in which freezing is distributed through a considerable verical depth. In contrast, cooling by melting is quite localized. In the case of the more stratiform precipitation, melting can result in a well-defined isothermal layer. The stability of this isothermal layer can inhibit the downward penetration of upper tropospheric winds to lower tropospheric levels.

4.6.7 Radiative Heating/Cooling

We shall see in the next chapter that cloud radiative heating and cooling rates are strongly modulated by the microphysical structure of a cloud. The reflection of incident solar radiation at the tops of clouds is dependent upon the concentration and size of cloud droplets or the concentration, size, and habit of ice crystals. The absorption and heating by solar radiation, in turn, is a function primarily of the integrated liquid-water path and secondarily of the size of the droplets. Likewise, the absorption of terrestrial radiation is related to the integrated liquid-water path. As a consequence, both the rates of solar and terrestrial radiative cooling are primarily related to the vertical distribution of condensate in a cloud. Because precipitation processes alter the vertical distribution of condensate, they also strongly impact upon longwave and shortwave radiative heating/cooling rates in clouds. The transformation of a cloud from the liquid phase to the ice phase can impact on the radiative properties of a cloud. Furthermore, the particular habit of ice crystal growth can affect the rate of absorption of solar radiation which, in turn, can alter the thermodynamic stability of the cloud system. We will examine these processes more quantitatively in the next chapter as well as in Chapters 7 and 11.

References

Auer, A. H., and John D. Marwitz (1969). Comments on the collection and analysis of freshly fallen hailstones. *J. Appl. Meteorol.* **8**, 303-304.

Aufdermauer, A. N., and D. A. Johnson (1972). Charge separation due to riming in an electric field. *Q. J. R. Meteorol. Soc.* **98**, 369-382.

Austin, P. H., M. B. Baker, A. M. Blyth, and J. B. Jensen (1985). Small-scale variability in warm continental cumulus clouds. *J. Atmos. Sci.* **42**, 1123-1138.

Baker, M. B., and J. Latham (1979). The evolution of droplet spectra and the rate of production of embryonic raindrops in small cumulus clouds. *J. Atmos. Sci.* **36**, 1612-1615.

Baker, M. B., R. G. Corbin, and J. Latham (1980). The influence of entrainment on the evolution of cloud droplet spectra: I. A model of inhomogeneous mixing. *J. Q. R. Meteorol. Soc.* **106**, 581-598.

Baker, M. B., R. E. Breidenthal, T. W. Choularton, and J. Latham (1984). The effects of turbulent mixing in clouds. *J. Atmos. Sci.* **41**, 299-304.

Bartlett, J. T. (1966). The growth of cloud droplets by coalescence. *Q. J. R. Meteorol. Soc.* **92**, 93-104.

Bartlett, J. T. (1970). The effect of revised collision efficiencies on the growth of cloud droplets by coalescence. *Q. J. R. Meteorol. Soc.* **96**, 730-738.

Bartlett, J. T., and P. R. Jonas (1972). On the dispersion of the sizes of droplets growing by condensation in turbulent clouds. *Q. J. R. Meteorol. Soc.* **98**, 150-164.

Beliaev, V. I. (1961). Size distribution of drops in a cloud during its condensation stage of development. *Izv. Akad. Nauk SSSR, Ser. Geofiz.* No. 8, 1209-1213.

Bennetts, D. A., and F. Rawlins (1981). Parameterization of the ice phase in a model of mid-latitude cumulonimbus convection and its influence on the simulation of cloud development. *Q. J. R. Meteorol. Soc.* **107**, 477–502.

Bergeron, T. (1935). On the physics of cloud and precipitation. *Proc. Assem. Int. Union Geodesy Geophys., 5th, Lisbon* pp. 156–178.

Berry, E. X. (1965). Cloud droplet growth by collection. Ph.D. Thesis, Univ. of Nevada.

Berry, E. X. (1967). Cloud droplet growth by collection. *J. Atmos. Sci.* **24**, 688–701.

Berry, E. X. (1968). Modifications of the warm rain process. *Proc. Nat. Conf. Weather Modif., 1st, Albany, N.Y.* pp. 81–88.

Berry, E. X., and R. L. Reinhardt (1974). An analysis of cloud drop growth by collection: Part I. Double distributions. *J. Atmos. Sci.* **31**, 1814–1824.

Bigg, E. K. (1953a). The supercooling of water. *Proc. Phys. Soc. London, Sect. B* **66**.

Bigg, E. K. (1953b). The formation of atmospheric ice crystals by the freezing of droplets. *Q. J. R. Meteorol. Soc.* **79**, 510–519.

Bigg, E. K. (1955). Ice-crystal counts and the freezing of water drops. *Q. J. R. Meteorol. Soc.* **81**, 478–479.

Blanchard, D., and A. T. Spencer (1970). Experiments on the generation of raindrop size distributions by breakup. *J. Atmos. Sci.* **27**, 101–108.

Blanchard, D. C. (1948). Observations of the behavior of water drops at terminal velocity in air. Occas. Rep., pp. 100–110. Proj. Cirrus, General Electric Res. Labs., Schenectady, N.Y.

Bleck, R. (1970). A fast, approximate method for integrating the stochastic coalescence equation. *J. Geophys. Res.* **75**, 5165–5171.

Braham, R. R. (1964). What is the role of ice in summer rain showers? *J. Atmos. Sci.* **21**, 640–645.

Braham, R. R. (1968). Meteorological bases for precipitation development. *Bull. Am. Meteorol. Soc.* **49**, 343–353.

Brazier-Smith, P. R., S. G. Jennings, and J. Latham (1972). The interaction of falling water drops: Coalescence. *Proc. R. Soc. London, Ser. A* **326**, 393–408.

Brazier-Smith, P. R., S. G. Jennings, and J. Latham (1973). Raindrop interactions and rainfall rates within clouds. *Q. J. R. Meteorol. Soc.* **99**, 260–272.

Broadwell, J. E., and R. E. Breidenthal (1982). A simple model of mixing and chemical reaction in a turbulent shear layer. *J. Fluid Mech.* **125**, 397–410.

Brownscombe, J. L., and P. Goldsmith (1972). On the possible production of sub-micron ice fragments during riming or the freezing droplets in free fall. *Proc. Int. Conf. Cloud Phys., R. Meteorol. Soc., London, August,* p. 27.

Brownscombe, J. L., and N. S. C. Thorndyke (1968). Freezing and shattering of water droplets in free fall. *Nature (London)* **220**, 687–689.

Byers, H. R. (1965). "Elements of Cloud Physics." Univ. of Chicago Press, Chicago, Illinois.

Carte, A. E. (1956). The freezing of water droplets. *Proc. Phys. Soc. London* **69**, 1028–1037.

Chen, S., and W. R. Cotton (1988). The simulation of a mesoscale convective system and its sensitivity to physical parameterizations. *J. Atmos. Sci.* **45**, 3897–3910.

Cheng, L., and M. English (1983). A relationship between hailstone concentration and size. *J. Atmos. Sci.* **40**, 204–213.

Chien, E. H., and M. Neiburger (1972). A numerical simulation of the gravitational coagulation process for cloud droplets. *J. Atmos. Sci.* **29**, 718–727.

Chisnell, R. F., and J. Latham (1976a). Ice multiplication in cumulus clouds. *Q. J. R. Meteorol. Soc.* **102**, 133–156.

Chisnell, R. F., and J. Latham (1976b). Comments on the paper by B. J. Mason, "Production of ice crystals by riming in slightly supercooled cumulus." *Q. J. R. Meteorol. Soc.* **102**, 713–715.

Clark, T. L. (1973). Numerical modeling of the dynamics and microphysics of warm cumulus convection. *J. Atmos. Sci.* **30**, 857–878.

Clark, T. L. (1976). Use of log-normal distributions for numerical calculations of condensation and collection. *J. Atmos. Sci.* **33**, 810–821.

Clark, T. L., and W. D. Hall (1983). A cloud physical parameterization method using movable basis functions: Stochastic coalescence parcel calculations. *J. Atmos. Sci.* **40**, 1709–1728.

Cotton, W. R. (1970). A numerical simulation of precipitation development in supercooled cumuli. Ph.D. Thesis, Pennsylvania State Univ.

Cotton, W. R. (1972a). Numerical simulation of precipitation development in supercooled cumuli, Part I. *Mon. Weather Rev.* **100**, 11, 757–763.

Cotton, W. R. (1972b). Numerical simulation of precipitation development in supercooled cumuli, Part II. *Mon. Weather Rev.* **100**, 11, 764–784.

Cotton, W. R., and N. R. Gokhale (1967). Collision, coalescence, and breakup of large water drops in a vertical wind tunnel. *J. Geophys. Res.* **72**, 16, 4041–4049.

Cotton, W. R., M. A. Stephens, T. Nehrkorn, and G. J. Tripoli (1982). The Colorado State University three-dimensional cloud/mesoscale model—1982. Part II: An ice-phase parameterization. *J. Rech. Atmos.* **16**, 295–320.

Cotton, W. R., G. J. Tripoli, R. M. Rauber, and E. A. Mulvihill (1986). Numerical simulation of the effects of varying ice crystal nucleation rates and aggregation processes on orographic snowfall. *J. Climate Appl. Meteorol.* **25**, 1658–1680.

Danielsen, E. F., R. Bleck, and D. A. Morris (1972). Hail growth by stochastic collection in a cumulus model. *J. Atmos. Sci.* **29**, 135–155.

Davis, M. H. (1964). Two charged spherical conductors in a uniform electric field: Forces and field strength. *Q. J. Mech. Appl. Math.* **17**, 499–511.

Davis, M. H. (1965). The effect of electric charges and fields on the collision of very small cloud drops. *Proc. Cloud Phys. Conf., Meteorol. Soc. Jpn., Tokyo–Sapporo* pp. 118–120.

de Almeida, F. C. (1975). On the effects of turbulent fluid motion in the collisional growth of aerosol particles. Res. Rep. 75-2, Dep. Meteorol., Univ. of Wisconsin, Madison.

de Almeida, F. C. (1976). The collisional problem of cloud droplets moving in a turbulent environment. *J. Atmos. Sci.* **33**, 1571–1578.

Douglas, R. H. (1964). Size spectra of Alberta hail. *Nat. Conf. Phys. Dyn. Clouds, Am. Meteorol. Soc., Chicago, Ill.* pp. 1–5

Dufuor, R., and L. Defay (1963). "Thermodynamic of Clouds." Academic Press, New York.

Dye, J. E., and P. V. Hobbs (1966). Effect of carbon dioxide on the shattering of freezing water drops. *Nature (London)* **209**, 464–466.

Dye, J. E., and P. V. Hobbs (1968). The influence of environmental parameters on the freezing and fragmentation of suspended water drops. *J. Atmos. Sci.* **25**, 82–96.

Farley, R. D., and C. S. Chen (1975). A detailed microphysical simulation of hydroscopic seeding on the warm rain process. *J. Appl. Meteorol.* **14**, 718–733.

Farley, R. D., and H. D. Orville (1986). Numerical simulation of cloud seeding experiments applied to an Alberta hailstorm. Rep. SDSMT/IAS/R-86/06, Inst. Atmos. Sci., South Dakota Sch. Mines Technol., Rapid City.

Federer, B., and A. Waldvogel (1975). Hail and raindrop size distributions from a Swiss multicell storm. *J. Appl. Meteorol.* **14**, 91–97.

Feingold, G., and Z. Levin (1986). The longitudinal fit to raindrop spectra from frontal convective clouds in Israel. *J. Climate Appl. Meteorol.* **25**, 1346–1363.

Findeisen, W. (1938). Die Kolloidmeteorologischen Vorgange der Niedershlagsbildung. *Meteorol. Z.* **55**, 121–133.

Fitzgerald, J. W. (1974). Effect of aerosol composition on cloud droplet size distribution: A numerical study. *J. Atmos. Sci.* **31**, 1358–1367.

Fletcher, N. H. (1962). "The Physics of Rainclouds." Cambridge Univ. Press, London.

Fukuta, N. (1972a). Growth theory of a population of droplets and the supersaturation in clouds. *Abstr. Vol., Int. Cloud Phys. Conf., ICCP, IAMAP, IUGG, R. Meteorol. Soc., WMO, London* pp. 147–148.

Fukuta, N. (1972b). Advances in organic ice nuclei generator technology. *J. Rech. Atmos.* Dessens Mem. Issue, Nos. 1-3, 155–164.

Fukuta, N., M. Kowa, and N.-H. Gong (1982). Determination of ice crystal growth parameters in a new supercooled cloud tunnel. *Prepr., Conf. Cloud Phys., Chicago, Ill.* pp. 325–328. Am. Meteorol. Soc., Boston, Massachusetts.

Fukuta, N., H.-H. Gong, and A.-S. Wang (1984). A microphysical origin of graupel and hail. *Proc. Int. Conf. Cloud Phys., 9th, Acad. Sci. USSR, Sov. Geophys. Comm., Acad. Sci. Eston. SSR, Inst. Astrophys. Atmos. Phys., USSR State Comm. Hydrometeorol. Control Nat. Environ., Cent. Aerol. Obs., Tallinn, USSR* pp. 257–260.

Gagin, A. (1971). Studies of the factors governing the colloidal stability of continental cumulus clouds. *Prepr., Int. Weather Modif. Conf., Canberra, Aust.*, pp. 5–11. Am. Meteorol. Soc., Boston, Massachusetts.

Gagin, A. (1972). Effect of supersaturation on the ice crystal production by natural aerosols. *J. Rech. Atmos.* **6**, 175–185.

Gillespie, D. T. (1975). Three models for the coalescence growth of cloud drops. *J. Atmos. Sci.* **32**, 600–607.

Gillespie, J. R. (1972). The stochastic coalescence model for cloud droplet growth. *J. Atmos. Sci.* **29**, 1496–1510.

Gillespie, J. R., and R. List (1976). Evolution of raindrop size distribution in steady state rainshafts. *Proc. Int. Cloud Phys. Conf., Boulder, Colo.* pp. 472–477. Am. Meteorol. Soc., Boston, Massachusetts.

Gokhale, N. R., and J. D. Spengler (1972). Freezing of freely suspended supercooled water drops by contact nucleation. *J. Appl. Meteorol.* **11**, 157–160.

Goldsmith, P., J. Goster, and C. Hume (1976). The ice phase in clouds. *Prepr., Int. Conf. Cloud Phys., Boulder, Colo.* pp. 163–167. Am. Meteorol. Soc., Boston, Massachusetts.

Grant, L. O. (1968). The role of the ice nuclei in the formation of precipitation. *Proc. Int. Conf. Am. Meteorol. Soc., Toronto* pp. 305–310.

Gunn, K. L. S., and J. S. Marshall (1958). Distribution with size of aggregate snowflakes. Sci. Rep. MW-20, pp. 9–32. MacDonald Phys. Lab., McGill Univ., Montreal.

Hallett, J. (1984). How snow crystals grow. *Am. Sci.* **72**, 582–589.

Hallett, J., and S. C. Mossop (1974). Production of secondary ice particles during the riming process. *Nature (London)* **249**, 26–28.

Hallgren, R. E., and C. L. Hosler (1960). Preliminary results on the aggregation of ice crystals. *Geophys. Monogr., Am. Geophys. Union* No. 5, 257–263.

Heymsfield, A. J. (1972). Ice crystal terminal velocities. *J. Atmos. Sci.* **29**, 1348–1357.

Heymsfield, A. J., and R. G. Knollenberg (1972). Properties of cirrus generating cells. *J. Atmos. Sci.* **29**, 1358–1366.

Heymsfield, A. J., D. N. Johnson, and J. E. Dye (1978). Observations of moist adiabatic ascent in northeast Colorado cumulus congestus clouds. *J. Atmos. Sci.* **35**, 1689–1703.

Hill, T. A., and T. W. Choularton (1985). An airborne study of the microphysical structure of cumulus clouds. *Q. J. R. Meteorol. Soc.* **111**, 517–544.

Hindman, E. E., II (1975). The nature of aerosol particles from a paper mill and their effects on clouds and precipitation. Ph.D. Thesis, Univ. of Washington.

Hobbs, P. V. (1969). Ice multiplication in clouds. *J. Atmos. Sci.* **26**, 315–318.

Hobbs, P. V. (1974). High concentrations of ice particles in a layer cloud. *Nature (London)* **251**, 694–696.

Hobbs, P. V., and D. A. Burrows (1966). The electrification of an ice sphere moving through natural clouds. *J. Atmos. Sci.* **23**, 757–763.

Hobbs, P. V., L. F. Radke, A. B. Fraser, J. D. Locatelli, C. E. Robertson, D. G. Atkinson, R. J. Farber, R. R. Weiss, and R. C. Easter (1972). Field observations and theoretical studies of clouds and precipitation over the Cascade Mountains and their modifications by artificial seeding (1971–72). Res. Rep. VII, Dep. Atmos. Sci., Univ. of Washington.

Hobbs, P. V., S. Chang, and J. Locatelli (1974). The dimensions and aggregation of ice crystals in natural clouds. *J. Geophys. Res.* **79**, 2199–2206.

Hobbs, P. V., D. A. Bowdle, and L. F. Radke (1977). Aerosol over the High Plains of the United States. Res. Rep. XII, Cloud Phys. Group, Univ. of Washington.

Hobbs, P. V., M. K. Politovich, D. A. Bowdle, and L. F. Radke (1978). Airborne studies of atmospheric aerosol in the High Plains and the structure of natural and artificially seeded clouds in eastern Montana. Rep. No. XIII, Dep. Atmos. Sci., Univ. of Washington.

Hocking, L. M. (1959). The collision efficiency of small drops. *Q. J. R. Meteorol. Soc.* **85**, 44–50.

Hocking, L. M., and P. R. Jonas (1970). The collision efficiency of small drops. *Q. J. R. Meteorol. Soc.* **96**, 722–729.

Holroyd, E. W. (1964). A suggested origin of conical graupel. *J. Appl. Meteorol.* **3**, 633–636.

Hosler, C. L., and R. E. Hallgren (1960). The aggregation of small ice crystals. *Discuss. Faraday Soc.* **30**, 200–208.

Houze, R. A., Jr. (1977). Structure and dynamics of a tropical squall-line system. *Mon. Weather Rev.* **15**, 1540–1567.

Howell, W. E. (1949). The growth of cloud drops in uniformly cooled air. *J. Meteorol.* **54**, 134–149.

Hsie, E.-Y., R. D. Farley, and H. D. Orville (1980). Numerical simulation of ice-phase convective cloud seeding. *J. Appl. Meteorol.* **19**, 950–977.

Huffman, P. J. (1973). Supersaturation spectra of AgI and natural ice nuclei. *J. Appl. Meteorol.* **12**, 1080–1082.

Huffman, P. J., and G. Vali (1973). The effect of vapor depletion on ice nucleus measurements with membrane filters. *J. Appl. Meteorol.* **12**, 1018–1024.

Isono, K. (1965). Variations of the ice nuclei concentration in the atmosphere and its effect on precipitation. *Proc. Int. Conf. Cloud Phys., Int. Assoc. Meteorol. Atmos. Phys., Tokyo* pp. 150–154.

Jaw, Jeou-Jang (1966). Statistical theory of precipitation process. *Tellus* **18**, 722–729.

Jayaweera, K. (1971). Calculations of ice crystal growth. *J. Atmos. Sci.* **28**, 728–736.

Jeffreys, H. (1916). Some problems of evaporation. *Philos. Mag.* **35**, 270–280.

Johnson, D. B. (1976). Ultragiant urban aerosol particles. *Science* **194**, 941–942.

Johnson, D. B. (1979). The role of coalescence nuclei in warm rain initiation. Ph.D. Thesis, Univ. of Chicago.

Johnson, D. B. (1980). The influence of cloud-base temperature and pressure on droplet concentration. *J. Atmos. Sci.* **37**, 2079–2085.

Johnson, D. B. (1982). The role of giant and ultragiant aerosol particles in warm rain initiation. *J. Atmos. Sci.* **39**, 448–460.

Jonas, P., and P. Goldsmith (1972). The collection efficiencies of small droplets falling through a sheared air flow. *J. Fluid Mech.* **52**, 593–608.

Jonas, P. R., and B. J. Mason (1982). Entrainment and the droplet spectrum in cumulus clouds. *Q. J. R. Meteorol. Soc.* **108**, 857–869.

Jones, R. F. (1960). Size distribution of ice crystals in cumulonimbus clouds. *Q. J. R. Meteorol. Soc.* **86**, 187–194.

Keller, V., and J. Hallett (1982). Influence of air velocity on the habit of ice crystal growth from the vapor. *J. Crystal Growth* **60**, 91–106.

Kessler, E., III (1969). On the distribution and continuity of water substance in atmospheric circulation. *Meteorol. Monogr.* **10**.

Klett, J. D., and M. H. Davis (1973). Theoretical collision efficiencies of cloud droplets at small Reynolds number. *J. Atmos. Sci.* **30**, 107–117.

Koenig, L. R. (1962). Ice in the summer atmosphere. Ph.D. Thesis, Univ. of Chicago.

Koenig, L. R. (1963). The glaciating behavior of small cumulonimbus clouds. *J. Atmos. Sci.* **20**, 29–47.

Koenig, L. R. (1965). Drop freezing through drop breakup. *J. Atmos. Sci.* **22**, 448–451.

Koenig, L. R. (1971). Numerical modeling of ice deposition. *J. Atmos. Sci.* **28**, 226–237.

Koenig, L. R. (1972). Parameterization of ice growth for numerical calculations of cloud dynamics. *Mon. Weather Rev.* **100**, 417–423.

Koenig, L. R. (1977). The rime-splintering hypothesis of cumulus glaciation examined using a field-of-flow cloud model. *Q. J. R. Meteorol. Soc.* **103**, 585–606.

Koenig, L. R., and F. W. Murray (1976). Ice-bearing cumulus cloud evolution: Numerical simulations and general comparison against observations. *J. Appl. Meteorol.* **15**, 747–762.

Kovetz, A., and B. Olund (1969). The effect of coalescence and condensation on rain formulation in a cloud of finite vertical extent. *J. Atmos. Sci.* **26**, 1060–1065.

Krasnogorskaya, N. V. (1965). Effect of electrical forces on the coalescence of particles of comparable sizes. *Izv. Acad. Sci. SSSR, Atmos. Ocean Phys. (Engl. Transl.)* **1**, 200–206.

Langmuir, I. (1948). The production of rain by a chain reaction in cumulus clouds at temperatures above freezing. *J. Meteorol.* **5**, 175–192.

Latham, J., and B. J. Mason (1961). Generation of electric charge associated with the formation of soft hail in thunderclouds. *Proc. R. Soc. London, Ser. A* **260**, 537–549.

Latham, J., and R. L. Reed (1977). Laboratory studies of the effects of mixing on the evolution of cloud droplet spectra. *Q. J. R. Meteorol. Soc.* **103**, 297–306.

Latham, J., and C. P. R. Saunders (1971). Experimental measurements of the collection efficiencies of ice crystals in electric fields. *Q. J. Roy. Meteorol. Soc.* **96**, 257–265.

Levin, L. M., and Y. S. Sedunov (1966). Stochastic condensation of drops and kinetics of cloud spectrum formation. *J. Rech. Atmos.* **2**, 425–432.

Levkov, L. (1971). Congelation de gouttes d'eau au contact de particules CuS. *J. Rech. Atmos.* **5**, 133–136.

Lin, Y.-L., R. D. Farley, and H. D. Orville (1980). The addition of a snow content field to a cloud model. Inst. Atmos. Sci., South Dakota Sch. Mines Technol., Rapid City.

Lin, Y.-L., R. D. Farley, and H. D. Orville (1983). Bulk parameterization of the snow field in a cloud model. *J. Climate Appl. Meteorol.* **22**, 1065–1092.

Lindblad, N. R., and R. G. Semonin (1963). Collision efficiency of cloud droplets in electric fields. *J. Geophys. Res.* **68**, 1051–1057.

Liu, J. Y., and H. D. Orville (1969). Numerical modeling of precipitation and cloud shadow effects on mountain-induced cumuli. *J. Atmos. Sci.* **26**, 1283–1298.

Locatelli, J. D., and P. Hobbs (1974). Fall speeds and masses of solid precipitation particles. *J. Geophys. Res.* **79**, 2185–2197.

Ludlam, F. H. (1951). The production of showers by the coalescence of cloud droplets. *Q. J. R. Meteorol. Soc.* **77**, 402–417.

Macklin, W. C. (1960). The production of ice splinters during riming. *Tellus* **3**, 30.

Macklin, W. C. (1961). Accretion in mixed clouds. *Q. J. R. Meteorol. Soc.* **87**, 413–424.

Macklin, W. C., and F. H. Ludlam (1961). The fallspeeds of hailstones. *Q. J. R. Meteorol. Soc.* **87**, 72–81.

Macklin, W. C., E. Strauch, and F. H. Ludlam (1960). The density of hailstones collected from a summer storm. *Q. J. R. Meteorol. Soc.* **3**, 12.

Magarvey, R. H., and J. W. Geldhart (1962). Drop collisions under conditions of free fall. *J. Atmos. Sci.* **19**, 107–113.

Magono, C. (1953). On the growth of snow flake and graupel. *Sci. Rep. Yokohama Nat. Univ., Sect. 7* **2**, 321–335.

Magono, C. (1954). Investigation of the size distribution of precipitation elements by the photographic paper method. *Sci. Rep. Yokohama Nat. Univ., Sect. 1* **3**, 41–51.

Magono, C., and C. W. Lee (1973). The vertical structure of snow clouds as revealed by "snow crystal sondes," Part II. *J. Meteorol. Soc. Jpn.* **51**, 176–190.

Manton, M. J. (1979). On the prediction of radiative cooling rates and fluxes in the troposphere. *J. Rech. Atmos.* **3**, 201–214.

Manton, M. J., and W. R. Cotton (1977). Parameterization of the atmospheric surface layer. *J. Atmos. Sci.* **34**, 331–334.

Marshall, J. S., and W. M. Palmer (1948). The distribution of raindrops with size. *J. Meteorol.* **5**, 165–166.

Marwitz, J. D. (1983). The kinematics of orographic airflow during Sierra storms. *J. Atmos. Sci.* **40**, 1218–1227.

Mason, B. J. (1956). On the melting of hailstones. *Q. J. R. Meteorol. Soc.* **82**, 209–216.

Mason, B. J. (1971). "The Physics of Clouds." Oxford Univ. Press (Clarendon), London.

Mason, B. J. (1975). Production of ice crystals by riming in slightly supercooled cumulus. *Q. J. R. Meteorol. Soc.* **101**, 675–679.

Mason, B. J., and P. R. Jonas (1974). The evolution of droplet spectra and large droplets by condensation in cumulus clouds. *Q. J. R. Meteorol. Soc.* **100**, 23–38.

Mason, B. J., and J. Maybank (1960). The fragmentation and electrification of freezing water drops. *Q. J. R. Meteorol. Soc.* **86**, 176–186.

Mazin, I. P. (1965). Toward a theory of the shaping of the particle size spectrum in clouds and precipitation. *Tr. Tsentr. Aerol. Obs.* **64**, 57–70.

Mazin, I. P. (1968). The stochastic condensation and its effect on the formation of cloud drop size distribution. *Proc. Int. Conf. Cloud Phys. Toronto* pp. 67–71.

McTaggert-Cowan, J. D., and R. List (1975). Collision and breakup of water drops at terminal velocity. *J. Atmos. Sci.* **32**, 1401–1411.

Mee, T. R., and D. M. Takeuchi (1968). Natural glaciation and particle size distribution in marine tropical cumuli. MRI Final Rep. MR 168, FR-823 Contract No. E22-30-68(N), Exp. Meteorol. Branch, ESSA, Univ. of Miami.

Mordy, W. A. (1959). Computations of the growth by condensation of a population of cloud droplets. *Tellus* **11**, 16–44.

Mossop, S. C., and J. Hallett (1974). Ice crystal concentration in cumulus clouds: Influence of the drop spectrum. *Science* **186**, 632–633.

Mossop, S. C., and A. Ono (1969). Measurements of ice crystal concentrations in clouds. *J. Atmos. Sci.* **26**, 130–137.

Mossop, S. C., A. Ono, and K. J. Heffernan (1967). Studies of ice crystals in natural clouds. *J. Rech. Atmos.* **3**, 45–64.

Mossop, S. C., R. E. Ruskin, and K. J. Heffernan (1968). Glaciation of a cumulus at approximately −4°C. *J. Atmos. Sci.* **25**, 889–899.

Mossop, S. C., A. Ono, and E. R. Wishart (1970). Ice particles in maritime clouds near Tasmania. *Q. J. R. Meteorol. Soc.* **96**, 487–508.

Mossop, S. C., R. E. Cottis, and B. M. Bartlett (1972). Ice crystal concentrations in cumulus and stratocumulus clouds. *Q. J. R. Meteorol. Soc.* **98**, 105–123.

Murray, F. W., and L. R. Koenig (1972). Numerical experiments on the relation between microphysics and dynamics in cumulus convection. *Mon. Weather Rev.* **100**, 717–732.

Nakaya, U., and T. Terada, Jr. (1935). Simultaneous observations of the mass falling velocity and form of individual snow crystals. *J. Fac. Sci., Hokkaido Imp. Univ., Ser. 2* No. 4.

Neiburger, M., and C. W. Chien (1960). Computations of the growth of cloud drops by condensation using an electronic digital computer. *Geophys. Monogr., Am. Geophys. Union* No. 5, 191–208.

Nelson, L. D. (1979). Observations and numerical simulations of precipitation mechanisms in natural and seeded convective clouds. Tech. Note No. 54, UCHI-CPL-79-54, Cloud Phys. Lab., Dep. Geophys. Sci., Univ. of Chicago.

Nelson, R. T., and N. R. Gokhale (1968). Concentration of giant particles below cloud base. *Prepr., Nat. Conf. Weather Modif., 1st, Albany, N.Y.* pp. 89–98. Am. Meteorol. Soc., Boston, Massachusetts.

Nickerson, E. C., J. M. Fritsch, C. F. Chappell, and D. R. Smith (1978). Numerical simulations of orographic and convective cloud systems. Annu. Rep. to Eng. Res. Cent., Bur. Reclam., U.S. Dep. Inter., Denver, Contract No. 8-07-83-V0017.

Nickerson, E. C., E. Richard, R. Rosset, and D. R. Smith (1986). The numerical simulation of clouds, rain, and airflow over the Vosges and Black Forest Mountains: A meso-β model with parameterized microphysics. *Mon. Weather Rev.* **114**, 398–414.

Ochs, H. T., III (1978). Moment-conserving techniques for warm cloud microphysical computations. Part II: Model testing and results. *J. Atmos. Sci.* **35**, 1959–1973.

Ochs, H. T., III, and C. S. Yao (1978). Moment-conserving techniques for warm cloud microphysical computations. Part I: Numerical techniques. *J. Atmos. Sci.* **35**, 1947–1958.

Ono, A. (1972). Evidence on the nature of ice crystal multiplication processes in natural cloud. *J. Rech. Atmos.* **6**, 399–408.

Orville, H. D., and F. J. Kopp (1977). Numerical simulation of the life history of a hailstorm. *J. Atmos. Sci.* **34**, 1596–1618.

Paluch, I. R., and C. A. Knight (1986). Does mixing promote cloud droplet growth? *J. Atmos. Sci.* **43**, 1994–1998.

Paluch, I. R., and C. A. Knight (1987). Reply to Telford's comment. *J. Atmos. Sci.* **44**, 2355–2356.

Passarelli, R. E., and R. C. Srivastava (1979). A new aspect of snowflake aggregation theory. *J. Atmos. Sci.* **36**, 484–493.

Peña, J. A., and C. L. Hosler (1971). Freezing of supercooled clouds induced by shock waves. *J. Appl. Meteorol.* **10**, 1350–1352.

Pitter, R. L., and H. R. Pruppacher (1973). A wind tunnel investigation of freezing of small water drops falling at terminal velocity in air. *Q. J. R. Meteorol. Soc.* **99**, 540–550.

Plumlee, H. R., and R. G. Semonin (1965). Cloud droplet collision efficiency in electric fields. *Tellus* **17**, 356–364.

Pruppacher, H. R., and J. D. Klett (1978). "Microphysics of Clouds and Precipitation." Reidel, Boston, Massachusetts.

Pruppacher, H. R., and R. J. Schlamp (1975). A wind tunnel investigation on ice multiplication by freezing of water drops falling at terminal velocity in air. *J. Geophys. Res.* **80**, 380–386.

Rau, W. (1950). Uber die Wirkungsweise der Gefrierkeime im unterkuhlten Wasser. *Z. Naturforsch., A* **5A**, 667–675.

Rauber, R. M. (1985). Physical structure of northern Colorado river basin cloud systems. Ph.D. Thesis, Atmos. Sci. Pap. No. 390, Dep. Atmos. Sci., Colorado State Univ.

Reinking, R. F. (1975). Formation of graupel. *J. Appl. Meteorol.* **14**, 745–754.

Rogers, D. C. (1974). The aggregation of natural ice crystals. M.S. Thesis, Dep. Atmos. Resour., Univ. of Wyoming.

Ryan, B. F. (1973). A numerical study of the nature of the glaciation process. *J. Atmos. Sci.* **30**, 824–834.

Sartor, D. (1960). Some electrostatic cloud droplet collision efficiences. *J. Geophys. Res.* **65**, 1953–1957.

Sasyo, Y. (1971). Study of the formation of precipitation by the aggregation of snow particles and the accretion of cloud droplets on snowflakes. *Pap. Meteorol. Geophys.* **22**, 69–142.

Schnell, R. C., and G. Vali (1976). Biogenic ice nuclei. Part I: Terrestrial and marine sources. *J. Atmos. Sci.* **33**, 1554–1564.

Scott, B. D., and P. V. Hobbs (1977). A theoretical study of the evolution of mixed-phase cumulus clouds. *J. Atmos. Sci.* **34**, 812–826.

Semonin, R. G., and H. R. Plumlee (1966). Collision efficiency of charged cloud droplets in electric fields. *J. Geophys. Res.* **71**, 4271–4278.

Simpson, J., and V. Wiggert (1969). Models of precipitating cumulus towers. *Mon. Weather Rev.* **97**, 471–489.

Simpson, J., and V. Wiggert (1971). Florida cumulus seeding experiment: Numerical model results. *Mon. Weather Rev.* **99**, 87–118.

Simpson, J., R. H. Simpson, D. A. Andrews, and M. A. Eaton (1965). Experimental cumulus dynamics. *Rev. Geophys.* **3**, 387–431.

Spilhaus, A. F. (1948). Drop size intensity and radar echo in rain. *J. Meteorol.* **5**, 161–164.

Squires, P. (1952). The growth of cloud drops by condensation. *Aust. J. Sci. Res., Ser. A* **5**, 59–86.

Squires, P. (1956). The microstructure of cumuli in maritime and continental air. *Tellus* **8**, 443–444.

Squires, P. (1958). The microstructure and colloidal stability of warm clouds. *Tellus* **10**, 256–271.

Srivastava, R. C. (1967). On the role of coalescence between raindrops in shaping their size distribution. *J. Atmos. Sci.* **24**, 287–292.

Srivastava, R. C. (1978). Parameterization of raindrop size distributions. *J. Atmos. Sci.* **35**, 108–117.

Stepanov, A. S. (1975). Condensational growth of cloud droplets in a turbulent medium taking into account diabatic effects in the approximation of the smallness of the water content. *Izv. Acad. Sci. USSR Atmos. Oceanic Phys.* (*Engl. Transl.*) **11**, 160–165.

Stepanov, A. S. (1976). Turbulence effect on cloud droplet spectrum during condensation. *Cloud Phys. Conf., Boulder, Colo.* pp. 27-32. Am. Meteorol. Soc., Boston, Massachusetts.

Subbarao, M. C., and P. Das (1975). Microphysical implications of precipitation formation in an adiabatic vertical current: Aerosol scavenging by enhanced nucleation. *Q. J. R. Meteorol. Soc.* **32**, 2338–2357.

Takahashi, C., and A. Yamashita (1969). Deformation and fragmentation of freezing water drops in free fall. *J. Meteorol. Soc. Jpn.* **47**, 431–436.

Takahashi, C., and A. Yamashita (1970). Shattering of frozen water drops in a supercooled cloud. *J. Meteorol. Soc. Jpn.* **48**, 373–376.

Takahashi, T. (1973). Numerical simulation of maritime warm cumulus. *J. Geophys. Res.* **20**, 6233–6247.

Takahashi, T. (1974). Numerical simulation of tropical shower. *J. Atmos. Sci.* **31**, 219–232.

Takahashi, T. (1976). Hail in an axisymmetric cloud model. *J. Atmos. Sci.* **33**, 1579–1601.

Takahashi, T. (1981). Warm rain development in a three-dimensional cloud model. *J. Atmos. Sci.* **38**, 1991–2013.

Telford, J. (1955). A new aspect of coalescence theory. *J. Meteorol.* **12**, 436–444.

Telford, J. W. (1987). Comment on "Does mixing promote cloud droplet growth?" *J. Atmos. Sci.* **44**, 2352–2354.

Telford, J. W., and S. K. Chai (1980). A new aspect of condensation theory. *Pure Appl. Geophys.* **118**, 720–742.

Telford, J. W., T. S. Keck, and S. K. Chai (1984). Entrainment at cloud tops and the droplet spectra. *J. Atmos. Sci.* **41**, 3170–3179.

Tripoli, G. (1989). A numerical study of an observed orogenic mesoscale convective system. Part 1: Simulated genesis and comparison with observations. *Mon. Weather Rev.* **117**, 269–300.

Tripoli, G. J., and W. R. Cotton (1980). A numerical investigation of several factors contributing to the observed variable intensity of deep convection over South Florida. *J. Appl. Meteorol.* **19**, 1037–1063.

Twomey, S. (1964). Statistical effect in the evolution of a distribution of cloud. *J. Meteorol.* **12**, 436.

Twomey, S. (1966). Computations of rain formation by coalescence. *J. Atmos. Sci.* **23**, 405–411.

Twomey, S. (1976). Computations of the absorption of solar radiation by clouds. *J. Atmos. Sci.* **33**, 1087–1091.

Twomey, S., and P. Squires (1959). The influence of cloud nucleus population on microstructure and stability of convective clouds. *Tellus* **11**, 408–411.

Twomey, S., and J. Warner (1967). Comparison of measurements of cloud droplets and cloud nuclei. *J. Atmos. Sci.* **24**, 702–703.

Vali, G., and E. J. Stansbury (1966). Time dependent characteristics of the heteorogeneous nucleation of ice. *Can. J. Phys.* **44**, 477–502.

Vardiman, L., and L. O. Grant (1972a). A case study of ice crystal multiplication by mechanical facturing. *Int. Cloud Phys. Conf., ICCP, IAMAP, IUGG, R. Meteorol. Soc., WMO, London* p. 22.

Vardiman, L., and L. O. Grant (1972b). Study of ice crystal concentrations in convective elements of winter orographic clouds. *Conf. Weather Modif., 3rd, Rapid City, S.D.* pp. 113–118. Am. Meteorol. Soc., Boston, Massachusetts.

Vonnegut, B., and C. E. Moore (1965). Nucleation of ice formation in supercooled clouds as the result of lightning. *J. Appl. Meteorol.* **4**, 640–642.

Warner, J. (1969). The micro-structure of cumulus clouds. I. General features of the droplet spectrum. *J. Atmos. Sci.* **26**, 1049–1059.

Warshaw, M. (1967). Cloud-droplet coalescence: Statistical foundations and a one-dimensional sedimentation model. *J. Atmos. Sci.* **24**, 278–286.

Wegener, A. (1911). Kerne derkristallbildung. *In* "Thermodynamik der Atmosphare" pp. 94–98. Barth, Leipzig.

Weinstein, A. I. (1970). A numerical model of cumulus dynamics and microphysics. *J. Atmos. Sci.* **27**, 246–255.

Weinstein, A. I., and L. G. Davis (1968). A parameterized numerical model of cumulus convection. Rep. II, NSF GA-777, 43, Natl. Sci. Found., Washington, D.C.

Wen, C.-S. (1966). Effects of the correlative time of the fluctuating force field on the random growth of cloud droplets. *Sci. Sin.* **15**, 870–879.

Wisner, C., H. D. Orville, and C. Myers (1972). A numerical model of a hail-bearing cloud. *J. Atmos. Sci.* **29**, 1160–1181.

Woodcock, A. H., R. A. Duce, and J. L. Moyers (1971). Salt particles and raindrops in Hawaii. *J. Atmos. Sci.* **28**, 1252–1257.

Yeh, J.-D., M. A. Fortune, and W. R. Cotton (1986). Microphysics of the stratified precipitation region of a mesoscale convective system. *Prepr., Conf. Cloud Phys., Snowmass, Colo.* pp. J151–J154. Am. Meteorol. Soc., Boston, Massachusetts.

Young, K. C. (1974a). The role of contact nucleation in ice phase initiation in clouds. *J. Atmos. Sci.* **31**, 768–776.

Young, K. C. (1974b). A numerical simulation of wintertime, orographic precipitation. I: Description of model microphysics and numerical techniques. *J. Atmos. Sci.* **31**, 1735–1748.

Young, K. C. (1974c). A numerical simulation of wintertime, orographic precipitation. II: Comparison of natural and AgI-seeded conditions. *J. Atmos. Sci.* **31**, 1749–1767.

Young, K. C. (1975). The evolution of drop spectra due to condensation, coalescence and breakup. *J. Atmos. Sci.* **32**, 965–973.

Chapter 5 Radiative Transfer in a Cloudy Atmosphere and Its Parameterization

5.1 Introduction

Cloud modelers have historically ignored the effects of radiative transfer. This is largely because the emphasis in cloud modeling has been on the simulation of individual convective clouds. For convective time scales of the order of 30 min to 1 h, radiative heating rates are of little importance. However, as cloud modeling has moved to the simulation of stratocumulus clouds, fogs, middle and high clouds, and to cloud processes on the mesoscale, where time scales are on the order of a day, cloud modelers have begun to consider radiative transfer processes. As with turbulent transport and cloud microphysical processes, cloud modelers must make a number of compromises in the design of a radiative transfer model; i.e., they must formulate the radiative transfer equations in a simplified parameterized form.

Similar to atmospheric motions that span a broad range of eddy sizes and cloud hydrometeors that span a broad range of particle sizes, atmospheric radiation covers a broad spectrum of radiation frequencies, wavelengths, or wave numbers. The sun, for example, emits radiation approximately as a blackbody having temperatures between 6000 and 5700 K, which peaks in intensity at a wavelength of 0.470 μm but which spans the range from less than 0.2 μm to greater than 1.8 μm (see Fig. 5.1). In contrast, the radiation energy emitted by the earth corresponds approximately to blackbody radiation at a temperature of 250 K. Thus, a combination of radiation emitted by the sun and the earth spans a range from less

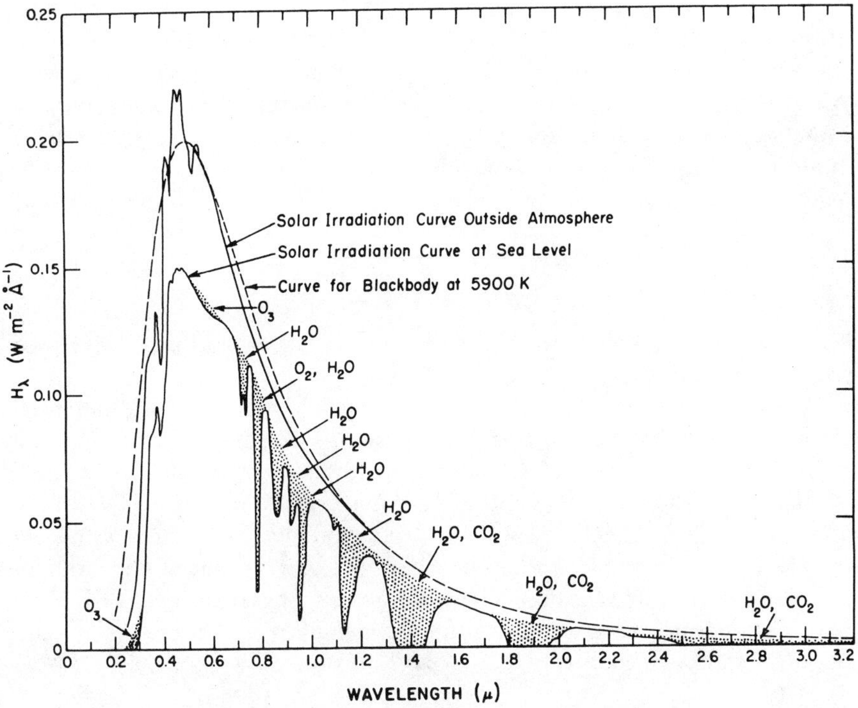

Fig. 5.1. Spectral distribution curves related to the sun; shaded areas indicate absorption, at sea level, due to the atmospheric constituents shown. [From Gast *et al.* (1965). Copyright © 1965 by McGraw-Hill. Reprinted by permission.]

than 0.2 μm to greater than 50 μm. However, the spectrum of radiation emitted by the sun and by the earth exhibits very little overlap. For this reason we refer to radiation emitted by the sun as *shortwave radiation* and radiation emitted by the earth and its atmosphere as *longwave radiation.* The two regions are separated arbitrarily at 4 μm. As we shall see later, this distinction allows some simplifications in the formulation of radiative transfer theories.

The important consideration as far as the dynamics of clouds is concerned is that the net heating rate at various levels in a cloud is a consequence of the attenuation of atmospheric radiation. The net heating rate is a function of the net radiative flux divergence,

$$\partial\theta/\partial t = (1/\rho_0 c_p)(\partial F_N/\partial z), \tag{5.1}$$

where F_N is the difference between downward and upward fluxes and has units of watts per square meter. In Eq. (5.1), for simplification, we consider only vertical flux divergences. In some cloud modeling problems, however, it may be desirable to consider horizontal radiative flux divergence as well. An example is a valley fog located between two radiating valley sides, or radiative fluxes passing through a field of cumuli. Nonetheless, for most cloud modeling applications, consideration of vertical radiative flux divergences is sufficient. The net vertical flux F_N is the difference between downward and upward fluxes, or

$$F_N = F\downarrow - F\uparrow. \tag{5.2}$$

The upward and downward fluxes, in turn, represent the fluxes integrated over all wavelengths.

A divergence of radiative flux is caused by a combination of differential extinction of radiation and thermal emission. Extinction of radiation is a result of absorption and of scattering of radiant energy. The absorptance A_λ represents the fraction of incoming radiation absorbed in a layer of the atmosphere. The reflectance Re_λ and transmittance Tr_λ simply represent the fraction of incoming radiation which is scattered out of the primary beam and transmitted through a layer of the atmosphere, respectively. Note that all three processes vary with the wavelength of radiation. As a result of conservation of energy,

$$A_\lambda + \mathrm{Re}_\lambda + \mathrm{Tr}_\lambda = 1. \tag{5.3}$$

Absorption, unlike scattering, results in a physical change in the medium. In the atmosphere this change is usually a change in internal energy or temperature. If the incident energy as well as the scattered and transmitted energy remain constant for a time, the internal energy of the system will remain unchanged. A radiative equilibrium is established in which as much radiation is being emitted as is being absorbed. However, in conditions of local thermodynamic equilibrium, the emitted radiation is in thermal equilibrium with the source level. If the atmosphere were a pure blackbody (i.e., $A_\lambda = 1$ for all wavelengths), then the total emitted radiative flux would be

$$F = \sigma T^4, \tag{5.4}$$

where σ is the Stefan–Boltzmann constant. According to Blevin and Brown (1971), $\sigma = (5.66961 \pm 0.0075) \times 10^{-8}\ \mathrm{W\,m^{-2}\,K^{-4}}$. Equation (5.4) can be obtained by integrating the Planck radiation distribution function over all wavelengths [see any basic radiation physics text or Paltridge and Platt (1976) for a definition of the Planck function]. The atmosphere does not, however, behave as a blackbody. Therefore, the amount of energy emitted by the absorbing atmosphere is given by

$$F = \varepsilon\sigma T^4, \tag{5.5}$$

where ε represents the emittance of the atmosphere. The emittance is the ratio of the flux emitted by a body to the flux emitted by a blackbody at the same temperature. We shall see later that the effective emittance of a cloudy atmosphere varies with the liquid-water content and particle spectra in clouds.

5.2 Absorptance, Reflectance, Transmittance, and Emittance in the Clear Atmosphere

As can be seen in Fig. 5.1, in a cloud-free atmosphere, the primary absorbers of shortwave radiation are ozone and water vapor. Aerosols also contribute to a lesser extent to absorption. At wavelengths shorter than 0.3 μm, oxygen and nitrogen absorb nearly all incoming solar radiation in the upper atmosphere. However, between 0.3 and 0.8 μm, little gaseous absorption occurs. Only weak absorption by ozone takes place in this spectral range. This is fortunate because these are the wavelengths in which the solar radiation peaks. At wavelengths of less than 0.8 μm, Rayleigh scattering of shortwave radiation back to space depletes the available flux. At longer wavelengths, absorption in various water vapor bands is quite pronounced. Some weak absorption by carbon dioxide and ozone also occurs at wavelengths greater than 0.8 μm, but water vapor is the predominant absorber.

Absorption of longwave radiation also occurs mainly in a series of bands. The principal absorbers of longwave or infrared (IR) radiation are water vapor, carbon dioxide, and ozone. Figures 5.1 and 5.2 illustrate the banded character of absorption in the IR region. Figure 5.2 represents the IR spectrum obtained by a scanning interferometer looking downward from a satellite over a desert region. Strong absorption by CO_2 at a wavelength band centered at 14.7 μm is shown by the emission of radiance at the temperature of 220 K. The stratosphere contributes mainly to the peak of this absorption band, with warmer tropospheric contributions occurring across the broader part of the absorption band. Water vapor absorption bands at 1.4, 1.9, 2.7, and 6.3 μm, and greater than 20 μm, cause emissions corresponding to midtropospheric temperatures. Little absorption is evident in the region called the "atmospheric window" between 8 and 14 μm. Here the radiance corresponds to the surface temperature of the desert except for a slight depression in magnitude due to departures of the emittance of sand from unity. A distinct ozone absorption band is evident in the region of 9.6 μm in the middle of the window. Although not very evident in the

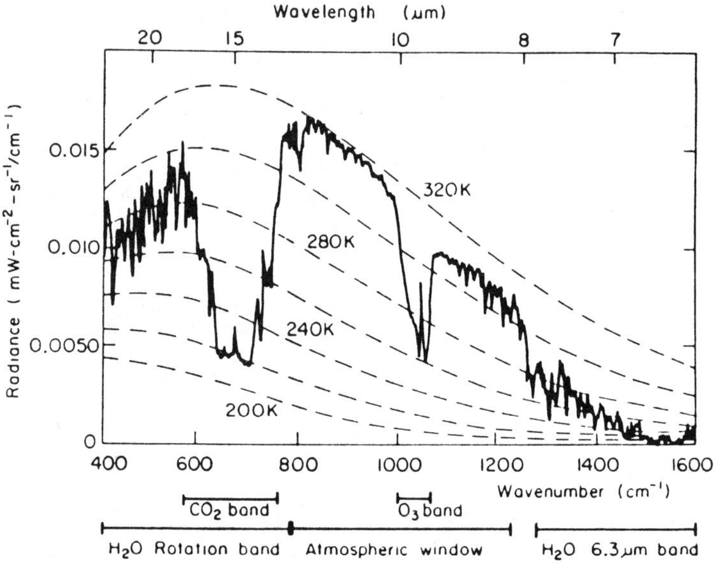

Fig. 5.2. Atmospheric spectrum obtained with a scanning interferometer on board the Nimbus 4 satellite. The interferometer viewed the earth vertically as the satellite was passing over the North African desert. [After Hanel *et al.* (1972), cited in Paltridge and Platt (1976).]

figure, weak continuous absorption also occurs across the "window." The intensity of the continuum absorption depends on water vapor pressure. This continuous absorption is due to the presence of clusters or dimers $(H_2O)_2$ of water vapor molecules.

A detailed knowledge of the absorption spectra in the IR region is essential to predicting the rate of cooling due to longwave radiative transfer. In the following sections, we first examine the interactions among cloud particles and radiation. Then we review some of the techniques for calculating radiative transfer in a cloudy atmosphere.

5.3 Shortwave Radiative Transfer in a Cloudy Atmosphere

The interaction of solar radiation incident upon a cloud is complicated by the fact that not only must we concern ourselves with the impact of a spectrum of radiative frequencies on cloud absorption, reflection, and transmission, but also with the consequences of a spectrum of droplets or ice crystals on radiative transfer. In the absence of emission, the azimuthally averaged radiative transfer equation appropriate to a cloudy medium is as follows:

$$\mu \frac{dI(\tau, \mu)}{d\tau} = -I(\tau, \mu) + \frac{\tilde{\omega}_0}{2} \int_{-1}^{+1} \bar{p}(\tau, \mu, \mu') I(\tau, \mu') \, d\mu'$$

$$+ \frac{S_0}{4\pi} \bar{p}(\tau, \mu, \mu_0) \, e^{-\tau/\mu_0}, \tag{5.6}$$

where τ is the optical depth, $\tilde{\omega}_0$ is the single-scattering albedo, $\bar{p}$ is the scattering phase function, and S_0 is the solar flux associated with a collimated beam incident on the cloud top. All parameters in Eq. (5.6) are functions of the frequency of radiation ν with the exception of μ and μ_0, where μ is a function of the cosine of the zenith angle. The quantity $I(\tau, \mu)$ is the radiance along an angle given by μ through a cloudy layer defined by the optical depth τ. The extinction optical thickness, τ, is the sum of droplet scattering, droplet absorption, and gaseous absorption. Gaseous absorption is primarily due to water vapor in both a clear and cloudy atmosphere. In fact, for some time it was thought that the major effect of clouds on shortwave radiation was to increase the optical path length of water vapor absorption as a result of multiple scattering by cloud droplets. We shall see that absorption by liquid droplets plays an important role in the overall absorption in clouds. Cloud droplet absorption is described in terms of τ_a, the droplet absorption optical thickness. If we define τ_s, the scattering optical thickness, and τ_g, the gaseous absorption optical thickness, then the extinction optical thickness is given by

$$\tau = \tau_s + \tau_a + \tau_g. \tag{5.7}$$

For any frequency one can calculate the optical thickness $\tau(\nu)$ and thus obtain a single-scattering albedo

$$\tilde{\omega}_0 = \tau_s / \tau. \tag{5.8}$$

Thus, $\tilde{\omega}_0 = 1$ for a nonabsorbing cloud and $\tilde{\omega}_0 = 0$ when scattering is negligible. The optical thickness varies only slightly with ν and the average size of a droplet. The single-scattering albedo, on the other hand, varies strongly with both frequency (or wavelength) and droplet size. Figure 5.3 illustrates the variation in single-scattering albedo as a function of wavelength for a maritime cloud with a droplet concentration of 25 cm^{-3} (solid curve) and a mildly continental cloud with a droplet concentration of 200 cm^{-3} (dashed curve) for a given temperature and a liquid-water content of 0.33 gm^{-3}. In a continental cloud, such as over the high plains of the United States, where droplet concentrations are approximately 700 cm^{-3}, or in polluted air masses, where droplet concentrations are approximately 2500 cm^{-3}, the single-scattering albedo would be much greater than in a maritime air mass.

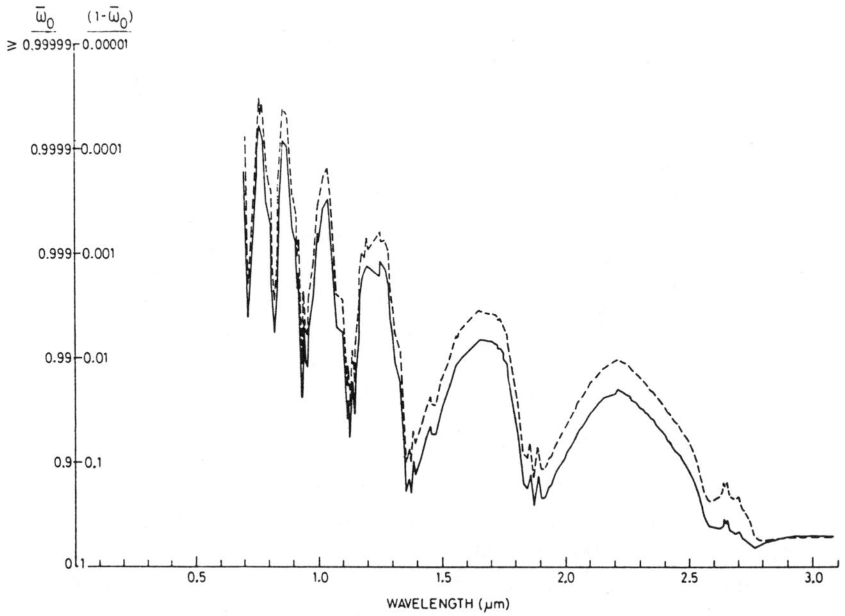

Fig. 5.3. Spectra of the single-scattering albedo ω_0 as a function of wavelength. [From Twomey (1976).]

This is because for a given liquid-water content, the higher droplet concentrations result in much smaller droplet sizes.

The scattering phase function $\bar{p}(\tau, \mu, \mu')$ in Eq. (5.6) characterizes the angular distribution of the scattered radiation field. For spherical droplets, this function exhibits a strong peak in the forward direction and produces rainbow and glory effects in the back direction. As noted by Joseph *et al.* (1976), for most applications the phase function can be conveniently expressed as

$$\bar{p}(\tau, \mu, \mu') = (1-g^2)/(1+g^2-2g\mu\mu'), \tag{5.9}$$

which was formulated by Henyey and Greenstein (1941). The parameter g in Eq. (5.9) is called the asymmetry factor, which is the average value of the cosine of the scattering angle (weighted according to energy scattered in the various directions). For symmetric scattering, which puts equal amounts of energy in the forward and backward directions, g is zero (e.g., Rayleigh scattering). As more energy is scattered in the forward hemisphere, g increases toward unity.

The single most important parameter defining the radiative properties of clouds is the optical thickness. Excluding the effects of water vapor, the formal definition of the optical thickness is

$$\tau = \int_0^z \int_0^\infty f(r) Q_e(x, n_\lambda) \pi r^2 \, dr \, dz, \tag{5.10}$$

where $Q_e(x, n_\lambda)$ is the extinction efficiency factor, and $f(r)$ is the spectral density of droplets of radius r. The first integral is over the cloud depth z. The second integral is over the cloud droplet radius r. Evaluation of Eq. (5.10) requires a knowledge of both the cloud droplet distribution and the behavior of Q_e. The extinction factor is defined as the ratio of the extinction to the cross-sectional areas of cloud droplets. For spherical cloud droplets, Q_e can be evaluated from Mie theory. It is a function of particle radius through the size parameter $x = 2\pi r/\lambda$ and the refractive index of the particle n_λ.

Figure 5.4 illustrates the variation of scattering efficiency factor for nonabsorbing water drops ($n_i = 0$) as calculated by Hansen and Travis (1974) using Mie theory. The extinction efficiency factor is the sum of the scattering and absorption efficiency factors. The variation in the absorption efficiency factor as a function of x also resembles Fig. 5.4. The efficiency factor was calculated assuming droplets were distributed in a γ-type size distribution function that is determined by two parameters: $a = r_e$, the effective radius defined later in Eq. (5.15), and a coefficient of dispersion (b) about the effective radius a. The coefficient of dispersion is a measure of the droplet spectral width, with $b = 0$ being a monodisperse distribution and larger values representing broader spectra. Typical observed droplet-size distributions in cumuli exhibit dispersion values slightly less than 0.2. Remember that the parameter x is a function of both the size of the droplets and the wavelength of electromagnetic radiation. Because the effective radius of most cloud droplet distributions is on the order of 10 μm, x is typically greater than 10 for the visible spectrum. For large values of x, Fig. 5.4 illustrates (1) a series of regularly spaced broad maxima and minima called the interference structure, which oscillates about the approximate value of 2, and (2) irregular fine structure called the ripple structure. For broader droplet spectra, this fine structure is smeared out, but the feature of Q_e approaching a limiting value of two for large x remains.

Because for large values of x (i.e., for shortwave radiation and typical cloud droplet-size distribution), Q_e tends to be an almost constant value of 2, Eq. (5.10) can be written as

$$\tau = 2\pi \int_0^z \int_0^\infty f(r) r^2 \, dr \, dz. \tag{5.11}$$

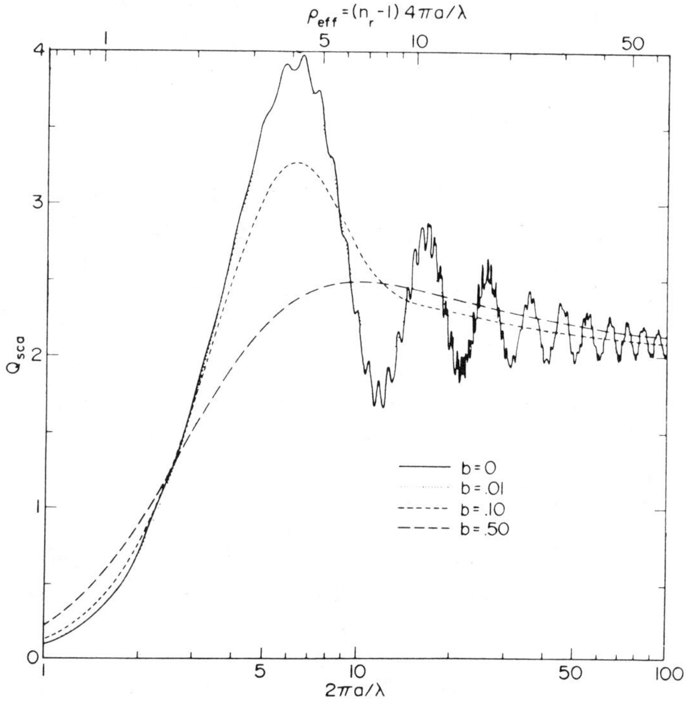

Fig. 5.4. Efficiency factor for scattering, Q_{sca}, as a function of the effective size parameter, $2\pi a/\lambda$. A γ-type size distribution is used with effective radius $a = r_e$ and coefficient of dispersion b. For the case $b = 0$, $2\pi a/\lambda = 2\pi r/\lambda \equiv x$. The refractive index is $n_r = 1.33$, $n_i = 0$. [From Hansen and Travis (1974). Copyright © 1974 by D. Reidel Publishing Company. Reprinted by permission.]

Thus Eq. (5.11) is only a function of the drop-size distribution and the depth of the cloud. To a good approximation we can write

$$\tau \simeq 2\pi N_c \bar{r}^2 h, \tag{5.12}$$

where N_c is the droplet concentration, $\bar{r}$ is the mean droplet radius, and h is the geometric thickness of a cloud layer. Using the notation in Chapter 4, we can express Eq. (5.12) in terms of the cloud droplet mixing ratio r_c as

$$\tau \simeq 2\pi(3\rho_0/\pi\rho_w)^{2/3} h N_c^{1/3} r_c^{2/3}, \tag{5.13}$$

where ρ_0 is the dry air density and ρ_w is the density of water. The cloud droplet optical thickness is therefore a function of the cloud properties (liquid-water mixing ratio and cloud depth) as well as droplet concentration.

The cloud droplet concentration, in turn, is largely determined by the aerosol content of the air mass or, in particular, the concentration of cloud condensation nuclei. For typical peak supersaturations in clouds, N_c varies from as low as 10–50 cm^{-3} in maritime clouds to 1000 cm^{-3} for continental clouds. The most important effect of changing the optical thickness of a cloud layer is that it changes the amount of reflected radiation and thereby alters the energy reaching the earth's surface and the atmosphere below a cloud layer. Twomey suggests that if the CCN concentration in the cleaner parts of the atmosphere, such as oceanic regions, were raised to continental atmospheric values, about 10% more energy would be reflected to space by relatively thin cloud layers. He also points out that an increase in cloud reflectivity by 10% is of more consequence than a similar increase in global cloudiness. This is because while an increase in cloudiness reduces the incoming solar energy flux, it also reduces the outgoing infrared radiation flux. Thus both cooling and heating effects occur when global cloudiness is increased. In contrast, an increase in cloud reflectance due to enhanced CCN concentrations does not appreciably affect infrared radiation but does reflect more incoming solar radiation, which results in a net cooling effect.

Stephens (1978a) has shown that another approximation to Eq. (5.10) is

$$\tau \simeq \tfrac{3}{2} W / r_e, \tag{5.14}$$

where W is the liquid-water path (g m^{-2}) and r_e is an effective radius defined as

$$r_e = \int_0^\infty f(r) r^3 \, dr \Big/ \int_0^\infty f(r) r^2 \, dr. \tag{5.15}$$

The liquid-water path represents the integrated liquid water through a cloud of depth h,

$$W = \int_0^h \rho_0 r_c \, dz. \tag{5.16}$$

5.3.1 *Absorption*

The greater the optical thickness of a cloud, the more incident solar radiation will be attenuated by it. Other factors being the same, an optically thick cloud reflects more solar energy and absorbs more incident energy. There is evidence that larger drops absorb energy more efficiently than smaller drops. As can be seen from Eq. (5.11), the optical thickness of a cloud is approximately twice the sum of the cross-sectional area of all drops in a column. Thus, a maritime cloud having fewer, but more numerous, large droplets absorbs more radiation than a continental cloud with more small droplets, and transmits proportionally more as well. The more numerous, smaller droplets in a continental cloud contribute to substantially greater cloud reflectivity. The calculations reported by Twomey (1976), Stephens (1978a), and Manton (1980), however, suggest that the differences in absorptance between nonprecipitating maritime and continental clouds is slight, the liquid-water path (W) having the dominating influence on cloud absorption.

Several investigators have estimated that precipitation-sized drops can appreciably increase cloud absorptance (Manton, 1980; Welch *et al.*, 1980; Wiscombe *et al.*, 1984). Because precipitation-sized drops form by collecting smaller cloud drops, they typically represent a miniscule percentage of the total drop concentration. Drizzle-sized drops have a concentration usually not exceeding $100\ m^{-3}$, compared to cloud droplet concentrations of the order of 10^8 to $10^9\ m^{-3}$. Thus precipitation-sized drops have little direct impact upon the reflectance of a cloud. On the other hand, because precipitation-sized drops can contribute significantly to the liquid-water content and, hence, to the liquid-water path, they can affect cloud absorption appreciably. Wiscombe *et al.* (1984) calculated that for a given liquid-water path, the absorptance is enhanced by 2–3% by the presence of large drops in a 4-km-deep cloud.

Recent calculations by Wiscombe and Welch (1986) in which more numerous precipitation-sized drops were considered (the distribution had an r_e of 216 μm) suggest that for large liquid-water paths and/or larger solar zenith angles, very large enhancements of absorptance occurred. They suggested that the presence of high concentrations of the large drops may explain measurements of cloud absorptance in the range 20–40% (Robinson, 1958; Drummond and Hickey, 1971; Reynolds *et al.*, 1975; Rozenberg *et al.*, 1974; Herman, 1977; Stephens *et al.*, 1978; Twomey and Cocks, 1982). The observed anomalously high values of absorptance have perplexed theoreticians, who predict a theoretical upper limit on cloud absorptance slightly in excess of 20% (Fritz, 1954, 1958; Wiscombe *et al.*, 1984). Other hypotheses include the finite dimensions of clouds and the role of aerosols in enhancing cloud absorptance. We shall examine these hypotheses in subsequent sections.

Welch *et al.* (1980) also noted that if the liquid water in the tops of middle and high clouds is distributed mainly in large drops (or ice particles) rather than on small cloud droplets, 10–23% reductions in cloud albedo (reflectance) can occur. Thus the facts that large drops exhibit very low reflectance and that they grow at the expense of cloud droplets suggest that the magnitude of cloud and planetary albedo is sensitive to the relative abundance of precipitation-sized particles in the upper levels of clouds.

5.3.2 Ice Clouds

The optical properties of ice clouds are complicated by the geometries of the ice particles, the uncertainties in ice crystal concentration, and their size spectra. We have seen in Chapter 4, that the habit of vapor-grown ice crystals varies with temperature and supersaturation with respect to ice. Moreover, as ice crystals grow by riming cloud droplets or aggregation, the geometrical and surface characteristics of the ice particles vary as do their optical properties. Crucial to any realistic assessment of the reflectance, transmittance, or absorptance of an ice or mixed-phase cloud system is the concentration of ice particles and their size spectra. Unfortunately, as we have pointed out in Chapter 4, we do not have a reliable way of diagnosing or predicting ice crystal concentration. At temperatures approaching 0°C, for example, the conventional approach of using ice nuclei concentrations to diagnose ice crystal concentrations may lead to an error of from three to five orders of magnitude. Equation (5.12) suggests that such an error in ice crystal concentrations will lead to an error of from three to five orders of magnitude in estimating the shortwave optical thickness (τ) for an ice cloud.

One further complication of ice clouds is that ice crystals generally are not randomly oriented in space. Instead, ice crystals fall preferentially with their major axis oriented horizontal (Ono, 1969; Jayaweera and Mason, 1965; Platt, 1978). In early studies of the multiple scattering properties of ice clouds, it was assumed that ice crystals were randomly oriented in three dimensions (e.g., Liou, 1972a, b). Stephens (1980a), however, examined the radiative properties of long cylinders exhibiting a preferential orientation in the horizontal plane. Stephens solved the radiative transfer equation for a horizontally infinite cloud layer. The hypothetical ice cloud was composed of 100-μm long cylinders that were 25 μm in radius, having a concentration of 0.05 cm^{-3} (50 $liter^{-1}$). Stephens performed calculations for (1) three-dimensionally randomly oriented crystals (3DRO), (2) crystals with their long axis randomly oriented in the horizontal plane only (2 DRO), and (3) spherical particles of equivalent radius (ES). The radius of an equivalent sphere was approximated by

$$r = (A/4\pi)^{0.5}, \tag{5.17}$$

where A is the surface area of the nonspherical particles. Figure 5.5 (a–c) illustrates the calculated 0.7- and 2.22-μm wavelength reflectance and the 2.22-μm absorptance as a function of cloud optical depth for the zenith angles 30° and 60° for the three crystal models discussed above. It is quite evident that the equivalent spherical model underestimates cloud albedo relative to either of the randomly oriented models. The cloud absorptance

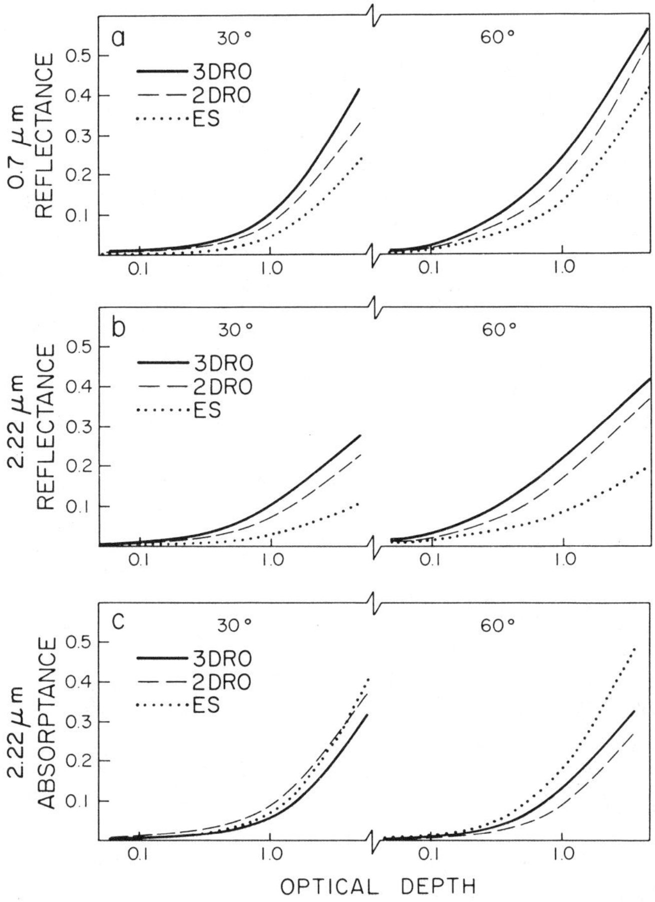

Fig. 5.5. The 0.7-μm (a) and 2.22-μm (b) reflectance and the 2.22-μm (c) absorptance as a function of cloud optical depth for the two solar zenith angles indicated and for the three different models of cloud microstructure. [From Stephens (1980a).]

for the horizontally oriented crystals is consistently less than the other two crystal models at the larger zenith angles shown here, particularly for the larger zenith angle. At small zenith angles (not shown), however, shortwave absorption by horizontally oriented crystals can be twice as large as the absorption for three-dimensionally oriented crystals.

Further evidence of the complexity of the shortwave radiative properties of ice clouds can be seen from the variety of optical phenomena that are frequently observed. Features such as halos, sundogs, and pillars result from the interaction of shortwave radiation with ice crystals having a particular crystal habit, spatial orientation, size spectra, and a growth history that often includes a relatively turbulent-free, slowly rising environment (Hallett, 1987; Tricker, 1970; Greenler, 1980).

5.4 Longwave Radiative Transfer in a Cloudy Atmosphere

In this discussion, we refer to longwave radiation as radiation emitted by the earth's surface or the atmosphere having wavelengths greater than about 4 μm. The effect of clouds on longwave radiation is quite different than it is for shortwave radiation. In the case of shortwave radiation, we find that cloud droplets are strong scatterers of incident radiation. Absorption of solar radiation by cloud droplets and ice crystals is small. By contrast, longwave radiation is strongly absorbed by cloud droplets in optically thick clouds. As much as 90% of incident longwave radiation can be absorbed in less than 50-m pathlengths in a cloud with high liquid-water content. Scattering of longwave radiation in clouds is secondary to absorption. Thus optically thick clouds are often considered to be blackbodies with respect to longwave radiation. Yamamoto *et al.* (1970) suggested that cumulonimbus clouds could be considered blackbodies after a pathlength of only 12 m. By contrast, the blackbody depth of thin cirrus ice clouds may be greater than several kilometers (Stephens, 1983), which is greater than the depths of those clouds. Thus, cirrus clouds, thin stratus, and many fogs do not behave as blackbodies over the infrared range.

We noted previously that in a cloud-free atmosphere, little gaseous absorption takes place between 8 and 14 μm, a band which is commonly referred to as the "atmospheric window." In a cloudy atmosphere, on the other hand, there are no spectral regions where absorption of longwave radiation is small. Clouds therefore have a major impact on the amount of longwave radiation emitted to space.

The behavior of the extinction efficiency Q_e is quite different in the infrared region than it is over visible wavelengths for cloud particles. The extinction efficiency varies similarly to the behavior of the scattering efficiency shown in Fig. 5.4. Thus, for small values of x, Q_e increases monotonically with x. Moreover, at wavelengths corresponding to the peak in the spectral density of terrestrial radiative flux ($10\ \mu m < \lambda < 20\ \mu m$), Q_e varies almost linearly with droplet radius for droplets less than 20 μm. Substitution of $Q_e = kr$ into Eq. (5.10) yields

$$\tau = \int_0^{\delta z} \int_0^{\infty} \pi k r^3 f(r)\, dr\, dz. \tag{5.18}$$

Because the liquid-water content is

$$\mathrm{LWC} = \int_0^\infty \frac{3\pi}{3\rho_l} r^3 f(r)\, dr, \tag{5.19}$$

Eq. (5.18) shows that the optical thickness over the infrared range is principally a function of the liquid-water content of a cloud and is not strongly dependent upon the details of the cloud droplet spectrum. This greatly simplifies the parameterization of longwave radiative transfer through clouds.

The calculations by Wiscombe and Welch (1986), however, suggest that predictions of infrared cooling rates can be significantly affected by the presence of drizzle or raindrops near the tops of optically thick clouds. Estimated cooling rates for a cloud containing cloud droplets only, and for one containing precipitation, may differ by a factor of 4 in the topmost 50 m of a cloud. The differences between cooling rates for the two cloud types are reduced substantially at greater penetration distances in the cloud, although for an 8-km-deep cloud there is still nearly a factor-of-2 difference in cooling rates between the two cloud types.

Paltridge and Platt (1976) noted that a commonly used approximation for estimating Q_e for complex-shaped crystals is to use the equivalent sphere approximation, Eq. (5.17). Mie theory can then be used to calculate the variation of Q_e as a function of x. The result is similar to Fig. 5.4, for water drops, with differences due to the variation of refractive indices between water and ice. Using the method of discrete space theory, Stephens (1980b) calculated the emittance and reflectance at a wavelength of 11 μm for three-dimensionally oriented crystals, crystals randomly oriented in the horizontal plane, and equivalent spheres. The calculated emittance did not vary among the three crystal models tested, nor did the calculated reflectance differ substantially between the two- and three-dimensionally oriented crystals. The reflectance calculated for equivalent spheres, on the other hand, was much less than either of the randomly oriented crystal models.

Although the longwave radiative reflectance is small, Stephens (1980a) calculated that it can have a significant impact on the flux profiles in the cloud and thus on cloud-heating profiles. This is true when the upward flux from the earth's surface is quite large. A longwave reflectance at cloud base of only a few percent can thus significantly affect the upwelling fluxes at colder cloud temperatures. This can substantially alter the strength of flux divergence and, hence, the rate of radiational cooling near cloud top. This effect is most pronounced in the tropics.

A commonly used concept in longwave radiation diagnostic studies as well as parameterizations is the effective emittance concept. Cox (1976) determined cloud emittance values from measurements of broadband radiative flux profiles through clouds. The emittance can thus be defined

$$\varepsilon(\uparrow) = \frac{F_{\mathrm{B}}(\uparrow) - F_{\mathrm{T}}(\uparrow)}{F_{\mathrm{B}}(\uparrow) - \sigma T_{\mathrm{T}}^4} \quad \text{for the upward irradiance,} \tag{5.20}$$

$$\varepsilon(\downarrow) = \frac{F_{\mathrm{B}}(\downarrow) - F_{\mathrm{T}}(\downarrow)}{\sigma T_{\mathrm{B}}^4 - F_{\mathrm{T}}(\downarrow)} \quad \text{for the downward irradiance.} \tag{5.21}$$

$F(\uparrow)$ and $F(\downarrow)$ refer to the upward and downward measured infrared irradiances, respectively. The subscripts T and B refer to the top and bottom of the cloud layer, respectively, and σ is the Stefan–Boltzmann constant. The definition of effective emittance combines the effects of reflection, emission, and transmission by cloud droplets as well as gas molecules. The effective emittance is therefore not a scalar but a directionally dependent vector, since the emissivity is dependent upon the particular path the radiation takes through the atmosphere. As noted by Stephens (1980a), when a cold cloud overlies a warm surface and reflects some longwave radiation, it can exhibit values of emittance considerably greater than unity.

We shall see below that the effective emittance concept is useful in the formulation of parameterizations of longwave radiative fluxes.

5.5 Radiative Influences on Cloud Particle Growth

In our discussion of cloud droplet growth (or evaporation) by vapor deposition, we did not mention the influence of radiative processes. Traditionally, cloud physicists have ignored the effects of radiation (Mason, 1971; Byers, 1965). It is generally argued that because the temperature difference between cloud droplets or ice crystals and their immediate surroundings is so small, nearly as much radiation is emitted from the cloud particle as is absorbed. This view is probably valid for a cloud particle that resides in the middle of an optically thick cloud. However, a cloud particle that resides at the top of a cloud layer or in an optically thin cloud such as a cirrus cloud essentially "sees" outer space, especially in the 8- to 12-μm spectral window. As a consequence, the surface temperature of the cloud particle will be cooler. As a result, the droplet or ice crystal will experience a higher supersaturation, or, in a subsaturated environment, a lesser subsaturation. This can be more readily seen by considering the rate of mass change due to vapor deposition or evaporation of an ice crystal or cloud droplet of mass M,

$$dM/dt = 4\pi C D f_1 f_2[\rho_{\mathrm{v}}(T_\infty) - \rho_{\mathrm{s}}(T_{\mathrm{s}})], \tag{5.22}$$

where C is the capacitance of an ice crystal defined in Chapter 4. For a spherical droplet or ice crystal, $C = r$. The coefficient D is the diffusivity of water vapor, f_1 is a factor that includes the accommodation coefficient for water molecules, f_2 is a ventilation function, $\rho_v(T_\infty)$ is the vapor density some distance from the particle surface, and $\rho_s(T_s)$ is the saturation vapor density with respect to ice or water at the surface temperature of the ice crystal. In order to estimate T_s, we normally assume that a cloud droplet or ice crystal is in thermal equilibrium,

$$L(dM/dt) + Q_r = 4\pi C K f_1 f_2 \Delta T, \tag{5.23}$$

where the first term on the LHS of Eq. (5.23) is the latent heat liberated in the growing cloud particle, L is either the latent heat of sublimation or vaporization, and the second term on the LHS is the net radiative heating of the particle. The RHS represents the rate of heat diffusion away from the particle, where K is the thermal diffusivity and ΔT is the temperature difference between a cloud particle and its environment. Roach (1976) and Barkstrom (1978) considered the radiative effects on cloud droplet growth while Stephens (1983) and Hallett (1987) considered radiative influences on ice crystal growth. Barkstrom showed that for optically thick clouds radiation can be important to droplet condensation for those droplets residing within 20 m of cloud top. For optically thin clouds such as fogs and thin stratus, cloud droplets throughout the cloud may be affected by radiation. For the case of thin ice crystal clouds such as some cirrus clouds, Stephens also concluded that (1) because radiation can enhance (suppress) particle growth (evaporation), radiative cooling at cloud top and warming at cloud base tend to broaden and narrow the spectrum, respectively and (2) the influence of radiation on the survival distance of falling ice particles is most significant in air having a relative humidity greater than 70%. At lower relative humidities evaporation is so strong that survival distances are altered little by radiation.

It is clear that radiative effects can be important to cloud particle growth and evaporation. Stephens also pointed out that the contribution of radiation to environmental heating prevails over latent heating when cirrus cloud supersaturations are small and when ice particle dimensions are large.

5.6 Radiative Characteristics of Clouds of Horizontally Finite Extent

Thus far we have considered only the radiative properties of horizontally infinite cloud layers. However, few cloud systems are horizontally homogeneous; a population of cumulus clouds is just one example. Even the tops of stratocumulus clouds are undulating, thus altering both the shortwave and longwave properties of those clouds.

Consideration of the finite geometries of clouds is quite complex. Looking out the window at a few cumulus clouds, we observe the complicated shapes these clouds assume, sometimes identifying cloud shapes with animals or other familiar objects. Needless to say, describing such complicated shapes with mathematical functions or computer algorithms can be quite difficult. Often the cloud shapes are approximated as simple cubes (McKee and Cox, 1974, 1976; Davis *et al.*, 1979a, b), while a few attempts have been made to simulate more complex shapes using superimposed sinusoidal functions (Takeuchi, 1986).

A variety of approaches have been used to examine the radiative properties of finite clouds. A particularly robust approach is the Monte Carlo simulation method, in which individual photons are followed from an assigned point of origin. A set of boundary conditions is defined for the geometrical limits of a cloud or population of clouds. Photons are then introduced in the cloud region. The direction in which the photon travels is specified and the photon is tracked until it interacts with a cloud boundary. At this point the type of interaction that takes place (i.e., scattering and/or absorption) must be ascertained.

A series of random interaction events is then simulated as the photon passes through the cloud or some other scattering/absorbing medium. As the photon progresses through the medium, the path of the photon is recorded, as is any other pertinent information affecting its behavior such as loss in energy due to absorption. Because scattering changes its direction, the photon can exit a cloud not only from its bottom, but also from the sides and top. The photon can then be followed through the cloud-free atmosphere, where further molecular absorption and scattering can be simulated. It should be noted that each photon's life history is only a single realization of what is assumed to be a random, statistical process. A large number of photons must be followed in order to obtain statistically meaningful results. Several shortcutting methods have been devised in order to reduce the number of photon life histories that must be simulated (Davis *et al.*, 1979a). Even with such cost-saving techniques, Davis *et al.* calculated some 20,000 photon life histories in their simulation of scattering in finite clouds. Determination of bulk radiative budgets for a layer using the Monte Carlo technique introduces many of the statistical uncertainties experienced by aircraft observers (Poellot and Cox, 1977).

The simple calculations performed by McKee and Cox (1974) for a nonabsorbing cloud revealed that radiation enters and leaves a finite cloud through its sides. The scattered radiation "leaking" through the sides of the clouds emerges in either the upward or downward hemisphere. The fraction of incident energy that emerges traveling in the upward hemisphere is defined as the directional reflectance (DR). In general, the directional reflectance for finite clouds is less than that for horizontally infinite clouds. Consideration of shortwave absorption, however, revealed a strong dependence on solar zenith angle, especially for optically thick clouds. At low zenith angles, Davis *et al.* (1979b) showed that the amount of absorbed solar radiation was greater in horizontally infinite clouds than in finite clouds. As the solar zenith angle increases, however, the exposure of the sides of the cloud to more radiant energy results in increasingly greater absorption, until at large zenith angles a finite cloud absorbs more solar energy than an infinite cloud. Davis *et al.* also estimated the total spectral radiative budget over the range 0.3–8.0 μm. The calculated time/space-averaged radiative budget indicated greater amounts of reflected and smaller amounts of transmitted radiative power for finite clouds than infinite clouds. The enhanced reflectivity of the finite clouds was due to the exposure of the vertical boundaries of the cloud to direct solar radiation. In a sense the increased exposure of the vertical cloud boundaries has an effect similar to an enhancement of the cloud fractional coverage. Davis *et al.* proposed an effective cloud cover that is greater than the geometrical cloud cover by a factor of $1+\tan\theta$, where θ is the solar zenith angle.

In general, it appears that the finite geometrical properties of clouds are more important to the bulk radiative properties than are variations in the cloud microstructure. This brings us to the question of when the radiative properties of a finite cloud approach those of an infinite cloud. Figure 5.6 illustrates the percentage of incident radiation that is absorbed and lost through the cloud sides as a function of the width-to-thickness ratio calculated by Welch *et al.* (1980). This shows that clouds that are 30 times wider than they are tall absorb incident radiation much like horizontally infinite clouds. Also the fraction of radiation lost through the sides of a cloud becomes quite small once a cloud is over 60 times wider than it is tall.

Welch and Zdunkowski (1981b) calculated broadband reflectance, transmittance, and absorptance as a function of the diameter of cumulus clouds and solar zenith angle using Monte Carlo simulations. Assuming cloud height to width ratios of one, and specified droplet size distributions and liquid-water paths, they calculated that the cloud absorptance increases with cloud diameter to a value of 19%. They also showed that the radiation exiting the cloud sides is distinctly anisotropic. For small cloud sizes, up to twice as much energy exiting the cloud sides is transmitted into the downward hemisphere than into the upward hemisphere. Clouds greater than 6 km in diameter, however, emit radiation from their sides nearly isotropically. As the solar zenith angle was increased from 0 to 60°, the

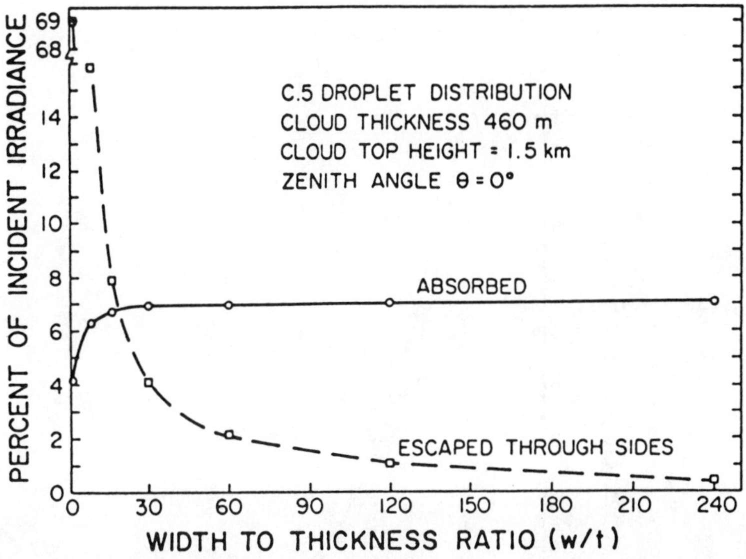

Fig. 5.6. Relationships between relative size of a finite cloud and absorption and energy escaping through the sides of the cloud element. [From Welch *et al.* (1980).]

computed cloud reflectance decreased from 73 to 48% while the cloud absorptance increased from 19 to 23%.

All the above-mentioned calculations assume that the radiation emanating from the sides of one cloud does not affect the radiative budget of a neighboring cloud. The radiative interaction of clouds has been studied using the Monte Carlo method (Aida, 1977) and by laboratory analogs to clouds (McKee *et al.*, 1983). Figure 5.7 illustrates the variation in the reflected radiation from finite clouds as a function of their separation distance. The parameter S is defined as the ratio of the distance between cloud centers D to the individual cloud diameter. At separation distances exceeding $S = 5$, very little cloud-to-cloud radiative interaction was observed in the laboratory simulation or in the Monte Carlo simulation experiments.

There also have been several recent studies of the longwave radiative characteristics of finite clouds (Liou and Ou, 1979; Harshvardhan *et al.*, 1981; Harshvardhan and Weinman, 1982; Ellingson, 1982). These studies reveal that when the height-to-width ratio of the cloud elements is on the order of one, it is quite unrealistic to model the net radiance over a cloud field by simply area weighting the radiances of the cloud sky and clear sky regions. Using the Monte Carlo technique for noninteracting cubic-shaped clouds, Harshvardhan *et al.* calculated differences in heating and cooling

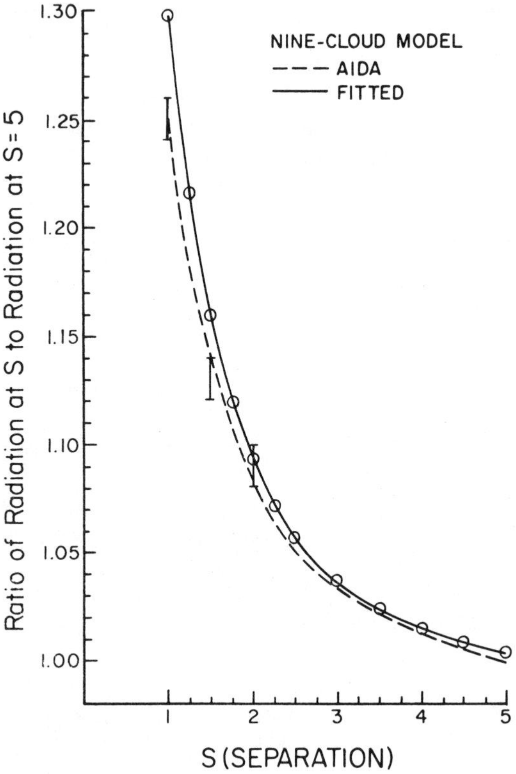

Fig. 5.7. Comparison between Aida (1977) and the experimental method for the nine-cloud model. [From McKee *et al.* (1983).]

rates between plane parallel clouds and finite clouds on the order of several tenths of a degree per hour (Fig. 5.8). Much of the difference in cooling rates is due to the infrared radiation that emits from the sides of the clouds. Assuming clouds radiate as blackbodies, Ellingson (1982) calculated the net radiances for finite cylindrically shaped clouds having variable radii and heights. His studies showed that cylindrical clouds result in more downward radiance at the surface (1–4%) and less escaping to the atmosphere (up to 8%) than from flat clouds. He also found that the subcloud layer experiences as much as 20% more heating from cylindrical clouds than from flat clouds.

Cloud heterogeneities may explain the anomalously high shortwave absorption that is often observed compared to theoretical predictions of absorptance. The above studies also suggest that cloud and mesoscale

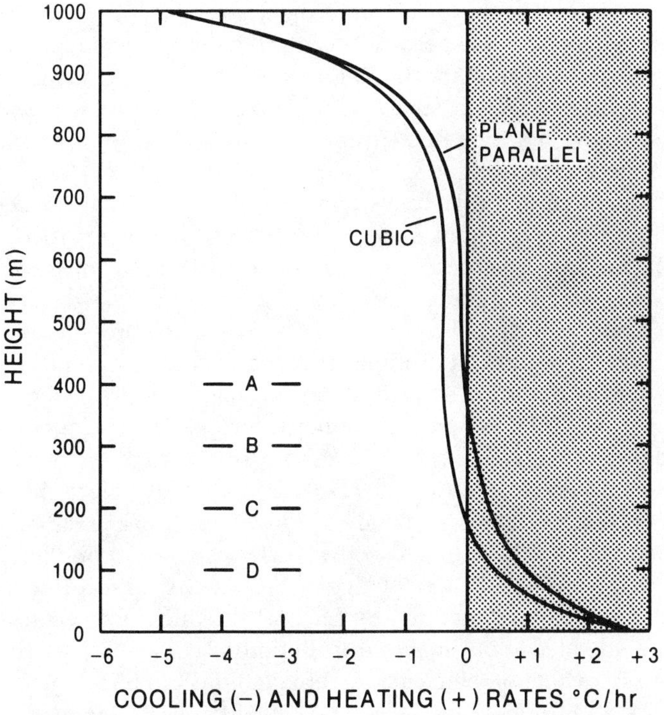

Fig. 5.8. Horizontally averaged heating (+) and cooling (−) rates in the 8- to 13.6-μm region within a 1-km cloud at −10°C overlying a black surface at 15°C. The cloud droplet-size distribution is a Deirmendjian C-1 type and has a concentration of 100 cm^{-3}. Stippled areas show heating. [From Harshvardhan *et al.* (1981).]

modelers should be concerned about shortwave radiation emitting from the sides of simulated clouds as well as the effects of horizontal radiative flux divergences on longwave radiative cooling rates. Clearly, the effects of radiative inhomogeneity in clouds must be formulated in a simple way for use in cloud and larger scale models.

5.7 Aerosol Effects on the Radiative Properties of Clouds

We have already seen that the concentration and the size spectra of cloud droplets have an important influence upon the shortwave radiative properties of clouds. The cloud droplet concentration, in turn, is largely a function of the concentration of cloud condensation nuclei activated in typical cloud supersaturations. The width of the droplet size spectra is to some extent also a consequence of the aerosol spectrum, although, as noted in Chapter 4, the width of the droplet spectrum is influenced by other cloud macroscopic parameters. We have also seen that the concentration and size spectra of ice crystals have a strong impact upon the radiative properties of clouds. Furthermore, the concentrations of aerosols active as ice nuclei are believed to play an important role in determining the ice crystal concentration, particularly for cold clouds such as altostratus and cirrus. Also, in those clouds having weak vertical motions, the size spectrum of ice crystals is largely determined by the competition for vapor among the ice crystals nucleated on ice nuclei. Thus the size spectra and chemical composition of aerosols have important controlling influences on the concentration of cloud droplets and ice crystals as well as on their size spectra, which, in turn, have important impacts upon the radiative properties of clouds.

In addition to affecting cloud radiative properties indirectly by influencing the cloud microstructure, aerosols can directly affect the radiative properties of clear as well as cloudy air. Assessment of the radiative effects of aerosols requires an estimate of the single scattering properties and, just as with cloud particles, a knowledge of the aerosol optical thickness. The optical thickness is determined by Eq. (5.10), which, like cloud droplets, requires a knowledge of the size distribution of the aerosol particles. Estimates of the extinction efficiency Q_e are complicated because aerosols are nonspherical, and, moreover, their variable chemical composition results in variability in their complex indices of refraction. Thus, the extinction efficiency varies with the source and life history of the aerosol. The life history is important because, as the aerosol ages, the particles coagulate with each other and form particles of mixed chemical composition. Furthermore, the extinction efficiency and the size spectra of the aerosol population change with relative humidity. At relative humidities greater than 70%, the hygroscopic aerosols take on water to become haze particles. As the relative humidity increases toward 100%, the aerosol particles swell in size and their complex indices of refraction change as the water-solution/particle mixture changes in relative amounts.

Computations of the radiative effects of natural dry aerosols suggest that polluted boundary layer air can result in shortwave radiative heating rates on the order of a few tenths of a degree to several degrees per hour (Braslau and Dave, 1975; Welch and Zdunkowski, 1976). Above the boundary layer, the aerosol shortwave radiative heating rates are much less. Even in the Saharan dust layer, heating rates are only of the order of 1-2°C per day (Carlson and Benjamin, 1980; Ackerman and Cox, 1982). Nonetheless, such heating rates strengthen the overlying inversion, which further concentrates the pollutants in the lower troposphere.

The impact of aerosols on infrared radiative transfer is usually less than it is for shortwave radiation. The effect of aerosols is greatest in the atmospheric window, where gaseous absorption of infrared radiation is least (Welch and Zdunkowski, 1976; Carlson and Benjamin, 1980; Ackerman *et al.*, 1976). Before longwave radiative cooling effects can be detected, rather large concentrations of aerosols through a deep layer must be present. Ackerman and Cox (1982) could not detect any significant change in longwave fluxes due to dust over Saudi Arabia. Carlson and Benjamin (1980) calculated that the Saharan dust layer can affect longwave radiative fluxes if the dust layer is sufficiently deep. Several researchers have found that longwave radiative cooling can be appreciable in polluted boundary layer air (Welch and Zdunkowski, 1976; Saito, 1981). Andreyev and Ivlev (1980) investigated the radiative properties of various organic and inorganic natural aerosols. They found that organic aerosols are typically less than 0.5 μm in radius and affect infrared radiation little, but have a significant impact on shortwave radiative fluxes. Infrared radiative fluxes were mainly affected by the presence of large ($r > 0.15$ μm) mineral substances. Andreyev and Ivlev's evaluations did not include exposure of the aerosols to increasing relative humidity. Some of the small inorganic aerosols may be activated as haze particles at higher relative humidities. As a result, once the small aerosols have swollen in size, they may alter longwave radiative fluxes appreciably. Welch and Zdunkowski calculated that net longwave radiative fluxes changed by more than 25% from a dry polluted boundary layer compared to a moist polluted boundary layer.

The effect of increasing humidity on the radiative properties of aerosols is most pronounced in the shortwave spectrum. The swelling of aerosol particles or activation of haze particles has an appreciable impact on local visibility (Kasten, 1969; Takeda *et al.*, 1986) and on the global albedo (Zdunkowski and Liou, 1976). Zdunkowski and Liou also calculate that the swelling of aerosol particles in humid atmospheres can alter the local albedo by as much as 5% relative to a dry atmosphere.

Aerosols can modify the radiative properties of clouds when they are mixed into a cloud system. The most hygroscopic of the aerosol particles participate in the nucleation of cloud droplets. Large and ultragiant aerosols (greater than 1 μm in radius) rapidly become wetted regardless of their chemical composition, and become engulfed in cloud droplets or raindrops as a result of hydrodynamic capture. No longer functioning as aerosols, they still influence the radiative properties of the cloud system. Smaller, submicrometer-sized aerosols, however, can remain outside of cloud droplets (interstitial). Those that are nonhygroscopic may remain dry with little change in size. Hygroscopic aerosols, while not being activated as cloud droplets, will swell in size in the saturated or supersaturated environment.

Thus, the interstitial aerosol population will resemble a cloud-free haze population with the exception of the removal of the largest, most hygroscopic components by nucleation scavenging and hydrodynamic capture of the particles greater than 1 μm. After a time, the remainder of the submicrometer particles may also be scavenged by cloud droplets. The major scavenging processes for submicrometer aerosols are Brownian diffusion and thermophoretic/diffusiophoretic scavenging. As noted in Chapter 4, Brownian diffusion is quite slow, and because thermophoresis predominates over diffusiophoresis, for a submicrometer-sized aerosol, phoretic scavenging does not enhance scavenging of submicrometer aerosols in a supersaturated cloud. In subsaturated regions of a cloud, however, evaporating cloud droplets can be very effective at scavenging submicrometer aerosols by phoretic processes. The droplet spectrum has to be broad enough to allow the largest droplets to survive evaporation in local subsaturated regions, however. This means that in the least dilute regions of convective storm updrafts, the time scale is short owing to intense updraft speeds, and because little evaporation occurs in the weakly mixed regions, many submicrometer particles will survive the ascent into the upper troposphere. At least for some time, submicrometer aerosol particles may remain interstitial and affect the radiative properties of the stratiform region of those clouds.

The presence of the ice phase also contributes to scavenging of interstitial aerosols. Slowly settling, branched snowflakes can be very effective removers of any remaining particles greater than 1 μm. Likewise, as ice crystals sublimate in sub-ice-saturated environments, they can be effective scavengers of submicrometer aerosols by phoretic processes. Furthermore, as we noted in Chapter 4, when ice crystals grow by vapor deposition, they do so at the expense of cloud droplets, causing their evaporation. Again, if the droplet spectrum is broad enough, the larger droplets may survive evaporation and contribute to the scavenging of submicrometer aerosols. Still, these processes are slow enough that the anvils or stratiform regions of deep convective systems, especially in polluted environments, contain a substantial interstitial aerosol population for some time. A significant interstitial aerosol population also exists in the tops of boundary layer stratocumuli and relatively young middle-tropospheric stratus clouds.

Assuming that the interstitial aerosol size spectrum is unchanged by the presence of cloud, Newiger and Bahnke (1981) calculated that aerosols can enhance absorption in a 4-km-thick horizontally uniform cloud by factors of 1.3 to 2.4. The absorptance did not exceed 30%, however, so their presence could not explain the observed so-called anomalous absorption.

The next question concerns how the scavenged aerosol particles affect the radiative properties of clouds. If the particles are soluble, they will dissolve once they have become embedded in water droplets for a time.

Only in very small droplets will the solution be concentrated enough to affect the refractive index of the droplet. If the droplet totally evaporates, however, they become aerosol particles again (perhaps altered in size and chemical composition) and affect radiative transfer again. If the particles, however, are insoluble, as are graphitic carbon or soot, they can remain embedded in droplets and alter the radiative properties of the droplets. If the insoluble particles are greater than 1 μm in radius, they could be scavenged by hydrodynamic capture. As noted previously, submicrometer-sized insoluble particles can be scavenged by Brownian or phoretic scavenging processes, or they can coagulate with hygroscopic particles and be removed by nucleation scavenging. Chylek *et al.* (1984) calculated the radiative properties of soot particles embedded in droplets. Assuming that

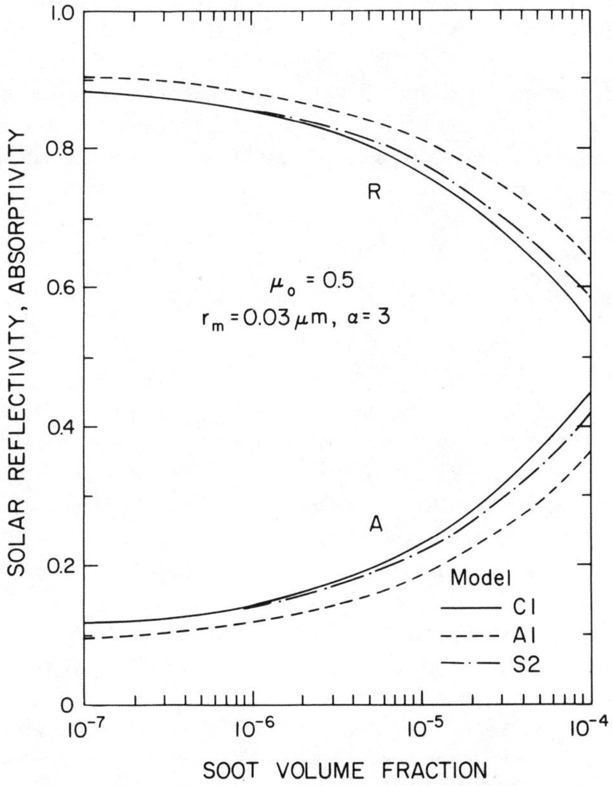

Fig. 5.9. Absorptance and reflectance for an optically thick cloud at $\lambda = 0.5$ μm as a function of soot volume fraction. The cosine of the solar zenith angle is taken to be 0.5. An amount of carbon between 5×10^{-6} and 1×10^{-5} by volume is required, depending on the cloud drop-size distribution, to obtain reflectance of 0.8 for thick clouds. [From Chylek *et al.* (1984).]

the soot particles were randomly distributed throughout the droplets, they showed that the absorption efficiency at short wavelengths of graphite carbon particles embedded in droplets is more than twice the efficiency of the same particles freely suspended in air, as long as the volume fraction of the particles in water is small. Figure 5.9 illustrates the calculated change in cloud reflectance and absorptance as a function of the volume fraction of graphite carbon for optically thick clouds having three different droplet size spectra. The cloud absorptance is substantially enhanced while the reflectance is reduced for soot volume fractions greater than 10^{-5}. Whether this amount of graphitic carbon occurs in cloud droplets in nature is unknown at this time. Certainly in the case of the hypothesized nuclear winter scenario (see, e.g., Pittock *et al.*, 1986) one would expect to find volume fractions of graphitic carbon in excess of those shown in Fig. 5.9.

What would be the effects of graphitic carbon particles attached to ice crystals? Again, little is known about the range of volume fractions of carbon that become attached to ice crystals. One would expect that graphitic carbon particles attached in sufficiently high numbers to ice crystals would decrease the reflectance of ice clouds substantially. There is certainly evidence that soot embedded in surface snow significantly affects the albedo of the snow surface (Warren, 1982).

5.8 Parameterization of Radiative Transfer in Clouds

5.8.1 Introduction

We have seen in previous sections that clouds have an important impact upon radiative transfer. Because divergence of radiative fluxes contributes to heating/cooling in a cloud system, radiative transfer processes can alter the thermodynamic stability of a cloud system and thereby contribute to the dynamics of the system. Because of the complexity of other cloud processes and their computational demands, cloud dynamicists and mesoscale modelers must seek simplifications in the formulation of radiative processes in models. Unfortunately, little work has been done with the specific aim of formulating radiation parameterization schemes that are suitable for cloud or mesoscale models in which specific data on cloud geometries, liquid-water contents, the partitioning of water among hydrometeor types (i.e. raindrops, ice crystals, and graupel), and hydrometeor-size spectra are explicitly modeled. Most of the research has been aimed at formulating radiative transfer parameterization schemes suitable for use in general circulation models. General circulation models typically do not have enough vertical resolution to make realistic estimates of the liquid-water path, let alone the other cloud properties such as hydrometeor type and spectra, and the finite dimensions of clouds.

5.8.2 Parameterization of Shortwave Radiation in Clouds

Any scheme designed to parameterize radiative processes in clouds must also take into account radiative transfer through the cloud-free atmosphere. For the sake of brevity, we will not review clear-air radiation parameterization schemes. The reader is referred to an excellent review on the topic by Stephens (1984b). We seek a radiation parameterization because the formal solution to the radiative transfer (5.6) is too complex and time-consuming for use in cloud and mesoscale models. One general approach to reducing the complexities associated with the solution to (5.6) is to introduce the so-called "two-stream approximation." In this approach the total radiation field is represented by two streams: one in the upward direction (↑) and one in the downward direction (↓). The radiation intensity $I(\tau, \mu)$ is then integrated over the upward and downward hemispheres to define the fluxes

$$F\uparrow\downarrow(\tau) = \int_0^1 \mu I(\tau, \pm\mu)\, d\mu, \tag{5.24}$$

which, as Eqs. (5.1) and (5.2) show, are the quantities needed to determine solar heating. Meador and Weaver (1980) showed that by assuming that I is dependent on μ, the hemispheric integral of Eq. (5.6) reduces to the standard ordinary differential equations of the form

$$dF\uparrow/d\tau = \gamma_1 F\uparrow - \gamma_2 F\downarrow + (F_0/4)\tilde{\omega}_0 \gamma_3\, e^{-\tau/\mu_0}, \tag{5.25}$$

$$dF\downarrow/d\tau = \gamma_2 F\uparrow - \gamma_1 F\downarrow + (F_0/4)\tilde{\omega}_0 \gamma_4\, e^{-\tau/\mu_0}. \tag{5.26}$$

The γ_i values are determined by the approximations used and are independent of τ. Solutions to Eqs. (5.25) and (5.26) can be obtained by standard techniques for specified boundary conditions. Meador and Weaver (1980) showed that several standard solution techniques, such as the Eddington approximation and the quadrature methods, could be transformed into the standard form of Eqs. (5.25) and (5.26) with the appropriate specification of the γ_i values. An illustration of a simple layered two-stream model is given in Fig. 5.10, where Re(n) and $A(n)$ represent the reflectances and absorptances of each layer, respectively.

As noted by Stephens (1984b), the basis of any parameterization based on the two-stream approximation rests with the determination of the cloud properties τ, $\tilde{\omega}$, and g in terms of modeled or specified microphysical and

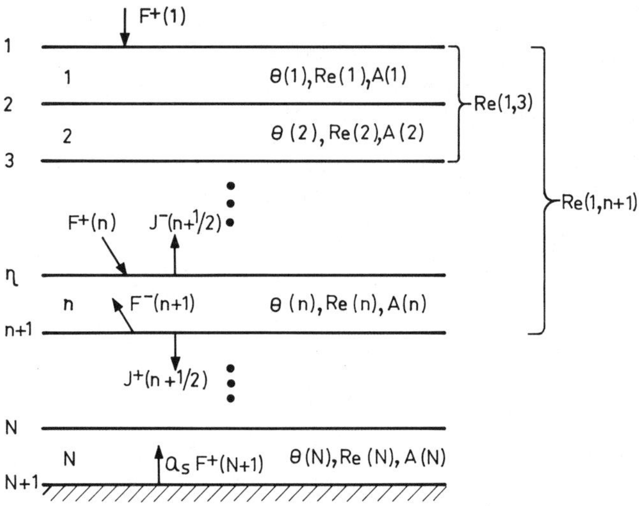

Fig. 5.10. The shortwave radiative transfer model; $F\downarrow$ and $F\uparrow$ denote the downward and upward flux. $\mathrm{Re}(n)$ and $A(n)$ represent, respectively, the reflectance and absorptance at the nth layer. $\mathrm{Re}(1, n+1)$ is the multiple reflectance from all layers above $(n+1)$th layer. [Adapted by Chen and Cotton (1983); from Stephens and Webster (1979).]

macrophysical variables. For spherical droplets, these properties can be obtained using Mie theory (see van de Hulst, 1957). If, however, the particle shape is complex, such as is the case for ice crystals or aerosols, solutions are not easily obtainable using Mie theory. In this case approximate solutions are necessary. For very large particles ($x = 2\pi r/\lambda \gg 1$), it has become customary to employ geometrical optics, in which the paths of individual rays traveling through a droplet or ice particle are traced. Rays passing through a particle and those not interacting with the particle are not allowed to interact or interfere with each other. Figure 5.11 illustrates an application of geometrical optics to the rainbow problem.

A somewhat less restrictive technique for obtaining the extinction and absorption efficiency factors is the so-called anomalous diffraction theory (van de Hulst, 1957; Ackerman and Stephens, 1987). Like geometrical optics, it is also valid for particles much larger than the wavelength of radiation ($x \gg 1$), and for which the index of refraction is $n \sim 1$, or what is often referred to as soft particles. The anomalous diffraction approximation is based on the premise that the extinction of light is primarily a result of the interference between the rays that pass through a particle and those rays that are not influenced by a particle (see Fig. 5.12).

5.8.2.a Cloud Optical Depth

Using Mie theory, or a suitable approximation to the interaction of radiation and droplets or ice particles, one can develop a parameterization of the cloud optical depth [Eq. (5.10)]. Stephens (1978b) used eight different cloud

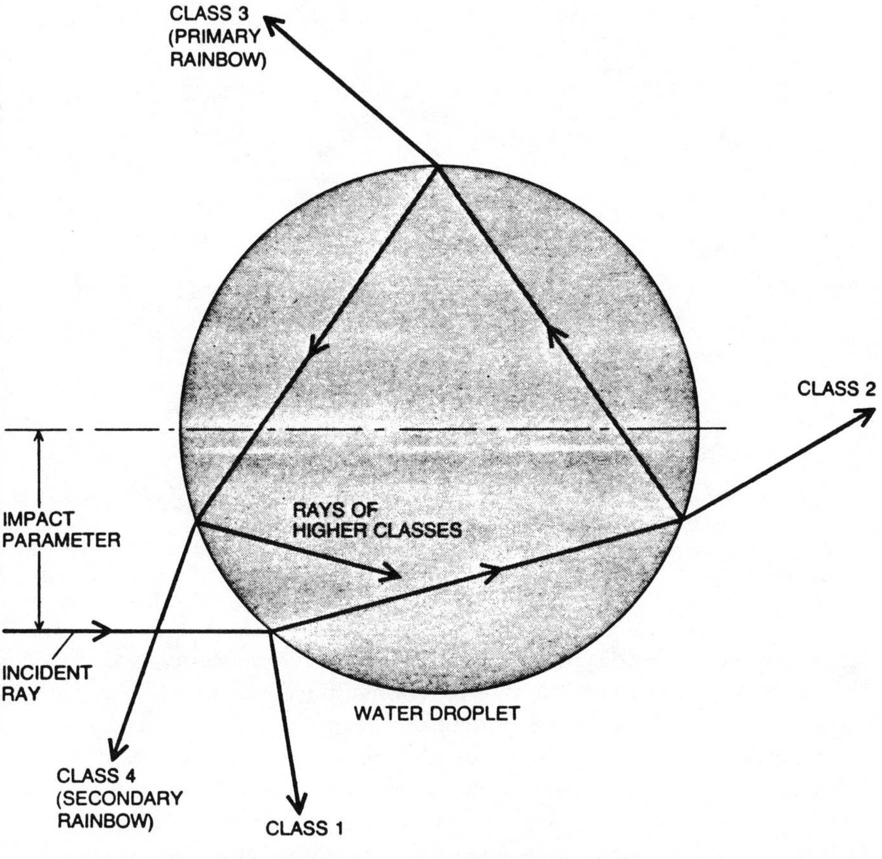

Fig. 5.11. Path of light through a droplet can be determined by applying the laws of geometrical optics. Each time the beam strikes the surface, part of the light is reflected and part is refracted. Rays reflected directly from the surface are labeled rays of Class 1; those transmitted directly through the droplet are designated Class 2. The Class 3 rays emerge after one internal reflection; it is these that give rise to the primary rainbow. The secondary bow is made up of Class 4 rays, which have undergone two internal reflections. For rays of each class, only one factor determines the value of the scattering angle. That factor is the impact parameter: the displacement of the incident ray from an axis that passes through the center of the droplet. [From Nussenzveig (1977). Copyright © 1977 by Scientific American, Inc. All rights reserved.]

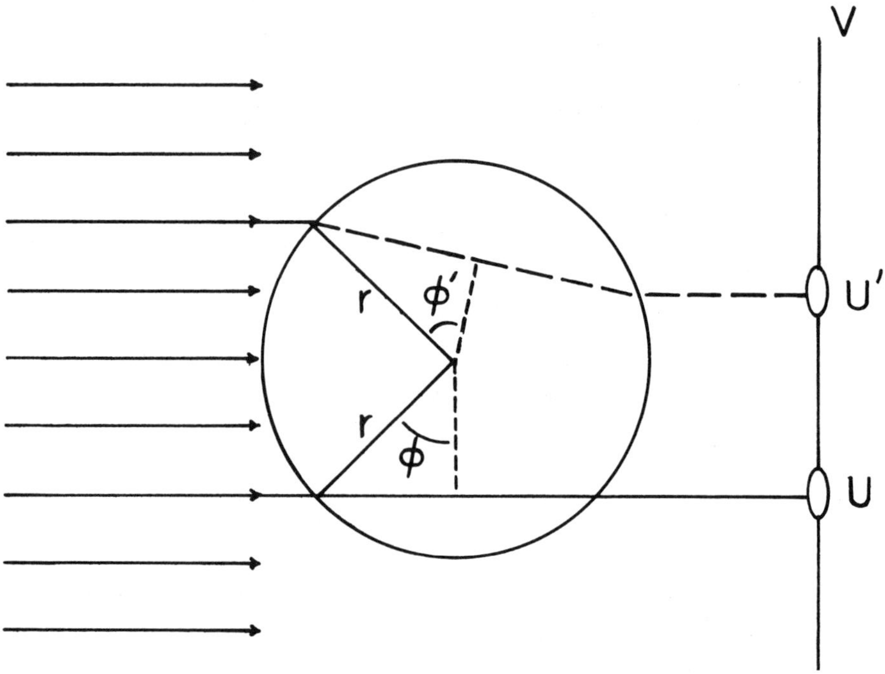

Fig. 5.12. Geometry of scattering by a large sphere with refractive index near 1. The solid ray passing through the sphere represents the anomalous diffraction theory (ADT), while the broken ray describes the ray path for the modified theory (MADT). [From Ackerman and Stephens (1987).]

droplet-size distributions to illustrate that the cloud optical depth as calculated with Mie theory is primarily a function of the cloud liquid-water path. He showed that τ could be approximated by Eq. (5.14). This is a useful approximation, because all one needs to do is calculate the liquid-water path, Eq. (5.16), and, given a climatologically derived effective radius, the optical thickness is easily determined. A cloud model with more explicit information on the cloud droplet-size distribution could also be used to calculate local values of r_e.

5.8.2.b Single-Scattering Albedo

The single-scattering albedo also has been parameterized in terms of the effective radius. Based on geometrical optics, Liou (1980) developed the following parameterization of single-scattering albedo

$$\tilde{\omega}_0 = 1 - 1.7k'r_e, \tag{5.27}$$

where k' is the complex part of the index of refraction ($n_\lambda = n_r - ik'$). Fouquart and Bonnel (1980) developed the expression

$$\tilde{\omega}_0 = 1 + \exp(-2k'r_e), \tag{5.28}$$

which is valid when liquid-water absorption is weak in the solar region. Using anomalous diffraction theory, van de Hulst (1957) derived a more general expression for $\tilde{\omega}_0$,

$$\tilde{\omega}_0 = 1 - \tfrac{1}{2}(\tfrac{4}{3}\rho \tan\Gamma - \rho^2 \tan^2\Gamma),$$

where $\rho = 2x(n_r - 1)$ and

$$\Gamma = \arc\tan(k'/n_r - 1). \tag{5.29}$$

Equation (5.28) is valid for small values of $4xk'$ and for $Q_e = 2$. Better estimates of $\tilde{\omega}_0$ can be obtained by using the tabulated values reported by Stephens *et al.* (1984). Stephens (1978b) showed that it is possible to remove the dependence of $\tilde{\omega}_0$ on r_e by empirically tuning $\tilde{\omega}_0$ as well as the integrated phase function parameter using accurate numerical solutions. Fouquart and Bonnel also used more accurate calculations to derive the spectrally averaged, single-scattering albedo

$$\tilde{\omega}_0 = 0.9989 - 0.0004\exp(-0.15\tau). \tag{5.30}$$

5.8.2.c Broadband Reflectance, Transmittance, and Absorptance

Stephens (1978b, 1984b) makes the case that the gross radiative properties of clouds can be modeled in terms of the liquid-water path and solar zenith angle. The effects of cloud microphysical variations, he claims, are likely to be smaller than variations in radiative properties due to fluctuations in liquid-water path. Stephens (1978b), therefore, divided the solar spectrum into two broadbands, one from 0.3 to 0.75 μm, where absorption by cloud droplets is small, and a second band from 0.75 to 4.0 μm, where absorption by cloud droplets is significant.

In the ultraviolet and visible region ($\lambda < 0.75$ μm), he parameterized the broadband reflectance, transmittance, and absorptance as follows:

$\lambda < 0.75\,\mu m$; nonabsorbing medium

$$\mathrm{Re}_1(\mu_0) = \frac{\beta_1(\mu_0)\tau/\mu_0}{1 + \beta_1(\mu_0)\tau/\mu_0}, \tag{5.31}$$

$$\mathrm{Tr}_1(\mu_0) = 1 - \mathrm{Re}_1(\mu_0), \tag{5.32}$$

$$A_1(\mu_0) = 0, \tag{5.33}$$

where τ is defined by Eq. (5.14).

In the near infrared region ($\lambda > 0.75\ \mu m$) they become as follows:

$\lambda > 0.75\ \mu m$; absorbing medium

$$\mathrm{Re}_2(\mu_0) = (u^2 - 1)[\exp(\tau_{\mathrm{eff}}) - \exp(-\tau_{\mathrm{eff}})]/R \tag{5.34}$$

$$\mathrm{Tr}_2(\mu_0) = 4u/R \tag{5.35}$$

$$A_2(\mu_0) = 1 - \mathrm{Re}_2(\mu_0) - \mathrm{Tr}_2(\mu_0) \tag{5.36}$$

where

$$u^2 = (1 - \omega_0 + 2\beta_2\omega_0)/(1 - \omega_0), \tag{5.37}$$

$$\tau_{\mathrm{eff}} = |(1 - \omega_0)(1 - \omega_0 + 2\beta_2\omega_0)|^{1/2}\tau/\mu_0, \tag{5.38}$$

$$R = (u + 1)^2 \exp(\tau_{\mathrm{eff}}) = (u - 1)^2 \exp(-\tau_{\mathrm{eff}}). \tag{5.39}$$

As noted previously, Stephens evaluated the single-scattering albedo ω_0 and the parameters β_1 and β_2, the back-scattered fraction of monodirectional incident radiation at the zenith angle μ_0, from detailed radiative spectral model calculations. Because the reflectance, absorptance, and transmittance are partitioned over two spectral regions, the totals of these are

$$\begin{aligned} \mathrm{Re} &= 0.517\mathrm{Re}_1 + 0.483\mathrm{Re}_2 \\ \mathrm{Tr} &= 0.517\mathrm{Tr}_1 + 0.483\mathrm{Tr}_2 \\ A &= 0.483A_2, \end{aligned} \tag{5.40}$$

where 51.7% of the reflectance, transmittance, and absorptance occurs for $\lambda < 0.75\ \mu m$ and 48.3% occurs for $\lambda > 0.75\ \mu m$. A disadvantage of this approach, however, is that we lose the ability to distinguish between the radiative properties of maritime versus continental clouds or clouds in polluted atmospheres. We also cannot consider the effects of raindrops on the enhancement of absorption, as noted previously.

5.8.2.d Radiative Transfer through a Multilayered Atmosphere

Our discussion thus far has only dealt with a single homogeneous cloud layer with fixed values of τ, $\tilde{\omega}_0$, and g. We now consider an atmosphere composed of a number of layers, each having different optical properties. One approach is to define an appropriate transmittance function in order to calculate the depletion of the directly transmitted solar beam as it travels through a multilayered atmosphere. Oliver *et al.* (1978) applied Beer's law to a broad radiation band by evaluating a transmission function $T_{\mathrm{r}}^2(z_1, z_2)$ through a layer of thickness $(z_2 - z_1)$ in a marine stratocumulus model. The transmission function was formulated as

$$T_{\mathrm{r}}^2(z_1, z_2) = \exp(-\alpha^{\mathrm{s}} W_{12}), \tag{5.41}$$

where α^s is a specific absorption coefficient for shortwave absorption by liquid drops and W_{12} represents the liquid-water path through the layer of thickness $z_2 - z_1$. They estimated α^s by integrating the spectral liquid-water absorption coefficient tabulated by Feigelson (1964) over the solar spectrum. They obtained a value of $\alpha^s = 0.0016\ \mathrm{m^2\ g^{-1}}$, which was independent of the droplet-size distribution. A disadvantage of this approach is that the effects of multiple scattering between layers cannot be simulated, and it is difficult to define a transmission function that correctly simulates the combined effects of gas absorption, droplet absorption, and droplet scattering in single and multiple layers.

A more precise and consistent approach is the so-called "adding" method. Grant and Hunt (1969a, b) considered an atmosphere composed of n homogeneous layers, each with their respective reflective (Re), transmissive (Tr), and absorptive (A) properties. For such an atmosphere, the reflectance $\mathrm{Re}(1, n+1)$ represents the combined multiple reflectance contributed by all layers above the $(n+1)$th layer. It may be defined as

$$\mathrm{Re}(1, n+1) = \mathrm{Re}(n) + \frac{\mathrm{Tr}{\downarrow}(n)\ \mathrm{Tr}{\uparrow}(n)\ \mathrm{Re}(1, n)}{1 - \mathrm{Re}(1, n)\ \mathrm{Re}(n)}, \tag{5.42}$$

where the reflection from a composite of all layers above the $(n+1)$th layer is obtained by adding the reflectance from two layers whose reflectance is $\mathrm{Re}(n)$ and $\mathrm{Re}(1,n)$.

The flux transmitted through the upper layer is represented by $V{\downarrow}(n+1/2)$ which may be computed as

$$V{\downarrow}(n+1/2) = \frac{\mathrm{Tr}{\downarrow}(n)\ V{\downarrow}(n-1/2)}{1 - \mathrm{Re}(1, n)\ \mathrm{Re}(n)}. \tag{5.43}$$

Similarly, the flux transmitted from the lower layer $V{\uparrow}(n+1/2)$ is calculated as

$$V{\uparrow}(n+1/2) = \frac{\mathrm{Re}(n)\ V{\downarrow}(n-1/2)}{1 - \mathrm{Re}(1, n)\ \mathrm{Re}(n)}. \tag{5.44}$$

The denominators in Eqs. (5.42), (5.43), and (5.44) account for multiple reflections between layers. As noted by Stephens (1984a), this factor is especially large when the reflection between layers is large. This occurs when a dense cloud overlaps a bright surface such as another cloud or a snow-covered surface.

To close such a layered model, boundary conditions must be supplied. At the upper boundary, one can assume that the reflectance is zero and the downward-transmitted radiation corresponds to the flux coming into the atmosphere, or

$$\mathrm{Re}(1, 1) = 0,$$

$$V\downarrow(1/2) = F\downarrow(1).$$

If the model top were placed at the tropopause, then $F\downarrow(1)$ would correspond to the downward flux from stratospheric levels and above.

At the lower boundary the upward flux is given by

$$F\uparrow(n+1) = a_s F\downarrow(n+1),$$

where a_s is the albedo of the earth's surface. The value of a_s varies depending on whether the surface is dry land, vegetated, snow covered, or a sea surface. It also varies depending on solar elevation. Typical values of a_s for solar elevations of 45° are 7% for a water surface, 20–35% for dry grass lands, 30–40% for sand, and 80–85% for fresh snow.

This layered model illustrates that the net radiative flux divergence is influenced by radiative transfer through all layers of the atmosphere above and below the level under consideration. To compute the transmittance and reflectance we note that Eq. (5.3) can be integrated over all wavelengths shorter than 4 μm to give

$$\mathrm{Re} + \mathrm{Tr} = 1 - A. \tag{5.45}$$

Thus, if a parameterization of absorptance and reflectance is available, one can compute the transmittance by using Eq. (5.45). To illustrate the effects of different parameterizations on radiative transfer calculations, let us estimate the absorption length L_s, which is a rough estimate of the distance solar radiation can penetrate into a cloud before it is totally absorbed, for two different parameterizations.

Oliver *et al.* (1978) defined absorption lengths for shortwave radiative transfer through clouds as

$$L_s = 1/(\alpha^s \rho_0 r_c), \tag{5.46}$$

where α^s is the absorption coefficient for shortwave radiation. Chen and Cotton (1983a, b) also computed equivalent absorption lengths for solar absorption using Stephens' (1978b) parameterization of solar absorption. The equivalent absorption coefficient derived from Stephens' parameterization is $\alpha^s = 0.0072\ \mathrm{m}^2\ \mathrm{g}^{-1}$, which may be compared with the value $\alpha^s = 0.0016\ \mathrm{m}^2\ \mathrm{g}^{-1}$ used by Oliver *et al.* (1978). Table 5.1 illustrates the variation of L_s as a function of liquid-water mixing ratios for the two parameterizations.

As can be seen from Table 5.1, depending on the particular parameterization of shortwave radiative transfer through clouds, the calculated penetrations of shortwave radiation through the cloud layer can differ markedly.

Table 5.1

Variation of L_s for Two Parameterizations

Liquid-water content r_c ($g\ kg^{-1}$)	Absorption depth (m) [Oliver *et al.* (1978)]	Absorption depth (m) [Stephens (1978b)]
0.1	5500	1222
0.2	2700	600
0.4	1350	300
0.8	700	155
1.0	500	111

This can dramatically change the predicted dynamic/thermodynamic response of the cloud to solar heating. In optically thin clouds, a parameterization having a long absorption length may give effectively no absorption of radiation, whereas one with a short L_s may estimate absorption of enough of the sun's rays to cause the breakup of the cloud. In optically thicker clouds, a large L_s will result in solar heating well within the interior of the cloud, while a short L_s will result in solar heating being localized near the cloud top. Solar heating near the cloud top may only slightly reduce the rate of cooling due to longwave radiation divergence, while solar heating in the cloud interior can enhance cloud buoyancy, thus having a greater impact upon the overall dynamics of the cloud. We believe that Stephens' parameterization is the more accurate of the two schemes.

5.8.2.e Partial Cloudiness

We have noted that the radiative properties of a population of finite clouds can differ substantially from that of a horizontally homogeneous cloud system. In the case of a region covered by partial cloudiness, the normal procedure is to weight the reflectance, transmittance, and absorptance calculated for a cloudy atmosphere by the cloud fractional coverage and the corresponding clear-air properties by the clear-air coverage. This ignores contributions from radiation emitted from the sides of clouds and the interaction of radiation among neighboring clouds. Welch and Zdunkowski (1981a) used second-order polynomials to parameterize cloud reflectance, transmittance, and absorptance which accounts for emission of radiation from the sides of finite clouds, but the radiative interaction among clouds has not yet been parameterized.

5.8.3 Parameterization of Longwave Radiative Transfer in Clouds

We have seen in Section 5.4 that clouds are effective absorbers of longwave radiation. A cloud of only modest liquid-water content of $0.2\ g\ kg^{-1}$ may absorb up to 90% of the upwelling infrared radiant energy within a depth of only 50 m. In contrast, an equivalent penetration distance for shortwave radiation is at least 600 m for a cloud having the same liquid-water content (Table 5.1). As a consequence, it is often assumed that optically thick clouds behave as blackbodies. Thus, all upwelling radiation is absorbed at the base of the cloud and is reemitted in the upward and downward directions with a flux equal to σT_b^4, where T_b is the cloud-base temperature. At the cloud top, the upward flux emitted by an assumed blackbody cloud will simply be σT_T^4, where T_T is the temperature at cloud top.

Many clouds, such as cirrus, stratus, and stratocumulus, as well as fogs, are not optically thick and, therefore, the blackbody approximation is a poor one. One must then seek alternate parameterizations of the longwave properties of such clouds.

Oliver *et al.* (1978) in an approach similar to that used for shortwave radiation parameterization, defined a transmission function T of the form

$$T_L(z_1, z_2) = \exp(-\alpha_T W_{12}), \tag{5.47}$$

where α_T represents an averaged liquid absorption coefficient for terrestrial radiation. Based on the work of Feigelson (1970), they indicated that α_T varies strongly with the drop-size distribution, ranging from 0.002 to 0.17 $m^2\,g^{-1}$ for drop spectra having an average drop radius ranging from 4.5 to 7 μm. The sensitivity of the transmission function to the droplet-size distribution is a disadvantage of this approach, because as we have seen above, the optical thickness in the infrared range is mainly dependent upon the liquid-water path and not on the details of the droplet spectrum.

By far the most popular approach to parameterizing longwave radiative transfer in optically thin or "gray" clouds is to use the effective emittance concept defined in Eqs. (5.20) and (5.21). If properly evaluated, the effective emittance combines the effects of reflection, transmission, and absorption by cloud droplets and gases. Stephens (1978b) parameterized the effective emittance as

$$\varepsilon\uparrow\downarrow = 1 - \exp(-a_0\uparrow\downarrow W), \tag{5.48}$$

where W is again the liquid-water path. He evaluated the coefficients for upward and downward emittance as

$$a_0\uparrow = 0.13 \quad m^2\,g^{-1} \qquad \text{and} \qquad a_0\downarrow = 0.158 \quad m^2\,g^{-1}.$$

The coefficients $a_0\uparrow$ and $a_0\downarrow$ correspond to a mass absorption coefficient, which may be compared to the values of α_T used by Oliver *et al.* (1978). Stephens, however, found little variation in $a_0\uparrow$ and $a_0\downarrow$ with changes in drop-size distributions. Assuming that absorption is the dominant longwave radiative process in clouds, the emittance form of the flux equations can then be used. For example, the fluxes through some level z in an isothermal cloud are

$$F\uparrow(z) = Z\uparrow(z_b)[1-\varepsilon(z_b, z)] + \varepsilon(z_b, z)\sigma T_c^4,$$
$$F\downarrow(z) = F\downarrow(z_t)[1-\varepsilon(z_t, z)] + \varepsilon(z_t, z)\sigma T_c^4, \qquad (5.49)$$

where z_b and z_t are the heights of cloud base and cloud top, respectively and T_c is the cloud temperature. The simplicity of this approach is that vertical profiles of longwave radiative fluxes can be obtained readily with only a knowledge of the variation of liquid-water path with height. Other recent examples of the use of the emittance approach can be found in Liou and Ou (1981) and Chylek and Ramaswamy (1982).

Thus far we have only considered a radiative path to consist of either cloud or clear air. Often one has to deal with a clear layer overlain by a cloudy layer, which in turn may have several clear and cloudy layers above it. A common approach in dealing with this problem is to use the mixed emittance concept introduced by Herman and Goody (1976). Thus the emittance (ε_m) along a radiative path composed of clear and cloudy regions may be formulated as

$$\varepsilon_m = 1-(1-\varepsilon_{\text{cloudy}})(1-\varepsilon_{\text{clear}}). \qquad (5.50)$$

Equation (5.50) is equivalent to stating that the transmittance through a mixed atmosphere is the product of the transmittance in cloud air and clean air. This should not be confused with the concept of partial cloudiness, in which only a portion of a grid volume or region of interest is covered by cloud. In the case of partial cloudiness, it is standard practice to calculate the fluxes by linearly weighting the fluxes by the corresponding clear and cloudy sky fractional coverages

$$F\uparrow(z) = (1-H)[F\uparrow(z)]_{\text{clear}} + H(F\uparrow(z)]_{\text{cloudy}}, \qquad (5.51)$$

where H is the cloud fractional coverage. As noted previously this approach does not consider the effects of the finite dimensions of clouds or the interaction of radiation among neighboring clouds.

5.9 Summary

In this chapter we have reviewed the interaction between the cloud microphysical and macrophysical structure and radiative transfer processes. We have also presented some of the concepts and approaches to parameterizing radiative transfer through clouds. In Part II of this book we will examine the effects of radiative processes on the dynamics and precipitation processes in several different cloud systems.

References

Ackerman, S. A., and S. K. Cox (1982). The Saudi Arabian heat low: Aerosol distributions and thermodynamic structure. *J. Geophys. Res.* **87**, 8991-9002.

Ackerman, S. A., and G. L. Stephens (1987). The absorption of solar radiation by cloud droplets: An application of anomalous diffraction theory. *J. Atmos. Sci.* **44**, 1574-1588.

Ackerman, T. P., K.-N. Liou, and C. B. Leovy (1976). Infrared radiative transfer in polluted atmospheres. *J. Appl. Meteorol.* **15**, 28-35.

Aida, M. (1977). Reflection of solar radiation from an array of cumuli. *J. Meteorol. Soc. Jpn.* **55**, 174-181.

Andreyev, S. D., and L. S. Ivlev (1980). Infrared radiation absorption by various atmospheric aerosol fractions. *Izv. Acad. Sci. USSR Atmos. Oceanic Phys.* (*Engl. Transl.*) **16**, 663-669.

Barkstrom, B. R. (1978). Some effects of 8-12 μm radiant energy transfer on the mass and heat budgets of cloud droplets. *J. Atmos. Sci.* **35**, 665-673.

Blevin, W. R., and W. J. Brown (1971). A precise measurement of the Stefan-Boltzmann constant. *Metrologia* **7**, 15-29.

Braslau, N., and J. V. Dave (1975). Atmospheric heating rates due to solar radiation for several aerosol-laden cloudy and cloud-free models. *J. Appl. Meteorol.* **14**, 396-399.

Byers, H. R. (1965). "Elements of Cloud Physics." Univ. of Chicago Press, Chicago, Illinois.

Carlson, T. N., and S. G. Benjamin (1980). Radiative heating rates for Saharan dust. *J. Atmos. Sci.* **37**, 193-213.

Chen, C., and W. R. Cotton (1983a). A one-dimensional simulation of the stratocumulus-capped mixed layer. *Boundary-Layer Meteorol.* **25**, 289-321.

Chan, C., and W. R. Cotton (1983b). Numerical experiments with a one-dimensional higher order turbulence model: Simulation of the Wangara Day 33 Case. *Boundary-Layer Meteorol.* **25**, 375-404.

Chylek, P., and V. Ramaswamy (1982). Simple approximation for infrared emissivity of water clouds. *J. Atmos. Sci.* **39**, 171-177.

Chylek, P., B. R. D. Gupta, N. C. Knight, and C. A. Knight (1984). Distribution of water in hailstones. *J. Climate Appl. Meteorol.* **23**, 1469-1472.

Cox, S. K. (1976). Observations of cloud infrared effective emissivity. *J. Atmos. Sci.* **33**, 287-289.

Davis, J. M., S. K. Cox, and T. B. McKee (1979a). Total shortwave radiative characteristics of absorbing finite clouds. *J. Atmos. Sci.* **36**, 508-518.

Davis, J. M., S. K. Cox, and T. B. McKee (1979b). Vertical and horizontal distribution of solar absorption in finite clouds. *J. Atmos. Sci.* **36**, 1976-1984.

Drummond, A. J., and J. R. Hickey (1971). Large-scale reflection and absorption of solar radiation by clouds as influencing earth radiation budgets: New aircraft measurements. *Proc. Int. Conf. Weather Modif., Canberra, Aust.* pp. 267-276. Am. Meteorol. Soc., Boston, Massachusetts.

Ellingson, R. G. (1982). On the effects of cumulus dimensions on longwave irradiance and heating rate calculations. *J. Atmos. Sci.* **39**, 886-896.

Feigelson, E. M. (1964). "Light and Heat Radiation in Stratus Clouds." Nauka, Moscow. (Engl. transl., Isr. Program Sci. Transl., 1966.)

Feigelson, E. M. (1970). "Radiant Heat Transfer in a Cloudy Atmosphere." Gidrometeorol., Leningrad. (Engl. transl., Isr. Program Sci. Transl., 1973.)

Fouquart, Y., and B. Bonnel (1980). Computations of solar heating of the earth's atmosphere: A new parameterization. *Atmos. Phys.* **53**, 35-62.

Fritz, S. (1954). Scattering of solar energy by clouds of "large drops." *J. Meteorol.* **11**, 291-300.

Fritz, S. (1958). Absorption and scattering of solar energy in clouds of "large water drops"—II. *J. Meteorol.* **15**, 51-58.

Gast, P. R., A. S. Jursa, J. Castelli, S. Basu, and J. Aarons (1965). Solar electromagnetic radiation. *In* "Handbook of Geophysics and Space Environments" (S. L. Valley, ed.), pp. 16-1-16-38. McGraw-Hill, New York.

Grant, I. P., and G. E. Hunt (1969a). Discrete space theory of radiative transfer. I. Fundamentals. *Proc. R. Soc. London, Ser. A* **313**, 183–197.

Grant, I. P., and G. E. Hunt (1969b). Discrete space theory of radiative transfer. II. Stability and non-negativity. *Proc. R. Soc. London, Ser. A* **313**, 199–216.

Greenler, R. G. (1980). "Rainbows, Halos and Glories." Cambridge Univ. Press, London.

Hallett, J. (1987). Faceted snow crystals. *J. Opt. Soc. Am., A* **4**, 581–588.

Hansen, J. E., and L. D. Travis (1974). Light scattering in planetary atmospheres. *Space Sci. Rev.* **16**, 527–610.

Harshvardhan, and J. A. Weinman (1982). Infrared radiative transfer through a regular array of cuboidal clouds. *J. Atmos. Sci.* **39**, 431–439.

Harshvardhan, J. A. Weinman, and R. Davies (1981). Transport of infrared radiation in cuboidal clouds. *J. Atmos. Sci.*, **38**, 2500–2513.

Henyey, L. G., and J. L. Greenstein (1941). Diffuse radiation in the galaxy. *Astrophys. J.* **112**, 445–463.

Herman, G. F. (1977). Solar radiation in summertime arctic stratus clouds. *J. Atmos. Sci.* **34**, 1423–1431.

Herman, G. F., and R. Goody (1976). Formation and persistence of summertime arctic stratus clouds. *J. Atmos. Sci.* **33**, 1537–1553.

Jayaweera, D. O., and B. J. Mason (1965). The behaviour of freely falling cylinders and cones in a viscous fluid. *J. Fluid Mech.* **22**, 709–720.

Joseph, J. H., W. J. Wiscombe, and J. A. Weinman (1976). The delta-Eddington approximation for radiative transfer. *J. Atmos. Sci.* **33**, 2452–2459.

Kasten, F. (1969). Visibility forecast in the phase of pre-condensation. *Tellus* **5**, 631–635.

Liou, K.-N. (1972a). Light scattering by ice clouds in the visible and infrared: A theoretical study. *J. Atmos. Sci.* **29**, 524–536.

Liou, K.-N. (1972b). Electromagnetic scattering by arbitrarily oriented ice cylinders. *Appl. Opt.* **2**, 667–674.

Liou, K.-N. (1980). "An Introduction to Atmospheric Radiation." Academic Press, New York.

Liou, K.-N., and S.-C. Ou (1979). Infrared radiative transfer in finite cloud layers. *Bull. Am. Meteorol. Soc.* **36**, 1985–1996.

Liou, K.-N., and S.-C. Ou (1981). Parameterization of infrared radiative transfer in cloudy atmospheres. *J. Atmos. Sci.* **38**, 2707–2716.

Manton, M. J. (1980). Computations of the effect of cloud properties on solar radiation. *J. Rech. Atmos.* **14**, 1–16.

Mason, B. J. (1971). "The Physics of Clouds," 2nd Ed. Oxford Univ. Press (Clarendon), Oxford.

McKee, T. B., and S. K. Cox (1974). Scattering of visible radiation by finite clouds. *J. Atmos. Sci.* **31**, 1885–1892.

McKee, T. B., and S. K. Cox (1976). Simulated radiance patterns for finite cubic clouds. *J. Atmos. Sci.* **33**, 2014–2020.

McKee, T. B., M. DeMaria, J. A. Kuenning, and S. K. Cox (1983). Comparison of Monte Carlo calculations with observations of light scattering in finite clouds. *J. Atmos. Sci.* **40**, 1016–1023.

Meador, W. E., and W. R. Weaver (1980). Two-stream approximations to radiative transfer in planetary atmospheres: A unified description of existing methods and a new improvement. *J. Atmos. Sci.* **37**, 630–643.

Newiger, M., and K. Bahnke (1981). Influence of cloud composition and cloud geometry on the absorption of solar radiation. *Contrib. Atmos. Phys.* **54**, 370–382.

Nussenzveig, H. M. (1977). The theory of the rainbow. *Sci. Am.* **236**, 116–127.

Oliver, D. A., W. S. Lewellen, and G. G. Williamson (1978). The interaction between turbulent and radiative transport in the development of fog and low-level stratus. *J. Atmos. Sci.* **35**, 301–316.

Ono, A. (1969). The shape and riming properties of ice crystals in natural clouds. *J. Atmos. Sci.* **26**, 138–147.

Paltridge, G. W., and Platt, C. M. R. (1976). "Radiative Processes in Meteorology and Climatology." Developments in Atmospheric Science, 5. Elsevier, New York.

Pittock, A. B., T. P. Ackerman, P. J. Crutzen, M. C. MacCracken, C. S. Shapiro, and R. P. Turco (1986). "Environmental Consequences of Nuclear War," Vol. 1.

Platt, C. M. R. (1978). Lidar backscatter from horizontal ice crystal plates. *J. Appl. Meteorol.* **17**, 482–488.

Poellot, M. R., and S. K. Cox (1977). Computer simulation of irradiance measurements from aircraft. *J. Appl. Meteorol.* **16**, 167–171.

Reynolds, D. W., T. H. Vonder Haar, and S. K. Cox (1975). The effect of solar radiation and absorption in the tropical atmosphere. *J. Appl. Meteorol.* **14**, 433–444.

Roach, W. T. R. (1976). On the effect of radiative exchange on the growth by condensation of a cloud or fog droplet. *Q. J. R. Meteorol. Soc.* **102**, 361.

Robinson, G. D. (1958). Some observations from aircraft of surface albedo and the albedo and absorption of cloud. *Arch. Meteorol. Geophys. Bioklimatol., Ser. B* **9**, 28–41.

Rozenberg, G. V., M. S. Malkevich, M. S. Malkova, and W. I. Syachov (1974). Determination of optical characteristics of clouds from measurement of reflected solar radiation by the cosmos 320 satellite. *Izv. Acad. Sci. USSR, Atmos. Oceanic Phys.* (*Engl. Transl.*) **10**, 14–24.

Saito, T. (1981). The relationship between the increase rate of downward longwave radiation by atmospheric pollution and the visibility. *J. Meteorol. Soc.* **59**, 254–261.

Stephens, G. L. (1978a). Radiation profiles in extended water clouds. I: Theory. *J. Atmos. Sci.* **35**, 2111–2122.

Stephens, G. L. (1978b). Radiation profiles in extended water clouds. II: Parameterization schemes. *J. Atmos. Sci.* **35**, 2123–2132.

Stephens, G. L. (1980a). Radiative properties of cirrus clouds in the infrared region. *J. Atmos. Sci.* **37**, 435–446.

Stephens, G. L. (1980b). Radiative transfer on a linear lattice: Application to anisotropic ice crystals. *J. Atmos. Sci.* **37**, 2095–2104.

Stephens, G. L. (1983). The influence of radiative transfer on the mass and heat budgets of ice crystals falling in the atmosphere. *Meteorol. Monogr.* **40**, 1729–1739.

Stephens, G. L. (1984a). Scattering of plane waves by soft obstacles: Anomalous diffraction theory for circular cylinders. *Appl. Opt.* **23**, 954–959.

Stephens, G. L. (1984b). The parameterization of radiation for numerical weather prediction and climate models. *Mon. Weather Rev.* **112**, 826–862.

Stephens, G. L., and P. J. Webster (1979). Sensitivity of radiative forcing to variable cloud and moisture. *J. Atmos. Sci.* **36**, 1542–1556.

Stephens, G. L., G. W. Paltridge, and C. M. R. Platt (1978). Radiation profiles in extended water clouds. III. Observations. *J. Atmos. Sci.* **35**, 2133–2141.

Stephens, G. L., S. Ackerman, and E. A. Smith (1984). A shortwave parameterization revised to improve cloud absorption. *Meteorol. Monogr.* **41**, 687–690.

Takeda, T., W. Pei-ming, and K. Okada (1986). Dependence of light scattering coefficient of aerosols on relative humidity in the atmosphere of Nagoya. *J. Meteorol. Soc. Jpn.* **64**, 957–966.

Takeuchi, Y. (1986). Effects of cloud shape on the light scattering. *J. Meteorol. Soc. Jpn.* **64**, 95–107.

Tricker, R. A. R. (1970). A note on the Lowitz and associated arcs. *Weather* **25**, 503.

Twomey, S. (1976). Computations of the absorption of solar radiation by clouds. *J. Atmos. Sci.* **33**, 1087–1091.

Twomey, S., and T. Cocks (1982). Spectral reflectance of clouds in the near-infrared: Comparisons of measurements and calculations. *J. Meteorol. Soc. Jpn.* **60**, 583–592.

van de Hulst, H. C. (1957). "Light Scattering by Small Particles." Wiley, New York.

Warren, S. G. (1982). Optical properties of snow. *Rev. Geophys. Space Phys.* **20**, 67–89.

Welch, R., and W. Zdunkowski (1976). A radiation model of the polluted atmospheric boundary layer. *J. Atmos. Sci.* **33**, 2170–2184.

Welch, R. M., and W. G. Zdunkowski (1981a). The radiative characteristics of noninteracting cumulus cloud fields. Part I: Parameterization for finite clouds. *Atmos. Phys.* **54**, 258–272.

Welch, R. M., S. K. Cox, and J. M. Davis (1980). "Solar Radiation and Clouds," Vol. 17, 96 pp. Meteorol. Monograph, Boston, Massachusetts.

Wiscombe, W. J., and R. Welch (1986). Reply. *J. Atmos. Sci.* **43**, 401–407.

Wiscombe, W. J., R. M. Welch, and W. D. Hall (1984). The effects of very large drops on cloud absorption. Part I: Parcel models. *J. Atmos. Sci.* **41**, 1336–1355.

Yamamoto, G., M. Tanaka, and S. Asano (1970). Radiative transfer in water clouds in the infrared region. *J. Atmos. Sci.* **27**, 282–292.

Zdunkowski, W. G., and K.-N. Liou (1976). Humidity effects on the radiative properties of a hazy atmosphere in the visible spectrum. *Tellus* **28**, 31–36.

Chapter 6 | Cumulus Parameterization and Diagnostic Studies of Convective Systems

6.1 Introduction

Because of the large magnitude of the energy transformations associated with the changes of phase of water in precipitating cumulus clouds, as well as the strong updrafts and downdrafts which often extend throughout much of the troposphere, it is not surprising that cumulus convection, especially deep, intense convection, can have an important effect on the dynamics and energetics of larger scale atmospheric systems. Riehl and Malkus (1958) showed the importance of cumulus convection in the heat balance of the tropical atmosphere. The tropical cyclone is perhaps the best example of a mesoscale atmospheric system that owes its existence to the release of latent heat in cumulus convection (Riehl and Malkus, 1961; Ooyama, 1964; Charney and Eliassen, 1964). However, many other atmospheric phenomena are affected strongly by the energy released in cumulus clouds and in the vertical convective transports of heat, moisture, and momentum. These include tropical squall lines (Zipser, 1977), tropical cloud clusters (Houze and Betts, 1981), mesoscale convective complexes (Fritsch and Maddox, 1981), extratropical squall lines (Ogura and Chen, 1977), cold fronts (Ogura and Portis, 1982), and extratropical cyclones (Tracton, 1973; Gyakum, 1983; Anthes *et al.*, 1983). In addition, the transport of momentum by cumulus clouds affects the general circulation through its contribution to the global angular momentum budget (Houze, 1973).

While the above examples demonstrate the importance of deep, intense cumulus convection on larger scale systems, on time scales of an hour or so up to several days, shallow, nonprecipitating cumulus clouds also affect much larger scales of motion, but usually over longer time scales. Stratocumulus decks, which cover vast areas of the eastern Pacific, for example, play a major role in the heat budget of the region. The cumulative effect of small cumulus clouds is important in maintaining the observed thermodynamic and moisture structure of the trade-wind regions (Betts, 1975). On a global scale, nonprecipitating clouds play a major role in determining the planet's radiation budget and hence its climate. Chapter 8 discusses the properties of nonprecipitating cumulus clouds and presents several models of these clouds. In this chapter the effects of deep, precipitating cumulus convection on their environment and the parameterization of those effects are discussed. The reader is referred to Chapters 9 and 10 for a more complete discussion of the physics and dynamics of precipitating, deep cumulus convection.

Because cumulus convection produces significant effects on larger scales of motion, it is necessary to account for these effects in a quantitative way in models of larger scale phenomena. Ideally, the horizontal resolution of the models would be sufficiently fine so that the individual clouds and their effects on the environment could be calculated directly. To model the clouds explicitly, however, would require horizontal and vertical grid sizes of between 100 and 1000 m, and such a high resolution covering the size of domain necessary for larger scale phenomena (such as extratropical cyclones) is far beyond present and foreseeable computational capability. Thus, an important problem in global numerical weather prediction, and in numerical simulation of phenomena such as fronts and tropical cyclones, is to estimate the physical effects of cumulus convection on the resolvable scales of motion by developing quantitative relationships between the cloud effects and the known parameters of the larger scale model. Relating the effects of the "subgrid-scale" cumulus clouds to the scales of motion resolved by the model is known as cumulus parameterization. As discussed by Ooyama (1971, 1982), the concept of cumulus parameterization in numerical models originated with studies of the tropical cyclone, perhaps because the importance of cumulus convection on the tropical cyclone was recognized early (Malkus and Riehl, 1960).

The concept of the parameterization of subgrid-scale effects in numerical models is not limited to cumulus convection, but includes other important physical processes such as the vertical transport of heat, moisture, and momentum by turbulent eddies in the planetary boundary layer, transfer of energy and momentum between the earth's surface and the atmosphere, and radiative heating and cooling by longwave and shortwave radiation and its interaction with clouds. The concept of parameterization (and subgrid-scale effects) for global models is discussed in the Global

Atmosphere Research Program (GARP) Publication Series No. 8 (World Meteorological Organization, 1972). According to this report, successful parameterization requires a number of steps:

(i) Identification of the process.
(ii) Determination of the importance of the process for the resolvable scales of motion.
(iii) Intensive studies of individual cases in order to establish the fact that the relevant physics and dynamics are adequately understood.
(iv) Formulation of quantitative rules for expressing the location, frequency of occurrence, and intensity of the subgrid-scale processes in terms of the resolvable scale.
(v) Formulation of quantitative rules for determining the grid-scale averages of the transports of mass, momentum, heat, and moisture and verification of these rules by direct observations.

All of the above requirements have been met to some degree by various cumulus parameterization schemes, although the understanding of the complicated physical interactions between the clouds and their environment and the verification of the different parameterizations are far from complete.

Before discussing specific cumulus parameterization schemes, it is necessary to show that the concept of cumulus parameterization is valid, i.e., that there is a physical basis, supported by observations, for developing such schemes. In order that successful parameterization of any small-scale process in numerical models be possible, it is necessary that there be a physical relationship between the resolvable and subgrid scales of motion. In particular, there must be large-scale parameters that control, or at least modulate, the behavior of the small-scale phenomena. For cumulus parameterization to be possible, it is necessary that cumulus convection be influenced by the large-scale dynamics and thermodynamics. The following section provides observational evidence that this necessary condition is met, i.e., cumulus clouds are strongly affected by the thermodynamic and moisture structures of their environment, as well as the horizontal and vertical velocity fields on much larger scales of motion.

6.2 Relationship between Cumulus Convection and Larger Scale Atmospheric Variables

6.2.1 Tropical Systems

One of the first observational studies that showed a strong dependence of deep cumulus convection on larger scale variables was by J. Malkus, H. Riehl, W. M. Gray, and C. Ronne in their photographic mapping of cloud conditions over the tropical Pacific. In a summary of this work, Malkus and Williams (1963) state "we found that penetrative cumulonimbus (tops > 35,000 ft) occurred only where low-level synoptic scale convergence prevailed and were snuffed out instantly when divergence took over." Malkus and Williams (1963) also noted that dynamic, rather than thermodynamic, factors were more crucial for cloud growth in the tropics, and that cumulonimbus clouds were characterized by marginal instability and low-level convergence, while very fair conditions are characterized by much stronger instability and divergence. This early observation, which was later confirmed by other studies (Matsumoto *et al.*, 1967; Cho and Ogura, 1974), forms the basis for one of the most popular cumulus parameterization schemes—the Kuo schemes (Kuo, 1965; 1974). In Kuo schemes, the total convective heating is proportional to the large-scale moisture convergence.

Detailed studies of tropical convective systems over the Marshall Islands in the Pacific Ocean and over the eastern Atlantic Ocean during the GARP Atlantic Tropical Experiment (GATE) have indicated a strong control of cumulus convection by the synoptic-scale flow. Reed and Recker (1971) formed a composite analysis of 18 tropical disturbances in the equatorial western Pacific. In the composite wave structures, the heaviest precipitation occurred in the trough region, where the highest relative humidity and the strongest upward motion were present (Fig. 6.1). The precipitation estimated from the large-scale moisture budget agreed closely with the observed precipitation, with the moisture convergence term being the largest contributor to the moisture budget.

Cho and Ogura (1974), in an analysis of the Reed–Recker composite wave, found a high correlation between the vertical mass flux in the deep clouds and the large-scale mass flux at 950 mbar. Yanai *et al.* (1976) showed that deep cumulus clouds in the Marshall Islands region were highly correlated with the large-scale vertical motion in the upper troposphere. Nitta (1978) also showed that deep cumulus convection over the GATE area was highly correlated with the large-scale vertical velocity at all levels, and that there was little or no lag between the time of maximum large-scale upward motion and the time of heaviest convective precipitation. The time of heaviest rain did lag behind the time of maximum PBL convergence, however, by about 6 h.

Krishnamurti *et al.* (1980, 1983) emphasized the rather remarkable result that the rainfall associated with tropical mesoscale convective systems, which propagate about twice the speed of the large-scale tropical wave, appears to be controlled by the large-scale wave dynamics. This strong relationship between the mesoscale convective rainfall and the large-scale variables suggests that parameterization of convective rainfall in the tropics is possible.

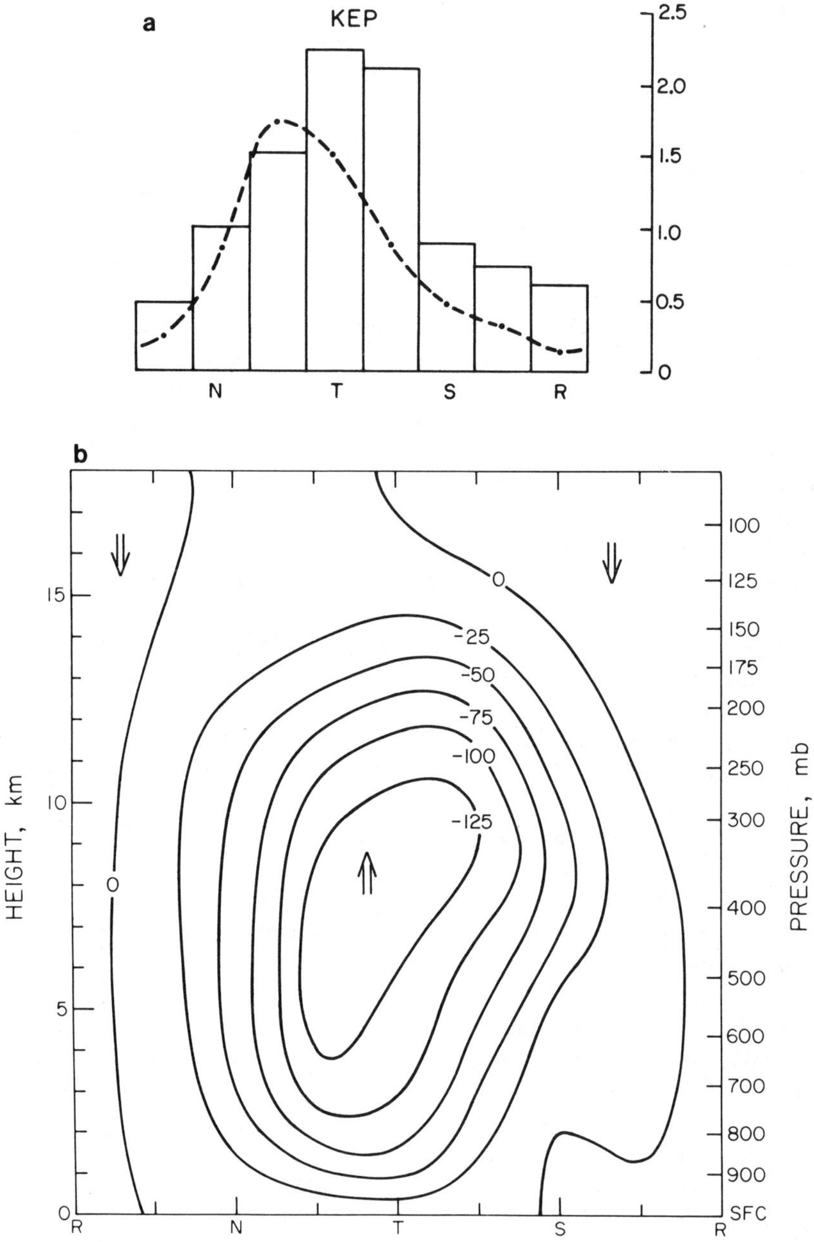

Fig. 6.1. (a) Observed average rainfall (cm day^{-1}) in various sectors of tropical wave in the equatorial West Pacific, with rainfall computed from moisture budget given by the dashed line. (b) Composite of vertical velocity ω (10^{-5} mbar s^{-1}) for 18 wave disturbances in the equatorial West Pacific. [From Reed and Recker (1971).]

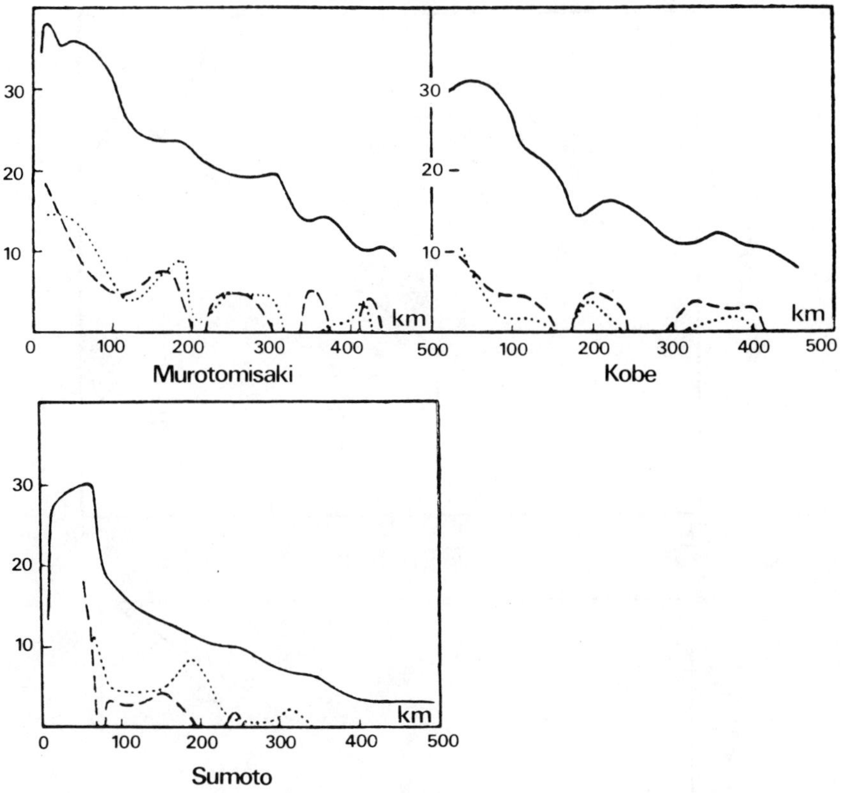

Fig. 6.2. Radial distributions of meteorological variables in a typhoon. Solid lines represent observed surface wind velocity in m s^{-1}, dotted lines represent the observed rate of precipitation in mm h^{-1}, and dashed lines represent the rate of precipitation computed from observed surface wind on the basis of Ekman pumping. [From Syono *et al.* (1951).]

The strong control of cumulus convection by the larger scales of motion in tropical cyclones has also been recognized for a long time. Syono *et al.* (1951) showed that the rate of precipitation in typhoons was related to the updrafts produced by frictional convergence in the PBL (so-called Ekman pumping) (Fig. 6.2). Later observational and modeling studies have confirmed the cooperative interaction between cumulus convection and the tropical cyclone through frictionally induced moisture convergence and enhanced evaporation in the PBL (see review in Anthes, 1982).

Ogura (1975) summarizes observational studies of precipitation in tropical waves and hurricanes. His results, shown in Fig. 6.3, indicate that there is a nearly linear correlation between the rainfall (or latent heating) rate and large-scale vertical velocity at the 950-mbar level over more than two orders of magnitude of rainfall rates.

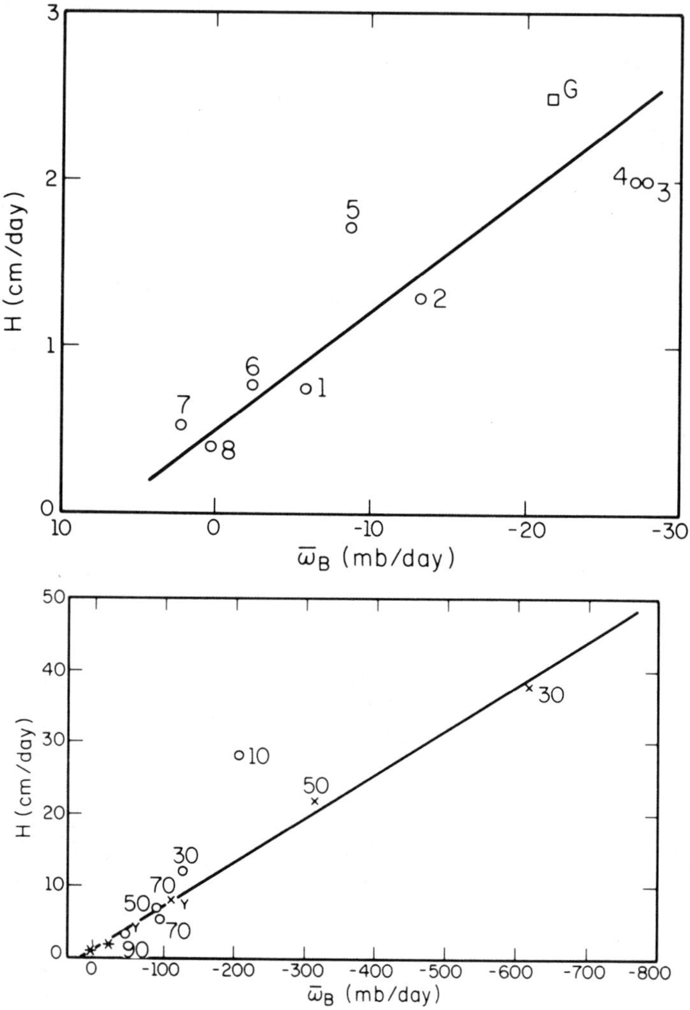

Fig. 6.3. Top: Net heating rate by cumulus clouds (H) in units of cm day^{-1} plotted against large-scale vertical velocity at the 950-mbar level ($\bar{\omega}_B$) for eight easterly wave categories. Bottom: H plotted against the larger scale vertical p-velocity ($\bar{\omega}_B$) at the 950-mbar level for several hurricanes. [From Ogura (1975).]

6.2.2 Extratropical Systems

The environment of convective systems in middle and high latitudes can differ considerably from the environment of tropical systems. In the extratropics, the synoptic-scale dynamic forcing can be stronger, the vertical shear of the horizontal wind often exceeds that of the tropics, and the lower relative and absolute humidities can enhance the role of cooling through evaporation of precipitation over that in the humid tropics. Even with these differences, however, there is abundant observational evidence that cumulus convection is often strongly controlled by large-scale processes in extratropical regions.

As in the tropics, the low-level convergence of mass and water vapor has been linked observationally to the occurrence of deep cumulus convection. Sasaki and Lewis (1970), Lewis (1971), and Hudson (1971) found close agreement between active convection and areas of mass and moisture convergence over the central United States. Lewis *et al.* (1974) indicated that a persistent mesoscale convergence area, with a horizontal scale of about 50 km, was present at about the 850-mbar level at least 2 h prior to the development of convective clouds. Fankhauser (1969, 1974) showed that convectively active areas associated with propagating squall lines were located inside mesoscale regions of convergence, with the magnitude of convergence reaching about $10^{-4}\ s^{-1}$. Convective precipitation is also related to surface convergence over a larger area (Ulanski and Garstang, 1978).

In contrast to tropical waves, in which there is an inverse relationship between the occurrence of convective rainfall and the magnitude of conditional or convective instability, there is a positive correlation between the magnitude of instability and amount of convective rainfall in the extratropics. Zawadzki and Ro (1978) and Zawadzki *et al.* (1981) found that a single thermodynamic parameter—the *static potential energy*—was highly correlated with maximum convective rainfall rates on both hourly and daily time scales (Fig. 6.4). The static potential energy, also called *parcel energy, available buoyant energy* (ABE), or *convective available potential energy* (CAPE), is essentially the "positive area" on a thermodynamic chart (the area between the environmental temperature sounding and the temperature of an undiluted parcel following a wet adiabat).

In the Zawadzki *et al.* (1981) study of convective rainfall near Montreal, Canada, the static potential energy parameter ABE explained about 60% of the storm-to-storm variability of mean and maximum rainfall rates. Although ABE was highly correlated with convective rainfall on days when convective rainfall occurred, there were a significant number of days that showed high values of ABE, but on which there was no convective rainfall. The occurrence of these nonstorm days in the presence of large values of ABE indicates that dynamical forcing, as well as thermodynamic instability, is important in initiating and maintaining extratropical convective systems.

The vertical shear of the horizontal wind can have a profound effect on extratropical convective systems. Fritsch (1975), summarizing work by J. Marwitz, showed that the precipitation efficiency (the ratio of rainout to

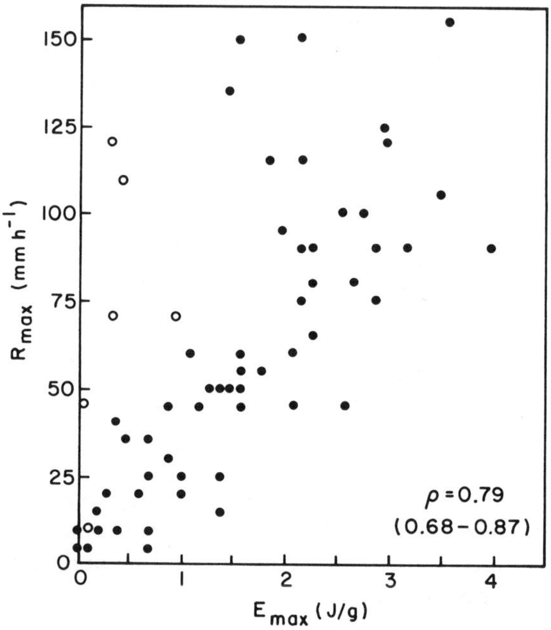

Fig. 6.4. Maximum rain rate (R_{max}) versus maximum parcel energy (E_{max}) for 67 storm days of summer seasons of 1969–1970. Indicated energies correspond to surface parcel or upper-air parcel, whichever was greater. Open circles indicate cases for which upper-air energy was greater than surface energy. Correlation coefficient ρ for the 61 remaining cases (darkened circles) and confidence limits, to the 5% significance level (in parentheses), are indicated. [From Zawadzki *et al.* (1981).]

water vapor inflow) is a strong function of vertical wind shear (see Chapter 9, Fig. 9.56). As discussed in Chapter 9, observational (Marwitz, 1972a, b, c) and theoretical (Moncrieff and Green, 1972; Weisman and Klemp, 1982) evidence indicates that the characteristics of severe, supercell thunderstorms are determined largely by the vertical wind shear.

6.2.3 Summary of Evidence for Large-Scale Controls over Cumulus Cloud Convection

As the above review indicates, there is abundant observational evidence, in both tropical and extratropical latitudes, that the synoptic and mesoscales of motion exert major controls over the formation and maintenance of cumulus convection, particularly deep, precipitating convection. Upward motion and convergence of water vapor are two large-scale parameters that are highly correlated with convective precipitation. In addition to these parameters, which are important in both tropical and extratropical systems, moist convection in middle latitudes is affected by the large-scale thermodynamic structure (degree of instability present) and by the vertical wind shear.

These relationships satisfy the necessary condition for parameterizing cumulus convection, i.e., that the cumulus convection be a strong function of the large-scale parameters. In the next section the mathematical framework is developed for cumulus parameterization. In subsequent sections the quantitative observational studies of the effects of cumulus convection on the larger scale environment, specific cumulus parameterization schemes, and the effect of some of these schemes on synoptic-scale and mesoscale numerical models are considered.

6.3 Mathematical Framework

For the concept of cumulus parameterization to be valid, there must be a distinct separation between the horizontal scale of the individual cumulus clouds and the larger scale of interest. For example, if the maximum diameter of a cumulonimbus cloud is about 10 km, the parameterization of the effects of a number of these clouds on their environment is most justifiable on scales 100 km or greater. In the following discussion, large-scale horizontal averages will refer to averages over regions at least as large as $(100 \times 100\ \text{km})$. For horizontal scales less than this, the concept of cumulus parameterization becomes muddy, since there is no clear distinction between resolvable, large-scale motions and the unresolvable, subgrid-scale cumulus clouds. In this case, it is more appropriate to adopt the generalized ensemble-averaging operator discussed in Chapter 3 and to resort to a different philosophical basis to the convective parameterization or closure problem.

6.3.1 Derivation of Area-Averaged Thermodynamic, Water Vapor, and Vorticity Equations

Most studies of the effect of cumulus convection on the large-scale environment have been concerned with temperature, water vapor, and vorticity. The equations for the time rate of change of these quantities, valid for a parcel of air in which these properties are homogeneous, are as follows:

$$\frac{\partial T}{\partial t}+\nabla\cdot\mathbf{V}T+\frac{\partial \omega T}{\partial p}-\frac{\omega RT}{c_{\mathrm{p}}p}=\frac{L}{c_{\mathrm{p}}}C^{*}+Q_{\mathrm{R}}, \tag{6.1}$$

$$\frac{\partial r_{\mathrm{v}}}{\partial t}+\nabla\cdot\mathbf{V}r_{\mathrm{v}}+\frac{\partial \omega r_{\mathrm{v}}}{\partial p}=-C^{*}, \tag{6.2}$$

$$\frac{\partial \zeta}{\partial t}+\nabla\cdot\mathbf{V}\zeta_{\mathrm{a}}+\omega\frac{\partial \zeta}{\partial p}+\hat{k}\cdot\nabla\omega\times\frac{\partial \mathbf{V}}{\partial p}=0, \tag{6.3}$$

where T is temperature, $\mathbf{V}$ is the horizontal vector velocity ($\mathbf{V} = u\hat{i} + v\hat{j}$), ω is the vertical velocity in pressure coordinates (dp/dt), R is the gas constant for dry air, L is the latent heat of condensation, c_p the specific heat at constant pressure for dry air, Q_R represents the temperature change due to radiative effects, r_v is the mixing ratio of water vapor, C^* is the net condensation rate (condensation minus evaporation, $c - e$), ζ is the vertical component of relative vorticity, $\zeta_a = \zeta + f$, f is the Coriolis parameter, and i, j, and k are the unit vectors in the west–east, south–north, and vertical directions, respectively. The above equations and the following discussion neglect the ice phase of water for simplicity.

Equations (6.1)–(6.3) are valid only for small parcels of air; they are not immediately applicable to the large-scale average properties of air associated with grid volumes in numerical models. (It should be remembered from Chapter 3 that a value of any variable defined at a grid point in a numerical model does not represent the point value of that variable in the atmosphere; instead it represents the average over an atmospheric volume of dimensions $\Delta x\, \Delta y\, \Delta z$, where Δx and Δy are the horizontal dimensions of the grid mesh and Δz is the thickness of the vertical layer of the model.)

To derive the appropriate averaged equations [Eqs. (6.1)–(6.3)], we adopt the horizontal averaging operator, Eq. (3.4), and decompose each variable onto a mean and fluctuating component, Eq. (3.17). Applying the Reynolds averaging procedure outlined in Chapter 3 yields the averaged equations

$$\frac{\partial \bar{r}_v}{\partial t} + \nabla \cdot \bar{\mathbf{V}}\bar{r}_v + \frac{\partial \bar{\omega}\bar{r}_v}{\partial p} = -\overline{C^*} - \nabla \cdot \overline{\mathbf{V}''r_v''} - \frac{\partial \overline{\omega'' r_v''}}{\partial p}. \tag{6.4}$$

Equation (6.4) is a commonly used expression in diagnostic studies and numerical models to estimate the time rate of change of the large-scale average mixing ratio. According to Eq. (6.4), $\bar{r}_v$ changes by the large-scale horizontal and vertical motions, the area-averaged condensation rate, and the effect of subgrid-scale motions [the horizontal and vertical eddy flux terms on the right side of Eq. (6.4)]. The left side of Eq. (6.4) is often called the "apparent moisture source" (Yanai *et al.*, 1973). For both modeling and diagnostic purposes, the *advective* form of Eq. (6.4) is often used:

$$\frac{\partial \bar{r}_v}{\partial t} + \bar{\mathbf{V}} \cdot \nabla \bar{r}_v + \bar{\omega}\frac{\partial \bar{r}_v}{\partial p} = -\overline{C^*} - \nabla \cdot \overline{\mathbf{V}''r_v''} - \frac{\partial \overline{\omega'' r_v''}}{\partial p}. \tag{6.5}$$

In deriving Eq. (6.4), we have assumed that correlations between mean and fluctuating variables and zero, i.e., $\overline{\bar{\mathbf{V}}r_v''} = \overline{\mathbf{V}''\bar{r}_v} = 0$. However, as we have noted in Chapter 3, they are not exactly zero, since the mean quantities vary slightly over the averaging area. The neglect of those terms becomes less valid as the averaging interval $\Delta x\, \Delta y$ is decreased to areas comparable with the area of convective systems such as squall lines or individual thunderstorms. At this time, unfortunately, there exists an imprecisely defined point at which we must abandon the grid-volume averaging concept and associated cumulus parameterization theory and employ the generalized ensemble-averaging operator and closure theory.

Following a procedure similar to deriving Eq. (6.4), approximate equations for the large-scale averaged temperature and vorticity may be obtained:

$$\frac{\partial \bar{T}}{\partial t}+\nabla \cdot \bar{\mathbf{V}}\bar{T}+\frac{\partial \bar{\omega}\bar{T}}{\partial p}-\frac{\bar{\omega}\bar{\alpha}}{c_{\mathrm{p}}}=\frac{L}{c_{\mathrm{p}}}\overline{C^*}+\bar{Q}_{\mathrm{R}}-\nabla \cdot \overline{\mathbf{V}''T''}-\frac{\partial \overline{\omega''T''}}{\partial p}-\frac{\overline{\omega''\alpha''}}{c_{\mathrm{p}}}, \quad (6.6)$$

where α in Eq. (6.6) is specific volume ($\alpha = RT/p$),

$$\frac{\partial \bar{\zeta}}{\partial t}+\nabla \cdot \bar{\mathbf{V}}\bar{\zeta}_{\mathrm{a}}+\bar{\omega}\frac{\partial \bar{\zeta}}{\partial p}+\hat{k}\cdot \nabla \bar{\omega} \times \frac{\partial \mathbf{V}}{\partial p}=-\nabla \cdot \overline{\mathbf{V}''\zeta''_{\mathrm{a}}}-\overline{\omega''\frac{\partial \zeta''}{\partial p}}-\overline{\hat{k}\cdot \nabla \omega'' \times \frac{\partial \mathbf{V}''}{\partial p}}. \quad (6.7)$$

Following Yanai *et al.* (1973), the left side of Eq. (6.6) is called the "apparent heat source" Q_1, while the left side of Eq. (6.7) is termed the "apparent vorticity source."

The apparent sources and sinks of water vapor, heat, and vorticity are often interpreted as the effects of cumulus convection. Cumulus parameterization schemes attempt to model the terms on the right side of the large-scale equations, Eqs. (6.4), (6.6), and (6.7), as a function of the known resolvable-scale variables on the left side of the equations in order to calculate the temporal rates of change of T, r_{v}, and ζ. In diagnostic studies, the apparent sources and sinks are estimated by calculating the left side of the equations from large-scale analyses.

The horizontal eddy flux terms ($\nabla \cdot \overline{\mathbf{V}''\phi''}$, where ϕ is any variable) in Eqs. (6.4), (6.6), and (6.7) have received little research attention and are generally assumed to be small compared to the vertical eddy flux terms. (An exception may be the vorticity equation, where Cho and Cheng (1980) have argued that the horizontal eddy flux of vorticity cannot be neglected compared to the vertical eddy flux.) Large-scale models do contain simple parameterizations of horizontal mixing; these typically take the form

$$\nabla \cdot \overline{\mathbf{V}''\phi''}=\nabla \cdot K_{\mathrm{H}}\nabla \bar{\phi}, \quad (6.8)$$

or

$$\nabla \cdot \overline{\mathbf{V}''\phi''}=-\Delta s^2\, \nabla^2 K_{\mathrm{H}}\nabla^2 \bar{\phi}, \quad (6.9)$$

where K_{H} is an eddy viscosity and Δs is the grid length (Williamson, 1978).

These terms are included primarily to damp small-scale waves (those with wavelengths close to the grid length of the model) rather than to model physical effects of horizontal mixing. In nonlinear numerical models, some form of dissipation of such short waves is necessary to prevent nonlinear numerical instability or the "pile-up" of energy cascading down to the grid scale (valid for truncated spectral models as well). In the rest of this chapter, the horizontal eddy flux terms will be neglected, in which case the equations for the temporal rate of change of the large-scale mixing ratio, temperature, and relative vorticity become (after dropping the overbars on the variables)

$$\frac{\partial r_{\mathrm{v}}}{\partial t}+\nabla \cdot \mathbf{V} r_{\mathrm{v}}+\frac{\partial \omega r_{\mathrm{v}}}{\partial p}=-C^{*}-\frac{\partial \overline{\omega'' r_{\mathrm{v}}''}}{\partial p}, \tag{6.10}$$

$$\frac{\partial T}{\partial t}+\nabla \cdot \mathbf{V} T+\frac{\partial \omega T}{\partial p}-\frac{\omega \alpha}{c_{\mathrm{p}}}=\frac{L}{c_{\mathrm{p}}} C^{*}+Q_{\mathrm{R}}-\frac{\partial \overline{\omega'' T''}}{\partial p}-\frac{\overline{\omega'' \alpha''}}{c_{\mathrm{p}}}, \tag{6.11}$$

$$\frac{\partial \zeta}{\partial t}+\nabla \cdot \mathbf{V} \zeta_{\mathrm{a}}+\omega \frac{\partial \zeta}{\partial p}+\hat{k} \cdot \nabla \omega \times \frac{\partial \mathbf{V}}{\partial p}=-\overline{\omega'' \frac{\partial \zeta''}{\partial p}}-\hat{k} \cdot \overline{\nabla \omega'' \times \frac{\partial \mathbf{V}''}{\partial p}}. \tag{6.12}$$

It is sometimes simpler to use potential temperature θ rather than temperature; in that case, the thermodynamic equation becomes

$$\frac{\partial \theta}{\partial t}+\nabla \cdot \mathbf{V} \theta+\frac{\partial \omega \theta}{\partial p}=\frac{L}{c_{\mathrm{p}}} \pi^{-1} C^{*}+\pi^{-1} Q_{\mathrm{R}}-\frac{\partial \overline{\omega'' \theta''}}{\partial p}, \tag{6.13}$$

where $\pi=(p/p_0)^{R/c_{\mathrm{p}}}$ and p_0 is a reference pressure (typically 1000 mbar). Another thermodynamic variable often used in diagnostic studies is *dry static energy*, $s=c_{\mathrm{p}}T+gz$; the equation describing the temporal change of s is

$$\frac{\partial s}{\partial t}+\nabla \cdot \mathbf{V} s+\frac{\partial \omega s}{\partial p}=L C^{*}+c_{\mathrm{p}} Q_{\mathrm{R}}-\frac{\partial \overline{\omega'' s''}}{\partial p}. \tag{6.14}$$

Before discussing quantitative estimates of the effects of cumulus clouds on the large-scale variables in later sections, we note some qualitative effects from Eqs. (6.10)–(6.14). In the thermodynamic equations, cumulus convection modifies the temperature through diabatic heating (cooling) due to condensation (evaporation)—the C^{*} term—and through vertical eddy fluxes associated with correlations between the temperature and vertical velocity. In a typical situation dominated by warm updrafts and cool downdrafts, the eddy flux term represents an upward transport of heat and a stabilization of the environment. Physically, this effect is most obvious when cool downdrafts reach the surface and stabilize the lower atmosphere.

In the continuity equation for water vapor, Eq. (6.10), the C^{*} term indicates that cumulus convection decreases (increases) the average mixing ratio when condensation (evaporation) prevails at a given level. The eddy flux term represents a net vertical transport of water vapor—upward in the typical case when moist updrafts and dry environmental subsidence are present.

In the vorticity equation, cumulus effects arise only through eddy terms. The first term on the right side of Eq. (6.12) represents the vertical eddy transport associated with correlations between vertical velocity and vorticity perturbations; the second term represents a twisting effect due to correlations between perturbations in the vertical velocity and vertical wind shear.

6.3.2 *Integral Constraints on the Large-Scale Equations*

A vertical integral of the large-scale thermodynamic and moisture equations yields information on the role of precipitation and evaporation as a lower boundary condition and provides important checks on observational budget studies. We first define the *moist static energy h*, which is conserved for both dry and wet adiabatic processes, as

$$h \equiv c_p T + gz + Lr_v = s + Lr_v. \tag{6.15}$$

Defining the apparent moisture sink Q_2 from Eq. (6.10)

$$Q_2 \equiv -(L/c_p)(\partial r_v/\partial t + \nabla \cdot \mathbf{V} r_v + \partial \omega r_v/\partial p), \tag{6.16}$$

and the apparent heat source Q_1 from Eq. (6.14)

$$c_p Q_1 \equiv \partial s/\partial t + \nabla \cdot \mathbf{V} s + \partial \omega s/\partial p, \tag{6.17}$$

and using Eqs. (6.10) and (6.14), we obtain

$$Q_1 - Q_2 - Q_R = -(1/c_p)(\partial \overline{\omega'' h''}/\partial p), \tag{6.18}$$

where the units of Q_1, Q_2, and Q_R are normally given in degrees Celsius per day. The vertical eddy transport F of total heat may be obtained by integrating Eq. (6.18) from p_t, the pressure at the top of the highest clouds where all cloud effects vanish,

$$F(p) \equiv -\frac{1}{g}\overline{\omega'' h''} = \frac{c_p}{g}\int_{p_t}^{p} (Q_1 - Q_2 - Q_R)\, dp, \tag{6.19}$$

where the units of F are W m^{-2}. Similarly, an integration of Eq. (6.14) from p_t to p_s, the surface pressure, yields

$$\frac{c_p}{g}\int_{p_t}^{p_s} (Q_1 - Q_R)\, dp = \frac{L}{g}\int_{p_t}^{p_s} C^*\, dp - \frac{1}{g}(\overline{\omega'' s''})_{p_s}. \tag{6.20}$$

The vertical integral of the net condensation minus evaporation rate equals the rate of precipitation P reaching the surface, while the eddy flux of dry static energy at the surface equals the sensible heat flux H_s there

$$\frac{c_p}{g}\int_{p_t}^{p_s}(Q_1 - Q_R)\,dp = LP + H_s, \tag{6.21}$$

where P is usually expressed in units of cm day^{-1} and H_s is expressed in W m^{-2}. A similar integration of Eq. (6.14) yields

$$\frac{c_p}{g}\int_{p_t}^{p_s} Q_2\,dp = L(P - E), \tag{6.22}$$

where E is the rate of evaporation at the surface. If the precipitation rate is known and the sensible heating and evaporation rates can be estimated from surface variables, Eqs. (6.21) and (6.22) serve as independent checks on the vertical integrals of $Q_1 - Q_R$ as estimated from observations.

6.3.3 Effects of a Cumulus Ensemble on the Large-Scale Variables

Cumulus parameterization would be considerably simpler if only one type of cumulus cloud existed; however, the apparent sources and sinks of temperature, water vapor, and vorticity are the result of many clouds of different sizes and with different thermodynamic and dynamic properties. Thus, the most general cumulus parameterization schemes (e.g., Ooyama, 1971; Arakawa and Schubert, 1974) consider a *spectrum* or *ensemble* of clouds and attempt to compute the net effect of the cloud ensemble by integrating over the spectrum of individual cloud effects. In this section, we present a model of a cumulus ensemble developed by Yanai *et al.* (1973). In this model, as in several others, cumulus clouds are classified according to the height of the cloud tops (see Section 6.3.5 and Fig. 6.6). If one assumes that the cloud-top height is uniquely determined by the lateral entrainment rate, this assumption is equivalent to classifying the clouds according to the entrainment rate. This classification is an important part of the Arakawa and Schubert (1974) cumulus parameterization scheme discussed later and the diagnostic cumulus ensemble models of Ogura and Cho (1973), Nitta (1975), and Johnson (1976).

The physical interpretation of the effects of cumulus convection on the environment is often facilitated by separating the large-scale average vertical motion $\bar{\omega}$ into the average vertical velocity in cumulus clouds ω_c and the vertical velocity $\tilde{\omega}$ in the environment of the clouds. These velocities, when written in pressure coordinates, are equivalent to vertical mass fluxes per unit area in the clouds and the environment of the clouds, i.e.,

$$\begin{aligned} \bar{M} &= -\bar{\omega}, \\ &= -a\omega_c - (1-a)\tilde{\omega}, \\ &= M_c + \tilde{M}. \end{aligned} \tag{6.23}$$

In Eq. (6.23), as in many studies, the mass flux has the same units as ω; to obtain the dimensions of mass flux per unit area, M should be defined as $M \equiv -\omega/g \approx \rho w$ (Yanai *et al.*, 1973). For the present, we will consider the presence of a single cloud type.

An important assumption in most cloud parameterizations is that the cumulus clouds occupy a small fraction, a, of the unit large-scale area. In this case, the total vertical mass flux in the unit area due to clouds and the total mass flux in the cloud environment are

$$M_c = -a\omega_c, \qquad \tilde{M} = -(1-a)\tilde{\omega}. \tag{6.24}$$

It is important to note that $\bar{M}$ and $\tilde{M}$ are not approximately equal and may even have opposite signs. In fact, observations of deep cumulus ensembles indicate that while $\bar{M}$ is usually positive (mean upward motion), $\tilde{M}$ is negative, representing compensating subsidence between active cumulus updrafts (Yanai *et al.*, 1973; Ogura and Cho, 1973; Nitta, 1977).

We next write the mean dry static energy s and mixing ratio r_v in terms of their representative (average) values over the cloud and the cloud environment,

$$s = as_c + (1-a)\tilde{s}, \tag{6.25a}$$

$$r_v = ar_{vc} + (1-a)\tilde{r}_v. \tag{6.25b}$$

Equations (6.24) and (6.25) may be used to obtain approximate expressions for the vertical eddy transports of dry static energy and water vapor due to the clouds

$$\overline{\omega'' s''} = a(1-a)(\omega_c - \tilde{\omega})(s_c - \tilde{s}), \tag{6.26a}$$

$$\overline{\omega'' r_v''} = a(1-a)(r_{vc} - \tilde{r}_v)(\omega_c - \tilde{\omega}). \tag{6.26b}$$

Making use of the assumptions that $a \ll 1$ and $|\tilde{\omega}| \ll |\omega_c|$, the eddy flux terms are approximately

$$\overline{\omega'' s''} \approx a\omega_c(s_c - \tilde{s}) = -M_c(s_c - \tilde{s}), \tag{6.27a}$$

$$\overline{\omega'' r_v''} \approx a\omega_c(r_{vc} - \tilde{r}_v) = -\tilde{M}_c(r_{vc} - \tilde{r}_v). \tag{6.27b}$$

Because $s_c - \tilde{s} = c_p(T_c - \tilde{T})$, Eq. (6.27) shows that the vertical eddy fluxes of dry static energy, or heat, and water vapor are proportional to the cloud vertical velocity and the temperature or moisture excess of the cloud.

The above relationships pertain to a single cloud type; in the general case, many clouds with different properties may exist. To generalize the derivation, we consider the cumulative effect of a number of clouds by summing over all clouds. Denoting the characteristics of each distinct cloud type by the index i, and assuming that $\sum_i a_i$ remains much less than 1, we have from Eq. (6.27)

$$-\overline{\omega'' s''} = \sum_i m_i (s_{ci} - \tilde{s}), \tag{6.28a}$$

$$-\overline{\omega'' r_v''} = \sum_i m_i (r_{vci} - \tilde{r}_v), \tag{6.28b}$$

and

$$M_c = \sum_i m_i. \tag{6.29}$$

From Eqs. (6.14), (6.17), and (6.28a), we have

$$c_p(Q_1 - Q_R) = LC^* - \frac{\partial \overline{\omega'' s''}}{\partial p} = L \sum_i C_i^* + \frac{\partial}{\partial p} \sum_i m_i (s_{ci} - \tilde{s}), \tag{6.30}$$

while Eqs. (6.10), (6.16), and (6.28b) imply

$$\frac{c_p}{L} Q_2 = C^* + \frac{\partial \overline{\omega'' r_v''}}{\partial p} = \sum_i C_i^* - \frac{\partial}{\partial p} \sum_i m_i (r_{vci} - \tilde{r}_v), \tag{6.31}$$

where C_i^* is the net condensation minus evaporation rate taking place in the ith cloud.

6.3.4 *Effects of Large-Scale Forcing on Cumulus Clouds*

As discussed earlier, many observational studies have indicated a significant effect of the large-scale dynamic and thermodynamic forcing on the development and characteristics of cumulus clouds. Soong and Ogura (1980) developed a two-dimensional numerical cloud model capable of investigating the properties of cumulus convection that develop in response to large-scale forcing. In their model, terms representing the temporal rates of change of potential temperature and water vapor mixing ratio due to large-scale processes (primarily horizontal and vertical advection) are specified in the thermodynamic and water vapor equations appropriate to the nonhydrostatic cloud model. The prescribed large-scale terms can be obtained from either observations or a large-scale model.

Soong and Ogura (1980) used their cloud model to investigate the response of shallow trade-wind cumulus clouds to various profiles of large-scale forcing. In this study, convection responded quickly to changes in the large-scale vertical motion by raising or lowering the height of the trade-wind inversion. Changes in the cloud properties, including condensation and evaporation, also changed in response to variations in the large-scale forcing in such a way as to maintain a balance between the large-scale and cloud effects.

Soong and Tao (1980) utilized the same cloud model to study the response of deep tropical cumulus clouds to observed large-scale forcing obtained from the Global Atmospheric Research Program Atlantic Tropical Experiment. In response to the specified large-scale vertical velocity, horizontal advection of temperature and moisture, and radiative cooling, the cloud properties, including the condensation and evaporation rates and the vertical fluxes of heat and moisture, were computed by the model. In these numerical simulations, the upward mass flux in the model clouds was about three times the specified large-scale vertical mass flux. Most of this excess was

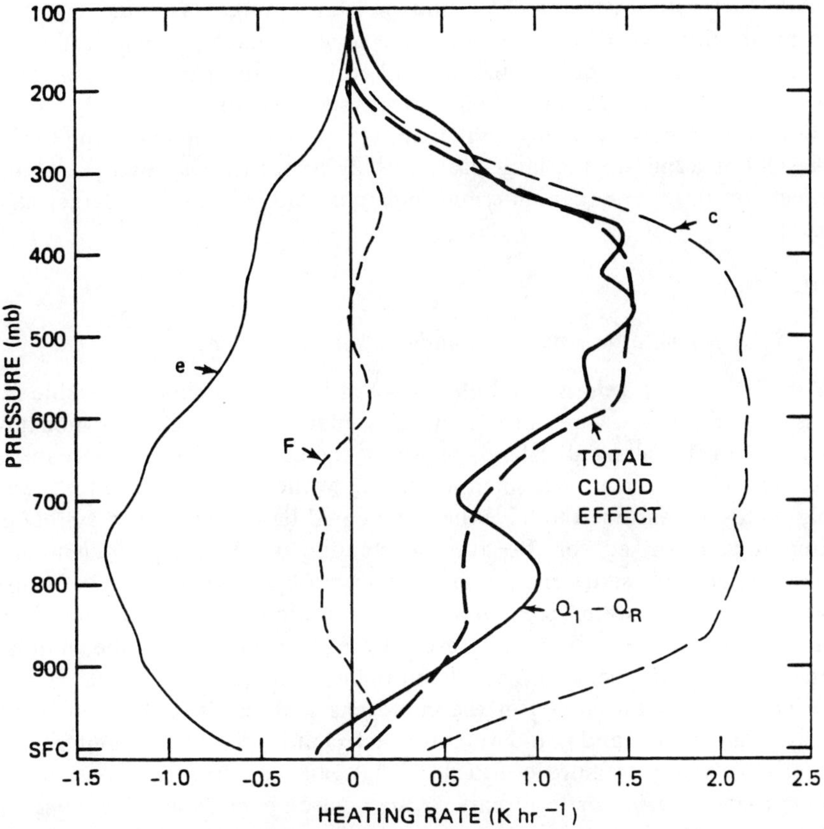

Fig. 6.5. Time-averaged heating rate by condensation (c), evaporation (e), and the vertical transport of sensible heat by clouds (F) predicted by a two-dimensional cumulus ensemble model. Sum of the three terms represents the total cloud heating effect. Cloud heating effect estimated from a large-scale heat budget over the period 0600–1200 GMT, 12 August 1974 is denoted by $Q_1 - Q_R$. [From Ogura (1984).]

compensated by downdrafts within the clouds. Warm updrafts were responsible for most of the upward flux of heat. However, both dry downdrafts and moist updrafts were important in producing the cloud-scale moisture flux.

Numerical experiments by Soong and Tao (1980) indicated that the properties of the clouds, such as their size and the heating and moistening profiles, adjusted to the large-scale forcing in a way that a near balance was maintained between the cloud-scale and large-scale processes. The properties of the model clouds were quite sensitive to small changes in the temperature and mixing ratio of the environment.

Ogura (1984) presents the response of the Soong and Ogura (1980) cloud model to prescribed large heat and moisture sources associated with a tropical rainband. Figure 6.5 shows the time-averaged vertical profiles of condensation, evaporation, and vertical eddy transport of heat associated with the model clouds in response to the specified profile $Q_1 - Q_R$. As in the earlier studies by Soong and Ogura (1980) and Soong and Tao (1980), the clouds adjust to the large-scale forcing in such a way that the cloud effects on the thermodynamic and moisture fields balance the large-scale effects.

6.3.5 A Model of a Tropical Cumulus Cloud Ensemble

Yanai *et al.* (1973) derived a bulk model of a cumulus cloud ensemble for use in a diagnostic study of the cumulus effects on the heat and moisture budgets of the Marshall Islands region of the equatorial Pacific Ocean. In this model, clouds are classified according to the height of their tops, and all clouds are assumed to have the same cloud base (Fig. 6.6). A system of equations is solved for the average cloud mass flux, entrainment and detrainment, dry static energy, moist static energy, mixing ratios of water vapor and liquid water, condensation, evaporation, and rainfall rates. Vertical profiles of these quantities describe the mean structure of the cumulus ensemble and the net cloud effects on the environment.

The basis of the model of the cloud ensemble is the balance of heat, mass, water vapor, and liquid water for each cloud type i and its environment (Fig. 6.6). At any pressure level, a cloud may gain mass from the environment (*mass entrainment*) or lose mass to the environment (*mass detrainment*). The mass entrainment and detrainment rates are denoted by ε_i and δ_i, respectively and increase or decrease the mass flux of a cloud at a level according to the continuity equation for cloud mass

$$\varepsilon_i - \delta_i + \partial m_i/\partial p = 0. \tag{6.32}$$

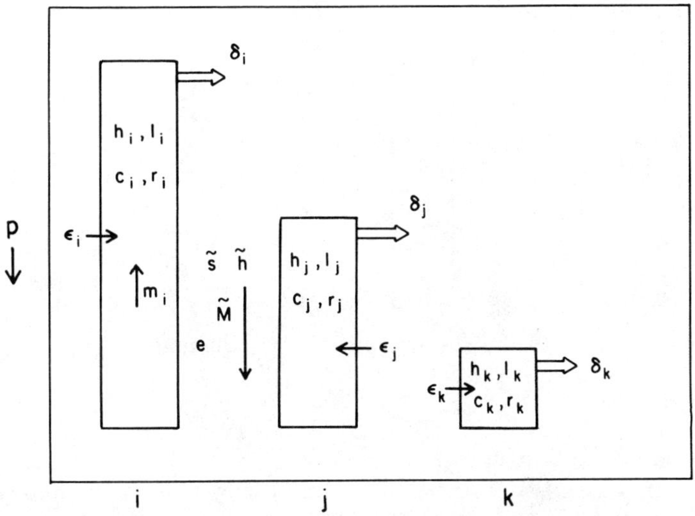

Fig. 6.6. Idealized model of ensemble of cumulus clouds of different types, classified according to their top heights. Symbols are defined in the text. [From Yanai *et al.* (1973).]

Entrainment brings in environmental air of different heat and water vapor contents than those of the cloud, while detrainment removes heat, water vapor, and liquid water from the cloud. These effects are accounted for in the continuity equation for
heat

$$\varepsilon_i \tilde{s} - \delta_i s_{ci} + \partial m_i s_{ci}/\partial p + Lc_i = 0, \tag{6.33}$$

water vapor

$$\varepsilon_i \tilde{r}_v - \delta_i r_{vci} + \partial m_i r_{vci}/\partial p - c_i = 0, \tag{6.34}$$

and *liquid water*

$$-\delta_i r_{li} + \partial m_i r_{li}/\partial p + c_i - \mathrm{PR}_i = 0, \tag{6.35}$$

where r_l is the mixing ratio of liquid water and PR_i is the rate of production of precipitating water, which is assumed to fall out of the cloud instantaneously. Detrained cloud water r_{li} is also assumed to evaporate instantly

$$e_i = \delta_i r_{li}. \tag{6.36}$$

The system of equations, Eqs. (6.32)-(6.36), is simplified by the following relations:

$$\begin{bmatrix} M_c \\ \varepsilon \\ \delta \\ c \\ \mathrm{PR} \\ e \end{bmatrix} = \sum_i \begin{bmatrix} m_i \\ \varepsilon_i \\ \delta_i \\ c_i \\ \mathrm{PR}_i \\ e_i \end{bmatrix} \tag{6.37}$$

Weighted averages of the cloud properties are defined as

$$\bar{\bar{h}}_c \equiv \sum_i m_i h_{ci} / M_c \tag{6.38}$$

with similar definitions for $\bar{\bar{s}}_c$, $\bar{\bar{r}}_{vc}$, and $\bar{\bar{r}}_l$.

Additional assumptions needed to close the system are as follows:

- Each cloud type detrains at the level of zero buoyancy ($T_{ci} = \tilde{T}$), unless the cloud reaches the tropopause, in which case $\delta = \partial M_c/\partial p$ is assumed.
- The liquid-water content detrained from each cloud type is equal to the weighted average of cloud liquid water

$$r_{li} = \bar{\bar{r}}_l. \tag{6.39}$$

- The rainfall rate is proportional to the average liquid-water content

$$\mathrm{PR} = K(p)\bar{\bar{r}}_l, \tag{6.40}$$

where the empirical constant of proportionality varies only with pressure.

With these assumptions, a closed set of 10 equations in the 10 unknowns (M_c, ε, δ, $\bar{\bar{h}}_c$, $\bar{\bar{s}}_c$, $\bar{\bar{r}}_{vc}$, $\bar{\bar{r}}_l$, c, PR, and e) is obtained:

$$c_p(Q_1 - Q_R) = L(c-e) + (\partial/\partial p)[M_c(\bar{\bar{s}}_c - \tilde{s})], \tag{6.41}$$

$$c_p Q_2 = L(c-e) - L(\partial/\partial p)[M_c(\bar{\bar{r}}_{vc} - \tilde{r}_v)], \tag{6.42}$$

$$\varepsilon - \delta + \partial M_c/\partial p = 0, \tag{6.43}$$

$$(\varepsilon - \delta)\tilde{s} + \partial M_c \bar{\bar{s}}_c/\partial p + Lc = 0, \tag{6.44}$$

$$\varepsilon \tilde{r}_v - \delta \tilde{r}_v^* + \partial M_c \bar{\bar{r}}_{vc}/\partial p - c = 0, \tag{6.45}$$

$$-\delta \bar{\bar{r}}_l + \partial M_c \bar{\bar{r}}_l/\partial p + c - \mathrm{PR} = 0, \tag{6.46}$$

$$\mathrm{PR} = K(p)\bar{\bar{r}}_l, \tag{6.47}$$

$$e = \delta \bar{\bar{r}}_l, \tag{6.48}$$

$$\bar{\bar{s}}_c - \tilde{s} = [1/(1+\gamma)](\bar{\bar{h}}_c - \tilde{h}^*), \tag{6.49}$$

$$L(\bar{\bar{r}}_{vc} - \tilde{r}_v^*) = [\gamma/(1+\gamma)](\bar{\bar{h}}_c - \tilde{h}^*), \tag{6.50}$$

where the asterisk refers to saturation and the saturation moist static energy of the environment h^* is defined as

$$\tilde{h}^* \equiv c_p\tilde{T} + gz + L\tilde{r}_v^*(p, \tilde{T}), \tag{6.51}$$

and

$$\gamma \equiv (L/c_p)(\partial r_v^*/\partial T)_p.$$

In diagnostic studies, synoptic-scale analyses are used to estimate $Q_1 - Q_R$, Q_2, $\tilde{s}$, and $\tilde{r}_v$. A slightly simplified version of the above set of 10 equations is solved by an iterative method for the 10 unknowns representing the cumulus ensemble.

To interpret the apparent heat source and the apparent moisture sink in this model, Eqs. (6.41) and (6.42) can be written, with the help of Eqs. (6.43), (6.44), and (6.45), as

$$c_p(Q_1 - Q_R) = -M_c(\partial\tilde{s}/\partial p) - Le, \tag{6.52}$$

$$c_pQ_2 = LM_c(\partial\tilde{r}_v/\partial p) - L\delta(\tilde{r}_v^* - r_v) - Le. \tag{6.53}$$

In this model, as in the Arakawa–Schubert cumulus parameterization scheme, the apparent heat source Q_1 equals radiative cooling, an adiabatic warming due to a compensating subsidence which is equal to the cloud mass flux M_c and evaporative cooling of detrained cloud water. The apparent moisture sink Q_2 is interpreted as drying due to compensating subsidence and moistening due to detrained cloud water vapor and evaporation of detrained cloud water.

The results of large-scale mass, heat, and moisture budgets and the cumulus effects on these budgets from Yanai *et al.* (1973) are summarized in Section 6.4. Similar equations and models were developed and used by Ogura and Cho (1973), Nitta (1975, 1977, 1978), Johnson (1976), and others in diagnostic studies. In addition, many of the physical concepts in this section are essential parts of the Arakawa–Schubert cumulus parameterization, discussed in Section 6.5.

6.4 Diagnostic Studies of the Effects of Cumulus Cloud Convection on the Environment

Observational studies which diagnose the effects of cumulus convection on the environment are important in designing and testing cumulus parameterization schemes. Three types of studies contribute to the understanding of cumulus effects. First, large-scale budgets of mass, heat, moisture, and vorticity provide estimates of the vertical profiles of the "apparent sources" of these quantities. Second, these large-scale budgets may be used with models of cumulus ensembles to diagnose the physical mechanisms by which clouds interact with the environment. These studies yield estimates of the vertical profiles of cloud properties, such as cloud-mass flux, entrainment, detrainment, condensation, and evaporation, as well as the mass flux in the environment of the clouds. Finally, the budget studies may be used to test cumulus parameterization schemes in a "semiprognostic" sense (Lord, 1982). In a semiprognostic test, the large-scale variables are used by the cumulus parameterization scheme to calculate the effects of the cumulus clouds on the large-scale variables for a single time step. The resulting vertical profiles of heating and moistening, as well as precipitation rates, may be compared with the diagnosed or observed variables for verification of the scheme. The semiprognostic tests differ from a complete test in that the large-scale variables are not allowed to evolve over a period of time and interact with the cumulus parameterization scheme in a prognostic fashion. A complete test would involve verification of the evolution of the large-scale system in response to the cumulus parameterization, as well as to other physical effects. This section summarizes diagnostic studies in the tropics and extratropics.

6.4.1 Diagnostic Studies in the Tropics

Three sets of data have been used in diagnostic studies of cumulus effects in the tropics. The early studies used surface, ship, and upper-air data from the Marshall Islands area during the period 15 April to 22 July 1956. This region lies in the equatorial Pacific (about 5–11°N and 160–170°E). A second data set, obtained from the Barbados Oceanographic and Meteorological Experiment (BOMEX) in 1969, was the basis for heat and moisture budgets in the Atlantic trade-wind region (Nitta and Esbensen, 1974; Nitta, 1975). The third widely used data set was obtained as part of the Global Atmospheric Research Program's Atlantic Tropical Experiment, which was carried out from June to September in 1974. The most commonly used data from GATE were the Phase 3 set (30 August–15 September 1974). In addition to these special data sets, several studies have utilized composite data of Pacific tropical waves (Reed and Recker, 1971; Cho and Ogura, 1974).

6.4.1.a Equatorial Pacific Studies

Reed and Recker (1971) formed a composite analysis, including a mass, heat, and moisture budget, of 18 synoptic-scale disturbances in the equatorial western Pacific. Results of significance to the cumulus param-

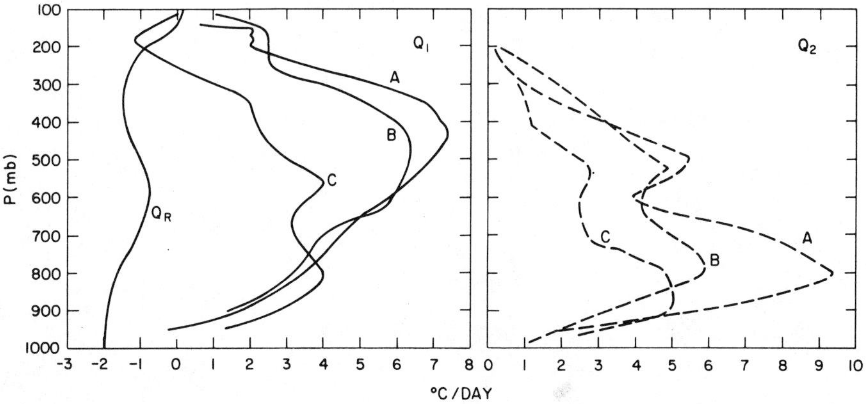

Fig. 6.7. Left: Vertical distributions of apparent heat source (Q_1) for (A) category 4 of easterly wave in Western Pacific (Cho and Ogura, 1974), (B) average cloud cluster over Marshall Islands (Yanai *et al.*, 1973), and (C) average profile over eastern Atlantic during Phase III of GATE (Nitta, 1978). Q_R is radiational heating rate estimated by Dopplick (1970). Right: Same as left side but for apparent moisture sink Q_2.

eterization problem included the following observations: (1) The observed precipitation was in close agreement with the synoptic-scale moisture convergence. (2) The large-scale atmosphere does not become absolutely unstable and readjust, as in some convective adjustment hypotheses; instead the vertical transport of heat and moisture takes place in convective updrafts and downdrafts within a conditionally unstable environment. (3) A substantial fraction of the total water vapor convergence occurs above the boundary layer.

Cho and Ogura (1974) used Reed and Recker's composite data to determine the apparent heat source and moisture sink, as well as the cloud-mass flux distribution in the eight portions of the composite wave. The Q_1 and Q_2 profiles for category 4, the trough axis, are shown in Fig. 6.7, together with other estimates from the tropics. Using the observed Q_1 and Q_2 profiles and a climatological radiational heating profile by Dopplick (1970) for Q_R, Cho and Ogura (1974) estimated the spectral distribution of the vertical mass flux associated with convective clouds at cloud base. Figure 6.8a shows the distribution of vertical mass flux at cloud base (m_B) as a function of cloud-top height for category 4 of the wave. A bimodal cloud distribution exists, with one group of clouds reaching the 300-mbar level and another population of shallow clouds not extending higher than about 850 mbar.

The vertical profiles of the vertical mass flux due to all clouds (M_c), the mass flux in the environment ($\tilde{M}$), and the large-scale average mass flux

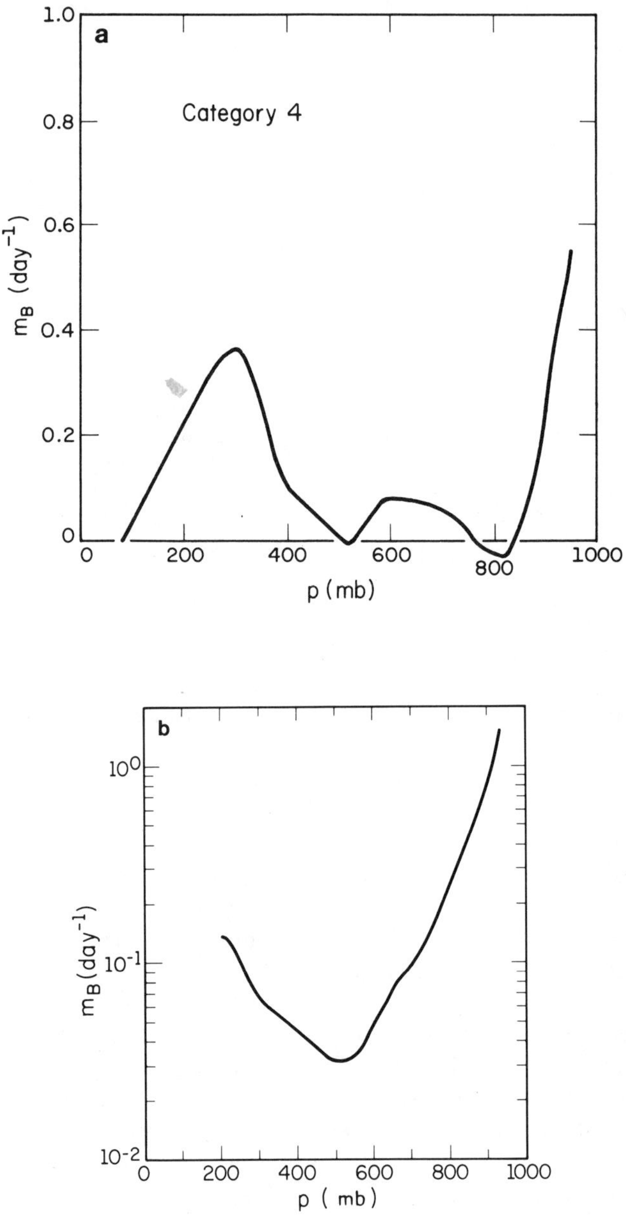

Fig. 6.8. (a) Spectral distribution of vertical mass flux at cloud base (m_B) as a function of cloud-top height (p) for category 4 of Pacific easterly waves. (b) Spectral distribution of vertical mass flux at cloud base as a function of cloud-top height for cloud clusters over Marshall Islands. [From Ogura and Cho (1973).]

($\bar{M}$) for category 4 are shown in Fig. 6.9. These profiles are typical of those found in convective regions in the tropics. They show that the cloud-mass flux exceeds the large-scale average mass flux, with the difference associated with compensating subsidence in the environment.

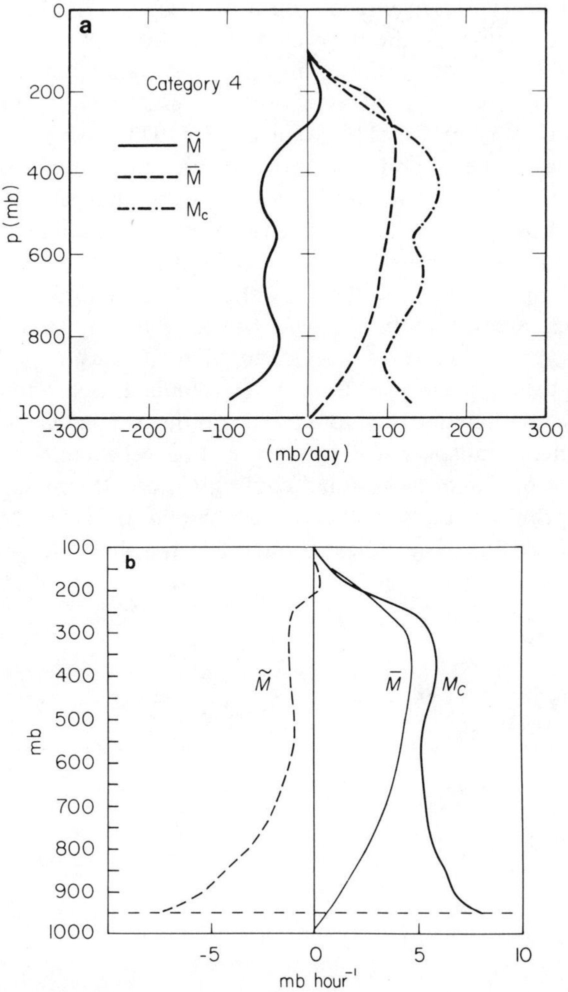

Fig. 6.9. (a) Vertical distributions of vertical mass flux due to all clouds (M_c), mass flux in environment ($\tilde{M}$), and large-scale mass flux ($\bar{M}$) for category 4 of Pacific easterly waves. [From Cho and Ogura (1974).] (b) Average cloud mass flux M_c, large-scale mass flux $\bar{M}$, and residual mass flux $\tilde{M}$ associated with cloud clusters over Marshall Islands. [From Yanai *et al.* (1973).]

The solutions for the vertical mass flux associated with different cloud types in the eight wave categories reveal two important points. First, while shallow clouds are present in all categories, deep clouds occur in significant numbers only near the trough axis. Second, although the total mass flux associated with all clouds is not strongly related to the low-level, large-scale convergence, the mass flux in the deep clouds increases linearly with an increase of low-level convergence. Both of these results indicate a strong synoptic-scale control on the deep cumulus clouds.

Important diagnostic studies using the Marshall Islands data include those by Nitta (1972), Yanai *et al.* (1973, 1976, 1982), Ogura and Cho (1973), and Chu *et al.* (1981). Yanai *et al.* (1973) computed large-scale budgets of mass, heat, and moisture for 366 individual cases over the Marshall Islands area. They used the large-scale analyses of temperature and mixing ratio to estimate the vertical profiles of apparent heat source (Q_1) and the apparent moisture sink (Q_2) from Eqs. (6.16) and (6.17). The ensemble averages of these profiles, together with profiles from other studies in the tropics, are shown in Fig. 6.7. In the Marshall Islands study, the averaged Q_1 profile shows a maximum of 6.4°C/day at 475 mbar. The maximum in the apparent moisture sink Q_2 is much lower (about 775 mbar). The difference is related to the radiational cooling profile Q_R and the vertical flux of eddy heat transport $F = -(\overline{\omega''h''})/g$ (Eqs. 6.18 and 6.19). Use of the climatological profile of radiational cooling Q_R by Dopplick (1970) yields the vertical profile of eddy heat transport shown in Fig. 6.10. This profile shows an upward flux of heat throughout the troposphere, a result of sensible

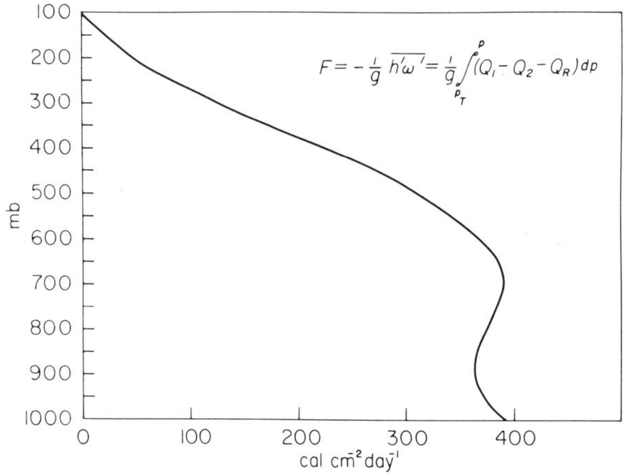

Fig. 6.10. Derived vertical eddy heat flux associated with cloud clusters over Marshall Islands. [From Yanai *et al.* (1973).]

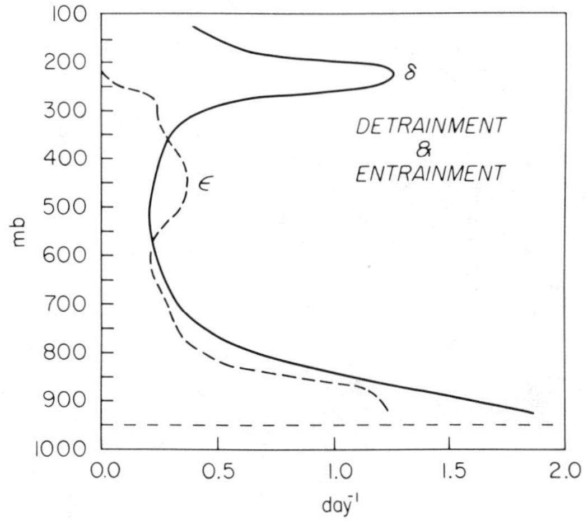

Fig. 6.11. Average detrainment δ and entrainment ε associated with cloud clusters over Marshall Islands. [From Yanai *et al.* (1973).]

and latent heat at the sea surface, turbulent eddies in the subcloud layer, and cumulus clouds in the free atmosphere.

The average cloud-mass flux M_c, together with the large-scale average mass flux $\bar{M}$ and the environmental mass flux M_c computed from the cloud ensemble model of Section 6.3.3, shows similar vertical variations to those associated with the composite easterly wave (Fig. 6.9). The diagnosed vertical profiles of entrainment and detrainment, averaged over all cloud types, are shown in Fig. 6.11. Because of the coexistence of many different clouds, both entrainment and detrainment occur at all levels. The greatest entrainment occurs in the lower levels and is associated with shallow clouds. The two maxima in the detrainment profile are associated with many shallow clouds that detrain just above cloud base and the deep clouds that detrain near the tropopause.

The heat balance diagnosed by the cumulus ensemble model is shown in Fig. 6.12. In this model, the heat source is associated mainly with diabatic warming due to compensating subsidence $[-M_c(\partial\tilde{s}/\partial p)]$ in the environment of the cloud. Warming of the environment due to detrainment of a warmer cloud occurs only at the uppermost cloud layer. Evaporative cooling associated with detrained cloud drops is most pronounced in the lower troposphere.

The moisture balance is shown in Fig. 6.13. The apparent moisture sink (Q_2) is determined largely by the drying associated with compensating subsidence $[-M_c(\partial\tilde{r}/\partial p)]$ in the middle and upper troposphere. In the

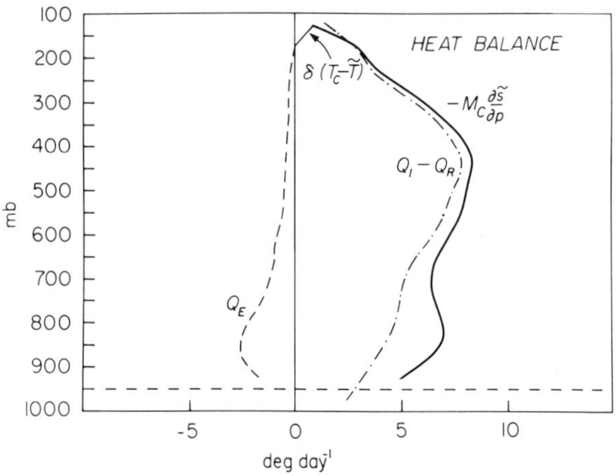

Fig. 6.12. Observed heat source $Q_1 - Q_R$, adiabatic heating by the compensating downward mass flux $-M_c(\partial\tilde{s}/\partial p)$, evaporative cooling $-Q_E = -Le$, and detrainment of heat $\delta(T_c - \tilde{T})$ associated with average cloud clusters over the Marshall Islands. [From Yanai *et al.* (1973).]

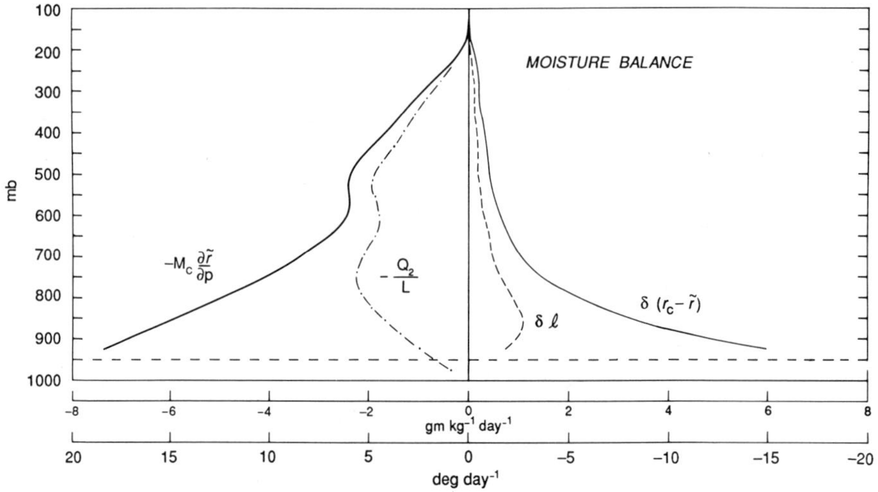

Fig. 6.13. Observed moisture source $-Q_2/L$, drying due to compensating sinking motion $-M_c(\partial\tilde{r}/\partial p)$, detrainment of water vapor $\delta(r_c - \tilde{r})$, and detrainment of liquid water δl, in units of $g\,kg^{-1}\,day^{-1}$ and in equivalent heating units ($°C\,day^{-1}$) associated with average cloud clusters over the Marshall Islands. [From Yanai *et al.* (1973).]

lower troposphere, this drying is offset to a considerable extent by evaporation of detrained cloud water and detrainment of moist cloud air.

Using vertical profiles of Q_1 and Q_2 as computed by Nitta (1972) from time-averaged data over the Marshall Islands (Fig. 6.7), Ogura and Cho (1973) applied a model of a cumulus ensemble to determine the properties of different types of clouds and their contributions to the changes of heat and moisture content of the large-scale environment. Unlike the cumulus ensemble model described by Yanai *et al.* (1973), which only estimated the averaged properties of clouds, the Ogura–Cho model solves for the properties of different types of clouds that comprise the cloud spectrum. A key element of this model is a one-dimensional cloud model in which the properties of a single cloud are uniquely determined by an entrainment parameter. Nitta (1975) developed a similar model of the cloud spectrum. As discussed by Yanai *et al.* (1976), the spectral model has several advantages over the bulk model. In addition to giving the bulk properties, it gives information about individual cloud types and the solution does not require iteration.

The spectral distribution of vertical mass flux at cloud base, plotted as a function of cloud top, is shown in Fig. 6.8b. A bimodal distribution, similar to that found in the trough region of the composite easterly wave is apparent; one group of deep clouds reaches the upper troposphere while another group of shallow clouds remains in the lower troposphere. The vertical distributions of the upward vertical mass flux due to all clouds, the downward mass flux between the clouds, and the large-scale mass flux are similar to those profiles derived by Yanai *et al.* (1973) using the bulk model.

The contributions of various processes to the large-scale heat and moisture budgets from the spectral cloud model are shown in Figs. 6.14 and 6.15. As in the budgets of the composite tropical wave (Figs. 6.12 and 6.13), the heat budget is dominated by warming in the compensating subsidence between clouds, with some cooling due to evaporation of detrained cloud water. The moisture budget represents a balance between drying in the environmental subsidence and moistening associated with detrained cloud liquid water and water vapor.

The vertical profiles of cloud environment temperature $(\bar{\bar{T}}_c - \tilde{T})$ and mixing ratio $(\bar{\bar{r}}_{vc} - \tilde{r}_v)$ differences for two entrainment rates λ, as diagnosed by the Ogura–Cho model, are presented in Figs. 6.16 and 6.17. The entrainment parameter is defined as

$$\lambda \equiv (1/m_i)(\partial m_i/\partial z), \tag{6.54}$$

where m_i is the mass flux in the ith cloud type. Also shown in Figs. 6.16 and 6.17 are the weighted average (over all clouds) profiles of temperature and mixing ratio excess, where the weighted average is defined in Eq. (6.38).

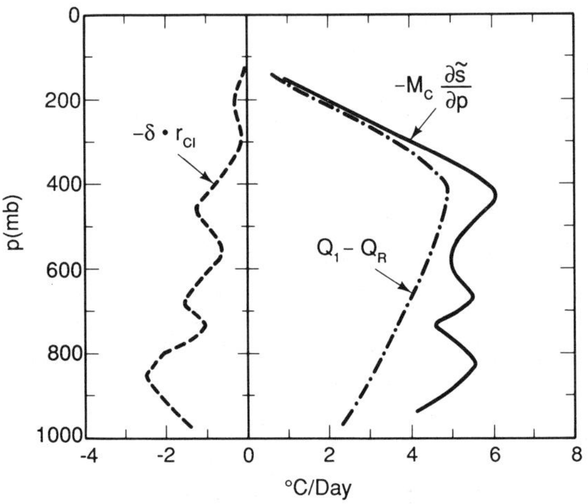

Fig. 6.14. Vertical distribution of contributions of various physical processes to large-scale heat balance of average cloud clusters over the Marshall Islands. Adiabatic heating due to compensating subsidence between clouds is represented by $-M_c(\partial\tilde{s}/\partial p)$; cooling due to evaporation of detrained cloud water is represented by δr_{cl}. [From Ogura and Cho (1973).]

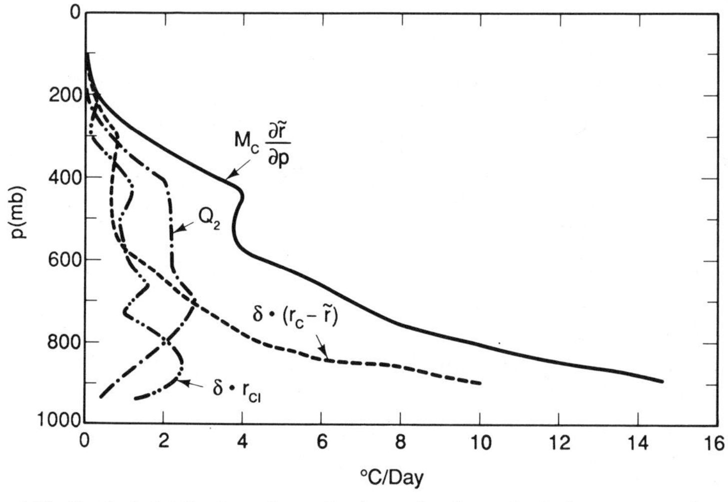

Fig. 6.15. Vertical distribution of contributions of various physical processes to large-scale moisture balance of average cloud clusters over the Marshall Islands. The drying effect due to compensating downdrafts between clouds is represented by $M_c(\partial\tilde{r}/\partial p)$, the moistening effect due to detrained cloud water and cloud vapor by δr_{cl} and $\delta(r_c - \tilde{r})$, and the apparent moisture sink by Q_2. [From Ogura and Cho (1973).]

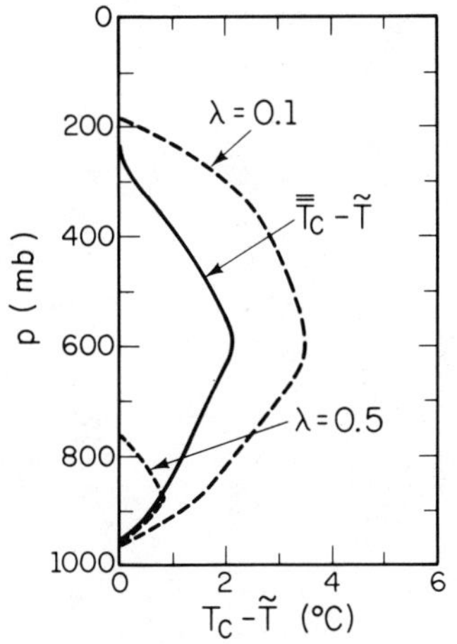

Fig. 6.16. Vertical distributions of excess temperature in clouds for two different entrainment rates (λ) as diagnosed by Ogura–Cho cloud model and excess temperature averaged over all clouds for average cloud clusters over the Marshall Islands. [From Ogura and Cho (1973).]

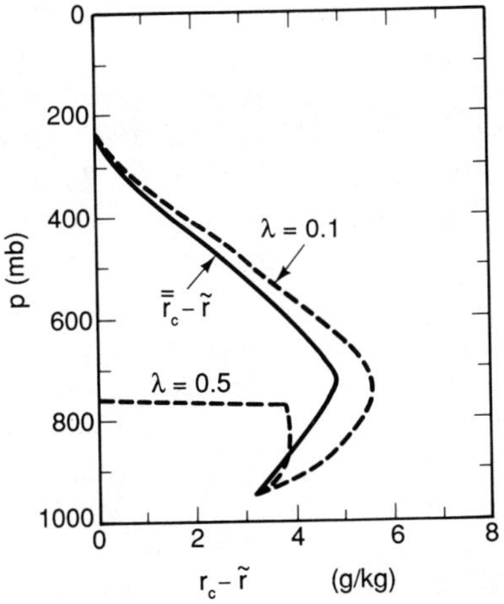

Fig. 6.17. Same as Fig. 6.16 except for excess water vapor mixing ratio. [From Ogura and Cho (1973).]

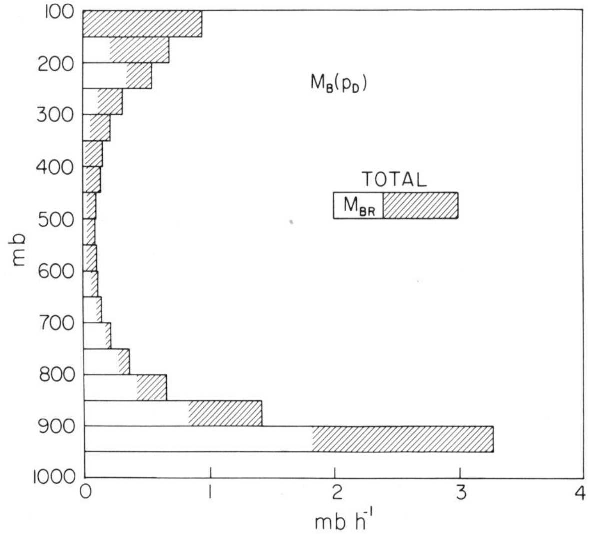

Fig. 6.18. Mean cloud-base mass flux M_B as a function of detrainment pressure p_D for average cloud cluster over the Marshall Islands. Unshaded portion of the bars shows the background mass flux induced by radiative cooling M_{BR}. [From Yanai *et al.* (1976).]

This average shows a maximum temperature excess of 2.2°C at 600 mbar and a maximum excess mixing ratio of 4.8 g kg^{-1} at 780 mbar.

In a further study of the Marshall Islands data, Yanai *et al.* (1976) used Nitta's (1975) spectral diagnostic model of a cumulus ensemble to estimate the spectrum of cloud-mass flux. The vertical profiles of average M_c, $\tilde{M}$, and $\bar{M}$ estimated from the spectral model agreed closely with those from the bulk method used by Yanai *et al.* (1973). The spectral distribution of cloud base mass flux M_B as a function of the detrainment pressure p_D is shown in Fig. 6.18. The unshaded portion of the bars indicates the contribution to the cloud-mass flux by the radiative cooling profile Q_R. It indicates that the bimodal character of the cloud distribution is, to a large extent, determined by the radiative cooling profile.

In order to examine the relationship between the mass fluxes associated with various types of clouds on the large-scale mass flux at different levels $\bar{M}$(p), Yanai *et al.* (1976) calculated the correlation coefficients between the mass flux at cloud base associated with each cloud type, $M_B(p_D)$ and $\bar{M}(p)$. The results, displayed in Fig. 6.19, indicate a positive correlation of the mass flux associated with deep clouds and the large-scale vertical motion. The deep clouds show the highest correlation (~0.6) with the mean vertical motion in the upper troposphere. Figure 6.19 also indicates that the shallow

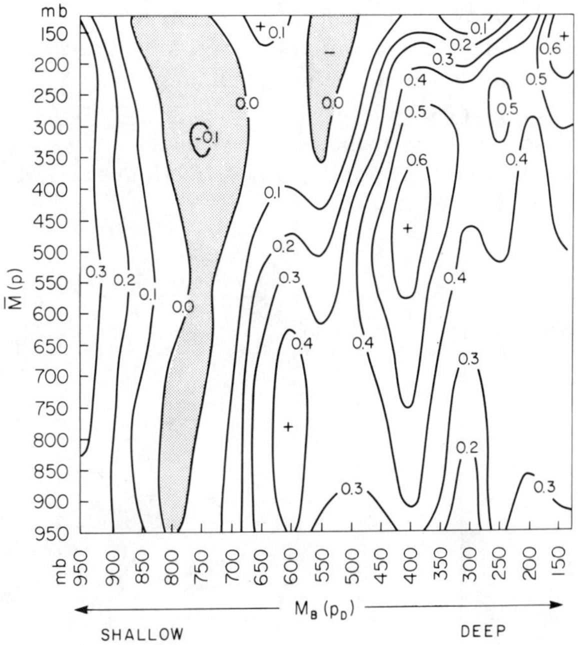

Fig. 6.19. Correlation coefficients between time series of large-scale mass flux $\bar{M}(\mathrm{p})$ at all levels and cloud-base mass flux M_B of various detraining pressures p_D. [From Yanai *et al.* (1976).]

clouds are not well correlated with the large-scale vertical motion, and indicates, as discussed in Chapter 8, that a successful parameterization of shallow cumulus is likely to be quite different from one for deep cumulus.

The heat and moisture budgets computed by various investigators using different methods from the time-averaged Marshall Islands data show consistency and are relatively easy to interpret. In contrast, the effect of cumulus convection on the large-scale wind field is more difficult to compute and interpret. If a momentum budget is used, it is necessary to compute the horizontal pressure-gradient force term, which is very difficult to estimate accurately. Therefore, most diagnostic studies compute a vorticity budget, Eq. (6.7), which eliminates the pressure-gradient force calculation but introduces the difficulty associated with computing second derivatives of the wind field and the complicated twisting term. Interpretation of the vorticity budgets and the effects of cumulus clouds are also more complicated than for the heat and moisture budgets because the vertical component of vorticity is not a conservative quantity.

In spite of the above difficulties, progress has been made in diagnosing the effects of cumulus convection on the large-scale vorticity field. Studies in the western Pacific have shown that cumulus convection generates a large apparent source of positive vorticity in the upper troposphere and an apparent sink in the lower troposphere (Williams and Gray, 1973; Reed and Johnson, 1974; Ruprecht and Gray, 1976; Hodur and Fein, 1977).

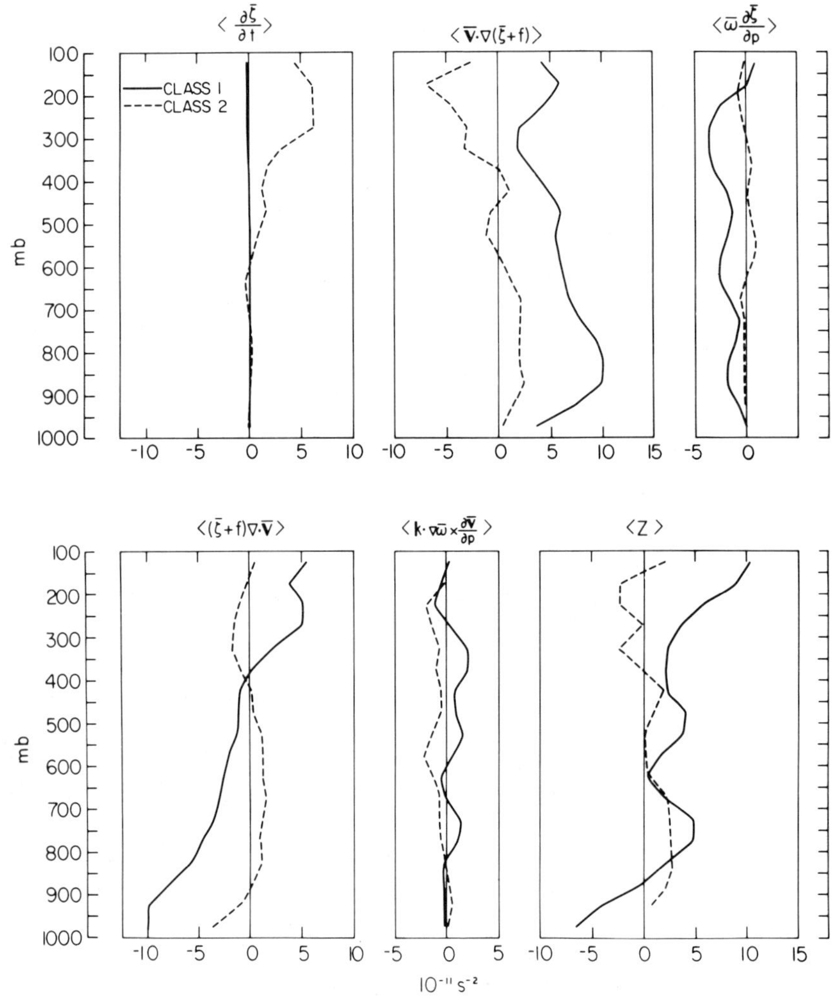

Fig. 6.20. Mean profiles of each large-scale term in the vorticity equation and budget residual Z for disturbed (solid curves) and undisturbed (dashed curves) cases over the Marshall Islands (units $10^{-11}\ s^{-2}$). [From Chu *et al.* (1981).]

The vertical profiles of the terms in the vorticity budget for convectively disturbed (upward motion between 300 and 450 mbar) and undisturbed (downward motion between 300 and 450 mbar) conditions over the Marshall Islands, computed by Chu *et al.* (1981), are shown in Fig. 6.20. During the period from 15 April to 22 July 1956, there were 352 analyses belonging to the disturbed class and 31 cases in the undisturbed category. The profiles are calculated by averaging the respective terms over all cases in the ensemble. Thus, if the average over all samples is denoted by the operator $\langle\ \rangle$, the mean horizontal advection of absolute vorticity is

$$\langle \mathbf{V} \cdot \nabla \bar{\zeta}_a \rangle = \frac{1}{N} \sum_{i=1}^{N} (\bar{\mathbf{V}} \cdot \nabla \bar{\zeta}_a)_i, \tag{6.55}$$

where i denotes the sample and the overbar denotes the large-scale average. Because of the nonlinearity of the terms such as vorticity advection, the ensemble mean of the terms does not equal the value of the terms computed from the mean or composite data set, e.g.,

$$\langle \bar{\mathbf{V}} \cdot \nabla \bar{\zeta}_a \rangle \neq \langle \bar{\mathbf{V}} \rangle \cdot \nabla \langle \bar{\zeta}_a \rangle. \tag{6.56}$$

However, as shown by a comparison of the vorticity advection, twisting, and residual terms computed by the two methods in Fig. 6.21, the ensemble

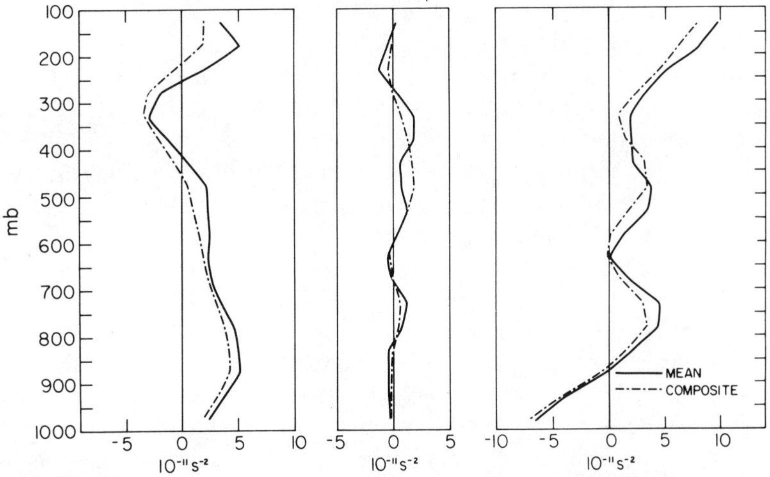

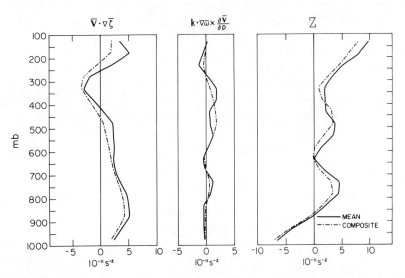

Fig. 6.21. Comparison between exact mean values (solid curves) and mean values obtained from composite data (dashed–dotted curves) for horizontal advection of relative vorticity (left), twisting term (middle), and budget residual Z (right) from the Marshall Islands data (units $10^{-11}\ \mathrm{s}^{-1}$). [From Chu *et al.* (1981).]

mean of the terms and the terms computed from the composite analysis are similar.

As indicated in Fig. 6.21, the largest terms in the vorticity budget for the disturbed cases are the horizontal advection term, which is positive throughout the troposphere, the divergence term, which is negative below 400 mbar and positive above, and the residual term Z, which is assumed to be dominated by cumulus effects. The Z profile indicates that the cumulus clouds create an apparent sink of vorticity in the lower troposphere and an apparent source of vorticity in the upper troposphere. Reed and Johnson (1974) and Yanai *et al.* (1982) interpret this profile as a result of upward transport of high values of relative vorticity in cumulus convection and detrainment of excess vorticity into the environment, as well as twisting and stretching of the large-scale vorticity by the compensating environmental subsidence. In a later study, Sui and Yanai (1984) found that detrainment was the main process in producing the upper-level source of vorticity.

6.4.1.b GATE Studies

The primary objective of GATE was to improve the basic understanding of tropical convection and its role in larger scale atmospheric circulations. Houze and Betts (1981) provide a review of the history of GATE and the observing systems used during the 1974 field experiment. Intensive diagnostic studies of the tropical convective systems observed during GATE followed the experiment (Reed *et al.*, 1977; Nitta, 1977; Cho *et al.*, 1979; Thompson *et al.*, 1979; Reeves *et al.*, 1979; Ogura *et al.*, 1979; Esbensen *et al.*, 1982; Tollerud and Esbensen, 1983).

Many of the properties of cumulus convection in the tropical Pacific were also found in the GATE data. For example, deep convection was strongly modulated by synoptic-scale disturbances, with deep convection occurring only during periods of large-scale upward motion. Also, there was close agreement between the observed precipitation and that diagnosed from the large-scale water vapor budget (Fig. 6.22).

Although the general appearance of the apparent heat source (Q_1) and apparent moisture sink (Q_2) profiles in the eastern Atlantic is similar to those of the western Pacific (Fig. 6.23a), there are significant differences. In particular, the Q_1 and Q_2 profiles show maxima at considerably lower levels in the Atlantic and smaller magnitudes in the upper troposphere. These differences are related to stronger low-level convergence (below 850 mbar) in the Atlantic. The differences in the Q_1 and Q_2 profiles between the Atlantic and Pacific imply differences in the vertical eddy heat flux, as shown in Fig. 6.23b.

An important advancement in the understanding of tropical convection since GATE has been the recognition that mesoscale downdrafts are important in the mass, water vapor, and heat budgets of tropical systems. These downdrafts, which have a characteristic horizontal scale of about 50 km, appear to be driven by evaporative cooling of liquid water and melting of ice (Zipser, 1977). Johnson (1980) extended the diagnostic spectral cloud model approach by including the effect of mesoscale downdrafts. In his model, both the cloud and the mesoscale downdrafts are related to the updraft mass flux by constants of proportionality. The constants are determined by matching the observed and diagnosed rainfall. Figure 6.24 shows the mass flux associated with the large-scale motion ($\bar{M}$), the convective-scale mass flux (without any moist downdrafts, with cumulus-scale downdrafts only, and with both cumulus-scale and mesoscale downdrafts), and the mass flux associated with dry air in the environment of the clouds. The main effect of including the moist downdrafts is to bring the net cloud-mass flux profile closer to that of the large-scale average. Nitta (1977) obtained a similar result.

Figure 6.25 shows the mean profiles of the apparent vorticity source (Z) over a 7°×7° array for undisturbed and disturbed periods of Phase III (31 August–18 September 1974) of GATE (Sui and Yanai, 1986). Similar profiles were obtained by Reeves *et al.* (1979). The small magnitudes of Z

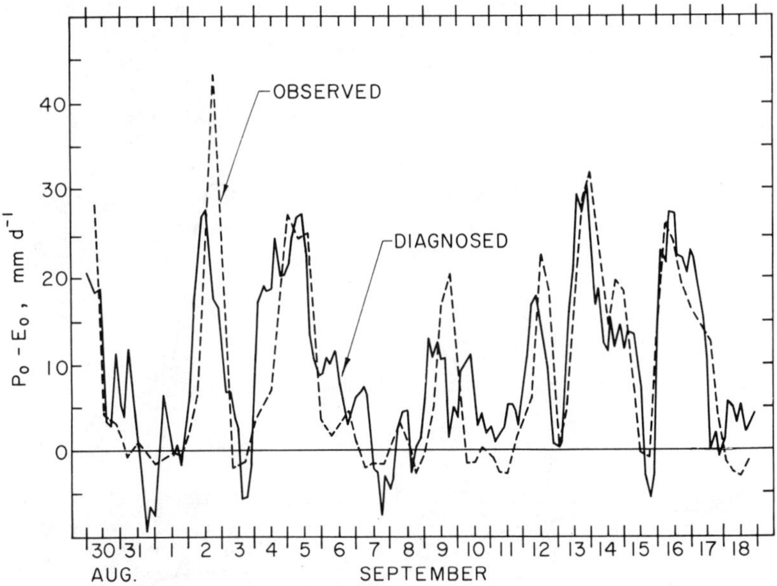

Fig. 6.22. Time series for GATE Phase III of diagnosed (solid curve) and observed (dashed curve) precipitation minus evaporation. [From Thompson *et al.* (1979).]

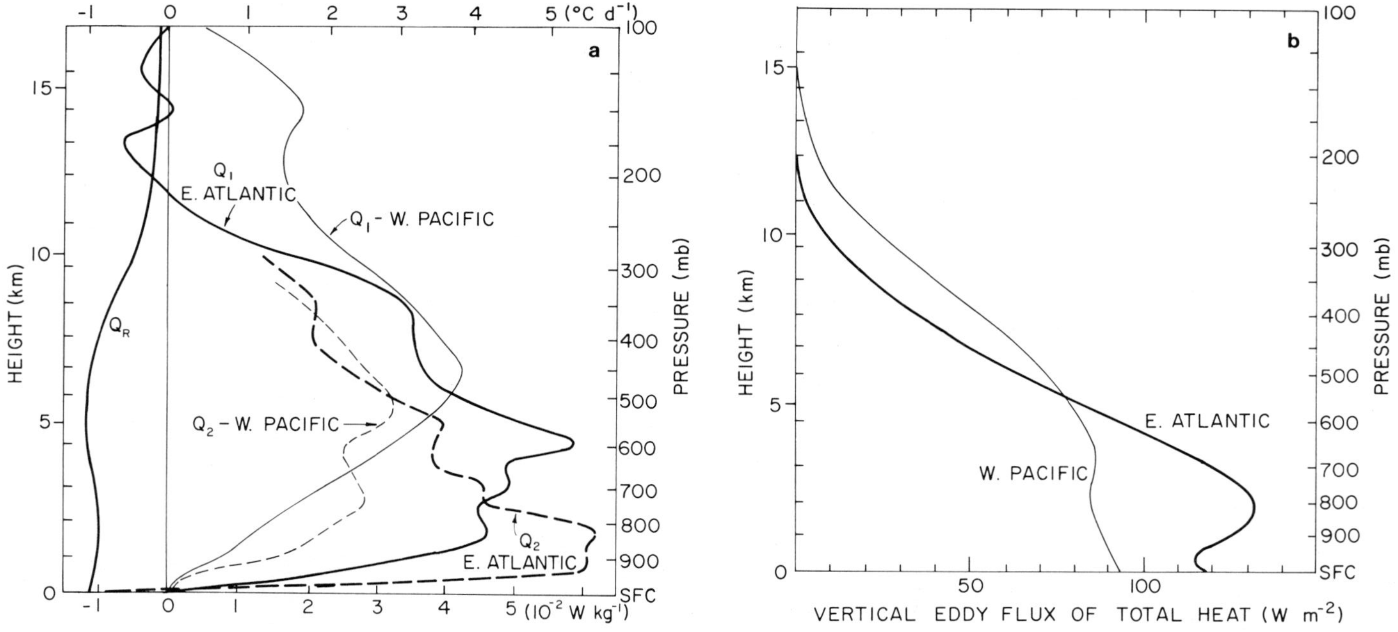

Fig. 6.23. Variation with height of apparent sensible heat source Q_1 and apparent latent heat sink Q_2; (a) mean radiational heating Q_R and (b) vertical eddy flux of moist static energy for GATE B-scale area and west Pacific triangle. [From Thompson *et al.* (1979).]

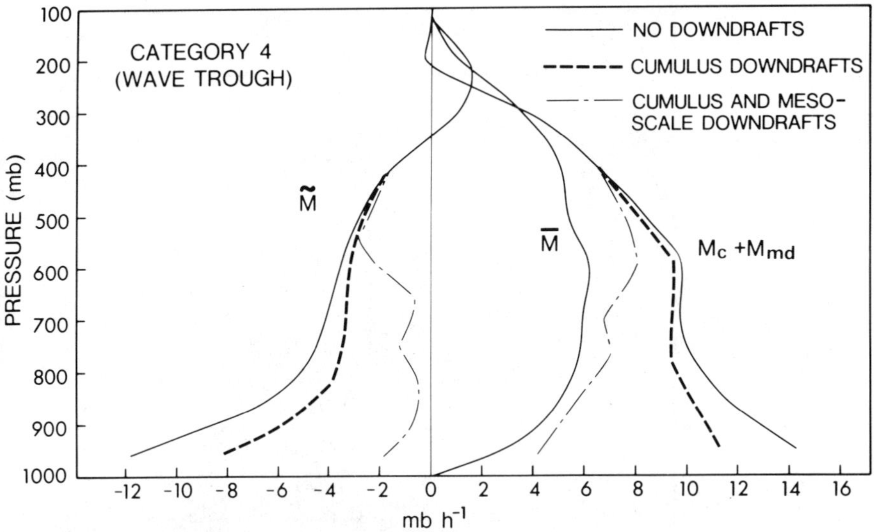

Fig. 6.24. Environmental mass flux $\tilde{M}$, mean mass flux $\bar{M}$, and net convective mass flux $M_c + M_{md}$ for wave trough in GATE Phase III for the case with and without downdrafts. [From Johnson (1980).]

in undisturbed periods compared to disturbed periods clearly indicate the importance of moist convection on the vorticity budget.

The apparent vorticity source in the upper troposphere during disturbed periods of GATE is consistent with the interpretation of Reed and Johnson (1974) and Yanai *et al.* (1982) that detrainment of vorticity from deep cumulus convection occurs at this level. The negative peak (vorticity sink) at about 400 mbar (Fig. 6.25) is a peculiar characteristic of the GATE area; it is not found in the Pacific data (Fig. 6.21). Sui and Yanai (1986) attribute this sink to cumulus-induced subsidence in a region of large-scale negative vertical shear of the relative vorticity. In the GATE region, a region of maximum positive vorticity exists at 650 mbar while strong anticyclonic vorticity is present at 200 mbar.

The negative values of Z (apparent vorticity sink) below 900 mbar in both the disturbed and undisturbed periods (Fig. 6.25) indicate the effect of turbulence in the PBL in destroying vorticity.

Given the vertical profiles of Z, Sui and Yanai (1986) estimated the effect of cumulus clouds on the rotational part of the horizontal wind. They showed that cumulus clouds, carrying weaker horizontal momentum from the lower troposphere, decelerated the flow aloft and tended to reduce the large-scale vertical shear—a down-gradient transport of horizontal momentum.

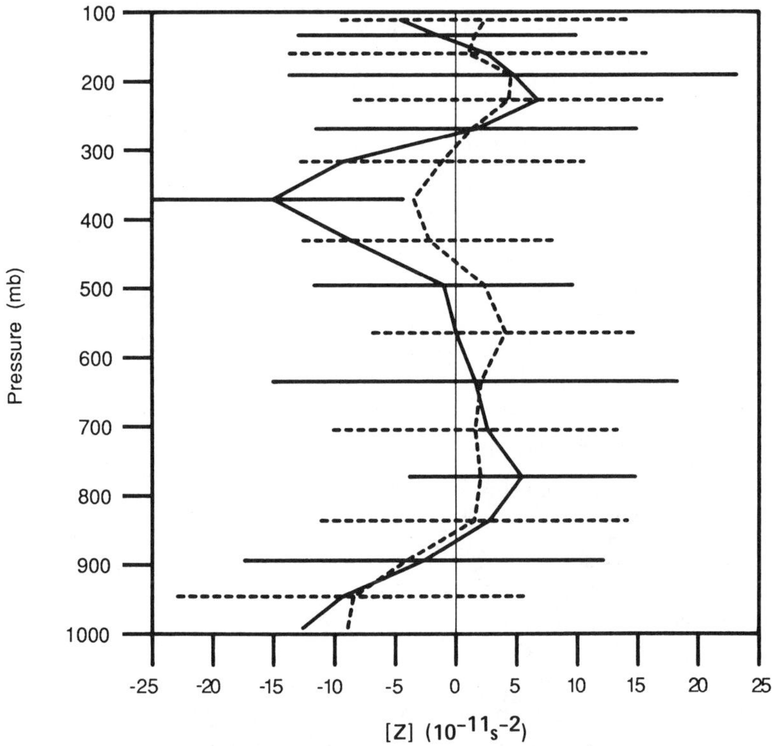

Fig. 6.25. Mean apparent vorticity source (10^{-11} s^{-2}) for disturbed (solid curves) and undisturbed (dashed curves) conditions of GATE Phase III with standard deviations at every other data level. [From Sui and Yanai (1986).]

The complexity of diagnosing the interactions between the cumulus clouds and the larger scale on the vorticity budget is also present when momentum budgets are considered, since the momentum equation contains the pressure-gradient term, which is difficult to calculate accurately. There is observational evidence that the vertical transport of horizontal momentum by cumulus convection can be either down-gradient (Lee, 1984; Sui and Yanai, 1986) or countergradient (LeMone, 1983). A striking example of countergradient transport is shown in Fig. 6.26. Figure 6.26a shows the average vertical profiles of u and v, and Fig. 6.26b shows the average profiles of the vertical flux of u and v momentum. Typical down-gradient parameterizations for momentum flux are written as

$$\overline{u''w''} = -K_{\mathrm{m}}(\partial\bar{u}/\partial z), \tag{6.57}$$

where K_m is a positive eddy exchange coefficient for momentum. As shown by Fig. 6.26, in the layer from 1 to 5 km, $\partial \bar{u}/\partial z$ is negative while $\overline{u''w''}$ is also negative, contrary to Eq. (6.57).

Asai (1970) showed that the sign of the vertical momentum flux by lines or rolls of cumulus convection depends on the orientation of the lines relative to the basic flow. When the bands are essentially parallel to the flow, the momentum flux is down-gradient. However, when the flow is perpendicular to the band, the momentum flux is countergradient. The observational studies of LeMone (1983) and Lee (1984) support this theory.

Many of the budget studies in the tropics have used time-averaged data, or composite data, in which the effects of random errors are largely removed. Computation of individual case studies is more difficult because random errors are not removed by averaging a large number of data. GATE, however,

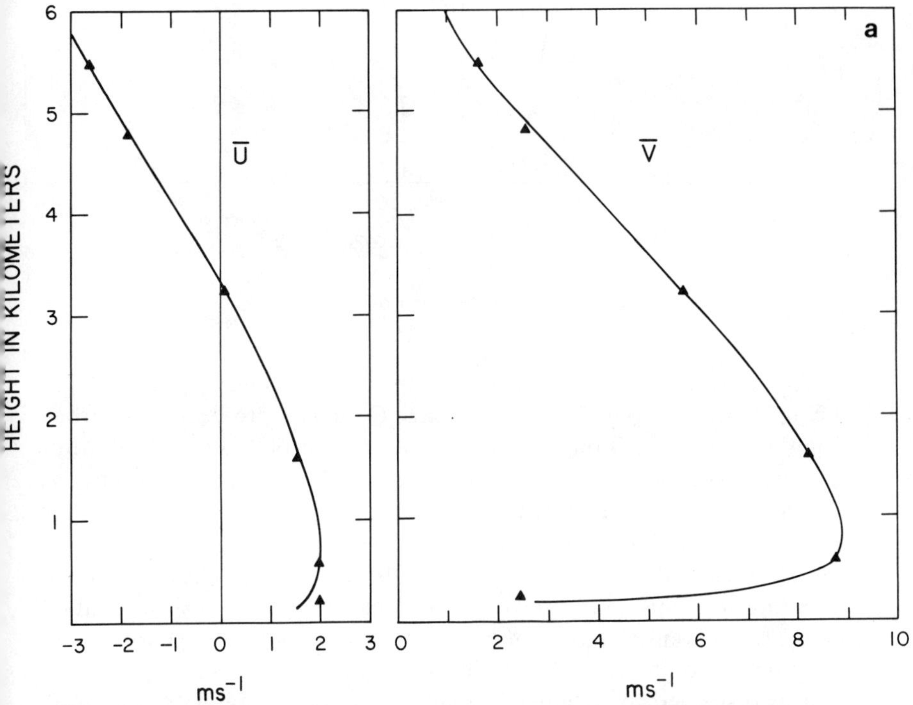

Fig. 6.26. (a) Vertical profiles of wind components $\bar{U}$ and $\bar{V}$, averaged in a 140-km east–west plane across north–south convective line of Day 257 of GATE. $\bar{U}$ is positive east, in direction of line motion; $\bar{V}$ is positive north. Values at each level are adjusted to 1330 GMT. (b) Profiles of vertical flux of horizontal momentum $\overline{\rho u'w'}$ and $\overline{\rho v'w'}$, in same coordinate system as for (a). Error bars denote standard deviations at each level. [From LeMone (1983).] (*Figure continues.*)

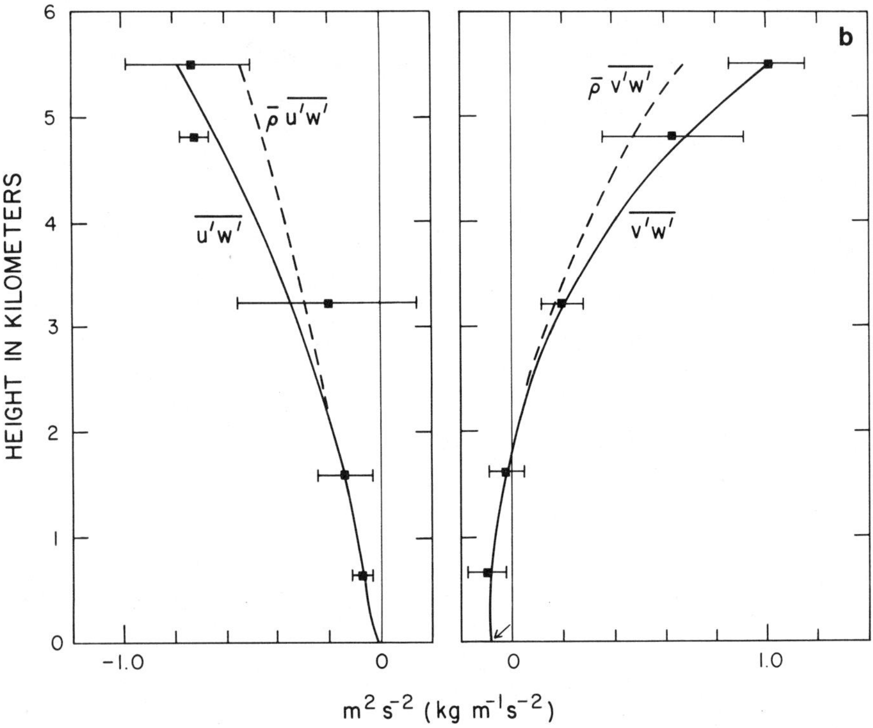

Fig. 6.26 (*continued*)

provided data sets of sufficient quality to calculate meaningful budgets. Figure 6.27, for example, shows the average Q_1 and Q_2 profiles for disturbed and undisturbed conditions during a 3-day period of GATE (0000 GMT 7 September to 0000 GMT 10 September 1974), as computed by Cho *et al.* (1979). During the undisturbed (light or no precipitation) portion of the 3-day period, cumulus clouds heated and moistened the troposphere below about 700 mbar and cooled and moistened the troposphere above 500 mbar. The cooling and moistening above 500 mbar could be caused by evaporation of cloud water transported from below; however, there is a possibility of large errors in computing a budget over short time periods.

During disturbed conditions, the cumulus clouds heat and dry the entire troposphere between 900 and 200 mbar. A difference in the Q_1 and Q_2 profiles for this individual case compared to the composite studies is that the Q_1 and Q_2 profiles are quite similar, with maxima at about 600 mbar.

In another case study covering the period 0600 GMT 5 September to 0000 GMT 7 September 1974, Nitta (1977) computed large-scale mass,

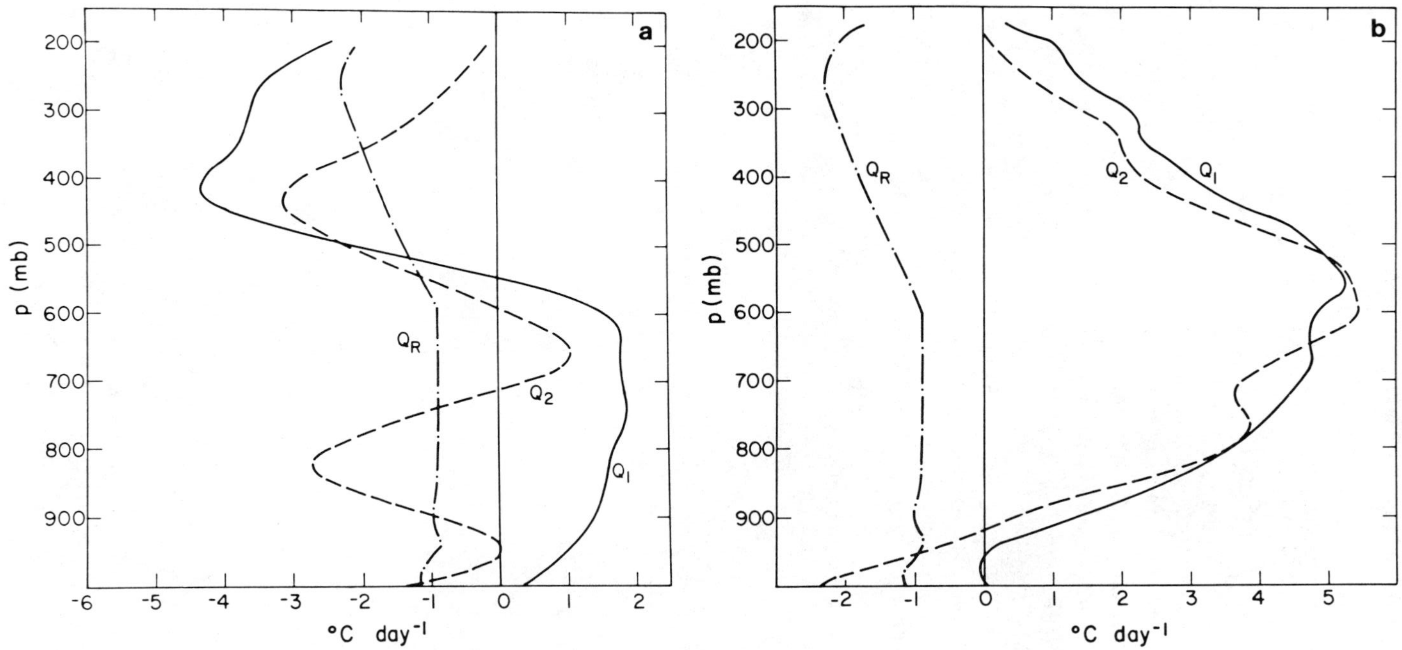

Fig. 6.27. (a) Apparent heat source Q_1, apparent moisture sink Q_2, and radiative heating rate Q_R profiles for undisturbed conditions over a 3-day period of GATE. (b) Same as (a) except for disturbed period. [From Cho *et al.* (1979).]

moisture, and heat budgets for three conditions, classified according to the degree of convective activity. In the first class, the cloud cluster developed and reached maturity. In the second class, the cloud cluster decayed, while in the third class there was no organized convective system. The Q_1, Q_2, and vertical eddy heat flux profiles for these three classes are shown in Figs. 6.28 and 6.29. In contrast to the case study by Cho *et al.* (1979), the Q_1 and Q_2 profiles have maxima at different levels during the mature stage, with Q_1 having a maximum at 400 mbar and Q_2 having a maximum at about 750 mbar. During the decaying stage, Q_1 shows an apparent heat sink in the middle troposphere. The Q_1 and Q_2 profiles during the undisturbed conditions show weak warming and drying throughout most of the troposphere. The vertical eddy heat flux due to cumulus and other unresolvable processes for the three periods indicates a strong upward heat flux during the developing and mature phase of the cloud cluster, a moderate upward flux during the undisturbed conditions, and weaker fluxes during the decaying stage. These individual case studies indicate a large variability in time of the effect of cumulus convective systems for individual tropical systems.

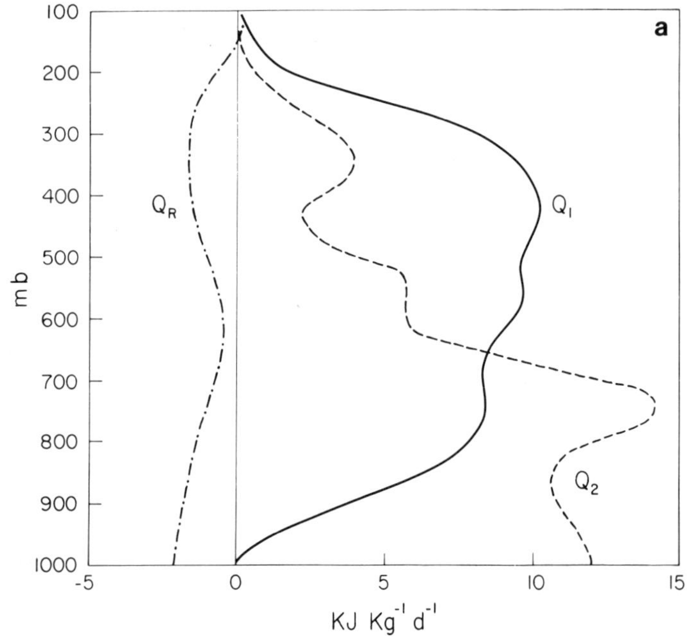

Fig. 6.28. Mean large-scale apparent heat source Q_1, apparent moisture sink Q_2, and radiation heating Q_R for three classes of convective activity during GATE. (a) Developing mature condition, (b) decaying conditions, and (c) no convection. [From Nitta (1977).] (*Figure continues.*)

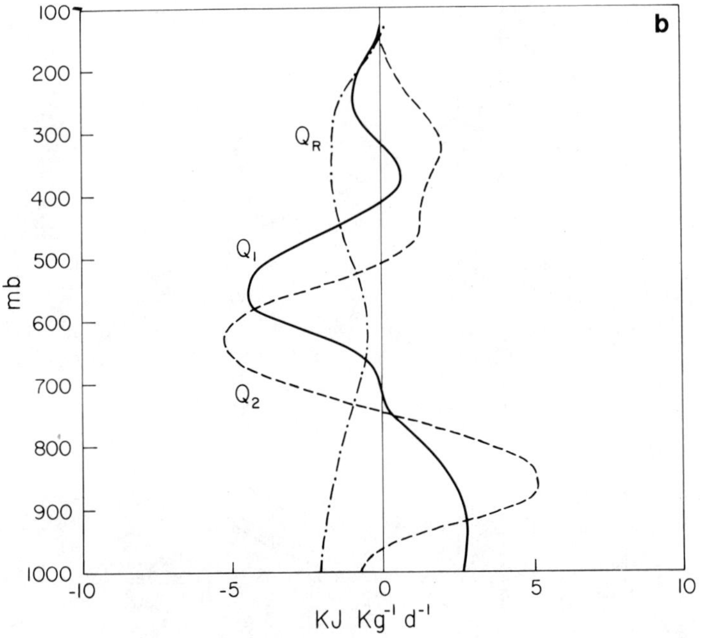

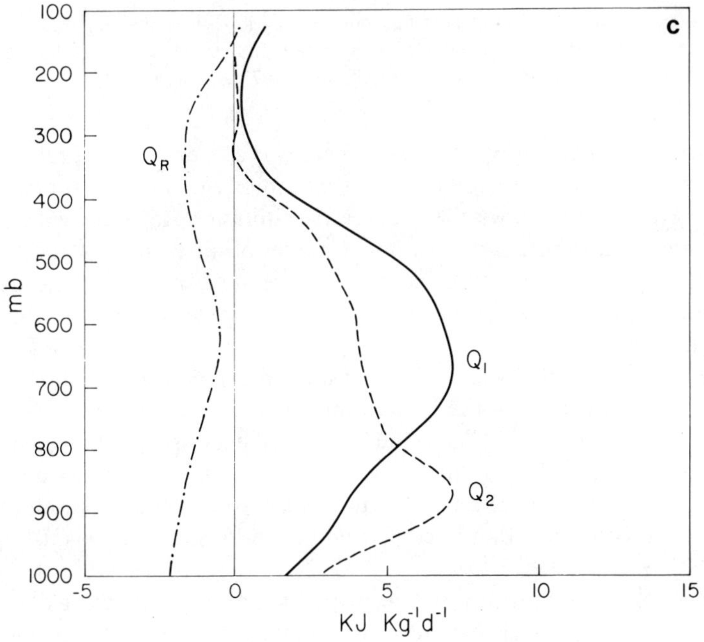

Fig. 6.28 (*continued*)

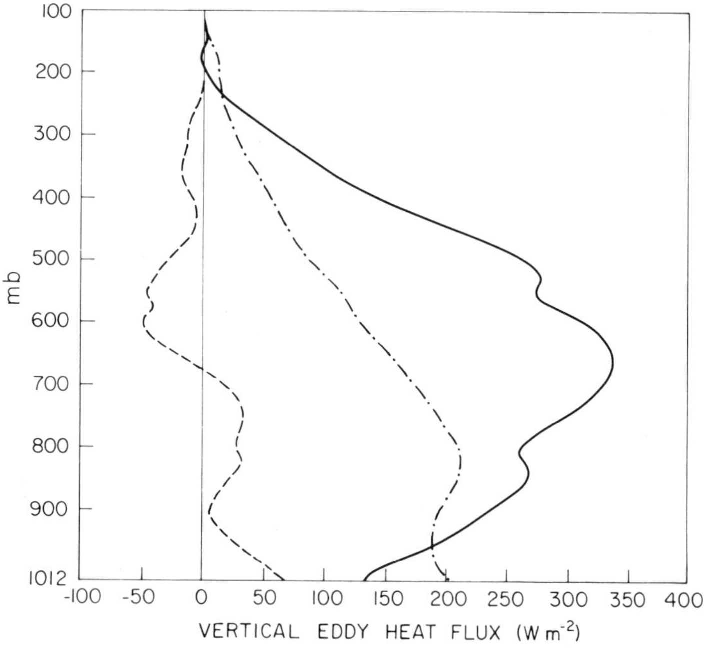

Fig. 6.29. Mean vertical total heat flux due to subgrid-scale eddies for three classes of convective activity during GATE, developing-mature (solid curve), decaying (dashed curve), and no convection (dash-dotted curve). [From Nitta (1977).]

As found in other tropical studies, there was a strong inverse relationship between convective precipitation and thermodynamic instability during GATE. Figure 6.30 shows the average precipitation, together with the net energy released in lifting an undiluted parcel of air from 1000 mbar to the level of zero buoyancy, as a function of the wave position. The heaviest precipitation occurs when the parcel energy, a measure of the convective instability, is at a minimum.

The convection in GATE in general was not as intense, nor did it extend as high as convection in the equatorial Pacific. The vertical eddy flux of total heat $[F = -1/g(\overline{h''\omega''})]$ and the associated heating rate showed maxima lower in the troposphere than in the Pacific (compare Figs. 6.31 and 6.10). The apparent heat source and moisture sink also showed maximum values at lower levels in the Atlantic compared to the Pacific (see Figs. 6.32, 6.33, and 6.23).

The strong relationship between the synoptic-scale waves and the cumulus effects in GATE can be seen in the cross sections of the eddy flux of heat, Q_1, Q_2, in Figs. 6.31–6.33; the cumulus activity is concentrated near, but slightly ahead of, the trough axis. Similar results were obtained by Nitta (1978).

6.4.2 *Diagnostic Studies in the Extratropics*

While considerably more attention has been directed over the past 25 years toward the study of tropical convection and its effect on the environment, compared to extratropical convection, there is considerable evidence that cumulus convection is controlled by synoptic-scale extratropical systems and can, in turn, modify these larger scale circulations. Ninomiya (1971a, b) studied the evolution of severe, tornado-producing thunderstorms in the midwestern United States and inferred a variety of feedbacks between the thunderstorms and the larger scale. In particular, vertical transport of momentum by the cumulus clouds apparently caused a low-level jet (LLJ) to form, and the convergence associated with this LLJ provided moisture to maintain the thunderstorm. Figure 6.34 shows the vertical profile of eddy heat transport in the convective area. The convergence of eddy heat flux in the upper troposphere, together with the net latent heating, helped create a warm core and an associated low-level pressure fall of 4–5 mbar. The upper-level warm core was also associated with the development of anticyclonic outflow in the upper troposphere over the convection. Later observational studies by Maddox *et al.* (1981) and Fritsch and Maddox

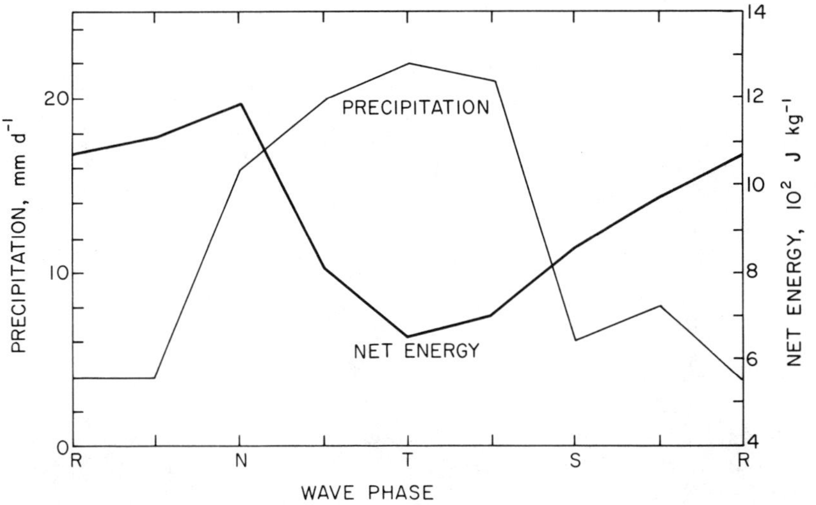

Fig. 6.30. Precipitation rate and net energy released in lifting undiluted parcel from 1000 mbar to the level of zero buoyancy as functions of wave position during GATE. [From Thompson *et al.* (1979).]

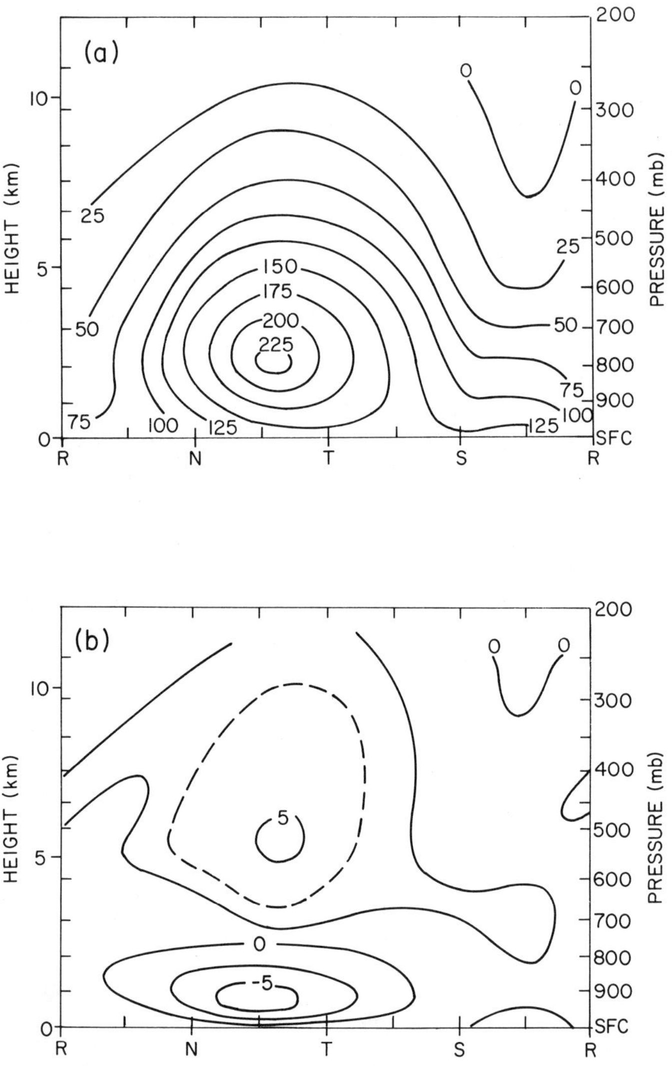

Fig. 6.31. (a) Vertical eddy flux of total heat (W m^{-2}), and (b) its convergence or cumulus heating (10^{-2} W kg^{-1} or approximately equivalent to °C day^{-1}) as functions of wave position during GATE. [From Thompson *et al.* (1979).]

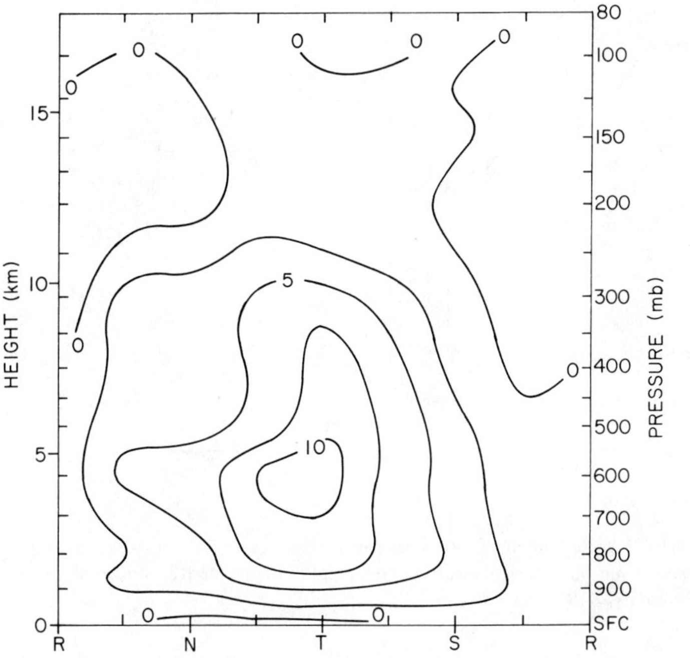

Fig. 6.32. Apparent sensible heat source Q_1 as function of wave position during GATE. Units: 10^{-2} W kg^{-1} (approximately equivalent to °C day^{-1}). [From Thompson *et al.* (1979).]

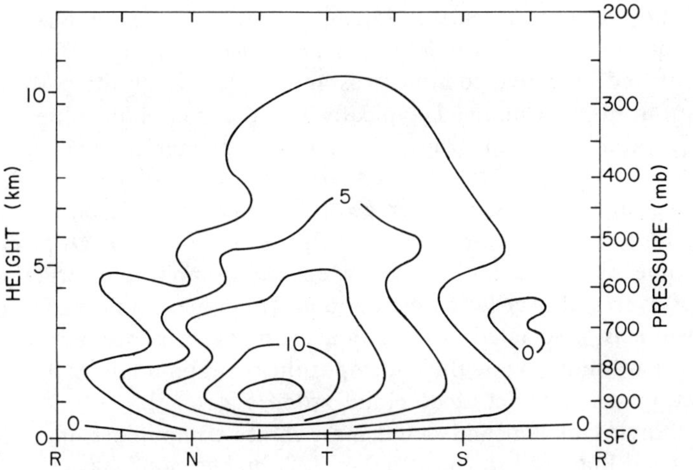

Fig. 6.33. Apparent latent heat sink Q_2 as a function of wave position during GATE. Units: 10^{-2} W kg^{-1}. [From Thompson *et al.* (1979).]

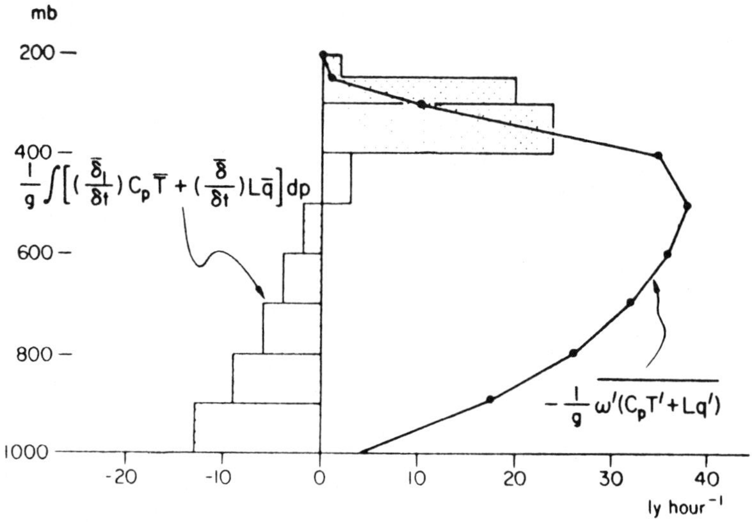

Fig. 6.34. Net increment of total heat energy for mean motion in each layer over a mature thunderstorm complex in midwestern United States and convective transfer of total heat energy. [From Ninomiya (1971a).]

(1981) confirmed the influence of moist convection on the development of anticyclonic perturbations in the upper troposphere (Fig. 6.35).

Lewis (1975) used the diagnostic spectral model of Ogura and Cho (1973) to estimate the convective effects on a mid-latitude squall line in Oklahoma on 8 June 1966. The calculated Q_1 and Q_2 profiles during the mature stage of the squall line are shown in Fig. 6.36. Because the fraction of the analyzed area covered by active cumulus was about 10%, while the corresponding fraction in tropical studies is typically 1%, the maximum values of the Q_1 and Q_2 profiles are an order of magnitude greater than in the tropical studies. The shape of the Q_1 and Q_2 profiles shows considerably more vertical oscillations than the average profiles in the tropics, possibly because of greater errors in the analyses, but also because Q_1 and Q_2 profiles over short time periods are likely to have more complicated structures than profiles averaged over long time periods. The upper-level maximum in the Q_1 profile is associated with compensating subsidence and associated adiabatic warming, while the midtropospheric minimum in Q_1 is associated with evaporation of detrained cloud water (Fig. 6.37).

The environmental and cloud mass fluxes computed from the 8 June 1966 squall line are shown in Fig. 6.38. A higher cloud base (~720 mbar) is present than is found in the tropics. The cloud-mass flux is nearly uniform from cloud base to a pressure of about 300 mbar, where rapid detrainment

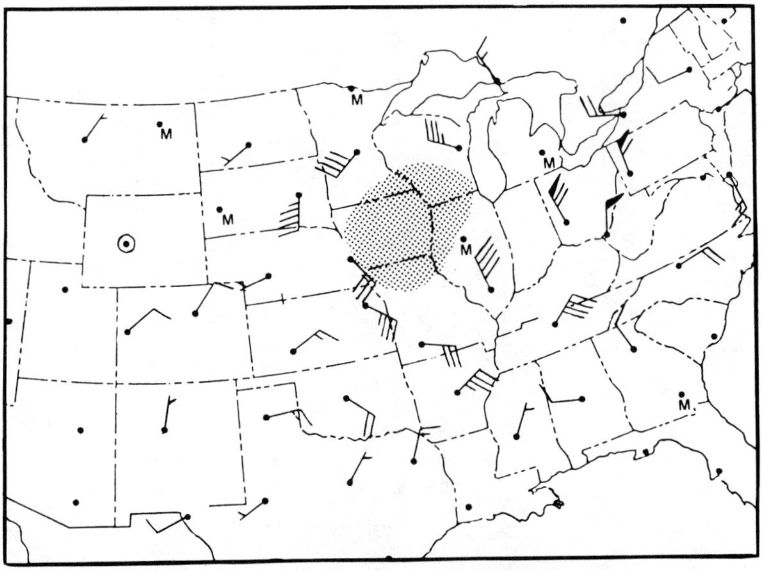

Fig. 6.35. Vector difference between observed flow at 200 mbar at 1200 GMT, 27 June 1979 and 200-mbar flow predicted by National Meteorological Center's Limited-Area Fine-Mesh Model (LFM), verifying at the same time. The difference is assumed to be mainly related to effects of latent heating in a mesoscale convective system which were not present in the LFM. [From Fritsch and Maddox (1981).]

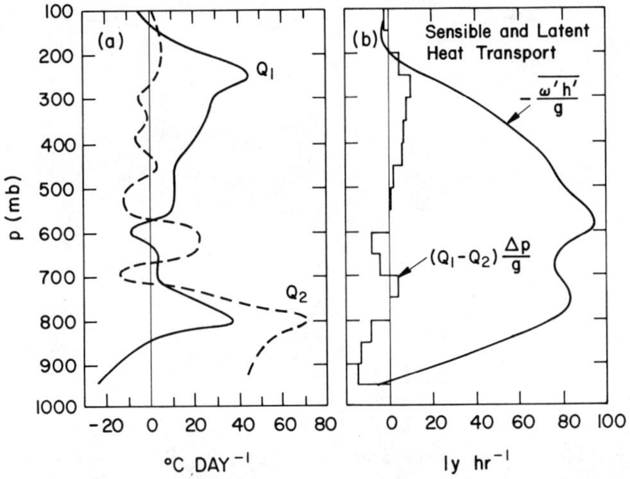

Fig. 6.36. (a) Area-averaged heat source (Q_1) and moisture sink (Q_2) during mature stage of an extratropical squall line; these quantities are expressed in units of heating rate °C day^{-1}. (b) Combined latent ($L\overline{\omega'q'}$) and sensible ($\overline{\omega's'}$) heat transport by small-scale motions. [From Lewis (1975).]

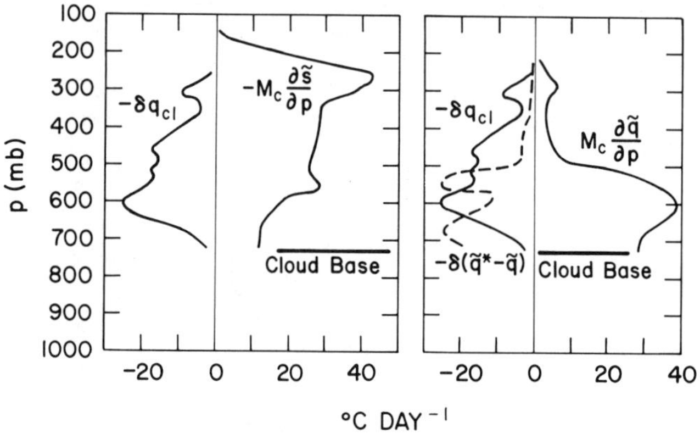

Fig. 6.37. Small-scale processes contributing to area-averaged heat and moisture budgets of an extratropical squall line. Detrainment of mass from cloud ensemble is indicated by δ, and mixing ratio of liquid water in clouds is denoted by q_{cl}. [From Lewis (1975).]

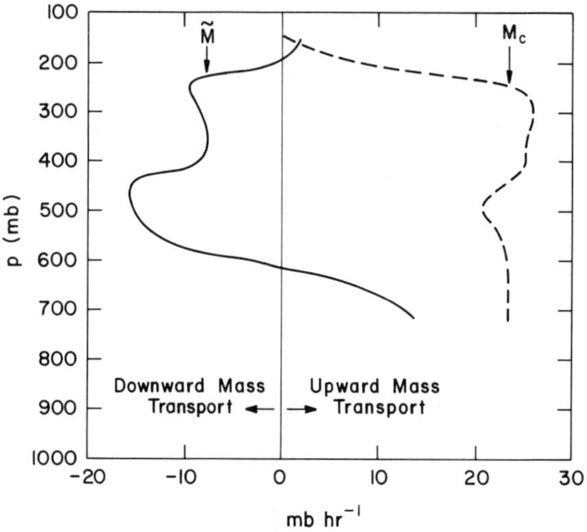

Fig. 6.38. Cumulative mass flux distribution within clouds M_c, and corresponding environmental mass flux $\tilde{M}$ associated with extratropical squall line. [From Lewis (1975).]

occurs. The cloud-mass flux spectrum (Fig. 6.39) shows that the cloud population is dominated by deep clouds that detrain between 200 and 300 mbar. The domination of the cloud spectrum by a single cloud type in this squall line case is a major difference from the bimodal distribution found in tropical systems.

In a later study of the 8 June 1966 squall line, Ogura and Chen (1977) showed that the squall line was triggered by a combination of low-level heating, which increased the depth of the PBL to the height of the lifting condensation level, and a region of mesoscale low-level convergence that existed at least 1.5 h *before* the development of the first clouds. A vertical cross section normal to the squall line of Q_1 and Q_2, averaged over a 3-h period, is shown in Figs. 6.40 and 6.41. Because the budget was estimated over a horizontal area much smaller than that used by Lewis (10×10 km versus 180×180 km), the maximum Q_1 in Fig. 6.40 is much greater than that diagnosed by Lewis (190°C/day versus 40°C/day). The vertical variation is similar, however, with a maximum in the upper troposphere, a minimum around 600 mbar, and a secondary maximum around 750 mbar.

Recently Lin (1986) performed a heat budget analysis from a composite sample of 135 cases of mesoscale convective complexes. The compositing technique was so designed that a crude depiction of the variation in heat and moisture budgets over the lifetime of a "composite" MCC was identified.

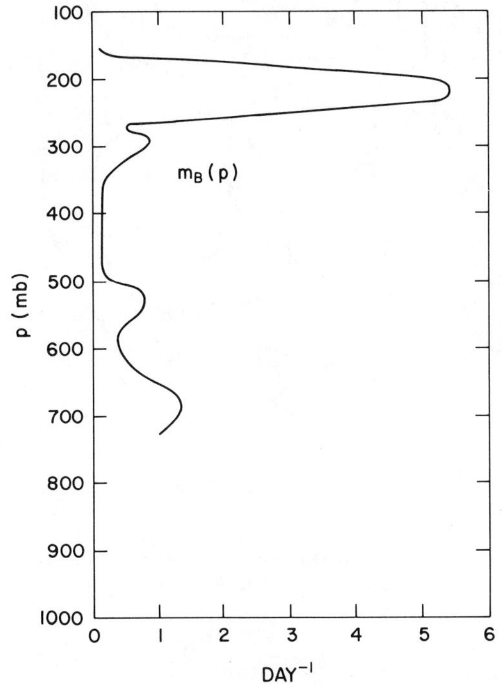

Fig. 6.39. Total mass flux into cloud bases, $M_c(p_B)$, associated with an extratropical squall line, partitioned according to detrainment level. Incremental area $m_B(p)\Delta p$ represents a fraction of $M_c(p_B)$ that is detrained in interval Δp. [From Lewis (1975).]

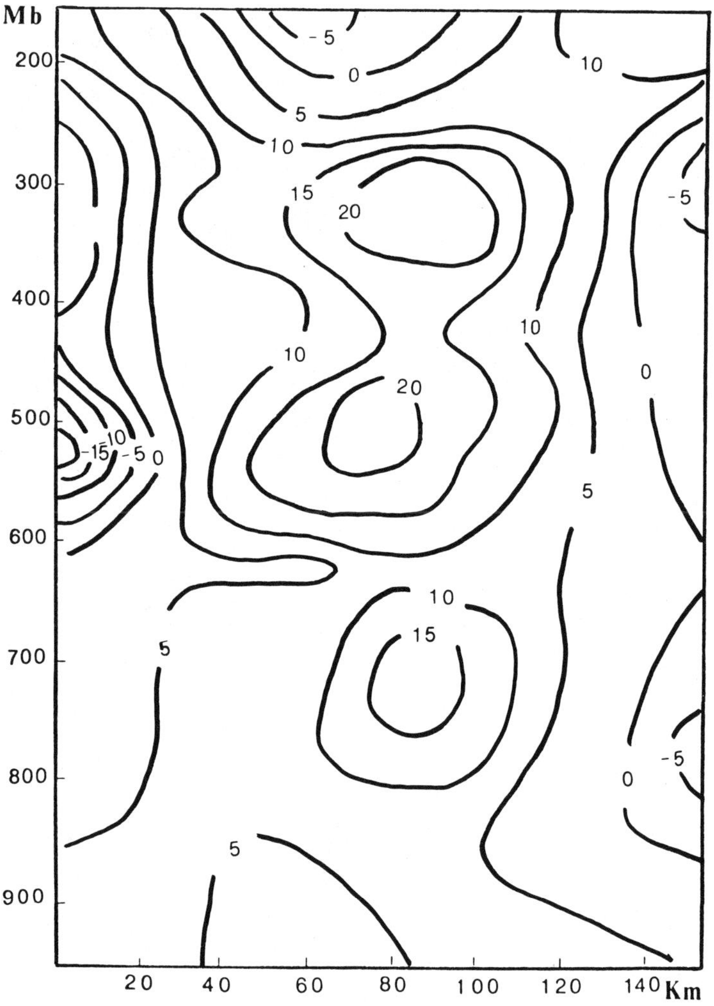

Fig. 6.40. Vertical cross section along line normal to an extratropical squall line of apparent heat source $[K\ (3\ h)^{-1}]$ averaged over the period from 1700 to 2000 CST, 8 June 1966. [From Ogura and Chen (1977).]

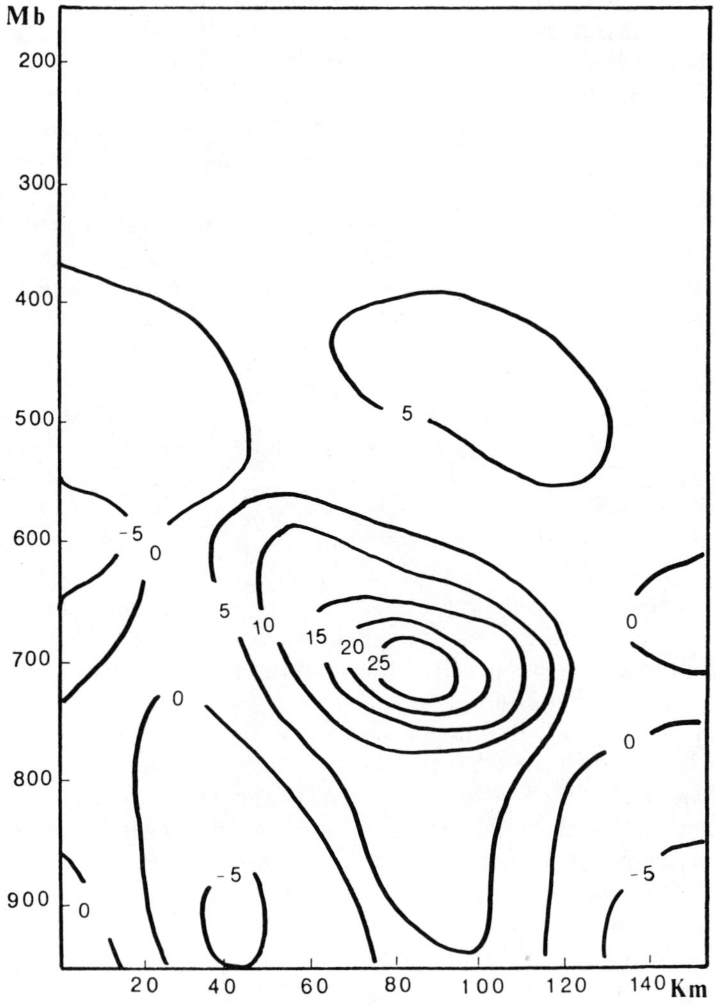

Fig. 6.41. As in Fig. 6.40, except for apparent moisture sink [g $kg^{-1}(3\ h)^{-1}$]. [From Ogura and Chen (1977).]

Figure 6.42 illustrates the computed temporal variation of Q_1 and Q_2, respectively. Early in the meso-α-scale convective complex (MCC) life cycle, low-level heating centered near the 850-mbar level prevails. As the system matures, however, the level of maximum heating ascends to near the 400-mbar level, with a maximum value of about 20 K/day. The magnitude is smaller than that obtained by Lewis for a squall line, probably because of the coarse resolution used in the compositing technique and to the inherent damping that is characteristic of compositing methods. Through most of the MCC life cycle, Lin infers cooling below 700 mbar, probably due to evaporative cooling in mesoscale and convective-scale downdrafts.

The computed temporal evolution of the apparent moisture sink Q_2 (Fig. 6.42b) shows a low-level sink of moisture to the MCC early in the system life cycle, with the moisture sink rising to the 700-mbar level as the system reaches maturity. The height of the maximum sink of moisture is surprisingly similar to that found in tropical studies. The moisture source at low-levels found by Lin at the mature stage suggests that precipitation exceeds evaporation at this stage.

In addition to the differences in shape and magnitude of the Q_1 and Q_2 profiles and the cloud mass flux distribution in the extratropical squall line case compared to those in the tropics, the storage terms ($\partial s/\partial t$ and $\partial r_v/\partial t$) are in general not negligible (compared to the horizontal and vertical advection terms) as they usually are in the tropics. McNab and Betts (1978), using data from the National Hail Research Experiment, computed average budgets of moist static energy and total water mixing ratio for four classes of convection (weak, developing, moderate, and precipitating) over the Great Plains of the United States. During periods of developing convection, they diagnosed a net loss of water vapor in the budget volume, although no precipitation was occurring. They ascribed this loss to an increase in cloud water in the region and showed that the storage of water in the increasing number and size of clouds was an important term in the water budget.

In the cases of developing convection studied by McNab and Betts (1978), the large-scale supply of water vapor exceeded the removal by precipitation, implying a positive storage of water in the volume. Other studies have indicated that precipitation may exceed the rate of supply of water into the volume, implying a drying, or negative storage. In a study of an Oklahoma squall line (21–22 May 1961), Fritsch *et al.* (1976) found that the instantaneous rate of consumption of water vapor by cumulus clouds exceeded that supplied by the synoptic-scale convergence by a factor of about 2. This result implies a large value of the storage term, since, from

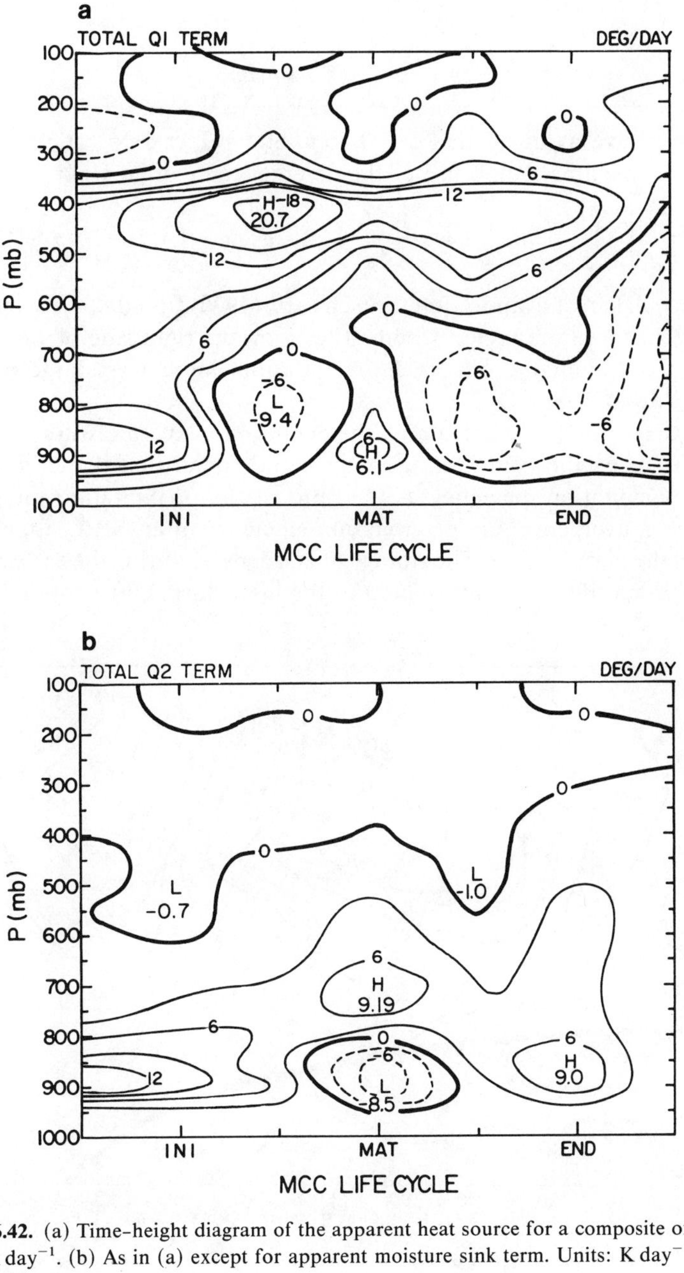

Fig. 6.42. (a) Time–height diagram of the apparent heat source for a composite of MCCs. Units: K day^{-1}. (b) As in (a) except for apparent moisture sink term. Units: K day^{-1}. [From Lin (1986).]

Eqs. (6.16) and (6.22),

$$\frac{1}{g}\int_{p_t}^{p_s}\frac{\partial r_v}{\partial t}\,dp=-\frac{1}{g}\int_{p_t}^{p_s}\nabla\cdot\mathbf{V}r_v\,dp-(P-E),\tag{6.58}$$

where we have assumed that $\omega=0$ at p_s and p_t. An integration over the synoptic-scale area A and use of the divergence theorem yield

$$\frac{1}{g}\int_A\int_{p_t}^{p_s}\frac{\partial r_v}{\partial t}\,dp\,dA=-\frac{1}{g}\int_{p_t}^{p_s}\oint v_n r_v\,ds\,dp-\int_A(P-E)\,dA.\tag{6.59}$$

With the neglect of evaporation, Fritsch *et al.* (1976) found that the moisture convergence into the region [the first term on the right side of Eq. (6.59)] supplied only half the observed precipitation, implying a large negative storage (drying) term.

Likewise Lin (1986) calculated the evolution of the moisture budget of an MCC using the compositing method and found that if one defined a storm precipitation efficiency as the ratio of the precipitation rate to the moisture convergence, the precipitation efficiency of an MCC approached 113% at the mature stage. The storage of condensate during the early portion of the system life cycle contributed to the high storm efficiency.

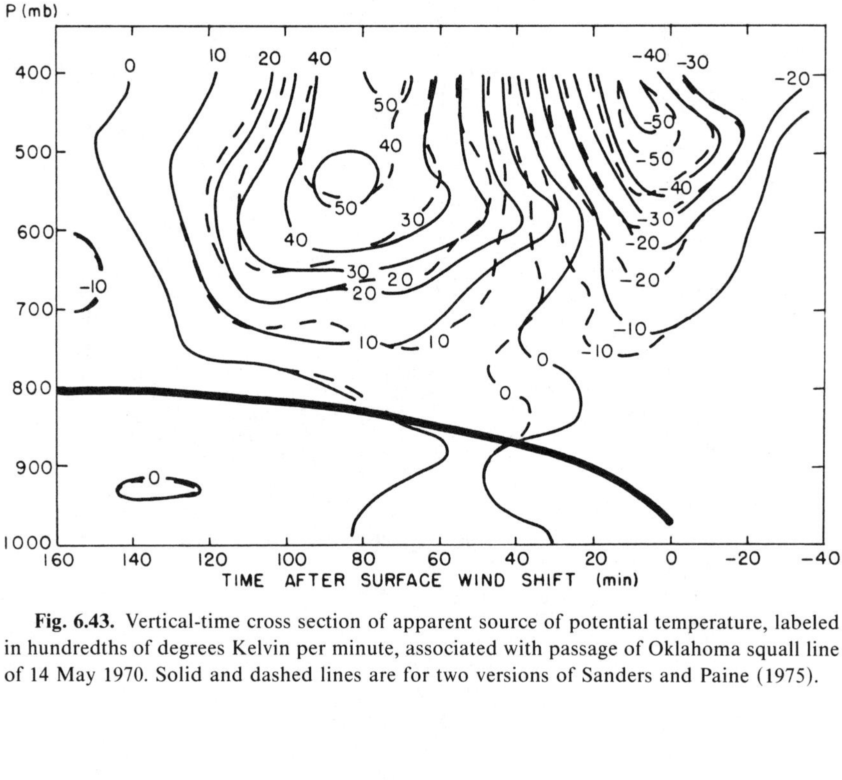

Fig. 6.43. Vertical-time cross section of apparent source of potential temperature, labeled in hundredths of degrees Kelvin per minute, associated with passage of Oklahoma squall line of 14 May 1970. Solid and dashed lines are for two versions of Sanders and Paine (1975).

In another analysis of an Oklahoma squall line (14 May 1970), Sanders and Paine (1975) analyzed an intense mesoscale updraft–downdraft pair. The downdraft was driven by strong cooling associated with evaporation of cloud and precipitation water. A time–height cross section through the squall of the apparent heat source (Fig. 6.43) indicates that the evaporative cooling centered over the surface wind shift and the convective heating behind the wind shift are about equal in magnitude. The strong effect of evaporation in this case, as well as in the 8 June case, is indicative of a general result that evaporation effects can be much larger in the extratropical continental regions, where the air in middle and lower levels can be considerably drier than in the tropics.

In a case study of a subtropical heavy rain event over north Florida (20–22 September 1969), Johnson (1976) used a spectral cloud model to diagnose the apparent heat source and moisture sinks. The Q_1 and Q_2 profiles (Fig. 6.44) are similar to the tropical Pacific profiles (Fig. 6.7), although the magnitude is about four times greater in the Florida case, because the rainfall rate over the smaller area of analysis is about four times greater. As in the tropical studies, a bimodal population of clouds existed in the Florida case.

6.4.3 *Use of Simulation Experiments in Diagnostic Studies*

A common problem in all observational studies of the effects of clouds on their environment is the estimation of the errors associated with the observations and the analysis procedures including vertical and horizontal interpolation error. One method of estimating the errors associated with various aspects of diagnostic studies is the use of observing systems simulation experiments (OSSEs). In an OSSE, a numerical model provides a high-resolution (spatial and temporal) data set which is internally consistent. If a realistic simulation of a particular phenomenon is obtained, the data are assumed perfect and can be used to test various analysis schemes. For example, the effect of errors in the observations, or sampling and analysis errors, can be studied by using modified subsets of the complete data set in budget calculations.

Kuo and Anthes (1984a) investigated the accuracy of diagnostic heat and moisture budgets using the AVE–SESAME 1979 special data set in a series of OSSE. A four-dimensional data set provided by a mesoscale model was used to simulate rawinsonde observations taken during the AVE–SESAME 1979 regional experiments, which were designed to study the interactions of severe moisture convective systems with the environment. Figure 6.45 shows the vertically integrated Q_1 and Q_2 at 12 h of the simulation (0000 GMT, 11 April 1979). At this time, the model simulated a region of

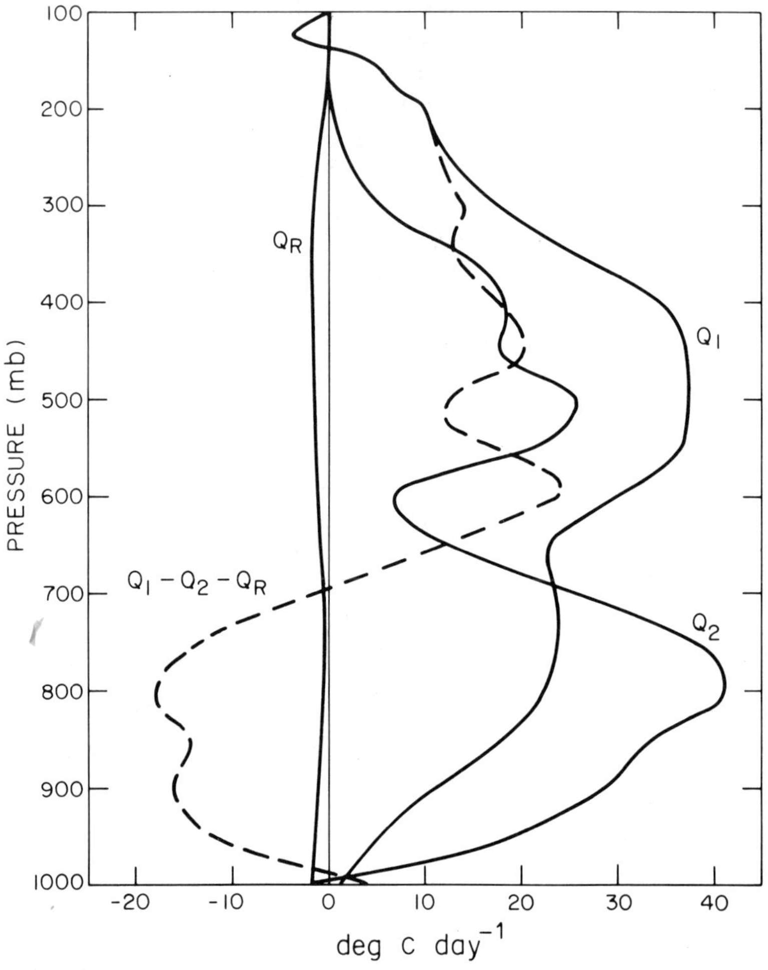

Fig. 6.44. Apparent heat source Q_1, apparent moisture sink Q_2, and radiative heating Q_R for a northern Florida case study. [From Johnson (1976).]

heavy convective rainfall over Oklahoma and Nebraska. The diagnosed Q_1 and Q_2 terms, using all the model data without superimposed error, were generally consistent with the model-simulated convective precipitation. However, there was some small-scale noise associated with the model equations horizontal diffusion terms, which were not accounted for in the budget equations. This noise nearly disappeared when the Q_1 and Q_2 terms were averaged over a synoptic-scale area (550 × 550 km), and the horizontal

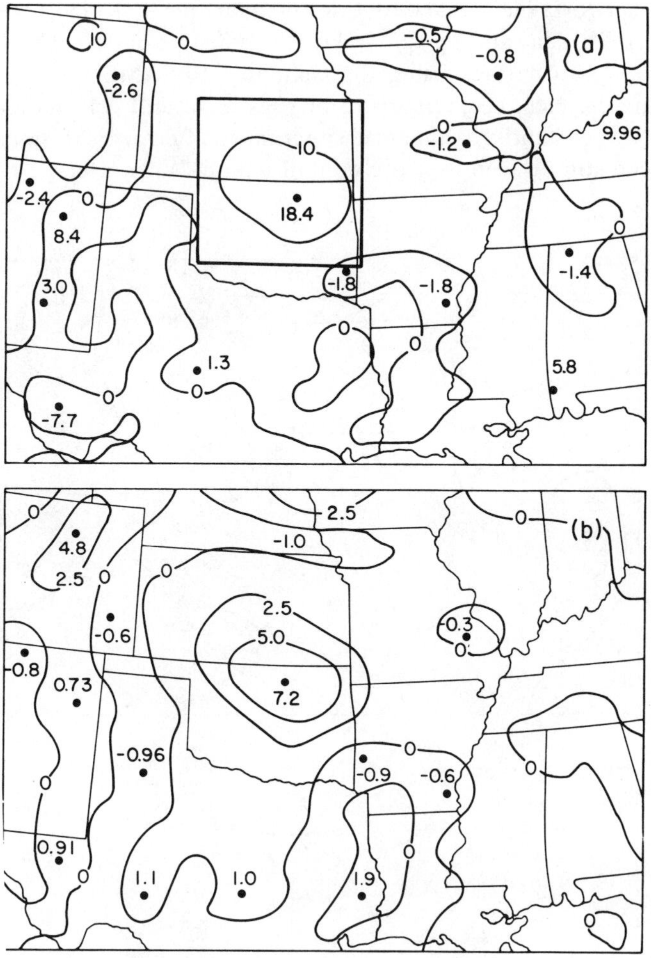

Fig. 6.45. Vertically integrated heat source Q_1 (a) and moisture sink Q_2 (b), computed from model output at 12 h of a simulation of heavy precipitation event over Oklahoma (0000 GMT, 11 April 1979). Units are °C day^{-1} and g kg^{-1} day^{-1}, respectively. [From Kuo and Anthes (1984a).]

average of the vertical profiles of Q_1 and Q_2 on this scale agreed very well with the prescribed model profile.

Kuo and Anthes (1984a) investigated the contribution to budget errors associated with observational frequency, observational density, objective analysis, vertical interpolation, and observational errors. When all sources of errors were present in a simulation of the actual AVE–SESAME 1979 regional network, the average root-mean-square error (RMSE) for the heat budget was 5°C/day and 2 g kg^{-1} d^{-1} for the moisture budget. The diagnosed Q_1 and Q_2 distributions using the simulated SESAME observations are shown in Fig. 6.46. A comparison of Figs. 6.46 and 6.45 indicates that, despite the substantial amount of superimposed errors, the signals in Q_1 and Q_2 can still be seen over the rainfall region. When average Q_1 and Q_2

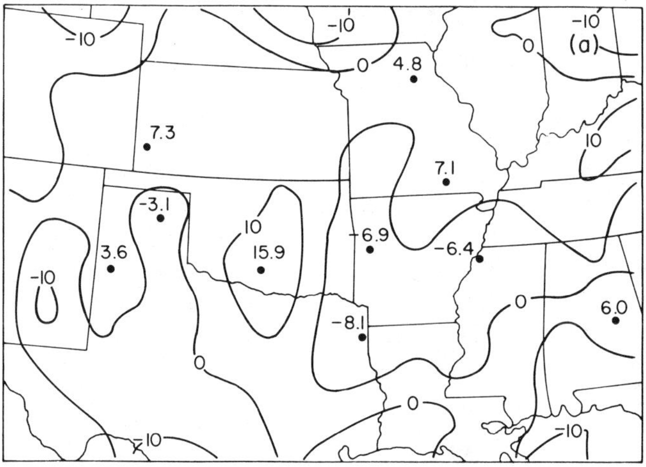

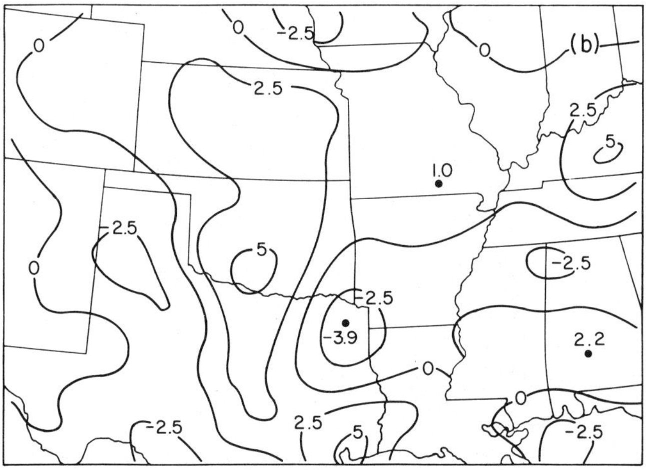

Fig. 6.46. As in Fig. 6.45, except from Observing Systems Simulation Experiment (OSSE) using a partial model data set with superimposed errors designed to simulate AVE-SESAME 1979 regional network. [From Kuo and Anthes (1984a).]

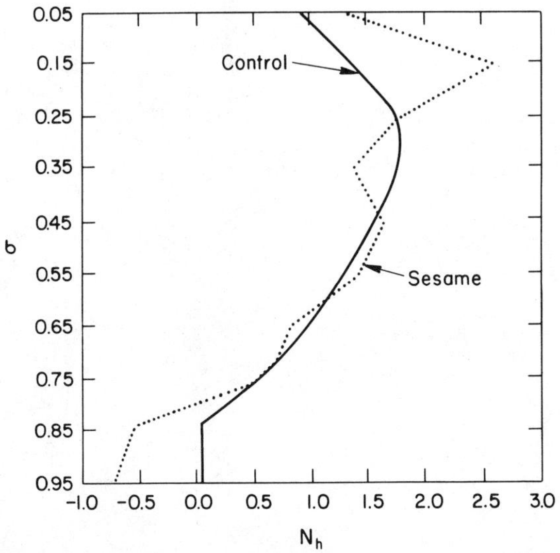

Fig. 6.47. Normalized convective heating profiles averaged over the boxed-in area of Fig. 6.45a for control model simulation and OSSE corresponding to AVE–SESAME 1979 regional network. [From Kuo and Anthes (1984a).]

profiles were computed over a 550×550 km box (box in Fig. 6.45a), the diagnosed profiles agreed fairly well with the control (Fig. 6.47).

The results from the above OSSE indicate that the accuracy and spatial and temporal density of the AVE–SESAME 1979 data sets were of sufficient quality to calculate meaningful budgets for moderate to strong convective systems (those with maximum values of Q_1 and Q_2 at least 10°C/day). The convective system of 10–11 April 1979 met this requirement during much of its life cycle. Figure 6.48 shows the time–height cross sections of terms in the water vapor budget, averaged over the box in Fig. 6.45a. The storage term ($\partial q/\partial t$, Fig. 6.48a) shows moistening of the large-scale environment before the squall line develops and drying as the squall line moves out of the budget area. Most of the large-scale moisture convergence (Fig. 6.48c) occurs below 700 mbar. However, the magnitude of the apparent moisture sink (Q_2, Fig. 6.48b) is substantial up to 450 mbar during the mature stage of the squall line. The large-scale vertical motion transports water vapor upward to support the water vapor consumption by the squall line. Vertical moisture flux convergence occurs above 800 mbar, with divergence below that level (Fig. 6.48d).

Time–height cross sections of terms in the heat budget are shown in Fig. 6.49. The form of the thermodynamic equation used in this budget is written

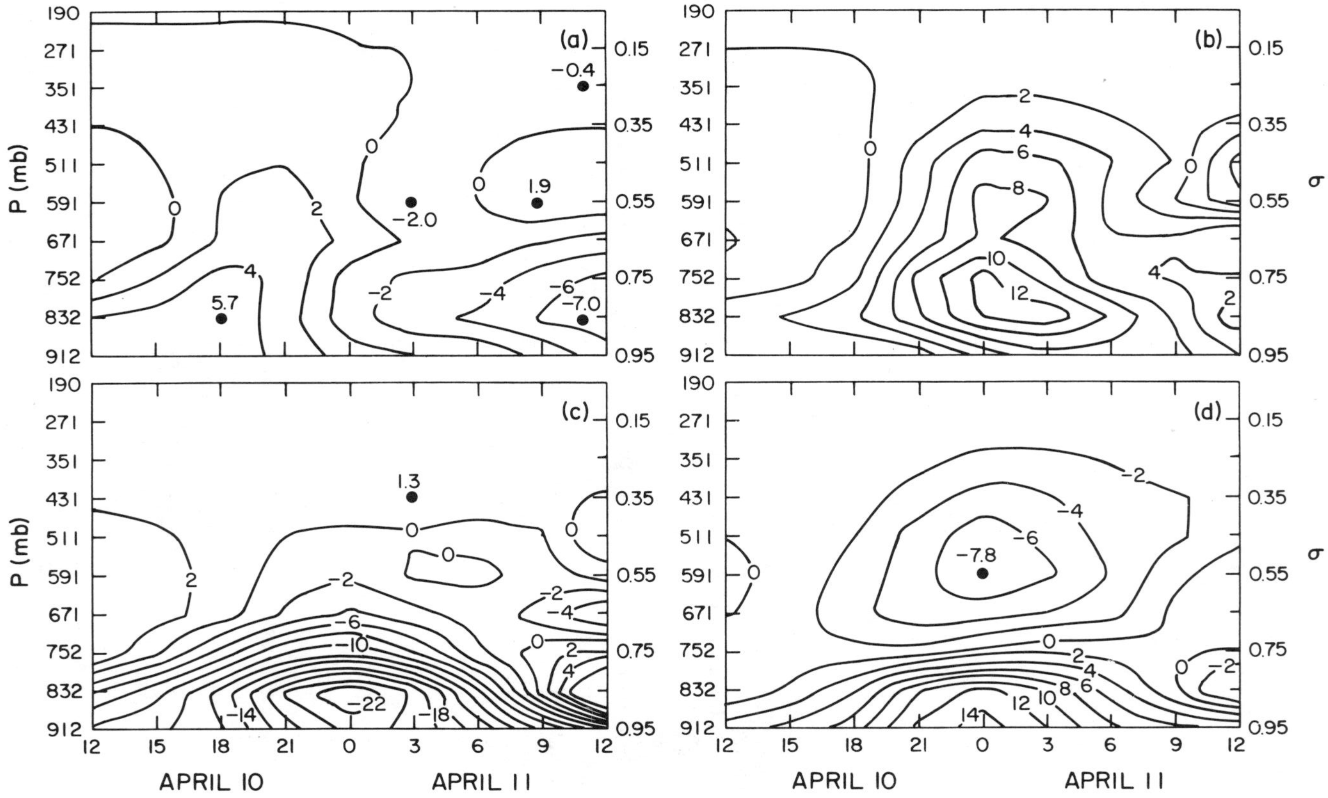

Fig. 6.48. Time–height section of terms in moisture budget of extratropical convective system of 10–11 April 1979, averaged over the boxed-in area of Fig. 6.45a. (a) Storage term, (b) moisture sink Q, (c) horizontal moisture flux convergence, and (d) vertical moisture flux convergence. Units are g kg^{-1} day^{-1}. [From Kuo and Anthes (1984b).]

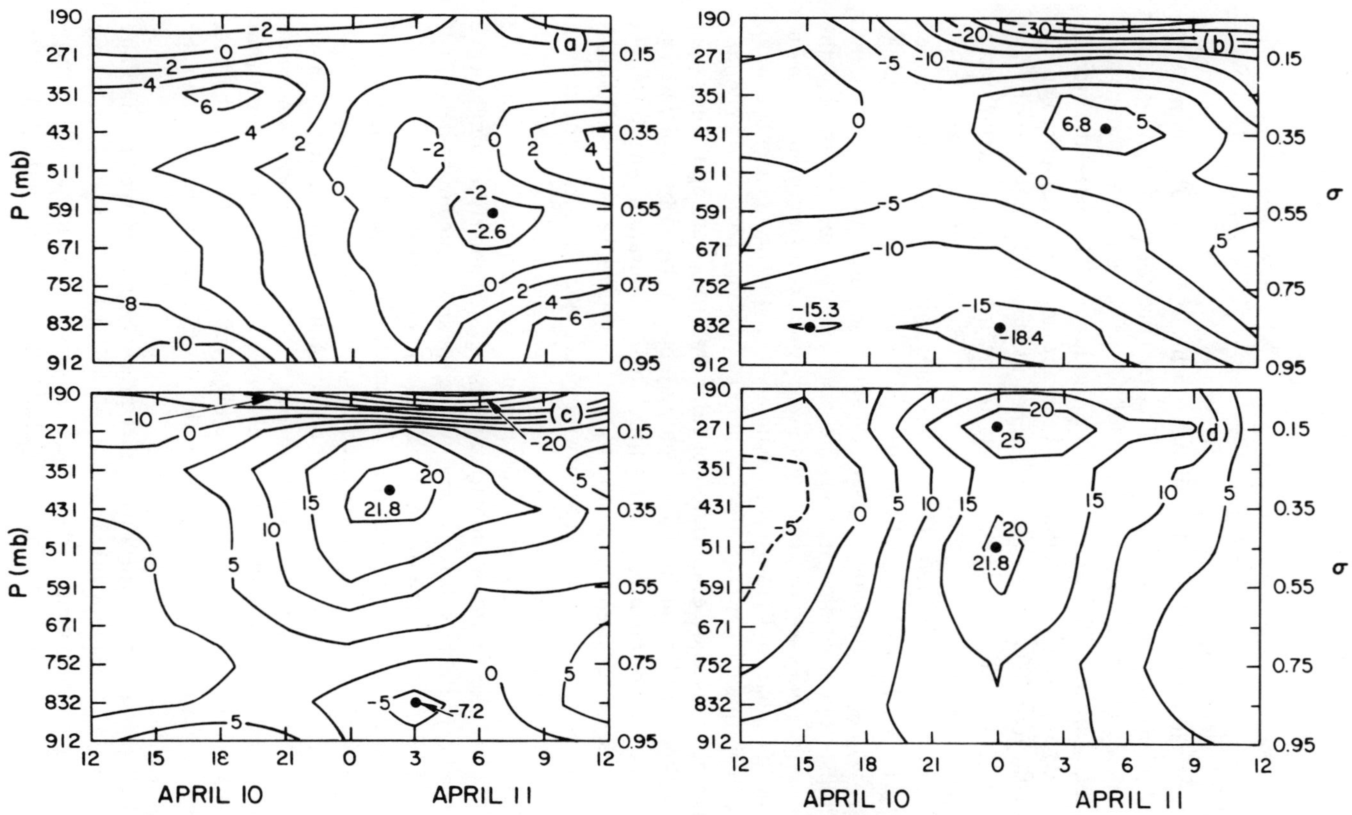

Fig. 6.49. As in 6.48, except for heat budget. (a) Storage term, (b) horizontal advection term, (c) heat source Q_1, and (d) sum of vertical advection and adiabatic cooling terms. [From Kuo and Anthes (1984b).]

in σ coordinates, where $\sigma=(p-p_t)/(p_s-p_t)$, p_s is surface pressure, and $p_t=100$ mbar:

$$Q_1-Q_R=\frac{\partial T}{\partial t}+\mathbf{V}\cdot\nabla T+\dot{\sigma}\frac{\partial T}{\partial \sigma}-\frac{RT\omega}{c_p p}. \tag{6.60}$$

The vertical advection and the adiabatic terms of Eq. (6.60) dominate the heat budgets and largely balance each other (Fig. 6.49d). Their difference is balanced by local storage, horizontal advection, and net condensation heating, with the adiabatic term being the largest. However, the storage term and horizontal advection term are not negligible compared to the adiabatic term. This is a basic difference of this and other extratropical budgets from the large-scale budgets in the tropics, where temperature changes are small.

The diabatic heating term (Fig. 6.49c) shows low-level warming between 1500–2100 GMT, 10 April, which corresponds to local times of 0800–1400. This warming is probably caused by solar heating. As the squall line develops, condensational heating increases. During the period that the squall line resides within the budget area, a maximum warming occurs at around 350 mbar, and a low-level maximum of cooling occurs near 800 mbar. Since the lifting condensation level is about 750 mbar, this low-level cooling is possibly caused by evaporation of rainfall. Strong cooling in the upper troposphere is also calculated, which may be produced by real physical processes (cloud-top radiation cooling, anvil evaporation, and turbulent mixing of stratospheric and tropospheric air because of overshooting cumulonimbus tops) in combination with inaccurate estimates of the vertical motion near the top level.

Figure 6.50 shows the observed rainfall rate, diagnosed rainfall rate (residual of the moisture budget), and integrated moisture convergence averaged over the box in Fig. 6.45a. As the squall line begins to develop at 1800 GMT, 11 April, the observed rainfall rate increases rapidly. It levels off at 0000 GMT, 11 April, and then decreases as the squall line moves out of the budget area. The area-averaged rainfall rate reaches a maximum of 4.5 cm day^{-1}. The diagnosed rainfall rate coincides well with the observed rainfall rate, indicating the high quality of the area-averaged moisture budget. In contrast, the rainfall rate diagnosed from the heat budget (not shown) did not agree as well with the observed rainfall, because the heat budget is more sensitive to errors in the vertical motion profile than the moisture budget.

The integrated moisture convergence precedes the observed rainfall rate by about 3 h. It exceeds the observed rainfall rate at the early stage of the squall line development. Later in the time period, the rainfall rate exceeds the large-scale supply of water vapor. This behavior is similar to that of the

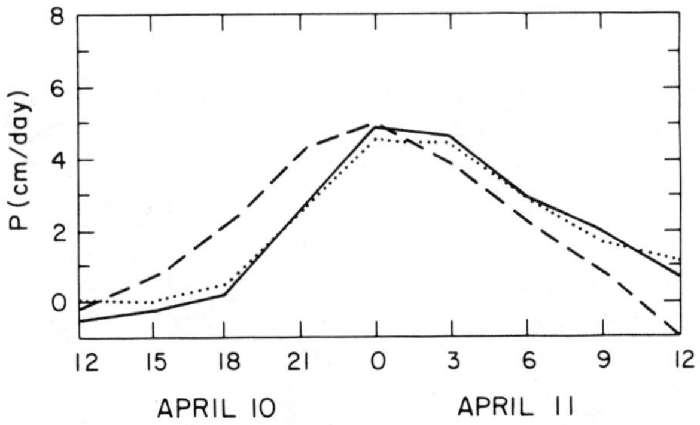

Fig. 6.50. Observed (dotted curve) and diagnosed (solid curve) rainfall rate and vertically integrated moisture convergence (dashed curve) averaged over the boxed-in area of Fig. 6.45a for the AVE-SESAME case. Units are cm day^{-1}. [From Kuo and Anthes (1984b).]

large-scale moisture storage and depletion in easterly waves, as discussed by Reed and Recker (1971), Cho (1976), and Thompson *et al.* (1979). These results suggest that the mean circulation on a scale of 550×550 km initiates the organized convective activity. Once convection has formed, a positive feedback undoubtedly occurs between the convective clouds and the mean circulation on this scale, with the mean flow providing the moisture for the cumulus convection while the energy release and redistribution in the convection enhance the mean circulation. This cooperative interaction is conceptually similar to the "conditional instability of the second kind" (CISK) mechanism of Ooyama (1964) and Charney and Eliassen (1964). However, a close balance between the large-scale, vertically integrated moisture convergence and the area-averaged rainfall rate does not exist at all times; after the convective rainfall rate reaches a maximum (at 0000 GMT, 11 April), the ratio of the large-scale moisture supply to the rainfall rate decreases from in excess of 100 to 0% over 12 h. Other studies (Fritsch *et al.*, 1976; Ogura and Cho, 1973; Lin, 1986) have shown that at particular times in the life cycle of convective systems the convective-scale consumption of water vapor can exceed the synoptic-scale supply.

6.4.4 *Summary of Observational Studies*

Numerous observational studies using tropical and extratropical data sets have indicated a strong control by the synoptic-scale motions on deep cumulus convection, with little synoptic-scale control on shallow clouds. Large-scale vertical motion and net horizontal convergence of water vapor are positively correlated with deep convection and convective rainfall rates.

The vertical profiles of the apparent heat source Q_1 are generally similar in the tropical and extratropical studies. Most studies show a maximum in Q_1 between 600 and 300 mbar. The Q_1 profiles diagnosed from time-averaged data are smoother than those computed from instantaneous data or from averages over short times. The Q_1 profiles in the middle latitudes frequently show less apparent heating in the lower troposphere than do those in the tropics because of higher cloud bases and greater evaporative cooling.

The vertical profiles of Q_2, the apparent moisture sink, also are similar in the tropics and middle latitudes, with a maximum sink present between 800 and 600 mbar, somewhat lower than the maxima in the Q_1 profiles.

The overall effect of cumulus clouds is to warm and dry most of the troposphere. This warming and drying occurs mainly through adiabatic compression in the subsiding environment of the clouds. However, detrainment of cloud water and transport of heat and moisture in moist downdrafts are also important terms in the heat and water budgets of the lower troposphere.

Cumulus clouds also modify the large-scale vorticity (and momentum) structure. In the tropics, cumulus convection generally produces an apparent sink of vorticity in the lower troposphere and an apparent source in the upper troposphere. Fewer quantitative studies have been made in the extratropics, but there is evidence of strong momentum transports by cumulus convection and a large environmental response in midlatitude convective systems. Both tropical and extratropical convective clouds are affected strongly by vertical wind shear.

The heat and moisture budgets associated with extratropical convective systems differ from those in the tropics in several ways. The storage terms ($\partial r_v/\partial t$ and $\partial T/\partial t$) are often as large as the other terms in the extratropics, whereas they are usually small in the tropics. The effect of evaporation can be larger in extratropical systems because of drier air and higher cloud bases. Finally, in addition to the importance of large-scale vertical motion and moisture convergence, moist convection in extratropical systems is positively correlated to static instability (unlike tropical convection). Altogether, the observational studies indicate a physical basis for cumulus parameterization, in spite of the complex nature of cumulus cloud, mesoscale, and synoptic-scale interactions.

6.5 Cumulus Parameterization Schemes

The problem of cumulus parameterization is to relate the convective condensation and transports of heat, moisture, and momentum by cumulus clouds, which cannot be explicitly resolved by the large-scale model, to the variables predicted by the model. There are two important aspects of cumulus parameterization. One is the modulation of convection by the large-scale forcing, which is related to the determination of total rainfall rate. The other is the feedback of cumulus convection to its environment, which is related to the vertical distribution of latent heating (condensation and evaporation, freezing, ice vapor deposition, and melting) in the clouds and the vertical transports of heat, moisture, and momentum. For parameterization to be possible, a relationship between cumulus convection and the large-scale circulation must exist. Since the scales of motion permitted in the model depend on the grid size, one might expect the relationships between the resolvable-scale and the unresolvable-scale circulations to vary with different model resolutions. A successful parameterization scheme requires knowledge of the mutual interaction between cumulus convection and its environment on different temporal and spatial scales.

This section considers several cumulus parameterization schemes that have been used in large-scale numerical models. These schemes are (1) moist convective adjustment schemes, (2) Kuo schemes, (3) the Arakawa-Schubert scheme, and (4) schemes based on local consumption of available buoyant energy.

6.5.1 *Moist Convective Adjustment*

Moist convective adjustment (MCA) schemes are among the simplest methods of parameterizing the effects of cumulus convection on the environment. In MCA parameterizations, it is assumed that there exists a critical temperature and moisture profile associated with the large-scale thermodynamic field. When the large-scale sounding becomes more unstable than this critical state, it is adjusted toward the critical, more stable state. This stabilization is assumed to be caused by cumulus convection.

A variety of MCA schemes have been proposed and tested in models (Manabe *et al.*, 1965; Miyakoda *et al.*, 1969; Krishnamurti and Moxim, 1971; Kurihara, 1973). In its most severe form, the so-called *hard convective adjustment*, an initial large-scale sounding in which $\partial\theta_e/\partial p > 0$, is adjusted so that θ_e or, equivalently, moist static energy h, is constant with height (Krishnamurti *et al.*, 1980). This adjustment is obtained by an iterative solution of four equations, the definition of h [Eq. (6.15)], the hydrostatic equation

$$\partial gz/\partial p = -(RT/p)(1+0.608r_v), \tag{6.61}$$

Tetens law

$$e_s = 6.11 \exp[a(T_s - 273.16)/(T_s - b)], \tag{6.62}$$

and the relation between saturation vapor pressure and mixing ratio

$$r_s = 0.622 e_s/(p - e_s). \tag{6.63}$$

In Eq. (6.62), the constants a and b are $a = 17.27$ and $b = 35.86$ for saturation over water, and $a = 21.87$ and $b = 7.66$ for saturation over ice. The precipitation rate P is given by

$$P = \frac{1}{g\Delta\tau} \int_{p_t}^{p_s} (r_v^i - r_v^f)\, dp, \tag{6.64}$$

where $\Delta\tau$ is a prescribed time scale associated with the lifetime of a deep cumulus cloud (~30 min), and i and f denote initial and adjusted values of r, respectively.

The hard MCA is simple to implement and conserves total moist static energy, since the average value of h in the adjusted sounding is equal to that of the initial sounding. However, it produces unrealistic modifications to the large-scale sounding by excessively cooling and drying the lower troposphere. Because of the removal of too much water vapor during the adjustment, the rainfall rates diagnosed by Eq. (6.64) are much too large. Krishnamurti *et al.* (1980) used GATE data over the period 1 September to 18 September 1974 in semiprognostic tests of the hard MCA scheme. The diagnosed rainfall rates from the hard MCA (Fig. 6.51) exceed the observed rainfall rates by more than an order of magnitude. Because of the unrealistic modification to the large-scale thermodynamic fields and the erroneous rainfall rates, the hard MCA is unsuitable for use in large-scale models.

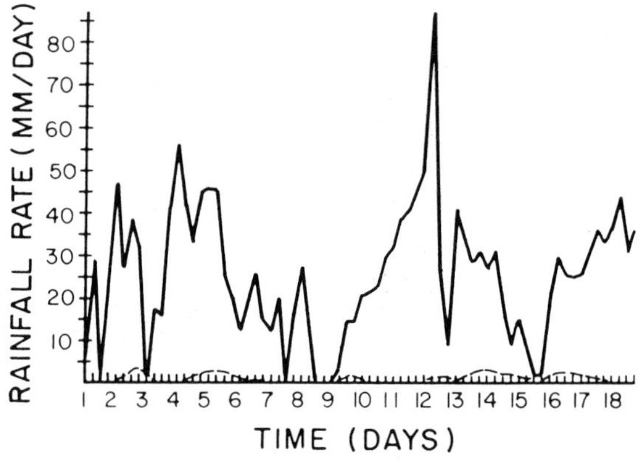

Fig. 6.51. Comparison of observed (dashed line) and predicted (solid line) rainfall rates (mm day^{-1}) using hard convective adjustment. Days 1–18 correspond to the third phase of GATE between 1 September and 18 September 1974. Data are for 6-h intervals beginning with 0000 GMT, 1 September. [From Krishnamurti *et al.* (1980).]

Because of the problems associated with the hard MCA schemes, several versions of MCA have been developed which produce much slower and more realistic adjustments to the large-scale sounding (Manabe *et al.*, 1965; Miyakoda *et al.*, 1969; Kurihara, 1973; Krishnamurti *et al.*, 1980). These methods are known as soft adjustment schemes. In one of the soft schemes, hard adjustment is assumed to occur over a small fraction a of the large-scale grid area; over the remaining $(1-a)$ fraction, the temperature and mixing ratio are assumed to remain constant at their initial value. The final adjusted large-scale values of T and r_v are given by

$$\begin{aligned} T^{f} &= aT_{c} + (1-a)T^{i}, \\ r_{v}^{f} &= ar_{vc} + (1-a)r_{v}^{i}, \end{aligned} \tag{6.65}$$

where T_c and r_{vc} are determined by a wet adiabat. The value of a is determined by a criterion based on mean relative humidity (RH) of the column. The relative humidity after adjustment RH^{f} is

$$\mathrm{RH}^{f} = r_{v}^{f}/r_{s}(T^{f}), \tag{6.66}$$

where $r_s(T^f)$ is the saturation mixing ratio at the adjusted temperature. The scheme is closed by an iterative search for a value of a which yields a vertical mean RH^{f} equal to that of a prescribed value $\overline{\mathrm{RH}}$. In semiprognostic tests of this soft MCA, Krishnamurti *et al.* (1980) found best agreement between calculated and observed rainfall rates when $\overline{\mathrm{RH}} = 82.4\%$ (Fig. 6.52).

As shown in Fig. 6.52, the soft MCA produces rainfall rates of the correct order of magnitude. However, because of the occurrence of the maximum instability before the time of maximum convection in the tropics, the soft

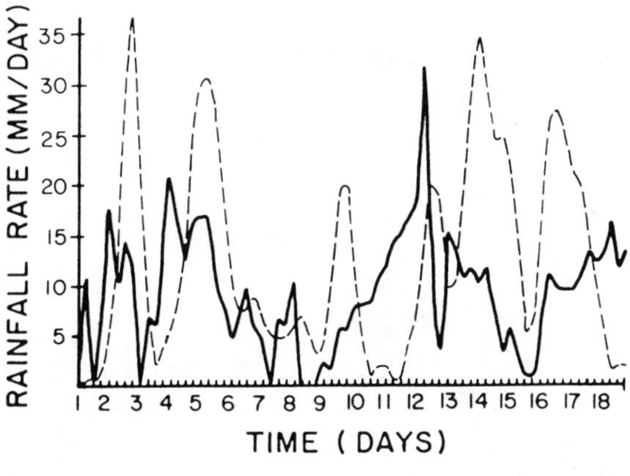

Fig. 6.52. As in Fig. 6.51, except for soft convective adjustment. [From Krishnamurti *et al.* (1980).]

MCA scheme shows a lag of 1 to 2 days between the computed and observed rainfall. This lag makes the soft MCA scheme undesirable for use in forecast models in which predicting the correct timing of rainfall is as important as predicting the correct amount. However, the soft MCA may be more suitable for climate models in which the timing of precipitation is less important.

6.5.2 Kuo Schemes

As reviewed earlier, there is a strong correlation between observed convective rainfall and total large-scale convergence of water vapor in a column. These observations suggest that large-scale water vapor convergence is a useful variable to parameterize the effects of convection in large-scale models, and many cumulus parameterizations have been based on a relationship between convective rainfall and large-scale moisture convergence. These schemes are called Kuo schemes here because of the early work by Kuo (1965). Versions of the Kuo (1965) scheme include those by Kuo (1974), Anthes (1977), Lian (1979), Krishnamurti *et al.* (1976, 1980, 1983), and Molinari (1982).

6.5.2.a Relation of Total Convective Heating and Moistening to Moisture Convergence

The use of moisture convergence to determine the rainfall rate is based on the large-scale water vapor budget, which may be written from Eqs. (6.10) and (6.16):

$$\frac{c_p}{L} Q_2 = -\left(\frac{\partial r_v}{\partial t} + \nabla \cdot r_v \mathbf{V} + \frac{\partial r_v \omega}{\partial p}\right) = \frac{\partial \overline{r_v'' \omega''}}{\partial p} + (c - e), \tag{6.67}$$

where c is the rate of condensation per unit mass of air, and e is the rate of evaporation.

A vertical integration of Eq. (6.67) from the surface pressure p_s to the top of the atmosphere ($p = 0$) yields

$$M_t + E = \frac{1}{g} \int_0^{p_s} (c - e)\, dp + S_{qv}, \tag{6.68}$$

where M_t is the vertically integrated horizontal moisture convergence

$$M_t \equiv -\frac{1}{g} \int_0^{p_s} \nabla \cdot r_v \mathbf{V}\, dp, \tag{6.69}$$

E is the surface evaporation rate

$$E \equiv -(1/g)(\overline{r_v'' \omega''})_{p_s}, \tag{6.70}$$

and S_{qv} is the storage rate of water vapor,

$$S_{qv} \equiv \frac{1}{g} \int_0^{p_s} \frac{\partial r_v}{\partial t} \, dp. \tag{6.71}$$

The relationship between the integrated net condensation and the precipitation rate can be obtained by considering the equation for cloud (liquid) water r_c

$$\frac{\partial r_c}{\partial t} + \nabla \cdot r_c \mathbf{V} + \frac{\partial r_c \omega}{\partial p} = -\frac{\partial \overline{r_c'' \omega''}}{\partial p} + c - e - \mathrm{CN}_{cr}, \tag{6.72}$$

where CN_{cr} is the rate of conversion of cloud water to precipitating water. A vertical integration of Eq. (6.72) from 0 to p_s yields

$$\frac{1}{g} \int_0^{p_s} (c - e) \, dp = P + S_{rl} - M_{t_l}, \tag{6.73}$$

where P is the precipitation rate, S_{rl} is the storage rate of liquid water and M_{t_l} is the vertical integral of the horizontal convergence of liquid water.

Substitution of Eq. (6.73) into Eq. (6.68) yields

$$M_t + M_{t_l} + E = P + S_{r_v} + S_{r_c}. \tag{6.74}$$

The budget represented by Eq. (6.74) indicates that the sources of water into a unit column are balanced by precipitation plus storage of vapor and liquid.

From Eq. (6.74), one observes that if the storage terms are small and the horizontal convergence of liquid water is small compared to the sum of the evaporation and the water vapor convergence, the net rainfall rate is equal to the large-scale moisture convergence plus the evaporation. Over large regions on a long temporal scale, this approximation is quite good. But, locally, on a short temporal scale, there can be substantial changes in the storage of water vapor and liquid water. In general, there is higher variability over midlatitudes than over the tropics. A key issue is then to relate the storage term to the large-scale variables (e.g., to determine what portion of the moisture convergence should go to storage and what portion should be removed as precipitation).

Kuo (1974) assumed that a fraction $(1 - b)$ of the total water vapor convergence M_t is condensed and precipitated, while the remaining fraction, b, is stored and acts to increase the humidity of the column (Fig. 6.53).

By the above assumption, the vertical integral of $c - e \equiv C^*$ is

$$\int_0^{p_s} C^* \, dp = (1 - b) g M_t. \tag{6.75}$$

The determination of b is discussed in a later section.

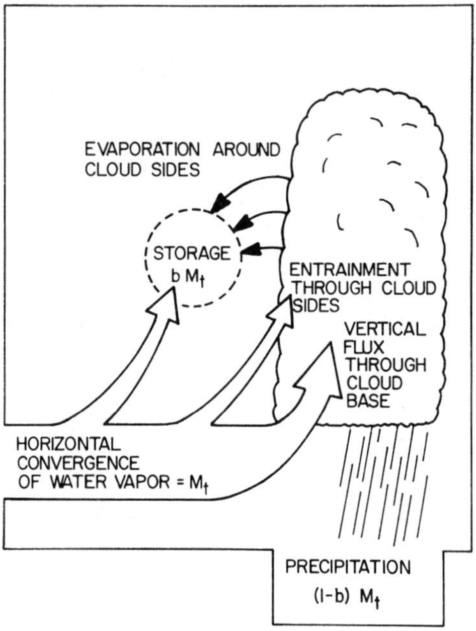

Fig. 6.53. Schematic diagram showing moisture cycle in a column which contains convection. [From Anthes (1977).]

6.5.2.b Vertical Partitioning of Apparent Heat Sources

After b is determined, the integral relation, Eq. (6.75), relates the total rainfall rate to the total moisture convergence. It is also necessary to specify or calculate the vertical distribution of C^*; in general, this distribution may be written

$$C^*(p) = [(1-b)gM_t/(p_b - p_t)]N(p), \tag{6.76}$$

where p_b and p_t are the pressures at cloud base and cloud top, respectively, and $N(p)$ is the vertical distribution function which obeys

$$\int_0^{p_s} N(p)\,dp = \int_{p_t}^{p_b} N(p)\,dp = p_b - p_t. \tag{6.77}$$

If Eq. (6.77) is satisfied, the total convective heating rate $Q_1 - Q_R$ in the column equals the latent energy condensed and removed as precipitation plus the surface sensible heat flux,

$$\frac{c_p}{g}\int_0^{p_s} (Q_1 - Q_R)\,dp = \frac{L}{g}\int_0^{p_s} C^*\,dp + H_s,$$
$$= L(1-b)M_t + H_s. \tag{6.78}$$

With the integral constraints represented by Eqs. (6.76)-(6.78), the remaining problem is to determine $N(p)$. As discussed earlier, this function has the same vertical distribution as does the cloud-scale net condensation. Anthes (1977) used the condensation rate C_c in a one-dimensional cloud model to estimate $N(p)$,

$$N(p) \equiv C_c/\langle C_c \rangle, \tag{6.79}$$

where the vertical averaging operator $\langle \ \rangle$ is defined by

$$\langle \ \rangle \equiv (p_b - p_t)^{-1} \int_{p_t}^{p_b} (\ ) \, dp. \tag{6.80}$$

Kuo (1965, 1974) assumed that $N(p)$ was determined by lateral mixing of warm-cloud air of temperature T_c with environmental air of temperature T, which gives

$$N(p) \equiv (T_c - T)/\langle T_c - T \rangle. \tag{6.81}$$

In most Kuo schemes, T_c is given by the wet adiabat associated with the equivalent potential temperature of a parcel of air originating near the surface.

Kuo and Anthes (1984c) compared the vertical profiles of convective heating computed from Eq. (6.79) with that given by Eq. (6.81) for an extratropical convective system. Figure 6.54 shows the observed and simulated convective heating profile by the Anthes and Kuo schemes for three different cloud radii without the inclusion of eddy sensible heat flux. The observed Q_1 is an average over an area 550×550 km (the box in Fig. 6.45a) and over a period of time (1800 GMT, 10 April 1970 to 1200 GMT, 11 April 1970) when the area-averaged rainfall exceeded 1 mm day^{-1}. When the cloud radius is small (Fig. 6.54a), the two formulas, by coincidence, give nearly identical profiles. Both simulate the maximum heating at a considerably lower level than the observed. As the cloud radius increases, the maxima in the two simulated convective heating profiles shift upward and hence become closer to the observed profile, with the maximum in Kuo's profile slightly higher than that of Anthes.

We next estimate the heating associated with the vertical divergence of the eddy heat flux associated with convection. In σ coordinates, the two terms associated with the eddy flux of sensible heat are

$$-\frac{\partial \overline{\dot{\sigma}'' T''}}{\partial \sigma} + \frac{R \overline{\omega'' T''}}{c_p (p^* \sigma + p_t)}. \tag{6.82}$$

The second term is usually one order of magnitude smaller than the first term and is neglected in the following discussion. For a small fraction of convective cloud cover, $a \ll 1$, and with the cloud vertical motion much

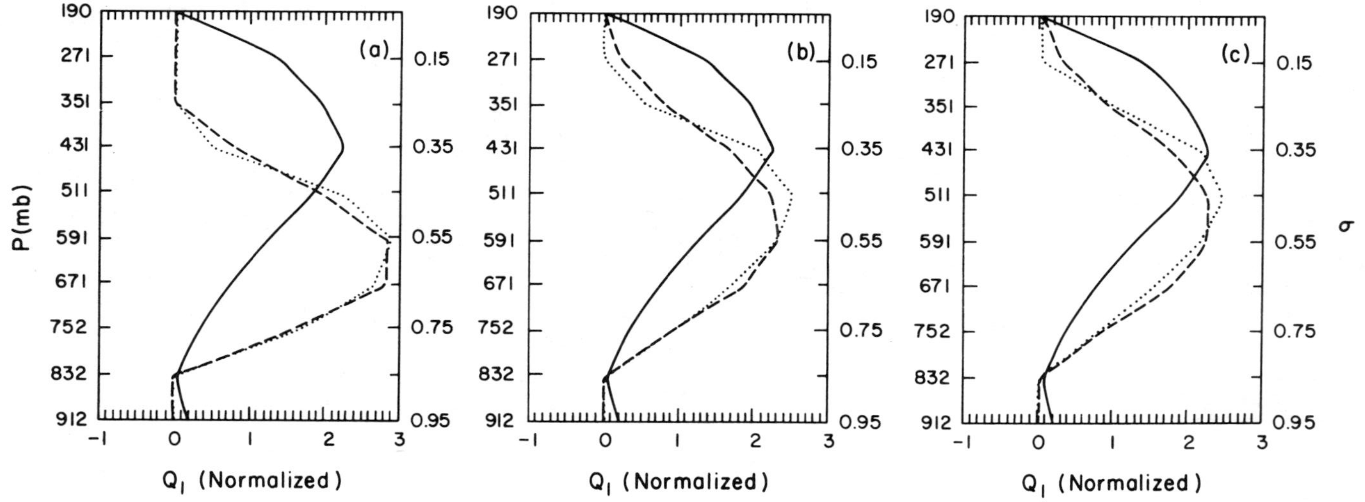

Fig. 6.54. Observed (solid curve) and simulated (normalized) Q_1 profiles by cumulus parameterization schemes of Kuo (1974; dotted curve) and Anthes (1977; dashed curve), averaged over the boxed-in area of Fig. 6.45a and for periods of the 10–11 April 1979 case when the observed rainfall rate was greater than 1 mm day^{-1}; for model cloud radii equal to (a) 1 km, (b) 3 km, and (c) 5 km, with no eddy sensible heat flux considered. [From Kuo and Anthes (1984c).]

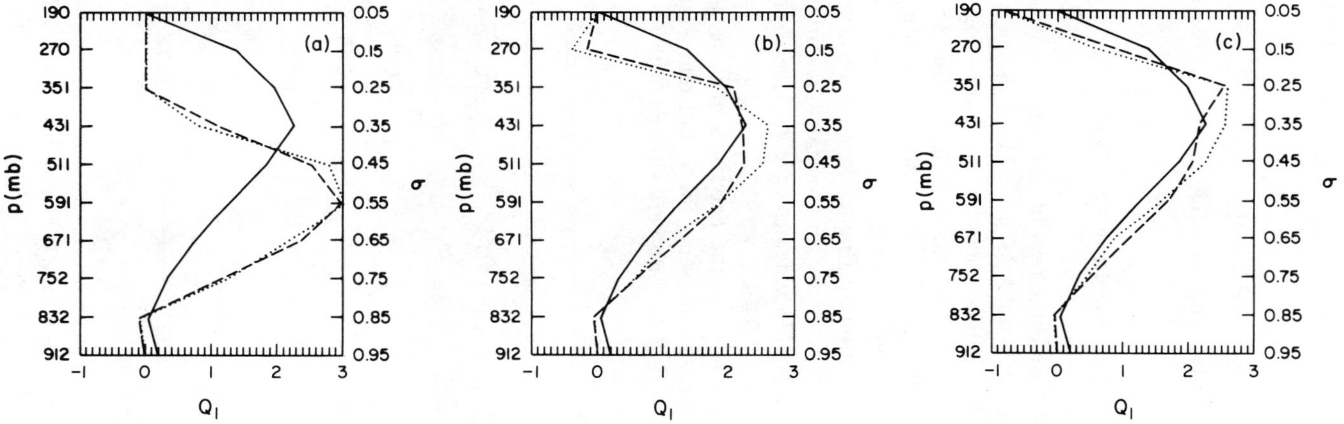

Fig. 6.55. As in Fig. 6.54, but with eddy sensible heat flux considered. [From Kuo and Anthes (1984c).]

larger than the large-scale vertical motion, they can be written as

$$\frac{\partial \overline{\dot{\sigma}''T''}}{\partial \sigma} = \frac{\partial}{\partial \sigma}\left[\frac{a}{1-a}\,\dot{\sigma}_c(T_c - T)\right]. \tag{6.83}$$

The fractional area of active convective cloud coverage can be computed as

$$a = P/P_c, \tag{6.84}$$

where P is the observed area-averaged rainfall rate and P_c is the rainfall rate of a single cloud.

The effect of the eddy sensible heat flux is to shift upward the maximum in the convective heating profiles (Fig. 6.55). With the increase of cloud radius, the effect of eddy sensible heat flux increases (Fig. 6.56). The rapid decrease of the cloud vertical motion near the cloud top for large clouds produces a strong eddy flux convergence and hence strong warming below the cloud top. The effect of eddy flux of sensible heat is not negligible compared to the condensation heating for large clouds. With the inclusion of eddy flux of sensible heat and with a large cloud radius, both the Kuo (1974) and the Anthes (1977) schemes give a close reproduction of the convective heating profile.

To consider an extreme case, the entrainment is set to zero. With no entrainment, a wet adiabat represents the cloud temperature and mixing ratio. The Q_1 profiles (Fig. 6.57) are very similar to those computed from

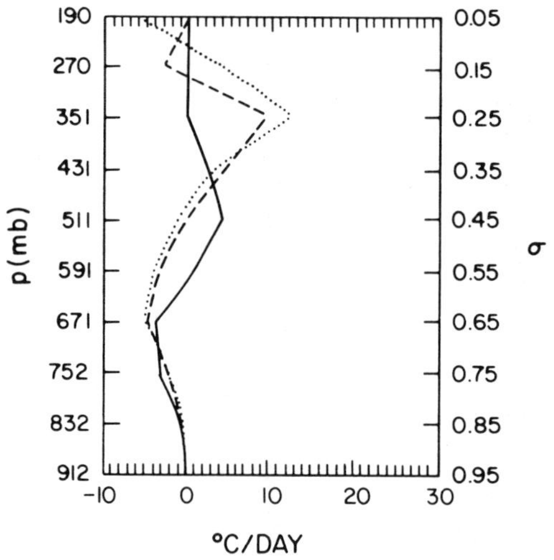

Fig. 6.56. Eddy sensible heat flux divergence (°C day^{-1}) for clouds with radii of 1 km (solid line), 3 km (dashed line), and 5 km (dotted line). [From Kuo and Anthes (1984c).]

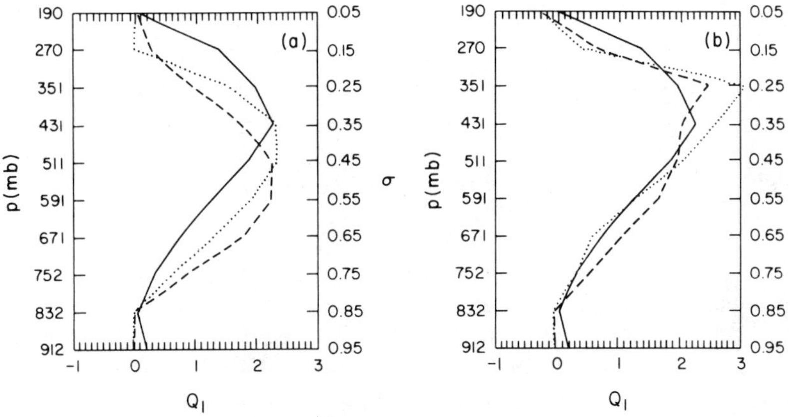

Fig. 6.57. Observed (solid line) and simulated Q_1 profiles (normalized) by cumulus parameterization schemes of Kuo (1974; dotted line) and Anthes (1977; dashed line) with no entrainment, averaged over the boxed-in area of Fig. 6.45a and over periods with the observed rainfall rate greater than 1 mm day^{-1}; (a) without eddy sensible heat flux, and (b) with eddy sensible heat flux. [From Kuo and Anthes (1984c).]

a large cloud radius with lateral entrainment and are very close to the observed profile. These results suggest that lateral entrainment is not important for strong convection of this type.

The above analysis indicates that both the Kuo (1974) and the Anthes (1977) schemes can reproduce the convective heating profile reasonably well when a large cloud radius is used (or when lateral entrainment is suppressed). It also suggests that the effect of eddy flux of sensible heat is quite important in midlatitude convection and should not be ignored in a cumulus parameterization scheme.

6.5.2.c Vertical Partitioning of Apparent Moisture Sinks

A fundamental assumption of Kuo schemes is that a fraction b of the total water vapor convergence is used to increase the humidity of the column. Therefore, in columns with convection the continuity equation for water vapor mixing ratio is written

$$\frac{\partial r_v}{\partial t} = \frac{gbM_t}{(p_b - p_t)} N_r(p) - \frac{\partial \overline{\omega'' r_v''}}{\partial p}, \tag{6.85}$$

where $N_r(p)$, the vertical distribution function for water vapor, satisfies an integral constraint such as Eq. (6.77). Various empirical functions have been proposed for $N_r(p)$; in the Kuo (1974) scheme,

$$N_r(p) = (r_{vc} - r_v)/\langle r_{vc} - r_v \rangle, \tag{6.86}$$

where r_{vc} is given by the wet adiabat used to compute T_c in Eq. (6.81). Anthes (1977) suggested that the convective moistening is a function of the relative humidity

$$N_r(p) = (100\% - \mathrm{RH})r_s(T)/\langle(100\% - \mathrm{RH})r_s(T)\rangle. \tag{6.87}$$

Equation (6.87) was later (Anthes *et al.*, 1982) further simplified to

$$N_r(p) = r_s/\langle r_s\rangle, \tag{6.88}$$

with the assumption that the relative humidity is fairly uniform in the vertical over areas of strong convection.

For all of these schemes to be valid, it is essential that the vertically integrated moisture convergence be greater than the rainfall rate, so that b is positive. However, observations show that at times the convective precipitation exceeds the total moisture convergence, implying a negative b (Fritsch *et al.*, 1976; Cho, 1976; Kuo and Anthes, 1984b; Lin, 1986).

One can revise the Kuo schemes by using the complete water vapor continuity equation [Eq. (6.4)] and using the condensation profile and eddy flux profile predicted by the cloud model. This will also enable the prediction of the convective drying (Q_2) profile consistent with that of the convective heating profile (Q_1). With the inclusion of eddy moisture flux, the normalized convective drying profile (Q_2) can be written as

$$N_r(\sigma) = \frac{ac(\sigma) + \partial\overline{\dot{\sigma}'' r_v''}/\partial\sigma}{\langle ac(\sigma) + \partial\overline{\dot{\sigma}'' r_v''}/\partial\sigma\rangle}. \tag{6.89}$$

The cloud coverage, a, can be computed following Eq. (6.84), and the effect of eddy moisture flux can be estimated by an equation analogous to Eq. (6.83).

The observed and simulated normalized Q_2 profiles by the revised scheme averaged over periods when the rainfall rate is larger than 1 mm day^{-1} (1800 GMT, 10 April to 1200 GMT, 11 April) are shown in Fig. 6.58. Most of the observed moisture sink (Q_2) is below 600 mbar, while most of the simulated condensation occurs in the middle troposphere. As the cloud radius is increased, the simulated condensation profile has its maximum at a slightly higher level. There are larger discrepancies between the simulated and observed Q_2 profiles when eddy moisture flux is ignored. The inclusion of eddy moisture flux shifts the simulated Q_2 profile downward and improves the scheme considerably. Better agreement is found when a larger cloud radius is used (Fig. 6.58b) or when entrainment is suppressed (not shown).

The convective eddy flux convergence of latent heat (moisture) is shown in Fig. 6.59. Its magnitude is compatible with that of the condensation heating. This again emphasizes the importance of eddy flux of moisture by cumulus convection.

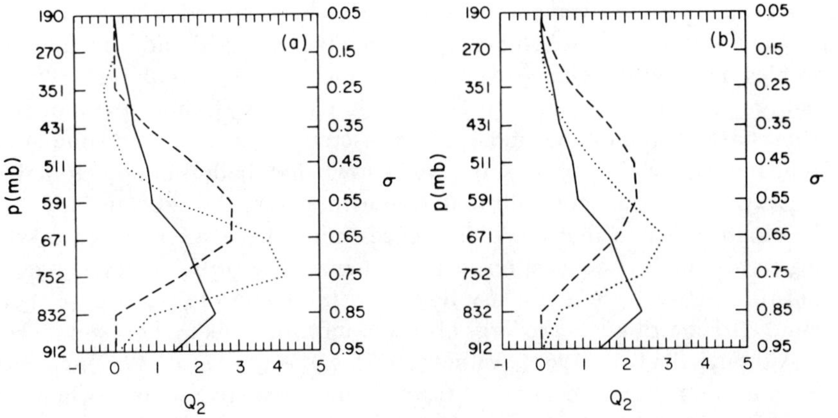

Fig. 6.58. Observed (solid) and simulated Q_2 profiles (normalized) by revised Anthes scheme with (dotted line) and without (dashed line) eddy moisture flux averaged over the boxed-in area of Fig. 6.45a and over periods when the observed rainfall rate was greater than 1 mm day^{-1}; (a) 1 km, and (b) 3 km; cloud radius is used in the steady-state cloud model. [From Kuo and Anthes (1984c).]

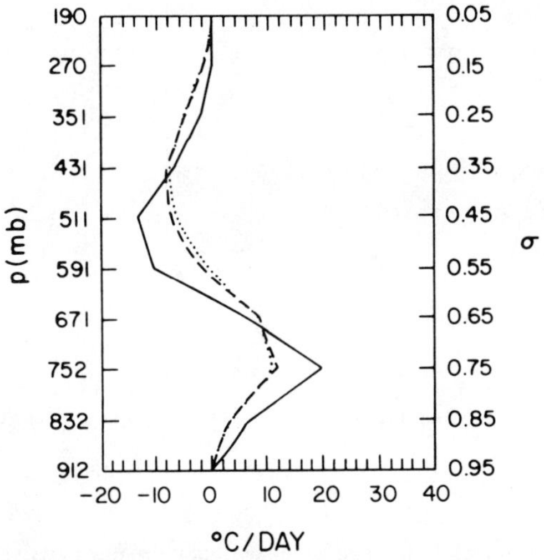

Fig. 6.59. Eddy latent heat flux convergence estimated by a steady-state cloud model with 1-km radius (solid line), 3-km radius (dashed line), and 1-km radius without entrainment (dotted line), averaged over the same area and period as in Fig. 6.58. [From Kuo and Anthes (1984c).]

Despite the improvement made by the inclusion of eddy moisture flux, substantial discrepancies still exist between the observed and simulated Q_2 profiles. The revised scheme overestimates the moisture sink above 800 mbar and underestimates the moisture sink below this level. Since the modeled cloud base is at about 830 mbar, these discrepancies cannot be eliminated by an increase of cloud radius. It is well known that shallow nonprecipitating convection can transport substantial moisture from the subcloud layer to the cloud layer, so that these differences might have been caused by the neglect of shallow convection in this scheme. The turbulent exchange of moisture between the planetary boundary layer and the free troposphere could also contribute to the large observed moisture sink in the lower layers.

Although the Kuo (1965), Anthes (1977), and Anthes *et al.* (1982) schemes are not valid after the mature stage of the convective system when the rainfall exceeds the moisture convergence (because b in these schemes cannot be negative), it is of interest to see how these schemes perform during the developing stage of the convective event when the observed b is positive. Figure 6.60 shows the normalized observed and simulated convective moistening profiles averaged over the developing period of the convective system (1500 GMT, 10 April to 2100 GMT, 10 April).

With the inclusion of eddy moisture flux, the Anthes *et al.* (1982) scheme shows some differences from the observed profile at small cloud radius (Fig. 6.60). However, when the cloud radius is increased (or when entrain-

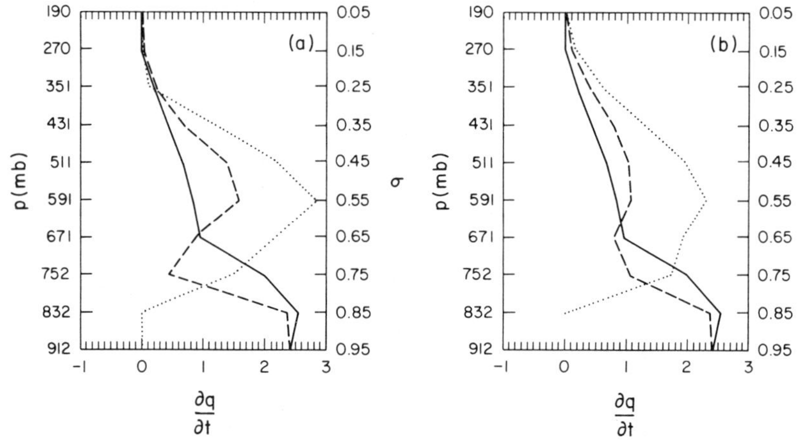

Fig. 6.60. Observed (solid line) and simulated normalized convective moistening ($\partial r_v/\partial t$) profiles by cumulus parameterization schemes of Kuo (1965; dotted line) and Anthes *et al.* (1982; dashed line) averaged over the boxed-in area of Fig. 6.45a and over the developing stage of the extratropical convective system for cloud radii equal to (a) 1 km and (b) 3 km, with inclusion of eddy moisture flux. [From Kuo and Anthes (1984c).]

ment is removed), the Anthes *et al.* (1982) scheme reproduces the general pattern of the convective moistening profile. The inclusion of convective eddy moisture flux does not improve the Kuo (1965) scheme very much. Since most of the convective moistening occurs in the lower troposphere, while $r_{vc} - r_v$ can only exist above the cloud base, these errors cannot be eliminated by either the inclusion of eddy sensible moisture flux or the variation of cloud radius.

The shape of the observed $(\partial r_v/\partial t)$ profile shows that most of the moistening exists in the lower troposphere. Since the moisture content also decreases with height, these results suggest that, during the developing stage, the net effect of the large-scale and convective redistribution of moisture is to increase the environmental relative humidity rather uniformly in the vertical.

6.5.2.d Determination of b

An important part of the Kuo schemes is the determination of the parameter b, the fraction of M_t used to moisten the column. Kuo (1965) partitioned the total rate of supply of moisture, I, into two parts, I_1 and I_2, where I_1 is the portion of I that is condensed and removed as precipitation, thereby increasing the temperature of the cloud column from T to T_c,

$$I_1 \equiv \frac{c_p}{L} \int_{p_t}^{p_b} (T_c - T)\, dp, \tag{6.90}$$

and I_2 is the part of I required to raise the mixing ratio of the cloud column from r_v to r_{vc},

$$I_2 \equiv \int_{p_t}^{p_b} (r_{vc} - r_v)\, dp. \tag{6.91}$$

After I is used to create a cloud of temperature T_c and mixing ratio r_{vc}, the large-scale temperature and moisture changes are computed by horizontal mixing of the cloud with its environment, with the total heating proportional to I_1 and the total moistening proportional to I_2. Although Kuo (1965) did not formally define a b parameter, an effective b for the 1965 scheme can be computed from the ratio of I_1 and I_2:

$$b \equiv \frac{I_2}{I_1 + I_2} = \frac{\langle r_{vc} - r_v \rangle}{\langle (c_p/L)(T_c - T) + r_{vc} - r_v \rangle}. \tag{6.92}$$

Kuo (1965) notes that, for a typical tropical sounding, I_1/I_2 is about 0.25, which yields a value of b of about 0.8. Thus, in the original Kuo scheme, most of the available moisture supply was used to moisten rather than heat the column, with the resulting underprediction of rainfall (Fig. 6.61). Observations (Kuo, 1974; Krishnamurti *et al.*, 1980) indicate b to be in the range 0.23–0.26.

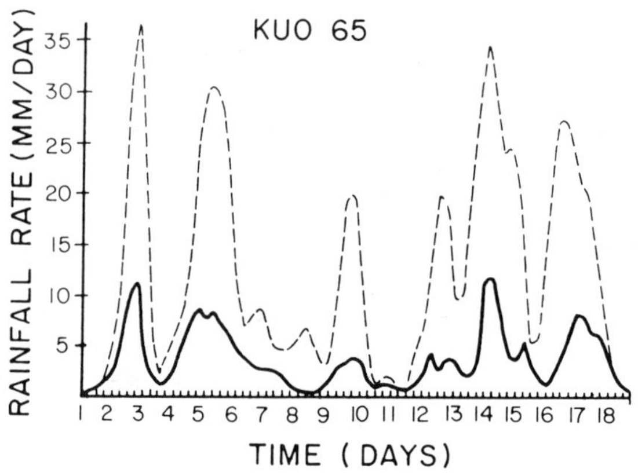

Fig. 6.61. Comparison between observed (dashed line) and predicted (solid line) rainfall rates for 1–18 September 1974, using the Kuo (1965) cumulus parameterization scheme. [From Krishnamurti *et al.* (1980).]

Anthes (1977) proposed that b could be related to the mean relative humidity in the column by an expression of the form

$$b = \left| \begin{matrix} \left[\dfrac{(1-\langle \mathrm{RH} \rangle)}{(1-\mathrm{RH_c})} \right]^n & \langle \mathrm{RH} \rangle \geq \mathrm{RH_c} \\ 1 & \langle \mathrm{RH} \rangle < \mathrm{RH_c} \end{matrix} \right|, \tag{6.93}$$

where $\mathrm{RH_c}$ is a critical value of relative humidity and n is a positive exponent of order 1 which may be empirically determined. In semiprognostic tests of the Anthes (1977) scheme, Kuo and Anthes (1984c) found the best agreement between observed and diagnosed rainfall rates when n is between 2 and 3 and $\mathrm{RH_c}$ is between 0.25 and 0.50.

Krishnamurti *et al.* (1980) partitioned the moisture convergence into horizontal and vertical advection and found that the vertical advection of mixing ratio coincided extremely well with observed rainfall rates during GATE Phase III; they wrote

$$b = -\frac{1}{gM_t} \int_0^{p_s} \mathbf{V} \cdot \nabla r_v \, dp. \tag{6.94}$$

Figure 6.62 shows the observed and diagnosed rainfall rates during GATE Phase III using a Kuo scheme with b determined by Eq. (6.94). Figure 6.63 shows the time variation of observed and diagnosed rainfall rates in an extratropical squall system (SESAME-I, 10–11 April 1979) using a Kuo scheme with b given by Eq. (6.93) with $n = 1.0$ and $\mathrm{RH_c} = 0.5$ and by Eq. (6.94). The success of Eq. (6.94) in diagnosing the observed rainfall rates

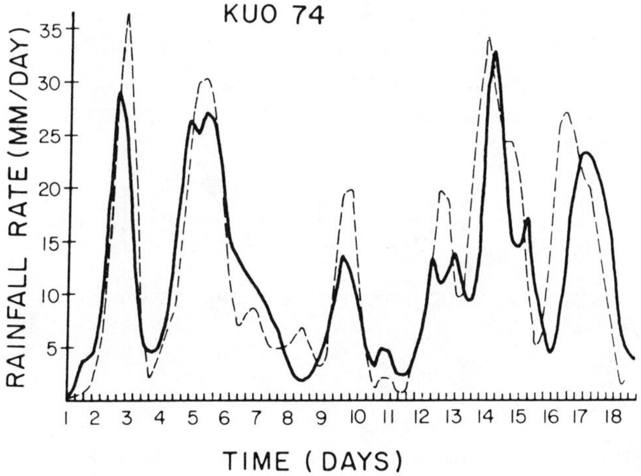

Fig. 6.62. As in Fig. 6.61, except for the Kuo (1974) scheme. [From Krishnamurti *et al.* (1980).]

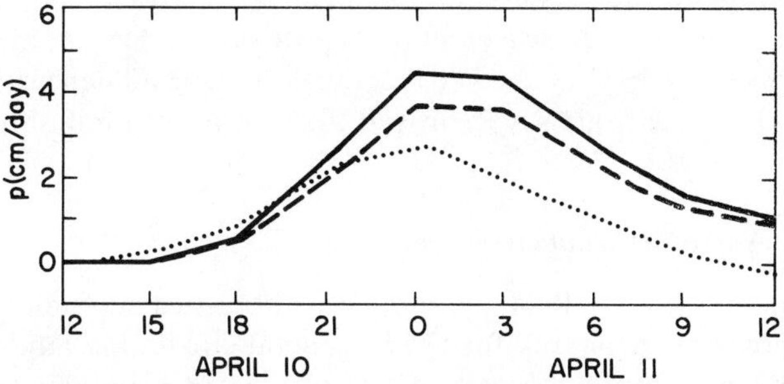

Fig. 6.63. Observed rainfall rate (solid line) and that simulated by Krishnamurti *et al.* (1980; dashed line) and Anthes (1977; dotted line) over the boxed-in area of Fig. 6.45a. [From Kuo and Anthes (1984c).]

in a tropical and extratropical system indicates that the net horizontal advection of moisture is not as closely related to precipitation as is the vertical advection, which is reasonable because a strong relationship between upward motion and precipitation is well known. In contrast, horizontal advection of water vapor of either sign may occur with little vertical motion or precipitation. Apparently, the horizontal advection of water vapor is more closely related to the storage term S_{rv} in Eq. (6.74).

Molinari (1982) compared the rate at which the standard subtropical sounding (Jordan, 1958) approached a moist neutral condition when a specified vertical motion and moisture supply was used with various formulations of b. Molinari found that the most realistic time evolution of the sounding occurred when a formulation of b was used that required the environmental temperature and mixing ratio to approach the cloud values at the same rate,

$$b = \frac{I+J}{I}\left[\frac{\langle r_{\mathrm{vc}} - r_{\mathrm{v}}\rangle}{\langle r_{\mathrm{vc}} - r_{\mathrm{v}}\rangle + (c_{\mathrm{p}}/L)\langle T_{\mathrm{c}} - T\rangle}\right], \tag{6.95}$$

where J is the net adiabatic temperature change in the column

$$J \equiv -\frac{1}{g}\int_{p_{\mathrm{t}}}^{p_{\mathrm{b}}} \frac{c_{\mathrm{p}}}{L}\frac{T}{\theta}\,\omega\frac{\partial\theta}{\partial p}\,dp. \tag{6.96}$$

In Molinari's tests, at the initial time $J \approx -0.8I$, so that the b parameter computed from Eq. (6.95) was about 0.2 times the effective b parameter of the original Kuo scheme. This reduction produced a value of b (0.13) much closer to observed values.

One deficiency of the Kuo schemes discussed above is that the effects of precipitating downdrafts are not included. Molinari and Corsetti (1985) incorporated a convective-scale and mesoscale downdraft parameterization in a Kuo-type scheme. They showed that the stabilizing influence of low-level cooling by downdrafts greatly improved the forecast of rainfall in one case study.

6.5.3 The Arakawa–Schubert Scheme

Arakawa and Schubert (1974) developed a sophisticated cumulus parameterization scheme, primarily for use in general circulation models, that is considerably more general than the Kuo schemes. In particular, a spectrum of cloud types is considered, the scheme is coupled with a model of the mixed layer, and the large-scale forcing function involves horizontal and vertical advection, radiation, and surface fluxes of heat and moisture (rather than only large-scale moisture convergence as in Kuo schemes).

The Arakawa–Schubert (AS) scheme assumes that an ensemble of cumulus clouds affects the environment in two major ways: (1) by inducing subsidence between the cloud which warms and dries the environment, and (2) through detrainment of saturated air which contains liquid water from the cloud top. Evaporation of the detrained cloud water causes a cooling and moistening of the environment. To determine these effects quantitatively, one must determine the vertical distribution of the total mass flux by the clouds, the detrainment of mass from the clouds, and the temperature, water vapor, and liquid-water contents of the detrained air.

The cloud mass flux is determined by assuming that a property of the large-scale atmosphere, called the *cloud-work function*, is in a quasiequilibrium state, determined by an approximate balance between large-scale processes and the effects of cumulus convection. The cloud-work function is a measure of the integrated buoyancy force in the clouds. A close balance between the large-scale and small-scale effects produces a series of quasi-balanced states from which the cloud-mass flux can be determined from the large-scale variables.

This section summarizes the AS cumulus parameterization scheme; because of its complexity, a complete derivation is not presented. The summary is divided into four parts: (1) derivation of the large-scale equations for heat and water vapor, which include cloud effects; (2) the spectral representation of the cumulus ensemble; (3) the cloud-work function and the quasiequilibrium assumption; and (4) tests of the AS scheme.

6.5.3.a Large-Scale Prognostic Equations for Heat and Moisture

The derivation of the prognostic equations for the large-scale average of dry static energy s and mixing ratio r_v is quite similar to that presented in Section 6.3.3 for diagnostic studies. Cumulus clouds, which are assumed to cover a small fraction of the large-scale area, are classified according to the height at which they detrain, and this height is assumed to be a unique function of an entrainment parameter λ. Budget equations for s, r_v, liquid-water r_l, and mass are written for each cloud type and s and r_v for the environment of the clouds, which is assumed to have horizontally uniform properties. Detrainment of saturated air containing liquid water is assumed to occur in a thin layer near cloud top. These assumptions and the budget equations lead to prognostic equations similar to the diagnostic equations, Eqs. (6.52) and (6.53):

$$\frac{\partial s}{\partial t} = -\mathbf{V} \cdot \nabla s - w\frac{\partial s}{\partial z} + Q_R + M_c\frac{\partial s}{\partial z} + D(\hat{s} - s + L\hat{r}_l), \tag{6.97}$$

$$\frac{\partial r_v}{\partial t} = -\mathbf{V} \cdot \nabla r_v - w\frac{\partial r_v}{\partial z} + M_c\frac{\partial r_v}{\partial z} + D(\hat{r}_v^* - r_v + \hat{r}_l). \tag{6.98}$$

The total cloud-mass flux M_c is given by Eq. (6.29); $\hat{s}$, $\hat{r}_v^*$, and $\hat{r}_l$ are the values of static energy, saturation mixing ratio, and liquid-water mixing ratio of all clouds which lose buoyancy at the same level z. Note that in this section the cloud-mass flux has the dimensions of w rather than ω as in Section 6.3. The total detrainment D is given by the sum of detrainment over all clouds

$$D \equiv \sum_i D_i, \tag{6.99}$$

where

$$D_i = \frac{M_{ci}}{\Delta z_{Di}} = \frac{a_i w_i}{\Delta z_{Di}}, \tag{6.100}$$

and a_i is the fraction of area covered by the ith cloud type, w_i is the vertical velocity in the ith cloud, and Δz_{Di} is the thickness of the detrainment layer for the ith cloud.

The cloud effects represented by the last two terms on the right side of Eqs. (6.97) and (6.98) are the same as those represented in the diagnostic equations, Eqs. (6.52) and (6.53), namely, adiabatic warming and drying associated with compensating subsidence and effects of detrained heat, water vapor, and liquid water from the cloud tops. In contrast to the diagnostic equations, however, the prognostic equations are expressed in terms of large-scale average variables (s and r_v) rather than the variables in the cloud environment ($\tilde{s}$ and $\tilde{r}_v$). As noted by Arakawa and Schubert, because the fraction of the area covered by cumulus clouds is small, prediction of the large-scale average is practically the same as prediction of the cloud environment thermodynamic variables. As indicated by Eqs. (6.97) and (6.98), the AS parameterization of cumulus effects requires determination of (1) the total mass flux in clouds $M_c(z)$, and (2) the total detrainment from the clouds and the mixing ratio of liquid water $\hat{r}_l(z)$ at the vanishing buoyancy level. The static energy $\hat{s}$ and saturation mixing ratio $\hat{r}_v^*$ are determined from the assumption of zero buoyancy at the detrainment level,

$$s_{v_i} = s_v, \tag{6.101a}$$

where s_v is the *virtual static energy* and is given by

$$s_v = s + c_p T(0.608 r_v - r_l). \tag{6.101b}$$

6.5.3.b Spectral Representation of the Cumulus Ensemble

In the AS cumulus parameterization scheme, a single parameter λ characterizes each cloud type, where λ is a fractional rate of mass entrainment which is constant with height (for a given cloud type). The total mass flux M_c is expressed as

$$M_c(z) = \int_0^{\lambda_D(z)} G(z, \lambda)\, d\lambda, \tag{6.102}$$

where $G(z, \lambda)\, d\lambda$ is the mass flux associated with clouds that have the parameter λ_i in the interval $(\lambda, \lambda + d\lambda)$ (Fig. 6.64),

$$G(z, \lambda)\, d\lambda = \sum_{\lambda_i} M_i(z). \tag{6.103}$$

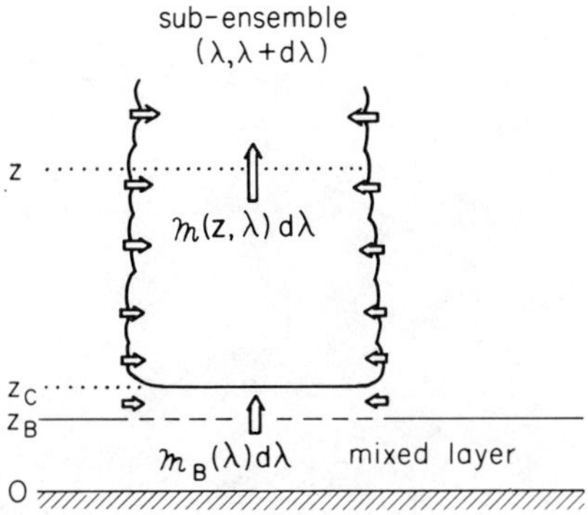

Fig. 6.64. Schematic diagram showing subensemble of type λ clouds and subcloud mixed layer. The updraft, which originates in the mixed layer, is unsaturated in region between z_C and z_B. [From Arakawa and Schubert (1974).]

This mass flux is written as a function of the mass flux at cloud base $G_B(\lambda)$

$$G(z, \lambda) \equiv G_B(\lambda)\eta(z, \lambda), \tag{6.104}$$

where the cloud base is assumed to be the height of the mixed layer (Fig. 6.64).

Budget equations, which include the effects of entrainment, are written for the mass, moist static energy and water substance $r_v + r_l$ for each subensemble; these yield the following equations:

mass

$$\partial\eta(z, \lambda)/\partial z = \lambda\eta(z, \lambda), \tag{6.105}$$

moist static energy

$$\partial h_c/\partial z = -\lambda(h_c - h), \tag{6.106}$$

and *water substance*

$$\partial(r_{vc} + r_l)/\partial z = -\lambda(r_{vc} - r_v + r_l) - \mathrm{CN}_{cr}, \tag{6.107}$$

where λ is the fractional rate of entrainment which determines each cloud type, and CN_{cr} represents the rate of conversion of cloud liquid water to precipitation. In Eqs. (6.106) and (6.107), h_c, r_{vc}, and r_l vary with height and λ, while the large-scale average variables h and r_v vary only with height.

The detrainment level (cloud top) of each cloud subensemble is determined by the level of vanishing buoyancy, determined by Eq. (6.101a). To find this level, the cloud profile of moist static energy is determined by integrating Eq. (6.106). This level, which is approximately the level at which

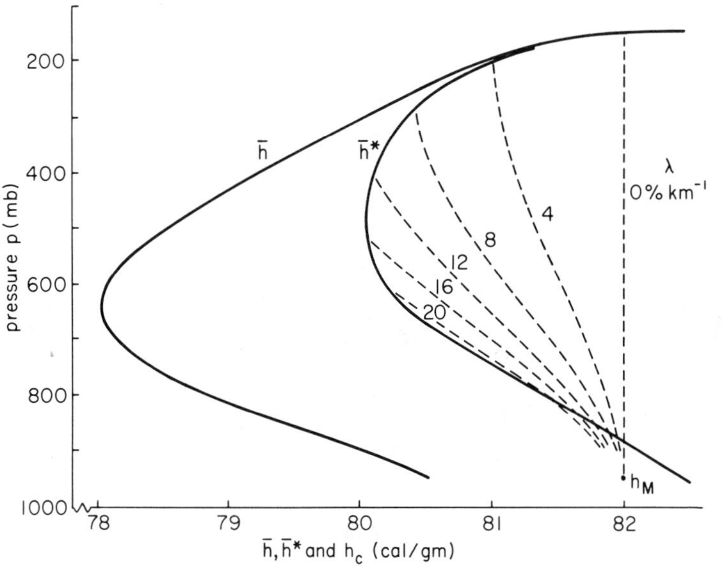

Fig. 6.65. Vertical profiles of $\bar{h}(p)$, $\bar{h}^*(p)$, and $h_c(p, \lambda)$; $h_c(p, \lambda)$ lines are dashed and are labeled with the value of λ in percent per kilometer. Profiles of $\bar{h}$ and $\bar{h}^*$ were obtained from Jordan's (1958) "mean hurricane season" sounding. The top p_B of the mixed layer is assumed to be 950 mbar; h_M is assumed to be 82 cal g^{-1}. [From Arakawa and Schubert (1974).]

$h_c(z, \lambda)$ intersects the saturation value of the moist static energy of the environment, $h^* \equiv s + Lr_v^*$, increases with decreasing entrainment rate (Fig. 6.65).

The remaining variables needed to complete the cumulus parameterization scheme are the total detrainment rate $D(z)$, and the variables at the cloud base level, z_B, $h_c(z_B, \lambda)$, $r_{vc}(z_B, \lambda)$, and the mass flux distribution at cloud base, $G_B(\lambda)$. The total detrainment $D(z)$ represents the detrainment by all clouds in the layer $z + dz$

$$D(z) = -G[z, \lambda_D(z)](d\lambda_D(z)/dz). \tag{6.108}$$

The height of the mixed layer z_B (and the base of all clouds) is determined by a prognostic equation which includes the effects of compensating subsidence between the cumulus clouds, large-scale horizontal advection and vertical motion, and surface turbulent fluxes of s and r_v,

$$\partial z_B/\partial t = -\mathbf{V}_B \cdot \nabla z_B - (M_B - \rho_B w_B) + kF_0/\Delta s_v, \tag{6.109}$$

where M_B is the vertical mass flux beneath the clouds, F_0 is the surface flux of virtual s, Δs_v is the jump in virtual s across z_B, k is a constant related to the entrainment of mass across z_B and is approximately 0.2, and the subscript B refers to the level of the top of the mixed layer.

The final variable required to complete the AS scheme is the cloud-base mass flux distribution, $G_B(\lambda)$. This variable is obtained from an assumption that a variable representing the total buoyancy associated with the cumulus ensemble, the *cloud-work function*, is in a quasiequilibrium state between the large-scale variables and the effects of the cumulus clouds.

6.5.3.c The Cloud-Work Function and Quasiequilibrium Assumption

The kinetic energy $K(\lambda)\,d\lambda$ per unit area of each subensemble of cumulus clouds which have fractional entrainment rates between λ and $\lambda + d\lambda$ changes through generation due to buoyancy forces and dissipation,

$$\frac{dK(\lambda)\,d\lambda}{dt} = [A(\lambda) - D(\lambda)]G_B(\lambda)\,d\lambda, \tag{6.110}$$

where $D(\lambda)$ is the cloud-scale dissipation rate per unit $G_B(\lambda)\,d\lambda$ and $A(\lambda)$ is the kinetic energy generation per unit $G_B(\lambda)\,d\lambda$. $A(\lambda)$ is called the cloud-work function and is an integral measure of the buoyancy force associated with clouds with λ in the range $\lambda + d\lambda$

$$A(\lambda) = \int_{z_B}^{z_D(\lambda)} \frac{g}{c_p T(z)}\,\eta(z, \lambda)[s_{vc}(z, \lambda) - s_v(z)]\,dz. \tag{6.111}$$

A positive value of A is necessary for the kinetic energy generation of clouds of type λ, therefore, $A(\lambda) > 0$ can be considered as a generalized criterion for moist convective instability. $A(\lambda) = 0$ for all λ indicates a neutral environment. Because $A(\lambda)$ depends only on the large-scale profiles of dry static energy and mixing ratio, including those in the mixed layer, $A(\lambda)$ and its time rate of change can be calculated in a prognostic model from the temporal derivatives of $s(z)$, $r_v(z)$, and z_B.

We have seen that instantaneous convective precipitation in the tropics is not closely related to the degree of instability; thus a given value of $A(\lambda)$ should not be expected to be closely related to the convective activity. Instead, the convective activity is related to the time rate of change of A, which is written as a sum of two terms, one due to large-scale processes $F(\lambda)$, and the other due to clouds

$$\frac{dA(\lambda)}{dt} = \left[\frac{dA(\lambda)}{dt}\right]_c + F(\lambda). \tag{6.112}$$

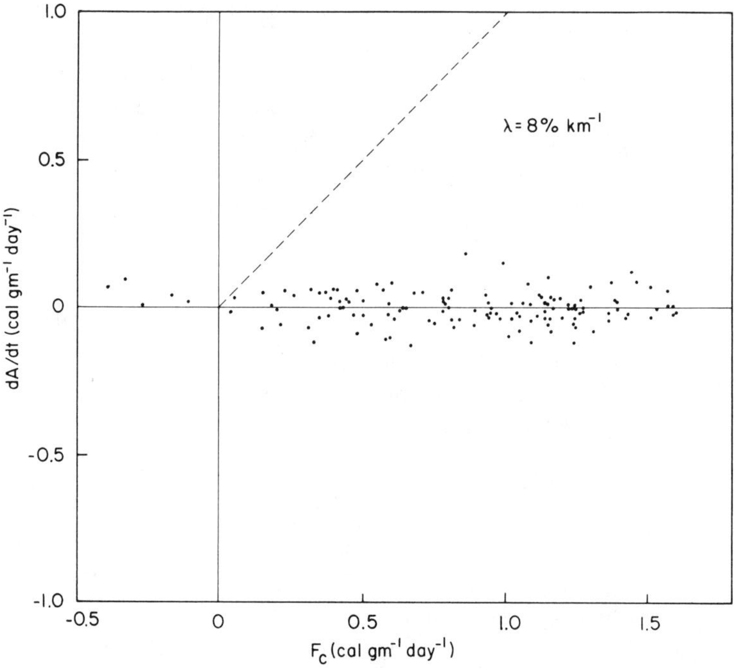

Fig. 6.66. Observational verification of quasiequilibrium assumption for cloud types $\lambda = 8\%\ \mathrm{km}^{-1}$ and $\lambda = 16\%\ \mathrm{km}^{-1}$, with the Marshall Islands data provided by Yanai *et al.* (1973). The abscissa is cloud-layer forcing $F_C(\lambda)$, and the ordinate is observed-time derivative of cloud-work function $dA(\lambda)/dt$. Top of mixed layer and cloud base are both assumed to be at 950 mbar. Zero buoyancy at cloud base is also assumed. Dashed line is the line along which $dA(\lambda)/dt = F_C(\lambda)$. [From Arakawa and Schubert (1974).] (*Figure continues.*)

In Eq. (6.112), the cloud terms depend linearly on $G_B(\lambda)$. The total contribution to the change of $A(\lambda)$ by all clouds is obtained by integrating over all λ

$$\left[\frac{dA(\lambda)}{dt}\right]_c = \int_0^{\lambda_{\max}} K(\lambda, \lambda')G_B(\lambda')\, d\lambda'. \tag{6.113}$$

The kernel $K(\lambda, \lambda')$, which depends on the cloud properties, including the mass flux at each level and the detrainment rate, is typically negative, which means that clouds act to stabilize the environment, mainly through adiabatic warming in compensating subsidence. The large-scale forcing function $F(\lambda)$ includes horizontal and vertical advection, sensible and latent heat fluxes from the surface, and radiative effects.

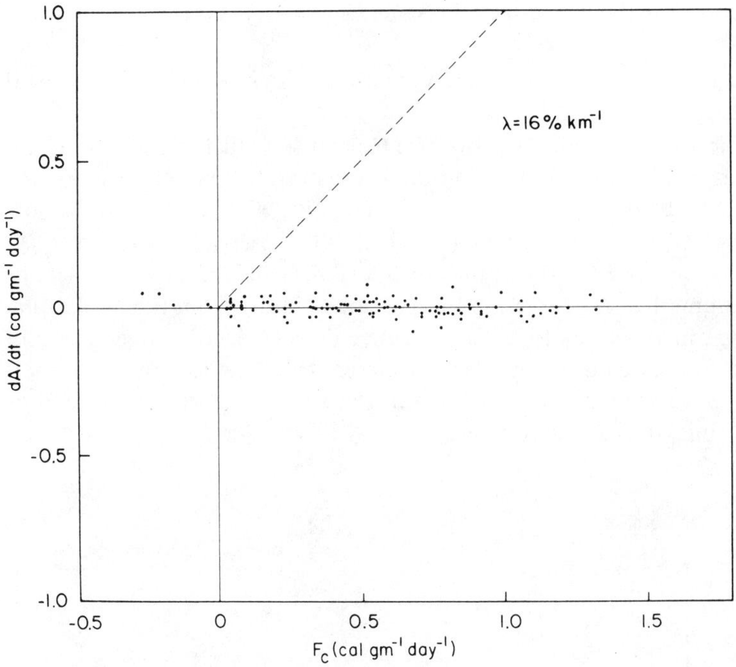

Fig. 6.66 (*continued*)

From Eqs. (6.112) and (6.113), $dA(\lambda)/dt$ can be written

$$\frac{dA(\lambda)}{dt} = \int_0^{\lambda_{\max}} K(\lambda, \lambda')G_B(\lambda')\, d\lambda + F(\lambda). \tag{6.114}$$

Observations indicate that the cumulus clouds react rapidly to the slower changes in the large-scale environment, so that the time rate of change of $A(\lambda)$ is small compared to the individual contributions by the cloud-scale and large-scale forcing functions in Eq. (6.114),

$$\left|\frac{dA(\lambda)}{dt}\right| \ll \left|\left[\frac{dA(\lambda)}{dt}\right]_c\right| \approx |F(\lambda)|. \tag{6.115}$$

Figure 6.66 shows observational verification of Eq. (6.115) using data from the Marshall Islands. The ordinate is $dA(\lambda)/dt$ and the abscissa is the contribution to the large-scale forcing $F(\lambda)$ due to processes above the mixed layer (F_c). Except for values of F_c close to 0, $dA(\lambda)/dt$ is much smaller than $F_c(\lambda)$. The assumption that Eq. (6.115) holds in general—the *quasiequilibrium assumption*—is used to obtain a diagnostic integral equation for the cloud-base mass flux from Eq. (6.114)

$$\int_0^{\lambda_{max}} K(\lambda, \lambda')G_B(\lambda')\, d\lambda' + F(\lambda) = 0. \tag{6.116}$$

The integral equation, Eq. (6.116), must hold for each cloud type for which $G_B(\lambda) > 0$. With additional constraints that clouds do not exist $[G_B(\lambda) = 0]$ if $A(\lambda)$ or $dA(\lambda)/dt$ is negative, Eq. (6.116) can be solved by an iterative scheme for $G_B(\lambda)$ given the large-scale forcing $F(\lambda)$. The expression for $F(\lambda)$ is complicated (Arakawa and Schubert, 1974); however, in a numerical model it can be estimated simply from the model-predicted change in $A(\lambda)$ due to all large-scale effects $(\Delta A/\Delta t)_{LS}$ over one time step. After first calculating this large-scale contribution to $dA(\lambda)/dt$, the cloud-scale processes can be calculated from $(\Delta A/\Delta t)_{LS}$, which will tend to bring $A(\lambda)$ back to its original value.

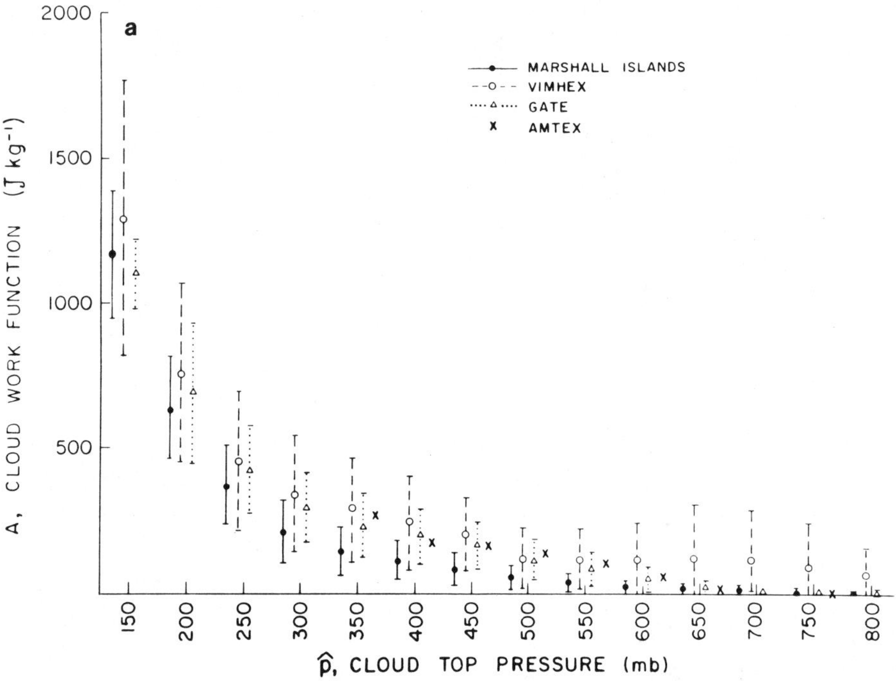

Fig. 6.67. (a) Mean values and standard deviations of cloud-work function versus cloud-top pressure $\hat{p}$ calculated from the Marshall Islands, VIMHEX, GATE, and AMTEX data sets. Error bars represent one standard deviation from the mean. (b) As in (a), except for the Marshall Islands undisturbed and disturbed cases, Jordan's sounding, and the composite typhoon data at $R = 0.7$, 2, and 4°. (c) As in (a), except for composite hurricane data at $R = 2$, 4, and 6°. The GATE mean values and the $R = 2°$ results are included for comparison. [From Lord and Arakawa (1980).] (*Figure continues.*)

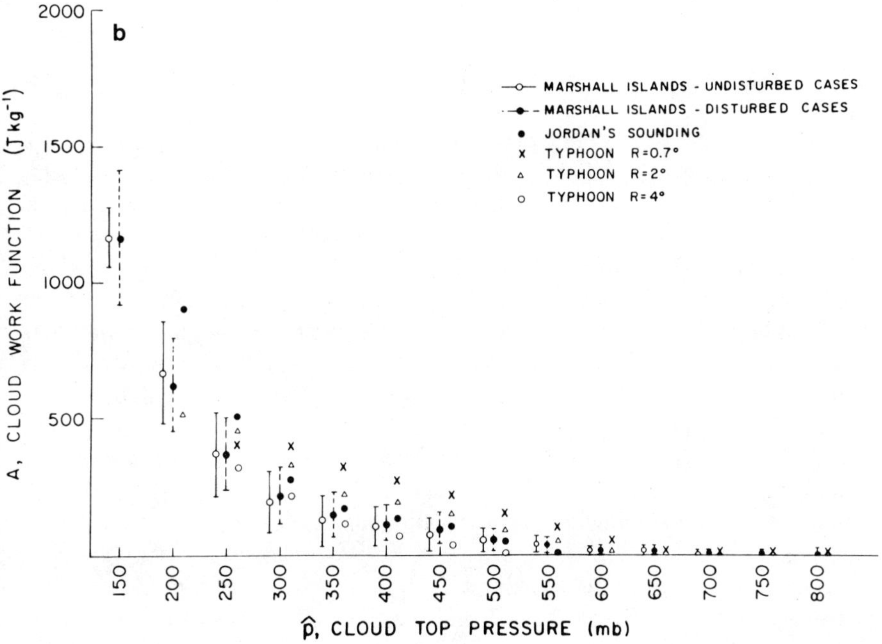

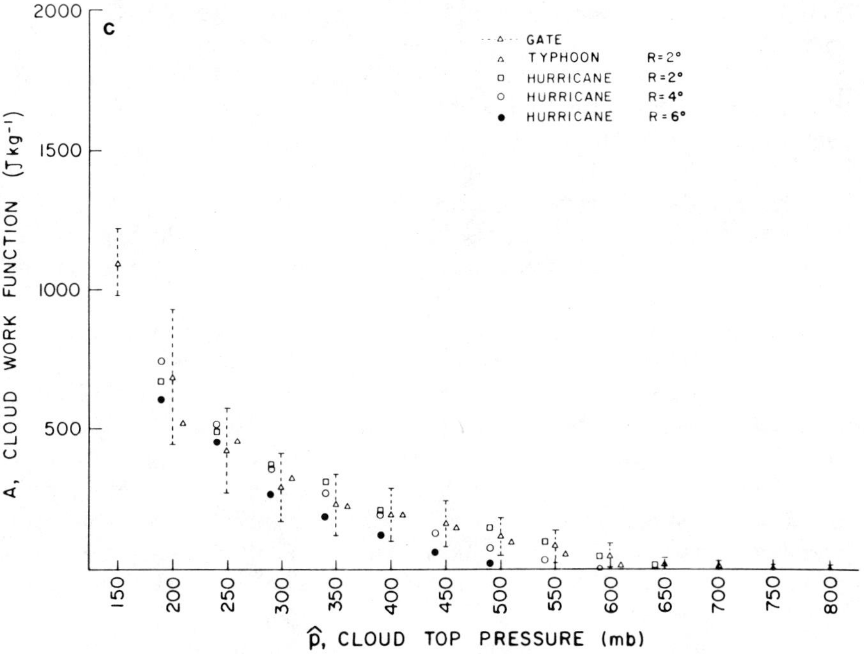

Fig. 6.67 *(continued)*

6.5.3.d *Verification of the Arakawa–Schubert Parameterization*

A complete verification of any cumulus parameterization scheme involves testing the scheme in a large-scale model and comparing the predictions of the model with the observed atmospheric structure at future times. It is useful, however, to verify individual parts of any scheme before attempting a complete verification. Figure 6.66, for example, supports the quasiequilibrium hypothesis of the AS scheme. This section summarizes additional verification studies of the AS scheme.

Lord and Arakawa (1980) consider the kinetic energy budget of a cumulus ensemble given by Eq. (6.110) and argue that the time scale associated with the dissipation term is much less than the time scale associated with the changes in the cloud-scale kinetic energy of the ensemble, which is equal to that of the large-scale processes. In this case, there is an approximate balance between $A(\lambda)$, which is a function only of the large-scale parameters, and $D(\lambda)$, which is mainly a function of cloud type. If this kinetic energy quasiequilibrium relation is valid, the cloud-work function should be closely correlated with cloud type and not a function of the synoptic situation or geographic location. To verify this assumption, Lord and Arakawa (1980) computed the cloud-work function $A(\lambda)$ for a variety of tropical locations and atmospheric conditions, including data sets from the Marshall Islands, the Venezuelan Meteorological and Hydrological Experiment (VIMHEX), GATE, AMTEX, mean West Indies sounding, and composite tropical cyclone soundings from the Pacific and West Indies. The cloud-work function for 17 possible cloud types, classified according to cloud top rather than the equivalent entrainment parameter λ, for the various data sets is plotted against cloud type in Fig. 6.67. The "error bars" represent one standard deviation from the mean. The similarity in cloud-work function versus cloud type, in spite of large individual difference in the temperature and moisture profiles between data sets, supports the hypothesis that the cloud-work function is uniquely related to cloud type in the tropics.

An important partial verification of cumulus parameterization schemes involves *semiprognostic tests*, in which observed estimates of large-scale advection, surface fluxes, and radiative heating are used in a cumulus parameterization scheme to estimate the cumulus properties, including precipitation, and the apparent heat source Q_1 and moisture sink Q_2 for successive observation times. The approach is termed semiprognostic because the computed cumulus effects do not influence the large-scale variables as they would in a prognostic model.

Lord (1982) used GATE data to test the AS scheme in a semiprognostic mode. Figure 6.68 shows the time series of precipitation rates from 1 to 18 September estimated by the AS parameterization P_{AS}, the observed moisture

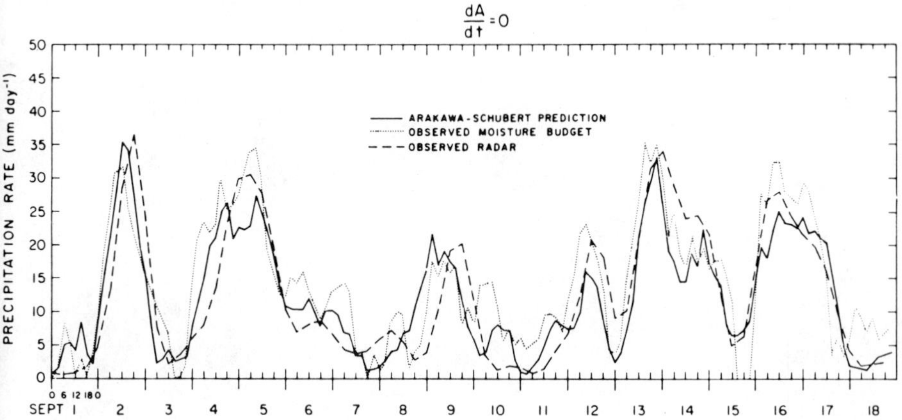

Fig. 6.68. Time series of precipitation rates (mm day^{-1}) from 1 to 18 September 1974, estimated by the Arakawa–Schubert scheme. [From Lord (1982).]

budget P_{Q2}, and observed radar measurements P_{RA}. There is generally good agreement among all three estimates, although radar estimates tend to lag behind P_{AS} and P_{Q2}. This lag is likely a result of the neglect of storage of liquid water in estimating P_{AS} and P_{Q2}.

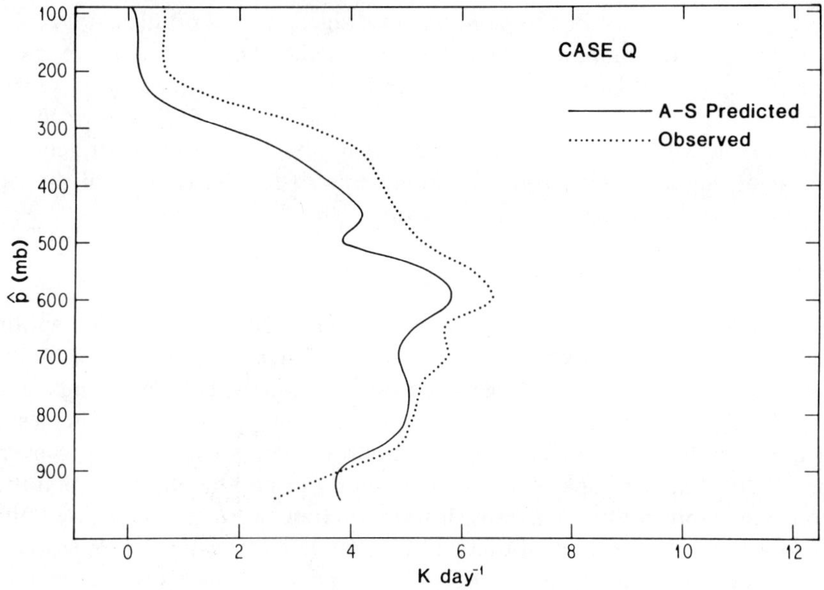

Fig. 6.69. Time-averaged calculated and observed $Q_1 - Q_R$. [From Lord (1982).]

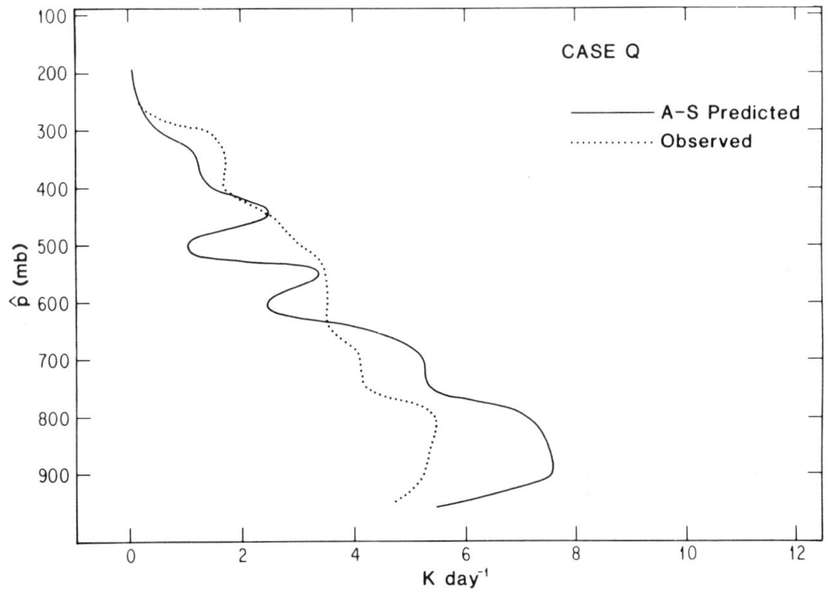

Fig. 6.70. Time-averaged calculated and observed Q_2. [From Lord (1982).]

Figure 6.69 shows the observed time-averaged vertical profiles of $Q_1 - Q_R$ and the profile computed from the AS scheme. The profiles have quite similar shapes, although the AS profile indicates a slight warm bias. The observed and calculated time-averaged Q_2 profiles (Fig. 6.70) show general agreement, with the greatest errors occurring below 750 mbar. In general, the semiprognostic tests provide strong support for the validity of the AS cumulus parameterization over the tropics.

6.5.4 Convective Parameterization on the Mesoscale

The Kuo schemes and the Arakawa–Schubert cumulus parameterization have the common property that the convective activity is controlled by the time rates of change of the large-scale variables (moisture convergence and the cloud-work function). Thus, the convection in both of these schemes is in quasiequilibrium with its large-scale environment. We have seen, however, on shorter time and space scales, particularly in middle latitudes, that the convective precipitation at any instant in time may greatly exceed the synoptic-scale supply, resulting in substantial changes in water storage. Also, as we consider the application of convective parameterization schemes in mesoscale models having resolution of less than 80–100 km, the assumption, that the convective scale and the resolvable scales are distinctly separate, becomes less and less defensible.

In seeking a convective parameterization scheme suitable for mesoscale models, we must first understand the different ways in which the atmosphere responds to heating (cumulus latent heating or other forms of heating) on different scales. To do so we must consider the relative roles of rotational stability and vertical static stability. A convenient parameter for identifying the scales at which rotational influences or inertial stability of a system become important is the Rossby radius of deformation λ_R (Rossby, 1938). Here we introduce the general definition of λ_R used by Frank (1983):

$$\lambda_R = \frac{NH}{(\zeta+f)^{1/2}(2VR^{-1}+f)^{1/2}}, \tag{6.117}$$

where N is the Brunt–Väisälä frequency, H is the scale height of the circulation, ζ is the relative vorticity, f is the Coriolis parameter, V is the rotational component of the wind, and R is the radius of curvature. The significance of λ_R is that it identifies the scale at which rotational influences or the inertial stability of a system become important. In linear CISK models, inertial stability is approximated by f^2. Schubert and Hack (1982) pointed out that this is only valid during the initial stages of cyclone development. As a tropical cyclone develops, for example, the relative vorticity and $2V/R$ become larger than f, which increases the inertial stability of the system. Thus if the scale of a disturbance exceeds λ_R, the system is nearly balanced such that the circulations evolve slowly, and vertical motions are largely controlled by those primary circulations. As shown by Schubert *et al.* (1980), a barotropic model with a single phase speed predicts that the mass field adjusts to the wind field when the scale of an initial disturbance is small compared to λ_R, and the wind field adjusts to the mass field when it is large compared to λ_R. This is illustrated in Fig. 6.71, which is adapted from Ooyama (1982) and Frank (1983). Thus in region III, the circulations are approximately in geostrophic balance and two-dimensional in nature. Convective systems smaller than λ_R represent unbalanced circulations in which convective heating excites gravity waves that propagate vertically and horizontally away from the convective disturbance. Thus instead of sustained local subsidence occurring just outside the region of convection, transient gravity waves are produced that can distribute the compensating motions well away from the region of convection. Region I includes the scale of convective clouds, at which deep convective overturning takes place. Vertical motions in region I are typically on the order of 10 m s^{-1}. In region II, by contrast, vertical motions are on the order of 1 m s^{-1} or less and are characterized by vertical displacements of a few kilometers.

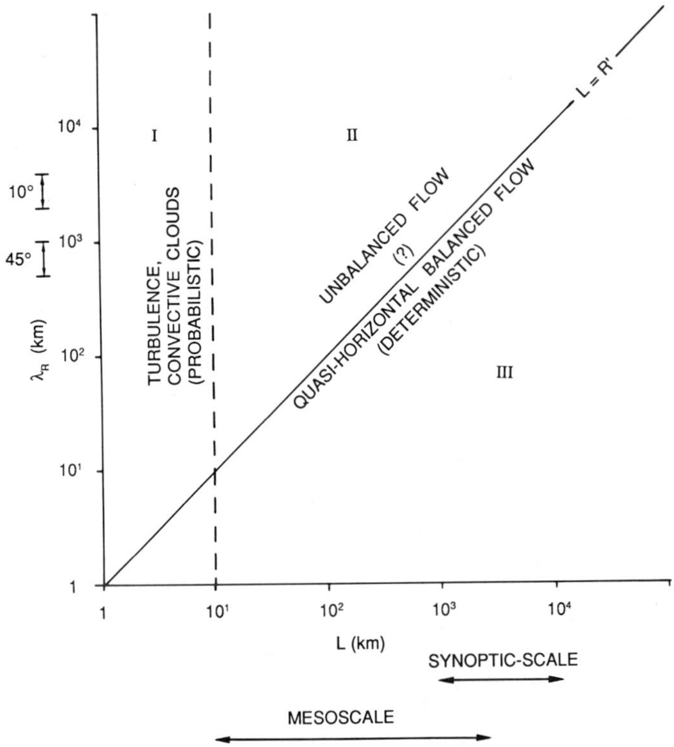

Fig. 6.71. Schematic of relationships between horizontal scale (L), the Rossby radius of deformation (λ_R), and modes of circulation. [After Ooyama (1982).] Typical values of λ_R for systems at latitudes 10° and 45° are indicated. [Adapted from Frank (1983).]

An important feature illustrated in Fig. 6.71 is that the geometric scale of a disturbance does not completely describe a system as being dynamically large or small. A disturbance having a radius of 700 km in the tropics may be dynamically small, while a similar-sized system in midlatitudes would be classified as dynamically large. Moreover, as a system develops relative vorticity, it can transform from a transient dynamically small system to a quasibalanced, dynamically large system without any significant change in geometric scale. Owing to the larger values of f in midlatitudes, however, small changes in relative vorticity or the rotational component of the wind results in little change in the magnitude of λ_R.

Tripoli *et al.* (1986) proposed that the Orlanski (1975) definition of mesoscales be modified so that the meso-α-scale represents scales of motion on the order of λ_R and greater or those scales where the atmospheric response to heating is dominated by the formation of balanced, persistent circulations. The meso-γ-scale is defined as the scale of individual deep convective clouds which typically range from 5 to 20 km. The meso-β-scale lies between λ_R and the cumulus scale and is characterized by transient gravity wave oscillations. The Kuo and Arakawa–Schubert-type parameterizations, in which a quasiequilibrium between convection and its large-scale environment is assumed, belong to the class of parameterization schemes suitable for the meso-α-scale and larger.

On the meso-β-scale and smaller, it is less obvious what the proper approach to parameterizing convection should be, since the convective response to transient meso-β-scale motions is not well understood.

In attempting to allow for the observed imbalances between the large-scale processes and the convection in mesoscale models, Kreitzberg and Perkey (1976) and Fritsch and Chappell (1980a) developed cumulus parameterization schemes that are based on the degree of potential instability present at a given time rather than large-scale rates of moisture convergence or destabilization. Because these schemes have not received the degree of use or verification that the Kuo and Arakawa–Schubert schemes have, we present only a summary of them here; the mathematical details are provided in the above references.

6.5.4.a The Kreitzberg–Perkey Scheme

In the Kreitzberg and Perkey (1976) scheme, a one-dimensional cumulus cloud model, is used to provide convective-scale tendencies to the larger scale model. In columns which support convection, a cloud base is selected which maximizes the cloud buoyancy. In contrast to most Kuo schemes and the AS scheme, cloud bases may occur at levels significantly higher than the PBL. After computation of the cloud properties, including cloud updraft, temperature, liquid water, and mixing ratio, the changes in the large-scale variables due to compensating subsidence between the clouds and mixing of cloud and environmental air are computed. These changes are converted to temporal tendencies for the large-scale variables by prorating the changes over 40 min.

In typical situations, the Kreitzberg–Perkey scheme produces warming and drying of the large-scale grid volume. Usually, one cloud is sufficient to remove the potential instability; however, if the modified sounding permits further convection, additional clouds are allowed.

The Kreitzberg–Perkey scheme has been used with some success in mesoscale numerical model simulations with real data (Chang *et al.*, 1981, 1982). The latent heating produced by the scheme is able to generate realistic perturbations to the temperature, geopotential, and wind field (Chang *et al.*, 1982).

6.5.4.b *The Fritsch–Chappell Scheme*

The cumulus parameterization scheme developed by Fritsch and Chappell (1980a) was designed especially for mesoscale models with horizontal resolutions of about 20 km. On this scale, convective clouds may occupy a substantial fraction of the model grid area. As in the Kreitzberg–Perkey scheme, a one-dimensional cloud model is used to calculate cloud properties. In contrast to the Kreitzberg–Perkey scheme, Fritsch and Chappell also consider moist downdrafts.

The basis for generating cumulus convection in the Fritsch–Chappell scheme is the presence of available buoyant energy, ABE, defined as the positive area on a thermodynamic chart when convection is possible

$$\mathrm{ABE} \equiv g \int_{\mathrm{LFC}}^{\mathrm{ETL}} \frac{T_c - T}{T}\, dz, \tag{6.118}$$

where LFC is the level of free convection and ETL is the level at which a parcel ascending wet adiabatically equals the environmental temperature (the equilibrium temperature level). Because convective rainfall is closely related to the amount of ABE in this scheme, the scheme should be less valid in the tropics where there is no strong relationship between convective precipitation and ABE.

Given the presence of ABE, moist updrafts and downdrafts are calculated with a one-dimensional cloud model. The fraction of updraft condensate evaporated in moist downdrafts is determined by an empirical relationship between the vertical wind shear and the precipitation efficiency. The adjusted large-scale average values of temperature, mixing ratio, and momentum are obtained from an area-weighted average of these properties in the environment, updraft and downdraft. The rate at which the stabilizing effects of the convection occur (removal of all ABE) is specified based on a time scale in the range of 30 min (the assumed lifetime of individual cells) to 1 h. The time scale in this range is based on the time required for the wind averaged throughout the cloud layer to advect the cloud across the mesoscale model grid.

In preliminary tests of this scheme using analytical initial conditions, Fritsch and Chappell (1980b) showed that the scheme was capable of generating convectively driven mesoscale pressure systems. Cooling by the parameterized moist downdrafts produced mesoscale highs in the lower troposphere, while warming by compensating subsidence was associated with mesoscale cyclogenesis.

One limitation of the Fritsch–Chappell scheme is that all the mass adjustments in response to deep convection must take place in a grid volume. Thus, all compensating subsidence due to convection must occur over relatively small volumes of air. As a result, this approach limits the smallest grid sizes that can be used in a model to be on the order of 20 km or greater. Otherwise one is faced with extreme stabilization of a grid volume when smaller grid sizes are employed.

6.5.4.c Explicitly Resolved Heating Schemes

Several modelers in recent years have tried to avoid the arbitrary nature of current convective parameterization schemes by explicitly calculating the effects of clouds and precipitation. Rosenthal (1978), for example, used an explicit representation of convective heating in a hydrostatic hurricane model with 20-km grid spacing. Precipitation processes and latent heating were modeled only on the explicitly resolved scales of motion. In spite of the absence of small-scale convective processes, Rosenthal obtained a reasonable simulation of hurricane amplification and structure. Yamasaki (1975, 1977) also simulated a tropical cyclone using a nonhydrostatic, axisymmetric model with a 400-m resolution in the inner region without a convective parameterization scheme. Yamasaki's success is not so surprising since he had sufficient resolution in the inner core region that motions on the cumulus-scale should be well represented. Rosenthal claims that an advantage of the explicitly resolved convective heating models is that they are able to "cross the line from noncooperative, non-CISK-scale interactions to the cooperative ones through the internal dynamics and energetics of the models. The CISK-type cooperation is not imposed."

Another example is Orlanski and Ross's (1984) simulation of a prefrontal squall line. They employed a hydrostatic, regional-scale model with 61.5-km grid spacing in which the explicitly resolved motions were the only source of convective heating and evaporative cooling. The only feature of their model that can be construed as a convective parameterization scheme is the introduction of a vertical eddy viscosity closure that is a function of a modified form of Richardson number. This has the effect of vertically mixing the model thermodynamic properties as well as momentum in regions of deep convection. The model simulated many large-scale features of observed squall lines (see Chapter 10 for further discussion).

Perhaps the reason that both Rosenthal and Orlanski and Ross were apparently successful in their simulations with coarse-resolution models without a convective parameterization is that in both systems the convection was strongly forced by the larger scales of motion. Rosenthal started with a weak-balanced vortex which was inertially stable and in which the convection was driven by low-level convergence established by a CISK-type mechanism. In Orlanski and Ross's case, the explicitly resolved convection was driven by convergence established by large-scale frontal baroclinicity. Orlanski and Ross did not have sufficient resolution to discern whether the convection was ever able to break away from the frontal forcing as in the case of a true prefrontal squall line. Support for this contention can be seen in Molinari and Corsetti's (1985) failure to simulate the more weakly forced MCC using only explicitly resolved heating. Thus the use of coarse-resolution models with explicitly resolved heating may be limited to applications where a preexisting, inertially stable circulation on scales greater than λ_R is the dominant feature driving the convective system.

It should also be noted that the resolution of the model, with either convective parameterization or explicitly resolved heating, has a major control on the scales to which convective heating is projected. That is, a coarse-resolution model is more likely to project convective heating onto scales greater than λ_R, and thereby contribute to the simulation of an inertially stable convective system, than is a fine-resolution model. As noted by Ooyama (1982), "we must be careful not to play the game with loaded dice."

Some further insight into the convective parameterization problem on the mesoscale can be obtained from the explicit simulations of orogenic mesoscale systems described by Tripoli (1986). As described more fully in Chapter 10, Tripoli simulated the genesis of mesoscale convective systems over the Colorado Rocky Mountains. The simulations were performed in two dimensions over an approximately 1000-km domain with 1.08-km resolution using explicit representations of precipitation physics and convective heating. Tripoli *et al.* (1986) then compared the results of simulations with the fine-resolution, explicit-physics model with a 14-km resolution model using a Fritsch and Chappell-type of parameterization scheme. The explicit-heating, fine-resolution model clearly demonstrated that there is no distinct scale separation between the meso-β-scale and the convective scale. Convective heating on the meso-γ-scale was clearly of larger amplitude, but the meso-β-scale transports and heating were of the same order of magnitude. In addition, the explicit-heating, fine-resolution model revealed that the convective systems continually produced transient gravity waves which propagated laterally and vertically out of the domain. The persistent, intense convective systems, however, remained locked to persistent planetary boundary layer circulations. In the simulations with parameterized convection, the convective systems often coupled with the transient gravity waves and moved with a phase speed of these waves much faster than either the explicitly simulated systems or observed systems.

Overall, these results suggest that the cumulus parameterization approach for models resolving meso-β-scale features is not well posed, whereas it is for models on the meso-α-scale and larger. Thus we must seek new ways for formulating convective parameterization schemes in models resolving features smaller than λ_R. It appears that fine-resolution models which explicitly resolve the details of convective-scale/mesoscale interactions are the most promising source of insight into those interactions.

With the continuing advances in the speed and memory of computers, the day is rapidly approaching when the interactions between cumulus convective systems and the larger scales of motion can be computed explicitly by models with high enough resolution to model convective-scale motion directly, while at the same time covering a large enough domain to resolve simultaneously the larger scales. In addition to providing fundamental insights into the interaction of these different scales of motion, including radiation and PBL effects, these future models will be useful in verifying the closure assumptions of the various cumulus parameterization schemes.

References

Anthes, R. A. (1977). A cumulus parameterization scheme utilizing a one-dimensional cloud model. *Mon. Weather Rev.* **105**, 270–286.

Anthes, R. A. (1982). Tropical cyclones—Their evolution, structure and effects. *Meteorol. Monogr.* **41**.

Anthes, R. A., Y.-H. Kuo, S. G. Benjamin, and Y.-F. Li (1982). The evolution of the mesoscale environment of severe local storms: Preliminary modeling results. *Mon. Weather Rev.* **110**, 1187–1213.

Anthes, R. A., Y.-H. Kuo, and J. R. Gyakum (1983). Numerical simulations of a case of explosive marine cyclogenesis. *Mon. Weather Rev.* **111**, 1174–1188.

Arakawa, A., and W. H. Schubert (1974). Interaction of a cumulus cloud ensemble with the large-scale environment. Part I. *J. Atmos. Sci.* **31**, 674–701.

Asai, T. (1970). Three-dimensional features of thermal convection in a plane Couette flow. *J. Meteorol. Soc. Jpn.* **48**, 18–29.

Betts, A. K. (1975). Parametric interpretation of trade-wind cumulus budget studies. *J. Atmos. Sci.* **32**, 1934–1945.

Chang, C. B., D. J. Perkey, and C. W. Kreitzberg (1981). A numerical case study of the squall line of 6 May 1975. *J. Atmos. Sci.* **38**, 1601–1615.

Chang, C. B., D. J. Perkey, and C. W. Kreitzberg (1982). A numerical case study of the effects of latent heating on a developing wave cyclone. *J. Atmos. Sci.* **39**, 1555–1570.

Charney, J. G., and A. Eliassen (1964). On the growth of the hurricane depression. *J. Atmos. Sci.* **21**, 68–75.

Cho, H. R. (1976). Effects of cumulus cloud activity on the large-scale moisture distribution as observed on Reed–Recker's composite easterly waves. *J. Atmos. Sci.* **33**, 1117–1119.

Cho, H. R., and L. Cheng (1980). Parameterization of horizontal transport of vorticity by cumulus convection. *J. Atmos. Sci.* **37**, 812–826.

Cho, H. R., and Y. Ogura (1974). A relationship between the cloud activity and the low-level convergence as observed in Reed–Recker's composite easterly waves. *J. Atmos. Sci.* **31**, 2058–2065.

Cho, H. R., R. M. Bloxam, and L. Cheng (1979). GATE A/B-scale budget analysis. *Atmos. Ocean* **17**, 60–76.

Chu, J.-H., M. Yanai, and C.-H. Sui (1981). Effects of cumulus convection on the vorticity field in the tropics. Part I: The large-scale budget. *J. Meteorol. Soc. Jpn.* **59**, 535–546.

Dopplick, T. G. (1970). Global radiative heating of the earth's atmosphere. Rep. No. 24, Planet. Circ. Proj., Dep. Meteorol., Mass. Inst. Technol.

Esbensen, S. K., E. I. Tollerud, and J.-H. Chu (1982). Cloud-cluster-scale circulations and the vorticity budget of synoptic-scale waves over the eastern Atlantic Intertropical Convergence Zone. *Mon. Weather Rev.* **110**, 1677–1692.

Fankhauser, J. C. (1969). Convective processes resolved by a mesoscale rawinsonde network. *J. Appl. Meteorol.* **8**, 778–798.

Fankhauser, J. C. (1974). The derivation of consistent field of wind and geopotential height from mesoscale rawinsonde data. *J. Appl. Meteorol.* **12**, 1330–1353.

Frank, W. M. (1983). The cumulus parameterization problem. *Mon. Weather Rev.* **111**, 1859–1871.

Fritsch, J. M. (1975). Cumulus dynamics: Local compensating subsidence and its implications for cumulus parameterization. *Pure Appl. Geophys.* **13**, 851–867.

Fritsch, J. M., and C. F. Chappell (1980a). Numerical prediction of convectively driven mesoscale pressure systems. Part I: Convective parameterization. *J. Atmos. Sci.* **37**, 1722–1733.

Fritsch, J. M., and C. F. Chappell (1980b). Numerical prediction of convectively driven mesoscale pressure systems. Part II: Mesoscale model. *J. Atmos. Sci.* **37**, 1734–1762.

Fritsch, J. M., and R. A. Maddox (1981). Convectively driven mesoscale weather systems aloft. Part I: Observations. *J. Appl. Meteorol.* **20**, 9–19.

Fritsch, J. M., C. F. Chappell, and L. R. Hoxit (1976). The use of large-scale budgets for convective parameterization. *Mon. Weather Rev.* **104**, 1408–1418.

Gyakum, J. R. (1983). On the evolution of the QE II storm. II. Dynamic and thermodynamic structure. *Mon. Weather Rev.* **111**, 1156–1173.

Hodur, R. M., and J. S. Fein (1977). A vorticity budget over the Marshall Islands during the spring and summer months. *Mon. Weather Rev.* **105**, 1521–1526.

Houze, R. A., Jr. (1973). A climatological study of vertical transports by cumulus-scale convection. *J. Atmos. Sci.* **30**, 1112–1123.

Houze, R. A., and A. K. Betts (1981). Convection in GATE. *Rev. Geophys. Space Phys.* **19**, 541–576.

Hudson, H. R. (1971). On the relationship between horizontal moisture convergence and convective cloud formation. *J. Appl. Meteorol.* **10**, 755–762.

Johnson, R. H. (1976). The role of convective-scale precipitation downdrafts in cumulus and synoptic-scale interactions. *J. Atmos. Sci.* **33**, 1890–1910.

Johnson, R. H. (1980). Diagnosis of convective and mesoscale motions during Phase III of GATE. *J. Atmos. Sci.* **37**, 733–753.

Jordan, C. L. (1958). Mean soundings for the West Indies area. *J. Meteorol.* **15**, 91–97.

Kreitzberg, C. W., and D. J. Perkey (1976). Release of potential instability: Part I. A sequential plume model within a hydrostatic primitive equation model. *J. Atmos. Sci.* **33**, 456–475.

Krishnamurti, T. N., and W. J. Moxim (1971). On parameterization of convective and nonconvective latent heat release. *J. Appl. Meteorol.* **10**, 3–13.

Krishnamurti, T. N., M. Kanamitsu, R. Godbole, C. B. Chang, F. Carr, and J. Chow (1976). Study of a monsoon depression (II). Dynamical structure. *J. Meteorol. Soc. Jpn.* **54**, 208–225.

Krishnamurti, T. N., Y. Ramanathan, H.-L. Pan, R. J. Pasch, and J. Molinari (1980). Cumulus parameterization and rainfall rates. I. *Mon. Weather Rev.* **108**, 465–472.

Krishnamurti, T. N., S. L.-Nam, and R. Pasch (1983). Cumulus parameterization and rainfall rates. II. *Mon. Weather Rev.* **111**, 815–828.

Kuo, H. L. (1965). On formation and intensification of tropical cyclones through latent heat release by cumulus convection. *J. Atmos. Sci.* **22**, 40–63.

Kuo, H. L. (1974). Further studies of the parameterization of the influence of cumulus convection on large-scale flow. *J. Atmos. Sci.* **31**, 1232–1240.

Kuo, Y.-H., and R. A. Anthes (1984a). Accuracy of diagnostic heat and moisture budgets using SESAME-79 field data as revealed by observing systems simulation experiments. *Mon. Weather Rev.* **112**, 1465–1481.

Kuo, Y.-H., and R. A. Anthes (1984b). Mesoscale budgets of heat and moisture in a convective system over the central United States. *Mon. Weather Rev.* **112**, 1482–1497.

Kuo, Y.-H., and R. A. Anthes (1984c). Semiprognostic tests of Kuo-type cumulus parameterization schemes in an extratropical convective system. *Mon. Weather Rev.* **112**, 1498–1509.

Kurihara, Y. (1973). A scheme of moist convective adjustment. *Mon. Weather Rev.* **101**, 547–553.

Lee, C.-S. (1984). The bulk effects of cumulus momentum transports in tropical cyclones. *J. Atmos. Sci.* **41**, 590–603.

LeMone, M. A. (1983). Momentum transport by a line of cumulonimbus. *J. Atmos. Sci.* **40**, 1815–1834.

Lewis, J. M. (1971). Variational subsynoptic analysis with applications to severe local storms. *Mon. Weather Rev.* **99**, 786–795.

Lewis, J. M. (1975). Tests of the Ogura-Cho model on a prefrontal squall line case. *Mon. Weather Rev.* **103**, 764–778.

Lewis, J. M., Y. Ogura, and L. Gidel (1974). Large-scale influences upon the generation of a mesoscale disturbance. *Mon. Weather Rev.* **102**, 545–560.

Lian, W.-J. (1979). Generalization of Kuo's parameterization of cumulus convection. *Pap. Meteorol. Res.* **2**, 101–115.

Lin, M.-S. (1986). The evolution and structure of meso-α-scale convective complexes by composite analysis. Ph.D. Thesis, Colorado State Univ.

Lord, S. J. (1982). Interaction of a cumulus cloud ensemble with the large-scale environment. Part III: Semi-prognostic test of the Arakawa-Schubert cumulus parameterization. *J. Atmos. Sci.* **39**, 88–103.

Lord, S. J., and A. Arakawa (1980). Interaction of a cumulus cloud ensemble with the large-scale environment. Part II. *J. Atmos. Sci.* **37**, 2677–2692.

Maddox, R. A., D. J. Perkey, and J. M. Fritsch (1981). Evolution of upper-tropospheric features during the development of a mesoscale convective complex. *J. Atmos. Sci.* **38**, 1664–1674.

Malkus, J. S., and H. Riehl (1960). On the dynamics and energy transformation in steady state hurricanes. *Tellus* **12**, 1–20.

Malkus, J. S., and R. T. Williams (1963). On the interaction between severe storms and large cumulus clouds. *Meteorol. Monogr.* **5**, 59–64.

Manabe, S., J. Smagorinsky, and R. F. Strickler (1965). Simulated climatology of a general circulation model with a hydrological cycle. *Mon. Weather Rev.* **93**, 769–798.

Marwitz, J. D. (1972a). The structure and motion of severe hailstorms. Part I: Supercell storms. *J. Appl. Meteorol.* **11**, 166–179.

Marwitz, J. D. (1972b). The structure and motion of severe hailstorms. Part II: Multi-cell storms. *J. Appl. Meteorol.* **11**, 180–188.

Marwitz, J. D. (1972c). The structure and motion of severe hailstorms. Part III: Severely sheared storms. *J. Appl. Meteorol.* **11**, 189–201.

Matsumoto, S., K. Ninomiya, and T. Akiyama (1967). Cumulus activities in relation to the meso-scale convergence field. *J. Meteorol. Soc. Jpn.* **45**, 292–305.

McNab, A. L., and A. K. Betts (1978). A mesoscale budget study of cumulus convection. *Mon. Weather Rev.* **106**, 1317–1331.

Miyakoda, K., J. Smagorinsky, R. F. Strickler, and G. D. Hembree (1969). Experimental extended predictions with a nine-level hemispheric model. *Mon. Weather Rev.* **97**, 1–76.

Molinari, J. (1982). A method for calculating the effects of deep cumulus convection in numerical models. *Mon. Weather Rev.* **11**, 1527–1534.

Molinari, J., and T. Corsetti (1985). Incorporation of cloud-scale and mesoscale downdrafts into a cumulus parameterization: Results of one- and three-dimensional integrations. *Mon. Weather Rev.* **113**, 485–501.

Moncrieff, M. W., and J. S. A. Green (1972). The propagation and transfer properties of steady convective overturning in shear. *Q. J. R. Meteorol. Soc.* **98**, 336–352.

Ninomiya, K. (1971a). Dynamical analysis of outflow from tornado-producing thunderstorms as revealed by ATS III pictures. *J. Appl. Meteorol.* **10**, 275–294.

Ninomiya, K. (1971b). Mesoscale modification of synoptical situations from thunderstorm development as revealed by ATS III and aerological data. *J. Appl. Meteorol.* **10**, 1103–1121.

Nitta, T. (1972). Energy budget of wave disturbances over the Marshall Islands during the years of 1956 and 1958. *J. Meteorol. Soc. Jpn.* **50**, 71–84.

Nitta, T. (1975). Observational determination of cloud mass flux distribution. *J. Atmos. Sci.* **32**, 73–91.

Nitta, T. (1977). Response of cumulus updraft and downdraft to GATE A/B-scale motion systems. *J. Atmos. Sci.* **34**, 1163–1186.

Nitta, T. (1978). A diagnostic study of interaction of cumulus updrafts and downdrafts with large-scale motions in GATE. *J. Meteorol. Soc. Jpn.* **56**, 232–242.

Nitta, T., and S. Esbensen (1974). Heat and moisture budget analyses using BOMEX data. *Mon. Weather Rev.* **102**, 17–28.

Ogura, Y. (1975). On the interaction between cumulus clouds and the larger scale environment. *Pure Appl. Geophys.* **13**, 869–889.

Ogura, Y. (1984). Response of cumulus clouds to large-scale forcing and cumulus parameterization. *FGGE Workshop, Natl. Res. Counc., Woods Hole, Mass.*

Ogura, Y., and Y. Chen (1977). A life history of an intense mesoscale convective storm in Oklahoma. *J. Atmos. Sci.* **34**, 1458–1476.

Ogura, Y., and H. R. Cho (1973). Diagnostic determination of cumulus cloud populations from observed large-scale variables. *J. Atmos. Sci.* **30**, 1276–1286.

Ogura, Y., and D. Portis (1982). Structure of the cold front observed in SESAME-AVE III and its comparison with the Hoskins–Bretherton frontogenesis model. *J. Atmos. Sci.* **39**, 2773–2792.

Ogura, Y., M.-T. Liou, J. Russell, and S. T. Soong (1979). On the formation of organized convective systems observed over the eastern Atlantic. *Mon. Weather Rev.* **107**, 426–441.

Ooyama, K. (1964). A dynamical model for the study of tropical cyclone development. *Geofis. Int.* **4**, 187–198.

Ooyama, K. (1971). A theory on parameterization of cumulus convection. *J. Meteorol. Soc. Jpn.* **39**(Spec. Issue), 744–756.

Ooyama, K. (1982). Conceptual evolution of the theory and modeling of the tropical cyclone. *J. Meteorol. Soc. Jpn.* **60**, 369–380.

Orlanski, I. (1975). A rational subdividion of scale for atmospheric processes. *Bull. Am. Meteorol. Soc.* **56**, 527–530.

Orlanski, I., and B. B. Ross (1984). The evolution of an observed cold front. Part II. Mesoscale dynamics. *J. Atmos. Sci.* **41**, 1669–1703.

Reed, R. J., and R. H. Johnson (1974). The vorticity budget of synoptic-scale wave disturbances in the tropical western Pacific. *J. Atmos. Sci.* **31**, 1784–1790.

Reed, R. J., and E. E. Recker (1971). Structure and properties of synoptic-scale wave disturbances in the equatorial western Pacific. *J. Atmos. Sci.* **28**, 1117–1133.

Reed, R. J., D. C. Norquist, and E. E. Recker (1977). The structure and properties of African wave disturbances as observed during Phase III of GATE. *Mon. Weather Rev.* **105**, 317–333.

Reeves, R. W., C. F. Ropelewski, and M. D. Hedlow (1979). On the relationship of the precipitation to variations in the kinematic variables during GATE. *Mon. Weather Rev.* **107**, 1154–1168.

Riehl, H., and J. S. Malkus (1958). On the heat balance in the equatorial trough zone. *Geophysica* **6**, 503–538.

Riehl, H., and J. S. Malkus (1961). Some aspects of hurricane Daisy, 1958. *Tellus* **13**, 181–213.

Rosenthal, S. L. (1978). Numerical simulation of tropical cyclone development with latent heat by the resolvable scales. I. Model description and preliminary results. *J. Atmos. Sci.* **35**, 258–271.

Rossby, C. G. (1938). Mutual adjustment of pressure and velocity distributions in certain simple current systems. *J. Mar. Res.* **2**, 239–263.

Ruprecht, E., and W. M. Gray (1976). Analysis of satellite-observed tropical cloud clusters: I. Wind and dynamic fields. *Tellus* **28**, 392–411.

Sanders, F., and R. J. Paine (1975). The structure and thermodynamics of an intense mesoscale convective storm in Oklahoma. *J. Atmos. Sci.* **32**, 1563–1579.

Sasaki, Y., and J. M. Lewis (1970). Numerical variational objective analysis of the planetary boundary layer in conjunction with squall line formation. *J. Meteorol. Soc. Jpn.* **48**, 381–399.

Schubert, W. H., and J. J. Hack (1982). Inertial stability and tropical cyclone development. *J. Atmos. Sci.* **39**, 1687–1697.

Schubert, W. H., J. J. Hack, P. L. Silva Dias, and S. R. Fulton (1980). Geostrophic adjustment in an axisymmetric vortex. *J. Atmos. Sci.* **37**, 1464–1484.

Soong, S.-T., and Y. Ogura (1980). Response of tradewind cumuli to large-scale processes. *J. Atmos. Sci.* **37**, 2035–2050.

Soong, S.-T., and W.-K. Tao (1980). Response of deep tropical cumulus clouds to mesoscale processes. *J. Atmos. Sci.* **37**, 2016–2034.

Sui, C.-H., and M. Yanai (1984). Vorticity budget of the GATE A/B area and its interpretation. *Postpr., Conf. Hurricanes Trop. Meteorol., 15th, Miami, Fla.* pp. 465–472. Am. Meteorol. Soc., Boston, Massachusetts.

Sui, C.-H., and M. Yanai (1986). Cumulus ensemble effects on the large-scale vorticity and momentum fields of GATE. Part I: Observational evidence. *J. Atmos. Sci.* **43**, 1618–1642.

Syono, Y., Y. Ogura, K. Gambo, and A. Kasahara (1951). On the negative vorticity in a typhoon. *J. Meteorol. Soc. Jpn.* **29**, 397–415.

Thompson, R. M., Jr., S. W. Payne, E. E. Recker, and R. J. Reed (1979). Structure and properties of synoptic-scale wave disturbances in the intertropical convergence zone of the eastern Atlantic. *J. Atmos. Sci.* **36**, 53–72.

Tollerud, E. I., and S. K. Esbensen (1983). An observational study of the upper-tropospheric vorticity fields in GATE cloud clusters. *Mon. Weather Rev.* **111**, 2161–2175.

Tracton, M. S. (1973). The role of cumulus convection in the development of extratropical cyclones. *Mon. Weather Rev.* **101**, 573–593.

Tripoli, G. (1986). A numerical investigation of an orogenic mesoscale convective system. Atmos. Pap. No. 401, Dep. Atmos. Sci., Colorado State Univ.

Tripoli, G., C. Tremback, and W. Cotton (1986). A comparison of a numerically simulated mesoscale convective system with explicit and parameterized convection. *Conf. Cloud Phys. Radar Meteorol., 23rd, Snowmass, Colo.* pp. J131–J134. Am. Meteorol. Soc., Boston, Massachusetts.

Ulanski, S. L., and M. Garstang (1978). The role of surface divergence and vorticity in the life cycle of convective rainfall. Part I: Observations and analysis. *J. Atmos. Sci.* **35**, 1047–1062.

Weisman, M. L., and J. B. Klemp (1982). The dependence of numerically simulated convective storms on vertical wind shear and buoyancy. *Mon. Weather Rev.* **110**, 504–520.

Williams, K. T., and W. M. Gray (1973). Statistical analysis of satellite-observed trade wind cloud clusters in the western North Pacific. *Tellus* **25**, 313–336.

Williamson, D. L. (1978). The relative importance of resolution, accuracy, and diffusion in short range forecasts with the NCAR global circulation model. *Mon. Weather Rev.* **106**, 69–88.

World Meteorological Organization (1972). Parameterization of sub-grid scale processes. GARP Publ. Ser. No. 8, WMO. (Available from Secretariat of WMO, Case Postale No. 1, CH-1211 Geneva 20, Switzerland.)

Yamasaki, M. (1975). A numerical experiment of the interaction between cumulus convection and larger scale motion. *Pap. Meteorol. Geophys.* **26**, 63–91.

Yamasaki, M. (1977). A preliminary experiment of the tropical cyclone without parameterizing the effects of cumulus convection. *J. Meteorol. Soc. Jpn.* **55**, 11–30.

Yanai, M., S. Esbensen, and J. Chu (1973). Determination of bulk properties of tropical cloud clusters from large-scale heat and moisture budget. *J. Atmos. Sci.* **30**, 611–627.

Yanai, M., J.-H. Chu, T. E. Stark, and T. Nitta (1976). Response of deep and shallow tropical maritime cumuli to large-scale processes. *J. Atmos. Sci.* **33**, 976–991.

Yanai, M., C.-H. Sui, and J.-H. Chu (1982). Effects of cumulus convection on the vorticity field in the tropics. Part I: Interpretation. *J. Meteorol. Soc. Jpn.* **60**, 411–423.

Zawadzki, I., and C. U. Ro (1978). Correlations between maximum rate of precipitation and mesoscale parameters. *J. Appl. Meteorol.* **17**, 1327–1334.

Zawadzki, I., E. Torlaschi, and R. Sauvageau (1981). The relationship between mesoscale thermodynamic variables and convective precipitation. *J. Atmos. Sci.* **38**, 1535–1540.

Zipser, E. J. (1977). Mesoscale and convective-scale downdrafts as distinct components of squall-line structure. *Mon. Weather Rev.* **105**, 1568–1589.

Part II | The Dynamics of Clouds

Chapter 7 | Fogs and Stratocumulus Clouds

7.1 Introduction

We begin our discussion of the dynamics of clouds and cloud systems by discussing what may be the least dynamic of all cloud phenomena, fog. Cloud dynamicists often classify fog as a micrometeorological phenomenon because it forms next to the earth in the atmospheric boundary layer, a domain traditionally covered by micrometeorologists. Lecturers in micrometeorology, however, often consider fog to be in the discipline of cloud dynamics or, perhaps, mesoscale meteorology. Fog does, in fact, span all these disciplines (as do many cloud systems); it occurs in the atmospheric boundary layer; it is a well-defined cloud in most instances, and it exhibits horizontal and temporal variability on scales normally thought to be the domain of mesoscale meteorology. We discuss fog as a cloud system that may be considered part of the general class of boundary layer stratiform clouds.

7.2 Types of Fog and Formation Mechanisms

Fog is normally categorized into four main types (see Willit, 1928; Byers, 1959; Jiusto, 1980):

A. Radiation fog
  (i) ground fog
  (ii) high inversion fog

   (iii) advection–radiation fog
   (iv) upslope fog
   (v) mountain–valley fog
B. Frontal fog
   (i) prefrontal (warm front)
   (ii) postfrontal (cold front)
   (iii) frontal passage
C. Advection (mixing) fog
   (i) sea fog
   (ii) tropical air fog
   (iii) land and sea-breeze fog
   (iv) steam fog ("Arctic sea smoke")
D. Other
   (i) ice fog
   (ii) snow fog

The physical mechanisms responsible for the formation of fog involve three primary processes: (1) cooling of air to its dewpoint, (2) addition of water vapor to the air, and (3) vertical mixing of moist air parcels having different temperatures.

The first mechanism generally explains radiation fogs, while the second causes frontal fogs and the third produces advection fog. A combination of all three mechanisms affects most fogs, though one mechanism may dominate.

To illustrate the processes further, consider the Clausius–Clapeyron diagram shown in Fig. 7.1. The formation of radiation fog may be illustrated by the line $(A_1 - A_2)$ where fog forms by the steady, isobaric cooling of air by contact with the cold ground and by radiative flux divergence in the moist or cloudy air until the air is cooled to saturation. According to Jiusto (1980), the line $A_1 - A_2'$ actually better represents radiational fog formation than does $A_1 - A_2$, because the water vapor content of the air is not fully conserved; some of the moisture is lost by dew deposition on the earth's surface. Thus, additional cooling of the air is needed for fog to form.

Frontal fogs often involve the addition of moisture by falling precipitation from relatively warm layers aloft into underlying cooler, subsaturated air. This is illustrated in Fig. 7.1 as line $B_1 - B_2$, where the addition of water vapor causes the mixing ratio in the air to exceed the saturation mixing ratio at point B_2.

Advection or mixing fogs were first described by Taylor (1917), who observed them from a whaling ship off the Grand Banks of Newfoundland. Advection fogs form when near-saturated air parcels of different temperatures mix vertically. Consider, for example, the formation of a sea fog

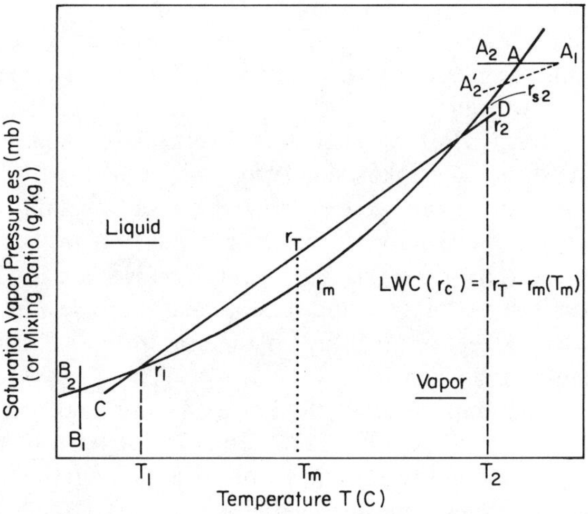

Fig. 7.1. Phase diagram and fog formation processes. [Adapted from Jiusto (1980).]

as warm, moist air flows over a cooler ocean surface. Suppose that the ocean surface has a temperature T_1, which is saturated at the mixing ratio r_1. Now a warm, moist air mass characterized by point D in Fig. 7.1, having a temperature T_2, flows over the cooler ocean surface. If the low-level air is well mixed in the vertical, the temperature will vary nearly linearly through the mixed layer. Likewise, the water vapor mixing ratio will vary linearly through the mixed layer along line $r_1 \rightarrow r_T \rightarrow r_2$ in Fig. 7.1. Because the saturation vapor pressure or the saturation mixing ratio varies nonlinearly along the line $r_1 \rightarrow r_m \rightarrow r_{s2}$, a region of the mixed layer becomes supersaturated with respect to water. A cloud thus forms which has a peak liquid-water content in the middle of the mixed layer equal to

$$r_c = r_T - r_m(T_m). \tag{7.1}$$

7.3 Radiation Fog Physics and Dynamics

7.3.1 The Role of Radiative Cooling

Radiation fog commences with strong radiative cooling of the earth's surface. According to Taylor (1917), nights with clear skies, light winds, and high relative humidities favor radiation fog. Radiation cools the earth's surface, which then cools the air close to the earth's surface by conduction. In addition, radiative flux divergence in the moist atmosphere is also important (Zdunkowski and Nielsen, 1969; Zdunkowski and Barr, 1972; Brown and Roach, 1976). Brown and Roach (1976) concluded that gaseous radiative cooling is necessary to account for fog formation on the observed time scale of a few hours.

Once the fog forms aloft, radiative flux divergence at the fog top (Fleagle *et al.*, 1952; Korb and Zdunkowski, 1970; Pilie *et al.*, 1975) increases the stability at and immediately above the fog top and destabilizes the lapse rate within and below the fog. The resultant vertical mixing of the cold foggy air with clear, nearly saturated air below causes the fog to propagate downward. Radiative cooling at the fog top also increases the liquid-water content and decreases the visibility in the fog, often contributing to its upward propagation as well.

Radiation is also important on the scale of individual droplets in fogs. As noted in Chapter 5, Roach (1976) and Barkstrom (1978) include a radiative transfer term in the equations for heat and water mass budgets of a spherical droplet. When combined with the Clausius–Clapeyron equation, they produce an equation for droplet growth. The resulting expression for the saturation vapor pressure at the droplet surface, including the effects of radiative loss, indicates droplet growth can occur in a slightly subsaturated environment. This process allows for fog formation with consequent radiation-induced changes in the stability of the entire cloudy layer, even though the environment may never become supersaturated with respect to water on the average. Brown (1980) and Mason (1982) conclude that the importance of the radiation term in the droplet growth equation varies with the concentration of activated cloud condensation nuclei. With high CCN concentrations, the radiative term has little influence on either the mean droplet size or liquid-water content. They suggest this is due to the fact that the resultant more numerous droplets have small values of absorption efficiencies. They conclude that radiative exchange between droplets and their environment will be greatest in clean fogs and maritime layer clouds, and will be less in heavily polluted air.

7.3.2 The Role of Dew

A number of investigators have emphasized the role of dew deposition to the formation of fog (see Wells, 1838; Geiger, 1965; Pilie *et al.*, 1975; Lala *et al.*, 1975; Brown and Roach, 1976). The deposition of dew at the surface is responsible for the development of a downward transport of moisture and the formation of a nocturnal dew-point inversion. Dew-point inversions have been observed to extend to between 40 and 200 m above the surface.

Some of the moisture supplied to dew deposition may come from the underlying soil. Most, however, is extracted from the overlying air mass.

Thus, Lala *et al.* (1975) and Brown and Roach (1976) view dew deposition as a "governor" on fog formation. For a given rate of radiative cooling, which drives the air toward saturation, if the dew deposition rate and accompanying downward transport of moisture is large, then fog formation may be inhibited. If the dew deposition at the surface is somewhat less, radiative cooling may be sufficient to initiate the formation of fog.

Pilie *et al.* (1975) also attribute the observed initial formation of fog aloft to the development of a dew-point inversion as a consequence of dew deposition. However, Jiusto (1980) has noted that many inland radiation fogs first develop at the surface and then build upward. The conditions under which fog formation occurs at the surface versus aloft are not well known.

From the time of fog formation and until sunrise, dew does not appear to serve any major function other than to maintain a saturated lower boundary (Pilie *et al.*, 1975). After sunrise, however, the surface temperature begins to rise and evaporation of dew commences. As the fog layer warms, a supply of water vapor is needed in order to maintain saturation. Pilie *et al.* (1975) estimate that the dew evaporation rate is sufficient to allow persistence of a fog for several hours longer than over a dew-free surface. Eventually solar heating causes the saturation vapor pressure to increase above the actual vapor pressure, in spite of the evaporation of dew. This leads to fog dissipation, first at the surface and then propagating upward. Thus originates the term "fog lifting."

7.3.3 The Role of Turbulence

Turbulent transport of heat and moisture plays an important role in the evolution of a fog. However, there is not a consensus as to whether the role of turbulence is primarily constructive, contributing to the formation of fog, or destructive, contributing primarily to the dissipation of fog. Brown and Roach (1976) conclude that turbulence inhibits the formation of radiation fog. They base their conclusion partially on a fog model in which turbulence is modeled with an eddy viscosity closure. The magnitude of eddy viscosity was specified by various profiles shown in Fig. 7.2. They conclude that numerical experiments with model II provide the most realistic results when compared to observations reported by Roach *et al.* (1976). Brown and Roach also experimented with a stability-dependent exchange coefficient formulation, which was derived using adiabatic similarity theory. As they point out, however, the existence of a constant-flux layer a few meters above the surface is questionable in such a stable environment as a radiation fog.

In a companion paper Roach *et al.* (1976) infer from observations that turbulence hinders fog formation. They formed this conclusion because

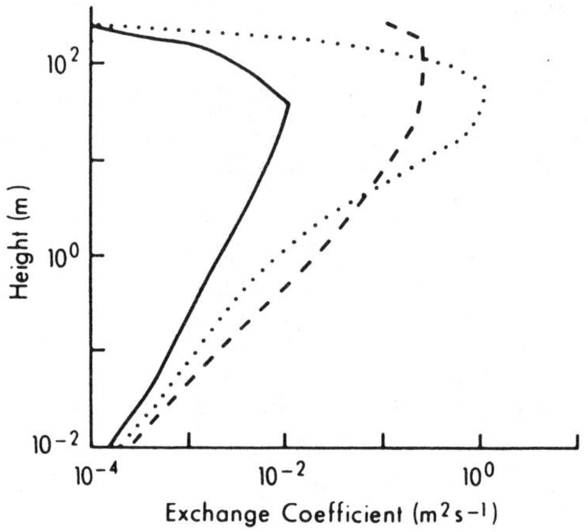

Fig. 7.2. Exchange coefficient regimes described in the text: dotted line, model I; dashed line, model II; solid line, model III. [From Brown and Roach (1976). Reproduced with the permission of the Controller of Her Britannic Majesty's Stationery Office.]

lulls in wind were accompanied by maximum cooling, and major lulls were accompanied by periods of significant fog development. Conversely, increases in wind (to $>2\ \mathrm{m\,s^{-1}}$) were associated with fog dispersal. They infer that as the wind speed decreases, turbulent transfer of moisture to the surface to form dew ceases. As a result, the moisture remains in the atmosphere, and as radiation cools the air, fog is formed. Alternately, at higher wind speeds, vertical mixing of drier air may inhibit fog formation.

Jiusto and Lala (1980) infer from observations that if radiative cooling and higher humidities extend to a greater depth, vertical turbulent mixing can contribute positively to the growth of the fog when it occurs in the presence of radiation cooling. Using various surface-layer turbulence exchange formulations, Welch *et al.* (1986) conclude that increased turbulence and reduced stability contribute to fog formation. They also suggest that fog intensification after sunrise is caused by increased turbulence generation and the resultant downward mixing of liquid in the upper part of the fog to the surface. Lala *et al.* (1982) suggest that turbulence in the early evening may inhibit fog, whereas later in the evening turbulent mixing can intensify fog.

Lala *et al.* (1975), Brown and Roach (1976), Zdunkowski and Barr (1972), and Welch *et al.* (1986) concur that the structure of fog and the occurrence and nonoccurrence of fog are strongly dependent upon the particular profile of eddy viscosity or turbulence models employed. Moreover, Brown and Roach (1976) conclude that a more realistic treatment of turbulence is needed, especially in the region beneath the fog top. They suggest that the top of deep fogs may behave somewhat like the ground, not only with respect to radiative processes, but also to turbulent transport. Comparison of more complicated turbulence models with observed fog structure shows that realistic simulations of fog life cycles can be obtained (Musson-Genon, 1987). This work further emphasizes the importance of turbulent transport to fog evolution.

7.3.4 The Role of Drop Settling

In Chapter 4 we discussed the partitioning of the droplet distribution in clouds into raindrops having significant terminal velocities and into cloud droplets which are assumed to have negligible fall velocities. In those models, cloud droplets are as large as 40–50 μm in radius. In fogs, however, only a few droplets exceed 20 μm in radius. Nonetheless, Brown and Roach (1976) have concluded that settling of droplets can play an important role in the evolution of fog structure. This reflects the fact that vertical velocities in fogs are quite small, much smaller than in cumulus clouds for which the models in Chapter 4 were developed.

Brown and Roach introduced drop settling into their model because, otherwise, unrealistically high liquid-water contents were predicted. Figure 7.3 illustrates the sensitivity of the model to several simple models of drop settling. Brown and Roach used the following feedback process to explain the sensitivity of the model to drop settling: the direct removal of liquid water by droplet settling causes a reduction of radiation cooling due to cloud droplets; lower radiation cooling, in turn, leads to a reduction in the rate of condensation and liquid-water contents. In a more detailed microphysical simulation of fog, Brown (1980) found that the simulated fog liquid-water content was sensitive to CCN concentrations. With lower concentrations of CCN, he found that the fog liquid-water content was about 20% less than with higher CCN concentrations; the difference was attributed to reduced drop settling in the high CCN case.

7.3.5 The Role of Vegetative Cover

Brown and Roach also investigated the role of surface vegetation on the rate of cooling at the surface. They argued that a grass surface will cool to a lower surface temperature than bare ground because of its small thermal capacity. Furthermore, grass will partially shield the soil from radiative loss. Thus, the air over a grass-covered surface will radiate to a colder surface than that of bare soil, and greater cooling of the air will occur.

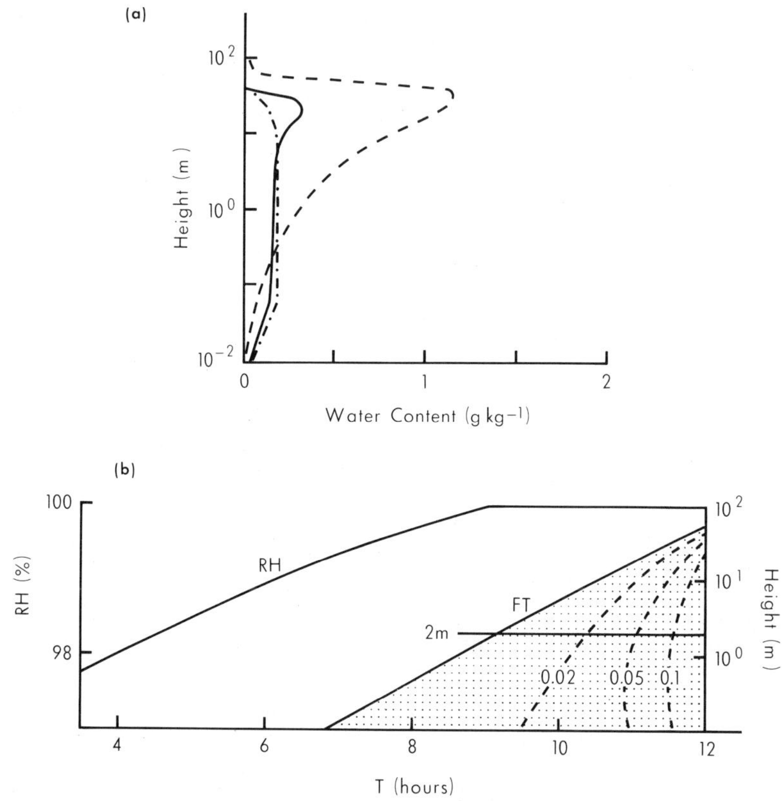

Fig. 7.3. (a) Liquid-water content profiles after 5 h of integration with gravitational droplet settling excluded (dashed line) and included using $\bar{v} = 6.25w$ (solid line) or droplet settling included using $\bar{v} = 10w$ (dot-dash line). (b) Plot of relative humidity (RH) at 2 m, height of fog top (FT), and heights of liquid-water mixing ratio isopleths (dashed lines, in units of g kg^{-1}) against time (T) since beginning of integration of model II. [From Brown and Roach (1976). Reproduced with the permission of the Controller of Her Britannic Majesty's Stationery Office.]

7.3.6 Advection of Cloud Cover

The advection of a cloud layer over a fog can alter the net divergence of longwave radiation from the top of the fog and thereby lead to an alteration in the fog structure or perhaps even the destruction of the fog. Saunders (1957) noted that a large fraction of fog cases cleared following the arrival of a cloud deck. He also observed a rise in temperature near the ground. Subsequently, Saunders (1960) observed that the net outgoing radiation at fog top decreased substantially compared to a clear sky, depending on the height and temperature of the overlying cloud layer. The lower the height of the overlying cloud layer, the greater was the reduction in net outgoing longwave radiation. Saunders also concluded that the observed rise in temperature near the ground following the advection of a cloud deck over an overlying fog was a consequence of the reduction of outgoing radiation of the fog top. The temperature rise was also due in part to sustained upward heat flux from the ground.

Because the upward heat flux from the ground is important in dissipating fog, another important controlling factor is the gradient in temperature between the soil and the air. If the soil–air temperature gradient is small, fog clearance is unlikely. If it is large, fog clearance is greatly enhanced when a cloud layer moves over the fog.

Brown and Roach (1976) simulated the effects of advection of cloud cover over a fog by increasing the downward longwave radiation flux incident at the model top boundary from 30 to 95 W m^{-2}. This value, they argued, was appropriate to a stratocumulus of a 600-m thickness and a base at 700 m in an atmosphere similar to one of the cases they observed. Figure 7.4 illustrates the model-predicted changes in temperature and liquid-water content through the fog layer. Over a period of 2 h the layer warmed and the liquid-water content decreased at all levels except near the top. They

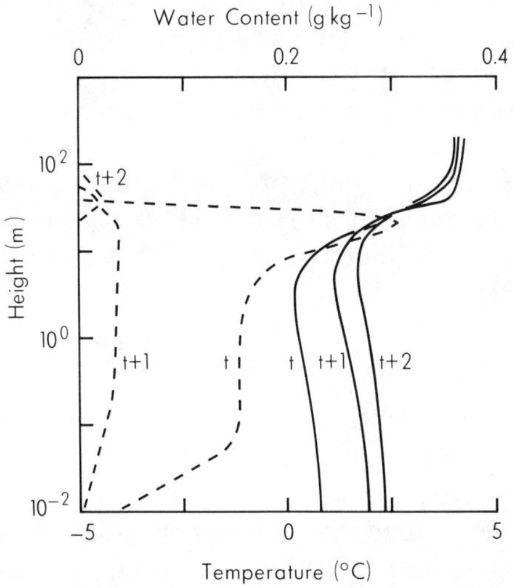

Fig. 7.4. Liquid-water content and temperature at times $t+x$ hours, where t is the time of advection of cloud cover: dashed lines, liquid-water content; solid lines, temperature. [From Brown and Roach (1976). Reproduced with the permission of the Controller of Her Britannic Majesty's Stationery Office.]

attributed the failure of the model to completely dissipate the fog to weaknesses in their turbulence model. Nonetheless, their model shows quite clearly that the overlying cloud deck reduces the net radiative cooling at the fog top to such an extent that the eddy transport of heat from the surface will drive the layer above water saturation.

7.3.7 Variability in Nocturnal Fogs

Once a nocturnal fog reaches maturity, the fog properties do not necessarily remain constant. Choularton *et al.* (1981) described periodic fluctuations in fog liquid-water content observed in Meppen, Federal Republic of Germany. They noted that the liquid water fluctuated with periods of 51, 31, and 96 s, respectively, for 15- to 20-min intervals. They speculated that the periodicities were due to the passage of Bernard cell-type convection past the observation site.

Longer period oscillations in fog liquid-water content of the order of 45 min to 1 h were found by Lala *et al.* (1982) in observations of fog in Albany, New York. The source of the periodicity was not determined, however. In their simulations of the Albany fog cases, Welch *et al.* (1986) simulated a series of pronounced oscillations in the fog parameters, including liquid-water content. The period of the oscillation found in the model was on the order of 30–40 min, although the length of the period varied considerably with the particular formation of the turbulence closure model. The oscillation was associated with a surge in eddy mixing, which promotes entrainment of warm, dry air from the inversion top into the fog layer, and an upward mixing of moist air from near the surface. The result is a rapid decrease in liquid-water content in the lower 20 m. Such longer period oscillations may be similar to those observed and modeled in stratocumulus clouds, as described later. Chen and Cotton (1987) attributed the oscillations in stratocumulus cloud simulations to interactions among radiation, turbulence, and droplet settling.

7.3.8 Dispersal of Fog by Solar Insolation

Brown and Roach (1976) investigated with a numerical model the dispersal of fog by solar heating. They introduced a simple shortwave radiative transfer model in which solar absorption in the fog was omitted. Fogs are so optically thin, they argued, that the solar heating rate is inconsequential relative to longwave radiative cooling and turbulent transport of heat. The primary feature modeled was the absorption of solar radiation at the earth's surface with about 30% of the incident solar beam reflected by the fog layer and 20% by the earth's surface. The predicted evolution of the liquid-water and temperature profiles after sunrise are shown in Fig. 7.5. A major

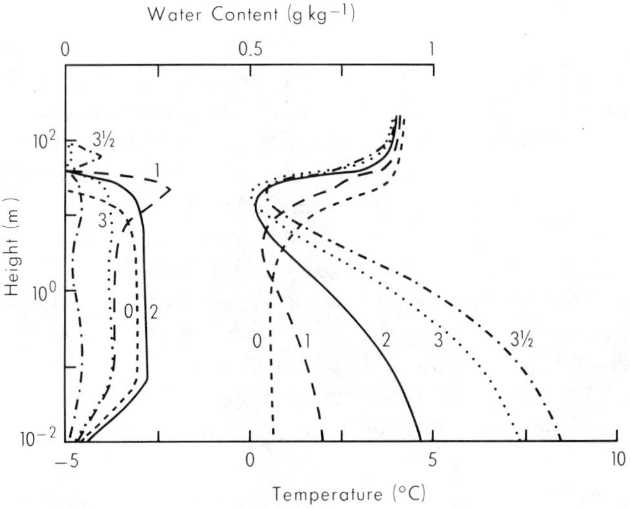

Fig. 7.5. Development of the model-predicted liquid-water and temperature profiles after sunrise: (- - -), sunrise; (— — —), 1 h; (——), 2 h; · · ·, 3 h; −·−·, 4 h. [From Brown and Roach (1976). Reproduced with the permission of the Controller of Her Britannic Majesty's Stationery Office.]

difference between the solar radiation case and the case with overlying cloud cover is that longwave radiative cooling is not diminished within the fog as a consequence of the sunrise. Thus, for a while longwave radiative cooling prevails over solar heating and the fog liquid-water content continues to rise after sunrise. After 3.5 h following sunrise, heat flux driven by solar heating of the surface prevails and the simulated fog completely dissipated except for levels above 22 m. This deficiency of the model was again attributed to the inadequacy of the turbulence model. Brown and Roach also felt that their observations did not support the simulated mode of dissipation, namely from the surface upward. They argued that in this case the fog appeared to clear from above. Thus, their neglect of solar absorption in the fog layer may not have been justified. Welch *et al.* (1986) conclude that fog intensification after sunrise is caused by enhanced mixing of upper-level liquid to the surface as a result of increased turbulence due to surface heating. Evaporation of dew and soil moisture also supplies additional moisture. Eventual dissipation of the fog occurs when the effect of surface heat flux on the relative humidity exceeds the effect of the supply of moisture from the surface or mixed downward from the fog top. Using a two-dimensional model, Forkel *et al.* (1987) predicted that fog dissipation occurs more quickly in a polluted atmosphere due to larger solar heating rates within the fog caused by the larger concentrations of aerosol.

7.4 Valley Fog

Valley fogs are also radiation fogs and are therefore regulated by all the physical processes discussed in the preceding section. In addition, valley fogs are influenced by organized slope circulations that develop in response to radiative cooling and drainage of cooler air down the slopes of the valley, where the cool air is pooled into the valley basins. Pilie *et al.* (1975a; Pilie, 1975) reported on an extensive study of the micrometeorological and microphysical characteristics of valley fogs near Elmira, New York. They interpreted their observations with respect to Defant's (1951) conceptual model of nocturnal mountain valley circulations shown in Fig. 7.6. According to their interpretation, the life cycle of a valley fog is as follows. Radiative cooling over the slope of the valley initiates a downslope wind and an upward return flow in the valley center region, as illustrated in Fig. 7.6a.

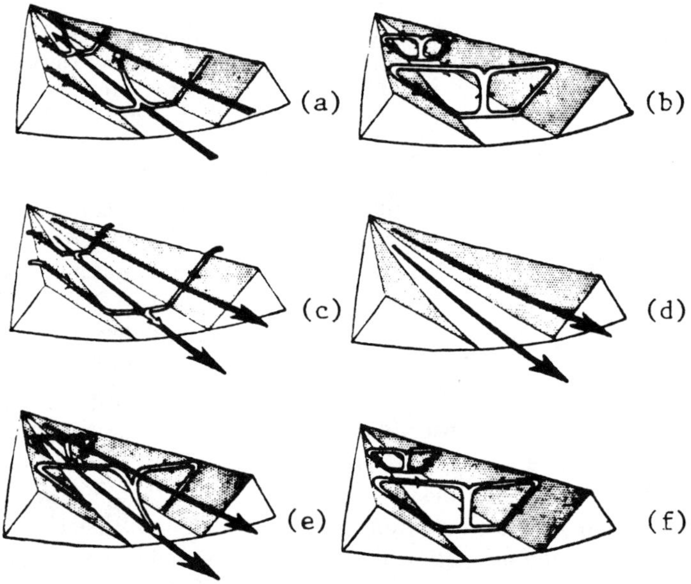

Fig. 7.6. (a) Downslope wind begins shortly after sunset before up-valley mountain wind dies. (b) In late evening, up-valley wind dies and only downslope wind and return flow at center of valley exist. (c) Return flow at center of valley ceases and down-valley mountain wind becomes established. (d) Late at night the downslope wind ceases and the down-valley mountain wind persists. (e) Sunrise; onset of upslope winds (white arrows), continuation of mountain wind (black arrows). Valley cold, plains warm. (f) Forenoon (about 0900); strong slope winds, transition from mountain wind to valley wind. Valley temperature same as plains. [From Defant (1951) and Pilie *et al.* (1975).]

This process contributes to the formation of a deep nocturnal inversion. Several hours after sunset the mountain wind is established by drainage of cool air down the axis of the valley. The speed of the wind is a maximum at levels of 40–200 m, with speeds of 2–4 m s^{-1}. As the mountain wind matures, downslope winds prevail throughout the valley (see Fig. 7.6c) Finally, in the late night hours, the mountain wind occupies the entire valley and continues until after sunrise (see Fig. 7.6d).

Pilie *et al.* (1975) then explain the formation of deep valley fogs as follows:

(1) Nocturnal radiation from the surface and subsequent turbulent heat transfer from air to ground which produces an initial low-level temperature inversion stimulates the downslope wind and upward return flow near the valley center. During the period, dew deposition at the cold surface creates the low-level dew-point inversion. The upward motion at the valley center carries the cool and somewhat dry air aloft to cause the inversion to deepen.

(2) Approximately 3 h before fog formation, the mountain wind forms, providing continuity for the downslope wind while restricting the upward motion of air near the valley center. Cooling is therefore restricted to low and middle levels of the valley, i.e., those levels in which fog will eventually form. The continuing downslope wind mixes with warmer air at midlevels in the valley and causes the cooling rate to maximize in that region. Through this period, the dew-point inversion persists. Temperature and dew point, therefore, converge at midlevels, and a thin layer of fog forms aloft.

The subsequent downward propagation of the fog is then envisaged to occur in a similar manner to that of a "pure" radiation fog as discussed in Section 7.3.

Pilie *et al.* (1975) also noted that on several occasions fog formed following sunrise. On these occasions they suggested that surface warming and dew evaporation following sunrise cause vertical mixing of moist low-level air with cooler air in the valley center aloft. This suggests that the vertical mixing mechanisms discussed in Section 7.1 and the adiabatic cooling of rising air parcels contribute to the fog formation.

Slope/mountain or valley circulations that develop in response to solar heating of exposed valley walls or hills, such as illustrated in e and f of Fig. 7.6, contribute to the eventual dissipation of the valley fog. As subsidence develops in the valley center in response to the slope circulation, the resultant adiabatic warming evaporates the fog. At the same time, slightly elevated hills in the valley center are the first regions where fog dissipates, because the optical depth of the fog is least over the tops of the hills. The solar insolation is able to penetrate the thinner fog layer, thus warming the higher ground. As the hills become exposed, slope circulations develop, with corresponding compensating sinking motions over the neighboring foggy lower terrain. Dissipation of the fog may then proceed quite rapidly. This is an area ripe for quantitative study that uses turbulence/radiation models coupled to mesoscale flow models.

7.5 Marine Fog

Marine fog differs from the radiation fogs we have discussed in that radiation does not rapidly alter the surface of the ocean as it does the land. Surface fluxes of moisture and heat are important in marine fog formation, but their fluxes are not affected significantly (on the time scale of the diurnal cycle) by variations in solar and terrestrial radiation. Another factor of importance to marine fog is that the air mass, often of maritime origin, contains fewer active cloud condensation nuclei. Giant sea salt nuclei also may represent a substantial fraction of the activated aerosol in the low supersaturations encountered in fogs. Compared to the generally more "continental" radiation fogs, marine fogs are more prone to drizzle formation with the potential for radiation/drop settling influences on the fog structure, as discussed in Section 7.3.4.

7.5.1 Fog Formation in a Turbulence-Dominated Marine Boundary Layer

Because the surface of the ocean provides a source of moisture, fog may form over the ocean surface without significant radiative cooling. That is, the fog formation mechanisms are purely a function of the surface fluxes of heat, moisture, and momentum and of the gradients of temperature, moisture, and wind in the overlying air mass. Oliver *et al.* (1978; hereafter referred to as OLW) examined the existence criteria for a turbulence-dominated fog in the surface layer. The surface layer is a thin ($\sim$10 m) layer next to the ground in which turbulent fluxes of water, momentum, and heat are assumed to be constant. It is also a region in which the Monin–Obukhov similarity theory, or adiabatic similarity theory, is applied. That is, the fluxes and mean gradients within the surface layer can be related to distance away from the surface layer and to surface fluxes of heat, moisture, and momentum.

In their analysis, OLW employ the conservative thermodynamic variable θ_s which they define as

$$\theta_s = T - T_0 + \Gamma z + (L_c/C_{pa}) r_v, \tag{7.2}$$

where Γ represents the dry adiabatic lapse rate ($\Gamma = g/C_{pa}$) and T_0 is the reference state temperature. The behavior of θ_s is similar to the behavior of θ_e.

They also define a virtual potential temperature θ_v as

$$\theta_v = T - T_0 + \Gamma z + (0.61 r_v - r_c) T_0. \tag{7.3}$$

To denote surface properties, we use a double zero subscript, so surface temperature is denoted as T_{00}. The surface friction velocity and non-dimensional surface heat and moisture fluxes are defined as

$$U_* = \overline{u''w''}^{1/2}|_{00}; \qquad \theta_{v*} = \overline{w''\theta_v''}|_{00}/U_*; \qquad r_* = \overline{w''r''}|_{00}/U_*. \tag{7.4}$$

In the surface layer the gradients of the mean variables can be expressed in terms of the diabatic profile functions: $\phi_u(\xi)$, $\phi_\theta(\xi)$, $\phi_r(\xi)$ as

$$\partial\bar{u}/\partial z = (u_*/kz)\phi_u(\xi), \tag{7.5}$$

$$\overline{\partial\theta_v}/\partial z = (\partial\theta_{v*}/kz)\phi_\theta(\xi), \tag{7.6}$$

$$\partial\bar{r}/\partial z = (r_*/kz)\phi_r(\xi), \tag{7.7}$$

where k is the von Karman constant and $\xi = z/L_m$. The parameter L_m is a generalized form of the Monin–Obukhov length defined as

$$L_m = -u_*^2 T_0/kg\theta_{v*}. \tag{7.8}$$

OLW generalized L to include the virtual heat flux from the ocean surface. The original definition of L by Obukhov (1946) included only the buoyancy effects of surface heat flux.

The behavior of the parameter $\xi = z/L_m$ is analogous to the behavior of the Richardson number, which is the ratio of the rate of buoyant production of turbulence to shear production. Thus, if $\xi < 0$, the surface is characterized by a positive surface buoyancy flux, and buoyant production of turbulence dominates for large negative values of ξ. In contrast, a stably stratified surface layer will be characterized by $\xi > 0$. A large and positive value of ξ implies that buoyancy is dominant over mechanical production of turbulence. Because buoyancy accelerations damp the generation of turbulence in a stable atmosphere, large positive values of ξ imply that turbulence will not be very active. However, small positive values of ξ imply that mechanically generated turbulence will dominate buoyant damping of turbulence.

In the absence of radiative effects in a surface-layer-bound fog, the turbulent correlations of θ_s and r are identical. The distributions of $r - r_{00}$ as well as their turbulent correlations are similar to one another and can be represented as a function of z/L_m and z/z_{00}, where z_{00} is the surface friction height.

According to OLW, similarity theory leads to the following theorem as noted by Taylor (1917):

> All fluid states in a surface layer in turbulent interaction must map onto a straight line in the $r - \theta_s$ plane which passes through the surface state (r_{00}, θ_{s00}) and has a slope determined by the surface humidity to heat flux ratio r^*/θ_{s*}.

Thus, OLW say that the existence criteria for a fog in the surface layer dominated by turbulence can be determined from Fig. 7.7, which is a generalization of Fig. 7.1. The surface state is denoted by the point "O" and the straight line OCDE represents the distribution of r and θ_s throughout the surface layer. The variation of the saturation mixing ratio with θ_s derived from Clausius–Clapyron equation is also shown in Fig. 7.7. (Note: OLW neglected variations in r_s with pressure for shallow surface-layer applications.) The tangent to the saturation curve at the surface is denoted β_{00} and defined as

$$\beta_{00} = (\partial r_s/\partial \theta_s)_p. \tag{7.9}$$

One can also relate β_{00} to the logarithmic derivative of saturation mixing ratio

$$\beta_T = (\partial \ln r_s/\partial \ln T)_p \tag{7.10}$$

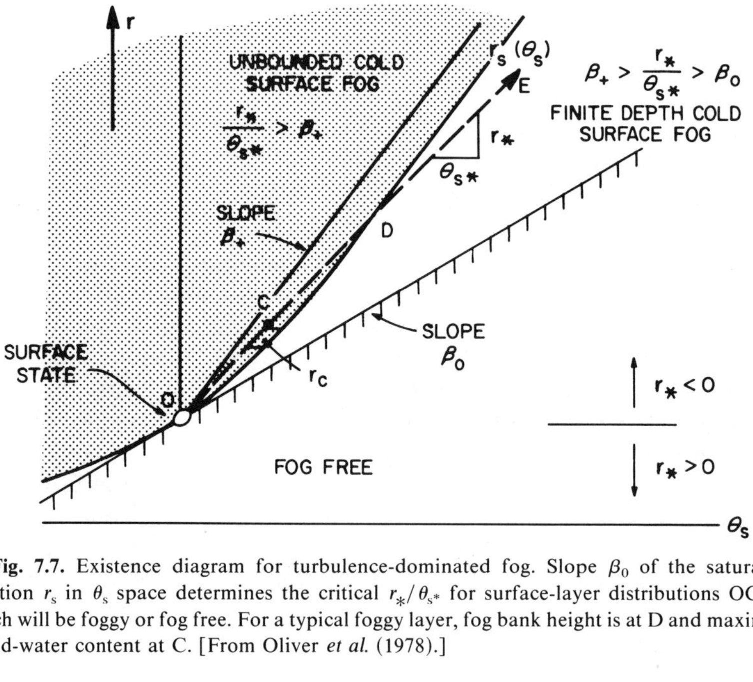

Fig. 7.7. Existence diagram for turbulence-dominated fog. Slope β_0 of the saturation function r_s in θ_s space determines the critical r_*/θ_{s*} for surface-layer distributions OCDE, which will be foggy or fog free. For a typical foggy layer, fog bank height is at D and maximum liquid-water content at C. [From Oliver *et al.* (1978).]

as

$$\beta_{00} = (r_s/\tilde{\mu} T_0)\beta_T, \tag{7.11}$$

where

$$\tilde{\mu} = 1 + (R_v/C_{pa}) r_s \beta_T^2. \tag{7.12}$$

For saturated conditions at the surface, any parcel trajectory will be a straight line emanating from the surface point r_{00}, θ_{s00}. Any parcel trajectory moving below the line β_0 will be a fog-free trajectory.

Thus, surface fog is *not possible* if

$$\theta_{s*} < 0, \qquad r_*/\theta_{s*} < \beta_0, \tag{7.13}$$

$$\theta_{s*} > 0, \qquad r_*/\theta_{s*} > \beta_0. \tag{7.14}$$

The saturation mixing ratio $r_s(\theta_s)$ becomes asymptotic to a straight line with slope $\beta_+ = r_{00}/\theta_{s00}$ for large θ_s (see upper part of diagram).

In the region between β_+ and β_0, surface fog may exist next to a cold surface and through the depth OCD on ray OCDE, with the top being at D. The maximum liquid-water content is at point C. Above the line β_+ cold surface fog exists, but it extends beyond the surface layer.

In the region $\theta_{s*} > 0$, warm surface fog exists. Such a fog may be bounded as OCDE or unbounded within the surface layer.

OLW showed that surface fog existence criteria could be assessed in terms of the surface flux ratio r_*/θ_{s*}, saturation curve at the surface:

$$(r_*/\theta_{s*})_{\text{crit}} = \beta_0. \tag{7.15}$$

According to OLW the existence criteria may be summarized:

> For a cold surface fog ($\theta_{s*} < 0$) a large humidity flux and small heat flux promote fog formation. Thus, dry nearly adiabatic air overrunning a cold water surface forms fog easily. On the other hand, for a warm surface fog ($\theta_{s*} > 0$), the flux ratio must be smaller in magnitude than the critical value. Hence, cold air with high relative humidity overrunning a warm surface will tend to fog.

OLW also examined conditions when turbulence will dominate over radiation in fog formation. They concluded that radiation plays a significant role in most surface fogs of depths greater than a few meters. Nonetheless, these existence criteria can be useful for diagnosing the initiation of a radiation/turbulence fog.

7.5.2 The Advection–Radiative Fog

In this section we are concerned with a deeper type of fog in which turbulence and radiative processes play an important role. Pilie *et al.* (1979; hereafter referred to as P79) describe a case of marine fog formation off the western coast of California on 30 August 1972. The fog occurred as air blew over cold water to a patch of relatively warm water. Figure 7.8 illustrates the changes in the low-level temperature profiles in the clear region over cool water, at the edge of the fog, and in the fog layer.

As can be inferred from Fig. 7.8, upwind of the fog, heat was being transferred from the warmer air to the cooler ocean surface. Within the fog layer, however, heat was being transferred from the warmer ocean surface to a "cooled" fog layer. P79 suggested that resultant mixing of the warm surface air over the warmer ocean and the near-saturated cool air produced the initial fog in accordance with the principles outlined in Sections 7.1 and 7.4 (i.e., Figs. 7.1 and 7.7). Radiative cooling of the top of the fog layer enhances the low-level instability and further promotes the turbulent transport of heat and moisture. The cool fog represents a sink of moisture which enhances evaporation from the warm sea surface. This is an excellent example of the interaction between turbulence and radiative processes. The fog grows progressively deeper in the downwind direction as vertical eddy transport provides the moisture for further fog formation and radiation cooling lifts the inversion and encourages further eddy transport.

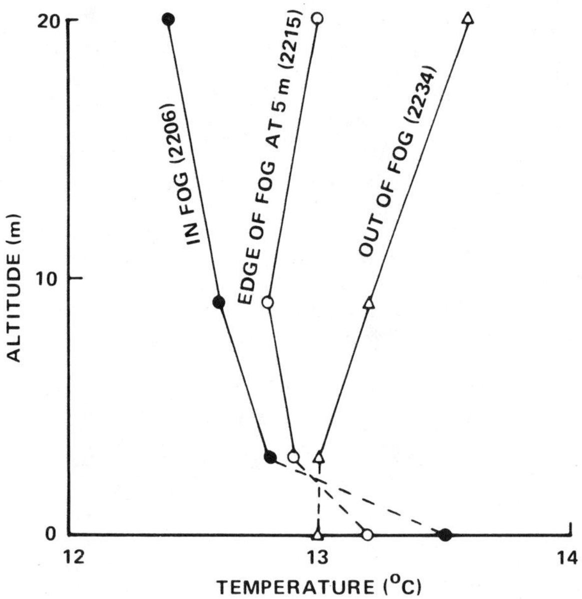

Fig. 7.8. Selected vertical profiles at indicated positions relative to fog edge, 30 August 1972. [From Pilie *et al.* (1979).]

OLW developed a sophisticated model of the growth of such an advective–radiative fog. Their model is a second-order transport turbulence model with radiative transfer processes included. The second-order transport model is closed with an eddy diffusion model on the third-order terms. The model also allows for cloud fractional coverage less than unity using a simple scheme. A simple shortwave and longwave radiative transfer model is employed which is based on the transmittance approach as described in Chapter 5. In their simulation of an advective–radiative fog with neglection of solar absorption, OLW simulated an air mass advecting over a warmer water surface by heating the surface by 0.5°C while the air mass moves about 3 km. The predicted changes in liquid-water content and temperature are shown in Fig. 7.9. The fog first forms at the surface and builds upward. When the fog has deepened to a depth of 100 m, radiative cooling becomes significant enough to generate a secondary maximum of liquid-water content.

7.5.3 Marine Fog Formation by a Stratus-Lowering Process

We shall reserve our discussion of the stratus-forming mechanisms until a later section. Before we leave the subject of fogs, however, we should note that on occasion fog forms as a result of the lowering of the base of stratus clouds. Anderson (1931) suggested that fog can form as a consequence of cloud-top radiation cooling, which causes instability in the cloud layer. This instability results in the downward development of the cloud base, in some cases to the surface. It has been noted by several authors that a condition necessary for stratus formation and lowering of the base requires that the inversion base must rise above the lifting condensation level.

P79 postulated the following conceptual model of the stratus-lowering process, shown schematically in Fig. 7.10. Net radiation from the stratus top causes rapid cooling that generates instability beneath the inversion. This causes a turbulent transport downward of cool air and cloud droplets. Evaporation of droplets beneath cloud base causes cooling and an increase in humidity. This leads to a further lowering of cloud base.

Additional insight into the stratus-lowering process can be derived from the numerical experiments described by OLW. They commence a simulation with a sounding such as shown in Fig. 7.11 derived from data reported by Mack *et al.* (1974). The height of the mixed layer z_i was varied from experiment to experiment. The experiments were all begun at 1900 local time. For z_i equal to 300 m, the cloud base propagated downward until 0500, where it reached 150 m above the surface. During the same period the top propagated upward to 450 m.

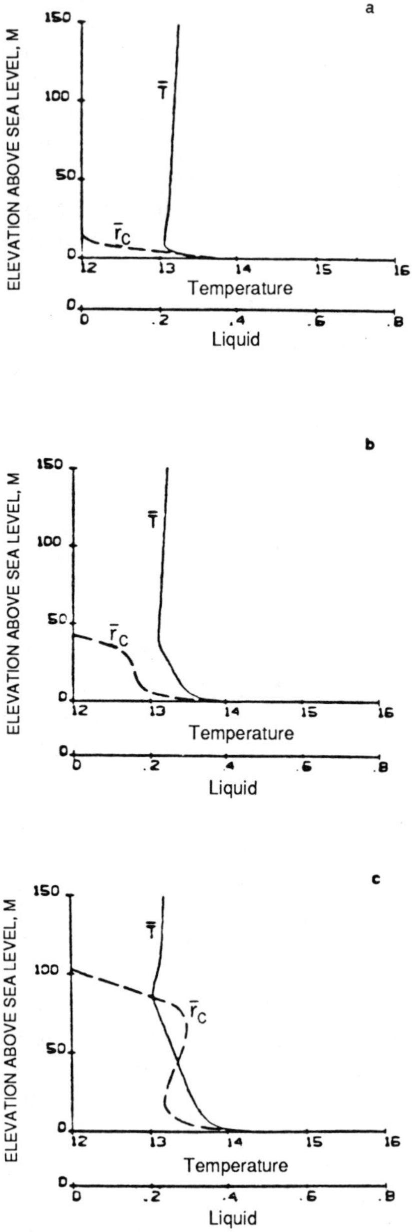

Fig. 7.9. Temperature and liquid-water content profiles in an advective–radiative fog at $x = 100$ m (a), $x = 1000$ m (b), and $x = 3000$ m (c). $\bar{T}$, temperature (°C); $\bar{r}_c$, liquid water (g/kg). [From Oliver *et al.* (1978).]

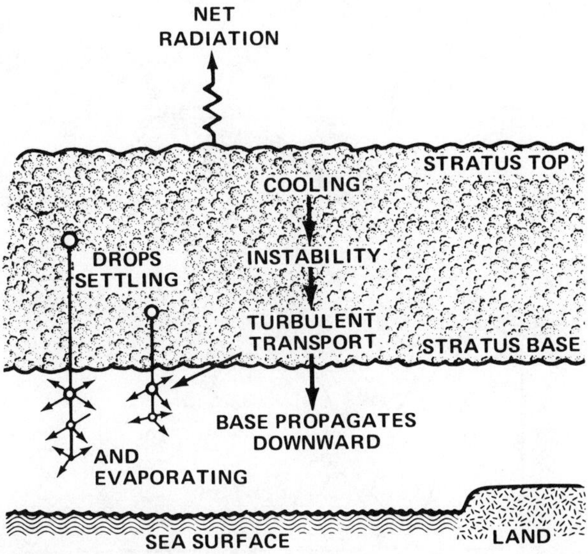

Fig. 7.10. Schematic representation of fog formation through the stratus-lowering process. [From Pilie *et al.* (1979).]

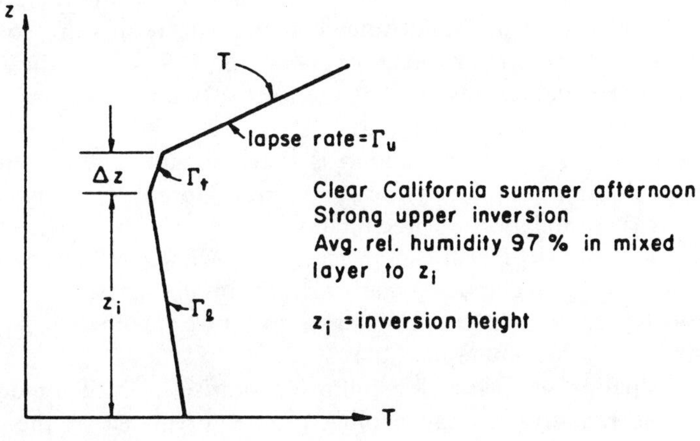

Fig. 7.11. Typical late afternoon temperature profile measured by Mack *et al.* (1974). [From Oliver *et al.* (1978).]

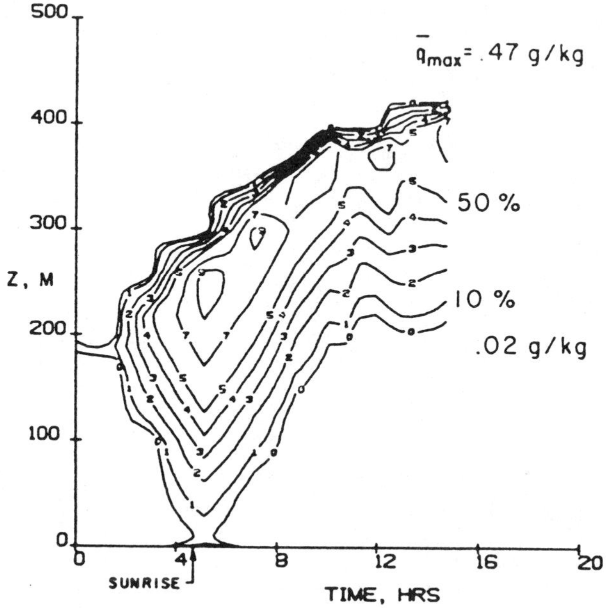

Fig. 7.12. Evolution of fog resulting from the lowering of stratus. Fog shows some tendency to form at surface just before downward-propagating stratus reaches the surface. [From Oliver *et al.* (1978).]

With z_i initially at 200 m, the stratus base descended to the surface by 0500, and there it remained until 0700. Figure 7.12 illustrates the predicted time–height evolution of liquid-water content. The results are consistent with observations reported by Leipper (1948) and P79 which indicated that stratus lowering did not occur until the later afternoon or early evening inversion height was 400 m or lower.

A feature of the OLW simulation is that the turbulent moisture flux profiles are positive at all levels and at all times. Moreover, the model does not contain a formulation of precipitation processes. Thus, the results of the model suggested by P79 are not in accordance with OLW's model results. Perhaps the conceptual model of marine stratocumuli suggested by Schubert *et al.* (1979a; see Fig. 7.18) is applicable to the stratus-lowering process. We shall discuss this model in a later section.

The dissipation of surface fog following sunrise in OLW's model is a result of the radiative heating/cooling profiles predicted by the models shown in Figs. 7.13 and 7.14. They note that direct solar heating is absorbed well into the interior of the fog, and that evaporation caused by solar heating occurs rather uniformly throughout the cloud because of the long absorption length for solar radiation (see Chapter 5). At the same time, infrared radiative cooling near cloud top causes instability and turbulence, which transports the shortwave warming throughout the cloud. As noted in Chapter 5, the magnitude of the solar absorption length is dependent upon the particular radiation parameterization used. It appears that OLW's parameterization exaggerates fog dissipation by solar heating. The drizzle process, however, may accelerate fog dissipation by removing the LWC near cloud top. This would allow shortwave radiation to penetrate more deeply into the interior of the fog where the radiation is absorbed more effectively due to the presence of large drops (see Chapter 5). This process should be studied quantitatively, however.

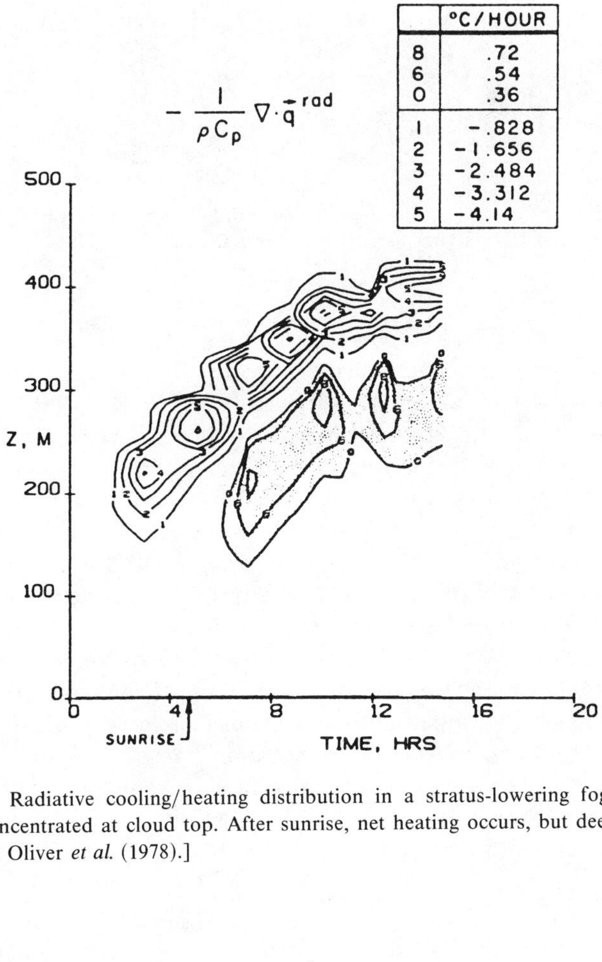

Fig. 7.13. Radiative cooling/heating distribution in a stratus-lowering fog. Maximum cooling is concentrated at cloud top. After sunrise, net heating occurs, but deep within the cloud. [From Oliver *et al.* (1978).]

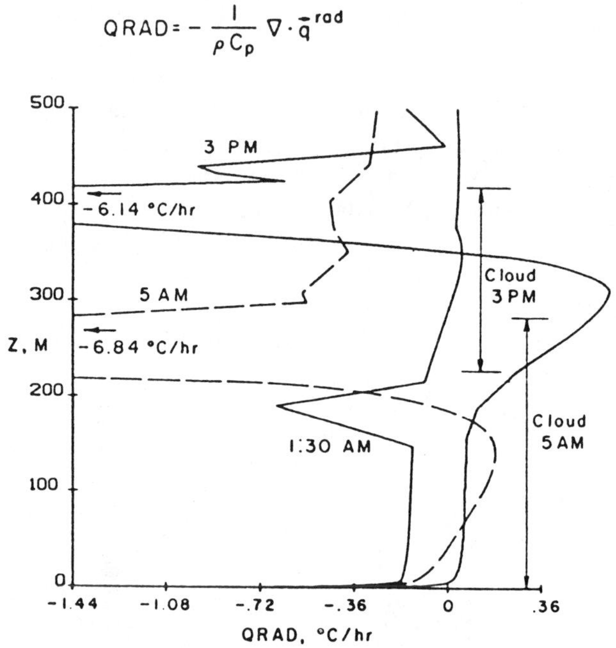

Fig. 7.14. Profiles of radiative cooling/heating at 0130 (stratus just forming), 0500 (stratus on the surface), and 1500 (stratus base lifted). [From Oliver *et al.* (1978).]

7.5.4 Fog Streets and Low-Level Convergence

Marine fog does not always have a horizontally homogeneous structure. P79 refer to observations of what they call *fog streets.* That is, the fog is organized in alternating parallel lines of foggy and clear air. Figure 7.15 illustrates the observed changes in visibility as the ship *Acania* cruised crosswind through a region characterized by fog streets. They noted that the individual fog patches range in width from 0.5 to 2 km. Along the direction of the wind, the fog patches appeared with the upwind edges of the fog touching the surface. At distances of several hundred meters to several kilometers from the upwind edge, the fog base lifted from the surface and persisted as a stratus deck aloft. An example of fog streets described by Walter and Overland (1984) is given in Chapter 8, Section 8.3.2. Mechanisms exhibiting a multiplicity of band scales responsible for the formation of such bands are also given in that section.

P79 also describe a case in which surface convergence appeared to be instrumental in fog formation. Upwind of the fog, the air was warmer than the sea, while within the fog it was cooler than the sea. Winds upwind of

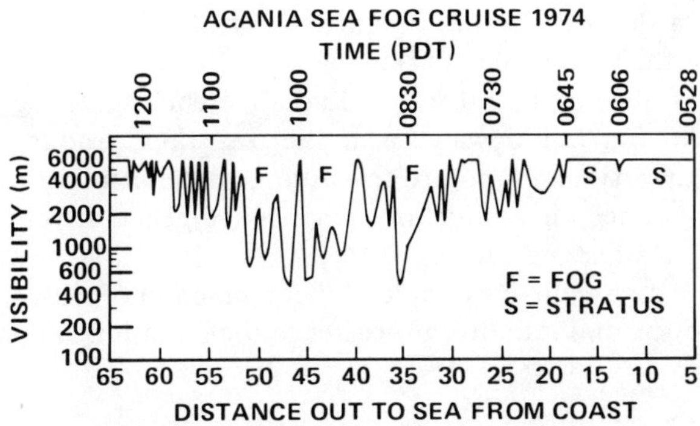

Fig. 7.15. Visibility data as a function of distance obtained while cruising crosswind through a series of fog patches, 23 August 1974. [From Pilie *et al.* (1979).]

the fog were out of the northwest at 4–8 m s^{-1}, while within the fog the speeds were consistently lower and, in some locations, were reversed in direction. Divergence values ranged from -0.7×10^{-4} to -2.7×10^{-4} s^{-1} and persisted for at least 20 h over a region of approximately 2500 km^2. P79 estimated average vertical velocities over the foggy region to be 1–2 cm s^{-1}. However, they suggested that organized patterns of updrafts and downdrafts

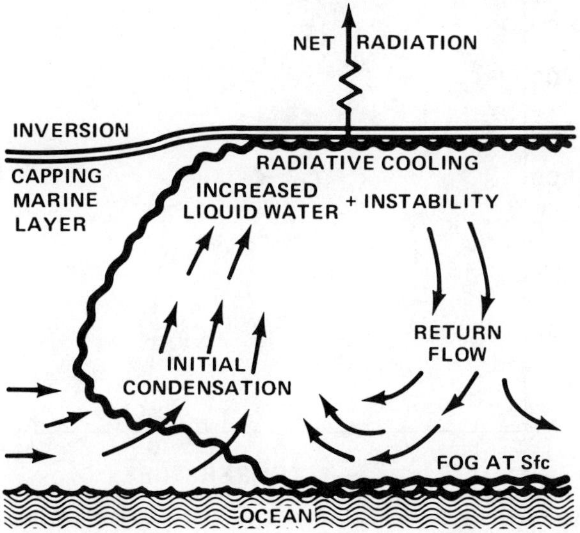

Fig. 7.16. Schematic representation of the vertical cross section of fog formed as a result of low-level convergence and radiative cooling. [From Pilie *et al.* (1979).]

existed within the area having significantly stronger magnitudes than the average. They thus postulated that adiabatic cooling within the local updrafts contributed to the fog liquid water. Radiative cooling at fog top further contributed to fog liquid water near the top, thus producing sufficient condensate to survive evaporative warming in descending drafts to the surface. Their conceptual model of such a convergence-driven, convective-radiation fog is illustrated in Fig. 7.16.

This case is an excellent example of the interaction of mesoscale circulations, turbulence, and radiative processes in the initiation and maintenance of marine fogs.

7.6 Stratocumulus Clouds

7.6.1 The Convective Boundary Layer

Low-level marine stratocumulus clouds occupy large portions of the eastern Pacific and eastern Atlantic oceans and small portions of the western Indian Ocean. Figure 7.17 illustrates a typical profile of temperature, moisture, and winds in an eastern North Pacific stratocumulus regime. The layer is charac-

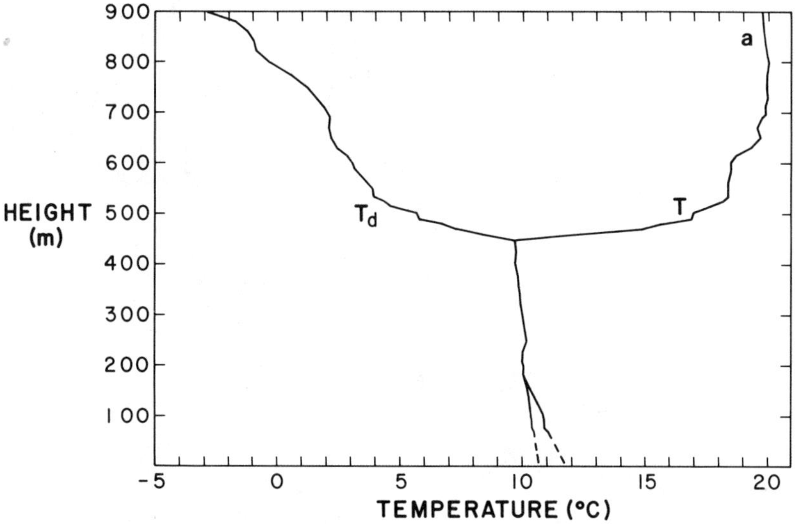

Fig. 7.17. Temperature, moisture, and wind data from an NCAR Electra sounding at 37.8°N, 125.0°W and between 1522 and 1526 GMT; (a) temperature and dew point; (b) dry static energy, moist static energy, and saturation static energy; (c) wind direction and speed. Dashed lines below 50 m are extrapolations. [From Schubert *et al.* (1979a).] (*Figure continues.*)

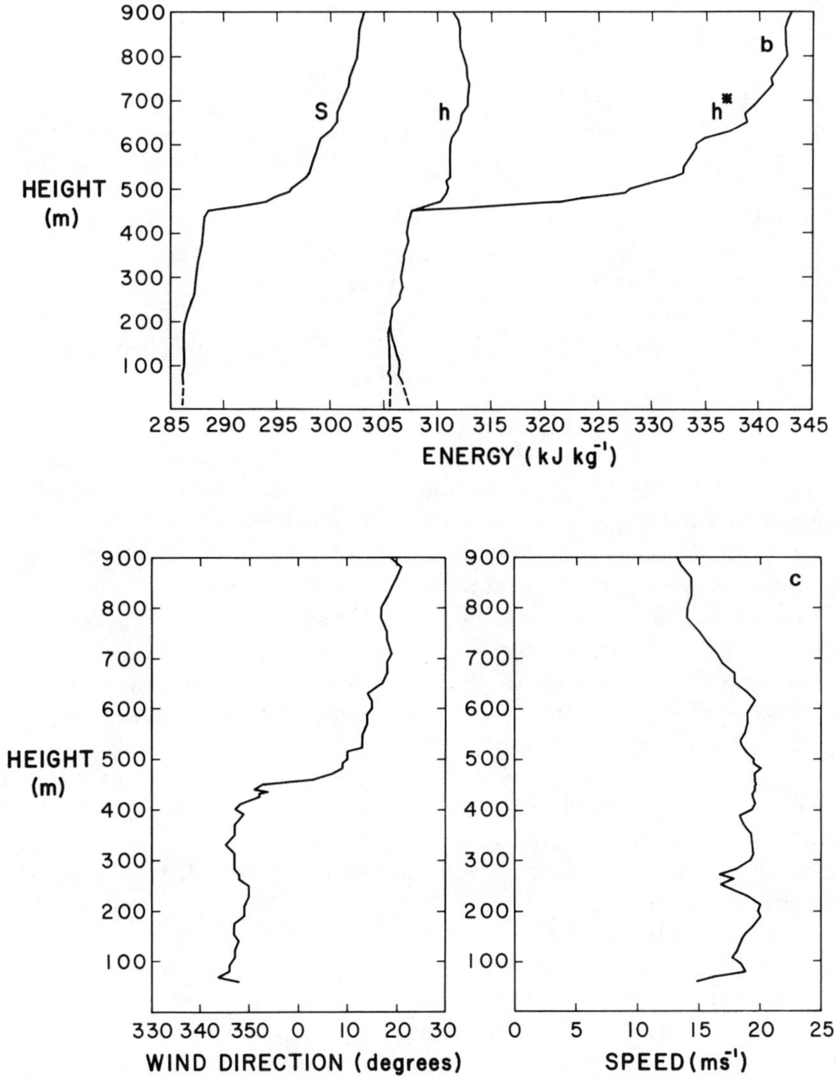

Fig. 7.17 (*continued*)

terized by a nearly adiabatic, well-mixed subcloud layer (dry static energy, $s = c_p T + gz$, is constant, or θ is constant) and by a wet adiabatic cloud layer (moist static energy, $h = s + Lr_T$, or θ_e, is constant), capped by a strong temperature inversion and drop in dew point. Winds through the entire mixed layer are nearly uniform. There are many questions that must be answered regarding the behavior and structure of stratocumulus clouds. Some of these are as follows:

(1) How does a stratocumulus layer maintain a steady depth against the effects of subsidence which cause a cloud layer to become shallower?
(2) What is the nature of entrainment into the top of a stratocumulus layer?
(3) What are the relative roles of buoyancy and shear in generating turbulent kinetic energy in a cloud layer?
(4) How important is cloud-top radiative cooling to the maintenance of a stratocumulus layer?
(5) What causes the breakup of a solid stratocumulus layer to form a field of cumuli?

In order to examine questions such as these as well as others, a hierarchy of models has been proposed to describe and predict the structure of marine stratocumulus clouds. These include one-dimensional layer-averaged or mixed-layer models, entity-type or plume models, higher ordered closure models, and large-eddy simulation models. Each of these modeling approaches offers certain advantages and disadvantages.

The primary advantage of layer-averaged models is their simplicity. They do not consume a great deal of computer time and, moreover, they yield a clear "signal" of the response of the simulated cloud layer to various physical processes. This is in contrast to higher ordered closure models or LES models, where the response of entrainment to changes in environmental parameters, for example, may involve a complex chain of events through triple correlation products or the ensemble-averaged effects of a large number of explicitly simulated convective cells. The primary disadvantage of the layer-averaged models is their lack of versatility. They are generally restricted to well-mixed layers. Departures from a well-mixed state violate the fundamental premise of such models. The presence of drizzle, for example, was found by Nicholls (1984) to violate the fundamental mixed-layer hypothesis in some cases. Moreover, extension of layer-averaged models to include the effects of wind shear or the presence of a broken cloud field requires parameterizations or additional modeling assumptions.

Entity-type models are a convenient vehicle for expressing concepts about the structure of a stratocumulus field. For example, the stratocumulus field can be viewed as being composed of a field of thermal-like elements or vertically elongated plumes. Justification for the particular form of entity structure hypothesized comes from observations. Entity-type models have the advantage that they provide a convenient framework for examining both a variety of cloud microphysical processes and the internal consistency of entrainment processes or the effects of radiative cooling.

Higher order closure models, when applied as ensemble-averaged models, are more general than layer-averaged models in their formulation, and as such are more versatile in principle. That is, stable and unstable cloud-capped boundary layers as well as boundary layers with shear are all possible atmospheric states which a higher ordered closure model may be able to simulate. Moreover, they can be applied to mesoscale problems where strict horizontal homogeneity is not present. Higher ordered closure models, however, are computationally demanding. Furthermore, important physical processes, such as entrainment, depend strongly on the details of the parameterization of poorly understood processes represented by terms such as triple-correlation terms and pressure–velocity correlations. Moreover, not unlike layer-averaged models, the determination of cloud fractional coverage requires the introduction of various ad hoc assumptions regarding the distribution functions of thermodynamic variables and/or water mixing ratios (see Chapter 3).

By far the most fundamental approach to modeling a cloud-capped boundary layer is the large-eddy simulation approach. The LES approach involves numerical integration of the equations of motion within a shallow boundary layer. LES offers the advantage of being able to simulate explicitly the detailed circulations and properties of the cloud layer. They exhibit far less sensitivity to the details of turbulent closure parameterizations than higher ordered closure models. Often the response of a LES model to changes in closure assumptions and/or changes in resolution is rather subtle, but nonetheless important to the overall simulation. LES models are generally quite versatile, being able to simulate both a convectively unstable solid stratocumulus deck and a trade-wind cumulus field. The major deficiency of LES models is that they are computationally demanding. They also generate large volumes of data which require considerable analysis, also at large computational expense. LES models are thus limited to exploring only a few physical questions and atmospheric states. Moreover, LES models can only simulate a limited range of scales of motion. Typically the largest scale resolved by a LES model is of the order of 5 km while the smallest is of the order of 50 m. This limitation becomes quite severe when the atmosphere is stably stratified and the dominant scales of motion may be on the order of 5 m. Observations also suggest that horizontal scales on the order of 5 m are important to cloud-top entrainment processes.

We will now examine the structure and characteristics of stratocumulus clouds using the various modeling approaches as a vehicle for describing them.

7.6.2 Layer-Averaged Models

The pioneering paper on this topic was published by Lilly (1968). He considered the effects of condensation and evaporation, large-scale vertical motion, and divergence of net radiation at the cloud top. The basic assumptions in his model are as follows:

(a) Below the base of the capping inversion, the boundary layer is well mixed, or uniform, in mean values of semiconservative properties such as total-moisture specific humidity (q_w) and wet-bulb potential temperature θ_w or equivalent-potential temperature θ_e.

(b) The capping inversion is of negligible thickness.

(c) Turbulence in the mixed layer is generated entirely by buoyant production (the effects of wind shear are ignored).

(d) The upper cloudy portion of the mixed layer is entirely saturated, and the buoyant flux of entrainment occurs entirely in the saturated air.

(e) There is no precipitation or drizzle.

(f) The divergence of net radiation occurs entirely within the capping inversion and not at all within the upper mixed layer.

(g) The jump in θ_w or θ_e across the capping inversion ($\Delta\theta_w$ or $\Delta\theta_e$) must be positive for parcel stability and maintenance of the cloud layer.

(h) The entrainment rate or growth rate relative to any mean subsidence of the mixed layer is bounded on the upper side by that which can be deduced if there were no dissipation of kinetic energy (Ball, 1960; maximum entrainment assumption), and on the lower side by the value that can be deduced if buoyancy flux were just zero at some height within the mixed layer and positive at all other heights within the mixed layer (minimum entrainment rate assumption).

Entrainment here is viewed solely as cloud-top entrainment. Lilly viewed the *maximum entrainment* assumption as occurring when the dissipation and transport terms in the vertically integrated TKE equation are small compared to opposing positive and negative contributions due to buoyancy generation of TKE. The minimum entrainment is seen to occur when dissipation is so strong that a region of negative heat flux cannot be supported.

Schubert (1976) extended Lilly's model by retaining assumptions (a)–(f) but estimated the entrainment rate as the weighted average of the maximum and minimum entrainment assumptions.

The most controversial aspect of Lilly's model is the relationship between radiative and convective fluxes and entrainment at the top of the mixed layer. Lilly assumed that the radiative cooling was confined to the cloud-top jump region so that radiative cooling did not appear in the mixed-layer heat budget. Deardorff (1976; hereafter referred to as D76) allowed only a fraction of the cloud-top radiation divergence to occur within the capping inversion and the remaining fraction to occur within the uppermost mixed layer just below the capping inversion. By having radiative cooling extend over a finite depth, enhanced TKE and cooling occurred within the mixed layer.

Kahn and Businger (1979) took an even stronger stand and argued that essentially all the cloud-top divergence of net radiation should be placed within the mixed layer and none within the capping inversion. They argued that all the cloudy air lies within the mixed layer and that the zone of longwave radiation cooling should extend below cloud top to the order of 100 m depth.

Schubert *et al.* (1979a) proposed a simple conceptual model of a stratocumulus cloud. In this model, shown in Fig. 7.18, air near the surface is accelerated toward the updraft and upward against negative buoyancy by lower pressure in the updraft near cloud base. If one computes the correlation of this pressure pattern with the convective-scale vertical motion field, one finds $(\overline{w''p''})_{z=z_c} < 0$. In other words, the work done on the subcloud layer by the cloud layer maintains the convection motions of the subcloud layer.

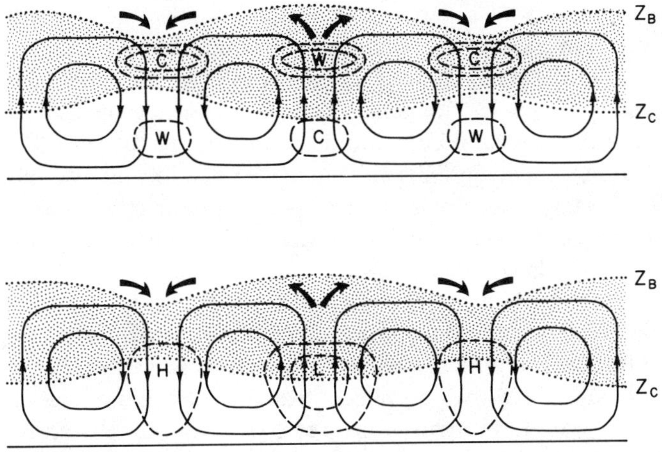

Fig. 7.18. Schematic depiction of the motion field of a convective element along with its associated cloud base, cloud top, temperature (top), and nonhydrostatic pressure (bottom) fields. Note that the updraft has positive buoyancy in the cloud layer and negative buoyancy below. To accelerate surface air upward into the updraft requires the nonhydrostatic pressure shown. Since $\overline{w'p'} < 0$ at cloud base, the cloud layer does work on the subcloud layer. [From Schubert *et al.* (1979a).]

This conceptual model may also help explain the stratus-lowering process discussed in Section 7.5.3. That is, as radiative cooling destabilizes the stratus layer, convective fluxes can propagate downward by building a lower pressure below the regions of active updrafts. Such a lowering process would not require drizzle to moisten the air, because convection would transport water vapor from the ocean surface to the low-pressure region near cloud base.

This brief summary does not come close to reviewing all the various approaches to layer-averaged models. Our purpose is to give the reader a basic understanding of some of the major concepts involved in their formulation.

7.6.3 Entity-Type Models

In an entity-type model, a stratocumulus field is viewed to be composed of a field of distinct convective elements which exhibit well-defined shapes, entrainment, or mixing laws, and so forth. The conceptual model illustrated in Fig. 7.18 is, in a sense, an entity-type model of a stratocumulus field in which the entities are regular thermal-like cells having comparable updraft and downdraft areas. Chai and Telford (1983) adopted a plume model to a stratocumulus field. They view a stratocumulus cloud to be composed of a population of plumelike updrafts and downdrafts. The ascending plumes originate at the top of a superadiabatic layer above the sea surface. The convective plume ascends, eventually becoming saturated, and gains buoyancy from latent heat release. Eventually the plume encounters a strong capping inversion which the rising plume cannot penetrate. The plume then turns over at the inversion base and descends toward the surface. Chai and Telford do not consider entrainment of air residing above the capping inversion into the descending plumes. They argue that entrainment effects happen rapidly and do not need to be modeled when seeking approximate time estimates. The descending plume is viewed to undergo horizontal motion at the bottom of its descent and reenter an adjacent rising plume. As the air reenters a plume, some surface air with surface properties will be mixed in. Thus, the plume becomes warmer and wetter. Because vapor is continuously added to the plumes near the sea surface, the cloud base descends and could eventually form sea fog. Only the erosion of the base of the capping inversion (a process called *encroachment*) by the rising plumes reduces the total moisture flux in the convective layer and inhibits the lowering of cloud base. Encroachment occurs when the whole layer is warmed up to the height of the inversion. When the air at greater heights is no longer buoyant relative to the air underneath, the air is incorporated into the boundary layer and the layer deepens. Mathematically the process is modeled by following the ascending and descending plumes which continually recycle air through the boundary layer. The plumes are assumed to remain well mixed internally and entrain air horizontally through the plume boundaries at rates proportional to the root-mean-square turbulent velocities. Chai and Telford also argue that radiative cooling is not necessary to model the time scales of overturning in the cloud layer (~15 min). While this may be valid from the perspective of the individual plumes, it is not necessarily valid with respect to time scales affecting the stability of the mixed layer as a whole (approximately several hours). Chai and Telford did not integrate through enough plume recycling to examine the importance of radiation to the longer term thermodynamic structure of the boundary layer.

7.6.4 Higher Order Closure Models

Higher order closure models involve the explicit prediction of turbulent kinetic energy, variances of the various quantities, and a number of covariances such as eddy fluxes. The general procedures involved in developing closure models are discussed in Chapter 3. Instead of specifying flux profiles through a mixed layer or entrainment rates at the top of a mixed layer, these quantities are explicitly predicted on a finite mesh. The validity of the predicted entrainment rates and the flux profiles depends on the vertical resolution of the model and on the closure assumptions. Examples of the application of closure models to the simulation of stratocumuli includes OLW's study of marine fog and the stratus-lowering process that we examined earlier. In addition, OLW examined the diurnal variation of a stratocumulus layer. Chen and Cotton (1983a) also examined the diurnal variation in a stratocumulus layer with a closure model; Bougeault (1985) studied the diurnal cycle of the marine stratocumulus layer with a higher order cloud model. We will discuss the results of more recent simulations of diurnal variations in a stratocumulus layer in another section. Chen and Cotton (1983b) and Moeng and Arakawa (1980) used higher order closure models to examine the onset of "entrainment instability" and the breakup of stratocumulus clouds. We will also examine this topic more fully in the next section. Thus, closure models have become an important vehicle for investigating the physics of stratocumulus layers. Interpretation of the results obtained with these models must be done cautiously, however.

7.6.5 Large-Eddy Simulation Models

The pioneering work in the application of LES models to stratocumulus-topped mixed layers was performed by Deardorff (1980b). He used a primitive equation model valid for shallow convection. Chosen in the cloud simulations was $\Delta y = \Delta x = \Delta z = 50$ m over a domain which was $2 \times 2 \times 2$ km. Explicit predictions were made for $\bar{u}$, $\bar{v}$, $\bar{w}$, $\bar{\theta}_1$, and $\bar{q}_w$ where the total specific humidity is

$$\bar{q}_w = q + q_1 \tag{7.16}$$

For eddies smaller than the grid scale, the model was closed with an eddy viscosity closure assumption in which the eddy viscosity was formulated to be a function of a predicted turbulent kinetic energy (TKE) [i.e., Eq. (3.50)]. The TKE was closed with a simple model of dissipation and with a down-gradient diffusion term on the triple-correlation terms and the pressure–velocity correlation terms.

It should be noted here that turbulence is relegated to scales less than 50 m. All larger eddy scales are explicitly resolved by the model. Deardorff carried out seven numerical experiments and then performed statistical analyses of the results predicted by the model in those cases. The first case was a clear-sky case based on Day 33 of the Australian Wangara boundary-layer experiment. The next two cases considered radiational cooling imposed at the top of the boundary layer, but they did not include any cloud effects. In the fourth case a stratocumulus deck was allowed to occupy the upper half of the mixed layer with cloud latent heat effects included but radiation excluded. The fifth case is the same as the fourth case except that the capping inversion was very weak.

The sixth case was the same as the fourth, except cloud-top radiative flux divergence was prescribed at the top grid point of the cloud layer. In the seventh case, everything was the same as in the sixth case except the cloud occupied almost the entire mixed layer.

Table 7.1 (from Deardorff, 1980b) summarizes the seven numerical experiments. The variables in this table are as follows: the average height of the mixed layer or cloud-top height is h, cloud base height is h_b, surface turbulence of fluxes of heat and moisture are $\langle w\theta \rangle_s$ and $\langle wq \rangle_s$, respectively, soil moisture fraction is w, and soil surface temperature and specific humidity are θ_s and q_{sfc}, respectively. We shall concentrate on cases 4 through 7.

Figure 7.19 illustrates the resultant mean profiles and eddy fluxes Deardorff analyzed from his 3-D model for case 4. Included in the figure are cloud fractional coverage σ_c. It should be noted that θ_1, which is conservative for dry and moist processes, is relatively constant throughout the entire mixed layer. Likewise, total specific humidity q_w is nearly constant throughout the mixed layer. As evidenced by the σ_c profile, the base of the cloud layer varies over a range of 100–150 m.

The simulated eddy heat flux $\langle \overline{w\theta_1} \rangle$ varies linearly from the surface up to the capping inversion. The buoyancy flux $\langle \overline{w\theta_v} \rangle$ also varies linearly through

Table 7.1

Descriptive Properties of the Seven Cases[a]

Case	Time period[b]	h (m)	h_b (m)	Phase change	Cloud-top radiative cooling	$\langle w\theta \rangle_s$ (mm s^{-1} K)	$\langle wq \rangle_s$ (mm s^{-1} g kg^{-1})	W	$\theta_s(K)$ (g kg^{-1})	q_{sfc}
1	13.37 to 13.43 h (1–3)	1300	—	No	No	198	23	0.10	300.2	5.0
2	13.31 to 13.50 h (10)	1200	—	No	Yes	99	15	0.10	289.0	3.85
3	14.24 to 14.78 h (10)	1220	—	No	Yes	3	5	0.10	283.5	3.46
4	14.35 to 14.80 h (10)	1400	780	Yes	No	89	10	0.09	286.2	5.82
5	16.32 h (1)	1795	990	Yes	No	38	6	0.09	284.8	5.70
6	14.44 to 14.98 h (10)	1160	620	Yes	Yes	73	22	0.30	285.7	6.85
7	16.52 to 16.90 h (10)	1550	200	Yes	Yes	27	7	0.70	284.2	8.02

[a] From Deardorff (1980b).

[b] Numbers in parentheses under the entries in the time period column refer to the number of elements or individual data sets that were averaged to obtain ensemble-average vertical profiles.

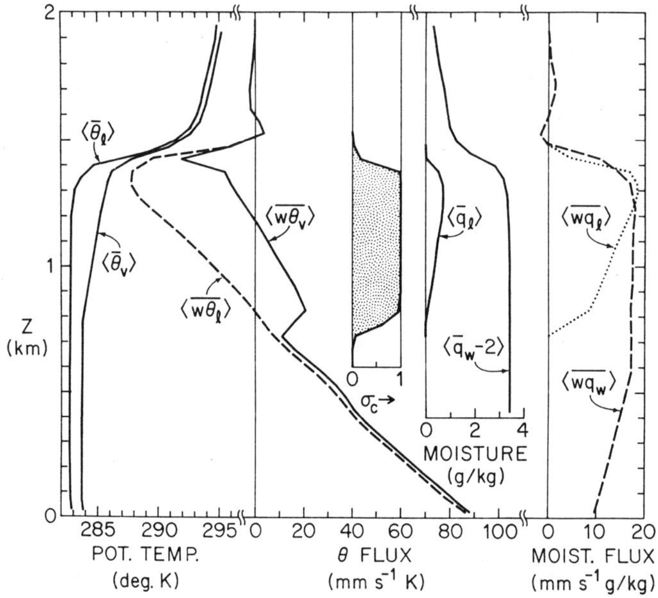

Fig. 7.19. Vertical profiles of mean liquid-water potential temperature (θ_l) and its vertical flux, virtual potential temperature and its vertical flux, cloud fraction (σ_c) at a given height (stippled area), total specific moisture (q_w) and its vertical flux, and liquid-water specific humidity (q_t) and its vertical flux, for case 4. Note that 2×10^{-3} is subtracted from ($\bar{q}_w$). [From Deardorff (1980b).]

the subcloud layer. In the cloud layer, however, the buoyancy flux departs significantly from linearity due to the effects of latent heat release. This is consistent with the models of Lilly (1968) and Schubert (1976). D76 noted that the negative buoyancy flux in the capping inversion was missed by their models because the inversion is of zero thickness. According to D76, above the solid cloud deck, q_l increases at the wet adiabatic rate up to the levels where cloud-top entrainment is active. Note that the vertical flux of q_l exceeds that of vapor (q) in the upper part of the cloud layer. D76 explained this by noting that

$$q_l'' = q_w'' - q_s'' = q_w'' - (\partial q_s/\partial T)T'' \tag{7.17}$$

where $q_w = q + q_l$, and that q_w'' and q_l'' correlate strongly with w. Deardorff notes that T'' correlates rather weakly with w because $\langle\overline{wT''}\rangle$ is closely related to the buoyancy flux, which is small in the upper part of the cloud layer. The total water flux remains large at these levels, giving rise to a liquid water flux $\langle\overline{w''q_l''}\rangle$ greater than $\langle\overline{w''q_w''}\rangle$.

Figures 7.20 and 7.21 illustrate the averaged profiles in Deardorff's simulation with radiational cooling introduced in the layer denoted ΔF. Figure 7.20 illustrates the case of a shallow cloud layer while Fig. 7.21 is a deeper cloud layer. Except for the slight dip in the $\langle w\theta_l \rangle$ profile, the cooling is confined to the capping inversion in support of Lilly's hypothesis. Otherwise, the results of this experiment differ little from case 4. The main effect of the deeper cloud layer is the increase in liquid-water content in the upper part of the layer.

Recently Moeng (1986) applied a LES model to the simulation of an unstable cloud-capped boundary layer. The model resolved features on the order of 60 m in the horizontal and 25 m in the vertical over a domain of $2.5 \times 2.5 \times 1$ km. An interesting result derived from her simulations and shown in Fig. 7.22 is the difference between a clear planetary boundary layer and a cloud-topped boundary layer (CTBL). Contours of virtual dry static energy in the dry case illustrate that the plumelike thermals have their origin near the earth's surface. By contrast, contours of liquid-water static energy in the cloud-topped boundary layer illustrate that the boundary layer

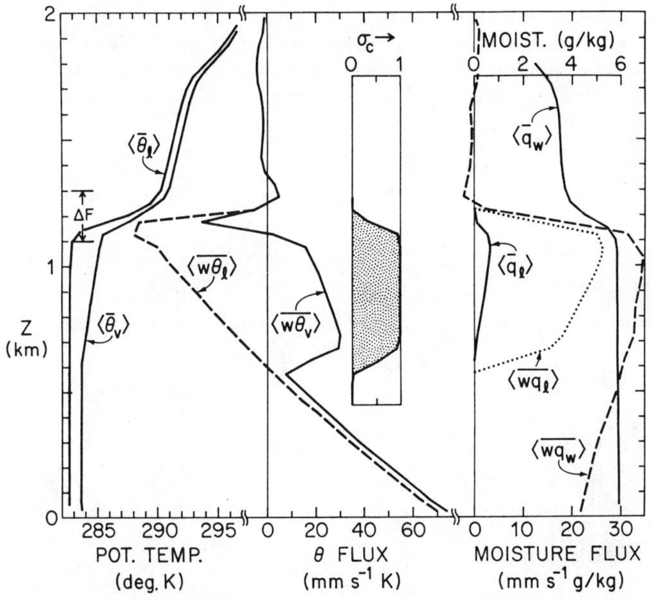

Fig. 7.20. Vertical profiles of mean liquid-water potential temperature and its vertical flux, virtual potential temperature and its vertical flux, cloud fraction, total specific moisture and its vertical flux, and specific liquid-water content and its vertical flux, for case 6. Layer of strong cloud-top radiative cooling indicated by arrows at left. [From Deardorff (1980b).]

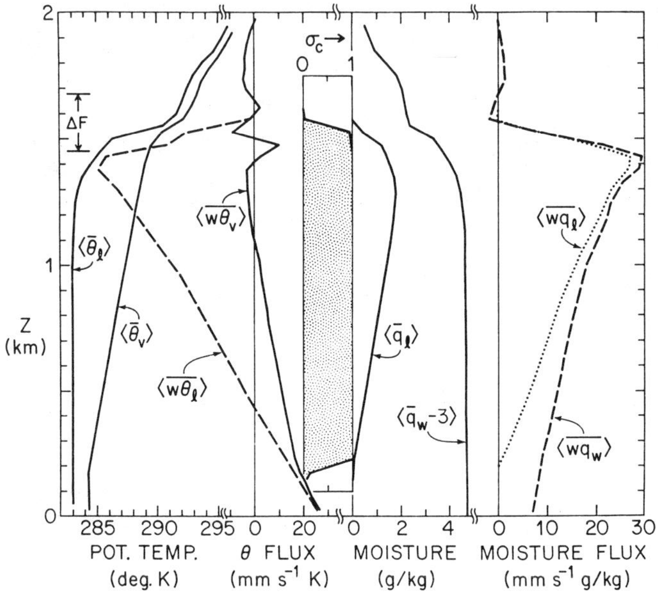

Fig. 7.21. Vertical profiles of mean liquid-water potential temperature and its vertical flux, virtual potential temperature and its vertical flux, cloud fraction, total specific moisture content and its vertical flux, and specific liquid-water content and its vertical flux, for case 7, the deep stratocumulus. Note that 3×10^{-3} is subtracted from $(\bar{q}_w)$. [From Deardorff (1980b).]

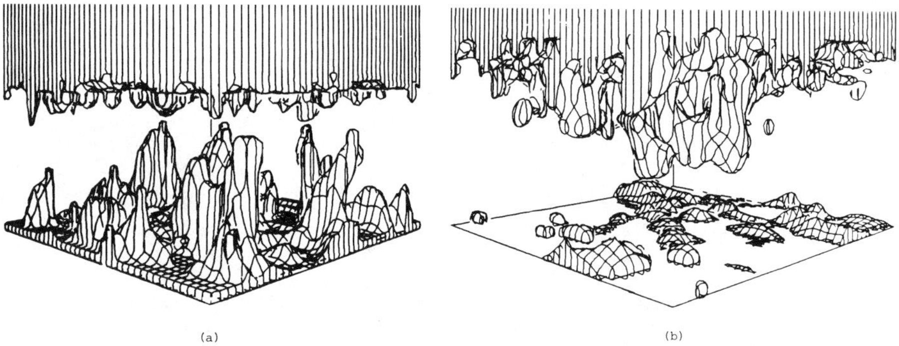

Fig. 7.22. Three-dimensional plots of a constant surface of (a) the virtual dry static energy from a clear convective PBL simulation and (b) the liquid-water static energy from a CTBL simulation. [From Moeng (1986).]

is dominated by descending plumelike convective elements. These descending elements obtain their negative buoyancy from radiative and evaporative cooling arising from entrainment of dry air at the capping inversion. Thus, the cloud-topped boundary layers are driven not by heating from below but by radiative and evaporative cooling near cloud top.

Further evidence of the dominance of descending plumelike thermals can be seen in Kuo's (1987) two-dimensional simulation of a cloud-capped boundary layer. Kuo used a numerical scheme which is spectral in both the horizontal and vertical dimensions (he used Fourier basis functions in the horizontal and Chebyshev polynomial expressions in the vertical). The model domain was 2.5×0.8 km with a resolution of nearly 10 m in each direction. Figure 7.23 illustrates the predicted velocity and liquid-water fields for a case with particularly strong entrainment. Radiative cooling is applied uniformly in the initial cloud-top region in this experiment. Note that the simulated cells are dome shaped in appearance with sharp gradients of liquid water on the cell sides. The sharp gradients are on a horizontal scale of 40 m and extend through the depth of the cloud layer. The maximum upward motion is at the center of the cell while the downward motion is near the side of the cell. It is interesting that the cloud bases in the downdraft regions are not necessarily higher than the updraft regions (contrary to Fig. 7.18). Thus, consistent with Chai and Telford's plume model, the downdraft does not appear to be mixing strongly with the dry inversion air. The horizontal scale of the cells are about 800 m, yielding an aspect ratio of about 2 for a 450-m-deep boundary layer. Kuo also found that horizontal variations in radiative cooling did not affect the boundary layer characteristics appreciably as long as the cooling is confined to the turbulent region. This is because local instabilities are rapidly eliminated by the vigorous turbulence. This is also consistent with Chai and Telford's hypothesis. Radiative effects are important, however, in that they contribute to the overall destabilization of the cloud-top layer in an average sense. Kuo's numerical experiments also illustrate that stratocumulus clouds can remain solid without breakup even in the presence of strong entrainment. We shall examine the implications of this modeling result more fully in the next section. While we must be a bit cautious about our interpretation of Kuo's two-dimensional simulations, nonetheless they do suggest that entrainment takes place in stratocumulus clouds in narrow descending plumelike elements that can extend through the depth of the cloud layer. The picture that emerges from these simulations is not too different from the plume concepts proposed by Telford and Chai. The updrafts, however, appear to be broad, thermal-like features rather than the narrow plumes Chai and Telford envision. We will now examine the implications of this concept of stratocumulus structure to the breakup of stratocumuli.

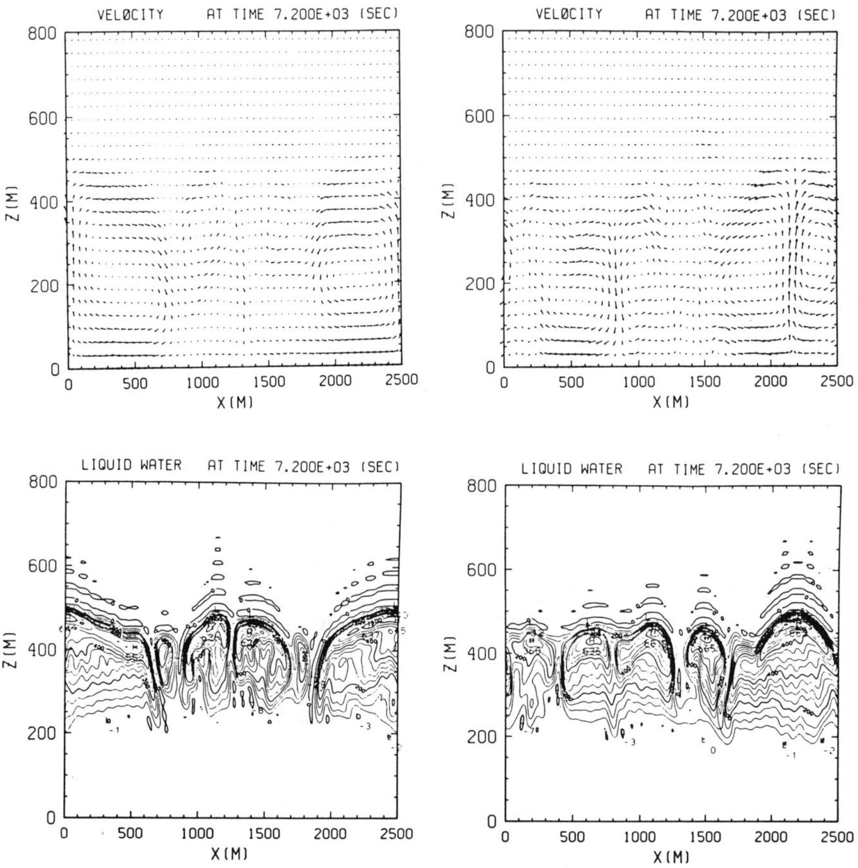

Fig. 7.23. The velocity and liquid-water mixing ratio in physical domain at 120 min for the U1 and S cases in the fixed radiation experiment. The liquid-water mixing ratio (in g kg^{-1}) is scaled by 10^3. The maximum velocity is 3.8 m s^{-1} for the U1 case and 3.1 m s^{-1} for the S case. [From Kuo (1987).]

7.6.6 Entrainment Instability

One of the most important questions that modelers of stratocumulus clouds must answer is, "at what point does a solid stratocumulus cloud layer break up into a broken cumulus field?" Lilly (1968) suggested that whenever the jump $\Delta\theta_e < 0$, where $\Delta\theta_e$ is the above cloud value of θ_e minus the in-cloud value, the cloud top will become unstable to entrainment. Lilly argues that

> If a parcel of the upper air is introduced into the cloud layer and mixed by turbulence, evaporation of cloud droplets into the dry parcel will reduce its

> temperature. If the mixed parcel reaches saturation at a colder temperature than that of the cloud top it will be negatively buoyant and can then penetrate freely into the cloud mass. In such a case the evaporation and penetration process will occur spontaneously and increase unstably until the cloud is evaporated.

Subsequently, Randall (1980) and Deardorff (1980a) pointed out that when water loading effects are considered, the criteria for cloud-top entrainment instability become $\Delta\theta_e < -1$ to -2 K.

Earlier Benoit (1976) predicted that the breakup of a stratocumulus would occur when $W_e/W_* > 0.06$, where W_e is the cloud-top velocity and

$$W_*^3 = 2.5\frac{q}{\theta}\int_0^h \overline{W''\theta_e''}\,dz,$$

is a generalized mixed-layer free convective velocity scale.

In Deardorff's (1980b) LES simulations described previously, he attempted to simulate cloud-top entrainment instability in his case 5 experiment (see Fig. 7.24).

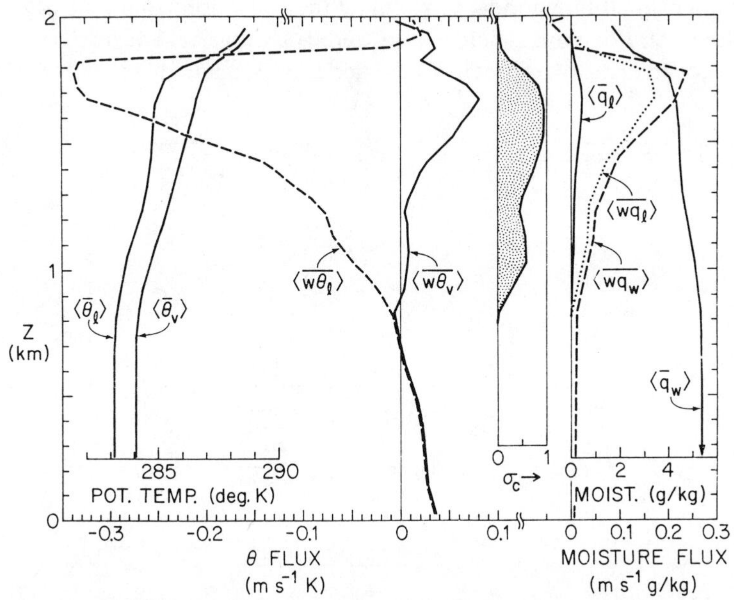

Fig. 7.24. Vertical profiles of mean liquid-water potential temperature and its vertical flux, virtual potential temperature and its vertical flux, cloud fraction at a given height (stippled area), total specific moisture and its vertical flux, and liquid-water specific humidity and its vertical flux, for case 5 of rapid entrainment. [From Deardorff (1980b).]

In this case the capping inversion was gradually cooled by about 5°C. Rapid entrainment ensued in which evaporation and water loading combined to produce penetrating downdrafts to the extent that cloud cover became less than 100%, particularly in the lower part of the cloud layer. The large negative eddy fluxes of θ_e are indicative of the rapid entrainment process. The resultant broken cloud layer is completely different in structure from the lower level, solid stratocumulus deck.

Chen and Cotton (1983b) attempted to replicate Deardorff's (1980b) 3-D simulations of entrainment instability with a 1-D, second-order turbulent transport model, following procedures outlined in Chapter 3. Figure 7.25 illustrates the profiles predicted by Chen and Cotton for Deardorff's rapid entrainment case (case 5). Comparison of Figures 7.24 and 7.25 illustrates that the two models predict similar eddy flux profiles although the magnitude of entrainment is larger in Chen and Cotton's experiment. In both cases following rapid entrainment, the cloud grows much higher and the cloud is no longer a solid deck. This clearly illustrates the extent to which the rapid entrainment process can alter the structure of a stratocumulus cloud layer.

Moeng and Arakawa (1980) applied a higher order closure model to a two-dimensional flow model to simulate the transition from a nearly 100% cloudiness stratocumulus field to a scattered cumulus field of small fractional cloud cover. The two-dimensional model was 3 km deep and extended

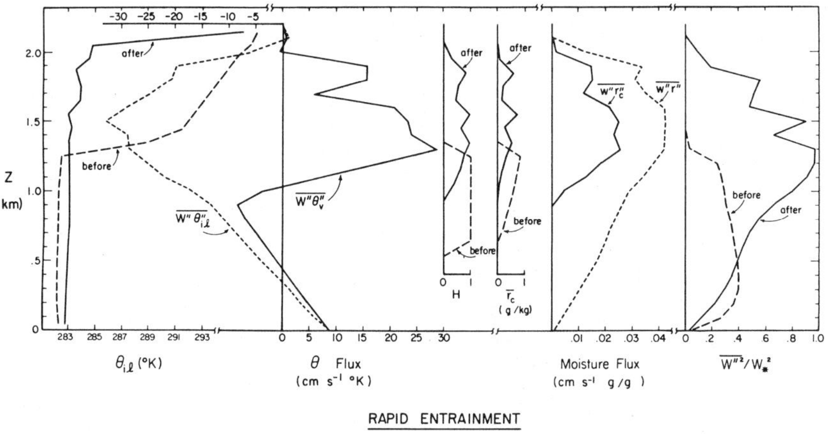

Fig. 7.25. vertical profiles of mean ice and liquid-water potential temperature ($\bar{\theta}_{il}$) and its vertical flux, vertical flux of vertical potential temperature ($\overline{w''\theta''_{il}}$, $\overline{w''\theta''_v}$), cloud fractional coverage (H), mean cloud water mixing ratio (r_c), and the normalized vertical velocity variance. The dashed lines represent the case before the onset of the rapid entrainment; solid lines represent the case after the onset of the rapid entrainment. [From Chen and Cotton (1983b).]

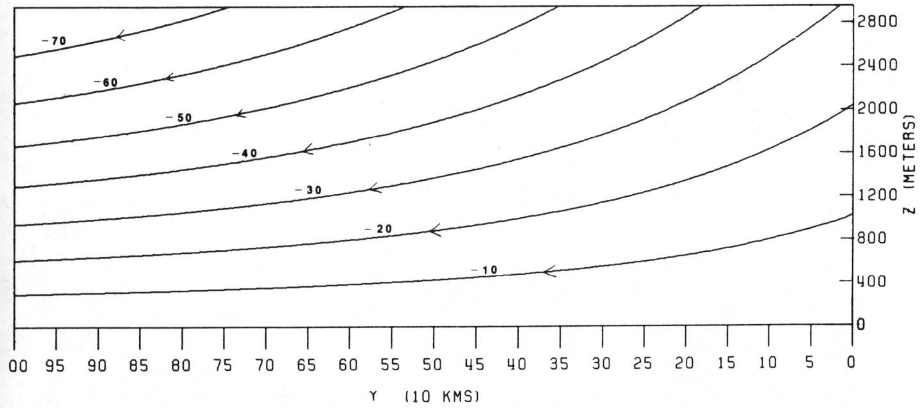

Fig. 7.26. The contour diagram of the initial streamfunction in the *y*–*z* domain. The right side of the diagram points to the northeast and the left to the southwest. Units are 100 $m^2\,s^{-1}$. [From Moeng and Arakawa (1980).]

horizontally over a 1000-km region extending from off the coast of California southwestward. The sea surface temperature increased equatorward at a constant gradient of 4 K $(1000\text{ km})^{-1}$. They imposed an initial circulation to represent the downward and equatorward branch of the Hadley cell. The circulation shown in Fig. 7.26 has a corresponding maximum subsidence of ~0.6 cm s^{-1} in the upper, northeast quadrant, decreasing downward and equatorward. The depicted liquid-water field shown in Fig. 7.27 illustrates that after 27 h, the cloud field breaks up into cumuli at about 600 km from the northeast boundary. Moeng and Arakawa conclude that entrainment instability ensues when cloudy air is advected into a region of relatively warm sea surface temperatures and relatively weak large-scale subsidence. Along the transect from northeast to southwest a general correspondence between Randall's and Deardorff's criteria for the onset of entrainment instability and the breakup of the simulated stratocumulus field was found. The jump in θ_e was positive in the northeast half of the domain and negative in the southwest half.

While Deardorff's LES model results and Chen and Cotton's and Moeng and Arakawa's higher order cloud model results suggest that the thermodynamic criteria for the onset of entrainment instability are valid, recent observations and LES model results suggest that the thermodynamic criteria may not be sufficient for predicting stratocumulus cloud breakup. Kuo (1987) compiled a number of observations of solid stratocumulus decks to demonstrate that the Randall–Deardorff thermodynamic criteria for entrainment instability are insufficient. Following arguments similar to that given

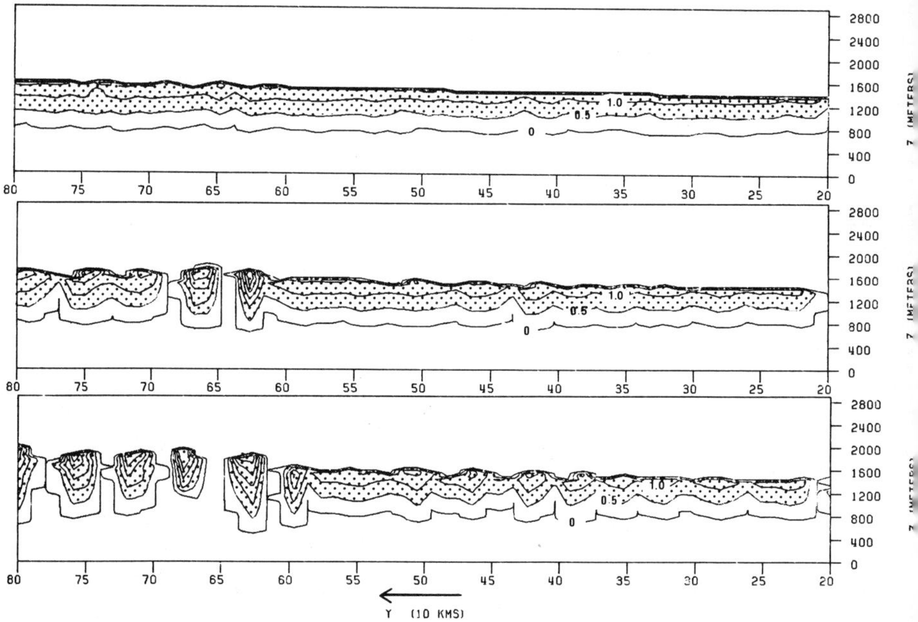

Fig. 7.27. The contour diagrams of the computed liquid-water mixing ratio in the y–z domain at hour 21 (top), hour 25 (middle), and hour 27 (bottom). The right sides of the diagrams point to the northeast and the left to the southwest. Units are g m^{-3}. [From Moeng and Arakawa (1980).]

by Albrecht *et al.* (1985) and Nicholls and Turton (1986), Kuo (1987) summarized the thermodynamic criteria for entrainment instability. Consider that X mass units of warm dry air just above the capping inversion have mixed with $1-X$ mass units of cool moist air just below the inversion. If the subscripts a and b denote parcels originating above the inversion and below the inversion, respectively, then the equivalent potential temperature θ_e and total mixing ratio r_T of the mixed parcel can be expressed

$$\theta_e = \theta_{e_b} + X\Delta\theta_e \tag{7.18}$$

and

$$r_T = r_b + X\Delta r. \tag{7.19}$$

The resultant virtual potential temperature of the mixed parcel can be expressed

$$\theta_v = \theta + \theta_0(0.608 r_v - r_l), \tag{7.20}$$

where r_v and r_l are the vapor and liquid-water mixing ratios of the mixture, respectively, and θ_0 is a reference potential temperature which he took to be 15°C. If we denote $\tilde{X}$ as the mass units of dry air required to evaporate all the liquid water of the mixed parcel, the buoyancy of the mixed parcel when $X \leq \tilde{X}$ can be expressed

$$\left(\frac{\theta_v - \theta_{vb}}{\theta_0}\right) g = X\left(\frac{C_p}{Lk}\Delta\theta_e - \Delta r\right) g, \qquad (7.21)$$

where $k \simeq 0.23$. If the term within parentheses on the RHS of Eq. (7.21) is negative, all mixtures with $0 < X \leq \tilde{X}$ will be negatively buoyant. The thermodynamic criteria for the onset of entrainment instability are

$$\Delta\theta_e < kL\Delta r / C_p. \qquad (7.22)$$

Figure 7.28 illustrates observations of solid-covered stratocumulus decks mapped relative to Lilly's ($\Delta\theta_e = 0$) and the Randall–Deardorff [Eq. (7.22)] thermodynamic criteria for the onset of entrainment instability. About two-thirds of the observations occur to the left of the Randall–Deardorff

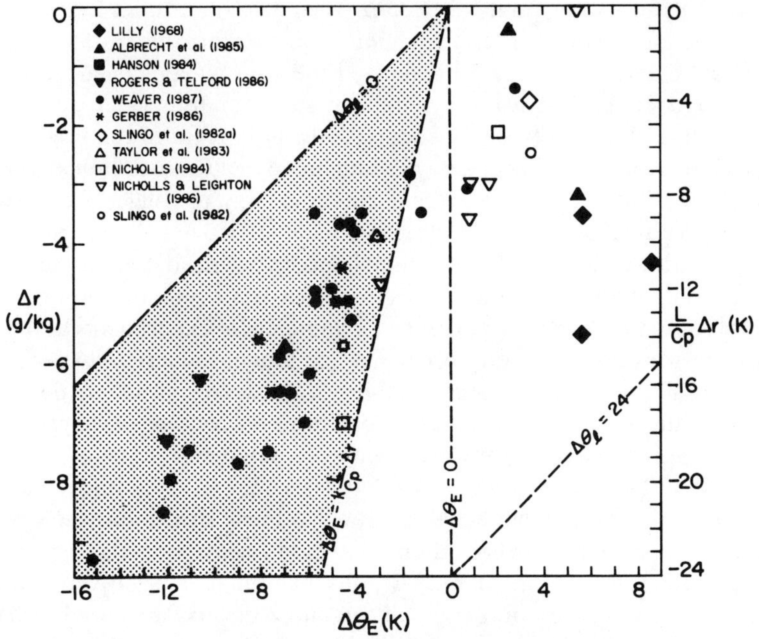

Fig. 7.28. The $\Delta\theta$, Δr plane, with the Lilly critical curve ($\Delta\theta_e = 0$) and the Randall–Deardorff critical curve ($\Delta\theta_e = k(L/C_p)\Delta r$). Observational data are indicated by the coded symbols, with solid symbols for subtropical cases and open symbols for midlatitude cases. About two-thirds of the observed solid stratocumulus clouds occur in the shaded region, where the thermodynamic theory predicts stratocumulus breakup. [Adapted from Kuo (1987).]

critical line, suggesting that the thermodynamic criteria for stratocumulus breakup are not sufficient. A similar conclusion was reached by Nicholls and Turton (1986), Hanson (1984), and Rogers and Telford (1986). As suggested by Nicholls and Turton (1986), Rogers and Telford (1986), and Kuo (1987), the amount of liquid water available in the tops of many stratocumuli may be so small that the maximum negative virtual temperature difference that can be generated will not allow deep downward penetration of dry air into the cloud interior. Kuo's (1987) two-dimensional LES-type simulations, however, suggested (see Fig. 7.23) that strong penetrative downdrafts may develop when Eq. (7.22) is satisfied, yet the cloud deck will not necessarily break up. Thus strong entrainment does not necessarily result in the breakup of the stratocumulus layer. Instead the ocean surface supplies the ascending buoyant cells with enough moisture to sustain the cloud layer against the drying effects of entrainment. As suggested by Randall (1984), under such conditions entrainment can lead to a deepening of the cloud layer rather than a breakup of the layer.

Tag and Payne (1987) performed three-dimensional LES numerical experiments of stratocumulus breakup. They also found that the existence of the thermodynamic criteria for cloud-top entrainment instability does not immediately result in the breakup of the cloud layer. As suggested by Rogers and Telford (1986), the simulated stratocumulus layer survived as a metastable cloud layer for periods as long as 100 min in the presence of an unstable cloud top. Tag and Payne concluded that stratocumulus breakup requires a critical level of turbulent kinetic energy to trigger the onset of breakup. Vertical velocities of the order of 0.1 to 0.3 m s^{-1} were necessary in the cases they simulated before cloud breakup occurred. Vertical velocities of such magnitudes, however, are common in stratocumulus clouds.

Why then did Deardorff's (1980b) LES numerical experiments and Chen and Cotton's (1983b) second-order closure model numerical experiments so readily simulate entrainment instability? The answer probably lies in the nature of the sounding Deardorff chose prior to cooling the layer above the capping inversion. Deardorff imposed a large surface buoyancy flux which Nicholls (1984) noted was six times larger in magnitude than the case he studied. This resulted in a vertical velocity distribution $\overline{w''^3}$ which was large and positive throughout the cloud and subcloud layers. By contrast, the case observed by Nicholls exhibited a negatively skewed vertical velocity distribution at cloud top and a positively skewed distribution at lower levels. It thus appears that Deardorff selected a case in which the strong surface heat flux yielded more than enough turbulent kinetic energy near the top of the stratocumulus layer, such that when cooling the layer above cloud top satisfied the Randall–Deardorff entrainment instability criteria, immediate breakup of the cloud layer ensued.

Clearly the processes involved in the breakup of stratocumulus clouds are complicated. Not only is the thermodynamic state of the cloud-top layer important, but so also is the magnitude of moisture and heat fluxes from the underlying sea surface, the turbulent kinetic energy in the bulk of the cloud layer, the liquid-water content, and, as suggested by Rogers and Telford (1986), perhaps even the drop-size spectrum. Moreover, Moeng and Arakawa (1980) and Kuo (1987) suggest that the magnitude of large-scale subsidence may also be important, with lesser subsidence favoring stratocumulus cloud breakup.

The attention that has been given in recent years to the processes involved in the breakup of stratocumulus clouds has also resulted in a greater focus on the nature of the entrainment processes at cloud top. Based on field observations, Mahrt and Paumier (1982) postulated a conceptual model of the entrainment process that differs little from the conceptual model of entrainment in isolated cumuli to be discussed in the following chapter. They observed that the coldest air near cloud top occurs on the downshear side of penetrative convective elements and on the upshear side of engulfed wisps or pockets of free-flow air. Mean vertical shear of the horizontal wind contributes to enhanced mixing in the downshear region of penetrative convective elements. Thus evaporative cooling contributes to the production of penetrative downdrafts on the downshear edge of the penetrating updrafts. If we also consider the observed and modeled flow about isolated penetrative towers described in Chapter 8, a schematic illustration of the interactions between penetrative updrafts, shear, and penetrative downdrafts is postulated as shown in Fig. 7.29. We will now examine the effects of vertical shear of the horizontal wind on the properties of stratocumulus clouds more fully.

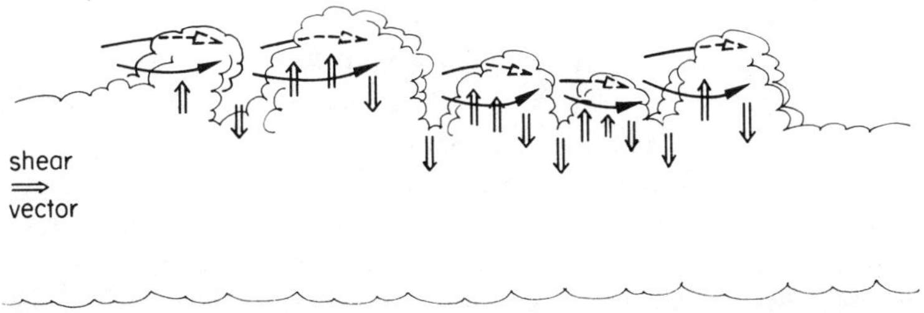

Fig. 7.29. Schematic model of a stratocumulus layer in sheared flow. Double arrows represent updrafts and downdrafts. Higher speed flow above the capping inversion is shown diverging around the emerging towers and entraining downshear of the updrafts. Evaporation in the downshear regions is shown to be coupled with the formation of downdrafts.

7.6.7 The Role of Vertical Shear of the Horizontal Wind

In our discussion of convectively unstable stratocumulus cloud layers, vertical shear of the horizontal wind is typically ignored as a contributing factor to the generation of kinetic energy in the cloud layer. This is particularly true of a number of the layer-averaged models such as Lilly (1968), Schubert (1976), and Deardorff (1976). The higher order closure models and 3-D large-eddy simulation models are inherently capable of including the effects of wind shear, though their application to the more stable cases with stronger wind shear has been quite limited. Brost *et al.* (1982a, b; hereafter referred to as BWL) presented the results of extensive analysis of several marine stratocumulus cases in which wind shear played an important role in the energetics of the cloud layer. Near the ocean surface, they characterized the cases as stable or slightly stable. In a horizontally homogeneous boundary layer, the time rate of change and advection of turbulent kinetic energy $\bar{e}$ are small; therefore Eq. (3.50) can be written as

$$0 = \underset{\text{(a)}}{-\rho_0 \overline{u''w''}\frac{\partial \bar{u}}{\partial z}} \underset{\text{(b)}}{- \rho_0 \overline{v''w''}\frac{\partial \bar{v}}{\partial z}} + \underset{\text{(c)}}{\rho_0\left(\frac{\overline{w''\alpha''_{\mathrm{m}}}}{\alpha_0} - \overline{w''r''_{\mathrm{w}}}\right)g}$$

$$\underset{\text{(d)}}{-\frac{\partial}{\partial z}(\overline{w''e})} + \underset{\text{(e)}}{\frac{\partial}{\partial z}(\overline{w''p''})} \underset{\text{(f)}}{- \rho_0 \varepsilon}. \tag{7.23}$$

Using a gust-probe-equipped aircraft, BWL measured the mechanical production terms (a and b), buoyancy production terms (c), and turbulent transport term (d). The dissipation rate ε was estimated from the average slopes of the individual spectra for u'', v'', and w''. The pressure transport term (e) was evaluated as a residual from Eq. (7.23). Figure 7.30 illustrates the profiles of the various terms in Eq. (7.23) which have been made nondimensional by multiplying them by $kz_i/\rho_0 u_*^3$. They have been calculated for several cases analyzed by BWL.

The outstanding feature in Fig. 7.30 is that shear production dominates buoyancy production of turbulent kinetic energy in almost every case. It is especially pronounced in case 17-2. Shear production can be expected to be large near the surface, but in more convectively unstable cases, buoyancy production is usually thought to be dominant throughout the cloud layer (see Deardorff, 1980b). This is only evident in case 17-1. Both shear production and buoyancy production (suppression) are quite large in the capping inversion for case 17-2. Dissipation shows the expected behavior of decreasing away from the earth's surface but occasionally shows a peak in the inversion, especially in case 17-2. Deardorff (1980b) also found a peak in ε near the inversion in his simulations of convectively unstable stratocumuli.

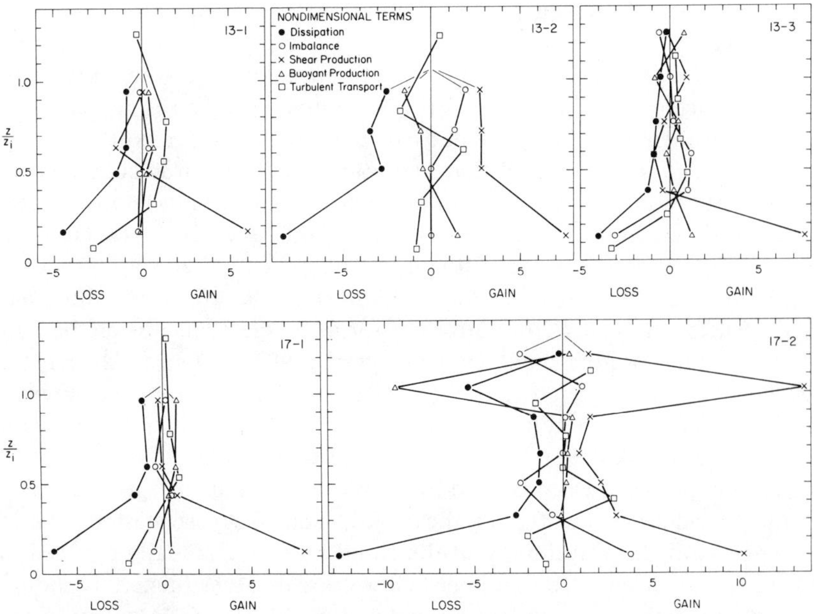

Fig. 7.30. Turbulent kinetic energy budgets. All terms are multiplied by kz_i/u_*^3. The terms as identified in the figure are dissipation ε, shear production $-\overline{u''w''}(\partial\bar{u}/\partial z) - \overline{v''w''}(\partial\bar{v}/\partial z)$, buoyant production $(\overline{w''\alpha_m}/\alpha_0 - \overline{w''r_w})g$, turbulent transport $-(1/\rho_0)(\partial/\partial z)(\overline{w''e})$, and pressure transport $-(\partial/\partial z)(\overline{w''p''}/\rho_0)$, which is calculated as a residual. [From Brost *et al.* (1982b).]

Deardorff found that pressure and turbulent transports act in concert near cloud base. They supply turbulent energy generated in the clear convective layer below. In the stable cases studied by BWL, pressure transport was quite small and not consistently related to turbulent transport.

Chen and Cotton (1987) applied the second-order closure model described by Chen and Cotton (1983a, b) to one of the cases described by Brost *et al.* (1982b). When the winds above the capping inversion were assumed to be in geostrophic balance, buoyant production of turbulence by radiative cooling in the solid cloud layer dominated shear production. It was noted, however, that a geostrophically balanced wind profile resulted in somewhat weaker shear than observed. When the wind shear was increased to observed magnitudes by imposing a supergeostrophic wind, shear production of turbulence became the dominant controlling factor and cloud-top radiative cooling was then balanced by entrainment warming. One consequence of imposing the enhanced shear was that the predicted liquid-water contents were reduced to about 0.1 g kg^{-1} which is comparable to observed values. With the weaker shear imposed by geostrophically balanced winds, however, predicted liquid-water contents ranged from 0.2 to 0.3 g kg^{-1}.

One consequence of the greater role of shear over buoyant production of turbulence is that the nature of the entrainment processes appears to be altered. In their observational study of nocturnal stratocumuli over land, Caughey *et al.* (1982) noted that acoustic sounder records suggested the absence of thermal plumes that could penetrate into the inversion layer and mix dry air well downward into the cloud and subcloud layer. However, the region near cloud top was fully turbulent and exhibited vertical velocity fluctuations of the order of 1 m s^{-1}, typical of the convective boundary layer. The length scale of the entraining turbulent elements was on the order of 10 m, which is quite small. Caughey *et al.* (1982) and also Caughey and Kitchen (1984) suggested the entrainment mechanism occurred intermittently, perhaps triggered by the formation of Kelvin–Helmholtz instability waves which break down into turbulence. Circumstantial evidence supporting their hypothesis was provided by the observation of stronger wind shear in the inversion at the top of the cloud layer. They suggested that the stronger wind shear at the top of the cloud layer occasionally reduced the local gradient Richardson number below critical values (generally thought to be 0.25) such that turbulent breakdown could occur. On occasion they observed the turbulent layer near cloud top to be much thicker, suggesting turbulent breakdown had occurred that resulted in locally weaker shear and the return of Richardson numbers at the top of the cloud layer to supercritical values. It was hypothesized that the turbulence would then decay, causing the interface layer near cloud top to thin and return to near laminar flow, at which point the shear would intensify.

When Chen and Cotton (1987) extended their closure model simulations to periods up to 6 h, they found that their model also predicted nighttime sporadic episodes of entrainment. With strong wind shear present as in Brost *et al.*'s observational study, a 15- to 20-min periodicity in cloud-top entrainment was simulated. When wind shear was removed from the initial conditions, sporadic episodes of entrainment also occurred, but with a 1-h or longer periodicity. Only when both vertical wind shear and drizzle formation were removed did the sporadic episodes of entrainment completely disappear. Thus, the simulated sporadic entrainment is controlled both by cloud-top radiative cooling, drizzle processes, and vertical wind shear with shear playing the dominant role. Chen and Cotton explained this phenomenon as follows.

The boundary layer large eddies are driven in part by the destabilization of the cloud layer by cloud-top radiative cooling. The undulation of the cloud top occurs when the boundary layer eddies are fully developed. The penetration of the convective elements into the capping inversion locally enhances the shear and thus aids in the breakdown of Kelvin–Helmholtz-like billows. The entrained air subsequently reduces the liquid-water content and lowers the rate of radiative cooling, which then stabilizes the cloud layer. As a result, the large-eddy circulation is quenched. The cycle begins to repeat itself due to the recovery of radiative cooling and increase of liquid-water content. Drizzle enhances the reduction in liquid-water content near cloud top by drop settling, and hence reduces the rate of radiative cooling, thus speeding up the process. Chen and Cotton called this process the Sporadic Cloud Radiative Instability Mechanism (SCRIM), in which radiative cooling, cloud water production, drizzle formation, and wind shear cooperatively interact to generate sporadic episodes of entrainment.

In their observation of a weaker, unstable, or stable cloud layer, Brost *et al.* (1982b) also found that the scale of the turbulent eddies contributing to cloud-top entrainment was only a few tens of meters.

These observations suggest a different link between cloud-top radiative cooling and entrainment in the stable or weakly convective stratocumulus cloud. It is generally believed that the local instability created by cloud-top radiative cooling is transmitted through the depth of the cloud layer and subcloud layer by strong vertical mixing caused by penetrative plumes. This is fundamental to the layer-averaged models such as Lilly's (1968), Schubert's (1976), and Stage and Businger's (1981a, b). It is also a predicted response in the higher order closure models of OLW and Chen and Cotton (1983a, b). Thus, radiative cooling causes enhanced mixed-layer turbulence in the form of penetrative plumes, which, in turn, causes greater rates of cloud-top entrainment.

In the case of more stable stratocumulus, it appears that the local instability caused by cloud-top radiative cooling generates small-scale turbulence. The small-scale turbulence interacts with local wind shear at the top of the radiatively cooled layer. This causes sporadic turbulent breakdown or shear-driven entrainment. In some cases, there does not appear to be any direct communication between radiation cooling at cloud top and the energetics of the entire depth of the cloud layer. As noted by Nicholls and Leighton (1986), almost all the cloud-top radiative cooling is balanced locally by entrainment. As a result there is no net generation of positive buoyancy and associated convective transport. Such clouds may be more properly called stratus rather than stratocumulus and exhibit properties more similar to altostratus and cirrus than to boundary layer stratus.

7.6.8 The Role of Drizzle

In several of the cases observed by Brost *et al.* (1982b), drizzle drops were present. The vertical water flux by drizzle represented in some cases a significant fraction of the water flux in the cloud layer. They suggested that drizzle can affect significantly the stability of a cloud layer by altering the vertical distribution of latent heating. That is, as drizzle settles from the top of the cloud layer, it removes water from that level which cannot be evaporated. As a consequence, drizzle contributes to a net latent heating in the upper part of the cloud layer. However, as the drizzle settles into the subcloud layer, it evaporates and causes evaporative cooling of the subcloud layer. In this process, the layer between the heating aloft and cooling below would be stabilized, whereas the shallow layers above the heating zones or below the evaporatively cooled zone would be destabilized. They hypothesized that this would tend to form two shallow unstable layers decoupled by an intermediate stable layer.

Anticipating a feedback loop among settling of drizzle, radiation, and turbulence as suggested by Brown and Roach (1976) for fogs, Chen and Cotton (1987) introduced a formulation for drizzle into the higher order closure model described by Chen and Cotton (1983b). The drizzle model included the autoconversion formulation given in Eq. (4.10) and the raindrop collection formula given in Eq. (4.28). Numerical experiments applying these formulas to stratocumulus must be considered exploratory sensitivity experiments only, because the theoretical foundations of these equations are not very appropriate for stratocumuli. To adjust to the different cloud environment, Chen and Cotton reduced the mean radius, R_m, in Eq. (4.28) from 0.27 mm, appropriate for deep convection, to 0.1 mm. Numerical experiments with these equations demonstrated a major change in their relative roles from that found in deep convective clouds. For example, in deep convective clouds, autoconversion acts mainly as a trigger to initiate rain. Once rain forms, however, raindrop accretion of cloud water [Eq. (4.28)] rapidly becomes the dominant contributor to the formation of precipitation. In the case of stratocumuli having weak updrafts, the rain formed by autoconversion rapidly settles out of the cloud. As a consequence, autoconversion continues to be the dominant contributor to drizzle or rain formation. This is to be expected, but unfortunately it places a great burden on what must be considered a very simple parameterization of the drizzle initiation process. Another effect of applying these formulas to stratocumuli can be seen in the collection equation, which is proportional to $\bar{r}_c\bar{r}_r + \overline{r_c'' r_r''}$. In deep convective clouds one can expect the product $\bar{r}_c\bar{r}_r$ to be the main contributor to rain formation by accretion. However, in stratocumuli, the correlation $\overline{r_c'' r_r''}$ is of the same order of magnitude as $\bar{r}_c\bar{r}_r$. Thus, the predominant turbulent nature of stratocumuli has a profound influence upon the microphysical processes in such clouds.

Caughey and Kitchen (1984) also calculated that with the weak vertical motions associated with the more stable stratocumulus clouds, the effects of radiative cooling on individual droplets (see discussion in Chapter 5) produced larger droplets near cloud top. The mixing of these drops into the cloud interior results in a broader droplet spectrum. The formation of such a broad droplet spectrum favors the production of drizzle. This again brings to our attention the different physical processes of importance in stratocumulus clouds with weak updrafts versus cumulonimbus clouds with strong updrafts.

Returning now to the coupling among drizzle, radiation, and turbulence, Chen and Cotton (1987) experimented with the sensitivity of the simulated stratocumulus properties to the drizzle process. Removal of water by drizzle lowered the maximum liquid-water content near cloud top by about 40%. Since the cloud-top longwave radiative cooling rate is determined by the liquid-water path, reduction in cloud-top liquid water reduced the radiative cooling rate, which further reduced the liquid-water production. Brown and Roach (1976) arrived at a similar result in their simulation of radiative fogs. In the case of stratocumulus clouds, Chen and Cotton predicted that the drizzle process reduced the cloud-top radiative cooling rate by 40°C day^{-1}. As a consequence of the reduced cloud-top radiative cooling, the cloud layer is stabilized somewhat and the rate of entrainment at cloud top is reduced when the drizzle process is present. In another example of a stratocumulus cloud containing significant drizzle, Nicholls (1984) calculated that the transport of water by gravitational settling of drops was comparable in magnitude to turbulent transport of water substance at all levels in and below the cloud layer. Nicholls suggested that evaporation of drizzle in the subcloud layer may have contributed to a separation in fluxes between the subcloud layer and cloud layer. Nicholls also suggested that other factors such as decreased surface buoyancy fluxes, decreased radiative flux divergence in the cloud layer, and increased entrainment of potentially warmer and drier air could also have contributed to a separation of fluxes between the cloud and subcloud layer. The importance of these other factors, especially cloud top entrainment, was illustrated in several other cases reported by Nicholls and Leighton (1986), where separation of subcloud and cloud layer fluxes was observed.

An interesting question addresses how a stratocumulus layer can be maintained against the drying effects of entrainment and drizzle removal of water when the subcloud becomes decoupled from the cloud layer, which thus inhibits the supply of water to the cloud layer. In some cases the decoupling of subcloud and cloud layer fluxes probably signals the demise of the cloud layer. Nicholls (1984) noted that in some cases in which there is a distinct separation of subcloud and cloud layer fluxes, cumulus clouds are observed to rise out of the subcloud layer into the stratocumulus deck. Nicholls suggested that separation of the cloud and subcloud layers is likely to lead to the buildup of conditional instability in the subcloud layer. In such a case cumulus clouds can resupply the cloud layer with moisture being lost by entrainment and settling of drizzle. Bougeault's (1985) simulation of such decoupled stratocumulus cloud layer with a higher order closure model also illustrated the role of cumuli in supplying the decoupled cloud layer with moisture.

Clearly, settling of drizzle and evaporation of drops can be important to the water budget and energetics of stratocumulus clouds. This is especially true of stratocumulus clouds occurring in the presence of weak surface fluxes. Its role in the more strongly forced cases has not been so clearly identified. It is also important to recognize that drizzle may be more important to the dynamics of nocturnal stratocumuli. This is because it is frequently observed (see Kraus, 1963, for example) that there is a well-defined nighttime maximum in maritime precipitation. Kraus found the largest diurnal amplitude to occur in midlatitude regions where the rain is mostly nonconvective. He speculated that absorption of solar radiation during the daytime resulted in less liquid-water production than at night. Thus, drizzle effects could be more important in the somewhat wetter and deeper, nocturnal stratocumulus clouds.

7.6.9 Role of Large-Scale Subsidence

It is generally recognized that large-scale subsidence plays an important role in establishing the environmental conditions favorable for formation of marine stratocumulus. That is, large-scale subsidence establishes the pronounced capping inversion, which serves as an upper lid to the atmospheric boundary layer and confines the moisture and heat fluxes from the ocean surface in a shallow layer. The overlying air mass is also dried out by the sinking motion. While subsidence may establish environmental conditions favorable for maintaining a solid stratus deck, too much subsidence may be responsible for the breakup of stratocumulus clouds. Roach *et al.* (1982) examined whether subsidence could account for the observed dispersal of a 300-m-thick cloud in 2–4 h at night. From the analysis of the vertical thermodynamic structure upwind of the cloud, they determined that a downward displacement of the air mass of 220 m in 2–4 h would result in the evaporation of the entire cloud. This corresponds to a subsidence rate of 2–4 cm s^{-1}. Subsidence rates of this magnitude cannot be directly measured but can be diagnosed by integrating the divergence of the horizontal wind field. They estimated that such a magnitude of subsidence was plausible, but the wind observations were too sparse to obtain a definitive mesoscale analysis of the local subsidence field. Thus, future observational studies of stratocumulus clouds should include the measurement of winds with sufficient accuracy and spatial resolution to calculate subsidence rates.

Recently, Chen and Cotton (1987) performed some sensitivity experiments with their higher order closure model to determine if large-scale subsidence played a more active role in the determination of cloud structure. In their first experiment they imposed a divergence of $5.0 \times 10^{-6}\ s^{-1}$, which produces a 0.25-cm s^{-1} subsidence rate at a height of 550 m. This corresponds to estimates for typical large-scale subsidence off the coast of California in the United States. If the intensity of the capping inversion is 10 K over a depth of 25 m, this subsidence rate will warm the capping inversion by 190 K per day. Thus a significant fraction of the longwave radiative cooling at cloud top is balanced by subsidence warming. As a consequence, the upward heat flux in the cloud layer and buoyancy production of turbulence are reduced. Relative to an environment with no vertical motion, this imposed sinking reduces the liquid-water content by 11.4%. If the subsidence is further increased by a factor of 2 (i.e., to 0.5 cm s^{-1} at 550 m), the liquid-water content is lowered to 45% of the value obtained in a zero-subsidence calculation. We shall see below that large-scale subsidence has an important influence on the response of the cloud layer to solar heating.

7.6.10 Diurnal Variations in Marine Stratocumuli

Using a second-order turbulent transport model, OLW simulated a diurnally varying stratocumulus cloud structure. As shown in Fig. 7.31, the stratus top rises and the cloud base lowers during the night and thickens until sunrise, when the stratus begins to dissipate due to solar insolation. As noted previously, however, the location of maximum solar heating well into the interior of the cloud layer shown in Fig. 7.32 may only be possible in clouds that are optically thin (i.e., contain little liquid-water content). The corresponding distribution of turbulent kinetic energy predicted by OLW is shown in Fig. 7.33. This shows a maximum in TKE in the interior of the cloud during the nighttime with a more vertically uniform profile of TKE from the surface to near cloud top during the daytime.

Bougeault's (1985) simulation of the diurnal cycle of a stratocumulus layer with a higher order closure model did not exhibit any significant variation in cloud-top height. The cloud base varied in height over several hundred meters, rising during daytime and descending during nighttime. During the afternoon a secondary stable zone was simulated just below cloud base. This reduces the supply of water vapor from the subcloud layer, thus reducing the liquid-water content. Bougeault found this feature to be transient; with longer daytime conditions a lower, solid, cloud deck subsequently formed. Transient cumulus clouds also enhanced the supply of water, yielding slightly higher values of liquid-water content.

Chen and Cotton (1987) also examined diurnal variations in the structure of marine stratocumuli. Similar to OLW, they found that absorption of shortwave radiation occurred deeper in the cloud layer than longwave radiation cooling. Shortwave heating reached a maximum about 100 m in the cloud interior, while longwave cooling was maximized only 25–30 m in the cloud interior. However, the difference in radiative depths is much less than predicted by OLW. As a consequence, Chen and Cotton found that in the absence of large-scale subsidence and a moist layer above the capping inversion, a net cloud-top radiative cooling remained after sunrise. Like Bougeault, they found the cloud top did not descend. Solar heating did, however, contribute to a rise in cloud base and a reduction in the maximum liquid-water content in the cloud layer. Only when large-scale subsidence was imposed did both the cloud top descend and the cloud base rise after sunrise. Rather strong large-scale subsidence (on the order of $10^{-5}\,s^{-1}$ divergence) was required to produce sufficient heating in conjunction with solar heating to offset longwave radiative cooling. Even so, it took over 4 h of solar heating to initiate a lowering of cloud top. The lowering process

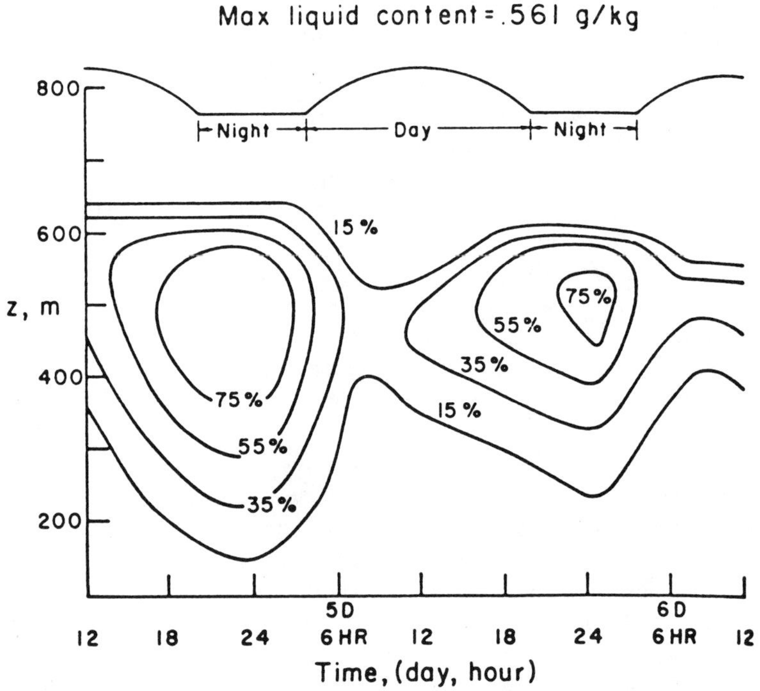

Fig. 7.31. Diurnal variation of liquid water in stratus cloud formed in a subsidence-capped boundary layer. [From Oliver *et al.* (1978).]

commenced by a reduction in cloud water content and cloud fractional coverage in the upper 100 m of the cloud layer. As a result of the reduced liquid-water path, the maximum of radiative cooling shifted to lower levels causing an abrupt collapse of the cloud top into a thinner cloud layer.

Twomey (1983) computed shortwave and longwave radiation heating profiles based on aircraft measurements of cloud microphysical parameters, temperature, and moisture through the depth of the marine boundary layer. He found the radiative warming prevailed near cloud top, with shortwave and longwave radiative influences being about equal and opposite. He

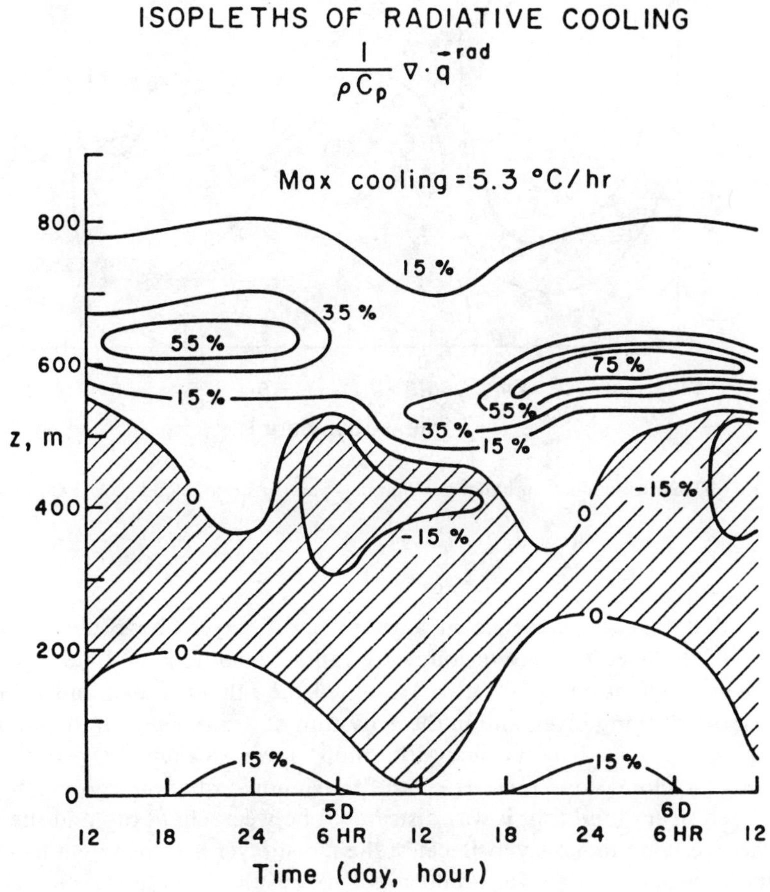

Fig. 7.32. Distribution of radiative cooling/heating in stratus cloud. Note that net heating occurs deep within the cloud interior. [From Oliver *et al.* (1978).]

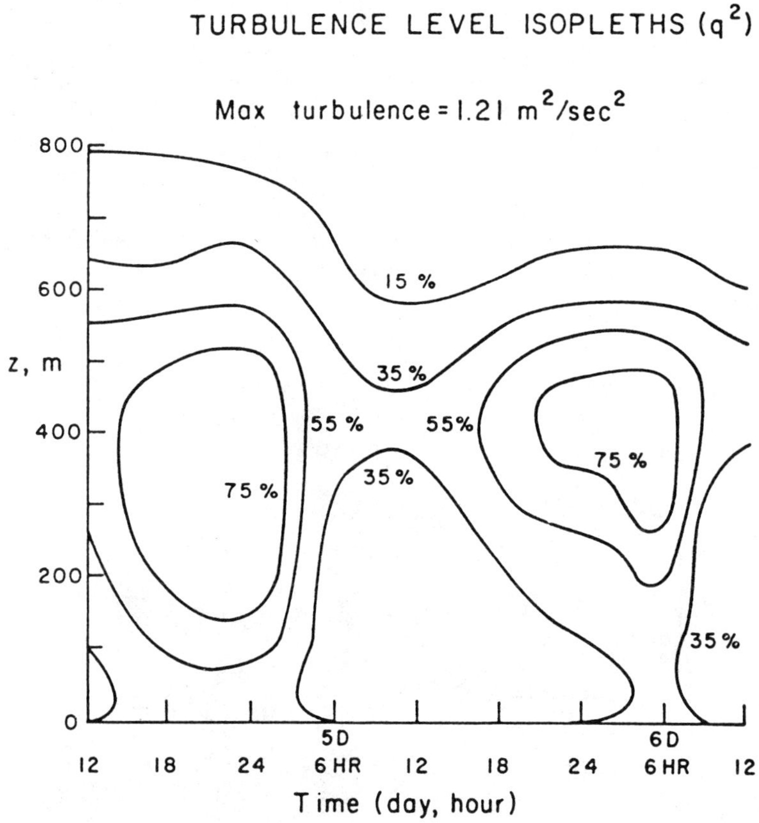

Fig. 7.33. Distribution of turbulence in diurnally varying stratus cloud. [From Oliver *et al.* (1978).]

concluded that the presence of a warm, moist layer above the capping inversion reduced the rate of longwave radiative cooling at cloud top.

Chen and Cotton (1987) also examined the influence of a moist layer above the capping inversion on the evolution of a simulated stratocumulus layer. As suggested by Twomey, the moist layer weakened the radiative cooling at cloud top. Instead of the maximum radiation cooling being localized near cloud top, it was distributed between cloud top and the top of the overlying moist layer. Because the moist layer is transparent to solar radiation, solar heating was unaffected. As a result, the cloud layer became far more responsive to solar heating, with cloud top descending and cloud base rising rapidly following sunrise. This is illustrated in Fig. 7.34.

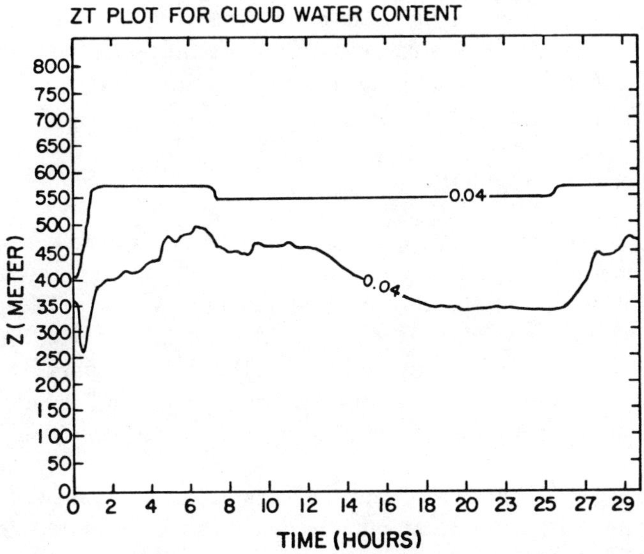

Fig. 7.34. The time–height contour of cloud water mixing ratio for the experiment with a moist layer above capping inversion. The magnitude of the maximum cloud water content is 0.138 g kg^{-1}. The units given are in g kg^{-1}. [From Chen and Cotton (1987).]

7.6.11 Influence of Mid- and High-Level Clouds

In our discussion of radiation fogs we noted that the advection of mid- and high-level clouds over a ground-based fog layer caused the dissipation of the fog by the reduction of longwave radiation cooling at fog top. One might expect a similar phenomenon to occur with marine stratocumulus clouds. Chen and Cotton (1987) simulated the response of a marine stratocumulus layer to upper level cloud layers with a second-order closure model. They imposed a cloud layer at a height of 5 km with a thickness of 50 m and a liquid-water content of 0.2 g kg^{-1} over the marine stratocumulus layer previously described. The upper-level cloud layer reduced the long-wave radiative cooling rate by 170°C day^{-1} and, as a consequence, lowered the maximum liquid-water content by 0.1 g kg^{-1}. In another experiment, Chen and Cotton introduced a cloud at a height of 3 km which was nearly 10°C warmer than the first overlying cloud layer. As a result of the warmer radiating temperature of the overlying cloud, longwave radiative cooling was reduced by about 232°C day^{-1}. The weaker longwave radiative cooling reduced the buoyancy production of turbulence and further reduced the cloud liquid-water content. While the response of a marine stratocumulus cloud to overlying cloud layers is more subtle than in a ground-based, radiative fog, the consequences are nonetheless quite important to the overall internal structure and dynamics of the stratocumulus layer.

7.7 Arctic Stratus Clouds

Low-level stratiform clouds are a common feature in the central Arctic region in the summertime. Monthly average cloud cover amounts are nearly 70% for the months of May through September (Tsay and Jayaweera, 1984). They form in the boundary layer and in the free atmosphere, typically at heights below 2000 m. They are normally rather tenuous clouds with thicknesses of a few hundred meters and are frequently found to be composed of two or more well-defined layers (Jayaweera and Ohtake, 1973; Herman, 1977).

According to Herman and Goody (1976) and Tsay and Jayaweera (1984), Arctic stratus clouds can form either as a convectively unstable cloud layer when cold polar air flows over a warmer sea surface or as a stable cloud layer when warm, moist air flows over a cold sea surface. Tsay and Jayaweera note that clouds that form in a stable air mass are relatively thin and have bases low enough to frequently reach the sea surface. They suggest that the formation of upper cloud layers is a result of horizontal advection of moist air undergoing weak ascent.

The most unique aspect of summertime Arctic stratus clouds is that they occur with a large solar zenith angle ($\sim 74°$), and at 80°N the sun remains above the horizon for 24 h day^{-1} from May through August. Diurnal variations in cloud properties can therefore be expected to be slight. Herman and Goody (1976) concluded from their model calculations that the layered structure of Arctic stratus clouds was due to the persistent solar heating in the cloud interior.

Recently Curry (1986) calculated shortwave and longwave radiative heating rates based on cloud liquid-water contents and droplet-size distributions observed in Arctic stratus. As seen in Fig. 7.35, owing to the large solar zenith angle, longwave radiative cooling predominates over shortwave radiative heating, and the cloud layers exhibit a net radiative cooling. This is particularly evident in the layered cloud (Deck 4) shown in Fig. 7.35. Of course, it is not essential that solar heating result in a net radiative warming for it to contribute to the dissipation of a layer. If radiative cooling maintains a cloud against the warming and drying effects of subsidence, then a reduced rate of net radiative cooling could lead to dissipation of a cloud layer. Curry *et al.* (1988) pointed out that the air above cloud top is often quite moist, thus the drying effects of subsidence are reduced.

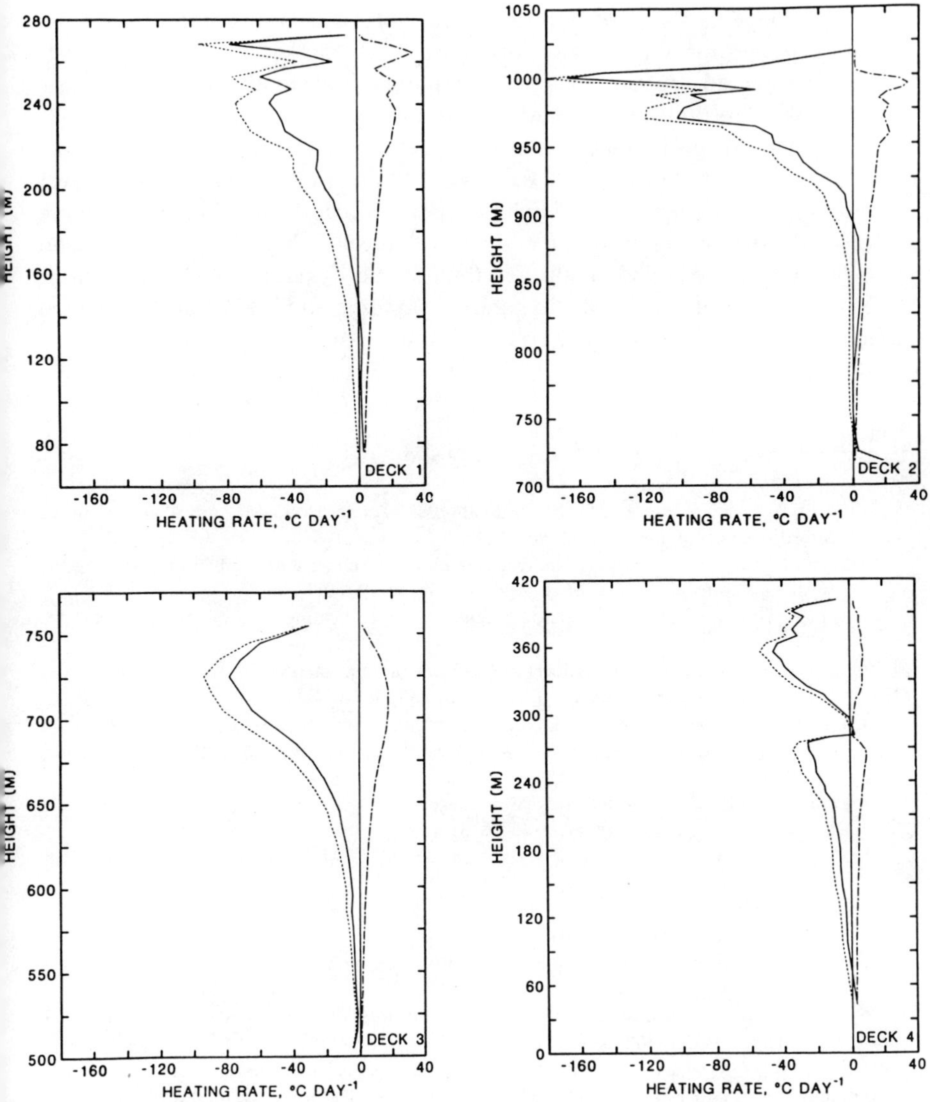

Fig. 7.35. Vertical profiles of heating rates: thermal radiation (dotted lines), solar radiation (dash–dot lines), and net radiation (solid lines). [From Curry (1986).]

As noted by Nicholls (1984), other factors such as drizzle and entrainment of warm, dry air into the tops of the cloud layer can contribute to a separation of cloud layers. Also Tsay and Jayaweera's (1984) analysis of the large-scale processes responsible for the formation of Arctic stratus clouds suggests that the layered structure may be a result of several simultaneous large-scale processes contributing to cloud formation.

In many respects the dynamics of layered Arctic stratus clouds appears to resemble the dynamics of weakly unstable stratocumuli observed over the United Kingdom (e.g., Nicholls, 1984; Nicholls and Leighton, 1986). In particular, an analysis of eddy fluxes reveals that many of the cloud layers are not coupled to surface fluxes. The dynamics of Arctic stratus, however, resemble more the dynamics of middle and high clouds discussed in Chapter 11 than ordinary stratocumulus.

References

Albrecht, B. A., R. S. Penc, and W. H. Schubert (1985). An observational study of cloud-topped mixed layers. *J. Atmos. Sci.* **42**, 800–822.

Anderson, J. B. (1931). Observations from air planes of cloud and fog conditions along the southern California Coast. *Mon. Weather Rev.* **59**, 264–270.

Ball, F. K. (1960). Control of inversion height by surface heating. *Q. J. R. Meteorol. Soc.* **86**, 483–494.

Barkstrom, B. R. (1978). Some effects of 8–12 μm radiant energy transfer on the mass and heat budgets of cloud droplets. *J. Atmos. Sci.* **35**, 665–673.

Benoit, R. (1976). A comprehensive parameterization of the atmospheric boundary layer for general circulation models. Ph.D. Thesis, McGill Univ. and Natl. Cent. Atmos. Res. Coop. Thesis No. 39.

Bougeault, P. (1985). The diurnal cycle of the marine stratocumulus layer: A higher-order model study. *J. Atmos. Sci.* **42**, 2826–2843.

Brost, R. A., D. H. Lenschow, and J. C. Wyngaard (1982a). Marine stratocumulus layers. Part I: Mean conditions. *J. Atmos. Sci.* **39**, 800–817.

Brost, R. A., J. C. Wyngaard, and D. H. Lenschow (1982b). Marine stratocumulus layers. Part II: Turbulence budgets. *J. Atmos. Sci.* **39**, 818–836.

Brown, R. (1980). A numerical study of radiation fog with an explicit formulation of the microphysics. *Q. J. R. Meteorol. Soc.* **106**, 781–802.

Brown, R., and W. T. Roach (1976). The physics of radiation fog: II—A numerical study. *Q. J. R. Meteorol. Soc.* **102**, 335–354.

Byers, H. R. (1959). "General Meteorology." McGraw-Hill, New York.

Caughey, S. J., and M. Kitchen (1984). Simultaneous measurements of the turbulent and microphysical structure of nocturnal stratocumulus cloud. *Q. J. R. Meteorol. Soc.* **110**, 13–34.

Caughey, S. J., B. A. Crease, and W. T. Roach (1982). A field study of nocturnal stratocumulus. II: Turbulence structure and entrainment. *Q. J. R. Meteorol. Soc.* **108**, 125–144.

Chai, S. K., and J. W. Telford (1983). Convection model for stratus cloud over a warm water surface. *Boundary-Layer Meteorol.* **26**, 25–49.

Chen, C., and W. R. Cotton (1983a). A one-dimensional simulation of the stratocumulus-capped mixed layer. *Boundary-Layer Meteorol.* **25**, 289–321.

Chen, C., and W. R. Cotton (1983b). Numerical experiments with a one-dimensional higher order turbulence model: Simulation of the Wangara day 33 case. *Boundary-Layer Meteorol.* **25**, 375–404.

Chen, C., and W. R. Cotton (1987). The physics of the marine stratocumulus-capped mixed layer. *J. Atmos. Sci.* **44**, 2951–2977.

Choularton, T. W., G. Fullarton, J. Latham, C. S. Mill, M. H. Smith, and I. M. Stromberg (1981). A field study of radiation fog in Meppen, West Germany. *Q. J. R. Meteorol. Soc.* **107**, 381–394.

Curry, J. A. (1986). Interactions among turbulence, radiation and microphysics in Arctic stratus clouds. *J. Atmos. Sci.* **43**, 90–106.

Curry, J. A., E. E. Ebert, and G. F. Herman (1988). Mean and turbulence structure of the summertime Arctic cloudy boundary layer. *Q. J. R. Meteorol. Soc.* **114**, 715–746

Deardorff, J. W. (1976). On the entrainment rate of a stratocumulus-topped mixed layer. *Q. J. R. Meteorol. Soc.* **102**, 563–582.

Deardorff, J. W. (1980a). Cloud top entrainment instability. *J. Atmos. Sci.* **37**, 131–147.

Deardorff, J. W. (1980b). Stratocumulus-capped mixed layers derived from a three-dimensional model. *Bound Layer Meteorol.* **18**, 495–527.

Defant, F. (1951). Local winds. *In* "Compendium of Meteorology" (T. F. Malone, ed.), pp. 655–672. Am. Meteorol. Soc., Boston, Massachusetts.

Fleagle, R. G., W. H. Parrott, and M. L. Barad (1952). Theory and effects of vertical temperature distribution in turbid air. *J. Meteorol.* **9**, 53–60.

Forkel, R., U. Sievers, and W. Zdunkowski (1987). Fog modelling with a new treatment of the chemical equilibrium condition. *Contrib. Atmos. Phys.* **60**, 340–360.

Geiger, R. (1965). "The Climate Near the Ground." Harvard Univ. Press, Cambridge, Massachusetts.

Hanson, H. P. (1984). On mixed-layer modeling of the stratocumulus-topped marine boundary layer. *J. Atmos. Sci.* **41**, 1226–1234.

Herman, G. F. (1977). Solar radiation in summertime Arctic stratus clouds. *Q. J. R. Meteorol. Soc.* **34**, 1423–1432.

Herman, G. F., and R. Goody (1976). Formation and persistence of summertime Arctic stratus clouds. *J. Atmos. Sci.* **33**, 1537–1553.

Jayaweera, K. O., and T. Ohtake (1973). Concentrations of ice crystals in Arctic stratus clouds. *J. Rech. Atmos.* **7**, 199–207.

Jiusto, J. E. (1980). Fog structure. *Invited Rev. Pap., Symp. Workshop Clouds: Their Form., Opt. Prop. Eff., IFAORS, Williamsburg, Va.*

Jiusto, J. E., and G. G. Lala (1980). Thermodynamics of radiation fog formation and dissipation—A case study. *Prep., Int. Cloud Phys. Conf., 8th, Clermont-Ferrand, Fr.* pp. 333–335.

Kahn, P. H., and J. A. Businger (1979). The effect of radiative flux divergence on entrainment of a saturated convective boundary layer. *Q. J. R. Meteorol. Soc.* **105**, 303–304.

Korb, G., and W. Zdunkowski (1970). Distribution of radiative energy in ground fog. *Tellus* **22**, 298–320.

Kraus, E. B. (1963). The diurnal precipitation change over the sea. *J. Atmos. Sci.* **20**, 551–556.

Kuo, H.-C. (1987). Dynamical modeling of marine boundary layer convection. Thesis, Dep. Atmos. Sci., Colorado State Univ.

Lala, G. G., E. Mandel, and J. E. Jiusto (1975). A numerical evaluation of radiation fog variables. *J. Atmos. Sci.* **32**, 720–728.

Lala, G. G., J. E. Jiusto, M. B. Meyer, and M. Kornfein (1982). Mechanisms of radiation fog formation on four consecutive nights. *Conf. Cloud Phys., Chicago, Ill.*

Leipper, D. F. (1948). Fog development at San Diego, California. *J. Mar. Res.* **7**, 337–346.

Lilly, D. K. (1968). Models of cloud-topped mixed layers under a strong inversion. *Q. J. R. Meteorol. Soc.* **94**, 292–309.

Mack, E. J., U. Katz, C. Rogers, and R. Pilie (1974). The microstructure of California coastal stratus and fog at sea. Rep. CJ-5405-M-1, Calspan Corp., Buffalo, New York.

Mahrt, L., and J. Paumier (1982). Cloud-top entrainment instability observed in AMTEX. *J. Atmos. Sci.* **39**, 622–634.

Mason, J. (1982). The physics of radiation fog. *J. Meteorol. Soc. Jpn.* **60**, 486–498.

Moeng, C.-H. (1986). Large-eddy simulation of a stratus-topped boundary layer. Part I: Structure and budgets. *J. Atmos. Sci.* **43**, 2886–2900.

Moeng, C.-H., and A. Arakawa (1980). A numerical study of a marine subtropical stratus cloud layer and its stability. *J. Atmos. Sci.* **37**, 2661–2676.

Musson-Genon, L. (1987). Numerical simulation of a fog event with a one-dimensional boundary layer model. *Mon. Weather Rev.* **115**, 592–607.

Nicholls, S. (1984). The dynamics of stratocumulus: Aircraft observations and comparisons with a mixed-layer model. *Q. J. R. Meteorol. Soc.* **110**, 783–820.

Nicholls, S., and J. Leighton (1986). An observational study of the structure of stratiform cloud sheets: Part I. Structure. *Q. J. R. Meteorol. Soc.* **112**, 431–460.

Nicholls, S., and J. D. Turton (1986). An observational study of the structure of stratiform cloud sheets. Part II. Entrainment. *Q. J. R. Meteorol. Soc.* **112**, 461–480.

Obukhov, A. M. (1946). Turbulence in an atmosphere with non-uniform temperature. *Tr. Inst. Teor. Geofiz., Akad. Nauk SSSR*, **1**, 95–115; Engl. transl., *Boundary-Layer Meteorol.* **2**, 7–29 (1971).

Oliver, D. A., W. S. Lewellen, and G. G. Williamson (1978). The interaction between turbulent and radiative transport in the development of fog and low-level stratus. *J. Atmos. Sci.* **35**, 301–316.

Pilie, R. J. (1975). The life cycle of valley fog. Part II: Fog microphysics. *J. Appl. Meteorol.* **14**, 364–374.

Pilie, R. J., E. J. Mack, W. C. Kocmond, W. J. Eadie, and C. W. Rogers (1975). The life cycle of valley fog, I, Micrometeorological characteristics. *J. Appl. Meteorol.* **14**, 357–363.

Pilie, R. J., E. J. Mack, C. W. Rogers, U. Katz, and W. C. Kocmond (1979). The formation of marine fog and the development of fog-stratus systems along the California coast. *J. Appl. Meteorol.* **18**, 1275–1286.

Randall, D. A. (1980). Entrainment into a stratocumulus layer with distributed radiative cooling. *J. Atmos. Sci.* **37**, 148–159.

Randall, D. A. (1984). Stratocumulus cloud deepening through entrainment. *Tellus* **36A**, 446–457.

Roach, W. T. (1976). On the effect of radiative exchange on the growth by condensation of a cloud or fog droplet. *Q. J. R. Meteorol. Soc.* **102**, 361–372.

Roach, W. T., R. Brown, S. J. Caughey, J. A. Garland, and C. J. Radings (1976). The physics of radiation fog: I—A field study. *Q. J. R. Meteorol. Soc.* **102**, 313–334.

Roach, W. T., R. Brown, S. J. Caughey, B. A. Crease, and A. Slingo (1982). A field study of nocturnal stratocumulus: I. Mean structure and budgets. *Q. J. R. Meteorol. Soc.* **108**, 103–123.

Rogers, D. P., and J. W. Telford (1986). Metastable stratus tops. *Q. J. R. Meteorol. Soc.* **112**, 481–500.

Saunders, P. M. (1957). The thermodynamics of saturated air: A contribution to the classical theory. *Q. J. R. Meteorol. Soc.* **83**, 342–350.

Saunders, P. M. (1961). An observational study of cumulus. *J. Atmos. Sci.* **18**, 451–467.

Schubert, W. H. (1976). Experiments with Lilly's cloud-topped model. *J. Atmos. Sci.* **33**, 436–446.

Schubert, W. H., J. S. Wakefield, E. J. Steiner, and S. K. Cox (1979a). Marine stratocumulus convection. Part I: Governing equations and horizontally homogeneous solutions. *J. Atmos. Sci.* **36**, 1286–1307.

Schubert, W. H., J. S. Wakefield, E. J. Steiner, and S. K. Cox (1979b). Marine stratocumulus convection. Part II: Horizontally homogeneous solutions. *J. Atmos. Sci.* **36**, 1308–1324.

Stage, S. A., and J. A. Businger (1981a). A model for entrainment into a cloud-topped marine boundary layer. Part I: Model description and application to a cold-air outbreak episode. *J. Atmos. Sci.* **38**, 2213–2229.

Stage, S. A., and J. A. Businger (1981b). A model for entrainment into a cloud-topped marine boundary layer. Part II: Discussion of model behavior and comparison with other models. *J. Atmos. Sci.* **38**, 2230–2242.

Tag, P. M., and S. W. Payne (1987). An examination of the breakup of marine stratus: A three-dimensional numerical investigation. *J. Atmos. Sci.* **44**, 208–223.

Taylor, G. I. (1917). The formation of fog and mist. *Q. J. R. Meteorol. Soc.* **43**, 241–268.

Tsay, S.-C., and K. Jayaweera (1984). Physical characteristics of arctic stratus clouds. *J. Climate Appl. Meteorol.* **23**, 584–596.

Twomey, S. (1983). Radiative effects in California stratus. *Contrib. Atmos. Phys.* **56**, 429–439.

Walter, B. A., jr., and J. E. Overland (1984). Observations of longitudinal rolls in a near neutral atmosphere. *Mon. Weather Rev.* **112**, 200–208.

Welch, R. M., M. G. Ravichandran, and S. K. Cox (1986). Prediction of quasi-periodic oscillations in radiation fogs. Part I: Comparison of simple similarity approaches. *J. Atmos. Sci.* **43**, 633–651.

Wells, W. C. (1838). An essay on dew and several appearances connected with it. *In* "On the Influence of Physical Agents on Life," by W. F. Edwards. Haswell, Barrington & Haswell, Philadelphia, Pennsylvania.

Willit, H. C. (1928). Fog and haze. *Mon. Weather Rev.* **56**, 435.

Zdunkowski, W. G., and A. E. Barr (1972). A radiative conductive model for the prediction of radiative fog. *Boundary-Layer Meteorol.* **2**, 152–177.

Zdunkowski, W. B., and B. C. Nielsen (1969). A preliminary prediction analysis of radiation fog. *Pure Appl. Geophys.* **75**, 278–299.

Chapter 8 | Cumulus Clouds

8.1 Introduction

Cumulus clouds take on a variety of forms and sizes ranging from nonprecipitating fair-weather cumuli to heavily precipitating thunderstorms. In this chapter we shall discuss the dynamic characteristics of cumuli, ranging from boundary layer cumuli to towering cumuli. Recently, Stull (1985) proposed a classification of fair-weather cumulus clouds according to their interaction with the atmospheric boundary layer. He considered three categories: forced, active, and passive clouds. Figure 8.1 illustrates the differences among the three cloud categories. *Forced cumulus clouds* form at the tops of boundary layer thermals that overshoot into the stable layer that caps the ABL. The thermals rise above the lifting condensation level but because they are unable to rise above the level of free convection (LFC), they remain negatively buoyant during overshoot. *Active* fair-weather cumulus clouds ascend above the LFC and therefore become positively buoyant. As a consequence of gaining positive buoyancy, they ascend to greater heights than forced cumuli, and develop circulations that depart from those characteristic of dry ABL thermals. *Passive clouds* are the decaying remnants of formerly active clouds. They can be readily identified by the absence of a flat base. Their importance is mainly due to the fact that decaying clouds may account for a significant fraction of the total cloud coverage (Albrecht, 1981). This is especially true on those days in which the free atmosphere is humid and cloud evaporation is slow. Thus, passive clouds can significantly affect the amount of radiative heating/cool-

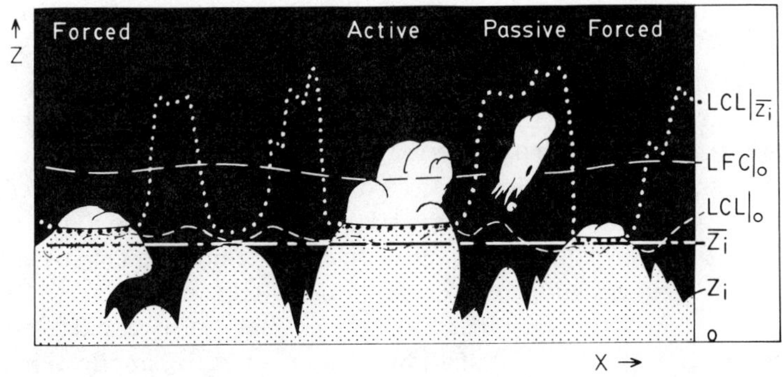

Fig. 8.1. Schematic of the relationship of cumulus clouds to various fair-weather mixed-layer (ML) characteristics. The lightly shaded region denotes ML air, the black region denotes free-atmosphere air, and the white regions denote clouds. The horizontal average ML depth is indicated by $\bar{Z}_i$, while Z_i is used for the local ML top. The dotted line shows the local lifting condensation level (LCL) for air measured at $\bar{Z}_i$; the short-dash line shows the local LCL for air measured in the surface layer; the long-dash line shows the local level of free convection (LFC) for air measured in the surface layer. Cloud classes are labeled above their respective sketches. [From Stull (1985).]

ing in the ABL. Over land, passive clouds mainly affect the ABL by shading the ground and thereby reducing surface heating. We begin this chapter by examining boundary layer or fair-weather cumuli as an ensemble of cloud elements that can be thought of as an extension of the cloud-free boundary layer. We then examine the organization of cumulus clouds. Finally, we focus on the properties of individual cumuli by examining the processes of entrainment in cumuli and the initiation and maintenance of convective-scale updrafts and downdrafts, and the interaction of clouds.

8.2 Boundary Layer Cumuli—An Ensemble View

Boundary layer cumuli are cumulus clouds whose vertical extent is limited by a pronounced capping inversion, often viewed as the top of the atmospheric boundary layer. As such, they do not markedly differ from stratocumulus clouds, except that the coverage of cumulus clouds remains considerably less than 100%. In fact, the breakup of a stratocumulus layer by entrainment instability or some other mechanism often results in the formation of a layer of boundary-layer-confined cumuli. Such clouds are generally nonprecipitating, and are usually referred to as fair-weather cumulus, cumulus humilus, or trade-wind cumulus. In some cases, warm-based, maritime boundary layer cumuli may precipitate, especially during high wind situations. Boundary layer cumuli are often composed of an ensemble of forced, active, and passive clouds. The active cells, however, do not ascend more than a few kilometers above the capping inversion.

There is some evidence that boundary-layer-confined cumuli differ significantly from more active, inversion-penetrating cumulus clouds such as towering cumuli or cumulonimbi. In a diagnostic study of tropical cumuli, Esbensen (1978) concluded that the time scale for boundary layer cumuli to reach a state of quasiequilibrium with the thermodynamic structure of their environment is significantly longer than the adjustment time for deeper clouds. He suggested it may therefore be necessary to parameterize separately the effects of deep and shallow cumuli for many applications. The observational study of the distribution of cloud cover by Plank (1969) also suggests that boundary layer cumuli and deep cumuli differ markedly. Plank noted that the morning populations of cumuli over Florida were essentially unimodal, whereas the distribution in the afternoon was bimodal. The morning cumuli were relatively small, so they were probably boundary-layer-confined cumuli. In the afternoon, the cloud populations could be considered to be a mix of boundary layer cumuli, towering cumuli, and cumulonimbi. We shall show later that the dominant mechanisms of entrainment in boundary layer cumuli and deep cumuli may also be quite different.

Since boundary layer cumuli may be an intimate component affecting the overall structure of the ABL, let us examine the characteristic thermodynamic structure of a cumulus-capped ABL. An example is the layered structure of the undisturbed trade-wind atmosphere. Similar structures are observed in middle latitudes in postfrontal air and in the early morning hours over land before the onset of deep convective disturbances. As summarized by Garstang and Betts (1974), the undisturbed trade-wind ABL can be characterized by five distinct layers (see Fig. 8.2):

(i) The *surface layer*, which may be adiabatic over water or superadiabatic over land and which may exhibit a decrease in specific humidity or mixing ratio. The surface layer is generally less than a few hundred meters in depth.

(ii) A *mixed layer*, which is nearly adiabatic throughout and is nearly constant in specific humidity. Its depth may extend to 900 mbar in the undisturbed trades and much higher over continental regions.

(iii) A *transition layer*, which is nearly isothermal but has sharply decreased moisture and a thickness less than 100 m. This layer separates the cloud layer above from the dry mixed layer below.

(iv) A *cloud layer*, which is conditionally unstable and extends from the transition layer to the base of the capping inversion and has a temperature gradient somewhat greater than wet adiabatic.

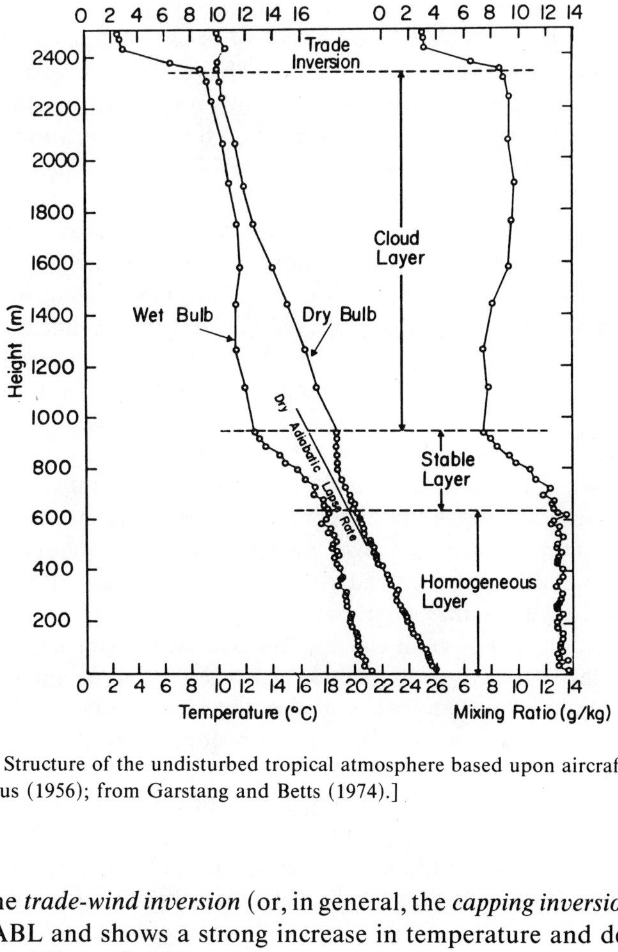

Fig. 8.2. Structure of the undisturbed tropical atmosphere based upon aircraft soundings. [After Malkus (1956); from Garstang and Betts (1974).]

(v) The *trade-wind inversion* (or, in general, the *capping inversion*), which caps the ABL and shows a strong increase in temperature and decrease in specific humidity with height.

The depths and structure of these layers are controlled by the fluxes of heat, moisture, and momentum at the surface, the strength of subsidence in the lower atmosphere, and the intensity of cumulus convection. In an analysis of data obtained from multiple-level aircraft flights during GATE, Nicholls and LeMone (1980) noted that a significant change in the distribution of heating and moistening throughout the depth of the mixed layer or subcloud layer was associated with the onset of cumulus convection. The virtual heat fluxes or buoyancy fluxes in the subcloud layer did not vary with the presence or absence of cumuli. Sommeria (1976) obtained a similar result in his 3-D large-eddy simulation of the trade-wind boundary layer.

There have been few attempts to obtain direct measurements of fluxes and energy budgets through the cloud layer because of two major problems. First, the cloud layer is too high for tethered balloon-borne sensors. Furthermore, most researchers agree that it is currently impossible to obtain reliable, high-frequency sampling of temperature and total water (water vapor plus condensed water) from high-speed airplanes. Direct (*in situ*) probes of air temperature either become wet or respond too slowly for accurate calculations of fluxes using eddy-correlation analyses. Likewise, total water content samplers either have too slow a response or are costly and unreliable.

The second problem arises from the marked difference in the character of eddies in a cumulus layer as opposed to those in the subcloud layer. In the cloud layer, the eddies composing the cumuli are clustered in widely spaced localized regions and separated by cloud-free, relatively quiescent regions. In order to obtain an adequate sampling of an ensemble of cloud and cloud-free eddies in the cloud layer, an aircraft must traverse a large volume of air. Unfortunately, cumuli are rarely horizontally homogeneous over long distances, so one can seldom obtain meaningful flux measurements by simply flying an airplane over long distances. Furthermore, because clusters of cumuli are rarely steady for periods of several hours, one cannot always obtain a meaningful ensemble average by flying an aircraft back and forth through a particular cluster. The expense of using multiple aircraft at different levels, and the present lack of appropriate cloud and temperature sensors, usually force us to resort to cloud models or diagnostic studies in which cloud-layer fluxes are inferred as a residual from total moisture and heat budget analyses.

Pennell and LeMone (1974) obtained some direct estimates of momentum fluxes and the kinetic energy budget of a trade-wind cloud layer under relatively high wind conditions. Using an aircraft equipped with a gust probe and an Inertial Navigation System, they obtained accurate measurements of the three wind components. Their measurements were, however, subject to the sampling problem noted above. Figure 8.3 illustrates the measured profiles of the standard deviations of the horizontal winds (σ_u and σ_v) and vertical velocity (σ_w) obtained through the depth of the boundary layer with winds as large as 16 m s^{-1} during a period of enhanced cloud activity. Below cloud base, the profiles of σ_u, σ_v, and σ_w are reasonably consistent with observations of the ABL under cloud-free conditions (Lenschow, 1970; Kaimal *et al.*, 1976). In particular, σ_u and σ_v exhibit a sharp decrease from a maximum value near the surface. The standard deviation of vertical velocity, however, exhibits a slight linear decay with height, whereas some other observations and laboratory models suggest (Willis and Deardorff, 1974; Nicholls and LeMone, 1980) that σ_w should exhibit a maximum in the subcloud layer at about one-half the mixed layer

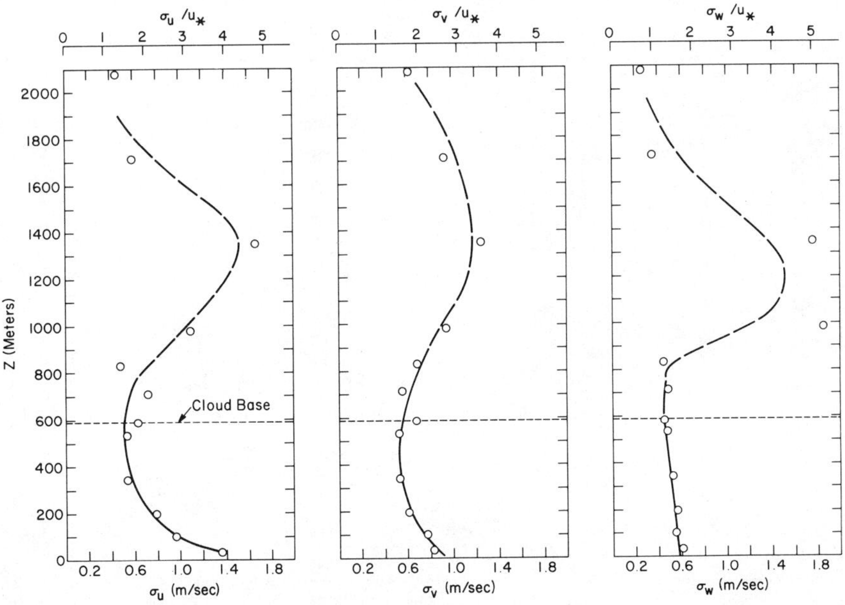

Fig. 8.3. Standard deviations of u', v', and w' as a function of height for the cloudy region ($u_* = 46$ cm s^{-1}). The sampling time is probably not sufficient to establish the profiles above cloud base. [From Pennell and LeMone (1974).]

depth. Above cloud base, where the release of latent heat is expected to be important, there is the suggestion that σ_u, σ_v, and σ_w reach a maximum. Likewise, the eddy flux of horizontal momentum $(-\overline{u'w'})$ shown in Fig. 8.4 exhibits a maximum in the cloud layer. Clearly, the vertical transport of momentum by clouds is important.

Using a three-dimensional large-eddy simulation model, Sommeria (1976) and Sommeria and LeMone (1978) simulated explicitly the turbulent structure of the trade-wind boundary layer, including the cloud layer. The model is an extension of Deardorff's (1972) boundary layer model to a nonprecipitating cloud layer. Using a grid resolution of 50 m, the model covered a horizontal domain of 2×2 km. The model data were compared with the observations during suppressed convective activity reported by Pennell and LeMone (1974). Figures 8.5 and 8.6 illustrate the simulated and observed profiles of the variances of the wind components u and w. The observed profiles are shown for raw variances and variances calculated after the wind field is filtered to remove wavelengths greater than 2.2 km. The model variances are shown for the case of weak cloud activity, which corresponds to the observed case, relatively strong cloud activity, and for

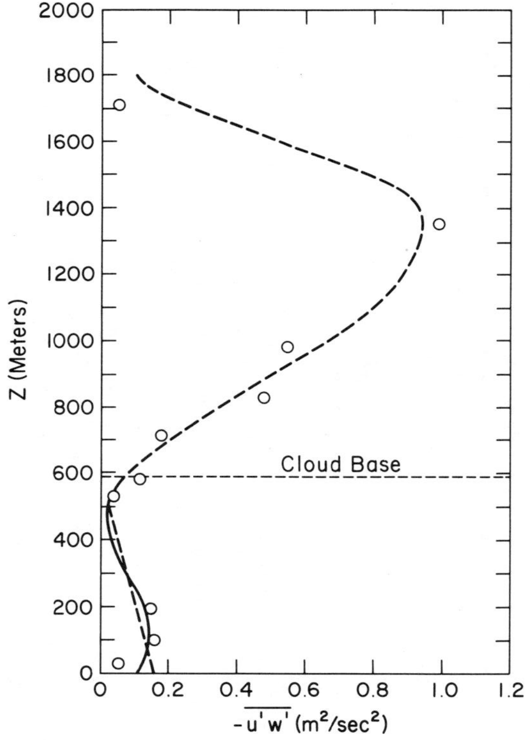

Fig. 8.4. Momentum flux as a function of height for the cloudy region. [From Pennell and LeMone (1974).]

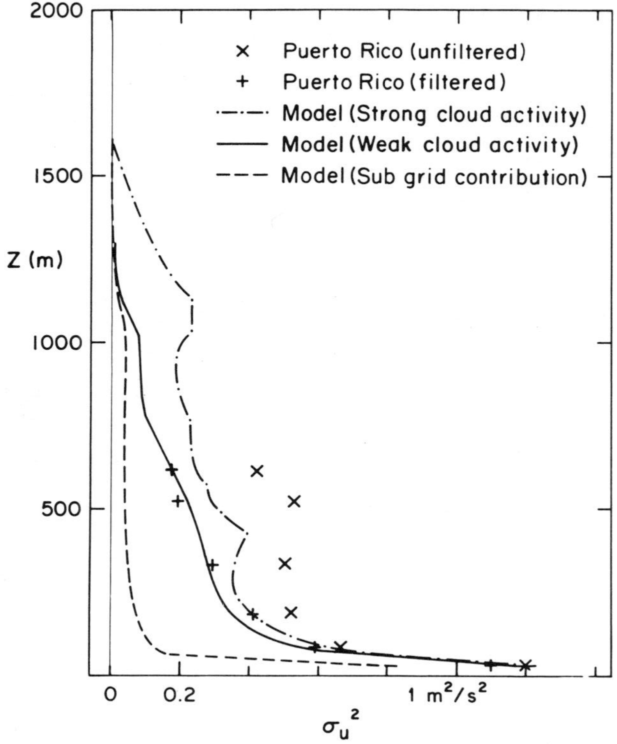

Fig. 8.5. Comparison of the variances of the *u* component of the velocity in the model and in the Puerto Rico experiment. [From Sommeria and LeMone (1978).]

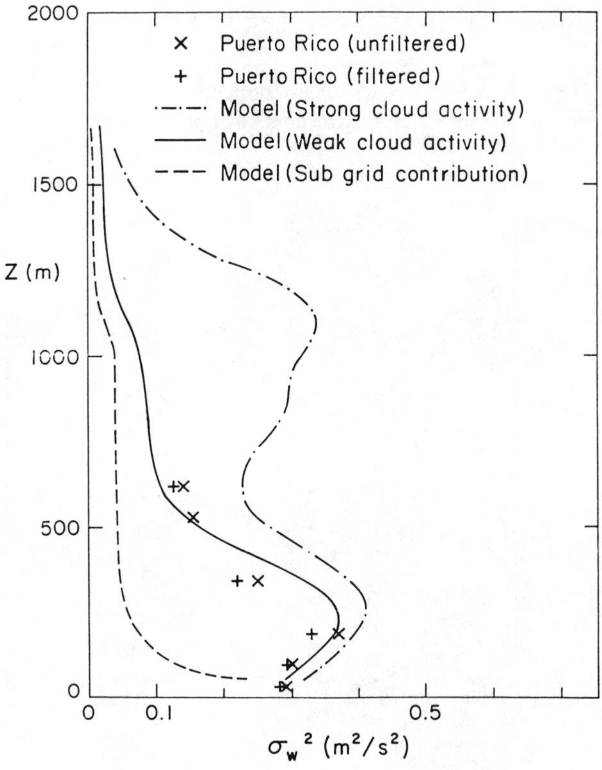

Fig. 8.6. Comparison of the variances of the vertical component of the velocity in the model and Puerto Rico experiment. [From Sommeria and LeMone (1978).]

the subgrid-scale contribution to the variances. The agreement between simulated and observed horizontal wind variances is quite good when compared to the filtered wind data. However, the unfiltered σ_u^2 are considerably greater than the observed, suggesting that contributions to the wind fluctuations on scales greater than 2 km is quite significant. Both the filtered and unfiltered observed σ_w^2 data are consistent with the simulated data. This suggests that eddy scales larger than 2 km are not as important to generating vertical velocity fluctuations. In the case of the simulated enhanced cloud activity, σ_w^2 exhibits a relative maximum in the cloud layer. This is consistent with the observed profile of σ_w^2 during enhanced cloud activity shown in Fig. 8.3, except that the observed values are quite a bit larger.

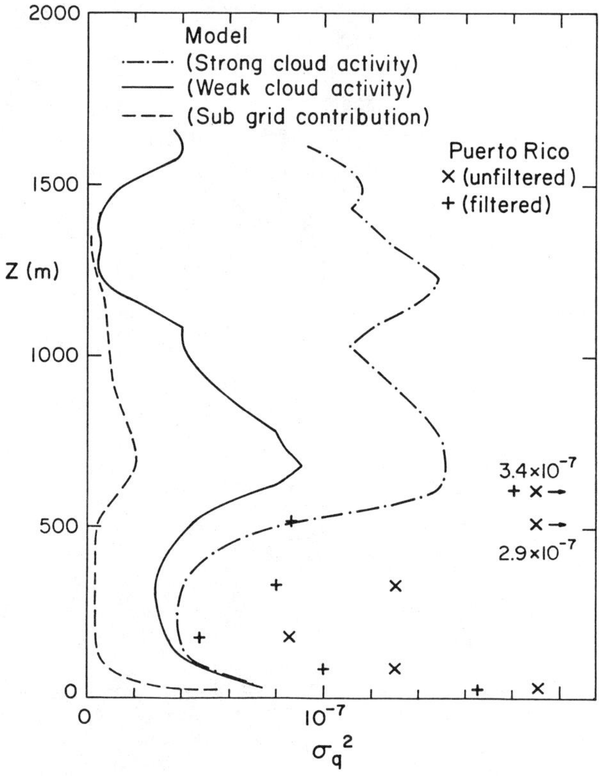

Fig. 8.7. Comparison of the variances of the specific humidity in the model and in the Puerto Rico experiment. Units are mass of water vapor per mass of air squared. [From Sommeria and LeMone (1978).]

As illustrated in Fig. 8.7, the model underestimates the variance in specific humidity. The authors suggested that this may be due to the neglect of anomalies in sea surface temperature which would generate local anomalies in surface evaporation rates. However, the errors are greatest relative to the unfiltered data; this suggests that scales larger than the domain of the model are contributing substantially to moisture fluctuations and perhaps to low-level moisture convergence.

The vertical transport of horizontal momentum simulated by the model shown in Fig. 8.8 agrees with observations in the subcloud layer. In the cloud layer, the observed fluxes are positive, especially under enhanced cloud activity (see Fig. 8.4), whereas the simulated fluxes are negative. The differences, however, may be due to the somewhat different wind profiles used in the model.

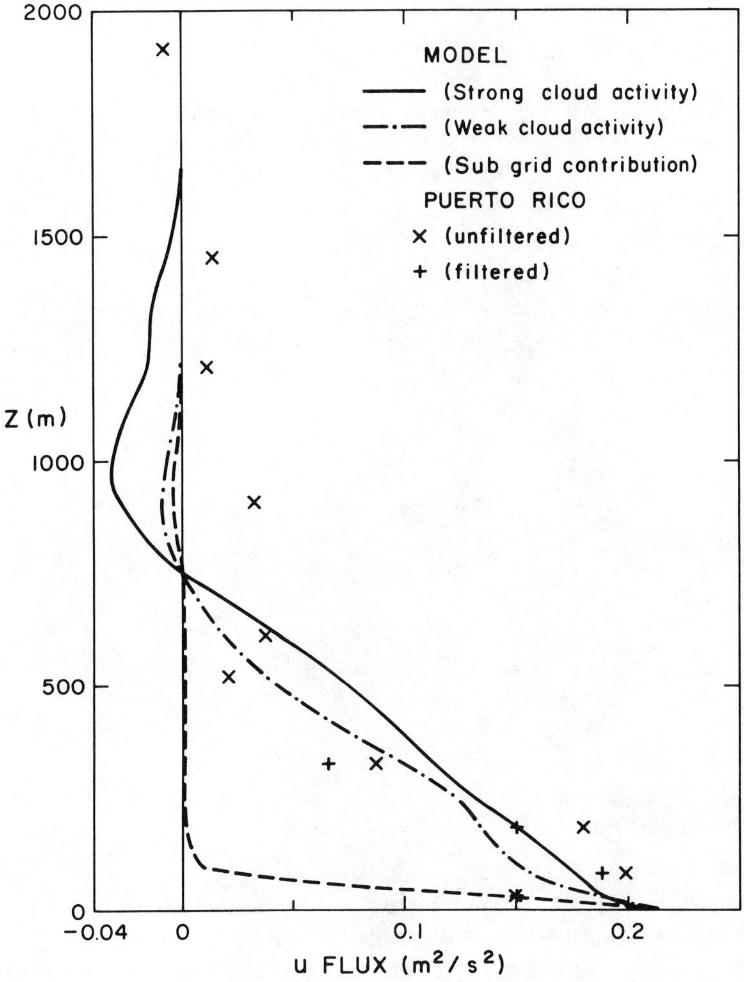

Fig. 8.8. Comparison of the flux of the *u* component of velocity in the model and in the Puerto Rico experiment. [From Sommeria and LeMone (1978).]

Figures 8.9 and 8.10 illustrate the observed and simulated moisture flux and the flux of virtual potential temperature. The observed and simulated fluxes of these quantities generally agree with observations in the subcloud layer. As convective activity increases, the moisture flux also increases substantially in the subcloud layer, whereas the buoyancy flux remains unchanged. In the cloud layer, however, both moisture and buoyancy fluxes are substantially modulated by the vigor of convective activity.

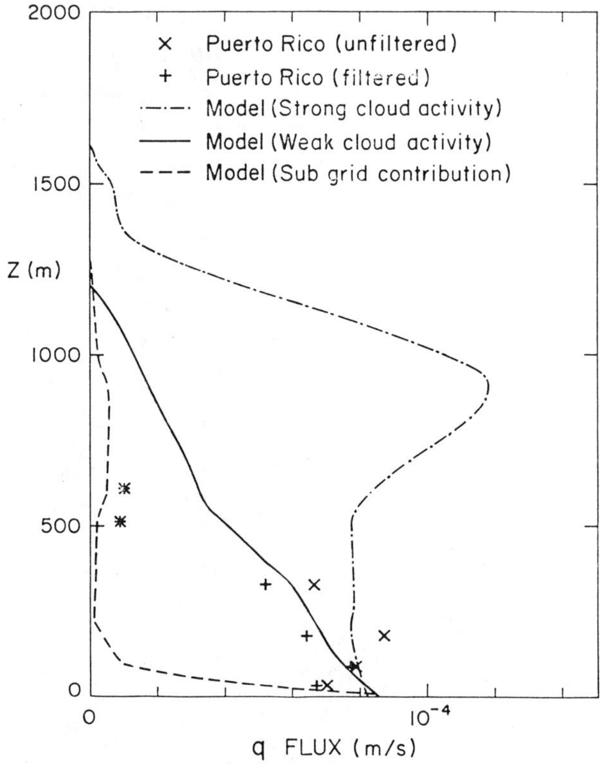

Fig. 8.9. Comparison of the flux of specific humidity in the model and in the Puerto Rico experiment. [From Sommeria and LeMone (1978).]

The various terms contributing to turbulent kinetic energy [see Eq. (3.50)] simulated by the model are illustrated in Fig. 8.11 for the case of strong cloud activity. The terms in the kinetic energy equation calculated from the somewhat more suppressed observed case are also shown in Fig. 8.11. As far as the kinetic energy budget is concerned, Sommeria and LeMone (1978) suggested that both the subcloud and the cloud layer could be divided into two layers. In the lower mixed layer up to approximately 200 m, shear production (and, secondarily, buoyancy) generates kinetic energy which is locally removed by viscous dissipation and vertical turbulent transport. In the upper mixed layer (up to about 700 m), turbulent transport of kinetic energy (and to a lesser extent, shear and buoyancy production) is balanced by dissipation. In the lower part of the cloud layer (approximately 700 to 1050 m), kinetic energy is produced by the buoyancy generated by condensation, or, secondarily, by shear and pressure transport. Local molecular

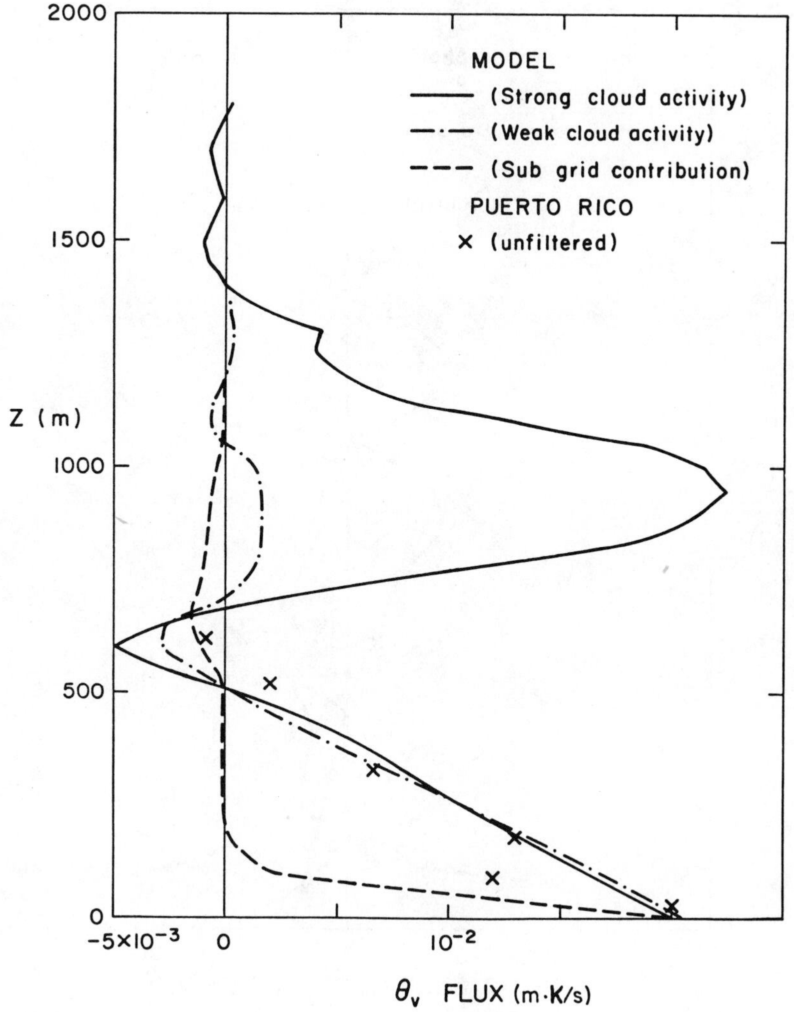

Fig. 8.10. Comparison of the flux of virtual potential temperature in the model and in the Puerto Rico experiment. [From Sommeria and LeMone (1978).]

dissipation and vertical turbulent transport share equally in balancing the production terms. Kinetic energy production in the upper part of the cloud layer is dominated by turbulent transport from lower levels along with smaller contributions from shear and buoyancy. Along with molecular viscosity, pressure transport plays a significant part in the dissipation of kinetic energy.

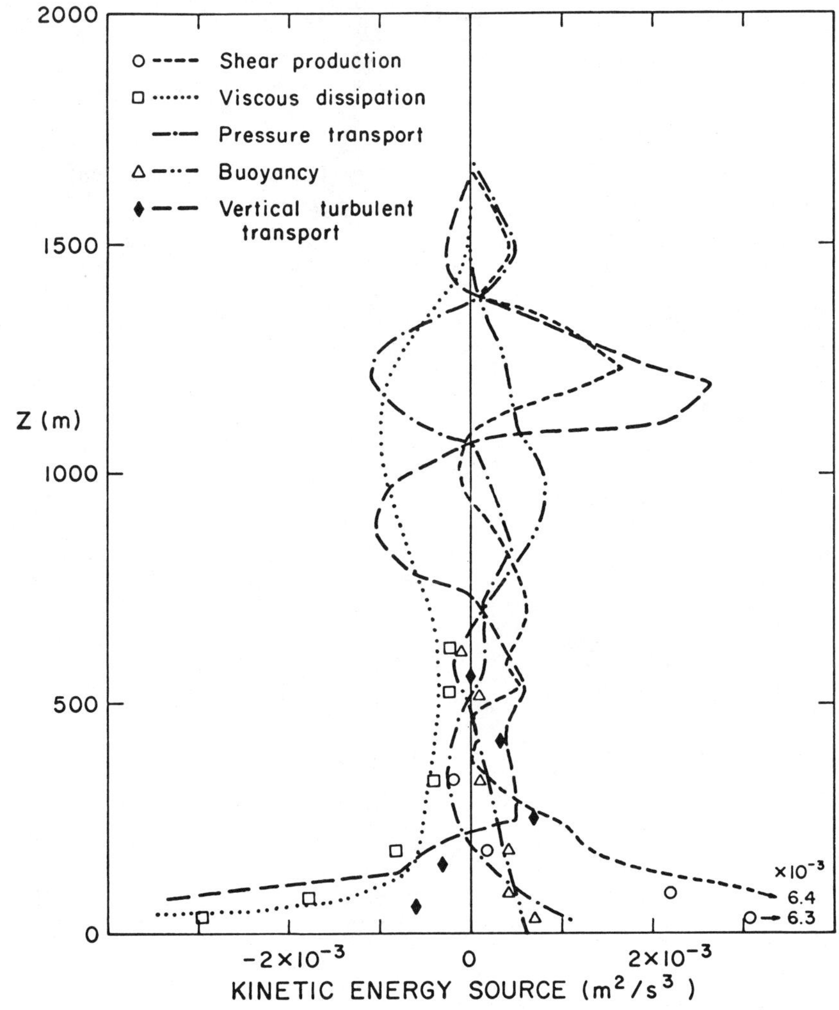

Fig. 8.11. Terms of the kinetic energy budget in the Puerto Rico experiment and the model. The model results are for the time of strong cloud activity, from model times 2.2 to 2.48 h. The signs denote observational data; the lines denote model results. [From Sommeria and LeMone (1978).]

Nicholls *et al.* (1982) applied Sommeria's model to a low-wind-speed case observed in the trade-wind region during GATE. Comparison with observations revealed that the model does quite well in predicting the observed heights of cloud base and cloud top, cloud cover, vertical fluxes of heat, moisture, and momentum, and vertical velocity variance. Again, the model underestimated the horizontal wind variances and the variances of moisture and temperature. The reason for the underestimate of moisture variance can be seen in Fig. 8.12, which illustrates a composite spectrum of the variations in specific humidity q, vertical velocity w, and vertical moisture flux. The interesting feature of the moisture spectrum is that it exhibits a peak at a wavelength of about 10 km. This implies that the largest contribution to the total moisture variance comes from mesoscale eddies having wavelengths on the order of 10 km. Nicholls and LeMone (1980) found that the horizontal wind variances also exhibited spectral peaks on wavelengths on the order of 10 km. However, as shown in b and c in Fig. 8.12, the composite vertical velocity spectrum and the cospectrum between w and q (which represents vertical moisture flux) exhibit peaks at wavelengths of about 700 m. The vertical eddy fluxes of u, v, and T also exhibit peaks at approximately 700 m. The results of this study and the earlier work of Sommeria and LeMone (1978) suggest that vertical velocity variance and vertical eddy fluxes scale with either (1) the boundary layer depth or (2) the depth of the subcloud layer. Total wind variances and total

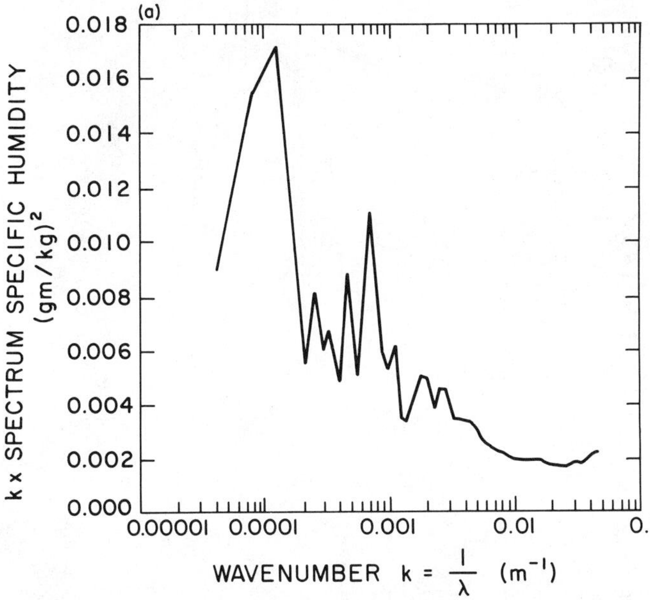

Fig. 8.12. (a) Frequency-weighted composite q-spectrum. The data were obtained from nine runs by the UK C130 at an altitude of 150 m. (b) As for (a), but for the w-spectrum. (c) As for (a), but for the wq-cospectrum. [From Nicholls *et al.* (1982).] (*Figure continues.*)

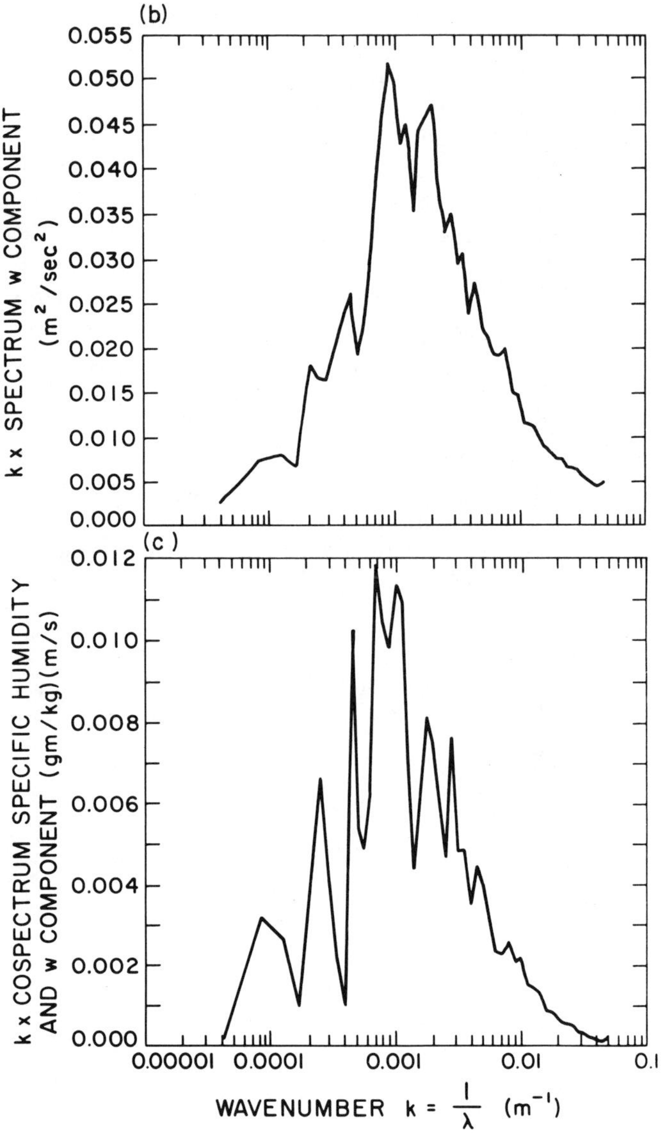

Fig. 8.12 (*continued*)

variances of such scalar properties as specific humidity (and perhaps temperature) scale with mesoscale eddies, whose wavelength contributing to the maximum variance of a quantity may vary depending on the meteorological situation. Thus, a large-eddy simulation model of limited horizontal extent such as Sommeria's appears to be quite useful for simulating eddies contributing to vertical fluxes.

If the horizontal wind fluctuations or variances in scalar properties are important to a particular problem, then the inability of a model to simulate mesoscale contributions to the total variances can lead to serious errors. One example is the prediction of cloud cover. Nicholls *et al.* (1982) noted that the model underpredicted cloud cover; where the observed coverage was between 5 and 10%, the modeled coverage was only 4 to 5%. Furthermore, the earlier work of Sommeria (1976) and Sommeria and LeMone (1978) indicated that the vertical fluxes of moisture from the sea surface were insufficient to maintain relatively steady moist convection. It is likely that this deficiency was caused not by an inadequate parameterization of surface fluxes but by a lack of resolution of mesoscale eddies contributing to low-level moisture convergence.

Despite these deficiencies, large-eddy simulation models can help us understand the behavior of boundary layer clouds. The models can guide us in formulating and testing simpler models of boundary layer clouds. However, the ultimate test of the skill of such models must be made by comparing them with the real atmosphere.

For example, Bougeault (1981a, b) used Sommeria's large-eddy simulation model to develop and test a one-dimensional higher order closure model of the trade-wind cumulus layer. He developed the basic turbulent transport equations of the model independently of Sommeria's model; these equations were an extension of Andre *et al.*'s (1976a, b, 1978) cloud-free, higher order closure boundary layer model. As noted in Chapter 3, Bougeault (1981a) used Sommeria's model to develop and test a scheme for diagnosing cloud fractional coverage and ensemble-averaged liquid-water contents. Subsequently, the 1-D model was tested against ensemble-averaged data simulated by Sommeria's model of clouds observed during the Puerto Rico Experiment (Pennell and LeMone, 1974; LeMone and Pennell, 1976). The 1-D and 3-D model results are consistent with each other. Figure 8.13 illustrates a comparison between the vertical profiles of vertical moisture flux simulated by the 1-D model and the ensemble-averaged moisture fluxes simulated by the 3-D model. Shown in Fig. 8.14 is a comparison of calculations of vertical velocity variance. In general, the agreement is good. Above cloud top, the 3-D model predicts significant contributions to $\overline{w'^2}$ due to gravity waves that are not present in the 1-D model. However, as shown in Fig. 8.14, these velocity fluctuations do *not* contribute to moisture fluxes.

It is interesting to note that the 1-D model exhibits considerable skill in simulating the ensemble characteristics of a field of boundary layer cumuli, in spite of the fact that the 1-D model does not consider lateral entrainment in the individual clouds. The 3-D model permits both vertical and horizontal

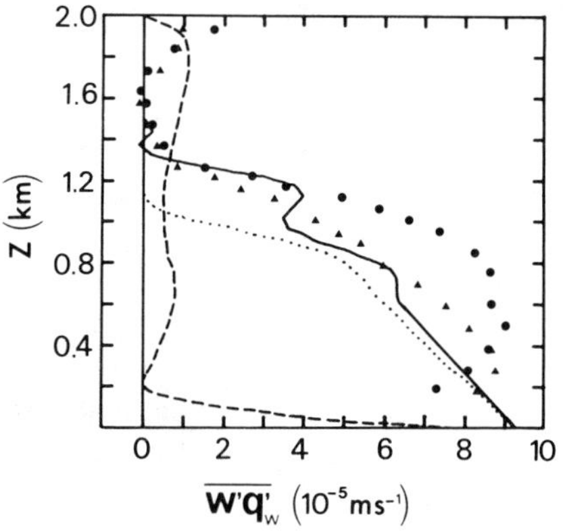

Fig. 8.13. Flux of total water content. Solid line, 1-D result at $t = 2.5$ h; dotted lines, 1-D result with Gaussian cloud scheme; triangles, 3-D results averaged from 2.28 to 2.54 h; circles, 3-D results averaged from 2.42 to 2.68 h; dashed lines, subgrid-scale contribution in the 3-D results. [From Bougeault (1981b).]

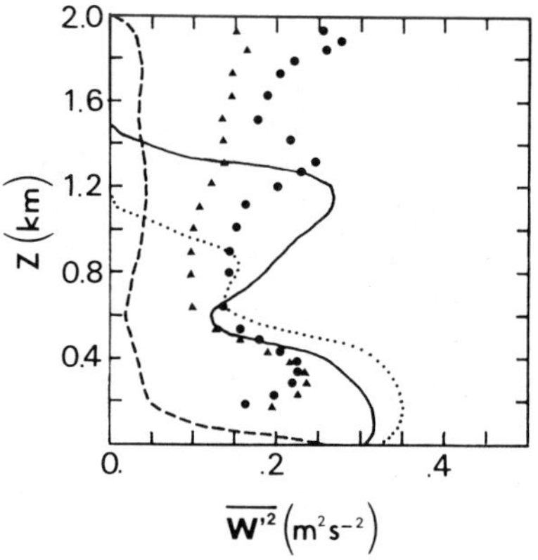

Fig. 8.14. Vertical component of the turbulent kinetic energy. Solid line, 1-D result at $t = 2.5$ h; dotted line, 1-D result with Gaussian cloud scheme; triangles, 3-D results averaged from 2.28 to 2.46 h; circles, 3-D results averaged from 2.42 to 2.68 h; dashed lines, subgrid-scale contribution in the 3-D results. [From Bougeault (1981b).]

mixing in the individual cumuli; by contrast, simple layer-averaged models of the trade-wind boundary layer such as Betts' (1975), Nitta's (1975), Esbensen's (1976), and Albrecht *et al.*'s (1979), which have been apparently successful in simulating the bulk properties of the cloud layer, consider explicitly lateral entrainment in cumuli. Beniston and Sommeria (1981), in fact, tested Betts' model against the data simulated by Sommeria's 3-D model, and found that the thermodynamic fluxes calculated by Betts' scheme were consistent with the 3-D model data.

In order to examine this apparent paradox further, let us look at Betts' model in greater detail. Betts formulated his model in terms of the thermodynamic quantities' moist and dry static energies.

The dry static energy s is defined as

$$s = c_p T + gz, \tag{8.1}$$

and moist static energy h is defined as

$$h = c_p T + gz + Lq. \tag{8.2}$$

In a water-saturated environment they may be defined as

$$S_L = c_p T + gz - Lq_L \tag{8.3}$$

and

$$H_s = c_p T + qz + Lq_s. \tag{8.4}$$

The behavior of the quantities s and h may be considered to be analogous to the behavior of θ and θ_l, respectively.

The fundamental premise in Betts' model is that the eddy fluxes of heat and moisture can be represented as the product of a single convective mass flux (ω^*) and a perturbation quantity derived from a single entraining cloud parcel (subscript c) rising through a known mean environment. Thus,

$$F_X = -\overline{\omega' x'} \equiv \omega^*(X_c - X_e), \tag{8.5}$$

where X represents h, q, or S_L. The convective mass flux represents the cloud area-averaged flux due to clouds having active updrafts with velocity w_c and having a fractional coverage σ. Thus,

$$\omega^* = \sigma \rho w_c.$$

In a manner similar to that used to parameterize the effect of cumulus clouds (Chapter 6), we write the budget equations for a saturated cloud layer in pressure coordinates,

$$Q_1 = \frac{\partial \bar{s}}{\partial t} + \mathbf{V} \cdot \nabla \bar{s} + \bar{\omega} \frac{\partial \bar{s}}{\partial p} = Q_R - \frac{\partial}{\partial p}(\overline{\omega' S_L'}), \tag{8.6}$$

$$Q_2 = L\left(\frac{\partial \bar{q}_v}{\partial t} + \mathbf{V} \cdot \nabla \bar{q}_v + \bar{\omega} \frac{\partial \bar{q}_v}{\partial p}\right) = -L \frac{\partial}{\partial p}(\overline{\omega' q_v'}), \tag{8.7}$$

$$Q_1 - Q_2 = \frac{\partial \bar{h}}{\partial t} + \mathbf{V} \cdot \nabla \bar{h} + \bar{\omega} \frac{\partial \bar{h}}{\partial p} = Q_R - \frac{\partial}{\partial p}(\overline{\omega' h'}), \tag{8.8}$$

where Q_R is the radiative heating and Q_1 and Q_2 are the apparent heat source and apparent water vapor sink, respectively. Substitution of Eq. (8.5) into Eq. (8.8), for example, gives

$$Q_1 - Q_2 - Q_R = \frac{\partial}{\partial p}[w^*(h_{sc} - \overline{h_e})]$$

$$= -\omega^* \frac{\partial h_e}{\partial p} + \omega^* \frac{\partial h_{sc}}{\partial p} + (h_{sc} - h_{\bar{e}}) \frac{\partial \omega^*}{\partial p}. \tag{8.9}$$

Betts defined the lateral entrainment rate for an ascending cloud parcel as λ (>0), where

$$\partial h_{sc}/\partial p = \lambda (h_{sc} - \bar{h}). \tag{8.10}$$

Substitution of Eq. (8.10) into Eq. (8.9) yields the budget equation

$$Q_1 - Q_2 - Q_R = -\omega^* \frac{\partial h_e}{\partial p} + \omega^*(h_{sc} - h_e)\left(\lambda + \frac{1}{\omega^*}\frac{\partial \omega^*}{\partial p}\right). \tag{8.11}$$

The term $\lambda + (1/\omega^*)(\partial\omega^*/\partial p)$ represents two sources, namely, entrainment of dry environmental air into clouds at the rate λ, and the net detrainment of cloud mass having properties differing from the environment. As noted by Betts, entrainment affects the properties of a cloud, while detrainment affects the environment on the time scale of the response of the environment. As a consequence, when Betts (1975) applied his model to the diagnostic interpretation of trade-wind cumulus budgets, he found that the diagnosed cloud transports were relatively insensitive to variations in entrainment rates. Even complete neglect of entrainment did not affect the results substantially. In contrast, the diagnosed detrainment rates were much larger and nearly independent of λ. In another diagnostic study, Esbensen (1978) concluded that detrainment dominates the mass budgets of shallow cumulus clouds. Thus, their diagnostic studies resolve the apparent paradox. The 1-D higher order transport model of Bougeault's (1981b) only calculates the net flux divergences (or detrainment rates) that affect the cloud environment. Since detrainment predominates over entrainment in determining the budgets of a shallow cumulus layer, a lack of consideration of entrainment in such higher order models does not appear to be a serious defect.

A consistent result from Sommeria's 3-D model, Bougeault's 1-D higher order transport model, and the diagnostic models developed by Betts and Esbensen is that boundary layer cumuli moisten the cloud layer. That is, the detrainment of water from clouds dominates the drying effect of cumulus-induced subsidence. Furthermore, the lower part of the cloud layer is warmed and the upper part cooled. The evaporative cooling at the top of the cloud layer and the radiative cooling maintain the trade inversion against the warming produced by large-scale subsidence. The moistened, destabilized cloud layer, however, favors the development of towering cumuli that can penetrate well above the capping inversion.

8.2.1 *Role of Radiation*

The importance of shortwave and longwave radiative transfer to the ensemble structure of a boundary layer cumulus layer has been investigated using 3-D, 1-D, and diagnostic models. These studies primarily focus on moist tropical or subtropical cumulus layers over the ocean. One can expect very important interactions between radiative transfer processes and the cloud layer over land, where surface heating/cooling can vary substantially over the diurnal cycle. However, there have been few quantitative investigations of the interactions among radiation, clouds, and the heating profiles over land.

Sommeria (1976) included longwave radiation cooling in his 3-D cloud model. The radiation model is based on the transmissivity approach of Yamamoto *et al.* (1970) and Sasamori (1972). In the clear air, radiative cooling is relatively uniform throughout the depth of the moist marine boundary layer. However, in the cloud layer, the cooling rate is a maximum at the level corresponding to the top of many small cumuli. The average cooling rate approaches 5°C day^{-1} at that level. In the cloud-free boundary layer, however, the cooling rate is about half that amount, i.e., 2.5°C day^{-1}. Since the cloud cover is about 5%, the cooling rate approaches 5°C h^{-1} at the tops of the individual cumuli.

Albrecht *et al.* (1979) included longwave radiational cooling in a time-dependent model of the trade-wind boundary layer. The model is an extension of Betts' (1976) model, in which the average values of mixing ratio and moist static energy in the subcloud and cloud layers are simulated. In addition, the slope of these quantities is calculated in the cloud layer, along with the heights of the transition layer and trade inversion. Simple treatments of longwave and shortwave radiative heating were included in their model. Based on detailed calculations with a broadband radiative transfer model, they assumed that under clear conditions the longwave cooling rate was uniformly distributed through the boundary layer at a rate determined from the detailed model. They found that clouds modified the vertical distribution of cooling, but did not change substantially the net cooling through the depth of the boundary layer. Under cloudy conditions, the clear-sky radiative cooling rate was confined to a thin layer at the top of the cloud layer.

A similar procedure was used to evaluate solar heating. Again, they used a detailed broadband model to evaluate clear-sky heating rates in the boundary layer. Under cloudy conditions, the clear-sky solar heating rates were confined to a thin layer at cloud top or the capping inversion. The radiative heating for the partly cloudy regions was modeled as a weighted average of the heating rates of the clear and cloudy regions, where the weighting factor was the fractional cloud cover. Calculations with their model further demonstrated that longwave radiative cooling and evaporation of cloud water at the inversion balances warming due to subsidence.

By implementing a simple variation in the solar heating function, Albrecht (1979) attempted to simulate the observed variation in the boundary layer depth found by Augstein *et al.* (1974) during ATEX. He was able to reproduce the phase of the observed variation, but the predicted amplitude was only 25% of the observed values. Other factors neglected in the model, such as diurnal variations of large-scale divergence (Nitta and Esbensen, 1974), magnitudes of solar absorption larger than modeled, and diurnal variations in cloud cover, may have been important in the observed case.

In a subsequent study, Brill and Albrecht (1982) modified the assumed vertical distribution of solar heating in the cloudy boundary layer. This change was motivated by studies by Welch *et al.* (1976), Grassl (1977), Tanaka *et al.* (1977), and Davis *et al.* (1979) which suggested that significant heating occurs below cloud top, particularly when the zenith angle is small or at larger zenith angles when the cloud fraction is small. Brill and Albrecht (1982) used the results of Davis *et al.* (1979) to parameterize the distribution of heating into a subcloud heating rate $\bar{Q}_M$, a lower half of the cloud-layer heating rate $\bar{Q}_{CB}$, and a top-half heating rate $\bar{Q}_{CT}$. In the top half of the cloud layer, a variable fraction α of the heating rate was placed in the inversion. Placing all the top-half heating in the inversion resulted in a much greater diurnal variation of the inversion than by placing it below. However, when all the top-half solar heating is distributed below the top of the inversion, the cloud layer is stabilized by warming the top half of the cloud layer more than the bottom half.

8.2.2 Cloud Coverage

As noted above, the amount of cloudiness has a significant impact upon the distribution of radiative heating in the boundary layer. In addition, the amount of cloud cover controls the net latent heating in the boundary layer and hence the resultant thermodynamic and dynamic impacts of the clouds on the environment.

The prediction or diagnosis of fractional cloud cover is important to a number of meteorological problems. For example, the problem of predicting conditions favorable for severe aircraft icing in supercooled boundary layer clouds depends upon both the liquid-water content and the fractional area of cloud coverage.

Unfortunately, our ability to predict cloud cover using the physical parameterization and resolution of current forecast models and observations is severely limited. LES models seem to adequately predict cloud cover induced by boundary layer eddies having a horizontal scale on the order of the boundary layer depth. However, we have noted previously that these models might underestimate total cloud cover. This, we indicated, may be because the eddies contributing the most to moisture variance have scales of 10 km or more; larger than the LES model domain. It is our concern that this same problem may apply equally to higher order closure boundary layer models. In Bougeault's (1981a) model, for example, cloud fractional coverage is diagnosed as a function of model-simulated variances and covariances of thermodynamic variables. A crucial aspect of this model is the assumption regarding the distribution function of the thermodynamic variables. Unfortunately, even if the question of the distribution function is fully resolved, there is no guarantee that the model will have a forecast application in cloud cover that is greater than LES models. Like the LES model, higher order closure models do not account for the role of mesoscale eddies, which may be contributing significantly to the determination of cloud coverage.

Albrecht (1981) approached the problem of predicting cloud cover in general circulations models somewhat differently. A fundamental premise in his scheme is that cloud cover depends principally on the time required for inactive (or passive) cumulus elements to decay. The basis of his assumption is the general observation that active cumulus elements occupy a fraction of a given large-scale ($\sim 100 \times 100$ km) area ($<3\%$), while total cloud cover may be an order of magnitude greater. He models the decay of the cloud as a simple diffusion process. The scheme then estimates the cloud cover as a function of the relative humidity near the top of the cloud layer and the liquid-water content produced by the active cloud elements. To estimate the liquid water content produced by the active cloud elements, Albrecht used the simple lateral entrainment cloud model employed by Betts (1975) and Albrecht *et al.* (1979) in their models of the trade-wind boundary layer. A comparison of predictions of the model with observations revealed that the scheme predicted cloud cover generally to within a factor of 2 of the observations. This is about the same level of accuracy obtained by Sommeria with a LES model. It is likely that this level of accuracy is sufficient for many meteorological applications.

Albrecht *et al.* (1979) demonstrated that in his scheme the simulated cloud cover was sensitive to sea surface temperature. In general, the simulated cloud amounts were a minimum for the maximum sea surface temperature tested of 22°C, with much larger cloud amounts at colder temperatures. The model was also sensitive to variations in sea surface temperatures at warmer temperatures.

Albrecht (1981) also investigated the sensitivity of the model to variations in radiative cooling. The results of those tests suggested that diurnal variations in the radiative budget of the boundary layer could lead to substantial variations in cloud cover.

In the follow-up study by Brill and Albrecht (1982), the interactions between a diurnally varying radiation budget and cloud cover were further investigated. Figure 8.15 illustrates a comparison between model predictions and diurnal variations of cloud cover observed during ATEX and GATE. Both the model and observations show a maximum of cloud cover in the morning. In comparison to ATEX data, the model predictions of cloud cover averaged over the entire day are only slightly less than the observations (0.54 for the model versus 0.57 for ATEX). The magnitude of the range of variations is 0.13 for the model and 0.18 for ATEX observations. However, the model predicts an afternoon minimum in cloudiness, whereas the ATEX data indicate sustained high fractional cloud cover. Brill and Albrecht (1982) suggest that this disagreement may occur because the ATEX data included

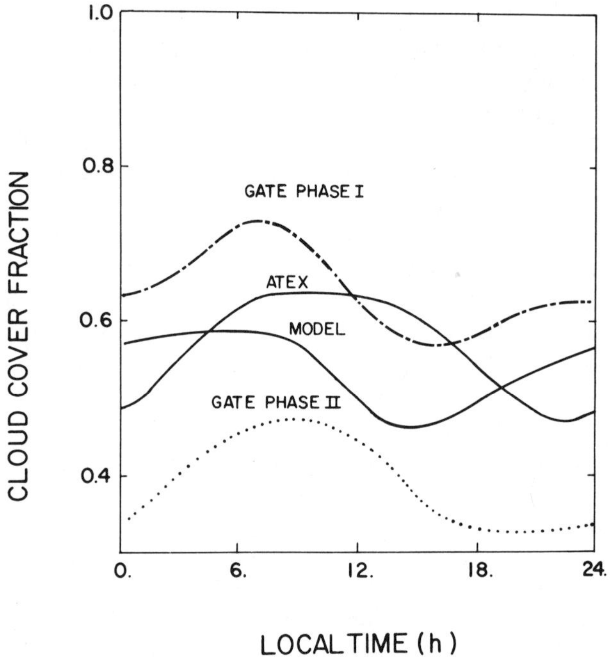

Fig. 8.15. A comparison of the model simulation with low-pass filtered cloud cover observations from ATEX, GATE Phase I (Endurer), and GATE Phase II (Charterer). [From Brill and Albrecht (1982).]

Table 8.1

Computed Fractional Coverage by σ_u, σ_0, and σ and the Observed Fractional Coverage (σ_s) for Four Kon Tur Situations[a]

Date	σ_u (%)	σ_0 (%)	σ_p (%)	σ (%)	σ_s (%)
29/09/1981	3	9	14	26	20
11/10/1981	6	16	69	91	87
13/10/1981	1.5	24	40	65.5	70
16/10/1981	1.5	20	35	56.5	70

[a] Taken from Becker (1986).

all clouds, whereas the model accounts for only boundary layer clouds. Comparison with GATE low-level cloud cover observations supports the model predictions of an afternoon of minimum cloudiness. Overall, the study by Brill and Albrecht (1982) suggests that the observed diurnal variations of marine trade-wind boundary layer features is due primarily to the diurnal variation of solar heating coupled with cloud cover variations and, secondarily, to the temporal variation of divergence associated with synoptic-scale motions.

Recently, Becker (1986) adapted Albrecht's (1981) method for determining fractional coverage of passive or decaying clouds to a simple scheme for parameterizing amounts of active cumulus clouds. He then tested the scheme against observations of cloud amount that occurred during the Kon Tur experiment (Becker, 1986). Table 8.1 compares model-predicted total cloud amounts (σ) to the observed fractional coverage (σ_s). Also listed are the model-predicted coverage for active updrafts (σ_u), the fractional coverage of active clouds not containing active updrafts (σ_0), and the fractional coverage of passive or decaying clouds σ_p. Table 8.1 shows that Becker's model predicted cloud amounts to within 30% of observed values for the four analyzed cases, with the greatest percentage of errors occurring with small cloud coverage. Furthermore, Table 8.1 shows the dominance of passive clouds to the overall cloud cover.

8.3 Organization of Cumuli

It is generally agreed that cumuli originate in boundary layer eddies. The association between subcloud eddies and cumulus clouds is most obvious over land, where thermals associated with mountains (Orville, 1965) or hot spots over flat terrain (Woodward, 1959) have been found to be the sources of cumuli. In the case of the marine boundary layer, however, the association between subcloud eddies and cumuli is much less distinct. In fact, early observations by Bunker *et al.* (1949) suggested that the origin of nonprecipitating trade cumuli was not in the subcloud layer. More recent observations using modern airborne instrumentation reported by LeMone and Pennell (1976) as well as 3-D modeling studies (Sommeria, 1976) indicate that cumuli have distinct (though not very intense) roots in the marine subcloud

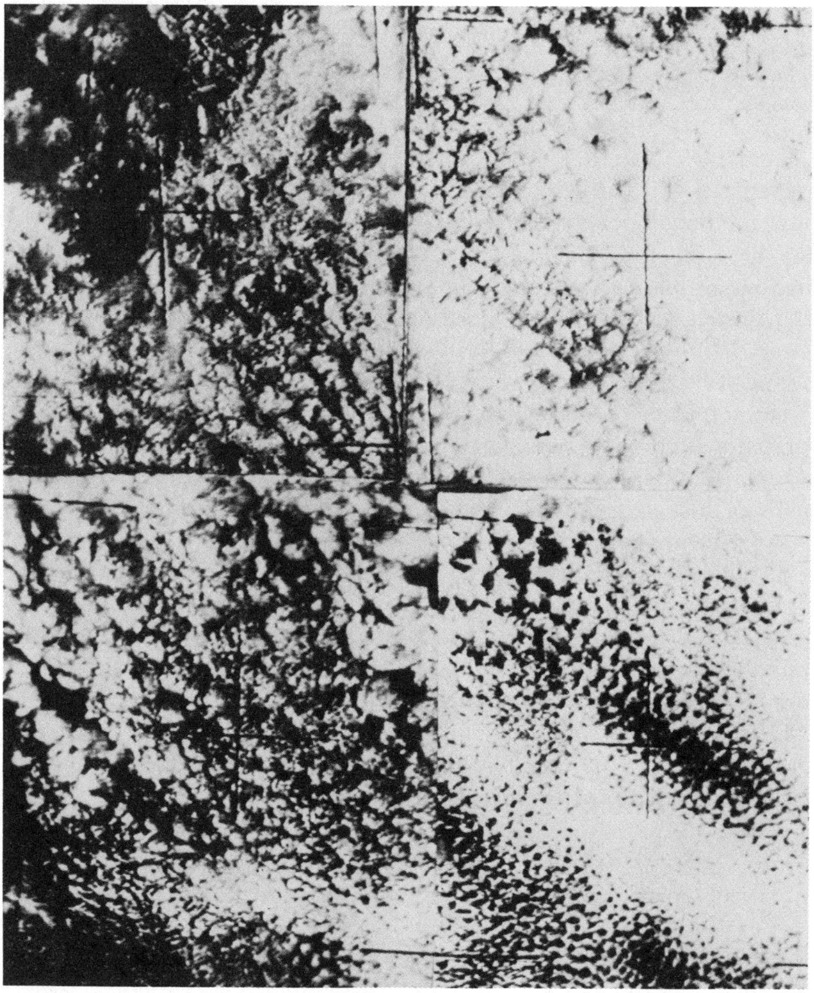

Fig. 8.16. Photograph of closed mesoscale cellular convection taken by COSMOS-144 at 0800 GMT, 20 April 1967. (Provided through the courtesy of Dr. V. A. Bugaev, Director, Hydrometeorological Research Centre of the USSR.) [From Agee and Chen (1973).]

layer. It is not surprising, then, that the organization of cumuli often reflects the organization of eddies in the subcloud layer. In fact, boundary layer cumuli are frequently used as a means of detecting subcloud eddies (Kuettner, 1959; Plank, 1966; Walter, 1980).

In recent years there has been a resurgence in interest in the study of factors affecting cloud organization. This largely results from the availability of satellites, which provide a perspective of cloud systems that has previously been unavailable. Furthermore, in recent years high-altitude aircraft have become available to the meteorological community. These observational platforms have demonstrated that there often exists an analogous behavior between the organization of atmospheric cloud systems and convective

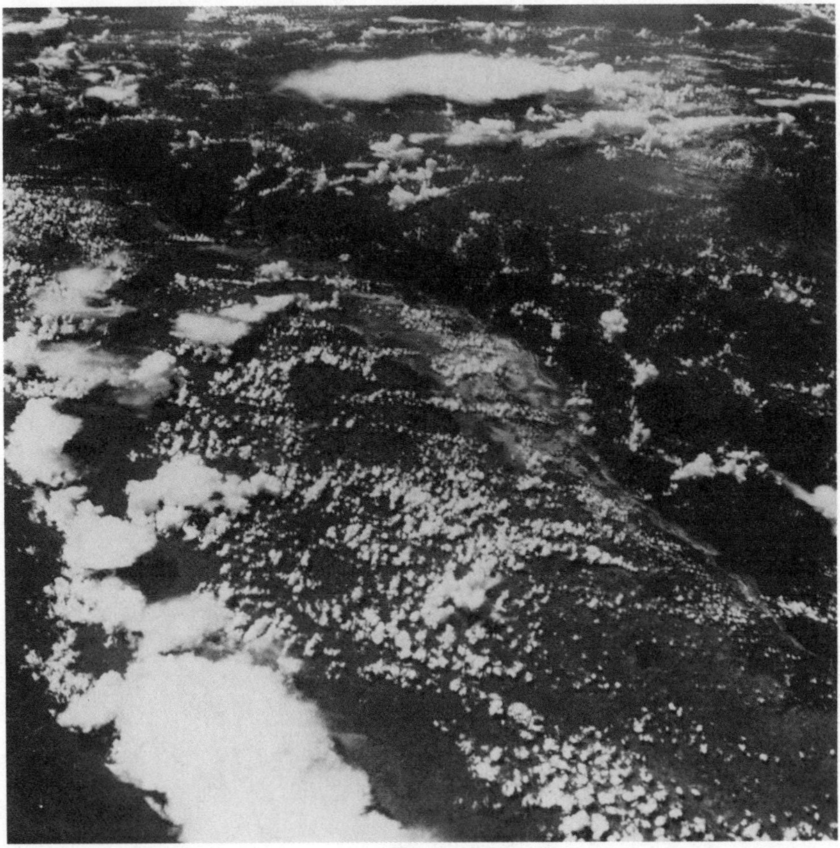

Fig. 8.17. Hexagonal "open" cells north of Cuba, photographed from Gemini 5 at 1827 GMT, 23 August 1965. Convective enhancement at the vertices of the hexagons can also be noted. [From Agee and Asai (1982).]

circulations observed in shallow laboratory fluid experiments. This observation inspired more laboratory studies of convection (e.g., Krishnamurti, 1968a, b; 1975b; Faller, 1965) as well as a proliferation of theoretical analyses of laboratory analogs to atmospheric flows (Kuettner, 1971; Krishnamurti, 1975a; Agee and Chen, 1973; Asai, 1970, 1972; Sun, 1978; Shirer, 1980). The latter studies can be traced to the classical convection theory developed by Rayleigh (1916).

Three different forms of convection that commonly appear in laboratory experiments as well as in the atmosphere have been identified. First, there is a form of cellular convection that occurs in laboratory fluid experiments that are initially at rest. One form of convection exhibits ascending motion

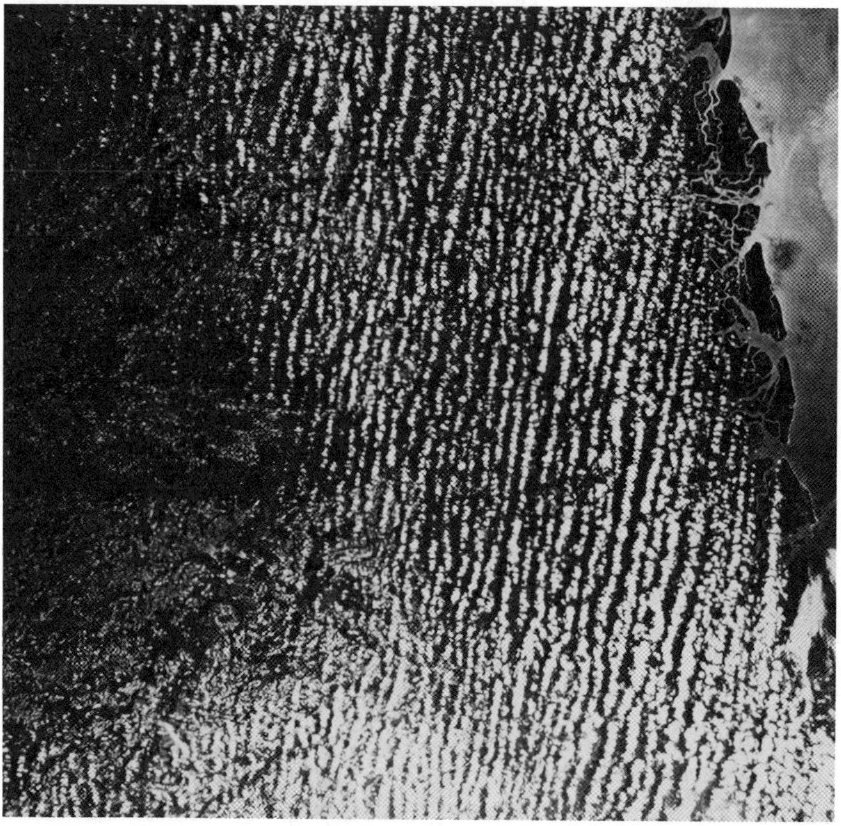

Fig. 8.18. Cloud streets over Georgia developing near the coastline in a southerly flow on 4 April 1968, as seen from Apollo 6. Maximum length of bands: over 100 km; spacing 2 to 2.5 km. (Photo from NASA.) [From Kuettner (1971).]

at the center of a cell and descending motion at the edge of the cell and is commonly referred to as *closed cellular* convection. Figure 8.16 provides an example of closed cellular convection in the upper right panel. *Open cellular* convection, in contrast, refers to cells that have downward motion at their centers and thin regions of ascent at their boundaries. Figure 8.17 and the lower right panel of Fig. 8.16 provide examples of open cellular convection in the atmosphere. The third form of convection is represented by cloud bands that are typically oriented parallel to the wind shear vector in the boundary layer or the average boundary layer wind. Wind shear is most important to cloud organization but the average boundary layer wind is usually in the same direction. Such clouds, often called *cloud streets*, are illustrated in Fig. 8.18. Because cloud streets are aligned parallel to the boundary layer wind shear, they are often called *longitudinal roll clouds*. Clouds are also observed to align in parallel bands that are perpendicular to the wind shear vector. Such clouds bands are called *transverse cloud bands*. Malkus and Riehl (1964) presented an example of a case in which both transverse and longitudinal cloud bands coexisted in the same environment. Figure 8.19 illustrates this case: to the left of a tropical wave trough only longitudinal roll clouds are present, whereas to the right of the trough, both longitudinal rolls aligned parallel to the low-level wind and transverse cloud bands aligned perpendicular to the low-level wind are evident.

Before we can examine the results obtained from laboratory experiments and analytic models, we must first become acquainted with a few nondimensional parameters common in these studies. Consider a fluid of finite

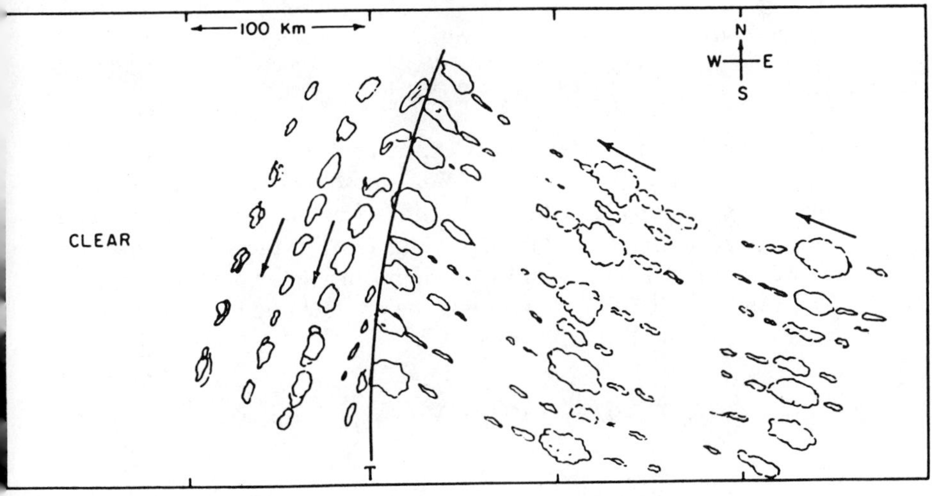

Fig. 8.19. Sketch of cloud observations near wave trough. [From Malkus and Riehl (1964).]

depth h which has a characteristic velocity U^* of the basic flow and an imposed temperature difference T^* between the top and bottom boundaries. We may then define a Reynolds number R_e which represents the ratio of inertial to viscous forces as

$$R_e = hU^*/\nu, \tag{8.12}$$

where ν represents the kinematic viscosity of the fluid.

A second parameter is the Rayleigh number (R_a) defined as

$$R_a = g\alpha h^3 T^*/K\gamma, \tag{8.13}$$

where α represents the coefficient of thermal expansion, T^* represents the temperature difference between the upper and lower boundaries of the fluid, and K represents the thermometric conductivity of the fluid. The Rayleigh number represents the ratio between the product of buoyancy forces and heat advection and the product of viscous forces and heat conduction in the fluid.

The third parameter is the Richardson number R_i defined as

$$R_i = g\alpha h T^*/U^{*2}. \tag{8.14}$$

It represents the ratio of work done against gravitational stability to energy transferred from mean to turbulent motion.

The fourth parameter is the Prandtl number (P_r) defined simply as

$$P_r = \nu/K. \tag{8.15}$$

The relationship among the four variables is

$$R_a = P_r R_i R_e^2 \tag{8.16}$$

Theoretical analysis of the laboratory experiments or their extension to the atmosphere normally begins with the Boussinesq form of the equations of motion and continuity equation (i.e., linearized in the thermodynamic variables) given in Chapter 2 [Eqs. (2.125)–(2.128)]. Typically, the governing equations are further linearized by decomposing the wind field into $U_i = U_{io} + U_i'$ and assuming $U_{30} = 0$. Ignoring the product of perturbations in the flow and rearranging Eq. (2.112) yields

$$\frac{\partial}{\partial t}(\rho_0 U') + U_0 \frac{\partial}{\partial x}(\rho_0 U') + V_0 \frac{\partial}{\partial y}(\rho_0 U') + \rho_0 w' \frac{\partial U_0}{\partial z}$$

$$= -\frac{\partial p'}{\partial x} + \gamma_0 \nabla^2 (\rho_0 U'), \tag{8.17}$$

$$\frac{\partial}{\partial t}(\rho_0 V') + U_0 \frac{\partial}{\partial x}(\rho_0 V') + V_0 \frac{\partial}{\partial y}(\rho_0 V') + \rho_0 w' \frac{\partial V_0}{\partial z}$$

$$= -\frac{\partial p'}{\partial y} + \gamma_0 \nabla^2(\rho_0 V'), \tag{8.18}$$

$$\frac{\partial}{\partial t}(\rho_0 W') + U_0 \frac{\partial}{\partial x}(\rho_0 W') + V_0 \frac{\partial}{\partial y}(\rho_0 W')$$

$$= -\frac{\partial p'}{\partial z} + g\gamma\rho_0\theta' + \gamma_0 \nabla^2(\rho_0 W'). \tag{8.19}$$

The system of equations, Eqs. (8.17)-(8.19), is fully linearized because the base state wind parameters U_0 and V_0 do not vary in time. Also, the equations are valid only for a nonrotating earth because the Coriolis force is ignored. When combined with a simplified thermodynamic energy equation (a linearized equation for θ') and the continuity equation, Eq. (2.114), we have a set of equations that can be transformed into non-dimensional variables and subjected to the linear analysis procedures outlined in Chapter 2. This transformation allows determination of the wavelengths of the most unstable waves and their phase speed of propagation. In addition to the simplification inherent in the linearization of the equations, this system is also limited to nonturbulent, viscous flow. Therefore, the system is quite suitable for the study of laboratory nonturbulent convection experiments such as Rayleigh's (1916) classical work. "The extension" to the atmosphere, however, requires the rather dubious assumption that the molecular viscosity γ_0 can be replaced by the much larger eddy viscosity K_m. Likewise, the molecular thermometric conductivity must be replaced by the eddy diffusivity for heat K_H. In most of the analytical studies mentioned below, the magnitudes of the eddy viscosities are assumed to be constant throughout the mixed layer or cloud layer. In some cases (e.g., Sun, 1978), the vertical and horizontal eddy diffusivities are assumed to be constant but of different magnitudes. In one study, Agee and Chen (1973) specified vertical profiles of eddy viscosity that were monotonically either increasing or decreasing. We must remember that the extension of the laboratory convection experiments (or their analytical model counterparts) to the atmosphere requires the assumption of an eddy viscosity whose behavior is independent of the properties of the flow field or, in other words, whose behavior is analogous to molecular viscosity. In Chapter 3 we noted that such an approximation is generally a poor one for simulating the fluxes and other properties of turbulent flow. However, as will be shown below, the qualitative results of the laboratory experiments and analytic models are quite realistic.

8.3.1 Cellular Convection

Using linear theory, Rayleigh determined the conditions required for the onset of convection. He determined that whenever R_a exceeded a critical value R_{ac},

$$R_a > 27\pi^4/4 = R_{ac}, \tag{8.20}$$

the layer would commence convective overturning. In addition, in the case where both boundaries were free, the unstable convective disturbance was found to have a characteristic scale or wavenumber K_c given by

$$K_c = \pi/\sqrt{2}. \tag{8.21}$$

Rayleigh's model, however, could not determine whether the convective cells were of an open or closed structure.

Using a linear model for an environment having a stagnant basic stage, Agee and Chen (1973) examined the influence of the assigned vertical profile of eddy viscosity on the structure of convective cells. They concluded that an unstable layer of cellular convection will have open (closed) cells if the layer is characterized by an eddy viscosity profile that is monotonically decreasing (increasing) with height. In an unstable boundary layer, the expected profile of eddy viscosity would have a maximum value above the surface layer and would decrease monotonically through the remainder of the layer. This corresponds most closely to the case of open cells predicted by Agee and Chen. A profile in which eddy viscosity increases with height through the real boundary layer is unlikely. Because both open and closed cells are observed over a relatively homogeneous region of the ocean, it is unlikely that the vertical profile of eddy viscosity is the major determination of atmospheric cellular structure.

Krishnamurti (1968a, b) proposed that changes in the mean temperature of a layer would determine the particular organization of convective cells. Subsequently, Krishnamurti (1975c) argued that for this effect to be the primary determining mechanism, the required degree of warming or cooling would be so large that cloud fields could not persist in a particular cellular configuration for several days as observed. Instead, she constructed a nonlinear steady-state model (Krishnamurti, 1975a) and tested it against a laboratory model (Krishnamurti, 1975b), in which the principal factor determining the mode of cellular convection was the direction of an imposed uniform vertical mass flux. If the environment exhibited large-scale sinking motion, Krishnamurti predicted that open hexagonal cells would be the predominant form of convection. Closed hexagonal cells should appear in regions of large-scale rising motion. For the case of no large-scale vertical motion, the predicted model is of the form of longitudinal rolls. All these solutions are valid for Rayleigh numbers near the critical value. Krishnamurti (1975c) then applied her theory to a cloud layer observed over the open Pacific ocean. Using an objectively analyzed vertical motion field, the model predicted open cells where open cells were observed by satellite imagery and closed cells where closed cells were observed. Thus, the theory satisfied this one crucial test.

The theory also predicted the wavelength of the cellular pattern at $R_a = R_{ac}$ to increase in proportion to γ^2, where

$$\gamma = W_0 h / K_H \tag{8.22}$$

and W_0 is the large-scale vertical velocity, h is the depth of the convective layer, and K_H is the eddy diffusivity for heat. While the predicted scales of cellular convective overlapped with the observed scales (10–30 km), the variability was too large to test adequately the theory. Note that in the actual atmosphere the value of R_a is 10–14 orders of magnitude larger than R_{ac}.

Agee (1987) reviewed recent studies of mesoscale cellular convection over the oceans. He presented a case of a cold-air outbreak off the east coast of the United States in which the organization of convection varied from cloud streets to open and closed cells. Figure 8.20 is a schematic illustration of the variation of the organization of convection with changes in the height of the capping inversion. Near the shoreline where the depth of the ABL is shallow, the preferred organization is longitudinal rolls. Due to strong surface heating from the warm sea surface, the ABL deepens and closed and open cells develop. In general, open cells are preferred where the inversion is lowered, presumably by descending motion, and closed cells are preferred where the inversion is higher, perhaps due to ascending motion consistent with Krishnamurti's theory. Agee notes, however, that in

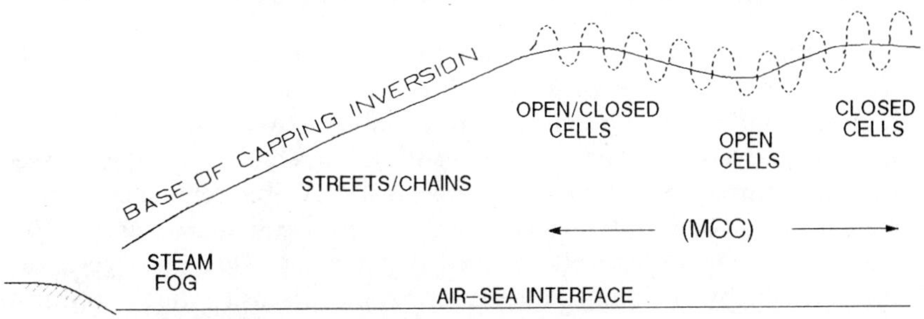

Fig. 8.20. Schematic of a vertical cross section through a type I cloud-topped boundary layer (CTBL) (from shoreline to open sea) and associated convective phenomena. Variation of convective depth between open and closed cell regions is noted, as well as for individual cell circulations (dashed line). [From Agee (1987).]

other cases both open and closed cells can be found in regions of large-scale sinking motion.

While there exists a qualitative similarity between laboratory convection experiments and atmospheric open and closed cellular convection, some substantial differences also exist. If we define λ as the diameter of a convective cell and h the depth of the convective layer, typical aspect ratios (λ/h) in laboratory experiments range from 2:1 to 4:1. In the atmosphere, however, the convective cells are extremely flat with aspect ratios of the order of 30:1. Fiedler (1985) noted that while individual cumulus cells display aspect ratios of 3:1, the atmospheric open and closed cellular convection is composed of many cumulus-scale elements and thus is distinct from both the cumulus-scale and laboratory convection experiments. Fiedler (1984) proposed a mesoscale instability mechanism related to entrainment into a cloud-topped mixed layer. He describes this mechanism, illustrated in Fig. 8.21, as follows:

> In general, thicker cloud cover at the top of a mixed layer tends to increase the rate of entrainment. Therefore entrainment is usually a process which tends to dissipate or stabilize fluctuations in cloud cover when the air above the mixed layer is drier than the mixed layer. However, if the temperature gradient in the stable layer is sufficiently great this tendency can be reversed. Regions with relatively more buoyant mixed-layer air will rise farther into the overlying stable air as compared with the less buoyant mixed-layer air and the density decrease across the inversion can intensify above the more buoyant regions. As a result the mass entrainment rate can be decreased above positive buoyancy fluctuations, and fluctuations in the mixing ratio can thereby be generated in phase with mesoscale buoyancy fluctuations. Mesoscale buoyancy fluctuations are generated by two mechanisms. First, the latent heat release associated with increased cloud thickness provides buoyancy. Second, although the mass entrainment rate decreases above the more buoyant regions, the production of buoyancy by entrainment can increase in those regions because the air entrained there is warmer than the air entrained into the less buoyant regions.

Fiedler then modeled the process with a Lilly-type mixed-layer model incorporated into a linearized two-dimensional flow model. The theory predicted that mesoscale entrainment instability (MEI) occurs at aspect ratios of about 10:1 and peaks at about 30:1. Van Delden (1985) expressed doubt about the applicability of Fiedler's theory to closed cellular convection where the cloud cover is typically only 10%, because Fiedler assumed 100% cloud cover. Van Delden also argued that Fiedler achieved large aspect ratios of MEI by assuming values of horizontal viscosity which appear to be too large for shallow convection. Fiedler (1985) argues, however, that there is a scientific basis for assuming horizontal viscosities of the order of 1000 $m^2 s^{-1}$. This discourse highlights how little is known about horizontal eddy mixing on the mesoscale and how important it is to predictions of

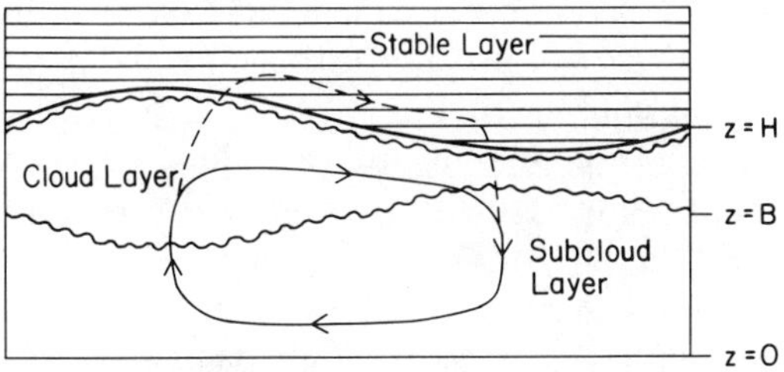

Fig. 8.21. Mesoscale entrainment instability. A positive buoyancy fluctuation rises into the stable layer and becomes adjacent to warmer air. A decreased mass entrainment flux occurs over the buoyancy fluctuation, resulting in the generation of a positive fluctuation in cloud thickness. The positive cloud thickness fluctuations in turn couple to the generation of the positive buoyancy fluctuations by latent heat release and by promoting the entrainment of buoyancy. Note that although less mass is entrained where the cloud is thicker, the air entrained is relatively warmer than that where the cloud is thinner. Therefore, entrainment can reinforce mesoscale buoyancy fluctuations as well as mesoscale humidity fluctuations. The level temperature stratification in the stable layer is shown. [From Fiedler (1984).]

cloud organization (at least with linearized models). It should also be noted that the mesoscale entrainment instability predicted by Fiedler is critically dependent upon the postulated relationship between cloud-top entrainment and environmental stability. Because, as we shall see in subsequent sections, entrainment is fundamentally a dynamic process, one should view the mesoscale entrainment instability with some skepticism.

8.3.2 Roll Convection

As mentioned above, it is commonly observed that boundary layer cumuli organize into longitudinal rolls or cloud streets. Clouds so organized are commonly observed under relatively high wind conditions (Kuettner, 1959, 1971). Krishnamurti's (1975a) nonlinear theory predicts that rolls are the preferred mode of convection when $R_a \simeq R_{ac}$ and the large-scale vertical motion is zero. Using free convection scaling arguments, Deardorff (1972, 1976) concluded that roll-type convection will predominate whenever $-h/L \leq 4.5$ to 25, where L is the Monin–Obukhov length defined as $L = U_*^3 \bar{T}/kg\overline{w''T''}$. For large values of $-h/L$, more or less random convection can be expected. LeMone (1973) observed rolls over the range $3 \leq -h/L \leq 10$. The precise value of $-h/L$ which separates the two regimes is still debatable and may vary depending on terrain and other factors. Because L increases with increasing wind speed, Deardorff's scaling arguments are at least qualitatively consistent with the observations that cloud streets are preferred under high wind conditions. Buoyancy is important to roll formation, however. Mason (1983) concluded from two-dimensional LES experiments, that when $-h/L > 0.3$, rolls become mainly shear driven but are too weak to be of consequence.

Kuettner (1959, 1971) also points out that cloud streets are preferred in an environment exhibiting a relatively unidirectional wind profile that is strongly curved, such as illustrated in Fig. 8.22. Kuettner also concludes that typically the cloud layer is convectively unstable. Other features of cloud streets: the lengths are 20 to 500 km, spacing between bands is 2 to 8 km, the depth of the boundary layer is 0.8 to 2 km, the width-to-height ratio varies from 2 to 4, and the vertical shear of the horizontal wind is 10^{-7} to 10^{-6} cm^{-1} s^{-2}. LeMone (1973) observed width-to-height ratios ranging from 2.2 to 6.5. The orientation of rolls is roughly parallel to the winds at the top of the convective layer. Kelley (1984) observed roll orientations from 2° to 15° to the right of the geostrophic wind at the top of the convective layer. Walter and Overland (1984), however, found fog streets to be oriented 16–18° to the left of the geostrophic wind in a near-neutral boundary layer. LeMone (1973) obtained a similar result.

Cloud streets, roll vortices, or longitudinal roll clouds have been the subject of intensive theoretical and laboratory investigations. Laboratory experiments and theoretical studies by Faller (1963) and Lilly (1966) have predicted the formation of roll vortices in a neutrally stratified, viscous boundary layer having a wind profile that varies both in direction and in speed following an Ekman profile (Blackadar, 1962). The mechanism of instability leading to roll vortex formation that they identified is generally referred to as *parallel instability*. It is dependent upon shear along the roll

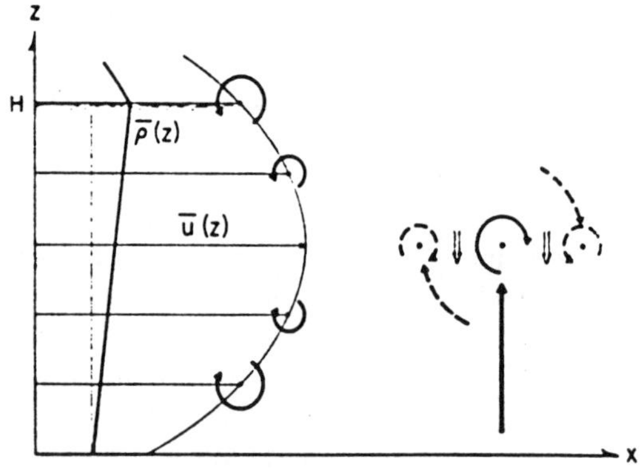

Fig. 8.22. Schematic presentation of basic flow and density profile (left), relative vorticity of vertically displaced convective element, and resulting restoring force (right). [From Kuettner (1971).]

axis, the earth's rotation (or Coriolis effect), and viscosity, but is independent of thermal instability.

A related mechanism for roll formation is commonly called *inflection point instability*. This mechanism of instability occurs in inviscid models in which the wind profile exhibits an inflection point or shear in the cross-roll direction (Brown, 1961; Barcilon, 1965; Faller, 1965). Lilly (1966) argued that in an unstable boundary layer, the inflection point instability mechanism would be suppressed, whereas the parallel instability mechanism would continue to operate. LeMone (1973) concluded that her observations of roll structure were consistent with the predictions of inflection point instability. She noted, however, that the neglect of buoyancy contributions may have contributed to an underestimate of the cross-roll component of the wind variance in those models.

Several investigators have used linear theory to examine roll vortex formation in a thermally unstable boundary with wind shear but no change in wind direction (Kuo, 1963; Kuettner, 1971; Asai, 1970, 1972). These studies show that the most unstable mode is the longitudinal mode in which the bands are oriented parallel to the wind shear and are stationary relative to the mean flow. However, in an unstable boundary layer in which both the speed and direction of the wind vary with height, Asai and Nakasuji (1973) found no preferred orientation of the rolls. Sun (1978) further extended the linear models to include latent heat effects. Following the wave–CISK hypothesis (Lindzen, 1974), latent heat was released in vertical columns in proportion to low-level moisture convergence, and evaporative cooling occurred when the low-level moisture was divergent. Sun found that the orientation and phase speed of propagation varied depending on whether or not buoyancy generated by latent heat release is dominant. If buoyancy dominated, the cloud bands were parallel to the wind shear and remained stationary relative to the mean wind. If the buoyancy force was weak and conversion of kinetic energy from the mean flow dominated, then the most unstable mode was a transverse mode (perpendicular to wind shear), and the cloud bands propagated relative to the mean wind.

Shirer (1980, 1982, 1986) also included latent heat effects in a nonlinear spectral model of cloud bands. In his model, latent heat is released in regions of upward motion above cloud base while air descends dry adiabatically. Because of the model's general formulation, it can be used to investigate all the previously described instability mechanisms believed to be important to cloud-street formation. Like Sun (1978), Shirer (1980) found that both longitudinal and transverse roll clouds could coexist in a given environment. However, the particular character of the wind profile was the major determining influence. If the wind direction was invariant with height, longitudinal rolls were the predominant mode. However, if the basic wind changed direction as well as speed, some wind profiles resulted in longitudinal rolls, some in transverse rolls, and others led to two or three coexisting bands of differing alignments.

The linear models mentioned above elucidated three instability mechanisms leading to cloud-street formation; inflection point instability, parallel instability, and thermal instability. Because the first two occur in a neutral atmosphere with shear, they can be referred to as dynamic instability mechanisms. Using a 3-D truncated spectral model, Shirer (1982, 1986) concluded that the thermal and parallel instability mechanisms are linked such that only one convective mode develops when both instability mechanisms are operating. This result is consistent with Lilly's earlier conclusions. As shown by Kelley (1984) and Shirer and Brummer (1986), Shirer's model yields good predictions of observed roll cloud orientation angles and wavelengths.

The emphasis in most of the theoretical studies as well as observational investigations is on roll structures having width-to-height ratios in the range 2 to 4. Occasionally, width-to-height ratios as large as 20 to 30 are reported (Ogura, 1985). In one case, Walter and Overland (1984) observed a multiplicity of band scales coexisting over a region. Band scales of 1.3–1.7, 5–6, 12–15, and 25–30 km were superimposed over a broad-scale fog bank. They suggested that the bands spaced at 12–15 km could be traced upstream to individual mountain peaks, which may have played a role in their genesis. They noted that upstream topographic influences on cloud street spacing has also been reported by Higuchi (1963), Asai (1966), and Tsuchiya and Fujita (1967) over the Sea of Japan. Walter and Overland also suggested that Fiedler's mesoscale entrainment instability mechanism or a resonant subharmonic to the basic boundary layer instability may also be causing the larger scale bands. It is also possible that the larger scale bands may be a result of symmetric instability (Emanuel, 1979). Symmetric instability theory involves the generalization of the linearized set of momentum equations, Eqs. (8.17)–(8.19), to a rotating, baroclinic atmosphere. If we define the vertical component of absolute vorticity as η and the gradient Richardson number R_i as $R_i = N^2/(\partial U/\partial Z)^2$, where N is the Brunt–Väisälä frequency, then the criterion for symmetric instability is $R_i < f/\eta$. Thus, symmetric instability will occur if the vertical shear is large or the absolute vorticity small. The resultant rolls are aligned parallel with the vertical shear vector.

It should be remembered that in all these laboratory and analytic studies of factors affecting cloud organization, idealized boundary layer models are assumed. In many of the analytic theories, a constant eddy viscosity model is specified that is analogous to molecular viscosity in the laboratory experiments. This assumption affects the quantitative results of the models.

Sun (1978), for example, assumed that the constant vertical eddy diffusivities for heat and momentum differed from the horizontal viscosities; this assumption improved his predictions of the spacing of the cloud bands. We emphasize that we must be cautious in our interpretations of the results of these analytic and laboratory studies when applying them quantitatively to the atmosphere.

8.3.3 The Role of Gravity Waves in Cloud Organization

In the studies of cloud organization examined thus far, the emphasis has been on instabilities in the boundary layer caused by shear and thermodynamic stratification. Recently Clark *et al.* (1986) and Kuettner *et al.* (1987) have argued that gravity waves in the stably stratified atmosphere above the atmospheric boundary layer play a role in cloud organization and spacing. They suggest that the gravity waves are excited by what they call "thermal forcing," in which boundary layer thermals deform the capping inversion or interface region, causing ripples which produce vertically propagating gravity waves. They also suggest that boundary layer eddies and cumulus clouds can act in an analogous way to obstacles in the flow in the presence of mean environmental shear. As we shall see in subsequent sections, as cumulus clouds penetrate into a sheared environment, pressure perturbations develop about the rising cloud towers with positive pressure anomalies on the upshear flanks of the updraft and at cloud top, and negative pressure perturbations on the downshear flanks of the updraft (see Fig. 8.36). The positive pressure perturbations tend to divert flow in the stably stratified environment over and around the updrafts of the convective towers. Thus the diversion of flow about the actively rising cumulus towers in a sheared environment qualitatively resembles flow about a solid obstacle. The resultant flow perturbations can excite gravity waves in the stably stratified environment, which can propagate vertically and horizontally. That boundary layer eddies or cumulus clouds can excite gravity waves is not too surprising. As Clark *et al.* and Kuettner *et al.* pointed out, sailplane pilots have known for some time that so-called thermal waves can frequently be found above the tops of cumulus clouds. What is surprising, however, is that Clark *et al.* showed, with a two-dimensional numerical prediction model, that the gravity waves excited by clouds and boundary layer thermals can feed back on the boundary layer eddies, causing a change in the spacing of cloud lines. They speculate that such tropospheric/boundary layer interactions can result in a weak resonant response that may at times involve the entire depth of the troposphere. This work introduces an entirely new perspective to the problem of cloud organization. It suggests that changes in vertical wind shear and thermodynamic stratification in the overlying stably stratified troposphere may be important to the organization of boundary-layer-confined cumulus clouds. Such changes in environmental shear and stratification will alter the vertical propagation of gravity waves, causing reflection and refraction of gravity wave energy into the boundary layer.

The numerical experiments reported by Clark *et al.* (1986) and Clark and Hauf (1986) suggest that not only is the interaction between cumulus clouds and the gravity wave field important to the spacing of cumulus clouds, but it is also important to the commonly observed phenomenon that new cloud turrent growth occurs preferentially on the upshear flank of a cloud system (e.g., Malkus, 1952). Clark and Hauf showed that in a sheared environment, as the cumulus clouds penetrate the stably stratified environment, gravity waves are excited and exhibit downward motion on the downshear side of the cloud tower and upward motion both on the upshear flank of the cloud and over the top of the turrent. This they attribute to the faster propagation of gravity waves upstream than the thermals, which are emitted from the boundary layer.

A clear implication of the Clark *et al.* and Clark and Hauf studies is that the dynamics of cumulus clouds cannot be modeled properly by confining

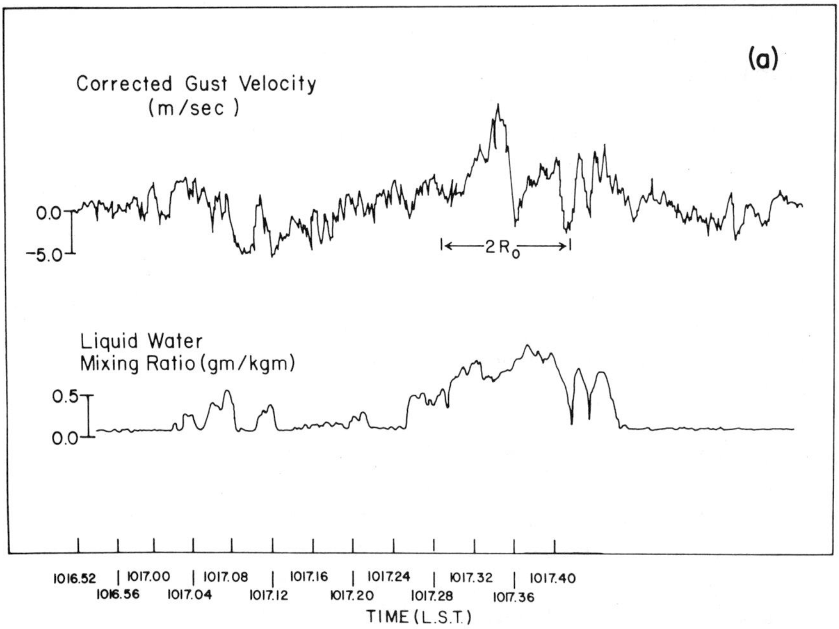

Fig. 8.23. Observed vertical velocity and cloud liquid-water content at 2.77 km (a), 2.34 km (b), 1.9 km (c), and 1.5 km (d). All heights mean sea level (MSL). [From Cotton (1975a).] (*Figure continues.*)

the cloud model to simulating motions only within the visible portion of the cloud. A cumulus cloud is clearly the product of interactions between the gravity wave field and convective thermals, which generally have their origins in the atmospheric boundary layer. We shall see in subsequent chapters that the interaction between gravity waves and convection is also important to the propagation and behavior of individual cumulonimbus clouds and mesoscale convective systems.

8.4 The Observed Structure of Individual Cumuli

In this section we focus mainly on observations of the liquid-water content and velocity structure of nonprecipitating or lightly precipitating cumuli. While the temperature structure or buoyancy structure of cumuli is of great importance, few observations of cloud temperatures are of sufficient accuracy to be more than order-of-magnitude estimates of cloud buoyancy.

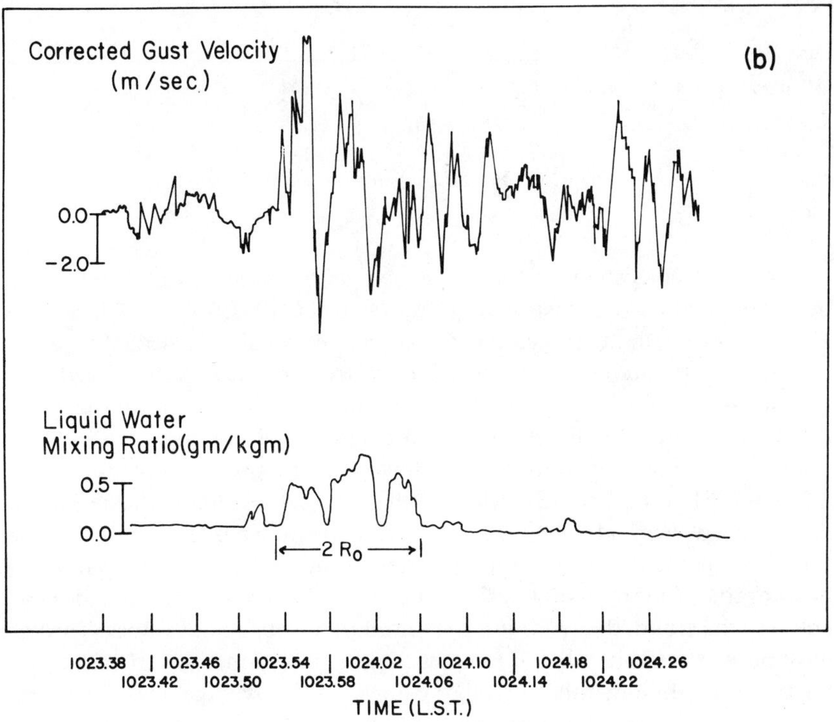

Fig. 8.23 (*continued*)

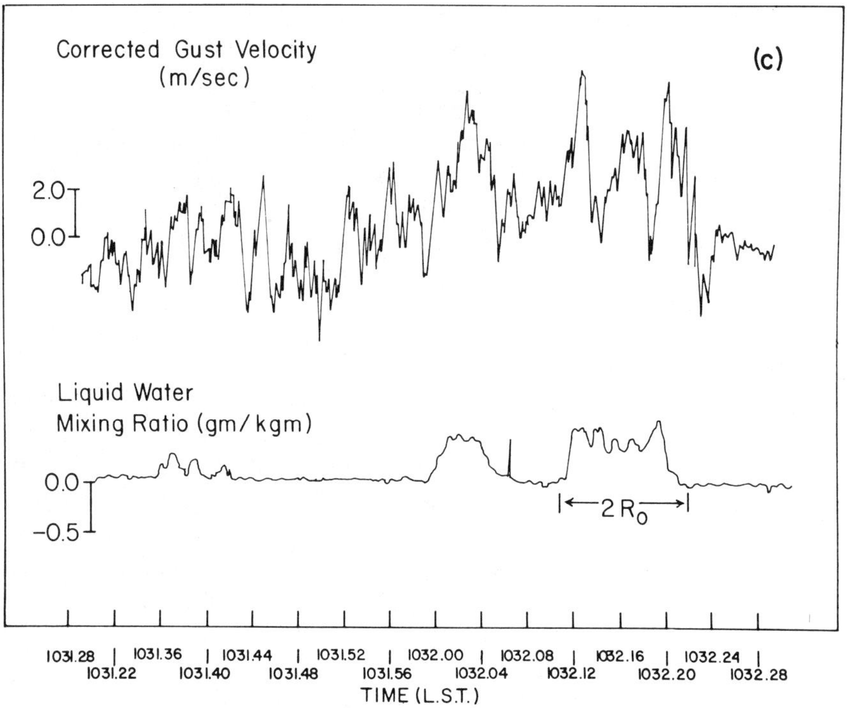

Fig. 8.23 *(continued)*

The most consistent and accurate sets of observations of cumuli have been obtained by the Australian group (Warner, 1955, 1970, 1978; Squires, 1958a, b) using an instrumented aircraft capable of simultaneously measuring cloud vertical velocities and cloud micro- and macrostructure. Figure 8.23 illustrates a set of observations obtained by the cloud physics group of CSIRO under the direction of J. Warner near Bundaberg, Queensland, Australia. The observations were performed in an aircraft descending from 2.77 km MSL to 1.5 km MSL. Most noteworthy in these figures is the extreme variability of vertical velocity as a cloud is transected. In some cases (see Fig. 8.23b), the vertical velocity varies from updrafts about 6 m s^{-1} to downdrafts of more than 4 m s^{-1} in less than 50 m across the cloud. Upon entering the cloud, the LWC jumps from zero to substantial values. Through most of the cloud, the LWC fluctuates slightly in response to vertical velocity fluctuations while maintaining high values, on the average. Only in regions of vigorous downdrafts does the LWC plunge to near-zero values in the cloud interior.

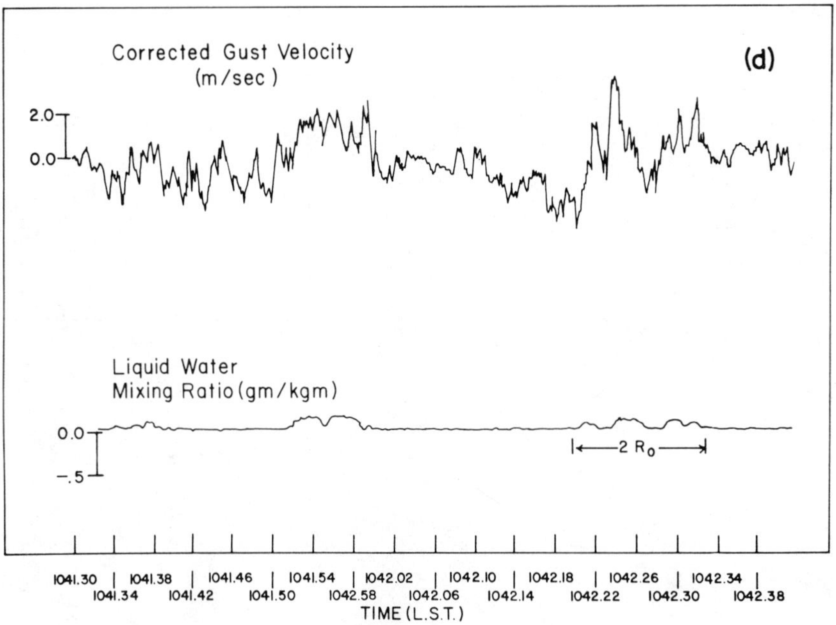

Fig. 8.23 (*continued*)

Warner (1955) examined the vertical variation in LWC from cloud base to near cloud top. Figure 8.24 illustrates that the peak LWC monotonically rises from cloud base to within 300 m of cloud top, where it rapidly falls to zero. If parcels of air entered the base of a cloud and then rose through the depth of the cloud without mixing with their environment, then the LWC would be given by

$$r_{\mathrm{la}} = r_{\mathrm{T}} - r_{\mathrm{s}}(T, P), \tag{8.23}$$

where r_{T} represents the total mixing ratio of the air that would be conserved in the absence of mixing, and $r_{\mathrm{s}}(T, P)$ represents the saturation mixing ratio that varies with pressure P and temperature T. For an unmixed parcel of air, the temperature T would be given by the rate of wet adiabatic cooling as the parcel ascends from cloud base. We define the quantity r_{la} as the adiabatic liquid-water mixing ratio. The ratio $(r_{\mathrm{l}}/r_{\mathrm{la}})$ represents the departure of the LWC from adiabatic values due to the effects of mixing. Only if the cloud-base temperature varies significantly from the estimated value or if precipitation allows water drops to settle to lower levels in a cloud will $r_{\mathrm{l}}/r_{\mathrm{la}}$ exceed unity. Figure 8.25 illustrates observed profiles of $r_{\mathrm{l}}/r_{\mathrm{la}}$ computed by Warner (1955), Squires (1958a), Ackerman (1959), and Skatskii (1965).

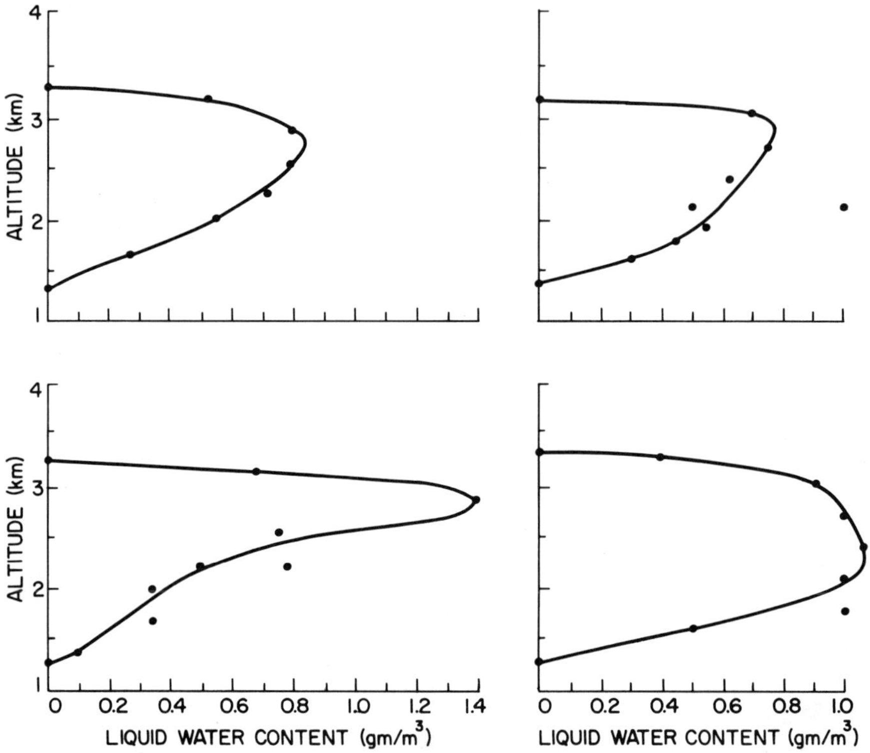

Fig. 8.24. Variation of peak water content with height. [From Warner (1955).]

Warner (1955) computed $\overline{r_{la}}$ as the average of the peaks in LWC shown in Fig. 8.23, whereas Squires (1958a) computed the average of a set of randomly distributed sampling points. Thus, Warner's (1955) averaging procedure results in slightly higher average values of LWC. Why Skatskii's average values of $\overline{r_l}/r_{la}$ are higher than Warner's is not known. The consistent feature of these observations is that the liquid-water content in cumuli departs substantially from adiabatic values within 500 m of cloud base. Furthermore, the average value of $\overline{r_l}/r_{la}$ rapidly approaches asymptotic values on the order 0.2. Squires (1958a) noted that the adiabatic LWC was attained in only 3% of his samples. The observations reported by Squires and Warner, however, were randomly oriented with respect to the wind shear vector. As we shall see, the asymmetry in cloud structure caused by wind shear can influence the likelihood of sampling near-adiabatic values of LWC.

As Fig. 8.23 shows, the most noteworthy feature of the structure of the vertical cross sections is its extreme variability. Warner (1970, 1977)

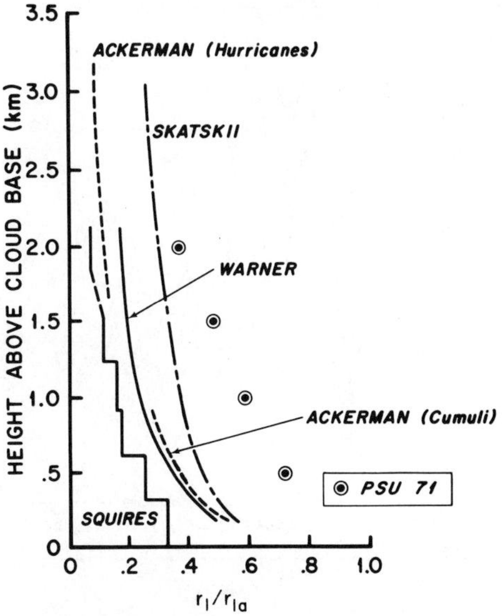

Fig. 8.25. The ratio of the mean liquid-water content at a given height above cloud base to the adiabatic value. The circular symbol represents calculated values of r_l/r_{la} using a lateral entrainment model. [From Cotton (1975a).]

examined the velocity structure of cumuli to determine updraft features that vary most consistently with height or with the lifetime of the cloud. First of all, Warner (1970) found that the vertical velocity averaged over the width of the cloud showed no consistent variations with height. The root-mean-square vertical velocity ω $(\omega = \overline{w''^2})^{1/2}$ averaged over the width of the cloud, however, exhibited a consistent, monotonic increase with height. He found that the root-mean-square velocity could be described by the regression equation

$$\omega = a_1 + b_1 z, \tag{8.24}$$

where z represents the height above cloud base in kilometers. The coefficient $a_1 = 1.26$ for observations over Coff's Harbour, Australia, and $a_1 = 1.13$ for observations over Bundaberg, Australia. Likewise b_1 varied from 0.76 over Coff's Harbour to 0.69 over Bundaberg.

The peak updrafts and downdrafts also showed consistent variations with height, both increasing with height above cloud base. The maximum updraft Warner encountered was 12.7 m s^{-1} at 2 km above the base of a 1.7-km-deep cloud. Warner (1970) noted that neither peak, average, nor root-mean-square velocity varied significantly with cloud width.

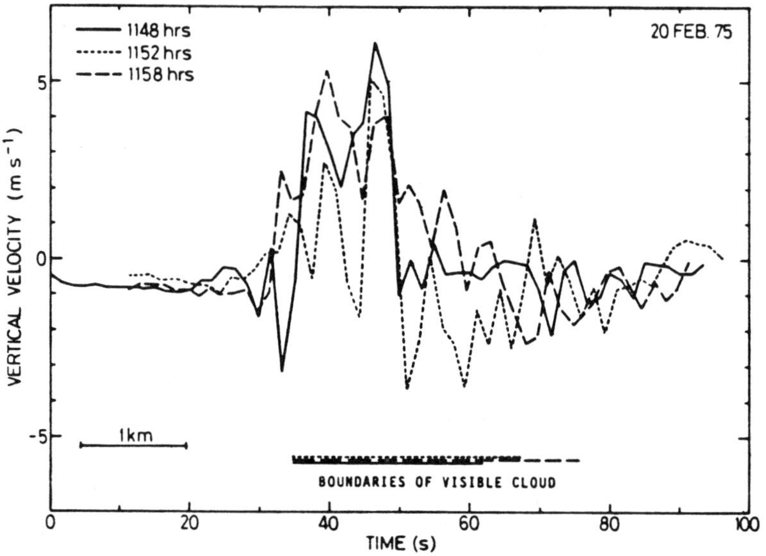

Fig. 8.26. Filtered velocity data from successive penetrations of a cumulus cloud showing the persistence of updraft structure near the upshear cloud boundary (left-hand side of diagram). [From Warner (1977).]

Examining the time variation of these properties, Warner (1977) found that the root-mean-square vertical velocity at a given level varied little throughout the lifetime of a cloud even for sampling periods as long as 20 min. Warner (1977) did find that a coherent, persistent updraft structure often occupies a small fraction of the cross-sectional area of the visible cloud. Figure 8.26 shows that a persistent updraft exists on the upshear side of a cloud for successive penetrations. Also, while fluctuations in vertical velocity on the upshear side of the cloud remain relatively small, they are more pronounced on the downshear side and extend a considerable distance from the cloud.

The appearance of an asymmetric structure of cumuli growing in an environment with wind shear was first noted by Malkus (1949) and was elaborated upon more fully by Heymsfield *et al.* (1978). Heymsfield *et al.* used an instrumented sailplane that allowed the simultaneous observation of cloud microstructure, temperature, and rise rate of a cloud (while spiraling upward in cloud updrafts). They used θ_e and r_l/r_{la} as two indicators of the degree of mixing in the cloud updrafts. Figure 8.27 illustrates an inferred vertical cross section of θ_e in a towering cumuli observed over northeastern Colorado. A core of constant θ_e can be seen nearly throughout the vertical

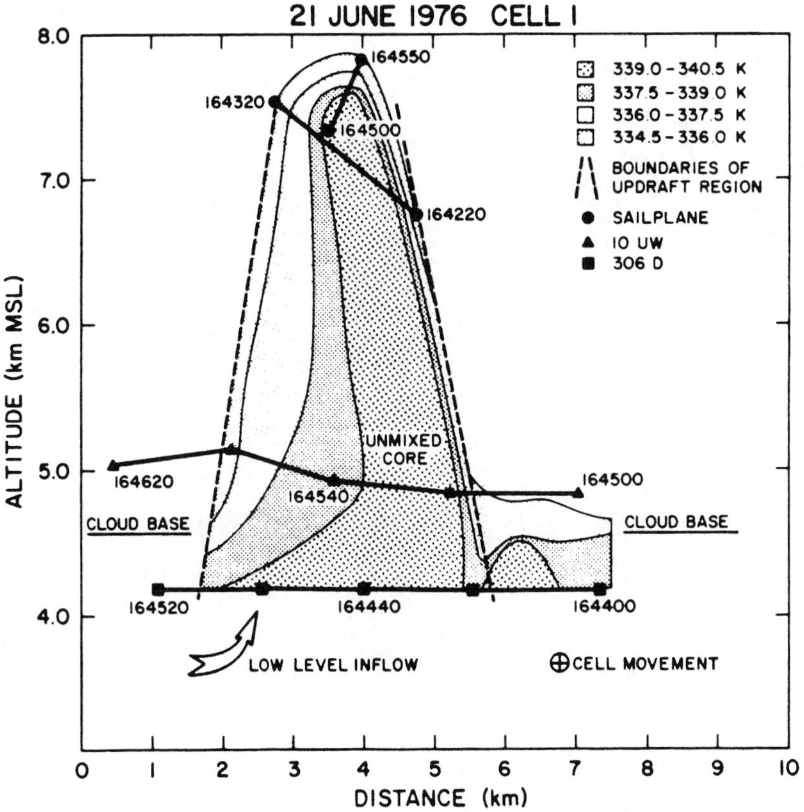

Fig. 8.27. Vertical section of θ_e data based on measurements from three aircraft plotted nearly perpendicular to the cell motion. Low-level inflow region is positioned on left side of cell. Times of aircraft penetrations are indicated. [From Heymsfield *et al.* (1978).]

extent of the cloud on its upshear flank. Over a much greater area of the cloud than the "unmixed" core, values of θ_e are smaller than those found at cloud base. Figure 8.28 illustrates a horizontal cross section of θ_e and r_l/r_{la} observed by the sailplane. A consistent pattern of an unmixed core on the upshear side of the cloud and vigorous mixing downshear can be seen in this figure. In another study, Ramond (1978) inferred from aircraft observations the pressure field about growing cumuli. He diagnosed regions of positive pressure anomalies upwind of actively growing cumuli. He suggested that the positive pressure anomaly forms a protective zone against entrainment, which provides a mechanism for forming upshear undiluted cores.

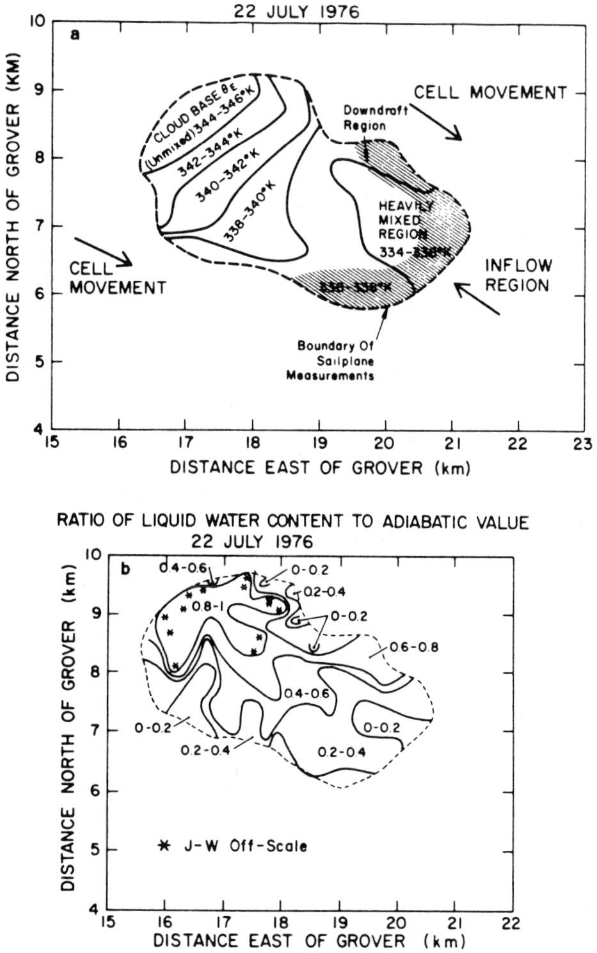

Fig. 8.28. Horizontal projection of data obtained from the sailplane on 22 July 1976 over an altitude range of 3.1 km. Data have been plotted relative to the cell motion. (a) τ_e data; (b) LWC/LWC_A data. [From Heymsfield *et al.* (1978).]

Using a tethered balloon system with multilevel turbulence probes, Kitchen and Caughey (1981) inferred a P-shaped circulation in shallow cumuli. Figures 8.29, 8.30, and 8.31 illustrate the inferred circulation field for three different cumuli. The feature of an updraft overturning at cloud top in the direction of the mean wind shear forming a reversed letter P pattern is consistent with the circulations inferred by Heymsfield *et al.* (1978) for larger towering cumuli. A feature of the circulation depicted in Figs.

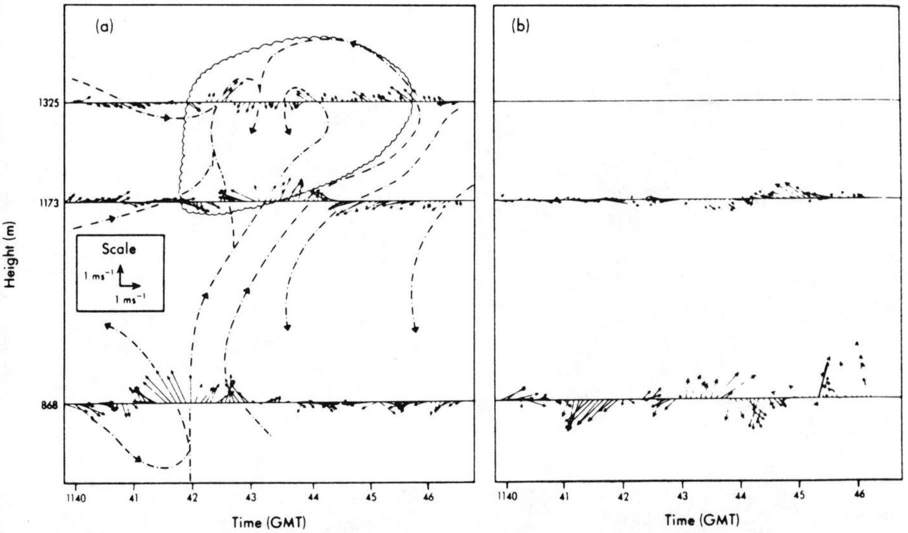

Fig. 8.29. Gust vectors (5 s) constructed from the u, v, and w wind components (see text) for cloud 1 in the vertical (a) and horizontal (b) planes. A suggested flow pattern (dot-dash lines with directional arrows) has an inclined updraft overturning at cloud top in the direction of the mean wind shear; this circulation assumes the shape of a reversed letter P. Schematic cloud boundaries (wavy lines) and possible horizontal rotations (large curved arrows) are marked. [From Kitchen and Caughey (1981).]

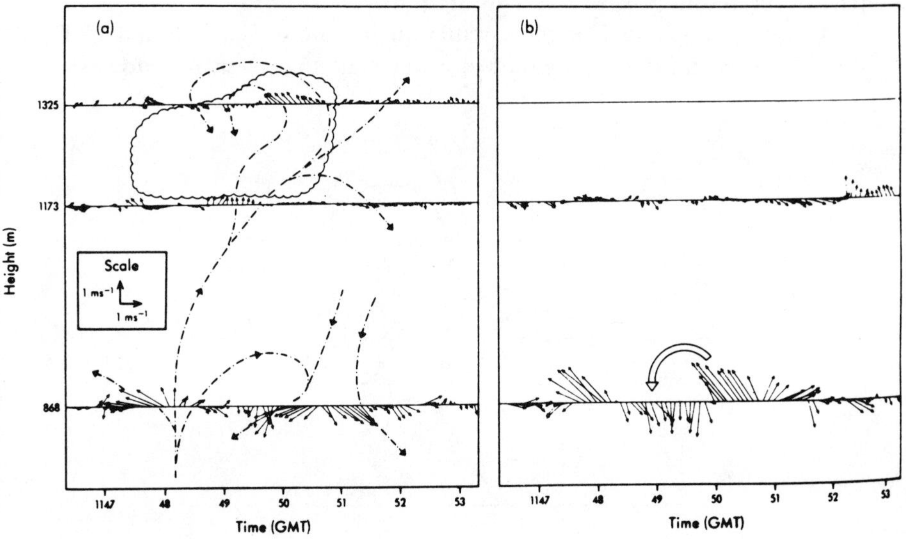

Fig. 8.30. As in Fig. 8.29, but for cloud 2, a weak dissolving cloud. [From Kitchen and Caughey (1981).]

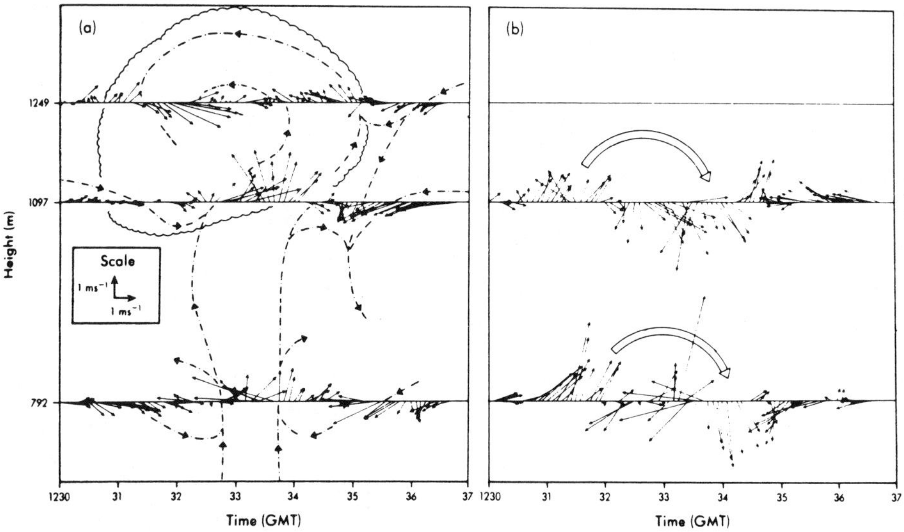

Fig. 8.31. As in Fig. 8.29, but for cloud 6, a larger cumulus cloud in the mature stage of development. [From Kitchen and Caughey (1981).]

8.29, 8.30, and 8.31 that was not inferred by Heymsfield *et al.* is the strong downdraft at the upwind boundary of the cloud. This, they suggested, is part of the return flow of the main updraft.

A conceptual model of the circulation in cumuli is illustrated in Fig. 8.32. Air rises through the cloud in an updraft that overturns downshear

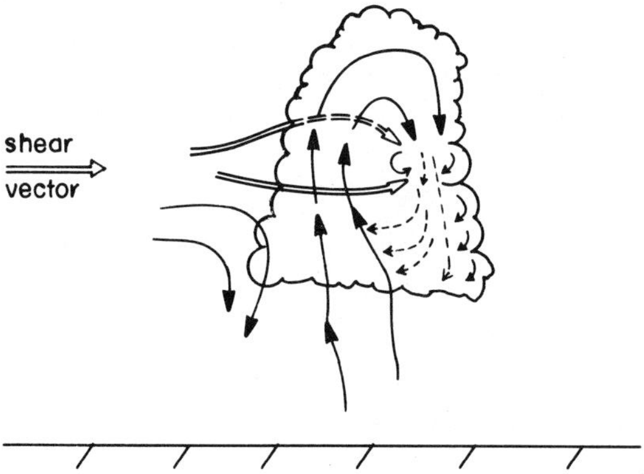

Fig. 8.32. Illustration of circulations in shallow cumuli. Dashed lines depict parcel motions following mixing in the downshear flank of the cloud.

near cloud top, and thus forms a downdraft. Also illustrated are downdrafts along the upshear flank of the cloud. Superimposed on the P-shaped circulation is a horizontal flow that diverges about the main updraft and converges in the descending branch of the P-shaped circulation.

8.5 Entrainment and Downdraft Initiation in Cumuli

As seen in the preceding section, there is considerable evidence that vigorous mixing takes place in cumuli leading to average LWCs well below adiabatic values. The mixing also leads to a vertical velocity structure noted for its variability rather than to a distinct organized structure on the scale of the cloud. Early concepts of entrainment of dry environmental air into clouds viewed entrainment as occurring primarily through the cloud sides. That is, clouds were thought to be composed of a principal buoyant updraft having a "jetlike" or "bubblelike" structure. Entrainment was thus seen to be generated by the shear between the principle updraft and a stagnant environment (Ludlam and Scorer, 1953; Malkus and Scorer, 1955; Woodward, 1959; Levine, 1959; Malkus, 1960; Schmidt, 1949; Stommel, 1947; Squires and Turner, 1962). The laterally entrained air was then thought to mix homogeneously across the width of the updraft. As a result, the thermodynamic properties of updraft air at any level in the cloud should be a mixture of the properties of air entering cloud base and air entrained into the updraft at all levels below that level. Also, one would expect that a distinct gradient in cloud properties should exist between the central interior of the cloud and near the cloud edge. Thus, vertical velocity should be greatest near the middle of the cloud, and the LWC should be highest in the middle. In Fig. 8.33, a and b illustrate schematic models of the "bubble" and "jet" concepts of lateral entrainment.

A corollary of the lateral entrainment theories is that the fractional rate of entrainment of environmental air into the updraft should vary inversely with the cloud radius (Malkus, 1960). Malkus (1960) formulated the rate of entrainment (μ_c) for the bubble model as

$$\mu_c = (1/M_c)(dM_c/dz) = b/R, \tag{8.25}$$

where M_c is the cloud mass, b is a dimensionless coefficient, and R is the radius of a cloud. Based on laboratory experiments (Turner, 1962, 1963), the rate of entrainment for a thermal is $b = 3\alpha = 0.6$, where α is the half-broadening angle of the rising bubble illustrated in Fig. 8.33a. In the case of a steady jet, Squires and Turner (1962) formulated the lateral entrainment rate (μ) as

$$\mu = (1/F_m)(dF_m/dz) = b/R, \tag{8.26}$$

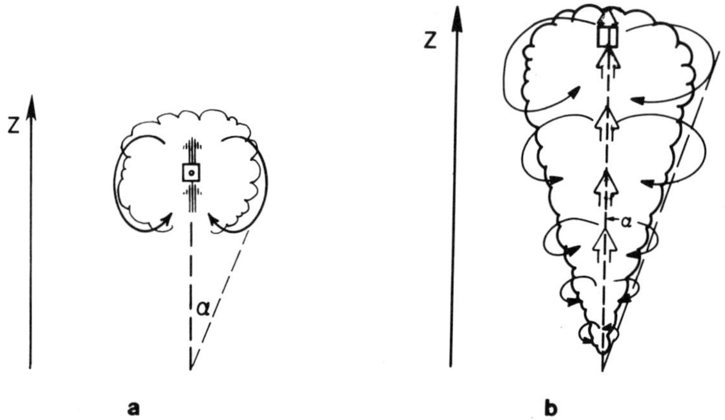

Fig. 8.33. (a) Schematic view of the "bubble" or "thermal" model of lateral entrainment in cumuli. (b) Schematic view of the "steady-state jet" model of lateral entrainment in cumuli.

where $b = 2\alpha = 0.2$, and again α is the half-broadening angle shown in Fig. 8.33b.

Aside from the differences in entrainment coefficients, these two models share much in common. Both models contain the same form of vertical rise rate equation (Morton *et al.*, 1956; Malkus and Williams, 1963), both view the entrainment process in cumuli to be primarily lateral, and both sets of equations can be integrated in a Lagrangian marching-type of solution technique. That is, one can vertically integrate the equations governing the rise rate and buoyancy by following the center of mass of a bubble or a characteristic parcel moving through a steady-state jetlike cloud (see a and b in Fig. 8.33). The lateral entrainment concepts have been generalized to models of a sequence of parcels released from the ground or the atmosphere (Danielsen *et al.*, 1972; Lopez, 1973) and to one-dimensional, time-dependent numerical models where the governing equations are cast in Eulerian form (Weinstein, 1970; Wisner *et al.*, 1972; Asai and Kasahara, 1967).

Lateral entrainment models have been used extensively in weather modification research, in diagnostic studies of cumuli, in cloud transports of pollutants and cloud chemistry, and in the formulation of cumulus parameterization schemes for use in larger scale models.

Perhaps the first challenge to the lateral entrainment concept came from Warner (1955), who observed that the ratio r_l/r_{la} at a given height above cloud base varied little with the horizontal dimension of a cloud. As illustrated in Fig. 8.23, Warner also noted that the LWC varies sharply across the cloud boundary, with the edges of the cloud being nearly as wet as the center. Subsequently, Squires (1958a) hypothesized that the observed structure of cumuli can be better explained by the entrainment of dry air into the tops of cumuli. The mechanism as envisaged is similar to the theory for entrainment instability in stratocumulus clouds described in Chapter 7. That is, tongues or plumes of dry environmental air engulfed into the tops of cumuli will cause the evaporation of neighboring cloudy air, thereby chilling the air and resulting in penetrating downdrafts. The downdrafts, bringing air having above-cloud-top-level environmental properties, will penetrate well into the interior of the cloud. A property of this theory is that the vigor of entrainment will increase in proportion to the liquid water produced by the cloud and by the dryness of the cloud environment.

A further deficiency of the lateral entrainment theory was pointed out by Warner (1970) and verified by Cotton (1975b). Warner showed that if the lateral entrainment rate was adjusted to enable the steady-state model to predict the observed cloud-top height, the LWC predicted by the model (or the profile of r_l/r_{la}) exceeded observed average peak values at all levels in the cloud. Figure 8.25 illustrates this result; the figure shows Cotton's (1975a) calculation of r_l/r_{la} exceeding observed values at all heights. In addition to the predicted magnitudes being too large, the predicted profile slope does not have the marked drop-off within the first few hundred meters above cloud base. Cotton (1975a) also showed that a time-dependent, one-dimensional model in which entrainment occurs principally laterally shares the same deficiency with predicted profiles of r_c/r_a, exceeding observed values.

Some support for the inverse radius dependence of lateral mixing was found in the observational study reported by McCarthy (1974). He found that the observed maximum temperature and liquid-water content in each cloud traverse was consistent with the inverse radius dependence of the fractional entrainment rate. However, the strength of the entrainment rate and inverse radius dependence increased as the criteria for cloud selection eliminated all but a few isolated cumuli growing in an undisturbed environment.

Paluch (1979) used the total mixing ratio r_T and wet equivalent potential temperature θ_q as tracers of cloud motions. The wet equivalent potential temperature is similar to θ_e except it is conservative for reversible adiabatic ascent and descent rather than pseudoadiabatic motions. The selected parameters have the property of mixing in a nearly linear way, which simplifies interpretation of the data. Paluch's analysis was applied to data collected by an instrumented sailplane ascending in developing cumuli in northeast Colorado. She concluded that air at a given observation level in an updraft could not have originated as a mixture of environmental air at that level or below and air originated near the earth's surface. Instead, most of the air entrained in updrafts originated several kilometers above the observation level near cloud-top height. Paluch's results have been supported by similar thermodynamic analyses reported by Boatman and Auer (1983), Jensen *et al.* (1985) and Blyth and Latham (1985), all of which showed that mixing occurs predominantly between two levels. Gardiner and Rogers (1987) argued that the clouds selected for analysis in those studies all were of limited vertical extent, and were capped by very stable, dry air which the clouds could not penetrate. Thus, the cloud-top remained at a given level for an extended period of the cloud lifetime. This enhances the likelihood that entrainment will take place from a single level.

Recently, Betts (1982a) applied the saturation-point analysis scheme to the analysis of mixing or entrainment processes in cumuli. The advantage of this scheme is that the conservative thermodynamic variables for moist and dry ascent and descent are fully specified by the single parameter, saturation level (SL).

To illustrate the application of saturation-point analysis to the diagnosis of mixing processes, consider Fig. 8.34. Air entering the cloud from the subcloud layer has a saturation point (SP) at C. If that air should now mix with air originating above cloud top having a SP at E, the resultant mixed parcel will have a SP along the mixing line CD'D″. As the cloud parcel ascends from cloud base p_B to any pressure level p, the amount of mixing determines how far the SP moves upward along the line CD'D″.

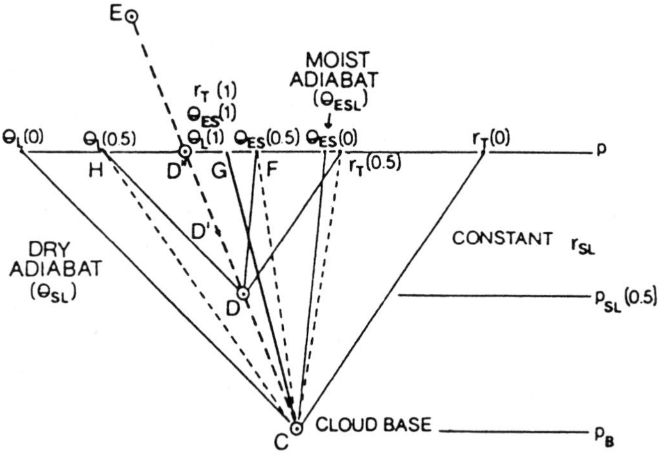

Fig. 8.34. Schematic tephigram showing relationship for cloud and clear-air mixing between change of saturation point and cloud parcel paths with respect to pressure. The heavy dashed line CE is the mixing line between cloud-base air and environmental air, with SP at E. The light solid lines are dry and moist adiabats and lines of constant r. The light dashed lines are cloud parcel paths for $\beta = 0.5$. Numbers in parentheses denote values of β. [From Betts (1982b).]

Using the transformation

$$d/dp = (dp_{SL}/dp)(d/dp_{SL}) = \beta(d/dp_{SL}), \quad (8.27)$$

the gradients following a parcel of pressure p can be written

$$d\theta_{SL}/dp = \beta(d\theta_{SL}/dp_{SL})_M, \quad (8.28a)$$

$$d\theta_{ESL}/dp = \beta(d\theta_{ESL}/dp_{SL})_M, \quad (8.28b)$$

$$dr_{SL}/dp = \beta(dr_{SL}/dp_{SL})_M, \quad (8.28c)$$

where the subscript M refers to gradients along a mixing line. If $\beta = 0$, no mixing occurs in the cloud and the SP remains at C in Fig. 8.34; $\beta = 1$ corresponds to mixing at a rate that corresponds to totally evaporating the water condensed as the parcel rises. The cloud parcel SP reaches D″ as the parcel reaches p. For $0 < \beta < 1$, partial evaporation of cloud water by mixing occurs. Thus, for example, for $\beta = 0.5$, the parcel path of θ_{ES} lies along the dashed line CF which is halfway between the mixing line and moist adiabat [$\theta_{ES}(0)$], while the parcel path for θ_L (line CH) lies halfway between the mixing line and the dry adiabat.

Betts (1982b) also estimated the ratio r_l/r_{la} as

$$r/r_{la} = (p_{SL} - p)/(p_B - p) = 1 - \beta, \quad (8.29)$$

since

$$\beta = (p_B - p_{SL})/(p_B - p) \quad (8.30)$$

and p_B is the cloud base pressure. For an ascending parcel to remain cloudy, $r_l > 0$ and $\beta < 1$.

For an ascending parcel to remain buoyant, however, its θ_{ES} and θ paths (neglecting virtual temperature correction) must lie to the right of the environment stratification illustrated as line CG in Fig. 8.34 and its SP point should not reach D′.

For the case illustrated,

$$\beta < \frac{d\theta_{ES}/dp}{(d\theta_{ESL}/dp_{SL})_M} \simeq 0.7, \quad (8.31)$$

or from Eq. (8.30) $r/r_{la} \simeq 0.3$. Thus, the requirement that a parcel remain buoyant constrains the range of permissible liquid-water content that a cloud can attain. This illustrative example considers only mixing between a rising cloud parcel and environmental air at one level. Betts (1982b) noted that this example is representative of actual cloud layers, since environmental SPs typically lie close to a well-defined mixing line.

Betts (1982b) applied the SP analysis technique to a case study analyzed by Cotton (1975a). The case was observed by Warner's group at Bundaberg,

Queensland, Australia. Airborne measurements of cloud liquid-water content, cloud temperature, and cloud-base and cloud-top heights were obtained. In agreement with earlier findings of Warner (1970), Cotton (1975a) found that the profile of r_l/r_{la} predicted with a one-dimensional lateral entrainment model exceeded observed values at all levels (see Fig. 8.25) when the lateral entrainment rate was set to give observed values of cloud-top heights. Betts (1982b) took a somewhat different approach by estimating the lateral entrainment rate that would give the best agreement with the observed LWC; he then examined the corresponding cloud buoyancy. Figure 8.35 shows the observed sounding and calculated SPs corresponding to each data level in the observed cloud. Also shown in Fig. 8.35 are the estimated cloud parcel temperatures, labeled L, corresponding to a parcel of air which laterally entrains environmental air and produces the LWC observed. The parcel temperature lies close to the environmental temperature. As noted by Betts, even with θ_v corrections, parcels which are characterized by lateral entrainment and corresponding profiles of r_l/r_{la} as observed are near neutral in buoyancy and cannot ascend to the observed cloud-top height.

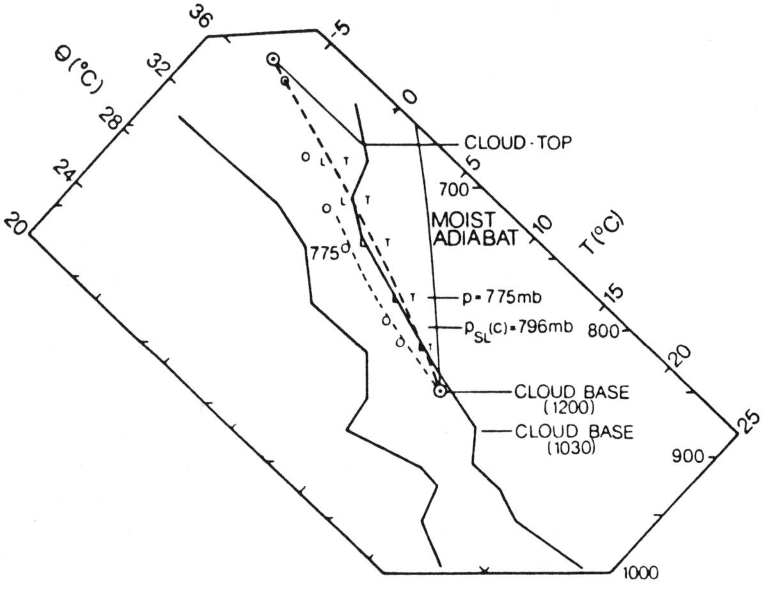

Fig. 8.35. Sounding at Bundaberg, Queensland, Australia, 10 November 1964 (solid lines). Heavy dashed line is mixing line with cloud-top entrainment, light dashed line is with lateral entrainment, and symbols L and T denote cloud parcel temperatures for lateral and cloud-top entrainment, respectively. [From Betts (1982b).]

However, if a parcel of air ascends from cloud base and mixes with cloud-top air, the parcel can have the same liquid-water content as with lateral entrainment, yet can have sufficient buoyancy to reach the observed cloud-top heights. Figure 8.35 illustrates the profile of saturation points due to cloud-top mixing (heavy dashed line) and the corresponding cloud temperatures (labeled T) calculated by Betts (1982b) for the case of cloud-top mixing.

Betts (1982b) also noted that Fig. 8.28 suggests that both lateral and cloud-top entrainment processes are important but are extreme cases. As mentioned previously, cloud-top entrainment results in buoyant updrafts. However, since the mixing line lies close to the temperature sounding, even continuous evaporative mixing will result in downdrafts which are near neutral buoyancy. Thus, cloud-top entrainment cannot account for both buoyant updrafts and negatively buoyant penetrative downdrafts such as observed in the case study described by Cotton (1975a). The mixing line for lateral entrainment, however, is to the left of the environmental temperature profile. Thus, negatively buoyant penetrative downdrafts are possible. Betts (1982b) concluded that a combination of lateral and cloud-top entrainment is likely in cumuli such that positively buoyant updrafts and negatively buoyant downdrafts can coexist. He suggested that the conceptual model of entrainment in cumuli in sheared flow as proposed by Heymsfield *et al.* (1978) and modeled numerically by Cotton and Tripoli (1979) and Tripoli and Cotton (1980) may be appropriate.

Using Paluch's $\theta_q - r_T$ thermodynamic analysis technique, Gardiner and Rogers (1987) concluded that the updrafts located on the upshear side of the cloud constantly mix with the environment as the air parcels rise. Air residing in the downshear portion of the cloud experiences considerable cloud-top mixing, which favors the formation of negatively buoyant downdrafts. To illustrate the asymmetric structure of cumulus clouds in shear flow, consider the vertical and horizontal cross section of a cloud simulated with the Colorado State University (CSU) three-dimensional cloud model shown in Figure 8.36. At the 4.9-km level, a relative high-pressure zone can be seen on the upshear side of the cloud (i.e., relative inflow side at that level), while on the downshear side a relative low exists. As inferred by Raymond (1978) from aircraft observations, the upshear side of the updraft remains protected against lateral entrainment, while the downshear side is readily accessible to the entrainment of dry environmental air. Thus, the downshear region of the simulated cloud is a favored region for the formation of negatively buoyant penetrative downdrafts. Betts (1982b) suggested that the air descending in these downdrafts might then mix with the updraft core lying on the upshear side of the cloud (see Fig. 8.32). This combination of processes would provide both negatively buoyant downdrafts and posi-

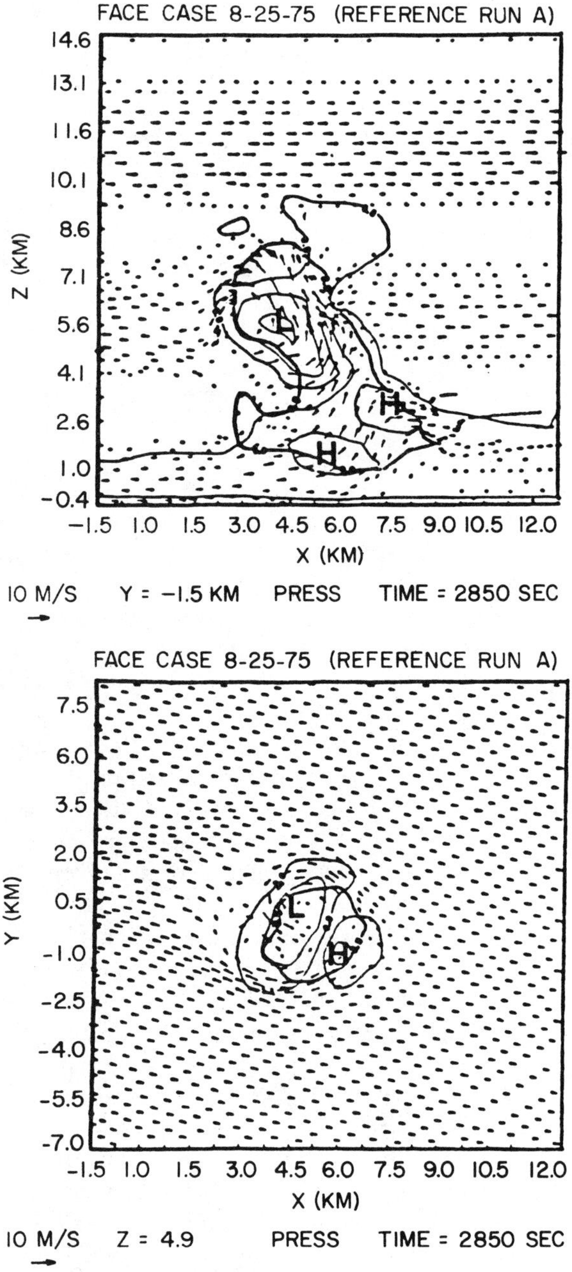

Fig. 8.36. Pressure analyses for natural cloud A. Contours are every 0.01 kPa, minima and maxima are denoted by L and H, respectively. [From Cotton *et al.* (1981).]

tively buoyant updrafts, with a bulk or net entrainment rate which is between the lateral and cloud-top models. The above-cited asymmetric entrainment process modeled with the three-dimensional cloud model (as well as that in Simpson *et al.* (1982) and others) that is illustrated in Figs. 8.32 and 8.36 can be thought of as a "large-eddy entrainment" process. That is, the scale of the horizontal and vertical eddies contributing substantially to lateral entrainment are of the dimensions of the cloud radius. The modeling studies reported by Tripoli and Cotton (1980) and observational studies of towering cumuli flanking cumulonimbi (Knupp and Cotton, 1982a, b) suggest that the entrainment rate is modulated by (i) the difference between the updraft horizontal momentum and environmental horizontal momentum, (ii) the updraft momentum flux and its vertical divergence, and (iii) the SP or θ_{ES} of the environment.

The high/low-pressure couplet illustrated in Fig. 8.36 is a result of the difference in horizontal momentum between the environmental air at a given level and the horizontal momentum carried by the updraft. Other things being the same, the greater the shear of the horizontal wind between the updraft source levels and cloud levels, the greater will be the tendency to generate "protected" updraft cores and vigorous lateral entrainment on the downshear side of the updraft. The interaction of a rising convective tower and environmental shear will be discussed further in Chapter 9.

The model sensitivity experiments reported by Tripoli and Cotton (1980) suggested that the vertical momentum flux carried by the updraft and its vertical divergence have a significant regulating influence on the intensity of large-eddy entrainment. If, for a given environmental stability in the cloud layer, the vertical momentum flux entering cloud base is relatively strong (weak), then the vertical divergence of updraft momentum flux will be weak (strong) or the rate of large-eddy entrainment will be weak (strong). Thus, local boundary features such as small hills or surface "hot spots" can have an important modulating influence upon the intensity of large-eddy entrainment in towering cumuli.

Finally, the drier the cloud environment (as evidenced by the minimum value of θ_{ES}, or the more the environmental SPs lie to the left of the cloud-top mixing line shown in Fig. 8.35), the greater will be the tendency for air entrained on the downshear side of the updraft to generate penetrative downdrafts. The modeling studies illustrated in Fig. 8.36 suggest that the penetrative downdrafts intensify the downshear relative pressure low, which further enhances the intensity of large-eddy entrainment. This positive-feedback loop between penetrative downdrafts and entrainment is analogous to "pulling the plug in a bathtub."

In the above discussion we referred to the entrainment process by eddies on the scale of a cloud tower radius as "large-eddy entrainment." Such eddies are readily observed by multiple-Doppler radar (e.g., Knupp and Cotton, 1982a, b) and are modeled by numerical prediction models. Entrainment by large eddies should be distinguished from entrainment by small eddies or turbulent eddies having horizontal scales of a few tens of meters.

As noted previously, Kitchen and Caughey (1981) observed the characteristics of eddies in boundary layer cumuli using multilevel sensors on a tethered balloon system. Using power spectrum analysis, they found that the kinetic energy of the cloud resided on two distinct scales, one being on the scale of the major updrafts and downdrafts (i.e., 0.5 km) and the second on a scale less than or equal to 10 m. They noted that the data suggested a shift in both peaks to shorter scales as the top is approached. They inferred that this was due to the influence of the capping inversion. Thus, one would anticipate that a small-eddy, cloud-top entrainment process would be more prevalent in boundary layer cumuli.

It should be noted that most of the three-dimensional modeling studies of cumulus convection involve the use of grid meshes that are only capable of resolving horizontal scales comparable to the scales of major updrafts and downdrafts (e.g., Sommeria, 1976; Beniston and Sommeria, 1981; Lipps, 1977; Cotton and Tripoli, 1978). These models rely on simple subgrid-scale parameterizations to represent the fluxes and energetics of small eddies on the order of 10 m in scale. Recently Klaassen and Clark (1985) attempted to simulate both a large-eddy and small-eddy entrainment process by using an interactive grid-nesting scheme in a two-dimensional numerical prediction model. The simulations were restricted to a no-shear environment, thereby preventing the interaction of the convective cell with ambient wind shear and the development of the asymmetric updraft/downdraft structure described previously. The use of the interactive grid-nesting scheme along with the restriction to two dimensions allowed them to achieve a spatial resolution of 5 m in the vicinity of the cloud. The restriction to two dimensions is an important limitation, because it may alter the dynamics of the entrainment process both qualitatively and quantitatively. Certainly the rate of air entrained into a cloud is underestimated. As a result, nonlinearity resulting from evaporation of cloud liquid-water will also be underestimated. Nonetheless, the two-dimensional simulations provide a first look at the role of small eddies in the entrainment process and the possibility that some inherent dynamic instabilities may exist which, in turn, trigger or drive the entrainment process. Figures 8.37 and 8.38 illustrate two cases of the simulated evolution of the entrainment process for clouds initiated by heating a region of air near the surface. Note that in both cases a relatively smooth cloud/no-cloud interface breaks down into several nodal surfaces in which environmental air becomes engulfed into the cloud

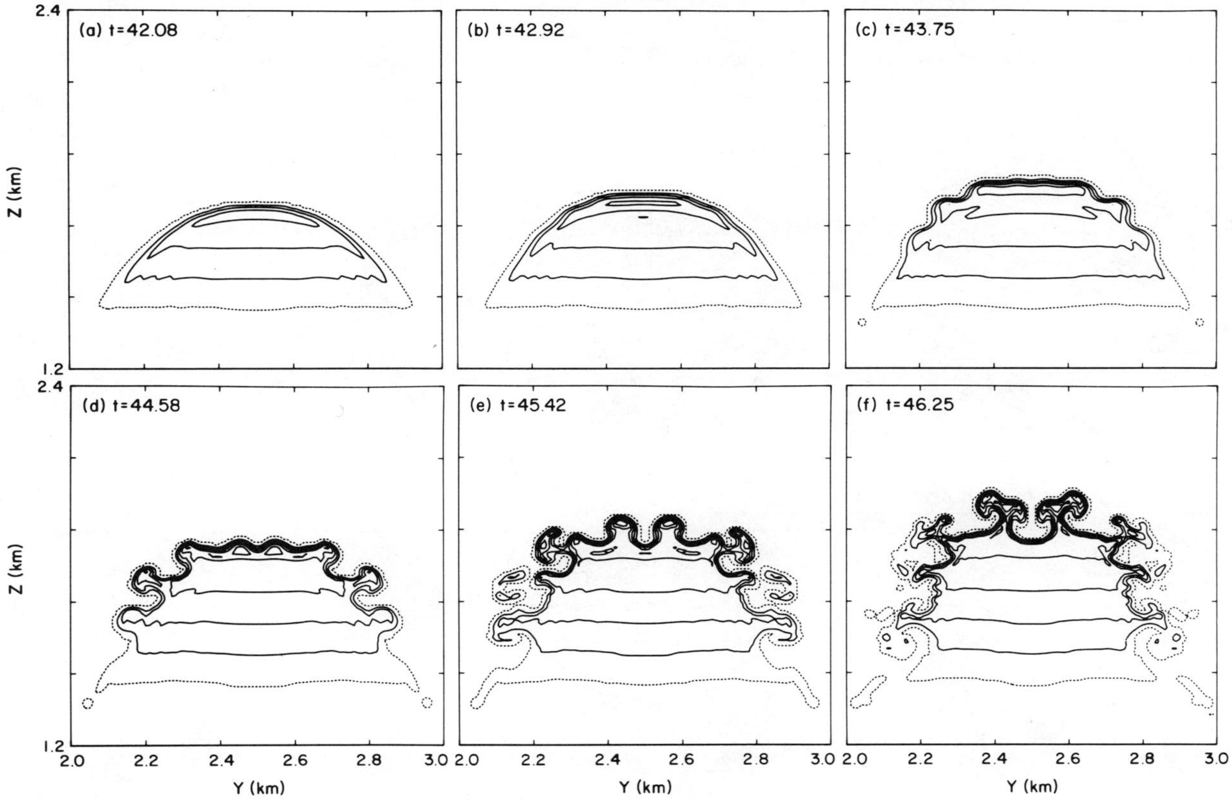

Fig. 8.37. Plots of the liquid-water mixing ratio, $q_c(y, z)$, for four different times from the 10-m-resolution Experiment 3. The contour interval is $0.2\ \text{g kg}^{-1}$. The times t shown for (a)–(f) are in minutes. [From Klaassen and Clark (1985).]

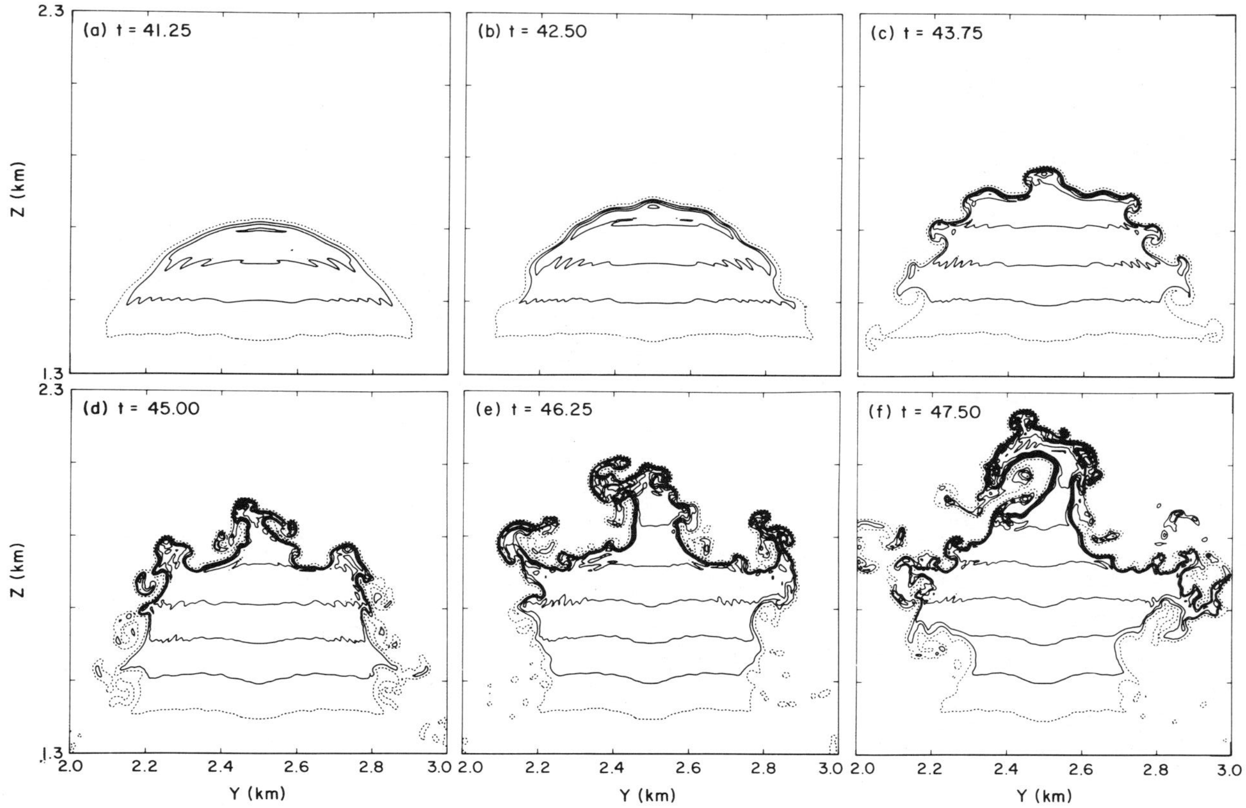

Fig. 8.38. Plots of the liquid-water mixing ratio, $q_c(y, z)$, for six different times from the 5-m-resolution Experiment 4. The contour interval is 0.2 g kg^{-1} and the 0.01 contour is shown as a dotted line. The times t shown for (a)–(f) are in minutes. [From Klaassen and Clark (1985).]

interior. Klaassen and Clark point out that the nodes first appear on the sides of the cloud and develop later in the central part of the cloud top. They conclude that the cloud-top nodal structure is a manifestation of an instability of the interface between the growing cloud and its environment. This instability, they suggest, selectively amplifies interfacial disturbances with a well-defined tangential length scale. Klaassen and Clark examined the role of two instability mechanisms in the development of the nodal structures: (1) Rayleigh–Taylor instabilities (e.g., Chandrasekhar, 1961), associated with positive vertical density differences across the cloud boundary; and (2) baroclinic instabilities associated with horizontal density gradients. They conclude that Rayleigh–Taylor instabilities are not important in the growth of the nodes because the nodes first appear on the sides of the cloud where vertical density differences are small. Horizontal density gradients are large near the cloud sides, which favors a baroclinic instability mechanism in the generation of the nodes. They also point out that local shear across the cloud boundary, not unlike the idealized thermal circulation shown in Fig. 8.33a, may also play a role in the development of the nodal instability.

In another numerical experiment, Klaassen and Clark set the small-scale eddy mixing to zero. An interesting result of this numerical experiment is that downdrafts still formed as a result of the formation of nodal instabilities and penetrated into the cloud interior. They suggest that evaporative cooling caused by the mixing of cloudy air with unsaturated downdraft air plays at most a secondary role in the penetration of downdrafts into the cloud interior. The entrainment process is thus viewed as a fundamentally inviscid, dynamic process. This is because mixing of cloudy air with the unsaturated air in the downdraft has two opposing effects: (1) the evaporation of cloud water cools the unsaturated parcel and accelerates its descent; and (2) cloudy air carries with it upward momentum, which, when mixed with downdraft air, decelerates the descent of the downdraft parcel. A similar result was obtained by Jonas and Mason (1982) in their simple two-thermal model of penetrating downdrafts.

The fact that entrainment is fundamentally a dynamical process is consistent with the earlier modeling results reported by Cotton and Tripoli (1978), although they emphasized the interaction of a cloud with environmental shear. They concluded that mixing and the resultant evaporative cooling further accelerates the entrainment process. Klaassen and Clark's (1985) and Jonas and Mason's (1982) modeling results suggest that eddy mixing actually retards the penetration of downdraft air into the cloud. T.L. Clark (personal communication) suggests that small-scale eddy mixing, modeled by an eddy viscosity parameterization, suppresses the production of vorticity along the cloud boundaries which causes entrainment.

It has also not been fully resolved under what conditions large eddies versus small eddies prevail in the entrainment process. One would anticipate that larger eddies take on greater importance as the scale of clouds increases to the size of towering cumuli. Clearly a great deal remains to be learned about the entrainment process in cumulus clouds.

8.6 The Role of Precipitation

In Chapter 4 we noted that the presence of condensed water affects the vertical accelerations in a cloud by generating a downward-directed drag force equivalent to the weight of the suspended water. Thus, one immediate consequence of a precipitation process is that it unloads an updraft from the weight of the condensed water. Several modeling studies (e.g., Simpson *et al.*, 1965; Simpson and Wiggert, 1969; Weinstein and Davis, 1968) have shown that as a result of precipitation unloading, a cumulus tower can penetrate to greater heights than its nonprecipitating counterpart. Additional modeling studies (Das, 1964; Takeda, 1965, 1966a, b; Srivastava, 1967) demonstrated that, in a nonshearing environment, the water which is removed from the upper parts of the updraft accumulates at lower levels, where it can eventually lead to the decay of the updraft.

In the case of a convective tower growing in a sheared environment, however, the precipitation may not settle into the updraft at lower levels. Instead, precipitation falls downshear of the updraft. Precipitation particles, therefore, typically settle in the region that, we have seen, is already preferred for downdraft formation as a consequence of the interaction of cloud vertical motions with environmental shear. Precipitation can thus enhance the negative buoyancy in that region. Moreover, as cloud water is depleted by evaporation in downdrafts, the settling of precipitation in downdrafts provides an additional supply of water to be evaporated in those regions. Thus, we see that there exists an intimate coupling between entrainment and precipitation processes in the initiation and maintenance of downdrafts in towering cumuli.

A number of modeling studies (Liu and Orville, 1969; Murray and Koenig, 1972; Yau, 1980) have shown that the thermodynamic consequences of the precipitation process are far more important than the water loading alterations. This is particularly true of the latter stages of the cloud lifetime, in which the evaporation of rainwater below the cloud is important in stimulating and intensifying downdrafts. As the downdraft air approaches the earth's surface, it spreads laterally. The evaporatively chilled, dense downdraft air can undercut surrounding moist, potentially warm (high θ_e)

low-level air, thereby lifting it to the lifting condensation level (or SP). In some cases, the lifting by the downdraft outflow may be great enough to lift the air to the level of free convection (i.e., the LFC, the height at which an air parcel becomes buoyant), thus initiating new convective towers. We shall see, however, that a distinguishing feature between a towering cumulus and a cumulonimbus cloud is that the downdraft outflow from a towering cumulus is frequently not vigorous enough to lift the low-level air to the LFC. Moreover, the profile of the horizontal wind or wind shear profile has a profound influence upon whether the precipitating downdraft from a towering cumulus can constructively or destructively influence the subsequent propagation of the cloud (i.e., create new convective towers).

Consider, for example, the three-dimensional simulations of towering cumuli over Florida reported by Levy (1982). Low-level flow in this case was westerly while midtropospheric flow was easterly. As a consequence, Levy found that the precipitation falling out of the cumulus tower fell into the relative inflow flank of the cloud updraft. As a result, the low-level updraft was quenched by the diverging rain-chilled air. This resulted in the dissipation of the parent cumulus tower. New cell growth occurred on the flanks of the diverging evaporatively cooled air, but these cells were relatively weak in intensity and shallow in depth. By contrast, when the low-level flow was altered from westerly to easterly, the evaporating rain did not cut off the low-level updraft inflow air; this caused a longer lived cloud circulation and a 32% increase in rainfall. Therefore, depending upon the environmental wind field, evaporation of precipitation can constructively or destructively contribute to the propagation and longevity of a towering cumulus cloud. As we shall see later, however, the volume of rainfall available for evaporation in the subcloud layer is also important to the subsequent propagation of convective towers. If the volume of rainfall from an individual cumulus tower or from several neighboring cumulus towers is large enough, the diverging rain-chilled air may initiate new cloud growth of sufficient depth to contribute further to the propagation of the cloud system, even in the case of westerly low-level flow illustrated above.

8.7 The Role of the Ice Phase

Since the early cloud-seeding experiments reported by Kraus and Squires (1947), wherein it was hypothesized that the observed "explosive growth" of cumuli was the result of artificially induced glaciation of the cloud, there has been intensive interest and research in the relationships between evolution of the ice phase and the dynamics of cumuli. As we have seen in previous sections, mixing or entrainment processes in cumuli act as a brake against cumulus growth (Malkus and Simpson, 1964). There often exists a delicate balance between the production of cloud buoyancy forces and their destruction by entrainment. Simple inspection of a thermodynamic diagram (e.g., skew T log P, or tephigram) reveals that the latent heat liberated during the growth of cloud droplets is quite large in the lower troposphere. However, in the middle to upper troposphere, the moist adiabat becomes more and more parallel to the dry adiabat; this illustrates that cloud buoyancy production by condensation growth of cloud droplets becomes less important. In contrast, at these same levels in a typical tropical and midlatitude environment, the formation of the ice phase becomes most active. Thus, the latent heat liberated during the freezing of supercooled droplets and the vapor deposition growth of ice particles can augment the diminishing cloud buoyancy production during condensation growth of cloud droplets. In some cases, the additional buoyancy may be sufficient to allow a cloud to penetrate weak stable layers or to survive the entrainment of relatively dry environmental air. The results of calculations of cloud growth with simple one-dimensional cloud models (Simpson *et al.*, 1965; Weinstein and Davis, 1968; Simpson and Wiggert, 1969; Cotton, 1972) suggests that the additional buoyancy liberated by the growth of the ice phase can, in some cases, promote the vertical growth of convective towers several kilometers above their nonglaciated counterparts.

The formation of frozen precipitation can also affect cloud circulations by the redistributions of total water or water loading caused by the differing terminal velocities and resultant settling of the various frozen species through updrafts and downdrafts. In some cold cloud-base continental cumuli, precipitation may only be initiated by ice-phase precipitation processes. Furthermore, the formation of rapidly falling hailstones can unload a vigorous updraft that is otherwise burdened by the weight of large quantities of slowly settling supercooled raindrops.

The buoyancy released during the formation of the ice phase is greatly dependent upon the prior rate of formation of liquid precipitation. For example, using a one-dimensional cloud model, Cotton (1972) found that the suppression of the warm-rain process could in some circumstances favor explosive cloud development. That is, suppose there exists a cumulus tower that is sufficiently vigorous (i.e., thermally buoyant) to overcome the negative buoyancy due to water loading. If the tower then penetrates into supercooled levels, the amount of condensed liquid water available for freezing is much greater than if the cloud had formed precipitation. It can easily be shown that the buoyancy gained by the freezing of condensed water exceeds the negative buoyancy contribution of the weight of the condensed water. Thus, if freezing commences before entrainment processes erode the cloud buoyancy, the additional condensed liquid water (relative to a raining cumuli) can contribute to explosive growth of the cloud tower.

However, the speed of glaciation of a cloud is also highly dependent upon the prior history of warm-rain processes. Several modeling studies (Cotton, 1972; Koenig and Murray, 1976; Scott and Hobbs, 1977) have shown that the coexistence of large, supercooled raindrops and small ice crystals nucleated by deposition, sorption, or Brownian contact nucleation favors the rapid conversion of a cloud from the liquid phase to the ice phase. This is because, in the absence of supercooled raindrops, small ice crystals first grow by vapor deposition until they become large enough to commence riming or accreting small cloud droplets. The riming process then proceeds relatively slowly until they have grown to millimeter-sized graupel particles. Thereafter, the conversion of the cloud to the ice phase can proceed relatively quickly. However, if supercooled raindrops are present, the slow-growth period can be circumvented. The large raindrops then quickly collide with small ice crystals; they immediately freeze to become frozen raindrops. The frozen raindrops can rapidly collect small supercooled cloud droplets, which further enhances the rate of conversion of a cloud to the ice phase. Secondary ice-crystal production by the rime-splinter mechanism (Hallett and Mossop, 1974; Mossop and Hallett, 1974) further accelerates the glaciation rate of the cloud. Several modeling studies (Chisnell and Latham, 1976a, b; Koenig, 1977; Lamb *et al.*, 1981) have shown that the presence of supercooled raindrops accelerates the cloud into a mature riming stage wherein large quantities of secondary ice crystals can be produced in the temperature range -3 to -8°C. The small secondary ice crystals collide with any remaining supercooled raindrops, causing them to freeze and further accelerate the glaciation process.

However, as noted by Keller and Sax (1981), in broad, sustained rapid-updraft regions, even when the criteria for rime-splinter secondary production are met, the secondary crystals and graupel will be swept upward and removed from the generation zone. Until the updraft weakens and graupel particles settle back into the generation zone, the positive-feedback aspect of the multiplication mechanism is broken. Therefore, the opportunities are greatest for rapid and complete glaciation of a single steady updraft if the updraft velocity is relatively weak. In contrast, Keller and Sax (1981) observed high concentrations of ice particles in the active updraft portion of a pulsating convective tower. They postulated that the graupel particles swept aloft in the first bubble of a pulsating convective tower settled downward into the secondary ice-particle production zone (-3 to -8°C), wherein they became incorporated into a new convective bubble and contributed to a prolific production of secondary ice crystals by the rime-splinter mechanism. This demonstrates that there exists a very intimate, nonlinear

Table 8.2

Summary of Dynamic Seeding Hypothesis Chain[a]

(1) Silver iodide is introduced at approximately the −10°C level in the cumulus clouds, i.e., in a region where there is believed to be a significant amount of supercooled water.
(2) This seeding results in conversion of water to ice, with resultant release of latent heat of fusion ($\sim$80 cal g^{-1}), producing increased buoyancy. Additional buoyancy is believed to be from the deposition of water vapor directly onto ice crystals, resulting from the fact that the saturation vapor pressure of ice is less than that of water.
(3) This buoyancy produces an increase in the updraft, which is transferred all the way down to the bottom of the cloud.
(4) This produces an increase in the inflow of moist air into the bottom of the cloud.
(5) This increased inflow of moisture eventually results in more rainfall.
(6) By appropriate seeding, neighboring clouds can be caused to merge.[b]
(7) The increased size of the merged cloud systems results in increased total rainfall.[b]

[a] From Simpson (1980).
[b] Steps 6 and 7 comprise the extension from single-cloud to area-wide effects.

coupling between buoyancy production by glaciation of a cloud and the evolution of the microstructure of the cloud, and the evolving cloud motion field.

Given that additional latent heat is liberated in a cloud associated with the glaciation process, the next question is whether the buoyancy that is gained aloft will be communicated to the moist, subcloud layer. If explosive growth of a convective tower above the 0°C isotherm is to have any significant impact upon the subsequent evolution of the cloud system, it must in some way alter the inflow of moist, potentially warm air (high-valued θ_e air) into the lower levels of the cloud system. The resolution of this question is essential to further advances in the technique of cloud modification by dynamic seeding (e.g., Simpson *et al.*, 1965; Simpson, 1980). Table 8.2 summarizes the hypothesized linkages which Simpson (1980) noted in 1978. The linkages occur between artificially triggered glaciation and enhanced inflow of moist air into the base of cumuli. A crucial step in the hypothesized chain of events is step 3, in which the buoyancy gained aloft produces an increase in the updraft, which is then transferred all the way down to the bottom of the cloud. Three interacting mechanisms are hypothesized to lead to the intensification of the low-level moist inflow into the cloud associated with glaciation aloft (Simpson, 1980):

(1) Pressure forces arising from the rapid warming and rise of the glaciated tower.

(2) Dynamical invigoration of lateral entrainment resulting in enhanced penetrative downdrafts.

(3) Downdraft invigoration by increased water loading and evaporation of precipitation.

Cunning and DeMaria (1981) argued in favor of the pressure communication mechanism. They cite observational evidence which indicated that rapid growth of the convective towers induced surface low pressure by hydrostatic and/or nonhydrostatic means. The surface convergence then increased in response to the low pressure, which resulted in increased convective growth and development prior to the detection of downdrafts on the surface. Lowering of surface pressure below simulated rapidly rising convective towers has also been described by Schlesinger (1980), Tripoli and Cotton (1980), Levy (1982), and Levy and Cotton (1984). In some instances, Tripoli and Cotton (1980) noted that the simulated low-level pressure deficits were strong enough to divert the boundary layer flow and downdraft outflow into the low-pressure region. This further enhanced the intensity of convection and resulted in a longer lasting cloud system. However, the cloud simulations reported by Schlesinger (1980) and Tripoli and Cotton (1980) were for an all water-phase cloud system. Thus, the simulated surface low pressure was a result of warming due to condensation growth of cloud droplets and the resultant rapid growth of the convective towers. This same process may have been the dominant contributor to the observed surface pressure lowering reported by Cunning and DeMaria (1981). They hypothesized that seeding-induced glaciation caused additional acceleration and warming within the updraft, which resulted in larger surface pressure deficits than those that would have occurred naturally.

Using a three-dimensional, time-dependent numerical prediction model, Levy (1982) and Levy and Cotton (1984) investigated the mechanism linking glaciation of the ice phase to the moist boundary layer. One of the cases they simulated was the observed case analyzed by Cunning and DeMaria (1981). In agreement with the findings of Cunning and DeMaria, they simulated surface pressure deficits on the order of 3.5 mbar. However, the contribution of glaciation aloft to the surface pressure deficits was small. They concluded that glaciation was weakly communicated to the surface with pressure falls of 1.0 mbar or less. Such small pressure anomalies had little influence upon the subsequent evolution of the cloud system. In the vicinity of the glaciated volume, however, the pressure lowered by as much as 1.4 mbar. Since this pressure response radiates roughly as a spherically symmetric disturbance, its magnitude diminishes in proportion to R^{-2}, where R is the radial distance from the glaciated cloud volume. What has not been examined, however, is the role of gravity waves in the stably stratified environment that may be excited by enhanced updraft velocities in the towering cumulus. Clark *et al.*'s (1986) numerical experiments suggest that the gravity waves can enhance the upshear growth of new convective towers. The gravity waves could also constructively and destructively interact with gravity waves emitted by neighboring convective towers, thus affecting their subsequent behavior.

The numerical experiments reported by Levy and Cotton (1984) also demonstrated that, while the temperature increased by 0.5 to 1.0°C in the glaciated cloud volume, just below the glaciation level, enhanced lateral entrainment resulted in a temperature decline of 0.1 to 0.3°C. The enhanced lateral entrainment was a direct response to pressure falls on the order of 1.4 mbar at and immediately below (within 500 m) the glaciated level. As a consequence of the entrainment-induced cooling, penetrative downdrafts were intensified by 1.4–2.0 m s^{-1}. Nonetheless, those intensified downdrafts rapidly lost most of their momentum several hundred meters below the glaciation level. Since the glaciation-induced downdrafts did not extend well into the moist subcloud air, they were not able to affect the resultant cloud propagation. The reason that the glaciation-induced downdrafts did not penetrate into the subcloud layer was that the entrainment was sufficiently vigorous to evaporate a large volume of the cloud. This resulted in a simulated cutoff tower similar to that often observed in natural clouds (Keller and Sax, 1981) and in seeded clouds (Simpson and Woodley, 1971). The initially negatively buoyant downdrafts had to descend in clear air, thereby warming dry adiabatically until they became neutrally or positively buoyant. In Levy and Cotton's numerical experiments, the negative buoyancy was lost in a penetration depth of only 1–1.5 km. A greater penetration depth could be expected only if the environment was more unstable or if the downdrafts had access to more liquid water. The entrainment-induced downdrafts could have better access to more liquid water only if heavy precipitation in the form of small raindrops fell from the glaciated levels of the cloud or if the cloud was composed of multiple towers, whereby the downdraft air could mix with liquid water imbedded in "fresh" cumulus towers.

Throughout the numerical experiments reported by Levy and Cotton (1984) and Tripoli and Cotton (1980), evaporation of rain in the subcloud layer caused the strongest thermal and pressure anomalies in the subcloud layer. While the high-pressure regions created by water loading and evaporation of precipitation in the subcloud layer were typically not strong enough to result in new cell development, they easily overrode any tendency for subcloud pressure lowering due to glaciation aloft or the influence of penetrative downdrafts. This led Levy and Cotton (1984) to conclude that the third mechanism (namely, downdraft invigoration by increased water loading and evaporation of precipitation in the subcloud layer) was the primary mechanism for communication of glaciation-induced explosive growth to the subcloud layer. Of course, this requires that glaciation aloft must result in increased precipitation, a condition that is by no means guaranteed.

In our discussion of the role of the ice phase in cumuli, we have not yet considered the influence of melting. Actually, there have been very few quantitative investigations of the role of melting in ordinary cumulus clouds. Using a one-dimensional time-dependent model, Wisner *et al.* (1972) considered the role of melting in a hail-bearing cloud. Since the updraft speeds and liquid-water contents predicted by the model are more characteristic of cumulonimbi than cumuli or towering cumuli, their results are not strictly applicable. However, their results indicate that if the mixing ratio of melting precipitation (i.e., hail) is on the order of 2 g kg^{-1}, melting can lead to significant low-level cooling. They predicted a virtual temperature deficit of about 2.4°C just above the surface in the rainshaft with melting; no such deficit was predicted when melting did not occur. A downdraft of the order of 8 m s^{-1} at 1 km above ground level (AGL) was predicted with melting; however, the downdraft was absent when melting was turned off. To some extent the constraint to one dimension exaggerated the effects of melting by forcing all precipitation downward onto the low-level updraft. However, it is clear that for towering cumuli in which the mixing ratio of ice-phase precipitation is of the order of magnitude of 1 g kg^{-1} or greater, cooling by melting can have a significant impact on the dynamics of cumulus cloud systems. We will examine the influence of melting in larger-scale cloud systems such as cumulonimbi and mesoscale convective systems more fully in subsequent chapters.

8.8 Cloud Merger and Larger Scale Convergence

A fundamental characteristic of the upscale growth of convective cloud systems is that individual convective elements merge together to form larger convective clouds. The merging of convective clouds has been identified observationally both visually and by radar (Byers and Braham, 1949) and from the analysis of aircraft-measured updraft structures (Malkus, 1954). The importance of the merger of convective clouds to convective precipitation was emphasized by Simpson *et al.* (1971), who noted that the merger of two moderately sized cumulonimbi produced a 10- to 20-fold increase of rainfall. In a subsequent study, Simpson *et al.* (1980) found that merged convective systems are responsible for 86% of the rainfall over an area under surveillance in south Florida, even though only 10% of the cells were merged.

Several different concepts of the merger process have been proposed. One view is that the merger process is a stochastic process in which larger aggregates of clouds arise spontaneously from the random clumping of smaller cloud elements. Ludlam and Scorer (1953), Lopez (1978), and Randall and Huffman (1980) represent several advocates of such an approach. The idea is that cumulus clouds leave behind, following their decay, a local environment which is more moist and more unstable (warm near cloud base and cool near cloud top). Thus, they provide an environment which is more favorable for further cloud growth. Given an initially random distribution of cloud elements, the locations where the greatest concentration of clouds initially occurred (by chance) would be able to support more vigorous, larger clouds. Thus, larger clumps of cumuli would be expected to arise in the more favored environment. Once a large clump of cumuli has formed, by virtue of its larger size and longer lifetime, the favored cumuli would be expected to aggregate randomly with smaller clumps. This would cause a still larger cloud system and a subsequently more favored environment for more cloud growth. This concept essentially views the merger process as a static process and ignores the dynamic properties of cumuli (i.e., their motion fields and pressure fields).

The role of cloud-induced pressure anomalies and circulations in the merger process has been emphasized by several investigators. In the numerical investigation of buoyant thermals, Wilkins *et al.* (1976) found that the overlapping buoyancy force fields between two neighboring thermals favors their coalescence. However, they also found that the velocity fields between the two rising thermals tend to interfere with each other and to suppress the circulation and velocity of rise. This causes a mutual repulsion between thermals. Orville *et al.* (1980) performed two-dimensional simulations of the interactions among precipitating cumuli. They varied both the spacing and the time of development of neighboring cumuli. They obtained results similar to Wilkins *et al.* (1976), in that two identical clouds initiated at the same time and height generated circulation fields and pressure fields between them which inhibited cloud merger. When two clouds were started at different times, or if one cloud was substantially stronger than the other and their spacing was less than about 7 km, cloud merger resulted. Figure 8.39 illustrates that for two clouds of differing strengths, a horizontal pressure gradient between the two clouds can occur which favors cloud merger. In the case of two identical clouds initiated at the same time, each cloud generates equal pressure perturbations with no favorable pressure gradient between them.

In contrast to Orville *et al.* (1980), Turpeinen (1982) found in three-dimensional numerical simulations of cloud interactions that differences in the timing and intensity of two neighboring cloud impulses did not promote

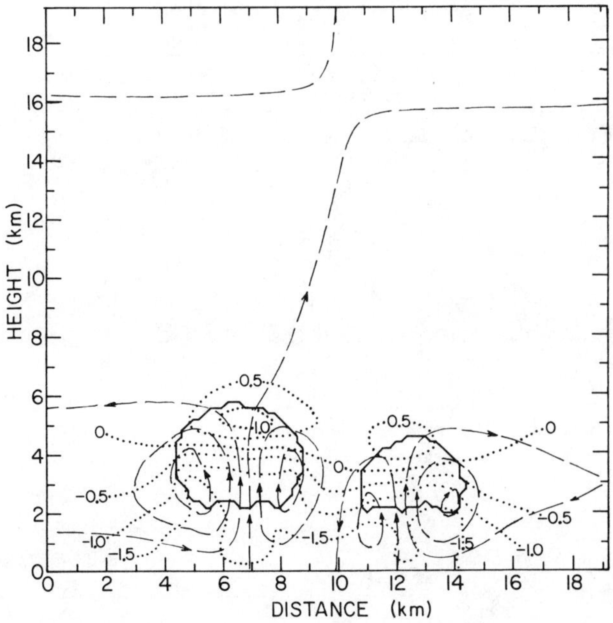

Fig. 8.39. Pressure perturbation pattern. The dotted lines are isobars at 0.05-kPa intervals; the solid lines outline the cloud (100% relative humidity) and the dashed lines are streamlines. [From Orville *et al.* (1980).]

cloud merger. Instead, they found that the maximum center-to-center separation between cloud impulses could not be greater than 3.6 km for cloud merger to occur. In agreement with Orville *et al.* (1980), Turpeinen found that the pressure perturbation field had the dominant influence on merger.

Using inferences drawn from radar observations of convective clouds, Simpson *et al.* (1980) postulated that the approach or collision of downdraft-induced gust fronts from adjacent cumulus clouds is the primary mechanism of shower merger. Figures 8.40 and 8.41 illustrate a conceptual model of the merger process in a light wind and weak shear case (Fig. 8.40) and in a moderate shear case (Fig. 8.41). Illustrated is the precursor "bridge" between cloud towers which, they noted, virtually always precedes radar echo merger. As the downdraft outflows approach and collide, new towers surge upward from the bridge filling the gap. In a sheared environment such as often occurs over Florida in the summer, east winds decrease in intensity with height and return to stronger northerlies just below the tropopause. In such an environment a young cloud (illustrated on the right side of Fig. 8.41) contains mostly updrafts and moves toward the west (to

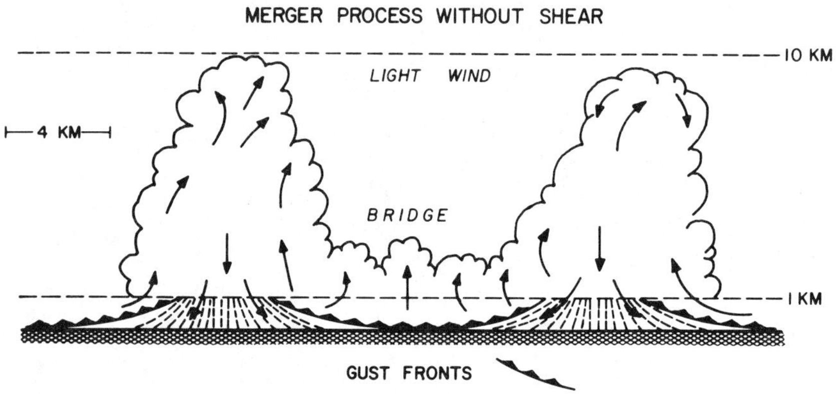

Fig. 8.40. Schematic illustration relating downdraft interaction to bridging and merger in case of light wind and weak shear. [From Simpson *et al.* (1980).]

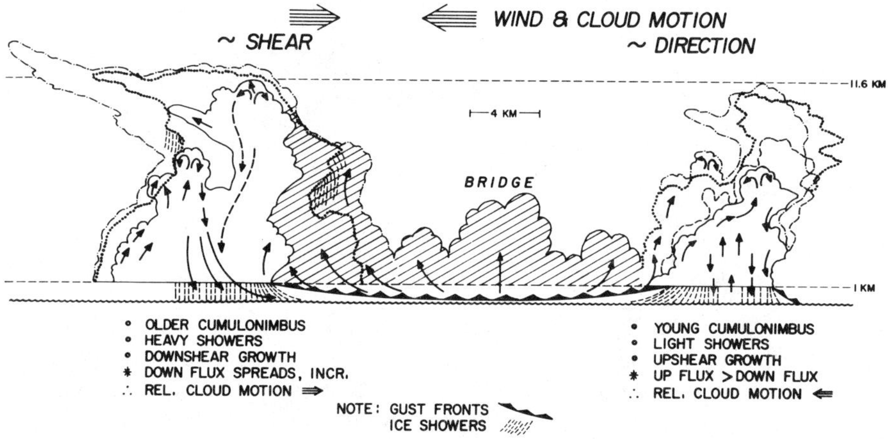

Fig. 8.41. Schematic illustration relating downdraft interaction to bridging and merger in case of moderate shear opposite to wind direction through most of the vertical extent of cloud layer. Younger cumulonimbus on right has predominant up motions and moves faster than the wind. Older cumulonimbus on left has predominant down motions and moves slower than wind, so clouds move and propagate toward each other. Interaction of downdrafts enhances bridge development. [From Simpson *et al.* (1980).]

the left) faster than the winds in the cloud layer. The gust front spreads out nearly symmetrically from the base of the young cloud. In an older cloud (illustrated on the left side of Fig. 8.41) downdrafts are predominant. Since the downdrafts transport slower moving easterlies downward, the relative motion of the older cloud is slower than the young cloud and of the wind in the subcloud layer. As a result, the downdraft spreads out mostly on the downshear (right side), where it collides with the ambient low-level easterly flow, setting off new towers. The new towers may then serve as the bridge for new cloud growth. Therefore, they hypothesize that in a sheared environment, relative motion and propagation (by the downdraft and outflows) play a major role in the merger process.

Further insight into the processes involved in merger can be gained by looking at the results of three-dimensional numerical simulations of cloud merger described by Turpeinen (1982). He performed two simulations of cloud merger, one with no ambient wind and the second with the wind field observed on Day 261 of GATE. Two cumulus clouds were initiated with impulses of radii 3.6 km and with a center-line separation distance of 3.2 km. As noted previously, this was the minimum separation distance for which merger was simulated. A warm-rain process was simulated in both cases. Figure 8.42 illustrates the sequence of events for a merger case for no environmental wind. As in Simpson *et al.*'s (1980) conceptual model (Fig. 8.40), a cloudy bridge can be seen as a precursor to the simulated merger event. A cloudy bridge prior to merger can also be seen in the simulation with moderate environmental wind shear (Fig. 8.43). Turpeinen attributed the merger event to accelerations caused by the perturbation pressure gradient. A schematic illustration of the simulated perturbation pressure field for the two cases is shown in Fig. 8.44. For the no-shear case, a symmetric pressure field is simulated with relative highs near cloud top and near the surface; the latter high is caused by the evaporatively chilled downdraft outflow. At midlevels, relative low-pressure cells were simulated in the cloudy air. However, between the clouds a relative low-pressure trough is found which is caused by subsidence warming in the clear air, as Orville *et al.* (1980) also found (see Fig. 8.39). High pressure occurs in the cloudy bridge region. This Turpeinen explains in terms of the vertical gradient of buoyancy with condensation warming at the base of the cloudy bridge and evaporative cooling at its top. The combined effects of the cloudy bridge and the clear-air subsidence establish a vertical pressure gradient which favors accelerated cloud growth. Turpeinen admitted that the cloudy bridge may have been a result of his overlapping initial impulses rather than the forced uplifting caused by the outflows from two neighboring clouds, as Simpson *et al.* (1980) inferred.

The process of environmental wind shear complicates the picture by displacing the central trough upshear so that it resides well within the upshear cell (see Fig. 8.44). There also exists a surface pressure high below the downshear cell only, since the upshear cell is not precipitating at the time of merging. The resulting pressure field favors the intensification and extended lifetime of the upshear cell. Because the merger in this case takes

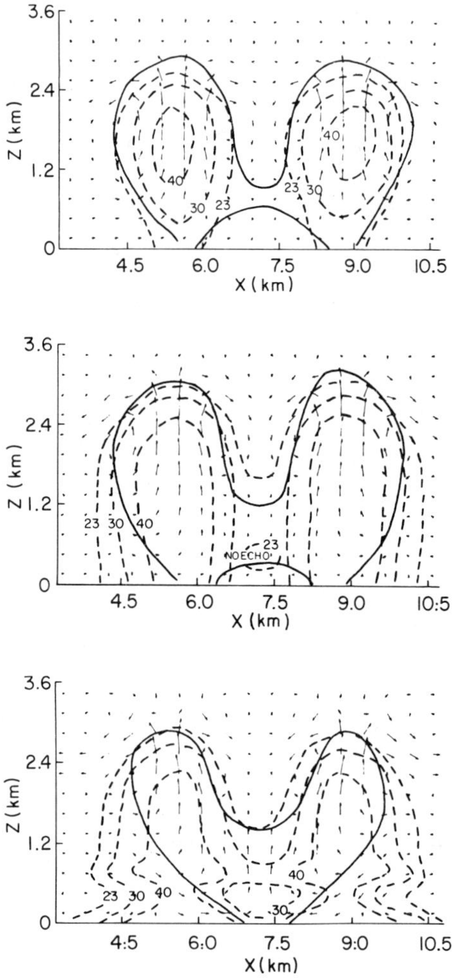

Fig. 8.42. Vertical sections of vector velocity (mean wind subtracted out), cloud water content (solid lines, 0.01 g kg^{-1}), and radar reflectivity factor (dotted lines, dBZ) for Run XIII at 20 (top), 25 (middle), and 30 min (bottom). [From Turpeinen (1982).]

place within the upshear cell, this differs from Simpson *et al.* (1980), who postulated that even in the presence of shear, merger should take place in the cloud bridge between the merging cells. Turpeinen suggested that the cases analyzed by Simpson *et al.* (1980) differed from his simulated cases in that mesoscale convergence could have played a role in the observed merger process.

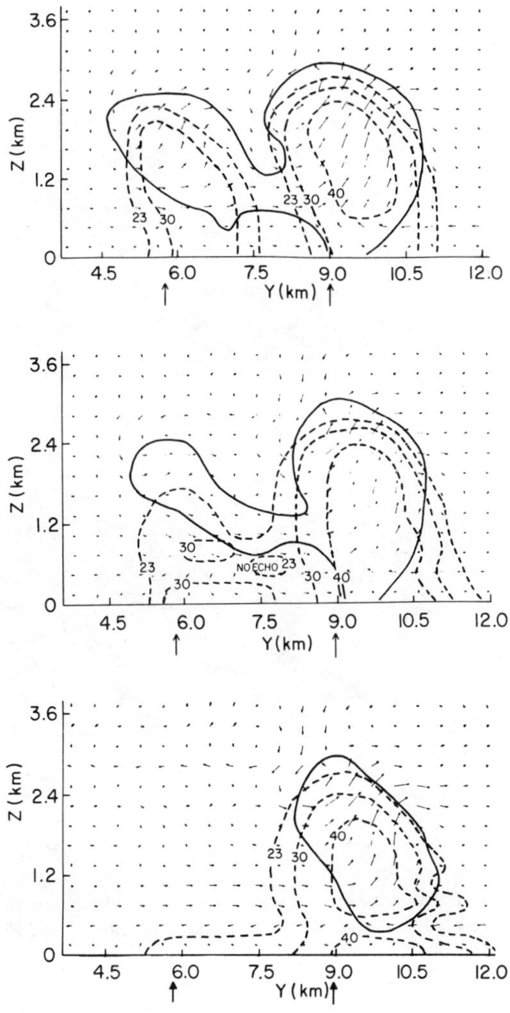

Fig. 8.43. Vertical sections of vector velocity (mean wind subtracted out), cloud water content (solid lines, 0.01 g kg^{-1}), and radar reflectivity factor [dotted lines, dBZ] for Run XII at 25 (top), 30 (middle), and 35 min (bottom). The downshear cell is to the left and the upshear cell is to the right. [From Turpeinen (1982).]

In summary, we can see that the cloud merger process can take place over a broad range of cloud scales, ranging from nonprecipitating cumuli to precipitating towering cumuli or cumulonimbi. The distribution of pressure around neighboring clouds clearly plays an important role in the merger

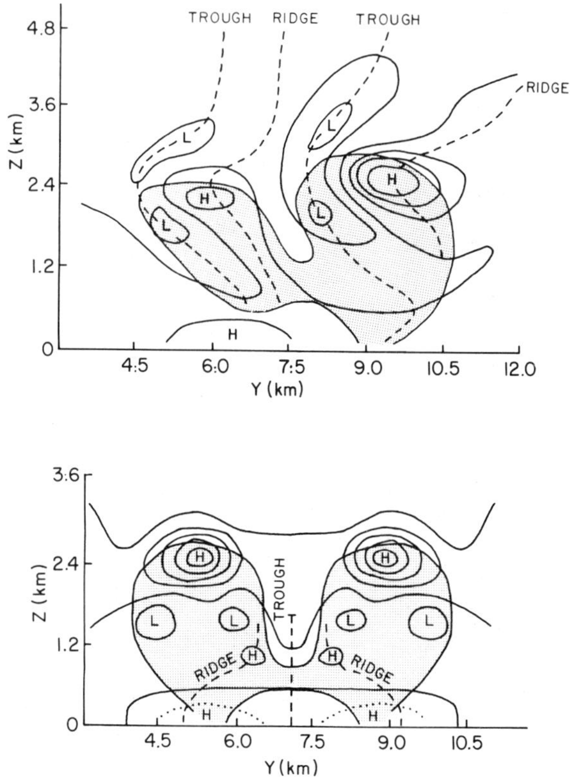

Fig. 8.44. Vertical section of nondimensional perturbation pressure (solid lines, in intervals of 1.5×10^{-5}) and cloud water content (shaded areas, $r_c \geq 0.01$ g kg^{-1}) with locations of troughs and ridges (dashed lines) for Runs XII (top) at 25 min and XIII (bottom) at 20 min. The dotted line indicates a perturbation pressure of 5×10^{-6}. In Run XII the downshear cell is the left and the upshear cell is to the right. [Adapted from Turpeinen (1982).]

process. The numerical experiments reported by Clark *et al.* (1986) suggest that the distribution of pressure about neighboring clouds could be, in part, a result of the constructive and destructive interference of gravity waves excited by each cloud. Just as colliding, low-level downdraft outflows can cause upward motion, so also can the collision of gravity waves emitted from neighboring clouds. Furthermore, the presence of precipitation accentuates the cloud-related pressure anomalies, especially in the subcloud layer. Thus, precipitating downdraft outflows and their interactions among neighboring clouds play a key role in the merger and upscale development of cloud systems.

Orville *et al.* (1980) also suggested that mesoscale convergence in the lower levels would increase the number of merger events because the effect of convergence is to move cells closer together. Simpson *et al.* (1980) provided circumstantial evidence that mesoscale convergence aids the merger process. They showed that many of the observed radar echo merger events coincided with the zones of convergence predicted by the three-dimensional sea-breeze model developed by Pielke (1974). This model did not contain any parameterization of cloud-scale processes. Thus, any degree of association between the predictions by the model and observed merger events could be attributed to the sea-breeze-generated convergence zones.

There are several reasons why one would expect areas of mesoscale convergence to be favored sites for cloud merger. As noted by Simpson *et al.* (1980), the horizontal influx of warm moist air provides a source for sustained buoyant ascent and the release of the slice method constraint (Bjerknes, 1938; Cressman, 1946), so that a larger fraction of the area can be filled by buoyant updrafts. This would also result in the more vigorous clouds being located closer together than in a divergent air mass.

Using a two-dimensional cloud model, Chen and Orville (1980) found that randomly initiated thermal eddies tended to merge together more frequently when a mesoscale convergence field was present. Also, the resultant clouds were broader and deeper in the presence of convergence than they were with no convergence or with divergence in the subcloud layer.

Tripoli and Cotton (1980) found that a three-dimensional cloud model predicted larger, more vigorous convective storms and greater precipitation intensities when low-level convergence was present. Also, Ulanski and Garstang (1978) presented observational evidence that surface convergence patterns nearly always precede the development of radar echoes for periods as long as 90 min over southern Florida. They suggested that "the most crucial factors in determining the total amount of rainfall produced by a given storm is the size of the area of surface convergence." Because larger precipitation rates result in greater subcloud evaporation rates, one would expect that the vigor of downdraft outflows from neighboring cumuli would be greater in an environment with low-level convergence, which favors the downdraft-related mechanisms of merger postulated by Simpson *et al.* (1980).

An extensive examination of the relationship between cloud merger and large-scale convergence has been performed by Tao and Simpson (1984) with a two-dimensional cloud model. They performed a total of 48 numerical experiments producing over 200 groups of tropical cloud systems. The result of those experiments showed that both convective activity and precipitation amounts increased with greater amounts of large-scale lifting (or low-level convergence). Likewise, they found a greater occurrence of cloud merger with increased intensity of large-scale lifting. The most favorable environmental conditions for cloud merger were (1) more unstable thermodynamic stratification and (2) stronger large-scale lifting.

In this chapter we have discussed the dynamics of cumuli, ranging in scale from shallow cumuli to towering cumuli to precipitating cumuli. Of these clouds, the shallow cumuli are least dependent upon mesoscale moisture convergence to sustain the cloud population. Such clouds can be sustained by boundary layer fluxes of heat and moisture. However, larger cumuli which transport large amounts of moisture, often through the depth of the troposphere, become dependent upon a supply of moisture and moist static energy which is greater than can be supplied by vertical eddy transport in the boundary layer. As we shall see in the next chapter, some cumulonimbi in a strongly sheared environment may be able to obtain sufficient moisture and moist static energy to sustain them by virtue of the convergence fields which they induce through downdraft outflows and cloud-induced vertical pressure gradients.

We have also seen in this chapter that boundary layer thermals and shallow cumulus clouds can excite gravity waves in the overlying, stably stratified environment, which, in turn, affect subsequent cloud organization and new cell growth. The merger of neighboring cumulus clouds and up-scale growth of cloud systems may also be partly a result of the interactions of gravity waves excited by neighboring convective clouds. We shall see in subsequent chapters that the interaction between cumulonimbus clouds and gravity waves may be responsible for propagation of thunderstorms and the further up-scale growth of cloud systems into mesoscale convective systems.

References

Ackerman, B. (1959). The variability of the water contents of tropical cumuli. *J. Meteor.* **16**, 191–198.

Agee, E. M. (1987). Mesoscale cellular convection over the oceans. *Dyn. Atmos. Oceans* **10**, 317–341.

Agee, E. M., and T. Asai, eds. (1982). Cloud dynamics: An introduction to shallow convective systems. *Proc. Symp. Gen. Assem. IAMAP, 3rd, Hamburg, West Germany*, 423 pp. Reidel, Boston, Massachusetts.

Agee, E. M., and T. S. Chen (1973). A model for investigating eddy viscosity effects on mesoscale cellular convection. *J. Atmos. Sci.* **30**, 180–189.

Albrecht, B. A. (1979). A model of the thermodynamic structure of the tradewind boundary layer: Part II. Applications. *J. Atmos. Sci.* **36**, 90–98.

Albrecht, B. A. (1981). Parameterization of trade-cumulus cloud amounts. *J. Atmos. Sci.* **38**, 97–105.

Albrecht, B. A., A. K. Betts, W. H. Schubert, and S. K. Cox (1979). A model of the thermodynamic structure of the trade-wind boundary layer, Part I. Theoretical formulation and sensitivity experiments. *J. Atmos. Sci.* **36**, 73–89.

Andre, J. C., G. DeMoor, P. Lacarrere, and R. DuVachat (1976a). Turbulence approximation for inhomogeneous flows. Part I: The clipping approximation. *J. Atmos. Sci.* **33**, 476–481.

Andre, J. C., G. DeMoor, P. Lacarrere, and R. DuVachat (1976b). Turbulence approximation for inhomogeneous flows. Part II: The numerical simulation of a penetration convection experiment. *J. Atmos. Sci.* **33**, 482–491.

Andre, J. C., G. DeMoor, P. Lacarrere, G. Therry, and R. DuVachat (1978). Modeling the 24-hour evolution of the mean and turbulent structures of the planetary boundary-layer model. *J. Atmos. Sci.* **35**, 1861–1883.

Asai, T. (1966). Cloud bands over the Japan Sea off the Hokuriku district during a cold air outburst. *Pap. Meteorol. Geophys.* **16**, 179–194.

Asai, T. (1970). Stability of a plane parallel flow with variable vertical shear and unstable stratification. *J. Meteorol. Soc. Jpn.* **48**, 129–139.

Asai, T. (1972). Thermal instability of a shear flow turning the direction with height. *J. Meteorol. Soc. Jpn.* **50**, 525–532.

Asai, T., and A. Kasahara (1967). A theoretical study of the compensating downward motions associated with cumulus clouds. *J. Atmos. Sci.* **24**, 487–497.

Asai, T., and I. Nakasuji (1973). On the stability of Ekman boundary layer flow with thermally unstable stratification. *J. Meteorol. Soc. Jpn.* **51**, 29–42.

Augstein, A., H. Schmidt, and F. Ostapoff (1974). The vertical structure of the atmospheric planetary boundary layer in undisturbed trade winds over the Atlantic ocean. *Boundary-Layer Meteorol.* **6**, 129–150.

Barcilon, V. (1965). Stability of non-divergent Ekman layers. *Tellus* **17**, 53–68.

Becker, P. (1986). A simple method for parameterizing cumulus cloud amount. *Contrib. Atmos. Phys.* **59**, 399–408.

Beniston, M. G., and G. Sommeria (1981). Use of a detailed planetary boundary layer model for parameterization purposes. *J. Atmos. Sci.* **38**, 780–797.

Betts, A. K. (1975). Parametric interpretation of trade-wind cumulus budget studies. *J. Atmos. Sci.* **32**, 1934.

Betts, A. K. (1976). Modeling subcloud layer structure and interaction with a shallow cumulus layer. *J. Atmos. Sci.* **33**, 2363–2382.

Betts, A. K. (1982a). Saturation point analysis of moist convective overturning. *J. Atmos. Sci.* **39**, 1484–1505.

Betts, A. K. (1982b). Cloud thermodynamic models in saturation point coordinates. *J. Atmos. Sci.* **39**, 2182–2191.

Bjerknes, J. (1938). Saturated ascent of air through dry-adiabatically descending environment. *Q. J. R. Meteorol. Soc.* **64**, 325–330.

Blackadar, A. K. (1962). The vertical distribution of wind and turbulence in a neutral atmosphere. *J. Geophys. Res.* **67**, 3095–3102.

Blyth, A. M., and J. Latham (1985). An airborne study of vertical structure and microphysical variability within a small cumulus. *Q. J. R. Meteorol. Soc.* **111**, 773–792.

Boatman, J. F., and A. H. Auer, Jr. (1983). The role of cloud top entrainment in cumulus clouds. *J. Atmos. Sci.* **40**, 1517–1534.

Bougeault, P. (1981a). Modeling the trade-wind cumulus boundary layer. Part I: Testing the ensemble cloud relations against numerical data. *J. Atmos. Sci.* **38**, 2414–2428.

Bougeault, P. (1981b). Modeling the trade-wind cumulus boundary layer. Part II: A high-order one-dimensional model. *J. Atmos. Sci.* **38**, 2429–2439.

Brill, K., and B. Albrecht (1982). Diurnal variation of the trade-wind boundary layer. *Mon. Weather Rev.* **110**, 601–613.

Brown, W. B. (1961). A stability criterion for three-dimensional laminar boundary layers. *In* "Boundary Layer and Flow Control" Vol. 2, pp. 913–923. Pergamon, New York.

Bunker, A. F., B. Haurwitz, J. S. Malkus, and H. Stommel (1949). Vertical distribution of temperature and humidity over the Caribbean sea. *Pap. Phys. Oceanogr. Meteorol.* **11.**

Byers, H. R., and R. R. Braham (1949). "The Thunderstorm." U.S. Weather Bur., Washington, D.C.

Chandrasekhar, S. (1961). "Hydrodynamic and Hydromagnetic Stability." Oxford Univ. Press (Clarendon), London.

Chen, C. H., and H. D. Orville (1980). Effects of mesoscale on cloud convection. *J. Appl. Meteorol.* **19**, 256–274.

Chisnell, R. F., and J. Latham (1976a). Ice multiplication in cumulus clouds. *Q. J. R. Meteorol. Soc.* **102**, 133–156.

Chisnell, R. F., and J. Latham (1976b). Comments on the paper by Mason, Production of ice crystals by riming in slightly supercooled cumulus. *Q. J. R. Meteorol. Soc.* **102**, 713–715.

Clark, T. L., and T. Hauf (1986). Upshear cumulus development: A result of boundary layer/free atmosphere interactions. *Prepr., Conf. Radar Meteorol. Conf. Cloud Phys., 23rd, Snowmass, Colo.* pp. J18–J21. Am. Meteorol. Soc., Boston, Massachusetts.

Clark, T. L., T. Hauf, and J. Kuettner (1986). Convectively forced internal gravity waves: Results from two-dimensional numerical experiments. *Q. J. R. Meteorol. Soc.* **112**, 899–925.

Cotton, W. R. (1972). Numerical simulation of precipitation development in supercooled cumuli. *Mon. Weather Rev.* **100**, 757–763.

Cotton, W. R. (1975a). On parameterization of turbulent transport in cumulus clouds. *J. Atmos. Sci.* **32**, 548–564.

Cotton, W. R. (1975b). Theoretical cumulus dynamics. *Rev. Geophys. Space Phys.* **13**, 419–448.

Cotton, W. R., and G. J. Tripoli (1978). Cumulus convection in shear flow—three-dimensional numerical experiment. *J. Atmos. Sci.* **35**, 1503–1521.

Cotton, W. R., and G. J. Tripoli (1979). Reply. *J. Atmos. Sci.* **36**, 1610–1611.

Cotton, W. R., T. Nehrkorn, and E. E. Hindman (1981). The dynamic response of Florida cumulus to seeding. Final Rep. to NOAA, Environ. Res. Lab., Weather Modif. Program Off., Grant No. 04-78-B01-29.

Cressman, G. P. (1946). The influence of the field of horizontal divergence on convective cloudiness. *J. Meteorol.* **3**, 85–88.

Cunning, J. B., and M. DeMaria (1981). Comment on "Downdrafts as linkages in dynamic cumulus seeding effects." *J. Appl. Meteorol.* **20**, 1081–1084.

Danielsen, E. F., R. Bleck, and D. A. Morris (1972). Hail growth by stochastic collection in a cumulus model. *J. Atmos. Sci.* **29**, 135–155.

Das, P. (1964). Role of condensed water in the life cycle of a convective cloud. *J. Atmos. Sci.* **21**, 404–418.

Davis, J. M., S. K. Cox, and T. B. McKee (1979). Vertical and horizontal distributions of solar absorption in finite clouds. *J. Atmos. Sci.* **36**, 1976–1984.

Deardorff, J. W. (1972). Numerical investigation of neutral and unstable planetary boundary layers. *J. Atmos. Sci.* **29**, 91–115.

Deardorff, J. W. (1976). On the entrainment rate of a stratocumulus topped mixed layer. *Q. J. R. Meteorol. Soc.* **102**, 563–583.

Emanuel, K. A. (1979). Inertial instability and mesoscale convective systems. Part I: Linear theory of inertial instability in rotating viscous fluids. *J. Atmos. Sci.* **36**, 2425–2449.

Esbensen, S. (1976). Thermodynamic effects of clouds in the trade wind planetary boundary layer. Ph.D. Thesis, Univ. of California at Los Angeles.

Esbensen, S. (1978). Bulk thermodynamic effects and properties of small tropical cumuli. *J. Atmos. Sci.* **35**, 826–387.

Faller, A. J. (1963). An experimental study of the instability of the laminar Ekman boundary layer. *J. Fluid Mech.* **15**, 560–576.

Faller, A. J. (1965). Large eddies in the atmospheric boundary layer and their possible role in the formation of cloud rows. *J. Atmos. Sci.* **22**, 176–184.

Fiedler, B. H. (1984). The mesoscale stability of entrainment into cloud-topped mixed layers. *J. Atmos. Sci.* **41**, 92–101.

Fiedler, B. H. (1985). Mesoscale cellular convection: Is it convection? *Tellus* **37A**, 163–175.

Gardiner, B. A., and D. P. Rogers (1987). On mixing processes in continental cumulus clouds. *J. Atmos. Sci.* **44**, 250–259.

Garstang, M., and A. K. Betts (1974). A review of the tropical boundary layer and cumulus convection: Structure, parameterization and modeling. *Bull. Am. Meteorol. Soc.* **55**, 1195–1205.

Grassl, H. (1977). Radiative effects of absorbing aerosol particles inside clouds. *In* "Radiation in the Atmosphere" (H. J. Bolle, ed.), pp. N180–N182. Science Press, Princeton, New Jersey.

Hallett, J., and S. C. Mossop (1974). Production of secondary ice particles during the riming process. *Nature (London)* **249**, 26–28.

Heymsfield, A. J., D. N. Johnson, and J. E. Dye (1978). Observations of moist adiabatic ascent in northeast Colorado cumulus congestus clouds. *J. Atmos. Sci.* **35**, 1689–1703.

Higuchi, K. (1963). The band structure of snowfalls. *J. Meteorol. Soc. Jpn.* **41**, 53–70.

Jensen, J. B., P. H. Austin, M. B. Baker, and A. M. Blyth (1985). Turbulent mixing, spectral evolution and dynamics in a warm cumulus cloud. *J. Atmos. Sci.* **42**, 173–192.

Jonas, P. R., and B. J. Mason (1982). Entrainment and the droplet spectrum in cumulus clouds. *Q. J. R. Meteorol. Soc.* **108**, 857–869.

Kaimal, J. C., J. C. Wyngaard, D. A. Haugen, O. R. Cote, Y. Izumi, S. J. Caughey, and C. J. Readings (1976). Turbulence structure in the convective boundary layer. *J. Atmos. Sci.* **33**, 2152–2169.

Keller, V. W., and R. I. Sax (1981). Microphysical development of a pulsating cumulus tower: A case study. *Q. J. R. Meteorol. Soc.* **107**, 679–697.

Kelley, R. D. (1984). Horizontal roll and boundary-layer interrelationships observed over Lake Michigan. *J. Atmos. Sci.* **41**, 1816–1826.

Kitchen, M., and S. J. Caughey (1981). Tethered-balloon observations of the structure of small cumulus clouds. *Q. J. R. Meteorol. Soc.*, **107**, 853–874.

Klaassen, G. P., and T. L. Clark (1985). Dynamics of the cloud-environment interface and entrainment in small cumuli: Two-dimensional simulations in the absence of ambient shear. *J. Atmos. Sci.* **42**, 2621–2642.

Knupp, K. R., and W. R. Cotton (1982a). An intense, quasi-steady thunderstorm over mountainous terrain. Part II: Doppler radar observations of the storm morphological structure. *J. Atmos. Sci.* **39**, 343–358.

Knupp, K. R., and W. R. Cotton (1982b). An intense, quasi-steady thunderstorm over mountainous terrain. Part III: Doppler radar observations of the turbulent structure. *J. Atmos. Sci.* **39**, 359–368.

Koenig, L. R. (1977). The rime-splintering hypothesis of cumulus glaciation examined using a field-of-flow cloud model. *Q. J. R. Meteorol. Soc.* **103**, 585–606.

Koenig, L. R., and F. W. Murray (1976). Ice-bearing cumulus cloud evolution: Numerical simulations and general comparison against observations. *J. Appl. Meteorol.* **15**, 747–762.

Kraus, E. B., and P. Squires (1947). Experiments on the stimulation of clouds to produce rain. *Nature (London)* **159**, 489.
Krishnamurti, R. (1968a). Finite amplitude convection with changing mean temperature. Part 1. Theory. *J. Fluid Mech.* **33**, 445–455.
Krishnamurti, R. (1968b). Finite amplitude convection with changing mean temperature. Part 2. An experimental test of the theory. *J. Fluid Mech.* **33**, 457–463.
Krishnamurti, R. (1975a). On cellular cloud patterns. Part 1: Mathematical model. *J. Atmos. Sci.* **32**, 1353–1363.
Krishnamurti, R. (1975b). On cellular cloud patterns. Part 2: Laboratory model. *J. Atmos. Sci.* **32**, 1364–1372.
Krishnamurti, R. (1975c). On cellular cloud patterns. Part 3: Applicability of the mathematical and laboratory models. *J. Atmos. Sci.* **32**, 1373–1383.
Kuettner, J. P. (1959). The band structure of the atmosphere. *Tellus* **11**, 267–294.
Kuettner, J. P. (1971). Cloud bands in the earth's atmosphere: Observations and theory. *Tellus* **23**, 404–425.
Kuettner, J. P., P. A. Hildebrand, and T. L. Clark (1987). Convection waves: Observations of gravity wave systems over convectively active boundary layers. *Q. J. R. Meteorol. Soc.* **113**, 445–467.
Kuo, H. L. (1963). Perturbations of plane couette flow in stratified fluid and origin of cloud streets. *Phy. Fluids* **6**, 195–211.
Lamb, D., J. Hallet, and R. I. Sax (1981). Mechanistic limitations to the release of latent heat during the natural and artificial glaciation of deep convective clouds. *Q. J. R. Meteorol. Soc.* **107**, 935–954.
LeMone, M. A. (1973). The structure and dynamics of horizontal roll vortices in the planetary boundary layer. *J. Atmos. Sci.* **30**, 1077–1091.
LeMone, M. A., and W. T. Pennell (1976). The relationship of trade-wind cumulus distributions to subcloud layer fluxes and structure. *Mon. Weather Rev.* **104**, 524–539.
Lenschow, D. H. (1970). Airplane measurements of planetary boundary structure. *J. Appl. Meteorol.* **9**, 874–884.
Levine, J. (1959). Spherical vortex theory of bubble-like motion in cumulus clouds. *J. Meteorol.* **16**, 653–662.
Levy, G. (1982). Communication mechanisms in dynamically seeded cumulus clouds. Atmos. Sci. Pap. No. 357, Dep. Atmos. Sci., Colorado State Univ.
Levy, G., and W. R. Cotton (1984). A numerical investigation of mechanisms linking glaciation of the ice-phase to the boundary layer. *J. Clim. Appl. Meteorol.* **23**, 1505–1519.
Lilly, D. K. (1966). On the stability of Ekman boundary flow. *J. Atmos. Sci.* **23**, 481–494.
Lindzen, R. S. (1974). Wave-CISK in the tropics. *J. Atmos. Sci.* **31**, 156–179.
Lipps, F. B. (1977). A study of turbulence parameterization in a cloud model. *J. Atmos. Sci.* **34**, 1751–1772.
Liu, J. Y., and H. D. Orville (1969). Numerical modeling of precipitation and cloud shadow effects on mountain-induced cumuli. *J. Atmos. Sci.* **26**, 1283–1298.
Lopez, R. E. (1973). A parametric model of cumulus convection. *J. Atmos. Sci.* **30**, 1354–1373.
Lopez, R. E. (1978). Internal structure and development processes of c-scale aggregates of cumulus clouds. *Mon. Weather Rev.* **106**, 1488–1494.
Ludlam, F. H., and R. S. Scorer (1953). Convection in the atmosphere. *Q. J. R. Meteorol. Soc.* **79**, 94–103.
Malkus, J. S. (1949). Effects of wind shear on some aspects of convection. *Trans. Am. Geophys. Union* **30**, No. 1.
Malkus, J. S. (1952). Recent advances in the study of convective clouds and their interaction with the environment. *Tellus* **2**, 71–87.

Malkus, J. S. (1954). Some results of a trade cumulus cloud investigation. *J. Meteorol.* **11**, 220–237.

Malkus, J. S. (1956). On the maintenance of the trade winds. *Tellus* **8**, 335–350.

Malkus, J. S. (1960). Penetrative convection and an application to hurricane cumulonimbus towers. *In* "Cumulus Dynamics" (C. F. Anderson, ed.), pp. 65–84. Pergamon, Oxford.

Malkus, J. S., and H. Riehl (1964). "Cloud Structure and Distributions Over the Tropical Pacific Ocean." Univ. of California Press, Berkeley.

Malkus, J. S., and R. S. Scorer (1955). The erosion of cumulus towers. *J. Meteorol.* **12**, 43–57.

Malkus, J. S., and R. H. Simpson (1964). Modification experiments on tropical cumulus clouds. *Science* **145**, 541–548.

Malkus, J. S., and R. T. Williams (1963). On the interaction between severe storms and large cumulus clouds. *Meteorol. Monogr.* **5**, 59–64.

Mason, P. J. (1983). On the influence of variations in Monin–Obukhov length on horizontal roll vortices in an inversion-capped planetary boundary layer. *Boundary-Layer Meteorol.* **27**, 43–68.

McCarthy, J. (1974). Field verification of the relationship between entrainment rate and cumulus cloud diameter. *J. Atmos. Sci.* **31**, 1028–1039.

Morton, B. R., Sir G. Taylor, F.R.S., and J. S. Turner (1956). Turbulent gravitational convection from maintained and instantaneous sources. *Proc. R. Soc. London, Ser. A* **235**, 1–23.

Mossop, S. C., and J. Hallett (1974). Ice crystal concentration in cumulus clouds: Influence of the drop spectrum. *Science* **186**, 632–634.

Murray, F. W., and L. R. Koenig (1972). Numerical experiments on the relation between microphysics and dynamics in cumulus convection. *Mon. Weather Rev.* **100**, 717–732.

Nicholls, S., and M. A. LeMone (1980). The fair weather boundary layer in GATE: The relationship of subcloud fluxes and structure to the distribution and enhancement of cumulus clouds. *J. Atmos. Sci.* **37**, 2051–2067.

Nicholls, S., M. A. LeMone, and G. Sommeria (1982). The simulation of a fair weather marine boundary layer in GATE using a three-dimensional model. *Q. J. R. Meteorol. Soc.* **108**, 167–190.

Nitta, T. (1975). Observational determination of cloud mass flux distributions. *J. Atmos. Sci.* **32**, 73–91.

Nitta, T., and S. Esbensen (1974). Heat and moisture budgets using BOMEX data. *Mon. Weather Rev.* **102**, 17–28.

Ogura, Y. (1985). Modeling studies of convection. *Adv. Geophys.* **28B**, 387–421.

Orville, H. D. (1965). A numerical study of the initiation of cumulus clouds over mountainous terrain. *J. Atmos. Sci.* **22**, 684–699.

Orville, Harold D., Y.-H. Kuo, R. D. Farley, and C. S. Hwang (1980). Numerical simulation of cloud interactions. *J. Rech. Atmos.* **14**, 499–516.

Paluch, I. R. (1979). The entrainment mechanism in Colorado cumuli. *J. Atmos. Sci.* **36**, 2467–2478.

Pennell, W. T., and M. A. LeMone (1974). An experimental study of turbulence structure in the fair-weather trade wind boundary layer. *J. Atmos. Sci.* **31**, 1308–1323.

Pielke, R. A. (1974). A three-dimensional numerical model of the sea breezes over South Florida. *Mon. Weather Rev.* **102**, 115–134.

Plank, V. G. (1966). Wind conditions in situations of pattern form and nonpattern form cumulus convection. *Tellus* **18**, 1–12.

Plank, V. G. (1969). The size distribution of cumulus clouds in representative Florida populations. *J. Appl. Meteorol.* **8**, 46–67.

Ramond, D. (1978). Pressure perturbations in deep convection. *J. Atmos. Sci.* **35**, 1704–1711.

Randall, D. A., and G. J. Huffman (1980). A stochastic model of cumulus clumping. *J. Atmos. Sci.* **37**, 2068–2078.

Rayleigh, Lord O. M. (1916). On convection currents in a horizontal layer of fluid when the higher temperature is on the underside. *Philos. Mag.* **32**, 529–546.

Sasamori, T. (1972). A linear harmonic analysis of atmospheric motion with radiative dissipation. *J. Meteorol. Soc. Jpn.* **50**, 505–518.

Schlesinger, R. E. (1980). A three-dimensional numerical model of an isolated thunderstorm. Part II: Dynamics of updraft splitting and mesovortex couplet evolution. *J. Atmos. Sci.* **37**, 395–420.

Schmidt, F. H. (1949). Some speculation on the resistance to motion of cumuliform clouds. *Meded. Ned. Meteorol. Inst. (b)* Deel 1, Mr. 8.

Scott, B. C., and P. V. Hobbs (1977). A theoretical study of the evolution of mixed-phase cumulus clouds. *J. Atmos. Sci.* **34**, 812–826.

Shirer, H. N. (1980). Bifurcation and stability in a model of moist convection in a shearing environment. *J. Atmos. Sci.* **37**, 1586–1602.

Shirer, H. N. (1982). Toward a unified theory of atmospheric convective instability. *In* "Cloud Dynamics" (E. M. Agee and T. Asai, eds.), pp. 163–177. Reidel, Dordrecht, Netherlands.

Shirer, H. N. (1986). On cloud street development in three dimensions. Parallel and Rayleigh instabilities. *Contrib. Atmos. Phys.* **59**, 126–149.

Shirer, H. N., and B. Brummer (1986). Cloud streets during Kon Tur: A comparison of parallel/thermal instability modes with observations. *Contrib. Atmos. Phys.* **59**, 150–161.

Simpson, J. (1980). Downdrafts as linkages in dynamic cumulus seeding effects. *J. Appl. Meteorol.* **19**, 477–487.

Simpson, J., and V. Wiggert (1969). Models of precipitating cumulus tower. *Mon. Weather Rev.* **97**, 471–489.

Simpson, J., and W. L. Woodley (1971). Seeding cumulus in Florida: New 1970 results. *Science* **172**, 117–126.

Simpson, J., R. H. Simpson, D. A. Andrews, and M. A. Eaton (1965). Experimental cumulus dynamics. *Rev. Geophys.* **3**, 387–431.

Simpson, J., W. L. Woodley, A. H. Miller, and G. F. Cotton (1971). Precipitation results of two randomized pyrotechnic cumulus seeding experiments. *J. Appl. Meteorol.* **10**, 526–544.

Simpson, J., N. E. Westcott, R. J. Clerman, and R. A. Pielke (1980). On cumulus mergers. *Arch. Meteorol. Geophys. Bioklimatol., Ser. A* **29**, 1–40.

Simpson, J., G. Van Helvoirt, and M. McCumber (1982). Three-dimensional simulations of cumulus congestus clouds on GATE Day 261. *J. Atmos. Sci.* **39**, 126–145.

Skatskii, V. I. (1965). Some results from experimental study of the liquid water content in cumulus clouds. *Izv. Acad. Sci. USSR, Atmos. Oceanic Phys. (Engl. Transl.)* **1**, 479–487.

Sommeria, G. (1976). Three-dimensional simulation of turbulent processes in an undisturbed trade wind boundary layer. *J. Atmos. Sci.* **33**, 216–241.

Sommeria, G., and M. A. LeMone (1978). Direct testing of a three-dimensional model of the planetary boundary layer against experimental data. *J. Atmos. Sci.* **35**, 25–39.

Squires, P. (1958a). Penetrative downdraughts in cumuli. *Tellus* **10**, 381–389.

Squires, P. (1958b). The microstructure and colloidal stability of warm clouds. *Tellus* **10**, 256–271.

Squires, P., and J. S. Turner (1962). An entraining jet model for cumulonimbus updraughts. *Tellus* **14**, 422–434.

Srivastava, R. C. (1967). A study of the effect of precipitation on cumulus dynamics. *J. Atmos. Sci.* **24**, 36–45.

Stommel, H. (1947). Entrainment of air into a cumulus cloud. Part I. *J. Appl. Meteorol.* **4**, 91–94.

Stull, R. B. (1985). A fair-weather cumulus cloud classification scheme for mixed-layer studies. *J. Clim. Appl. Meteorol.* **24**, 49–56.

Sun, W. Y. (1978). Stability analysis of deep cloud streets. *J. Atmos. Sci.* **35**, 466–483.

Takeda, T. (1965). The downdraft in convective shower-cloud under the vertical wind shear and its significance for the maintenance of convective system. *J. Meteorol. Soc. Jpn.* **43**, 302–309.

Takeda, T. (1966a). Effect of the prevailing wind with vertical shear on the convective cloud accompanied with heavy rainfall. *J. Meteorol. Soc. Jpn.* **44**, 129–143.

Takeda, T. (1966b). The downdraft in the convective cloud and raindrops: A numerical computation. *J. Meteorol. Soc. Jpn.* **44**, 1–11.

Tanaka, M., S. Asano, and G. Yamamoto (1977). Transfer of solar radiation through water clouds. *In* "Radiation in the Atmosphere" (H.-J. Bolle, ed.), pp. 177–179. Science Press, Princeton, New Jersey.

Tao, W.-K., and J. Simpson (1984). Cloud interactions and merging: Numerical simulations. *J. Atmos. Sci.* **41**, 2901–2917.

Tripoli, G. J., and W. R. Cotton (1980). A numerical investigation of several factors contributing to the observed variable intensity of deep convection over South Florida. *J. Appl. Meteorol.* **19**, 1037–1063.

Tsuchiya, K., and T. Fujita (1967). A satellite meteorological study of evaporation and cloud formation over the western Pacific under the influence of the winter monsoon. *J. Meteorol. Soc. Jpn.* **45**, 232–250.

Turner, J. S. (1962). "Starting plumes" in neutral surroundings. *J. Fluid Mech.* **13**, 356–368.

Turner, J. S. (1963). Model experiments relating to thermals with increasing buoyancy. *Q. J. R. Meteorol. Soc.* **89**, 62–74.

Turpeinen, O. (1982). Cloud interactions and merging on Day 261 of GATE. *Mon. Weather Rev.* **110**, 1238–1254.

Ulanski, S., and M. Garstang (1978a). The role of surface divergence and vorticity in the lifecycle of convective rainfall, Part I: Observation and analysis. *J. Atmos. Sci.* **35**, 1047–1062.

Ulanski, S., and M. Garstang (1978b). The role of surface divergence and vorticity in the life cycle of convective rainfall. Part II: Descriptive model. *J. Atmos. Sci.* **35**, 1063–1069.

van Delden, A. (1985). On the preferred mode of cumulus convection. *Contrib. Atmos. Phys.* **58**, 202–219.

Walter, B. A. (1980). Wintertime observations of roll clouds over the Bering Sea. *Mon. Weather Rev.* **108**, 2024–2031.

Walter, B. A., Jr., and J. E. Overland (1984). Observations of longitudinal rolls in a near neutral atmosphere. *Mon. Weather Rev.* **112**, 200–208.

Warner, J. (1955). The water content of cumuliform cloud. *Tellus* **7**, 449–457.

Warner, J. (1970). The microstructure of cumulus cloud. Part III. The nature of the updraft. *J. Atmos. Sci.* **27**, 682–688.

Warner, J. (1977). Time variation of updraft and water content in small cumulus clouds. *J. Atmos. Sci.* **34**, 1306–1312.

Warner, J. (1978). Physical aspects of the design of PEP. Rep. No. 9, pp. 49–64. Precipitation Enhancement Proj., Geneva Weather Modif. Programme, World Meteorol. Organ.

Weinstein, A. T. (1970). A numerical model of cumulus dynamics and microphysics. *J. Atmos. Sci.* **27**, 246–255.

Weinstein, A. T., and L. G. Davis (1968). A parameterized numerical model of cumulus convection. Rep. II, GA-777, 43, Natl. Sci. Found., Washington, D.C.

Welch, R., J. F. Geleyn, G. Korb, and W. Zdunkowski (1976). Radiative transfer of solar radiation in model clouds. *Contrib. Atmos. Phys.* **49**, 128–146.

Wilkins, E. M., Y. K. Sasaki, G. E. Gerber, and W. H. Chaplin, Jr. (1976). Numerical simulation of the lateral interactions between buoyant clouds. *J. Atmos. Sci.* **33**, 1321–1329.

Willis, G. E., and J. W. Deardorff (1974). A laboratory model of the unstable planetary boundary layer. *J. Atmos. Sci.* **31**, 1297–1307.

Wisner, C., H. D. Orville, and C. Myers (1972). A numerical model of a hail-bearing cloud. *J. Atmos. Sci.* **29**, 1160–1181.

Woodward, E. B. (1959). The motion in and around isolated thermals. *Q. J. R. Meteorol. Soc.* **85**, 144–151.

Yamamoto, G., M. Tanaka, and S. Asano (1970). Radiative transfer in water clouds in the infrared region. *J. Atmos. Sci.* **27**, 282–292.

Yau, M. K. (1980). A two-cylinder model of cumulus cells and its application in computing cumulus transports. *J. Atmos. Sci.* **37**, 488–494.

Chapter 9 Cumulonimbus Clouds and Severe Convective Storms

9.1 Introduction

The cumulonimbus cloud, or thunderstorm, is a convective cloud or cloud system that produces rainfall and lightning. It often produces large hail, severe wind gusts, tornadoes, and heavy rainfall. Many regions of the earth depend almost totally upon cumulonimbus clouds for rainfall. Cumulonimbus clouds also play an important role in the global energetics and the general circulation of the atmosphere by efficiently transporting moisture and sensible and latent heat into the upper portions of the troposphere and lower stratosphere. They also affect the radiative budgets of the troposphere. Moreover, cumulonimbus clouds influence tropospheric air quality and the chemistry of precipitation.

We begin this chapter by reviewing descriptive models of thunderstorms, and then we attempt to identify several storm types. This will provide the reader with a perspective on the variable nature of thunderstorms and an introduction to the terminology that is typically used in discussing thunderstorms.

9.2 Descriptive Storm Models and Storm Types

As defined by Byers and Braham (1949) and Browning (1977), the fundamental building block of a cumulonimbus cloud is the "cell." Normally identified by radar as a relatively intense volume of precipitation or a local, relative maximum in reflectivity, the cell can also be described as a region of relatively strong updrafts having spatial and temporal coherency. These updrafts give rise to local regions of intense precipitation, which may not be exactly colocated with the updrafts.

Byers and Braham (1949) identified three stages in the evolution of an ordinary cumulonimbus cloud: the *cumulus stage*, the *mature stage*, and the *dissipating stage*. During the *cumulus stage*, cloud towers, mainly with updrafts, characterize the system. The characteristics of the cumulus stage parallel those described in Chapter 8, except that the horizontal scale of the updrafts is generally larger than typical cumuli. As illustrated in Fig. 9.1a, the cumulus stage is characterized by one or more towering cumulus clouds that are fed by moisture convergence in the boundary layer. While updrafts prevail during this stage, penetrative downdrafts near cloud top and on the downshear flank of the cumuli can occur. Also during this stage, precipitation may form in the upper portion of the cumuli, but significant rainfall in the subcloud layer is unlikely.

The merger of the cumulus elements into a larger-scale convective system characterizes the transition to the *mature stage*. As noted in Chapter 8, the merger process is frequently associated with the collision of downdraft-induced gust fronts from adjacent cumulus clouds. Thus, the onset of

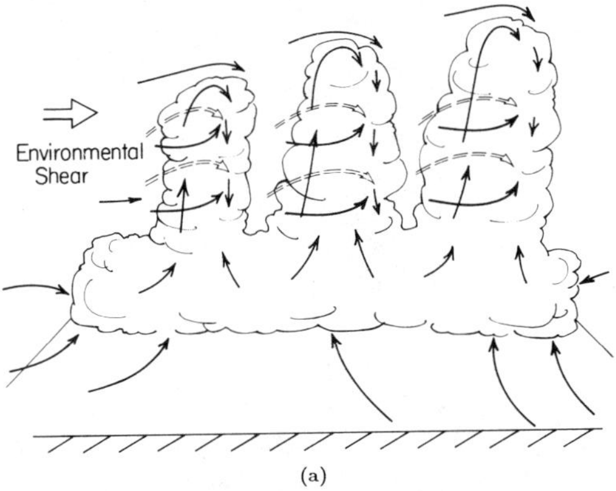

Fig. 9.1. Schematic model of the life cycle of an ordinary thunderstorm. (a) The cumulus stage is characterized by one or more towers fed by low-level convergence of moist air. Air motions are primarily upward, with some lateral and cloud-top entrainment depicted. (b) The mature stage is characterized by both updrafts and downdrafts and rainfall. Evaporative cooling at low levels forms a cold pool and a gust front that advances, lifting warm, moist, unstable air. An anvil at upper levels begins to form. (c) The dissipating stage is characterized by downdrafts and diminishing convective rainfall. Stratiform rainfall from the anvil cloud is also common. The gust front advances ahead of the storm, preventing air from being lifted at the gust front into the convective storm. (*Figure continues.*)

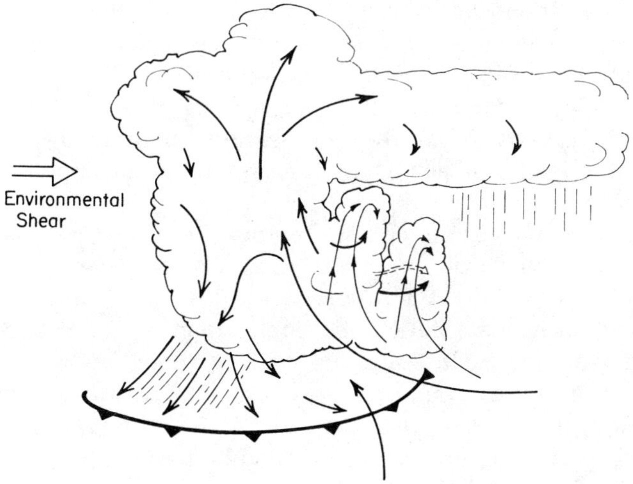

(b)

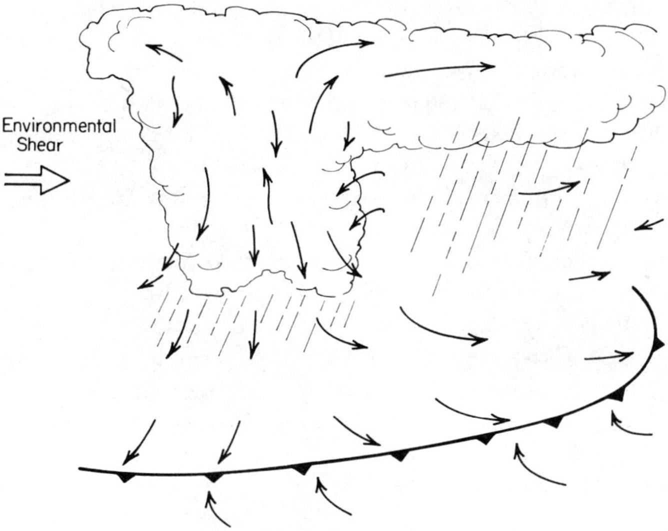

(c)

Fig. 9.1 (*continued*)

precipitation into the subcloud layer is also characteristic of the transition from the cumulus to the mature phase. As illustrated in Fig. 9.1b, both updrafts and downdrafts characterize the mature phase. Updrafts may extend through the depth of the troposphere. Divergence of the updrafts just below the tropopause results in the formation of the anvil cloud, and a cloud dome is often present. A cloud dome signifies updraft air overshooting into the stable stratosphere. Near the ground, the diverging downdraft air, chilled by the evaporation and melting of precipitation, spreads out to form a gust front. This front forces warm, moist air ahead of it, thus feeding the updrafts of new cumuli. Heavy localized rain showers also characterize this stage.

Downdrafts characterize the lower portions during the *dissipating stage* of a cumulonimbus. However, local pockets of convective updrafts can remain, as shown in Fig. 9.1c, especially in the upper half of the cloud. Entrainment through the sides of the cloud and turbulence also occur. The cloud dome, often visible during the mature phase, is absent. Near the ground, the diverging, evaporatively chilled air feeds the gust front, and the front advances far away from the cloud; thus air lifted by the gust front can no longer feed the storm updrafts. Light but steady, stratiform precipitation prevails during the dissipating stages.

There have been numerous attempts to classify thunderstorms into various storm types (Browning, 1977; Chisholm, 1973; Marwitz, 1972a, b, c; Weisman and Klemp, 1984; Foote, 1984). Browning (1977) used the label "ordinary" to refer to a thunderstorm that undergoes the three stages in evolution in a period of 45–60 min and in which the mature stage lasts for only 15–30 min. He distinguishes these from a more vigorous type of convection that is often referred to as a "*supercell.*"

Large size and intensity distinguish supercell storms, where the updraft and downdraft circulations coexist in a nearly steady-state form for periods of 30 min or longer. The model of a supercell thunderstorm has undergone a series of refinements and changes over the years (Browning and Ludlam, 1960, 1962; Browning and Donaldson, 1963; Browning, 1965, 1977; Marwitz, 1972a, b, c; Chisholm, 1973; Browning and Foote, 1976; Lemon and Doswell, 1979; Rotunno and Klemp, 1985). Researchers often refer to supercell storms as severe right (SR) storms because the major low-level inflow is on the right flank of the storm relative to its direction of motion. Therefore the SR storm propagates to the right of the mean tropospheric winds.

Figure 9.2 shows a schematic model of a supercell storm moving toward the east. The model illustrates a broad, intense updraft entering the southeast flank of the storm; the updraft rises vertically and then curves anticyclonically in the anvil outflow region. Figure 9.2 also illustrates a midlevel downdraft that originates on the forward flank of the storm, curves to the north, and exits on the rear flank of the storm. Some supercell models (e.g., Lemon and Doswell, 1979) exhibit downdrafts originating in both the forward and rear flanks of the storm. We shall discuss the origin of such downdrafts in subsequent sections.

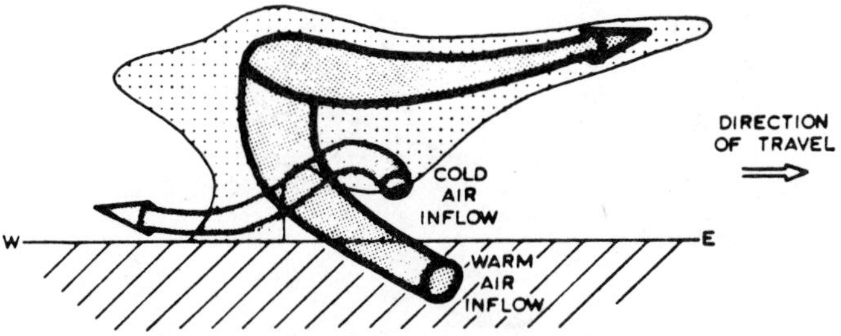

Fig. 9.2. Model showing the airflow within a three-dimensional severe right storm traveling to the right of the tropospheric winds. The extent of precipitation is lightly stippled and the updraft and downdraft circulations are shown more heavily stippled. Air is shown entering and leaving the updraft with a component into the plane of the diagram. However, the principal difference of this organization is that cold-air inflow, entering from outside the plane of the vertical section, produces a downdraft ahead of the updraft rather than behind it. [From Browning (1968).]

Rotating thunderstorms (or *mesocyclones*), typically associated with tornado-producing storms, and radar-echo characteristics also help identify quasisteady supercell storms. Figure 9.3 illustrates plan views (a) and a vertical cross section (b) through a supercell storm. Particularly noteworthy is the "hook echo" that wraps around a so-called "vault" (Browning and Ludlam, 1960, 1962) or "bounded weak-echo region" (BWER) (Chisholm and Renick, 1972). Browning hypothesized that the vault, or BWER, is caused by air rising so rapidly in strong updrafts that insufficient time is available for the formation of radar-detectable precipitation elements. Weisman and Klemp (1984) have found in their numerical experiments that the BWER resides on the gradient of strong updrafts. They suggested that both the rotational character of the updraft and the updraft strength are important in producing the features of the BWER. Figure 9.3b shows that an intense precipitation region on the storm's rear flank, and an overhanging precipitating region on the storm's forward flank, bound the echo-free vault.

Byers and Braham (1949), Browning (1977), Marwitz (1972b), and Chisholm and Renick (1972), among others, have identified another form of severe storm called the *multicell storm*, which, as its name implies, is typically composed of two to four cells. At any time, some cells may be in the cumulus stage, others in the mature stage, and others in the dissipating stage of the life cycle of a cumulonimbus cell. Figure 9.4 shows schematic horizontal and vertical radar depictions of a multicell storm existing at various times in the evolution of the storm system. New cumulus towers typically form on the right flank of the storm complex (i.e., cells at time 0 in Fig. 9.4). Some researchers have referred to the flanking cumulus towers as "feeder" clouds (Dennis *et al.*, 1970) because the new cells move into

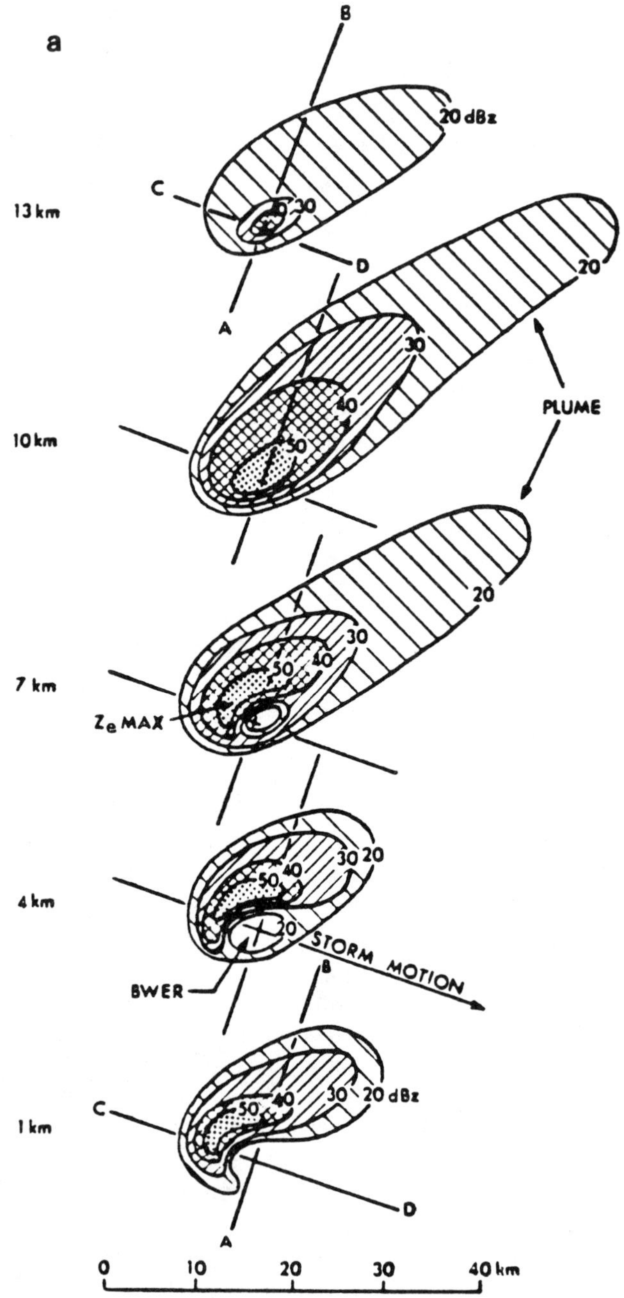

Fig. 9.3 (*Figure continues.*)

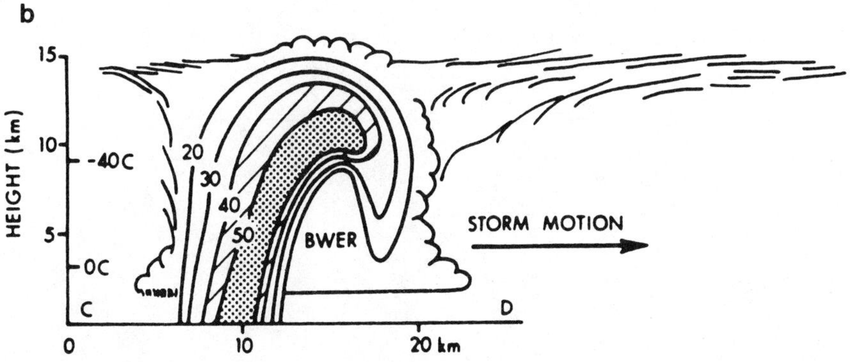

Fig. 9.3. (a) Schematic horizontal sections showing the radar structure of a unicellular supercell storm at altitudes of 1, 4, 7, 10, and 13 km AGL. Reflectivity contours are labeled in dBZ. Note the indentation on the right front quadrant of the storm at 1 km which appears as a weak echo vault (or BWER, bounded weak echo region, as it is labeled here) at 4 and 7 km. On the left rear side of the vault is a reflectivity maximum extending from the top of the vault to the ground. (b) Schematic vertical section through a unicellular supercell storm in the plane of storm motion [along CD in (a)]. Note the reflectivity maximum, referred to elsewhere as the hail cascade, which is situated on the (left) rear flank of the vault (or BWER, as it is labeled here). The overhanging region of echo bounding the other side of the vault is referred to as the embryo curtain, where it is shown to be due to millimeter-sized particles, some of which are recycled across the main updraft to grow into large hailstones. [From Chisholm and Renick (1972).]

the "parent" storm complex and merge with the parent cell. Others, such as Browning (1977), refer to the flanking line of cumuli as "daughter" cells, wherein the new cells do not merge with the parent circulation but grow rapidly to become the new storm center. The older parent cell then begins to decay. According to Browning, new cells typically form at intervals of 5–10 min and exhibit characteristic lifetimes of 20–30 min. Generally, multicell thunderstorms are somewhat less intense than supercell storms. As seen in Fig. 9.4, they frequently exhibit radar features such as hook echoes and weak-echo regions (WERs), although generally the WER is not fully bounded in the form of an echo-free vault. Occasionally we observe short-lived BWERs or echo-free vaults in multicelled storms.

Browning (1977) distinguished between a typical multicell storm and a supercell by the visual appearance of daughter cells (Fig. 9.5). However, some scientists argue that a supercell storm is nothing more than a multicell storm in which the daughter cells are embedded within the forward overhanging anvil cloud and precipitation. In some cases multicell storms evolve into supercell storms (e.g., Vasiloff *et al.*, 1986; Knupp and Cotton, 1982a, b). Moreover, there is some evidence of a continuum of storm types, ranging

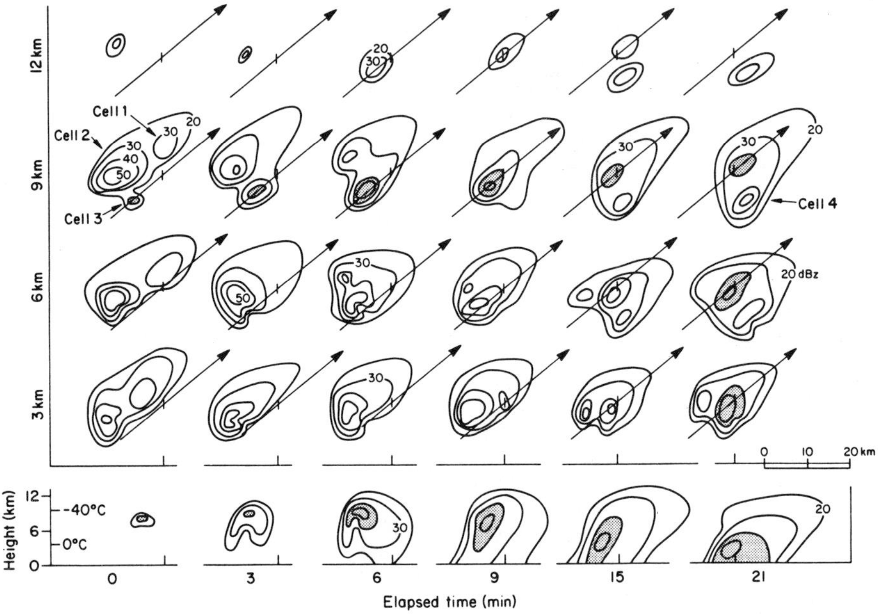

Fig. 9.4. Schematic horizontal and vertical radar sections for an ordinary multicell storm at various stages during its evolution showing reflectivity contours at 10-dBZ intervals. Horizontal sections are illustrated for four altitudes (3, 6, 9, and 12 km AGL) at six different times. The arrow superimposed on each section depicts the direction of cell motion and is also a geographical reference line for the vertical sections at the botton of the figure. Cell 3 is shaded to emphasize the history of an individual cell. [From Chisholm and Renick (1972).]

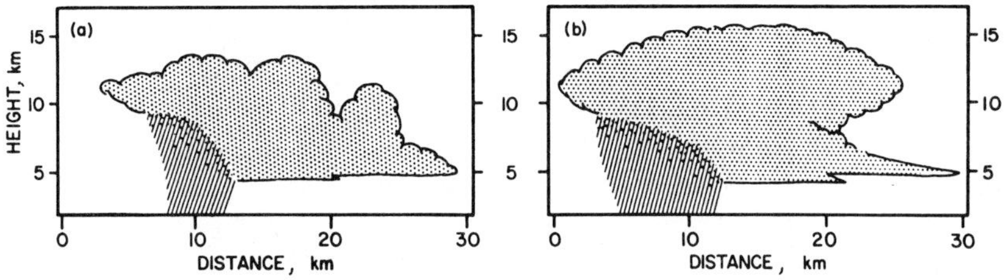

Fig. 9.5. Schematic diagrams illustrating the distinction between the visual appearance of (a) an ordinary multicell storm with growing daughter clouds and (b) a unicellular supercell storm with no discrete daughter clouds. Inflow toward the updraft is from the right in both cases; the precipitation falls out on the left. [From Browning (1977).]

from the lesser organized multicell storms, to organized multicell storms, to the steady supercell storm (e.g., Foote and Wade, 1982). Vasiloff *et al.* (1986) suggest that the distance between successive updraft cells L relative to updraft diameter D, can be used to identify storm type. When $L > D$, storms appear as multicell storms, and when $L \ll D$ they resemble supercell storms. When $L/D < 1$, they suggest that weak evolving multicell storms should prevail in which individual updraft perturbations associated with cells are embedded within a larger scale region of background updraft.

Weisman and Klemp (1984) propose a more dynamic classification of supercell versus ordinary multicell storms. They differentiate between supercell and multicell storms by noting such features as dynamically induced low-level pressure minima, vertical pressure gradients that enhance updrafts, degree of correlation between updraft and vertical vorticity, and propagation characteristics of the storm. They also argue that strong shear over the lowest 6 km, or a proper range of bulk Richardson number R_i (defined as $R_i = \text{Buoyancy}/\frac{1}{2}\Delta\bar{u}^2$, where $\Delta\bar{u}$ represents a difference between environmental wind speeds at low and midlevels), is a necessary condition for supercell formation. We shall discuss these factors more fully in subsequent sections.

Occasionally thunderstorms organize into clusters of cumulonimbi called mesoscale convective systems (MCSs) that have maximum dimensions of 100 km or more (see Chapter 10). Often MCSs form major lines of thunderstorms called *squall lines*, in which the cells align in a direction perpendicular to the direction of movement of the storm system (Newton, 1963). Typically the thunderstorm building blocks of the squall-line system go through a multicellular life cycle with new cells forming on the southern flank of the squall-line system. One or several supercell storms may comprise the squall-line building blocks. More circular mesoscale convective systems, as viewed from satellites, are referred to as mesoscale convective complexes (Maddox, 1980) in midlatitudes, or as cloud clusters in the tropics. Occasionally MCCs have been observed to contain simultaneously smaller squall lines and multicellular thunderstorms (Wetzel *et al.*, 1983).

9.3 Updrafts and Turbulence in Cumulonimbi

One of the fundamental properties of a cumulonimbus cloud is the intense, deep updraft associated with the cloud system. Many factors govern the intensity of updrafts in convective clouds. This can readily be seen by applying Eq. (3.22) to the updraft velocity $\bar{w}$ averaged across its width.

Neglecting Coriolis effects, Eq. (3.22) becomes

$$\frac{d\bar{w}}{dt} = \underset{(1)}{-\frac{1}{\rho_0}\frac{\partial p'}{\partial z}} + \underset{(2)}{\left(\frac{\theta_v'}{\theta_0} - \frac{c_v}{c_p}\frac{p'}{p_0} - r_w'\right)g} - \underset{(3)}{\frac{1}{\rho_0}\frac{\partial}{\partial x_j}(\rho_0 \overline{w'' u_j''})} + \underset{(4)}{\text{viscous terms.}} \quad (9.1)$$

Thus we see that the mean updraft speed is controlled by (1) local vertical pressure gradients; (2) buoyancy due to virtual temperature anomalies, pressure anomalies, and the drag or loading due to the presence of liquid or frozen water; (3) turbulent Reynolds stresses; and (4) viscous diffusion and dissipation. For high-Reynolds-number flows characteristic of cumulonimbi, viscosity has a negligible influence on mean updraft speeds. The three remaining terms, however, all exert an important influence on the magnitude and scale of updrafts in cumulonimbi, although their relative importance varies with location and the lifetime of the cloud system.

Turbulence also distinguishes cumulonimbus clouds. One can apply Eq. (3.50) to form a turbulent kinetic energy equation averaged across the widths of updrafts (or downdrafts) of cumulonimbi,

$$\frac{d\bar{e}}{dt} = \underset{(a)}{-\rho_0 \overline{u_i'' u_j''}\frac{\partial \bar{u}_i}{\partial x_j}} + \underset{(b)}{\left(\frac{\overline{u_i''\theta_v''}}{\theta_0} - \frac{c_v}{c_p}\frac{\overline{u_i'' p''}}{p_0} - \overline{u_i'' r_w''}\right) g\delta_{i3}}$$

$$\underset{(c)}{-\frac{\partial}{\partial x_j}(\overline{e u_j''})} - \underset{(d)}{\frac{\partial}{\partial x_j}(\overline{u_j'' p''})} - \underset{(e)}{\rho_0 \varepsilon}, \quad (9.2)$$

where (a) represents the mechanical or shear production of turbulent kinetic energy, (b) represents buoyant production of TKE, (c) represents the transport of kinetic energy by turbulence, (d) represents the diffusion of TKE by pressure–velocity correlations, and (e) is the rate of molecular dissipation of TKE.

We next examine observational and modeling studies that shed light upon the characteristics of updrafts and turbulence in cumulonimbi.

9.3.1 Updraft Magnitudes and Profiles

As can be seen from Eq. (9.1), a significant force in determining the magnitude of updrafts in cumulonimbi is the average buoyancy across the updraft width. The buoyancy, in turn, is regulated by the stability of the environment and by the amount of turbulent mixing between the updraft and the dry environment. As a rough estimate of the effects of environmental stability on updraft intensities, it is common practice to assume that a parcel of air lifted from the surface will rise dry adiabatically to the lifting condensation level and thereafter rise wet adiabatically until the air parcel attains substantial negative buoyancy. Ignoring all mixing processes and pressure-gradient influences, we can then use Eq. (9.1) to estimate the vertical profile of updraft velocity. Thus, for a hailstorm environment, such as that shown in Fig. 9.6, which exhibits a maximum buoyancy of 2.5–3°C at 3.5 km above ground level (AGL), Chisholm (1973) calculated the corresponding vertical velocity and liquid-water content profiles shown in Fig. 9.7. The maximum vertical velocity is 25 m s^{-1} near 6 km AGL.

Considerably more intense updraft velocities can be expected in supercell thunderstorms. Miller *et al.* (1988) diagnosed updraft magnitudes in excess of 40 m s^{-1} in a supercell storm observed near Miles City, Montana (Fig. 9.8). At the other extreme, in a disturbed tropical environment, such as a hurricane, the latent heat released by large numbers of cumulonimbi drives the environmental sounding to nearly wet adiabatic, resulting in maximum parcel buoyancies on the order of 1°C or less. As a result, typical updraft velocities are less than 6 m s^{-1}. Figure 9.9 shows this in the updraft data compiled by Zipser and LeMone (1980) and Jorgensen (1984). By contrast, observations made during the thunderstorm project of ordinary thunderstorms over Ohio and Florida in the United States exhibit typical updraft

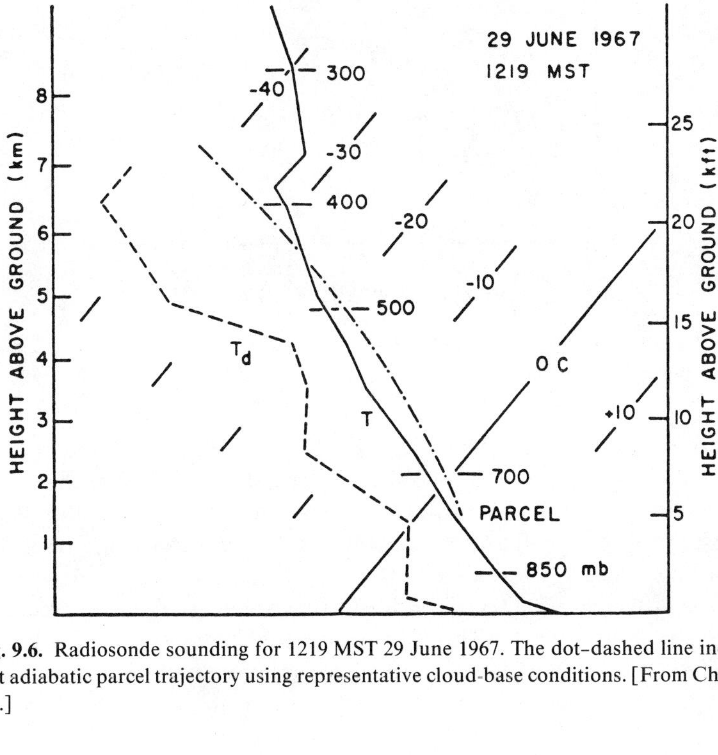

Fig. 9.6. Radiosonde sounding for 1219 MST 29 June 1967. The dot-dashed line indicates a moist adiabatic parcel trajectory using representative cloud-base conditions. [From Chisholm (1973).]

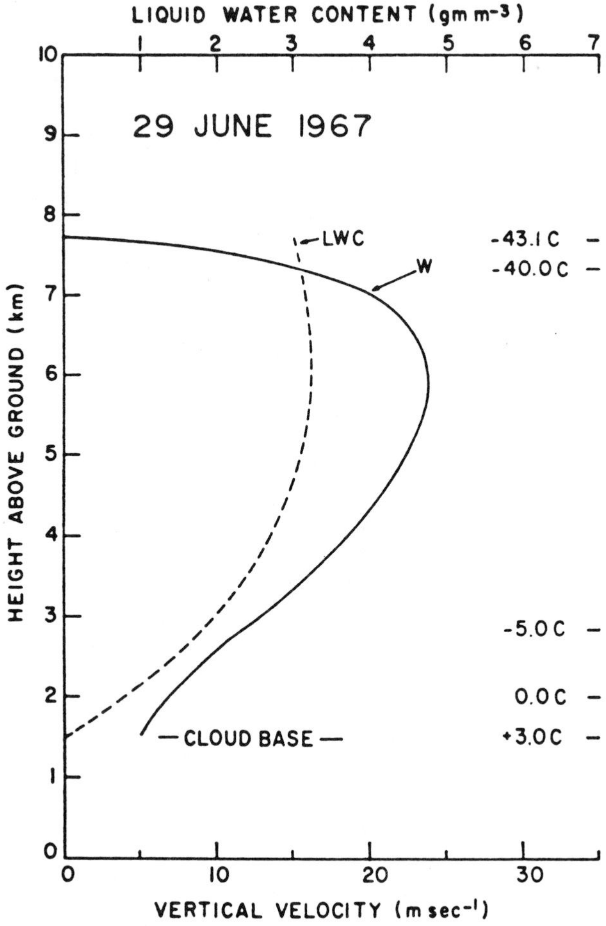

Fig. 9.7. Vertical velocity (W) and water content (LWC) profiles for 29 June 1967. Note the maximum W value near 25 m s^{-1} and the storm top at 7.8 km AGL. [From Chisholm (1973).]

magnitudes that exceed 10 m s^{-1}. Table 9.1 lists some observed maximum and average updraft speeds for various types of cumulonimbi. Not surprisingly, severe midlatitude cumulonimbi over land exhibit the highest updraft speeds; these exceed 40 m s^{-1}. Also, the width of updrafts varies from about 2 km to more than 10 km.

Above the freezing level, updraft speeds may greatly exceed values attained in lower levels. This is because the largest virtual temperature anomalies usually occur in severe thunderstorm environments above the freezing level. Moreover, the additional latent heat released by the freezing

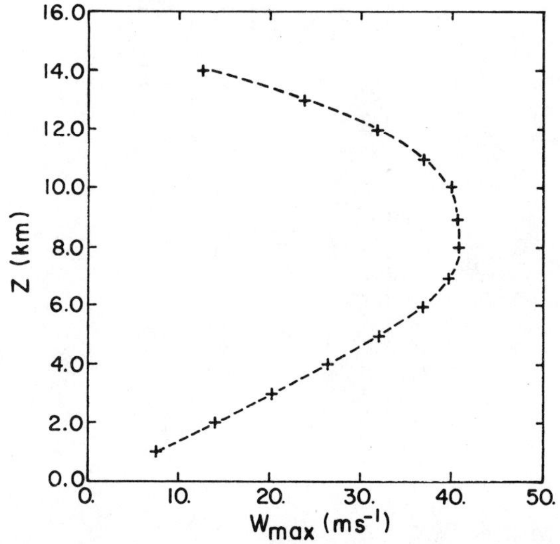

Fig. 9.8. Vertical profile of maximum updraft speed versus height obtained from multiple-Doppler radar from a supercell storm observed during the 1981 CCOPE near Miles City, Montana. The maximum value of updraft speed through the entire depth was 40.88 m s^{-1}. [Provided by L. Jay Miller for the storm described in Miller *et al.* (1988).]

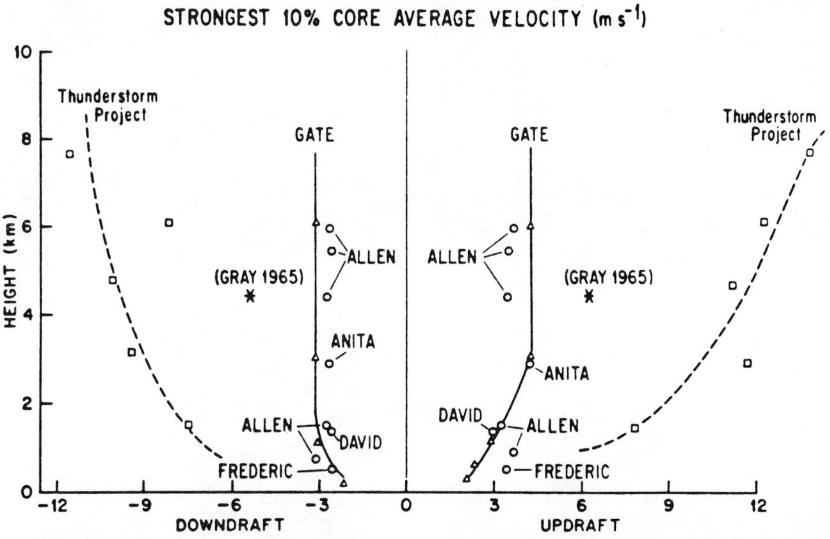

Fig. 9.9. Strongest 10% level average vertical velocity in updraft and downdraft cores as a function of height. GATE values are denoted by triangles, Thunderstorm Project values by squares, and hurricane values by circles and storm name. Observations made by Gray (1965) are also indicated. [From Jorgensen (1984).]

Table 9.1

Summary of Draft Magnitudes and Widths as Measured by Penetrating Aircraft[a]

Location	Reference	Penetration/ clouds	Height of penetration (km MSL)	Updraft speed (max/mean) ($m\ s^{-1}$)	Updraft width (max/mean) (km)	Downdraft speed (max/mean) ($m\ s^{-1}$)	Downdraft width (max/mean) (km)	Cloud type
Florida	Wiggert *et al.* (1982)	35/15	—	31.5/16.8[b]	—	23.5/6.8[b]	—	Precipitating towering cumulus and cumulonimbus
Illinois	Wiggert *et al.* (1982)	16/16	—	20.5/7.2[b]	—	7.8/3.5[c]	—	Precipitating towering cumulus and cumulonimbus
Hurricanes	Jorgensen *et al.* (1985)	—	0.5–6.1	Average 6[b] median 1.5	—	Average 5 median 1.5	—	Four hurricanes, inner core and outer bands
Tropical Atlantic	LeMone and Zipser (1980)	—	0.15–8.0	14/2.9[c]	7/~1.8[c]	7/~1.8[c]	7/~1[c]	Precipitating and non-precipitating cumulus congestus to cumulonimbus
Florida, Ohio	Byers and Braham (1949)	1363/76	2–8	26/7	11.5/1.5	24/5	7/1.2	Precipitating cumulonimbus

NE Colorado	Musil *et al.* (1973)	2/1	~6	18/12[b]	6/3.7	10/~6[b]	4/2.5	Precipitating cumulonimbus
NE Colorado	Musil *et al.* (1976)	1/1	~7	18/—	10/—	7/—	4/—	Precipitating cumulonimbus
NE Colorado	Musil *et al.* (1977)	108/24	5–7	40/10–15[b]	10/2–3[b]	20/5–10	8/2.5	Precipitating cumulonimbus, some intense towering cumuli
NE Colorado	Sand (1976)	7/3	5–7	18/11[b]	9/3.4[b]	11/7[b]	3/1.8[b]	Precipitating cumulonimbus
Colorado, Oklahoma	Sinclair (1973)	—	9.2–9.8	26/—	15/—	10/—	12/—	Precipitating cumulonimbus
Oklahoma	Sinclair (1979)	—	4.5–6.0	20/—	—	>20/—	—	Precipitating cumulonimbus
NE Colorado, Colorado	Heymsfield and Musil (1982)	3/1	7	26/17[b]	7.5/4.5[b]	14/12[b]	3.6/3.2	Precipitating cumulonimbus, hail
Oklahoma	Heymsfield and Hjelmfelt (1981)	1/1	6–7	40/26[b]	6.5/4[b]	~10–20/8.5[b]	~6/~2	Squall line
Eastern Montana	Musil *et al.* (1982)	1/1	6–7	40/—	9.5/—	20/—	9/1	Severe storm

[a] From Knupp and Cotton (1985).
[b] Mean of maximum gusts.
[c] Median values.

of supercooled water drops and by the vapor deposition growth of ice crystals further contributes to cloud buoyancy. Owing to the small differences between saturation mixing ratios and environmental mixing ratios at colder temperatures, entrainment has a less inhibiting influence on updraft intensities than it does in the lower troposphere. This was clearly illustrated in early one-dimensional modeling studies (Simpson *et al.*, 1965; Simpson and Wiggert, 1969; Cotton, 1972). Figure 9.10 illustrates how different rates of activation of the ice phase affect updraft speeds aloft. Clearly, the ice phase can accelerate updrafts greatly in the upper troposphere. Vertically pointing Doppler radar-estimated updraft speeds reported by Battan (1975) and shown in Fig. 9.11 also show updraft speeds reaching maximum values in the upper troposphere. Battan inferred updraft speeds greater than 24 m s^{-1} at the 10.5-km level.

Updrafts in cumulonimbi are not regulated solely by buoyancy [i.e., term (2) in Eq. (9.1)]. Analyses of updrafts below cloud base and in weak-echo

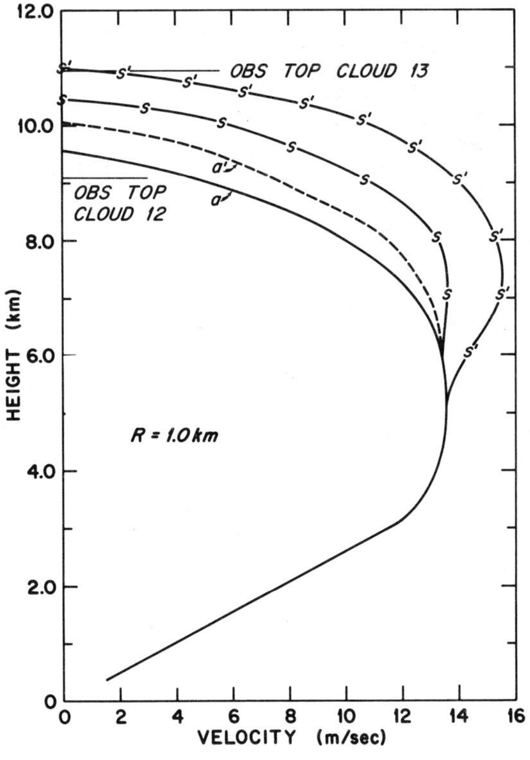

Fig. 9.10. Predicted cloud vertical velocity as a function of height for 27 May 1968, EML case study. [From Cotton (1972).]

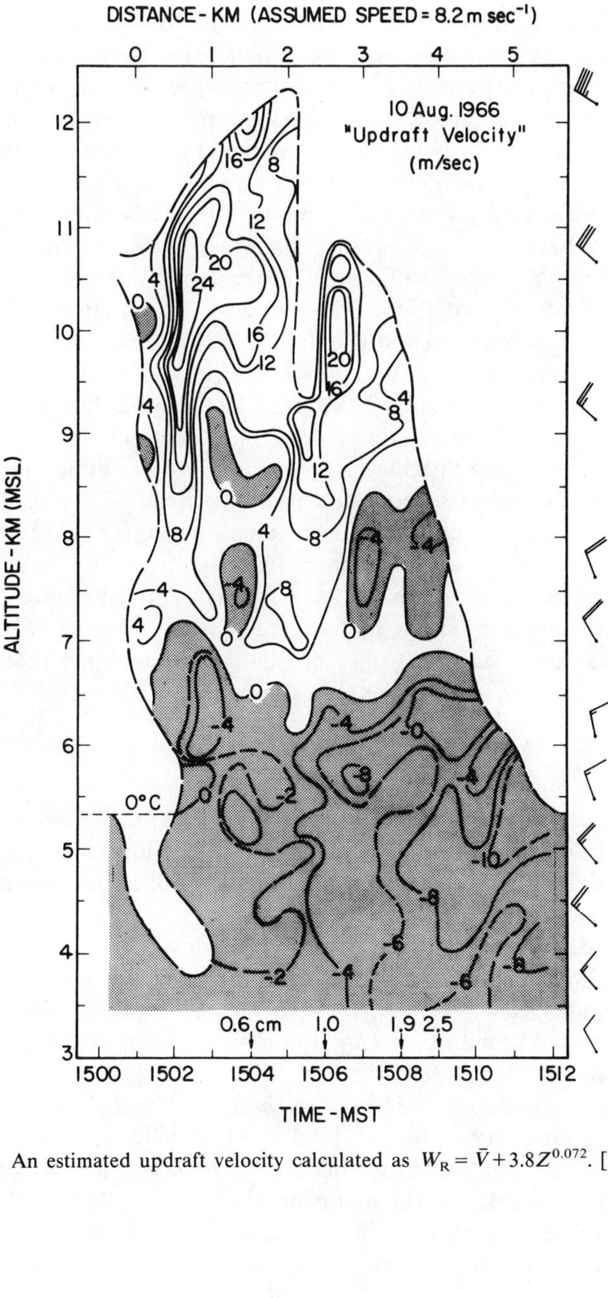

Fig. 9.11. An estimated updraft velocity calculated as $W_R = \bar{V} + 3.8Z^{0.072}$. [From Battan (1975).]

regions of severe cumulonimbi (Auer and Marwitz, 1968; Marwitz, 1973; Grandia and Marwitz, 1975; Ellrod and Marwitz, 1976) have revealed that updrafts below cloud base are frequently negatively buoyant. Researchers inferred that the vertical pressure-gradient force [term (1) in Eq. (9.1)] accelerates air near the surface upward to cloud base. Three-dimensional thunderstorm models and linear theory support this inference. Weisman and Klemp (1984) concluded from numerical experiments that, with a clockwise-curved wind hodograph, the updraft on the right flank of the storm system is locally induced by the lowering of pressure caused by the dynamic interaction of the storm with environmental wind shear. This dynamic forcing is often sufficiently strong to lift some of the negatively buoyant low-level air into the updraft circulation (Schlesinger, 1975, 1978, 1980; Rotunno and Klemp, 1982, 1985). Schlesinger (1978) also noted that a significant pressure high is simulated near the tops of growing cumulonimbi. We will examine the processes responsible for lowering of pressures in a cloud more fully in subsequent sections.

Marwitz (1973) and Sulakvelidze *et al.* (1967) also found that the height of the maximum updrafts were often quite low in the WER, i.e., 10–25 m s^{-1} at some 1–4 km above cloud base. Presumably water loading and entrainment of low-valued θ_e air could have weakened updraft strengths at levels above the updraft maximum in the WER. However, because researchers could not reliably estimate w by tracking chaff or balloons in supercooled clouds, these observations do not preclude the possibility that a primary updraft maximum existed at heights above the freezing level.

9.3.2 Turbulence

The observations reported by Marwitz (1973), Grandia and Marwitz (1975), and Ellrod and Marwitz (1976) indicate that the updraft air entering the base of cumulonimbi is smooth and relatively free of turbulence and remains so through a significant depth of the WER. This is consistent with the observation that the updrafts are negatively buoyant and accelerated by the vertical pressure-gradient force; negative buoyancy suppresses the production of turbulence.

The turbulent structure of cumulonimbi has been observed by aircraft (Steiner and Rhyne, 1962; Marwitz, 1973; Grandia and Marwitz, 1975; Ellrod and Marwitz, 1976) and Doppler radar (Frisch and Clifford, 1974; Frisch and Strauch, 1976; Battan, 1975, 1980; Donaldson and Wexler, 1969; Battan and Theiss, 1973). Turbulence levels can be estimated with a Doppler radar using two methods. The first method involves estimates of the energy dissipation rate [Eq. (9.2), term (e)] from the variance of the Doppler spectra obtained over a given radar pulse volume. The estimate of the energy dissipation rate assumes that the turbulence is homogeneous and isotropic over the radar pulse volume. This estimate of turbulence can be analogous to a subgrid-scale estimate of turbulence in a large-eddy simulation model. The radar pulse volume is a conical section having typical dimensions of 150 m in pulse length and 350 m in transverse beam width. The second method of estimating turbulence levels by Doppler radar is from the variability in mean radial Doppler velocities V_R from radar range gate to range gate ($\Delta V_r/\Delta R$) or by directly calculating spectral energy using point estimates of V_r. This estimate is analogous to the explicitly predicted turbulence at grid points in a LES model.

Knupp and Cotton (1982b) synthesized both estimates of turbulence variations in a cumulonimbus cloud with multiple-Doppler estimated mean flow fields and radar reflectivities. The storm system they analyzed was a quasisteady storm that moved to the left of the mean cloud-layer environmental winds. Figure 9.12 illustrates a conceptual model of the storm updraft and downdraft circulations as inferred from the Doppler analyses. Major features are a primary updraft located in the northwest (downshear) storm

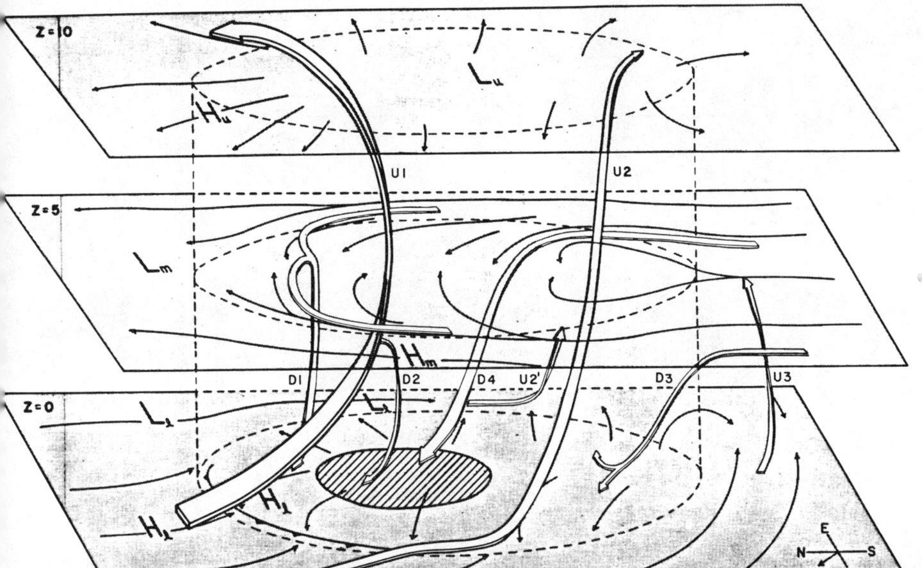

Fig. 9.12. Conceptual model of the flow patterns within C11 during its intense quasisteady stage. Streamlines depict airflow (storm relative) in the given horizontal planes. The arrowed ribbons represent updraft and downdraft circulations. Each H and L denotes regions of strong and weak flow, respectively, at lower (l), middle (m), and upper (u) levels. The hatched region denotes heavy rain. [From Knupp and Cotton (1982a).]

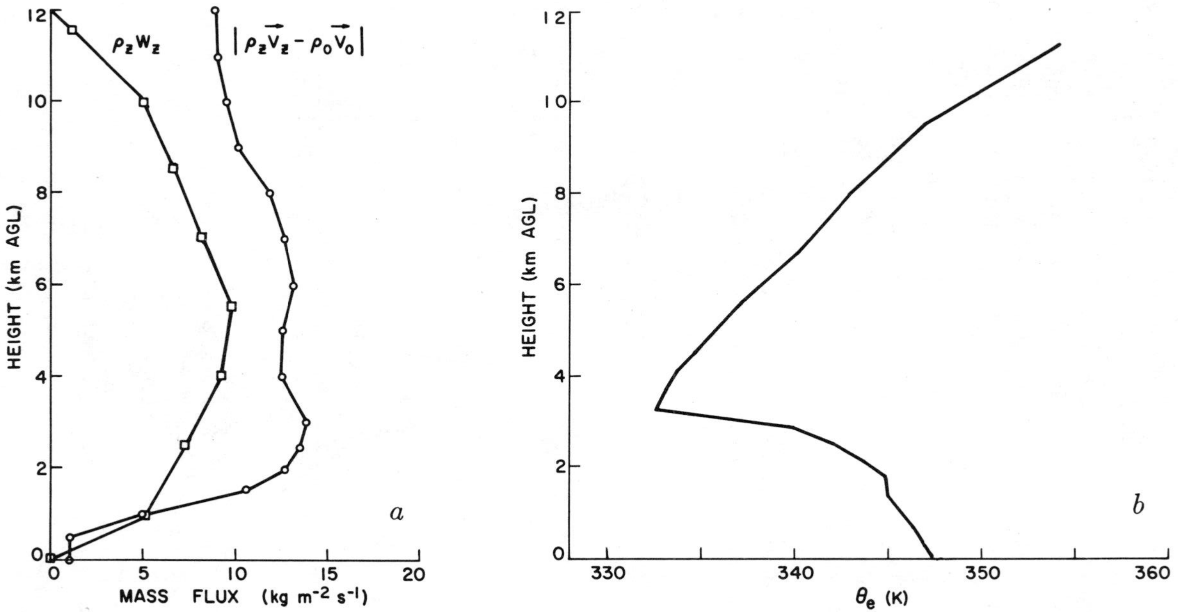

Fig. 9.13. Vertical profiles of (a) the magnitude of the difference between low-level momentum ($\rho_0 V_0$) and environmental momentum at cloud levels $\rho_z V_z$ and $|\rho_0 V_0 - \rho_z V_z|$, and the maximum updraft mass flux ($\rho_z w_z$) at 1927 MDT; (b) environmental equivalent potential temperature (θ_e). Add 3 km for height above MSL. [From Knupp and Cotton (1982b).]

quadrant with peak speeds of approximately 25 m s^{-1} and a secondary weaker updraft (10–20 m s^{-1}) located in the southern (upshear) quadrant. The most significant downdrafts were situated in the southeast–southwest storm quadrants, where relative inflow of low-valued θ_e air produced evaporation of cloud and precipitation particles. Figure 9.13b shows that a pronounced minimum in θ_e between 3 and 4 km AGL characterized the environment. Furthermore, the wind hodograph shown in Fig. 9.14 exhibits strong shear at low levels near 4 km MSL. Figure 9.15 shows that the largest dissipation (ε) magnitudes at low levels (1 km AGL) were in the eastern storm quadrant, where the air mass was confluent and a downdraft was present. Within the primary updraft inflow sector (northwest quadrant), turbulence levels were quite low. Figures 9.16 and 9.17 illustrate vertical cross sections of ε, $\Delta V_R/\Delta R$, mean wind vectors, and radar reflectivity as well as horizontal sections at 4.0 and 7.0 km AGL. At 4.0 km, turbulence was most substantial along the western and southwestern regions of the

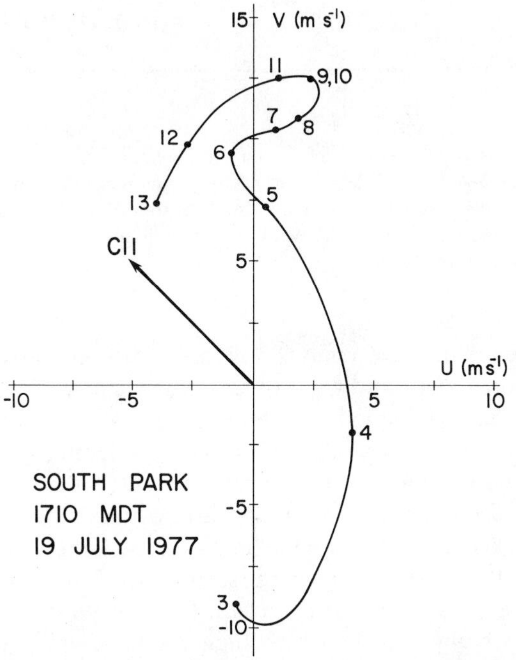

Fig. 9.14. Environmental wind hodograph derived from the 1710 MDT South Park sounding. Winds above 11 km MSL are probably influenced by anvil outflow from storms to the east. Numbers adjacent to hodograph curve denote height (kilometers MSL). The motion of storm C11 is depicted by the arrow. [From Knupp and Cotton (1982a).]

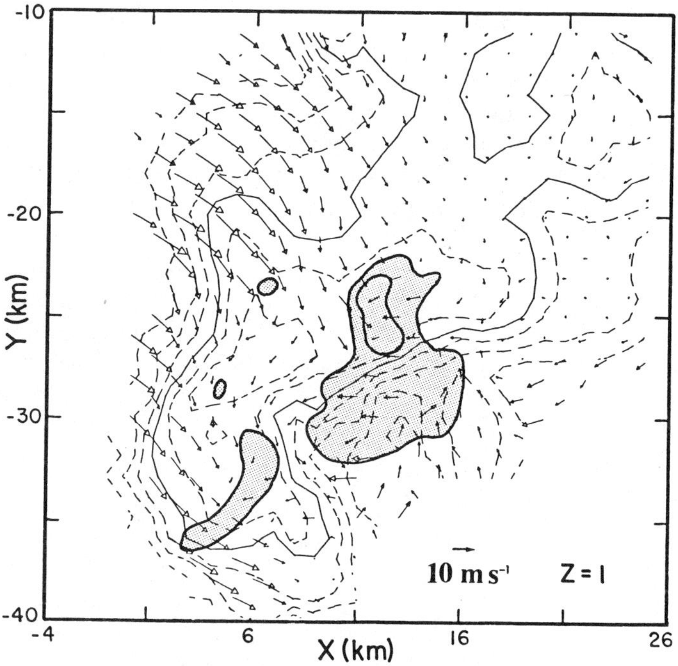

Fig. 9.15. Patterns of reflectivity factor, mean wind vectors (storm relative), and dissipation rate estimates at 1 km AGL at 1907 MDT. Reflectivity contours (thin solid and dashed lines) are drawn every 5 dBZ, with 30 dBZ denoted by thick solid lines. Dissipation rate contours are drawn at 0.02 and 0.04 $m^2\, s^{-1}$. The NOAA-D radar is located at the coordinate origin. [From Knupp and Cotton (1982b).]

storm. Turbulence in these regions appears to be generated by the inflow of dry environmental air and wind shear in the vicinity of the updraft. Both buoyant and shear production of turbulence appear to be important. Strong shear production is evident in the region (labeled SH in Fig. 9.16d) in which the horizontal flow carried by the updraft encounters the midlevel southerly flow diverging around the primary updraft. While the most intense levels of turbulence exist between 4 and 6 km AGL, the largest areal coverage of turbulence is in the upper portion of the storm in the form of small turbulence eddies, because large values of ε prevail. In the 4- to 6-km layer, $\Delta V_r/\Delta R$ has the largest areal coverage, which suggests that entrainment first occurs in large eddies that then cascade to smaller scales as they are carried aloft in the updrafts. Moreover, the turbulence is initiated at levels just below the 4-km level, where the minimum values in environmental θ_e reside (see Fig. 9.13b). At the same level, the greatest difference exists between the

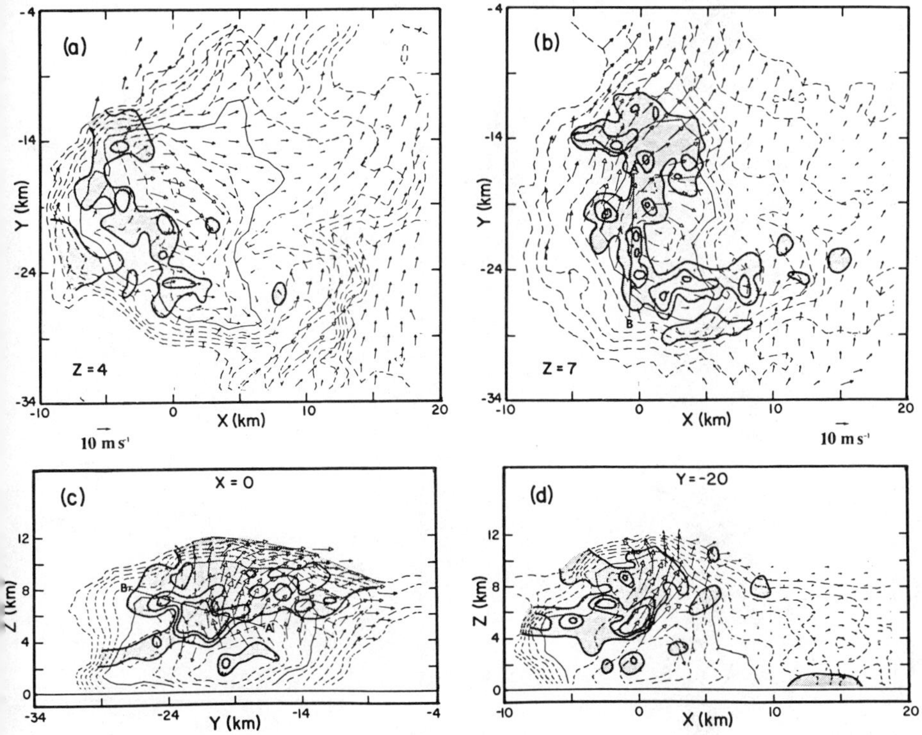

Fig. 9.16. Patterns of reflectivity factor, mean wind vectors (storm relative), and dissipation rate estimates at 1927 MDT for selected horizontal and vertical planes. Contours of ε are drawn for $\varepsilon = 0.02$, 0.04, and 0.07 $m^2 s^{-3}$. The NOAA-D radar is located at the coordinate origin. [From Knupp and Cotton (1982b).]

horizontal momentum carried by the updraft $(\rho_z \mathbf{V}_z)$ and the environmental momentum $(\rho_0 \mathbf{V}_0)$ (see Fig. 9.13a). The differences in horizontal momentum create horizontal shears and also support shear and buoyant production of turbulence as well as the generation of entrainment in the downshear flank of the cloud, as discussed in Chapter 8. Also shown in Fig. 9.13a is the vertical profile of updraft mass flux $(\rho_z w_z)$. Large values of $(\rho_z w_z)$ imply the existence of substantial lateral shear between the updraft and a relatively quiescent environment. Moreover, considering mass continuity, strong vertical divergence of $(\rho_z w_z)$ implies that a strong horizontal convergence (or dynamic entrainment) in the updraft must be present. Thus, above 4 km AGL, buoyancy and shear production of turbulence can be induced by a variety of cloud-environment interactions. In general, the intensity and spatial distribution of turbulence will vary depending on many of the

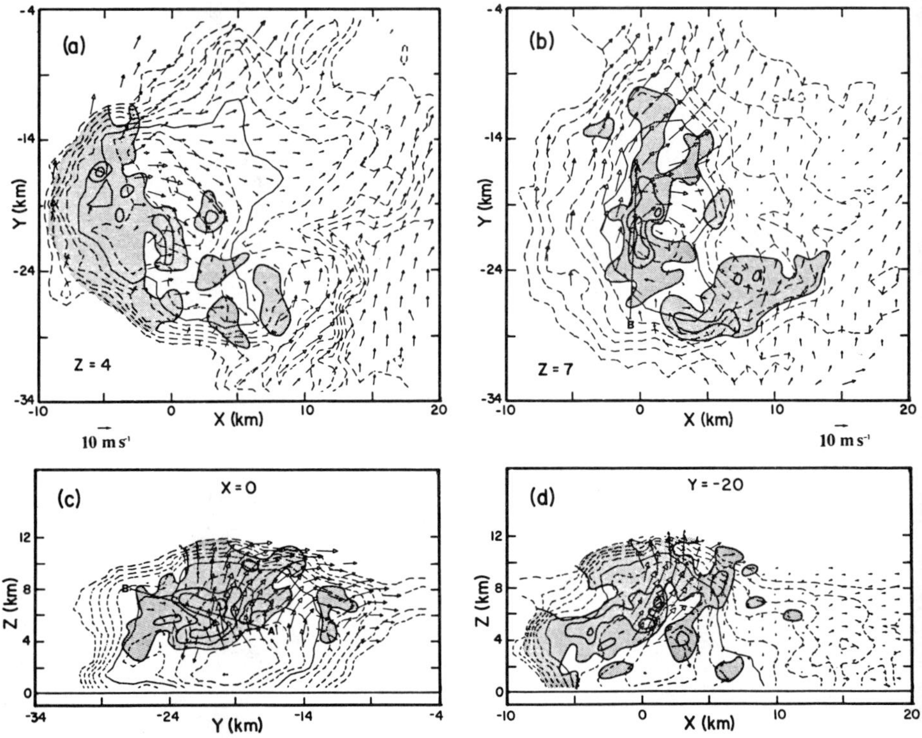

Fig. 9.17. Patterns of reflectivity factor, mean wind vectors (storm relative), and radial velocity differences ($\Delta V_r/\Delta R$) at 1927 MDT. $\Delta V_r/\Delta R$ contours are drawn at 5×10^{-3}, 10×10^{-3}, and $15\times10^{-3}\ s^{-1}$. The NOAA-D radar is located at the coordinate origin. [From Knupp and Cotton (1982b).]

characteristics of the storm environment we described above. As we shall see, the same environmental characteristics that enhance entrainment into the storm updrafts and turbulence generation also affect the genesis and intensity of storm downdrafts.

9.4 Downdrafts: Origin and Intensity

The same forces that affect updrafts initiate, maintain, or dissipate downdrafts in cumulonimbi. Thus Eq. (9.1) applies equally well to estimating downdraft speeds. Vertical pressure-gradient force, buoyancy, and turbulent Reynolds stresses initiate, maintain, or dissipate downdrafts. Similarly, Eq. (9.2) can describe the rate of change of turbulent kinetic energy in downdrafts. As in updrafts, both buoyant and shear production of turbulence are important. The relative importance of the various terms in Eqs. (9.1) and (9.2), however, may differ in downdrafts as compared to updrafts. Water loading, for example, may play a critical role in the initiation and maintenance of downdrafts because precipitation is more likely to settle in downdraft regions.

Downdrafts within cumulonimbi exhibit a wide spectrum of magnitudes and sizes. Vertical velocity data tabulated from the Thunderstorm Project flights (Byers and Braham, 1949) reveal median downdraft speeds and widths of 5–6 m s^{-1} and 1.2 km, respectively. More recent data acquired from intense northeastern Colorado cumulonimbi, as summarized by Musil *et al.* (1977), show a respective mean maximum downdraft speed and width of 8 m s^{-1} and 2.5 km. Maximum measured downdraft gusts and widths have exceeded 20 m s^{-1} and 7 km in several cases shown in Table 9.1, including the above two references. While these peak values are typically measured at and above midlevels, downdrafts of similar size and magnitude may also exist at low levels, because measurements within and near low-level precipitation cores have been avoided. Sinclair (1973, 1979) has reported frequent occurrences of downdrafts at middle to upper cloud levels in both clear and cloudy portions of intense cumulonimbi. Figure 9.18 (Sinclair, 1973) gives an example of the variability of updraft/downdraft structure in cumulonimbi and of clear-air downdrafts with magnitudes of several meters per second and approximately 10-km widths bordering an active updraft. Temperatures within this clear-air downdraft were about 2°C warmer than adjacent environmental air, in contrast to relatively cold temperatures (−2°C) within the stronger cloud interior downdraft. Other indirect evidence

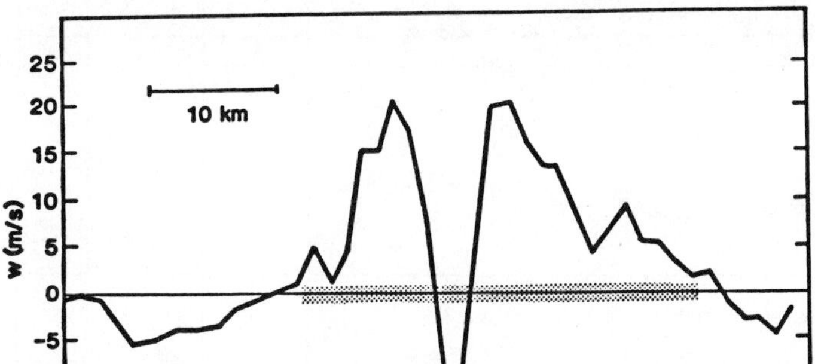

Fig. 9.18. Aircraft-measured vertical velocity time series at the 9.8-km level of an intense convective storm over northeastern Colorado on 21 June 1972. Cloudy regions are represented by stippling. [Adapted from Sinclair (1973).]

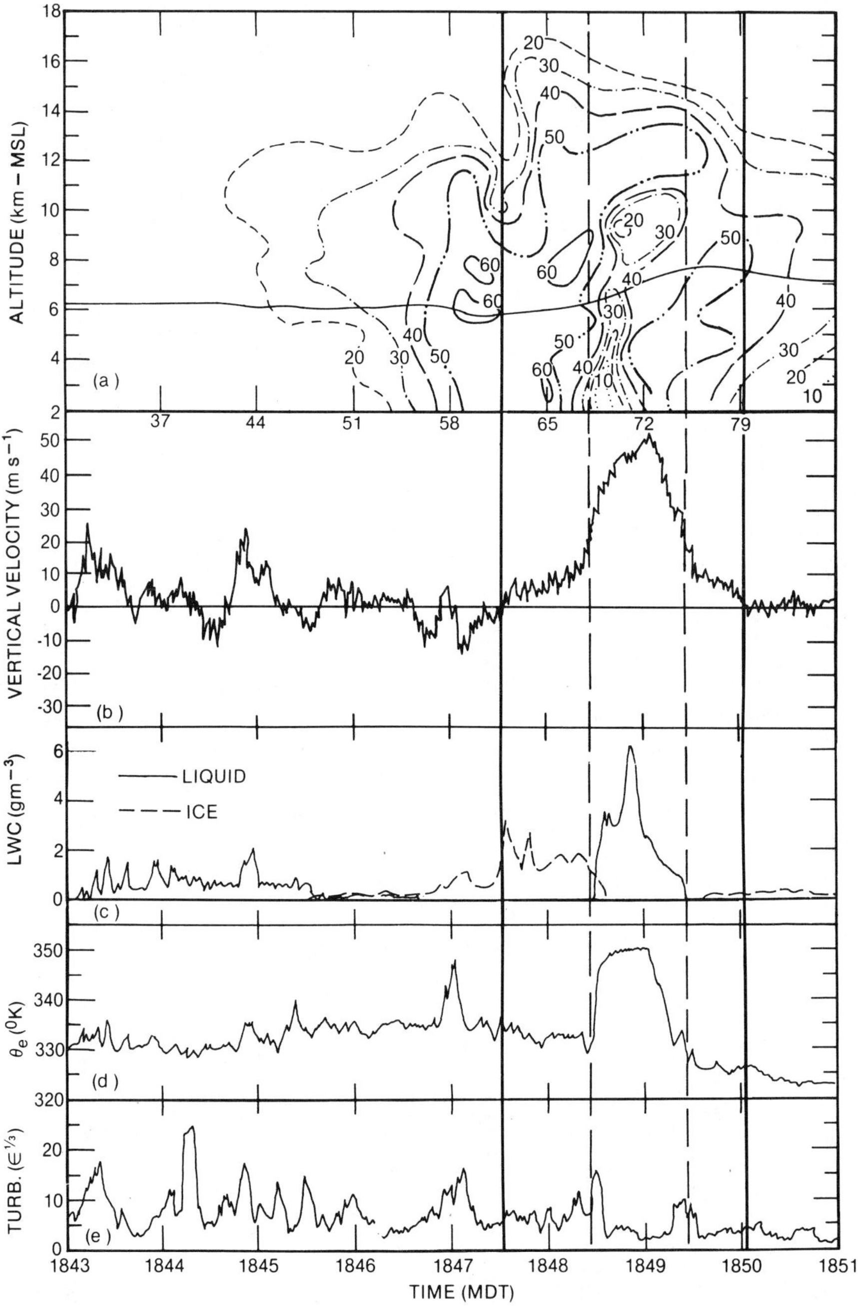

ALTITUDE (km – MSL)
(a)
37
44
51
58
65
72
79
VERTICAL VELOCITY (m s^{-1})
(b)
LWC (gm^{-3})
LIQUID
ICE
(c)
θ_e (^{0}K)
(d)
TURB. ($\in^{1/3}$)
(e)
1843
1844
1845
1846
1847
1848
1849
1850
1851
TIME (MDT)

supporting the presence of clear-air downdrafts or larger regions of weaker subsidence adjacent to precipitating convection is summarized in Fritsch (1975) and in Hoxit *et al.* (1976).

Figure 9.19, from Musil *et al.* (1986), illustrates an extreme case of an extensive and strong downdraft measured near the core of a large, severe storm in southeastern Montana. In this case, extensive downdrafts 5–10 km wide with amplitudes of 10–20 m s^{-1} flanked an intense 40-m s^{-1} updraft at 6–7 km MSL. Somewhat smaller but intense downdrafts and updrafts were encountered on the upshear flank.

Relative to midlatitude continental cumulonimbi, tropical maritime cumulonimbi contain much weaker updrafts and downdrafts. LeMone and Zipser (1980) summarized vertical motions measured within GATE convective clouds, and they found a median value of 1.8 m s^{-1} with maximum downdraft speeds of 10 m s^{-1} in rare instances. A summary of these measurements and a comparison to Thunderstorm Project data are shown in Fig. 9.9. This shows that GATE draft magnitudes are one-third to one-half the Thunderstorm Project draft magnitudes. Figure 9.9 also shows composite draft profiles Jorgensen (1984) obtained through hurricane convective bands and inner cores that are similar to GATE profiles.

Also shown in Fig. 9.9 is an increase with height of both updraft and downdraft magnitudes for GATE, Thunderstorm Project, and hurricane drafts, a behavior similar to that measured in nonprecipitating cumulus congestus. Such a pattern may be biased due to the previously mentioned lack of penetrations through low-level precipitation cores, particularly over continents. Some measurements within midlatitude precipitating convection indicate that low-level downdrafts associated with precipitation may attain intense magnitudes. For example, Rodi *et al.* (1983) measured 15-m s^{-1} peak downdrafts within light precipitation beneath the bases of cumulus congestus clouds forming above deep, dry mixed layers in Colorado.

Vertically pointing Doppler (VPD) radar observations provide further evidence that strong downdrafts of continental cumulonimbi exist at low levels. Battan (1975, 1980) has presented the most comprehensive set (four cases) of VPD observations. These are supplemented by additional VPD

Fig. 9.19. Summary of T-28 data during penetration 1 between 1843 and 1851 MDT: (a) vertical section of reflectivity along the penetration path of the T-28, including line showing T-28 altitude; radar contours are in dBZ and an approximate horizontal scale in kilometers is indicated; (b) vertical air velocity from T-28 measurements; (c) liquid-water and ice mass concentrations; (d) equivalent potential temperatures; and (e) intensity of turbulence. The dashed vertical lines represent the approximate boundary of the WER, while the solid vertical lines outline the total region of updraft associated with the WER. [From Musil *et al.* (1986).]

data contained in Battan and Theiss (1970), Strauch and Merrem (1976), Wilson and Fujita (1979), and Mueller and Hildebrand (1983). Figure 9.11 typifies one of those presented in Battan's (1975, 1980) examples. The VPD observations generally reveal vertically continuous, large-scale downdrafts (6–12 m s^{-1} maximum) in the lowest 3–4 km. Pockets of downdraft that typify the middle to upper levels represent small-scale drafts commonly measured by aircraft (e.g., Fig. 9.17). The observations of Wilson and Fujita (1979) show considerable variability in small-scale intense updrafts and downdrafts near the radar-echo top. The magnitude of such near-cloud-top (overshooting) downdrafts may approach 40 m s^{-1}, as Fujita (1974) determined from airborne photogrammetric analyses of intense cumulonimbi. Cloud-top-overshooting downdrafts differ from those driven by evaporative cooling of cloud and precipitation at lower to middle levels (<6 km). Figure 9.20, taken from Newton (1966), portrays a qualitative differentiation between these two downdraft types. Here, a representative environmental sounding illustrates buoyancy within updrafts and downdrafts whose parcel paths on the thermodynamic diagram in Fig. 9.20a are depicted by long dashed lines. Differences between the updraft parcel curves A, B, and U arise from differences in assumed rates of lateral mixing of updraft air and environmental air. The vertical velocity profiles in Fig. 9.20b were obtained

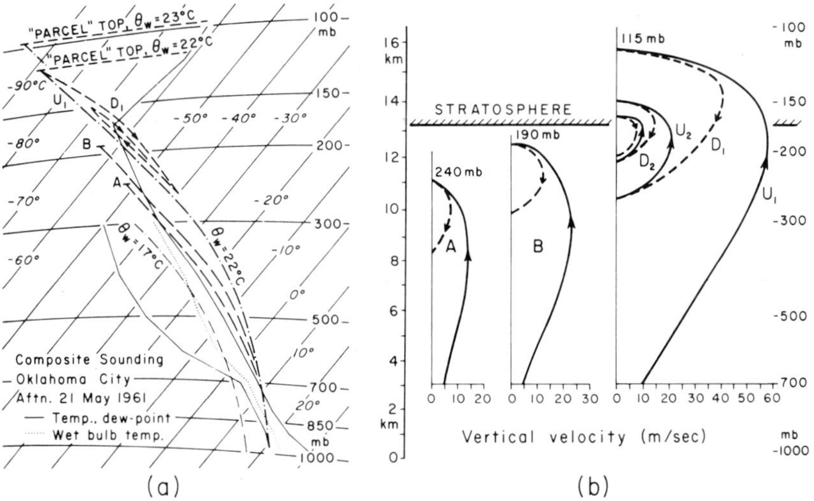

Fig. 9.20. (a) Representative sounding at Oklahoma City, Oklahoma, afternoon of 21 May 1961 (thin solid lines, temperature and dew point; dotted line, wet-bulb temperature). (b) Vertical velocities in updrafts (solid lines) and downdrafts (dashed lines), under different assumptions. [From Newton (1966).]

by integrating the vertical equation of motion, including buoyancy forces, modified by entrainment of momentum. Curves U and D illustrate updraft ascent to heights exceeding (overshooting) the parcel equilibrium level, beyond which rising air quickly becomes negatively buoyant. Very strong downward forces from negative buoyancy then lead to downdrafts that are negatively buoyant only in the upper levels. These initial overshooting downdrafts similarly overshoot their level of neutral buoyancy, leading to subsequent decaying buoyancy oscillations U_2, D_2.

In an analysis of tropical convective systems, Zipser (1969, 1977) distinguished 1- to 5-m s^{-1} downdrafts of scale ~1 km associated with active convective cloud cores from weaker, 0.1- to 0.5-m s^{-1} mesoscale downdrafts of scale 10–100 km associated with an extensive anvil cloud trailing the active convection. Other studies on both squall lines (Houze, 1977; Ogura and Liou, 1980) and cumulonimbus cloud clusters (Leary and Houze, 1979a) show similar scale separation between convective-scale and mesoscale downdrafts. Mesoscale downdrafts will be discussed more fully in Chapter 10.

Similar spatial-scale variations greater than one order of magnitude appear in convective downdrafts. As discussed previously, direct aircraft observations revealed nonprecipitating convective cloud downdrafts no greater than ~500 m, in contrast to ~10-km-wide downdrafts occasionally measured within precipitating cumulonimbi. Indirect observations suggest a similar range of scales in intense downdrafts (downbursts). From their inspection of surface damage patterns that exhibited scales from a few hundred meters to greater than 10 km, Fujita and Wakimoto (1981) and Forbes and Wakimoto (1983) inferred a wide spectrum of downburst sizes. Fujita (1981) also found short time scales (~5 min) of low-level outflow wind associated with small downbursts.

A number of investigators have inferred downdraft source levels by analyzing thermodynamic tracers such as equivalent potential temperature (θ_e), wet-bulb potential temperature (θ_w), or moist static energy ($h = c_p T + Lq + gz$), all of which are conserved approximately for dry and moist adiabatic processes assuming no mixing or ice-phase change. Vertical profiles of θ_e in the environment of cumulonimbi typically show a minimum near 500–600 mbar. Mal and Desai (1938), Normand (1946), and Newton (1950) were among the first to apply this principle in inferring that cold downdraft air measured at the surface originated several kilometers above.

Other investigators have subsequently indicated that low-valued midlevel θ_e air often reaches the surface within downdrafts. Using an analysis of θ_e, Zipser (1969) inferred that midlevel air near the level of minimum θ_e descended approximately 500 mbar to the surface behind a tropical squall line. Similar inferences concerning the origin of downdraft air are made using thermodynamic analysis in many other cases, for example, in midlatitude convective storms (Newton and Newton, 1959; Foote and Fankhauser, 1973; Fankhauser, 1976; Lemon, 1976; Barnes, 1978a, b; Ogura and Liou, 1980). In other less intense cases, downdrafts apparently originate just above cloud base, significantly below the level of minimum θ_e. Betts (1976) estimated that downdraft air descended about 100 mbar from just above the cloud base of Venezuelan storms. Barnes and Garstang (1982) and Johnson and Nicholls (1983) inferred downdraft source levels near 700–750 mbar for precipitating tropical convection of moderate intensity.

Many analyses indicate that low-level downdrafts are closely associated with precipitation falling beneath cloud base from convective clouds of weak to severe intensity. Byers and Braham (1949) demonstrated a close association between downdrafts and surface rainfall. They inferred that downdrafts were initiated by precipitation loading and were maintained by evaporation of cloud and precipitation. Other striking examples showing this relationship can be seen in the surface mesonet analyses of Foote and Fankhauser (1973), Fankhauser (1976, 1982), Holle and Maier (1980), and Wade and Foote (1982). Finally, Barnes and Garstang (1982) established a positive correlation between areal precipitation rate and downdraft transport of mass and low static energy (h) into the boundary layer. They determined that precipitation rates needed to exceed a threshold of $\sim 2\ \mathrm{mm\ h^{-1}}$ (averaged over an area of $\sim 16\ \mathrm{km^2}$) before air with low values of h was transported into the subcloud layer.

Figure 9.21, derived from a detailed case study of a northeastern Colorado hail storm (Fankhauser, 1976; Browning *et al.*, 1976), shows the colocation of heavy precipitation, downdraft, and low-valued θ_e air. In other cases, the lowest valued θ_e air is located just upshear of the downdraft and precipitation core (e.g., Barnes, 1978a, b; Nelson, 1977; Lemon, 1976). The multiple-Doppler radar presentations in Kropfli and Miller (1976), Ray *et al.* (1981), Foote and Frank (1983), and Wilson *et al.* (1984), among others, also illustrate that low-level downdrafts are either located within or along the upshear edge of heaviest low-level precipitation. An example shown in Fig. 9.22 (from Klemp *et al.*, 1981) depicts flow patterns derived from a multiple-Doppler radar analysis and a comparative three-dimensional cloud model simulation of a tornadic thunderstorm. In this case low-level downdrafts with magnitudes up to $10\ \mathrm{m\ s^{-1}}$ were located within and just upshear of the precipitation core (see also Ray *et al.*, 1981, for additional details of this case). Also note that downdraft regions located in the far eastern flank at midlevels (4 and 7 km) are significantly damped in the same relative locations at low levels. Downdraft air parcel trajectories analyzed from both observations and model results indicated that very little of the air reaching the surface originated above 3 km AGL.

The relationship between subcloud precipitation and downdrafts appears to be especially strong in cases where precipitating cumulus congestus and cumulonimbi form above deep, dry boundary layers in the western United States. Braham (1952) speculated that subcloud evaporation of a significant fraction of precipitation was of primary importance in dry regions. In his study of cold mesoscale outflows, Fujita (1959) indicated that rises in pressure beneath convective systems, which are proportional to the total cooling from precipitation evaporation at low levels, were a function of the boundary layer dryness and total surface rainfall. Storm systems forming in relatively dry areas such as western Texas also produced rises in surface pressure essentially equal to those over southern and midwestern states where significantly more rain fell.

Krumm (1954) and MacDonald (1976) estimated that precipitation evaporation within a deep dry adiabatic layer alone could account for strong surface winds frequently experienced near high-based (3–4 km AGL) convective clouds. Brown *et al.* (1982) recently examined this problem further and found that even relatively shallow clouds producing virga and no lightning were capable of generating surface outflow winds in excess of

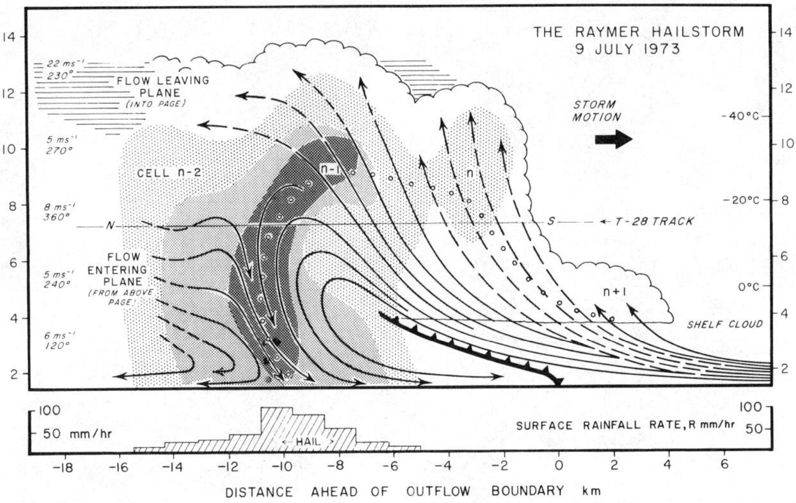

Fig. 9.21. Schematic model from a northeastern Colorado storm case study. Storm relative airflow is composited from aircraft, Doppler radar, and surface mesonet measurements. Light stippling represents cloud; successively darker stippling denotes radar reflectivity of 35, 45, and 50 dBZ. Measurements of rain, θ_e, and wind (the component in the plane of the figure), from a surface station over which the storm passed, are shown in the lower portion. The maximum wind vector is $11 \mathrm{~m~s^{-1}}$. [Adapted from Browning *et al.* (1976) and Fankhauser (1976).]

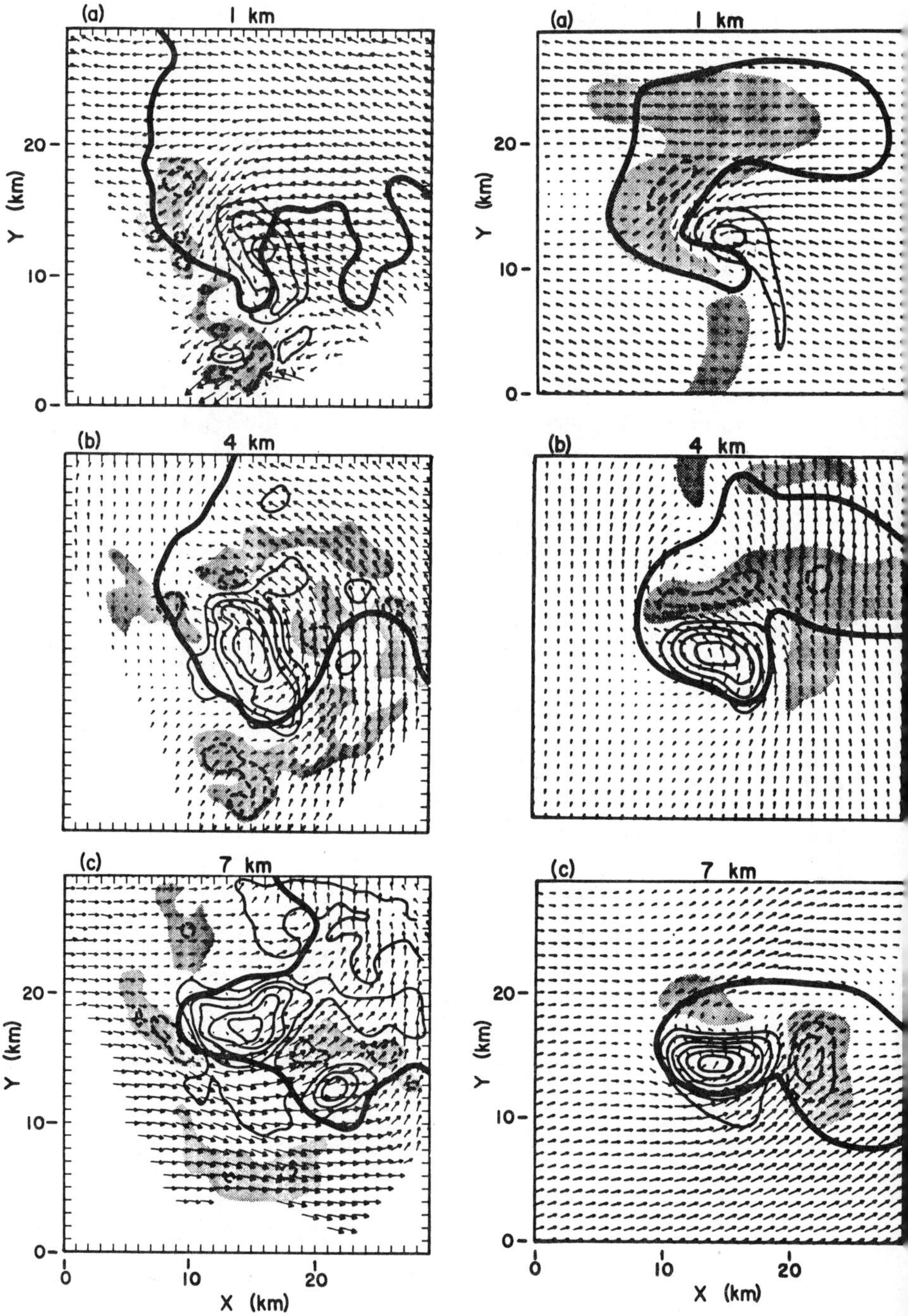

(a)
1 km
(b)
4 km
(c)
7 km
Y (km)
X (km)
0
10
20
(a)
1 km
(b)
4 km
(c)
7 km
Y (km)
X (km)
0
10
20

30 m s^{-1}. They constructed a composite sounding which depicts a dry adiabatic layer (3–4 g kg^{-1} mean mixing ratio) from the surface up to 500 mbar. Brown *et al.* hypothesized that weak updrafts within such clouds produced small precipitation particles that evaporate more readily than larger particles usually found in more intense cumulonimbi. Recently, *in situ* measurements beneath clouds of this type in Colorado (Rodi *et al.*, 1983) have revealed a striking correlation between small precipitation particles (most less than 1 mm in diameter) and very strong downdrafts, or downbursts (up to 15 m s^{-1}), 3 km below the base of lightly precipitating convection.

Numerous modeling studies have demonstrated the importance of the drop-size distribution on downdraft speeds. Hookings (1965) calculated that with other factors remaining unchanged, more vigorous downdrafts were produced for (i) smaller droplet sizes, (ii) greater liquid-water content, and (iii) lower initial humidity at downdraft origin. Kamburova and Ludlam (1966) and Das and Subba Rao (1972) showed that for specified downdraft speeds, the rates of evaporation and associated cooling strongly depend on precipitation size and intensity. More recently Knupp (1985) demonstrated in a two-dimensional cloud simulation that quartering the characteristic precipitation size for both rain and graupel particles increased maximum downdraft magnitudes over 2 m s^{-1}, enhanced low-level cooling by 3.5°C, and increased downdraft depth by 0.8 km. The dependence of downdraft intensity on precipitation size is relatively easy to understand, because evaporation occurs at the surface of raindrops. Because the surface-to-volume ratio is greatest for smaller drops, a downdraft containing a given liquid-water content will expose a greater surface area to evaporation if the drops are small than if they are large.

Although subcloud downdrafts within mixed boundary layers are often associated with precipitation, not all precipitation shafts generate strong downdrafts. A good example appears in Knight (1981), in which light precipitation from a long-lived cumulonimbus cloud fell into a deep, relatively dry mixed layer and failed to generate downdrafts and significant outflow. Limited precipitation observations indicated the presence of sparse,

---

Fig. 9.22. Horizontal cross sections of flow patterns from a four-Doppler radar analysis at 1833 CST (left column) and from a three-dimensional cloud model after 2-h simulation (right column) initiated with a sounding considered representative of the observed storm's environment. Updrafts (solid lines) and downdrafts (dashed lines) are contoured at 5-m s^{-1} increments, with values less than −1 m s^{-1} shaded. Wind vectors are scaled such that one grid interval represents 20 m s^{-1}. The heavy solid line outlines the 0.5-g kg^{-1} rainwater contour for the model output (left) and the 30-dBZ radar reflectivity contour for the Doppler analysis (right). [From Klemp *et al.* (1981).]

large particles, which, having a slow net evaporation rate, may explain the lack of significant cooling and downdraft activity.

Occasionally, low-level downdrafts in severe storms appear to assume a two-celled pattern: one associated with heavy precipitation as described above and another located on the upshear flank within lighter precipitation. The schematic in Fig. 9.23 illustrates the relative locations of what Lemon and Doswell (1979) term a forward-flank downdraft (FFD) located within the precipitation core downshear, and a colder rear-flank downdraft (RFD) within lighter precipitation on the upshear storm flank. Lemon and Doswell speculate that the RFD is initially dynamically forced by perturbation pressure gradients on the upshear flank at high levels (7–10 km), and is then maintained by loading and evaporation of anvil precipitation at middle to lower levels. The pressure gradients were assumed to be generated by high pressure typically present within the upshear flank of updrafts in which perturbation pressure increases with height up to midlevels. Other possible mechanisms will be discussed in subsequent sections.

Although the existence of the RFD as distinct from the FFD is weakly supported by some surface mesonet analyses (Lemon, 1976; Barnes,

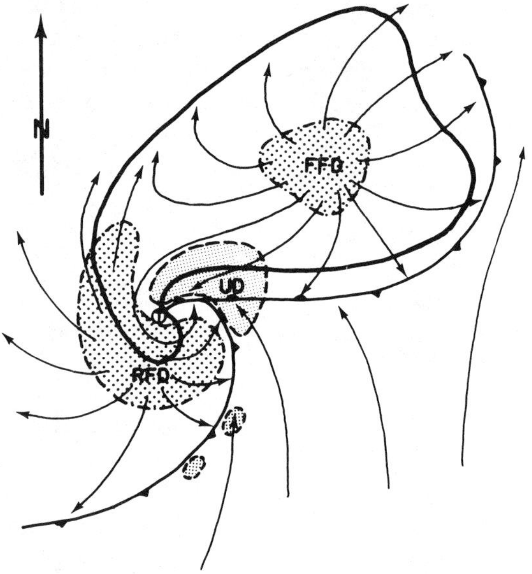

Fig. 9.23. Schematic plan view of surface features associated with a tornadic thunderstorm. Gust fronts are depicted by barbed frontal symbols. Low level positions of draft features are denoted by stippled areas, where UD is updraft, RFD is rear-flanking downdraft, and FFD is forward-flanking downdraft. Streamlines denote storm-relative flow. [From Lemon and Doswell (1979).]

1978a, b), the RFD structure and dynamics are unclear. Many Doppler studies show horizontal continuity in downdrafts associated with precipitation and extending toward the upshear flank. Some investigations indicate that minimum-valued θ_e air lies within downdraft cores (see Fig. 9.21), while in others the θ_e minimum lies on the upshear edge of the downdraft core.

Knupp (1985), in a comprehensive analysis of low-level downdrafts in high-plains convective storms over the United States, included the synthesis of a number of cases observed by multiple-Doppler radar, two- and three-dimensional numerical simulations of the observed convective storms, model sensitivity studies, and the use of simple diagnostic models. Figure 9.24a illustrates a conceptual model of several predominant downdraft flow branches that would occur in an environment having moderate shear, such as is illustrated in Fig. 9.24c. Figure 9.24c illustrates a sounding that has a planetary boundary layer depth of approximately 2 km, a conditionally unstable layer between 2 and 4 km, and an absolutely stable structure above 4 km. A projection of the downdraft branches onto an east–west plane is illustrated in Fig. 9.24b. Also shown in Fig. 9.24b are processes affecting downdraft intensities and the relationship between the sounding and those branches.

Knupp referred to those branches originating above the PBL as "mid-level" and those originating within the PBL he called "up–down." Originating within the updraft inflow sector, the up–down branch may rise up to 4 km before descending within the precipitation-laden primary downdraft region. Initially an upward-directed pressure gradient assists the lifting of the parcel until latent heating produces positive buoyancy along the upper portion of its path. Cooling produced by melting of precipitation along with "loading" by precipitation may provide sufficient negative buoyancy to initiate its transition to the down segment of this branch. Thus the up–down branch typically occurs during the mature storm phases. Both Doppler radar analyses and cloud-model results indicate that mixing occurs near the summit between moist air of this branch and drier, lower valued θ_e air flowing along the midlevel branch b (Fig. 9.24a). Such mixing produces subsaturated, intermediate-valued θ_e air, which promotes increased evaporation rates along the descending portion of the up–down branch. The sudden decrease in buoyancy associated with the enhanced evaporation rates, along with water loading and, perhaps, a relative movement of this trajectory away from the region of favorable upward-directed pressure gradient forces, all contribute to the fast descent rates along this branch.

The midlevel downdraft branches (b and c) originate within a 2- to 4-km AGL layer above the PBL in the southwestern quadrant of the storm. These branches are most pronounced during the developing downdraft stages.

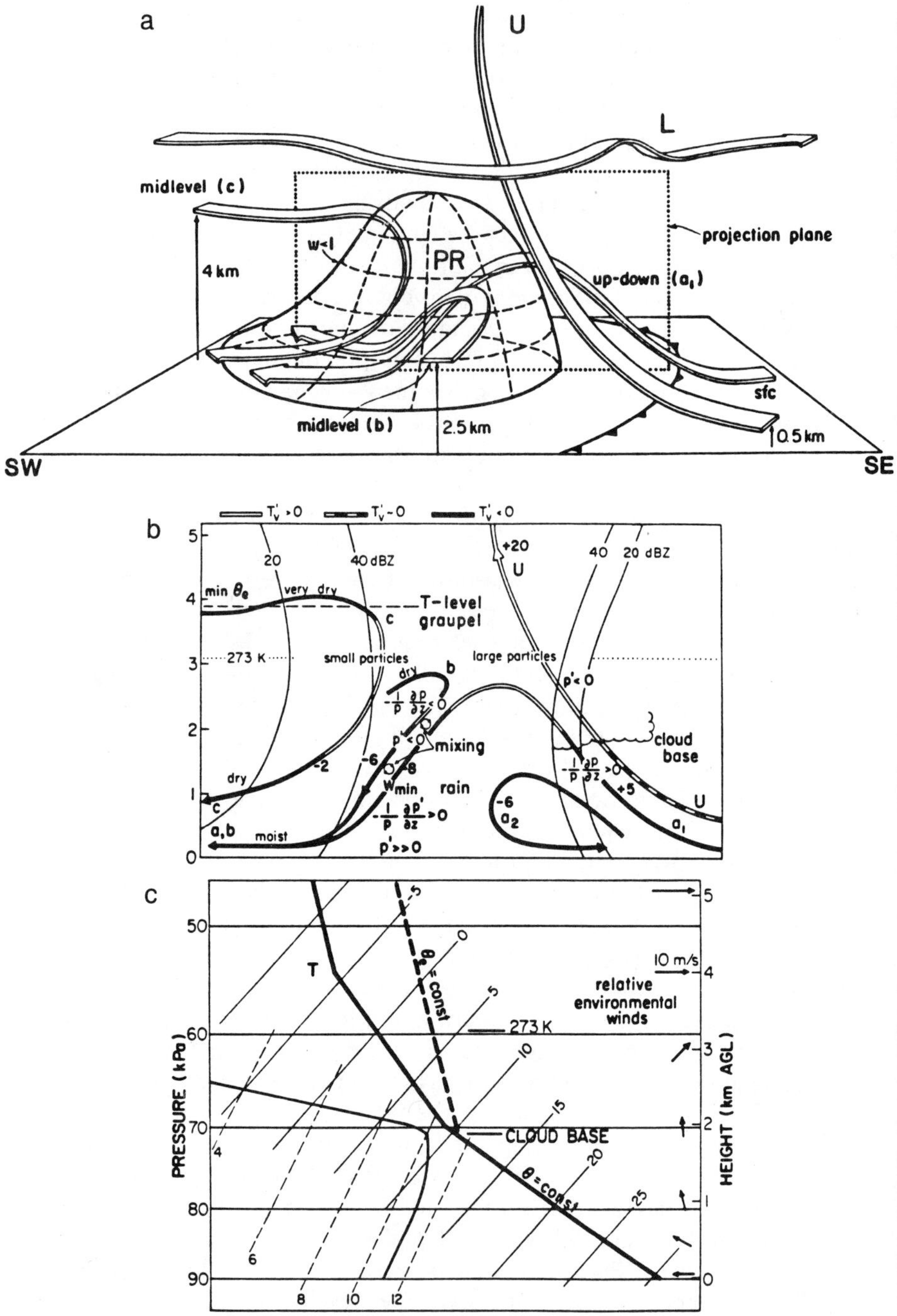
a
U
L
midlevel (c)
projection plane
w<1
PR
4 km
up-down (a₁)
sfc
midlevel (b)
2.5 km
0.5 km
SW
SE
b
T'v > 0
T'v ~ 0
T'v < 0
20
40 dBZ
+20
U
40
20 dBZ
min θe
very dry
T-level
graupel
c
273 K
small particles
large particles
dry
b
p' < 0
mixing
cloud
base
dry
-2
-6
-8
Wmin
rain
-6
a2
+5
U
a1
c
a,b
moist
p' >> 0
c
T
θe = const
10 m/s
relative
environmental
winds
273 K
CLOUD BASE
θ = const
PRESSURE (kPa)
HEIGHT (km AGL)

Observations and model results indicate that both branches exhibit a period of weak ascent before descending within precipitation. Air with low-valued θ_e is transported to near the up-down branch where mixing occurs (trajectory b) and along trajectory c to the surface. In agreement with earlier diagnostic modeling studies of Kamburova and Ludlam (1966), Knupp found that descent rates along midlevel branches depend on static stability of environmental air at the level of origin of the branch, with greater stability favoring slower descent rates. Downdraft branches such as c typically descend at rates of 1–3 m s^{-1} outside heaviest precipitation and reach the surface some distance behind the convective precipitation cores. This accounts for the lowest valued θ_e air at the surface being observed along the upshear sector of the storm (Lemon, 1976; Barnes, 1978a, b; Ray *et al.*, 1981; Knupp, 1985). Knupp noted that the thermodynamic and kinematic properties along branch c resemble those found within mesoscale downdrafts (Zipser, 1977; Leary and Houze, 1979b).

Figure 9.24b indicates that as the precipitation-laden air along downdraft branches (b) and the up-down branch enters the subcloud layer, evaporation of precipitation accelerates the air downward and creates a downward-directed pressure-gradient force. This pressure-gradient force further enhances the downward acceleration of air and encourages the mixing between it and branch b containing lower valued θ_e air. This example demonstrates that the vertical pressure-gradient force contributes significantly to downdraft intensities and to the volume of downdraft air displaced into the subcloud layer.

An interesting finding by Knupp was the importance of melting to total cooling along downdraft trajectories. He found that melting can account for 10–60% of the total cooling along certain downdraft trajectories. The largest contribution by melting occurred along the up-down trajectories in which parcels often pass through the melting zone twice or reside within the melting zone for considerable time at the apex of the up-down path. Precipitation evaporation accounts for the remainder of the total cooling and is typically greatest within drier air along midlevel trajectories. The contribution of melting to total downdraft cooling becomes greater in clouds having relatively moist subcloud layers and low cloud-base heights such as exist in the maritime tropics.

Fig. 9.24. (a) Schematic illustrating primary relative flow branches comprising the low-level precipitation-associated downdraft located along the upshear flank with respect to the updraft. (b) Projection of primary downdraft flow branches onto a vertical east–west plane. Physical processes along each branch are portrayed. (c) Sounding illustrating the relationship of downdraft properties to environmental structure. [From Knupp (1985).]

We can see from preceding sections that the intensity of updrafts and downdrafts in cumulonimbi is quite variable both in space and in time. Furthermore, the organization of updrafts and downdrafts in cumulonimbi is fundamental in determining the particular character of the storm system. We will now examine the downdraft outflows near the surface.

9.5 Low-Level Outflows and Gust Fronts

As the downdraft air approaches the surface, it diverges and forms a gust front that commonly produces significant convergence along its leading edge. Gust front properties and associated flows have been studied from analyses of observations (Charba, 1974; Goff, 1976; Wakimoto, 1982; Fankhauser, 1982; Sinclair and Purdom, 1983), laboratory experiments (Simpson, 1969; Simpson and Britter, 1980), and numerical modeling studies (Mitchell and Hovermale, 1977; Thorpe *et al.*, 1980; Droegemeier and Wilhelmson, 1985a, b). Figure 9.25 illustrates some gust-front features as composited from observational and numerical studies. Major features include the nose, or elevated leading edge, the head, which marks the greatest height of the advancing system, and the turbulent wake. The gust-front system is usually 0.5–2 km deep, depending on strength and distance from downdraft source. Greatest vertical motions range from several to ~10 m s^{-1} near the upper portion of the front edge (e.g., Matthews, 1981; Wakimoto, 1982; Fankhauser, 1982), above which arc clouds or deep convection may form. Downward motion characterizes the turbulent wake zone behind the head.

Wakimoto (1982) described four stages in the evolution of a gust front. During Stage I evaporatively cooled downdraft air begins to diverge near the surface. During Stage II an advancing gust front forms; it exhibits a roll structure at its leading edge. The mature stage is similar to that illustrated in Fig. 9.25. During the dissipating stage (Stage IV), the gust front is no

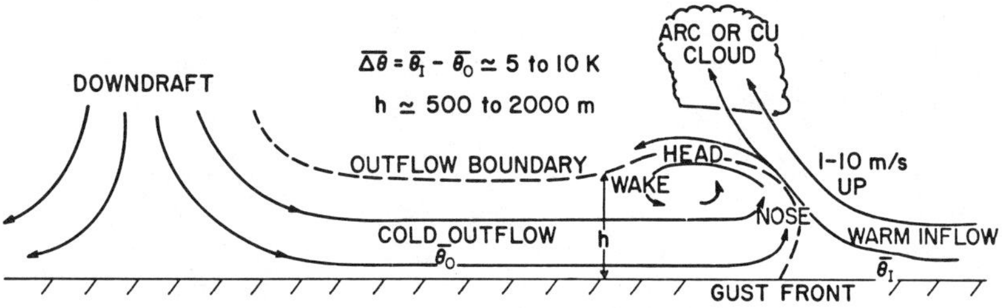

Fig. 9.25. Schematic structure of a gust front. [Adapted from Goff (1976), Fankhauser (1982), and Wakimoto (1982).]

longer fed by evaporatively chilled downdraft air, and the advancing gust front shrinks in vertical extent.

Movement speeds of the advancing front relative to ambient flow is often found to be close to the speed of density currents, where if we approximate the density difference across the gust front as $\Delta\rho/\rho_0 \simeq \overline{\Delta\theta}/\theta_0$, we find

$$c = k[gh(\overline{\Delta\theta}/\theta_0)]^{1/2}, \tag{9.3}$$

where $\overline{\Delta\theta}$ is the average potential temperature deficit over the depth h of the outflow, θ_0 is the mean potential temperature, and k is the Froude number, which is a constant that is determined empirically or theoretically. Typical values are about 0.8, ranging from 0.7 to 1.1 (Wakimoto, 1982). The typical speed of the gust front is 10 m s^{-1}, although speeds in excess of 20 m s^{-1} have been observed (Wakimoto, 1982; Goff, 1975).

By introducing the equation of state, Seitter (1987) showed that the density difference across the gust front can be estimated from the surface hydrostatic pressure difference across the gust front, $\Delta p = gh\Delta\rho$, thus

$$c = k(\Delta p/\rho_0)^{1/2}, \tag{9.4}$$

where k is the Froude number and ρ_0 is the density of air. An advantage of Eq. (9.4) is that only surface pressure need be measured to estimate the speed of propagation of a gust front. Moreover, Nicholls (1987) found in two-dimensional numerical simulations of tropical squall lines that whenever clouds occurred above the gust front, latent heat release and associated warming as well as water loading at levels above the depth of the outflow altered the surface pressure difference. This, in turn, altered the propagation speed of the gust front. These affects are not depicted in Eq. (9.3). Thus, both Seitter (1987) and Nicholls (1987) found that Eq. (9.4) yields a more consistent prediction of observed and modeled gust-front speeds than does Eq. (9.3), in which only temperature perturbations in the cold pool affect gust-front movement.

Equation (9.4) is valid only when there is no ambient wind ahead of the gust front. Seitter noted that simply adding the wind component $\bar{U}$ parallel to the gust front motion to Eq. (9.4) does not yield satisfactory agreement with observations. Based on laboratory simulations of density currents reported by Simpson and Britter (1980), Seitter suggested that $\bar{U}$ must be multiplied by the factor 0.62 before being added to Eq. (9.4). Note that $\bar{U}$ is the wind component parallel to the gust-front motion that is averaged over the depth of the head of the gust front and is positive in the direction of the gust-front motion.

While outflow θ profiles may assume constant mixed-layer values around the active turbulent portions, the coldest air may become stratified within the lower 500 m of less vigorous outflow air (e.g., Betts, 1984). The vertical motion over gust fronts is one of the primary mechanisms for triggering new clouds (e.g., Purdom, 1982). Gust-front interactions with other mesoscale boundaries such as cold fronts, dry lines, sea-breeze fronts, or other gust fronts are important in the formation of convective clouds (e.g., Purdom, 1976, 1979; Sinclair and Purdom, 1983).

Purdom (1976) and Weaver and Nelson (1982) observed that new cloud growth is particularly explosive within regions where two outflow boundaries collide. The relative importance of such boundaries has recently been established by Purdom and Marcus (1982), who found that outflow interactions were present in 73% of all cumulonimbi initiated over the southeast United States during the 1979 summer months. Surface analyses presented in Cooper *et al.* (1982) also suggest that, in several cases examined, convergence produced by outflow from precipitating convection intensified and expanded cumulonimbi activity over southern Florida.

Droegemeier and Wilhelmson (1985a, b) simulated cloud development along intersecting thunderstorm outflow boundaries using the Klemp and Wilhelmson (1978a) three-dimensional cloud model. In the no-wind simulations (or those with weak wind shear), as the outflows from two convective storms collide, two symmetric vertical velocity maxima are generated on the ends of the intersecting outflow boundaries. The structure of the convergence field driving the vertical velocity is shown schematically in Fig. 9.26. As the two outflows approach within several kilometers of each other, a zone of lifting is created between them. As the gust fronts come closer, air in the region labeled A is squeezed out laterally and vertically. The largest accelerations are horizontal because vertical accelerations require work against gravity. As a result, two regions of maximum convergence form on either side of the intersection point, as Fig. 9.26b shows (the center of A). If the low-level air is moist and the level of free convection is low enough, two clouds form over the two zones of maximum convergence.

Droegemeier and Wilhelmson (1985b) also simulated the effects of a unidirectional wind shear of varying depths and intensities on outflow interactions. The shear vector was aligned perpendicular to the center line of two initial outflow-producing clouds (i.e., oriented vertically in Fig. 9.26). The model results showed that in strong shear, the upshear member of the pair of clouds that formed in the maximum convergence zones became the stronger cell. This was because the upshear member had a head start on development and because the downdraft from the upshear member tilted downshear, thus suppressing the other cloud.

Downdraft outflows appear to exert primary influences on cumulonimbus maintenance and motion. Results of several three-dimensional cloud modeling studies indicate that convergence and associated uplift along downdraft-driven gust fronts may produce quasistationary storms (Miller, 1978),

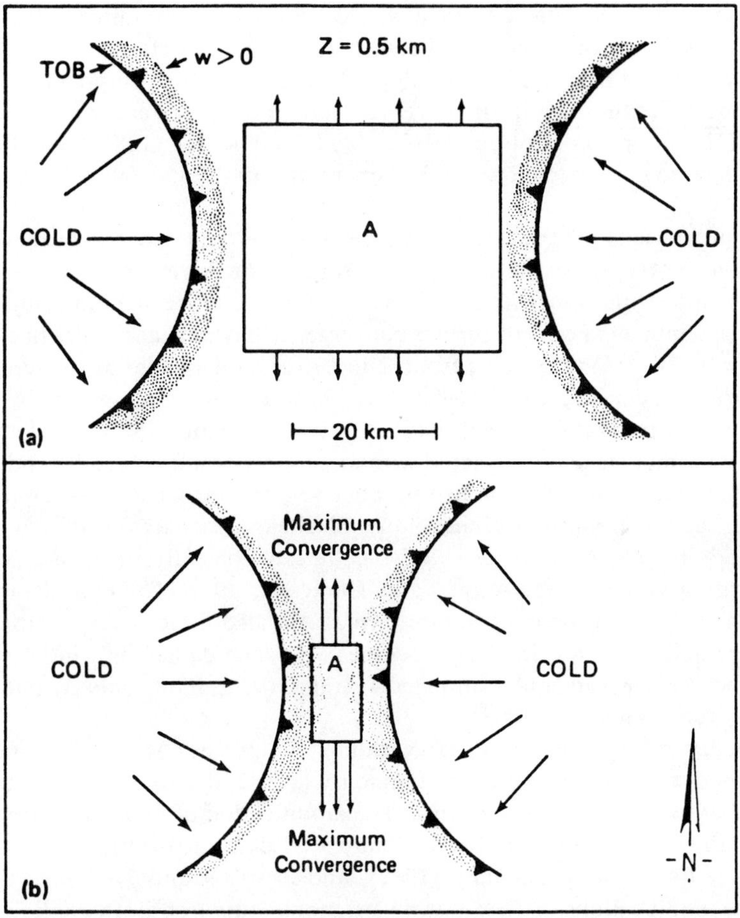

Fig. 9.26. Schematic diagram for the case of no wind shear in the model environment showing a plan view of (a) two outflows aproximately 40 km apart and moving toward each other. The gust front is indicated by the bold solid lines with barbs, and the regions of upward motion along the gust fronts are stippled. An arbitrary area A is shown by the box, and the small arrows indicate horizontal flow out of A as the outflows approach each other. (b) The outflows are now approximately 10 km apart. The regions of maximum horizontal convergence due to the rapid flow out of A from mass continuity are also indicated. [From Droegemeier and Wilhelmson (1985a).]

upshear propagation of tropical squall lines (Moncrieff and Miller, 1976), and storm regeneration (Thorpe and Miller, 1978; Klemp *et al.*, 1981). Moncrieff and Miller and Thorpe and Miller inferred that steady storm behavior requires zero relative motion between a storm and its gust front.

This condition is satisfied when the low-level updraft inflow equals the outflow velocity of air behind the gust front. The importance of gust-front convergence in maintaining cumulonimbi and controlling storm motion has also been emphasized in the observational studies of Weaver and Nelson (1982), Knupp and Cotton (1982a), and Fankhauser (1982), and in the cloud model investigations of Klemp *et al.* (1981) and Wilhelmson and Chen (1982).

Fankhauser (1982) determined from mesonet and radar data that new cells generated above the surface gust front attained greater intensities than those appearing above the outflow air some distance behind. Objective analyses indicated cells of surface convergence having magnitudes in excess of $3 \times 10^{-3}\ s^{-1}$ along the gust front. Numerical model experiments reported by Wilhelmson and Chen (1982) have produced similar results. In their study, the development of five cells over a 4-h simulation was related to outflow enhancement associated with mature precipitating cell downdrafts.

Downdraft outflows can also produce severe weather such as damaging winds and extreme wind shears. Wind shear associated with low-level downdraft divergence patterns is sometimes sufficiently extreme to cause aircraft accidents (e.g., Fujita and Caracena, 1977; National Research Council, 1983). Damaging surface winds are also directly responsible for deaths and numerous injuries. In some cases such damaging wind systems can be widespread and prolonged (Fujita, 1978; Fujita and Wakimoto, 1981; Johns and Hirt, 1983).

Downdraft outflows affect larger atmospheric scales by transforming the thermodynamic structure of the boundary layer and through vertical transports of mass, heat, and moisture. The modification of boundary layers by downdrafts has been described by many, e.g., Betts (1976), Ogura and Liou (1980), Barnes and Garstang (1982), and Johnson and Nicholls (1983). These modifications include downward momentum transport, cooling, and drying, which lead to significant reduction in near-surface moist static energy. Over tropical oceans, the cooling and dry effects of boundary layer modifications may persist for 2–12 h, and may result in an augmentation of surface-sensible and latent heat fluxes by an order of magnitude (Johnson and Nicholls, 1983). Due to the stabilizing effect, convection is markedly suppressed in these modified regions, or wakes, which cover about 30% of the total area (Gaynor and Mandics, 1978). Over land where rainfall augmentation of soil moisture is also important, relatively cool areas often remain for 12 h or more due to modification of boundary layer thermodynamics and surface energy budgets.

Clearly, downdraft outflows and gust fronts play an important role in the behavior of convective storms. We will now examine theoretical models of storm propagation which include these as well as other mechanisms of storm propagation and movement.

9.6 Theories of Storm Movement and Propagation

Storm movement and propagation can be classified into three different mechanisms: (i) translation or advection, (ii) forced propagation, and (iii) autopropagation. Translation or advection is the process whereby a storm is blown along by the mean wind as it evolves through its lifetime. Forced propagation refers to the sustained regeneration of a convective storm by some external forcing mechanism usually larger in scale than the convective storm. Examples of external forcing mechanisms are fronts and rainband convergence associated with midlatitude cyclones, sea-breeze fronts, convergence associated with mountains, rainband convergence in tropical cyclones, frontal boundaries produced by the outflows from decayed convective storms, and convergence associated with gravity waves excited by external forcing mechanisms. Often the systems providing the forced propagation have lifetimes considerably longer than individual thunderstorms and are only mildly modulated by the presence of the storms.

Autopropagation refers to the process in which a thunderstorm can regenerate itself or cause the generation of similar storm elements (cells) within the same general system. Examples of autopropagation mechanisms are downdraft forcing and gust fronts, updraft forcing via warming aloft (causing enhanced inflow due to vertical pressure gradients), development of vertical pressure gradients due to storm rotation, and the triggering of gravity waves by a thunderstorm which generates areas of enhanced low-level convergence.

Many convective storm systems are affected by all three mechanisms of movement and propagation for at least some part of their lifetime. Some storm systems are primarily affected by autopropagation mechanisms for a major portion of their lifetime. An example of the latter is the supercell thunderstorm. Squall-line thunderstorms can form by both forced propagation and autopropagation. The severe, prefrontal squall-line thunderstorm system (Newton, 1963), once initiated, is a classic example of an autopropagating system. Other squall-line thunderstorm systems are coupled with fronts throughout their lifetime. Such frontal squall lines are influenced primarily by forced propagation.

In this section we focus our attention on the theories of autopropagation of thunderstorms. We break these down into two classes: (1) the convective overturning models of Moncrieff and Green (1972; hereafter referred to as MG), Moncrieff and Miller (1976; hereafter referred to as MM), Moncrieff (1981; hereafter referred to as M), and Thorpe *et al.* (1982; hereafter referred to as TMM), and (2) the wave–CISK models of Lindzen (1974), Stevens and Lindzen (1978), Raymond (1975, 1976, 1983; hereafter referred to as R75, R76, R83, respectively), Davies (1979), and Silva-Dias *et al.* (1984; hereafter referred to as SBS). These models represent two fundamentally different concepts of the propagation of cumulonimbi, although they also exhibit some common features. We then examine theories of updraft splitting and storm propagation.

9.6.1 Moncrieff–Green–Miller–Thorpe (MGMT) models

The MGMT models are comprehensive theories of steady, convective, overturning storms. Turbulent diffusion and dissipation are ignored in the models, which can be considered models of ensemble-averaged flow fields in steady convective storms. A major uncertainty, however, is the extent to which the neglect of turbulent diffusion and dissipation affects the updraft and downdraft velocities, displacement heights of the inflow and outflow streamlines, and the propagation speeds of the storm.

A common assumption used in the MGMT models is that the flow is steady in a reference frame that moves at the translation speed of the thunderstorm system. Each of the MGMT models involves a "characteristic" updraft/downdraft streamline pattern. Miller and Moncrieff (1983) summarized the formulation of the MGMT models as follows.

For steady-state flow, an energy equation per unit mass can be written as

$$\frac{D}{Dt}\left(\frac{1}{2}V^2+\frac{p'}{\bar{\rho}}\right)=g\frac{\theta'}{\bar{\theta}}w, \tag{9.5}$$

where $V^2=u_1^2+u_2^2+u_3^2$. (The notation used here is the same as in Chapters 2 and 3.) The left-hand side of Eq. (9.5) represents the change in total energy moving along a steady streamline, or trajectory, and the right-hand side represents the production of energy by buoyancy forces.

For steady flow it can be shown that

$$w\frac{\theta'}{\bar{\theta}}=\frac{D}{Dt}\int_{z_0}^{z}\frac{\theta'}{\bar{\theta}}\,dz, \tag{9.6}$$

where the integrand is evaluated along streamlines and $z-z_0$ represents the displacement of a parcel from its inflow level z_0. Substitution of Eq. (9.6) into Eq. (9.5) results in

$$\frac{1}{2}v^2+\frac{p'}{\bar{\rho}}-\int_{z_0}^{z}g\frac{\theta'}{\bar{\theta}}\,dz=C_1(\psi), \tag{9.7}$$

which shows that the energy along a streamline in steady, inviscid flow is conserved. Assuming $p'/\bar{\rho}=0$ at $z=z_0$ results in $C_1(\psi)=\frac{1}{2}V_0^2$, the inflow specific kinetic energy.

The thermodynamic energy equation can be written in the form

$$(D/Dt)(\theta'/\bar{\theta})+WB=Q/\bar{T}, \tag{9.8}$$

where B is the static stability. It is assumed that the source of entropy Q is of the form $W\Gamma(z)$, where Γ is the lapse rate along a steady updraft or downdraft streamline. The term Γ can be specified as being either the dry or wet adiabatic lapse rates. Similar to Eq. (9.6) we find that

$$w(\Gamma - B) = \frac{D}{Dt}\int_{z_0}^{z} (\Gamma - B)\, dz. \tag{9.9}$$

Thus, Eq. (9.8) becomes

$$\frac{\theta'}{\bar{\theta}} - \int_{z_0}^{z} (\Gamma - B)\, dz = C_2(\psi). \tag{9.10}$$

Assuming $\theta' = 0$ at $z = z_0$ yields $C_2(\psi) = 0$. Equation (9.10) shows that if Γ is the wet adiabatic lapse rate, $\theta'/\bar{\theta}$ is caused by the integrated release of latent heat by condensation or evaporation as one follows a streamline along updrafts or downdrafts from their source heights.

An equation for the y-component of vorticity for two-dimensional flow in the (x, z) plane is:

$$\frac{D}{Dt}(\eta) + \eta\left(\frac{\partial u}{\partial x} + \frac{\partial w}{\partial z}\right) + g\frac{\partial}{\partial x}\left(\frac{\theta'}{\bar{\theta}}\right) = 0, \tag{9.11}$$

where $\eta = \partial u/\partial z - \partial w/\partial x$.

The first term on the left-hand side of Eq. (9.11) is the rate of change of vorticity along a streamline, the second term is the generation of vorticity due to vertical variations in density, and the third term is the generation of vorticity by horizontal gradients in buoyancy.

MG showed that by using a streamfunction (ψ) and the mass continuity equation, the following quantity is conserved along streamlines:

$$\frac{\eta}{\rho} - \int_{z_0}^{z} \left(\frac{\partial}{\partial \psi}\right)_z \left(\frac{\theta'}{\bar{\theta}}\right) dz = C_3(\psi), \tag{9.12}$$

where $C_3(\psi)$ equals the inflow vorticity divided by the inflow density.

The horizontal wind speed u along a streamline can be obtained (M) from the conservation relation

$$u - \int_{z_0}^{z} \left(\frac{\partial}{\partial \psi}\right)_z \frac{\delta p}{\rho}\, dz = C_4(\psi), \tag{9.13}$$

where $C_4(\psi)$ represents the inflow wind speed.

Equations (9.7), (9.10), (9.12), and (9.13) represent the conservation equations along streamlines of a steady-state system of energy, entropy, y-component of vorticity, and momentum production.

Moncrieff (1981) then derived a general displacement equation relating the inflow and outflow heights along streamlines. Assuming that the lateral dimensions of the outflow and inflow streamtubes are equal, his result becomes

$$\left(\frac{dz_0}{dz_1}\right)^2 = 1 - \frac{\Delta p}{1/2u_0^2} + \frac{g(\Gamma - B)}{1/2u_0^2}\left|\frac{(z_1 - z_0)^2}{2} - \int_{z_*}^{z_1} (z_1 - z_0)\, dz_1\right|, \quad (9.14)$$

where z_1 represents the height of the outflow level and Δp is the pressure perturbation at the outflow reference level $z = z_*$. A simplification made in the MGMT model is that the density ρ, the static stability B, and the parcel lapse rate Γ are constant throughout the convective layer.

The above set of equations has been applied to a family of "characteristic" flow models.

9.6.1.a The Steering-Level Model

An analytic model of steady, midlatitude squall-line convection was first described by MG and more fully by Moncrieff (1978). The theory is designed to simulate storm systems which move at a speed corresponding to some level in the atmosphere, or steering level. Figure 9.27 illustrates the revised hypothesized flow structure (Moncrieff, 1981) of such a steady storm system. The flow is two-dimensional with low-level inflow air tilting downstream as it rises through the troposphere and flowing out ahead of the storm at the anvil outflow level. The most unrealistic feature of this model is that the downdraft originates at the tropopause level and then descends through the depth of the troposphere.

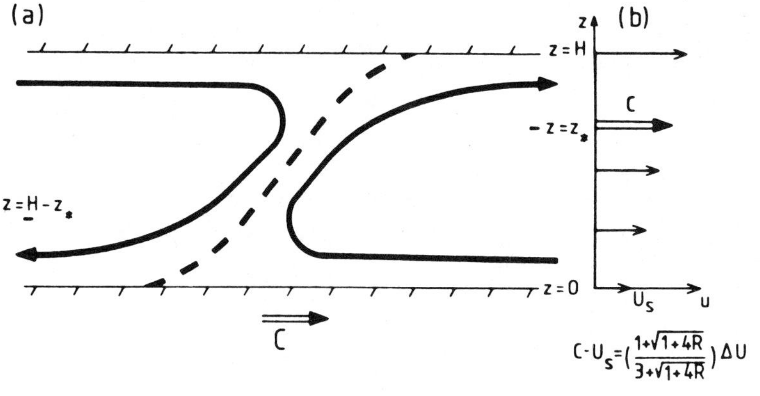

Fig. 9.27. (a) Schematic of the relative flow in the steering-level model. The level $z = z_*$ is the height at which the travel speed equals the environmental wind speed. (b) The relationship between the travel speed C and the undisturbed flow relative to the ground. [From Moncrieff (1981).]

If the vertical shear is constant and the undisturbed change in wind speed between low levels and the top of the troposphere (Δu) is large, MG and M showed that the speed of propagation of the storm system can be determined from the parameter R, where

$$R = \mathrm{CAPE}/\tfrac{1}{2}(\Delta u)^2 \qquad (9.15)$$

is a type of Richardson number.

The convective available potential energy (CAPE) is defined as

$$\mathrm{CAPE} = \int_{-H/2}^{H/2} g\, \delta\phi_{\mathrm{p}}\, dz, \qquad (9.16)$$

where H is the depth of convection and $\delta\phi_{\mathrm{p}}$ is the log potential temperature difference between parcel and undisturbed environment.

The storm speed relative to the surface wind u_{s} is given by

$$c - u_{\mathrm{s}} = \Delta u[1 + (1 + 4R)^{1/2}]/[3 + (1 + 4R)^{1/2}]. \qquad (9.17)$$

Equation (9.13) shows that horizontal momentum is transferred upward against the mean velocity gradient and the vertical shear in the troposphere is increased with stronger flow aloft and weaker flow at low levels. MG showed that steady, steering-level-type convection can exist only within the range $-\frac{1}{2} \leq R \leq 1$.

MG showed that the predicted propagation speed and steering-level height agreed with several supercell and squall-line storms in which the value of R was small, or equivalently if

$$R_{\mathrm{i}} = -RH/z_{*} \qquad (9.18)$$

was small, where z_{*} is the steering level. However, the model did poorly in predicting the propagation speed of a severe right-moving storm system, which they suggested was a result of the model's neglect of three-dimensional characteristics.

9.6.1.b Steady-Upshear Tilted, Two-Dimensional Storm Model

There are two features of the propagating storm model that are unrealistic. First of all, we noted that it is unrealistic for a storm downdraft to extend from the tropopause to the surface. Second, since the updraft tilts downshear, Moncrieff (1978) noted that precipitation falling out of the updraft will settle into the inflow flank of the storm, thus cutting off the supply of moisture into the storm and preventing steady convection. TMM showed by numerical experiment that with constant shear, convection lasted only about 1 h although the behavior of the convection during its mature stage was well represented by MG's dynamical solutions.

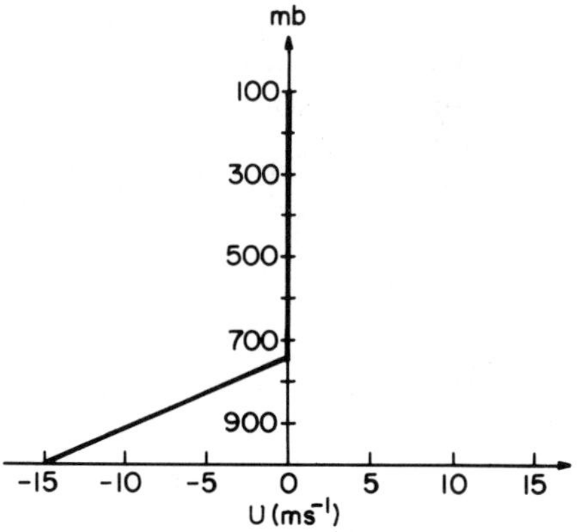

Fig. 9.28. Profile of the initial flow used in Thorpe *et al.*'s (1982) simulations.

TMM also found by two-dimensional numerical experiments that the most efficient and steady storms could be produced when there was strong low-level shear and weak mid- and upper-level shear. Figure 9.28 illustrates the optimum shear profile they examined.

Using the numerical model results as guidance, TMM then developed an analytic model of steady, two-dimensional flow based on the idealized model illustrated in Fig. 9.29. Features of this model are an upshear-tilted updraft, a downshear-tilted updraft, and a shallow downdraft on the upshear flank. Equations (9.7), (9.10), (9.12), (9.13), (9.14), and (9.17) can be used to calculate the kinetic energy, cloud buoyancy, vorticity, horizontal wind speed, and displacement heights along each of the hypothesized streamlines

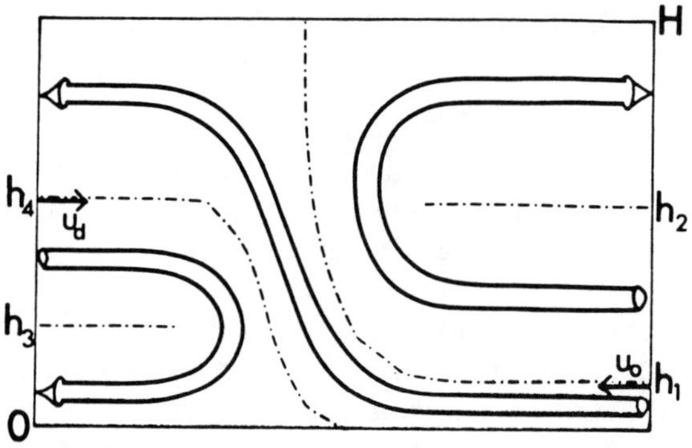

Fig. 9.29. Idealized model with reference heights labeled. [From Thorpe *et al.* (1982).]

and the propagation speed of the storm. As in the MM tropical squall-line model described in the next section, TMM found it necessary to define a closure hypothesis in the model in order to obtain a unique solution. Using the total energy equation and requiring the interface between the updraft and downdraft to be oriented to allow rain to fall from the updraft to the downdraft, they required that

$$U_d/U_0 < 1,$$
$$E \leq 1$$
$$F \leq 1,$$
$$H - h_4 \geq h_1, \tag{9.19}$$

where U_d is the downdraft and U_0 is the low-level inflow speed. The first condition simply states that the downdraft must be energetically less important than the updraft. The second condition shows that since

$$E = \Delta p/(\tfrac{1}{2}\rho U_0^2) \tag{9.20}$$

must be of the order of unity, the cross-storm pressure-gradient Δp is fundamental to maintaining steady, two-dimensional flow. The constraint that

$$F = U_0/(\mathrm{CAPE})^{1/2} \simeq 1 \tag{9.21}$$

requires that the low-level inflow speed U_0 be quite large. Note that F and the convective Richardson number are related by the identity $F - 1/\sqrt{R}$. Remember that CAPE, given by Eq. (9.16), represents the buoyant production of energy that can be thought of as the positive area on a thermodynamic diagram. It is thus proportional to maximum updraft strength. TMM estimated this condition corresponds to a wind speed of the order of 20 m s^{-1}. The final constraint is simply a limitation on the geometry of the flow as illustrated in Fig. 9.29. The main result is that steady, two-dimensional flow is only possible in the presence of strong low-level shear with weak shear aloft.

9.6.1.c The Tropical Squall Line or Propagating Storm Model

A common characteristic of tropical squall lines is that the storm system moves faster than the undisturbed flow at all levels. As a result, the low-level inflow occurs on the forward flank of the system and outflow takes place on the rear flank. No relative flow reversal exists in such a system.

A steady convective model of such a storm system was developed by MM. Figure 9.30 illustrates the hypothesized storm structure. As can be seen from Fig. 9.30, the hypothesized flow field is inherently three

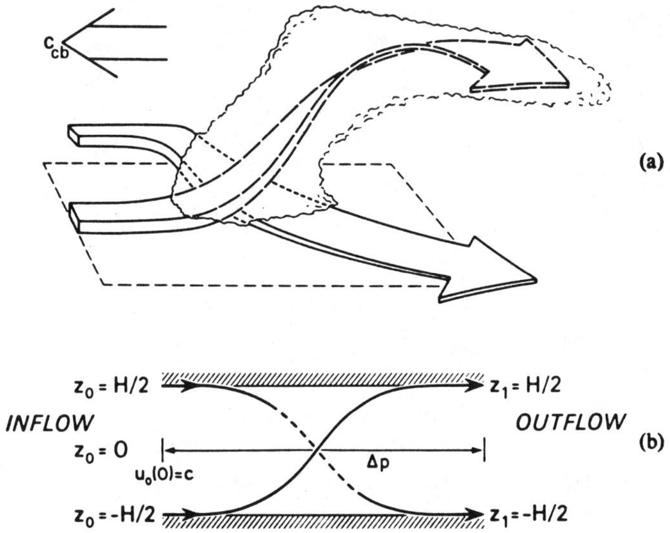

Fig. 9.30. (a) Schema of a propagating cumulonimbus cloud. (b) Limiting relative streamlines in the analytic model. On inflow, the velocity profile $U_0 = u_0 - c$, the static stability, and the parcel lapse rate are specified as a function of the inflow level z_0, but c is calculated. On outflow, for each streamline passing through z_0, the values of $z_1(z_0)$, $U_1(z_0)$, and $\phi_1(z_0)$ are calculated, where $-H/2 \leq z_0 \leq H/2$. [From Moncrieff and Miller (1976).]

dimensional. Equations (9.7), (9.10), (9.12), (9.13), (9.14), and (9.17) can again be used to calculate the kinetic energy, cloud buoyancy, vorticity, horizontal wind speed, and displacement heights along each of the hypothesized streamlines as well as the propagation speed of the storm system. As noted by Moncrieff (1981), inflow and outflow on opposite sides of the storm requires that a net pressure gradient must exist across the storm system. Thus, a unique solution to the above equations can only be obtained if the cross-storm pressure gradient is specified by some form of closure assumption. MM hypothesized that given an inflow velocity profile, and the mass conservation and energy constraints along a streamline, there exists an outflow mass and velocity distribution that optimizes the upward buoyancy flux and release of kinetic energy. Identifying

$$E = \Delta p / (\tfrac{1}{2}\rho U_0^2) \tag{9.22}$$

as the mechanical efficiency in converting available potential energy to mean-flow kinetic energy, MM assume that E is optimized in a steady storm system, or that $E = 1.0$.

An important prediction of the MM model is that the propagation speed of the storm system can be determined mainly from the difference between parcel and environmental lapse rates, or

$$C_M \simeq 0.32(\text{CAPE})^{1/2}, \tag{9.23}$$

where C_M represents the propagation speed of the convective system relative to midlevel winds.

Another result from this model is that if the inflow shear is too large, then steady overturning is not possible. The limiting shear is given by

$$\partial u_0/\partial z = 2C_M/H. \tag{9.24}$$

As noted previously, a common feature of the MGMT models is that steady convection acts to increase the vertical shear in the troposphere. For the case of propagating tropical squalls, this is illustrated in Fig. 9.31, where the "before and after" winds are plotted for different values of R [Eq. (9.15)]. In each case, steady convection is predicted to increase the westerly momentum at the surface and aloft while decreasing westerly momentum in middle levels.

A distinct advantage of the MGMT models over simple parcel models is that they recognize the role of horizontal pressure-gradient forces on the storm dynamics and storm propagation. As we shall see, MGMT models have this feature in common with the wave-CISK models. Moreover, the MGMT models simultaneously conserve mass, energy, momentum, and vorticity along steady streamlines of a convective system.

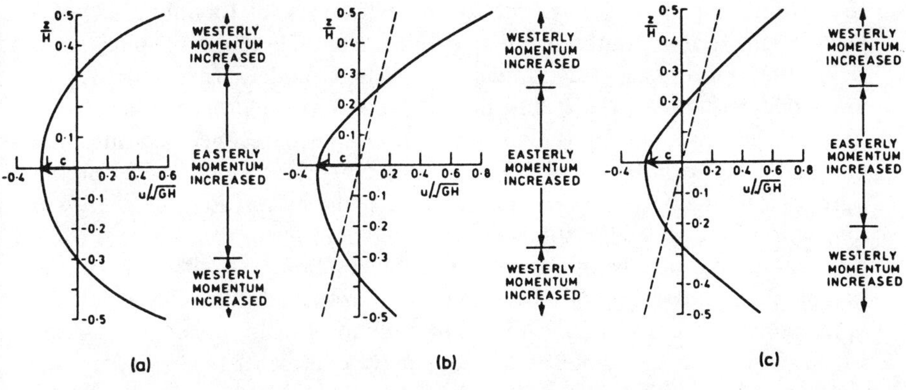

Fig. 9.31. The modification of the flow by convection. The modified and unmodified profiles are represented by full and broken lines, respectively. (a) Unsheared incompressible flow ($R = \infty$); (b) sheared incompressible flow ($R = 4$); (c) sheared compressible flow ($R = 4$). [From Moncrieff and Miller (1976).]

9.6.2 Wave-CISK Models

The concept of wave-CISK (conditional instability of the second kind) is a simple one. From the wave-CISK perspective, a convective system is viewed as a symbiotic interplay among the convective elements and a wave disturbance in the stably stratified, cloud-free environment. The convective elements provide the heating and, perhaps, momentum transports to initiate and drive the wave disturbance, while the wave provides the large-scale, moisture supply to the clouds. As seen in Chapter 8, even shallow cumulus clouds excite gravity waves in the stably stratified troposphere. The gravity waves, in turn, appear to exert an important influence on the resultant organization of clouds and propagation of clouds. In this section we examine the possible role of gravity waves on the propagation of thunderstorms by using simple linear models to describe the interaction between deep convection and gravity waves.

The wave-CISK concept is a derivative of the original CISK concept pioneered by Charney and Eliassen (1964) and Ooyama (1964) in its application to hurricanes. In those studies, cumulus convection was perceived to be driven by boundary layer frictional convergence associated with a tropical depression. Latent heat release and warming by convection then led to a deepening of the cyclone with a corresponding increase in boundary layer convergence.

Wave-CISK differs from CISK in that the forcing mechanism for cumulus convection is the convergence field associated with inviscid gravity-wave motions rather than boundary layer convergence. As in CISK, cumulus convection is introduced as a heat source with a specified distribution in the vertical. Figure 9.32 illustrates a schematic view of Raymond's (1975) wave-CISK representation of a convective system. There is no explicit scale separation specified in Raymond's wave-CISK models. Instead, a separation occurs between moist and dry processes. Moist processes are not a part of the solutions of the model; rather, they are specified in some parameterized form. The parameterized moist processes then serve as forcing functions to the set of linearized equations of motion. Predicted growth rates and phase speeds of the normal mode solutions to dry, inviscid wave motion are obtained as an eigenvalue problem. Omitted in the wave-CISK scheme is the possibility of moist, mesoscale ascent and descent, which are known to occur in tropical cyclones and other mesoscale convective systems such as squall lines and mesoscale convective complexes (see Chapter 10).

Wave-CISK models have been primarily developed and applied to larger scale tropical wave disturbances such as easterly waves (Yamasaki, 1969, 1971; Hayashi, 1970, 1971a, b, c; Lindzen, 1974; Kuo, 1965; Chang, 1976; Stark, 1976; Stevens *et al.*, 1977). However, Raymond

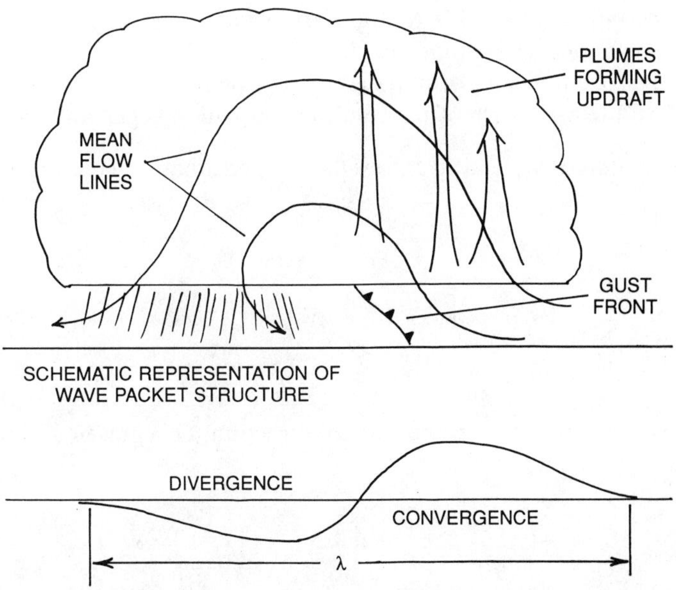

Fig. 9.32. Schematic diagram of the hypothesized relationship between a packet of gravity waves and the associated convective storm, subject to the limitations of a two-dimensional drawing. The wave packet consists of one convergent and one divergent region with a dominant wavelength λ comparable to the diameter of the storm. Convective plumes develop in the convergent region, passing into the divergent region as they decay and produce rain. [From Raymond (1975).]

(1975, 1976, 1983, 1984), Silva-Dias *et al.* (1984), Emanuel (1982), and Nehrkorn (1985) have applied wave-CISK theory to convective disturbances on the scale of supercell thunderstorms and of tropical and midlatitude squall lines.

To illustrate the formulation of wave-CISK models, we follow the notation of Silva-Dias *et al.* (1984). We begin with the equations of motion [Eq. (2.126)], the thermodynamic energy equation [Eq. (2.37)], and the continuity equation [Eq. (2.128)] as given in Chapter 2. The analysis is similar to that described in Chapter 2 for formulating the anelastic, Boussinesq set of equations of motion linearized in the thermodynamic variables.

(i) Scale separation is assumed between the horizontally averaged basic state (subscript zero) and deviations from it (primed variables) that contain mesoscale and small-scale contributions.

(ii) The effects of earth's rotation are usually neglected in mesoscale wave–CISK models (Raymond, 1975, 1976; Silva-Dias *et al.*, 1984). Larger scale models are typically formulated on a β-plane.

(iii) The anelastic form of the continuity equation is generally assumed.

These assumptions lead to the system of perturbation equations

$$\frac{\partial \mathbf{v}'}{\partial t} + \mathbf{v}_0 \cdot \nabla \mathbf{v}' + \mathbf{v}' \cdot \nabla \mathbf{v}_0 + \nabla \frac{p'}{\rho_0} - \frac{\theta'}{\theta_0} g\mathbf{k} = \psi \tag{9.25}$$

$$\frac{\partial \theta'}{\partial t} + \mathbf{v}_0 \cdot \nabla \theta' + \mathbf{v}' \cdot \nabla \theta_0 = Q' + \psi_\theta \tag{9.26}$$

$$\nabla \cdot (\rho_0 \mathbf{v}') = 0, \tag{9.27}$$

where $\mathbf{k}$ is the unit vector in the vertical direction, Q represents a specified heating function, and

$$-\psi = \left(v' \cdot \nabla v' + \frac{\rho'}{\rho_0^2} \nabla p' \right), \qquad -\psi_\theta = (v' \cdot \nabla \theta').$$

Note that all the nonlinear terms are embodied in ψ and ψ_θ.

Many theories introduce an averaging operator or filter to separate the convective scales from the mesoscale. Silva-Dias *et al.* (1984) introduce the formal averaging operator

$$\langle \mathbf{v}'(x, y, z, t) \rangle = \frac{1}{\Delta x \, \Delta y} \int_{x-\Delta x/2}^{x+\Delta x/2} \int_{y-\Delta x/2}^{y+\Delta x/2} v'(x', y', z, t) \, dx' \, dy', \tag{9.28}$$

where $v' = \langle v' \rangle + v''$.

This form of averaging operator corresponds to the grid-volume average described in Chapter 3. The separation of cumulus-scale motions from mesoscale wave motions is acceptable provided the spatial structure of the cumulus clouds is distinctly different from the mesoscale wave motions. Unfortunately, in the presence of zero or weak shear, conventional wave–CISK models predict that the most unstable waves have a scale corresponding with the largest wavenumber or smallest scales resolved by the truncated spectral model. In other words, a scale separation between the explicitly resolved waves and the cumulus scale is not guaranteed. Raymond (1983) does not introduce a formal averaging operator but instead distinguishes between the convective scale and mesoscale waves by considering moist processes only on the convective scale. He thus introduces a form of top-hat averaging or filtering between the moist convection and the mesoscale.

If we define (i) a horizontally homogeneous basic state, (ii) some form of filter or averaging operator between the cumulus and mesoscale wave motions, and (iii) linearize Eqs. (9.25) and (9.26) by neglecting correlations between mesoscale variables, then we can decompose the mesoscale variables into their Fourier components [assuming a single-valued complex frequency $\omega(\mathbf{k})$]:

$$\langle \mathbf{v}'(x, y, z, t)\rangle = \int_{-\infty}^{+\infty}\int \hat{v}(\mathbf{k}, z) \exp[i(\mathbf{k}\cdot\mathbf{r}-\omega t)]\, dk_x\, dk_y, \qquad (9.29)$$

where here

$$\mathbf{k} = k_x i + k_y j, \qquad \mathbf{r} = xi + yj.$$

The preceding assumptions then reduce the above equations to the set of linear, homogeneous equations:

$$-i(\omega - k\cdot v_0)\hat{u} + \frac{du_0}{dz}\cdot\hat{w} + \frac{ik_x\hat{p}}{\rho_0} = \psi_x, \qquad (9.30)$$

$$-i(\omega - k\cdot v_0)\hat{v} + \frac{dv_0}{dz}\cdot\hat{w} + \frac{ik_y\hat{p}}{\rho_0} = \psi_y, \qquad (9.31)$$

$$-i(\omega - k\cdot v_0)\hat{w} + \frac{d}{dz}\frac{\hat{p}}{\rho_0} - g\frac{\hat{\theta}}{\theta_0} = \psi_z, \qquad (9.32)$$

$$-i(\omega - k\cdot v_0)\hat{\theta} + \frac{d\theta_0}{dz}\hat{w} = Q + \psi_\theta, \qquad (9.33)$$

$$ik\cdot\mathbf{v}_0 + \frac{1}{\rho_0}\frac{d}{dz}(\rho_0\hat{w}) = 0. \qquad (9.34)$$

The procedure then is to (1) define a basic state, (2) choose a parameterization scheme for the right-hand side of Eqs. (9.30)–(9.33), (3) impose boundary conditions, (4) impose initial conditions, and (5) solve the resulting linear homogeneous system as an eigenvalue problem. In contrast to Silva-Dias *et al.*, most wave–CISK models are formulated as hydrostatic models. Because the most nonhydrostatic part of the convective system is parameterized in wave–CISK models, the inclusion of nonhydrostatic contributions to the pressure field should not make a great deal of difference in the solutions. Recent wave–CISK models have also included the effects of the earth's rotation in their formulation.

We will not concentrate here on the various solution techniques, boundary conditions, and so forth, but we will focus on the formulation of and effects of the various parameterizations (ψ_x, ψ_y, ψ_z, Q, and ψ_θ) and model predictions for certain basic states. As noted by Raymond (1983), it is the parameterization of these terms that is the Achilles heel of wave–CISK models. That is, the model predictions generally depend critically upon the details of the parameterization embedded in those terms. Particularly important is the parameterization of cumulus convective heating and fluxes. As seen in Chapter 6, this is a problem that is by no means fully solved, especially for unbalanced disturbances such as we are examining here.

In most wave-CISK models, ψ_x, ψ_y, and ψ_z are ignored, and the heating function Q and heat advection terms ψ_θ are combined into a simple heating function Q or a simple cloud model representation. The classical wave-CISK hypothesis is that the heating function is proportional to the low-level convergence or vertical velocity at cloud base. The major variations are in the form of the vertical distribution of the heating function. Raymond (1975, 1976), for example, distributed the heating uniformly between cloud base and cloud top. Silva-Dias *et al.* (1984) varied the heating rate nonlinearly as a function of height. The heating profile had a maximum at a prescribed level in the troposphere. In Silva-Dias's study the predicted growth rates, group velocities, and phase velocities were quite sensitive to the height of maximum heating. Nehrkorn (1985) also found his model predictions to be sensitive to the specified heating profile and the height of the maximum heating.

Marroquin and Raymond (1982) departed from the traditional wave-CISK formulation by prescribing a heating function proportional to the local vertical velocity. Such a heating function is similar to that used in the Moncrieff models. This heating function allows the heating to remain in phase with the updrafts even though the updraft may tilt with height. Again, this form of heating function significantly altered the predicted storm movement from Raymond's (1975, 1976) original models.

A common assumption in early wave-CISK models is that cooling occurs where there is downward motion at cloud base. Several of the more recent models, however, remove this restriction. Moreover, most wave-CISK models predict convective heating (cooling) whenever there are updrafts (downdrafts) at cloud base, regardless of the amount of moist static energy available. Raymond (1984) introduced a parameterization that turned off the heating function whenever the moist static energy was depleted by more than 10%.

The fundamental weakness in wave-CISK theory is that, in the presence of zero or weak shear, the scale of the waves exhibiting the maximum growth rate are the smallest scales resolved by the model. This means that the most unstable waves correspond to the cumulus scale or the scale that is being parameterized.

For larger scale applications, Davies (1979) introduced a time lag between low-level convergence and the cloud response producing convective heating. By introducing an adjustment time of the order of 12 h, the instability of inertial gravity waves was suppressed, thus providing a scale selection mechanism in which larger scale waves exhibited the greatest instability. Raymond (1983) suggested that such a lagged model is not appropriate for mesoscale applications. He presented a lagged heating model in which convective "bubbles" respond instantaneously to low-level convergence at the level of free convection but take a certain amount of time to rise through the atmosphere. He introduced this form of lagged updraft into a modified wave-CISK model that was cast in mass flux form. Unfortunately, this form of lagged updraft removed the shortwave cutoff found by Davies. Thus, scale selection is very sensitive to the particular form of the lagged updraft model.

Wet but unsaturated downdrafts are known to play an important role in thunderstorm propagation. However, early wave-CISK models did not contain any explicit model for unsaturated, low-level downdrafts. Two recent models have included some form of low-level downdraft parameterization. The simplest approach, taken by Nehrkorn (1985), was to introduce a vertical profile of heating that includes low-level cooling. Nehrkorn found that his baroclinic wave-CISK model predicted that the wavelength of maximum growth shifted toward shorter wavelengths as low-level cooling was intensified. A far more complicated downdraft model was introduced by Raymond (1983). Downdraft fluxes were parameterized to be a function of the specified precipitation efficiencies of the convective cell. He also introduced a time lag between the onset of downdrafts and low-level convergence. An interesting consequence of introducing a lagged downdraft is that a new mode appeared in his solutions. In the previous wave-CISK solutions, only propagating modes appear. However, with a lagged downdraft, a mode is present that is stationary with zero wind or moves with the mean wind, and appears with the propagating mode. If the downdraft mass flux exceeds a critical value, Raymond found that this advection mode exhibits a growth rate much larger than the normal propagating mode. Furthermore, Raymond indicates that the growth rate of this mode is almost independent of wavelength.

Several modelers have introduced simple parameterizations of the effects of mixing of momentum between updrafts or downdrafts and a quiescent environment (Schneider and Lindzen, 1976; Stevens *et al.*, 1977; Raymond, 1984). Stevens *et al.* concluded that momentum mixing was only important for larger scale waves. Squall lines have such a short time scale compared to the time scale of momentum mixing that they are not affected by its parameterization. Raymond (1984) also concluded that the effect of momentum mixing on squall-line propagation is negligible.

Early wave-CISK models were formulated in a barotropic atmosphere; the effects of shear were not considered. Recent models of tropical and midlatitude squall lines and mesoscale convective systems (Emanuel, 1982;

Silva-Dias *et al.*, 1984; Raymond, 1983; Nehrkorn, 1985) have included vertical shear of the horizontal wind. Shear is found to contribute to the scale selection of an unstable mode on the mesoscale. Silva-Dias *et al.* emphasized the importance of a low-level jet in scale selection. In Raymond's (1983) model, stronger shear suppressed the normal wave-CISK propagating modes. Only the upshear fundamental mode was enhanced by shear. Nehrkorn (1985) found that the wavenumber and growth rate of the fastest growing mode are proportional to shear. Inclusion of the Coriolis force with shear also affects wave-CISK model predictions (Emanuel, 1982; Nehrkorn, 1985), further reinforcing scale selection. Nehrkorn noted that Coriolis turning contributes significantly to larger scale waves of the order of several thousand kilometers, while convective heating or normal wave-CISK prevails at shorter wavelengths.

In summary, the variety of wave-CISK models formulated in recent years exhibits a considerable degree of sensitivity to the details of the convective parameterizations and to the characteristics of the environmental thermodynamic stability and wind profiles. In the following section we compare wave-CISK models with the MGMT models and with observations.

9.6.3 Comparison of Wave-CISK and MGMT Models with Observations

Both the MGMT and wave-CISK models are extreme simplifications or idealizations of the behavior of wet convective systems. The formulations of both models have aspects in common with theoretical models of stably stratified flow over mountain barriers (see Chapter 12 for similarities between mountain wave theory and these models). Therefore, wave solutions are inherent in both approaches, as is the role of horizontal pressure gradients on storm propagation. The assumption of steady state is a severe limitation of the MGMT models, as is the assumption of linearity in the wave-CISK models. Comparison of model predictions with observations has generally been limited to a few case studies. Unfortunately, because the MGMT models require the updraft/downdraft profiles to be specified and the wave-CISK models are sensitive to the details of the convective parameterizations, each new development in model formulation requires a completely new evaluation of the models.

Betts *et al.* (1976) compared the MM model predictions of tropical squall propagation speeds with observations during VIMHEX 1972. They found propagation speeds to be fairly well represented by the model. In a more extensive study, Fernandez and Thorpe (1979) compared the MM model and Raymond's (1975, 1976) versions of the wave-CISK model with 15 tropical squall-line cases observed over tropical Venezuela, the eastern

Atlantic, and West Africa. Comparison of wave-CISK model phase speeds and directions with observations is somewhat arbitrary, because the model solutions are waves with different phase speeds. The standard and most objective procedure compares the phase speed and direction of the most unstable mode (the mode exhibiting the largest growth rate) with observed storm propagation. However, Raymond (1975) argued from scale analysis that the longest wave ($k = 0$) in the model domain should be compared with observed storm movement. Subsequently Raymond (1976) abandoned this argument and reconstructed the predicted spectra to allow comparison among the modes giving the "best" prediction of observed storm movement. Recently D.J. Raymond (personal communication) suggested that the second harmonic (the wavelength exhibiting the second largest growth rate) best agrees with observed storm propagation. Fernandez and Thorpe (1979) followed Raymond's (1976) suggestion and compared observed speeds with the closest predicted phase speed with the model. Much of the arbitrariness is not present in the MGMT models, because they do not predict a spectrum of solutions; moreover, they only predict storm speed and not direction.

Fernandez and Thorpe found that the wave-CISK model did well in predicting the propagation velocities of five out of eight Venezuelan storms and poorly on one, while a unique propagation speed could not be assessed in the remaining two cases. The model gave consistently poor predictions on all west African storms and in eastern Atlantic squall lines. They argued that the weakness in the wave-CISK model in the eastern Atlantic squall lines may be associated with the assumption that convection breaks into plumes at cloud base level, whereas they suggested that plumelike convection is not observed in eastern Atlantic squalls up to a much higher level. This is consistent with the author's (WRC) observation during GATE that cellular convection was not pronounced until levels as high as 500 mbar. Fernandez and Thorpe adjusted a parameter that they associated with the level of formation of plumelike convection and found that by increasing the height of plume convection initiation, the agreement between predicted and observed storm movement improved.

Fernandez and Thorpe's study was restricted to only tropical squall lines; therefore, only the MM model was compared with observations. It should be remembered that the propagation speed predicted by the MM model is a function of the midlevel wind speed U_m and CAPE. Fernandez and Thorpe noted that a better prediction was obtained with MM's model when the mean value of the component of wind along the direction of motion was used rather than the observed midlevel wind. They found that the MM model slightly overpredicted the propagation speeds of the Venezuelan and east Atlantic storms. However, the speeds of the west African storms were significantly overestimated. In contrast, the wave-CISK model underestimated storm movement.

Comparison of observed and predicted storm movement in middle latitudes with the MGMT and wave-CISK models has mainly been limited to a few studies. MG showed that their model propagation speeds compared well with the Wokingham supercell storm and an Indian squall line, while their model slightly overpredicted the movement of a severe storm in the Indian premonsoonal environment. The latter two cases were not in a midlatitude environment. They found that the model underpredicted storm movement by 2.5 m s^{-1} in a midlatitude case with a large negative R_i and overpredicted storm movement (by 11 m s^{-1}) for a severe right-moving storm in the United States.

Raymond (1976) compared his wave-CISK model predictions of storm movement to two isolated, midlatitude severe storms. One case was an isolated supercell storm in which the predicted storm movement was 260° at 15 m s^{-1}, while the observed movement was 275° at 14 m s^{-1}. The second case was a severe storm system that multiplied from a single cell via the splitting process. The model predicted that the left-moving, split cell moved 187° at 14 m s^{-1}, compared to 178° at 17.5 m s^{-1}. The right-moving cell was predicted to move from 250° at 10 m s^{-1} compared with the observed movement 257° at 11.2 m s^{-1}. Thus, for these two cases, the model performed remarkably well. Moreover, it exhibited an ability to simulate storm splitting, at least for the case where the wind hodograph is not precisely planar. (We shall talk more about the splitting process in the next section.) More recently, Raymond (1984) compared his wave-CISK model in mass flux form with a time-lagged downdraft parameterization with observed features of a tropical and a midlatitude squall line. He showed that the speed of propagation of the simulated storms agreed with observations. In addition, the simulated streamline patterns shared many common features with the observed flow structure of midlatitude squall lines.

When comparing his wave-CISK model predictions with observed storm propagation, Nehrkorn (1985) arrived at a markedly different conclusion. As did Raymond, he found that the wave-CISK model predicted major observed circulation features such as the upstream tilt of the major updraft. Furthermore, the predicted surface pressure field with decreasing pressure in advance of the squall line and rising pressure behind is reasonable. However, he concluded that the predicted wavelength of the squall line was some two to three times greater than observed. The largest discrepancy occurred, however, between the predicted and observed storm speed and direction. Instead of comparing his model predictions with a few selected case studies, Nehrkorn compared his model predictions with the squall-line climatological data presented by Bluestein and Jain (1985) and Wyss and

Emanuel (1988). The comparison with a climatology of storms removes the possibility that agreement between model predictions and observations is limited to only a few cases. Nehrkorn found that the model could predict an orientation angle of the squall line relative to mean tropospheric winds in reasonable accord with Bluestein and Jain's analyses when a heating profile with upper-level heating and low-level cooling is used. Even with this heating profile, he found that the predicted orientation angle of the most unstable mode disagreed with Wyss and Emanuel's analyses by 25°. Even more disturbing is the fact that when the observed orientation angles are predicted within 5°, the predicted phase speeds are in error by a factor of 2.4 to 2.9. Nehrkorn did not determine whether the failure of his model in reproducing observed storm movement was due to (1) weaknesses in his cumulus parameterization, such as the lack of time-lagged downdrafts; (2) failure of the CISK concept (i.e., low-level convergence forces upper-level convective heating); or (3) nonlinear wave interactions. Clearly this problem must be resolved before wave–CISK can be considered a viable model of convective storm propagation.

In summary, we have reviewed two fundamental theories of convective storm propagation. Both modeling approaches have been helpful in clarifying and isolating the fundamental processes affecting convective storm autopropagation. Moreover, both theories have been tested against observations and have been shown to have deficiencies. Before we leave the subject of storm propagation, however, we will discuss the physics of storm splitting and its relationship to storm propagation. The cell splitting process is not only a mode of cell propagation; it also appears to be important to the development of rotating storms. While Raymond's (1976) wave–CISK model did represent a splitting process, it did not elucidate the physics behind the process.

9.6.4 *Updraft Splitting and Storm Propagation*

The process of storm splitting was first recognized from the behavior of radar reflectivity fields (Fujita and Grandoso, 1968; Achtemeier, 1975). More recently it has been identified by multiple-Doppler radar as a process of a splitting of the updraft in convective storms (Bluestein and Sohl, 1979; Knupp and Cotton, 1982a). Three-dimensional numerical models have been used extensively to analyze the splitting process (Wilhelmson and Klemp, 1978, 1981; Klemp and Wilhelmson, 1978a, b; Thorpe and Miller, 1978; Schlesinger, 1980; Clark, 1979; Rotunno and Klemp, 1982, 1985; Weisman and Klemp, 1984; Tripoli and Cotton, 1986).

Klemp and Wilhelmson (1978a) argued that downward drag due to strong rainwater loading in the cloud interior contributes to updraft splitting.

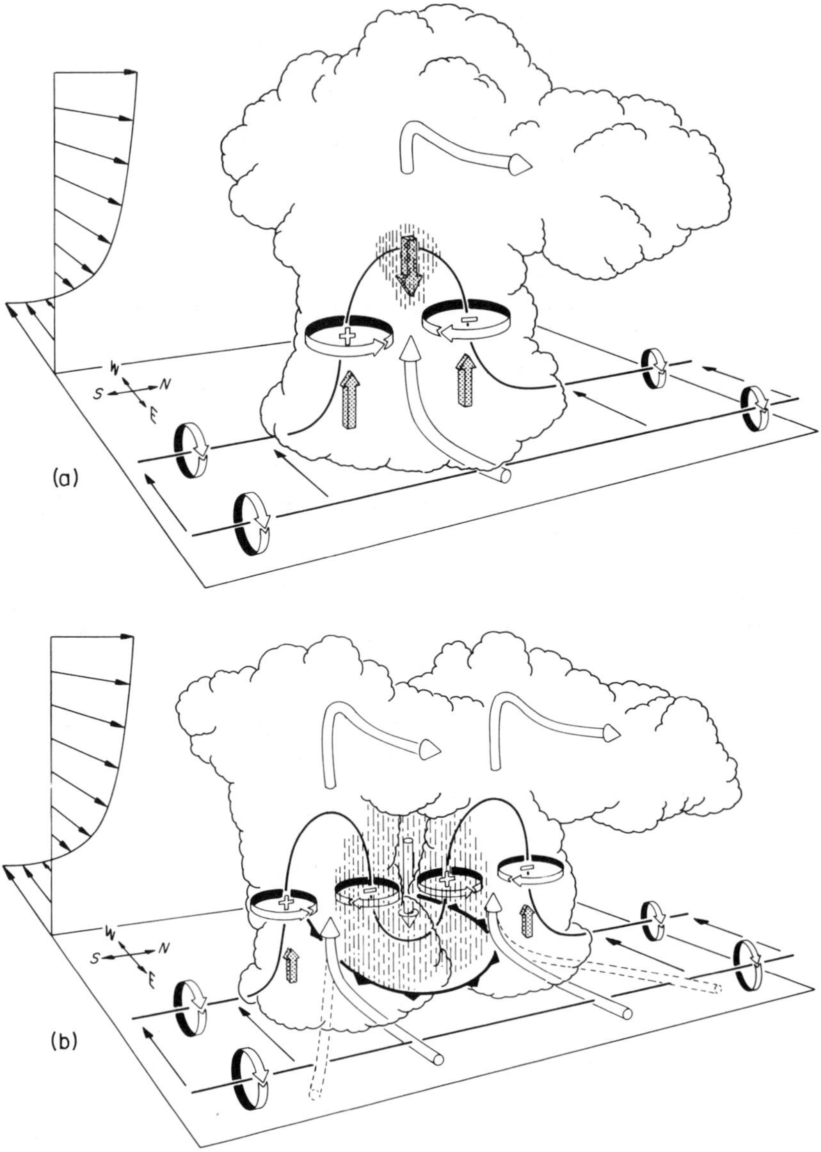

Fig. 9.33. Schematic depicting how a typical vortex tube contained within (westerly) environmental shear is deformed as it interacts with a convective cell (viewed from the southeast). Cylindrical arrows show the direction of cloud-relative airflow, and heavy solid lines represent vortex lines with the sense of rotation indicated by circular arrows. Shaded arrows represent the forcing influences that promote new updraft and downdraft growth. Vertical dashed lines denote regions of precipitation. (a) Initial stage: vortex tube loops into

Schlesinger (1980) concluded that the dynamic pressure perturbations associated with the entrainment of potentially cool air outside the storm also contributed strongly to storm splitting. Clark (1979) also concluded that the entrainment process contributed to storm splitting. He argued that the entrainment process and, hence, storm splitting were very sensitive to the details of the often unrealistic model initial conditions.

Vertical shear of the horizontal wind is clearly important in the splitting process. Thorpe and Miller (1978) concluded that strong shear contributes to a large downshear slope of the updraft core with rain falling into the inflow region, resulting in a split updraft cell. Klemp and Wilhelmson's (1978a) calculations indicate that strong low-level vertical shear or the presence of a low-level jet favors the updraft splitting process.

The directional shear of the horizontal wind has been found to be important in determining which of the two split-cell members predominates. If the wind shear is unidirectional (i.e., shows no change in direction with height) such as illustrated in Fig. 9.33, Wilhelmson and Klemp (1981) found that a vortex pair develops in a horizontal plane in which positive vertical vorticity occurs on the right side of the updraft relative to the shear vector. On the left side of the updraft, negative vorticity occurs. This vortex pair is essentially responsible for the production of entrainment on the down-shear flank of towering cumulus clouds as discussed in Chapter 8. Rotunno (1981) showed that the vortex pair is a direct consequence of the tilting of vortex tubes by rising convective updrafts. The vortex tubes, shown in Fig. 9.33a, are produced by the environmental shear. The downshear downdraft induced by entrainment and rainwater loading subsequently splits the original cell into two cells, one dominated by cyclonic vorticity at low levels propagating to the right of the shear vector; a second cell propagating to the left of the shear vector is dominated by anticyclonic vorticity. These two storms, illustrated in Fig. 9.33b, possess mirror-image symmetry. Furthermore, these storms possess properties similar to Browning's (1964) and Browning and Wexler's (1968) conceptual model of severe right (SR)- and left (SL)-moving storms.

Klemp and Wilhelmson (1978b) found that the relative strengths of the SR- and SL-moving storms varied with directional changes of the shear vector. They found that if the wind shear vector veers with height (i.e., turns clockwise in the northern hemisphere), the development of a cyclonic,

---

the vertical as it is swept into the updraft. (b) Splitting stage: downdraft forming between the splitting updraft cells tilts vortex tubes downward, producing two vortex pairs. The barbed line at the surface marks the boundary of the cold air spreading out beneath the storm. [Adapted by Klemp (1987), from Rotunno (1981). Copyright © 1987 by Annual Reviews, Inc. Reproduced with permission.]

right-moving storm is favored. In contrast, if the wind shear vector backs with height (i.e., turns counterclockwise in the northern hemisphere), then an anticyclonic, left-moving storm prevails. Rotunno and Klemp (1982) developed a simple linear model to explain these results. Figure 9.34 is a conceptual model illustrating the variations in pressure gradient with changes in the direction of the shear vector. Figure 9.34a illustrates the case of unidirectional shear flow where symmetric vortex pairs develop in the updraft with relative high-pressure upshear of the updraft and relative low-pressure downshear. With no change in the shear vector with height, the relative highs and lows are vertically aligned. Thus, the storm circulations develop symmetrically about the shear vector. When the shear vector changes direction with height, such symmetry is lost, however. Figure 9.34b illustrates the case when a shear vector turns clockwise with height. This illustrates that a relative high forms to the south and a low to the north at low levels, and at a higher level a high is situated to the north and a low to the south. This creates a favorable vertical pressure gradient for updraft intensification on the southern side of the updraft and an unfavorable one on the northern side. Therefore, such a wind hodograph favors the SR-moving storm.

Weisman and Klemp (1984) performed a detailed analysis of the pressure fields and convergence patterns simulated by a three-dimensional cloud model for wind profiles in which the shear vector turns clockwise through 180° over the lowest 5 km. As noted previously, such profiles favor the formation of the SR storm system. As in earlier simulations, splitting of the updraft of the initial cell occurred which resulted in right-flank and left-flank updraft cells. Two distinct mechanisms forced the low-level convergence fields supporting the updrafts. In the case of the left-flank updraft, the strong convergence was driven by a cold pool of air spreading against an incoming flow of potentially unstable environmental air. The convergence is thus forced by high pressure behind the gust front. This effect is strongest where the storm relative inflow and downdraft outflow most directly oppose each other. The simulated left-flank cells were multicellular, shorter lived, and less intense than the right-flank cells. However, the left-flank cells produced the most rainfall.

In the case of the right-flank cell, the low-level convergence field was locally induced by the lowering of pressure caused by the dynamic interaction of the storm with environmental wind shear, similar to that illustrated in Fig. 9.34b. The resultant mesolow produced low-level convergence of moist air into the updraft in regions where convergence caused by the spreading cold outflow is relatively weak. The strength of such a dynamically induced mesolow increases with higher values of environmental wind shear. Weisman and Klemp found that a bulk Richardson number,

$$R_{\mathrm{i}} = B/\tfrac{1}{2}(\Delta\bar{u})^2, \qquad (9.35)$$

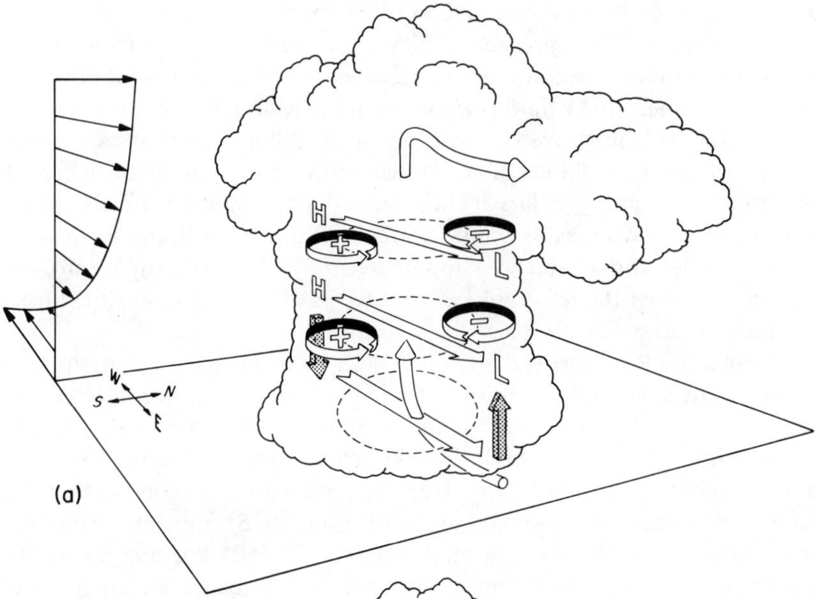

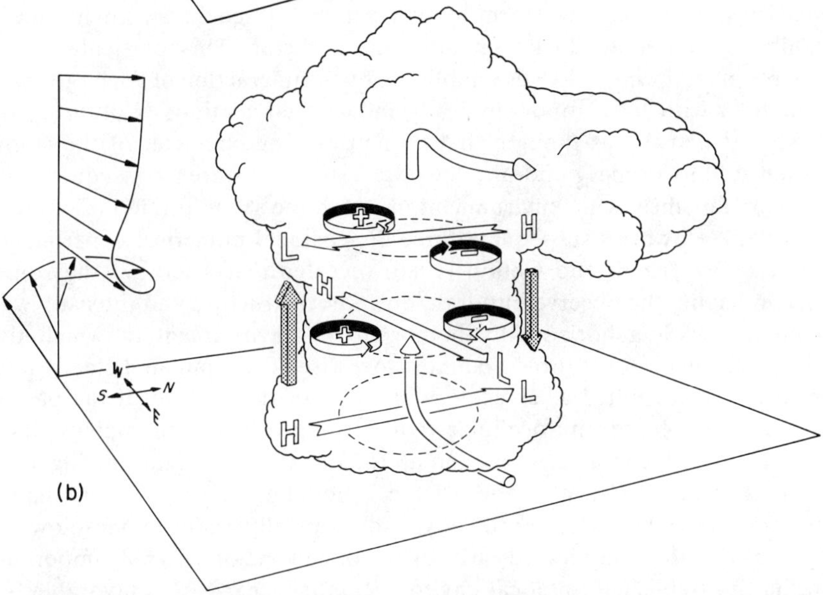

Fig. 9.34. Schematic illustrating the pressure and vertical vorticity perturbations arising as an updraft interacts with an environmental wind shear that (a) does not change direction with height and (b) turns clockwise with height. The high (H) to low (L) horizontal pressure gradients parallel to the shear vectors (flat arrows) are labeled along with the preferred location of cyclonic (+) and anticyclonic (−) vorticity. The shaded arrows depict the orientation of the resulting vertical pressure gradients. [Adapted by Klemp (1987), from Rotunno and Klemp (1982). Copyright © 1987 by Annual Reviews, Inc. Reproduced with permission.]

where B represents the convective available potential energy in the storm's environment and $\Delta\bar{u}$ represents a difference between environmental wind speeds at low levels and midlevels, was a useful parameter for determining when such a mesolow would prevail. As in the MG definition, the Richardson number is a measure of the ratio of available potential to available kinetic energy. Low values of R_i correspond to an environment where the available kinetic energy is large. They found that the right-flank cells increase in strength with decreasing values of R_i until the shear becomes so strong that the initial convection was suppressed ($R_i \sim 15$). The right-flank cells predominate over the left-flank cells (in terms of updraft intensities) for R_i less than about 6.0.

As noted by Rotunno and Klemp (1982), the majority of severe tornadic storms over the United States are SR-moving storms. This is consistent with Maddox's (1976) climatology of tornadic storms which reveals that they occur in a wind field in which the wind shear vector between the surface and 700 mbar turns clockwise. However, occasionally counterclockwise wind shear vectors are observed along with intense SL-moving storms. One such storm, described by Knupp and Cotton (1982a, b), was an intense, quasisteady left-moving storm system which formed in an environment exhibiting a counterclockwise wind shear vector. This particular local environmental wind field was established by the interaction of the large-scale flow field with local topographically induced circulations (Cotton *et al.*, 1982). At first it was thought that the left-moving character of the storm system and its primary updraft/downdraft structure were a consequence of the complex mesoscale environment in which the storm formed (Cotton *et al.*, 1982). However, subsequent three-dimensional numerical experiments reported by Tripoli and Cotton (1986) revealed that a left-moving storm system having the observed updraft/downdraft structure and intensity was simulated with a horizontally homogeneous environment in which the low-level shear vector turned counterclockwise. The simulated steady left-moving storm formed as a result of splitting of the updraft of the parent cell. A less intense, multicellular right-moving cell was simulated which also exhibited characteristics similar to an observed right-moving cell. Consistent with Weisman and Klemp's modeling results, the simulated left-moving cell was supported by a dynamically induced mesolow. It appears that the complicated early history of convection was only important in that it established the local environmental shear that was favorable for the dominance of the left-moving cell.

In summary, several mechanisms are involved in the propagation of convective storms. Which mechanism prevails depends strongly upon the environmental winds and stability. Updraft rotation also affects storm propagation. We now examine how updraft rotation affects storm dynamics as well as the formation of tornadoes.

9.7 Mesocyclones and Tornadoes

In this section we concentrate on the processes involved in the generation of thunderstorm rotation and tornado-scale vorticities. The emphasis, however, is on the generation of rotation on the thunderstorm scale, with only a brief discussion of structure and vorticity in tornadoes. We begin by defining the various parameters and equations that are used to describe and measure rotation in thunderstorms and tornadoes. We then discuss the larger scale conditions most favorable for generating tornado-producing storms then discuss the processes involved in developing rotating thunderstorms (mesocyclones). We conclude with a brief summary of the characteristics of tornadoes and models of tornadoes.

9.7.1 Vorticity and Circulation Equations

One measure of rotation frequently used in meteorology is the vorticity which represents the local and instantaneous rate of rotation of the system. In the case of solid rotation, vorticity is twice the angular velocity. To obtain the vorticity of a system, one takes the curl or vector cross product of the equations of motion. As shown by Dutton (1976), this results in the general vorticity equation of the form:

$$d\boldsymbol{\omega}/dt = \overset{\text{(a)}}{(\boldsymbol{\omega}\cdot\nabla)\mathbf{V}} - \overset{\text{(b)}}{\boldsymbol{\omega}\nabla\cdot\mathbf{V}} - \overset{\text{(c)}}{\nabla\alpha\times\nabla p} + \overset{\text{(d)}}{\gamma\nabla^2\boldsymbol{\xi} + \nabla\gamma\times[\nabla^2\mathbf{V} + \nabla(\nabla\cdot\mathbf{V})]}, \tag{9.36}$$

where the absolute vorticity $\boldsymbol{\omega} = \boldsymbol{\xi} + 2\mathbf{r}$, $\boldsymbol{\xi}$ is the relative vorticity, and $\mathbf{r}$ is the earth's rotation rate. Ignoring the effects of viscosity, Eq. (9.36) can be decomposed into its scalar components as follows:

$$\frac{d}{dt}\begin{bmatrix}\xi\\ \eta\\ \zeta\end{bmatrix} = (\boldsymbol{\omega}\cdot\nabla)\begin{bmatrix}u\\ v\\ w\end{bmatrix} - \begin{bmatrix}\xi\\ \eta\\ \zeta\end{bmatrix}\nabla\cdot\mathbf{V} - \begin{bmatrix}\partial\alpha/\partial y & \partial p/\partial z & - & \partial p/\partial y & \partial\alpha/\partial z\\ \partial\alpha/\partial z & \partial p/\partial x & - & \partial\alpha/\partial x & \partial p/\partial z\\ \partial\alpha/\partial x & \partial p/\partial y & - & \partial\alpha/\partial y & \partial p/\partial x\end{bmatrix}. \tag{9.37}$$

Terms (a) and (b) of Eqs. (9.36) and (9.37) represent the so-called tilting and convergence terms, respectively. These terms act together to concentrate and transfer vorticity from one plane to another. An updraft, for example, can tilt a horizontally oriented vortex tube into a vertically oriented one. Term (b), the convergence term, represents the fluid analog to the angular acceleration of a solid rotating system due to change in the radius of rotation.

Term (c) is called the solenoidal term and it represents the production of vorticity by baroclinicity. It acts predominantly in the horizontal, where large-scale baroclinic fronts or convective-scale gust fronts can generate vorticity about a horizontal axis.

For a Boussinesq system (see Chapter 2), Eq. (9.37) reduces to

$$\frac{d}{dt}\begin{bmatrix}\xi\\ \eta\\ \zeta\end{bmatrix} = \underset{\text{(a)}}{(\boldsymbol{\omega}\cdot\boldsymbol{\nabla})\begin{bmatrix}u\\ v\\ w\end{bmatrix}} - \underset{\text{(b)}}{\begin{bmatrix}\xi\\ \eta\\ \zeta\end{bmatrix}\boldsymbol{\nabla}\cdot\mathbf{V}} + \underset{\text{(c)}}{\begin{bmatrix}g\,(\partial/\partial y)(\alpha'/\alpha_0)\\ -g\,(\partial/\partial x)(\alpha'/\alpha_0)\\ 0\end{bmatrix}}, \tag{9.38}$$

where the solenoidal term affects only the horizontal components of vorticity.

Another useful quantity for investigating the rotational properties of thunderstorms is *equivalent potential vorticity*, defined as

$$\omega_{\theta_e} = \alpha_0 \boldsymbol{\nabla}\theta_e \cdot \boldsymbol{\omega}. \tag{9.39}$$

Rotunno and Klemp (1985) extended the concept of potential vorticity (see Dutton, 1976, p. 382) to a cloud system. Introducing the anelastic continuity equation (see Chapter 2) in Eq. (9.38) and defining $B = \alpha' g/\alpha_0$, we have

$$\frac{1}{\alpha_0}\frac{d(\alpha_0\boldsymbol{\omega})}{dt} = (\boldsymbol{\omega}\cdot\nabla)\mathbf{V} + \boldsymbol{\nabla}\times(B\mathbf{k}). \tag{9.40}$$

Making use of the identity

$$\boldsymbol{\omega}\cdot\frac{d}{dt}\boldsymbol{\nabla}\theta_e = \boldsymbol{\omega}\cdot\nabla\frac{d\theta_e}{dt} - \nabla\theta_e\cdot[(\boldsymbol{\omega}\cdot\nabla)\mathbf{V}]$$

in the left-hand side of Eq. (9.40) yields

$$\alpha\nabla\theta_e\left[\frac{1}{\alpha_0}\frac{d(\alpha_0\boldsymbol{\omega})}{dt} - \boldsymbol{\omega}\cdot\nabla\mathbf{V}\right] = \frac{d}{dt}(\alpha_0\nabla\theta_e - \boldsymbol{\omega}) - \alpha\boldsymbol{\omega}\cdot\nabla\frac{d\theta_e}{dt}, \tag{9.41}$$

or

$$\frac{d}{dt}(\alpha_0\nabla\theta_e\cdot\boldsymbol{\omega}) - \alpha_0\boldsymbol{\omega}\cdot\nabla\frac{d\theta_e}{dt} = -\alpha_0\nabla\theta_e\cdot\nabla\times(B\mathbf{k}). \tag{9.42}$$

Assuming that cloud updrafts and downdrafts are wet adiabatic, $d\theta_e/dt = 0$, then

$$\frac{d}{dt}(\omega_{\theta_e}) = \frac{d}{dt}(\alpha_0\nabla\theta_e\cdot\boldsymbol{\omega}) = -\alpha_0\nabla\theta_e\cdot\nabla\times(B\mathbf{k}) \tag{9.43}$$

Ignoring the effects of water loading, in saturated air $B = B(\theta_e)$. Therefore, the right-hand side is zero and ω_{θ_e} is conserved. In unsaturated downdrafts, Rotunno and Klemp argue by scale analysis that the right-hand side is small so that ω_{θ_e} is nearly conserved. Thus, ω_{θ_e} has the property of being nearly conserved along updraft and downdraft trajectories in convective storms. This useful property can help us understand the intimate relationship between the vorticity and thermodynamics of a convective storm.

Another parameter useful for studying storm rotation is the circulation $C(t)$ defined as

$$C(t) = \oint \mathbf{V} \cdot d\mathbf{l}, \tag{9.44}$$

where the integration is performed around a closed material surface. The change in circulation around the material surface for an inviscid, Boussinesq system is given by

$$dC/dt = \oint B\mathbf{k} \cdot d\mathbf{l}. \tag{9.45}$$

Recently, Lilly (1986a,b) has advocated the use of the "helicity" concept in studying thunderstorm rotation. The helicity H is defined as

$$H = \mathbf{V} \cdot \boldsymbol{\omega}, \tag{9.46}$$

which is essentially a covariance of the velocity and vorticity vectors. Lilly also defined relative helicity r_{H} as

$$r_{\mathrm{H}} = H/V\omega, \tag{9.47}$$

where r_{H} ranges in magnitude between $+1$ and -1.

We will subsequently use these principles in our examination of factors contributing to thunderstorm rotation and tornado formation.

9.7.2 Large-Scale Conditions Associated with Tornadoes

In this section we examine the large-scale conditions associated with the formation of tornadoes. As we shall see, forecasting tornado-producing storms involves forecasting intense storms that are also capable of producing hail, flash floods, and strong, nonrotational surface winds. Few severe storms produce only one form of severe event. However, we can identify conditions which favor the formation of the most intense tornadic outbreaks. These conditions usually occur in the springtime associated with vigorous baroclinically driven midlatitude cyclonic storms. Figure 9.35 illustrates the synoptic-scale conditions often associated with major tornado outbreaks. Pertinent features are the wave cyclone with its attendant warm front (WF) and cold front (CF), along with the low-level jet (LJ) and upper-level jet

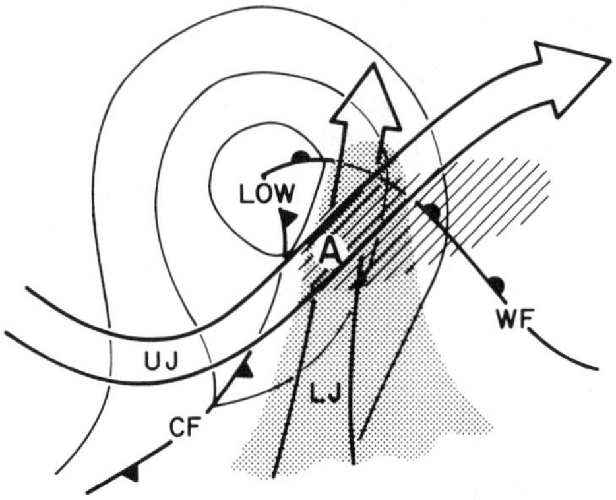

Fig. 9.35. Idealized sketch of a situation especially favorable for development of severe thunderstorms. Thin lines denote sea-level isobars around a low-pressure center, CF and WF indicating cold and warm fronts at the ground. Broad arrows represent jet streams in low (LJ) and upper (UJ) levels of the troposphere. Region rich in water vapor at low levels is stippled. Severe storms are most likely to originate near A and, during the ensuing 6–12 h as the cyclone moves eastward and the severe storms move generally eastward at a faster speed, affect the hatched region (outside which thunderstorms may also occur that will probably be less intense). [From Newton *et al.* (1978).]

(UJ) streams. Beebe and Bates (1955) noted that the region labeled A where the two jet streams cross is the most favored region for the formation of tornado-producing storms. This region is often characterized by the following conditions:

(1) Strong low-level advection of heat and moist air, providing an adequate supply of moist static energy.
(2) A conditionally unstable atmosphere through a deep layer.
(3) A strong capping inversion, which inhibits the widespread outbreak of thunderstorms.
(4) Moderate to strong vertical shear of the horizontal winds.

Also necessary for severe storm occurrence is some mechanism for breaking the capping inversion. Some of the major mechanisms are frontal lifting, surface heating, drylines, gravity waves, and lifting due to terrain effects.

It has also been shown that major tornadic outbreaks are associated with jet streaks (speed maxima) in the upper-level jet stream. Riehl *et al.* (1952) concluded that jet streaks were associated with divergence fields which produce upward motion in the left front of the jet exit zone and in the right rear of the entrance zone of the jet streak. Subsequently, Uccellini and Johnson (1979) showed that the UJ and LJ jets are often coupled by mass adjustments associated with the propagation of geostrophically unbalanced flows associated with the jet streaks. Thus Kloth and Davies-Jones (1980) found that the left-exit and right-entrance zones are favored zones for upward motion and tornado occurrences.

The warm sector of the wave cyclone is often a region of strong positive vorticity advection (PVA), which, as discussed in Chapter 11, is associated with upward motion and is therefore a very important parameter in severe-storm forecasting (Koscielski, 1965; Miller, 1967). However, Maddox and Doswell (1982) showed several examples of tornadic outbreaks that were not associated with intense cyclonic storms and their attendant PVA and UJ/LJ relationships. These were significant events; they had as many as 15 tornado reports and considerable loss of life and property damage. They noted that the most consistent large-scale feature associated with such storms was a pronounced low-level thermal advection field. It should be remembered that the omega equation (see Chapter 11) shows that for quasigeostrophic motion, thermal advection contributes directly to large-scale vertical motion. Thus, Maddox and Doswell argue that in the absence of significant PVA and jet-streak dynamics, low-level warm advection in regions of strong conditional instability signals the presence of large-scale upward motion that can trigger significant severe weather occurrences.

Tornadoes are also generally associated with tropical cyclones and can account for a significant fraction of the damage and loss of life in such storms. Novlan and Gray (1974) found that 25% of the hurricanes in the United States spawned tornadoes, with 10 as the average number of tornadoes per storm. This number excludes hurricane Beulah in 1967, which produced 141 tornadoes. Fujita *et al.* (1972) found the average number of tornadoes associated with typhoons in Japan to be somewhat less, that is, 2.3. Studies of the location of tornadoes relative to storm motion in the United States (Malkin and Galway, 1953; Smith, 1965) and Japan (Fujita *et al.*, 1972) suggest that the right-forward sector of the storm is the most frequent location of the tornadoes. Orton (1970) and Novlan and Gray (1974) argued that the frame of reference relative to true north was more meaningful than one relative to storm motion. The latter found that the sector from 350 to 120° and 60–250 nautical miles from the storm center was a preferred location for tornadoes. They also noted that hurricane-spawned tornadoes were not associated with particularly unstable soundings; arguing that strong low-level shear in the hurricane environment was much more important in generating tornadoes than were unstable soundings. Furthermore, a hurricane coming ashore and filling rapidly is likely to develop a cold core and an associated strong low-level vertical wind shear.

In one case studied by Fujita *et al.* (1972), a tornado was associated with a rotating thunderstorm (also referred to as a mesocyclone or tornado cyclone) embedded in the tropical cyclone. Tornadoes are also associated with mesoscale convective complexes (see Chapter 10), a storm system that often develops in weakly baroclinic atmospheres.

In summary, we find that the major tornadic outbreaks occur generally with vigorous midlatitude and tropical cyclonic disturbances. However, any time the environment is able to support an intense thunderstorm system, tornadoes are possible, along with other forms of severe weather such as flash floods, hail, and severe winds, so forecasting tornadoes remains a major challenge to meteorologists. For a more extensive discussion about forecasting tornadoes and other severe storms the reader is referred to Schaeffer (1986).

9.7.3 Rotating Thunderstorms or Mesocyclones

In this section we will examine the characteristics and dynamics of rotating thunderstorms, also known as mesocyclones. If a rotating thunderstorm produces a tornado, it is also referred to as a tornado-cyclone. Most severe tornadoes are associated with rotating thunderstorms.

The basic conceptual model of a rotating thunderstorm builds on the supercell storm model as discussed in Section 9.1. Most, if not all, supercell storms are also rotating thunderstorms. However, one must not be misled into thinking that only supercell storms are rotating storms. There is considerable evidence that the mature cell in multicellular storms often exhibits the rotational characteristics of the supercell storm.

A comprehensive conceptual model of a rotating thunderstorm has been derived by Lemon and Doswell (1979) from the synthesis of data obtained for aircraft, Doppler radar, and surface mesoscale networks, as well as from visual observations. Figure 9.23 illustrates the surface features of the mesocyclone at the time of tornado formation. Pertinent features are the locations of the storm updraft (UD), and the forward-flanking downdraft (FFD) and rearward flanking downdraft (RFD). The storm system resembles an occluding midlatitude cyclonic storm. The most intense tornadoes (T) are often located near the tip of the "occlusion" close to the updraft/downdraft interface. The authors emphasize the importance of the rear-flanking downdraft in establishing the occlusion and the descent of the mesocyclone from midlevels. Often associated with the descending mesocylone is a local region of intense horizontal shear, which can be detected by Doppler radar. This local horizontal shear region is referred to as the tornado vortex signature (TVS), and is often found to precede tornado touchdown at the surface by tens of minutes.

A three-dimensional depiction of the temporal evolution of Lemon and Doswell's model storm is shown in Fig. 9.36. The evolution of a tornado-producing storm is depicted in four separate stages.

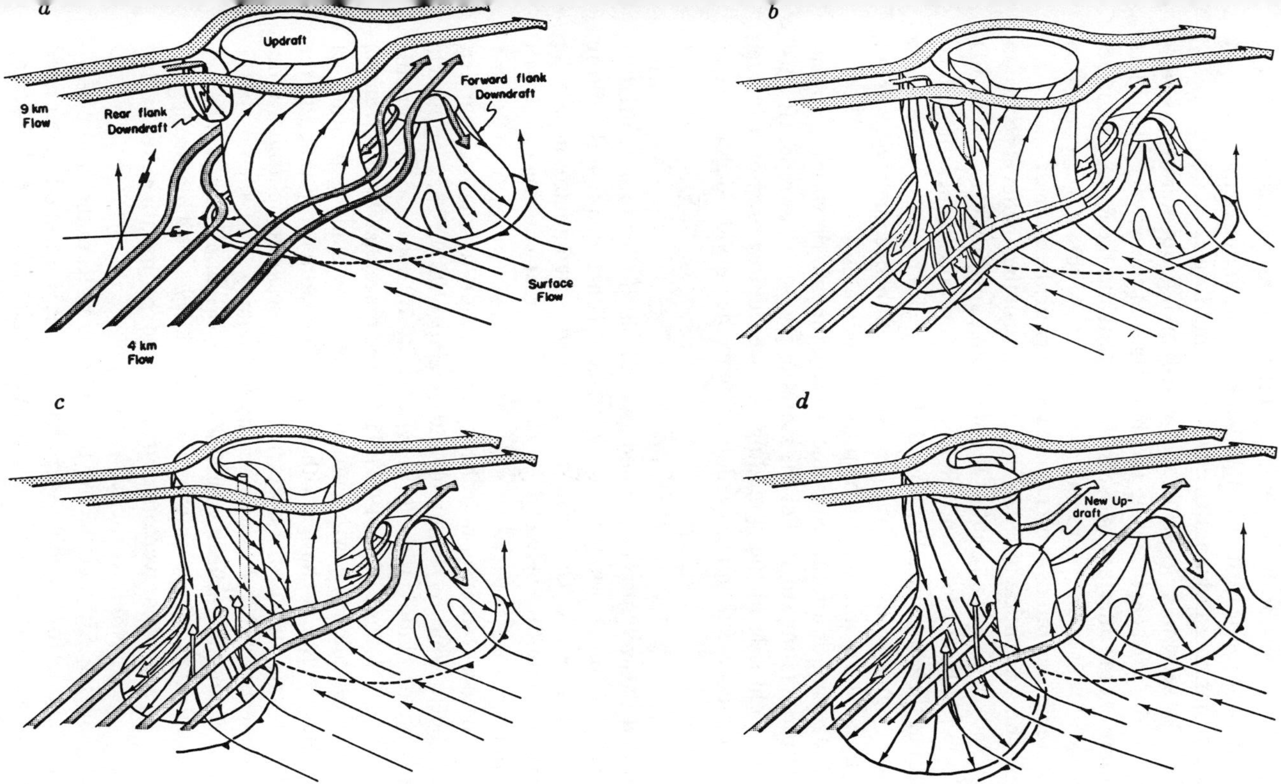

Fig. 9.36. Schematic three-dimensional depiction of evolution of the drafts, tornado, and mesocyclone in an evolving supercell storm. The stippled flow line suggesting descent of air from the 9-km stagnation point has been omitted from (c) and (d), for simplicity. Fine stippling denotes the TVS. Flow lines throughout the figure are storm relative and conceptual only, and are not intended to represent flux, streamlines, or trajectories. Conventional frontal symbols are used to denote outflow boundaries at the surface, as in Fig. 9.23. Salient features are labeled on the figure. [From Lemon and Doswell (1979).]

Stage 1: At this stage the storm has a strong rotating updraft with a forward-flanking downdraft in the precipitation region. A gust front at the surface forms at the forward and right flanks of the radar echoes. Only an incipient rear-flanking downdraft is evident at midlevels.

Stage 2: As seen in Fig. 9.36b, the rear-flanking downdraft has now descended to the surface, initiating the first stages of the wave–cyclone-like features there. At the interface between the updraft and the RFD, a midlevel vortex or TVS forms at the rear flank of the storm. At this stage, a bounded weak-echo region or radar hook echo characteristically forms.

Stage 3: This stage is identified by the formation of the low-level occlusion (see Fig. 9.36c), which is a result of the spreading gust front from the RFD catching up with the outflow from the FFD. The tornado reaches the surface near the tip of the occlusion. The bounded weak-echo region or radar hook echo region begins to fill or wrap up. Precipitation becomes widespread throughout the storm and the major tornado dissipates.

Stage 4: The major updraft continues to weaken and downdrafts spread throughout the rotating storm. For the storm to persist, a new updraft must form on the right side of the spreading surface outflow as shown in Fig. 9.36d. The sequence is then repeated, leading to the periodic occurrence of tornadoes.

Lemon and Doswell also performed an analysis of the vorticity production in such a rotating storm [see Eq. (9.37)] and concluded that the tilting term [term (a)] prevails. They also noted that the solenoidal term may be significant at the interface between the cold, dense, rear-flanking downdraft and the updraft.

A more extensive analysis of vorticity production in rotating thunderstorms has been done by Rotunno and Klemp (1985). They applied a three-dimensional cloud model to a case study of a tornado-producing storm described by Klemp *et al.* (1981). They used the actual thermodynamic sounding, but modified the wind profile to have the same magnitude of shear without directional change with height. The model responded to the initial thermal perturbation in a manner similar to the storm-splitting studies described in Section 9.5.4. As described by Wilhelmson and Klemp (1981), a downdraft forms downshear which splits the original updraft into two cells, one dominated by cyclonic vorticity at midlevels, which propagates to the right of the shear vector, and a second propagating to the left dominated by anticyclonic vorticity.

Focusing on the SR member of the split pair, Rotunno and Klemp integrated the model for 150 min. As shown in Fig. 9.37, the model simulated a surface occlusion similar to the conceptual model developed by Lemon and Doswell (1979). Note the well-defined rotation at both upper and lower

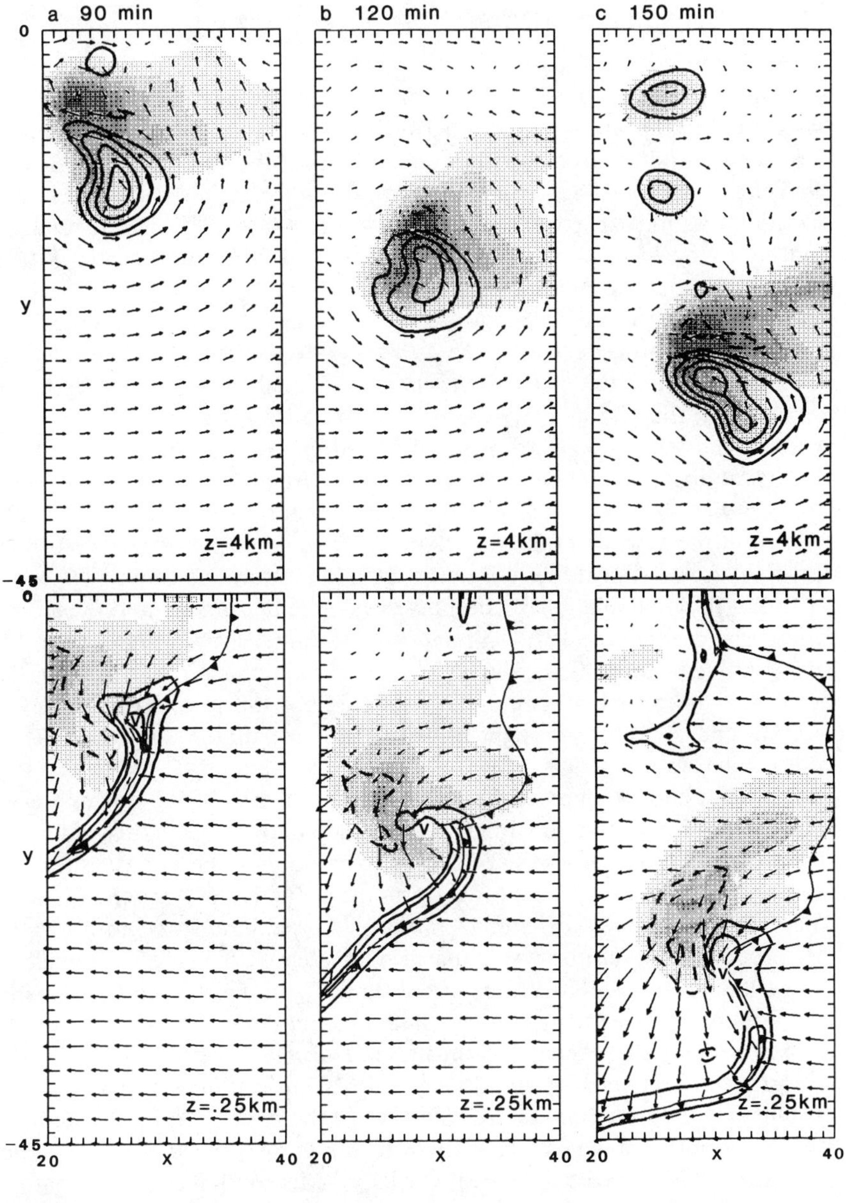

Fig. 9.37. Time sequence of the flow at $z = 4$ and 0.25 km over a 20×45-km portion of the computational domain at (a) 90 min, (b) 120 min, and (c) 150 min. The contour interval for w is 5 m s^{-1} at $z = 4$ km and is 1 m s^{-1} at 0.25 km with the zero lines omitted. The rainwater field is shaded in 2-g kg^{-1} increments beginning with 1 g kg^{-1}; horizontal wind vectors are represented as one grid length = 10 m s^{-1} plotted at every other grid point. The locations of the low-level maxima of vertical vorticity are indicated by the v. [From Rotunno and Klemp (1985).]

levels. At the surface, the maximum vorticity is located at the updraft center at the position marked by a v at 90 min. By 150 min the maximum vorticity has shifted to the sharp gradient between updraft and downdraft, but now it is on the updraft side. At this time (150 min) the updraft strength has weakened sharply. The reduction in updraft strength occurs when the low-level rotation exceeds the midlevel rotation. Associated with the rotating updraft is a cyclostrophic reduction in pressure at the updraft center. When the low-level portion of the updraft rotates faster than the midlevel portion, an adverse vertical pressure gradient is established that retards the updraft strength. This process, called the "vortex valve" effect by Lemon *et al.* (1975), explained the decline in updraft strength as low-level rotation increased in the Union City, Oklahoma, tornado. Figure 9.38a illustrates schematically the downward-directed pressure gradient force.

Several observations have shown that the tornado forms in the zone of strong horizontal gradient of vertical velocity that resides between the main updraft of the storm and the rear-flanking downdraft. The existence of the rear-flanking downdraft has been attributed to evaporative cooling (Barnes, 1978a) or to dynamical interaction between the environment and the storm (Lemon and Doswell, 1979). Klemp and Rotunno (1983) argue that the rear-flanking downdraft is induced by intense rotation near the ground. The downward or adverse pressure gradient noted previously with the low-level rotating updraft is also thought to drive the rear-flanking downdraft. This is illustrated in Fig. 9.38b.

Rotunno and Klemp showed that the development of the midlevel rotation was fundamentally different than that at low levels in the simulated storm. At midlevels the primary source of rotation is the vertical shear of the horizontal wind, which is tilted [term (a) of Eq. (9.36)] into the vertical. They used the concept of equivalent vorticity [see Eq. (9.42)] to illustrate this process. It should be noted that if equivalent vorticity is conserved, a vortex line must lie on a surface of constant θ_e. In their numerical experiments there is no vertical vorticity initially; thus, the vortex lines are all aligned horizontally. Figure 9.39 illustrates a view looking west of the 336-K θ_e surface after 10 min of simulation. In this illustration the shear vector producing the vortex lines is directed toward the viewer. When the high-θ_e air is transported upward in the updraft, the vortex line is tilted into the vertical-producing positive vertical vorticity on its south flank and negative on its north flank (not shown). At 40 and 60 min, a downdraft forms in the downshear precipitating flank of the cloud, producing a depression in the θ_e surface on the northeast side. Because the vortex line in this region tilts downward, negative vorticity is produced. As the updraft splits, the updraft maximum is displaced southeastward into the location of positive production of vertical vorticity. Thus, the updraft maximum and the region of

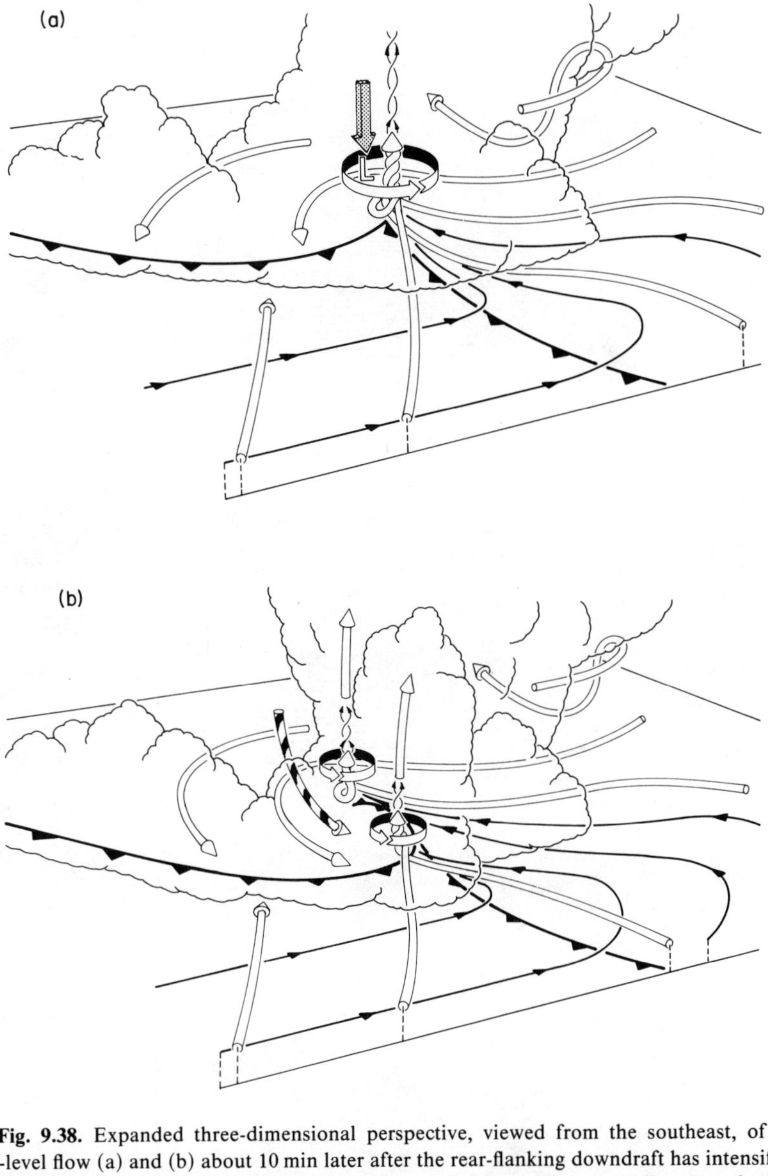

Fig. 9.38. Expanded three-dimensional perspective, viewed from the southeast, of the low-level flow (a) and (b) about 10 min later after the rear-flanking downdraft has intensified. The cylindrical arrows depict the flow in and around the storm. The vector direction of vortex lines are indicated by arrows along the lines. The sense of rotation is indicated by the circular ribbon arrows. The heavy barbed line works the boundary of the cold air beneath the storm. The shaded arrow in (a) represents the rotationally induced vertical pressure gradient, and the striped arrow in (b) denotes the rear-flanking downdraft. [From Klemp (1987). Copyright © 1987 by Annual Reviews, Inc. Reproduced with permission.]

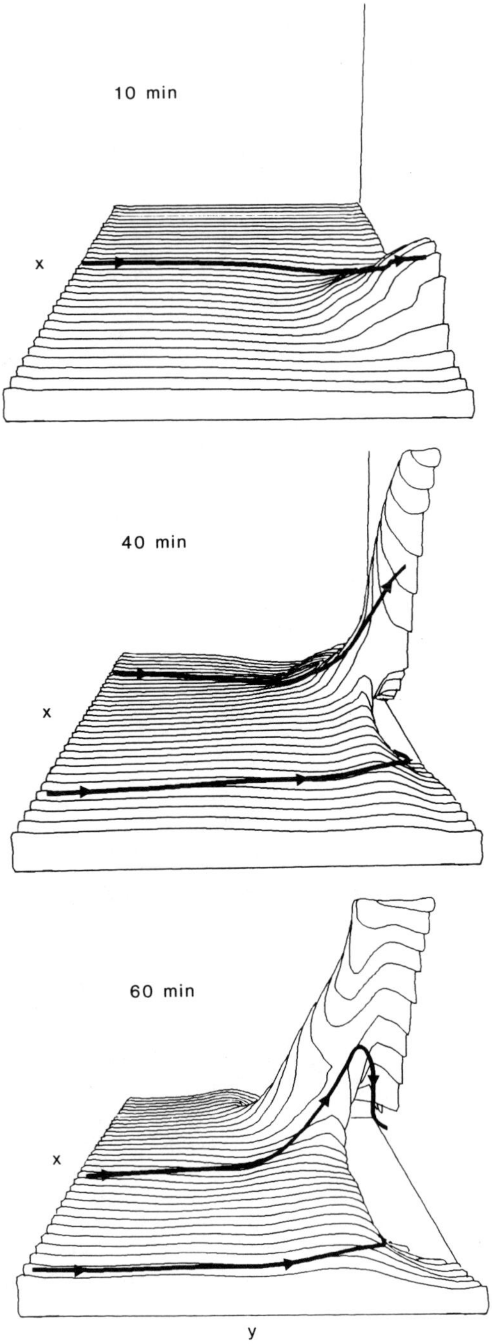
10 min
x
40 min
x
60 min
x
y

positive vorticity tend to come into phase. Also shown at 60 min is a vortex line that does not enter the updraft, but instead moves from south to north and then turns to the southwest as it enters the gust front. This illustrates the mixing of environmental vorticity along the gust front.

This mixing of baroclinically produced vorticity and environmental vorticity is a feature which—Rotunno and Klemp argue—is distinct from vorticity production at middle levels. To further investigate the production of low-level vorticity, Rotunno and Klemp analyzed the various terms contributing to vorticity production in a Boussinesq system, Eq. (9.38). They found that a major contribution to the production of horizontal vorticity at low levels was the baroclinicity along the gust front [term (c) in Eq. (9.38)]. Figure 9.40 illustrates the 331-K θ_e surface as viewed looking northwest. Here an environmental vortex line (i.e., produced by the ambient shear) moves from the south but turns southwestward as it encounters the cold-air boundary of the gust front. As the vortex line travels along the gust front, it mixes with low-valued θ_e downdraft air, where baroclinic production of vorticity is also occurring. Of course, mixing between the two air masses destroys conservation of equivalent vorticity, so this analysis becomes invalid. Nonetheless, we can see from Fig. 9.40 that the vortex line on the 331-K θ_e surface, and presumably its intermingled gust-front counterpart, eventually encounter the updraft where it is tilted into positive vertical vorticity (see also Fig. 9.38b).

Because the concept of equivalent vorticity was not useful in the baroclinic zone of the gust front, Rotunno and Klemp applied the circulation equation, Eq. (9.44), to the gust-front region to investigate further the process involved in the generation of vorticity or circulation in that region. That analysis showed the importance of the cold air in generating the low-level circulation. It also showed that the circulation produced along the gust front is a consequence of the vorticity production in all the air parcels that move along the front. This includes the mixing of air parcels possessing ambient vorticity and air parcels in which vorticity is produced baroclinically.

Fig. 9.39. Three-dimensional perspective (the view is westward) showing the contour $\theta_e = 336$ K surface on a $30 \times 20 \times 9$-km portion of the computational domain. The upward surging of the high-θ_e air reflects the updraft region. Vortex lines, denoted by heavy lines, lie approximately on the constant-θ_e surface owing to near conservation of the initially zero equivalent potential vorticity in the model. These vortex lines tilt upward on the south flank, producing a positive vertical component of the vorticity vector. As the updraft propagates to the south, low-θ_e air descends on the north flank. The vortex lines, adhering to the surface, tilt downward, producing a negative component of the vorticity vector on the north flank. The vertical scale is exaggerated by a factor of 2. [From Rotunno and Klemp (1985).]

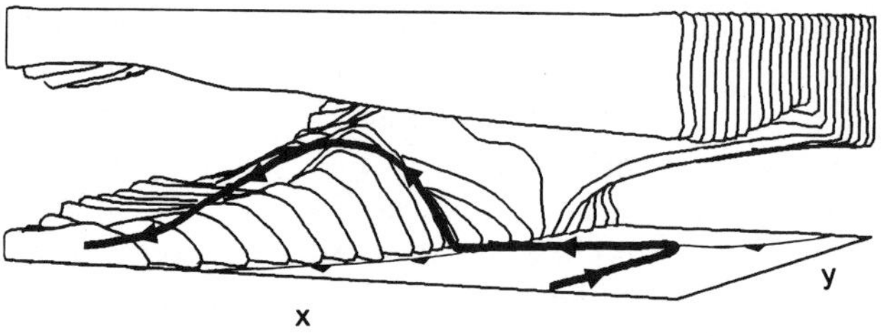

Fig. 9.40. Three-dimensional perspective (looking to the northwest) of the surface which encloses all values of θ_e below 331 K, together with the vortex line (heavy line) which passes through the location of maximum vertical vorticity at $t = 90$ min on a $14 \times 14 \times 4$-km portion of the computational domain. The vertical scale is exaggerated by a factor of 2. The cold frontal boundary indicates the location of the $-1°C$ perturbation surface isotherm. [From Rotunno and Klemp (1985).]

An obvious question concerns the relationship between the midlevel rotation and the low-level rotation of the storm. Rotunno and Klemp argue that the primary importance of the midlevel rotation is that it affects the transport of potentially cold air to the forward and left flanks of the storm. Upon being evaporatively chilled by rain, it descends and forms a cold pool on the left and forward flank. This location is the right place for baroclinically produced horizontal vorticity of the proper sign to become tilted into positive vertical vorticity upon encountering updrafts. Figure 9.38 illustrates schematically Rotunno and Klemp's concept of the relationship between midlevel mesocyclone rotation and low-level tornadic rotation.

Rotunno and Klemp's view of a more passive role of the mesocyclone in tornado genesis differs markedly from the view derived by Lemon and Doswell (1979) and Brandes (1981, 1984) from the analysis of Doppler radar data. From their viewpoint, rotation first begins at middle levels (5–10 km AGL) and then builds upward and downward. They argue that stretching and tilting terms concentrate vertical vorticity in the main updraft, leading to the formation of the tornado vortex aloft which then builds down to the surface.

Brandes (1984) shows that the major tilting occurs in the mesocyclone several kilometers away from the actual tornado location. A weakness in Klemp and Rotunno's argument is that little tilting of the baroclinically produced horizontal vorticity can occur near the ground where vertical motion is nearly zero. If a tornado derives its vorticity from near-surface tilting of baroclinically produced vorticity, how then can a tornado's vortex lines remain essentially vertical down close to the ground, turning horizontal in the friction layer? Clearly, better resolution is needed in the tornado-producing mesocyclone than is currently possible with Doppler radar or numerical prediction models to identify the linkage between the mesocyclone and tornado.

Whether or not there exists a more direct link between the upper-level vortex and the low-level vortex remains to be determined. The emphasis, however, is on the parent mesocyclone as the location of tornado formation. In particular, Lemon and Doswell (1979) argue that the location most favorable for tornado formation is at the cusp of the occlusion in the mesocyclone. Certainly the cases best documented by multiple-Doppler radar have so far indicated this to be the favored location of tornado formation. Bates (1968), however, reported on a number of airborne observations of tornadoes below towering cumulus congestus, located along the flanking line of a severe thunderstorm system. He noted that the tornadoes are frequently located as much as 36 km from the parent thunderstorm in a region free of precipitation over the life cycle of a tornado.

The processes involved in the formation of anticyclonically rotating tornadoes also need further study. Several cases of anticyclonically rotating tornadoes have been well documented (Burgess, 1976; Fujita, 1977; Brown and Knupp, 1980). In these cases the tornado did not form on the left side of the storm or in a left-moving storm as one might suspect, but instead it formed in the right rear flank of storms moving to the right or with the upper-level steering winds. Generally, such anticyclonic tornadoes form in association with other cyclonically rotating tornadoes.

We have seen that rotation in thunderstorms, in addition to being responsible in some way for the formation of tornadoes, is also responsible for (or at least enhances) vertical pressure gradients that affect the main updraft/downdraft strengths and organization of the storm system. Using the concepts of helicity, Levich and Tzvetkov (1984) and Lilly (1986b) argue that a helical storm experiences reduced diffusive and dissipative effects compared to a nonrotating storm system. Thus rotating storms are longer lived and energetically more efficient. Lilly also notes that the inherent inertial stability of a helical storm simplifies the modeling problem by allowing the use of coarser grid resolution than in nonrotating storms and lowering the model's sensitivity to the grid closure schemes. Lilly also speculates that the helical storm should be a more predictable storm system than nonrotating storms.

9.7.4 Tornado Features

As can be seen from the preceding section, there is considerable experimental as well as numerical modeling evidence that tornadoes form in a highly baroclinic environment of mesocyclones. The extent to which this baroclinicity is present in or affects the tornado circulation itself remains largely unknown. There is some evidence from the analysis of debris and cloud-tag trajectories (Golden and Purcell, 1978) that the circulation in tornadoes is strongly asymmetric in both the rotational and vertical wind components. Most theoretical, numerical, and laboratory models of tornadoes do not include the effects of this baroclinicity nor asymmetric characteristics of the tornado circulation. It remains to be seen how well the axisymmetric concepts of tornado rotation relate to the asymmetric, baroclinic of the parent rotating thunderstorm. Clearly the biggest gap in our understanding of tornado formation occurs on scales between the mesocyclone and the tornado vortex. Currently, Doppler radars and numerical models resolve scales of motion of the order of 1 km and greater. In contrast, the scale of a tornado is normally only a few hundred meters and rarely more than 1 km. The important scales of motion within the tornado itself are on the order of tens of meters. Thus, to model or observe a tornado, scales on the order of tens of meters must be resolved.

We will briefly review some of the theories and concepts of the dynamics of tornado vorticities in this section. For more complete reviews the reader is referred to Morton (1966), Davies-Jones and Kessler (1974), Davies-Jones (1982), Lewellen (1976), and Snow (1982).

Current concepts governing the dynamics of tornadoes have been derived from a few photogrammetric studies of tornado motions (Fujita, 1960; Hoecker, 1960a, b; Golden and Purcell, 1978), some axisymmetric, numerical model simulations and analytic models, and axisymmetric, barotropic laboratory vortex experiments. To illustrate the characteristics of the flow field in steady, fully developed tornadoes, we shall partition the flow into five regions, following Morton (1970), Lewellen (1976), and Snow (1982) as shown in Fig. 9.41. Some of the characteristics of these regimes described by Snow (1982) are given below.

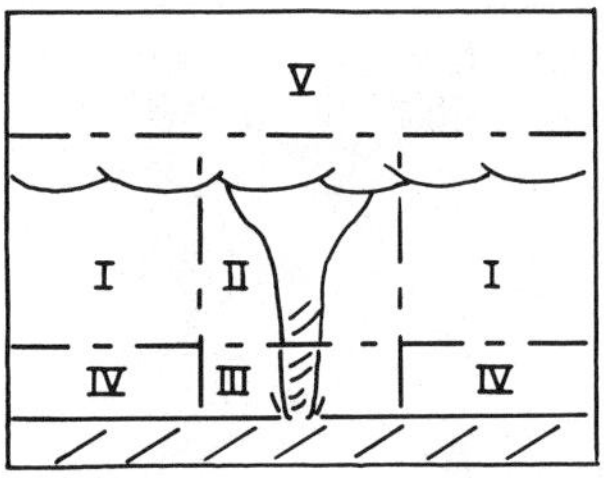

Fig. 9.41. Sketch of an idealized tornado vortex showing the five regions of the flow discussed in the text: region I, outer flow; region II, core; region III, corner; region IV, inflow layer; and region V, convective plume. [From Snow (1982). Copyright by the American Geophysical Union.]

9.7.4.a *Region I: The Outer Flow*

The flow in region I responds to the concentrated vorticity in the core (region II) and to the positive buoyancy and vertical pressure gradients associated with the cloud aloft (region V). The flow in this region is expected to approximately conserve its angular momentum. The flow therefore spins faster as it approaches the central core.

9.7.4.b *Region II: The Core*

The core region surrounds the central axis and extends outward to the radius of maximum tangential winds. It often contains a visible condensation funnel which extends from cloud base. Dust and debris raised from the surface may further outline the region. Owing to the high rotation speeds in this region, the flow is in approximate cyclostrophic balance. Cyclostrophic balance and a radial increase of angular momentum suppress radial motions or entrainment. Centrifugal effects also contribute to a lowering of the central pressure below that occurring at the same level in the outer-flow region. These conditions have led to the application of a variety of simplified flow models in this region (Lewellen, 1976).

9.7.4.c *Region III: The Corner*

This region represents that part of the boundary layer where the flow changes from primarily horizontal flow (region IV) to upward motion into the core. It is the least understood region of tornadoes. Wilson and Rotunno (1982) suggest that the outer boundary of this region should be identified with the radius of maximum tangential winds. Vertical pressure-gradient forces are responsible for deflecting the flow from the horizontal into the vertical.

9.7.4.d *Region IV: Boundary Layer Flow*

As the low-level inflow air interacts with the earth's surface, a turbulent boundary layer is created of the order of tens of meters in depth, and occasionally, in the largest tornadoes, a few hundred meters. The dynamics of this layer has been discussed by Rotunno (1980), Baker (1981), and Wilson (1981). These studies suggest that the cyclostrophic balance that is characteristic of region II is disrupted by frictional effects. As a result the flow is retarded and a net inward-directed force is present. This results in an inward acceleration of the flow toward the core, with the rate of inflow being limited by the eddy stresses near the surface and inertial forces. Figure 9.42 is a schematic illustration of several vortex lines at different levels in the boundary layer synthesized by Snow (1982). The schematic also includes an illustration of the corner region.

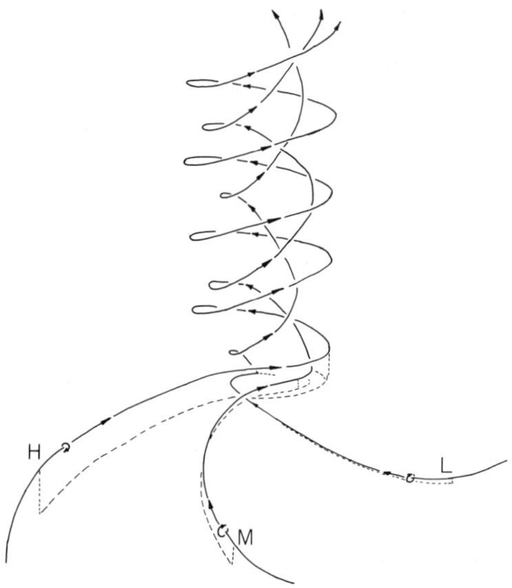

Fig. 9.42. Schematic interpreting the near-surface vorticity distribution. Initially, all vortex lines spiral in the same manner. As the radial flow accelerates, the vortex lines at the top of the boundary layer (H) are turned to spiral opposite to those at lower levels. As radius decreases, the inflection point (change in direction of spiraling) moves inward and downward, affecting next those at midlevels (M). Lines at the very bottom (L) are not modified. The figure is based on work by Rotunno (1980). [From Snow (1982). Copyright by the American Geophysical Union.]

9.7.4.e Region V: The Rotating Updraft

This region is the storm-scale rotating updraft region discussed in Section 9.7.4. It is also the region where vertically oriented vortex lines in the tornado spread laterally outward, eventually becoming horizontal. As mentioned earlier, many unanswered questions remain with regard to the interactions between this region and the tornado-scale regions above.

9.7.5 Summary

Finally, tornadoes do not necessarily consist of a single vortex. Fujita (1970) and Agee *et al.* (1977) describe secondary vortices or suction spots existing within the main tornado circulation. Fujita (1970) has noted that much of the damage associated with tornadoes occurs within cycloidal streaks within the main track. Photographs of tornadoes reported by Agee *et al.* (1977) show multiple vortices that apparently rotate around the tornado axis and have diameters considerably less than the parent tornado. Investigations into the structure and causes of multiple vortices have involved laboratory experiments (Ward, 1972; Church *et al.*, 1979), analytic theory (Gall and Staley, 1981; Gall, 1983), and numerical simulations (Rotunno, 1981; Rotunno and Lilly, 1981; Walko and Gall, 1984). Figure 9.43 is a schematic drawing of three secondary vortices by Snow (1982). The above-mentioned studies suggest that the multiple vortices are created by a shearing instability

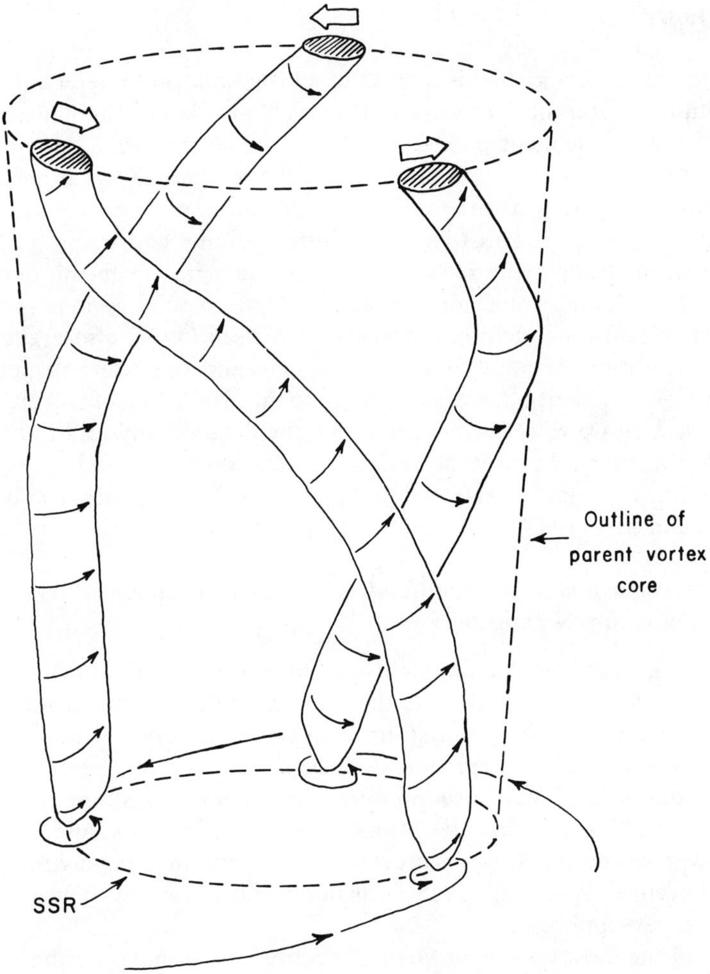

Fig. 9.43. Sketch of a vortex containing three subsidiary vortices. These are in rapid rotation about their spiral axes, while moving about the central core. [From Snow (1982). Copyright by the American Geophysical Union.]

caused by the shear in both the tangential and axial components of the flow of an initially symmetric vortex.

As one can see, much remains to be learned about the dynamics of tornadoes. Owing to the range of scales from the parent rotating storm to the secondary vortices (a few tens of meters) and to their sporadic nature, they remain a major challenge to theoreticians and experimentalists alike.

9.8 Hailstorms

The formation of large hail is a result of a broad range of interacting scales of motion and physical processes. However, the size of the hailstones is also strongly dependent on the microstructure of the storm system. This includes the source, location, and size of the ice particles (hail embryos) suitable for initiation of the rapid stages of hail particle growth; the time needed to grow large hailstones relative to the time constraints provided by the storm system; the nature of hail particle growth in the optimum temperature, liquid-water content, and updraft velocity regions; and the rapidity of hailstone melting processes. Danielsen (1977) also argues that the concentrations of cloud condensation nuclei and ice nuclei and, perhaps, the aerosol size distribution are important to the production of hail embryos. In this section we will not concentrate on the detailed physics of hailstone growth processes. For recent reviews of the properties and growth of hailstones, the reader is referred to Mason (1971), Pruppacher and Klett (1980), and List (1982).

9.8.1 Synoptic and Mesoscale Conditions Suitable for the Formation of Hailstones

Forecasting hailstorms is similar to the forecasting of the occurrence of severe convective storms. Beyond that, it is difficult to distinguish between a hail-producing storm and a storm that produces severe winds and tornadoes. This is not too surprising, because often a severe convective storm that produces large hail also produces tornadoes and strong winds. In general, flash-flood storms prevail in conditions of low wind and low wind shear, whereas most hailstorms prevail in stronger wind shear environments. The Cheyenne, Wyoming, 1985 flash flood and hailstorm, however, represents an exception.

One of the most important factors affecting the formation of hailstorms is the thermodynamic instability of the atmosphere. The more unstable the atmosphere, the more likely that thunderstorms will form with updraft strengths capable of supporting large hailstones. Most of the hail forecast techniques and indices use this concept as the basis of the forecast scheme (Humphreys, 1928; Showalter, 1953; Fawbush and Miller, 1953; Foster and Bates, 1956; Galway, 1956; Boyden, 1963; Miller, 1972; Zverev, 1972; Haagenson and Danielsen, 1972; Maxwell, 1974). The most sophisticated of the thermodynamic stability forecast schemes use simple one-dimensional cloud models to predict maximum updraft velocity and cloud temperature (Fawbush and Miller, 1953; Miller, 1972; Maxwell, 1974; Zverev, 1972; Foster and Bates, 1956; Haagenson and Danielsen, 1972). Danielsen (1977) used such a model to predict hail growth. Renick and Maxwell (1977) summarized the hail forecast scheme using the nomogram applied to Alberta, Canada, thunderstorms shown in Fig. 9.44. Due to deficiencies in the one-dimensional model, perhaps the neglect of vertical pressure-gradient forces, the peak updraft velocity predicted by the model was adjusted to a lower value. Using observations as a basis, English (1973) adjusted the height of maximum updraft velocity to cloud-base height plus 0.61 times cloud-top height. With these adjustments, the nomogram is then used to predict the likely size of hailstones that will form. For example, if the model-predicted maximum W is between 35 and 44 m s^{-1} and the temperature (at the adjusted height) is between about -32 and -38°C, then hail between 33 and 52 mm is forecast.

Hailstorms are frequently associated with an upper-level jet stream and strong wind shear (Ludlam, 1963; Schleusener, 1962; Modahl, 1969; Das, 1962; Frisby, 1962; Fawbush and Miller, 1953). Often the wind veers with height from a low-level southeasterly jet to a midlevel to upper-level westerly jet in the northern hemisphere. Wind shear aids the formation of a sustained updraft/downdraft couplet and a long-lived self-propagating storm system. However, damaging hail can also form in shorter lived, intense single cells that develop in relatively weakly sheared environments (Battan, 1964; Battan and Theiss, 1966; Renick, 1971). Thus, strong wind shear is not a necessary condition for the formation of damaging hail.

The height of the melting level is also important in determining the amount, if any, of hail that will reach the surface. Foote (1984) illustrated the importance of melting by modifying the temperature and humidity profiles for a Colorado hailstorm to representative profiles of Alberta, Canada, and southern Arizona. For Colorado conditions, he calculated that 74% of the hail falling through the 0°C level melts before reaching the surface. For Alberta the amount was 42%, while it was 90% for southern Arizona. This result is consistent with the observation that hailstorms are more frequent at higher latitudes. In fact, G.B. Foote (personal communication) has noted that any convective storm with a radar-echo top over 8 km above ground level in Alberta is likely to produce significant hailfall. This probably explains the usefulness of the height at which the wet-bulb tem-

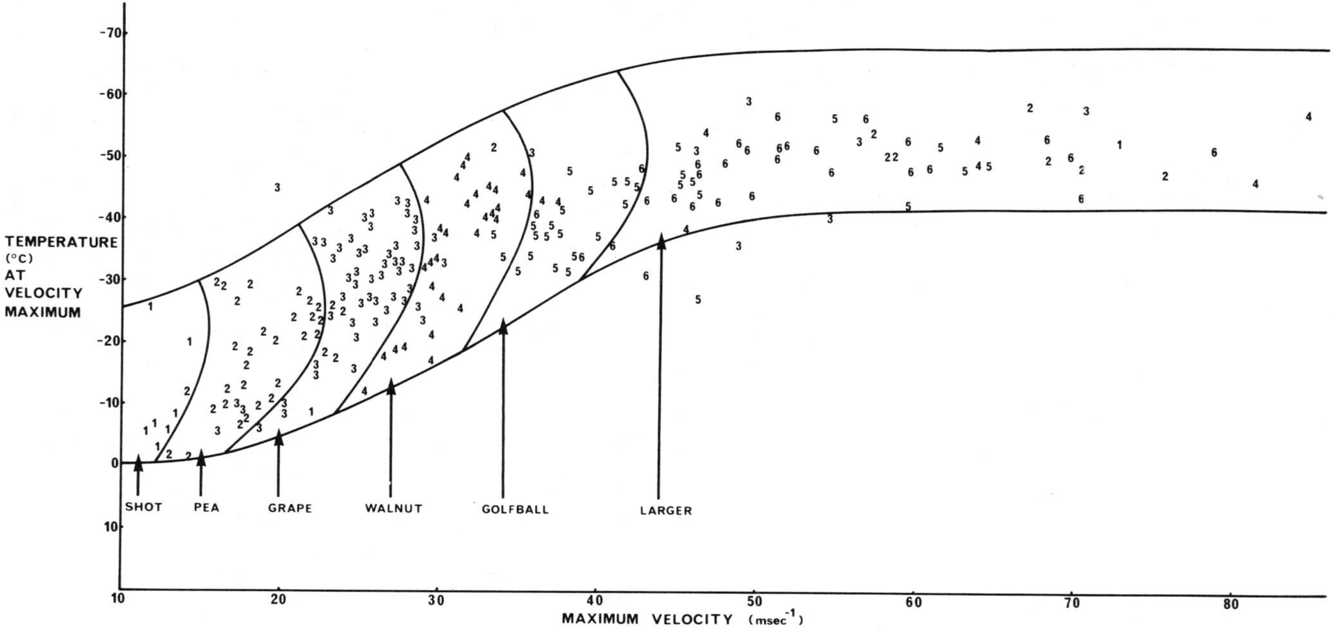

Fig. 9.44. Results and proposed nomogram for forecasting maximum hail size from model diagnostic reruns of 1969–1973 hail days. Hail size categories are (1) shot, 1–3 mm; (2) pea, 4–12 mm; (3) grape, 13–20 mm; (4) walnut, 21–30 mm; (5) golfball, 33–52 mm; (6) larger than golfball. [From Renick and Maxwell (1977).]

perature is 0°C (often referred to as the wet-bulb zero height, or WBZ) as a predictor for hail formation (Miller, 1972). Large hail is generally associated with a WBZ between 2.1 and 2.7 km, while only small hailstones are typically found when the WBZ is below 1.5 km or above 3.4 km. Miller and McGinley (1977) argue that the WBZ represents the minimum depth that a hailstone will fall without melting appreciably, after allowing for evaporative cooling of the stone. One would, therefore, expect that there should not be a lower limit on the WBZ height. However, Morgan (1970) found the WBZ is correlated with environmental low-level moisture in Italy's Po Valley. Thus, if the WBZ is too close to the ground, environmental moisture is not great enough to support intense convection.

Danielsen (1977) argues that the distributions of cloud condensation nuclei and ice nuclei of an air mass are also important to the formation of hail, and that variations in these quantities contribute to the difficulty in forecasting hail. As noted in Chapter 4, cloud-base temperature also influences the microphysical evolution of the storm system. Thus, if the cloud-base temperature is relatively warm, warm-cloud collision and coalescence processes are more likely to prevail. A vigorous warm-cloud precipitation process below the freezing level is likely to deprive the hail generation levels of supercooled liquid water as well as to introduce a large number of hail embryos that can compete for the available water at freezing temperatures. Thus, both the greater prominence of warm-rain processes in warm-based cumulonimbi as well as the greater depths available for melting of hailstones may contribute to infrequent observations of hailfalls in the tropics.

The climatology of the frequency of hailstorms suggests that thermally driven mesoscale circulations such as mountain ridge/valley circulations and sea-breeze and land-breeze circulations aid in the formation of hailstones. This is dramatically illustrated in the July climatology of hailfalls shown in Fig. 9.45. To the east of the central Rocky Mountains, three predominant regions of frequent hailfall can be seen over southern Wyoming, central Colorado, and northern New Mexico. Each of these hail/frequency maxima correspond to elevated ridges that extend into the High Plains. Also, the maxima near Rapid City, South Dakota, correspond to the location of the Black Hills, and the maxima near Cody, Wyoming, correspond to the Big Horn Mountains. Presumably ridge/valley circulations and flow about the local hills establish low-level convergence zones that enhance the formation of the most intense, hail-producing storms. Likewise, the maxima along the shores of the Great Lakes may be associated with sea-breeze and land-breeze circulations in those regions.

It is clear from the foregoing discussion that a large range of scales of motion and a number of physical processes contribute to the formation of significant hailfalls. Using the 30 July 1979 hailstorm over Fort Collins,

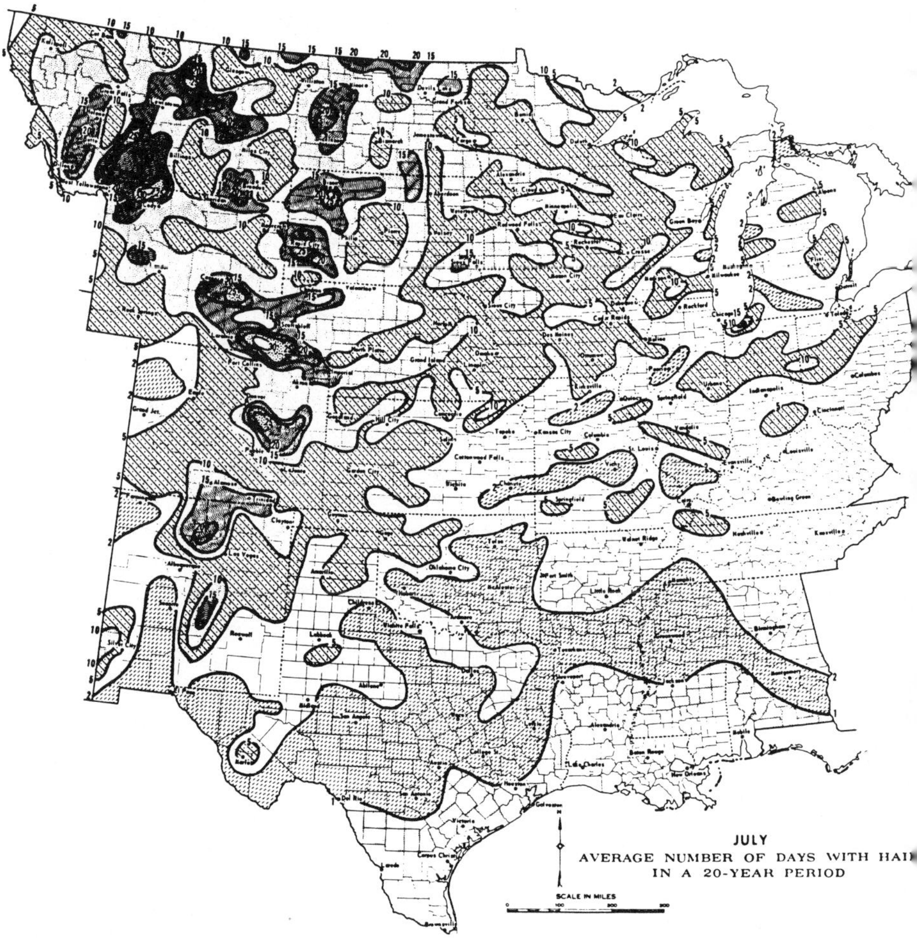

Fig. 9.45. Average number of days with hail in July in a 20-yr period. [From Stout and Changnon (1968).]

Colorado, as an example, Fritsch and Rogers (1981) argued that successful prediction of hailstorms requires the quantitative assessment of a multitude of smaller scale or mesoscale meteorological processes. They argue that quantification of these processes will require advances in (1) observational capability and (2) increased computer power.

9.8.2 Models of Hailstorms and Hail Formation Processes

In this section we will not attempt to review all of the extensive literature on models and concepts of hail growth. We will focus instead on a few key papers illustrating the differing concepts of hailstorm structure and hailstone growth. Nor will we concentrate on the detailed particle-growth physics such as the mathematical models of hailstone growth by collection or accretion, wet and dry growth, and melting processes. The reader is instead referred to the brief summary of microphysical processes in Chapter 4 as well as to the articles mentioned in the introduction to this section.

There is a great diversity of views, concepts, and models of the formation and growth of hail. Some of the diversity is due to the evolution in our ability to observe quantitatively convective storms as well as to model them numerically. Early concepts and models were essentially one-dimensional. As we obtained the ability to observe convective storms with multiple-Doppler radar and aircraft, and as our numerical models evolved from one dimension to two dimensions (see review in Orville, 1978) and more recently three dimensions (Xu, 1983), the models of hail growth have evolved into complex interactions between particle-growth physics and a time-evolving three-dimensional flow structure of the storm system. Some of the diversity in concepts may result from geographical variations in the structure of hailstorm motion fields, water contents, and microstructure. Because only a few hailstorms have been simulated numerically in three dimensions and because combined multiple-Doppler radar and aircraft observations of hailstorms have been limited to a few experimental cases and geographical areas, it is possible that other storm types and hail growth models may apply to storms that have not yet been observed with modern instrumentation or modeled on computers.

9.8.2.a The Soviet Hail Model

The Soviet hail model is described by Sulakvelidze *et al.* (1967). We describe this model in some detail, since it formed the basis of the Soviet hail-suppression techniques (Sulakvelidze *et al.*, 1967) as well as being the model tested in the United States in the National Hail Research Project (Foote and Knight, 1979). Because this model was developed prior to modern multiple-Doppler radar and multidimensional numerical models, it is vague in its depiction of the flow structure of a hailstorm. In this model, the updraft of the storm must be evolving slowly in time and nearly erect in its vertical extent. Moreover, the updraft velocity should increase with height and exhibit a maximum (W_m) somewhere in the warmer portion (−5 to −15°C) of the supercooled liquid-water zone. Another aspect of the Soviet model is that a warm-rain, collision, and coalescence process takes place at levels below the level of maximum velocity. This allows the formation of large, supercooled raindrops whose terminal velocities are comparable to W_m. Thus, the growing raindrops will ascend to near W_m where they will be trapped in what the Soviets call an "accumulation zone." In this region the liquid-water content of the raindrops will increase to greater than 10-15 g m^{-3}. For this large water content to remain suspended, the temperature excess (ΔT) in the updraft must exceed 3-5°C.

If some of the large supercooled drops reach a temperature range of −15 to −22°C and freeze by immersion freezing, or if smaller ice particles form by some type of primary and secondary nucleation and are rapidly swept up by the supercooled raindrops, millimeter-sized hail embryos form in a water-rich environment. The Soviets calculate that hail will grow by accretion of cloud droplets from 0.1 to 1-3 cm in just 2-4 min as the hailstone settles from the accumulation zone to cloud base. This model is in accord with radar observations of a rapidly developing reflectivity maximum aloft, which then descends rapidly to the surface. A crucial aspect of the Soviet model is that the warm-rain process be present to feed the accumulation zone. If the warm-rain process takes place too quickly, so much water will be depleted in the warmer portions of the cloud that an insufficient amount will be available to accumulate in the supercooled zones, thus limiting hail growth. We suggest that this is one factor contributing to the low frequency of hail occurrence in the tropics. However, if a warm-rain process is not present, supercooled water cannot accumulate in a relatively narrow zone and the rapid generation of hail will not occur. If the Soviet model is valid, it should apply in clouds whose bases are moderately warm (10-15°C), where the air mass exhibits moderate CCN concentrations (e.g., 300-600 cm^{-3}), and in multicellular storms whose peak updrafts are relatively low in height and low in magnitude ($W_m \sim$ 12-20 m s^{-1}).

In the Soviet hail model, supercooled raindrops serve as millimeter-sized embryos for the growth of hailstones. Many hailstorms exhibit steady updrafts of the magnitude of 25-50 m s^{-1}. With such strong updrafts, the time available for broadening the droplet spectrum to precipitation-size raindrops is extremely short. Danielsen *et al.* (1972) calculated that the initial droplet spectrum had to be quite broad before hail could form and grow entirely within such strong updrafts. Presumably, this motivated Danielsen's (1977) hypothesis that details of the air mass aerosol population are important to hail growth. Low concentrations of CCN or large concentrations of ultragiant aerosol particles can result in rapid initial broadening of the cloud droplet spectrum and enhance the formation of supercooled raindrops as hailstone embryos. In the case of hailstorms over the High Plains of the United States and Canada, there is considerable evidence that only 20% of the hailstone embryos are frozen raindrops (e.g., Knight *et al.*, 1974). Many of those may actually be ice particles that have melted and have been swept up by the vigorous updrafts. Thus, for cold cloud-base, continental hailstorms, the millimeter-sized embryos for hailstone growth generally form by primary and/or secondary nucleation of ice crystals, followed by slow growth by vapor deposition, riming, and aggregation. The time required for the growth of millimeter-sized particles in 35- to 40-m s^{-1} steady updrafts is longer than the time necessary to eject such particles in the anvil region of the cloud system. The hail growth process in such storms is thus viewed as a multiple-staged process in which hail embryos first form in relatively weak updrafts and then are transported into the more vigorous, water-rich updrafts. Such a multistaged process depends on the detailed temporal and spatial characteristics of the storm circulations. We will thus examine the hail growth processes in several storm types, namely, multicelled storms, supercell storms, and a more hybrid storm system that Foote and Wade (1982) have labeled "organized" multicelled storms. We shall see that there exists a continuum of storm types, ranging from lesser organized multicell storms to the classical steady, dominant updraft/downdraft model of a supercell storm.

At one extreme of the continuum resides the classic multicell storm illustrated in Fig. 9.4. Individual cells in the storm system evolve through the ordinary storm life cycle of growth stage and mature stage, followed by decay stage. Using the "daughter" cell concept proposed by Browning (1977), millimeter-sized hail embryos first form during the cumulus stage in flanking cumulus towers. As the towers grow into the mature stage, these daughter cells become the "parent" cell of the storm complex. The hail embryos grow rapidly in the water-rich environment of the upper portions of the evolving cell and then descend through the storm system as the cell evolves into its dissipating stage.

9.8.2.b Ordinary Multicell Storms

Battan's (1975) single-Doppler radar observations of a multicellular hail storm reveal a structure quite different from that of the Soviet model. Instead of a quasisteady updraft and an accumulation zone, his observations reveal that the cloud system is composed of a series of turbulent thermals that are some 1–2 km in diameter. Figure 9.11 illustrates that some of the thermals exhibit updraft speeds greater than 24 m s^{-1}. Battan hypothesized that hail embryos form in the pulsating updrafts and are carried aloft while growing into hailstones. The larger hailstones may fall out of the thermal and settle through weak updrafts or downdrafts before reaching the ground. In some cases, the descending hailstones may encounter several vigorous updraft cells ascending and descending accordingly as they eventually find their way to the ground. Battan's concept of the hail growth process is certainly more complex than the Soviet model and, as we shall see, the models for hail growth in more organized multicellular storms and supercell storms.

9.8.2.c Organized Multicellular Storms

To illustrate the concept of hail growth in an organized multicellular storm, we will use the 22 July 1976 case study observed during the National Hail Research Experiment (NHRE). This storm has been described extensively (Foote and Wade, 1982; Heymsfield and Musil, 1982; Foote and Frank, 1983; Jameson and Heymsfield, 1980; Heymsfield *et al.*, 1980; Foote, 1984). According to Foote and Wade (1982) the early history of the storm system was characterized as an organized multicellular storm system in which new cells periodically formed on the right flank (relative to the storm's direction of motion) of the older parent cells. Following the merger of several weaker cells, an intense cell formed which exhibited a persistent intense updraft. During the latter portion of the storm's lifetime, this steady flow persisted while a series of reflectivity and updraft perturbations could be seen superimposed on the steady circulation. Figure 9.46 illustrates the persistent updraft/downdraft structure derived from multiple-Doppler radar data.

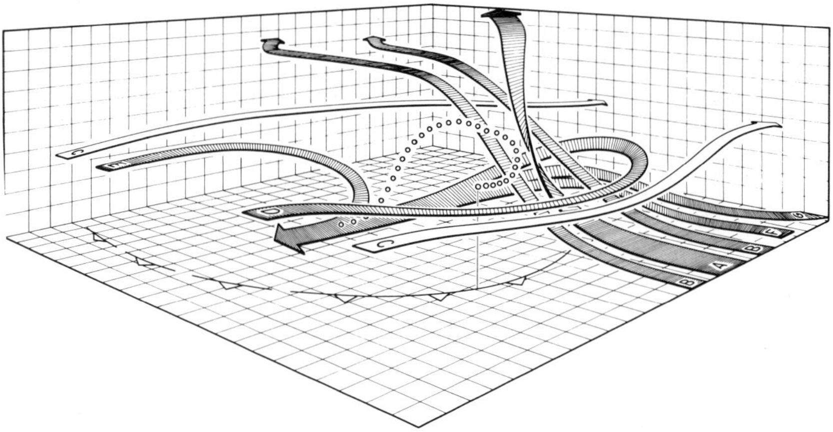

Fig. 9.46. Major components of the airflow in the Westplains storm. The strong updraft is depicted by the ribbon labeled A, which starts in the low levels to the south-southeast of the storm, rises sharply in the storm interior, and leaves the storm toward the northeast to form the anvil outflow. On the flanks of the strong updraft the air rises more slowly and penetrates farther to the rear of the storm before also turning to the northeast. In the middle levels there is a tendency for the westerly environmental flow to be diverted around the sides of the storm (streamlines labeled C), but some air also enters the storm (streamlines D and E) and contributes to the downdraft. A contribution to the downdraft flux is also made by air originally in the low levels to the southeast and east of the storm (streamlines F and G), which then rises several kilometers before turning downward in the vicinity of the echo core. The various streamlines are depicted relative to the storm, which is moving toward the south-southeast as shown, rather than relative to the ground. The small circles indicate the possible trajectory of a hailstone. [From Foote and Frank (1983).]

Major updraft components are labeled A, B, F, and G. Updraft A is the dominant updraft feature; it originates at low levels in the south–southeast, rises abruptly as it is lifted over the gust front, then passes through cloud base, ascends through cloud levels, and turns to form the anvil streaming to the northeast. On the flanks of A the updraft streamlines labeled B tilt more to the back of the storm and enter the lower part of the anvil. Updraft streamlines F and G can be identified with the "up–down," downdraft circulation identified by Knupp (1985).

At midlevels, the flow is from the west. Streamlines labeled C encounter positive pressure anomalies on the upshear side of the updraft and are diverted around the storm. Streamline D is also diverted around the storm, but along the east side of the storm it encounters lower pressure, causing it to divert close to the updraft, where precipitation loading, evaporation, and melting cause it to descend as a major component of the storm downdraft and surface outflow. The flow along streamline E is substantially weaker than along D and enters the rear of the storm to become a rear-flanking

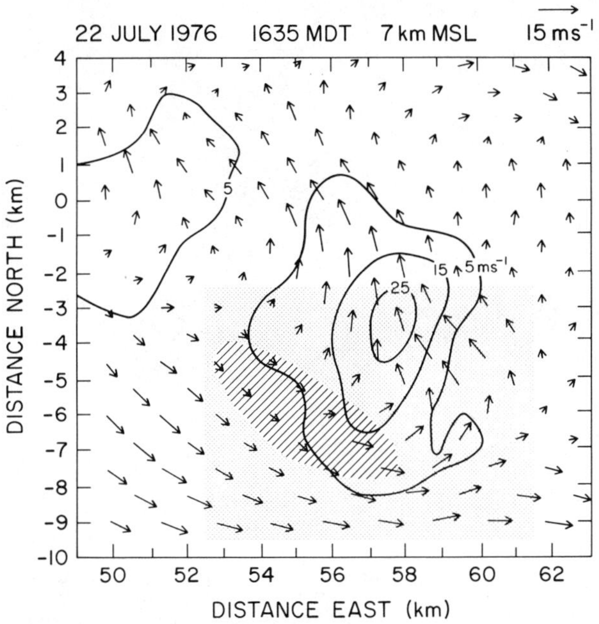

Fig. 9.47. Embryo starting positions relative to the airflow at 7 km MSL. Embryos of various sizes are inserted into the flow over a 1-km grid defined within the shaded region. Horizontal wind vectors are shown (scale in upper right), and updraft contours are indicated. The largest hail originated within the hatched region, where embryos were in the best position to be transported into the updraft core. [From Foote (1984).]

downdraft. Extension of Knupp's (1985) analysis of thunderstorm downdrafts to this case suggests that air along this streamline is forced to descend by a downward pressure-gradient force as well as by negative buoyancy created by melting and mixing of low-valued θ_e air with cloudy air.

Particularly important to the formation of hail is the fact that the updraft streamlines are horizontally convergent. Frank and Foote (1982) found that the convergence in the updrafts extends from the surface to the 6- to 7-km levels. Foote (1984) simulated hail growth in a model of the Westplains, Colorado, storm by releasing hail embryos of 0.1, 0.2, 0.4, 0.6, and 0.8 cm in diameter in the observed flow fields. The particles were released at altitudes of 5, 6, 7, and 8 km MSL in the shaded region shown in Fig. 9.47. He then computed the growth of the hail embryos into hailstones by continuous accretion using the growth equations developed by Paluch (1978). Foote found that with the right combination of initial particle size, injection altitude, and particle drag law, hail could originate at any position in the shaded region except near the southern corners. Figures 9.48 and 9.49 illustrate a plan view and vertical cross section of three selected

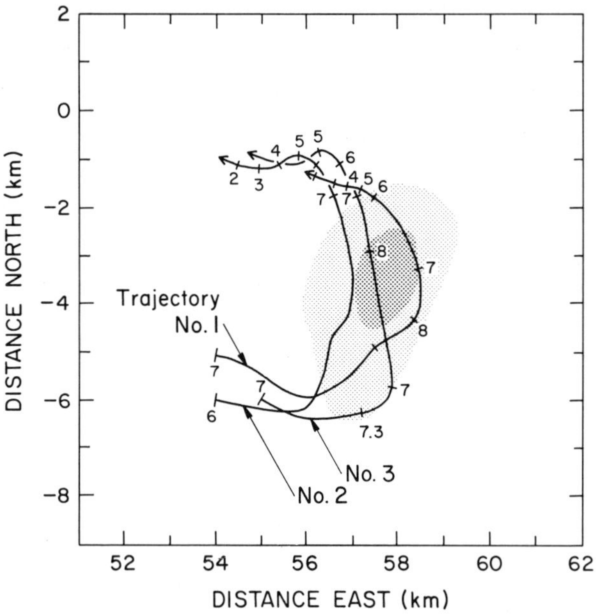

Fig. 9.48. Plan view of trajectories 1, 2, and 3 relative to the 15- and 25-m s^{-1} updraft contours as in Fig. 9.47. These are examples of trajectories that produced the largest hail. Heights are indicated alongside the tracks in units of kilometers above mean sea level. [From Foote (1984).]

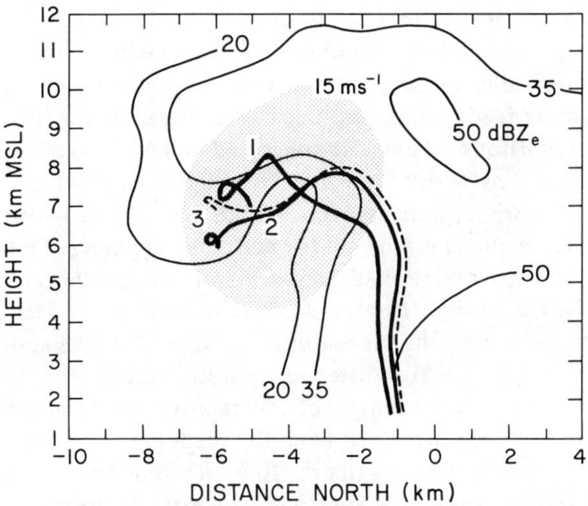

Fig. 9.49. Trajectories 1, 2, and 3 projected onto a north–south vertical section. Reflectivities are shown at three levels of intensity for a north–south plane passing through the updraft maximum. The region of updraft exceeding 15 m s^{-1} in the same plane is also indicated. After entering the updraft from the west, the hailstones grow while making a simple traverse to the north, falling out in the vicinity of the radar echo core. [From Foote (1984).]

hailstone trajectories that produced the largest simulated hailstones. Note that the embryos enter the strong updraft region and then turn northward and move along the long dimension of the updraft. It is important to hail growth in the early period that the particle fall speed and updraft speed are nearly balanced so that the hailstone moves less than 2.0 km vertically. During this period the hailstone accretes cloud liquid water. Subsequently, it passes to the north of the updraft core where it falls to the ground. Foote noted that the hailstones falling into strong downdrafts are more likely to fall to the ground as large hail, since they spend less time in warm air. He calculated that a 1.0 cm hailstone particle above the melting level can lose 73% of its mass due to melting.

Overall, Foote concluded that (1) hail is grown during a single pass through the sloping updraft passing generally from south to north; (2) embryos of a single size can produce a wide range of hail sizes, and the range of sizes is insensitive to initial embryos size; (3) most of the hailstone growth occurs in the temperature range −10 to −25°C; and (4) larger hail tends to originate from embryos that find themselves in a region on the southwest side of the large updraft.

Not fully resolved is the question of the origin of the embryos of the hailstones. Heymsfield *et al.* (1980) argued that hail embryos originate in feeder cells which were either small-scale, embedded updrafts and flanking cumulus towers or upwind mature cells. They concluded that aggregates of dendrites and smaller single dendrites were the predominant type of particle found within the feeder cells, and these were the embryos for most graupel particles and hailstones. They also suggested that embryo production would be enhanced if new turrets injected large aggregates which settle in the precipitation debris region associated with the forward overhang of the mature cell. Once the embryos are formed, they are carried into the mature cell by the storm circulation described by Foote and Frank (1983) and Foote (1984). Heymsfield *et al.*, also concluded that the feeder cells have to be relatively close to the parent updraft to be effective suppliers of hail embryos. The actual effective distance separating the feeder from the parent updraft depends on the strength of the relative winds and the height at which the embryos are detrained from the feeder cells. As we shall see, the relationship of feeder cells to hail growth is an important difference between hail growth in multicellular storms and in supercell storms.

9.8.2.d Supercell Storms

Earlier in this chapter we identified the supercell thunderstorm as a quasisteady storm system exhibiting a persistent, primary updraft/downdraft circulation that generally travels to the right of the mean tropospheric winds in the northern hemisphere. Such a storm system is characterized by a persistent, bounded weak echo region or echo-free vault. Supercell hailstorms have been described by Browning and Ludlam (1962), Browning and Donaldson (1963), Browning (1962, 1965), Marwitz (1972a), Chisholm (1973), and Chisholm and Renick (1972). While supercell storms are rather infrequent in terms of the total number of thunderstorms or even hail-producing storms, they generally produce the largest hailstones and hailswaths which are quite long and wide. Therefore, they are a major contributor to hail damage (Summers, 1972).

In the context of hail growth, the best example of a supercell storm is the conceptual model developed by Browning and Foote (1976). Their model was derived from calibrated radar and multiple aircraft observations of the Fleming, Colorado, hailstorm, which occurred on 21 June 1972 over northeast Colorado. Figure 9.50 illustrates the evolution of the radar reflectivity field and hailswath for the storm system. Note that the hailswath was approximately 300 km long and 15–20 km wide, with baseball-sized hail falling near the town of Fleming. Figures 9.51 and 9.52 illustrate horizontal maps of the radar reflectivity fields and a vertical cross section along a northwest to southeast line through the bounded weak echo region during the time large hail began reaching the ground. Figure 9.51a shows the weak echo region which is bounded by the large forward overhang or embryo curtain on its southeast quadrant and by the main hail precipitation region on its northern quadrant.

The wind field at midlevels observed by aircraft (see Fig. 9.53) suggests a flow field similar to that obtained by Doppler radar in the organized multicellular storm discussed previously. The flow diverges about the main updraft region and converges downshear of it. Figure 9.51 illustrates that superimposed on the quasisteady storm structure were a number of local reflectivity maxima, which Browning and Foote labeled "hot spots." They did not regard the hot spots as being particularly important in themselves, but only used them as tracers of air motions. The hot spots were observed to form along the western flank of the storm and more cyclonically around the main embryo curtain.

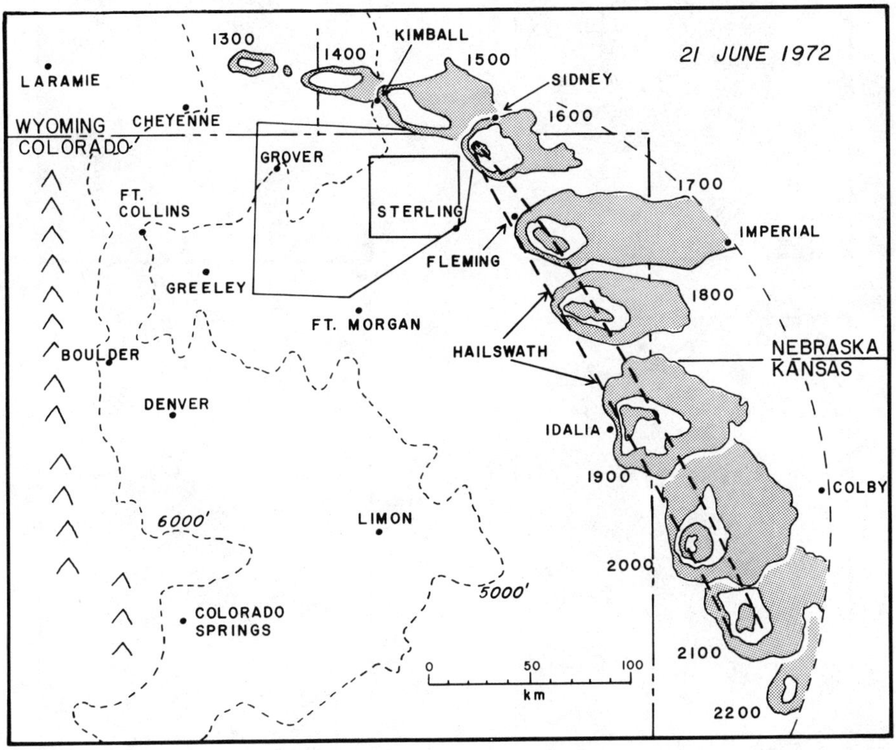

Fig. 9.50. Hourly positions of the Fleming hailstorm as determined by the NWS Limon radar (CHILL radar data used, 1300–1500 MDT). The approximate limits of the hailswath are indicated by the bold dashed line. Continuity of the swath is not well established, but the total extent is. Special rawinsonde sites were located near the towns of Grover, Ft. Morgan, Sterling, and Kimball. Contour intervals are roughly 12 dB above 20 dBZ. [From Browning and Foote (1976).]

Browning and Foote visualized hail growth in such a storm system as a three-stage process, as shown schematically on Fig. 9.54. During stage 1, hail embryos form in a relatively narrow region on the edge of the main updraft, where speeds are typically $10 \mathrm{~m~s^{-1}}$, allowing time for growth to

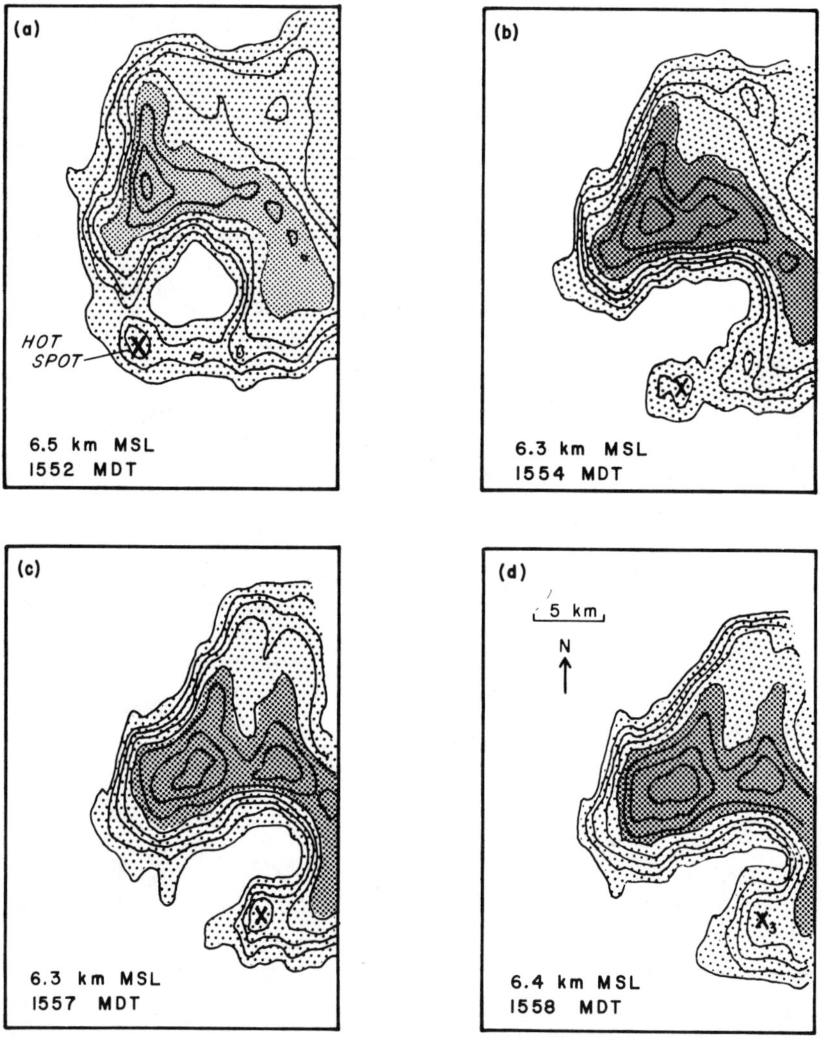

Fig. 9.51. Quasihorizontal sections at four altitudes showing three-dimensional pattern of radar reflectivity for the Fleming storm at 1552–1553 MDT. Reflectivity contours are at 5-dBZ intervals. Areas in excess of 30 and 50 dBZ, respectively, are stippled thinly and thickly; X is a fiducial mark. [From Browning and Foote (1976).] (*Figure continues.*)

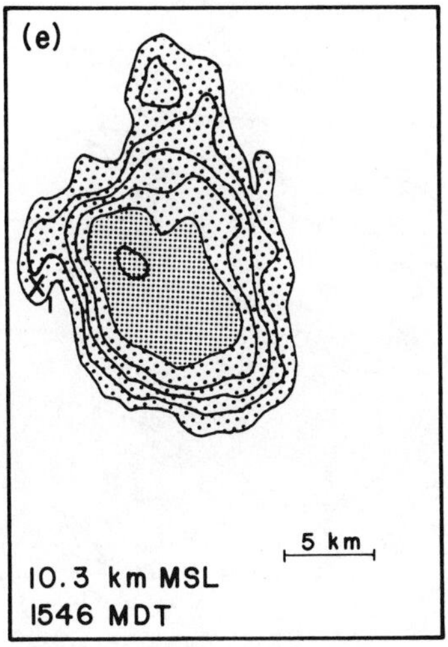

Fig. 9.51 (*continued*)

millimeter size. Those particles forming on the western edge of the updraft have a good chance of entering the embryo curtain region, and follow the trajectory labeled 1 in Fig. 9.54b. The particles labeled trajectory 0 ascend through the core of the updraft and do not have sufficient time to grow to large size. They form the weak echo region and are exhausted into the anvil outflow.

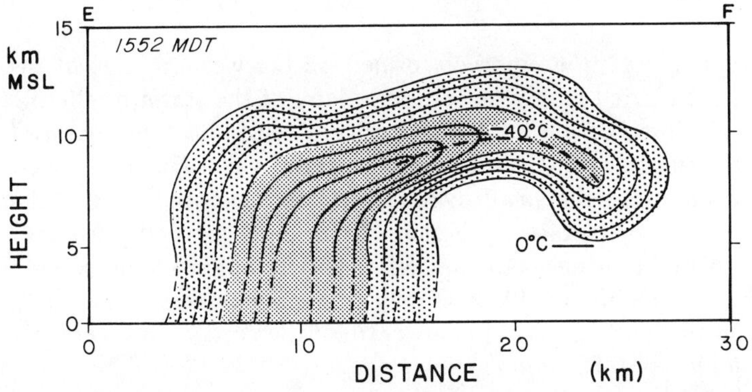

Fig. 9.52. Pattern of radar reflectivity for the Fleming storm in a vertical section along a NW to SE line through the bounded weak echo region in Fig. 9.51. Contours and shading are as in Fig. 9.51. The resolution of this figure, as in Fig. 9.51, is limited by the 1° beamwidth and by the 1-s time integration while scanning in azimuth at $1° s^{-1}$. The Grover radar was located about 95 km west of the storm. [From Browning and Foote (1976).]

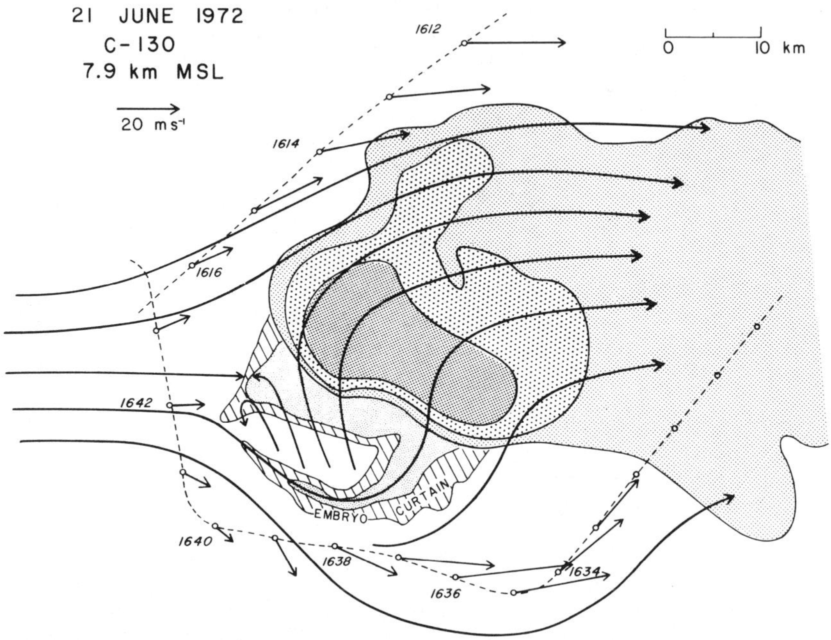

Fig. 9.53. Winds measured by the C-130 aircraft as it flew around the Fleming storm at 7.9 km MSL (track and winds shown relative to the storm). The radar data consist of a low-level PPI from the CHILL system, and a horizontal composite from the DC-6 radar showing the overhanging echo at an altitude of 7.5 km. The streamline analysis emphasizes the blocking flow with a forward stagnation point. North is toward the top of the figure. [From Browning and Foote (1976).]

During stage 2, the embryos formed on the western edge of the main updraft are carried along the southern flank of the storm by the diverging flow field. Some of the large hail embryos settle into the region of weak updrafts that characterizes the embryo curtain. The trajectory labeled 2 also illustrates particles that, as Browning and Foote hypothesize, find their way into the embryo curtain by erosion of updraft air and particles as the environmental flow meets the updraft air on its western flank. Some further growth of the embryos is likely as they descend in the embryo curtain. Some of the particles settle out of the lower tip of the curtain and reenter the foot of the main updraft, commencing stage 3.

Stage 3 represents the mature and final stage of hail growth in which the hailstones experience nearly adiabatic liquid-water contents during their ascent in the main updraft. Similar to Browning (1963) and English (1973), Browning and Foote visualize the growth of hail from embryos as a single up-and-down cycle. Those embryos reentering the main updraft at the lowest

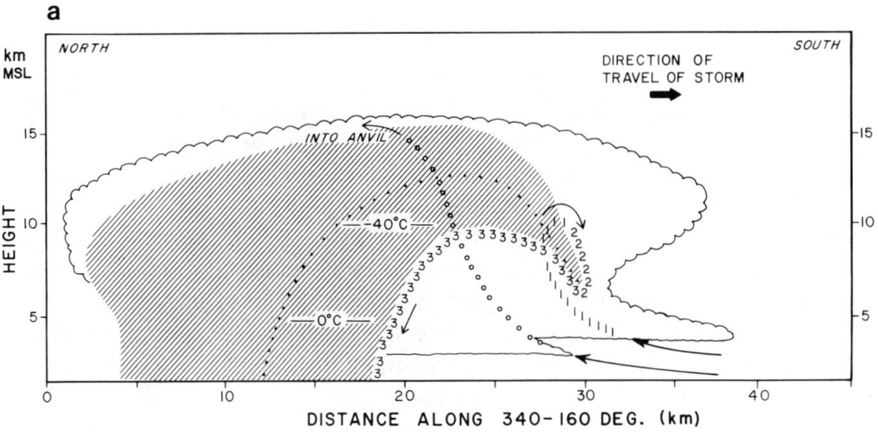

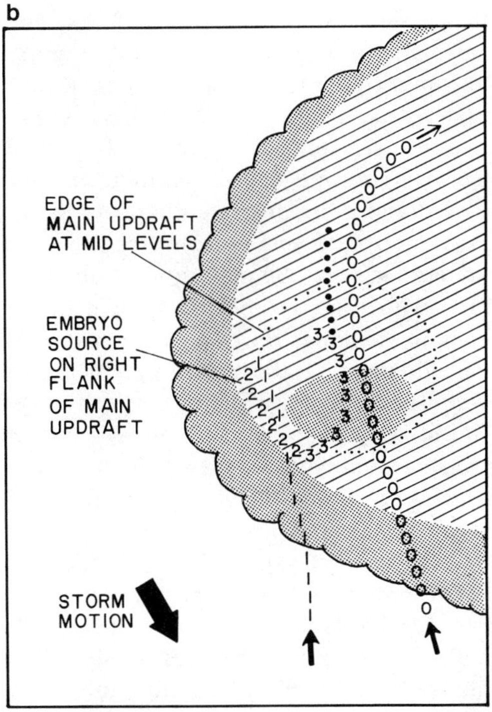

Fig. 9.54. (a) and (b) Schematic model of hailstone trajectories within a supercell storm based upon the airflow model inferred by Browning and Foote (1976). (a) Hail trajectories in a vertical section along the direction of travel in the storm; (b) these same trajectories in plan view. Trajectories 1, 2, and 3 represent the three stages in the growth of large hailstones discussed in the text. The transition from stage 2 to 3 corresponds to the reentry of a hailstone embryo into the main updraft prior to a final up-and-down trajectory during which the hailstone may grow large, especially if it grows close to the boundary of the vault. Other, slightly less favored, hailstones will grow a little farther away from the edge of the vault and will follow trajectories resembling the dotted trajectory. Cloud particles growing "from scratch" within the updraft core are carried rapidly up and out into the anvil along trajectory 0 before they can attain precipitation size. [From Browning and Foote (1976).]

levels, where the updraft is weakest, are likely to have their fall speed nearly balanced by the updraft speed. As a result of their slow rise rate, they will have plenty of time to accrete the abundant liquid water. Browning and Foote (1976) and Browning *et al.* (1963) emphasize that the requirement that the particle fall speed nearly balance the updraft speed accounts for the infrequent occurrence of large hail. The most favored particles remain near the lower boundary of the embryo curtain as they slowly ascend and rapidly accrete cloud droplets. Eventually, they reach the apex of their ascent where, according to Atlas (1966), they spend a large amount of time accreting cloud droplets in a region called the "balance level." Their fall velocity will eventually become large enough to overcome the intense updraft speeds and/or they will settle into the downdraft region as they travel to the northern flank of the storm in a tilted updraft.

Nelson (1983) computed hail growth in flow fields derived by multiple-Doppler radar in a supercell storm. Like Browning and Foote, he calculated that major hail growth occurred in a single up-and-down trajectory. His calculations suggested that a factor more important for hail growth than maximum updraft speed is the presence of a broad region of moderate updraft (20–40 m s^{-1}), because hail cannot grow in the strongest part of the updraft. He concludes that the flow field of the storm, not the lack of embryos, is a major modulating factor in hail growth.

A major point of contention with the Browning and Foote model is the assumption that embryos are formed at the edges of the main updraft or are eroded from the main updraft by the interaction between the environmental flow and the updraft. A competing hypothesis is that supercell storms contain flanking, towering cumulus elements or feeder clouds that are embedded within the overall supercell precipitation field. Krauss and Marwitz (1984) provide evidence for such embedded feeder clouds in a supercell storm. They hypothesize that the feeder cells serve as a source region for hailstone embryoes, as they do in the conceptual models of multicell storms. The supercell storm observed by Nelson (1983) also contained local updraft regions that could be interpreted as embedded feeder cells. One could argue that the supercell storms observed by Krauss and Marwitz (1984) and Nelson (1983) represent different positions in the continuum of storm types residing between the classic supercell storm of Browning and Foote and the organized multicell storm described previously. Alternatively, one could interpret the hot spots described by Browning and Foote as actual embedded feeder cells that represent the primary sources for embryos to the main updraft.

We have seen that hailstorms are a result of a complex series of scale interactions, ranging from the synoptic scale down to the scale of hailstones and their embryos. We now examine the processes involved in the formation of rainfall in cumulonimbus clouds.

9.9 Rainfall from Cumulonimbus Clouds

In many parts of the world, rainfall from cumulonimbus clouds is the dominant contributor to crop-growing season rainfall and to the supply of water for people and livestock. Simultaneously, cumulonimbi can produce heavy rainfall and flash floods, which can kill people and livestock and cause millions of dollars in property damage and losses.

The amount of rainfall produced by cumulonimbi depends on the organization and structure of weather systems over a broad range of scales, ranging from the synoptic scales of motion, through the mesoscale, and down to the scale of individual cumulonimbi. This environmental organization can result in rainfall from cumulonimbi ranging from less than 1 cm for ordinary storms (Byers and Braham, 1949) to 38 cm for the more severe flash-flood-producing storms (Maddox *et al.*, 1978).

It is not surprising that the rainfall from cumulonimbus clouds depends strongly on the moisture content of the air mass near the surface as well as the moisture content of the air through the depth of the troposphere. The flash-flood events over the United States analyzed by Maddox *et al.* (1979) exhibited typical surface mixing ratios of 10 g kg^{-1} to greater than 14 g kg^{-1} and precipitable water contents ranging from 2.58 to 4.16 cm.

Often, heavy rainfall occurs during periods when the large-scale flow pattern appears innocuous, typically exhibiting a large-scale ridge pattern (Maddox *et al.*, 1979). They noted that in many flash floods a shortwave trough at the 500-mbar level was evident to the west or northwest (upstream) of the heaviest rain areas. The heavy rain events over the western United States, however, were often more directly associated with a shortwave trough at the 500-mbar level. This results from the more complicated interactions between large-scale weather disturbances and the complex topography of that region.

A common characteristic of the flow patterns at the 850-mbar level associated with heavy convective rainfall over the United States was the presence of a pronounced low-level jet. The LLJ is effective in fueling convective storms with the low-level moisture.

Maddox *et al.* (1979) noted that there exist several characteristic surface weather patterns associated with flash floods over the United States. Some events are associated with active cyclonic storms which exhibit slowly moving north- and south-oriented fronts. Others are associated with a nearly stationary surface front oriented in an east–west direction. Still other flash-flood events were not associated with a synoptic-scale surface front but were linked to a nearly stationary thunderstorm outflow boundary. Presumably such thunderstorm outflow boundaries could be associated with large complexes of thunderstorms or mesoscale convective systems described by

Maddox (1980), Bosart and Sanders (1981), and Wetzel *et al.* (1982). We will describe these systems more fully in Chapter 10.

The mesoscale processes associated with heavy thunderstorm rainfall events are variable and complex. Often the heavy rainfall is associated with the colocation of thermally driven mesoscale circulations such as sea-breeze fronts (Cotton *et al.*, 1974), mountain ridge/valley circulations (Maddox *et al.*, 1978) upper-level large-scale troughs, or surface fronts. In some cases slope circulations associated with small hills or urban heat-island circulations are contributing factors to flash-flood events (Miller, 1978). Some are associated with MCCs that survive for several days and locally interact with complex terrain to yield heavy rainfall (Bosart and Sanders, 1981).

The presence of weak winds aloft and weak-to-moderate vertical shear of the horizontal wind is often a characteristic of heavy convective rainfall events (Maddox *et al.*, 1979). Weak winds aloft result in reduced storm motion and more localized heavy rainfall, compared to strong winds aloft in which rain is spread over a large area as the storm moves with stronger winds. The role of weak-to-moderate wind shear in producing heavy convective rainfall is more complicated. Figure 9.55 illustrates the variation of precipitation efficiency with vertical shear of the horizontal winds for High Plains thunderstorms over North America as compiled by Marwitz (1972d). Precipitation efficiency is defined as the ratio of the measured precipitation rate at the surface to the water vapor flux though the base of a cloud system. The figure suggests that cumulonimbi residing in high wind shears have low precipitation efficiencies, whereas clouds existing in an environment of low wind shear exhibit high precipitation efficiencies.

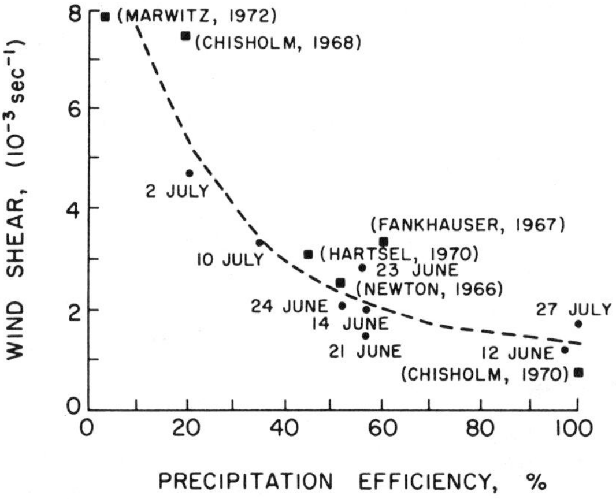

Fig. 9.55. Scatter diagram of wind shear versus precipitation efficiency for 14 thunderstorms which occurred on the High Plains of North America. [From Marwitz (1972d).]

To understand this result, let us consider the water budget of a typical cumulonimbus cloud. The primary source of water for a cumulonimbus cloud is the flow of water vapor into the base of the cloud. As the air ascends and cools, the vapor is converted into liquid cloud droplets and some is converted into liquid or frozen precipitation. A portion of the water rapidly falls out as surface rainfall, while some of the water is injected into the anvil portion of the cloud, where it eventually evaporates or slowly settles out as steady precipitation. Some of the water is evaporated from the sides of the cloud due to entrainment processes. As the cloud decays, some of the cloud water and smaller precipitation elements also evaporate. Another portion evaporates in the dry subcloud layer in low-level downdrafts. As an indication of the relative contribution of the various water sinks for a severe squall-line thunderstorm, Newton (1966) estimated that between 45 and 53% of the water vapor entering the updraft reaches the ground as precipitation, while about 40% evaporates in downdrafts and about 10% is injected into the anvil portion of the cloud. As noted by Fujita (1959), the loss of water by subcloud evaporation is appreciable and increases with the height of cloud base. In one example, he showed that the amount of rain that evaporated as it descended from a cloud-base height of approximately 3.0 km equaled the amount of rain reaching the ground.

As we have seen, vertical shear of the horizontal wind increases the rates of entrainment into the cloud system and aids the organization of the storm into a vigorous updraft/downdraft couplet. With larger values of wind shear, one would expect greater water losses in downdrafts and greater rates of transport of water into the upper troposphere, thus lowering the precipitation efficiency of storms.

In contrast, wind shear may increase the storm-relative inflow of warm moist air into the storm system, and sustain the cloud lifetime such that, even with reduced precipitation efficiencies, greater amounts of precipitation are produced than in low-shear environments. Wind shear, however, is clearly not the only important factor influencing the precipitation efficiency of a cumulonimbus. For thunderstorms observed over Florida and Ohio during the Thunderstorm Project, Braham (1952) estimated a precipitation efficiency of only 10%. He attributed the low precipitation efficiency of these storms to the loss of water by entrainment, which is greater in smaller, ordinary cumulonimbi.

Not only is the magnitude of wind shear important to the efficiency of precipitation production, but its directional variation is also critical to rainfall production. Miller (1978) simulated a localized heavy rainfall event over London, called the Hampstead storm. The model for this study was a three-dimensional, nonhydrostatic numerical model described by Miller and Pearce (1974). An important feature of the environment of the storm was that the wind veered with height. Veering of environmental winds is also a characteristic of the flash-flood events analyzed by Maddox *et al.* (1979). Miller demonstrated that the veering wind profile was important in producing a localized heavy rainfall event.

The storm simulated by Miller was a multicellular storm system in which new cells repeatedly formed on the southeast flank of the spreading low-level outflow. It was on this flank that the spreading outflow most directly opposed the low-level winds. It is commonly observed that flash-flood-producing storms are multicellular (e.g., Caracena *et al.*, 1979; Dennis *et al.*, 1973; Hoxit *et al.*, 1978). This is consistent with Weisman and Klemp's (1982) model results which suggested that multicellular storms prevail in relatively low-sheared environments in which the bulk Richardson number exceeds 50. Miller noted that part of the mature cell and the decaying older cell merged to form an elongated raining anvil. The elongated rainfall pattern created a similarly elongated mesohigh that, in turn, assisted in the persistence and regeneration of the storm system as a whole. The mesoscale circulation of the storm as characterized by the low-level inflow, anvil outflow, and downdraft outflow was quite persistent. Figure 9.56 illustrates a conceptual model of the interlocking nature of the persistent updraft/downdraft circulation. The advance of the gust front toward the southeast was nearly matched by the prevailing southeasterly flow, so the

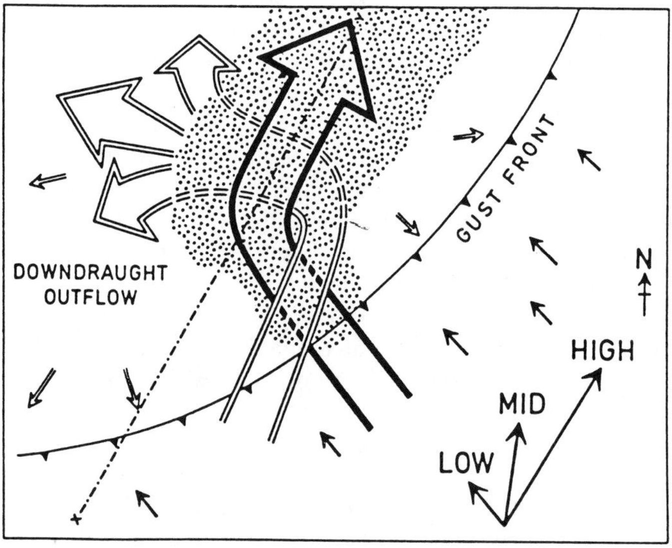

Fig. 9.56. A schematic of the primary features of the storm model deduced from the simulation. [From Miller (1978).]

gust front remained nearly stationary relative to the ground. Consequently, new cell development occurred repeatedly at the point of maximum convergence between the outflow and inflow at a nearly constant geographical position.

Miller demonstrated further that the veering of the winds through 90° between the surface and 400 mbar was instrumental in maximizing the convergence of air approaching the storm. He did so by repeating the simulation with middle-level and upper-level winds rotated to a southeasterly direction but with no speed change. The storm system elongated toward the northwest and hence in the direction of the low-level relative flow. As a result, the basic stationary character of the storm circulation was lost. Figure 9.57 is a schematic diagram of the results of the two numerical experiments. Miller's results are consistent with the rule of thumb that the precipitation from a storm system increases with the horizontal area of the storm. The elongated precipitation/convergence pattern would be expected to enhance the total volume of rainfall from the storm system. Miller's simulation is an ideal illustration of the importance of the speed and direction of environmental winds on the rainfall production of a storm system.

Not only are synoptic-scale, mesoscale, and storm-scale features important to rainfall production, but the microphysical characteristics of the storm are also important. Generally, heavy rain-producing storms are warm-based storms in an environment with a high mixing ratio, which afford the opportunity for warm-rain processes to be very active. If the air mass is maritime (low in CCN) or is cleansed by earlier precipitation scavenging of aerosol, then the opportunity for dominance of the warm-rain process results in the precipitation process being concentrated at low levels. This was apparently true for the Big Thompson Storm (Caracena *et al.*, 1979).

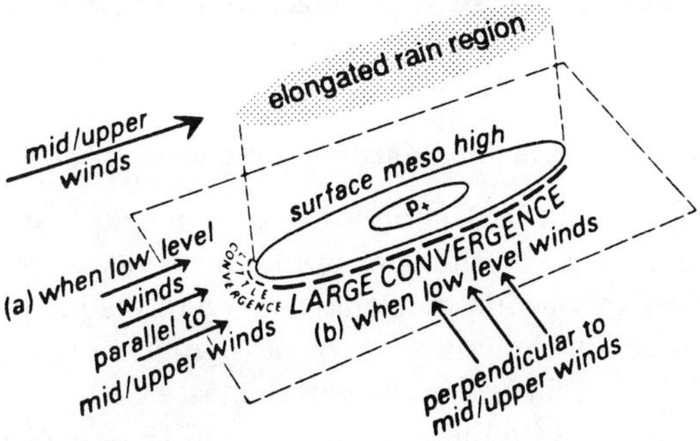

Fig. 9.57. A schematic of the proposed mechanism by which vector shear assists the regeneration or maintenance of a storm system. [From Miller (1978).]

Modeling studies indicate that a cloud system in which the warm-rain process is predominant is a more efficient rain-producing cloud than is a cloud system dominated by ice processes (Levy and Cotton, 1985; Tripoli and Cotton, 1982). As the ice phase becomes more predominant in a large convective storm, more total water is thrust upward into the anvil region.

This brings us to the question of the relationship between the production of hail versus rainfall in cumulonimbi. As we have seen, the storms producing the largest hailstones occur in strongly sheared environments. Thus, in general, we should not expect that the storm systems producing the largest hailstones are also heavy rain-producing storms. However, there are many storm systems that produce large quantities of smaller yet damaging hailstones. Such storms occur in less severely sheared environments and can thus be prolific rain producers as well. For example, calculations by Crow *et al.* (1976) suggest that for northeastern Colorado hailstorms, the hail contribution to the total precipitation mass is typically less than 4%. Rainfall is the dominant form of precipitation even in some of the most severe hailstorms. Dennis (1980) has noted that cloud seeders are concerned about whether seeding clouds to reduce hail increases, decreases, or has no effect upon rainfall production by those storm systems. When one considers the small fraction of the water budget that hail represents, it is not surprising that this problem has not been resolved at this time.

9.10 Thunderstorm Electrification and Storm Dynamics

In this section we review thunderstorm electrification processes with a focus on the possible impact of storm electrical processes on storm dynamics and the impact of storm dynamics on charge separation processes. We begin by discussing the possible role of storm electrification processes on the storm thermodynamics and dynamics.

9.10.1 Influence of Storm Electrification on Cloud Dynamics

It was briefly mentioned in Section 9.7 that Vonnegut (1960) hypothesized that lightning discharges in the core of the tornado vortex would generate sufficient heating to become a significant energy source in driving the tornado. In order for lightning to be a significant energy source, the frequency of lightning flashes in the vortex must be very high, on the order of 1000 km^{-2} min^{-1}. While there are occasional reports of electrical glows in and near tornadoes (Vaughan and Vonnegut, 1976), there is little confirmation evidence of vigorous lightning activity such as strong radio sferics (Davies-Jones and Golden, 1975a), or other evidence of electromagnetic disturbances near tornadoes (Zrnic, 1976). As noted by Davies-Jones (1982), the importance of electrical heating in tornadoes is by no means fully resolved; the pros and cons of the theory are still being debated (Davies-Jones and Golden, 1975a, b, c; Vonnegut, 1975; Colgate, 1975; Watkins *et al.*, 1978).

Electrical heating, however, is not the only way in which cloud electrification processes can affect storm dynamics. It is possible that the electrical fields and high space charge densities in thunderstorms could accelerate air parcels directly. However, numerical calculations (Chiu, 1978) and simple order-of-magnitude estimates (Vonnegut, 1963) suggest that such electrical forces are small in comparison to buoyancy accelerations. Cloud electrification may also affect the dynamics of clouds by altering the terminal velocities of precipitation elements. Numerical calculations by Levin and Ziv (1974) and Chiu (1978) suggest that when the electrical field strengths approach breakdown potential, the terminal velocities of precipitation particles can be appreciably modified. The precipitation particles are then levitated by the electric fields. The alteration of fall velocities of precipitation particles then influences the distribution of water substances; the distribution, in turn, alters the buoyancy of the cloud by changing its water loading. The numerical experiments by Chiu (1978) suggest that these effects can significantly alter the subsequent dynamics of a cloud. Rawlins (1982) found in a three-dimensional cloud model that levitation of hail had a negligible influence on the early development of electrification but had a modest influence at large field strengths, suggesting that levitation altered the water distribution of the cloud.

Schonland (1950) and Levin and Ziv (1974) suggested that the cessation of levitation of precipitation elements immediately following a lightning discharge results in the occurrence of the so-called rain gush. Moore *et al.* (1964) observed as much as a 10-fold increase in precipitation content of a storm following a lightning discharge. If levitation of precipitation particles is important, then one should observe substantial changes in particle motions by vertically pointing Doppler radar immediately following lightning discharges. Williams and Lhermitte (1983) attempted to examine such a response in Florida thunderstorms. They observed infrequent Doppler velocity changes at high levels in the cloud where the radar reflectivities and precipitation particle sizes were small. In general, however, they found little correspondence between velocity changes of the precipitation and lightning discharges in regions of high reflectivity. This suggests that levitation does occur, at least on smaller particles, but there is little evidence that it is strong enough to substantially alter the motions of larger precipitation particles.

Cloud electrification processes can also alter the dynamics of clouds in a more subtle way. There is considerable evidence suggesting that the presence of strong electric fields and charged drops can enhance the collection efficiency among cloud and precipitation elements (Sartor and Miller, 1965; Davis, 1961, 1964a, b; Lindblad and Semonin, 1963; Semonin and Plumlee, 1966; Plumlee and Semonin, 1965; Schlamp *et al.*, 1976, 1979; Latham, 1969; Saunders and Wahab, 1975). As a result of enhanced collection processes, larger, faster falling precipitation elements will more readily form, resulting in a redistribution in condensed water with its subsequent impact upon cloud buoyancy and dynamics. Indeed, Moore *et al.* (1964) suggested that the rain gush is a result of increased coalescence of precipitation elements in the strong electric fields prior to the lightning stroke. The Doppler radar observations reported by Williams and Lhermitte (1983) are consistent with Moore's hypothesis. They noted that gradual changes in downward particle velocity were well correlated with electric field changes. One would expect a more gradual response in particle motions by enhanced collection than by levitation effects.

In conclusion, there is considerable evidence that cloud electrification processes can influence the dynamics of clouds. It appears, however, that these effects are localized, and they occur at electric field strengths approaching breakdown potential.

9.10.2 Influence of Cloud Dynamics on Cloud-Charging Processes

To begin our discussion of the influence of cloud dynamics on cloud-charging processes, we shall follow the lead of Mason (1971) by first listing the characteristics of a satisfactory charge separation theory. In so doing we will update Mason's characteristics with the results of recent observations. The updated characteristics are as follows:

(1) The average duration of precipitation and electrical activity from a single thunderstorm cell is about 30 min.

(2) The electric field strength destroyed in a lightning flash is about 3-4 kV cm^{-1}; the breakdown field in clear air is much higher (30 kV cm^{-1}).

(3) In a large, extensive cumulonimbus cloud, this charge is generated and separated in a volume bounded by the −5 and −40°C levels and having a radius of approximately 2 km.

(4) Negative charges are usually centered between the −10 and −20°C levels with the positive charge several kilometers above, and a secondary pocket of positive charge is occasionally found near cloud base in precipitation; the center of negative space charge may be somewhat lower in mesoscale systems, closer to the freezing level.

(5) The charge generation and separation processes are closely associated with the development of precipitation, although the space charge center appears to be displaced both vertically and horizontally from the main precipitation core.

(6) Sufficient charge must be generated and separated to supply the first lightning flash within about 20 min of the appearance of precipitation particles of radar-detectable size.

The charge generation theories consistent with most of these characteristics can be classified as being either a precipitation-related theory or a convection theory. The relative merits of these two classes of charge generation theory have been debated extensively by Mason (1976) and Moore (1976). Here we will briefly review the basic concepts and their relationship to the dynamic structure of the cloud.

9.10.2.a Convection Charging Theory

The convection theory is intimately coupled with the overall dynamics of the cloud system. Advocates of this theory of charge separation include Grenet (1947), Vonnegut (1955, 1963), and Wagner and Telford (1981). According to this theory, a normal fair-weather electric field establishes a net concentration of positive ions in the lower troposphere. As convective updrafts form, they carry the positive space charges into the cloud layer, causing the cloud initially to be positively charged. As the cloud penetrates to higher levels in the troposphere, it encounters air in which the mobility of free ions (or conductivity of the air) increases with increasing height. These ions are produced in the ionosphere or at heights above 6 km by cosmic radiation. The rising positively charged cumulus preferentially attracts the negative free ions, causing the cloud top to become negatively charged. Instead of neutralizing the positive space charge in the cloud, Vonnegut hypothesizes that convective downdrafts transport the negative ions to the lower part of the cloud while updrafts carry the positive ions to the upper part of the cloud. It is hypothesized that the resulting buildup of positive space charge near the earth's surface then causes preferential point discharge of positive ions, which are then transported into the cloud by updrafts. The resulting increase of positive charge in the cloud enhances the flow of negative ions to cloud top, leading to an exponentially increasing cloud polarity. Figure 9.58 illustrates the process as hypothesized by Vonnegut.

Chiu and Klett (1976) attempted to simulate the convective charging theory using a simple steady-state, axisymmetric cloud model developed by Gutman (1963, 1967). They found that the theory could not produce sufficient cloud charging to induce a flux of positive space charge near the

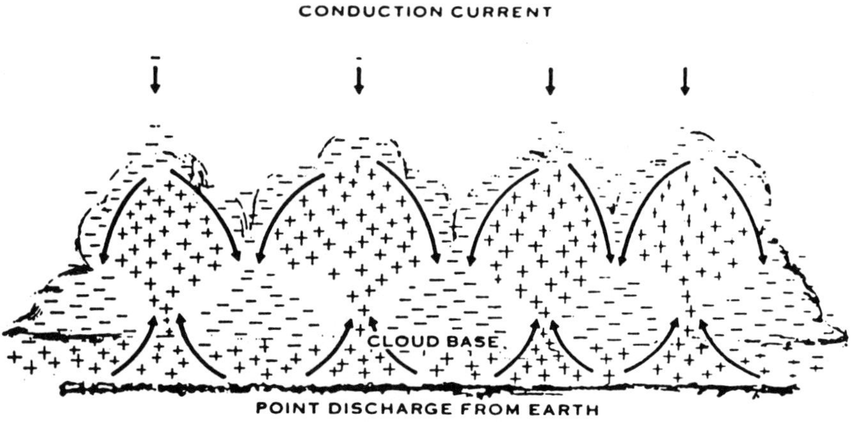

Fig. 9.58. Schematic representation of a group of thunderstorm cells illustrating how the electrification process might be maintained by convection. According to this representation, the negative charge attracted to the top of the cloud is carried to the lower part of the cloud by downdrafts while positive charge created by point discharge at the ground is carried to the upper part of the cloud by updrafts. [From Vonnegut (1963).]

earth's surface by point discharge. In fact, when their simulated cloud had a base height at observed levels, the cloud exhibited weak, negatively charged cores and relatively weak upper layers of positive charge. Only when the cloud-base height was lowered to a few tens of meters above the ground did a weak charging of opposite polarity develop.

Wagner and Telford (1981) modified the convective charging theory by adapting it to recent concepts of the dynamics structure of nonprecipitating cumuli discussed in Chapter 8. First, they note that a sharp interface exists at the leading edge of downdraft and updraft thermals. This interface represents a zone of sharp transition between the rapidly rising or sinking air and the quiescent environment. They also note that updrafts generally carry numerous small cloud droplets, whereas downdrafts consist of a factor of 2 or so fewer cloud droplets. As a consequence of fewer drops in downdrafts, absorption of ions will be less, yielding larger concentrations of free ions and greater conductivity than updrafts. In a fair-weather electric field, they postulate that positive ions will become concentrated at the sharp leading edge of the downdraft, while negative ions will reside preferentially in its interior. Figure 9.59 illustrates the process. In this model the positive ions at the leading edge of the downdraft thermal are swept aside into the main body of the cloud where they are carried aloft. The lack of a sharp interface in the trailing edge of the plume inhibits the loss of negative ions in that region. This leads to downdrafts being preferentially negatively

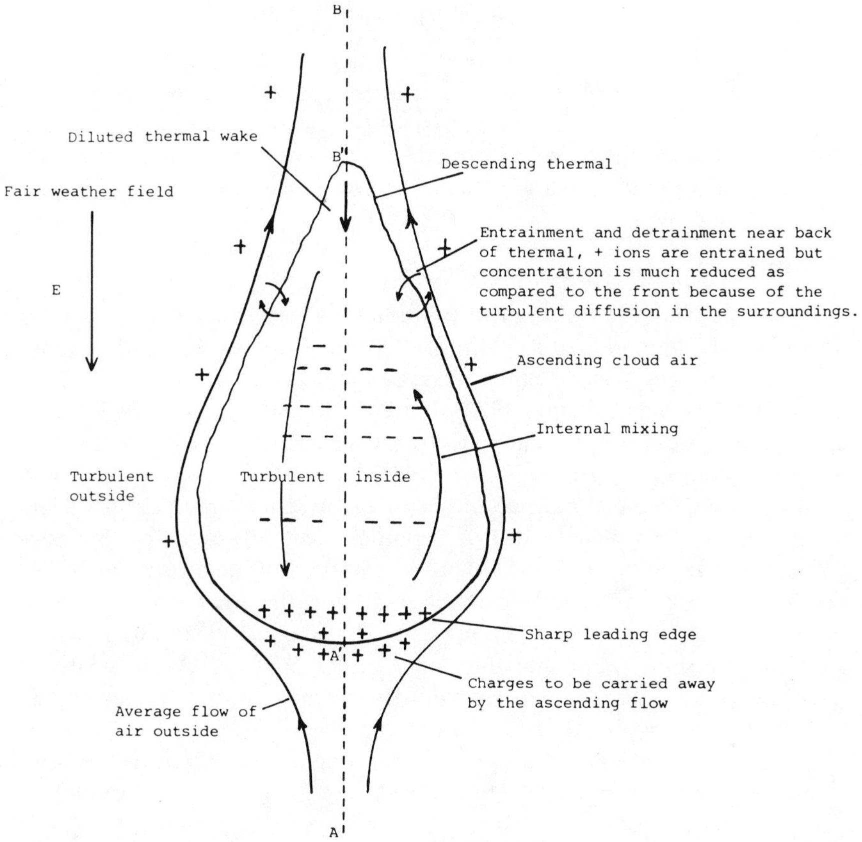

Fig. 9.59. The descending thermal, with sharp leading edge, contains far fewer particles than that in the ascending cloud, and the ion concentration in the thermal is higher than in the updraft. Due to the sharp leading edge of the descending thermal, the ion concentration has its maximum right behind the leading edge. Thus, when a downward field, which can be the fair-weather field, is applied, the ions are separated into positive and negative regions, where some of the positive ions will move out of the leading edge of the thermal and be carried away by the ascending flow. This process will then leave the thermal negatively charged. [From Wagner and Telford (1981).]

charged. The positive ions lost by the downdraft become readily attached to the numerous cloud droplets in the updrafts. The ice phase accelerates the process because a few ice crystals in the downdraft readily deplete the cloud droplets, further increasing the mobility of the ions in the downdraft. Aside from some order-of-magnitude estimates, this extended convection theory has not been examined in a comprehensive cloud model.

9.10.2.b Precipitation Charging Theories

The precipitation charging theories do not depend directly or solely on the convective motions of the cloud for charge separation. However, they do depend indirectly upon the dynamic structure of the cloud for vertical and horizontal redistribution of precipitation elements. We will examine the induction-precipitation charging theory and several noninduction-precipitation charging theories.

9.10.2.c Induction Charging Theory

The particle-charging mechanism by induction has enjoyed a long list of advocates (Elster and Geital, 1913; Muller-Hillebrande, 1954, 1955; Sartor, 1961; Latham and Mason, 1962; Mason, 1968, 1976; Scott and Levin, 1975; Levin, 1976; Colgate *et al.*, 1977; Illingworth and Latham, 1977; Chiu, 1978). The basic concept is that in the presence of a fair-weather field, cloud and precipitation elements become polarized, as shown in Fig. 9.60, with the lower part of the cloud particle being positively charged and its upper part being negatively charged. It is hypothesized that when a precipitation particle and a cloud droplet or small ice crystal collide and rebound, the larger particle becomes negatively charged and the smaller one becomes positively charged. The resultant positively charged small precipitation elements are then swept into the upper portions of the cloud, while the larger, negatively charged particles settle in the lower portions. This process leads to a cloud polarization with a positive space charge in the upper part of the cloud and negative space charge residing on the larger precipitation elements in the lower part of the cloud. Using greatly different models of particle physics and cloud dynamics, both Levin (1976) and Chiu (1978) conclude that particle charging by induction involving collision between liquid droplets can only develop an electric field strength of breakdown

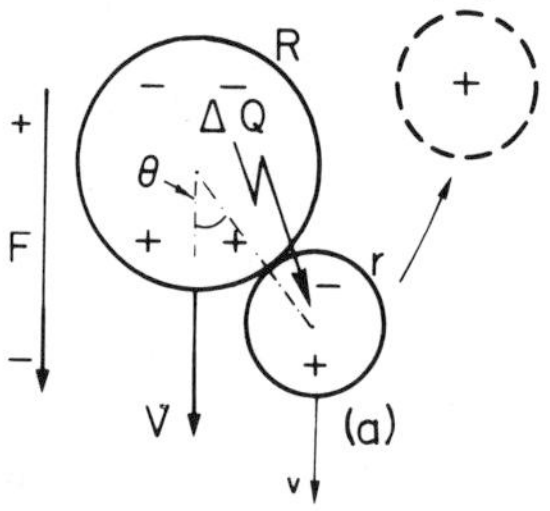

Fig. 9.60. Sequence of events during the polarization charging of cloud hydrometeors. (a) During contact; (b) after separation. [From Levin and Ziv (1974). Copyright by the American Geophysical Union.]

potential within a typical cloud lifetime of the order of 1000 s. Illingworth and Latham (1977), however, concluded that liquid–liquid induction charging is not capable of thunderstorm electrification. They concluded that collision between ice crystals and hail pellets is a far more powerful induction-charging mechanism. The induction mechanism needs a high frequency of collision and rebound between cloud particles. This condition is more likely to be satisfied for ice–ice collisions than for water–water or water–ice collisions. However, as noted by Gaskell (1979), the relaxation times to conduct charge through ice is relatively long, possibly too long for significant charge transfer. Also, when Rawlins examined the ice–ice induction mechanism in a cloud model, he found that if multiple collisions of each ice particle with more than one hail particle were included, a breakdown field was not simulated.

9.10.2.d Noninduction Charging of Graupel and Hail Particles

There has been considerable circumstantial evidence over the years that the presence of graupel is linked to the separation of charge in thunderstorms. As a consequence, there have been a number of hypotheses and laboratory experiments aimed at explaining its role. Some of these hypotheses, such as the thermoelectric effect (Latham and Mason, 1961), are now thought to have too small a charging rate to be of significance (Marshall *et al.*, 1978). A number of laboratory studies have exhibited a complex variation in charging rates with temperature and relative humidity. The laboratory experiments by Takahashi (1978), for example, suggested that at temperatures colder than −10°C, the sign of charging during riming depended on temperature and liquid-water content. At temperatures warmer than −10°C, positive charge occurred regardless of the liquid-water content. Hallett and Saunders (1979) examined the role of secondary ice-particle production during riming as well as rime-particle collision with ice crystals on rime deposit charging. The experiments were carried out at −4°C, where secondary multiplication has been found to maximize. They noted that the rimed surface became positively charged both by secondary ice-particle production and by colliding with artificially nucleated crystals. However, if the supply of moisture was turned off and the saturation ratio decreased, the sign of charging reversed to produce negative charging. They concluded that the local saturation ratio is a controlling factor on charge transfer. They hypothesize, therefore, that graupel particles will become positively charged in updrafts due to secondary ice-particle production and secondarily by collisions with ice crystals. Moreover, in downdrafts that become subsaturated or are in near equilibrium, the graupel particles will become negatively charged through collisions with ice crystals. In cold cloud-base continental clouds, where coalescence processes are inhibited, secondary particle charging at temperatures colder than −10°C could occur if graupel riming takes place vigorously enough to warm the surface temperature of the graupel to near −4°C. Both Latham (1981) and Rawlins (1982) argue that the rate of charging during secondary production is too weak and that the mechanism is too restrictive in the range of conditions under which it can operate. However, it is interesting that the polarity of charging is dependent upon the updraft/downdraft structure of the cloud. Even if secondary production of ice crystals is not directly responsible for cloud charging, the consequences of production of large concentrations of ice crystals at warm temperatures can have a significant bearing on charge generation by ice–ice collisions.

The process of noninduction charging resulting from collisions between ice particles and graupel has been studied by a number of investigators through the years (Reynolds *et al.*, 1957; Marshall *et al.*, 1978; Buser and Aufdermaur, 1977; Caranti and Illingworth, 1980; Gaskell, 1979; Hallett and Saunders, 1979; Illingworth and Latham, 1977; Jayaratne *et al.*, 1983; Baker *et al.*, 1987). Buser and Aufdermaur (1977) and Caranti and Illingworth (1980) propose that the charging is a function of the surface properties of ice. They argue that a difference between the contact potentials of ice formed in different ways drives the transfer of charge. An advantage of this hypothesis over the ice–ice induction hypothesis is that charge is rapidly transferred between charged states at the interface between rimed graupel/hail and vapor-grown ice crystals. In contrast to induction charging, where short contact times limit the amount of bulk charge that can be conducted across the interface, charge transfer by this process is quite rapid. It should be noted that this process requires a particular mix of ice-particle types. The process operates most effectively when there is a relatively high concentration of vapor-grown crystals that can collide with large, rapidly falling rimed ice particles such as graupel or hail. Rawlins (1982) examined this process quantitatively in a three-dimensional cloud model and concluded it was capable of producing lightning 20 min after the formation of precipitation. In his calculations, however, he estimated crystal concentrations using the Fletcher (1962) ice-nuclei equation, which is known to underestimate the concentration of vapor-grown crystals at warm temperatures. The use of the Hallett–Mossop secondary ice-particle production process would further accelerate the process.

9.10.2.e Examination of Charging Theories in Relation to Recent Field Observations

We will now examine in greater detail the results of recent field observations of thunderstorm charging, air motions, and precipitation development in terms of their consistency with the above theories.

First, we will discuss the relationship between the locations and presence of precipitation and the charge centers of the storm. The observations reported by Krehbiel *et al.* (1983) in New Mexico indicate that lightning discharges formed throughout the precipitating region of the storm; the lightning discharges also appeared to be bounded by the precipitation region. In one case, they noted that the lightning preferentially discharged in regions where the precipitation was strongest. Williams (1985), however, summarized a number of observations of lightning source regions and the locations of precipitation; he concluded that in many cases the regions of intense precipitation are not regions of maximum space charge density. There certainly is little evidence that the main charge centers are associated with regions exhibiting high radar reflectivity (>45–50 dBZ). These results are not consistent with numerical models such as Rawlins's, in which the maximum space charge density created by ice–ice interactions is colocated with the maximum precipitation. However, this may be an artifact of the model parameterizations because, as noted by Latham (1981), a few large particles can contribute significantly to the reflectivity, but a large number of smaller precipitation elements having a larger net surface area can contribute substantially to the space charge density. Thus the radar-charge location studies do not fully support the precipitation charging theories, nor do they completely invalidate them.

Recent observations continue to support earlier findings that the center of negative charge is located in the supercooled region of the cloud. Krehbiel *et al.* (1983), for example, found that the main source of negative charge for cloud-to-ground lightning strokes was the region between −10 and −25°C that generally coincided with the presence of precipitation. Taylor (1980) also found the main center of lightning activity to be associated with the supercooled cloud layer, but at slightly warmer temperatures between −5 and −20°C. Krehbiel *et al.* (1983) found that the main center of negative charge associated with lightning activity lies between −10 and −20°C, in warm-based, maritime cumuli over Florida, in cold-based, continental cumuli over New Mexico, and in cold-based, relatively shallow cumuli over Japan; this finding strongly supports an ice-phase charging process. Figure 9.61 illustrates the results of their study. It is also interesting to note that Chauzy *et al.* (1985) found the main center of negative charge to be closer to 0°C in the stratified portion of a tropical squall line. This suggests that the temperature at which the negative space charge is centered is a function of the updraft speeds or the characteristics of precipitation, both of which will differ in more local convective storms versus the more stratified mesoscale systems. In summary, the observations of the locations of space charge support the ice-phase precipitation mechanisms rather than a convection theory.

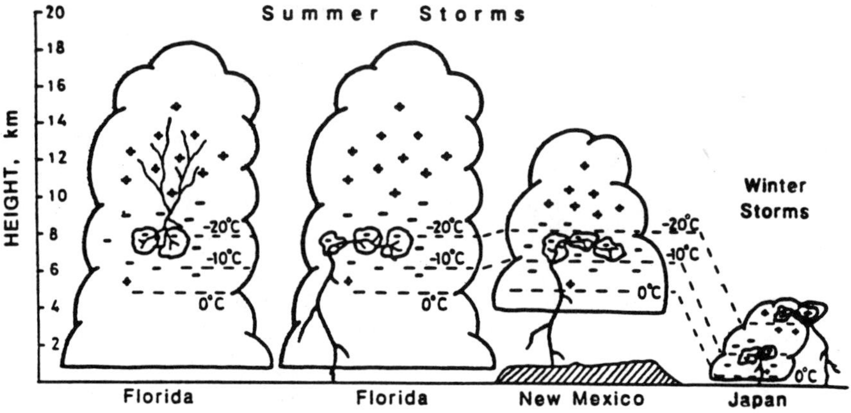

Fig. 9.61. Schematic diagram illustrating ground flash charge source levels and development in Florida and New Mexico summer thunderstorms and in Japanese winter storms. [From Krehbiel *et al.* (1983). Original artwork prepared by Marx Brook. Copyright by A. Deepak Publishing.]

All the above theories either directly or indirectly depend upon cloud air motions for charge separation. Lhermitte and Krehbiel (1979) have synthesized Doppler radar-derived air motions with remotely sensed locations of electrical discharges. They found that the onset of vigorous electrical activity coincided with the development of a vigorous updraft ($>20\ \mathrm{m\ s^{-1}}$) which penetrated the -10 to -15°C region. The initial lightning sources formed an umbrella-like shield over the well-organized updraft in a region where the updraft merges with the external downdraft. The sources of lightning occur around the periphery of the high-reflectivity region near the -10°C level and are well separated from the high-precipitation core. These observations appear to support more strongly the convection charging theories than the precipitation theories.

In another study, however, Lhermitte and Williams (1985) found that the main center of negative charge associated with lightning was less than 1 km above a high-reflectivity region where the particle vertical motions were near zero (a balance level or accumulation zone). They attributed the relative stability of the location of the negative charge center to the persistence of the balance level at the same location through much of the storm lifetime. The negative charge center corresponded to the -15°C level and was linked to, though not colocated with, a precipitation core; this fact supports the idea of an ice-related precipitation-charging mechanism.

In yet another study, Rust *et al.* (1981) found that the source locations for lightning strokes occurred in regions of weaker updrafts ($<10\ \mathrm{m\ s^{-1}}$), often adjacent to downdrafts. Again, few of the lightning sources were in the cores of high reflectivity. It is interesting to note that they found a good correlation between increases in updraft, reflectivity, and lightning flashing rates. An interesting finding by MacGorman *et al.* (1980) is that lightning usually occurred in or near regions of cyclonic shear, often associated with updrafts.

Winn *et al.* (1978) reported on the results of flying an instrumented balloon into a thunderstorm. They found changes in the electric field associated with changes in horizontal winds and in encounters with updrafts and downdrafts. In one instance they found an organized and persistent downdraft along the lower surface of the anvil that carried negative charges.

Observations in winter thunderstorms over Japan reported by Brook *et al.* (1982) suggest that vertical shear of the horizontal wind is strongly correlated with the fraction of cloud-to-ground lightning strokes that carry positive charge. They suggested that the vertical wind shear horizontally displaces the positive space charge region in the upper part of the cloud from the lower negative charge region. This would allow the development of a strong potential gradient between cloud top and the earth's surface. They speculated that in the absence of shear, the positive discharges occur principally as cloud-to-cloud strokes.

Furthermore, Ray *et al.* (1987) found that lightning tended to be downshear of the main storm updraft and radar reflectivity core in supercell storms, whereas it was concentrated in the updraft and reflectivity core of multicell storms. Because supercell storms prevail in more strongly sheared environments, their finding is consistent with Brook *et al.*'s and MacGorman's (1978) suggestion that the positive space charge is advected downwind of the change-generating region in the updraft in strongly sheared environments.

Overall measurements of air motions, along with electrical measurements in thunderstorms, provide partial support for both the convection theories and the ice-phase-related precipitation theories, although they do not completely confirm either of them.

We conclude this section by noting a few observations suggesting that different intensities of lightning activity occur with different storm types. Pakiam and Maybank (1975) examined the electrical activity of multicellular and supercell hail-producing thunderstorms over Alberta, Canada. While the number of cases they studied was limited, they noted the following relationships with storm type:

(1) If the storm was of an ordinary multicellular type with limited depth (6–7.5 km MSL in Alberta), then rain and small hail occurred with a low frequency of lightning flashes.

(2) If the thermodynamic instability was greater, the storms became better organized multicellular storms with high cloud tops (7.5–12 km MSL in Alberta); rain and hail and the lightning frequency increased appreciably.

(3) With even greater instability and higher shear, organized multicellular storms with tops well into the stratosphere (12+ km MSL in Alberta) prevailed. In these storms, large hail forms, and the frequency of lightning depends on the number and proximity of thunderstorm cells. A system composed of five cells may produce 35 flashes min^{-1}, a number of which are intracloud. An isolated cell produces only 2–3 flashes min^{-1}.

(4) In one case of a severe, hail-producing supercell storm, the frequency of flashes was only 2–3 min^{-1}. They concluded that such a storm is only a single electrical cell resulting in a low frequency of lightning flashes.

It is not surprising that the results of their study suggest that the frequency of lightning discharges does not increase with updraft speeds, but is primarily a function of the depth and number of multicellular convective elements in a storm system. Although Pakiam and Maybank's study is limited, because it considers only a few cases for a single geographical region, it nevertheless suggests that storm organization has an important control on the rate of thunderstorm discharge.

Further support of the concept that the frequency of lightning discharges is a function of the depth and number of multicellular convective elements is Goodman and MacGorman's (1986) study of cloud-to-ground lightning activity in mesoscale convective complexes. As will be shown in Chapter 10, the MCC represents a very well-organized mesoscale convective system which is composed of numerous cumulonimbus cells. Goodman and MacGorman showed that MCCs produce maximum ground lightning strikes of 54 min^{-1} averaged over 1 h or a sustained lightning frequency in excess of 17 min^{-1} for nine consecutive hours! Goodman and MacGorman noted that such a high, sustained lightning frequency over the lifetime of an MCC suggests that the passage of a single MCC over a given location can produce 25% of the estimated mean annual strike density for that site. Furthermore they found that the ratio of ground discharges in MCCs to ordinary thunderstorms observed in Florida is 4:1, whereas for severe or multicell storms in the High Plains of the United States the ratio is in excess of 20:1. It is interesting that Goodman and MacGorman (1986) found that the peak flash rate in MCCs occurs at the time that McAnelly and Cotton (1985) found that the storm exhibits its coldest cloud-top temperatures. This corresponds to the time that the MCC achieves its peak rainfall rates. It also corresponds to the time that the MCC is composed of numerous, convectively active cells.

In summary, we have seen that the electrical activity of a storm varies with the organization of the storm system. On the scale of an individual cell, charge separation appears to be linked to local regions of updraft/downdraft shear and to regions of horizontal shear of the horizontal wind. The cloud-charging mechanisms are most active when ice-phase precipitation processes are prevalent. However, the centers of charge do not coincide with the regions of highest radar reflectivity or precipitation content. The observational evidence thus far does not conclusively support any one of the current charging theories. However, there is evidence supporting the convection theories as well as an ice-phase-related precipitation theory. At this time, the most one can say is that the observational and modeling evidence leans toward a graupel-ice precipitation theory. Furthermore, the most prolific producers of cloud-to-ground lightning are thunderstorm systems organized on the mesoscale that are composed of a number of multicellular convective elements. We will examine the characteristics of mesoscale convective systems more fully in Chapter 10.

References

Achtemeier, G. L. (1975). Doppler velocity and reflectivity morphology of a severe left-moving split thunderstorm. *Prepr., Radar Meteorol. Conf., 16th, Houston, Tex.* pp. 93–98. Am. Meteorol. Soc., Boston, Massachusetts.

Agee, E. M., J. T. Snow, F. S. Nickerson, P. R. Clare, C. R. Church, and L. A. Schaal (1977). An observational study of the West Lafayette, Indiana, tornado of 20 March 1976. *Mon. Weather Rev.* **105**, 893–907.

Atlas, D. (1966). The balance level in convective storms. *J. Atmos. Sci.* **23**, 635–651.

Auer, A. H., Jr., and J. D. Marwitz (1968). Estimates of air and moisture flux into hailstorms on the high plains. *J. Appl. Meteorol.* **7**, 196–198.

Baker, B., M. B. Baker, E. R. Jayaratne, J. Latham, and C. P. R. Saunders (1987). The influence of diffusional growth rates on the charge transfer accompanying rebounding collisions between ice crystals and soft hailstones. *Q. J. R. Meteorol. Soc.* **113**, 1193–1215.

Baker, G. L. (1981). Boundary layers in laminar vortex flows. Ph.D. Thesis, Purdue Univ., West Lafayette, Indiana.

Barnes, S. L. (1978a). Oklahoma thunderstorms on 29–30 April 1970. Part I: Morphology of a tornadic storm. *Mon. Weather Rev.* **106**, 673–684.

Barnes, S. L. (1978b). Oklahoma thunderstorms on 29–30 April 1970. Part II: Radar-observed merger of twin hook echoes. *Mon. Weather Rev.* **106**, 685–696.

Barnes, G., and M. Garstang (1982). Subcloud layer energetics of precipitating convection. *Mon. Weather Rev.* **110**, 102–117.

Bates, F. C. (1968). A theory and model of the tornado. *Proc. Int. Conf. Cloud Phys., Toronto.* pp. 559–563. Am. Meteorol. Soc., Boston, Massachusetts.

Battan, L. J. (1964). Some observations of vertical velocities and precipitation sizes in a thunderstorm. *J. Appl. Meteorol.* **3**, 415–420.

Battan, L. J. (1975). Doppler radar observations of a hailstorm. *J. Appl. Meteorol.* **14**, 98–108.

Battan, L. J. (1980). Observations of two Colorado thunderstorms by means of a zenith-pointing Doppler radar. *J. Appl. Meteorol.* **19**, 580–592.

Battan, L. J., and J. B. Theiss (1966). Observations of vertical motions and particle sizes in a thunderstorm. *J. Atmos. Sci.* **23**, 78–87.

Battan, L. J., and J. B. Theiss (1970). Measurements of vertical velocities in convective clouds by means of a pulsed Doppler radar. *J. Atmos. Sci.* **27**, 293–298.

Battan, L. J., and J. B. Theiss (1973). Observations of vertical motion and particle sizes in a thunderstorm. *J. Atmos. Sci.* **23**, 78–87.

Beebe, R. G., and F. C. Bates (1955). A mechanism for assisting in the release of convective instability. *Mon. Weather Rev.* **83**, 1–10.

Betts, A. K. (1976). The thermodynamic transformation of the tropical subcloud layer by precipitation and downdrafts. *J. Atmos. Sci.* **33**, 1008–1020.

Betts, A. K. (1984). Boundary layer thermodynamics of a High Plains severe storm. *Mon. Weather Rev.* **112**, 2199–2211.

Betts, A. K., R. W. Grover, and M. W. Moncrieff (1976). Structure and motion of tropical squall-lines over Venezuela. *Q. J. R. Meteorol. Soc.* **102**, 395–404.

Bluestein, H. B., and M. H. Jain (1985). The formation of mesoscale lines of precipitation: Severe squall lines in Oklahoma during the spring. *J. Atmos. Sci.* **42**, 1711–1732.

Bluestein, H. B., and C. J. Sohl (1979). Some observations of a splitting severe thunderstorm. *Mon. Weather Rev.* **107**, 861–873.

Bosart, L. F., and Frederick Sanders (1981). The Johnstown Flood of July 1977: A long-lived convective system. *J. Atmos. Sci.* **38**, 1616–1642.

Boyden, C. J. (1963). A simple instability index for use as a synoptic parameter. *Meteorol. Mag.* **92**, 198–210.

Braham, R. R. (1952). The water and energy budgets of the thunderstorm and their relation to thunderstorm development. *J. Meteorol.* **9**, 227–242.

Brandes, E. A. (1981). Fine structure of the Del City–Edmond tornadic mesocirculation. *Mon. Weather Rev.* **109**, 635–647.

Brandes, E. A. (1984). Vertical vorticity generation and mesocyclone sustenance in tornadic thunderstorms: The observational evidence. *Mon. Weather Rev.* **112**, 2253–2269.

Brook, M., M. Nakano, P. Krehbiel, and T. Takeuti (1982). The electrical structure of the Hokuriku winter thunderstorms. *J. Geophys. Res.* **87**, 1207–1215.

Brown, J. M., and K. R. Knupp (1980). The Iowa cyclonic–anticyclonic tornado pair and its parent thunderstorm. *Mon. Weather Rev.* **108**, 1626–1646.

Brown, J. M., K. R. Knupp, and R. Caracena (1982). Destructive winds from shallow, high-based cumuli. *Prepr., Conf. Severe Local Storms, 12th.* pp. 272–275. Am. Meteorol. Soc., Boston, Massachusetts.

Browning, K. A. (1962). Cellular structure of convective storms. *Meteorol. Mag.* **91**, 341–350.

Browning, K. A. (1963). The growth of large hail within a steady updraft. *Q. J. R. Meteorol. Soc.* **89**, 490–506.

Browning, K. A. (1964). Airflow and precipitation trajectories within severe local storms which travel to the right of the winds. *J. Atmos. Sci.* **21**, 634–639.

Browning, K. A. (1965). Some inferences about the updraft within a severe local storm. *J. Atmos. Sci.* **22**, 669–677.

Browning, K. A. (1968). The organization of severe local storms. *Weather* **23**, 429–434.

Browning, K. A. (1977). The structure and mechanisms of hailstorms. *Meteorol. Monogr.* **16**(38), 1–43.

Browning, K. A., and R. J. Donaldson, Jr. (1963). Airflow and structure of a tornadic storm. *J. Atmos. Sci.* **20**, 533–545.

Browning, K. A., and G. B. Foote (1976). Airflow and hail growth in supercell storms and some implications for hail suppression. *Q. J. R. Meteorol. Soc.* **102**, 499–533.

Browning, K. A., and F. H. Ludlam (1960). Radar analysis of a hailstorm. Tech. Note No. 5, Dep. Meteorol., Imperial College, London.

Browning, K. A., and F. H. Ludlam (1962). Airflow in convective storms. *Q. J. R. Meteorol. Soc.* **88**, 117–135.

Browning, K. A., and R. Wexler (1968). The determination of kinematic properties of a wind field using Doppler radar. *J. Appl. Meteorol.* **7**, 105–113.

Browning, K. A., F. H. Ludlam, and W. R. Macklin (1963). The density and structure of hailstone. *Q. J. R. Meteorol. Soc.* **89**, 75–84.

Browning, K. A., J. C. Fankhauser, J.-P. Chalon, P. J. Eccles, R. G. Strauch, F. H. Merrem, D. J. Musil, E. L. May, and W. R. Sand (1976). Structure of an evolving hailstorm. Part V: Synthesis and implications for hail growth and suppression. *Mon. Weather Rev.* **104**, 603–610.

Burgess, D. W. (1976). Anticyclonic tornado. *Weatherwise* **29**.

Buser, O., and A. N. Aufdermaur (1977). Electrification by collisions of ice particles on ice or metal targets. *In* "Electrical Processes in Atmospheres" (H. Dolezalek and R. Reiter, eds.), pp. 294–301. Steinkopff Verlag.

Byers, H. R., and R. R. Braham (1949). "The Thunderstorm." U.S. Weather Bur., Washington, D.C.

Caracena, F., R. A. Maddox, L. R. Hoxit, and C. F. Chappell (1979). Forecasting likelihood of microbursts along the front range of Colorado—Results of the JAWS project. *Prepr., Conf. Severe Local Storms, 13th* pp. 262–264. Am. Meteorol. Soc., Boston, Massachusetts.

Caranti, J. M., and A. J. Illingworth (1980). Surface potentials of ice and thunderstorm charge separation. *Nature (London)* **284**, 44–46.

Chang, C.-P. (1976). Vertical structure of tropical waves maintained by internally-induced cumulus heating. *J. Atmos. Sci.* **33**, 729–739.

Charba, J. (1974). Application of gravity current model to analysis of squall line gust front. *Mon. Weather Rev.* **102**, 140–156.

Charney, J. G., and A. Eliassen (1964). On the growth of the hurricane depression. *J. Atmos. Sci.* **21**, 68–75.

Chauzy, S., M. Chong, A. Delannoy, and S. Despiau (1985). The June 22 tropical squall line observed during COPT 81 experiment: Electrical signature associated with dynamical structure and precipitation. *J. Geophys. Res.* **90**, 6091–6098.

Chisholm, A. J. (1973). Radar case studies and airflow-models. Part I, Alberta hailstorms. *Meteorol. Monogr.* **36**, 1–36.

Chisholm, A. J., and J. H. Renick (1972). The kinematics of multicell and supercell Alberta hailstorms, Alberta Hail Studies, 1972. Res. Counc. Alberta Hail Stud. Rep. No. 72-2, pp. 24–31.

Chiu, C.-S. (1978). Numerical study of cloud electrification in an axisymmetric, time-dependent cloud model. *J. Geophys. Res.* **83**, 5025–5049.

Chiu, C.-S., and J. D. Klett (1976). Convective electrification of clouds. *J. Geophys. Res.* **81**, 1111–1124.

Church, C. R., J. T. Snow, G. L. Baker, and E. M. Agee (1979). Characteristics of tornado-like vortices as a function of swirl radio: A laboratory investigation. *J. Atmos. Sci.* **36**, 1755–1776.

Clark, T. L. (1979). Numerical simulations with a three-dimensional cloud model: Lateral boundary condition experiments and multicellular storm simulations. *J. Atmos. Sci.* **36**, 2191–2215.

Colgate, S. (1975). Comment on "On the relation of electrical activity to tornadoes" by R. P. Davies-Jones and J. H. Golden. *J. Geophys. Res.* **80**, 4556.

Colgate, S. A., Z. Levin, and A. G. Petschek (1977). Interpretation of thunderstorm charging by the polarization-induction mechanism. *J. Atmos. Sci.* **34**, 1433–1443.

Cooper, H. J., M. Garstang, and J. Simpson (1982). The diurnal interaction between convection and peninsular-scale forcing over South Florida. *Mon. Weather Rev.* **110**, 486–503.

Cotton, W. R. (1972). Numerical simulation of precipitation development in supercooled cumuli, Part I. *Mon. Weather Rev.* **100**, 757–763.

Cotton, W. R., P. T. Gannon, and R. A. Pielke (1974). Numerical experiments of the influence of the mesoscale circulation on the cumulus scale. *Conf. Cloud Phys., Tucson, Ariz.* pp. 424–429. Am. Meteorol. Soc., Boston, Massachusetts.

Cotton, W. R., R. L. George, and K. R. Knupp (1982). An intense, quasi-steady thunderstorm over mountainous terrain. Part I: Evolution of the storm-initiating mesoscale circulation. *J. Atmos. Sci.* **39**, 328–342.

Crow, E. L., P. W. Summers, A. B. Long, C. A. Knight, G. B. Foote, and J. E. Dye (1976). "Experimental Results and Overall Summary." Vol. I. Final Rep., Natl. Hail Res. Exp. Randomized Seeding Exp. 1972–1974. Natl. Cent. Atmos. Res., Boulder, Colorado.

Danielsen, E. F. (1977). Inherent difficulties in hail probability prediction. *Meteorol. Monogr.* **38**, 135–143.

Danielsen, E. F., R. Bleck, and D. Morris (1972). Hail growth by stochastic collection in a cumulus model. *J. Atmos. Sci.* **29**, 133–155.

Das, P. (1962). Influence of wind shear on the growth of hail. *J. Atmos. Sci.* **19**, 407–414.

Das, P., and M. C. Subba Rao (1972). The unsaturated downdraft. *Indian J. Meteorol. Geophys.* **23**, 135–144.

Davies, H. C. (1979). Phase-lagged wave-CISK. *Q. J. R. Meteorol. Soc.* **105**, 325–353.

Davies-Jones, R. P. (1982). Tornado dynamics. *In* "Thunderstorms: A Social, Scientific, and Techological Documentary" (E. Kessler, ed.), Vol. 2. U.S. Dep. Commer., Washington, D.C.

Davies-Jones, R. P., and J. H. Golden (1975a). On the relation of electrical activity to tornadoes. *J. Geophys. Res.* **80**, 1614–1616.

Davies-Jones, R. P., and J. H. Golden (1975b). Reply. *J. Geophys. Res.* **80**, 4557–4558.

Davies-Jones, R. P., and J. H. Golden (1975c). Reply. *J. Geophys. Res.* **80**, 4561–4562.

Davies-Jones, R. P., and E. Kessler (1974). Tornadoes. *In* "Weather Climate Modification" (W. N. Hess, ed.), pp. 552–595. Wiley, New York.

Davis, M. H. (1961). The forces between conducting spheres in a uniform electric field. RM-2607. Rand Corp., 1700 Main St., Santa Monica, California 90406.

Davis, M. H. (1964a). Two charged spherical conductors in a uniform electric field: Forces and field strength. RM-3860-PR. Rand Corp.

Davis, M. H. (1964b). Two charged spherical conductors in a uniform electric field: Forces and field strength. *Q. J. Mech. Appl. Math.* **17**, 499–511.

Dennis, A. S. (1980). "Weather Modification by Cloud Seeding." Academic Press, New York.

Dennis, A. S., C. A. Schock, and A. Koscielski (1970). Characteristics of hailstorms of Western South Dakota. *J. Appl. Meteorol.* **9**, 127–135.

Dennis, A. S., R. A. Schleusener, J. H. Hirsch, and Alexander Koscielski (1973). Meteorology of the Black Hills Flood of 1972. Rep. No. 73-4, Inst. Atmos. Sci., South Dakota Sch. Mines Technol., Rapid City.

Donaldson, R. J., Jr., and R. Wexler (1969). Flight hazards in thunderstorms determined by Doppler velocity variance. *J. Appl. Meteorol.* **8**, 128–133.

Droegemeier, K. K., and R. B. Wilhelmson (1985a). Three-dimensional numerical modeling of convection produced by interacting thunderstorm outflows: Part I. Control simulation and low-level moisture variations. *J. Atmos. Sci.* **42**, 2381–2403.

Droegemeier, K. K., and R. B. Wilhelmson (1985b). Three-dimensional numerical modeling of convection produced by interacting thunderstorm outflows: Part II. Variations in vertical wind shear. *J. Atmos. Sci.* **42**, 2404–2414.

Dutton, J. A. (1976). "The Ceaseless Wind." McGraw-Hill, New York.

Ellrod, G. P., and J. D. Marwitz (1976). Structure and interaction in the subcloud region of thunderstorms. *J. Appl. Meteorol.* **15**, 1083–1091.

Elster, J., and H. Geitel (1913). Zur Influenztheorie der Niederschlagselek-trizitat. *Phys. Z.* **14**, 1287.

Emanuel, K. A. (1982). Inertial instability and mesoscale convective systems. Part II: Symmetric CISK in a baroclinic flow. *J. Atmos. Sci.* **39**, 1080–1097.

English, M. (1973). Alberta hailstorms. Part II: Growth of large hail in the storm. *Meteorol. Monogr.* **36**, 37–98.

Fankhauser, J. C. (1976). Structure of an evolving hailstorm, Part II: Thermodynamic structure and airflow in the near environment. *Mon. Weather Rev.* **104**, 576–587.

Fankhauser, J. C. (1982). The 22 June 1976 case study: Large-scale influences, radar echo structure and mesoscale circulations. *In* "Hailstorms of the Central High Plains" (C. A. Knight and P. Squires, eds.), Vol. 2, pp. 1-33. Colorado Assoc. Univ. Press, Boulder.

Fawbush, E. F., and R. Miller (1953). A method for forecasting hailstone size at the earth's surface. *Bull. Am. Meteorol. Soc.* **34**, 235–244.

Fernandez, W., and A. J. Thorpe (1979). An evaluation of theories of storm motion using observations of tropical convective systems. *Mon. Weather Rev.* **107**, 1306–1319.

Fletcher, N. H. (1962). "The Physics of Rainclouds." Cambridge Univ. Press, London.

Foote, G. B. (1984). A study of hail growth utilizing observed storm condition. *J. Clim. Appl. Meteorol.* **23**, 84–101.

Foote, G. B., and J. C. Fankhauser (1973). Airflow and moisture budget beneath a Northeast Colorado hailstorm. *J. Appl. Meteorol.* **12**, 1330–1353.

Foote, G. B., and H. W. Frank (1983). Case study of a hailstorm in Colorado. Part III: Airflow from triple Doppler measurements. *J. Atmos. Sci.* **40**, 686–707.

Foote, G. B., and C. A. Knight (1979). Results of a randomized hail suppression experiment in Northeast Colorado. Part I: Design and conduct of the experiment. *J. Appl. Meteorol.* **18**, 1526–1537.

Foote, G. B., and C. G. Wade (1982). Case study of a hailstorm in Colorado. Part I: Radar echo structure and evolution. *J. Atmos. Sci.* **39**, 2828–2846.

Forbes, G. S., and R. M. Wakimoto (1983). A concentrated outbreak of tornadoes, downbursts and microbursts, and implications regarding vortex classification. *Mon. Weather Rev.* **110**, 220–235.

Foster, D. S., and F. Bates (1956). A hail size forecasting technique. *Bull. Am. Meteorol. Soc.* **37**, 135–141.

Frank, H. W., and G. B. Foote (1982). The 22 July 1976 case study: Storm airflow, updraft structure, and mass flux from triple-Doppler measurements. *In* "Hailstorms of the Central High Plains" (C. A. Knight and P. Squires, eds.), Vol. 2, pp. 131–162. Colorado Assoc. Univ. Press, Boulder.

Frisby, E. M. (1962). Relationship of ground hail damage patterns to features of the synoptic map in the Upper Great Plains of the United States. *J. Appl. Meteorol.* **1**, 348–352.

Frisch, A. S., and S. F. Clifford (1974). A study of convection capped by a stable layer using Doppler radar and acoustic echo sounders. *J. Atmos. Sci.* **31**, 1622–1628.

Frisch, A. S., and R. G. Strauch (1976). Doppler-radar measurements of turbulence kinetic energy dissipation rates in a northeastern Colorado convective storm. *J. Appl. Meteorol.* **15**, 1012–1017.

Fritsch, J. M. (1975). Cumulus dynamics: Local compensating subsidence and its implications for cumulus parameterization. *Pure Appl. Geophys.* **113**, 851–867.

Fritsch, J. M., and D. M. Rodgers (1981). The Fort Collins Hailstorm—An example of the short-term forecast enigma. *Bull. Am. Meteorol. Soc.* **62**, 1560–1569.

Fujita, T. T. (1959). Precipitation and cold air production in mesoscale thunderstorm systems. *J. Meteorol.* **16**, 454–466.

Fujita, T. T. (1960). A detailed analysis of the Fargo tornadoes of June 20, 1957. USWB Res. Pap. No. 42, Chicago, Illinois.

Fujita, T. T. (1970). The Lubbock tornadoes. A study of suction spots. *Weatherwise* **23**, 161–173.

Fujita, T. T. (1974). Overshooting thunderheads observed from ATS and Learjet. Satellite Mesometeorol. Res. Pap. No. 117, Dep. Geophys. Sci., Univ. of Chicago.

Fujita, T. T. (1977). Anticyclonic tornadoes. *Weatherwise* **30**, 51–64.

Fujita, T. T. (1978). Manual of downburst identification for project NIMROD. Satellite Mesometeorol. Res. Pap. No. 156, Dep. Geophys. Sci., Univ. of Chicago.

Fujita, T. (1981). Tornadoes and downbursts in the context of generalized planetary scales. *J. Atmos. Sci.* **38**, 1511–1534.

Fujita, T. T., and F. Caracena (1977). An analysis of three weather-related aircraft accidents. *Bull. Am. Meteorol. Soc.* **58**, 1164–1181.

Fujita, T., and H. Grandoso (1968). Split of a thunderstorm into anticyclonic and cyclonic storms and their motion as determined from numerical model experiments. *J. Atmos. Sci.* **25**, 416–439.

Fujita, T., and R. M. Wakimoto (1981). Five scales of airflow associated with a series of downbursts on 16 July 1980. *Mon. Weather Rev.* **109**, 1438–1456.

Fujita, T., K. Watanabe, K. Tsuchiya, and M. Schimada (1972). Typhoon-associated tornadoes in Japan and new evidence of suction vortices in a tornado near Tokyo. *J. Meteorol. Soc. Jpn.* **50**, 431–453.

Gall, R. (1983). A linear analysis of the multiple vortex phenomenon in simulated tornadoes. *J. Atmos. Sci.* **40**, 2010–2024.

Gall, R., and D. O. Staley (1981). Nonlinear barotropic instability in a tornado vortex. *Abstr., Conf. Atmos. Waves Stab., 3rd. Am. Meteorol. Soc., Boston, Mass.*

Galway, J. G. (1956). The lifted index as a predictor of latent instability. *Bull. Am. Meteorol. Soc.* **37**, 528–529.

Gaskell, W. (1979). Field and laboratory studies of precipitation charge. Ph.D. Thesis, Univ. of Manchester.

Gaynor, J. E., and P. A. Mandics (1978). Analysis of the tropical marine boundary layer during GATE using acoustic sounder data. *Mon. Weather Rev.* **106**, 223–232.

Goff, R. C. (1975). Thunderstorm outflow kinematics and dynamics. NOAA Tech. Memo. ERL NSSL-75, Natl. Severe Storms Lab., Norman, Oklahoma.

Goff, R. C. (1976). Vertical structure of thunderstorm outflow. *Mon. Weather Rev.* **104**, 1429–1440.

Golden, J. H., and D. Purcell (1978). Airflow characteristics around the Union City Tornado. *Mon. Weather Rev.* **106**, 22–28.

Goodman, S. J., and D. R. MacGorman (1986). Cloud-to-ground lightning activity in mesoscale convective complexes. *Mon. Weather Rev.* **114**, 2320–2328.

Grandia, K. L., and J. D. Marwitz (1975). Observational investigations of entrainment within the weak echo region. *Mon. Weather Rev.* **103**, 227–234.

Gray, W. M. (1965). Calculation of cumulus vertical draft velocities in hurricanes from aircraft observations. *J. Appl. Meteorol.* **4**, 47–53.

Grenet, G. (1947). Essai d'explication de la charge electrique des nuage d'orages. *Ann. Geophys.* **3**, 306–307.

Gutman, L. N. (1963). Stationary axially symmetric model of a cumulus cloud. *Dokl. Akad. Nauk SSSR* **150**, No. 1.

Gutman, L. N. (1967). Calculation of the velocity of ascending currents in a stationary convective cloud. *In* "Formation of Precipitation and Modification of Hail Processes" (E. K. Fedorov, ed.), p. 12. Isr. Program Sci. Transl., Jerusalem.

Haagenson, P. L., and E. Danielsen (1972). Operational steady-state model. NCAR Intern. Rep., Natl. Cent. Atmos. Res., Boulder, Colorado.

Hallett, J., and C. P. R. Saunders (1979). Charge separation associated with secondary ice crystal production. *J. Atmos. Sci.* **36**, 2230–2235.

Hayashi, Y. (1970). A theory of large-scale equatorial waves generated by condensation heat and accelerating the zonal wind. *J. Meteorol. Soc. Jpn.* **48**, 140–160.

Hayashi, Y. (1971a). Instability of large-scale equatorial waves with a frequency-dependent CISK parameter. *Q. J. R. Meteorol. Soc.* **49**, 59–62.

Hayashi, Y. (1971b). Instability of large-scale equatorial waves under the radiation condition. *Q. J. R. Meteorol. Soc.* **49**, 316–319.

Hayashi, Y. (1971c). Large-scale equatorial waves destabilized by convective heating in the presence of surface friction. *Q. J. R. Meteorol. Soc.* **49**, 450–457.

Heymsfield, A. J., and M. R. Hjelmfelt (1981). Dynamical and microphysical observations in two Oklahoma squall lines, II. In situ measurements. *Prepr., Conf. Radar Meteorol., 20th*, pp. 60–65. Am. Meteorol. Soc., Boston, Massachusetts.

Heymsfield, A. J., and D. J. Musil (1982). Case study of a hailstorm in Colorado. Part II: Particle growth processes in mid-levels deduced from *in-situ* measurements. *J. Atmos. Sci.* **39**, 2847–2866.

Heymsfield, A. J., A. R. Jameson, and H. W. Frank (1980). Hail growth mechanisms in a Colorado storm. Part II: Hail formation processes. *J. Atmos. Sci.* **37**, 1779–1807.

Hoecker, W. H. (1960a). Windspeed and air flow patterns in the Dallas tornado of April 2, 1967. *Mon. Weather Rev.* **88**, 167–180.

Hoecker, W. H. (1960b). The dimensional and rotational characteristics of the tornadoes and their parent cloud systems. USWB Res. Pap. No. 41, pp. 53–112. Washington, D.C.

Holle, R. L., and M. Maier (1980). Tornado formation from downdraft interaction in the FACE mesonetwork. *Mon. Weather Rev.* **108**, 1010–1028.

Hookings, G. A. (1965). Precipitation maintained downdrafts. *J. Appl. Meteorol.* **4**, 190–195.

Houze, R. A., Jr. (1977). Structure and dynamics of a tropical squall-line system. *Mon. Weather Rev.* **105**, 1541–1567.

Hoxit, L. R., C. F. Chappell, and J. M. Fritsch (1976). Formation of mesolows or pressure troughs in advance of cumulonimbus clouds. *Mon. Weather Rev.* **104**, 1419–1428.

Hoxit, L. R., R. A. Maddox, C. F. Chappell, F. L. Zuckerberg, H. M. Mogil, I. Jones, D. R. Greene, R. E. Saffle, and R. A. Scofield (1978). Meteorological analysis of the Johnstown, Pennsylvania, flash flood, 19–20 July 1977. NOAA Tech. Rep. ERL 401-APCL 43.

Humphreys, W. J. (1928). The uprush of air necessary to sustain the hailstone. *Mon. Weather Rev.* **56**, 314.

Illingworth, A. J., and J. Latham (1977). Calculations of electric field structure and charge distributions in thunderstorms. *Q. J. R. Meteorol. Soc.* **103**, 281–295.

Jameson, A. R., and A. J. Heymsfield (1980). Hail growth mechanisms in a Colorado storm. Part I: Dual-wavelength radar observations. *J. Atmos. Sci.* **37**, 1763–1778.

Jayaratne, E. R., C. P. R. Saunders, and J. Hallett (1983). Laboratory studies of the charging of soft-hail during ice crystal interactions. *Q. J. R. Meteorol. Soc.* **109**, 609–630.

Johns, R. H., and W. D. Hirt (1983). The derecho—A severe weather producing convective system. *Prepr., Conf. Severe Local Storms, 13th*, pp. 178–181. Am. Meteorol. Soc., Boston, Massachusetts.

Johnson, R. H., and M. Nicholls (1983). A compositive analysis of the boundary layer accompanying a tropical squall line. *Mon. Weather Rev.* **111**, 308–319.

Jorgensen, D. P. (1984). Mesoscale and convective-scale characteristics of mature hurricanes. Ph.D. Thesis, Colorado State Univ.

Jorgensen, D. P., E. J. Zipser, and M. A. Le Mone (1985). Vertical motions in intense hurricanes. *J. Atmos. Sci.* **42**, 839–856.

Kamburova, P. L., and F. H. Ludlam (1966). Rainfall evaporation in thunderstorm downdrafts. *Q. J. R. Meteorol. Soc.* **92**, 510–518.

Klemp, J. B. (1987). Dynamics of tornadic thunderstorms. *Annu. Rev. Fluid Mech.* **19**, 369–402.

Klemp, J. B., and R. Rotunno (1983). A study of the tornadic region within a supercell thunderstorm. *J. Atmos. Sci.* **40**, 359–377.

Klemp, J. B., and R. B. Wilhelmson (1978a). The simulation of three-dimensional convective storm dynamics. *J. Atmos. Sci.* **35**, 1070–1096.

Klemp, J. B., and R. B. Wilhelmson (1978b). Simulations of right- and left-moving storms produced through storm splitting. *J. Atmos. Sci.* **35**, 1097–1110.

Klemp, J. B., R. B. Wilhelmson, and P. Ray (1981). Observed and numerically simulated structure of a mature supercell thunderstorm. *J. Atmos. Sci.* **38**, 1558–1580.

Kloth, C. M., and R. P. Davies-Jones (1980). The relationship of the 300-mb jet stream to tornado occurrence. NOAA Tech. Memo. ERL NSSL-88, Natl. Severe Storm Lab., Norman, Oklahoma.

Knight, C. A. (1981). Case studies on convective storms, Case study 9: 13 June 1974: Mature storm study. A small, isolated, "steady state" convective storm. NCAR Tech. Note NCAR/TN-163+STR.

Knight, C. A., N. C. Knight, J. E. Dye, and V. Toutenhoofd (1974). The mechanism of precipitation formation in Northeastern Colorado cumulus. I. Observations of the precipitation itself. *J. Atmos. Sci.* **31**, 2142–2147.

Knupp, K. R. (1985). Precipitation convective downdraft structure: A synthesis of observations and modeling. Ph.D. Thesis, Dep. Atmos. Sci., Colorado State Univ.

Knupp, K. R., and W. R. Cotton (1982a). An intense, quasi-steady thunderstorm over mountainous terrain. Part II: Doppler radar observations of the storm morphological structure. *J. Atmos. Sci.* **39**, 343–358.

Knupp, K. R., and W. R. Cotton (1982b). An intense, quasi-steady thunderstorm over mountainous terrain—Part III: Doppler radar observations of the turbulence structure. *J. Atmos. Sci.* **39**, 359–368.

Knupp, K. R., and W. R. Cotton (1985). Convective cloud downdraft structure: An interpretive study. *Rev. Geophys. Space Phys.* **23**, 183–215.

Koscielski, A. (1965). 110 tornado forecasts and reasons why they did or did not verify. Unpublished manuscript, U.S. Weather Bur. (NSSFC, Rm. 1728, Federal Bldg., 601 E. 12th St., Kansas City, Missouri 64106).

Krauss, T. W., and J. D. Marwitz (1984). Precipitation processes within an Alberta supercell hailstorm. *J. Atmos. Sci.* **41**, 1025–1034.

Krehbiel, P. R., M. Brook, R. L. Lhermitte, and C. L. Lennon (1983). Lightning charge structure in thunderstorms. *In* "Proceedings in Atmospheric Electricity" (Lothar H. Ruhnke and John Latham, eds.), pp. 408–410. A. Deepak Publ., Hampton, Virginia.

Kropfli, R. A., and L. J. Miller (1976). Kinematic structure and flux quantities in a convective storm from dual-Doppler radar observation. *J. Atmos. Sci.* **33**, 520–529.

Krumm, W. R. (1954). On the cause of downdrafts from dry thunderstorms over the plateau area of the United States. *Bull. Am. Meteorol. Soc.* **35**, 122–126.

Kuo, H. L. (1965). On the formation and intensification of tropical cyclones through latent heat release by cumulus convection. *J. Atmos. Sci.* **22**, 40–63.

Latham, J. (1969). Experimental studies of the effect of electric fields on the growth of cloud particles. *Q. J. R. Meteorol. Soc.* **95**, 349–361.

Latham, J. (1981). The electrification of thunderstorms. *Q. J. R. Meteorol. Soc.* **107**, 277–298.

Latham, J., and B. J. Mason (1961). Generation of electric charge associated with the formation of soft hail in thunderclouds. *Proc. R. Soc. London, Ser. A* **260**, 537–549.

Latham, J., and B. J. Mason (1962). Electrical charging of hail pellets in a polarizing field. *Proc. R. Soc. London, Ser. A* **266**, 387–401.

Leary, C. A., and R. A. Houze, Jr. (1979a). The structure and evolution of convection in a tropical cloud cluster. *J. Atmos. Sci.* **36**, 437–457.

Leary, C. A., and R. A. Houze, Jr. (1979b). Melting and evaporation of hydrometers in precipitation from the anvil clouds of deep tropical convection. *J. Atmos. Sci.* **36**, 669–679.

Lemon, L. R. (1976). The flanking line, a severe thunderstorm intensification source. *J. Atmos. Sci.* **33**, 686–694.

Lemon, L. R., and C. A. Doswell, III (1979). Severe thunderstorm evolution and mesocyclone structure as related to tornadogenesis. *Mon. Weather Rev.* **107**, 1184–1197.

Lemon, L. R., D. W. Burgess, and R. A. Brown (1975). Tornado production and storm sustenance. *Prepr., Conf. Severe Local Storms, 9th, Norman, Okla.* pp. 100–104. Am. Meteorol. Soc., Boston, Massachusetts.

LeMone, M. A., and E. J. Zipser (1980). Cumulonimbus vertical velocity events in GATE. Part I: Diameter, intensity and mass flux. *J. Atmos. Sci.* **37**, 2444–2457.

Levich, E., and E. Tzvetkov (1984). Helical cyclogenesis. *Phys. Lett.* **100A**, 53–56.

Levin, Z. (1976). A refined charge distribution in a stochastic electrical model of an infinite cloud. *J. Atmos. Sci.* **33**, 1756–1762.

Levin, Z., and A. Ziv (1974). The electrification of thunderclouds and the rain gush. *J. Geophys. Res.* **79**, 2699–2704.

Levy, G., and W. R. Cotton (1985). A numerical investigation of mechanisms linking glaciation of the ice-phase to the boundary layer. *J. Clim. Appl. Meteorol.* **23**, 1505–1519.

Lewellen, W. S. (1976). Theoretical models of the tornado vortex. *In* "Proceedings of Symposium on Tornados" (R. E. Peterson, ed.), pp. 107–143. Texas Tech. Univ., Lubbock.

Lhermitte, R., and P. R. Krehbiel (1979). Doppler radar and radio observations of thunderstorms. *IEEE Trans. Geosci. Electron.* **GE-17**, 162–171.

Lhermitte, R., and E. Williams (1985). Thunderstorm electrification: A case study. *J. Geophys. Res.* **90**, 6071–6078.

Lilly, D. K. (1986a). The structure, energetics and propagation of rotating convective storms. Part I: Energy exchange with the mean flow. *J. Atmos. Sci.* **43**, 113–125.

Lilly, D. K. (1986b). The structure, energetics and propagation of rotating convective storms. Part II: Helicity and storm stabilization. *J. Atmos. Sci.* **43**, 126–140.

Lindblad, N. R., and R. G. Semonin (1963). Collision efficiency of cloud droplets in electric fields. *J. Geophys. Res.* **68**, 1051–1057.

Lindzen, R. S. (1974). Wave-CISK in the tropics. *J. Atmos. Sci.* **31**, 156–179.

List, R. (1982). Properties and growth of hailstones. *In* "Thunderstorms: A Social, Scientific, and Technological Documentary, Vol. 2, Thunderstorm Morphology and Dynamics," pp. 409–445. U.S. Dep. Commer., Washington, D.C.

Ludlam, F. H. (1963). Severe local storms: A review. *Severe Local Storms, Meteorol. Monogr.* No. 27, 1–30.

MacDonald, A. E. (1976). Gusty surface winds and high level thunderstorms. Natl. Weather Serv. West. Region Tech. Attachment No. 76–14.

MacGorman, D. R. (1978). Lightning location in a storm with strong wind shear. Ph.D. Thesis, Rice Univ., Houston, Texas.

MacGorman, D. R., W. L. Taylor, and A. A. Few (1980). Lightning location from acoustic and VHF techniques relative to storm structure from 10-cm radar. *Int. Conf. Atmos. Electr., 6th, Manchester, Engl.* Sess. XI-15. Am. Meteorol. Soc., Boston, Massachusetts.

Maddox, R. A. (1976). An evaluation of tornado proximity wind and stability data. *Mon. Weather Rev.* **104**, 133–142.

Maddox, R. A. (1980). Mesoscale convective complexes. *Bull. Am. Meteorol. Soc.* **61**, 1374–1387.

Maddox, R. A., and C. A. Doswell (1982). An examination of jet stream configurations, 500 mb vorticity advection and low-level thermal advection patterns during extended periods of intense convection. *Mon. Weather Rev.* **110**, 184–197.

Maddox, R. A., C. F. Chappell, L. R. Hoxit, and F. Caracena (1978). Comparison of meteorological aspects of the Big Thompson and Rapid City flash floods. *Mon. Weather Rev.* **106**, 375–389.

Maddox, R. A., C. F. Chappell, and L. R. Hoxit (1979). Synoptic and meso-α scale aspects of flash flood events. *Bull. Am. Meteorol. Soc.* **60**, 115–123.

Mal, S., and N. Desai (1938). The mechanism of thundery conditions at Karachi. *Q. J. R. Meteorol. Soc.* **64**, 525–537.

Malkin, W., and J. G. Galway (1953). Tornadoes associated with hurricanes. *Mon. Weather Rev.* **81**, 299–303.

Marroquin, A., and D. J. Raymond (1982). A linearized convective overturning model for prediction of thunderstorm movement. *J. Atmos. Sci.* **39**, 146–151.

Marshall, B. J. P., J. Latham, and C. P. R. Saunders (1978). A laboratory study of charge transfer accompanying the collision of ice crystals with a simulated hailstone. *Q. J. R. Meteorol. Soc.* **104**, 163–178.

Marwitz, J. D. (1972a). The structure and motion of severe hailstorms. Part I: Supercell storms. *J. Appl. Meteorol.* **11**, 166–179.

Marwitz, J. D. (1972b). The structure and motion of severe hailstorms. Part II: Multicell storms. *J. Appl. Meteorol.* **11**, 180–188.

Marwitz, J. D. (1972c). The structure and motion of severe hailstorms. Part III: Severely sheared storms. *J. Appl. Meteorol.* **11**, 189–201.

Marwitz, J. D. (1972d). Precipitation efficiency of thunderstorms on the high plains. *J. Rech. Atmos.* **6**, 367–370.

Marwitz, J. D. (1973). Trajectories within the weak echo regions of hailstorms. *J. Appl. Meteorol.* **12**, 1174–1182.

Mason, B. J. (1968). The generation of electric charges and fields in precipitating clouds. *Proc. Int. Conf. Cloud Phys., Toronto* pp. 657–662. Am. Meteorol. Soc., Boston, Massachusetts.

Mason, B. J. (1971). "The Physics of Clouds," 2nd ed. Oxford Univ. Press (Clarendon), London.

Mason, B. J. (1976). In reply to a critique of precipitation theories of thunderstorm electrification by C. B. Moore. *Q. J. R. Meteorol. Soc.* **102**, 219–225.

Matthews, D. A. (1981). Observations of a cloud are triggered by thunderstorm outflow. *Mon. Weather Rev.* **109**, 2140–2157.

Maxwell, J. B. (1974). Unpublished LMA diagnostic results. Atmos. Environ. Serv., Toronto.

McAnelly, R. A., and W. R. Cotton (1985). The precipitation lifecycle of mesoscale convective complexes. *Prepr., Conf. Hydrometeorol., 6th, Indianapolis, Indiana* pp. 197–204. Am. Meteorol. Soc., Boston, Massachusetts.

Miller, L. J., J. D. Tuttle, and C. K. Knight (1988). Airflow and hail growth in a severe northern High Plains supercell. *J. Atmos. Sci.* **45**, 736–762.

Miller, M. J. (1978). The Hampstead storm: A numerical simulation of a quasi-stationary cumulonimbus system. *Q. J. R. Meteorol. Soc.* **104**, 413–427.

Miller, M. J., and M. W. Moncrieff (1983). Dynamics and simulation of organized deep convection. *Proc. NATO Adv. Study Inst. Mesoscale Meteorol.—Theor., Obs. Models, Bonas, Fr., 1982* pp. 451–496. Reidel, Dordrecht, Netherlands.

Miller, M. J., and R. P. Pearce (1974). A three-dimensional primitive equation model of cumulonimbus convection. *Q. J. R. Meteorol. Soc.* **100**, 133–154.

Miller, R. C. (1967). Notes on analysis and severe storm forecasting procedures of the Military Weather Warning Center. Air Weather Serv. Tech. Rep. No. 200, Scott AFB, Illinois.

Miller, R. C. (1972). Notes on analysis and severe storm forecasting procedures of the Air Force Global Weather Central. Air Weather Serv. Tech. Rep. No. 200, Scott AFB, Illinois.

Miller, R. C., and J. A. McGinley (1977). Response to "Inherent Difficulties in Hail Probability Prediction" and "Forecasting Hailfall in Alberta." *Meteorol. Monogr.* **38**, 153–154.

Mitchell, K. E., and J. B. Hovermale (1977). A numerical investigation of the severe thunderstorm gust front. *Mon. Weather Rev.* **105**, 657–675.

Modahl, A. C. (1969). The influence of vertical wind shear on hailstorm development and structure. Pap. No. 137, Dep. Atmos. Sci., Colorado State Univ.

Moncrieff, M. W. (1978). The dynamical structure of two-dimensional steady convection in constant vertical shear. *Q. J. R. Meteorol. Soc.* **104**, 543–567.

Moncrieff, M. W. (1981). A theory of organized steady convection and its transport properties. *Q. J. R. Meteorol. Soc.* **107**, 29–50.

Moncrieff, M. W., and J. S. A. Green (1972). The propagation and transfer properties of steady convective overturning in shear. *Q. J. R. Meteorol. Soc.* **98**, 336–352.

Moncrieff, M. W., and M. J. Miller (1976). The dynamics and simulation of tropical cumulonimbus and squall lines. *Q. J. R. Meteorol. Soc.* **102**, 373–394.

Moore, C. B. (1976). Reply (to B. J. Mason). *Q. J. R. Meteorol. Soc.* **102**, 225–240.

Moore, C. B., B. Vonnegut, E. A. Vrablik, and D. A. McCraig (1964). Gushes of rain and hail after lightning. *J. Atmos. Sci.* **21**, 646.

Morgan, G. M., Jr. (1970). An examination of the wet bulb zero as a hail forecasting parameter in the Po Valley, Italy. *J. Appl. Meteorol.* **9**, 537–540.

Morton, B. R. (1966). Geophysical vortices. *Prog. Aeronaut. Sci.* **7**, 145–193.

Morton, B. R. (1970). The physics of firewhirls. *Fire Res. Abstr. Rev.* **12**, 1–19.

Mueller, C. K., and P. H. Hildebrand (1983). The structure of a microburst: As observed by ground-based and airborne Doppler radar. *Prepr., Conf. Radar Meteorol., 21st*, pp. 602–608. Am. Meteorol. Soc., Boston, Massachusetts.

Muller-Hillebrande, D. (1954). Charge generation in thunderstorms by collisions of ice crystals with graupel falling through a vertical field. *Tellus* **6**, 367–381.

Muller-Hillebrande, D. (1955). Zur Frage des Ursprunges der Gewitterelektrizitat. *Ark. Geofys.* **2**, 395.

Musil, D. J., W. R. Sand, and R. A. Schleusener (1973). Analysis of data from T-28 aircraft penetrations of a Colorado hailstorm. *J. Appl. Meteorol.* **12**, 1364–1370.

Musil, D. J., E. L. May, P. L. Smith, and W. R. Sand (1976). Structure of an evolving hailstorm, IV. Internal structure from penetrating aircraft. *Mon. Weather Rev.* **104**, 596–602.

Musil, D. J., P. L. Smith, J. R. Miller, J. H. Killinger, and J. L. Halvorson (1977). Characteristics of vertical volocities observed in T-28 penetrations of hailstorms. *Prepr., Conf. Weather Modif., 6th*, pp. 161–169. Am. Meteorol. Soc., Boston, Massachusetts.

Musil, D. J., A. J. Heymsfield, and P. L. Smith (1982). Characteristics of the weak echo region in an intense High Plains thunderstorm as determined by penetrating aircraft. *Prepr., Conf. Cloud Phys.*, pp. 535–538. Am. Meteorol. Soc., Boston, Massachusetts.

Musil, D. J., A. J. Heymsfield, and P. L. Smith (1986). Microphysical characteristics of a well-developed weak echo region in a High Plains supercell thunderstorm. *J. Clim. Appl. Meteorol.* **25**, 1037–1051.

National Research Council (1983). "Low-Altitude Wind Shear and its Hazard to Aviation." Natl. Acad. Press, Washington, D.C.

Nehrkorn, Thomas (1985). Wave-CISK in a baroclinic basic state. Ph.D. Thesis, Mass. Inst. Technol.

Nelson, S. P. (1977). Rear flank downdraft: A hailstorm intensification mechanism. *Prepr., Conf. Severe Local Storms, 10th, Am. Meteorol. Soc., Boston, Mass.* pp. 521–525.

Nelson, S. P. (1983). The influence of storm flow structure on hail growth. *J. Atmos. Sci.* **40**, 1965–1983.

Newton, C. W. (1950). Structure and mechanism of the prefrontal squall line. *J. Meteorol.* **7**, 210–222.

Newton, C. W. (1963). Dynamics of severe convective storms. *Meteorol. Monogr.* No. 27, 33–58.

Newton, C. W. (1966). Circulations in large sheared cumulonimbus. *Tellus* **18**, 699–712.

Newton, C. W., and H. R. Newton (1959). Dynamical interactions between large convective clouds and environment with vertical shear. *J. Meteorol.* **16**, 483–496.

Newton, C. W., R. C. Miller, E. R. Fosse, D. R. Booker, and P. McManamon (1978). Severe thunderstorms: Their nature and their effects on society. *Interdiscip. Sci. Rev.* **3**, 71–85.

Nicholls, M. (1987). A numerical investigation of tropical squall lines. Ph.D. Thesis, Dep. Atmos. Sci., Colorado State Univ.

Normand, C. W. B. (1946). Energy in the atmosphere. *Q. J. R. Meteorol. Soc.* **72**, 145–167.

Novlan, D. J., and W. M. Gray (1974). Hurricane-spawned tornadoes. *Mon. Weather Rev.* **102**, 476–488.

Ogura, Y., and M. T. Liou (1980). The structure of a mid-latitude squall line: A case study. *J. Atmos. Sci.* **37**, 553–567.

Ooyama, K. (1964). A dynamical model for the study of tropical cyclone development. *Geofis. Int.* **4**, 187–198.

Orton, R. (1970). Tornadoes associated with Hurricane Beulah on September 19–23, 1967. *Mon. Weather Rev.* **98**, 541–547.

Orville, H. D. (1978). A review of hailstone–hailstorm numerical simulations. *Meteorol. Monogr.* No. 38, 49–61.

Pakiam, J. E., and J. Maybank (1975). The electrical characteristics of some severe hailstorms in Alberta, Canada. *J. Meteorol. Soc. Jpn.* **53**, 363–383.

Paluch, I. R. (1978). Size sorting of hail in an three-dimensional updraft and implications for hail suppression. *J. Appl. Meteorol.* **17**, 763–777.

Plumlee, H. R., and R. G. Semonin (1965). Cloud droplet collision efficiency in electric fields. *Tellus* **17**, 356–363.

Pruppacher, H. R., and J. D. Klett (1980). "Microphysics of Clouds and Precipitation." Reidel, Dordrecht, Netherlands.

Purdom, J. F. W. (1976). Some uses of high-resolution GOES imagery in the mesoscale forecasting of convection and its behavior. *Mon. Weather Rev.* **104**, 1474–1483.

Purdom, J. F. W. (1979). The development and evolution of deep convection. *Prepr., Conf. Severe Local Storms, 11th, Kansas City, Mo.* pp. 143–150. Am. Meteorol. Soc., Boston, Massachusetts.

Purdom, J. F. W. (1982). Subjective interpretation of geostationary satellite data for nowcasting. *In* "Nowcasting" (Keith Browning, ed.), pp. 149–156. Academic Press, New York.

Purdom, J. F. W., and K. Marcus (1982). Thunderstorm triggered mechanisms over the Southeast United States. *Prepr., Conf. Severe Local Storms, 12th, San Antonio, Tex.* pp. 487–488. Am. Meteorol. Soc., Boston, Massachusetts.

Rawlins, F. (1982). A numerical study of thunderstorm electrification using a three-dimensional model incorporating the ice phase. *Q. J. R. Meteorol. Soc.* **108**, 779–800.

Ray, P. S., B. C. Johnson, K. W. Johnson, J. S. Bradberry, J. J. Stephens, K. K. Wagner, R. B. Wilhelmson, and J. B. Klemp (1981). The morphology of several tornadic storms 20 May 1977. *J. Atmos. Sci.* **38**, 1643–1663.

Ray, P. S., D. R. MacGorman, and W. D. Rust (1987). Lightning location relative to storm structure in a supercell storm and a multicell storm. *J. Geophys. Res.* **92**, 5713–5724.

Raymond, D. J. (1975). A model for predicting the movement of continuously propagating convective storms. *J. Atmos. Sci.* **32**, 1308–1317.

Raymond, D. J. (1976). Wave-CISK and convective mesosystems. *J. Atmos. Sci.* **33**, 2392–2398.

Raymond, D. J. (1983). A wave-CISK in mass flux form. *J. Atmos. Sci.* **40**, 2561–2572.

Raymond, D. J. (1984). A wave-CISK model of squall lines. *J. Atmos. Sci.* **40**, 1946–1958.

Renick, J. H. (1971). Radar reflectivity profiles of individual cells in a persistent multicellular Alberta hailstorm. *Prepr., Conf. Severe Local Storms, 7th, Kansas City, Mo.* pp. 63–70. Am. Meteorol. Soc., Boston, Massachusetts.

Renick, J. H., and J. B. Maxwell (1977). Forecasting hailfall in Alberta. *Meteorol. Monogr.* **38**, 145–151.

Reynolds, S. E., M. Brook, and M. F. Gourley (1957). Thunderstorm charge separation. *J. Meteorol.* **14**, 426–436.

Riehl, H., J. Badner, J. E. Hovde, N. E. LaSeur, L. L. Means, W. C. Palmer, M. J. Schroeder, and L. W. Snellman (1952). Forecasting in the middle latitudes. *Meteorol. Monogr.* **5**.

Rodi, A. R., K. L. Elmore, and W. P. Mahoney (1983). Aircraft and Doppler air motion comparisons in a JAWS microburst. *Prepr., Conf. Radar Meteorol., 21st*, pp. 624–629. Am. Meteorol. Soc., Boston, Massachusetts.

Rotunno, R. (1980). Vorticity dynamics of convective swirling boundary layer. *J. Fluid Mech.* **97**, 623–640.

Rotunno, R. (1981). On the evolution of thunderstorm rotation. *Mon. Weather Rev.* **109**, 577–586.

Rotunno, R., and J. B. Klemp (1982). The influence of shear-induced pressure gradients on thunderstorm motion. *Mon. Weather Rev.* **110**, 136–151.

Rotunno, R., and J. B. Klemp (1985). On the rotation and propagation of simulated supercell thunderstorms. *J. Atmos. Sci.* **42**, 271–292.

Rotunno, R., and D. K. Lilly (1981). A numerical model pertaining to the multiple vortex phenomenon. Contract Rep. NUREG/CR-1840, U.S. Nucl. Regul. Comm., Washington, D.C.

Rust, W. D., W. L. Taylor, and D. R. MacGorman (1981). Research on electrical properties of severe thunderstorms in the Great Plains. *Bull. Am. Meteorol. Soc.* **62**, 1286–1293.

Sand, W. R. (1976). Observations in hailstorms using the T-28 aircraft system. *J. Appl. Meteorol.* **15**, 641–650.

Sartor, J. D., and J. S. Miller (1965). Relative cloud droplet trajectory computation. *Proc. Int. Cloud Phys. Conf.*, pp. 108–112. Meteorol. Soc. Japan, Tokyo/Sapporo.

Sartor, J. D. (1961). Calculations of cloud electrification based on a general charge separation mechanism. *J. Geophys. Res.* **66**, 831–843.

Saunders, C. P. R., and N. M. A. Wahab (1975). Influence of electric fields on the aggregation of ice crystals. *J. Meteorol. Soc. Jpn.* **53**, 121–126.

Schaeffer, J. T. (1986). Severe thunderstorm forecasting: A historical perspective. *Weather Forecasting* **1**, 164–189.

Schlamp, R. J., S. N. Grover, H. R. Pruppacher, and A. E. Hamielec (1976). A numerical investigation of the effect of electric charges and vertical external fields on the collision efficiency of cloud drops. *J. Atmos. Sci.* **33**, 1747–1755.

Schlamp, R. J., S. N. Grover, H. R. Pruppacher, and A. E. Hamielec (1979). A numerical investigation of the effect of electric charges and vertical external electric fields on the collision efficiency of cloud drops: Part II. *J. Atmos. Sci.* **36**, 339–349.

Schlesinger, R. E. (1975). A three-dimensional numerical model of an isolated deep convective cloud: Preliminary results. *J. Atmos. Sci.* **32**, 934–957.

Schlesinger, R. E. (1978). A three-dimensional numerical model of an isolated thunderstorm. Part I. Comparative experiments for variable ambient wind shear. *J. Atmos. Sci.* **35**, 690–713.

Schlesinger, R. E. (1980). A three-dimensional numerical model of an isolated thunderstorm. Part II: Dynamics of updraft splitting and mesovortex couplet evolution. *J. Atmos. Sci.* **37**, 395–420.

Schleusener, R. A. (1962). On the relation of the latitude and strength of the 500 millibar west wind along 110 degrees west longitude and the occurrence of hail in the lee of the Rocky Mountains. Atmos. Sci. Tech. Pap. No. 26, Civ. Eng. Sect., Colorado State Univ.

Schneider, E. K., and R. S. Lindzen (1976). A discussion of the parameterization of momentum exchange by cumulus convection. *J. Geophys. Res.* **81**, 3158–3161.

Schonland, B. F. J. (1950). "The Flight of Thunderbolts," pp. 150–151. Oxford Univ. Press, New York.

Scott, W. D., and Z. Levin (1975). A stochastic electric model of an infinite cloud: Charge generation and precipitation development. *J. Atmos. Sci.* **32**, 1814–1828.

Seitter, K. L. (1987). A numerical study of atmospheric density current motion including the effects of condensation. *J. Atmos. Sci.* **43**, 3068–3076.

Semonin, R. G., and H. R. Plumlee (1966). Collision efficiency of charged cloud droplets in electric fields. *J. Geophys. Res.* **71**, 4271–4278.

Showalter, A. K. (1953). Synoptic conditions associated with tornadoes. U.S. Weather Bureau Hydrometeorol. Sect. Report.

Silva-Dias, M. F., A. K. Betts, and D. E. Stevens (1984). A linear spectral model of tropical mesoscale systems: Sensitivity studies. *J. Atmos. Sci.* **41**, 1704–1716.

Simpson, J. E. (1969). A comparison between laboratory and atmospheric density currents. *Q. J. R. Meteorol. Soc.* **95**, 758–765.

Simpson, J. E., and R. E. Britter (1980). A laboratory model of an atmospheric mesofront. *Q. J. R. Meteorol. Soc.* **106**, 485–500.

Simpson, J., and V. Wiggert (1969). 1968 Florida seeding experiment: Numerical mode results. *Mon. Weather Rev.* **97**, 471–489.

Simpson, J., R. H. Simpson, D. A. Andrews, and M. A. Eaton (1965). Experimental cumulus dynamics. *Rev. Geophys.* **3**, 387–431.

Sinclair, P. C. (1973). Severe storm velocity and temperature structure deduced from penetrating aircraft. *Prepr., Conf. Severe Local Storms, 8th*, pp. 25–31. Am. Meteorol. Soc., Boston, Massachusetts.

Sinclair, P. C. (1979). Velocity and temperature structure near and within severe storms. *Conf. Severe Local Storms, 11th, Colorado State Univ.*

Sinclair, P. C., and J. F. W. Purdom (1983). The genesis and development of deep convective storms. Final Rep. for NOAA Grant NA80AA-D-00056.

Smith, J. S. (1965). The hurricane-tornado. *Mon. Weather Rev.* **93**, 453–459.

Snow, J. T. (1982). A review of recent advances in tornado vortex dynamics. *Rev. Geophys. Space Phys.* **20**, 953–964.

Stark, T. E. (1976). Wave-CISK and cumulus parameterization. *J. Atmos. Sci.* **33**, 2383–2391.

Steiner, R., and R. H. Rhyne (1962). Some measured characteristics of severe storm turbulence. U.S. Weather Bur., Natl. Severe Storms Proj. Rep. No. 10.

Stevens, D. E., and R. S. Lindzen (1978). Tropical wave-CISK with a moisture budget and cumulus friction. *J. Atmos. Sci.* **35**, 940–961.

Stevens, D. E., R. S. Lindzen, and L. J. Shapiro (1977). A new model of tropical waves incorporating momentum mixing by cumulus convection. *Dyn. Atmos. Oceans* **1**, 365–425.

Stout, G. E., and S. A. Changnon, Jr. (1968). Climatology of hail in the central United States. CHIAA Res. Rep. No. 38, prepared for Crop Hail Insurance Actuarial Association.

Strauch, R. G., and F. H. Merrem (1976). Structure of an evolving hailstorm. Part III: Internal structure from Doppler radar. *Mon. Weather Rev.* **104**, 588–595.

Sulakvelidze, G. K., N. S. Bibilashvili, and V. F. Lapcheva (1967). "Formation of Precipitation and Modification of Hail Processes." Isr. Program Sci. Transl., Jerusalem.

Summers, P. W. (1972). Project Hailstop: A review of accomplishments to date. *In* "Alberta Hail Studies 1972," Hail Stud. Rep. 72-2, pp. 47–53. Res. Counc. Alberta.

Takahashi, T. (1978). Riming electrification as a charge generation mechanism in thunderstorms. *J. Atmos. Sci.* **35**, 1536–1548.

Taylor, W. L. (1980). Lightning location and progression using VHF space-time mapping technique. *Int. Conf. Atmos. Electr., 6th, Manchester, Engl.* Sess. V-1. Am. Meteorol. Soc., Boston, Massachusetts.

Thorpe, A. J., and M. J. Miller (1978). Numerical simulations showing the role of downdraft in cumulonimbus motion and splitting. *Q. J. R. Meteorol. Soc.* **104**, 873–893.

Thorpe, A. J., M. J. Miller, and M. W. Moncrieff (1980). Dynamical models of two-dimensional downdrafts. *Q. J. R. Meteorol. Soc.* **106**, 463–484.

Thorpe, A. J., M. J. Miller, and M. W. Moncrieff (1982). Two-dimensional convection in non-constant shear: A model of mid-latitude squall lines. *Q. J. R. Meteorol. Soc.* **108**, 739–762.

Tripoli, G. J., and W. R. Cotton (1982). The Colorado State University three-dimensional cloud/mesoscale model—1982. Part I: General theoretical framework and sensitivity experiments. *J. Rech. Atmos.* **16**, 185–220.

Tripoli, G. J., and W. R. Cotton (1986). An intense quasi-steady thunderstorm over mountainous terrain. Part IV: Three-dimensional numerical simulation. *J. Atmos. Sci.* **43**, 894–912.

Uccellini, L. W., and D. R. Johnson (1979). The coupling of upper and lower tropospheric jet streaks and implications for the development of severe convective storms. *Mon. Weather Rev.* **107**, 682–703.

Vasiloff, S. V., E. A. Brandes, and R. P. Davies-Jones (1986). An investigation of the transition from multicell to supercell storms. *J. Clim. Appl. Meteorol.* **25**, 1022–1036.

Vaughan, D. H., and B. Vonnegut (1976). Luminous electrical phenomena associated with nocturnal tornadoes in Huntsville, Alabama, 3 April 1974. *Bull. Am. Meteorol. Soc.* **57**, 1220–1224.

Vonnegut, B. (1955). Possible mechanism for the formation of thunderstorm electricity. *In* "Conference on Atmospheric Electricity," Res. Pap. No. 42, pp. 169–181. Geophys. Res. Dir., Air Force Cambridge Res. Cent., Bedford, Massachusetts.

Vonnegut, B. (1960). Electrical theory of tornadoes. *J. Geophys. Res.* **65**, 203–212.

Vonnegut, B. (1963). Some facts and speculations concerning the origin and role of thunderstorm electricity. *Meteorol. Monogr.* **5**, 224–241.

Vonnegut, B. (1975). Comment on "On the relation of electrical activity to tornadoes" by R. P. Davies-Jones and J. H. Golden. *J. Geophys. Res.* **80**, 4559–4560.

Wade, C. G., and G. B. Foote (1982). The 22 July 1976 case study: Low-level airflow and mesoscale influences. *In* "Hailstorms of the Central High Plains" (C. A. Knight and P. Squires, eds.), Vol. 2, pp. 115–130. Colorado Assoc. Univ. Press, Boulder.

Wagner, P. B., and J. W. Telford (1981). Cloud dynamics and an electric charge separation mechanism in convective clouds. *J. Rech. Atmos.* **15**, 97–120.

Wakimoto, R. M. (1982). The life cycle of thunderstorm gust fronts as viewed with Doppler radar and rawinsonde data. *Mon. Weather Rev.* **110**, 1050–1082.

Walko, R., and R. Gall (1984). A two-dimensional linear stability analysis of the multiple vortex phenomenon. *J. Atmos. Sci.* **41**, 3456–3471.

Ward, N. B. (1972). The exploration of certain features of tornado dynamics using a laboratory model. *J. Atmos. Sci.* **29**, 1194–1204.

Watkins, D. C., J. D. Cobine, and B. Vonnegut (1978). Electric discharges inside tornadoes. *Science* **199**, 171–174.

Weaver, J. F., and S. P. Nelson (1982). Multiscale aspects of thunderstorm gust fronts and their effects on subsequent storm development. *Mon. Weather Rev.* **110**, 707–718.

Weisman, M., and J. Klemp (1982). The dependence of numerically simulated convective storms on vertical wind shear and buoyancy. *Mon. Weather Rev.* **110**, 504–520.

Weisman, M. L., and J. B. Klemp (1984). The structure and classification of numerically simulated convective storms in directionally varying wind shears. *Mon. Weather Rev.* **112**, 2479–2498.

Wetzel, P. J., W. R. Cotton, and R. L. McAnelly (1982). The dynamic structure of the mesoscale convective complex—Some case studies. *Conf. Severe Local Storms, 12th*, pp. 265–268. Am. Meteorol. Soc., Boston, Massachusetts.

Wetzel, P. J., W. R. Cotton, and R. L. McAnelly (1983). A long-lived mesoscale convective complex. Part II: Morphology of the mature complex. *Mon. Weather Rev.* **111**, 1919–1937.

Wiggert, V., R. I. Sax, and R. L. Holle (1982). On the modification potential of Illinois summertime convective clouds, with comparison to Florida and FACE observations. *J. Appl. Meteorol.* **21**, 1293–1322.

Wilhelmson, R. B., and C. S. Chen (1982). A simulation of the development of successive cells along a cold outflow boundary. *J. Atmos. Sci.* **39**, 1466–1483.

Wilhelmson, R. B., and J. Klemp (1978). A numerical study of storm splitting that leads to long-lived storms. *J. Atmos. Sci.* **35**, 1974–1986.

Wilhelmson, R. B., and J. Klemp (1981). A three-dimensional numerical simulation of splitting severe storms on 3 April 1964. *J. Atmos. Sci.* **38**, 1581–1600.

Williams, E. R. (1985). Large-scale charge separation in thunderclouds. *J. Geophys. Res.* **90**, 6013–6025.

Williams, E. R., and R. Lhermitte (1983). Radar test of the precipitation hypothesis for thunderstorm electrification. *J. Geophys. Res.* **88**, 10,984–10,992.

Wilson, J. W., and T. Fujita (1979). Vertical cross section through a rotating thunderstorm by Doppler radar. *Prepr., Conf. Severe Local Storms, 11th*, pp. 447–452, Am. Meteorol. Soc., Boston, Massachusetts.

Wilson, J. W., R. D. Roberts, C. Kessinger, and J. McCarthy (1984). Microburst wind structure and evolution of Doppler radar for airport wind shear detection. *J. Clim. Appl. Meteorol.* **23**, 898–915.

Wilson, T. (1981). Vortex boundary layer dynamics. M.S. Thesis, Univ. of California, Davis.

Wilson, T., and R. Rotunno (1982). Numerical simulation of a laminar vortex flow. *Proc. Int. Conf. Comput. Methods Exp. Meas.* pp. 203–215. Springer-Verlag, Berlin.

Winn, W. P., C. B. Moore, C. R. Holmes, and L. G. Byerly, III (1978). A thunderstorm on July 16, 1975, over Langmuir Laboratory: A case study. *J. Geophys. Res.* **83**, 3080–3092.

Wyss, J., and K. A. Emanuel (1988). The Pre-STORM environment of mid-latitude pre-frontal squall lines. *Mon. Weather Rev.* **116**, 790–794.

Xu, J.-L. (1983). Hail growth in a three-dimensional cloud model. *J. Atmos. Sci.* **40**, 185–203.

Yamasaki, M. (1969). Large-scale disturbances in the conditional unstable atmosphere in low latitudes. *Pap. Meteorol. Geophys.* **20**, 298–336.

Yamasaki, M. (1971). A further study of wave disturbances in the conditionally unstable model tropics. *J. Meteorol. Soc. Jpn.* **49**, 391–415.

Zipser, E. J. (1969). The role of organized unsaturated convective downdrafts in the structure and rapid decay of an equatorial disturbance. *J. Appl. Meteorol.* **8**, 799–814.

Zipser, E. J. (1977). Mesoscale and convective-scale downdrafts as distinct components of squall-line structure. *Mon. Weather Rev.* **105**, 1568–1589.

Zipser, E. J., and M. A. LeMone (1980). Cumulonimbus vertical velocity events in GATE. Part II: Synthesis and model core structure. *J. Atmos. Sci.* **37**, 2458–2469.

Zrnic, D. S. (1976). Magnetometer data acquired during nearby tornado occurrences. *J. Geophys. Res.* **81**, 5410–5412.

Zverev, A. S., ed. (1972). "Practical Work in Synoptic Meteorology," pp. 225–252. Hydrometeorol. Publ. House, Leningrad.

Chapter 10 | Mesoscale Convective Systems

10.1 Introduction

In this chapter we discuss the dynamics and characteristics of precipitating mesoscale convective systems, focusing on the mesoscale features of the earth's dominant precipitating convective systems: tropical and midlatitude squall lines, tropical and midlatitude cloud clusters, including mesoscale convective complexes, and the mesoscale structure of tropical cyclones. We first look at tropical mesoscale convective systems, discuss midlatitude mesoscale convective systems, and then focus on the mesoscale structure of tropical cyclones.

10.2 Mesoscale Convective Systems

The term MCS describes a deep convective system that is considerably larger than an individual thunderstorm and that is often marked by an extensive middle to upper tropospheric stratiform–anvil cloud of several hundred kilometers in horizontal dimension. The cloud systems typically have lifetimes of 6 to 12 h and, on some occasions, the stratiform–anvil portion of the system can survive for several days. In some regions of the world, including many portions of the oceanic tropics and subtropics and even parts of the High Plains of the United States, MCSs are the dominant contributor to the annual precipitation. Thunderstorms embedded within MCSs are often the source of flood-producing rainfalls, strong, damaging winds, and, in midlatitudes, violent storms producing tornadoes and hail. We first investigate tropical mesoscale systems, such as squall and nonsquall clusters, and then their midlatitude cousins, squall lines, and MCCs.

10.2.1 Large-Scale Conditions Associated with Tropical Mesoscale Systems

In a composite study of tropical mesoscale systems, McBride and Gray (1980) found that the dominant forcing of such systems in the western Pacific and the eastern Atlantic oceans is the convergence associated with the Intertropical Convergence Zone (ITCZ). Frank (1978) found that large-scale convergence precedes by several hours the formation of MCSs observed in the Global Atmospheric Research Program's Atlantic Tropical Experiment (GATE) area. He suggested that the large-scale convergence causes a buildup of middle-level moisture, which allows the convection to survive the drying influences of entrainment. The large-scale convergence is not limited to the ITCZ in the tropics, however; regional circulations such as land and sea breezes (Houze *et al.*, 1981) can also be a source of sustained large-scale mesoscale systems forced by convergence.

McBride and Gray (1980) determined that easterly waves in the tropics are an important forcing mechanism of MCSs but not as important as large-scale convergence in the ITCZ. Frank (1978) viewed easterly waves as a triggering mechanism for MCSs. As noted by Reed and Recker (1971), an easterly wave exhibits a characteristic vertical profile of divergence, with convergence at low levels, nondivergence between 500 and 300 mbar and strong divergence above 175 mbar. They suggested that the divergence aloft may result from the outflow produced by thunderstorms. Thus, the vertical motions associated with easterly waves, which are of the order of a few centimeters per second, establish an environment favorable for deep convection. Motions of such weak magnitudes, however, cannot be considered triggers for tropical clusters.

Payne and McGarry (1977) found that the genesis of cloud clusters occurred fairly uniformly relative to the trough axis of easterly waves, but large, long-lived clusters were generally located ahead of the 700-mbar trough axis. The longer-lived clusters moved along with, but at a speed slightly less than, the wave disturbance. Once they fell behind the wave trough axis, they eventually dissipated. Payne and McGarry found that squall clusters in the GATE area formed in the afternoon over land somewhat to the west of the trough axis and north of the 700-mbar disturbance. Because of their faster propagation speeds, squall clusters advanced westward through the wave disturbance and terminated just east of the 800-mbar ridge.

Of course, for squall and nonsquall clusters, conditional instability in the lower troposphere and some sort of large-scale forcing mechanism such as easterly wave troughs are essential. Frank (1978) noted, however, that strong low-level vertical shear of the horizontal wind is critical to the organization of convection into a squall-line configuration. Although he could not define precisely a threshold value of zonal shear, the mean easterly shear from 950 to 650 mbar was 6 m s^{-1} in the cluster cases, whereas it was 13 m s^{-1} in the squall-line cases. In the West African disturbances, Fortune (1980) observed that the African easterly jet located near 600 mbar may provide the critical shear for squall-line formation.

In comparing the environments of fast- and slow-moving squall lines, Barnes and Sieckman (1984) defined slow-moving lines as having speeds $V_L < 3$ m s^{-1} and fast-moving lines as having speeds $V_L > 7$ m s^{-1}. They indicated that the fast-moving lines are synonymous with the squall-line systems described here. Consistent with previous studies, the environment of squall lines was characterized by greater shear in the lower troposphere. In fast-moving lines, most of the lower tropospheric shear existed in the wind component (V_N) perpendicular to the leading edge of the line. By contrast, most of the lower tropospheric shear resided in the wind component (V_T) parallel to the line in slow-moving lines. The magnitude of shear of V_N in fast-moving lines was nearly double the shear of V_T in slow-moving lines. The squall-line environment also exhibited a more pronounced minimum of θ_e at about 650 mbar than did slow-moving lines.

10.2.2 Characteristics of Tropical Squall Clusters

The tropical squall cluster or squall line is identifiable by a line of vigorous convective cells from 100 to several hundred kilometers along its major axis. At the surface, the passage of the squall line is noted for a distinct roll cloud followed by a sudden wind squall of 12–25 m s^{-1} (Hamilton and Archbold, 1945). Immediately behind the surface squall, a heavy downpour sets in, producing as much as 30 mm of rainfall in 30 min. In the tropics, the heavy downpour is followed by several hours of relatively steady and uniform rainfall from the stratiform anvil of the system. A low-level horizontal radar display of an approaching squall line (Fig. 10.1) is earmarked by a convex-shaped line of convective radar echoes that advances at speeds of 15–20 m s^{-1}. Often the squall advances as a continuous disturbance, but occasionally it moves as a family of squall systems (Fortune, 1980) that may propagate for several days (Fig. 10.2).

The strengths of convective-scale updrafts and downdrafts in tropical squall lines are similar to those in tropical cumulonimbus (Fig. 9.19, Chapter 9). In the terminology of Betts (1976), the convective-scale updrafts strip

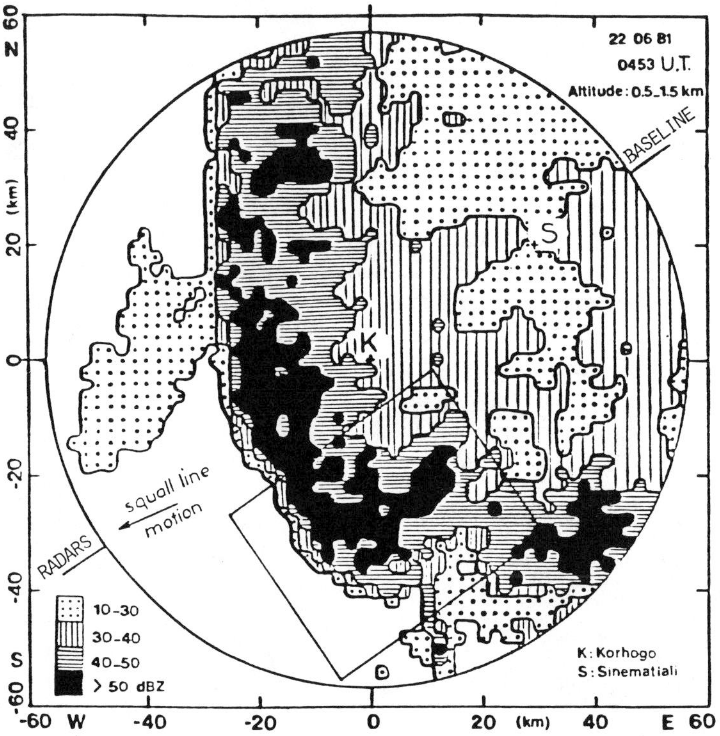

Fig. 10.1. Radar reflectivity contours (dBZ) at low levels (0.5–1.5 km) deduced from the Korhogo radar scan. Rectangular mesh south of Korhogo (K) represents the region of dual-Doppler radar observations. [From Chauzy *et al.* (1985). Copyright by the American Geophysical Union.]

away the high moist-static-energy air in the subcloud layer and carry it in buoyant updrafts into the middle and upper troposphere. At the same time, convective-scale downdrafts bring low moist-static-energy air downward from the middle troposphere into the boundary layer. In the maritime tropics, the clouds are warm based and exhibit low cloud droplet concentrations. Because the updraft strengths are relatively weak, peak radar-echo intensities reside at low levels (Szoke and Zipser, 1986), thus favoring a low-level source to the downdrafts. The evaporative cooling and melting of raindrops result in cool air brought down to the surface in convective-scale downdrafts, which are significant in forming the cold pool of air immediately behind the gust front. The convective-scale downdrafts and the cold pool play an important role in the propagation of tropical squall lines, which behave much like a density current (Section 9.5).

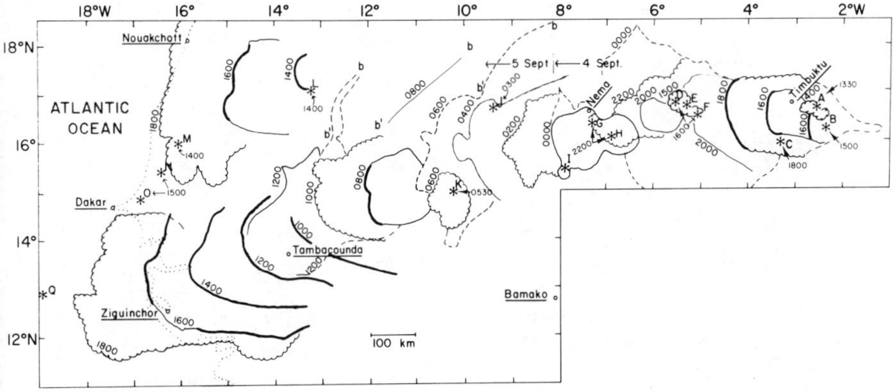

Fig. 10.2. Advance of the squall-line disturbance on 4 and 5 September 1974 from its origins to the Atlantic. Asterisks mark the points of origin of squall-line elements A–Q, with hour of origin indicated. Alternating scalloped and thin lines are the leading edge of the anvil cloud at 2-h intervals. Thick solid curves mark the position of the arc front on the visible pictures. Dashed lines outline the anvil every 6 h. The series of arcs emanating to the west of point K is the principal squall line. Line b–b′ is a long-lived but dormant arc of middle cloud. [From Fortune (1980).]

Figure 10.3 displays the locations and intensities of convective-scale updrafts and downdrafts inferred by dual-Doppler radar in a West African squall line (Chong *et al.*, 1987). The strong updraft at the leading edge of the line is fed by warm boundary layer air. Maximum updrafts speeds of 13 m s^{-1} occur at the 2.5-km level. Heavy precipitation forms in this tilted updraft (Fig. 10.4), as evidenced by high radar reflectivity below it. Between 1.0 and 1.5 km, the updraft is continuous along the leading edge of the line. Above 2.0 km, the updraft splits into two parts (Fig. 10.3), with the most intense core in the northern region. Associated with the split updrafts is an intrusion of midtropospheric rear-to-front flow; a similar feature is common in midlatitude squall lines.

Behind the leading updraft line is a band of convective-scale downdrafts, most evident at high levels. Maximum downdraft speeds of 4 m s^{-1} are observed and are similar in magnitude to those observed from aircraft by Zipser and LeMone (1980). Figure 10.5 illustrates vertical cross sections of the corresponding horizontal winds relative to the squall-line motion derived from dual-Doppler radars. The horizontal flow is marked by (1) a front-to-rear flow at all levels at the system's leading edge and (2) below 3 km a rear-to-front flow behind the leading squall line.

Accompanying the convective-scale updrafts and downdrafts is a region of weaker upward and downward motions and associated stratiform cloudi-

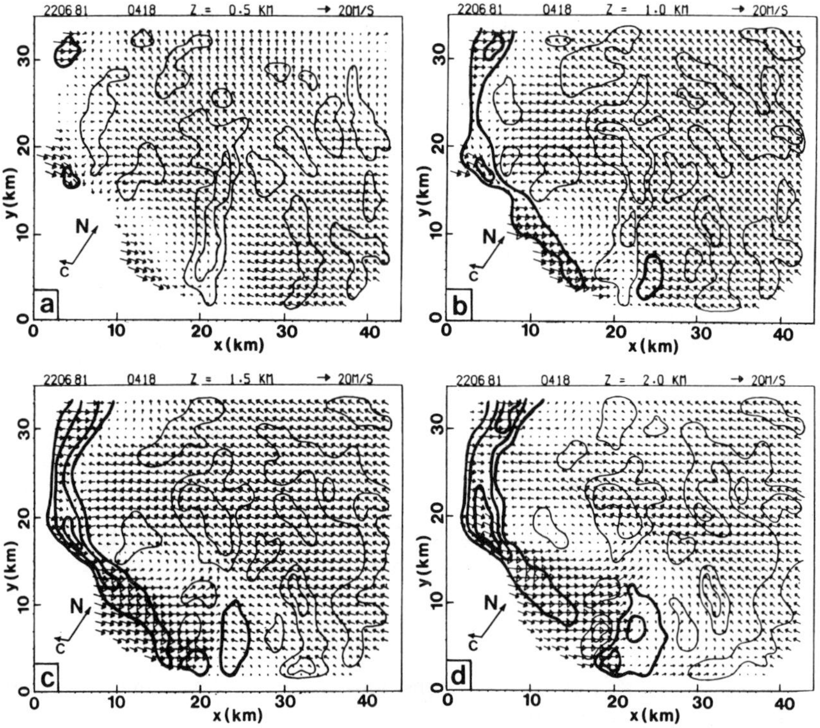

Fig. 10.3. Convective-scale updraft (heavy lines) and downdraft (light lines) contours superimposed on the horizontal flows at the altitudes of 0.5, 1.0, 1.5, 2.0, 2.5, 3.0, 3.5, and 4 km, observed at 0418 UTC, 22 June 1981. Contours correspond to 3, 6, and 9 $m\ s^{-1}$ for updrafts and to -0.5 and $-2\ m\ s^{-1}$ for downdrafts. [From Chong *et al.* (1987).] (*Figure continues.*)

ness and precipitation. The intense convective precipitation just behind the surface squall (Fig. 10.4) is followed by weaker stratified precipitation extending some 120 km behind the leading squall. A well-defined radar bright band near the melting level from 90 to 120 km is typical of stratiform precipitation in a variety of weather systems, including midlatitude cyclones (Houze, 1981). At temperatures slightly colder than 0°C, the bright band results from vigorous aggregation of ice crystals creating sizeable aggregates with large radar cross sections. At temperatures warmer than 0°C, the aggregates begin to melt, yet retaining their large aggregate shape. The wetted surface of the aggregate yields a dielectric constant similar to water instead of ice (Battan, 1973), further enhancing the radar reflectivity. At still warmer temperatures, the aggregates melt fully and break up into smaller drops, thus lowering the magnitude of back-scattered radar energy.

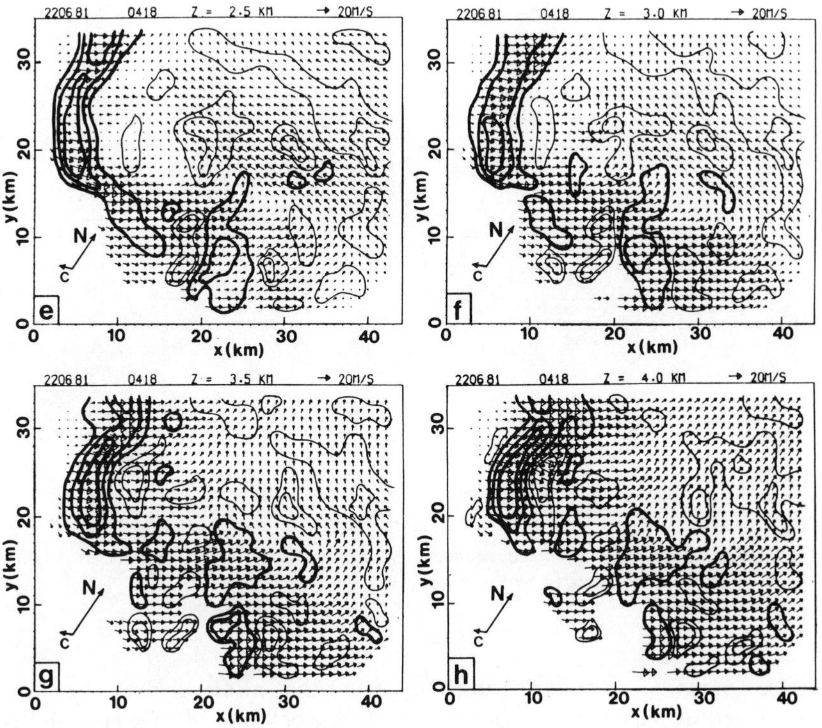

Fig. 10.3 (*continued*)

The greater fall velocity of the raindrops formed from melted snowflakes causes a divergence in particle concentrations below the melting level, further contributing to a reduction in radar reflectivity.

In the tropics, the stratiform region generally trails the squall line, but on some occasions a region of stratiform cloud and precipitation extends ahead of the surface squall. Houze and Rappaport (1984) described a system exhibiting a stratiform cloud and precipitation that trailed the surface squall and extended in advance of it. Figure 10.6 shows a vertical radar cross section through the storm at the time the precipitation ahead of the surface squall was at a maximum. Houze and Rappaport (1984) concluded that the prefrontal stratiform precipitation resulted from forward advection of hydrometeors by the relative flow across the squall line at upper levels. They could not detect a mesoscale updraft in that region. In the trailing stratiform region, they hypothesized that the precipitation was maintained by condensation produced by mesoscale ascent as well as by advection of condensate from another convective region to the south of the storm. R.A.

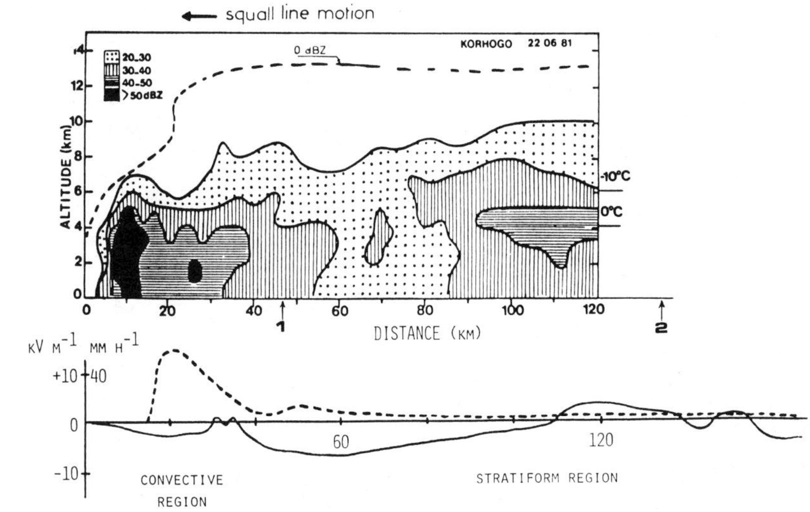

Fig. 10.4. Vertical section of radar reflectivity contours along the propagation axis. [From Chauzy *et al.* (1985). Copyright by the American Geophysical Union.]

Houze (personal communication) suggested that midlevel detrainment of slowly settling ice particles from the squall-line cells could have contributed to the trailing stratiform precipitation. From the fact that this squall moved more slowly than others observed during GATE, they concluded that the leading stratiform cloud and precipitation were a result of relative horizontal flow through the squall line at upper levels transporting condensate ahead of the surface squall. The contribution of stratiform precipitation to the total was 42% in this case. In two storms with only trailing stratiform rain areas, Houze (1977) and Gamache and Houze (1983) found that the stratiform contribution to total rainfall was 40 and 49%, respectively.

The above observations show that mesoscale upward and downward motions are associated with the stratiform precipitation field. Generally, the base of the stratiform cloud is at or above the melting level. From the observations of low values of θ_e in the upper part of the boundary layer in the wake of the disturbance, Zipser (1969) inferred that a mesoscale downdraft, driven by evaporation of rain, exists below the melting level.

Zipser (1977) showed that warming and drying associated with the subsaturated descent beneath and behind the trailing anvil resulted in a so-called onion or diamond shape to the temperature and dew-point profiles (Fig. 10.7). A maximum spread in the temperature and dew-point profiles occurred between 900 and 960 mbar.

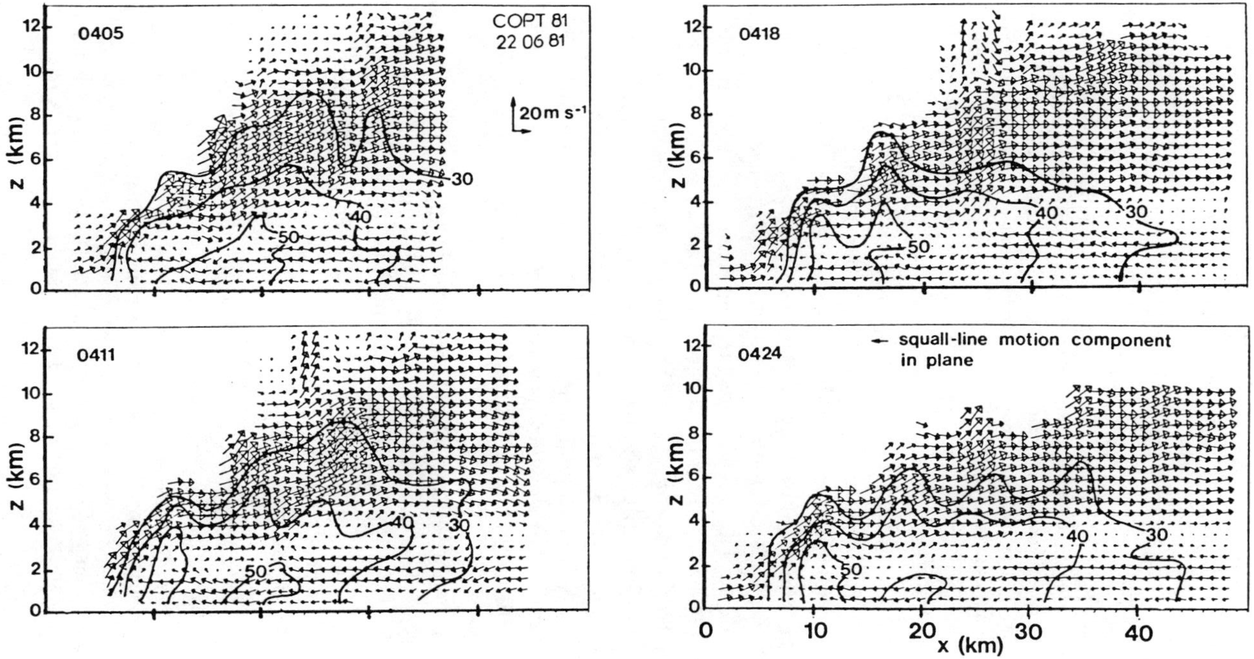

Fig. 10.5. Relative wind fields in vertical cross sections through the convective part of the squall line along $y - 18$ km, deduced from coplane sequence analyses at 0405, 0411, 0418, and 0424 UTC. The x-axis, parallel to the radar baseline, is very close to the system's direction of propagation. Vectors representing 20-m s^{-1} horizontal and vertical velocities are shown; note the different scaling for the horizontal and vertical wind components. Contours of the reflectivity factor (dBZ) are also indicated. [From Chong *et al.* (1987).]

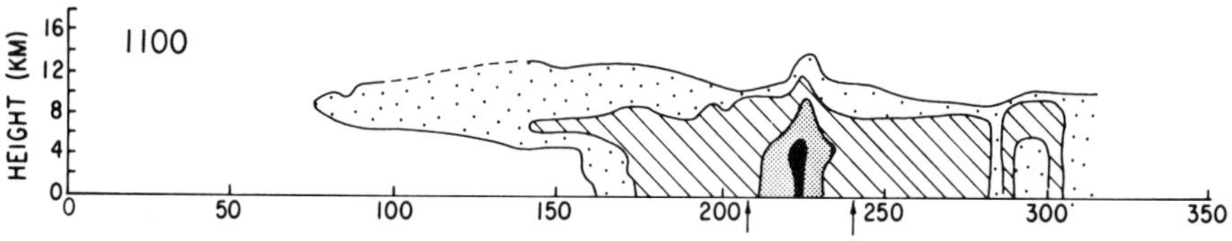

Fig. 10.6. Evolution of the vertical structure of squall system radar echoes. Cross sections are along the propagation direction, with motion from right to left. Note the stratiform anvil cloud extending well in advance of the main convective cores. Shading thresholds are for the minimum detectable 24, 34, and 44 dBZ. Small arrows denote front and back edges of convective zone. [From Houze and Rappapart (1984).]

Further support for this hypothesis was provided with a two-dimensional numerical model by Brown (1979) who found that evaporative cooling of precipitation accounted for most of the mesoscale descent below the anvil. Brown's model predicted subsidence with peak strengths of about 10 cm s^{-1}, comparable to Zipser's estimate of mesoscale subsidence of 5–25 cm s^{-1} magnitude at 500 m. The subsidence may be driven, in part, by vertical pressure gradients arising from the divergence of air in the cold pool in the subcloud layer (Miller and Betts, 1977). Using a three-dimensional cloud model, Miller and Betts inferred that evaporation of rain from the stratified anvil cloud was not essential to maintenance of the mesoscale unsaturated downdraft. Instead, they argued that only evaporation of rain in nearly saturated convective-scale downdrafts was needed to create low-level divergence, which, in turn, dynamically forced mesoscale descent below the anvil cloud.

Neither Brown's nor Miller's model contained ice-phase thermodynamic processes. Leary and Houze (1979b), however, calculated that melting of ice can amount to several degrees per hour, and they suggested that melting may cooperate with evaporation in both maintaining and initiating mesoscale descent. We discuss further in the section on midlatitude MCSs the significance of ice-phase processes.

Fig. 10.7. Characteristic soundings in postsquall regions. (a) 5 September 1974 by NCAR *Queen Air*, mostly in rain; (b) 1630 GMT, 12 September 1974, for the *Fay* (GATE position 28A, about 200 km behind leading edge and 50 km behind trailing precipitation); (c) 1804 GMT, 12 September 1974, for the *Oceanographer* (GATE position 4, about 250 km behind leading edge and 100 km behind trailing precipitation); (d) 1130 GMT, 12 September 1974, for the *Poryv* (GATE position 10, about 350 km behind leading edge and 174 km behind trailing precipitation); (e) 200 AST, 28 August 1969, Anaco, Venezuela (about 50 km behind trailing precipitation); (f) aircraft sounding from Fig. 10.6; (g) Barbados sounding from Fig. 10.6 (no T_d available). [From Zipser (1977).]

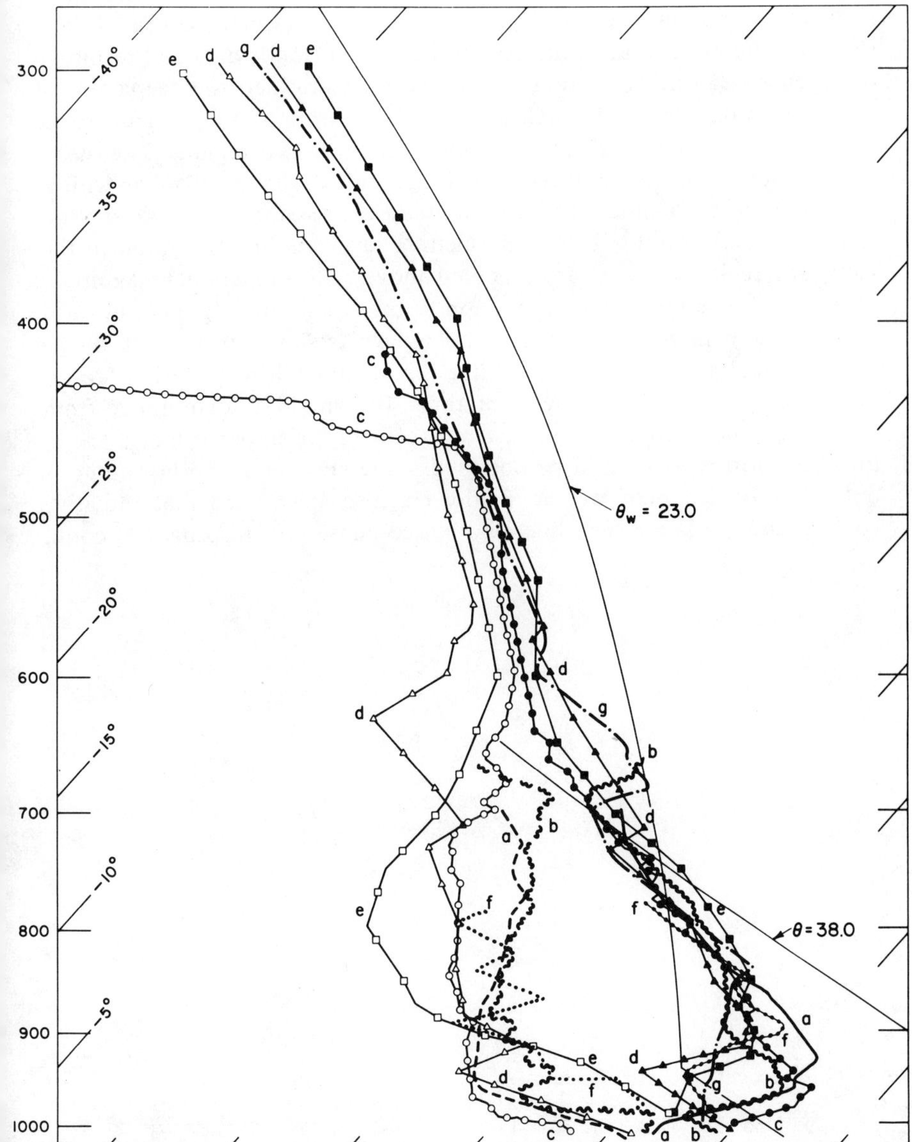
300
400
500
600
700
800
900
1000
-40°
-35°
-30°
-25°
-20°
-15°
-10°
-5°
0°
5°
10°
15°
20°
25°
30°
θ_w = 23.0
θ = 38.0
a
b
c
d
e
f
g

At heights above the melting level, a number of researchers (e.g., Houze, 1977; Zipser, 1977) speculated that organized mesoscale ascent within the stratiform region must occur. Gamache and Houze (1982) used radar data to composite horizontal winds derived from rawinsonde data and resultant divergence and vertical motion estimates for the stratified and deep convective regions of a tropical squall line observed during GATE. Figure 10.8a illustrates the calculated vertical motion for these two regions. The mean vertical motion in the stratiform anvil is upward above 650 mbar with a maximum near 300 mbar. The magnitude of the maximum upward motion is 15 cm s^{-1}. At heights below the melting level, the mean motion in the stratiform region is downward, as noted above, with a maximum magnitude of 6 cm s^{-1}. The mean vertical motion in the squall-line region is upward at all levels with a maximum near 650 mbar. As shown in Fig. 10.8b, the mean vertical motion in the squall-line region at low levels is the net result of strong upward and downward motions. Brown (1979) determined from his model calculations that persistent ascent in the upper troposphere in the stratiform region could be obtained if the environmental lapse rate is only slightly less than wet adiabatic. He also speculated that radiative cooling and the latent heat liberated by ice-phase growth processes could

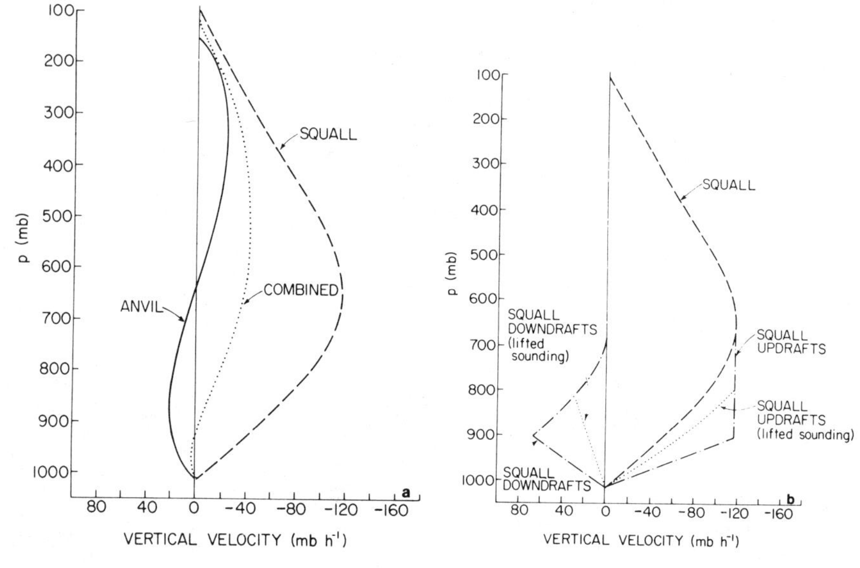

Fig. 10.8. Average vertical velocity (ω) for the squall line (dashed), anvil (solid), and combined (dotted) regions (a), and for the squall-line updrafts and downdrafts in either lifted or unlifted presquall environments (b). [Adapted from Gamache and Houze (1982).]

allow upward motion in the upper troposphere, even in an environment that is slightly stable for water-saturated motions.

More direct confirmation of the presence of mesoscale vertical motions in the trailing stratiform region has been obtained by Chong *et al.* (1987) using single-Doppler radar velocity azimuth display (VAD) techniques. They found mesoscale ascent above 620 mbar, or just below the 0°C isotherm, and mesoscale descent below that level. The magnitudes of the inferred mesoscale updrafts were 35–45 cm s^{-1}, and the downdrafts were 15–30 cm s^{-1}. These are considerably larger than the 6- to 15-cm s^{-1} mesoscale vertical motions calculated by Gamache and Houze. The differences are presumably because the Doppler-radar-inferred vertical motions were averaged over a 60-km-diameter VAD circle, whereas Gamache and Houze's calculations were based on more widely spaced rawinsonde data.

To demonstrate perturbations in horizontal wind fields about tropical squall lines, we examine the composite wind patterns derived by Gamache and Houze (1982) from a case observed during GATE. Figure 10.9 depicts the derived surface-wind patterns. Ahead of the squall line is confluence where a line of cumulus clouds is observed. The squall line is clearly defined by a line of convergence, behind which strong surface divergence exists in

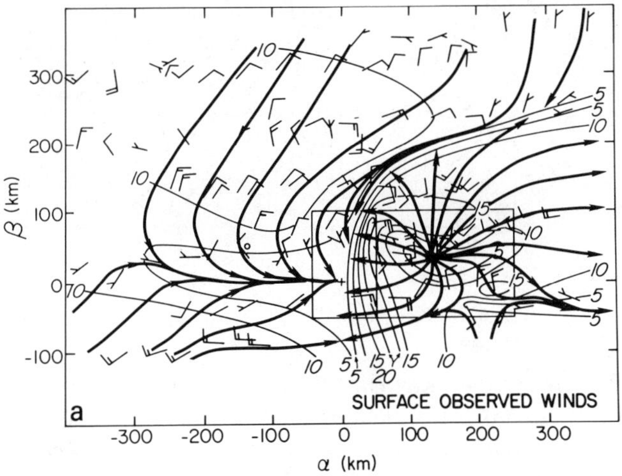

Fig. 10.9. Composite observed surface winds with (a) streamline and isotach analysis. Cross indicates squall-line center and origin of the α–β coordinate system. Solid contours indicate wind speed in knots. Full wind barb is for 10 kt (or 5 m s^{-1}) and half barb is for 5 kt (or 2.5 m s^{-1}). (b) Composite of surface relative winds, streamlines, and isotachs as in (a). (c) Divergence pattern for surface composite wind in units of 10^{-5} s^{-1}. [From Gamache and Houze (1982).] (*Figure continues.*)

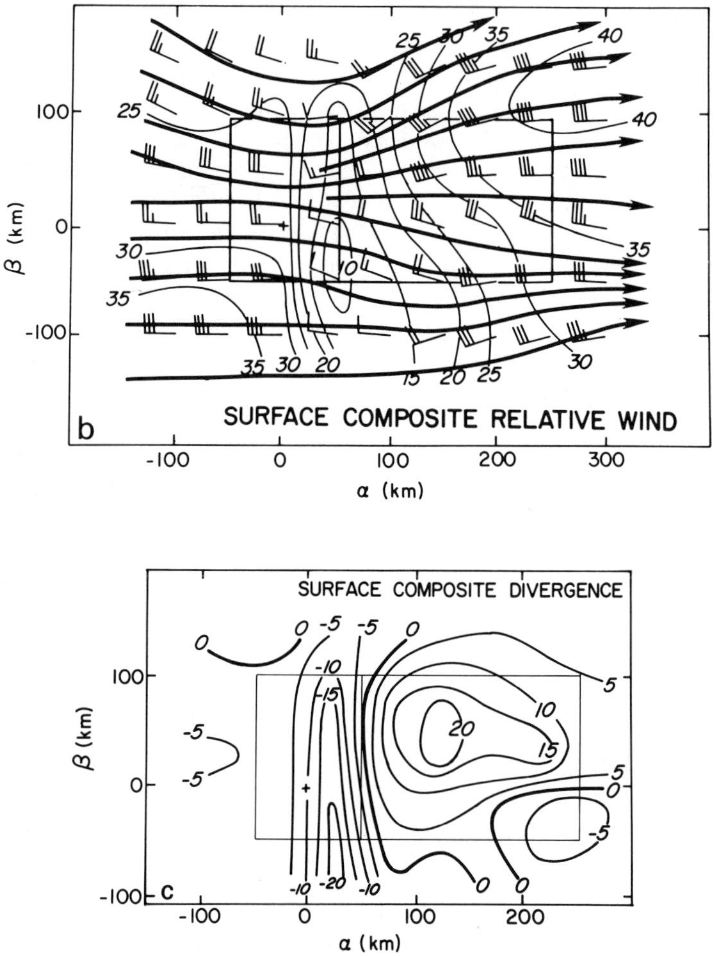

Fig. 10.9 (*continued*)

the cold pool. A relative flow (i.e., flow relative to the advancing system) of 12–15 m s^{-1} streams into the storm in advance of the squall line (Fig. 10.9b). The flow slows to a near calm at the squall line where it ascends into the convective updrafts (e.g., Fig. 10.5). Figure 10.10 shows at 850 mbar a strong jet located in the storm's left quadrant and terminating within the squall-line region. The storm-relative flow in Fig. 10.10b exhibits flow entering the storm in advance of the squall and primarily from its left flank. The flow converging into the squall line either ascends in convective-scale updrafts or descends in downdrafts.

The outstanding flow feature at 650 mbar is the pronounced cyclonic vortex in the relative flow (Fig. 10.11). Associated with this vortex is a region of convergence at the base of the stratiform layer. The 650-mbar level in this case separates the region of mesoscale ascent above and

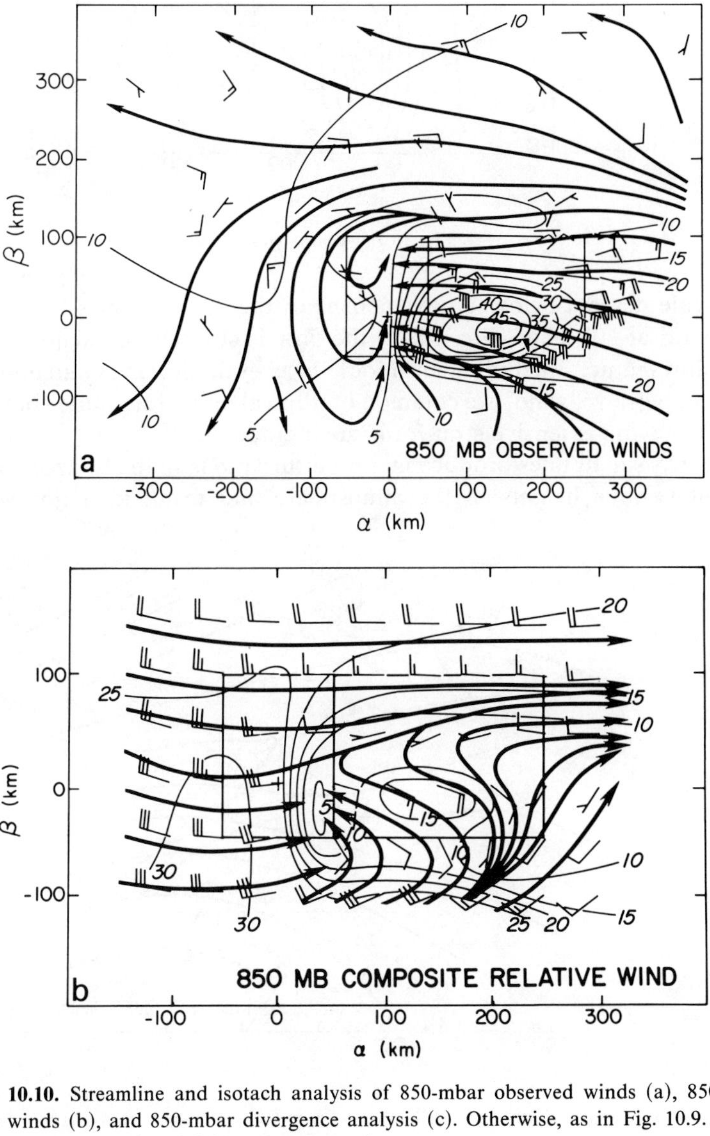

Fig. 10.10. Streamline and isotach analysis of 850-mbar observed winds (a), 850-mbar relative winds (b), and 850-mbar divergence analysis (c). Otherwise, as in Fig. 10.9. [From Gamache and Houze (1982).] (*Figure continues.*)

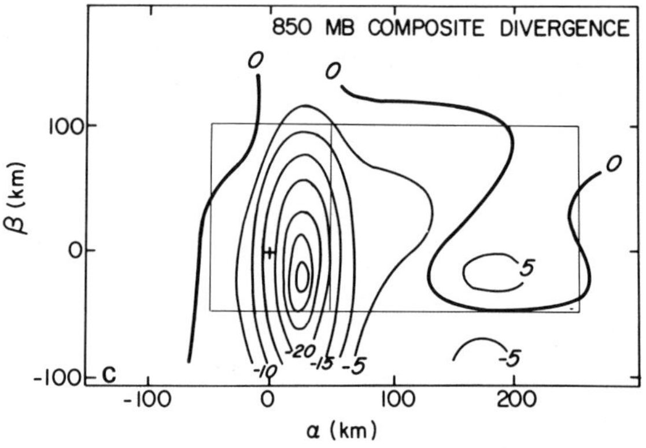

Fig. 10.10 (*continued*)

mesoscale descent below. Above 650 mbar, the flow becomes increasingly divergent; at 200 mbar (Fig. 10.12) the flow is strongly diffluent. Many of these flow features, including the middle-level cyclonic vortex and diffluence of the upper-level wind, are common in tropical and midlatitude convective systems having extensive stratiform-anvil clouds.

Clearly, squall lines produce major perturbations in the horizontal wind field at various heights in the atmosphere and result in major vertical

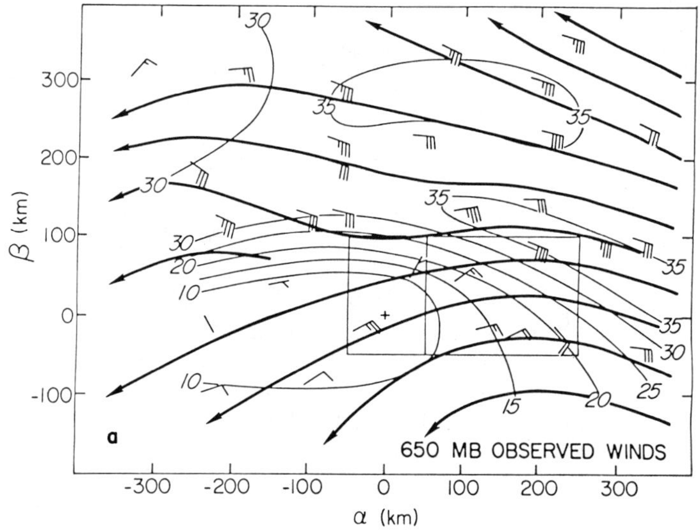

Fig. 10.11. As in Fig. 10.10, for 650 mbar. [From Gamache and Houze (1982).] (*Figure continues.*)

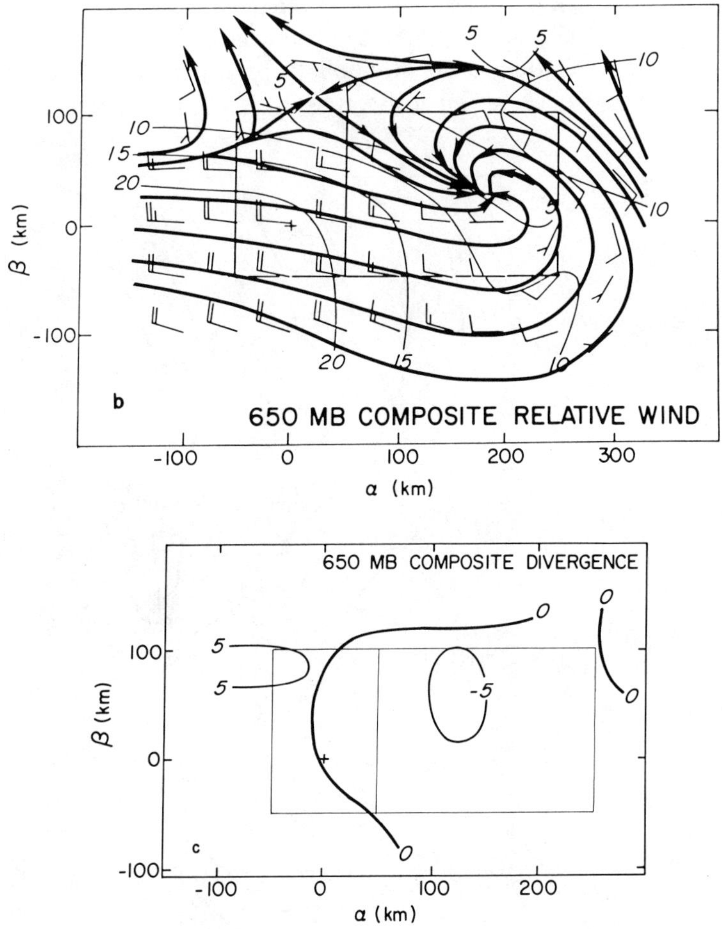

Fig. 10.11 *(continued)*

redistributions of horizontal momentum. Using a three-dimensional numerical prediction model and an analytic model, Moncrieff and Miller (1976) predicted that tropical squall-line convection would increase the upper- and lower-layer shears in the east–west direction or in the plane perpendicular to the direction of line propagation. Based on the numerical model results, they predicted that the convection would decrease the shear in the north–south plane or the plane perpendicular to the direction of line propagation. As discussed in Chapter 6, LeMone (1983) and LeMone *et al.* (1984) have extensively analyzed how squall lines redistribute horizontal momentum in the vertical. With aircraft data, they found that, in well-organized squall-line systems, the convective systems increased momentum aloft in the *direction opposite* to their motion and increased momentum at low levels in the *direction of* their motion, a result similarly predicted by Moncrieff and Miller. Such behavior is inconsistent with the downgradient mixing

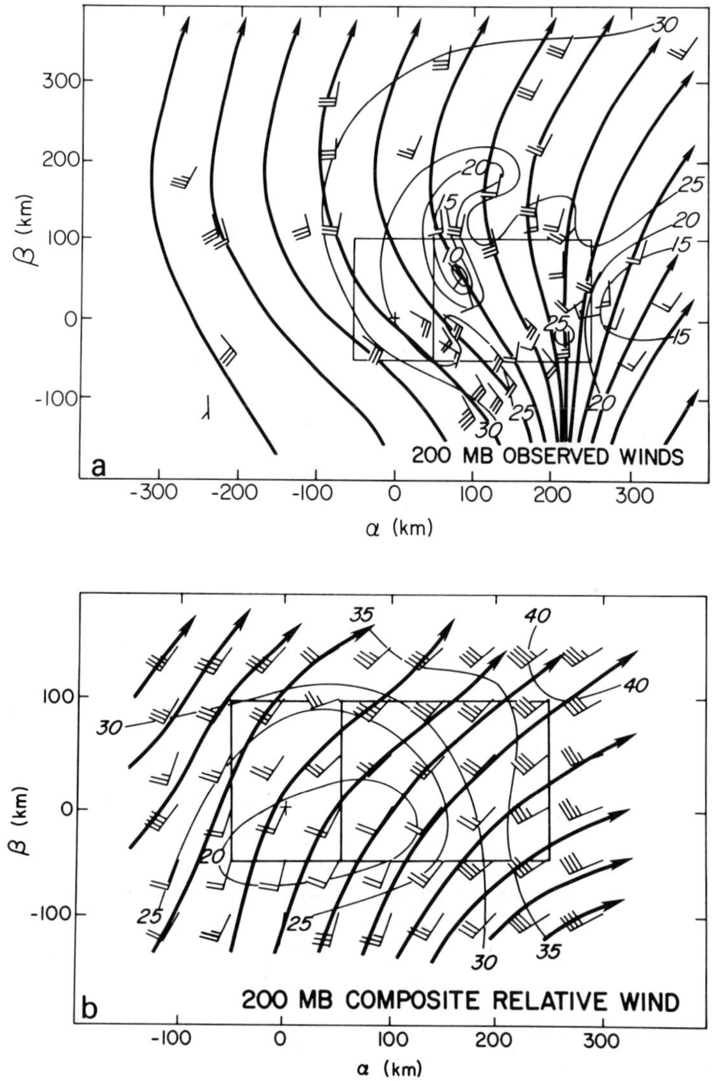

Fig. 10.12. As in Fig. 10.10, for 200 mbar. [From Gamache and Houze (1982).] (*Figure continues.*)

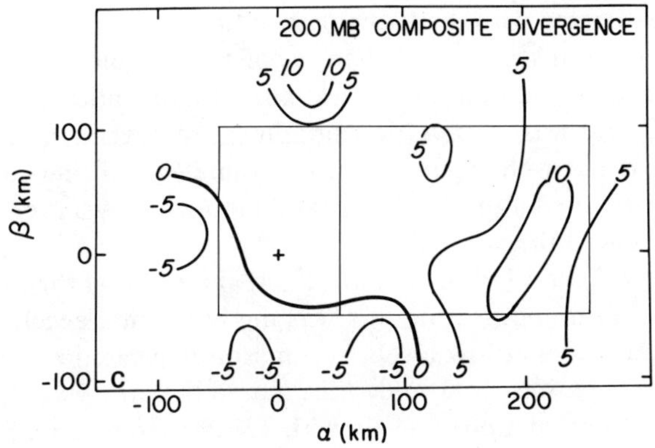

Fig. 10.12 *(continued)*

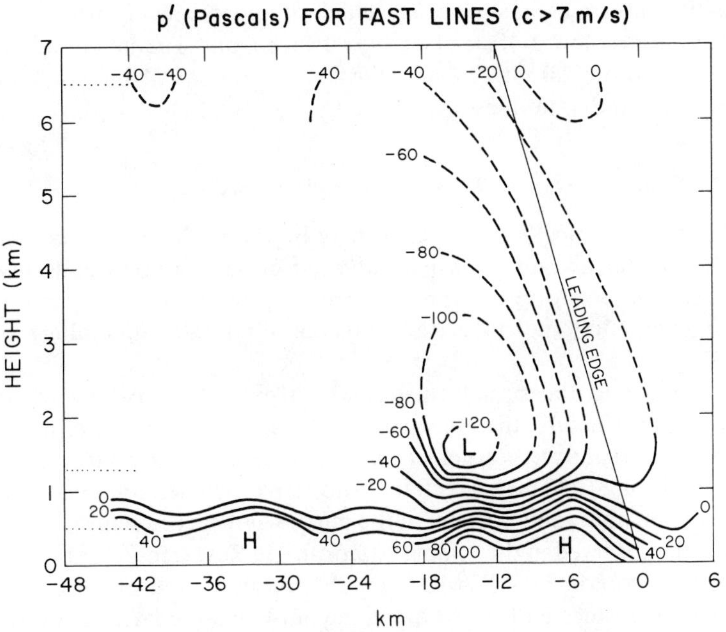

Fig. 10.13. Composite pressure deviation p' (pascals) relative to inflow environment, for fast lines based on Days 225 and 255; slant line indicates average leading edge in time and space, in coordinate system moving with lines. Vertical:horizontal exaggeration, 6:1. Dotted lines correspond to flight levels or averages of level close in altitude. [From LeMone *et al.* (1984).]

hypothesis associated with the K-theories of turbulent mixing. (See Chapter 3 for an extensive discussion of theories of momentum mixing.) LeMone *et al.* (1984) also found that the vertical flux of momentum aligned parallel to the squall line was always downgradient. In several cases where the convection was not well-organized in two-dimensional lines or where isolated towers prevailed aloft, the horizontal momentum was vertically mixed and thus decreased the wind shear.

LeMone (1983) and LeMone *et al.* (1984) showed that the primary cause of the generation of momentum flux was the horizontal acceleration of the air entering the storm at low levels by a mesolow located at the rear of the storm. At the same time, buoyancy accelerated the air upward, resulting in air leaving the line at upper levels with greater front-to-rear momentum than it had upon entering the storm. Downdraft air entering the rear of the updraft at midlevels was also accelerated into the mesolow, and, after descending, the air deposited enhanced flow in the direction of storm movement, thus contributing to negative momentum flux. Figure 10.13 illustrates the inferred pressure distribution across a fast-moving squall line. The main features are a surface mesohigh of approximately 1-mbar magnitude and a mesolow aloft trailing the surface front. The magnitude of the elevated mesolow was −1.2 mbar in the fast-moving squall, but much weaker in slow and moderate lines.

10.2.3 Characteristics of Tropical Nonsquall Clusters

Although the tropical squall cluster may be the most spectacular form of the MCS as far as rain intensity and wind gusts, the nonsquall cluster is the most frequent form and probably plays a more important role in the overall energetics, moisture transports, and rainfall climatology of the tropics.

In contrast to the squall cluster, which exhibits a distinct mesoscale precipitation structure, the nonsquall cluster exhibits a variety of mesoscale precipitation features. Warner *et al.* (1980) described a nonsquall cluster observed during GATE in which the convective cells were aligned in bands roughly 9 km apart and parallel to the mean wind shear vector (Fig. 10.14). In a cluster observed in the Global Atmospheric Research Program's Winter Monsoon Experiment (WMONEX), Churchill and Houze (1984a) described a cluster that formed along the outflow boundaries of two neighboring clusters. The convective cells (Fig. 10.15) do not exhibit any preferred alignment or organization. On occasion, nonsquall clusters are organized relative to a low-level cyclone (Smith *et al.*, 1975; Zipser and Gautier, 1978); a radar composite obtained from one such system is shown in Fig. 10.16. In some cases, a squall cluster is a substructural feature of a larger scale nonsquall cluster exhibiting a variety of convective organizations (Leary and Houze, 1979a). The only precipitation feature common to all of these cloud clusters is the large stratiform-anvil cloud and precipitation region surrounding the convective precipitation features. In fact, the stratiform region of nonsquall clusters differs little from that observed in squall clusters.

The life cycle of nonsquall clusters is characterized by a formative stage in which convective cells prevail, a mature stage in which convective cells and stratified precipitation features coexist and contribute nearly equally to observed rainfall, and a dissipating stage in which the stratiform precipitation prevails and slowly dissipates. Figure 10.17 portrays the evolution of convective cores, area coverage, and precipitation rates for stratiform and

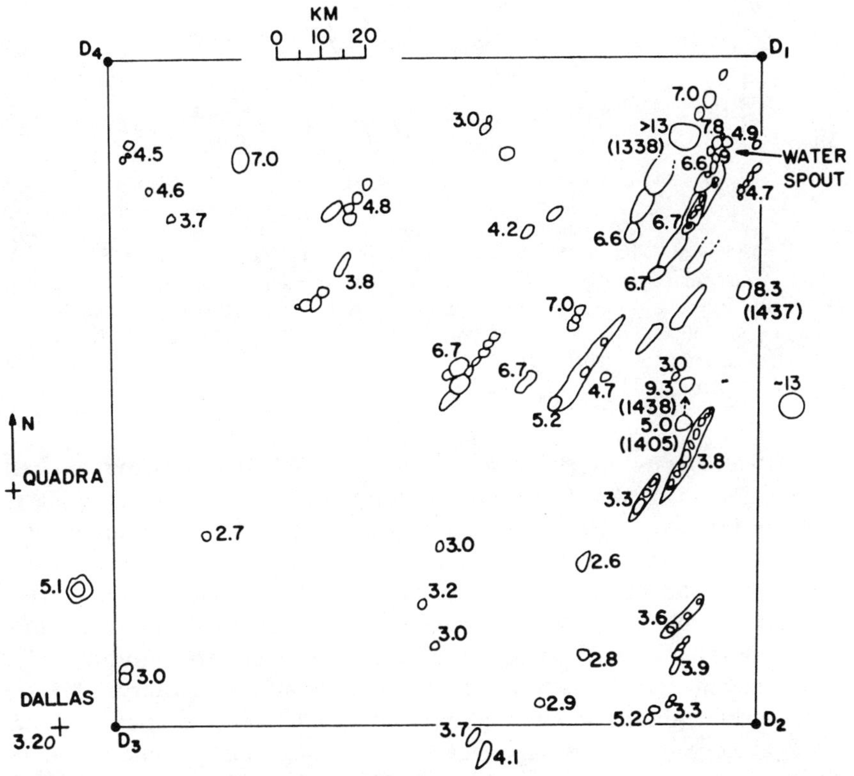

Fig. 10.14. Active cloud tops (outlined by contours) in the height interval 2.5 to >13 km, during the first box circuit by the aircraft, 1300–1445 GMT. Numbers are heights (kilometers) above the sea and times of measurement. The corners D_1 to D_4 define the box circuit between latitudes 8°34′ and 9°56′N, and longitudes 21°2′ and 22°23′W. Clouds are aligned parallel to the mean wind shear vector. [From Warner *et al.* (1980).]

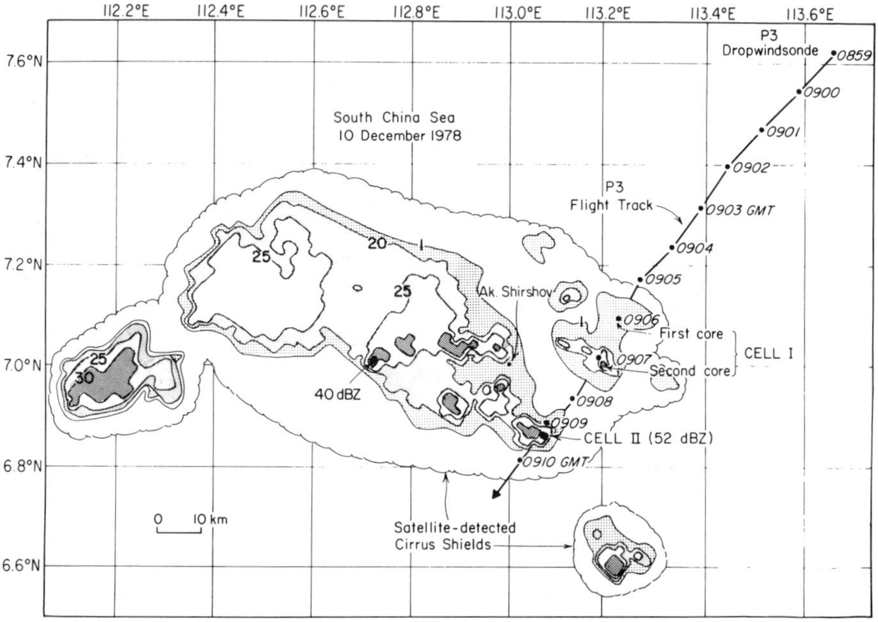

Fig. 10.15. Radar and satellite depiction of cluster A. The scalloped outline indicates the approximate boundary of the cloud shield. The contours indicate precipitation at 1, 20, 25, 30, and 40 dBZ detected by the P3's lower fuselage radar at 0908 GMT. The flight track of the P3 is annotated with time (GMT) to indicate the aircraft position. The two convective cells penetrated by the P3 are labeled I and II. The flight level was 7.8 km. [From Churchill and Houze (1984a).]

convective regions of a cluster observed during WMONEX (Churchill and Houze, 1984a). It shows the dominance of convective precipitation early in the system's life cycle, followed by the dominance of stratiform precipitation rates later in the system's life cycle. For this cluster, 46% of the rain detected by radar fell as stratiform rain over the system's lifetime. This contribution to the total rainfall by the stratiform component is comparable to the 30–50% values found in other tropical clusters (Houze, 1977; Gamache and Houze, 1983; Houze and Rappaport, 1984; Leary, 1984).

In general, the relatively weak updraft speeds observed in WMONEX cloud clusters (Churchill and Houze, 1984a) were comparable to values determined in clusters observed in GATE by LeMone and Zipser (1980) (Fig. 9.9). Churchill and Houze (1984a) did observe one convective core updraft of 17 m s^{-1} that exceeded any of the updraft intensities reported by LeMone and Zipser. However, Churchill and Houze's observations were at 400 mbar, or approximately 8 km MSL, which is higher than any of the

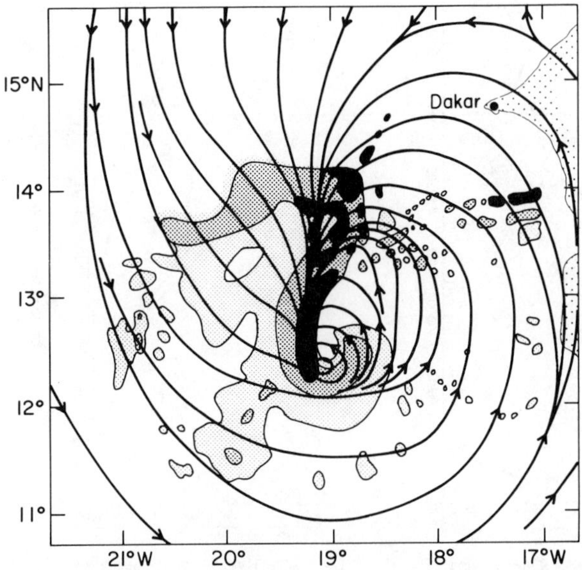

Fig. 10.16. Radar composite, 990-mbar streamlines superimposed, for 1400 GMT, 15 July 1974. Three levels of reflectivity are indicated, determined subjectively. The northwest portion of the echo is not enclosed, as it is believed to extend beyond the limits shown. [From Zipser and Gautier (1978).]

observations reported by LeMone and Zipser. R.A. Houze (personal communication) noted that updraft speeds exceeding 15 m s^{-1} were not uncommon in the Equatorial Mesoscale Experiment (EMEX) flights at 4–5 km. Moreover, Dudhia and Moncrieff (1987) predicted updraft velocities on the order of 17 m s^{-1} at 10 km in their two-dimensional simulation of convective cells observed in GATE nonsquall clusters.

Johnson (1982) estimated the mesoscale vertical motions from rawinsonde data in nonsquall clusters observed in WMONEX. Figure 10.18 illustrates the calculated mesoscale vertical velocity for WMONEX along with Gamache and Houze's (1982) estimates for a GATE squall line and Ogura and Liou's (1980) estimates for an Oklahoma squall line. Both the WMONEX cluster and the GATE squall line exhibit peak upward motions of 15–20 cm s^{-1} near 250 mbar. The WMONEX peak mesoscale ascent is somewhat weaker and at a slightly higher level than the GATE squall line. The mesoscale downdraft is lower in height and stronger in magnitude in the GATE squall line. It is not clear if this is because of differences between squall lines and nonsquall clusters or regional differences in soundings. A much sharper distinction can be seen in the Oklahoma squall line, which

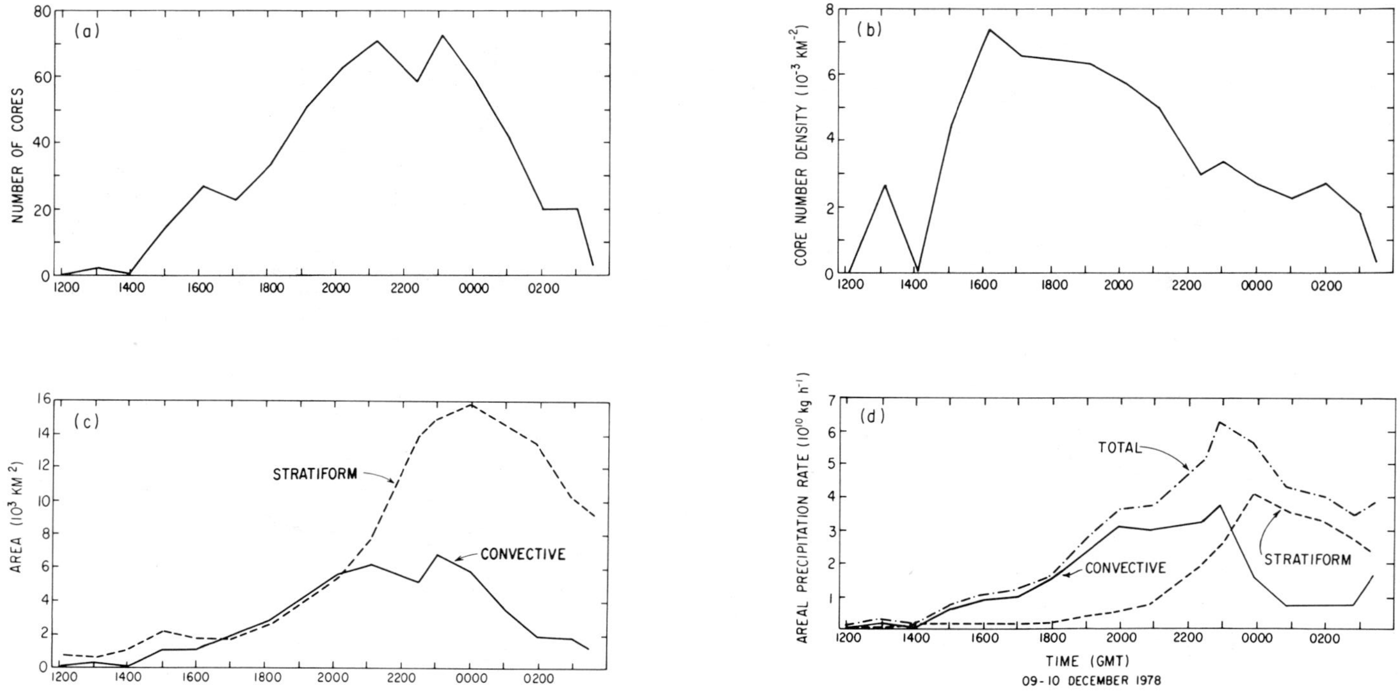

Fig. 10.17. (a) Number of convective cores in cluster B. (b) Number density of convective cores in cluster B, computed by dividing the values shown in (a) by the total area covered by precipitation at 3 km shown in (c). (c) Area covered by objectively determined stratiform (dashed curve) and convective precipitation (solid curve) at 3 km, defined by a 1-dBZ threshold. (d) Area-integrated precipitation rate for cluster B at 3 km. The top curve (dot–dashed line) represents the sum of the convective (solid line) and stratiform (dashed line) components of precipitation. [From Churchill and Houze (1984a).]

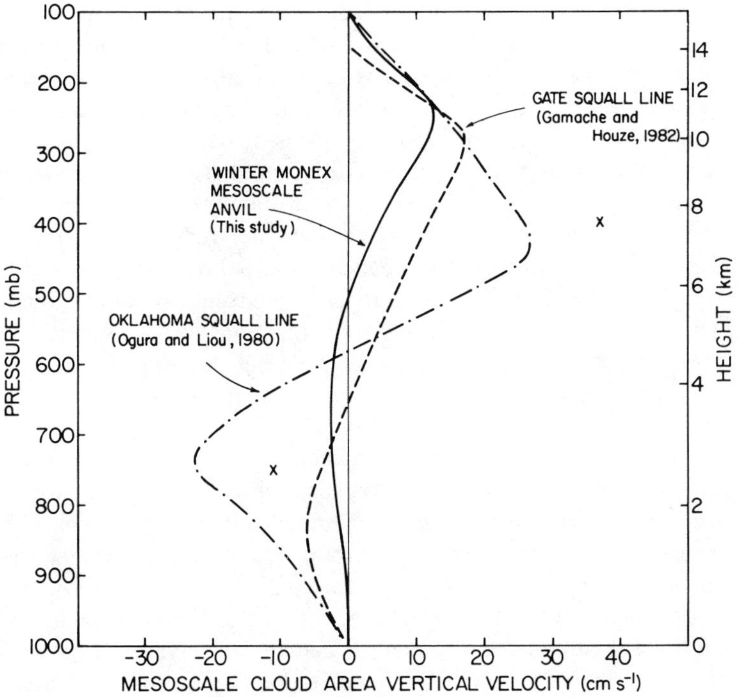

Fig. 10.18. Mesoscale cloud area vertical velocity for winter MONEX mesoscale anvil, GATE squall line, and Oklahoma squall line. Each X indicates extreme values for mesoscale updraft and downdraft in the modeling study of Brown (1979; S1 case at 9 h). [From Johnson (1982).]

displays higher amplitude vertical velocities and updraft maxima much lower in height near 450 mbar.

Analysis of the cloud microphysical structure of the stratified region of nonsquall clusters observed in WMONEX revealed a structure consistent with Johnson's diagnosed vertical motion profile. In the upper portion of the stratified region, Houze and Churchill (1984) observed that the identifiable ice particles were primarily branched, vapor-grown crystals and aggregates. The presence of branched, vapor-grown crystals at temperatures colder than −20°C suggests the crystals must have grown in a water-saturated environment in regions of mesoscale ascent weak enough to allow the crystals to settle to flight level. Variations in crystal concentrations and habits suggested that the stratiform cloud was not horizontally uniform (Churchill and Houze, 1984a). They frequently encountered regions of enhanced concentrations 30–100 km across and noted that the nonuniformity could have been associated with convective overturning within the stratiform cloud deck. It is not clear if the nonuniformity is cellular, as one might expect in a stratocumulus layer, if it existed in mesoscale bands, or if it is simply the remains of the upper parts of decaying convective cells.

Houze and Churchill (1984) noted a difference between stratiform regions exhibiting higher radar reflectivities (greater than 20 dBZ) and those having reflectivities between 1 and 20 dBZ. The higher reflectivity regions were typical near active convection and consisted of debris from convective cells and particles grown in the more stratiform environment. In the low reflectivity stratiform regions, the particle concentrations were lower, and vapor-grown crystals and aggregates prevailed among the identifiable particles. This is consistent with precipitation formation in relatively weak upward motions.

Tropical squall clusters produce noticeable perturbations in the horizontal wind and vertical motion fields. Not surprisingly, nonsquall clusters also produce significant perturbations in the wind fields. There are two approaches in analyzing such perturbations. One involves the analysis of perturbations from individual cases; an example is the case analyzed by Leary (1979) using GATE data. The advantage to this approach is that if the wind observations are sufficiently dense and have enough temporal resolution, a clear signal of the perturbations in winds created by the disturbance emerges. However, because one is dealing with only a single case study, it is difficult to generalize about the behavior of a given type of weather system. A second approach is to composite or average data from a set of observations of similar disturbances (e.g., Williams and Gray, 1973; Frank, 1978; McBride and Gray, 1980). An advantage is that if the data have insufficient resolution to depict adequately perturbations created by a single disturbance, a composite analysis of a number of well-defined similar systems can present a clear, meaningful representation of the perturbations generated by such a weather system. A disadvantage of the composite method is that it smooths out fine-scale features or features such as temperature inversions that may vary in height from one storm system to another.

Tollerud and Esbensen (1985) performed a composite analysis of the largest nonsquall clusters observed during GATE. A striking feature of the composite wind field is the absence of any strong wind shear in the cloud layer between 700 and 250 mbar. This sharply contrasts with the composite winds determined from squall clusters (Frank, 1978), where a midlevel jet between 600 and 700 mbar is characteristic. Tollerud and Esbensen noted that the small shear is a general feature of the environment because it is present before and after the cluster moves through the observing network.

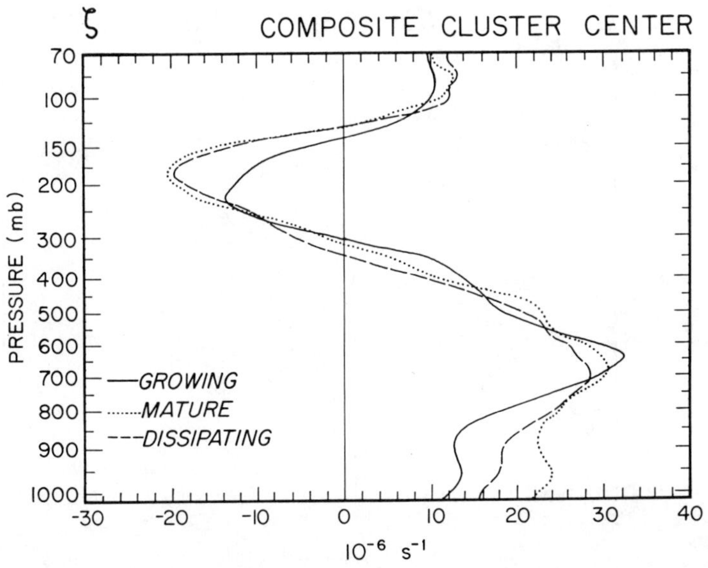

Fig. 10.19. Vertical component ζ of relative vorticity at the center of the composite nonsquall cluster for three life-cycle stages. [From Tollerud and Esbensen (1985).]

Tollerud and Esbensen calculated the composite average vertical component of vorticity for developing, mature, and dissipating clusters. Figure 10.19 shows that the maximum cyclonic vorticity at 700 mbar changes little throughout the life cycle of the cluster. This feature is more a property of the environment than a perturbation created by the cluster. At low levels, however, the initial positive vorticity is amplified considerably as the system matures. Likewise, the initial anticyclonic vorticity at 200 mbar intensifies as the system matures. The strengthened negative vorticity at the outflow level is maintained through the dissipating stages of the system. This is also a characteristic of composite midlatitude MCCs, but as Fig. 10.20 shows, not of horizontal divergence at the outflow level. The divergence exhibits a pronounced maximum at the mature stage of the system but returns to its growing-stage magnitude by the dissipation stage. This latter feature was not observed in the composite structure of midlatitude MCCs (Lin, 1986). The vorticity field clearly retains a longer "memory" of the presence of the system. Another feature of the divergence profile is that the convergence at low levels is a maximum early in the system's life cycle and weakens through the mature and dissipating stages. Weak divergence near 500 mbar is evident during the mature stage, while weak convergence near 400 mbar reaches a maximum in the dissipating stage.

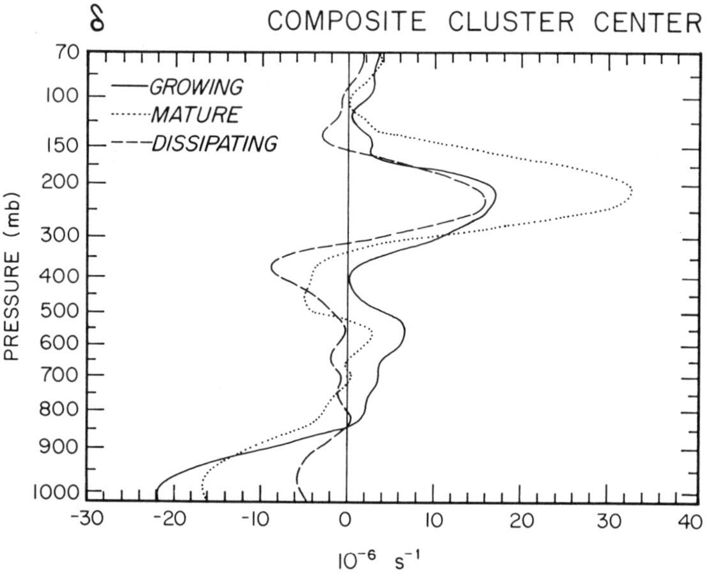

Fig. 10.20. As in Fig. 10.19, except for horizontal divergence δ. [From Tollerud and Esbensen (1985).]

The corresponding composite vertical motion is shown in Fig. 10.21. Upward vertical velocity is diagnosed at all levels throughout the life cycle of the cluster. The peak upward motion occurs just below 300 mbar during the mature stage of the system. The composite vertical motion in Fig. 10.21 should not be confused with Johnson's estimate of vertical motions in the stratiform region of tropical clusters (Fig. 10.18). The composite vertical motion represents the average of convective-scale, mesoscale, and synoptic-scale vertical motions averaged over a number of cloud clusters. The peak composite upward motion at the mature stage is about 100 mbar lower than Johnson's estimated peak motion in the stratiform region of a WMONEX cluster. This is probably because the tropopause is higher in the WMONEX region. Also, Lin's (1986) composite vertical motion field for MCCs, discussed later, exhibits a maximum at the dissipation stage of the system rather than at the mature stage.

In performing three-dimensional numerical simulations of nonsquall convection observed in a GATE tropical cluster, Dudhia and Moncrieff (1987) found that the simulated growth of a mesoscale system depended on the presence of sustained large-scale ascent rates on the magnitude of 1-2 cm s^{-1}. Such a magnitude of lifting is typical of the values observed in the ITCZ, and is consistent with the findings of McBride and Gray (1980)

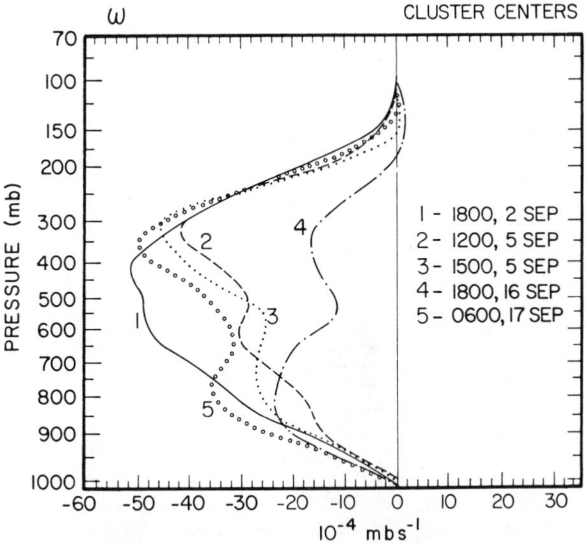

Fig. 10.21. Vertical velocity ω at the center of individual clusters during their mature stage. Dates and times (in GMT) for each profile are given. [From Tollerud and Esbensen (1985).]

based upon composite data analysis. Dudhia and Moncrieff (1987) concluded that the effects of large-scale ascent are most important during the growth stage of the convective bands, because the large-scale ascent restores convective available potential energy and the supply of low-level moisture, thereby counteracting the stabilizing influence of environmental subsidence.

An interesting feature of Dudhia and Moncrieff's simulation is that the clouds were aligned in bands parallel to the low-level shear vector similar to that shown in Fig. 10.14. They concluded that the large shear from the strong low-level easterly jet was crucial in influencing the alignment and produced an elongated rain area which, in turn, caused an east–west elongated cold pool. The gust front along the southern boundary of the cold pool was inhibited from moving southward by the strong low-level northerly shear. This kept the rain area and gust front close together, and the cold pool was continuously supplied with downdraft air in a line parallel to the gust front.

Dudhia and Moncrieff's simulation provides insight into the processes involved in the transformation from a line of convective cells to an organized mesoscale system with mesoscale ascent aloft. Figure 10.22 depicts the time evolution of simulated mean vertical motion. After 180 min, the level of the maximum mean upward motion rises from 850 to 550 mbar. This transition corresponds to the greater vertical penetration of convective updrafts, result-

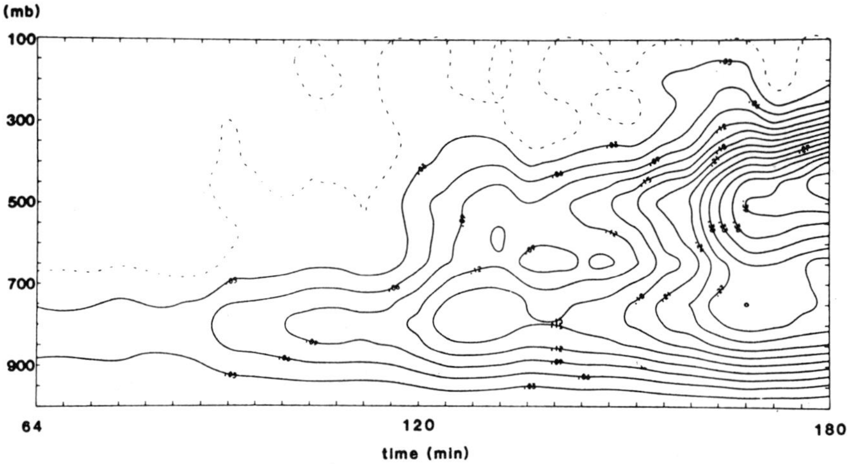

Fig. 10.22. A time–pressure plot of the horizontally averaged vertical velocity ($m\,s^{-1}$) between 64 and 180 min. The spatial average is taken over a 10-km slab in the north–south direction. [From Dudhia and Moncrieff (1987).]

ing in condensation becoming larger than evaporation above 600 mbar. The height of the midlevel jet appears to be crucial in causing this abrupt transition. Early in the system's life cycle, shear associated with the midlevel jet causes the vigorous mixing and detrainment of towers reaching 600 mbar. Convective heating, therefore, is concentrated at low levels. Once the towers penetrate the jet level, they can reach heights of 10 km with vigorous updraft speeds on the order of 15 $m\,s^{-1}$, resulting in an upward shift of the level of convective heating and mean vertical motion.

Dudhia and Moncrieff's simulations reveal some characteristics of momentum transport by convective bands aligned parallel to the lower tropospheric shear. Their simulations predicted that vertical transport of horizontal momentum is downgradient (i.e., it results in a simple mixing of horizontal momentum). They argue that squall lines have a much greater effect on the mean shear profile because they are aligned perpendicular to the shear. Thus, the kinetic energy associated with the mean shear is available to them, causing large eddy-momentum transports. Because the convective bands in nonsquall clusters are aligned parallel to the shear, most of the midlevel jet air deviates around the band and, consequently, cannot transport large-scale momentum as effectively as squall lines. Some of the jet air in the bands is transported downward in downdrafts driven by vertical pressure gradients, rain evaporation, and drag. These downdrafts transport easterly momentum downward, causing a mixing of momentum. A complex interplay exists among environmental shear profiles, the organization of convection, and the resultant vertical transport of horizontal momentum.

Evidence suggests that radiative effects are important to the life cycle of tropical clusters. Oceanic, tropical, deep convection exhibits an early morning maximum and early evening minimum in heavy rainfall, which Gray and Jacobson (1977) attribute to organized weather systems such as cloud clusters. They hypothesize that the diurnal variations result from the following mechanism:

> The atmosphere surrounding organized cloud regions adjusts to its large radiational cooling at night through extra subsidence. This extra night time subsidence increases low-level convergence into the adjacent cloud regions. During the day solar heating reduces tropospheric radiative loss. Clear region subsidence warming and cloud region low-level convergence are substantially reduced.

We will examine other radiative influences on MCSs in our analysis of midlatitude MCSs.

10.2.4 Large-Scale Conditions Associated with Midlatitude Squall Lines and Other MCSs

The large-scale environment of prefrontal squall lines has the features of the warm sector of a midlatitude cyclonic storm and a surface cold front. Figure 9.36 shows schematically a cyclonic storm with its low-level and upper-level jets that provide the necessary vertical shear of the horizontal wind for squall-line maintenance. Differential thermal advection caused by the warm, low-level jet and, in some cases, cold advection by the upper-level jet are destabilizing influences in the warm sector of the storm. Moreover, the moist air advected by the low-level jet is a source of conditional instability. In a developing springtime cyclone to the lee of the Rocky Mountains in the United States, lower tropospheric to midtropospheric air frequently originates over the warm, dry Mexican plateau. As it is swept northeastward by the evolving cyclonic storm, it encounters warm moist air originating over the Gulf of Mexico. The horizontal contrast in moisture forms the "dry line," a well-known source region for severe weather. As the warm dry air migrating from the Mexican plateau moves over the slightly cooler, Gulf of Mexico air, a pronounced capping inversion is created which inhibits the consumption of moist static energy by small convective elements (Carlson *et al.*, 1980). Combination of all these factors creates an environment capable of sustaining long-lived squall-line convective systems.

Miller (1959) emphasized the significance of the dry-air intrusions between 850 and 700 mbar, not only in sustaining severe convection, but also as an important trigger for initiating those systems. We examine storm-trigger mechanisms more fully in Section 10.4.

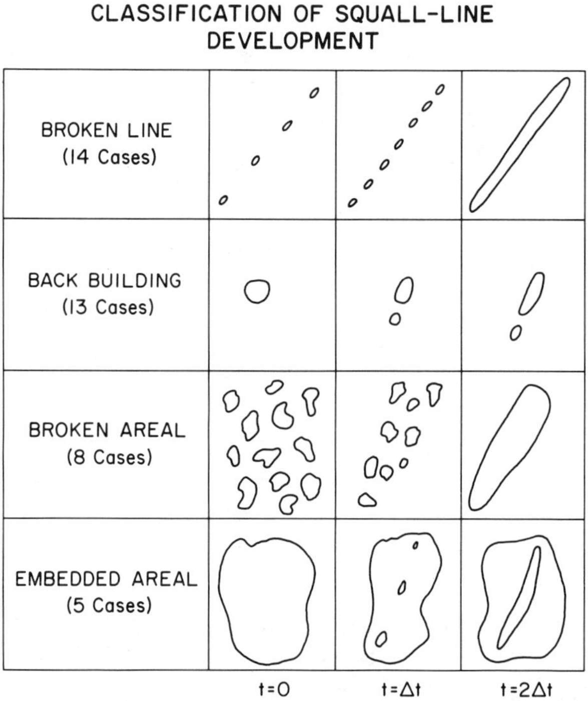

Fig. 10.23. Idealized depiction of squall-line formation. [From Bluestein and Jain (1985).]

Vertical shear of the horizontal wind is a key environmental factor influencing squall-line formation. In our discussion of the theories of storm propagation in Chapter 9 (Section 9.6), we noted that the Moncrieff and Green (1972) model predicted storm-propagation speed to be a function of a parameter that was proportional to a Richardson number averaged over the depth of the troposphere. Furthermore, Weisman and Klemp (1982) found from numerical experiments that a parameter proportional to the Richardson number could discriminate between environments supporting supercell storms versus multicellular storms. Bluestein and Jain (1985) identified four different classes of severe squall-line development: broken lines, back-building lines, broken areal lines, and embedded areal lines, as illustrated in Fig. 10.23. They defined the Richardson number R_i as

$$R_i = \frac{\text{CAPE}}{\frac{1}{2}[(\bar{U}_6 - \bar{U}_{0.5})^2 + (\bar{V}_6 - \bar{V}_{0.5})^2]}, \tag{10.1}$$

where $\bar{U}_6$ and $\bar{U}_{0.5}$ represent pressure-weighted means over the lowest 6 and 0.5 km. They found that the average R_i for broken-line storms is 111 which, as found by Weisman and Klemp (1982), favors multicell storms. The average R_i for back-building lines is 32, similar to the value Weisman and Klemp found for supercell storms. Bluestein and Jain argue that the low values of R_i for back-building lines are a result of strong vertical shear in their environment. Thus, wind shear influences the particular mode of organization of squall lines and their speed of propagation. Because isolated supercells and back-building lines exist in the same R_i regimes, R_i cannot distinguish between environments supporting one of these storm types. Bluestein *et al.* (1987) extended Bluestein and Jain's analysis to nonsevere squall lines. The principal difference between the environments of severe and nonsevere squall lines is that the convective available potential energy in the nonsevere squall-line environment is about 60% of that found in severe squall-line environments. Vertical shear is less in the nonsevere squall-line environments and, therefore, the bulk Richardson numbers are larger.

Another form of MCS frequently observed in middle latitudes is the MCC, whose characteristics are listed in Table 10.1 (Maddox, 1980). In a composite of 10 MCC cases, Maddox (1983) found that the pre-MCC environment, as determined by 0000 GMT data taken just before typical storm maturity, is characterized by abundant moisture through a deep layer and weak vertical shear of the horizontal wind. Lin (1986) found that typical environmental values of R_i for MCCs range in the hundreds where multicellular storms prevail.

Figure 10.24 illustrates the environment of the MCC genesis region as determined by Maddox (1983). At low levels, MCCs are associated with a nearly stationary surface front oriented in a general east–west direction. The low-level flow (south of the front) is marked by warm advection. At

Table 10.1

Mesoscale Convective Complex Definition

Size:	(A) Cloud shield with IR temperature $\leq -32°C$, must have an area $\geq 100{,}000\ km^2$
	(B) Interior cold cloud region with temperature $\leq -52°C$, must have an area $\geq 50{,}000\ km^2$
Initiate:	Size definitions A and B are first satisfied
Duration:	Size definitions A and B must be met for a period ≥ 6 h
Maximum extent:	Continuous cold cloud shield (IR temperature $\leq -32°C$) reaches maximum size
Shape:	Eccentricity (minor axis/major axis) ≥ 0.7 at time of maximum extent
Terminate:	Size definitions A and B no longer satisfied

[a] From Maddox (1980).

850 mbar, a pronounced low-level jet feeds warm, moist air into the MCC precursor environment. The systems reach maturity at the time that the diurnally varying low-level jet reaches its maximum intensity over the High Plains of the United States. At 700 mbar, a weak short wave is evident with warm-air advection over the generating region. In several case studies, McAnelly and Cotton (1986) found that individual meso-β-scale thunderstorm elements followed the contours of the 700-mbar height fields and merged into a meso-α-scale MCC system in the confluence zone of the height field. Maddox noted a weak shortwave trough at 500 mbar west of the generating zone of the systems. This shortwave trough progresses eastward throughout the storm's life cycle and is evidenced by larger amplitude perturbations in the 500-mbar height and temperature fields as the storm system matures.

In some cases, MCCs form in episodes on several consecutive days. Wetzel *et al.* (1983) analyzed an 8-day episode in early August 1977 in

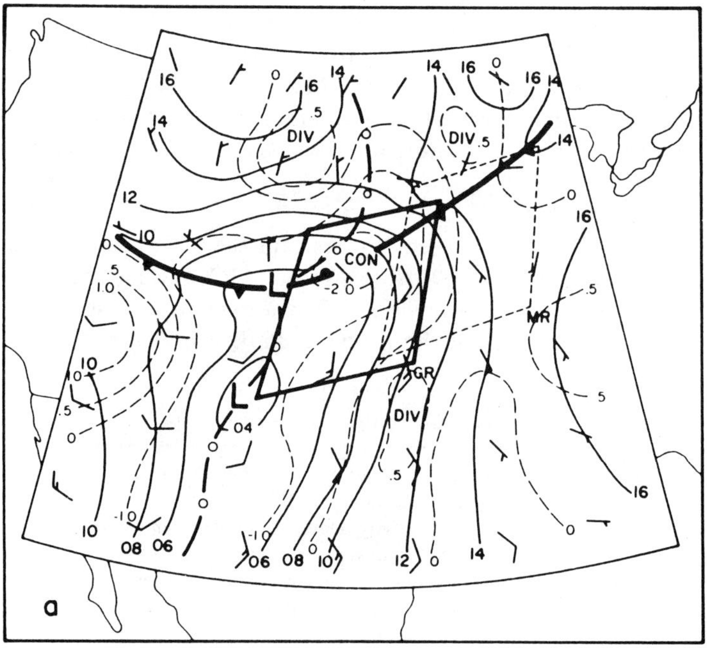

Fig. 10.24. (a) Analysis of surface features prior to MCC development. Surface winds are plotted at every other grid point (full barb $=5\ \mathrm{m\ s^{-1}}$). Isobars of pressure reduced to sea level are shown as solid lines, with surface divergence ($\times 10^{-5}\ \mathrm{s^{-1}}$) shown as dashed lines. (b) Analysis of surface features prior to MCC development. Isopleths of mixing ratio ($\mathrm{g\ kg^{-1}}$) are solid lines and isotherms (°C) are dashed. (c) Analysis of the 850-mbar level prior to MCC development. Heights (meters) are heavy solid contours, isotherms (δeC) are dashed, and

which one or more MCCs formed. In analyzing the large-scale conditions that prevailed throughout the period as well as shortly before and after the episode, Culverwell (1982) showed that two large-scale circulation features combine to establish a deep, moist, conditionally unstable atmosphere over the High Plains of the United States. At low levels over the High Plains, the Atlantic subtropical high establishes itself far enough westward to drive warm, moist air northward over the High Plains. This circulation favors the development of a vigorous nocturnal low-level jet laden with moist air. The generally southerly flow associated with the Atlantic subtropical high is a factor in retarding the southward penetration of weak synoptic-scale surface fronts and aids the establishment of an east–west-oriented quasistationary front across the central High Plains.

At the same time, Culverwell ascertained that the midlevel moistening of the air mass was a consequence of the development of the southwest monsoon (Rasmussen, 1967; Hales, 1972; Brenner, 1974). This circulation

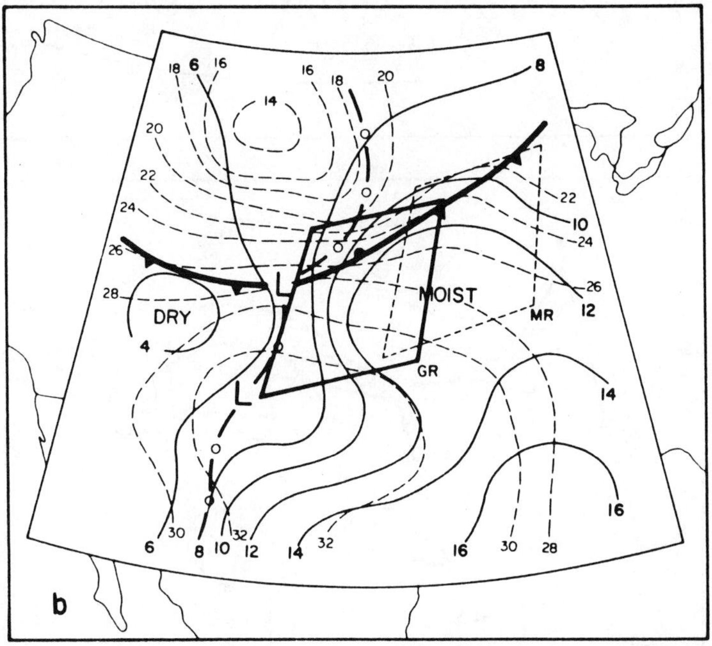

mixing ratio ($g\,kg^{-1}$) is indicated by light solid contours. Winds (full barb = $5\,m\,s^{-1}$) are plotted at every other grid point and dark arrow shows axis of maximum winds. The cross-hatched region indicates terrain elevations above the 850-mbar level. (d) As in (c), but for the 700-mbar level prior to MCC development. (e) As in (c), but for the 500-mbar level prior to MCC development. (f) As in (c), but for the 200-mbar level prior to MCC development (note that wind flat = $25\,m\,s^{-1}$). [From Maddox (1983).] (*Figure continues.*)

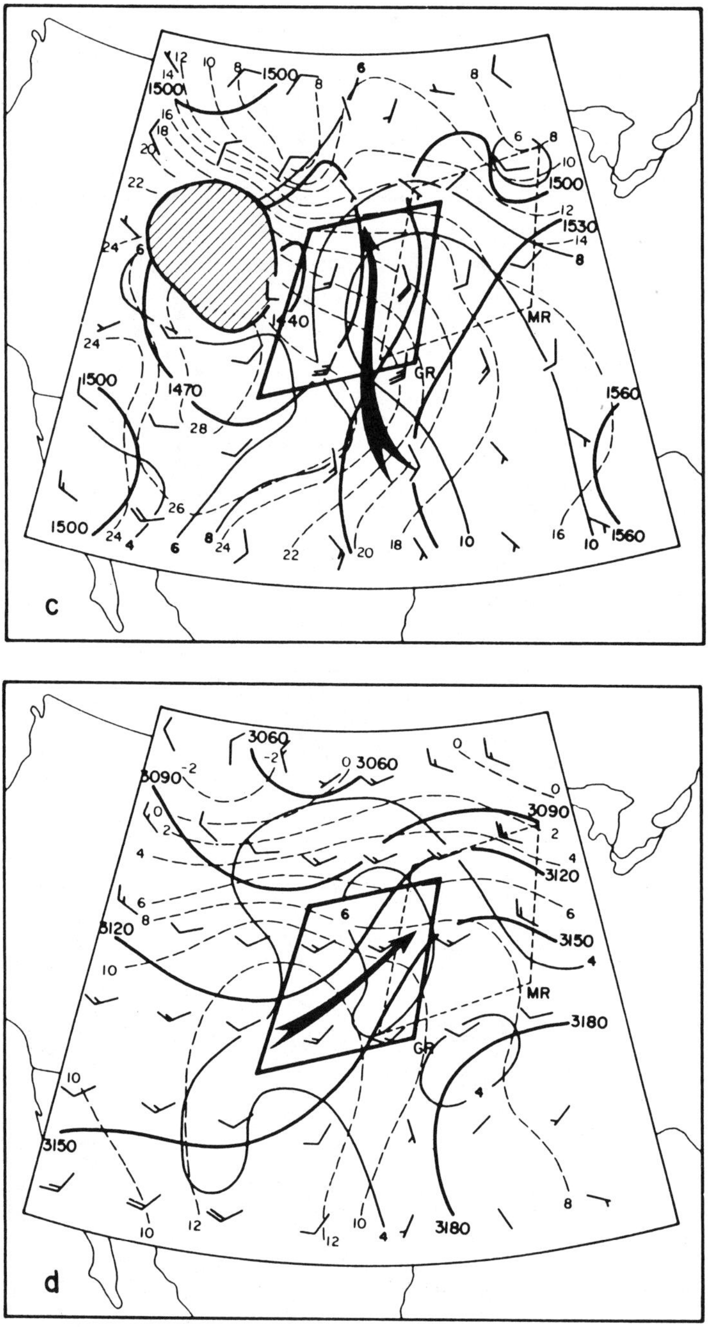

Fig. 10.24 (*continued*)

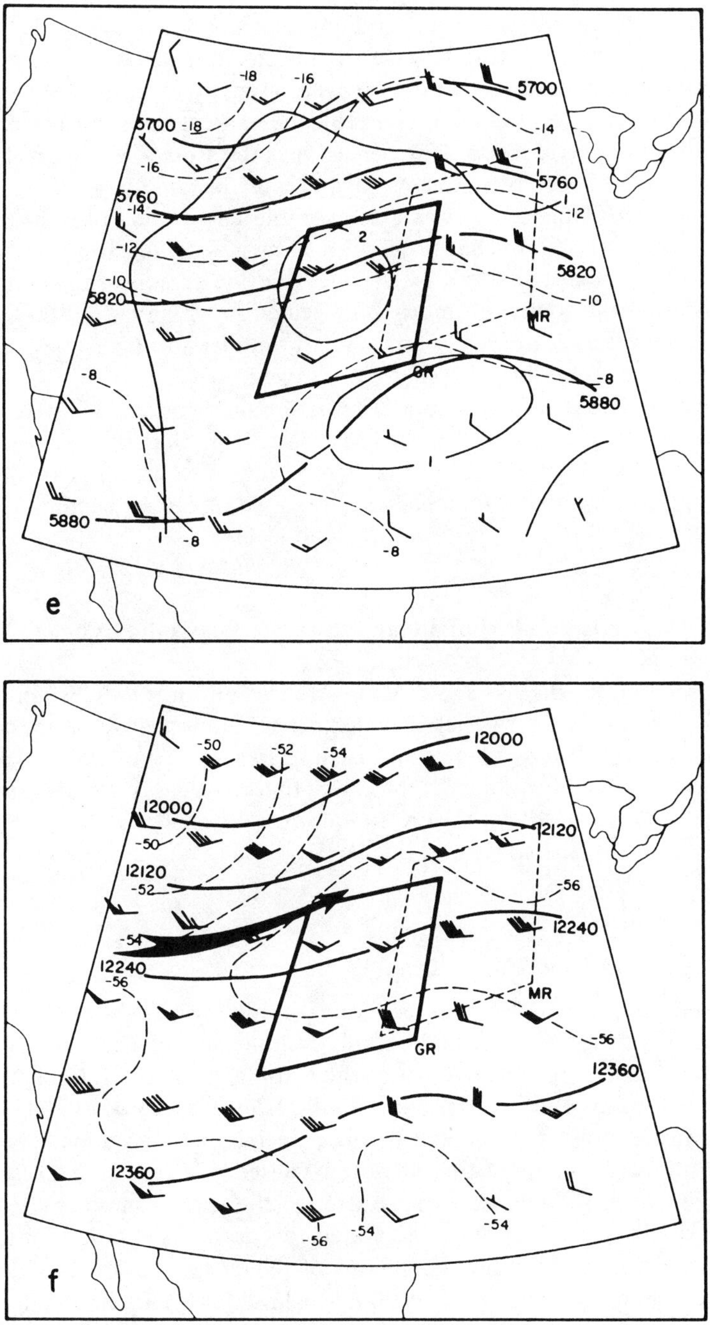

Fig. 10.24 (*continued*)

pattern, which usually reaches maximum intensity between 15 July and 15 August, transports moisture-laden air from the Gulf of California and the southern Pacific Ocean northeastward over the central Colorado Mountains. Deep convection over the mountains aids the vertical transport of moist air into midtropospheric levels as it passes over the Rocky Mountain barrier (Section 10.4). These two streams of moist air become vertically aligned over the High Plains and provide a favorable environment for sustaining MCC-scale convection. That both sources of moistening were needed to sustain MCCs was evidenced by the fact that the episode ended with the weakening of the southwest monsoon circulation which cut off the supply of middle-level moisture. Only a few small thunderstorm clusters could be sustained over Texas in the days that followed.

The main distinguishing feature we can identify between an environment capable of sustaining squall lines versus MCCs is that MCCs prevail in low-shear environments with deep moisture, whereas squall lines are favored in strongly sheared environments with a tongue of dry air at midlevels.

10.3 Characteristics of Midlatitude Mesoscale Convective Systems

In this section, we discuss squall lines, MCCs, and other MCSs and some factors involved in their genesis. A distinct separation of these mesoscale system types is not possible because each shares many common properties with the other classifications. To the extent that it is possible to discriminate among the classes of MCSs, the prefrontal squall line and the MCC represent opposing ends of the spectrum.

10.3.1 Squall Lines

10.3.1.a Prefrontal Squall Lines

The prefrontal squall line forms along or in advance of a cold front associated with a vigorous, midlatitude cyclonic storm. The line forms in the warm sector of the cyclone where the warm conveyor-belt (Chapter 11) air slowly ascends ahead of the surface cold front. This low-level jet carries with it warm, moist, high-valued θ_e air, especially in springtime. The low-level air is overrun by warm, dry, low-valued θ_e air that has recently descended from the middle troposphere. In the springtime to the east of the Rocky Mountains in the United States, this normally explosive situation is further intensified by the interactions of the cyclonic storm circulations with the regional geography. Thus, the low-level jet is enriched with moist air originating over the Gulf of Mexico. The air immediately above the moist surface air often has its origins over the warm, dry Mexican plateau, yielding a strong inversion which caps intense convective storms until a triggering mechanism permits their sudden outbreak (Carlson *et al.*, 1980). Moreover, the vertical wind profile in the warm sector of the cyclone yields a wind shear profile that is favorable for long-lived thunderstorm systems. All of these features combine to form an environment capable of sustaining some of the most violent outbreaks of severe thunderstorms on earth. Fortunately, only rarely does an intense midlatitude cyclone form where it has access to this very explosive combination of air masses. This rarity also means that dense observational programs equipped with modern instrumentation seldom observe a truly severe prefrontal squall line. Because the scale of the system ranges from the individual thunderstorm to the cyclonic storm, the prefrontal squall presents a challenge to modelers. Most observations of these storms are based on routine radar, satellite, and rawinsonde measurements. The models have been limited to two dimensions or coarse-resolution, three-dimensional, regional-scale models.

A classic prefrontal squall line relative to its parent cyclonic storm is shown in Fig. 10.25. The squall line extends to the north and south of the warm front and, once formed, it can move as much as 1000 km ahead of the advancing front. As shown in Fig. 10.26, while the line moves eastward, individual thunderstorms form at the southern end of the line and move northeastward. Occasionally, the individual thunderstorm elements are long-lived rotating supercell storms, but more frequently they are severe multicellular types, exhibiting lifetimes on the order of an hour.

A partly schematic model of a prefrontal squall-line storm system developed by Newton (1966) is shown in Fig. 10.27. Noteworthy is the upshear-tilted updraft that rises from the subcloud layer and overshoots the tropopause, penetrating the stratosphere for several thousand meters. As shown in the schematic, the downdraft has its origin centered about 500 mbar and transports low-valued θ_e air downward to the surface. The evaporatively chilled air advances forward, undercutting the inflow air and creating a lifting mechanism for sustained propagation of the storm system. The downdraft outflow also spreads to the rear of the storm system. Knupp's (1985) analysis of thunderstorm downdrafts and more recent three-dimensional squall-line simulations (Rotunno *et al.*, 1988) suggest that a significant contributor to the cold pool of air behind the squall front is updraft air that participates in Knupp's so-called up–down downdraft branch. It is likely that this downdraft branch (not portrayed in Fig. 10.27) contributes to the most vigorous downdrafts immediately behind the squall front. The more slowly descending air trailing the squall front contributes to the large volume of low-valued θ_e air trailing behind the storm.

The schematic also illustrates an upper-level anvil region extending some

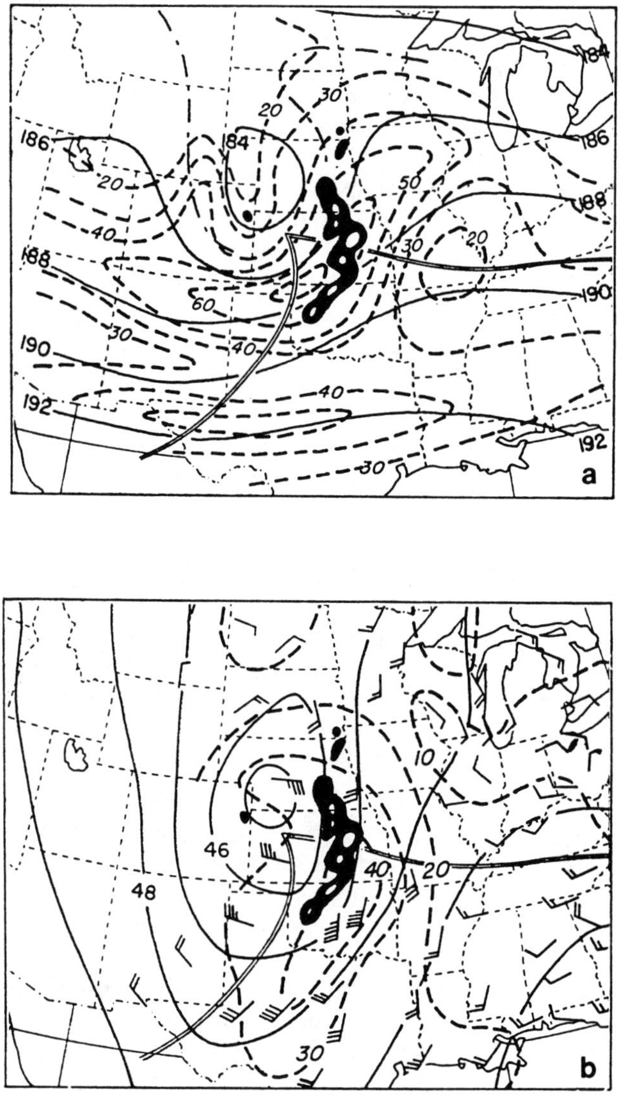

Fig. 10.25. (a) 500-mbar and (b) 850-mbar charts, 2100 CST, 20 May 1949. Solid lines, contours (hundreds of feet); dashed lines, isotachs (knots); thin double lines, surface fronts. Blacked-in areas, rainfall in excess of 0.20 inch h^{-1}; inner light areas, 0.50 inch or more. [From Newton and Newton (1959).]

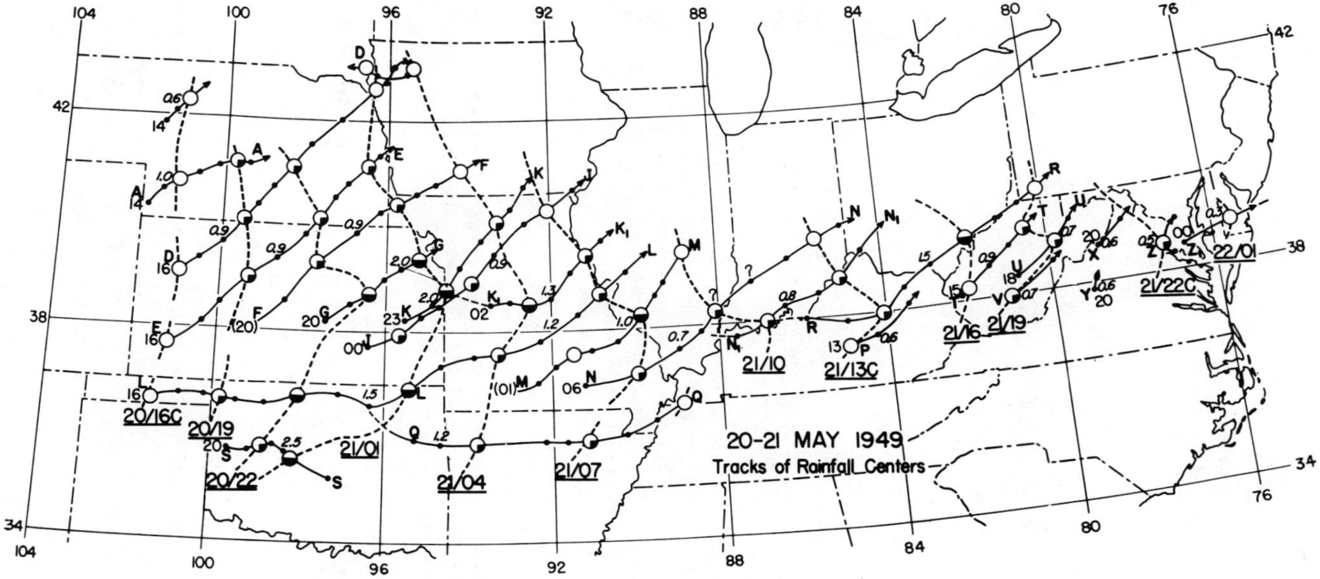

Fig. 10.26. Tracks of individual rainstorms, squall line of 20–21 May 1949. Dashed lines connect rainfall centers observed at same time (date/hour below); dots on tracks show positions at successive hours. In circles at 3-h intervals, intensity: open circle, maximum 1-h rainfall less than 0.50 inch; quarter filled, 0.50 inch or more; half filled, 1.00 inch or more per hour. [From Newton and Newton (1959).]

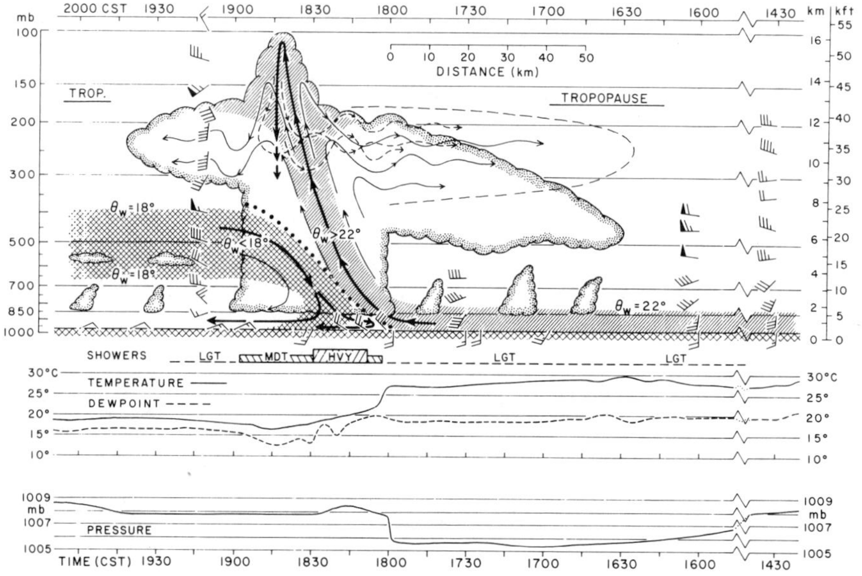

Fig. 10.27. Partly schematic cross section through squall line of 21 May 1961, as it passed Oklahoma City (fivefold vertical exaggeration). Winds plotted with directions relative to north at top of figure. Prior to squall-line passage, there were other thunderstorms in the vicinity; these have been omitted for clarity (the balloon released at 1730 oscillated near 650 mbar for 14 min, probably having encountered the downdraft of one of the earlier storms; during its further ascent winds were unmeasurable because of low-elevation angles). Hatching indicates depth of air with θ_w in excess of 22°C ahead of squall line and its probable extent in updraft and upper portion of cloud. Cross-hatching indicates extent of air with θ_w less than 18°C, based on sounding behind storm and on surface observations. A similar layer at closely corresponding heights was present before squall-line passage; this is omitted for the sake of clarity. At bottom, periods of precipitation, and temperature, dew point, and sea level pressure traces (somewhat smoothed) are shown. Heavy arrows, axes of main drafts; light arrows, relative streamlines, dashed where air emanates from core of stratospheric tower. At right, long dashes suggest outline of mass of air plume originating in storm and spreading out essentially horizontally; the radar-detected cloud plume, at lower elevations, consists of small precipitation particles which have partly fallen out of the air plume. [From Newton (1966).]

50 to 60 km in advance of the surface squall and a smaller anvil region located upshear. Light precipitation falls from the stratiform regions, with moderate to heavy precipitation below the intense convective regions. As recognized by Ludlam (1963), the upshear tilt of the updraft of severe prefrontal squall lines is an important dynamic property of the storm system and is responsible for the long lifetimes of the squall-line convective cells. As a consequence of the upshear-tilted updraft, precipitation settles out of

the updraft, unloading the updraft from the burden of the condensed water substance. At the same time, the precipitation contributes to downdraft initiation by water loading and supplies the downdraft with water to sustain evaporative cooling. This results in an updraft and downdraft couplet in which the downdraft-induced gust front lifts air feeding the updraft and enhances the updraft circulation. The updraft and downdraft circulations operate constructively to create an efficient convective machine.

The mechanisms responsible for the upshear tilt of the updraft are still debated in the literature. Bates (1961) and Newton (1966) argued that the updraft conserves its easterly momentum as it rises and thereby moves in a storm-relative sense toward the rear of the storm as it encounters environmental air of increasing westerly strength. Seitter and Kuo (1983) argued that Moncrieff's (1981) investigations of two-dimensional nonprecipitating cloud circulations in shear flow do not support the Bates and Newton concepts. Moncrieff showed that with deep tropospheric shear, the updraft tilted downshear (Fig. 9.28), contrary to the conservation of horizontal momentum arguments. Only when the shear was confined to below 2.5 km could Thorpe *et al.* (1982) obtain the steady upshear-tilted updraft circulation shown in Fig. 9.30. Seitter and Kuo (1983) argued that water loading by precipitation would erode the downdraft side of the updraft because of the large quantities of liquid water present. They argued that a balance between the water-loading mechanism and environmental shear maintains an updraft and downdraft couplet at a stable upshear slope. Their mechanism, however, seems to require an initial upshear tilt of the updraft. When they performed calculations without water loading, an upshear-tilting updraft was obtained that was driven by the horizontal spread of the cold pool caused by the evaporation of precipitation. The cold pool apparently undercut the low-level inflow, initiating the upshear-tilted updraft.

Further clarification of the role of the cold pool in inducing an upshear-tilted updraft has been obtained in the two-dimensional numerical experiments by Thorpe *et al.* (1982) and the two- and three-dimensional numerical simulations by Rotunno *et al.* (1988). The upshear tilt of the convective system signals the increasing dominance of the cold pool and its interaction with lower tropospheric shear. Thorpe *et al.* (1982) argued that the cold pool acts as an obstacle to the flow. At the interface between the outflow from the cold pool and the storm-relative inflow, steady convergence is established which prolongs the cell's life. If the storm-relative inflow is weak, $|u_0| \geq 15\ \mathrm{m\ s^{-1}}$ (e.g., the low-level shear is weak), the gust front is not stationary, and localized boundary layer convergence is not maintained and the convection dies. These results are consistent with earlier two-dimensional simulations reported by Takeda (1971), who concluded that a jet at low levels, particularly at 2.5 km AGL, is most favorable for producing a "long-lasting" cloud system. Rotunno *et al.* (1988) interpreted the cold-pool and relative-inflow interaction in terms of the opposing shears present on each side of the cold-pool and inflow interface. They noted that the circulation associated with a cold pool is characterized by a negative horizontal component of vorticity at the cold-pool and inflow interface. In contrast a strong relative inflow, particularly one with a low-level jet, is characterized by a positive vorticity of the low-level shear. They argued that the optimum situation for maintaining steady, upshear-tilted convection is one in which the negative vorticity associated with the cold pool and the positive vorticity associated with the inflow are approximately balanced. Because cold pools are necessarily shallow, shear can promote convection only when restricted to low levels. This conclusion is consistent with Bluestein and Jain's (1985) composite study, which revealed that squall lines, in general, are oriented along the shear vector in the lowest level.

Unfortunately, this leads to a dilemma—the environment of severe prefrontal squall lines is characterized by strong shear through the depth of the troposphere, and yet the above numerical simulations, as well as two-dimensional numerical simulations reported by Hane (1973) and Seitter and Kuo (1983), among others, suggested that a steady upshear-tilted updraft circulation cannot be maintained in an environment with strong shear through the depth of the troposphere. Hane suggested that this is an artifact of two-dimensional models that cannot allow the strong upper-level flows to divert around the rising updraft air. Nonetheless, Hane's simulations did show that strong shear leads to more intense, longer lasting, and broader circulations, compared to cases with weak or moderate shear. This, he argued, results from the faster translation speeds of the storm system in strong shear, which enhances the relative inflow of warm, moist air (large CAPE) into the storm updrafts.

Following a suggestion by Lilly (1979), Rotunno *et al.* (1988) performed a three-dimensional numerical simulation in which shear was introduced at a 45° angle relative to the squall lines through a 5-km layer, yielding a shear of $30\,\mathrm{m\,s^{-1}}$ through the 5-km depth. The resulting squall line (Fig. 10.28) is composed of a line of rotating supercell-like storms. Lilly (1979) speculated that each supercell propagating to the right of the shear vector would remain uninhibited by other cells, whereas leftward members of a split-cell pair would move back into the cold air behind the line and decay. The Rotunno *et al.* simulation generally supports Lilly's hypothesis. Earlier simulations of supercell storms have revealed that they thrive on deep tropospheric shear (e.g., Rotunno and Klemp, 1985). The updraft rotation induces a vertical pressure gradient that preserves strong inflow into the storm's updraft (Chapter 9). Lifting by the cold pool is relegated to a minor role in the supercell. The implication then is that severe prefrontal squall

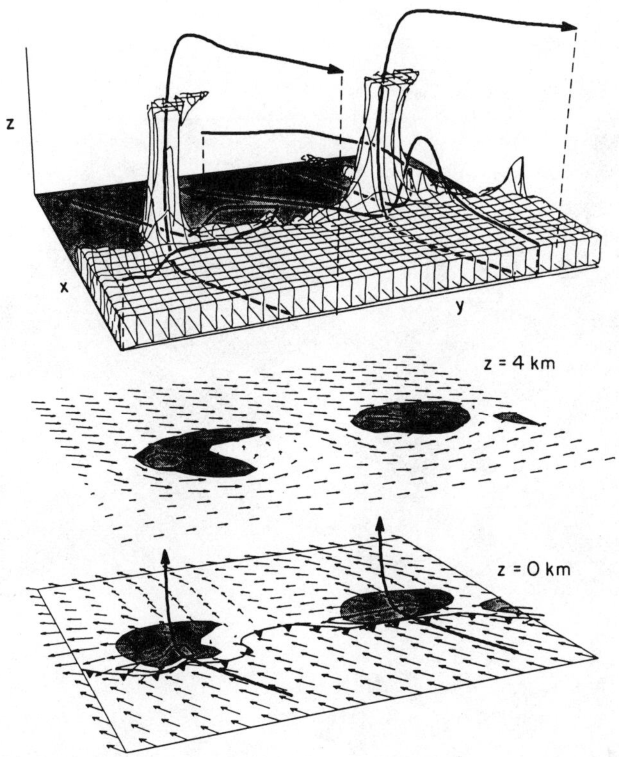

Fig. 10.28. Top: a three-dimensional perspective view of the $\theta_e = 335$ K surface for a portion ($56 \le x \le 116$ km, $14 \le y \le 82$ km) of a line containing two supercells. Below there is the $z = 4$ km horizontal plane exhibiting line-relative flow vectors at every other grid point (a length of two grid intervals $= 20\ \mathrm{m\ s^{-1}}$); the shaded regions encompass places where rainwater exceeds $0.1\ \mathrm{g\ kg^{-1}}$; the circular contour encompasses updraft greater than $10\ \mathrm{m\ s^{-1}}$. The flow in the horizontal plane at the surface is denoted similarly, except the updraft contour (at $z = 350$ m) encompasses values greater than $1\ \mathrm{m\ s^{-1}}$ and the barbed line denotes the cold-air boundary defined by the -1 K perturbation. [From Rotunno *et al.* (1988).]

lines can be maintained as a steady line of thunderstorms in the presence of strong, deep shear because such squall lines are composed of rotating, supercell-like storm elements. We shall later present a case of a more ordinary squall line existing in the presence of strong deep shear. This system is composed of a supercell-like element and numerous, shorter lived, ordinary thunderstorm cells, which suggests that a steady squall-line system can be maintained in the presence of strong, deep shear with both ordinary cells and supercells coexisting.

An example of an extensive prefrontal squall line has been discussed by

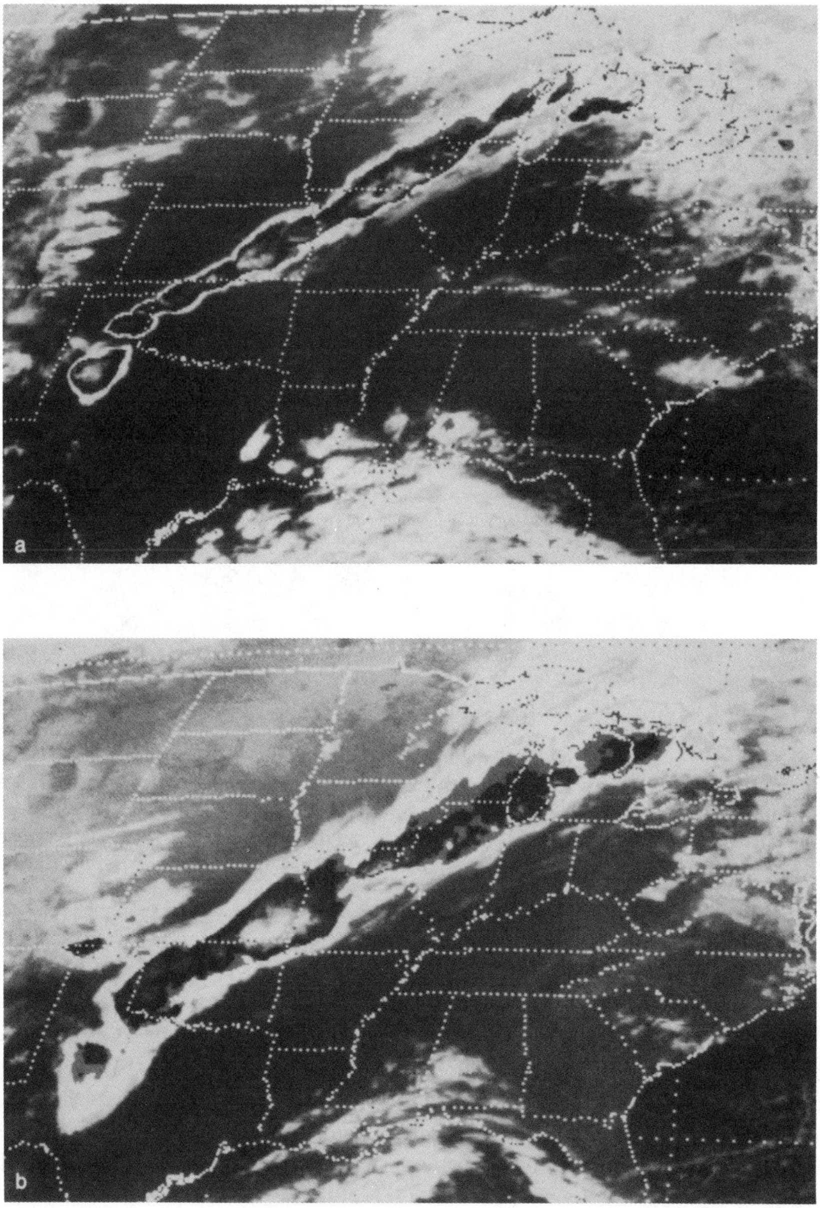

Fig. 10.29. (a) Satellite infrared image showing the squall line at 2000 CDT fairly early in life. (b) As in (a) but at 2330 CDT, showing increase in width of cloud shield. [From Srivastava *et al.* (1986).]

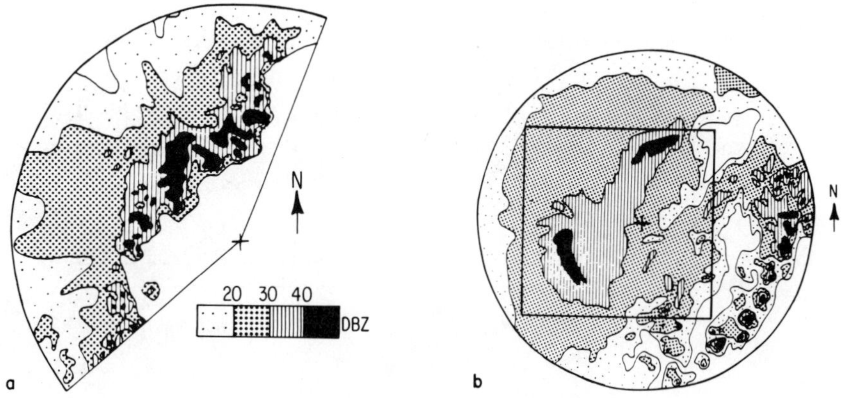

Fig. 10.30. PPI display of CP3 radar reflectivity factors. Maximum range from CP3 radar (+) is 108 km. (a) For 2215 CDT, the elevation angle is 4.5°. (b) For 2346 CDT, the elevation angle is 3.5°. The maxima in reflectivity factor at about 50-km range to the north and southwest are associated with the melting band. The square is the 120 × 120-km area of multiple-Doppler radar analysis. (Note scale change between panels a and b.) [From Srivastava *et al.* (1986).]

Srivastava *et al.* (1986) (hereafter referred to as SML). They performed single- and multiple-Doppler radar analyses of a prefrontal squall line that passed through the Northern Illinois Meteorological Research on Downbursts (NIMROD) experimental area (Fujita, 1978, 1981). The squall line was a large system extending from Illinois to the Texas panhandle (Fig. 10.29). Radar reflectivity PPI displays of the storm system (Fig. 10.30) depict a line of intense convective cells formed a trailing anvil or stratiform cloud early in the observations. An extensive trailing stratiform region can be seen 1.5 h later, separated from the squall-line convection by a band of weak echo. SML suggested that a factor in the formation of the stratiform region was the general southeastward propagation of the squall line combined with the northeastward motion of the individual storm cells. This propagation favored the merger of anvil debris from decaying cells with active anvils.

As in tropical stratiform regions, the trailing anvil exhibited a characteristic bright band near the melting level. Vertical fall speed estimates with the single-Doppler radar VAD technique suggested that from 11 to 6.5 km (temperatures colder than −12°C) fall speeds increased from 1 to 2 m s^{-1}, typical of individual snow particles being transformed to small aggregates. Between 6.5 and 4.2 km (−10 to 0°C), particle fall speeds and radar reflectivities did not change appreciably, suggesting the presence of large aggregates that did not significantly change fall speed with size. The inferred microstructure is similar to that found in the stratiform region of tropical

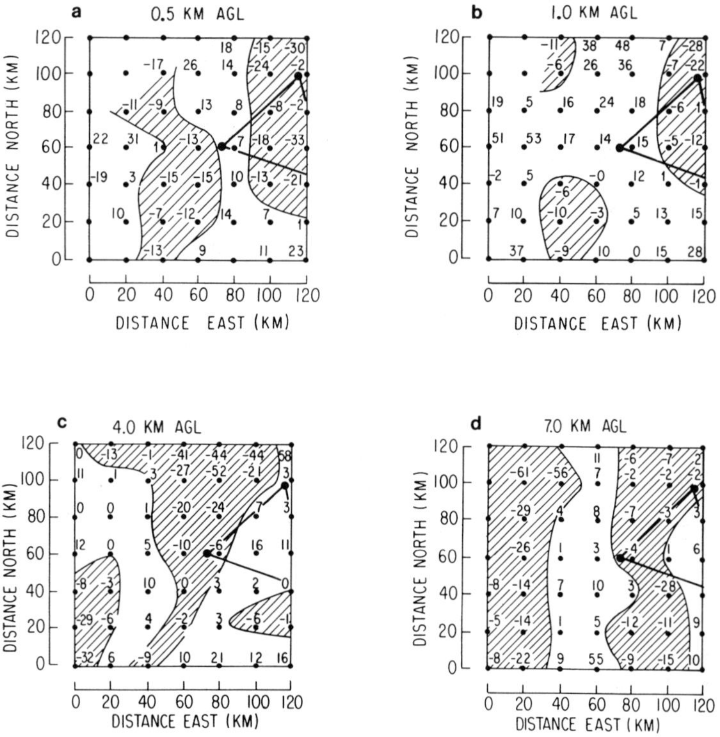

Fig. 10.31. Divergence of the horizontal wind ($10^{-5}\ s^{-1}$) from the MDOP analysis at (a) 0.5 km, (b) 1.0 km, (c) 4.0 km, and (d) 7.0 km AGL (add 0.226 km to get heights above sea level). Convergence regions are hatched. The Doppler triangle is indicated. Figure is based on data collected during the time period 2345–2354 CDT. [From Srivastava *et al.* (1986).]

nonsquall clusters (Churchill and Houze, 1984a). Variations in crystal concentrations and habit suggested that the stratiform cloud was not uniform in the horizontal. SML used multiple-Doppler radar analysis to show a banded structure in the stratiform region, as one might expect from gravity waves in stably stratified flow (see Fig. 10.31).

SML also inferred divergence and mesoscale motions in the stratiform region of the system using single-Doppler radar analysis. The divergence profile (Fig. 10.32) exhibits low-level convergence with divergence between 1.0 and 2.5 km and a zone of convergence between 2.5 and 5.0 km. The maximum of convergence in this layer is at 3.5 km in the melting layer. A deep layer of convergence exists between 6.0 and 10.0 km, capped by a layer of divergence near the tropopause. Except for the region of strong surface convergence, the divergence profiles are similar to those obtained

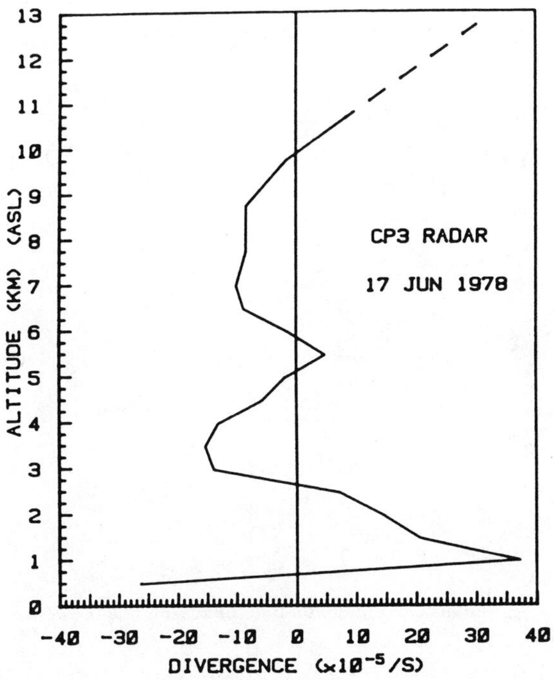

Fig. 10.32. Vertical profile of divergence of the horizontal wind from the EVAD analysis. The dashed line is an extrapolation of the profile to the satellite-estimated cloud top. The data for the EVAD analysis were acquired during 2345–2354 CDT. [From Srivastava *et al.* (1986).]

from the analysis of rawinsonde data in tropical systems and in the midlatitude system described by Ogura and Liou (1980) (discussed later). SML argued that the intense surface convergence below 270 m is real because it could be seen in the multiple-Doppler analysis as well as in the surface-wind data. Its absence in other mesoscale systems may result from the deficiencies of rawinsonde analysis that requires system-relative steadiness and the operator's ability to track balloons at the lowest levels.

The inferred vertical motion field from the divergence profile described above is shown in Fig. 10.33. The three estimates shown depend on the boundary conditions applied to the divergence profile. SML argued that curve C is the most consistent profile; it exhibits upward motion near the surface and a layer of descent from 1.1 to 4.5 km. The 4.5-km level marks the division between unsaturated and saturated air. The region of near-zero vertical motion above 4.5 km corresponds to a stable layer associated with the melting level. Above 6.5 km, a deep layer of ascent is inferred that is

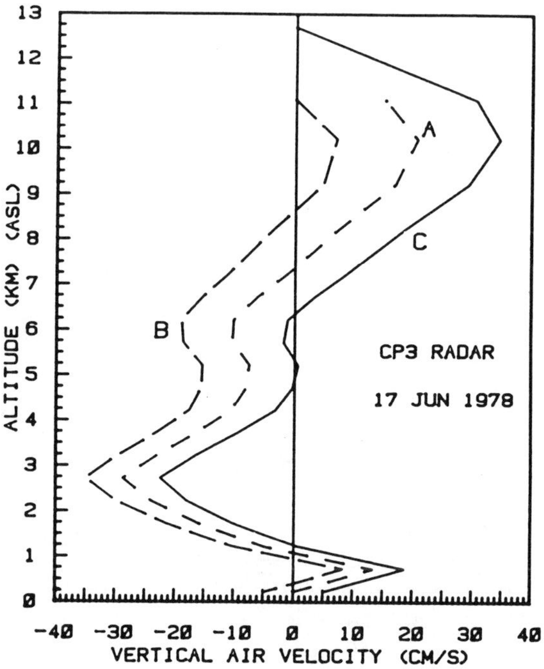

Fig. 10.33. Profiles of vertical air velocity, ω, from integration of the divergence in Fig. 10.32. Curve A assumes $\omega = 0$ at the ground (226 m ASL) and uses the solid divergence profile of Fig. 10.32. Curves B and C assume $\omega = 0$ at the radar and satellite-estimated cloud tops, respectively. Curve B uses the solid divergence curve, while curve C uses the dashed extrapolation also. [From Srivastava *et al.* (1986).]

similar to that found in the stratiform region of tropical and midlatitude MCSs.

Another interesting feature of SML's study is their analysis of the wind field distribution transverse to the squall line. Figure 10.34a illustrates the wind field obtained from multiple-Doppler radar analysis and Fig. 10.34b illustrates the corresponding radar-reflectivity field. Assuming that the instantaneous wind represents a system-relative, two-dimensional, stationary flow field and that horizontal momentum is conserved, they interpreted the analysis as follows: (1) above 3 km, air enters the forward portion of the stratiform region and ascends to above 6.5 km, where it exits the anvil; (2) between 3 and 6.5 km, air enters the rear of the stratiform region and descends toward the front of the system, where it exits in a layer extending from the ground to 3 km; and (3) surface air ahead of the old convective line ascends in the stratiform region in a layer that deepens to 3 km to the

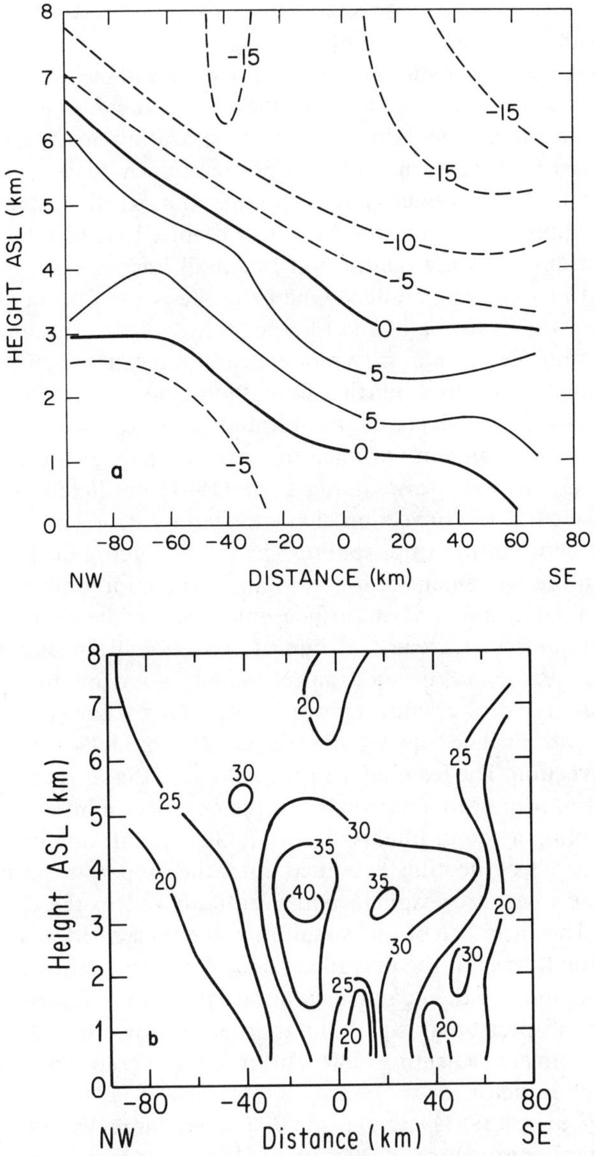

Fig. 10.34. (a) Cross section of wind component transverse to the squall line and relative to it. The contours are labeled in m s^{-1}. Positive numbers indicate winds toward the southeast. The cross section is along the northwest–southeast diagonal of the square in Fig. 10.30b. The projection of the CP3 radar on the diagonal is the origin of the distance coordinate (see text). (b) Cross section of reflectivity factor (dBZ) distribution along the same diagonal as in (a). [From Srivastava *et al.* (1986).]

rear of the system. A new convective line is at an abscissa value of about 90 km, outside the range of the plot.

Until now, we have focused on the storm-scale and meso-β-scale characteristics of prefrontal squall lines. In concert with the dramatic storm-scale features of prefrontal squall lines is a line organization on the meso-α-scale. Detailed observational documentation of the meso-α-scale organization of severe prefrontal squall lines is not available, but excellent documentation of nonprefrontal squall-line structure is available. Evidence for the simultaneous existence of the organization of squall lines on the meso-α-scale is suggested in modeling studies. One example is the Chang *et al.* (1981) simulation of the prefrontal squall line of 6 May 1975. The finest mesh in their simulations was 35 km; such model resolution cannot resolve explicitly the deep, moist convection. Furthermore, their convective parameterization does not include the effects of unsaturated moist downdrafts, which are believed to be of major importance to storm propagation (see discussion in Section 9.5). Nonetheless, Chang *et al.* (1981) predicted the formation of an extensive line of convection characterized by strong mesoscale ascent through the depth of the troposphere. The initial forcing of the system was attributed to the movement of dry cold air in the middle troposphere over northward-moving moist, warm surface air as a part of the attendant cyclonic storm circulation. This created a zone of conditionally unstable air. Low-level thermal advection, in addition to creating a narrow tongue of warm air, contributed to surface convergence. Thus, a narrow zone of conditionally unstable air with surface convergence triggered the model's parameterization of deep convection. The resultant heating enhanced the surface convergence, which, in turn, produced a narrowing of the convection into a 100-km-wide band resembling a squall line. As the simulated squall line intensified, the level of convective heating extended into the upper troposphere. This stabilized the lower troposphere and eventually led to the dissipation of the system. The short life of this squall line, they reasoned, may be characteristic of squall lines in the occluded zone of cyclonic storms. It may also reflect the inability of the coarse-resolution model to develop the observed upshear tilt of prefrontal squall lines, thereby lowering the stabilizing influence of convective heating aloft. The absence of convective-scale downdrafts may be a factor too.

Chang *et al.*'s. (1981) simulation displayed large values of cyclonic vorticity (maximum values of $20 \times 10^{-5}\,\mathrm{s}^{-1}$) similar to observed values of like storms. Convergence was the major contributor to the low-level vorticity production in the model simulation.

An even more dramatic example of the meso-α-scale characteristics of prefrontal squall lines is Orlanski and Ross's (1984) simulation. The model used in their study, described by Ross and Orlanski (1982), is hydrostatic with a horizontal grid spacing of 61.5 km. In spite of the coarse grid spacing, they did not represent the effects of deep moist convection by a convective parameterization. Instead, they relied on the explicitly resolved motions to release conditional instability and to provide convective heating and evaporative cooling. The model contains neither an explicit precipitation scheme nor evaporation of rain; only cooling by evaporation of explicitly predicted cloud water is permitted. Thus, a subsaturated, precipitating mesoscale downdraft is not simulated. The only feature of their model that can be construed as a convective parameterization scheme is the introduction of a vertical eddy-viscosity closure that is a function of a modified form of Richardson number. This has the effect of vertically mixing the model's thermodynamic properties as well as momentum in regions of deep convection.

Orlanski and Ross applied this model to the case of a cold front which occurred on 2 May 1967 over the United States and which produced a severe prefrontal squall line. Despite insufficient resolution to resolve explicitly the details of deep convection and despite the lack of a convective parameterization scheme, they were able to simulate a squall-line-like feature which moved ahead of the cold front and contained many mesoscale features observed in midlatitude squall lines. Because detailed mesoscale analyses of prefrontal squall lines were not available, they compared their model predictions with Ogura and Liou's (1980) analysis of an ordinary midlatitude squall line. Some features that the model and observations have in common include the fields of divergence. In both cases, maximum values of convergence occur at midlevels approximately 120 km to the rear of the surface maximum but the level of the midlevel convergence maximum is about 150–200 mbar higher in the observed case. Ogura and Liou argued that the midlevel convergence maximum was linked to divergence beneath this midlevel feature caused by water loading and evaporation of falling rain. The Orlanski and Ross model had no mechanism for rainwater evaporation and contained only a modest effect of water loading due to cloud water. Even so, a structure of midlevel convergence and lower-level divergence out of phase with surface convergence was simulated. They used a simple two-layer linear model to strengthen their hypothesis that the midlevel convergence maximum is driven by a net positive buoyancy from latent heat release above the level of free convection. They argued that the low-level convergence maximum may occur as forced convection since it occurs in the stable layer below the cloud layer.

Associated with the simulated divergence field was an updraft field that resembles the observed profile, including a midlevel updraft above a low-level downdraft displaced 120 km behind the low-level updraft. The observed low-level downdraft is considerably more intense, presumably because of water loading and melting, and evaporation of falling rain.

Both the model and observations exhibit strong cyclonic vorticity at low levels, but Ogura and Liou's analysis displays strong cyclonic vorticity at middle levels which they attribute to vortex stretching as a result of the midlevel convergence. Orlanski and Ross's simulation contained a strong anticyclone at middle levels to the rear of the surface cyclone. In their case, the anticyclone results from an enhancement of the horizontal shear in the southwesterly jet stream due to the simulated convection. Therefore, the differences in midlevel vorticity fields may result from differences in environmental wind fields associated with the prefrontal environment compared to the environment of ordinary squall lines.

Orlanski and Ross also noted that the model excited an apparent gravity-wave response in the stable air mass above and surrounding the cloud region similar to that found experimentally by Cerasoli (1979). As the simulated, explicitly resolved convection grew and decayed, gravity waves were produced. Orlanski and Ross pointed out that if a convective system is to survive in a stably stratified environment, it must be in dynamic balance with the surrounding stably stratified atmosphere. Otherwise, the penetrative convection within the system will produce a field of dispersive gravity waves that will carry away much of the energy of the system and destructively interfere with the organization of the system. If a system is to maintain its organized structure, it must maintain not only its low-level supply of moist static energy or CAPE which sustains convection, but it must also have a physical structure and propagation speed consistent with a solitary wave (as opposed to a family of waves) that can be maintained in the surrounding stable environment. Such solitary waves exist for many times their intrinsic period because of their nondispersive characteristics. Two requirements must be satisfied for this compatibility to occur—the speed of the convective system should be similar to the group velocity of a solitary wave, and the aspect ratio H/L of the horizontal to vertical scales of the system should be compatible with such a nondispersive wave.

The second condition—that the waves be nondispersive—defines the horizontal scale for such storm systems. If the wavelength is too long, Coriolis effects will prevail and the waves will propagate as dispersive inertial-gravity waves. If the wavelength is too short, nonhydrostatic effects will create dispersive waves. Orlanski and Ross specified the following limits on the aspect ratio H/L so that the gravity waves will be nondispersive in a rotating system:

$$f/N \ll H/L \ll 1, \tag{10.2}$$

where N is the Brunt-Väisälä frequency. For characteristic values of f, N, and H, the horizontal scale of most mesoscale systems must lie between 10 and 1000 km. Orlanski and Ross also stated that the horizontal scale of most mesoscale systems lie within the range $H \ll L \ll \lambda_R$, where λ_R is the Rossby radius of deformation NH/f. (A more general definition of λ_R is presented in Chapter 6 and Section 10.3.2.)

Another example of the mesoscale wave character of prefrontal squall lines is the degree of success that wave–CISK models have in simulating the characteristics of squall lines. Wave-CISK models have been successful in predicting squall-line propagation speeds as well as the upshear tilt of the mesoscale updraft profile (Section 9.6, Chapter 9). Their success in predicting observed features, however, depends strongly on the details of the convective parameterization. Moreover, Nehrkorn (1985) found that his wave–CISK model could not predict simultaneously the observed squall-line orientation angle relative to the mean tropospheric winds and the phase speed of the system. This suggests that although wave–CISK models may represent the basic wave nature of squall-line convection, other features such as nonlinear wave–wave interactions or the solitary wave character of actual squall lines may not be well represented by the linear wave–CISK models.

The severe prefrontal, midlatitude squall-line system has dual characteristics. It exhibits intense thunderstorm-scale line organization often observed by quantitative radar and simulated by numerical prediction models and steady-state analytical models such as Moncrieff's (1978) convective overturning model (Section 9.6, Chapter 9). At the same time, the system exhibits mesoscale wave features such as those evident in the analysis of less severe squall lines and modeled by mesoscale numerical models and, to some extent, wave–CISK-type models. The prefrontal squall-line storm best exemplifies the symbiotic nature of interaction between the thunderstorm scale and the mesoscale. At the same time, the prefrontal squall line is triggered by and generally at the mercy of its larger scale cyclonic storm environment and many of its mesoscale and storm-scale features are determined by the dynamic and thermodynamic structure of this environment.

10.3.1.b Ordinary Midlatitude Squall Lines

We have characterized the severe, prefrontal squall line as representing one extreme of the overall family of squall-line mesoscale systems existing in the midlatitudes. More commonly observed is a continuum of squall-line-type systems that are shorter in their along-line component and typically have shorter lifetimes than severe prefrontal squalls. They too may produce severe weather, but the number of severe events is less. We refer to such squall lines as ordinary squall lines. Ordinary squall lines may be associated with weak cyclonic storms and stationary fronts, or they may have no obvious relationship to a cyclonic storm system. Many ordinary midlatitude squall lines exhibit properties similar to tropical squall-line systems and they can be one of the meso-β-scale components of a MCC, just as tropical squall lines can be a substructure of tropical clusters.

An ordinary squall-line system studied in detail is the 22 May 1976 squall line that passed through the National Severe Storms Network in central Oklahoma. Ogura and Liou (1980) examined the storm system from the meso-β-scale perspective using data obtained from a special rawinsonde network and hourly surface-station analysis. With single- and dual-Doppler radar analysis, Smull and Houze (1985, 1987a) obtained a detailed analysis of the flow structure and radar-reflectivity fields of the storm system. An infrared satellite analysis of the storm top indicated that the system met Maddox's criteria for an MCC shown in Table 10.1. However, it appears from Smull and Houze's data, that the system did not meet Maddox's criterion that the system maintain the specified areal extent for at least 6 h. The ability of a system to maintain this feature for an extended period has important dynamic implications.

Ogura and Liou's analysis revealed a storm structure (Fig. 10.35a–f) in which the east–west component of flow relative to the squall line included a strong easterly inflow at low levels and an equally strong outflow to the rear of the squall line centered at 300 mbar. At 200 mbar, a relative maximum in outflow ahead of the surface squall and midlevel rear inflow jet are evident. The v-component of flow reveals strong southerly flow at low levels; the flow appears to be conserved as it rises in convective-scale and mesoscale updrafts to the 200- to 300-mbar level. Northerly flow prevails at lower levels behind the squall front. The corresponding divergence field in Fig. 10.35c is characterized by strong surface convergence ahead of the squall line and even stronger convergence at 550 mbar 120 km behind the surface convergence. Ogura and Liou argued that, since positive temperature anomalies to the rear of the squall line at upper levels were rather weak in this case, the convergence at middle levels was dynamically rather than thermodynamically induced. This is contrary to Orlanski and Ross's conclusion in their prefrontal squall-line simulation. The analysis also revealed divergence at low levels in the rain area behind the squall line and at upper levels. The vertical velocity shown in Fig. 10.35d exhibits generally upward motion extending from low levels over the squall line and tilting upshear to 400 mbar. Pronounced downward motion is evident behind the squall line in the rain-chilled region extending downward from 550 mbar.

The streamlines in Fig. 10.35e illustrate a deep inflow layer from the surface to near 400 mbar. Air entering the rear of the storm system at middle levels feeds the downdraft as in Newton's and Srivastava *et al.*'s analysis of prefrontal squall lines. In contrast to Newton's analysis, however, air coming from the front of the squall line also feeds the downdraft. We suspect that the absence of air feeding the downdraft in front of the squall line in Newton's analysis is more a reflection of differences in analysis

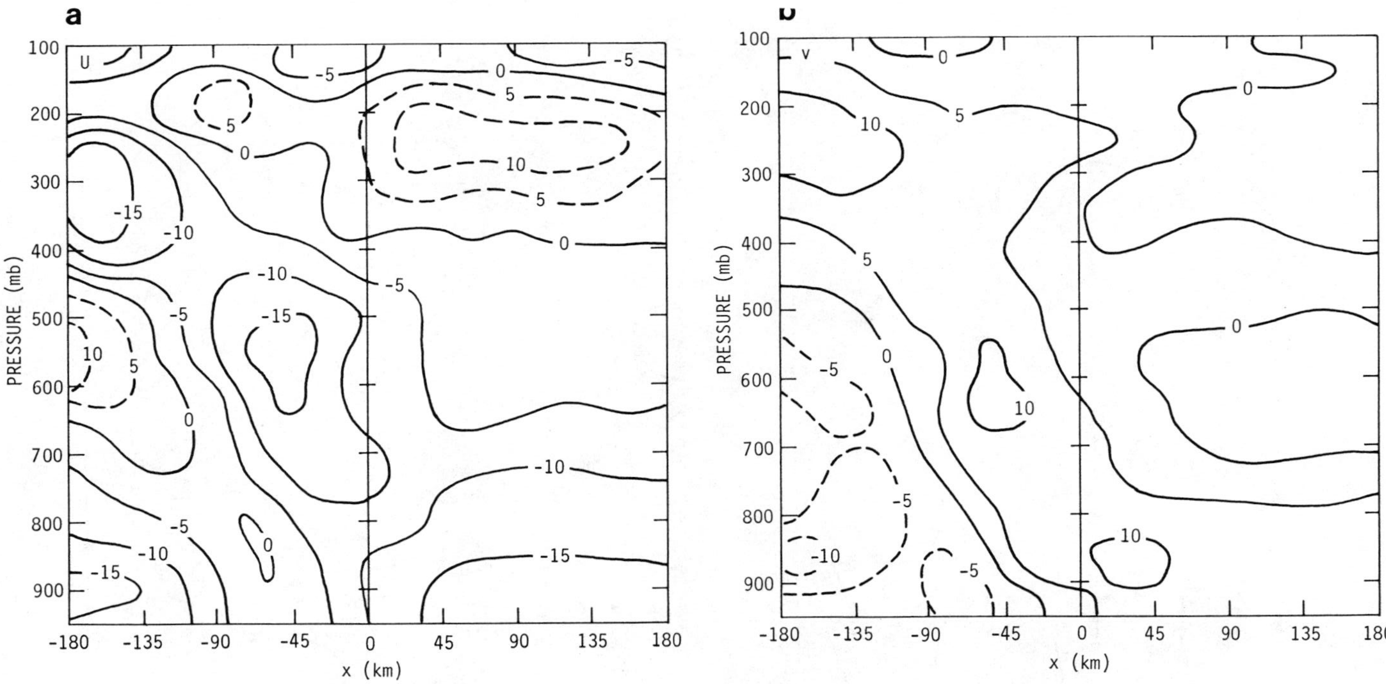

Fig. 10.35. (a) East–west component of wind ($m\ s^{-1}$) relative to the moving squall line in the vertical plane, with x representing distance ahead of the leading edge of the squall line. (b) As in (a), except for north–south component on wind ($m\ s^{-1}$). (c) As in (a), except for divergence ($10^{-5}\ s^{-1}$). (d) As in (a), except for vertical velocity ω (10^{-3} mbar s^{-1}). (e) As in (a), except for streamlines determined from u and ω. (f) As in (a), except for the vertical component of vorticity ($10^{-5}\ s^{-1}$). [From Ogura and Liou (1980).] (*Figure continues.*)

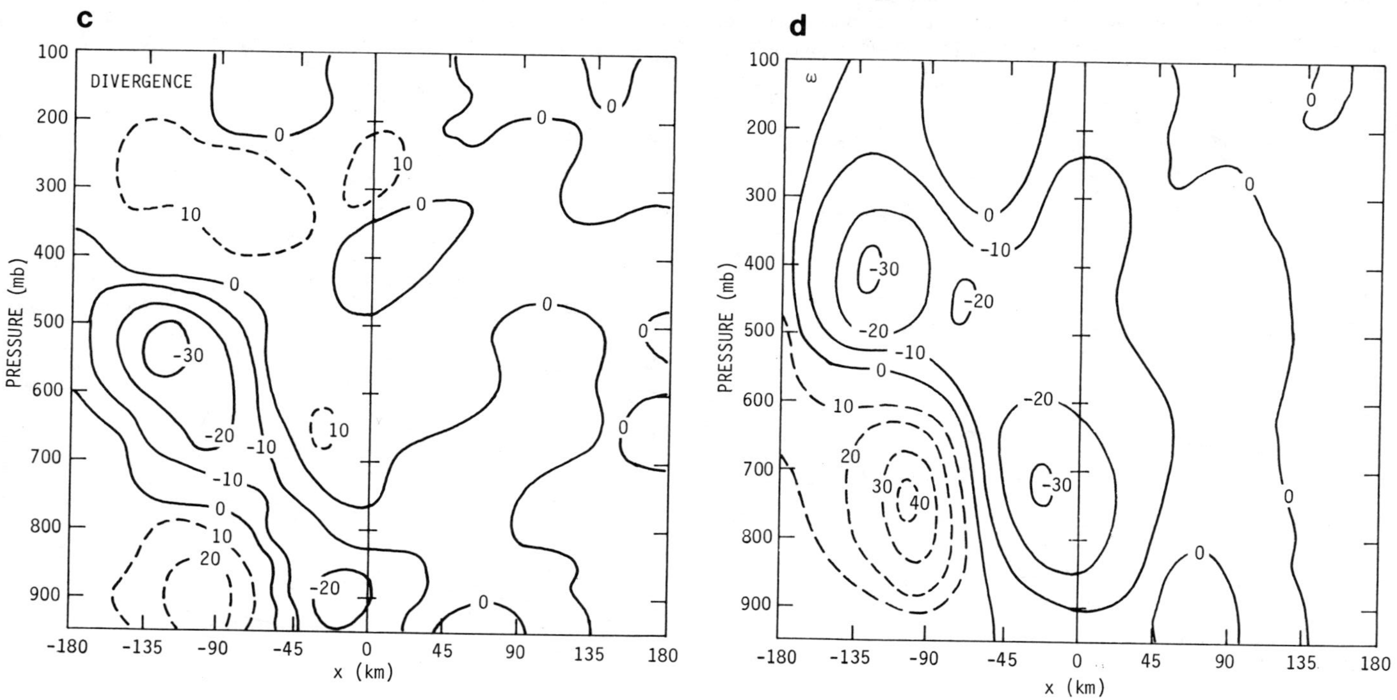

Fig. 10.35 (*continued*)

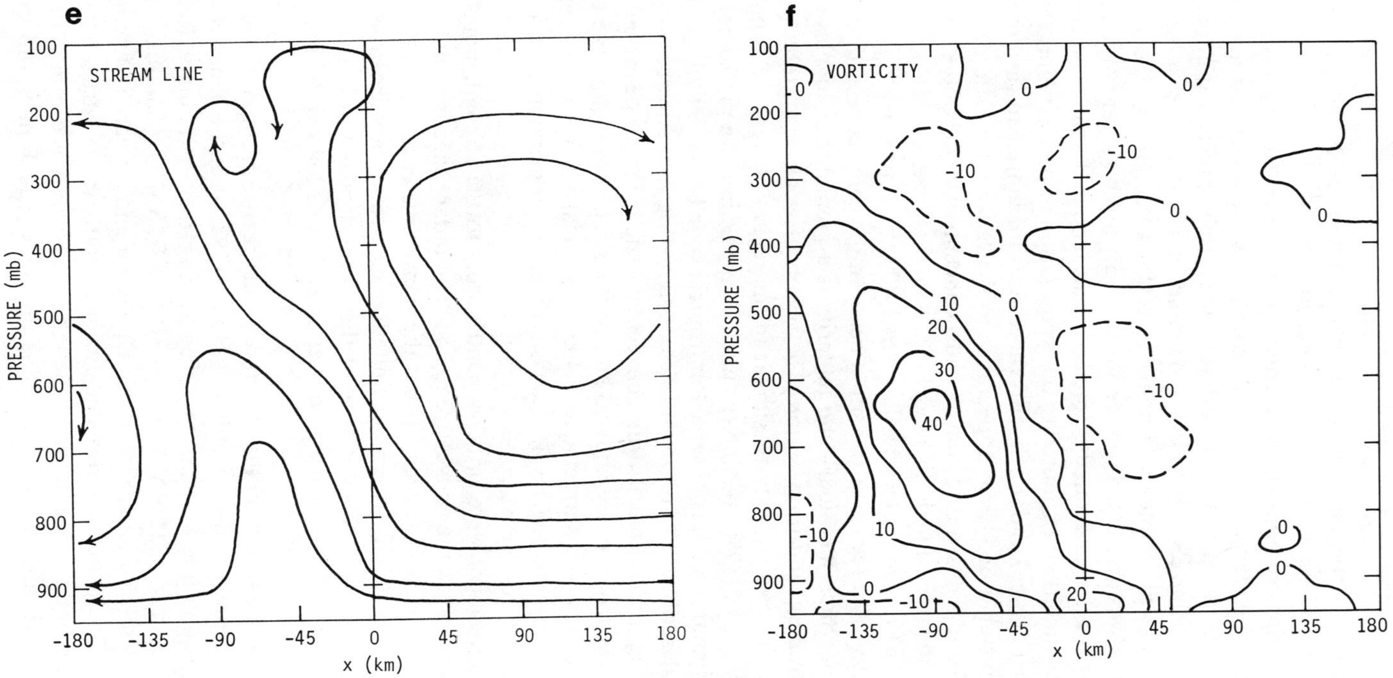

Fig. 10.35 (*continued*)

procedures and/or data resolution than a difference in the flow structure between prefrontal and nonprefrontal squall lines.

The vertical component of vorticity field in Fig. 10.35f shows cyclonic vorticity at low levels at the squall line, tilting rearward to a maximum at 650 mbar. As in other MCSs, anticyclonic vorticity prevails at upper levels. Smull and Houze's (1985) single-Doppler analysis of the storm system revealed that the midlevel cyclonic vortex was centered in the stratiform region and was associated with an influx of dry air to the rear of the system. The influx of dry air created a notch of weak echo in the radar precipitation pattern, which they attributed to the evaporation of precipitation elements. The notch of weak radar echo advanced forward in time toward the squall front, causing a bow-echo appearance to the radar-echo pattern.

Smull and Houze (1985) developed a conceptual model of the squall-line system and later refined this model (Smull and Houze, 1987a). Their model (Fig. 10.36) includes (1) a leading anvil, (2) the convective region of the squall line, (3) a region of transition between the convective and stratiform regions in which the radar-echo intensity is a minimum at low levels, and (4) the trailing stratiform region. The region of weak echo, referred to as the transition region, may be similar to the region of weak echo noted by Srivastava *et al.* (1986) behind the intense convection of a prefrontal squall line. A similar feature is observed in tropical squall lines. Smull and Houze (1987a) described a "front-to-rear" jet originating in the lower portion of the leading anvil cloud. In the convective region, the jet deepened to fill much of the troposphere above 3 km. In the stratiform region, the core of the front-to-rear jet contracted to 3–4 km in vertical extent and was centered just above the melting level. Analysis of eddy momentum fluxes in the vicinity of the convective cores suggested that convective downdrafts in the upper troposphere were primary contributors to the strengthening of the front-to-rear jet in the convective region. The radar-reflectivity fields in the maximum jet region suggested that the convective cells were also highly effective in expelling frozen precipitation elements rearward toward the anvil region. On the western edge of the storm depicted in Fig. 10.36 is a midlevel rear-to-front jet which descends and turns toward the rear of the storm at low levels.

Further elaboration on the structure of the rear-to-front midlevel inflow jet has been presented by Smull and Houze (1987b) in the analysis of two ordinary squall-line systems observed during the Preliminary Regional Experiment for STORM-Central (PRE-STORM) (Cunning, 1986). (STORM is the National Stormscale Operation and Research Meteorology Program.) Their analysis revealed further details in the structure of the rear-to-front inflow jet and established that this feature is common to many different squall-line systems, including prefrontal squall lines and tropical

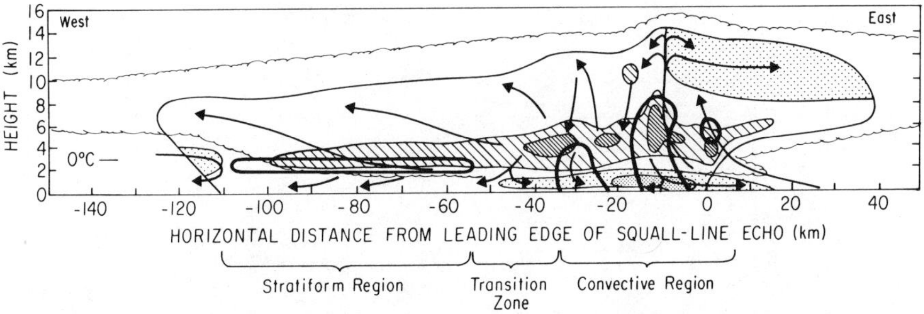

Fig. 10.36. Conceptual model of the mature 22 May 1976 Oklahoma squall-line system viewed in a vertical cross section oriented normal to the convective line. System motion is from left to right (i.e., eastward) at $15 \mathrm{~m} \mathrm{~s}^{-1}$. Outermost scalloped line marks extent of cloud. Outermost solid contour marks boundary of detectable precipitation echo, while heavy solid lines enclosed more intense echo features. Stippling indicates regions of system-relative horizontal wind directed from rear to front (left to right); darker stippling represents stronger flow. Elsewhere within echo, relative flow is from front to rear (right to left). Maximum front-to-rear flow at middle and upper levels is shown by hatching; darker hatching denotes embedded speed maxima. Thin streamlines show two-dimensional projection of relative flow determined from the dual-Doppler analysis and the composite rawinsonde analysis of Ogura and Liou (1980). The 0°C level in the trailing stratiform region is indicated at echo's rear. [From Smull and Houze (1987b).]

squall lines. Detailed Doppler radar analysis showed that the rear inflow jet occupied a continuous channel extending from middle levels at the back edge of the stratiform region to lower levels of the leading convective region. Near the leading edge of the squall line, the rear-to-front flow appeared to merge with convective-scale downdrafts. Smull and Houze suggested that the mixing between the rear inflow jet and convective-scale downdrafts enhanced the intensity of the leading gust front by contributing to enhanced evaporation.

Analysis of PRE-STORM data by Johnson and Hamilton (1988) has clarified some of the surface-pressure features accompanying an ordinary midlatitude squall line which exhibited trailing stratiform precipitation. Figure 10.37 is a conceptual model of those features, along with a vertical cross section of the system-relative flow fields. The major surface features are as follows:

(a) *The presquall mesolow.* In advance of the squall line is a local region of relative low pressure. Hoxit *et al.* (1976) attributed this feature to convectively induced subsidence warming in the middle to upper troposphere in advance of squall lines.

(b) *The squall mesohigh.* To the rear of the squall line in the region of heavy precipitation and convective-scale downdrafts is a pronounced

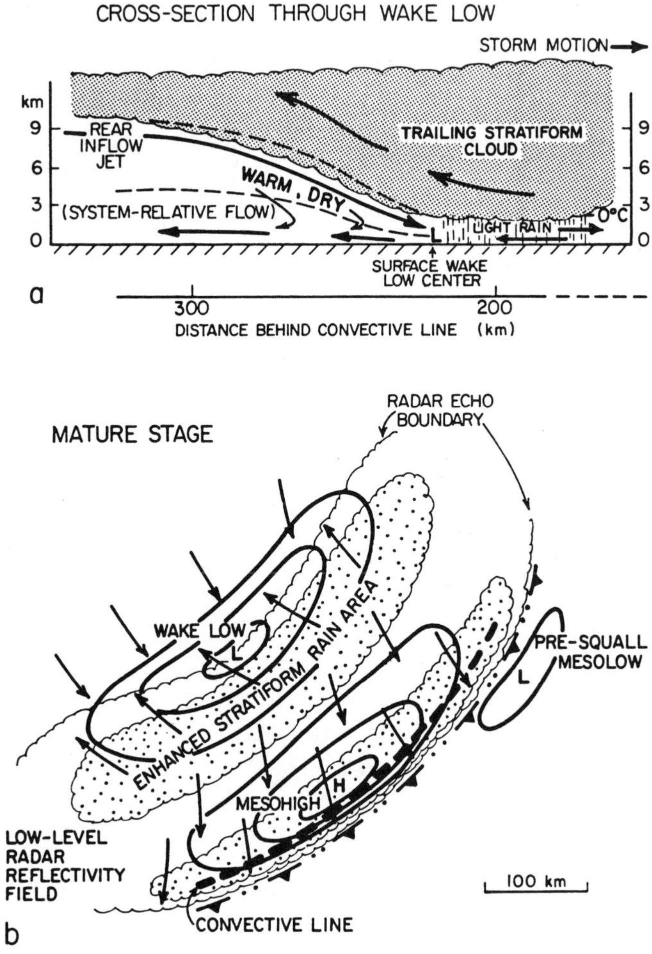

Fig. 10.37. Schematic cross-section through wake low (a) and surface-pressure and wind fields and precipitation distribution during squall-line mature stage (b). Dashed line in (a) denotes zero relative wind. Arrows indicate streamlines, not trajectories. [From Johnson and Hamilton (1988).]

mesohigh. This feature was clearly defined in the "thunderstorm project" (Byers and Braham, 1949) and has been attributed to cooling by evaporation (Sawyer, 1946; Fujita, 1959), as well as the weight of suspended water (Sanders and Emanuel, 1977) and cooling by melting. Fujita suggested that the surface mesohigh may be enhanced by nonhydrostatic pressure contributions caused by the impact of downdrafts on the ground.

(c) *The wake low.* Behind the mesohigh and much of the stratiform rain area is a region of low surface pressure which Fujita (1955) referred to as the wake low. This trailing low-pressure region has also been identified by Brunk (1953), Pedgley (1962), and Williams (1952, 1963). Williams (1963) showed that subsidence warming can account for the lower pressure in that region. Zipser (1977) observed the "onion-shaped" soundings behind tropical squall lines (Fig. 10.7) in this same region of a squall line. As shown in the vertical cross section of Fig. 10.37, Johnson and Hamilton (1988) argue that the wake low is a surface manifestation of the descending rear inflow jet. The warming is maximized at the back edge of the precipitation area. Evaporative cooling is insufficient here to offset adiabatic warming.

A class of midlatitude squall lines exists that differs little in structure from their tropical cousins. Both exhibit a line of ordinary thunderstorm cells along the squall front, and the pronounced front-to-rear flow that rises behind the squall line is prevalent in tropical and midlatitude systems. Middle-level, rear-to-front flow is evident in most systems. As noted by Smull and Houze (1987b) for midlatitude systems, its strength and height are quite variable from one system to another. Both systems exhibit intense convective rainfall along the leading edge of the squall line, followed by a transition region of relatively lower radar reflectivity and by a region of stratiform rainfall that varies in horizontal extent. A radar bright band near the melting level in the stratiform region is frequently present in tropical and midlatitude squall lines.

Still, many extratropical squall lines exhibit significant differences from tropical squall lines. One such squall line has been analyzed extensively by Schmidt and Cotton (1988) using multiple-Doppler radar. Figure 10.38 is a conceptual model of this squall-line system, which could be classified as a fast-moving squall line because its speed is $21\ \mathrm{m\,s^{-1}}$. Depicted in the figure are a line of ordinary thunderstorm cells immediately behind the squall front and a more dominant cell labeled G1, which is a quasisteady, rotating cell resembling a supercell, but which is more properly labeled a weak evolution storm of the type described by Foote and Frank (1983). That such a supercell-like storm should be present is not surprising, since the vertical wind shear is strong through a 6-km layer ($\sim 8\times10^{-3}\ \mathrm{s^{-1}}$) and the bulk Richardson number is 31, which Weisman and Klemp (1984) show is conducive to supercell-like storms. The presence of this supercell-like storm distinguishes this squall-line system from tropical squall-line storms and many ordinary midlatitude squall lines. Moreover, in Fig. 10.38, this quasisteady cell labeled G1 coexisted with a more transient line of ordinary cells labeled F2. These more transient cells formed repeatedly along the squall front and, as they aged, trailed farther behind the squall line into

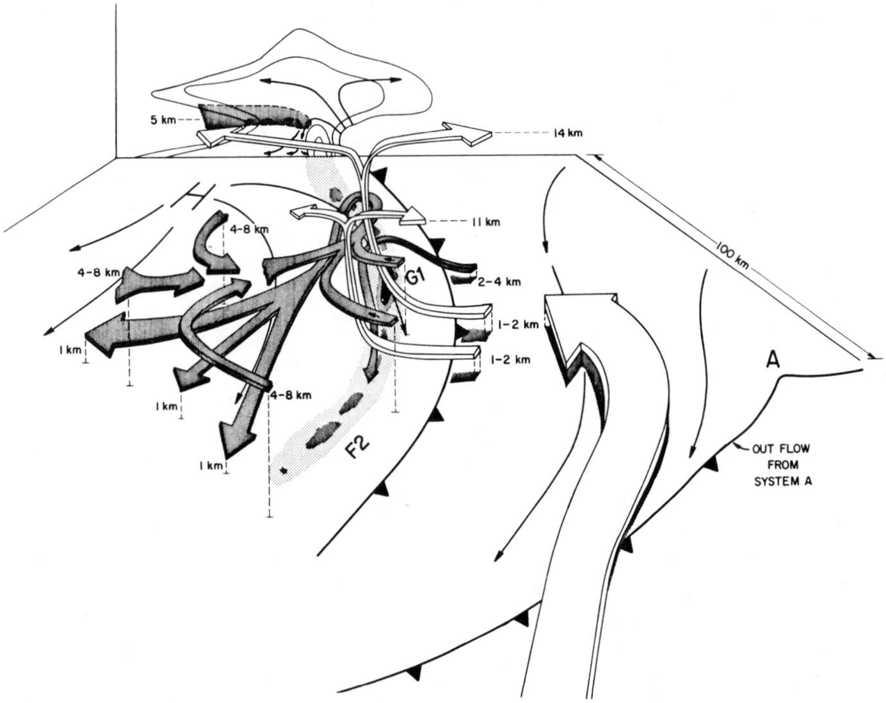

Fig. 10.38. Schematic depiction summarizing the 2-D and 3-D flow features discussed in the text for the 2 August 1981 CCOPE squall line showing mesoscale outflow boundaries, surface streamlines (thin arrows), convective reflectivity structure (stippled), overriding flow (bold arrow), and storm-relative flow (thin ribbons). The vertical cross section corresponds to a representative depiction of the storm core G1 and shows reflectivity (thin solid lines), schematic storm-relative flow (thin arrows), and location of the midlevel upshear inflow (shaded). The bold A refers to an isolated supercell system, G1 and F2 represent the cell groups along the squall line, and the H represents the location of the surface mesohigh (upper left). Labeling of the storm-relative flow ribbons refers to height above ground level. [From Schmidt and Cotton (1989).]

the stratiform region. The aged cells were then replaced by new cells forming along the squall front, similar to the cases Fujita (1978) described as a bow echo. The quasisteady cell G1 maintained a pivotal position with regard to the behavior of the more transient cell groups as well as the overall flow structure of the squall-line system. The fact that supercell and ordinary cells coexist in the same environment suggests that the value of R_i does not yield, both a necessary and sufficient condition for the existence of supercell storms. Other conditions, such as total available energy over a domain, influence of one storm on another, and localized subcloud convergence, may be factors.

A two-dimensional depiction of the flow through the G1 cell (Fig. 10.38) illustrates that this squall line resembles the general class of ordinary squall lines discussed previously, including a front-to-rear flow that rises behind the line and a middle-level, rear-to-front flow that feeds into the convective region. Trajectory analysis of the Doppler-radar-derived flow fields suggests, however, that the rear-to-front flow also interacts with a middle-level, front-to-rear flow and together feeds the convective downdraft region. The front-to-rear branch of downdraft air may be a consequence of relative flow about the dominant cell G1 and/or to the rapid propagation of the squall-line system. This says that a two-dimensional depiction of a squall line can be a misleading representation of the flow structure of the storm system.

Another major property of this squall-line system is that the low-level inflow feeding the convective updrafts is decoupled from the surface. That is, the inflow depicted in Fig. 10.38 is actually gliding over the surface-based cool outflow from an isolated supercell storm labeled A in the figure. This may have been responsible, in part, for the rapid movement of the squall line and the types of convective cells along the line.

The squall system studied by Schmidt and Cotton (1988) is noteworthy because the anvil clouds associated with it merged with the anvil from the supercell labeled A in Fig. 10.38, as well as with the anvil from another cluster of thunderstorms, to form a MCC. Moreover, the resulting MCC, including the squall line during the formative stages of the MCC, produced a nearly continuous band of severe winds covering a four-state area from Montana to Minnesota. Such a severe, straight-line, wind-producing storm system is called a "derecho" by Hinrichs (1888) and Johns and Hirt (1983, 1985). Johns and Hirt note that derecho-type systems produce severe-damage swaths of 100 km or more in width and extending 500–1000 km in length. As in the case here, the derecho-producing storm system generally exhibits a bow-echo (Fujita, 1978) radar structure. Schmidt and Cotton (1988) found that, for 165 MCC events, between 25 and 30% of the cases produced severe, derecho-type winds. A distinction between the environmental conditions favorable for derechos versus the more normal MCCs has not been identified, however. Johns and Hirt suggest that stronger wind shear and thermodynamic stability may be involved.

10.3.2 Mesoscale Convective Complexes

As shown in Table 10.1, Maddox arrived at his definition of MCCs from the characteristics of the infrared satellite imagery of a midlatitude MCSs. This definition is based on the areal extent and depth, as determined by the temperature of its capping stratiform cloud shield. Maddox discriminated between the longer squall lines and the more circular clusters by requiring that the cloud shield have an eccentricity (ratio of minor axis to major axis) greater than or equal to 0.7. As a result of the areal extent of the cloud shield required of MCCs, smaller MCSs are eliminated from his definition; Maddox required that an MCC maintain its size characteristics for at least 6 h, thus discriminating the MCC from more transient MCSs and enabling examination through standard synoptic-data sources. This criterion has implications for the overall dynamic properties of MCCs. Maddox noted that MCCs are large and intense enough that the thermal and wind perturbations created by the systems are minimally resolved by the standard rawinsonde spacing in the United States (typical spacing is 450 km). The MCC is an extreme example of a continuum of MCS types, just as the supercell represents an extreme type of thunderstorm. The prefrontal squall line represents still another extreme form of MCS.

Most of our understanding of MCCs is derived from case studies or composite analyses of operational rawinsonde data, along with satellite imagery and operational radar and surface meteorological data. Occasionally, more detailed observations have been made as a result of the chance passage of an MCC through a dense special network designed to study other systems.

The large-scale environment characterizing the MCC genesis region is described in Section 10.2.4. The early stages of convective organization often take place in the form of meso-β-scale thunderstorm systems outside the genesis region defined by Maddox (1983). These systems move into the region characterized in Fig. 10.24. Figure 10.39 contains such a sequence of events as seen by satellite and radar. The stratiform region reaches its fullest areal extent by 0900 GMT (0300 CST). MCCs typically reach full maturity between 0100 and 0500 local standard time and are characterized as nocturnal thunderstorm systems. In later sections, we discuss the factors in MCC genesis as well as diurnal influences on MCSs.

Maddox (1983) described the mature stage of an MCC by compositing standard rawinsonde soundings taken at 1200 GMT, which is close to the average time of maximum areal extent of MCCs over the central United States. Figure 10.40 illustrates Maddox's composite analysis for a mature MCC from the surface to 200 mbar and should be compared with the prestorm conditions in Fig. 10.24. At 1200 GMT, the surface front has moved southward, and the surface divergence pattern suggests a weak mesohigh and outflow boundary on the southern side of the front. The surface air mass has been moistened considerably from moist downdrafts and rainfall evaporation. At 850 mbar the low-level jet has increased slightly in speed and veered to a more southwesterly direction. Associated with the low-level jet is a region of higher moisture content.

The flow at 700 mbar has strengthened ($>5\ \mathrm{m\ s^{-1}}$ increase) to a west-southwesterly jet over the mature MCC region. The 700-mbar shortwave trough and ridge pattern has moved eastward with the speed of the system. The shortwave trough is also evident at 500 mbar. Maddox emphasized the fact that the entire composite MCC life cycle is linked to the eastward progression of a meso-α-scale shortwave trough having a wavelength of nearly twice the MCC diameter (1500 km).

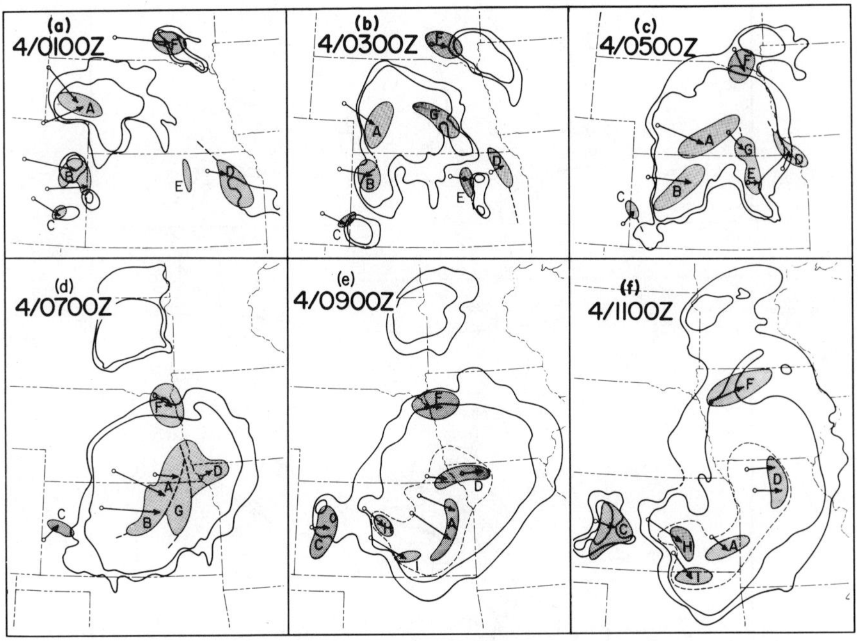

Fig. 10.39. Schematic IR satellite and radar analysis at 2-h intervals, from 0100 to 1100 GMT, 4 August 1977, for the western MCC #1. The anvil cloud shields are indicated by the −32 and −53°C IR contours (outer and inner solid lines, respectively), remapped from satellite images at the labeled times. Darkly shaded regions (identified by letters) denote significant radar-observed, meso-β-scale convective features at about 25 min after the indicated whole hour, with the vectors showing their previous 2-h movements. The dashed line segments extending from the meso-β convective features indicate flanking axes of weaker convection. In the more developed MCC stages, in (e) and (f), the light-shaded area within the dashed envelope indicates weaker, more uniform and widespread echo. [From McAnelly and Cotton (1986).]

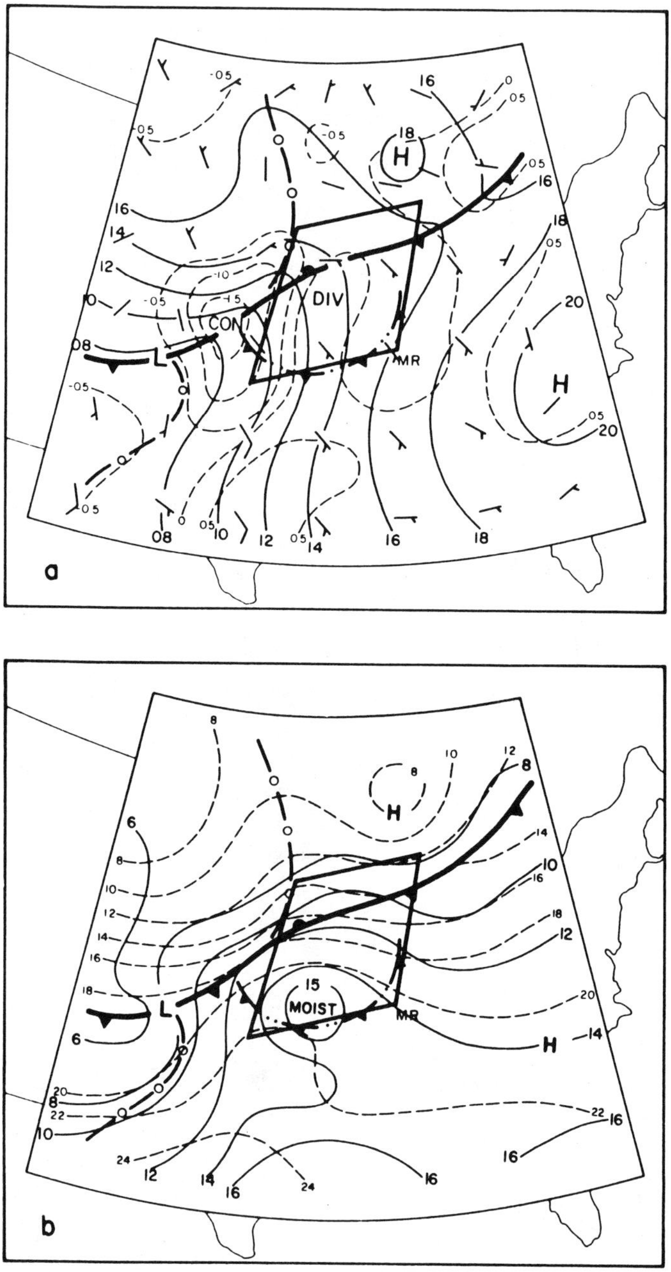

Fig. 10.40. Same as Fig. 10.24, except at the time of maximum MCC. [From Maddox (1983).] (*Figure continues.*)

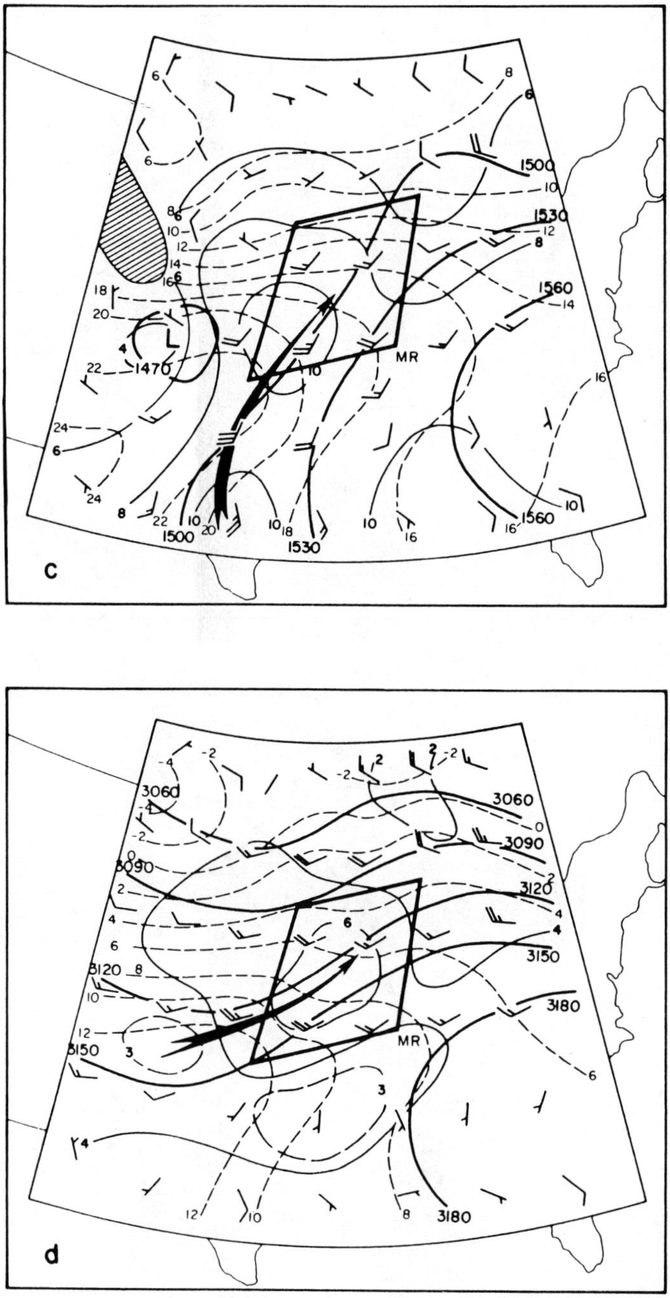

Fig. 10.40 (*continued*)

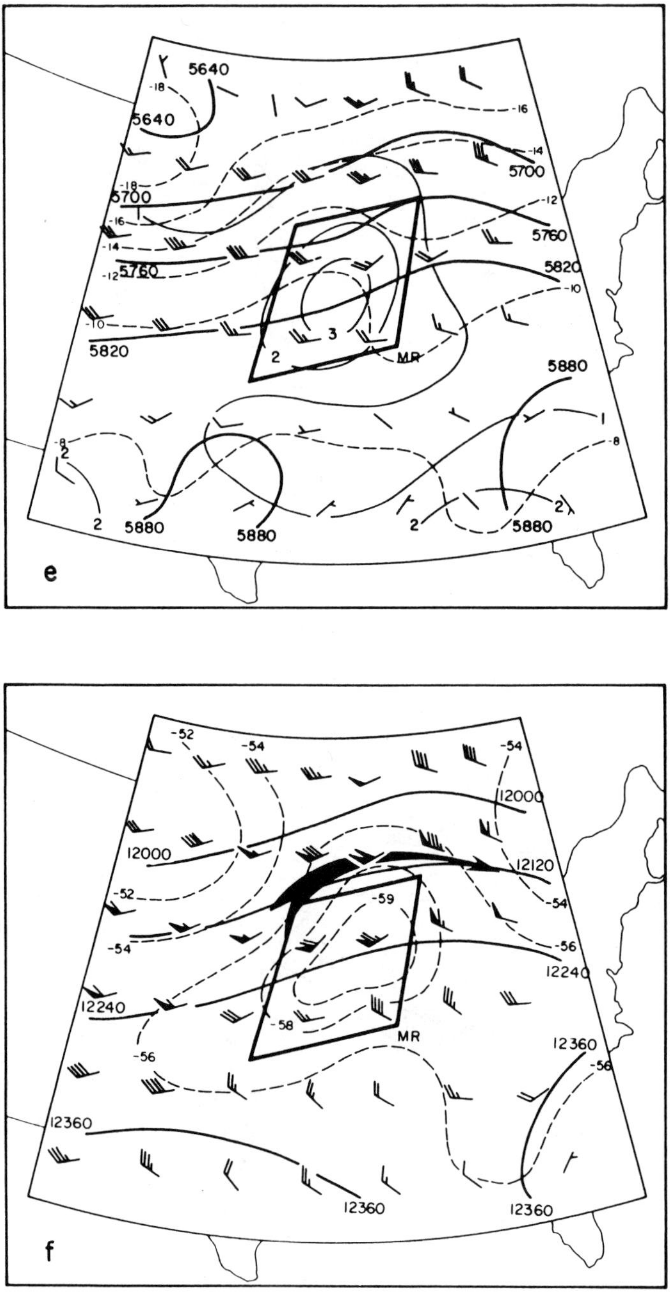

Fig. 10.40 *(continued)*

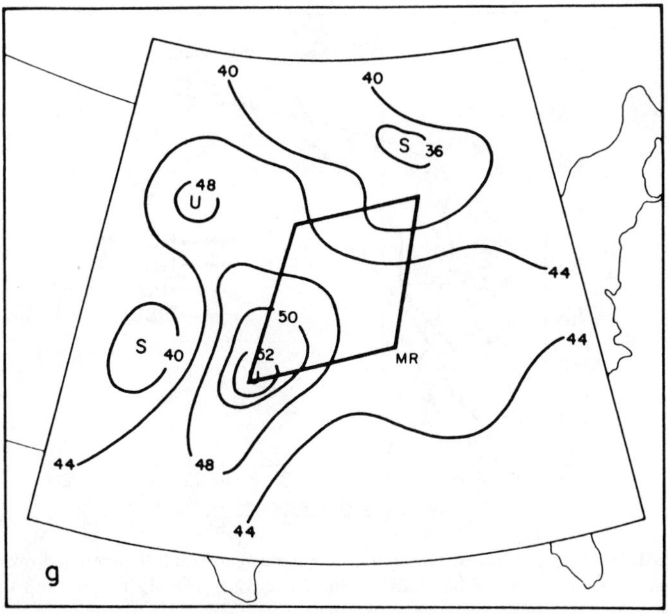

Fig. 10.40 (*continued*)

The flow at 200 mbar is characterized by a pronounced jet streak to the north of the MCC. The anticyclonically curved jet streak exhibits speeds $>15\ \mathrm{m\ s^{-1}}$, greater than any observed wind at 200 mbar at 0000 GMT. Lin (1986) and Cotton *et al.* (1989) extended Maddox's 10-case composite analysis by considering over 128 MCC cases. Lin also attempted to define better the MCC life cycle by stratifying MCC life cycles into a total of seven time bins, each of a 2-h duration, centered at the time of MCC maximum maturity, as identified by infrared satellite imagery. MCCs show considerable variability, so at the time the operational 1200 GMT soundings are taken, systems may be anywhere in their life cycle from maturity to 2 to 4 h before or after maturity. Likewise, some systems may be in the early stages of growth or within several hours of maturity at the time the 0000 GMT soundings are taken and some may be in the final stages of decay when the 1200 GMT soundings are taken. Thus, a composite model of an MCC life cycle can be obtained, even though high-time resolution soundings may not be available.

Figure 10.41 shows the divergence profile associated with Lin's (1986) composite data set. Early in its life cycle, the MCC is characterized by a layer of convergence up to 750 mbar and divergence aloft. By the time of

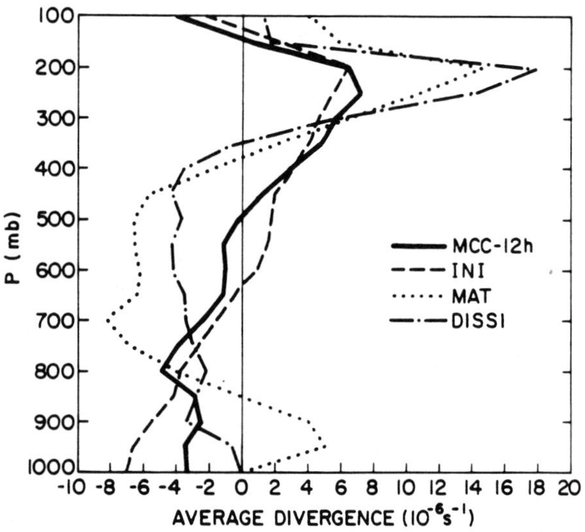

Fig. 10.41. Vertical profiles of horizontal divergence (horizontally averaged over the 3×3 central grid points at 50-mbar intervals) at the MCC 12-h, initial, mature, and dissipation stages. Units: 10^{-6} s^{-1}. [From Cotton *et al.* (1989).]

MCC maturity (6 h later), the layer of convergence has deepened to 400 mbar, while the layer of divergence aloft has strengthened with its center at 200 mbar. A low-level region of divergence is evident below 900 mbar, presumably because of the development of a mesohigh as a result of evaporation of precipitation. At the time of MCC dissipation (6 h beyond MCC maturity), the layer of midtropospheric convergence has weakened and so has the low-level convergence zone. The region of upper tropospheric divergence reaches its maximum intensity very close to the time of system decay.

The corresponding composite vertical motion in Fig. 10.42 shows that the strongest upward motion is centered near 650 mbar early in the system's life cycle, similar to tropical clusters. Later in the life cycle, the upward motion maximum shifts upward to 350 mbar. Similar upper tropospheric maxima of upward motion were seen in the composite analysis, as well as in individual case studies of mature nonsquall tropical clusters. In Lin's composite analysis, the maximum vertical motion at 350 mbar remains strong through the dissipating stage of the system rather than rapidly weakening after the mature stage, as occurs in tropical clusters (Fig. 10.21). Another feature not evident in the composite models of tropical clusters is the deep layer of downward motion below 700 mbar at the time of system

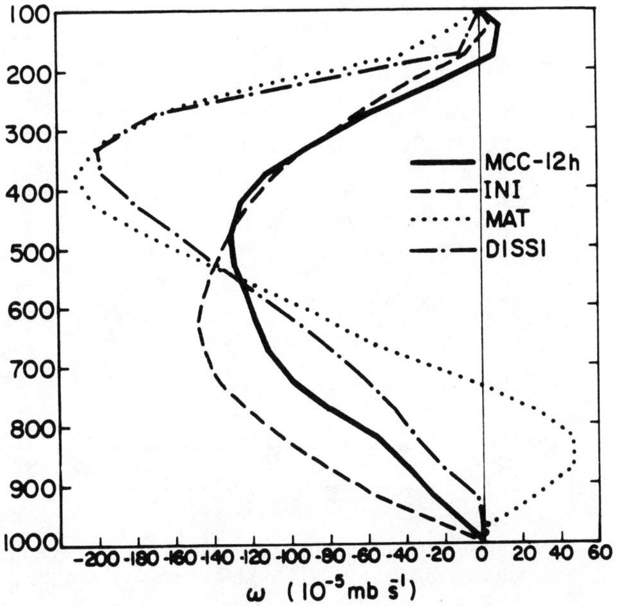

Fig. 10.42. Vertical profiles of vertical velocity (ω) at the MCC 12-h, initial, mature, and dissipation stages, calculated by integrating the corresponding divergence profiles in Fig. 10.41. They represent average ω over the 3×3 central grid-point region (4.4×10^5 km^2). Units: 10^{-5} mbar s^{-1}. [From Cotton *et al.* (1989).]

maturity. This is probably the result of higher cloud bases and drier subcloud layers in midlatitude, continental MCCs which strengthens low-level downdrafts and makes them more resolvable in rawinsonde networks.

The composite vorticity and the vorticity change are shown in Figs. 10.43 and 10.44, respectively. The environment of the MCC is characterized by weak cyclonic vorticity below 700 mbar throughout the MCC life cycle, with a maximum at 900 mbar at maturity. Likewise, above 700 mbar, the environment is characterized by anticyclonic flow, with a maximum at 200 mbar toward the end of the system's life cycle. Because the average vorticity of the MCC prestorm environment can mask changes in vorticity by the MCC, the change in vorticity was calculated. A region of cyclonic vorticity increase between 450 and 600 mbar is evident in the latter stages of the system's life cycle (Fig. 10.44). Cyclonic vorticity increases at 800 to 900 mbar, while anticyclonic vorticity increases near 200 mbar. The maximum in anticyclonic vorticity at 200 mbar occurs at the decay stage of the composite system. This corresponds to the time at which vertical motion is also a maximum.

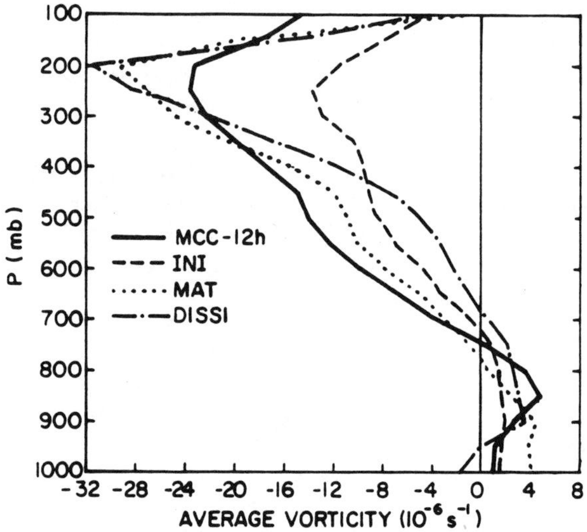

Fig. 10.43. Vertical profiles of relative vorticity (horizontally averaged over the 3×3 central grid points at 50-mbar intervals) at the MCC 12-h, initial, mature, and dissipation stages. Units: $10^{-6}\,s^{-1}$. [From Cotton *et al.* (1989).]

The midlevel cyclonic vorticity maxima in the latter stages of the MCC are probably associated with the convergence that reaches a maximum at middle levels at the time of system maturity. A similar midlevel positive vorticity center was analyzed by Bosart and Sanders (1981) in their investigation of the Johnstown Flood MCC. They speculated that the positive vorticity center resulted from interactions between a low-level cold pool, which intensified a preexisting lower tropospheric jet. In the analysis of one MCC, Leary and Rappaport (1987) found that the stratiform precipitation region was organized in a set of curved bands that were aligned with cyclonic inflow at 500 mbar. A similar alignment of bands was found in tropical clusters (Fig. 10.16). Figure 10.45 shows the cyclonically curved bands in a decaying MCC in which the absence of a cirrus shield permits a view of the low- and middle-level clouds.

Analysis of hourly surface precipitation associated with MCCs reveals a well-defined precipitation life cycle that is consistent with the composite kinematic fields and satellite appearance. McAnelly and Cotton (1986) determined the composite precipitation life cycle for an episode of MCCs observed in 1977, and, more recently, the composite precipitation life cycle for MCCs using the same set of cases that Lin used in his composite study. Figure 10.46 shows that the average precipitation rates increase in intensity through the developmental stages of the MCC and reach their peak at the

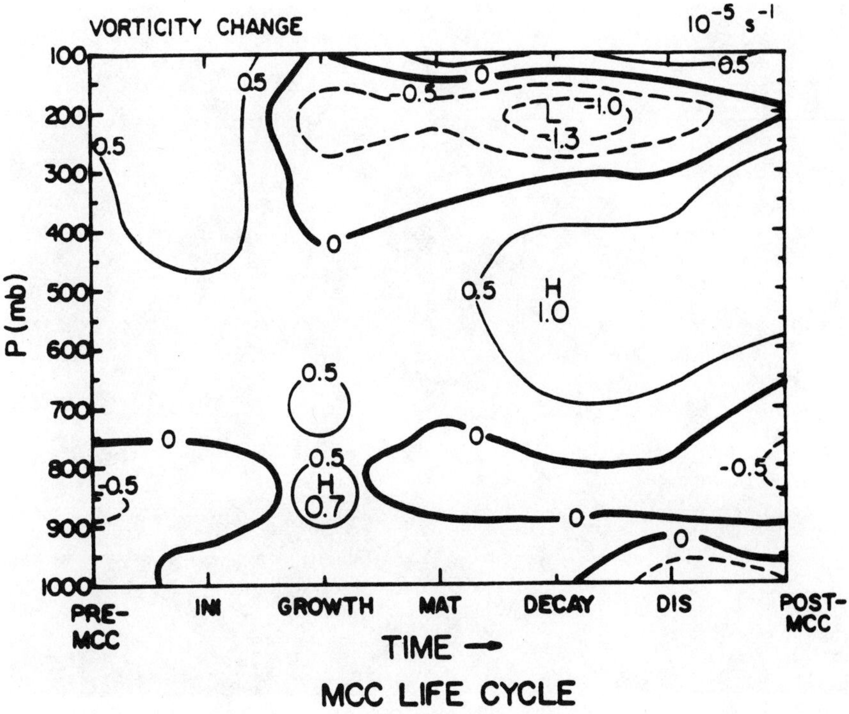

Fig. 10.44. Time–height plot of the difference in relative vorticity (horizontally averaged over the 3 × 3 central grid points at each 50-mbar level and subperiod) from its corresponding value at the MCC 12-h stage. Positive values indicate a more cyclonic (or less anticyclonic) vorticity than at the MCC 12-h stage. Units: 10^{-5} s^{-1}. [From Cotton *et al.* (1989).]

time the MCC exhibits its maximum cloud shield. During this same period, the rainfall rates are intense and convective in scale. The volumetric rainfall reaches a maximum, however, during the mature stage of the system when its meso-α-scale circulation is best organized. Later in the MCC life cycle, the rain area continues to expand, but the precipitation rate becomes lighter. This is consistent with the weakening mesoscale circulation in Lin's composite kinematic fields. As discussed in Chapter 6, Section 6.4.2, Lin (1986) calculated that the precipitation efficiency of the MCC actually exceeded 100% at the mature stage. This occurs because storage of condensate in the stratiform–anvil cloud during the early portion of the MCC life cycle contributed to a precipitation amount that exceeded the instantaneous convergence of water vapor in the storm system.

McAnelly and Cotton (1986) found that the precipitation was largely confined to the meso-β-scale convective features embedded in the systems.

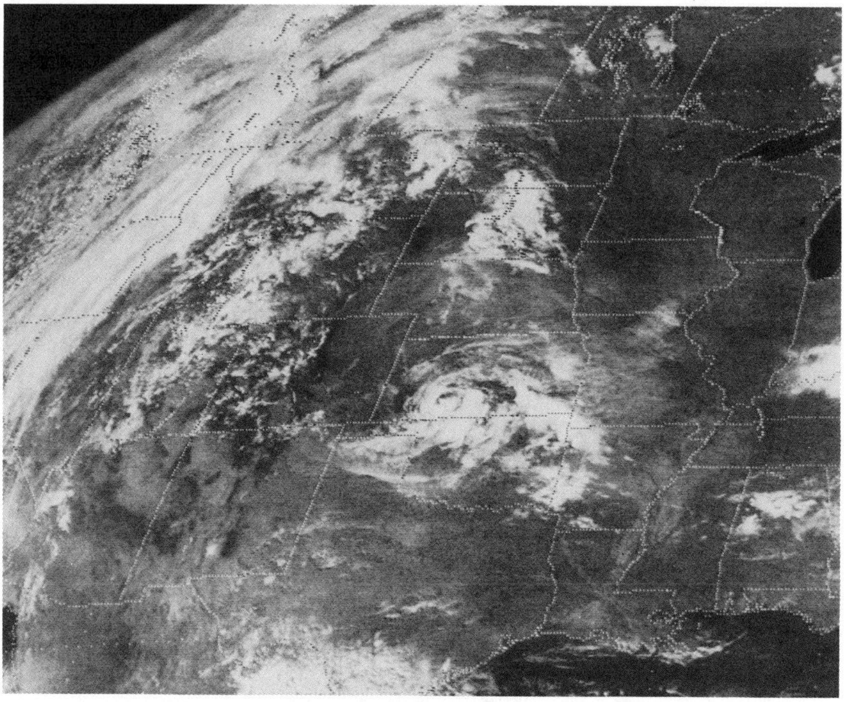

Fig. 10.45. Visible satellite photograph (1530 GMT, 14 May 1984) of decaying MCC showing circulation remaining long after active convection has ceased.

This was true of systems forming in the western High Plains region of the United States. Thus, in contrast to tropical clusters where as much as 40% or more of the rainfall is stratiform in nature, the stratiform rainfall in some western United States systems evaporates before it reaches the surface. Other midlatitude MCSs, such as the squall line studied by Johnson and Hamilton (1988), exhibit stratiform rainfall contributions as large as 39% of the total rainfall, similar to tropical systems. One would expect the contribution to total rainfall by the stratiform clouds to increase as systems form or propagate into moister subcloud regions such as in the southeastern United States.

As in nonsquall tropical clusters, the thunderstorm-scale organization of MCCs is variable. In Fig. 10.39, the meso-β-scale organization of some MCCs reflects the organization of individual cells during the early genesis stages of the system. In other cases, the convection is organized in a more bow-echo-shaped squall line of convection (Schmidt and Cotton, 1988; Smull and Houze, 1987a). In still others, the convection appears to be more

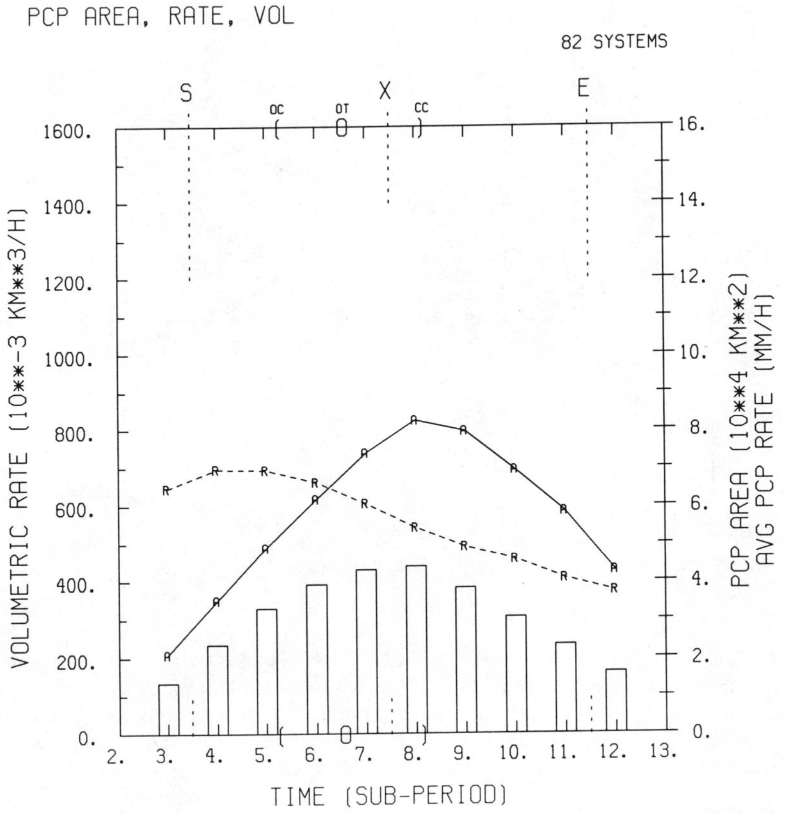

Fig. 10.46. Time evolution of MCC precipitation characteristics, averaged over 82 systems. The time scale represents a normalized life cycle (based on satellite-observed cloud shield size), where the MCC is defined to start, maximize, and end at times 3.5, 7.5 and 11.5, respectively. Solid curve shows area of active precipitation, and dashed curve shows average rainfall rate within that area. Their product gives the volumetric rainfall rate of the MCC, depicted by the bars.

randomly organized (Fig. 10.47). Maddox (1981) noted that the more intense convection in MCCs tended to be oriented parallel to the midlevel steering winds rather than perpendicular to them as in prefrontal squall lines. This organization is rather similar to that simulated and described by Dudhia and Moncrieff (1987) in nonsquall tropical clusters. Also, the veering of the wind profile from low-level southerlies to westerlies aloft is similar to the profile identified by Miller (1978; see Chapter 9) as being favorable for flash-flood situations. Miller (1978) showed that such a wind profile caused the precipitation to stream out perpendicular to the low-level southerly

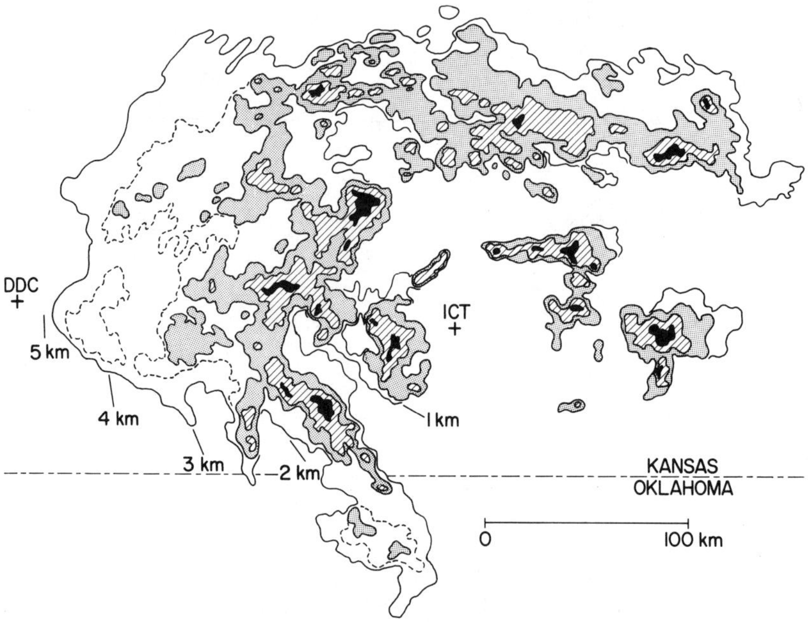

Fig. 10.47. Large-scale radar reflectivity structure of the PRE-STORM MCC at 0916 GMT, as seen from the digitized NWS radar at Wichita, Kansas (ICT). Contours indicate regions exceeding 10 (clear), 25 (dashed contour), 30 (stippled), 43 (hatched), and 48 (black) dBZ. Arcs along southern edge of echo between the ICT and Dodge City (DDC) indicate height intersections of this 0.6° PPI. [From Fortune and McAnelly (1986).]

flow, resulting in a mesohigh and associated gust front which were oriented perpendicular to the low-level flow. The resultant convergence field created a situation in which steady, heavily precipitating convection is favored.

As the MCC reaches the mature and dissipating stages of its life cycle, intense cellular convection weakens. This transition can take place quite abruptly, on a time scale on the order of an hour. Before the transition occurs, radar reflectivities are on the order of 35–50 dBZ. After the transition, radar reflectivities decrease to typical peak values of 10–25 dBZ (see Fig. 10.47). Not only does the intensity of the echoes change abruptly, but so does the pattern or distribution of echoes. Before the transition, the precipitation is concentrated in two or three intense convective cells. After the transition, numerous regions of moderate reflectivity can be identified within the prevailing stratiform rain pattern. The mechanisms responsible for such a transition in storm dynamics are not well understood, but we shall examine one possible mechanism in a later section.

While severe weather is most prevalent during the early formative stages of an MCC, occasionally, sporadic tornadoes and other severe-weather events occur late in the life cycle of a MCC. These events occur along the southern flank of the system where the low-level mesohigh encounters the southerly low-level jet. Pockets of very unstable air can be encountered in this region, even in the early morning hours, leading to a flare-up of severe convection.

Information about the stratiform region of MCCs is just now coming forth. Yeh *et al.* (1986) found the stratiform region similar to that found in tropical clusters and prefrontal squall lines. Generally, the stratiform region is bounded by the 0°C isotherm at its base and by the tropopause at its top. Aggregates of ice crystals comprise the dominant form of precipitation to heights corresponding to temperatures as cold as −14°C. Cloud liquid-water contents are low, on the order of $0.3\ \mathrm{g\,m^{-3}}$. Typical water contents of precipitation particles are on the order of $0.6\ \mathrm{g\,m^{-3}}$, although maxima on the order of $3\ \mathrm{g\,m^{-3}}$ were found near −3°C. The precipitation fields are inhomogeneous, as in Churchill and Houze (1984a, b), although the structure of the inhomogeneities has not been determined.

A variety of models have been developed and adapted to the simulation of MCCs. Maddox *et al.* (1981) used the Kreitzberg and Perkey (1976, 1977) mesoscale model to see if the 200-mbar jet streak in Maddox's composite study of mature MCCs results from deep convection associated with MCCs or from baroclinic processes independent of the convection (e.g., Wilson, 1976; Uccellini and Johnson, 1979). Numerical experiments with and without deep convection demonstrated that the 200-mbar jet streak and associated divergence were primarily caused by convective storms. Subsequently, Fritsch and Brown (1981) applied the Fritsch and Chappell (1980a, b) mesoscale model to determine the mechanisms responsible for the generation of the mesohigh aloft and the creation of the cold cloud tops seen by satellite infrared imagery associated with MCCs. They examined two hypotheses: (1) overshooting penetrative convective towers "detrain" cold turbulent air into the upper troposphere or lower stratosphere, and (2) the parameterized convective heating with a maximum at 350–400 mbar produces mesoscale ascent that extends above the level of heating, resulting in cooling by adiabatic expansion. Numerical experiments with and without detrainment from convective towers revealed that adiabatic cooling due to mesoscale ascent alone can account for the observed cooling aloft. Cooling by detrainment from convective towers actually reduces the cooling effect by mesoscale ascent, because mesoscale subsidence warming occurs in response to the convective updrafts.

In another study, Perkey and Maddox (1985) applied the Kreitzberg-Perkey model to the simulation of an observed MCC event. Simulations with and without cumulus heating revealed that convective heat release induced the formation of a relative upper-level high, which, in turn, increased upper-level divergence. The upper-level divergence caused an uncompensated removal of mass at upper levels; this resulted in the hydrostatic lowering of pressure at low levels. In response to the lowering of pressure, low-level convergence and cyclonic vorticity were created. Their experiments suggest that, because no such response was simulated in the experiments without convection, the observed circulation features associated with MCCs, including the low-level cyclone and upper-level anticyclone, are a direct response to deep convection.

Xu and Clark (1984) applied a wave–CISK model to the simulation of midlatitude MCSs. They emphasized that a time-lagged, nonequilibrium cloud model was important to scale selection of the most unstable modes on the mesoscale. Sensitivity experiments revealed that the model was sensitive to the gross features of the imposed convective heating profile, especially the magnitude and level of the maximum heating in the upper troposphere, lower tropospheric cooling, and cooling just above the tropopause. The model was insensitive to thin layers of heating and cooling, however.

Two independent groups applied different hydrostatic mesoscale models to the simulation of the MCC involved in the formation of the Johnstown flood. Molinari and Corsetti's (1985) model simulated many observed large-scale features of the MCC storm system, as reported by Bosart and Sanders (1981). Molinari and Corsetti's model also simulated the composite features reported by Maddox (1983), including the upper tropospheric anticyclone and a low-level cyclonic eddy. The model produced only weak perturbations in the wind field at midtropospheric levels and they emphasized that the parameterization of downdrafts was instrumental in the simulation of observed rainfall amounts and a realistic life-cycle behavior. Without downdrafts, the intensity of the vertical circulation was overestimated; without convective heating, the circulation never developed.

Zhang and Fritsch (1986) employed a two-way nested grid version of the Pennsylvania State University/National Center for Atmospheric Research (NCAR) regional model with Fritsch and Chappell's (1980a) convective parameterization scheme, including moist downdrafts. Use of the nested grid scheme allowed them to depict better the meso-β-scale features of the MCC, including an observed squall line and other meso-β-scale features, such as surface mesohighs and lows as well as upper-level troughs and ridges. They also simulated a warm-core mesovortex in response to the mesolow in the low troposphere to midtroposphere.

In summary, the MCC represents a large mesoscale convective system that exhibits a lifetime in excess of 6 h. Convection within the MCC takes on a variety of forms and, therefore, is not a distinguishing property of these systems. The presence of a stratiform–anvil cloud of considerable areal extent, however, is a feature of all MCCs. The upper levels of MCCs are characterized by divergent anticyclonic flow while the flow at midlevels is convergent and cyclonic. At the surface, the MCC is dominated by a divergent mesohigh of large areal extent. As the system reaches maturity, deep convection becomes a less dominant feature of the MCC, and, instead, stratiform rainfall dominates. From the above features, we conclude that the MCC represents an inertially stable form of the MCS that is nearly geostrophically balanced. Remember from Chapter 6 that a parameter that crudely identifies the scale at which the inertial stability of a system becomes important is the Rossby radius of deformation λ_R, where

$$\lambda_R = \frac{NH}{(\zeta + f)^{1/2}(2VR^{-1} + f)^{1/2}}. \qquad (10.3)$$

The parameter N is the Brunt–Väisälä frequency, H the scale height of the circulation, ζ the vertical component of relative vorticity, f the Coriolis parameter, and V the tangential component of the wind at the radius of curvature R. It is often difficult to evaluate the depth H over which the parameters controlling λ_R should be calculated. Moreover, once the atmosphere becomes saturated, N should be evaluated for a saturated atmosphere (Chapter 12). Cotton *et al.* (1989) computed λ_R for the set of 128 cases in their composite study. The calculated value of λ_R was about 300 km, and it varied by only 4% over the life cycle of the composite MCC, because in midlatitudes λ_R is dominated by the earth's rotation, or f^{-1}. This is in contrast to tropical cyclones (Schubert *et al.*, 1980), where the inertial stability is significantly modified by the intensification of the cyclonic circulation during its life cycle.

Using the area outlined by the −32°C isotherm at the mature stage of an MCC as an estimate of the "dynamic" radius R_L of such systems, Cotton *et al.* computed R_L to be 322 km or slightly greater than λ_R. Thus, a more dynamically based definition of an MCC is "a mature MCC represents an inertially stable mesoscale convective system which is nearly geostrophically balanced and whose horizontal scale is comparable to or greater than λ_R."

This is consistent with Maddox's satellite-based definition of an MCC, especially his requirements of large areal extent and long duration. It is also consistent with the observation that deep convection becomes less intense as the system reaches maturity and the flow becomes increasingly geostrophically balanced and nondivergent. Of course, the ageostrophic component of the circulation sustains the low-level convergence and vertical motion through the depth of the system. Low-level convergence is driven, in part, by the baroclinicity generated by the intersection of the evaporatively chilled mesohigh with the warm, moist, low-level jet present in the MCC environment. It is also possible that geostrophic imbalance in the upper troposphere caused by a convectively induced jet streak maintains ascending motion in the upper troposphere even as convection decays. This may explain why Lin (1986) found that the upward motion in his composite sample of midlatitude MCCs reached a maximum at the dissipation stage of the system, whereas Tollerud and Esbensen (1985) found the maximum upward motion in tropical clusters occurred at the mature stage of the system.

This definition distinguishes the MCC from the shorter-lived, smaller MCSs in middle latitudes and is a basis for comparing tropical mesoscale systems with midlatitude MCCs. Thus, an MCS residing at tropical latitudes must either be much larger in areal extent than a typical extratropical MCC, or it must exhibit considerable rotation.

10.3.3 Other Mesoscale Convective Systems

Among the variety of MCSs, many are not large enough to be classified as MCCs or do not exhibit a pronounced squall-line structure. These systems range in scale from slightly larger than a multicellular thunderstorm system to slightly smaller than MCCs. One such system analyzed by Knupp and Cotton (1986) had horizontal dimensions on the order of 100 km, much smaller than a typical MCC. Nonetheless, this system exhibited many features in common with its larger-scale cousins in both midlatitudes and the tropics. These included (1) a large stratiform cloud region at levels above the melting level, (2) mesoscale updrafts above and downdrafts below the melting level, (3) convective-scale updrafts and downdrafts of typical speeds of +10 and $-5\ \mathrm{m\ s^{-1}}$, respectively, being weaker than more isolated convection, (4) strong divergence of flow near the tropopause and near the surface in the rain-chilled region, and (5) a midlevel inflow jet. In the system studied by Knupp and Cotton (1986), the midlevel inflow jet appeared to be a response to the formation of convective-scale downdrafts. They suggested that reduced perturbation pressure associated with the formation of downdrafts in the lower and middle troposphere accelerated air into the downdrafts, creating the midlevel jet. Knupp and Cotton also noted that the dissipation of updrafts in the upshear flank of the cluster removed the positive perturbation pressure associated with those updrafts and made the upshear flank of the storm system more permeable to the prevailing flow. It is still not clear if this relationship between midlevel jets and convective-scale downdrafts prevails in all mesoscale systems or if a different relationship prevails in the larger-scale systems. For example, the numerical and analytical studies by Orlanski and Ross (1984) and Chen's (1986) numerical experiments described below suggest that midlevel convergence and jets may be a response to heating aloft. In that case, convective-scale and mesoscale downdrafts respond to the influx of low-valued θ_e air rather than vice versa.

Knupp and Cotton noted a symbiotic coupling between upshear convective cells and the stratiform region immediately downshear. The upshear convection created a supply of condensate to the downshear stratiform region, while the stratiform region produced the mesoscale downdraft. Outflow from the mesoscale downdraft, in turn, created a low-level mesohigh which maintained or intensified upshear convection.

Chen (1986) applied a two-dimensional, nonhydrostatic version of the Colorado State University Regional Atmospheric Modeling System (RAMS) to the simulation of the stratiform region of a midlatitude MCS. The numerical experiments were designed to be similar to Brown's (1979) simulation of tropical clusters, except that the role of the ice-phase and radiative heating and cooling processes was also considered. Whereas Brown employed a convective parameterization to drive the mesoscale circulation, Chen explicitly resolved precipitating deep convection with a 1.5-km horizontal grid covering a 300-km domain. The model was initialized with a composite sounding taken on 15 July 1984 during the Airborne Investigation of Mesoscale Convective Systems (AIMCS) experiment. An MCS formed on that day along a nearly stationary surface cold front and reached MCC proportions at the time of maturity, but it barely satisfied Maddox's criteria for duration and circular appearance of the stratiform anvil. Because the simulations were performed without the effects of the earth's rotation ($f = 0$), they do not actually represent an MCC according to our previous definition. Chen first attempted to simulate the MCC by setting up the two-dimensional plane of simulation perpendicular to the surface cold front. Vigorous sustained convection necessary to feed a stratiform cloud region could not be produced, however. To obtain steady, deep convection, the plane had to be oriented parallel to the surface front in the plane of the observed low-level jet. The jet maximum of 20 m s^{-1} was observed from the southwest at 850 mbar. These experiments illustrated the importance of the low-level jet to sustaining deep convection. The control simulation, which contained ice-phase precipitation processes as well as longwave radiative cooling, exhibited many features in common with observed MCSs, including (1) organized mesoscale ascent above the melting level and descent below, with attendant warming and drying, (2) a low-level evaporatively cooled mesohigh, (3) a mesohigh and associated divergence near the tropopause, and (4) a midlevel rear-to-front inflow centered near the melting level.

The strength of the rear-to-front flow, though relatively weak, was modulated by radiative heating and cooling. Radiative cooling at the top of the stratiform layer and radiative warming at its base destabilized the stratiform cloud region which, in turn, strengthened the mesoscale and convective-scale updraft and downdraft circulations. The rear-to-front flow was a manifestation of midlevel convergence created by the enhanced convective heating above the melting level in response to radiative destabilization and to the formation of a mesoscale downdraft by water loading and evaporation of precipitation. The former process is similar to Orlanski and Ross's proposed mechanism for the formation of midlevel convergence in prefrontal squall lines.

Chen found that additional heating associated with the ice phase was significant in organizing the mesoscale circulations. When the ice phase was turned off, a deep stratiform cloud with its sustained mesoscale circulations could not be obtained with the initial conditions imposed on the control experiment. Instead of a steady mean vertical velocity in the upper troposphere, a pulsating mean vertical velocity was more characteristic of the experiment without the ice phase.

When cooling by melting was turned off, but latent heating from freezing and sublimation growth of ice particles was retained, the strengths of the midlevel, rear-to-front flow and the mesoscale downdraft were essentially unchanged. The removal of cooling by melting strengthened the mesolow pressure region at the base of the melting level which, in turn, strengthened the updraft circulation and influx of moist air into the cloud system.

The fact that the midlevel, rear-to-front flow and associated midlevel convergence were essentially unaltered by the neglect of melting suggests that one of two possible processes compensated for its removal. First, latent heating associated with the condensation, freezing, and ice-vapor deposition of the enhanced vertical flux of moisture into the upper cloud levels could have generated additional convergence in the manner proposed by Orlanski and Ross (1984). Second, the enhanced vertical flux of moist air resulted in greater condensate formation and water loading. The enhanced water loading could have compensated for the absence of cooling by melting, thereby maintaining a downdraft of intensity similar to that with melting present. The midlevel convergence and rear-to-front flow could have been maintained by the convergence driven by a mesoscale downdraft, which was driven by intensified water loading. Once dry air enters the cloudy region, evaporation of cloud droplets and precipitation particles intensifies the mesoscale downdraft. At lower levels, the convective-scale "up–down" downdraft component, identified by Knupp (1985), feeds the low-level mesohigh with evaporatively chilled air. This drives low-level divergence, which can reinforce a dynamically driven downdraft, as suggested by Miller and Betts (1977).

Midlevel convergence and associated rear-to-front midlevel flow or jet in MCSs can be induced either by heating aloft or by cooling in the lower troposphere. Once an MCS organizes, both heating aloft and cooling at lower levels contribute to the intensity of midlevel inflow into the storm system. The strength of the midlevel inflow is modulated by the stability in the upper and lower troposphere, the dryness of the midlevel air, vertical shear of the horizontal wind, and diabatic processes such as radiative heating and cooling, ice-phase heating aloft, and evaporative and melting cooling in the lower troposphere.

Considering the conclusions from analysis of tropical MCSs, the insensitivity of the mesoscale downdraft to cooling by melting found by Chen is surprising. Leary and Houze (1979b) suggested that cooling from melting of ice particles is instrumental in the initiation and maintenance of mesoscale downdrafts. Lord *et al.* (1984) arrived at a similar conclusion in their examination of the effects of ice-phase microphysics on the dynamics of a simulated tropical cyclone. The widespread horizontal distribution of cooling as a result of melting can initiate and maintain mesoscale downdrafts. These downdrafts contribute to the formation of multiple convective rings in their axisymmetric model which subsequently modifies the development of the tropical cyclone. These studies focus on an important difference between maritime tropical and many extratropical MCSs—that is, in the maritime tropics, the base of convective clouds is low, often 600 to 700 m above the surface, and the layer between the stratiform cloud base and the surface is relatively moist. In many extratropical MCSs, especially in the western portion of the High Plains of the United States, the base of convective clouds is generally quite high, often 3.0 km above ground level and only 0.5 km below the base of the stratiform layer. As a result, evaporation of precipitation in the subcloud layer is much greater than in the comparable tropical MCS. The important difference between tropical and extratropical convective systems was discussed by Anthes *et al.* (1982). Melting is relegated to lesser importance in the higher cloud-base, drier subcloud layers characteristic of the extratropical MCS. Other factors make many extratropical MCSs less sensitive to melting. One is that the conditional instability is often much greater over the heated continental regions. As a result, convective-scale updrafts and downdrafts can be more dominant relative to mesoscale updrafts and downdrafts than in comparable oceanic tropical MCS. An excellent example is the ordinary squall line described by Schmidt and Cotton (1988) which, we note, was dominated by a single rotating cell.

10.4 Genesis of Mesoscale Convective Systems

In earlier sections, we discussed the large-scale conditions associated with MCSs. At the very least, the large-scale conditions establish an environment thermodynamically and dynamically ripe for their development. In some instances, the large-scale environment not only establishes the setting favorable for MCS genesis, but it can also supply the trigger to develop the MCS and organize it into its mature configuration. This is true for some midlatitude prefrontal squall lines (Ross, 1987). In many tropical and midlatitude MCSs, however, their genesis involves a complex interplay among the larger scale environment, mesoscale circulations driven by differential heating associated with variations in terrain elevation, land–sea temperature contrasts or other geographically related forcings, and deep cumulonimbus convection.

The first stages of MCS genesis are associated with the development of low-level moisture convergence whose source can be a variety of forms, including the interaction of land-breeze circulations over tropical islands with the prevailing flow (Houze *et al.*, 1981), orographic circulations (Cotton *et al.*, 1983; Tripoli, 1986), convergence along the interface between air masses along a dry line (Ross, 1987) or along frontal boundaries (Ogura and Chen, 1977; Miller and Sanders, 1980; Orlanski and Ross, 1984), symmetric instability in the warm sector of a midlatitude cyclone (Ogura *et al.*, 1982), convergence in the intertropical convergence zone (McBride and Gray, 1980), or outflows from previous MCSs (Fortune, 1980; Wetzel *et al.*, 1983).

Whatever the source of low-level moisture convergence, the first stages of MCS formation are accompanied by the development of deep cumulonimbi in close proximity. Often the first signature of MCS genesis is the merger of anvil clouds emanating from the neighboring cumulonimbi (Leary and Houze, 1979a; McAnelly and Cotton, 1986). At the same time, a cold pool of air at low levels of meso-β-scale proportions forms as a consequence of the evaporation of precipitation (Leary and Houze, 1979a; Maddox, 1981). It is not clear which of these processes is more instrumental to the upscale growth of a convective system.

Evidence suggests that, with the merger of anvil clouds and the formation of elevated stratiform precipitation, an upward shift in mean vertical motion of about 200 mbar occurs (Houze, 1982; Johnson, 1982; Gamache and Houze, 1982; Chen, 1986; Dudhia and Moncrieff, 1987) as well as an associated upward shift of the level of maximum heating (Esbensen and Wang, 1984; Lin, 1986). CISK (Yamasaki, 1968; Ooyama, 1969; Koss, 1976) and wave–CISK (Silva-Dias *et al.*, 1984; Nehrkorn, 1985) models have repeatedly demonstrated a pronounced sensitivity to the level of maximum heating in MCS development. Using a wave–CISK model with a downdraft parameterization, Raymond (1987) concluded that an intensity threshold in convection may control the level and magnitude of convective heating. He suggested that, in some environments, convective heating must reach a threshold intensity before the convective system can survive the consequences of strong evaporation in the dry layers at midlevels.

In their numerical simulation of midlatitude cyclone development, Anthes and Keyser (1979) pointed out that greater upper-level heating stabilizes the atmosphere, therefore an upward shift in the level of heating would weaken MCSs. Using a combination of numerical and analytical models, to investigate the influence of vertical heating profiles on tropical cyclone intensity, Hack and Schubert (1986) found a larger percentage of the kinetic energy projected onto larger horizontal scales of motion (i.e., those near λ_R) when the heating maximum is shifted into the upper troposphere. They discovered that a vortex generated by upper-level heating is quite deep and relatively diffuse in the lower troposphere, while a vortex produced by lower-level heating is shallower and more tightly organized. Thus, an upward shift in convective heating associated with the merger of anvils and development of stratiform precipitation may weaken the overall convective intensity of an embryonic MCS and at the same time favor its upscale growth to mesoscale proportions. Weakening of convective intensity is characteristic of the development of mature midlatitude MCCs.

In preceding sections, observations and conceptual, analytic, and numerical models have shown that low-level cooling from evaporation and melting in convective-scale and mesoscale downdrafts is important to the initiation and maintenance of MCSs. In squall lines, the evaporatively chilled air forms a gust front, which then propagates similarly to a density current and contributes to the continued regeneration of the convective system. In moister, more weakly sheared environments, gust-front propagation may not be as important as the consequences of forming a low-level mesohigh. The numerical experiments by Zhang and Fritsch (1986) and Song (1986), as well as the advective wave–CISK models of Raymond (1984, 1987), suggest that the evaporatively cooled air induces a mesoscale baroclinicity which establishes a low-level pressure gradient that can function like land and sea temperature contrasts in driving sea-breeze fronts. For a given environment, a threshold rain volume or, more properly, an evaporatively cooled volume would be needed to develop a mesohigh of large enough amplitude and scale to generate a self-sustaining mesoscale system.

Large-scale thermodynamic stability, adequate moisture content, wind shear, some mechanism to generate low-level convergence and convective feedbacks through the development of elevated convective heating in the upper troposphere, and low-level cooling by evaporation of precipitation are instrumental to the genesis of MCCs. Moreover, large-scale deep tropospheric lifting as driven by shortwave troughs (Maddox, 1981), jet streaks (Uccellini and Johnson, 1979; Matthews, 1983), easterly waves, or the intertropical convergence zone (McBride and Gray, 1980; Frank, 1983;

Dudhia and Moncrieff, 1987) are also key in sustaining deep convection long enough to transform a system from the cumulonimbus scale to the mesoscale. Other factors of consequence to MCC genesis are revealed in the numerical experiments reported by Tripoli (1986).

Tripoli's (1986) study focused on the genesis of MCSs over and to the lee of the Rocky Mountains in the United States. An example of such an eastward, propagating system is shown in Fig. 10.39. As noted by Cotton *et al.* (1983) and Wetzel *et al.* (1983), this was just one of many such systems propagating away from the mountains and participating in the formation of an MCC. To investigate this phenomenon, Tripoli applied the Colorado State University RAMS to a 1000-km domain extending from the western slopes of the Rocky Mountains eastward into central Kansas. Three-dimensional coarse-mesh simulations (~14-km grid spacing), with and without a convective parameterization, and two-dimensional, fine-mesh simulations (~1-km grid spacing) were tried. The two-dimensional, fine-mesh simulations yielded the most realistic simulations. The model was able to simulate explicitly precipitating deep convection, including the effects of the ice phase and shortwave and longwave radiative heating and cooling, along with a mesoscale response to those processes. The model contained a complete surface-energy budget that allowed simulation of a diurnal cycle. Simulations were run for 24 h commencing at midnight the day before the MCS formed.

With the model simulations for guidance, Tripoli developed a conceptual model of six stages in the life cycle of the evolving MCS. Figures 10.48–10.52 illustrate five of the six stages. Stage 1, not shown, begins in the morning hours as the nocturnal downslope drainage flow is replaced by an upslope flow driven by heating of the elevated terrain. About 60 km east of the ridgetop, a convergence zone forms as a result of the interaction between the mountains' and plains' thermally driven upslope circulations and the ambient flow over the mountain. The mountain lee wave flow inflects upward at the convergence zone. The meridional flow is sheared anticyclonically slightly above ridgetop early in the day; this residual effect of the nocturnal regime accelerates the development of upslope flows. Cumulus convection at this stage can be characterized as cumulus and towering cumulus. During this and subsequent stages, the mountains' and plains' circulations exert a strong stabilizing influence over the eastern plains. The shallow upslope flow cools the low-level air, while sinking motion above the upslope flow causes warming and strengthens the formation of a strong capping inversion.

The commencement of stage 2 (shown in Fig. 10.48) is indicated by deep convection that first forms over the mountain ridgetop as the prevailing westerly flow, augmented by the slope circulation, advects moisture into the higher elevations, where it is lifted by the terrain to trigger cumulonimbi.

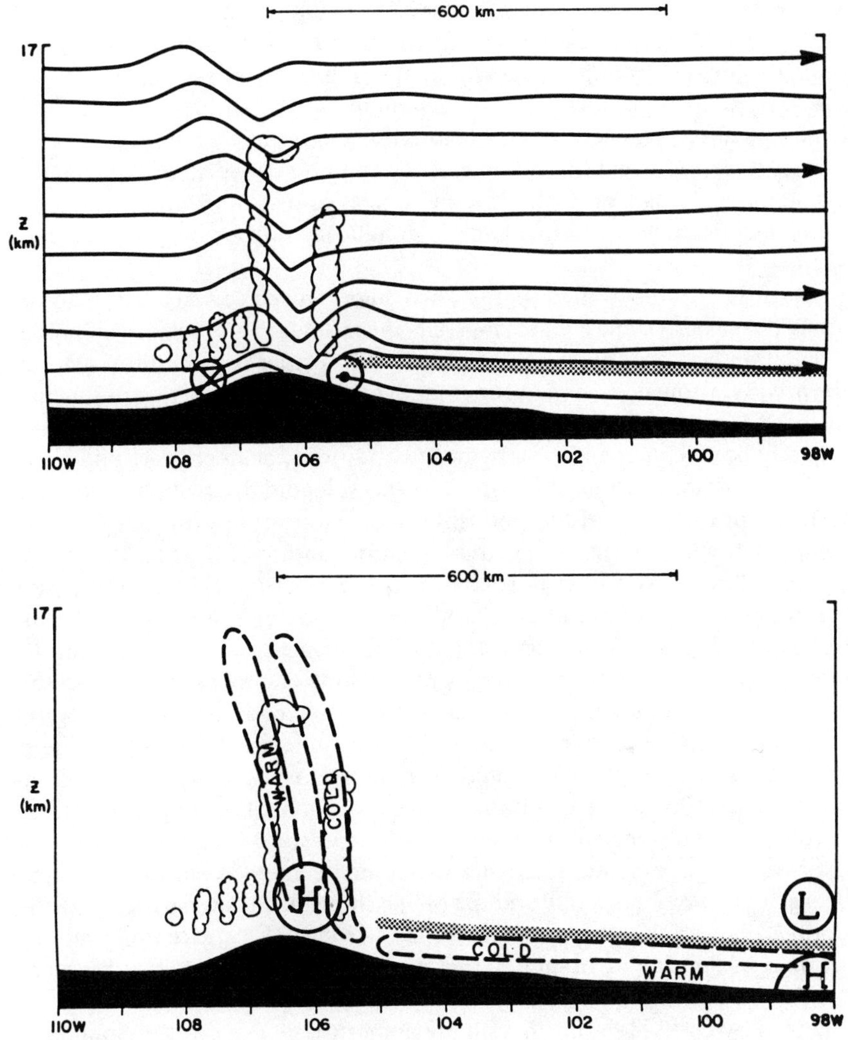

Fig. 10.48. Conceptual model showing flow field and position of convective elements at the time deep convection forms (stage 2). Axes and topography are as described. The fields are derived from 86-km running averages of actual model-predicted variables. The surface topography is depicted by the black shading. The vertical axis is height in kilometers above mean sea level and the horizontal axis is west longitude. The stippled line represents the position of the plains inversion. Regions of cloud are indicated. Top: The flow field with ground-relative streamlines. Circles depict flow perturbation normal to plane. Bottom: The pressure and temperature response. Pressure centers are depicted by solid closed contours and temperature by dashed contours. The length scale of 600 km ($2L_R$) is indicated. [From Tripoli (1986).]

As the deep convection moves eastward with the prevailing westerly flow, low-level western slope moisture is carried along with it. When the eastward-migrating cumulonimbus system approaches the convergence zone between the eastern upslope flow and the mountain wave, all elements favoring explosive development of intense convection come together, including moisture, surface convergence, and preexisting cumulus convection. As explosive development of an ensemble of mesoscale convection ensues, moisture advection from the eastern plains is enhanced, further invigorating convection.

Stage 3 (Fig. 10.49) thus begins when an organized convective line forms along the eastern slopes. Until this time, the mountain and plains solenoidal circulation was confined to below 5 km MSL. Under dry conditions, when deep convection does not occur, the solenoid circulation remains below 5 km MSL and slowly migrates eastward in the presence of prevailing westerly flow. Under sufficiently moist conditions, deep convection organized in a mesoscale line develops and the solenoidal circulation deepens to 12 km or the depth of the troposphere. In association with the deepened solenoidal cell, a pronounced thunderstorm outflow forms in the 10- to 12-km layer. Geostrophic adjustment to the persistent circulation forms anticyclonic shear of meridional flow aloft and cyclonic shear at low levels.

Stage 4 (Fig. 10.50) begins when the convective system moves from the mountains over the eastern plains, where a zone of suppression is encountered; the zone of suppressed convection is a consequence of several topographically related phenomena. At the juncture between the steep mountain slope and the more gentle sloping plains, two solenoidal circulations interact. Associated with the heated steep mountain slope, an intense deep solenoidal circulation develops with its upward branch focused over the mountain peaks and its downward branch centered over the transition between the mountain slope and the plains slope. Associated with the gentle sloping plains is another solenoidal circulation with its ascending branch just east of the sinking branch of the mountain solenoid and its subsidence branch located much farther east. [Figure 10.53 illustrates of these cells, obtained by Dirks (1969) with a simple numerical model.] As the embryonic MCS moves onto the plains, it encounters greater subsidence as well as surface divergence as the low-level air is forced up the mountain slope or away from the system core. Because of the greater subsidence in the system core, upward motion within the system core collapses.

The upper-level anticyclone and low-level cyclone persist, however, because they slowly adjust to Coriolis accelerations (see Fig. 10.45, for an example). In a sense, they represent a "flywheel" effect similar to that discussed by Emanuel (1982) with respect to the generalized CISK concept. The collapse of the system also initiates a deep gravity wave 150–200 km in horizontal wavelength which propagates laterally eastward and westward. The westward-propagating mode triggers weak convection, where it encounters moisture resupplied to the mountain convergence zone. The eastward-propagating mode fails to trigger deep convection, however, because the plains inversion prevents it from tapping the low-level moisture. More importantly, the collapsing core, augmented by precipitation loading and

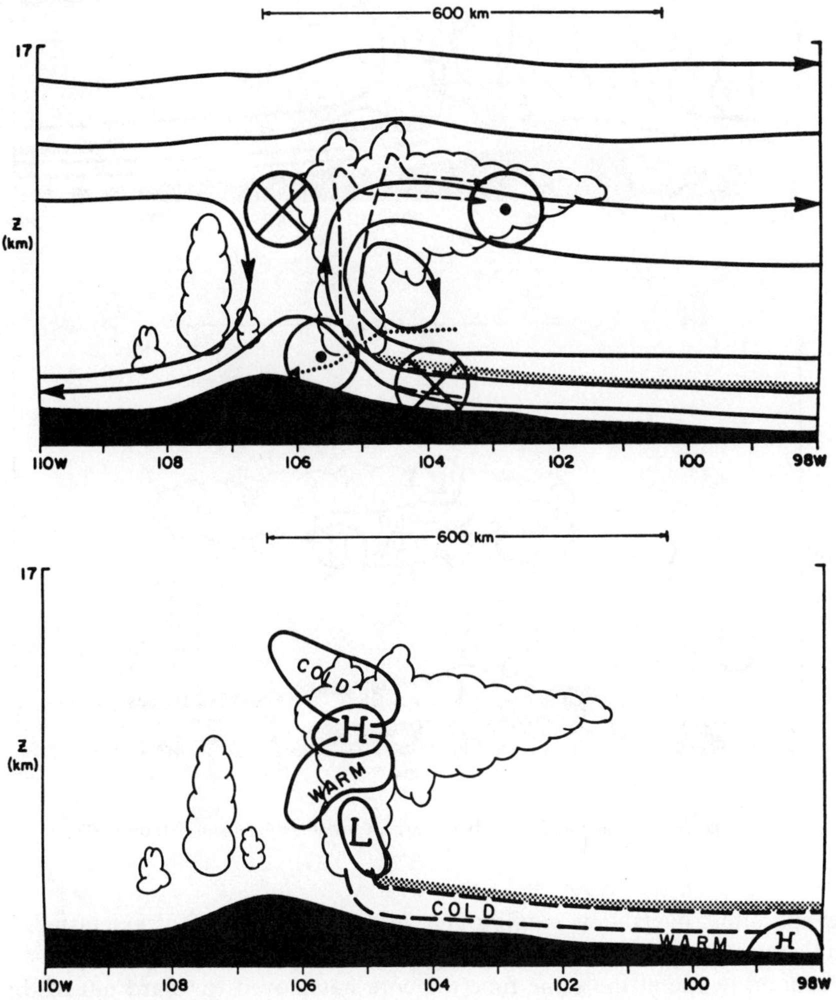

Fig. 10.49. Same as Figure 10.48, except for stage 3. Also, flow is storm-relative flow. Individual parcel paths are given by dashed (updraft) and dotted (downdraft) lines. [From Tripoli (1986).]

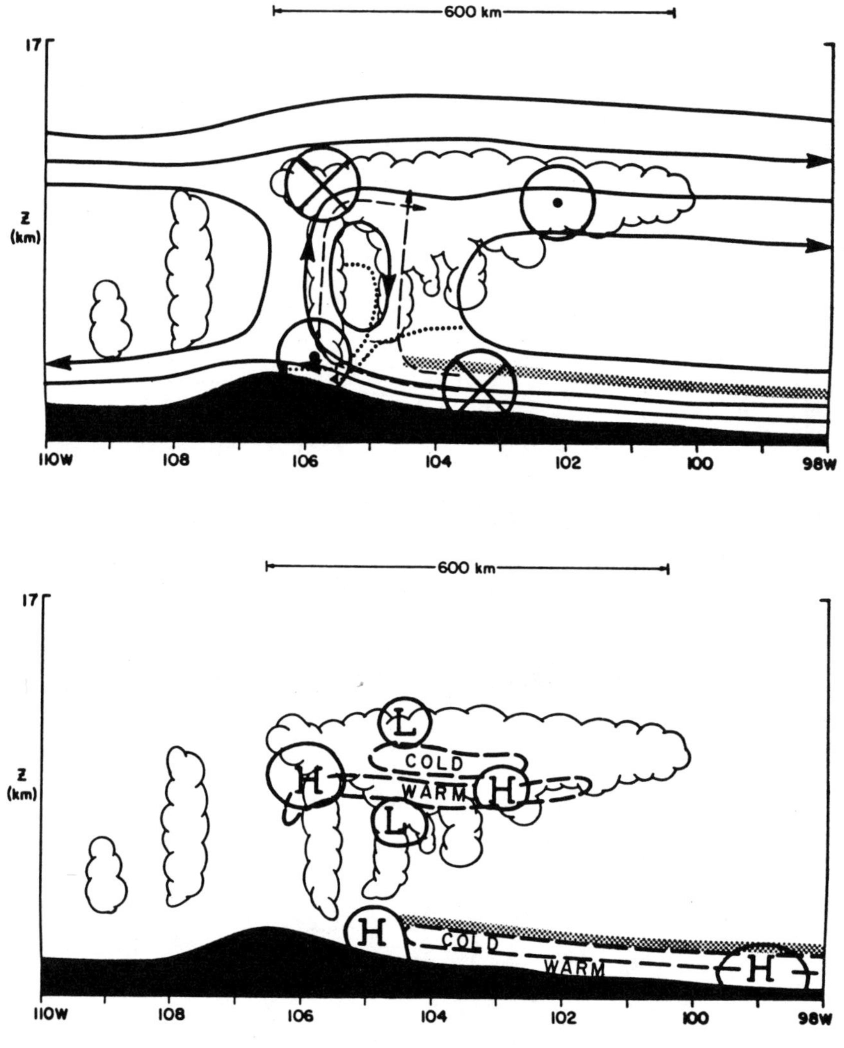

Fig. 10.50. Same as Figure 10.48, except for stage 4. [From Tripoli (1986).]

evaporation, overshoots equilibrium, and after condensate is exhausted, it adiabatically warms. The warmed core causes the upward motion to rebound, while at the same time the core has moved eastward out of the zone of suppression, where it encounters the influx of moisture-rich air from beneath the plains inversion. The convective system, therefore, intensifies and matures to MCS proportions.

At this juncture, the system enters stage 5 (Fig. 10.51). Until now, the system evolution is very site specific and is a strong function of the particular properties of the Rocky Mountains' thermally driven slope/plains circulations and the availability of moisture unique to the synoptic-scale circulations of that region. This is not to say that other locations along the north–south extent of the Rocky Mountain barrier spanning Alberta,

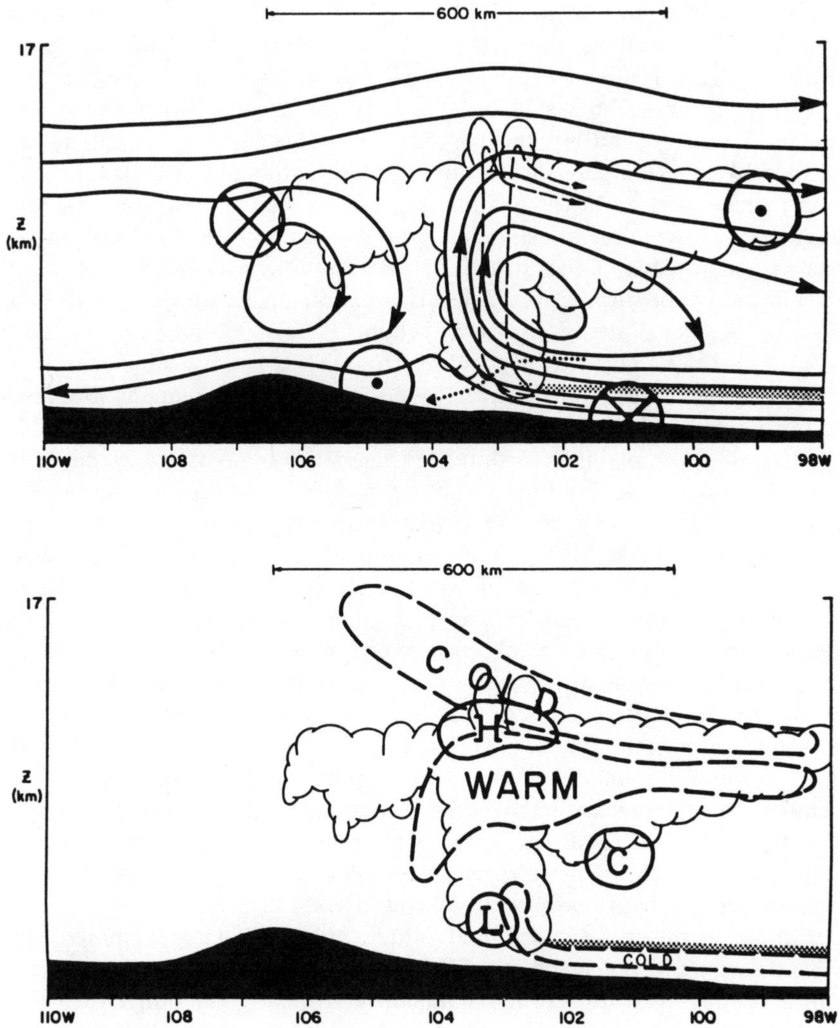

Fig. 10.51. Same as Figure 10.48, except for stage 5. [From Tripoli (1986).]

Canada, to the New Mexico mountains may not also be favorable for such a development scenario. Similar "orogenic" forcing of MCSs may occur along other major mountain ranges (see, e.g., Velasco and Fritsch, 1987). Upon entering stage 5, the convective system enters a stage of its life cycle that is more generic in its MCS characteristics. The simulated MCS is not steady; it undergoes repeated cycles of growth, overdevelopment, and weakening, although the system does not collapse as it did at the end of stage 4. The cycles resemble the breakdown of an unstable advective CISK mode described by Raymond (1984). Each breakdown cycle initiates oppositely propagating transient gravity waves that expand radially from the system core. Unlike Raymond's model, the convection does not remain linked to the transient waves because they are usually isolated from the low-level unstable air by the capping inversion. As the system moves eastward, the mean-core circulation gradually strengthens as the kinetic energy generated by latent heating is partially retained while the system undergoes geostrophic adjustment. That only a minor fraction of the kinetic energy generated by latent heat is retained as a geostrophically adjusted circulation is not surprising, since this is characteristic of systems smaller than λ_R. Instead, as demonstrated by Schubert *et al.* (1980), the vast majority of energy is radiated vertically and horizontally as gravity-wave energy.

Throughout stage 5, the convective core of the MCS remains localized at the western edge of the plains inversion. Short-lived convective cells residing over the plains inversion are triggered by gravity waves emitting from the system core, but they do not intensify because they cannot tap the moist air below it. The western boundary of the plains inversion is continually eroded by the MCS core circulation. Here, adiabatic cooling associated with the mesoscale ascent and turbulent mixing by the convection itself destroy the inversion interface. Precipitation is advected eastward where its evaporation cools the air above the inversion and destabilizes the air on the westernmost edge of the inversion. Overall, the system moves at about 10 m s^{-1}, which corresponds to the speed of the upper tropospheric wind. The solenoidal circulation of the mature MCS resembles the observationally derived squall-line model of Ogura and Liou (1980), as well as the more detailed flow features depicted by Smull and Houze (1985, 1987a, b). A rear midlevel inflow jet is absent in this simulation, however. This may be a consequence of the moist air mass at middle levels which would inhibit evaporative cooling in mesoscale downdrafts.

Similar to the tropical and midlatitude MCSs we have examined, the system forms a deep stratiform cloud aloft. A large part of the stratiform cloud extends in advance of the main convective core. The radiative effects of the anvil become more important as darkness approaches. Prior to this period, heating at cloud top from absorption of solar radiation largely offsets longwave radiation cooling at the cloud top. As night approaches, longwave radiative cooling at cloud top and heating at the base of the stratiform layer lead to further destabilization of the stratiform layer. The intensity of the MCS, therefore, increases with peak vertical velocity being simulated at 2000 MST. Another consequence of the greater dominance of longwave radiative flux divergence to the radiative budget of the MCS is that the Brunt–Väisälä frequency is lowered in magnitude in the stratiform layer, particularly near cloud top. As a result, the phase speed of propagation of gravity waves is reduced, causing a partial refraction of the vertically propagating gravity wave modes. A greater fraction of the vertically propagating gravity wave energy becomes trapped beneath the stratiform cloud top. As the stratiform cloud gains greater prominence, so does the intensity of the anticyclonic shear of the meridional winds aloft and the cyclonic shear at low levels.

Stage 6 of the system (Fig. 10.52) represents the transformation from an unsteady MCS having a scale less than the Rossby radius of deformation to the more geostrophically balanced MCC system greater in scale than λ_R. Of the stages of the conceptual model derived from Tripoli's numerical simulation, this is the most speculative because the observed systems are strongly influenced by the diurnally varying low-level jet (LLJ) over the High Plains of the United States. Tripoli's two-dimensional simulations were unable to simulate properly the MCS response to the LLJ for the following reasons: (1) The eastern boundary of the model domain was near the climatological centroid of the southerly LLJ over the High Plains. (2) The LLJ is not only a function of regional sloping terrain and diurnally varying boundary layer effects, but it is also dependent upon the position and strength of the subtropical high. (3) The horizontal advection of heat and moisture by the southerly LLJ cannot be easily simulated in an east-west-oriented two-dimensional model domain.

Nonetheless, Tripoli's simulations suggest that some significant transformations take place in the genesis of an MCC from an ordinary MCS as nighttime approaches. The surface begins to cool, and low-level convective available potential energy is reduced. Furthermore, as a low-level nocturnal inversion forms, convective updrafts no longer draw on surface air and, instead, begin to draw on air residing above the nocturnal inversion. At this point, the nocturnal LLJ intensifies and fuels the convective updrafts of observed systems. Without the thermal and moisture advection associated with the LLJ in Tripoli's simulation, the simulated MCS begins to weaken. Second, as a consequence of the greater trapping of internal gravity-wave energy, gravity waves propagating away from the system core have greater amplitude. With increased moisture available to them above the inversion, the higher amplitude gravity waves can excite new convective elements

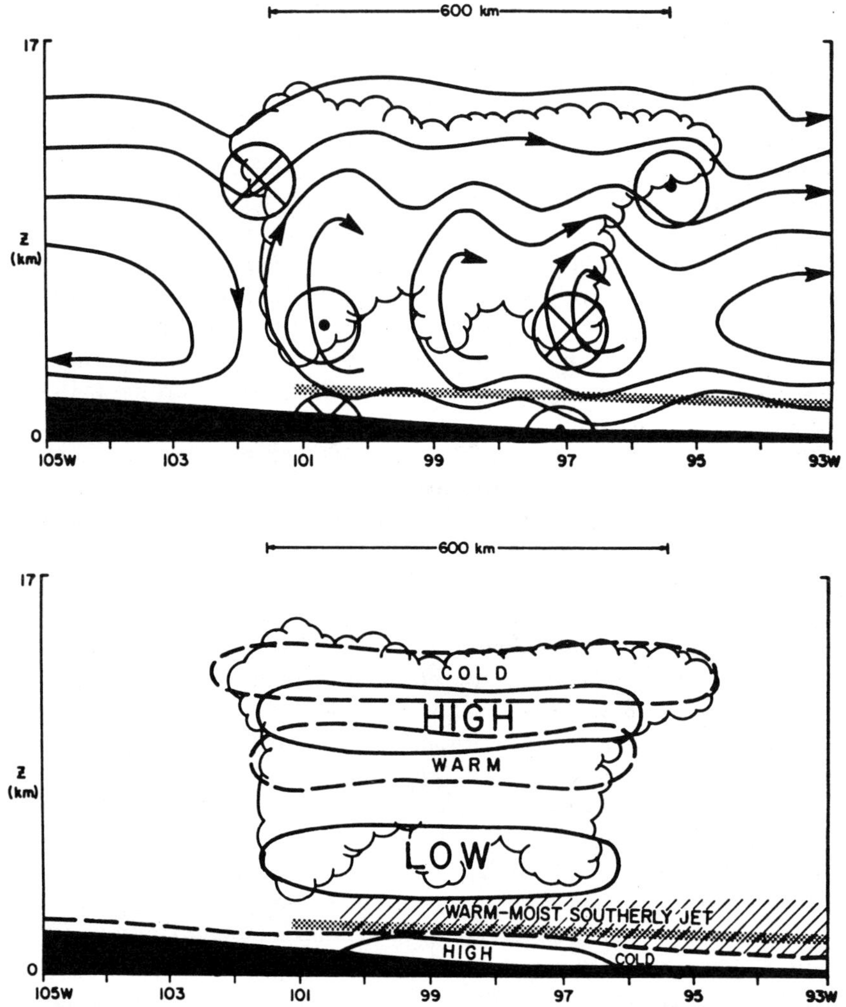

Fig. 10.52. Same as Figure 10.48, except for stage 6. Also, the region of the low-level southerly jet is hatched. [From Tripoli (1986).]

away from the main convective core. Consistent with the observations, convection becomes more widespread, though weaker in intensity, beneath the stratiform cloud. The resultant more widespread convection favors the projection of a greater fraction of the kinetic energy generated by latent-heat release onto a scale comparable to or greater than the Rossby radius of deformation. At the same time, the system is developing greater anticyclonic shear aloft and cyclonic shear at low levels. Thus, in accordance with Eq. (10.3), λ_R shrinks, favoring the system to become geostrophically balanced. It is perhaps not fortuitous that Maddox (1981) required that an MCC maintain a large anvil area for at least 6 h; this longer duration distinguishes the unbalanced MCS circulation from the more balanced MCC.

Tripoli's simulations illustrate dramatically that the genesis of an MCS and, subsequently, the more balanced MCC involves a complex series of steps. Thunderstorms are initiated in an organized low-level convergence field. If the supply of CAPE is great enough, the anvils emitted by the neighboring cumulonimbi may merge to form a stratiform cloud layer extending from the melting level to the tropopause. This will result in the elevation of the level of heating to the upper troposphere. At the same time, downdraft outflows from the neighboring storms merge into a mesohigh at low levels. In a sense, the embryonic MCS creates its own baroclinicity. Both processes favor the upscale growth of a convective system to mesoscale proportions. As the system organizes on the mesoscale, anticyclonic shear develops aloft and cyclonic shear forms at low levels, and a greater fraction of the kinetic energy generated by latent-heat release is projected into a geostrophically balanced flow. Greater trapping of gravity-wave energy by a radiatively destabilized stratiform–anvil cloud in the nocturnal regime further contributes to the projection of kinetic energy generated by latent heating onto the more balanced meso-α-scale.

Of course, more often than not, one or more of these processes do not take place and the upscale growth of a convective system into MCS or MCC proportions is curtailed. It is not essential that an MCS evolve step by step in the way we have just outlined. Instead, the large-scale environment can provide the vertical motion field, cyclonic and anticyclonic shears, and conditional instability, which can immediately organize convection on the meso-β-scale and completely circumvent the first stages of MCS genesis. It is entirely possible that some large-scale environments will support the immediate organization of a convective system onto the more balanced meso-α-scale, favoring the nearly spontaneous development of an MCC.

Why are MCCs more prevalent at night? An answer to this question is not fully resolved at this time. One possibility is that the time scale required for MCC genesis from afternoon convection (see Fig. 10.53 for idealized daytime circulation), either orogenic or not, naturally places the time at which an MCC matures after local midnight. The greater dominance of cloud-top radiative cooling after sunset favors the organization of a system on the meso-α-scale. Furthermore, Maddox's (1983) composite study revealed the importance of the low-level jet to the occurrence of MCCs. The LLJ strengthens after sunset and its orientation produces strong convergence with an eastward-moving, thunderstorm-generated mesohigh. Toth

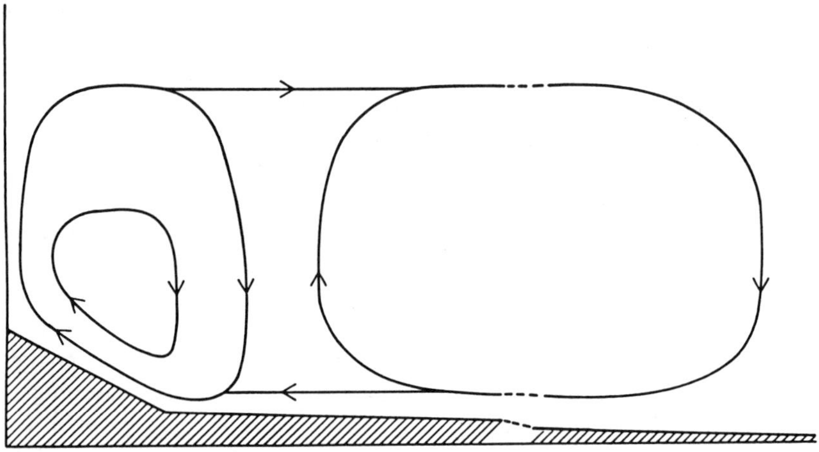

Fig. 10.53. Schematic illustration of the idealized Rocky Mountains–Great Plains daytime circulation. [From Dirks (1969).]

(1987) examined the behavior of shallow cold pools, including MCS-generated cold pools, during the day and at night. Using observational analysis and numerical experimentation, Toth showed that frontogenesis along the leading edge of a cold pool was weakened by the presence of afternoon convection. During the night, when dry boundary layer convection is absent, frontogenesis along the boundary of the cold pool strengthened and the frontal boundary propagated faster. Thus, the weakening of boundary layer convection at night may strengthen the baroclinicity along the frontal surface of a MCS-generated mesohigh and further enhance the convergence between the mesohigh frontal boundary and the LLJ.

10.5 Tropical Cyclones

10.5.1 General Characteristics of Tropical Cyclones

The dynamical behavior of an atmospheric circulation system can be characterized by the scaling parameter λ_R, the Rossby radius of deformation [Eq. (10.3)]. A substitution of typical values of the parameters in Eq. (10.3) for tropical cyclones and extratropical cyclones (averaged over the entire cyclone) shows that, for the largest scale of these cyclonic disturbances, the atmosphere behaves as a two-dimensional, quasibalanced fluid (Fig. 6.72). Because of the enormous spatial variation of λ_R within tropical and extratropical cyclones, a variety of clouds and mesoscale-precipitating phenomena are embedded within these systems. Many of these are described in other chapters of this book. In this section we describe the physical processes and properties of the environment that determine the types of mesoscale structures forming within larger scale tropical cyclonic storms.

Unlike extratropical cyclones (discussed in Chapter 11), tropical cyclones are embedded in an air mass with rather uniform properties. The large-scale aspects of tropical storms are described by Frank (1977) and Anthes (1982). Tropical cyclones occur over all of the tropical oceans except the south Atlantic, where water temperatures are too low for their formation. The strong (over 1 m s^{-1}) upward motion near the storm center and the extremely high water-vapor content (mixing ratios greater than 20 g kg^{-1}) of the low-level air feeding the storm produce deep and extensive clouds and heavy precipitation. Typical rainfall rates averaged over the inner 200 km of tropical cyclones are 10 cm day^{-1}, although maximum amounts can exceed 50 cm day^{-1}.

Anthes (1982) reviewed observational and theoretical aspects of tropical cyclones, and the reader is referred to this review for further details. A summary of these aspects related to the cloud structure is given here. Figure 10.54 is a schematic of the three-dimensional structure of a mature tropical cyclone. Near the surface, warm moist air spirals inward toward the center

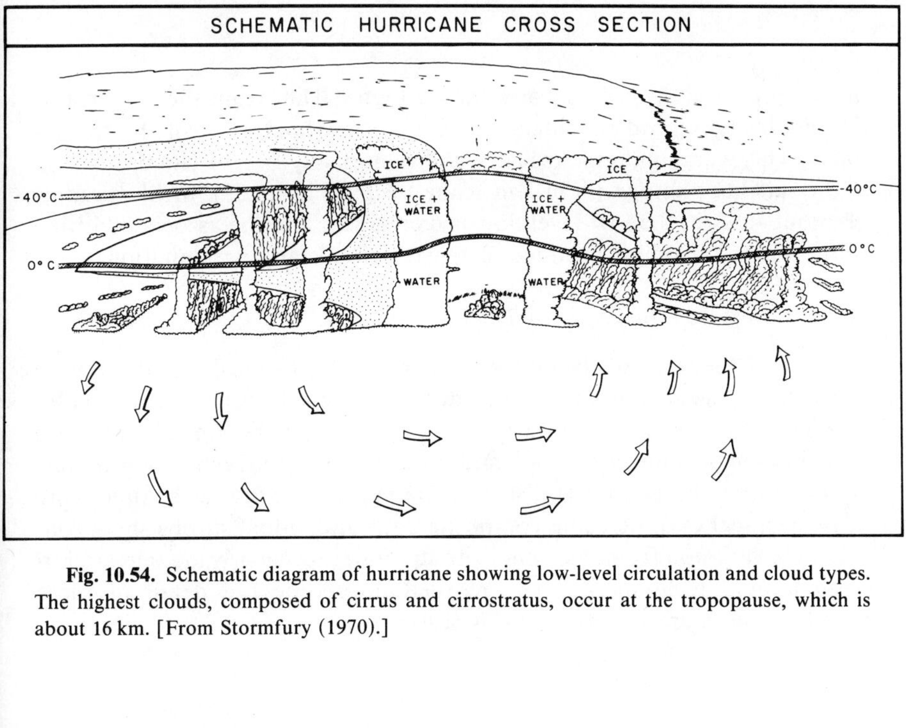

Fig. 10.54. Schematic diagram of hurricane showing low-level circulation and cloud types. The highest clouds, composed of cirrus and cirrostratus, occur at the tropopause, which is about 16 km. [From Stormfury (1970).]

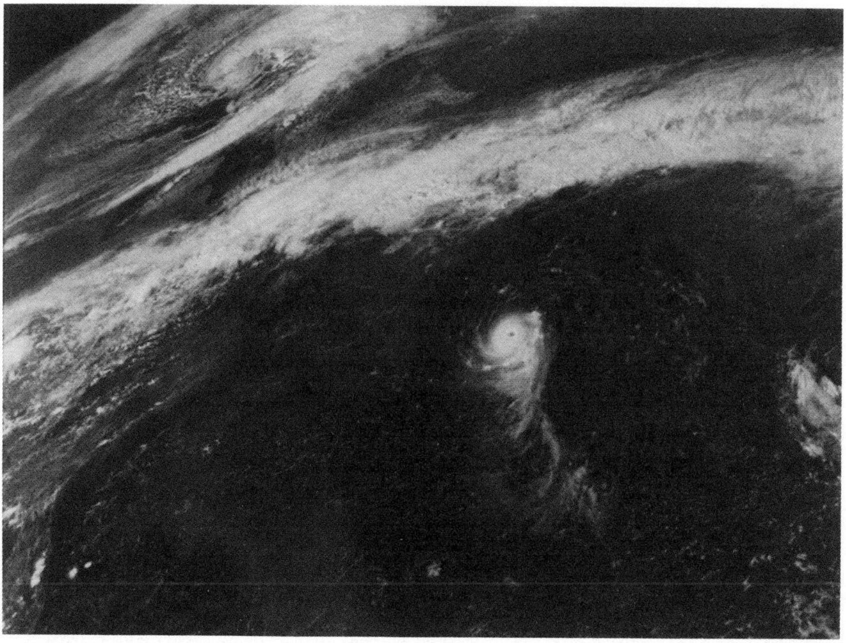

Fig. 10.55. Satellite photograph of Hurricane Becky, 1800 GMT, 20 August 1974. [From Anthes (1982).]

of low pressure. At radii greater than about 400 km from the center, this flow is divergent and subsidence extends throughout most of the troposphere. This warm sinking air is dry and remarkably free of clouds, as seen in the satellite photograph of Hurricane Becky (Fig. 10.55). Inside a radius of about 400 km, the low-level flow is convergent and the associated lifting of the warm humid air produces extensive clouds and precipitation.

10.5.2 Mesoscale Structure of Tropical Cyclones

In spite of their relatively uniform environment, tropical storms contain important mesoscale features, including the eyewall, a generally circular ring of intense convection surrounding the eye; a region of stratiform precipitation outside the eyewall; and spiral bands of convection that assume various forms. Frank (1984) discusses the mean structure of the inner core of the tropical cyclone using composite data. Individual storms show considerable variation from the composite structure; recent advances in aircraft observational techniques have provided unprecedented detail of these features, which are reviewed in this section.

Figure 10.56 shows visible and infrared satellite imagery of four mature hurricanes. All show a well-developed eye and bands of clouds that spiral around and into the eye. It is difficult to distinguish on the satellite photographs between the eyewall convection and the stratiform and other convective clouds near the center of the hurricanes. In Fig. 10.56, the inner side of the eyewall is highlighted by its high reflectivity in the photographs of Anita (a), Frederic (c), and Allen (e). A dramatic photograph of the inner edge of Hurricane Allen's eyewall, taken from the NOAA-42 aircraft, is shown in Fig. 10.57.

Radar observations from C-band radars aboard research aircraft show quite clearly the eyewall convection, stratiform regions, and convective

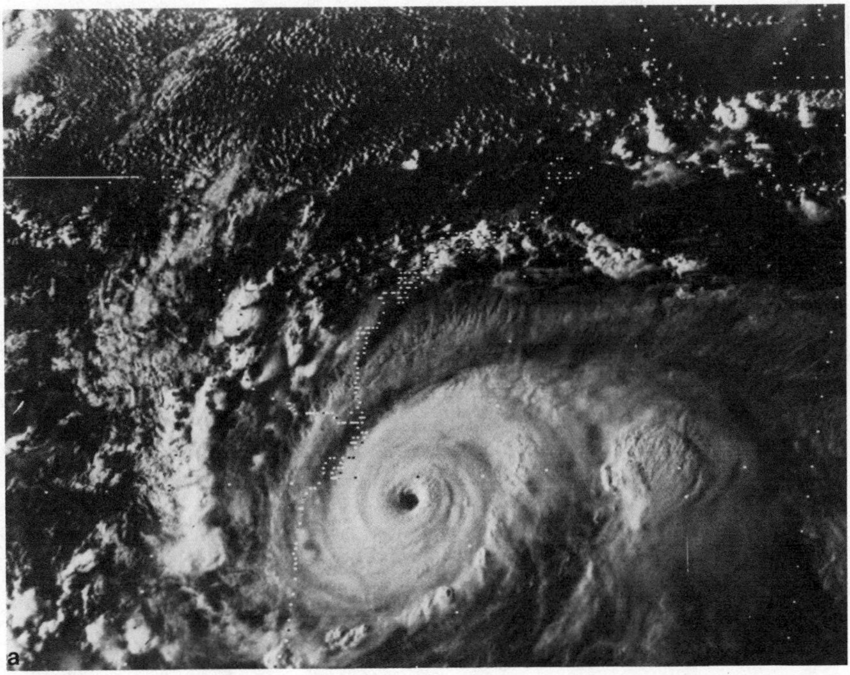

Fig. 10.56. Satellite imagery of the four hurricanes near the time of flights by the research aircraft: (a) GOES-2 visible view of Hurricane Anita at 2300 GMT, 1 September 1977, in the western Gulf of Mexico. (b) SMS-2 IR picture of Hurricane David at 1530 GMT, 30 August 1979, south of Puerto Rico. (c) SMS-2 visible picture of Hurricane Frederic at 1300 GMT, 12 September 1979, south of Mobile, Alabama. (d) SMS-2 IR view of Hurricane Allen at 1430 GMT, 5 August 1980, south of Hispaniola. (e) TIROS-N visible image of Hurricane Allen at 2115 GMT, 8 August 1980, in the central Gulf of Mexico. [From Jorgensen (1984a).] (*Figure continues.*)

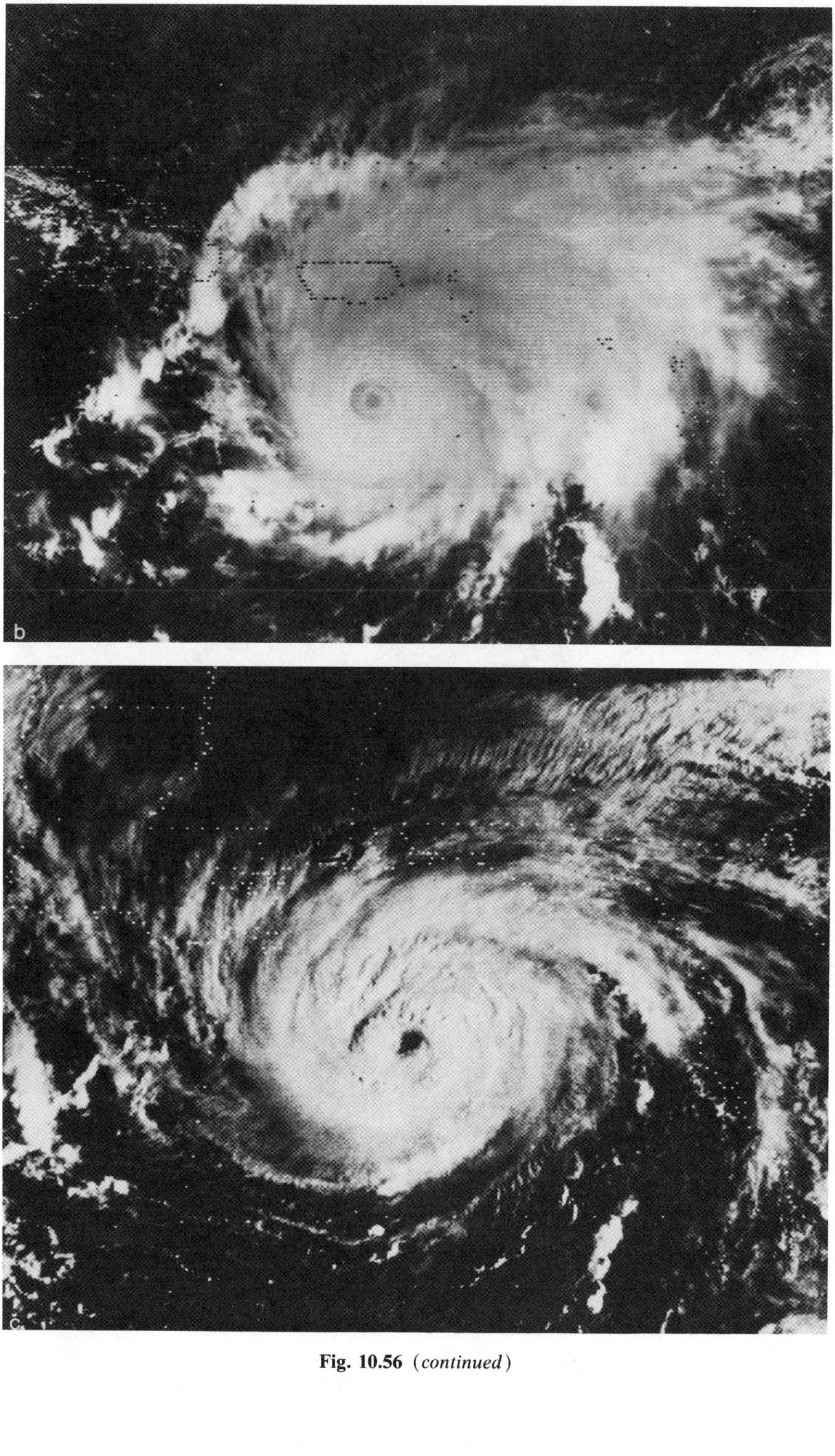

Fig. 10.56 (*continued*)

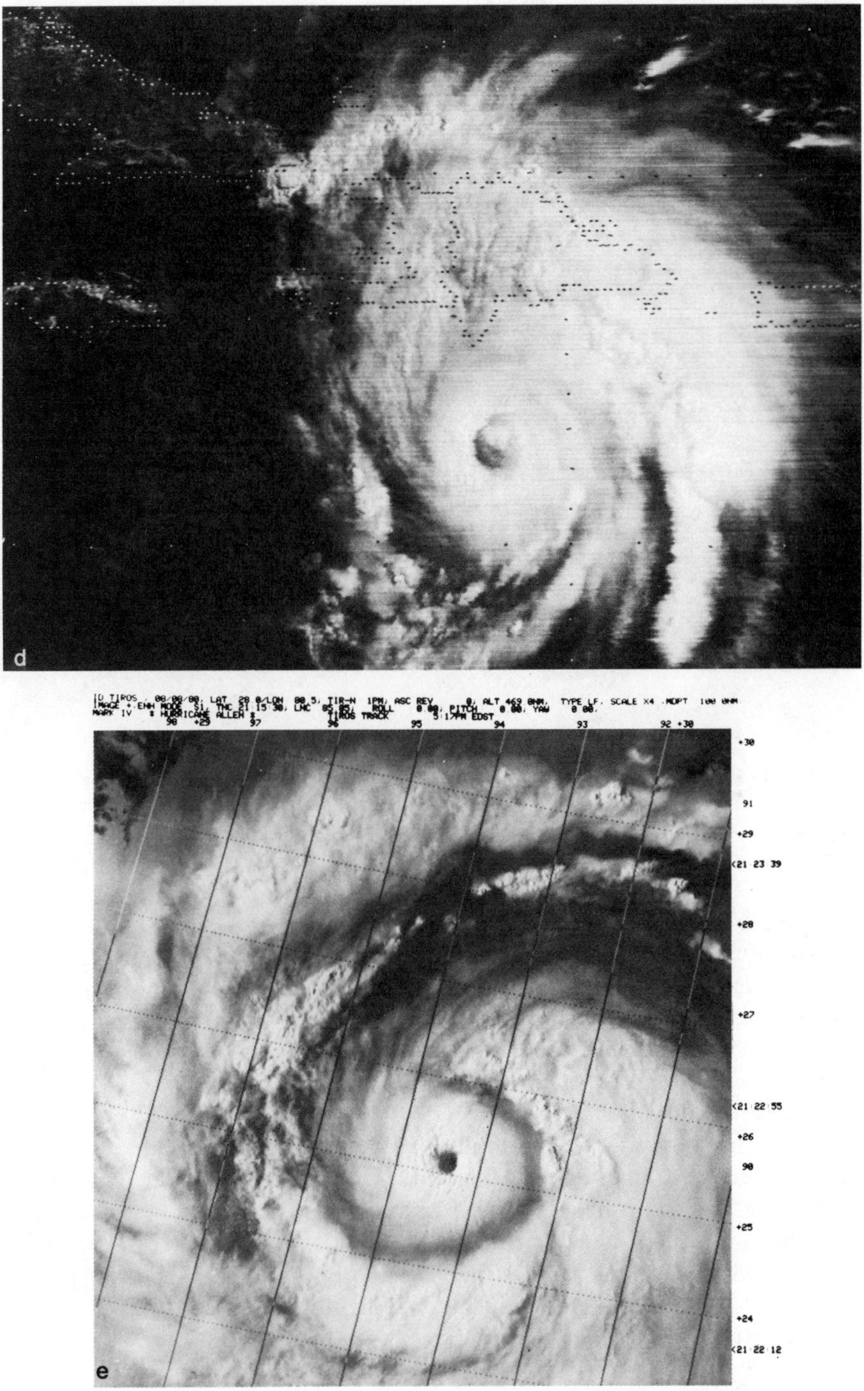

Fig. 10.56 (*continued*)

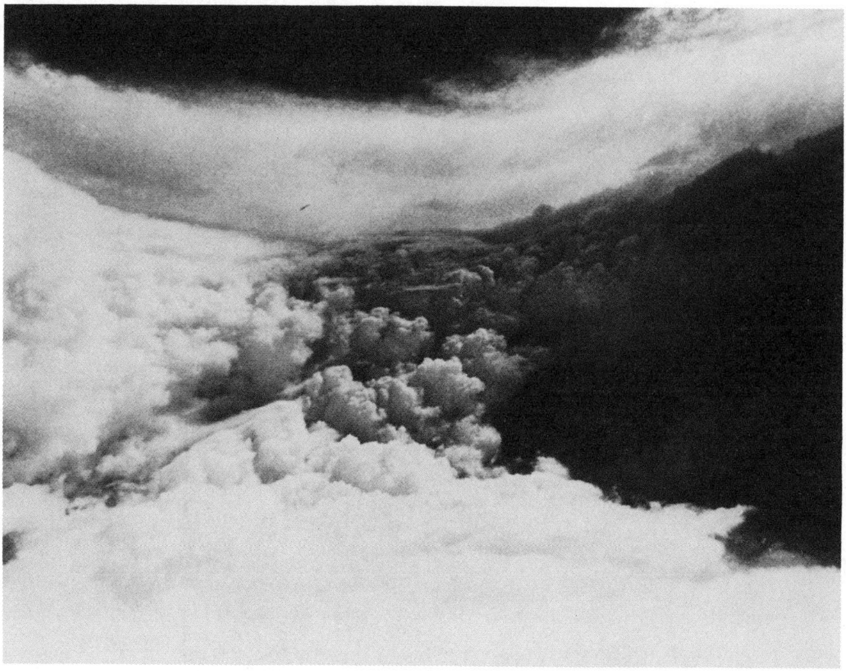

Fig. 10.57. Photograph from NOAA-42 of the inner edge of Allen's eyewall during a flight on 7 August 1980, near the time of record minimum pressure (89.9 kPa) recorded by an aircraft in an Atlantic hurricane. Photo courtesy of C. B. Emanuel. [From Jorgensen (1984b).]

bands. For example, Fig. 10.58 is a vertical cross section of reflectivity for Hurricane Frederic (Jorgensen, 1984a). This cross section, typical of other mature hurricanes studied by Jorgensen, exhibits an intense (greater than 40 dBZ) circular eyewall between 30 and 44 km from the eye. The eyewall slopes away from the storm center with height, making an angle to the horizontal of about 30°.

The other regions of highest reflectivity in Fig. 10.58 beyond the eyewall are convective rainbands embedded in a region of lower reflectivity corresponding to stratiform clouds and rainfall. The stratiform nature of this rainfall is indicated clearly by the bright band of reflectivity corresponding to the melting level at about 5 km. These bright bands are evident in the ranges 74–88 and 102–125 km in Fig. 10.58. Although the rainfall intensity in the stratiform regions is considerably less than that in the convective regions, the area covered by the stratiform rainfall is much greater, so that

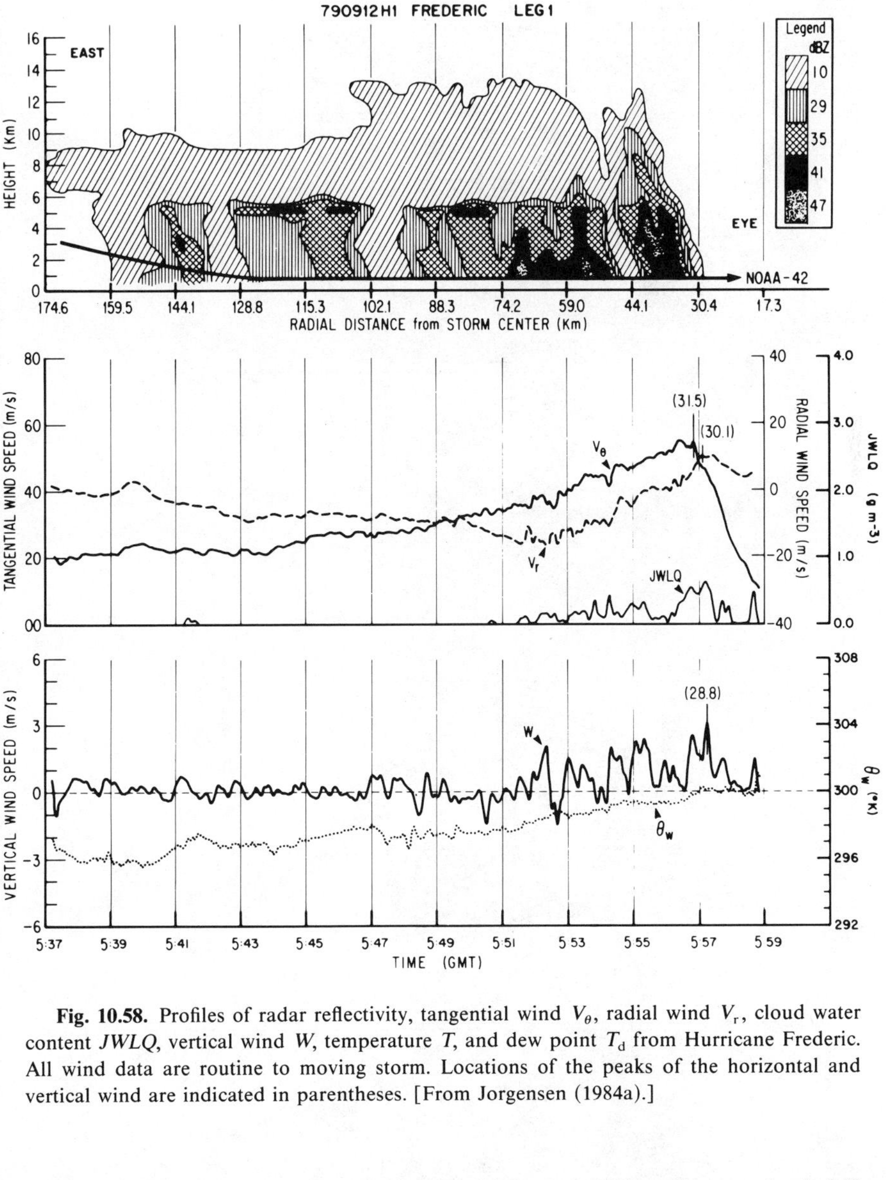

Fig. 10.58. Profiles of radar reflectivity, tangential wind V_θ, radial wind V_r, cloud water content *JWLQ*, vertical wind *W*, temperature *T*, and dew point T_d from Hurricane Frederic. All wind data are routine to moving storm. Locations of the peaks of the horizontal and vertical wind are indicated in parentheses. [From Jorgensen (1984a).]

stratiform precipitation accounts for approximately 60% of the total rainfall volume in the storm (Jorgensen, 1984a).

The three-dimensional structure of the eyewall region of Hurricane Allen appears in Fig. 10.59 (Jorgensen, 1984b). The eyewall is a nearly circular region of moderate (38–43 dBZ) reflectivity. Outside the eyewall are cells

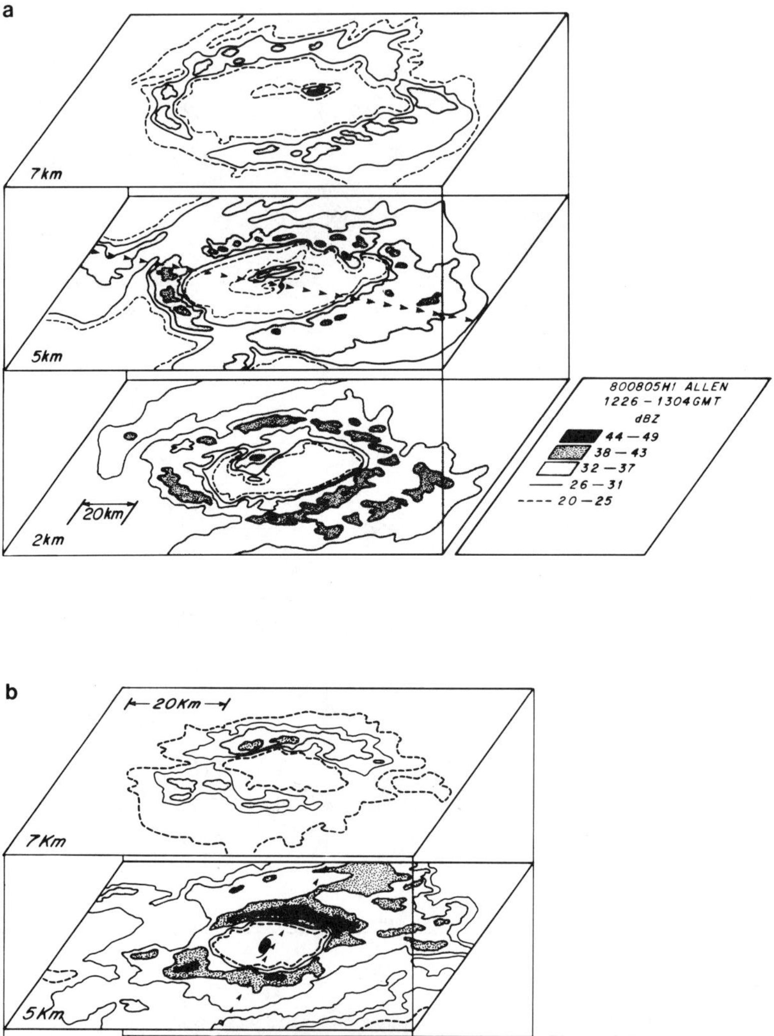

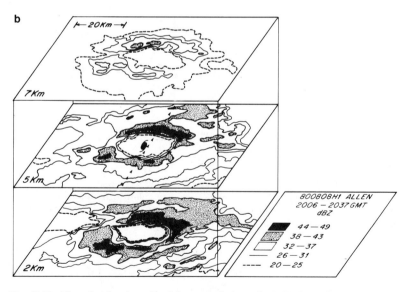

Fig. 10.59. Three-level radar reflectivity analysis from the NOAA-42 X-band tail radar of Hurricane Allen's inner region on (a) 5 August and (b) 8 August 1980. Aircraft track (line with arrow) and storm circulation (hurricane symbol) are indicated on the middle plane. [From Jorgensen (1984b).]

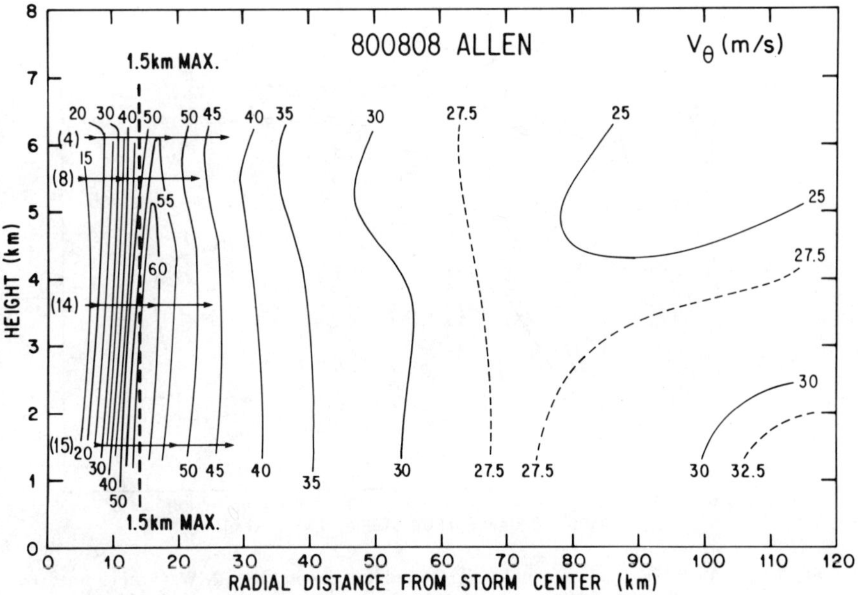

Fig. 10.60. Composite cross section of tangential wind (m s^{-1}) relative to the radius of maximum wind (RMW) from the 8 August Hurricane Allen flight leg data. Aircraft altitudes are indicated by the solid arrows; the number of individual legs is given in parentheses. Vertical dashed line denotes the location of the 1.5-km maximum. The period of this composite is 1832–2359 GMT. [From Jorgensen (1984b).]

of convection that move with the strong winds in this region and are relatively short lived.

The eyewall generally lies inside the radii of the maximum updraft, the maximum tangential wind, the maximum low-level convergence of the radial wind, and the highest equivalent potential temperature. Figure 10.60 is a composite vertical cross section of the tangential wind for Hurricane Allen. A narrow band of maximum (greater than 60 m s^{-1}) wind lies between 10 and 20 km from the center and slopes outward with height at the same angle as the maximum radar reflectivity. Extremely large cyclonic shear radially inward from the radius of maximum wind (RMW) is associated with relative vorticity values in excess of 3.0×10^{-3} s^{-1}.

Figure 10.61 is a composite vertical cross section of equivalent potential temperature for Allen. Noteworthy are the increase of θ_e with decreasing radial distance, which implies a source of sensible and latent energy from the sea, and the band of maximum θ_e associated with the eyewall convection. Instantaneous analyses of θ_e in the eyewall often reveal a nearly constant

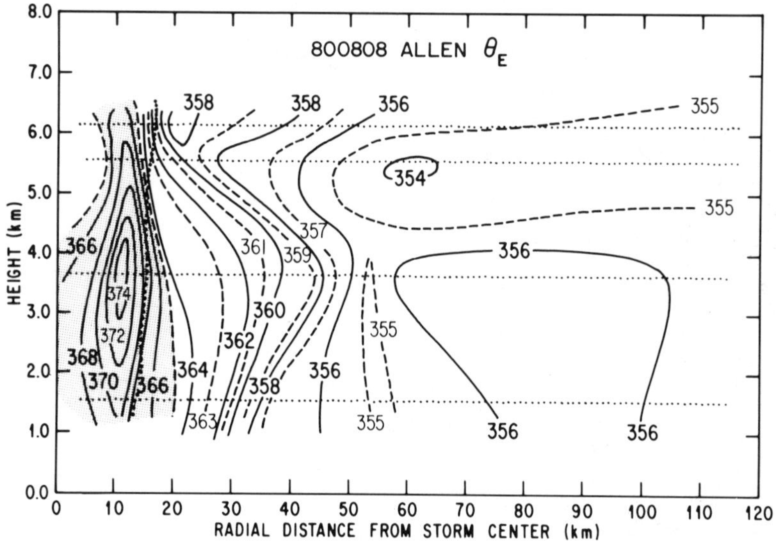

Fig. 10.61. As in Fig. 10.60, except for θ_e. Dotted line denotes the RMW. [From Jorgensen (1984b).]

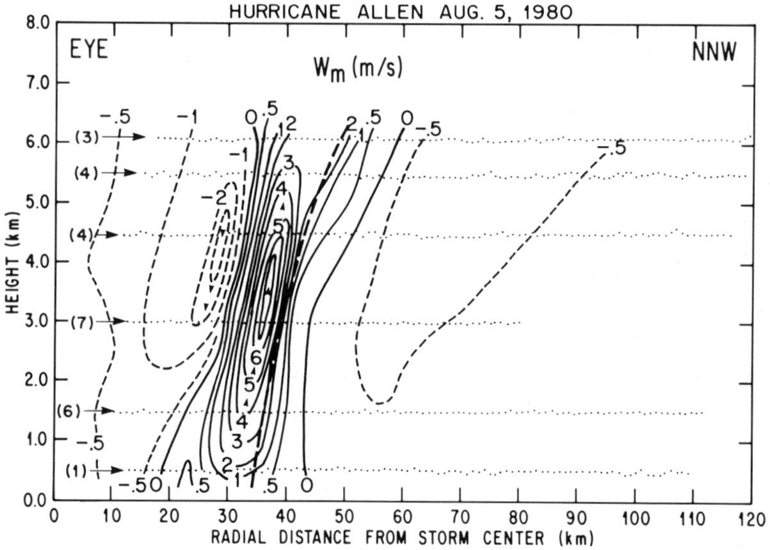

Fig. 10.62. Composite cross section of mesoscale vertical velocity W_m (m s^{-1}) computed by integrating the two-dimensional continuity equation and using the composite radial divergence analysis on 5 August 1980. [From Jorgensen (1984b).]

value of θ_e from the boundary layer to the upper troposphere, indicating nearly undilute ascent of low-level air.

An estimate of the mesoscale (horizontal scales of motion between approximately 10 and 100 km) vertical velocity in the eyewall region can be made by integrating the mass continuity equation using the divergence of the measured radial wind components under the assumption of axial symmetry. The composite vertical cross section of vertical motion for Hurricane Allen (three days earlier than the previous analyses) is shown in Fig. 10.62. Mean upward motion exceeding 7 m s^{-1} occurs in the sloping core of the eyewall. Subsidence occurring in the eye itself is responsible for the warm, dry air in the eye. Temperatures in the middle troposphere of the eye are typically 10 K warmer than the temperatures in the eyewall and the relative humidity is often as low as 50% (Anthes, 1982).

Using multiple aircraft observations, Jorgensen (1984b) constructed the schematic vertical cross section through the eyewall region (Fig. 10.63). In

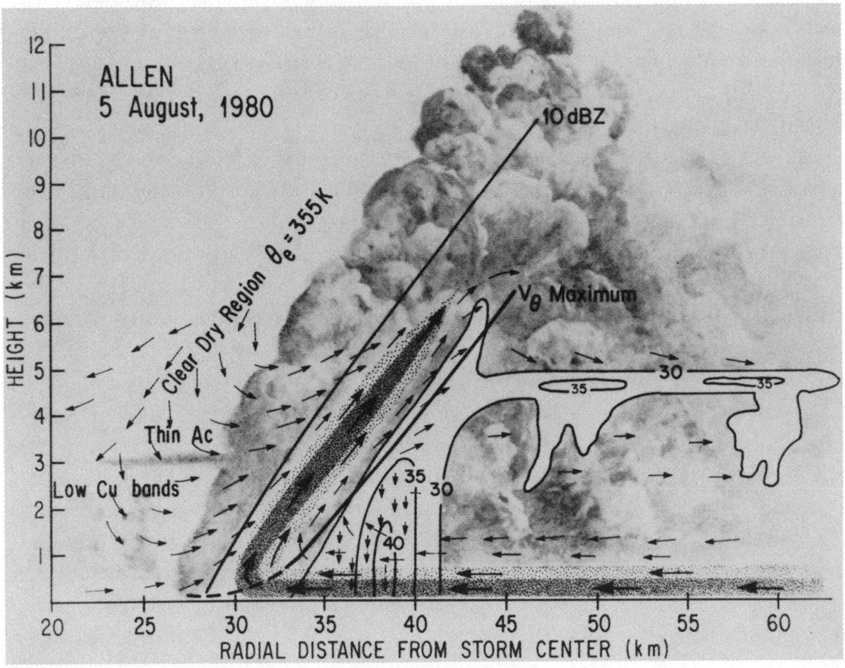

Fig. 10.63. Schematic cross section depicting the locations of the clouds and precipitation, RMW, and radial-vertical airflow through the eyewall of Hurricane Allen on 5 August 1980. The slope of the cloudy region on the inside edge of the eyewall is based on radar minimum detectable signal analysis, aircraft altimeter readings, handheld photography, and observer notes. Darker shaded regions denote the location of the largest radial and vertical velocity. [From Jorgensen (1984b).]

the outer regions of the storm, air flows radially inward in the lowest 2 km. The inflowing air rises abruptly in the eyewall in a radial band between 30 and 35 km from the center. The eyewall tilts radially outward with height at an angle of about 30° to the horizontal. Inside the eyewall, air sinks and moves outward. The strongest radar reflectivity is located several kilometers outside the radius of maximum winds, while the maximum updraft lies several kilometers inward from the RMW. The local regions of high reflectivity (bright bands) near the 5-km height outward from 45 km represent the melting of stratiform precipitation.

Beyond the eyewall region of the hurricane, the predominant mesoscale features of the tropical cyclone are spiral bands of clouds and rainfall. The structure of these bands is highly variable, but they may be classified into two basic types—propagating wavelike bands associated with internal gravity waves and groups of bands that remain in a nearly fixed location relative to the storm center, termed "stationary band complex" (SBC) by Willoughby *et al.* (1984). The predominant tangential wave number of the SBC is one and is usually located on the east side of the storm in a boundary zone between the inner-core region of the storm, where air trajectories form closed paths, and an outer envelope, where trajectories are not closed and air flows through the storm system.

Barnes *et al.* (1983) and Barnes and Stossmeister (1986) analyzed aircraft observations to describe the mesoscale thermodynamic, kinematic, and reflectivity structure of two hurricane rainbands. The reflectivity observations showed that the bands contained both convective and stratiform rainfall. Low-level air entering the band from the upwind (outer) end was convectively unstable and thus ascended in convective clouds. On the downwind end of the bands the environment was much more stable and the precipitation mostly stratiform.

Figure 10.64 shows a composite vertical cross section of θ_e through Hurricane Floyd [1981] (Barnes *et al.*, 1983). The notable features are the convective instability outside the band, where θ_e of the subcloud layer air exceeds 354 K, and the much more stable air inside the band, where θ_e in the subcloud layer is as much as 12 K lower than the air outside the band. The dramatic decrease of θ_e from 90 to 70 km (across the rainband) was a result of convective-scale vertical motions in the band, with convective downdrafts bringing air with lower θ_e to the surface while updrafts brought air with higher θ_e to the middle troposphere. Considerable mixing of the updraft and downdraft air occurred; there was no evidence of undilute ascent as is often observed in the eyewall.

Barnes *et al.* (1983) speculate that the air with low θ_e inside the bands was an exhaust product of convection. The static stability of this air was unfavorable for convection, and if it reached the eyewall before it was

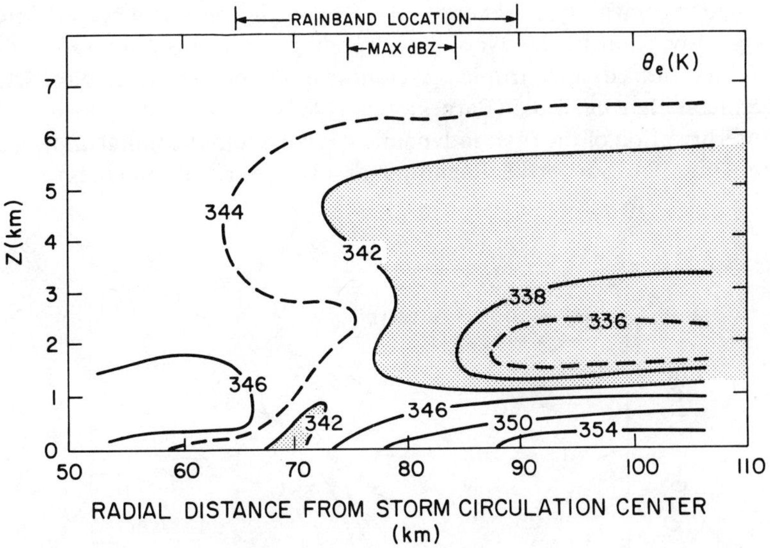

Fig. 10.64. Composite field of equivalent potential temperature (θ_e) as a function of radial distance from storm circulation center and height. Values of θ_e less than 342 are shaded. [From Barnes *et al.* (1983).]

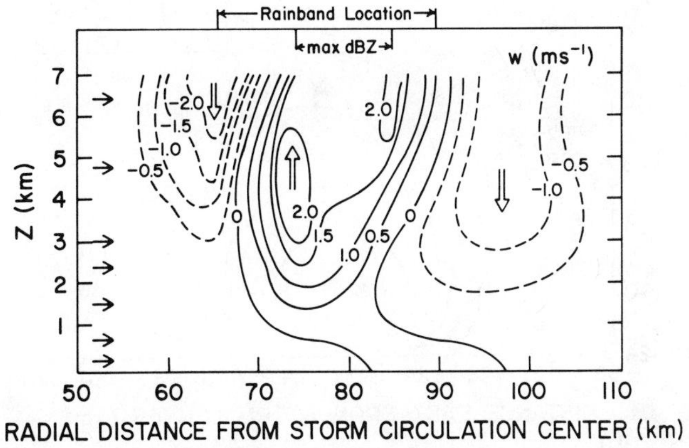

Fig. 10.65. Mesoscale vertical velocity derived from integration of two-dimensional continuity equation. Contours are every 0.5 m s^{-1}. Arrows on the left side of the figure show aircraft levels. Downward vertical velocities above 4 km and inward from 65 km from the circulation center are considered unreliable. [From Barnes *et al.* (1983).]

modified significantly by energy fluxes from the ocean, it would inhibit strong convection in the eyewall and perhaps reduce the intensity of the hurricane. Indeed, most tropical cyclones with strong convective rainbands have incomplete eyewalls (Barnes *et al.*, 1983).

Interpretation of the thermodynamic structure and the radial and vertical winds (Fig. 10.65 shows the diagnosed mesoscale vertical velocities) allowed

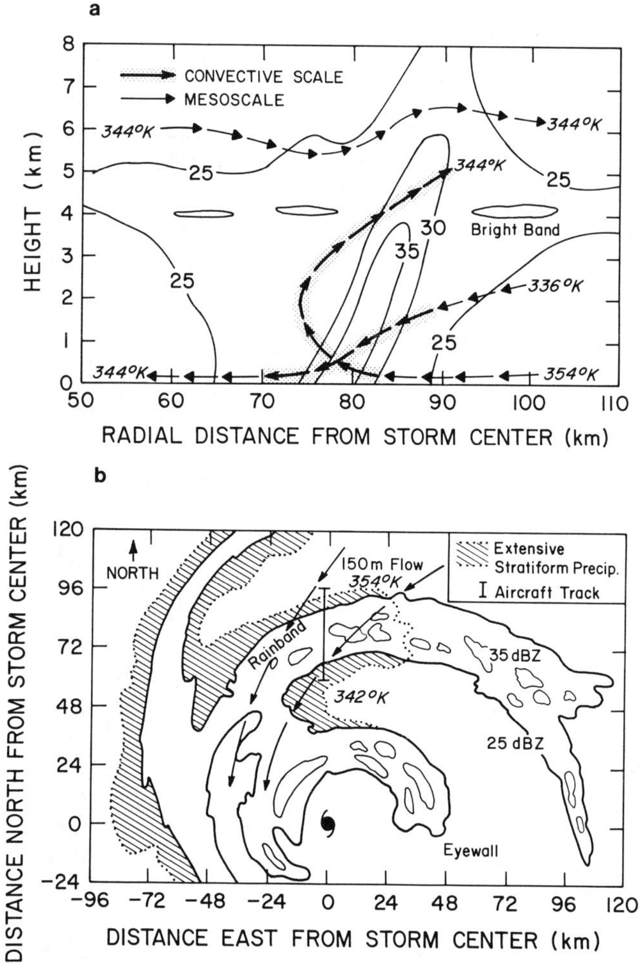

Fig. 10.66. (a) A schematic of the rainband in r and z coordinates. Reflectivity, θ_e, mesoscale (arrows), and convective-scale motions are shown. (b) Rainband in x and y coordinates. Aircraft track, reflectivities, cells, stratiform precipitation, and 150-m flow and θ_e values are shown. [From Barnes *et al.* (1983).]

Barnes *et al.* (1983) to construct the schematic diagrams of the vertical and horizontal structure of the rainband in Fig. 10.66; the 25-dBZ contour in Fig. 10.66b defines the rainband. The plain arrow represents the mesoscale flow (approximately 10 km averaged) and the arrows on the stippled background represent convective-scale updrafts and downdrafts. These convective-scale motions transport air with low values of θ_e downward from the middle troposphere into the boundary layer on the inside of the band, where it continues to move inward toward the storm center. Strong rainbands of this type could have a significant damping effect on the intensity of tropical storms by causing relatively cool stable air to enter the eyewall.

The structure of hurricane rainbands is extremely variable. Barnes and Stossmeister (1986) analyzed a dissipating rainband in the 1981 Hurricane Irene; this rainband showed a structure considerably different than did the rainband in Hurricane Floyd. Only weak convective activity and, consequently, little or no decrease of θ_e across the band were evident. Differences in the wind structure of the band were also present. While the rainband in Floyd showed only a weak maximum in the tangential wind component in the band (Fig. 10.67), the Irene band showed a strong local maximum in its center (Fig. 10.68).

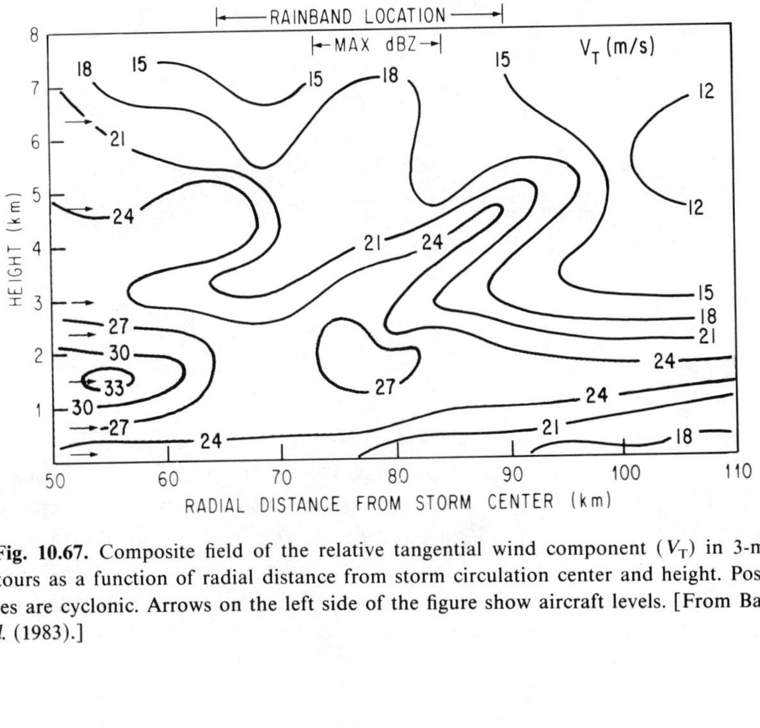

Fig. 10.67. Composite field of the relative tangential wind component (V_T) in 3-m s^{-1} contours as a function of radial distance from storm circulation center and height. Positive values are cyclonic. Arrows on the left side of the figure show aircraft levels. [From Barnes *et al.* (1983).]

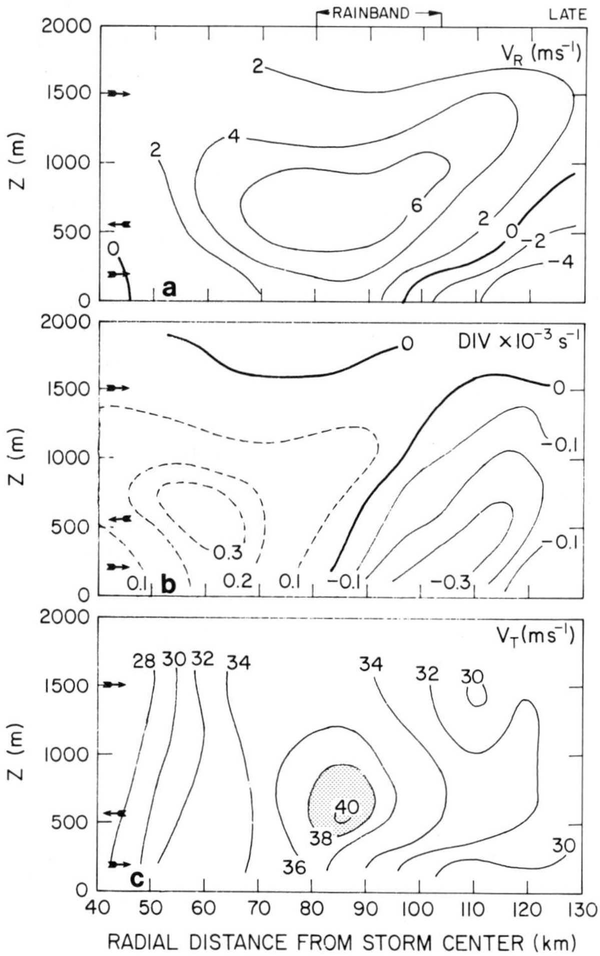

Fig. 10.68. Composite vertical cross section of radial velocity V_R, horizontal divergence DIV, and tangential wind V_T during the decaying stage of Hurricane Irene [1981]. [From Barnes and Stossmeister (1986).]

The rainbands in Hurricanes Floyd and Irene occurred in different locations with respect to the moving storm centers, and the environment of the two bands had different thermodynamic and dynamic characteristics. The rainband in Floyd was located ahead of the moving storm and was fed by energy-rich air that had not been processed by the hurricane. In contrast, the Irene rainband was located in more stable air to the rear of the storm center in a region of mesoscale subsidence associated with the eyewall of the hurricane. These two examples indicate that some rainbands may inhibit the strength of the eyewall by modifying the subcloud air, while in other cases, the eyewall circulation may suppress the rainband circulation by adversely affecting the static stability and the mean vertical motion.

In summary, the mesoscale structure of tropical cyclones is determined by the high moisture content of the lower troposphere, the convective and conditional instability present in the tropical air mass, and the dynamical structure of the tropical cyclone's large-scale circulation, which produces a large-scale inflow and convergence of air in the planetary boundary layer. These characteristics of the environment favor the ring of eyewall convection and the convective rainbands.

References

Anthes, R. A. (1982). Tropical cyclones—Their evolution, structure, and effects. *Meteorol. Monogr.* **41**.

Anthes, R. A., and D. Keyser (1979). Tests of a fine-mesh model over Europe and the United States. *Mon. Weather Rev.* **107**, 963–984.

Anthes, R. A., H. D. Orville, and D. J. Raymond (1982). Mathematical modeling of convection. *In* "Thunderstorms: A Social, Scientific, and Technological Documentary, Vol. 2, Thunderstorm Morphology and Dynamics," pp. 495–579. U.S. Dep. Commer., Washington, D.C.

Barnes, G. M., and K. Sieckman (1984). The environment of fast- and slow-moving tropical mesoscale convective cloud lines. *Mon. Weather Rev.* **112**, 1782–1794.

Barnes, G. M., and G. J. Stossmeister (1986). The structure and decay of a rainband in Hurricane Irene (1981). *Mon. Weather Rev.* **114**, 2590–2601.

Barnes, G. M., E. J. Zipser, D. Jorgensen, and F. Marks, Jr. (1983). Mesoscale and convective structure of a hurricane rainband. *J. Atmos. Sci.* **40**, 2125–2137.

Bates, F. C. (1961). An airborne observation of a tornado. Univ. Kansas Cent. Res. Eng. Sci., Weather Res. Rep. No. 1.

Battan, L. S. (1973). "Radar Observation of the Atmosphere." Univ. of Chicago Press, Chicago, Illinois.

Betts, A. K. (1976). The thermodynamic transformation of the tropical subcloud layer by precipitation and downdrafts. *J. Atmos. Sci.* **33**, 1008–1020.

Bluestein, H. B., and M. H. Jain (1985). The formation of mesoscale lines of precipitation: Severe squall lines in Oklahoma during the spring. *J. Atmos. Sci.* **42**, 1711–1732.

Bluestein, H. B., M. H. Jain, and G. T. Marx (1987). Formation of mesoscale lines of precipitation: Non-severe squall lines in Oklahoma during the spring. *Conf. Mesoscale Processes, 3rd, Vancouver, B.C.* pp. 198–199. Am. Meteorol. Soc., Boston, Massachusetts.

Bosart, Lance F., and Frederick Sanders (1981). The Johnstown Flood of July 1977: A long-lived convective system. *J. Atmos. Sci.* **38**, 1616–1642.

Brenner, I. S. (1974). A surge of maritime tropical air—Gulf of California to the southwestern United States. *Mon. Weather Rev.* **102**, 375–389.

Brown, J. M. (1979). Mesoscale unsaturated downdrafts driven by rainfall evaporation: A numerical study. *J. Atmos. Sci.* **36**, 313–338.

Brunk, I. W. (1953). Squall lines. *Bull. Am. Meteorol. Soc.* **34**, 1–9.

Byers, H. R., and R. R. Braham (1949). "The Thunderstorm." U.S. Weather Bur., Washington, D.C.

Carlson, T. N., R. A. Anthes, M. Schwartz, S. G. Benjamin, and D. G. Baldwin (1980). Analysis and prediction of severe storms environment. *Bull. Am. Meteorol. Soc.* **61**, 1018–1032.

Cerasoli, C. P. (1979). Experiments on buoyant parcel motion and the generation of internal gravity waves. *J. Fluid Mech.* **86**, 247–271.

Chang, C. B., D. J. Perkey, and C. W. Kreitzberg (1981). A numerical case study of the squall line of 6 May 1975. *J. Atmos. Sci.* **38**, 1601–1615.

Chauzy, S., M. Chong, A. Delannoy, and S. Despiau (1985). The June 22 tropical squall line observed during COPT 81 experiment: Electrical signature associated with dynamical structure and precipitation. *J. Geophys. Res.* **90**, 6091–6098.

Chen, S. (1986). Simulation of the stratiform region of a mesoscale convective system. M.S. Thesis, Dep. Atmos. Sci., Colorado State Univ.

Chong, M., P. Amayenc, G. Scialom, and J. Testud (1987). A tropical squall line observed during the COPT 81 experiment in West Africa. Part I: Kinematic structure inferred from dual-doppler radar data. *Mon. Weather Rev.* **115**, 670–694.

Churchill, D. D., and R. A. Houze, Jr. (1984a). Development and structure of winter monsoon cloud clusters on 10 December 1978. *J. Atmos. Sci.* **41**, 933–960.

Churchill, D. D., and R. A. Houze, Jr. (1984b). Mesoscale updraft magnitude and cloud-ice content deduced from the ice budget of the stratiform region of a tropical cloud cluster. *J. Atmos. Sci.* **41**, 1717–1725.

Cotton, W. R., R. L. George, P. J. Wetzel, and R. L. McAnelly (1983). A long-lived mesoscale convective complex. Part I: The mountain-generated component. *Mon. Weather Rev.* **111**, 1893–1918.

Cotton, W. R., M. S. Lin, R. L. McAnelly, and C. J. Tremback (1989). A composite model of mesoscale convective complexes. *Mon. Weather Rev.* **117**, 765–783.

Culverwell, A. H. (1982). An analysis of moisture sources and circulation fields associated with an MCC episode. M.S. Thesis, Dep. Atmos. Sci., Colorado State Univ.

Cunning, J. B. (1986). The Oklahoma-Kansas preliminary regional experiment for STORM-Central. *Bull. Am. Meteorol. Soc.* **67**, 1478–1486.

Dirks, R. (1969). A theoretical investigation of convective patterns in the lee of the Colorado Rockies. Atmos. Sci. Pap. No. 154, Dep. Atmos. Sci., Colorado State Univ.

Dudhia, J., and M. W. Moncrieff (1987). A numerical simulation of quasi-stationary tropical convection bands. *Q. J. R. Meteorol. Soc.* **113**, 929–967.

Emanuel, K. A. (1982). Inertial instability and mesoscale convective systems. Part II: Symmetric CISK in a baroclinic flow. *J. Atmos. Sci.* **39**, 1080–1097.

Esbensen, S. K., and J.-T. Wang (1984). Heat budget analysis and the synoptic environment of GATE cloud clusters. *Prepr., Conf. Hurricanes Trop. Meteorol., 15th*, pp. 455–460. Am. Meteorol. Soc., Miami, Fla.

Foote, G. B., and H. W. Frank (1983). Case study of a hailstorm in Colorado. Part III: Airflow from triple-doppler measurements. *J. Atmos. Sci.* **40**, 686–707.

Fortune, M. A. (1980). Properties of African squall lines inferred from time-lapse satellite imagery. *Mon. Weather Rev.* **108**, 153–168.

Fortune, M. A., and R. L. McAnelly (1986). The evolution of two mesoscale convective complexes with different patterns of convective organization. *Prepr., Conf. Radar Meteorol., 23rd, Snowmass, Colo.* pp. J175–J178. Am. Meteorol. Soc., Boston, Massachusetts.

Frank, W. M. (1977). The structure and energetics of the tropical cyclone. I. Storm structure. *Mon. Weather Rev.* **105**, 1119–1135.

Frank, W. M. (1978). The life cycles of GATE convective systems. *J. Atmos. Sci.* **35**, 1256–1264.

Frank, W. M. (1983). The cumulus parameterization problem. *Mon. Weather Rev.* **111**, 1859–1871.

Frank, W. M. (1984). A composite analysis of the core of a mature hurricane. *Mon. Weather Rev.* **112**, 2401-2420.

Fritsch, J. M., and J. M. Brown (1981). On the generation of convectively driven mesohighs aloft. *Mon. Weather Rev.* **110**, 1554-1563.

Fritsch, J. M., and C. F. Chappell (1980a). Numerical prediction of convectively driven mesoscale pressure systems. Part I: Convective parameterization. *J. Atmos. Sci.* **37**, 1722-1733.

Fritsch, J. M., and C. F. Chappell (1980b). Numerical prediction of convectively driven mesoscale pressure systems. Part II: Mesoscale model. *J. Atmos. Sci.* **37**, 1734-1762.

Fujita, T. T. (1955). Results of detailed synoptic studies of squall lines. *Tellus* **7**, 405-434.

Fujita, T. T. (1959). Precipitation and cold air production in mesoscale thunderstorm systems. *J. Meteorol.* **16**, 454-466.

Fujita, T. T. (1978). Manual of downburst identification for project NIMROD. Satellite Mesometeorol. Res. Pap. No. 156, Dep. Geophys. Sci., Univ. of Chicago [NTIS No. N78-30771/7GI].

Fujita, T. (1981). Tornadoes and downbursts in the context of generalized planetary scales. *J. Atmos. Sci.* **38**, 1511-1534.

Gamache, J. F., and R. A. Houze, Jr. (1982). Mesoscale air motions associated with a tropical squall line. *Mon. Weather Rev.* **110**, 118-135.

Gamache, J. F., and R. A. Houze, Jr. (1983). The water budget of a mesoscale convective system in the tropics. *J. Atmos. Sci.* **40**, 1835-1850.

Gray, W. M., and R. W. Jacobson, Jr. (1977). Diurnal variation of deep cumulus convection. *Mon. Weather Rev.* **105**, 1171-1188.

Hack, J. J., and W. H. Schubert (1986). Nonlinear response of atmospheric vortices to heating by organized cumulus convection. *J. Atmos. Sci.* **43**, 1559-1573.

Hales, J. E., Jr. (1972). Surges of maritime tropical air northward over the Gulf of California. *Mon. Weather Rev.* **100**, 298-306.

Hamilton, R. A., and J. W. Archbold (1945). Meteorology of Nigeria and adjacent territory. *Q. J. R. Meteorol. Soc.* **71**, 231-265.

Hane, C. E. (1973). The squall line thunderstorm: Numerical experimentation. *J. Atmos. Sci.* **30**, 1672-1690.

Hinrichs, G. (1888). Tornadoes and derechos. *Am. Meteorol. J.* **5**, 306-317, 341-349.

Houze, R. A. (1977). Structure and dynamics of a tropical squall-line system. *Mon. Weather Rev.* **105**, 1540-1567.

Houze, R. A., Jr. (1981). Structure of atmospheric precipitation systems—A global survey. *Radio Sci.* **16**, 671-689.

Houze, R. A., Jr. (1982). Cloud clusters and large-scale vertical motions in the tropics. *J. Meteorol. Soc. Jpn.* **60**, 396-410.

Houze, R. A., Jr., and D. D. Churchill (1984). Microphysical structure of winter monsoon cloud clusters. *J. Atmos. Sci.* **41**, 3405-3411.

Houze, R. A., Jr., and E. N. Rappaport (1984). Air motions and precipitation structure of an early summer squall line over the eastern tropical Atlantic. *J. Atmos. Sci.* **41**, 553-574.

Houze, R. A., Jr., S. G. Geotis, F. D. Marks, Jr., and A. K. West (1981). Winter monsoon convection in the vicinity of North Borneo. Part I: Structure and time variation of the clouds and precipitation. *Mon. Weather Rev.* **109**, 1595-1614.

Hoxit, L. R., C. F. Chappell, and J. M. Fritsch (1976). Formation of mesolows or pressure troughs in advance of cumulonimbus clouds. *Mon. Weather Rev.* **104**, 1419-1428.

Johns, R. H., and W. D. Hirt (1983). The derecho—A severe weather producing convective system. *Prep., Conf. Severe Local Storms, 13th*, pp. 178-181. Am. Meteorol. Soc., Boston, Massachusetts.

Johns, R. H., and W. D. Hirt (1985). The Derecho of 19–20 July 1983—A case study. *Natl. Weather Dig.* **10**, 17–32.

Johnson, R. H. (1982). Vertical motion of near-equatorial winter monsoon convection. *J. Meteorol. Soc. Jpn.* **60**, 682–690.

Johnson, R. H., and P. J. Hamilton (1988). The relationship of surface pressure features to the precipitation and air flow structure of an intense midlatitude squall line. *Mon. Weather Rev.* **116**, 1444–1472.

Jorgensen, D. P. (1984a). Mesoscale and convective-scale characteristics of mature hurricanes. Part I. General observations by research aircraft. *J. Atmos. Sci.* **41**, 1268–1285.

Jorgensen, D. P. (1984b). Mesoscale and convective-scale characteristics of mature hurricanes. Part II. Inner core structure of hurricane Allen (1980). *J. Atmos. Sci.* **41**, 1287–1311.

Knupp, K. R. (1985). Precipitation convective downdraft structure: A synthesis of observations and modeling. Ph.D. Thesis, Dep. Atmos. Sci., Colorado State Univ.

Knupp, K. R., and W. R. Cotton (1986). Internal structure of a small mesoscale system. *Mon. Weather Rev.* **115**, 629–645.

Koss, W. J. (1976). Linear stability of CISK-induced disturbances: Fourier component eigenvalue analysis. *J. Atmos. Sci.* **33**, 1195–1222.

Kreitzberg, C. W., and D. J. Perkey (1976). Release of potential instability: Part I. A sequential plume model within a hydrostatic primitive equation model. *J. Atmos. Sci.* **33**, 456–475.

Kreitzberg, C. W., and D. J. Perkey (1977). Release of potential instability: Part II. The mechanism of convective/mesoscale interaction. *J. Atmos. Sci.* **34**, 1569–1595.

Leary, C. A. (1979). Behavior of the wind field in the vicinity of a cloud cluster in the Intertropical Convergence Zone. *J. Atmos. Sci.* **36**, 631–639.

Leary, C. A. (1984). Precipitation structure of the cloud clusters in a tropical easterly wave. *Mon. Weather Rev.* **112**, 313–325.

Leary, C. A., and R. A. Houze, Jr. (1979a). The structure and evolution of convection in a tropical cloud cluster. *J. Atmos. Sci.* **36**, 437–457.

Leary, C. A., and R. A. Houze, Jr. (1979b). Melting and evaporation of hydrometeors in precipitation from the anvil clouds of deep tropical convection. *J. Atmos. Sci.* **36**, 669–679.

Leary, C. A., and E. N. Rappaport (1987). The life cycle and internal structure of a mesoscale convective complex. *Mon. Weather Rev.* **115**, 1503–1527.

LeMone, M. A. (1983). Momentum transport by a line of cumulonimbus. *J. Atmos. Sci.* **40**, 1815–1834.

LeMone, M. A., and E. J. Zipser (1980). Cumulonimbus vertical velocity events in GATE. Part I: Diameter, intensity and mass flux. *J. Atmos. Sci.* **37**, 2444–2457.

LeMone, M. A., G. M. Barnes, and E. J. Zipser (1984). Momentum flux by lines of cumulonimbus over the tropical oceans. *J. Atmos. Sci.* **41**, 1914–1932.

Lilly, D. K. (1979). The dynamical structure and evolution of thunderstorms and squall lines. *Annu. Rev. Earth Planet. Sci.* **7**, 117–161.

Lin, M.-S. (1986). The evolution and structure of composite meso-α-scale convective complexes. Ph.D. Thesis, Colorado State Univ.

Lord, S. J., H. E. Willoughby, and J. M. Piotrowicz (1984). Role of a parameterized ice-phase microphysics in an axisymmetric, nonhydrostatic tropical cyclone model. *J. Atmos. Sci.* **41**, 2836–2848.

Ludlam, F. H. (1963). Severe local storms: A review. *Meteorol. Monogr.* **5**(27), 1–30.

Maddox, R. A. (1980). Mesoscale convective complexes. *Bull. Am. Meteorol. Soc.* **61**, 1374–1387.

Maddox, R. A. (1981). The structure and life-cycle of midlatitude mesoscale convective complexes. Atmos. Sci. Pap. No. 336, Dep. Atmos. Sci., Colorado State Univ.

Maddox, R. A. (1983). Large-scale meteorological conditions associated with midlatitude, mesoscale convective complexes. *Mon. Weather. Rev.* **111**, 1475–1493.

Maddox, R., D. J. Perkey, and J. M. Fritsch (1981). Evolution of upper tropospheric features during the development of a mesoscale convective complex. *J. Atmos. Sci.* **38**, 1664–1674.

Matthews, David A. (1983). Analysis and classification of mesoscale clouds and precipitation systems. Ph.D. Thesis, Dep. Atmos. Sci., Colorado State Univ.

McAnelly, R. L., and W. R. Cotton (1986). Meso-beta-scale characteristics of an episode of meso-alpha-scale convective complexes. *Mon. Weather Rev.* **114**, 1740–1770.

McBride, J. L., and W. M. Gray (1980). Mass divergence in tropical weather systems. I: Diurnal variations. *Q. J. R. Meteorol. Soc.* **106**, 501–516.

Miller, D. A., and F. Sanders (1980). Mesoscale conditions for the severe conditions of 3 April 1974 in the east-central United States. *J. Atmos. Sci.* **37**, 1041–1055.

Miller, M. J. (1978). The Hampstead storm: A numerical simulation of a quasi-stationary cumulonimbus system. *Q. J. R. Meteorol. Soc.* **104**, 413–427.

Miller, M. J., and A. K. Betts (1977). Travelling convective storms over Venezuela. *Mon. Weather Rev.* **105**, 833–848.

Miller, R. C. (1959). Tornado-producing synoptic patterns. *Bull. Am. Meteorol. Soc.* **40**, 465–472.

Molinari, J., and T. Corsetti (1985). Incorporation of cloud-scale and mesoscale downdrafts into a cumulus parameterization: Results of one- and three-dimensional integrations. *Mon. Weather Rev.* **113**, 485–501.

Moncrieff, M. W. (1978). The dynamical structure of two-dimensional steady convection in constant vertical shear. *Q. J. R. Meteorol. Soc.* **104**, 543–567.

Moncrieff, M. W. (1981). A theory of organized steady convection and its transport properties. *Q. J. R. Meteorol. Soc.* **107**, 29–50.

Moncrieff, M. W., and J. S. A. Green (1972). The propagation and transfer properties of steady convective overturning in shear. *Q. J. R. Meteorol. Soc.* **98**, 336–352.

Moncrieff, M. W., and M. J. Miller (1976). The dynamics and simulation of tropical cumulonimbus and squall lines. *Q. J. R. Meteorol. Soc.* **102**, 373–394.

Nehrkorn, T. (1985). Wave-CISK in a baroclinic basic state. Ph.D. Thesis, Mass. Inst. Technol.

Newton, C. W. (1966). Circulations in large sheared cumulonimbus. *Tellus* **18**, 699–712.

Newton, C. W., and H. R. Newton (1959). Dynamical interactions between large convective clouds and environment with vertical shear. *J. Meteorol.* **16**, 483–496.

Ogura, Y., and Y.-L. Chen (1977). A life history of an intense mesoscale convective storm in Oklahoma. *J. Atmos. Sci.* **34**, 1458–1476.

Ogura, Y., and M.-T. Liou (1980). The structure of a midlatitude squall line: A case study. *J. Atmos. Sci.* **37**, 553–567.

Ogura, Y., H.-M. Juang, K.-S. Zhang, and S.-T. Soong (1982). Possible triggering mechanisms for severe storms in SESAME-AVE IV (9–10 May 1979). *Bull. Am. Meteorol. Soc.* **63**, 503–515.

Ooyama, K. (1969). Numerical simulation of the life cycle of tropical cyclones. *J. Atmos. Sci.* **26**, 3–40.

Orlanski, I., and B. B. Ross (1984). The evolution of an observed cold front. Part II. Mesoscale dynamics. *J. Atmos. Sci.* **41**, 1669–1703.

Payne, S. W., and M. M. McGarry (1977). The relationship of satellite inferred convective activity to easterly waves over west Africa and the adjacent ocean during Phase III of GATE. *Mon. Weather Rev.* **105**, 413–420.

Pedgley, D. E. (1962). A meso-synoptic analysis of the thunderstorms on 28 August 1958. *Br. Meteorol. Off., Geophys. Mem.* No. 106.

Perkey, D. J., and R. A. Maddox (1985). A numerical investigation of a mesoscale convective system. *Mon. Weather Rev.* **113**, 553–566.

Rasmussen, E. M. (1967). Atmospheric water vapor transport and the water balance of North America: Part 1. Characteristics of the water vapor field. *Mon. Weather Rev.* **95**, 403–426.

Raymond, D. J. (1984). A wave-CISK model of squall lines. *J. Atmos. Sci.* **40**, 1946–1958.
Raymond, D. J. (1987). A forced gravity-wave model of self-organizing convection. *J. Atmos. Sci.* **44**, 3528–3543.
Reed, R. J., and Recker, E. E. (1971). Structure and properties of synoptic-scale wave disturbances in the equatorial western Pacific. *J. Atmos. Sci.* **28**, 1117–1133.
Ross, B. B. (1987). The role of low-level convergence and latent heating in a simulation of observed squall line formation. *Mon. Weather Rev.* **115**, 2298–2321.
Ross, B. B., and I. Orlanski (1982). The evolution of an observed cold front. Part I. Numerical simulation. *J. Atmos. Sci.* **39**, 296–327.
Rotunno, R., and J. B. Klemp (1985). On the rotation and propagation of simulated supercell thunderstorms. *J. Atmos. Sci.* **42**, 271–292.
Rotunno, R., J. B. Klemp, and M. L. Weisman (1988). A theory for strong, long-lived squall lines. *J. Atmos. Sci.* **45**, 463–485.
Sanders, F., and K. A. Emanuel (1977). The momentum budget and temporal evolution of a mesoscale convective system. *J. Atmos. Sci.* **34**, 322–330.
Sawyer, J. S. (1946). Cooling by rain as a cause of the pressure rise in convective squalls. *Q. J. R. Meteorol. Soc.* **72**, 168.
Schmidt, J. M., and W. R. Cotton (1989). A High Plains squall line associated with severe surface winds. *J. Atmos. Sci.* **46**, 281–302.
Schubert, W. H., J. J. Hack, P. L. Silva-Dias, and S. R. Fulton (1980). Geostrophic adjustment in an axisymmetric vortex. *J. Atmos. Sci.* **37**, 1464–1484.
Seitter, K. L., and H. L. Kuo (1983). The dynamical structure of squall-line type thunderstorms. *J. Atmos. Sci.* **40**, 2831–2854.
Silva-Dias, M. F., A. K. Betts, and D. E. Stevens (1984). A linear spectral model of tropical mesoscale systems: Sensitivity studies. *J. Atmos. Sci.* **41**, 1704–1716.
Smith, C. L., E. J. Zipser, S. M. Daggupaty, and L. Sapp (1975). An experiment in tropical mesoscale analysis: Part I. *Mon. Weather Rev.* **103**, 878–903.
Smull, B. F., and R. A. Houze (1985). A midlatitude squall line with a trailing region of stratiform rain: Radar and satellite observations. *Mon. Weather Rev.* **113**, 117–133.
Smull, B. F., and R. A. Houze (1987a). Dual-Doppler radar analysis of a mid-latitude squall line with a trailing region of stratiform rain. *J. Atmos. Sci.* **44**, 2128–2148.
Smull, B. F., and R. A. Houze (1987b). Rear inflow in squall lines with trailing stratiform precipitation. *Mon. Weather Rev.* **115**, 2869–2889.
Song, J.-L. (1986). A numerical investigation of Florida's sea breeze—cumulonimbus interactions. Ph.D. Thesis, Dep. Atmos. Sci. Colorado State Univ.
Srivastava, R. C., T. J. Matejka, and T. J. Lorello (1986). Doppler radar study of the trailing anvil region associated with a squall line. *J. Atmos. Sci.* **43**, 356–377.
Stormfury (1970). Project Stormfury Annu. Rep. 1969, Natl. Hurricane Res. Lab., NOAA, Coral Gables, Florida.
Szoke, E. J., and E. J. Zipser (1986). A radar study of convective cells in mesoscale systems in GATE. Part I: Vertical profile statistics and comparison with hurricanes. *J. Atmos. Sci.* **43**, 182–218.
Takeda, T. (1971). Numerical simulation of a precipitating convective cloud: The formation of a "long-lasting" cloud. *J. Atmos. Sci.* **28**, 350–376.
Thorpe, A. J., M. J. Miller, and M. W. Moncrieff (1982). Two-dimensional convection in non-constant shear: A model of mid-latitude squall lines. *Q. J. R. Meteorol. Soc.* **108**, 739–762.
Tollerud, E. I., and S. K. Esbensen (1985). A composite life cycle of nonsquall mesoscale convective systems over the tropical ocean. Part I: Kinematic fields. *J. Atmos. Sci.* **42**, 823–837.

Toth, J. J. (1987). Interaction of shallow cold surges with topography on scales of 100–1000 km. Ph.D. Thesis, Dep. Atmos. Sci. Colorado State Univ.

Tripoli, G. J. (1986). A numerical investigation of an orogenic mesoscale convective system. Ph.D. Thesis, Dep. Atmos. Sci. Colorado State Univ.

Uccellini, L. W., and D. R. Johnson (1979). The coupling of upper and lower tropospheric jet streaks and implications for the development of severe convective storms. *Mon. Weather Rev.* **107**, 682–703.

Velasco, I., and J. M. Fritsch (1987). Mesoscale convective complexes in the Americas. *J. Geophys. Res.* **92**, 9591–9613.

Warner, C., J. Simpson, G. van Helvoirt, D. W. Martin, D. Suchman, and G. L. Austin (1980). Deep convection on day 261 of GATE. *Mon. Weather Rev.* **108**, 169–194.

Weisman, M., and J. Klemp (1982). The dependence of numerically simulated convective storms on vertical wind shear and buoyancy. *Mon. Weather Rev.* **110**, 504–520.

Weisman, M. L., and J. B. Klemp (1984). The structure and classification of numerically simulated convective storms in directionally varying wind shears. *Mon. Weather Rev.* **112**, 2479–2498.

Wetzel, P. J., W. R. Cotton, and R. L. McAnelly (1983). A long-lived mesoscale convective complex. Part II: Evolution and structure of the mature complex. *Mon. Weather Rev.* **111**, 1919–1937.

Williams, D. T. (1952). Pressure wave observations in the central midwest. *Mon. Weather Rev.* **81**, 278–298.

Williams, D. T. (1963). The thunderstorm wake of May 4, 1961. Natl. Severe Storms Proj. Rep. No. 18, U.S. Dep. Commer., Washington, D.C. [NTIS PB 168223].

Williams, K. T., and W. M. Gray (1973). Statistical analysis of satellite-observed trade wind cloud clusters in the western North Pacific. *Tellus* **25**, 313–336.

Willoughby, H. E., F. D. Marks, Jr., and R. J. Feinberg (1984). Stationary and moving convective bands in hurricanes. *J. Atmos. Sci.* **41**, 3189–3211.

Wilson, G. S. (1976). Large-scale vertical motion calculations in the AVE IV experiment. *Geophys. Res. Lett.* **3**, 735–738.

Xu, Qin, and J. H. E. Clark (1984). Wave CISK and mesoscale convective systems. *J. Atmos. Sci.* **41**, 2089–2107.

Yamasaki, M. (1968). Numerical simulation of tropical cyclone development with the use of primitive equations. *J. Meteorol. Soc. Jpn.* **46**, 178–201.

Yeh, J.-D., M. A. Fortune, and W. R. Cotton (1986). Microphysics of the stratified precipitation region of a mesoscale convective system. *Conf. Cloud Phys. Radar Meteorol., 23rd, Snowmass, Colo.* pp. J151–J154. Am. Meteorol. Soc., Boston, Massachusetts.

Zhang, D.-L., and J. M. Fritsch (1986). Numerical simulation of the meso-β-scale structure and evolution of the 1977 Johnstown Flood. Part I: Model description and verification. *J. Atmos. Sci.* **43**, 1913–1943.

Zipser, E. J. (1969). The role of organized unsaturated convective downdrafts in the structure and rapid decay of an equatorial disturbance. *J. Appl. Meteorol.* **8**, 799–814.

Zipser, E. J. (1977). Mesoscale and convective-scale downdrafts as distinct components of squall-line structure. *Mon. Weather Rev.* **105**, 1568–1589.

Zipser, E. J., and C. Gautier (1978). Mesoscale events within a GATE tropical depression. *Mon. Weather Rev.* **106**, 789–805.

Zipser, E. J., and M. A. LeMone (1980). Cumulonimbus vertical velocity events in GATE. Part II: Synthesis and model core structure. *J. Atmos. Sci.* **37**, 2458–2469.

Chapter 11 The Mesoscale Structure of Extratropical Cyclones and Middle and High Clouds

11.1 Introduction

In Chapters 6 and 10 we considered how the dynamical behavior of an atmospheric circulation system can be characterized by the scaling parameter λ_R, the Rossby radius of deformation [Eq. (10.3)]. A substitution of typical values of the parameters in Eq. (10.3) for tropical and extratropical cyclones (averaged over the entire cyclone) shows that, for the largest scale of these cyclonic disturbances, the atmosphere behaves as a two-dimensional, quasibalanced fluid (Table 11.1 and Fig. 6.71). An enormous spatial variation of λ_R exists within both tropical and extratropical cyclones and, thus, a variety of clouds and mesoscale precipitating phenomena are embedded within these systems. In this chapter we describe the physical processes and properties of the environment that determine the types of mesoscale structures that form within extratropical cyclones and the characteristics of the associated clouds.

11.2 Large-Scale Processes that Determine Mesoscale Features

Mesoscale cloud and precipitation features are initiated by two mechanisms—forcing on the mesoscale by inhomogeneities in the surface (such as terrain features) and instabilities in the larger scale environment. Terrain-forced features are discussed in Chapter 12; here we discuss the large-scale

Table 11.1

Horizontal Scale and Rossby Radius of Deformation (λ_R) for the Large-Scale Structure of Extratropical and Tropical Cyclones

	L (km)	H (km)	N (s^{-1})	f (s^{-1})	ζ (s^{-1})	$2V/R$ (s^{-1})	λ_R (km)
Extratropical cyclone	10,000	8	0.01	10^{-4}	f	f	400
Tropical cyclone	5,000	8	0.01	0.5×10^{-4}	$2f$	$2f$	500

processes that produce an environment that is stable or unstable with respect to mesoscale cloud and precipitation systems.

We have seen that the type of cloud and precipitation system is determined by six factors: (1) water vapor content of air (both relative and absolute humidity), (2) temperature, (3) aerosol types and amounts, (4) static stability, (5) vertical motion, and (6) vertical shear of the horizontal wind. Because these atmospheric properties vary greatly throughout extratropical cyclones, these storms contain a rich variety of clouds and mesoscale precipitating systems. After a brief description of the physical processes determining the above parameters on large scales of motion (100 km and greater), we discuss their variation in extratropical cyclones which, in turn, determines the variation of clouds and precipitation in these systems.

11.2.1 Water Vapor Content

Chapters 2 and 4 contain the detailed equations that determine the temporal variation of water vapor as well as liquid water and ice. For convenience, we repeat the continuity equation for the water vapor mixing ratio appropriate for large-scale models,

$$\frac{\partial r_v}{\partial t} = -\mathbf{V} \cdot \nabla r_v - w\frac{\partial r_v}{\partial z} + E - C + F_{r_v}, \qquad (11.1)$$

where E represents evaporation (from the surface or from precipitation), C represents condensation (including sublimation), and F_{r_v} represents unresolvable (subgrid-scale) transports. Over short time periods and for the large scales of motion considered here, horizontal and vertical transports are the major processes contributing to the change of water vapor. Evaporation from the surface is important for long time periods (greater than 12 h) over water and land surfaces that are moister and warmer than the air immediately above the surface. Condensation is a key removal mechanism in precipitating systems, whereas evaporation from cloud water and precipitation water can be locally significant over short time periods. Subgrid-scale (turbulent) transports are greatest in the unstable planetary boundary layer where water vapor evaporated from the surface is transported upward.

Equation (11.1) describes the temporal variation of the mass of water vapor in a unit mass of dry air, or the mixing ratio. The *relative* humidity is determined by the mixing ratio and the temperature.

11.2.2 Temperature

Detailed forms of the thermodynamic equation are presented in Chapter 2. A simplified form appropriate for interpreting the temporal variation of the large-scale temperature is

$$\frac{\partial \theta}{\partial t} = -\mathbf{V} \cdot \nabla \theta - w \frac{\partial \theta}{\partial z} + \frac{\theta}{c_{\mathrm{p}} T} Q + F_{\theta}, \tag{11.2}$$

where θ is potential temperature and Q is the net grid scale averaged diabatic heating. The first term represents horizontal advection; second term represents vertical advection, which in a statically stable atmosphere produces cooling with upward motion and warming with subsiding motion. The diabatic heating term Q represents latent heating and cooling effects associated with condensation or evaporation and radiation. In precipitation systems, there is a close balance between vertical advection (adiabatic cooling) and diabatic heating. The last term F_θ is the subgrid-scale transport of heat and includes the effect of sensible heating from the surface and the upward transport in the PBL. On the large scale above the PBL and in the absence of precipitation, horizontal and vertical advection are the largest terms. In the daytime, heated PBL, the last term, dominates.

11.2.3 Aerosol Types and Amounts

Condensation does not usually occur at relative humidities of exactly 100% (Chapter 4). The presence and type of cloud depend on the amount and distribution of aerosols in the atmosphere, in particular, cloud condensation nuclei and ice nuclei. The effects of these aerosols are discussed in Chapter 4 and are mentioned here for completeness. They are not explicitly considered in most large-scale models at present because, on the large scale, their effects are thought to be relatively small compared to other factors determining cloud type and precipitation, notably water vapor content, temperature, and vertical motion.

11.2.4 Static Stability

Once condensation occurs, the subsequent evolution of the cloud and precipitation system depends on the mean vertical motion and the static stability. The static stability of the large-scale environment is classified as *absolutely unstable, conditionally unstable*, or *absolutely stable*, depending upon the relationship of the actual lapse rate γ,

$$\gamma \equiv -\partial T/\partial z, \tag{11.3}$$

the dry adiabatic lapse rate γ_d, and the wet adiabatic lapse rate γ_w (see Chapter 2 for definitions of the dry and wet adiabatic lapse rates).

These relationships are as follows:

$$\begin{aligned} &\text{Absolutely unstable:} && \gamma > \gamma_d \\ &\text{Conditionally unstable:} && \gamma_d > \gamma > \gamma_w \\ &\text{Absolutely stable:} && \gamma_w > \gamma \end{aligned} \tag{11.4}$$

Conditional instability refers to the lifting of a parcel of air through its environment. *Convective instability,* in contrast, refers to the lifting of an entire layer of air. The condition for convective instability is that equivalent potential temperature decrease throughout the layer,

$$\text{Convective instability:} \quad \partial \theta_e/\partial z < 0. \tag{11.5}$$

It is possible for a layer of air to be convectively unstable but conditionally stable. Convective instability is often associated with the large-scale environment prior to the development of severe convective storms and tornadoes, when warm, dry air overlies warm, moist air, with an associated rapid decrease of θ_e with height. When the layer is lifted, it rapidly becomes unstable and favorable for severe thunderstorm development.

When the mean vertical motion is near zero and the atmosphere is conditionally stable, fogs or layered clouds occur (Chapter 7). When the mean vertical velocity is upward (typically a few centimeters per second) and the atmosphere is conditionally stable, deep layers of nonconvective (stratiform) clouds are produced. Under conditionally unstable conditions, convective clouds and precipitation can occur, even with near-zero mean (large-scale) vertical velocities or even weak subsidence.

If we define a static stability parameter $\gamma_\theta \equiv \partial\theta/\partial z$, a simple equation describing the temporal variation of the large-scale static stability can be derived from Eq. (11.2):

$$\frac{\partial \gamma_\theta}{\partial t} = \frac{-\partial \mathbf{V} \cdot \nabla \theta}{\partial z} - \frac{\partial w \gamma_\theta}{\partial z} - \frac{\partial}{\partial z}\left(\frac{\theta}{c_p T} Q\right) + \frac{\partial F_\theta}{\partial z}. \tag{11.6}$$

The first term on the right side of Eq. (11.6) represents differential horizontal advection of potential temperature; for example, cold advection overlying warm advection contributes to destabilization. The second term represents the effect of vertical stretching of a column, i.e., if γ_θ is constant in the vertical, an increase of upward motion with height (stretching) represents destabilization. This process is effective in producing or destroying temperature inversions. The third term represents the effect of differential heating in the vertical. A decrease of diabatic heating with height, as occurs above the region of maximum latent heating associated with cumulus convection, for example, destabilizes the environment. Another example of this process is radiative cooling near the tops of layered clouds, which destabilizes this region. The final term represents the vertical variation of turbulent heat fluxes and is largest in the heated PBL.

Conditional and convective instabilities do not consider the effects of rotation, but rotation affects the stability of fluid motions. Bennetts and Hoskins (1979) and Emanuel (1979, 1982, 1983a, b, c, d) discuss the combined effect of rotation and static stability in a theory of *conditional symmetric instability.* To summarize this theory, we consider a mean zonal flow U which increases with height and is in geostrophic and thermal wind balance. The absolute angular momentum M, defined by

$$M = U - fy, \tag{11.7}$$

is conserved approximately in the absence of friction. Thus, parcels of air, when lifted, will move along surfaces of constant M, which in this example slope upward toward the north. If a parcel moves upward along a surface of constant M and becomes warmer than its environment because of the release of latent heat, the atmosphere is in a state of *conditional symmetric instability.* A conditionally stable atmosphere may possess conditional symmetric instability. Moist convection arising from this instability is called *moist slantwise convection.* Conditional symmetric instability is thought to be the process with produces rainbands in extratropical cyclones.

11.2.5 Vertical Motion

Along with moisture content and static stability, vertical motion is one of the most important properties of the large-scale environment that determine the presence and type of clouds and precipitation systems. Large-scale upward motion favors clouds and precipitation because it cools the air toward saturation as well as destabilizes the air. In addition, the low-level convergence associated with rising motion in the middle troposphere is associated with moisture convergence in all but the driest air masses. In contrast, sinking air becomes more stable, relative humidity decreases, and low-level moisture divergence usually results.

A useful diagnostic equation for isolating the large-scale physical processes associated with vertical motion is the quasigeostrophic omega equation, derived from the vorticity equation and the first law of thermodynamics, using the assumption that the vorticity and the large-scale wind are in quasigeostrophic balance (see Holton, 1979, for a derivation). A convenient form of the omega equation with pressure as the vertical coordinate can be written as

$$\nabla^2\omega + \frac{f_0^2}{\sigma}\frac{\partial^2\omega}{\partial p^2} = \frac{f}{\sigma}\frac{\partial}{\partial p}[\mathbf{V}_g \cdot \nabla(\zeta + f)] + \frac{R}{p\sigma}\nabla^2\mathbf{V}_g \cdot \nabla T$$

$$-\frac{R}{c_p p\sigma}\nabla^2 Q - \frac{f_0}{\sigma}\frac{\partial F_\zeta}{\partial p}, \tag{11.8}$$

where the static stability parameter σ is given by

$$\sigma = -(1/\rho\theta)(\partial\theta/\partial p), \tag{11.9}$$

and F_ζ represents the contribution of subgrid-scale effects (friction) to the temporal rate of change of the vertical component of the relative vorticity (ζ).

To interpret Eq. (11.8), wave forms may be assumed for the horizontal and vertical variation of ω (Holton, 1979),

$$\omega \approx \sin[\pi(p/p_0)]\sin(kx)\sin(ly), \tag{11.10}$$

where k and l are horizontal wave numbers and p_0 is a reference pressure. With the assumption Eq. (11.10), the left side of Eq. (11.8) is

$$\left(\nabla^2 + \frac{f_0^2}{\sigma}\frac{\partial^2}{\partial p^2}\right)\omega \approx -\left[(k^2+l^2) + \frac{1}{\sigma}\left(\frac{f_0\pi}{p_0}\right)^2\right]\omega. \tag{11.11}$$

With the use of Eq. (11.11), the omega equation, Eq. (11.8), can be interpreted as follows. The first term on the right side represents the vertical derivative of the vorticity advection. Positive vorticity advection that increases with height contributes to upward vertical motion (negative omega). The second term on the right side represents the Laplacian of the horizontal temperature advection. In regions of maximum warm-air advection, this term is positive and contributes to upward motion. The third term on the right is the Laplacian of diabatic heating; regions of maximum diabatic heating are associated with upward motion.

The last term in Eq. (11.8) represents the effect of subgrid-scale motions (turbulence). In the PBL, its effect may be estimated by assuming a quadratic stress law for the frictional terms in the equations of motion,

$$\frac{\partial u}{\partial t} = \cdots - C_D|\mathbf{V}|u/h,$$
$$\frac{\partial v}{\partial t} = \cdots - C_D|\mathbf{V}|v/h, \qquad (11.12)$$

where C_D is the drag coefficient, u and v are the mean horizontal wind components in the PBL, and h is the depth of the PBL. With the frictional terms represented by Eq. (11.12), the linearized form of F_ζ is:

$$F_\zeta(p_s) \approx -K\zeta(p_s), \qquad (11.13)$$

where K is a mean value of $C_D|\mathbf{V}|/h$. Thus, using Eqs. (11.8), (11.11), and (11.13), the vertical velocity near the top of the PBL is approximately

$$\omega \propto \frac{\partial F_s}{\partial p} \propto F_\zeta(p_s), \qquad (11.14)$$

where we have used the fact that F_ζ vanishes near the top of the PBL. From (11.13) and (11.14) we see that surface friction induces upward motion in cyclonic systems and downward motion in anticyclonic systems. This effect is sometimes called "Ekman pumping."

11.2.6 Vertical Shear of the Horizontal Wind

A fifth property of the environment in the development of some types of clouds and convective systems is the vertical shear of the horizontal wind. In the PBL, the wind shear organizes fair-weather cumulus clouds into bands, rolls, rings, and streets (Chapter 8). In addition, strong vertical wind shear is a major factor in determining the organization and structure of cumulonimbus clouds. Wind shear affects the entrainment rate, the strength, movement, precipitation efficiency, and lifetime of convective clouds and storms (Section 9.5.4). Wind shear is also a factor in the splitting of severe thunderstorms and the development of rotating storms and hail- and tornado-producing thunderstorms.

The development of wind shear in the large-scale environment is closely tied to the development of baroclinicity, since the thermal wind balance is approximately satisfied for these scales of motion. Figure 11.1 shows a horizontal cross section of a strong baroclinic zone and associated wind shear (Shapiro *et al.*, 1984). Baroclinicity on these scales is produced primarily by two mechanisms—frontogenetic processes and differential heating associated with latent heat release. Frontogenetic processes are important in extratropical cyclone systems. As reviewed by Keyser and Shapiro (1986), confluence and deformation in the large-scale environment

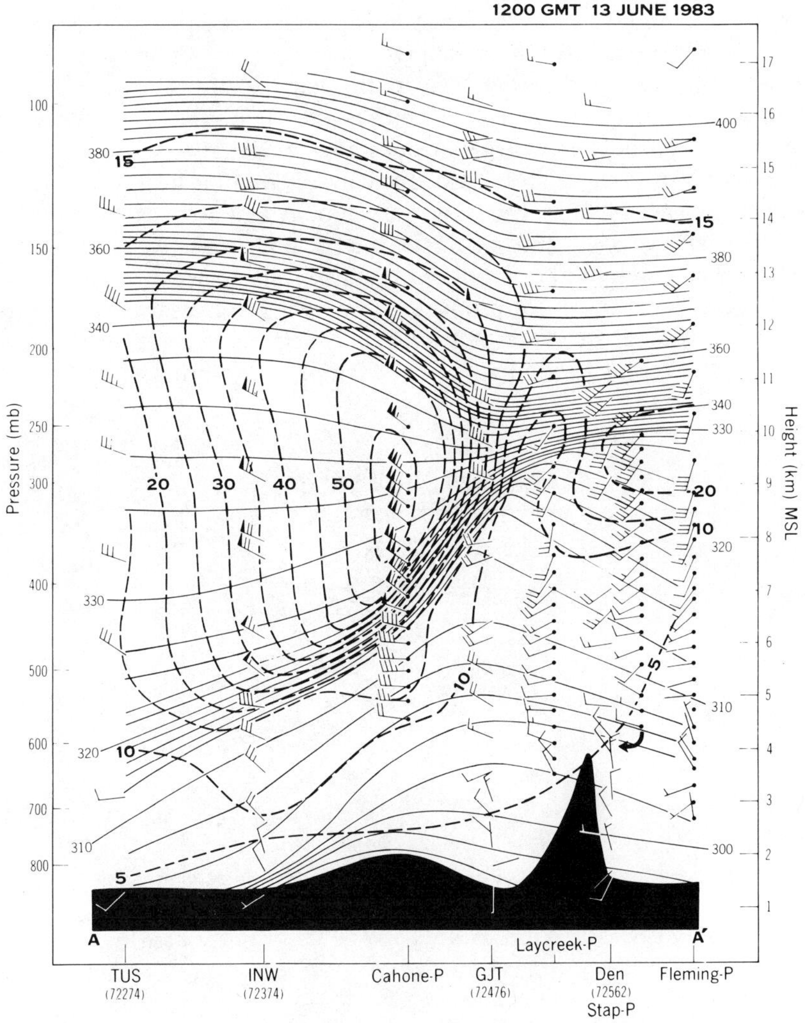

Fig. 11.1. Cross-sectional analysis of wind speed (m s^{-1}, dashed lines) and potential temperature (K, solid lines) at 1200 GMT, 13 June 1983 along a SW–NE line from Tucson, Arizona to Fleming, Colorado. Analysis is a composite of rawinsonde winds and radar wind profiles. Profiles soundings are designated by the letter P at the horizontal axis. Flag = 25 m s^{-1}; full barb = 5 m s^{-1}; half barb = 2.5 m s^{-1}. [From Shapiro *et al.* (1984).]

and an ageostrophic response of the atmosphere to thermal-wind imbalances produced by the changing baroclinicity are key elements of the frontogenesis process. Figure 11.2 shows the transverse ageostrophic circulation associated with frontogenesis in a numerical model. This vertical circulation, in addition to playing an essential role in the frontogenesis process and the development of wind shear, destabilizes the environment on the warm, moist side of fronts and triggers clouds and precipitation systems.

Baroclinicity can also be produced by differential heating associated with the release of latent heat. The baroclinicity and associated wind shear in tropical cyclones and in some mesoscale convective systems and extratropical cyclones are produced by this mechanism.

Another mechanism for producing low-level wind shear is surface friction. The general decrease of frictional effects with height in the lower troposphere results in wind shear. For example, the diurnal variation in the

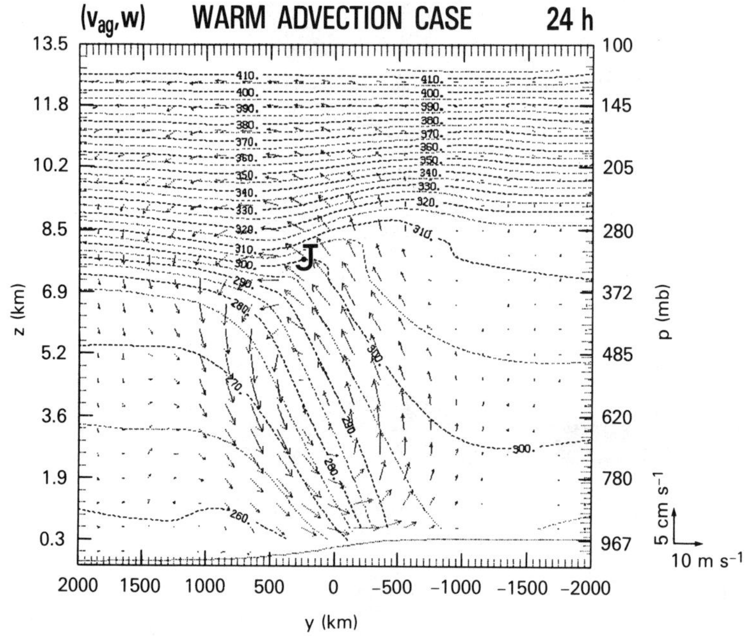

Fig. 11.2. Cross section of transverse ageostrophic circulation (v_{ag}, w) and potential temperature (dashed lines, contour interval 5 K) after a 24-h integration of a two-dimensional primitive-equation model of frontogenesis (Keyser and Pecnick, 1985a, b) due to confluence in the presence of advection. Location of upper-level jet in along-front-velocity component is indicated by J; magnitudes of components of transverse ageostrophic circulation are represented by vector scales on lower right margins of figure. [From Keyser and Shapiro (1986).]

depth of the PBL and the intensity of turbulent mixing of momentum can lead to the development of low-level jets (Blackadar, 1957; Bonner, 1966, 1968). The horizontal convergence and wind shear associated with these jets can affect significantly the development of convective storms.

11.3 Mesoscale Structure of Extratropical Cyclones

Driven inexorably by differential radiative heating between high and low latitudes, the middle-latitude atmosphere is characterized by large-scale horizontal temperature gradients and, through the thermal wind relationship, westerly winds that normally increase with height throughout the troposphere. The poleward decrease of temperature is rarely uniform, and, instead, is usually concentrated in relatively narrow baroclinic zones or fronts. These baroclinic zones become unstable with respect to wavelike perturbations, and the result is the development of cyclones in the baroclinic zone. The wavelength of maximum instability depends on the static stability and horizontal temperature gradient (Staley and Gall, 1977); on the average it is around 3000 km.

As cyclones develop, cold air is carried southward to the rear of the cyclone while warm air is carried northward. (For convenience, we refer to the Northern Hemisphere in this discussion.) Confluence and deformation associated with the developing circulation produce increasing horizontal temperature gradients, in narrow bands, i.e., warm and cold fronts. Temperature changes associated with horizontal advection and vertical motions destroy thermal wind balance, and the resulting ageostrophic motions produce organized regions of divergence, convergence, and associated vertical motions.

Extratropical cyclones dominate the large-scale variability of the weather of middle latitudes. Figure 11.3 shows visible satellite imagery of a mature cyclone over North America. The circulation of this cyclone covers most of North America and adjacent waters east of the Rocky Mountains. Produced by baroclinic instability, these cyclones set up the environment to host a variety of mesoscale cloud and precipitation systems.

Many aspects of the mesoscale structure of extratropical cyclones, discussed in detail in this text, include fogs and stratocumulus clouds (Chapter 7), cumulus clouds (Chapter 8), severe thunderstorms and tornadoes (Chapter 9), squall lines and other mesoscale convective systems (Chapter 10), and middle- and high-level cloud systems (later in this chapter). In this section, we interpret the formation of these features in different regions of the extratropical cyclone in terms of the physical processes and characteristics of the environment discussed in the preceding section.

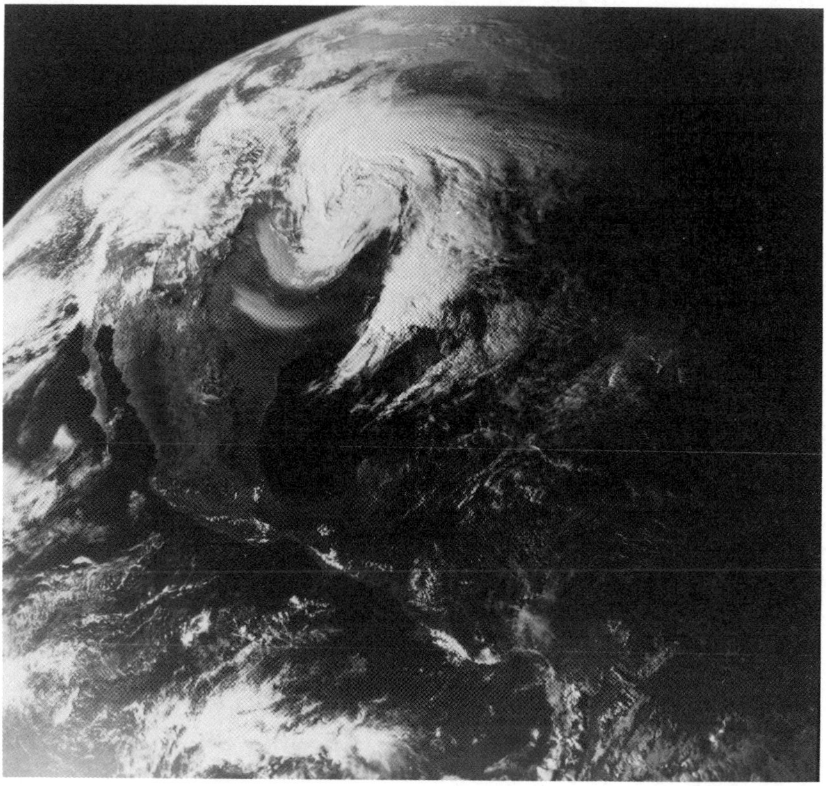

Fig. 11.3. Visible satellite photograph of occluding cyclone over North America at 2200 GMT, 23 February 1977.

Atkinson (1981) summarizes the history of research leading to our knowledge of the mesoscale precipitation structure of extratropical storms. Austin (1960) and Austin and Houze (1972) used radar observations in New England to reveal the universal existence of organized mesoscale precipitation systems in all parts of the extratropical cyclone. Hobbs and his collaborators were leaders in aircraft and radar investigations of these systems in the vicinity of the Pacific Northwest coast (Houze and Hobbs, 1982). In the United Kingdom, Browning and his collaborators observed similar mesoscale structures and related the precipitation features to the circulation of the extratropical cyclone. Browning (1971) introduced the concept of the "conveyor belt," which Harrold (1973) developed further. The warm conveyor belt is a warm, moist airstream which ascends relative to the warm front and which is at least as important in the distribution of precipitation as the warm front itself. Carlson (1980) has extended these ideas to a conceptual model of the extratropical cyclone. Carlson's model (Fig. 11.4) consists of three streams of air. A *warm conveyor belt* enters the cyclone from the southeast, rises in the warm sector and over the surface warm front, and exits the system to the northeast ahead of an upper-level trough.

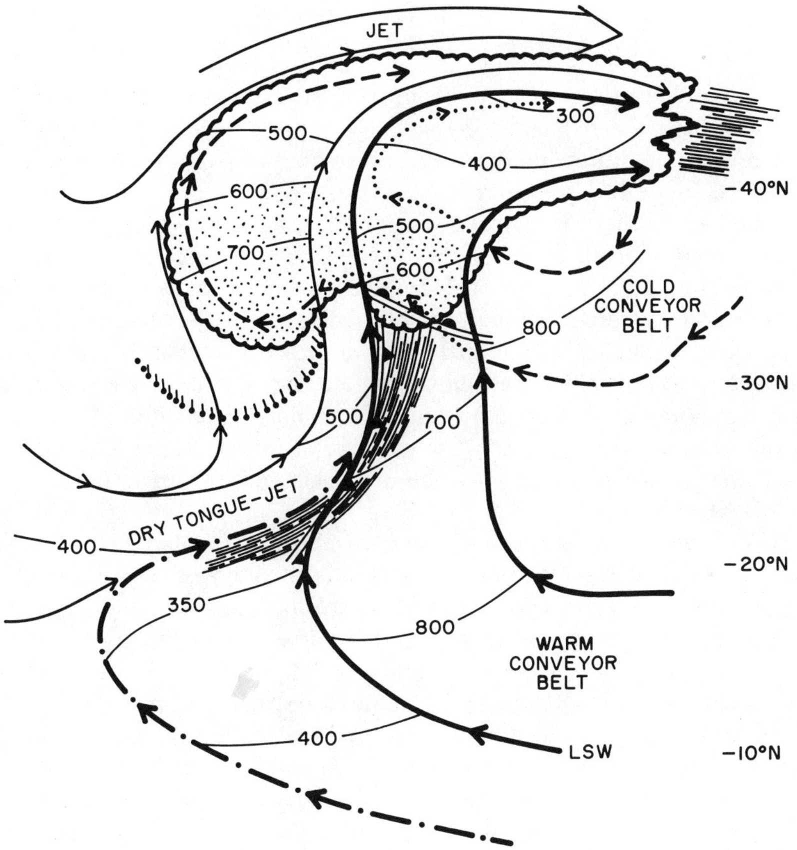

Fig. 11.4. Schematic composite of airflow through middle-latitude cyclone. Heavy solid streamlines depict airflow at top of the warm conveyor belt. Dashed lines represent cold conveyor belt (dotted when it lies beneath the warm conveyor belt or dry airstream). Dot-dashed line represents air originating at middle levels in tropics. Thin solid streamlines pertain to dry air which originates at upper levels west of the trough. Thin solid lines denote heights of the airstreams (millibars) and are approximately normal to the direction of respective air motion (isobars are omitted for cold conveyor belt where it lies beneath warm conveyor belt or beneath jet-stream flow). The region of dense upper- and middle-level layer cloud is represented by scalloping and sustained precipitation is represented by stippling. Streaks denote thin cirrus. The edge of low-level stratus is shown by curved border of small dots with tails. The major upper tropospheric jet streams are labeled JET. The limiting streamline for warm conveyor belt is labeled LSW. See text for further explanation. [From Carlson (1980).]

A *cold conveyor belt* approaches the cyclone in the low levels from the northeast, rises near the center of the low, and merges with the warm conveyor belt as it leaves the system. The third stream, the *dry stream*, originates in the upper troposphere upstream of the cyclone and flows into the system from the northwest.

11.3.1 Warm Conveyor Belt

The warm conveyor belt is a stream of relatively warm, moist air that, in the Northern Hemisphere, originates in the low levels of the southeast quadrant of the storm and flows northward and westward toward the center of the cyclone. Because isentropic surfaces slope upward to the north, this air rises as it flows through the warm sector and above the surface warm front. Interpreted in terms of the quasigeostrophic omega equation [Eq. (11.8)] on constant-pressure surfaces, this air, which is ahead of an advancing upper-level trough, is usually associated with positive vorticity advection and warm advection. If the air in the warm conveyor belt is convectively unstable, the lifting will destabilize the air and lead to convective clouds and precipitation systems, including rainbands, squall lines, and severe thunderstorms. These systems are often most intense along and ahead of the cold front where the upward motion associated with the frontal dynamics is the greatest.

If the air in the warm conveyor belt is absolutely stable, the lifting will produce extensive layers of stratiform clouds, including nimbostratus. Often parts of the warm conveyor belt, usually the low levels in the warm sector, will be convectively unstable, while other parts, usually the upper levels in the cool sector north of the surface warm front, will be absolutely stable. Thus, convective clouds and precipitation systems are most likely in the warm sector, while stratiform clouds and precipitation are predominant in the cool sector.

A vertical cross section through an idealized extratropical cyclone (Fig. 11.5) illustrates typical cloud types in various regions of the cyclone. This classic model, developed by the Norwegians over 60 years ago, remains a useful, though simplified, conceptual model. Ahead of the cold front, where the warm conveyor belt air is moist and convectively unstable, rising motion initiates deep cumulus convection and thunderstorms. This lifting is often strongest just ahead of the front, where ageostrophic vertical circulation associated with the front and frictional convergence at the surface cold front produces maximum updraft speeds.

Diabatic heating is important in modifying the air in the warm conveyor belt. In the low levels, surface heating destabilizes the air when sufficient solar radiation is available. Fair-weather cumulus clouds populate this

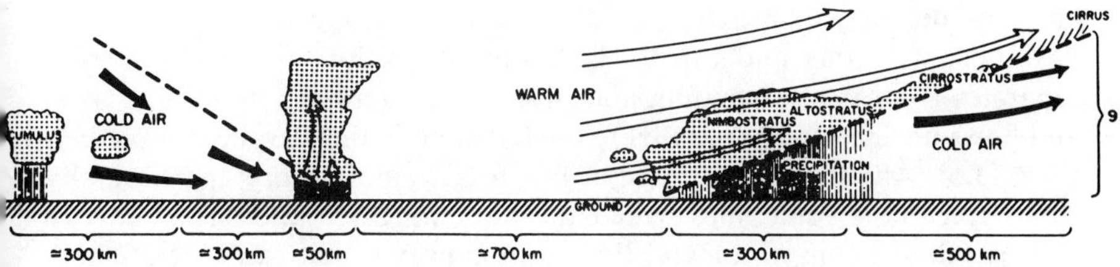

Fig. 11.5. Idealized vertical cross section through a midlatitude cyclone, according to a Norwegian model. (Vertical scale is stretched by a factor of about 30 compared to horizontal scale.) [From Houze and Hobbs (1982).]

region during the daytime. When either convective or nonconvective precipitation occurs in the warm conveyor belt, latent heat release reduces the adiabatic cooling of the rising air and represents a significant energy source to the cyclone. In addition, the diabatic heating associated with the latent heat release transports angular momentum upward from the lower troposphere, increasing the rate of upper-level cyclogenesis (Johnson *et al.*, 1976).

The air in the warm conveyor belt eventually reaches the upper troposphere and turns anticyclonically toward the northeast, representing the southwesterly flow ahead of the upper-level trough on a constant-pressure surface. By this time, the air is usually absolutely stable, so the clouds are extensive sheets of altostratus and cirrostratus. Far ahead of the cyclone, the leading edge of the warm conveyor belt is heralded by thin cirrus or cirrostratus clouds (Fig. 11.5).

The left (west) edge of the warm conveyor belt is identified by the streamline labeled LSW (limiting streamline for the warm conveyor belt) in Fig. 11.4. The air following and immediately to the right of this streamline originated farthest south and most closely approaches the cold front. Because of its origin, this air is the warmest, moistest, and most unstable air in the cyclone system. As it approaches the cold front, it also experiences the greatest rate of lifting. Because of the strong baroclinicity in the frontal region, strong wind shear is present as well. The combination of the moist convectively unstable air, rapid lifting, and strong wind shear makes this region favorable for the severe mesoscale convective phenomena discussed in Chapters 9 and 10.

11.3.2 The Dry Stream

The air to the immediate left of the LSW streamline (Fig. 11.4) is dry and represents the right (eastern) edge of the dry stream. This air originated in the upper troposphere or lower stratosphere west of the upper-level trough and has descended into the middle and lower troposphere. As the air descends, it warms and dries. The northern portion of the dry stream separates from the descending anticyclonic flow, crosses the trough axis, and flows northeastward parallel to the left edge of the warm conveyor belt (Fig. 11.6). Although this stream begins to ascend, it is dry and normally cloud free. It is sometimes called the "dry tongue," and the boundary between this stream of air and the warm conveyor belt can be extremely sharp, resulting in a surface feature frequently observed over the High Plains of the United States called a "dryline." The satellite photograph (Fig. 11.3) shows a dry tongue and a sharp boundary between the dry stream and the warm conveyor belt.

The southern portion of the dry stream continues to descend, warm and dry as it approaches the surface. Strong surface winds associated with the lower portions of this sinking airstream produce significant blowing dust

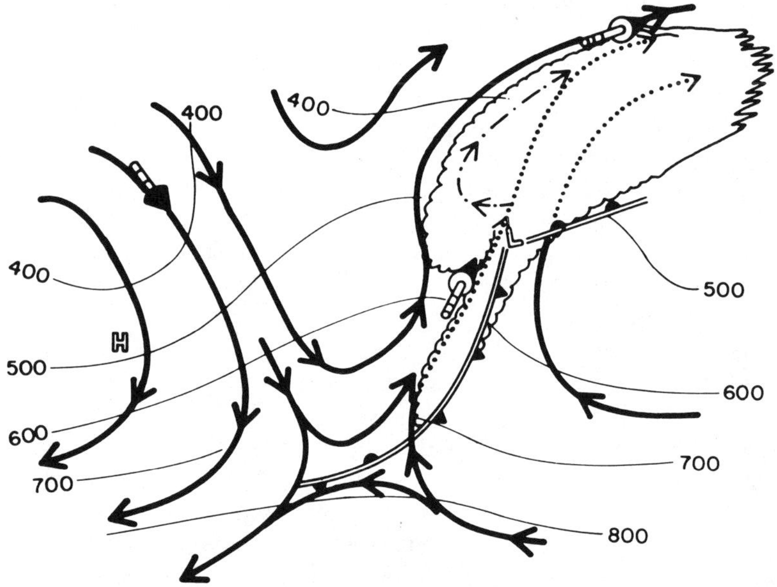

Fig. 11.6. Schematic of relative flow on a dry isentropic surface. Thin solid lines are isobars; these are discontinued in regions of saturation where flow is shown as dotted lines for warm conveyor belt and as a dot–dash line for cold conveyor belt. Dry airstream originates at high levels in northwest and descends toward trough axis. There it splits into two branches, one branch descending into low levels west of trough and the other flowing around trough to ascend in a narrow stream over the western extension of cloud shield (scalloped border). Symbols L and H, respectively, refer to locations of surface high- and low-pressure centers. Fronts are depicted in conventional symbols. Jet maxima are represented by solid arrowhead segments. [From Carlson (1980).]

in arid regions (Fig. 11.3), where a large plume of dust originates over the southwestern United States south of the cyclone and streams more than 500 km eastward. These dust storms are favored by strong surface heating, which causes a deep (sometimes up to 3 km) boundary layer in the southwest quadrant of the cyclone.

11.3.3 The Cold Conveyor Belt

The stream of air called the cold conveyor belt originates to the northeast of the cyclone in the descending air associated with the previous anticyclone. This air flows westward beneath the warm conveyor belt. Over flat terrain, the most pronounced lifting of the cold conveyor belt occurs just north of the cyclone center and under the southwesterly jet aloft. This usually stable air rises, turns anticyclonically, and flows parallel to the upper regions of the warm conveyor belt. Saturation is favored by evaporation of precipitation falling from the warm conveyor belt aloft, and thick fog and stratus clouds occur in this airstream north of the surface warm front.

Along the eastern slopes of the Rocky Mountains in the United States, the cold conveyor belt experiences significant lifting as it follows the rising terrain, and these "upslope" conditions are responsible for almost all of the nonconvective precipitation in this region (Boatman and Reinking, 1984; see also Chapter 12).

The cool cyclonically flowing low-level air to the rear of the extratropical cyclone is frequently moist enough that the frictionally induced rising motion produces an extensive layer of stratocumulus or fair-weather cumulus clouds at the top of the PBL. The cyclone shown in Fig. 11.3 contains a region of stratocumulus clouds to the south and southwest of the cyclone center and behind the western edge of the dry tongue.

The conceptual model of the extratropical cyclone discussed above is modified somewhat for cyclones over the ocean. Two major differences are the greater availability of moisture and the smoother, more homogeneous surface. The latter allows more rapid development and ultimately more intense cyclones. Rapid development can be pronounced when the sea surface is relatively warm, thus destabilizing lower levels and leading to explosively deepening cyclones (Sanders and Gyakum, 1980; Roebber, 1984; Rogers and Bosart, 1986; Nuss and Anthes, 1987). In addition, the more homogeneous surface favors the development of organized mesoscale convective clouds in the PBL (Chapter 8).

Figure 11.7 shows an occluding extratropical cyclone over the North Pacific. As the upper-level trough deepens and becomes a closed cyclone, the low- and middle-level clouds are wrapped around the cyclone center. A region of organized, cellular cumulus convection is present in the unstable

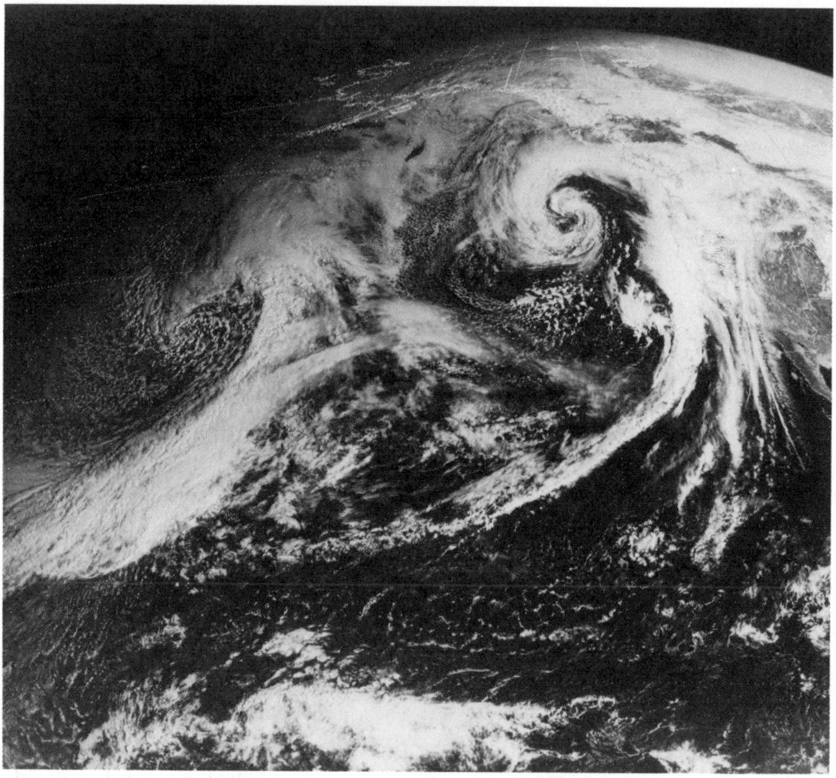

Fig. 11.7. Visible satellite photograph of occluding extratropical cyclone over North Pacific. Note cellular cumulus convection south of cyclone center and small cloud system southeast of center behind cold-frontal band of clouds. This system is an incipient mesoscale cyclone. Time is 1845 GMT, 27 February 1980.

PBL south of the cyclone center. Bands of convective clouds occur along the cold front and in the convectively unstable warm sector ahead of the front. A variety of rainbands occurs with such Pacific storms (Houze and Hobbs, 1982).

Figure 11.8 shows the location of six types of bands with respect to the cold and warm fronts in a typical cyclone. Warm frontal bands, depicted in cross-sectional form in Fig. 11.9, are oriented parallel to the surface warm front where a deep layer of ascending warm, moist air exists. Warm sector bands (50 km wide) are located ahead of and parallel to the cold front. Wide (50-km) cold frontal bands are oriented parallel to the cold front and either straddle the front or occur behind the front. In occlusions, they can be associated with the cold front aloft. A narrow (5-km) cold frontal band at the surface cold front is probably caused by intense frictionally induced convergence at the front (Keyser and Anthes, 1982). Prefrontal cold-surge bands are associated with surges of cold air ahead of the cold front. Finally, lines of convective clouds that form in the unstable cold air behind the front are called postfrontal bands.

The seeding of supercooled water by ice crystals falling from above, as indicated in Fig. 11.9, enhances precipitation. The region where the ice crystals are generated is the seeder zone. The remaining precipitation is produced in the zone below, termed the *feeder* zone because of the strong advective supply of moisture by the cyclone-scale circulation. The *seeder*-

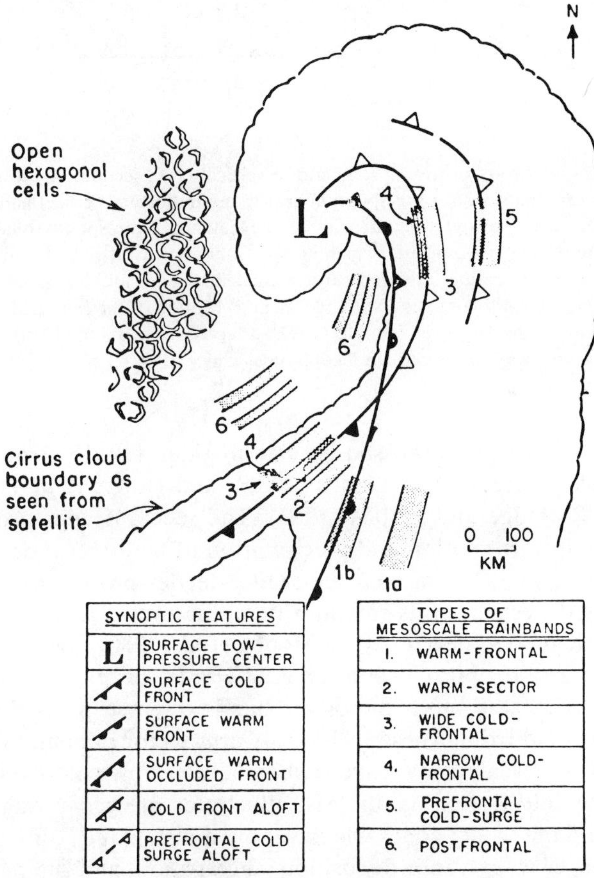

Fig. 11.8. Schematic depiction of types of rainbands observed in extratropical cyclones. [From Houze and Hobbs (1982); adapted from Hobbs (1978). Copyright by the American Geophysical Union.]

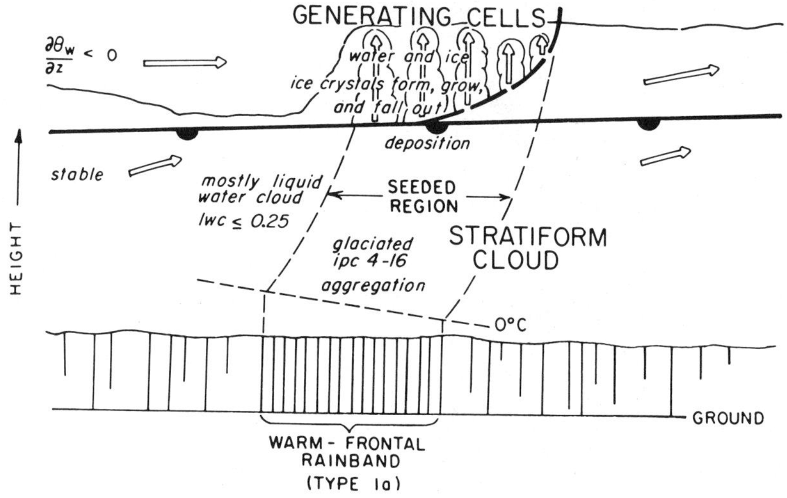

Fig. 11.9. Model of a warm-frontal rainband in vertical cross section. Structure of clouds and predominant mechanisms for precipitation growth are indicated. Vertical hatching below cloud bases represents precipitation: density of the hatching corresponds qualitatively to precipitation rate. Heavy broken line branching out from front is a warm-frontal zone with convective ascent in generating cells. Ice-particle concentrations (ipc) are given in numbers per liter; cloud liquid-water contents (lwc) are in g m^{-3}. Motion of the rainband in the figure is from left to right. [From Houze and Hobbs (1982); adapted from Hobbs (1978) and Matejka *et al.* (1978). Copyright by the American Geophysical Union.]

feeder process has been discussed by Cunningham (1951), Browning and Harrold (1969), Herzegh and Hobbs (1980), and Houze *et al.* (1981) and modeled by Rutledge and Hobbs (1983). The seeder–feeder process also enhances the upward motion and precipitation of the other types of bands depicted in Fig. 11.8. We discuss the seeder–feeder process more fully in Chapter 12 with respect to its role in orographic precipitation.

A schematic model, developed by Hobbs *et al.* (1980), depicts the band structure of an extratropical cyclone (Fig. 11.10). In this model, the heaviest precipitation is located in six mesoscale bands oriented parallel or nearly parallel to the cold front. These include a warm-sector rainband consisting of a series of mesoscale convective subbands, a narrow cold-frontal band, and four wide cold-frontal bands. Also illustrated are generating zones of ice crystals and the feeder zones which supply moist, cloudy air from which seeder crystals grow by vapor deposition, aggregation, and riming of cloud droplets. Hobbs *et al.* (1980) estimated that about 20% of the precipitation in the wide cold-frontal rainbands originates in the seeder zones and ~80% originates in the feeder zones.

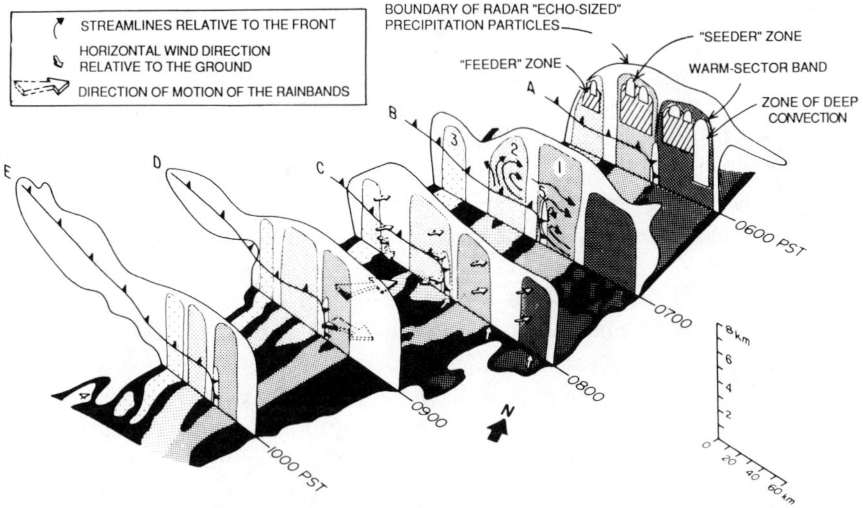

Fig. 11.10. Schematic summary of structure and organization of clouds and precipitation and airflow associated with a typical cold front. Outline of the radar-echo pattern in a vertical plane normal to the front is shown for five different times. Various rainbands (1–5 and the warm-frontal band) are indicated by different shadings. Various features of the structure are highlighted in each vertical section. (A) Locations of "feeder–seeder" zones and regions of deeper convection; (B) streamlines of airflow relative to the front; (C) horizontal winds; and (D) motions of wide and narrow cold-frontal rainbands. [From Hobbs *et al.* (1980).]

Using triple-Doppler radar observations, Carbone (1982) obtained a detailed analysis of the kinematic and thermodynamic structure of a narrow cold-frontal band of intense precipitation. As this band passed through the Central Valley of California, it produced a variety of severe mesoscale weather conditions, including strong winds, electrical activity, and tornadoes.

The Doppler radar observations showed a nearly two-dimensional updraft of magnitude 15–20 m s^{-1}; this updraft was associated with a gravity current that propagated ahead of the cold front toward a strong prefrontal low-level jet. Convective available potential energy was small in this case. The vertical shear, which occurred in a direction nearly parallel to the band, and symmetric instability (Emanuel, 1979) were likely important factors in the development of this band. Diabatic cooling associated with the melting of ice was important in maintaining the density contrast across the gravity current.

The mechanisms responsible for the variety of bands in extratropical cyclones are still under debate. This is partly because of the limitations of the theoretical models and the variation of the dominating mechanisms from storm to storm and even over portions of a storm sector. This is particularly true of the warm sector of extratropical cyclones. In the more severe cyclonic storms, the warm sector can be conditionally unstable, leading to the formation of prefrontal squall lines. Mechanisms such as symmetric instability (Ogura *et al.*, 1982) have been proposed as being responsible for triggering the linear structure of the squall lines.

In other extratropical cyclones, the warm sector may be weakly unstable to deep convection and at the same time unstable to conditional symmetric instability (CSI). Bennetts and Ryder (1984), for example, compared the predictions of CSI and symmetric wave-CISK (Emanuel, 1982; or see Chapter 9) to the observed banded structure in the warm sector of an extratropical cyclone. The primary energy source for symmetric instability is the kinetic energy of the basic flow, while conditional symmetric instability receives additional energy from latent heat release. Symmetric wave-CISK, in contrast, derives its energy principally from latent heat release, while additional energy is supplied by the mean flow (e.g., vertical wind shear). Both theories predicted banded rolls, but CSI predicted that the rolls move with the mean wind while the wave-CISK theory predicted that the rolls exhibit a propagation velocity relative to the mean flow of $6\ \mathrm{m\ s^{-1}}$. The observations indicated that the bands moved less than $2\ \mathrm{m\ s^{-1}}$ relative to the mean flow. This suggests that CSI theory was more applicable to this case than the wave-CISK theory. As noted in Chapter 9, however, a number of variations in the parameterizations used in wave-CISK theory can account for the faster propagation of Emanuel's model. Raymond's (1983) advective wave-CISK model, for example, predicts a propagation velocity comparable to the mean wind speed. Thus, in the warm sector of many extratropical cyclones, both CSI and wave-CISK may occur, the differences being so slight that variations in the details of the parameterizations used in the models are greater than the fundamental differences between the two theories.

In still another extratropical cyclone, Parsons and Hobbs (1983) found that CSI appeared to be the stronger candidate for explaining the observed warm-sector rainbands. Parsons and Hobbs (1983) also concluded that the CSI theory was consistent with the observed structure of the wide cold-frontal rainbands in the same case.

Thus the simple linear models discussed above can explain some, but not all, of the factors responsible for the banded structures observed in the extratropical cyclones. Further insight into the factors responsible for the banded structure of extratropical cyclones can be obtained from numerical prediction models which simulate explicitly clouds and precipitation processes.

Hsie *et al.* (1984), for example, used an explicit model for clouds and precipitation to study the diabatic effects of condensation and evaporation on an idealized model of a cold front. The prognostic equations for r_v, r_c, and r_r were added to the dry model of Keyser and Anthes (1982), which considered the effects of boundary layer processes on the adiabatic inviscid frontal model of Hoskins and Bretherton (1972).

Without moisture, frontogenesis occurs in the two-dimensional model as a result of geostrophic shearing deformation. A jet of upward motion occurs just ahead of the surface cold front (SCF) (Fig. 11.11), a result of deformational and frictional processes (Baldwin *et al.*, 1984). A northerly jet develops at the top of the PBL behind the SCF, and a low-level southerly jet develops in the warm sector ahead of the SCF.

The moist simulation contains some distinctive features not present in the dry simulation. The vertical velocity shows a banded structure in the warm sector (Fig. 11.12). The first band is associated with the frictional convergence around the pressure minimum at the SCF, and it is stronger than the updraft in the dry case (Fig. 11.11). The second and third bands are associated with moist convection; these bands form in a region of conditional symmetric instability. The horizontal wavelength of 200–300 km

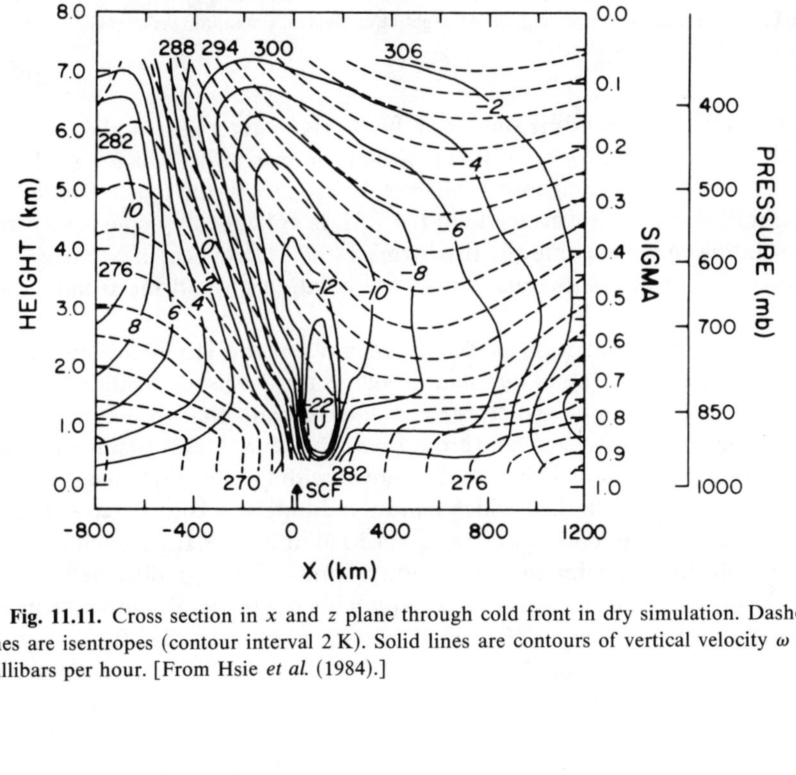

Fig. 11.11. Cross section in x and z plane through cold front in dry simulation. Dashed lines are isentropes (contour interval 2 K). Solid lines are contours of vertical velocity ω in millibars per hour. [From Hsie *et al.* (1984).]

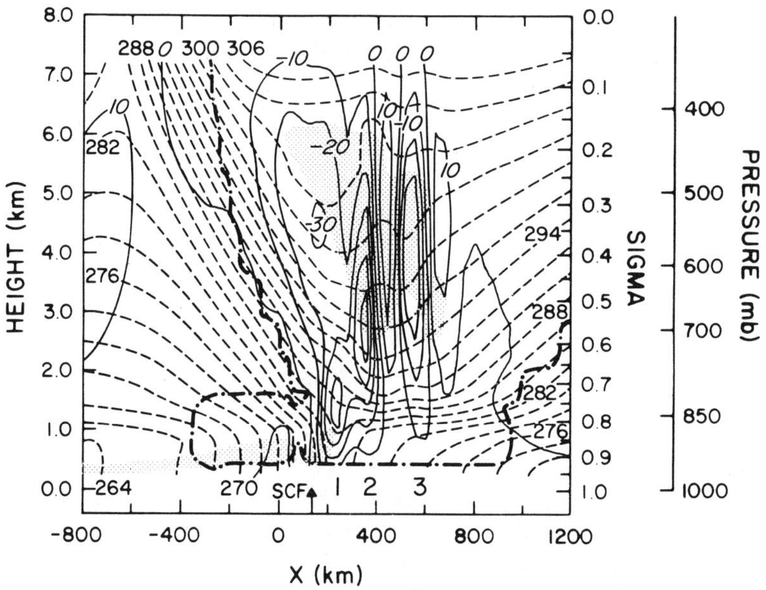

Fig. 11.12. As in Fig. 11.11, but for moist simulation. Cloud boundary is indicated by a thick dash-dot line; position of surface cold front (SCF) is indicated by vertical arrow at about 160 km. Shaded areas denote regions of negative equivalent potential vorticity. [From Hsie *et al.* (1984).]

of the convective bands is similar to the shorter wavelength mode predicted by Emanuel (1982) for conditional symmetric instability. Mesoscale downdrafts also exist. The ageostrophic flow is stronger in the moist simulation. The inflow from the warm side in the PBL is enhanced by convection and the outflow in the upper levels also increases. The latent heat of condensation produces a stronger, thermally direct circulation around the frontal zone in the moist simulation.

Initially, the formation of the cloud is from the large-scale upgliding component of the frontal circulation. Later, this large-scale motion breaks down into the banded structure. The narrow band that forms in the low levels close to the surface cold front (band 1 in Fig. 11.12) is similar to the "narrow cold-frontal rainband" that moves with the same speed as the SCF (Hobbs *et al.*, 1980). The wider band straddling the cold front at higher levels (centered at $x = 180$ km and $z = 4.5$ km in Fig. 11.12) is similar to the wide cold-frontal rainband. Two other bands resemble observed warm-sector bands. Both the warm-sector and wide cold-frontal rainbands move at a speed faster than the SCF. Hobbs *et al.* (1980) described the evolution of observed warm-sector bands (Fig. 11.10) as follows: "The warm-sector rainband displayed a different evolution and motion: sub-band (b) formed near the surface cold front, and sub-band (a) developed ahead of sub-band (b). Both sub-bands advanced relative to the front. A third, incipient sub-band subsequently formed and intensified ahead of sub-band (a)." The behavior of the bands in the moist simulation agrees qualitatively with the above description.

In addition to producing the rainbands and a stronger low-level jet, the effects of latent heat on the frontal circulations are summarized as follows:

(1) Latent heating produces a stronger horizontal potential temperature gradient across the front, especially in the middle and upper levels. The gradient in the low levels is relatively unaffected.

(2) Convection results in a stronger static stability across the front and a weaker static stability in the convective region.

(3) Both the upper- and low-level jets are stronger when latent heating is present. The horizontal wind shear (or relative vorticity) is stronger in the moist simulation, especially in the low levels.

(4) Convection intensifies the ageostrophic circulation around the frontal zone. The circulations in the warm sector appear to be dominated by convection.

(5) Moist convection increases the speed of the SCF and reduces the slope of the front.

In summary, for mesoscale models with high horizontal resolution, it is possible to resolve explicitly cloud and precipitation water by including prognostic equations for these variables in the model. Relatively simple representations of cloud microphysical processes account for the major effects of latent heating of condensation, evaporational cooling, and growth and fallout of precipitation. Such explicit models of clouds and precipitation, though still highly simplified, avoid many of the assumptions needed in the parameterizations for large-scale models.

11.3.4 Polar Lows

Besides the banded structures associated with extratropical cyclones, other mesoscale features include mesoscale vortices; one such mesoscale vortex is referred to as a *polar low.* A small region of organized convective clouds to the southeast of the major cyclone system and in the cold air behind the cold front is shown in Fig. 11.7. This system is an example of an incipient mesoscale cyclonic system that frequently occurs in polar airstreams behind or poleward of cold fronts in the north Pacific, North Sea, north Atlantic (particularly south of Iceland), and other locations where cold air flows over warm water. The systems are called polar lows. Although often only a trough of low pressure occurs at the surface, some polar lows develop closed cyclonic circulations, as shown by the satellite photograph of a polar low between Iceland and the Faroe Isles in Fig. 11.13 (Rasmussen, 1981). These small-scale vortices, which resemble tropical cyclones in their scale and cloud structure, are accompanied by high winds and convective precipitation (Fig. 11.14; Rasmussen, 1979).

Reed (1979) analyzed two polar low cases in detail, while Mullen (1979) analyzed a composite of 22 cases. The polar lows developed within deep baroclinic zones on the low-pressure side of well-developed jet streams in

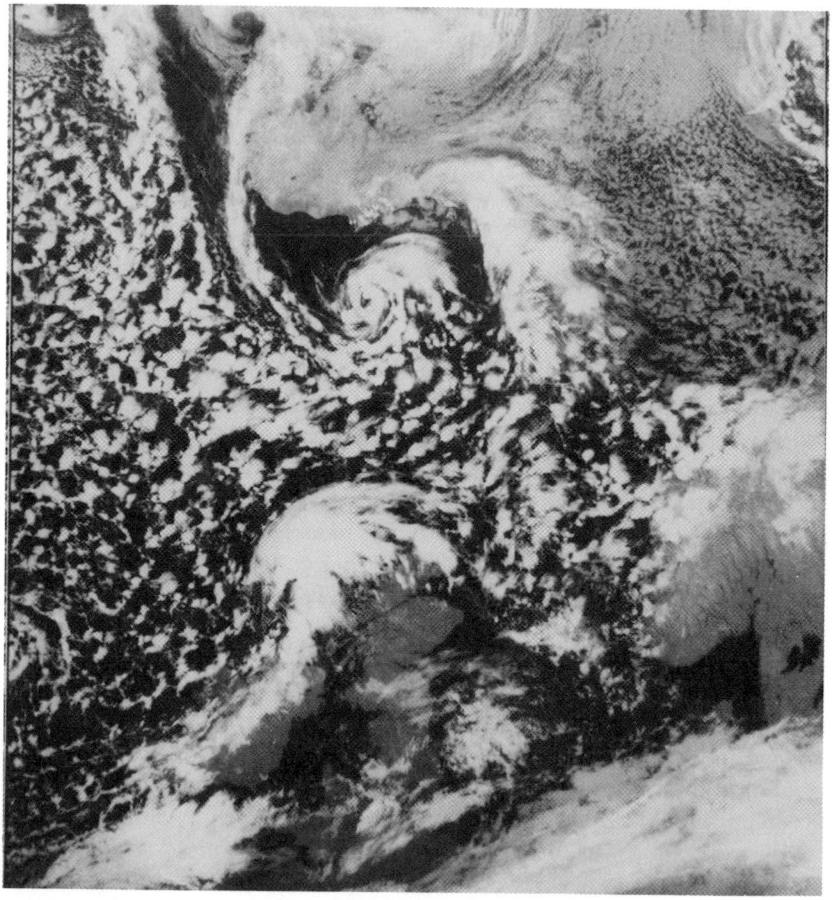

Fig. 11.13. NOAA-5 VHRR infrared satellite image for 1932 GMT, 24 November 1978, showing a polar low just south of Iceland. Faroe Isles are located just south of cyclone, which shows a cloud-free center. [From Rasmussen (1981).]

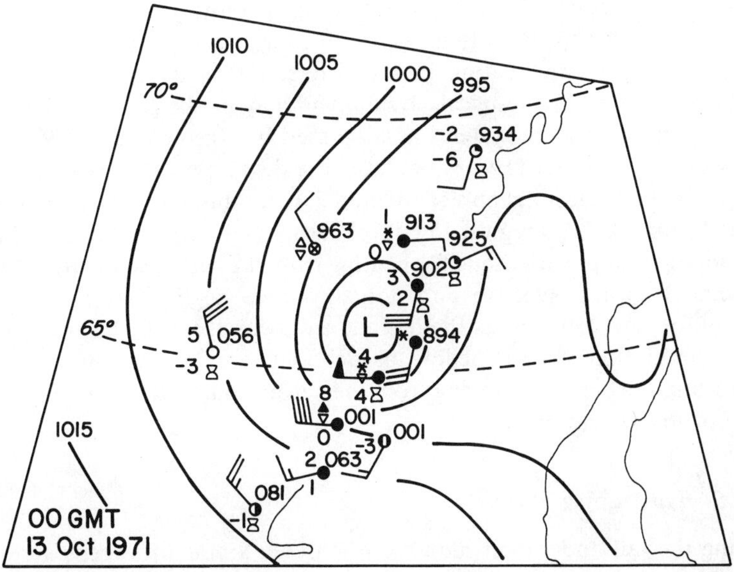

Fig. 11.14. A well-developed polar low off Norwegian coast at 0000 GMT, 13 October 1971. Surface data are plotted according to normal convention. Each full wind barb represents 10 kt. Maximum wind south of center is 50 kt. Temperature and dewpoint are in °C. Isobars are labeled in millibars. [From Rasmussen (1979).]

regions of strong cyclonic shear. The lower troposphere was conditionally unstable and strongly heated from below by the ocean in the early stages of development, which was 2–6 K warmer than the air. These conditions were also present in a case study by Rasmussen (1985).

Reed and Mullen conclude that the formation of polar lows is probably a result of baroclinic instability in the presence of low static stability in the lower troposphere as cold air flowing around the large-scale cyclone is heated by the warmer ocean surface. The small static stability explains the mesoscale size of the system (Staley and Gall, 1977).

Businger (1985) analyzed a composite of 42 cases of well-developed polar lows over the Norwegian and Barents seas and found large-scale conditions similar to those observed by Reed, Mullen, and Rasmussen. In particular, the lows developed in a strong baroclinic region of very low static stability under a region of positive vorticity advection. His study also showed an outbreak of deep convection at the time of rapid deepening, which suggested that latent heating may play a role in the deepening process.

The conditionally unstable air mass in which polar lows occur favors the development of convective clouds. Locatelli *et al.* (1982) analyzed the mesoscale structure of three polar lows and found convective rainbands and cells similar to those that occur with large-scale Pacific extratropical cyclones. The latent heating in the convection is likely to enhance the intensification of the mesoscale system and may be the primary mechanism for deepening of some systems as discussed by Rasmussen (1979, 1981).

Sardie and Warner (1985) simulated the development of two polar lows with a three-dimensional numerical model. In a simulation of a polar cyclone in the Denmark Strait region, low-level baroclinicity was sufficient to initiate development of the low. Sensible heating from the surface and latent heating associated with convective and nonconvective heating were essential to sustain the development as the low moved away from the baroclinic zone. In the simulation of a Pacific low, baroclinicity and latent heating were also important, but sensible heating from the surface had little effect on the time scale of the development.

11.3.5 Lake-Effect Storms

During the fall and winter months, when cold arctic air sweeps across the Great Lakes, local, often heavy, snowstorms occur along the lee shores. Most of the heavy snowfall does not occur during the passage of cold fronts, but several hours afterward. In some of the more severe storms, snowfall accumulations of more than 75 cm day^{-1} is not uncommon (Wiggin, 1950; Sykes, 1966) and snowfall rates of 30 cm h^{-1} have been reported (E.S.S.A., 1966). An important feature of these storms is their persistence for several days over limited and sharply defined regions. According to Davis *et al.* (1968), lake-effect snow may result from either loosely organized snow flurries, covering a broad area of the lee shores, or from highly-organized and nearly stationary cloud bands. The former type of lake-effect snowfall can occur quite frequently [as often as 1 out of 5 days from November through January along Lake Erie (McVehil and Peace, 1965)], while the latter type of disturbance requires a more unique set of atmospheric conditions occurring as infrequently as once per year.

Braham and Kelly (1982), Forbes and Merritt (1984), and Hjelmfelt (1988) have identified four general types of lake-effect storms over Lake Michigan. Shown schematically in Fig. 11.15 are (a) cloud streets or transverse bands, which occur during strong winds and strong static stability similar to those described in Chapter 7; (b) shoreline parallel bands with a well-developed land breeze on the lee shore (Braham, 1983), which occur with moderate wind speeds and weak stability; (c) midlake bands with low-level convergence centered over the lake (Passarelli and Braham, 1981), which occur with weak synoptic-scale pressure gradients such as the center of an anticyclone; and (d) mesoscale vortices with a well-developed cyclonic

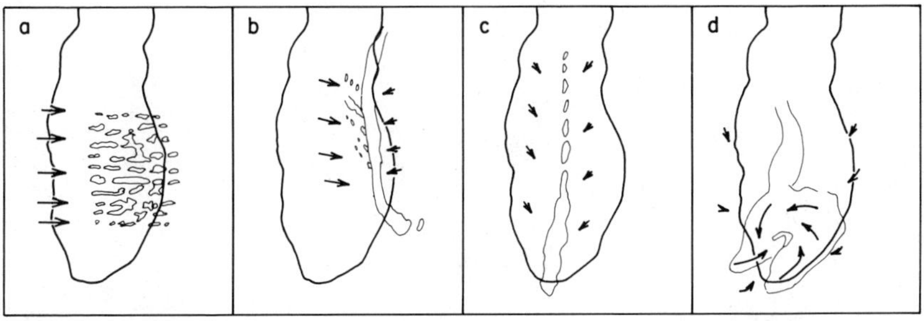

Fig. 11.15. Schematic of four morphological types of lake-effect snow storms over Lake Michigan as defined from observations: (a) broad area coverage, (b) shoreline band, (c) midlake band, and (d) mesoscale vortex. [From Hjelmfelt (1988).]

flow in the boundary layer, usually occurring with weak surface-pressure gradients and a ridge of high pressure centered over the lake or to the west of the region (Peace and Sykes, 1966; Forbes and Merritt, 1984; Pease *et al.*, 1988). The lake vortices appear to be a weak version of the polar lows described above.

The most intense snowstorms are the shoreline parallel bands which, according to Sheridan (1941), Wiggin (1950), and Rothrock (1969), occur following the passage of a cold front when the following conditions are met:

(1) The difference between the lake surface temperature and the 850-mbar temperature exceeds 13°C.

(2) An onshore wind with an overwater fetch of more than 100 km is present.

(3) Over water, low-level wind speeds are moderate to strong.

(4) The height of the boundary layer capping inversion exceeds 1000 m.

A conceptual model of a major snowstorm over Lake Erie developed by Davis *et al.* (1968) is shown in Fig. 11.16. Illustrated is a stratocumulus cloud that deepens over the lake as it nears the lee shore. Corresponding to the deepening cloud layer is an upward displacement of the capping inversion. The faster falling, heavily rimed snow crystals such as graupel particles precipitate just onshore. This is also the location of the heaviest snowfall amounts. Lightly rimed dendrites and plates are carried further inland. Aggregates of snow crystals having terminal velocities between graupel and unrimed single crystals settle somewhere in between. Table 11.2 (Juisto, 1967) summarizes both the microphysical and mesoscale features of lake-effect snowstorms over Lake Erie.

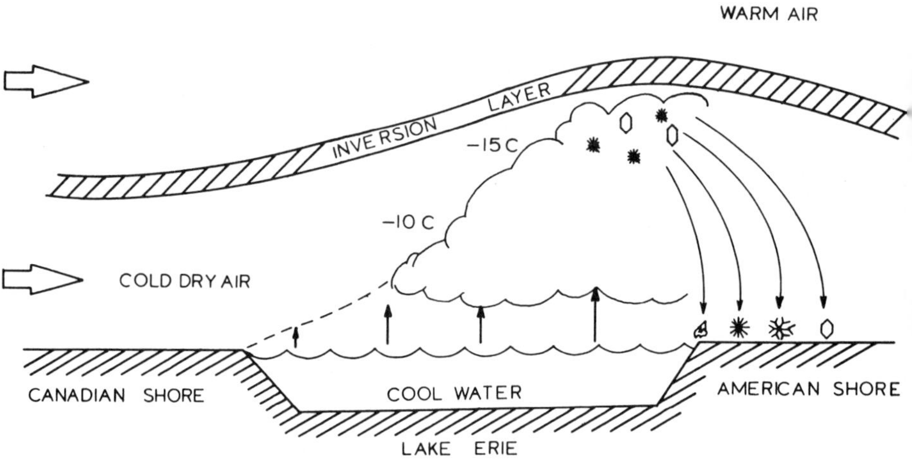

Fig. 11.16. Schematic cross section of lake-effect snow. [From Davis *et al.* (1968).]

A number of numerical models have been applied to the simulation of lake-effect storms. Using a simple mixed-layer model, Lavoie (1972) demonstrated the importance of wind speed and direction to the strength of the mesoscale disturbance, with the longest fetch of air across the lake surface producing the strongest disturbance. Both Lavoie (1972) and Ellenton and Danard (1979) showed that the heat and moisture fluxes from the lake were the principal causes of precipitation, while shoreline convergence and orographic lift played secondary roles. Likewise, Hsu (1987) showed that the coupling between surface heat fluxes and winds were responsible for the formation of patterns of low-level convergence that generated precipitation. Hjelmfelt and Braham (1983) simulated the lee-shore parallel band over Lake Michigan (see Fig. 11.15c), including the formation of a surface mesolow beneath the major cloud band. Pease *et al.* (1988) and Hjelmfelt (1988) showed that strong surface heating and weak winds from the north were conducive to the formation of a mesovortex at the southern end of Lake Michigan.

In summary, three major streams of air are involved in the three-dimensional circulation of extratropical cyclones. Because of their different origins and the variety of dynamical and physical processes that they experience in their passage through the cyclone system, the characteristics of the environment vary greatly with time and space. These environments, with different temperatures, moisture content, vertical motion, static stability, and wind shear, host the enormous variety of mesoscale phenomena and clouds present in extratropical cyclones.

Table 11.2

Characteristics of Lake-Effect Snow Bands[a]

Characteristics	Single	Multiple
A. Mesoscale features		
1. Precipitation	Heavy snowfall (>10 cm)	Modest snowfall (<10 cm)
2. Location/orientation	Over lake or along south shore	NW–SE bands, south shore
3. Winds aloft	SW to W/15–20 $m\ s^{-1}$	NW
4. Air temperature—low level	−15 to −10°C	−15 to −5°C
5. Cells		
a. Movement	With prevailing wind	With prevailing wind
b. Echo persistence	Long (>1 h)	Short (<1 h)
c. Entire band "movement"	At angle to prevailing wind	—
6. Convergence pattern	Sharp convergence line beneath storm	—
7. Band dimensions (horizontal)	9–18 km wide; 90–180 km long	1.8–5.5 km wide; 36–90 km long
B. Cloud structure		
1. Vertical extent	2–2.5 km	1 km
2. Bases/tops	0.5 km/3 km	1 km/2 km
3. Temperature	−15 to −30°C	0.45- to 1.8-km-wide cloud streets over water fusing into continuous sheet inland
C. Cloud microphysics[b]		
1. Updrafts—average	<2 $m\ s^{-1}$	<1 $m\ s^{-1}$
2. LWC	0.5–2 $g\ m^{-3}$	0.2–1 $g\ m^{-3}$
3. Droplets		
a. Size (average radius)	5 μm	—
b. Concentration	300–500 cm^{-3}	—
4. Natural ice nucleus concentration	1–10 $liter^{-1}$	1–10 $liter^{-1}$
5. Ice-to-water ratio (initial)	10^{-4} to 10^{-6}	—
6. Ice-crystal size /concentration	1–4 mm/1–10 $liter^{-1}$	—

[a] From Juisto (1967).

[b] Tentative data obtained from sparse aircraft and ground measurements, or deduced from calculations where appropriate.

11.4 Middle- and High-Level Clouds

11.4.1 Introduction

On an annual average, clouds cover between 55 and 60% of the earth (Matveev, 1984), and much of this cloud cover consists of vast sheets of middle (altostratus and altocumulus) and high (cirrus, cirrostratus, and cirrocumulus) clouds (Fig. 11.17). In association with organized tropical and extratropical cyclonic storm systems, these clouds produce significant precipitation over many regions. In addition, they play a vital role in the planet's energy budget through the release of latent heats of vaporization and sublimation and through their profound effect on shortwave and long-wave radiation.

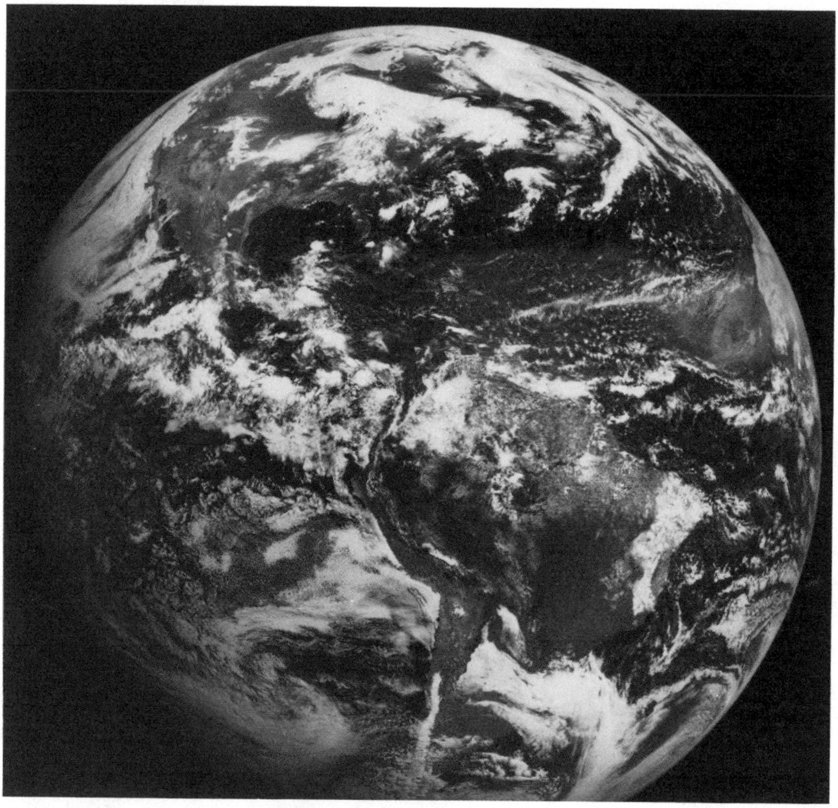

Fig. 11.17. GOES visible satellite photograph at 1500 GMT, 18 January 1977.

The physical properties of middle- and high-level clouds and their association with extratropical weather systems are discussed in this section, as are mathematical models of these clouds in which the physics and dynamics are modeled explicitly. While these high-resolution models are useful theoretical tools in understanding the formation and dissipation of clouds, they are inappropriate in large-scale forecast or climate models. In these models, it is necessary to account for the effects of middle and high clouds, as well as cumulus convection, on large-scale weather and climate systems through simplified, parameterized models. Several such parameterizations are discussed.

Middle-level clouds refer to altocumulus and altostratus clouds, composed entirely of liquid water, regardless of their elevation. While the elevation of middle-level clouds may vary considerably with season and latitude, a typical elevation in middle latitudes is 3 km, or about 700 mbar. High-level clouds refer to cirrus (including cirrostratus and cirrocumulus) clouds which are composed mostly of ice. Again, the elevation of cirrus clouds may vary considerably; a typical height is 10 km, or 250 mbar.

Middle- and high-level clouds frequently occur in layers or sheets of great horizontal dimensions (hundreds or even thousands of kilometers). Because their horizontal dimensions are much greater than their vertical extent, which is typically a kilometer, they are called stratiform or layered clouds. Although this terminology suggests a statically stable lapse rate, thin layers of conditional or convective instability often are present, and these instabilities are conducive to the formation of small-scale convection embedded in the cloud layer.

11.4.2 Descriptive Aspects of Middle- and High-Level Clouds

Middle- and high-level clouds form when the lifting of moist air, often on large scales in association with cyclonic storms, adiabatically cools the air to its dew-point temperature. Even at temperatures well below freezing, the clouds almost always form first as water droplets, with the initial condensation occurring on the usually abundant *cloud condensation nuclei.* Direct transformation from water vapor to ice crystals (sublimation) is possible, but rarely occurs because of the scarcity of *sublimation nuclei* in the atmosphere.

The evolution of the cloud after the formation of the first cloud drops depends on two factors—the temperature of the cloud and the vertical motion. At temperatures below freezing, ice processes are likely to be important because of the relative abundance of *freezing nuclei* in the atmosphere. These nuclei trigger the freezing of supercooled water to ice. Once ice appears, the difference in saturation vapor pressure over ice and water

(Chapter 4) favors the rapid growth of ice crystals at the expense of the water drops (the Bergeron–Findeisen process; see Section 4.3).

If the upward vertical velocities are sufficiently strong ($10\ \mathrm{cm\ s^{-1}}$ or greater) and persist for enough time to cause further cooling of the layer, the water drops or ice crystals will grow to sizes wherein fall velocities relative to the air motion become significant ($10\ \mathrm{cm\ s^{-1}}$ or greater) and precipitation will occur. The physical processes that produce this growth are collision and coalescence in all water clouds, the Bergeron–Findeisen process, and the riming of supercooled cloud droplets and aggregation among ice crystals in clouds containing both ice and water.

For weaker upward velocities (e.g., $1\text{–}2\ \mathrm{cm\ s^{-1}}$), the temperature change resulting from adiabatic expansion is reduced and the temperature changes

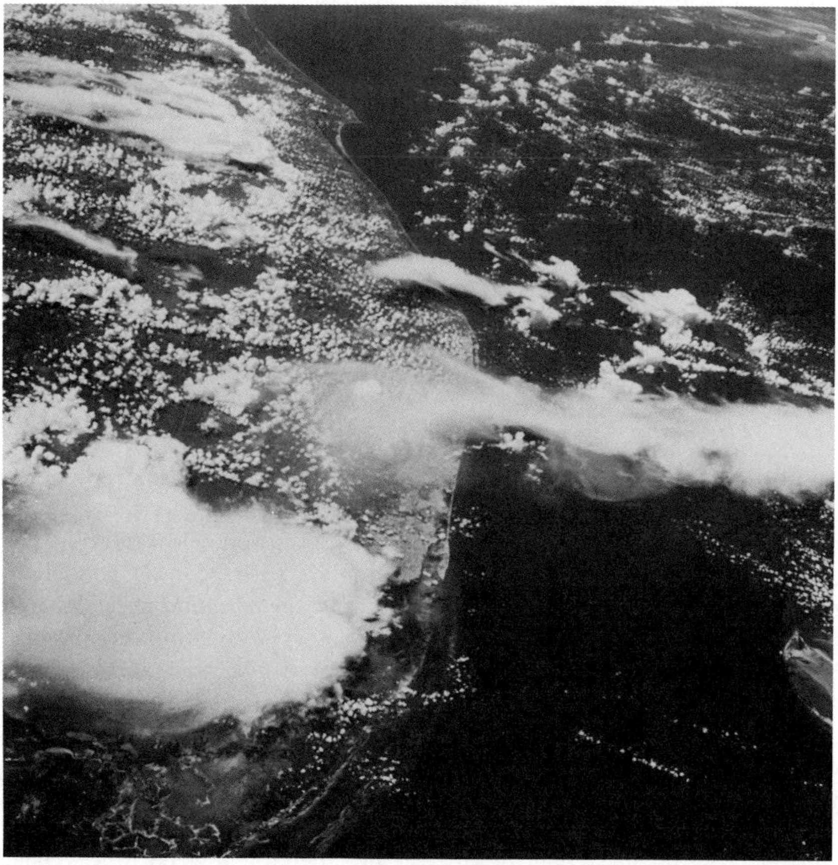

Fig. 11.18. Satellite photograph of cumulus convection over Florida peninsula and adjacent waters. A thunderstorm in south Florida is producing a massive shield of cirrus clouds.

from radiative effects become significant. Cooling at cloud top from the divergence of the vertical flux of longwave radiation and warming at cloud base as a result of the convergence of the vertical flux of longwave radiation destabilize the cloud layer and lead to internal convective circulations in the cloud layer. This destabilization is aided by the release of latent heat near cloud base and evaporative cooling near cloud top. During the day, absorption of solar radiation contributes to a heating of the cloud.

The type of middle or high cloud is strongly affected by the static stability. Altostratus clouds are produced by the lifting of a layer of air in which the lapse rate is less than the wet adiabatic lapse rate. For layers in which the lapse rate exceeds γ_w, vertical convection results in altocumulus clouds. For high clouds, absolutely stable layers are associated with cirrostratus clouds, while cirrus and cirrocumulus clouds occur when the lapse rate exceeds γ_w. Alternating regions of upward- and downward-moving air associated with gravity waves in a stable cloud layer can result in a banded structure in the cloud layer.

In addition to the mechanism of cloud formation associated with the large-scale lifting of air, several other mechanisms produce middle- and high-level clouds. Cumulus convection produces significant cirrus clouds when their tops encounter a stable layer (frequently the tropopause). Cirrus anvils may spread horizontally outward and cover areas many times the size of the convective updraft (Fig. 11.18). The anvils from neighboring

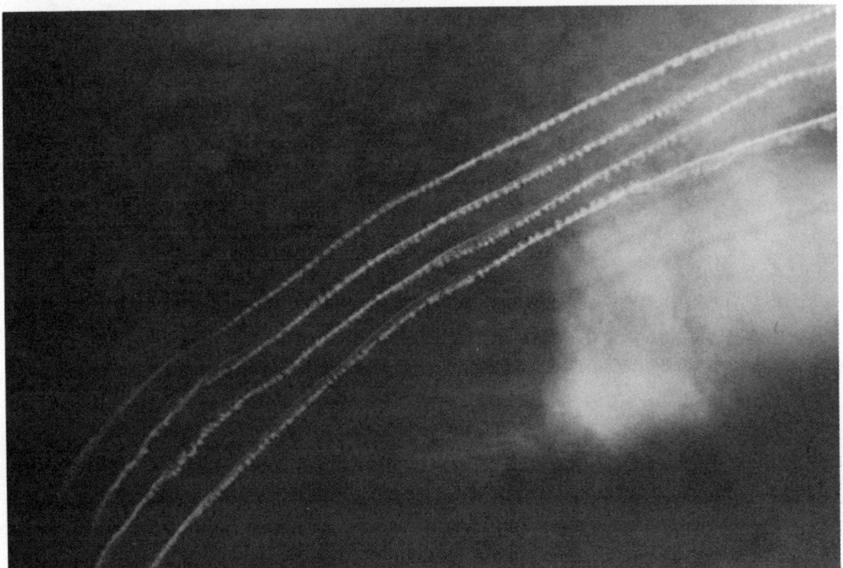

Fig. 11.19. Photograph of condensation trails. (Photo by R. Anthes.)

cumulonimbus clouds can merge into an extensive stratiform layer in the middle and upper troposphere in many MCSs (Chapter 10). Air flowing over mountains may generate orographic clouds of water or ice extending to great heights in the troposphere. Cirrus clouds sometimes form in the vicinity of the jet stream in association with small-scale vertical circulations that develop around the jet. Condensation trails (contrails) from jet aircraft can produce cirrus clouds (Fig. 11.19). Unlike the mechanism of adiabatic cooling associated with the other processes that produce middle and high clouds, contrails are produced through the mixing of warm, moist exhaust air with colder, drier environmental air. This mechanism, made possible by the exponential increase of saturation vapor pressure with temperature, is the same mechanism that produces steam fogs when cold, dry air flows over warmer water (Section 7.1 and Fig. 7.1). In corridors of high aircraft density, individual contrails can merge to form a thin blanket of cirrus-type clouds.

11.4.3 Properties of the Environment of Middle and High Clouds

Starr and Cox (1980) examined temperature, moisture, and wind soundings associated with more than 3600 cloud cases to determine the characteristics of the environment of middle and high clouds. Knowledge of these characteristics is useful to develop methods of parameterizing clouds in large-scale models as a function of the large-scale thermodynamic and dynamic variables predicted by the model. The parameters examined by Starr and Cox included static stability, defined by

$$\sigma \equiv \partial\theta/\partial z, \tag{11.15}$$

vertical wind shear,

$$S \equiv |\partial \mathbf{V}/\partial z|, \tag{11.16}$$

and Richardson number,

$$R_{\mathrm{i}} = (g/\theta)(\sigma/S^2), \tag{11.17}$$

in addition to the mean temperature and relative humidity in the cloud layer.

A common property of most cloud observations was that the lapse rate was rarely wet adiabatic. Other properties of the environment showed considerable variation from case to case, indicating the difficulty of parameterizing the cloud effects in large-scale models. There were statistical differences between the characteristics of thick cloud layers (defined as extending through a depth of greater than 50 mbar) and thin clouds (50-mbar thickness or less). In general, thick clouds were associated with frontal circulations and cyclonic storms and thin clouds were not. The general characteristics of thick and thin clouds are summarized in Table 11.3.

Table 11.3

Characteristics of Thick and Thin Middle- and High-Level Clouds as Deduced from Radiosonde Ascents[a]

A. Thick clouds (depth greater than 50 mbar)	
Static stability:	Decreases with height from subcloud layer to above cloud layer, with a typical range of 5 K km^{-1} in the subcloud layer to 3 K km^{-1} near the cloud top
Vertical wind shear:	Usually positive (increasing wind speed with height); magnitude ranging from about 4 m s^{-1} km^{-1} in summer to 6.5 m s^{-1} km^{-1} in winter
Richardson number:	Average values of R_i throughout cloud layer 12–18; very small percentage (about 15%) of soundings showed R_i less than 1.0
B. Thin clouds (depth less than or equal to 50 mbar)	
Static stability:	Quite variable, but on average greatest stability (5.5 K km^{-1}) above cloud layer and least stability (3.5 K km^{-1}) above cloud layer and least stability (3.5 K km^{-1}) in and below cloud layer; stable layer on top may correspond to tropopause
Wind shear:	No significant difference from thick clouds
Richardson number:	Average values slightly greater (18–24) than for thick clouds

[a] Adapted from Starr and Cox (1980).

Two conclusions may be drawn from the summary of average cloud conditions in Table 11.3. First, the layers are generally statically stable in an absolute sense, with lapse rates less than the saturation-adiabatic rates with respect to either water or ice. The fact that thick clouds are generally above stable layers indicates that they are associated with upper-level fronts. In contrast, thin clouds are usually below stable layers, which indicates that they are located below the tropopause.

A second conclusion from Table 11.3 is that shear-induced turbulence (Kelvin–Helmholtz instability) is not likely in the vicinity of most middle- and high-level cloud systems. A necessary condition for Kelvin–Helmholtz instability is that the Richardson number be less than 1/4; as shown in Table 11.3, the average values of R_i are much greater than this value. The coarse vertical resolution used to evaluate R_i in this study may bias the estimates of R_i toward higher values; it is possible that considerably smaller values of R_i exist locally in thin layers where strong destabilization associated with radiative effects is important. An exception to this conclusion occurs in shallow layers in the vicinity of jet streams, where mesoscale vortices produce turbulence and extensive bands of cirrus (Section 11.4.5).

11.4.4 Microphysical Properties of Middle- and High-Level Clouds

Aircraft observations of middle- and high-level clouds have revealed much about their structure and microphysical characteristics.

As summarized in Table 11.4, cirrus clouds contain ice crystals of typical length 0.5 mm and in concentrations that vary widely from 10^4 to 10^6 m^{-3}. The terminal velocity of the ice crystals is about 0.5 m s^{-1}, which is of the same order as the updraft speed. The water content of cirrus clouds is typically 0.2 g m^{-3}. Observations have shown that ice-water content is

Table 11.4

Microphysical Characteristics of Middle and High Cirrus Clouds

Property or variable	Value	Reference
Concentration of ice crystals (m^{-3})	10^5–10^6	Braham and Spyers-Duran (1967)
	2×10^5–5×10^5	Heymsfield (1975a)
	1.0×10^4–2.5×10^4	Heymsfield and Knollenberg (1972)
	6.0×10^5–38.0×10^5	Ryan *et al.* (1972)
	2.0×10^4–8.0×10^4	Houze *et al.* (1981)
	1.0×10^4–5.5×10^5	Churchill and Houze (1984)
Length of crystals	Up to 0.17 mm	Braham and Spyers-Duran (1967)
	0.6–1.0 mm	Heymsfield and Knollenberg (1972)
	0.35–0.9 mm	Heymsfield (1975a)
Terminal velocity of crystals	Typically 50 cm s^{-1}; max 120 cm s^{-1}	Heymsfield (1975a)
Ice-water content	0.15–0.25 g m^{-3}	Heymsfield and Knollenberg (1972)
	0.15–0.30 g m^{-3}	Heymsfield (1975a)
	0.10–0.50 g m^{-3}	Rosinski *et al.* (1970)
Precipitation rate	0.5–0.7 mm h^{-1}	Heymsfield and Knollenberg (1972)
	0.1–1.0 mm h^{-1}	Heymsfield (1975a)
Updraft velocity	1.0–1.5 m s^{-1} (in cirrus uncinus)	Heymsfield (1975a)
	2.0–10.0 cm s^{-1} (warm front overrunning)	Heymsfield (1977)
	20 cm s^{-1} (warm front occlusion)	Heymsfield (1977)
	25–50 cm s^{-1} (closed low aloft)	Heymsfield (1977)
Lifetime of individual cloud	15–25 min	Heymsfield (1975a)

strongly dependent on the vertical velocity and the temperature. Figure 11.20 shows plots of observed ice-water contents (IWC) versus temperature for vertical velocities ranging from 1 to 50 cm s^{-1}. The observations were made under various synoptic situations, including warm-frontal overrunning, warm-frontal occlusion, closed low aloft, and jet stream cloudiness (Heymsfield, 1977). The data shown in Fig. 11.20 indicate, for a given updraft speed (W), a nearly exponential increase of water content with increasing temperature, from low values of less than 10^{-3} g m^{-3} at temperatures below −50°C to maximum values around 0.3 g m^{-3} at temperatures in the range 0 to −10°C. At a given temperature (e.g., −30°C), the water content increases from 10^{-2} g m^{-3} for an updraft of 0.01 m s^{-1} to 2.0×10^{-1} g m^{-3} for an updraft speed of 0.5 m s^{-1}. The empirical equation fitting the data in Fig. 11.20 is

$$\mathrm{IWC} = 0.072\, W^{0.78} \exp[-0.01\, W^{0.186}(-T)^{1.59\, W^{-0.04}}]. \qquad (11.18)$$

The total ice-crystal concentration also shows a strong dependence on temperature and vertical velocity. As shown in Fig. 11.21, the concentration varies from less than 10^{-1} liter^{-1} (100 m^{-3}) for weak (0.01 m s^{-1}) updrafts and temperatures below −50°C to 100 liter^{-1} (10^5 m^{-3}) for temperatures around −10°C and updraft speeds of 0.5 m s^{-1}.

Observations also indicate a relationship between crystal size and ice-water content. Figure 11.22 is a plot of observed mean and maximum crystal lengths as a function of ice-water content. The maximum lengths, in particular, show an increase with increasing IWC, from about 0.5 mm at 10^{-4} g m^{-3} to 5 mm at 1 g m^{-3}.

A property of precipitating cloud systems having practical consequences is the close relationship between precipitation rate R and ice-water content. Figure 11.23 shows this strong positive correlation, which is well represented by

$$R = 3.6(\mathrm{IWC})^{1.17}, \qquad (11.19a)$$

where IWC is expressed in g m^{-3} and R is in mm h^{-1}. A very similar expression was derived independently by Sekhon and Srivastava (1970),

$$R = 5.0(\mathrm{IWC})^{1.16}. \qquad (11.19b)$$

This empirical relationship is useful because radar reflectivity Z is a measure of IWC (Fig. 11.24),

$$Z = 750(\mathrm{IWC})^{1.98}, \qquad (11.20)$$

where Z is expressed in units of mm^6 m^{-3}. Equations (11.19) and (11.20), or other similar empirical expressions, can be used with radar measurements to estimate precipitation rates from cirrus clouds.

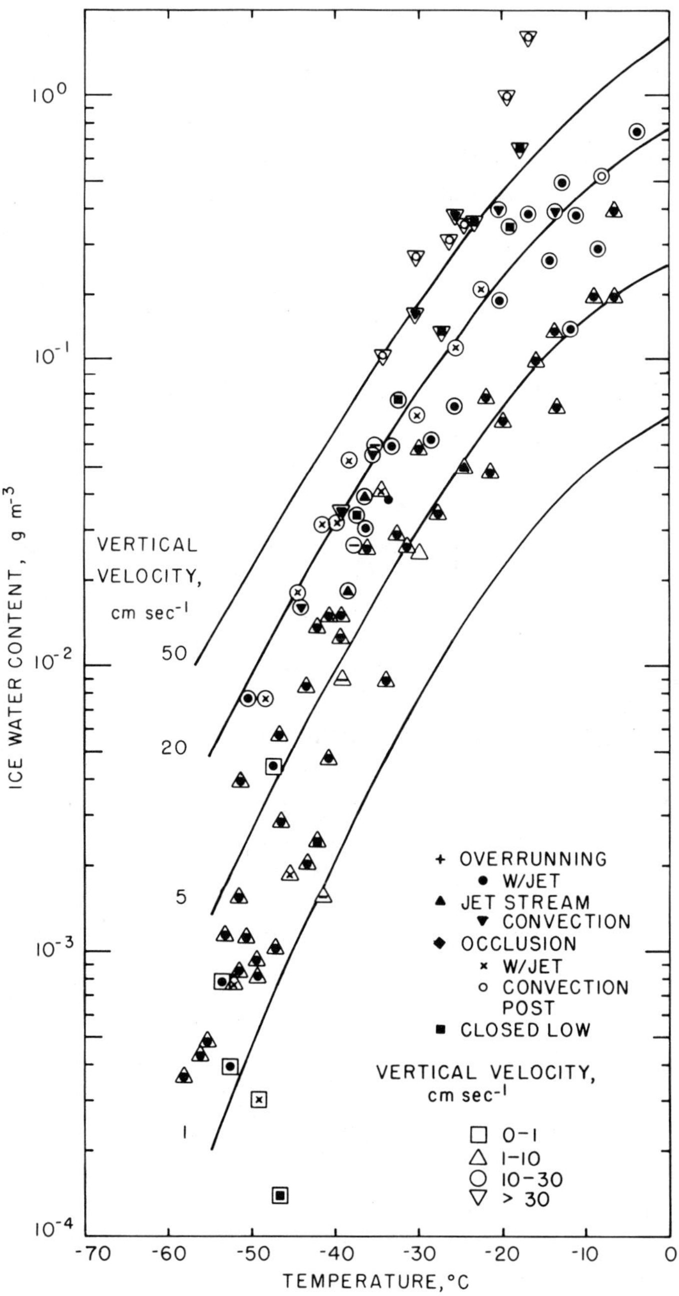

Fig. 11.20. Ice content (g m^{-3}) plotted against temperature and parameterized in terms of vertical air velocity: small, solid inner symbols, synoptic type; larger, open outer symbols, vertical velocity range. [From Heymsfield (1977).]

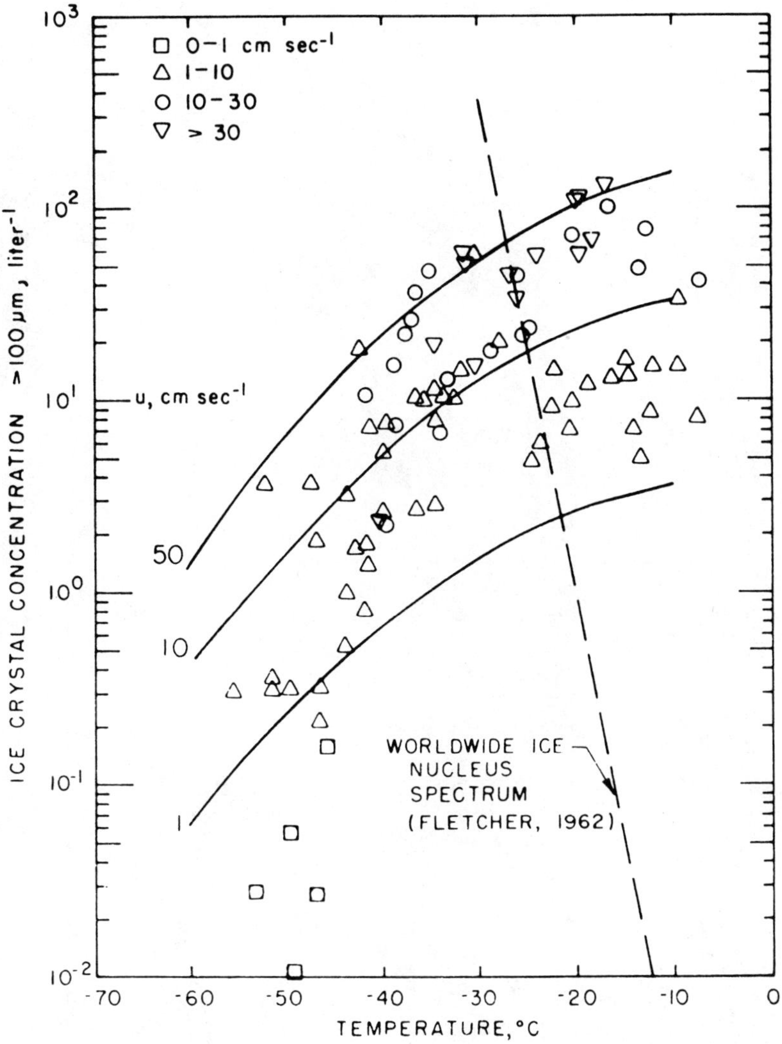

Fig. 11.21. Total ice-crystal concentration (>100 μm) plotted against temperature and parameterized in terms of air velocity. Worldwide ice-nucleus spectrum is plotted. [From Heymsfield (1977).]

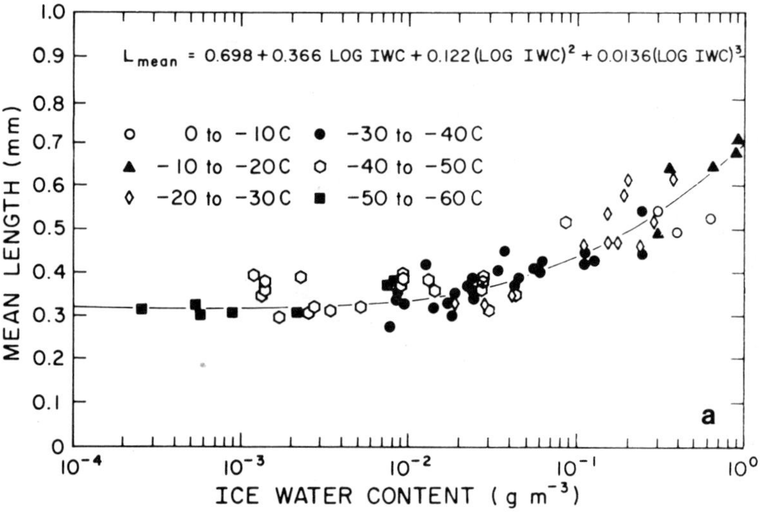

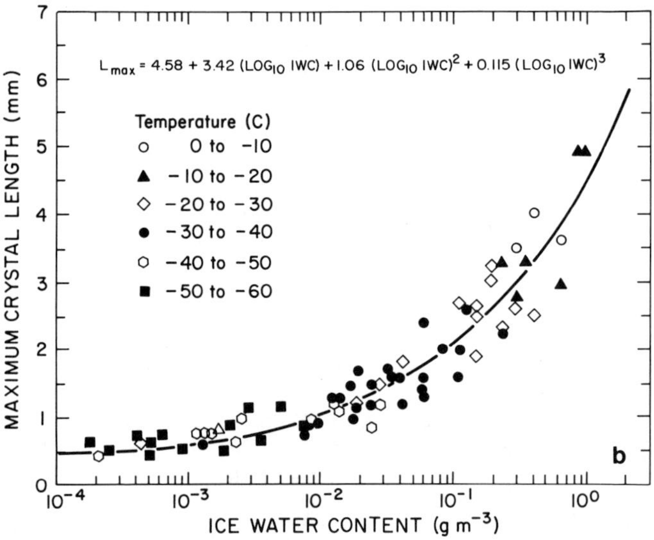

Fig. 11.22. Mean (a) and maximum (b) crystal lengths plotted against ice-water content. Maximum crystal length corresponds to particles in concentrations of 1 m^{-3} per millimeter size class. Sampling temperature is indicated. [From Heymsfield (1977).]

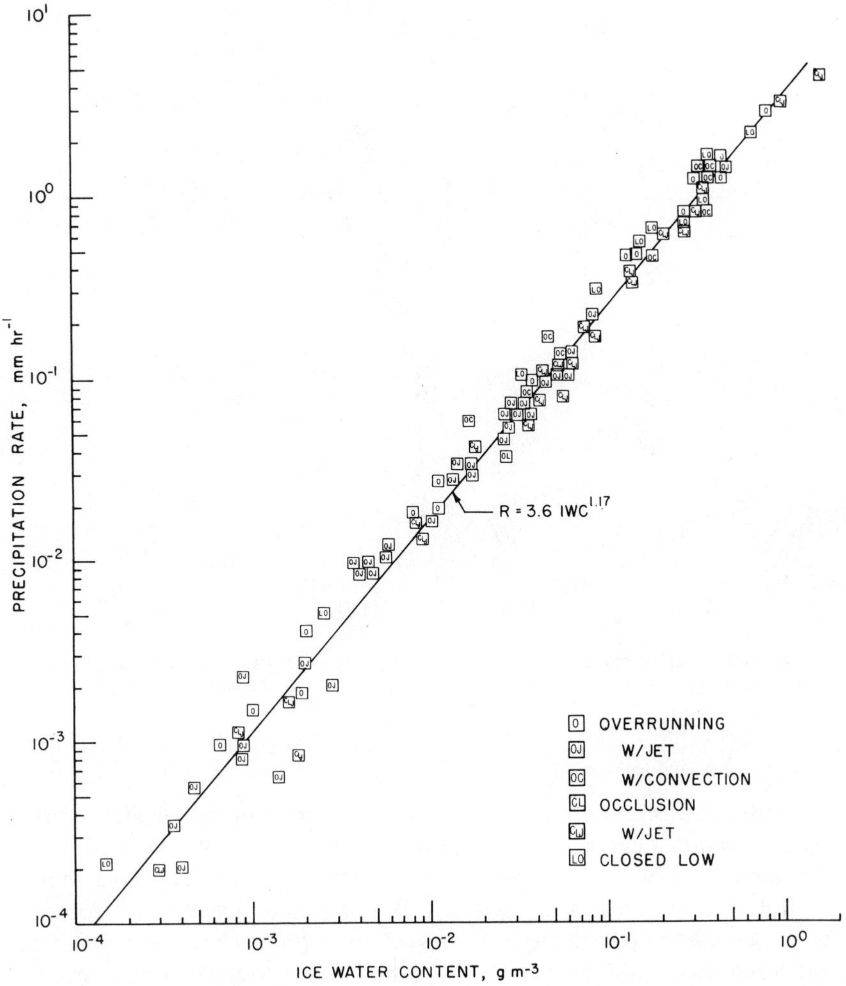

Fig. 11.23. Calculated precipitation rate plotted against ice content in synoptic systems sampled. Best-fit line through data points is indicated. [From Heymsfield (1977).]

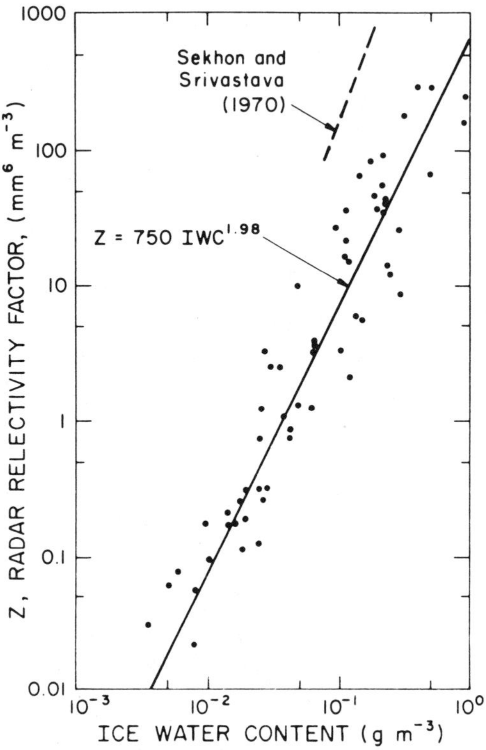

Fig. 11.24. Calculated radar reflectivity factor plotted against ice content. Best-fit line and Sekhon and Srivastava (1970) curve for aggregate snow are shown for comparison. [From Heymsfield (1977).]

Cirrus clouds assume a variety of forms, depending on the mean vertical velocity, wind shear, relative humidity, and static stability. Dense layers of cirrostratus occur under conditions of gentle, uniform upward motion, saturated air, and high static stability. With less stability and in the presence of weaker mean upward motion, convection may form cirrus uncinus, dense patches of cirrus which produce ice crystals large enough to acquire appreciable terminal velocities. These falling crystals, in the presence of wind shear, may form trails of considerable length (Fig. 11.25). Because of the high-saturation relative humidity with respect to ice compared to that of water, the crystals may survive falls of more than 5 km through clear air (Braham and Spyers-Duran, 1967; Hall and Pruppacher, 1976). Sometimes these ice crystals "seed" middle-level supercooled water clouds, and the associated release of latent heat of fusion stimulates the growth of the clouds at the

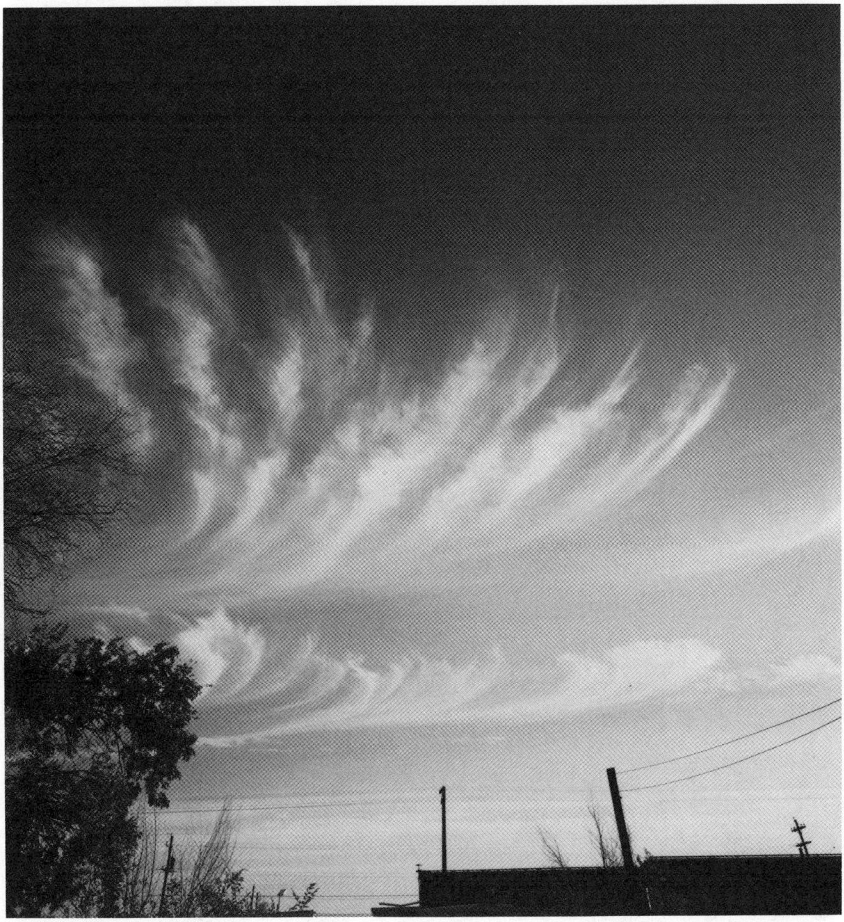

Fig. 11.25. Photograph of cirrus uncinus and fall streaks. (NCAR photograph.)

lower level. Braham (1967) observed the development of a "major precipitation system just at the point where ice-crystal trails from cirrus uncinus dipped into a region of disordered supercooled middle-level cloudiness."

Heymsfield (1975a, b) studied with aircraft observations the dynamics and microphysics of cirrus uncinus clouds. A conceptual model for cases of positive wind shear (west wind increasing with height) throughout the cloud system is presented in Fig. 11.26a. According to this model, clouds are initiated in an updraft, which develops in a layer with nearly dry adiabatic lapse rate. The updraft velocity in this convective head is 1.0–1.5 m s^{-1},

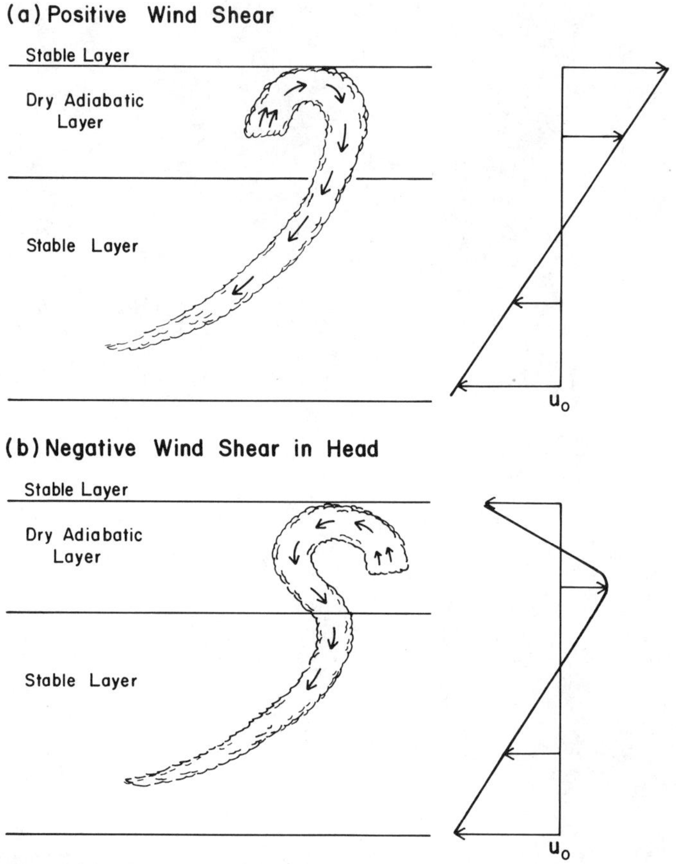

Fig. 11.26. Schematic west–east vertical cross sections illustrating development of cirrus uncinus with fall streaks in two types of wind shear. (a) Positive wind shear (west wind increasing with height) and (b) negative wind shear in generating region (head) of cloud.

considerably larger than average updraft velocities in cirrus clouds. As the ice crystals rise, they grow and move downshear relative to their point of origin. When their size and terminal velocity increase and/or they move out of the convective updraft, they fall and form a trail extending upshear of the generating point. A downdraft of magnitude 50 cm s^{-1} has been observed in the trail portion of the head (Heymsfield, 1975a).

Cirrus uncinus clouds may form in layers with a variety of wind shears. When the head forms in a region of negative wind shear, the head curves the opposite way (Fig. 11.26b). The particles grow and are carried downshear from the generating region until they become large enough to fall out of the updraft. In a relative sense, they then move back toward their point of origin until they reach the region of positive shear, when they again move upshear away from the point of origin. The resulting cloud resembles a reversed question mark. An almost infinite variety of trails can be produced depending on the wind shear.

11.4.5 Models of Middle- and High-Level Clouds

Numerical models of middle- and high-level clouds are useful to determine the quantitative roles of dynamics, microphysics, and radiation in the life cycle of these clouds. They can also be useful in studying the potential for modifying clouds and precipitation through cloud seeding. Because most of the physical processes in middle- and high-level clouds require only one horizontal dimension, they can be studied by two-dimensional (x and z) numerical models. In this section, we summarize the results obtained with the cirrus model developed by Starr and Cox (1985a). Although their model was developed primarily for the study of cirrus clouds, with minor modifications it may also be used to model altocumulus and altostratus clouds. The model is two-dimensional (x and z), time dependent, and Eulerian with a 100-m resolution in the horizontal. The physical parameterizations include explicit predictions of phase changes of water, radiation, fallout of precipitation, and large-scale vertical motion. Details of the model equations and parameterizations are given in Starr and Cox (1985a).

11.4.5.a Cirrus Simulation

The initial sounding for the control simulation of cirrus clouds by Starr and Cox (1985a) (Fig. 11.27) is characterized by a weakly stable lapse rate $\gamma = \gamma_m - 1.5°\mathrm{C\ km^{-1}}$ from 5.5 km (the lower boundary of the model) to 7.0 km, where γ_m is the saturation adiabatic lapse rate. From 7.0 to 7.5 km, the lapse rate is equal to γ_m and the relative humidity with respect to ice is 115%. Above 7.5 km, the lapse rate is again stable ($\gamma = \gamma_m - 4.0°\mathrm{C\ km^{-1}}$). A mean upward velocity of 2.0 cm $\mathrm{s^{-1}}$ is prescribed; this value is typical of synoptic-scale ascent in cirrus clouds.

The development of cirrus clouds occurs when initial perturbations are added to the potential temperature field. Buoyancy thus generated in the supersaturated layer immediately initiates sublimation of vapor to ice, and the release of latent heat of fusion provides energy for the perturbation to grow. Because of the large amount of available latent energy, the first 10–20 min are characterized by a rapid growth of cloud water and perturbation kinetic energy. Following this period of rapid consumption of available energy, the cloud ice water declines and reaches a quasisteady state after

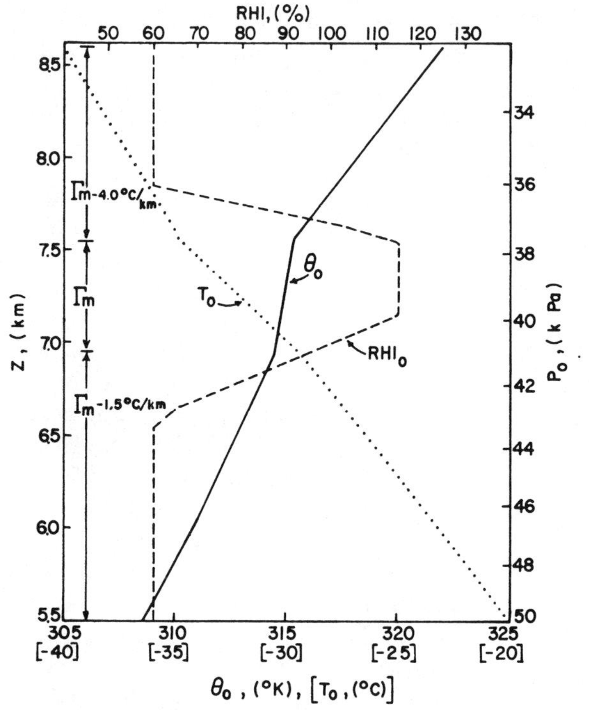

Fig. 11.27. Initially specified basic-state thermodynamic and moisture structure for a cirrus cloud simulation. [From Starr and Cox (1985a).]

about 40 min. Figure 11.28 shows vertical profiles of ice water at 10, 20, 30, 40, and 60 min. In the steady state, a near balance exists between the formation and destruction of ice water. The average value of liquid water of about 5 mg m^{-3} and the maximum value of 0.01 g m^{-3} at 60 min agree well with observations.

The horizontally and temporally averaged potential temperature tendency resulting from phase changes of water and radiative effects over three time periods is shown in Fig. 11.29. Because the terminal velocity of ice crystals becomes considerably larger than w_0 ($V_t \sim 80$ cm s^{-1} at 10 min), ice falls out of the cloud and sublimates to vapor in the dry subcloud layer. The cooling associated with this sublimation is shown by the minimum in the Q_c profile below cloud base in Fig. 11.29. The region of positive Q_c in Fig. 11.29 represents the *generator region* of net ice formation.

The heating profiles from infrared and shortwave radiation (Fig. 11.29) are comparable in magnitude to Q_c at all stages of the cloud, indicating the importance of radiation. Strong absorption of shortwave radiation near the cloud top is offset by infrared cooling of the same magnitude. The net radiation profile in the steady-state cloud shows weak heating near the top of the cloud layer, net cooling of about 2°C day^{-1} throughout most of the cloud, and weak heating (0.2°C day^{-1}) below the cloud.

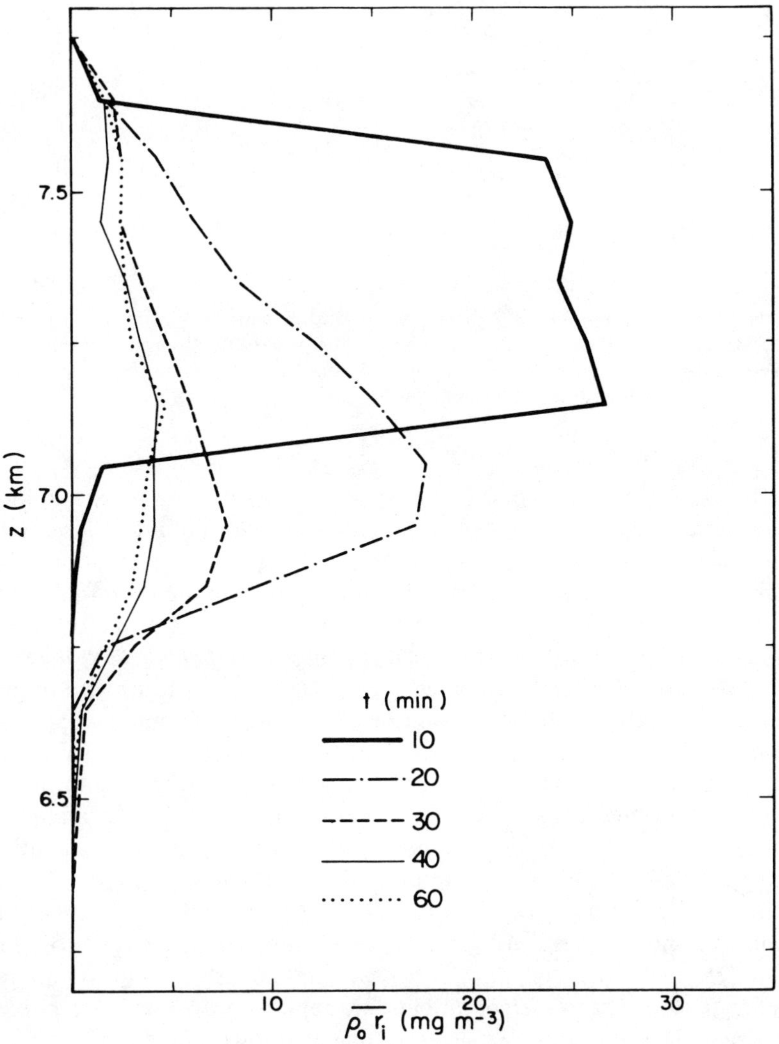

Fig. 11.28. Vertical profiles of horizontally averaged ice content ($\overline{\rho_0 r_i}$) at various times during simulation of a thin cirrus cloud layer. [From Starr and Cox (1985a).]

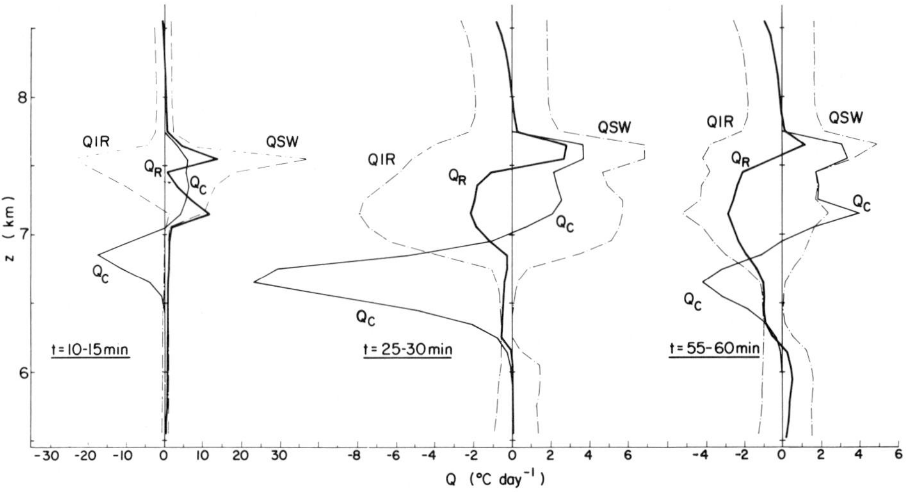

Fig. 11.29. Vertical profiles of horizontally averaged potential temperature tendencies due to phase changes of water ($\overline{Q_C}$), infrared radiative processes ($\overline{QIR}$), absorption of solar energy ($\overline{QSW}$), and net radiative processes ($\overline{Q_R}$) over various time periods during a simulation of thin cirrus cloud layer. [From Starr and Cox (1985a).]

The above simulation of cirrus clouds indicates the importance of large-scale upward motion, latent heating, radiation, and fallout of ice crystals in a quasisteady-state daytime cirrus cloud. In additional simulations, Starr and Cox (1985b) tested the sensitivity of the model to the magnitude of the mean vertical velocity, the relative fall velocity, and the radiation. Figure 11.30 shows the vertical profiles of water contents of ice crystals r_i at various times for mean vertical velocities of −2, 0, 2, and 10 cm s^{-1}. For $w_0 =$ −2 cm s^{-1} (sinking motion), a cloud forms (owing to the initial supersaturation), but the cloud is short lived compared to the other cases. The simulation with $w_0 = 0$ is similar to the control ($w_0 = 2.0$ cm s^{-1}) during the first half hour, but r_i is significantly less at 60 min. Without a mean upward motion, the initial water vapor supply is exhausted, the ice crystals fall out of the domain, and the convection gradually decays.

In the simulation with the strong sustained updraft of 10 cm s^{-1}, the cloud becomes thicker during the entire 60 min. The peak value of cloud ice-water content at 60 min is by far the largest of the simulations (30 mg m^{-3}), owing to the continuous and rapid supply of water vapor into the cloud. The peak in r_i occurs at a somewhat lower elevation than in the other simulations, a result of the combined effects of ice falling from above and ongoing ice formation at this level. The maximum value of r_i of

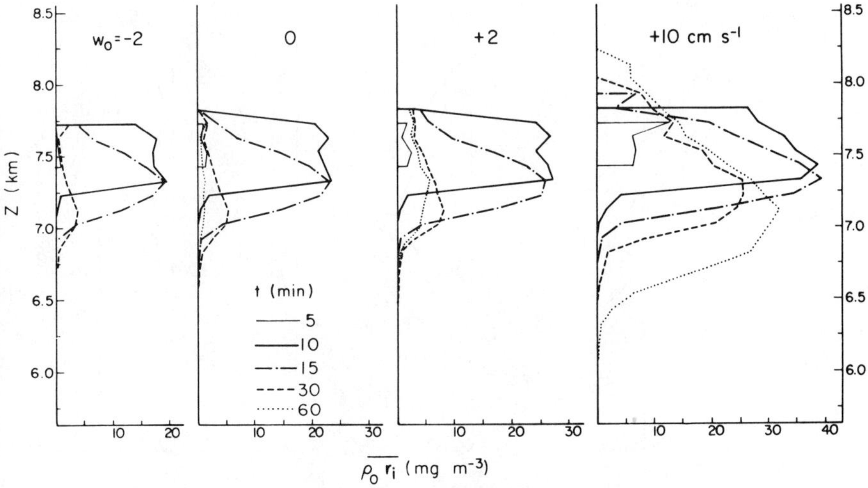

Fig. 11.30. Vertical profiles of horizontally averaged ice content ($\overline{\rho_0 r_i}$) at various times during simulations of cirrus clouds at various specified values of large-scale vertical velocity (w_0). [From Starr and Cox (1985b).]

30 mg m^{-3} is consistent with observations of 10–100 mg m^{-3} under conditions of strongly forced ($w_0 \sim$ 10–30 cm s^{-1}) thick cirrostratus clouds (Heymsfield, 1977).

Because the ice-particle relative fall speeds are large compared to w_0, depletion of cirrus clouds by fallout is of consequence. The sensitivity of the cloud model to 20% higher or lower values of V_t is demonstrated in Fig. 11.31. The greatest difference occurs after the generation of maximum cloud ice-water content (20–30 min). In the near steady state, when cloud ice-water contents are smaller (5 mg m^{-3}), the sensitivity to V_t is less.

The sensitivity of the model to radiation was considered by comparing the control daytime simulation, in which shortwave radiation was present, to a nighttime simulation in which only infrared radiative effects were neglected entirely. Figure 11.32 shows the temporal evolution of the domain-average ice-water content for the three simulations. Although the night simulation shows about 20% more cloud water in the quasisteady state, the sensitivity of the model to radiation is less than the sensitivity to the fall velocity and the mean vertical motion. The vertical profiles of the net radiative heating rates for the day and night cases, together with Q_c, are shown in Fig. 11.33.

Although the domain-averaged condensed water substance is similar in the three cases with quite different radiative heating profiles, differences in

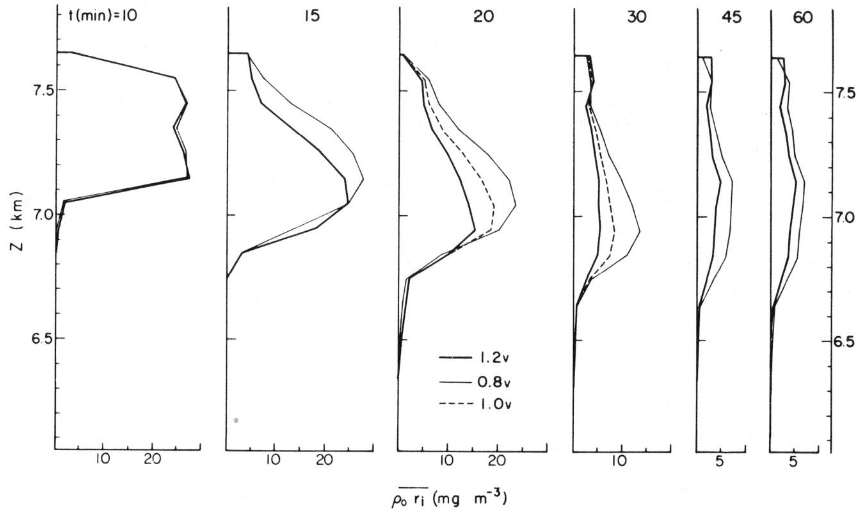

Fig. 11.31. Vertical profiles of horizontally averaged ice content ($\overline{\rho_0 r_i}$) at various times during simulations of cirrus clouds, differing only in diagnostic relationship used to evaluate ice relative fall speed (V_t). [From Starr and Cox (1985b).]

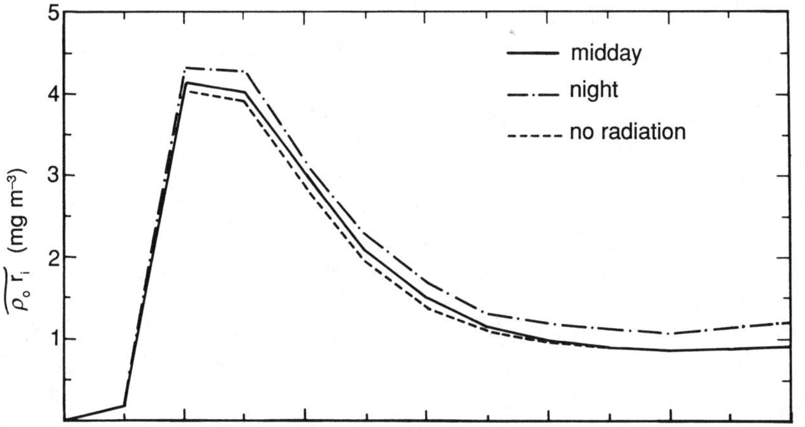

Fig. 11.32. Time-dependent behavior of domain-averaged ice content ($\rho_0 r_i$) during cirrus cloud simulations under nighttime and midday conditions and where radiative processes are neglected. [From Starr and Cox (1985b).]

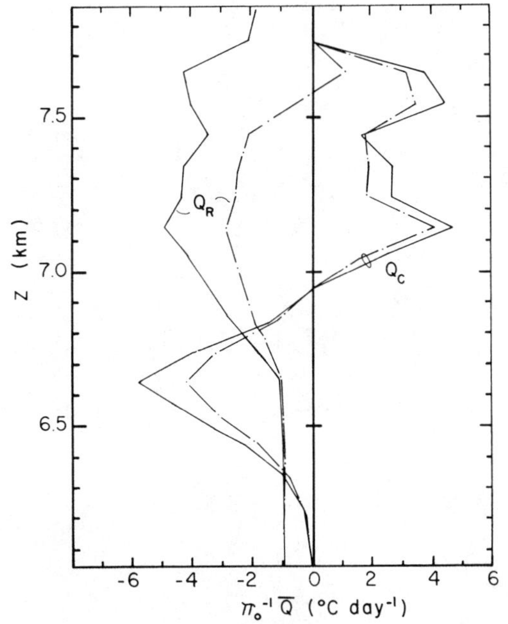

Fig. 11.33. Vertical profiles of horizontally averaged potential temperature tendency due to phase changes of water ($\overline{Q_C}$) and radiative processes ($\overline{Q_R}$) over the time period $t = 55$–60 min during cirrus cloud simulations under nighttime (solid lines) and midday (dash-dot lines) conditions. [From Starr and Cox (1985b).]

the structure of the cirrus clouds exist in the three simulations. Figure 11.34 shows the cloud water contents at 30 min for the night and day cirrus cases. During the day, the cirrus are more cellular and the maxima of cloud water in the convective cells are greater. These differences are related to the feedback among infrared cooling, shortwave heating, and buoyancy in convective cells. Infrared cooling increases with increasing cloud water and partially offsets the warming from the latent heat of fusion. Infrared radiation acts as a negative feedback mechanism to decrease the buoyancy in convective cells. In contrast, absorption of solar radiation increases with cloud water content, and the additional heating enhances the convection. Other factors being the same, these results indicate that daytime cirrus should have a more cellular, convective structure than night cirrus.

11.4.5.b Altostratus Simulation

The same basic model framework used to simulate cirrus clouds can be applied to model clouds other than cirrus, with appropriate modification

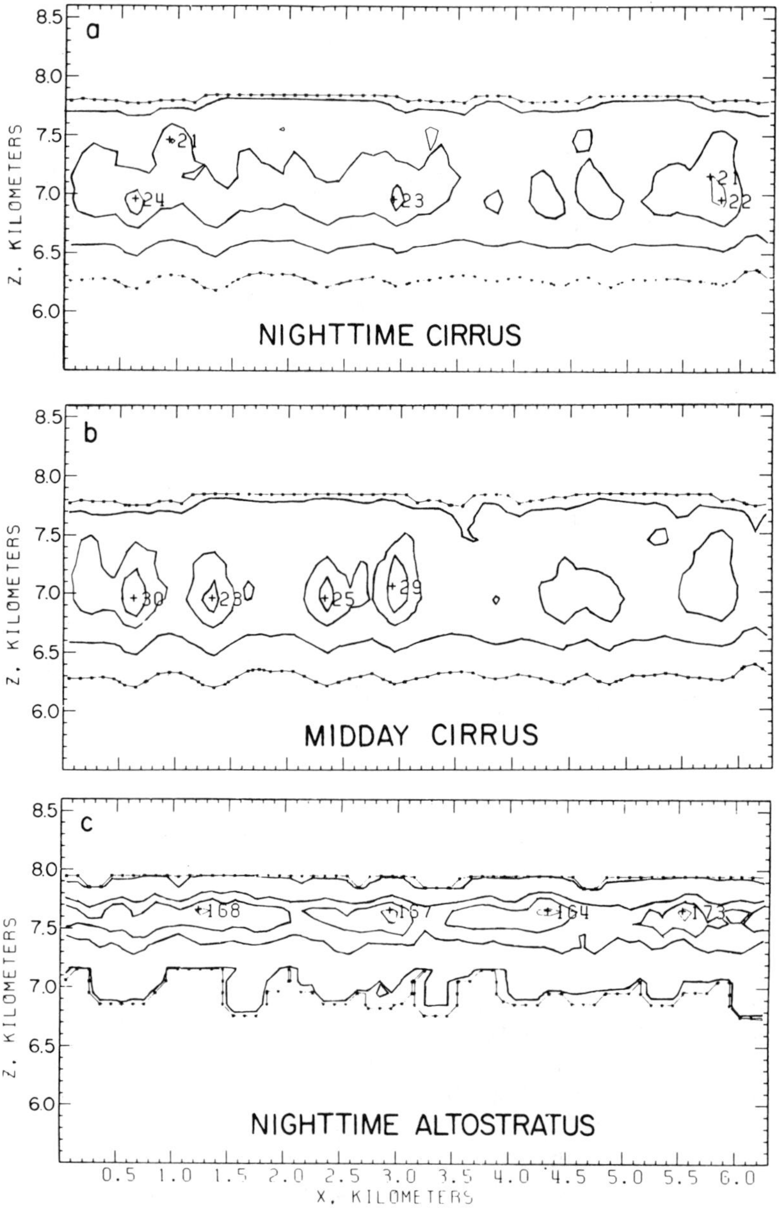

Fig. 11.34. Contour plot of field of cloud or ice mixing ratio at time $t = 30$ min during simulation of (a) cirrus at night, (b) cirrus at midday, and (c) altostratus at night. Contour levels correspond to $r_i = 10^{-3}$ (*), 1, 10, and 20 $\mu g\ g^{-1}$ on panels (a) and (b) and $r_c = 10^{-3}$ (*), 1, 50, 100, and 150 $\mu g\ g^{-1}$ on panel (c). [From Starr and Cox (1985b).]

of the physical parameterizations. To model all water altostratus clouds, Starr and Cox (1985b) modified the terminal velocity to a constant value of 0.9 cm s^{-1}, much smaller than the typical value of 100 cm s^{-1} of cirrus cloud ice crystals. The radiation parameterization is modified so that the upward- and downward-directed emittances (Chapter 5) increase more rapidly with cloud water path. Thus, an altostratus cloud will appear blacker than a cirrus cloud for the same total condensed water paths. The phase change of water is also modified; condensation occurs whenever the relative humidity (RH) exceeds 100%, where RH is the relative humidity with respect to liquid water. Evaporation occurs whenever RH is less than 100% and liquid water is present. Growth of cloud drops to precipitation-sized drops is not considered (an assumption that limits the model to simulations of weakly forced altostratus clouds over short time periods) and so fallout of cloud drops does not occur.

Figure 11.35 depicts the vertical profiles of domain-averaged cloud water content for the altostratus simulation at 15, 30, and 60 min. Notable differences are present from the cirrus simulation (compare with Fig. 11.28). In particular, the total cloud water content increases throughout the period of the altostratus simulation rather than peaking early as in the cirrus simulation. At 60 min, there is about five times as much liquid water as ice water. This increase is also evident in the vertical cross section of cloud water content at 30 min (Fig. 11.34c). This difference is mainly due to the absence of fallout of water drops in the altostratus case. The shape of the vertical profiles of liquid versus ice water is also quite different. The profiles in the altostratus case increase sharply with height, and the level of maximum water increases throughout the period. Again, this difference can be

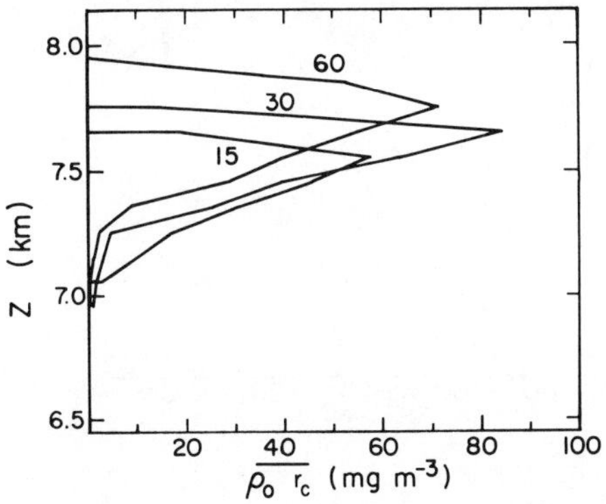

Fig. 11.35. Vertical profiles of horizontally averaged liquid water content ($\overline{\rho_0 r_c}$) at various times during a simulation of an altostratus cloud at night. [From Starr and Cox (1985b).]

explained by the lack of fallout in the altostratus simulation; the cloud drops are carried upward continuously by the basic-state vertical velocity.

The vertical profiles of infrared radiative cooling and condensation heating are shown in Fig. 11.36. Compared to the cirrus case (Fig. 11.29), the diabatic terms are increased by an order of magnitude. Strong infrared cooling occurs in the upper part of the cloud and strong infrared warming in the lower portion. This differential heating in the vertical represents a significant destabilization process. A maximum condensation heating of about 40°C day^{-1} occurs in the upper part of the cloud; evaporative cooling occurs in a thin layer near cloud top and over a thicker layer near cloud base.

In summary, the neglect of fallout of cloud drops in the altostratus simulation produces large differences in cloud evolution and structure compared to the cirrus simulation. Larger liquid-water contents and stronger convective cells occur in the altostratus cloud. Although the water-loading effect is greater in the altostratus convection, this drag is more than offset

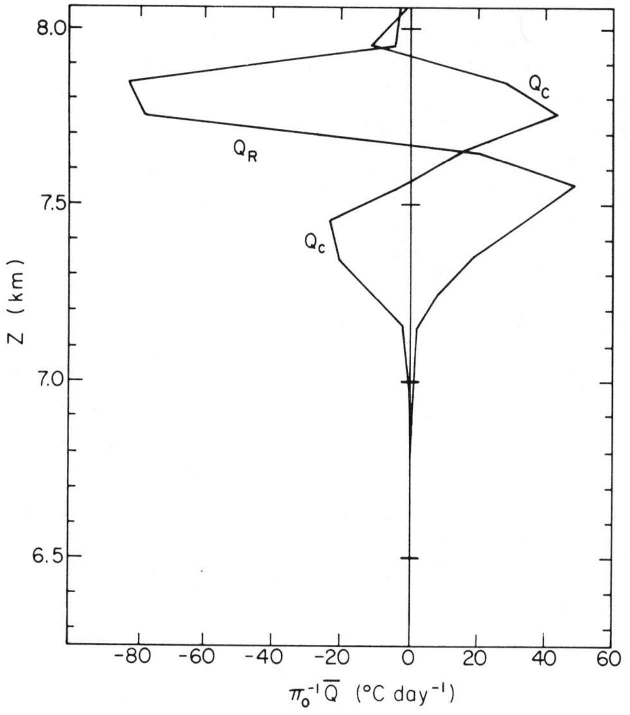

Fig. 11.36. Vertical profiles of horizontally averaged potential temperature tendency due to phase changes of water ($\overline{Q_C}$) and radiative processes ($\overline{Q_R}$) over the time period $t = 55$–60 min during simulation of an altostratus cloud at night. [From Starr and Cox (1985b).]

by the increased latent heating. In the cirrus convection, fallout of ice crystals and sublimation below the convective cell produce cooling which weakens the convection. As a result of the more vigorous convection in the altostratus, updrafts penetrate into the stable overlying air and detrained cloud water evaporates in this region. The higher water content in the altostratus cloud allows downdrafts to penetrate relatively far into the subcloud layer where evaporation occurs. The greater water content associated with the altostratus case and the higher emittance of water versus ice lead to a much greater destabilization by radiative cooling in the altostratus case.

11.4.5.c Application of Mixed-Layer Models to Middle and High Clouds

Lilly (1988) proposed the application of radiative–convective mixed-layer models to middle- and high-level clouds. These one-dimensional, layer-averaged models have been applied with reasonable success to the simulation of boundary-layer stratocumulus clouds. The basic concepts of mixed-layer models and the various closure approximations employed in their formulation are discussed in Chapter 7. Lilly extended the mixed-layer concept to elevated cloud layers that are bounded above and below by stable nonturbulent flow. In contrast to boundary layer clouds, cirrus clouds cannot be considered blackbodies and, therefore, radiative flux divergences are distributed throughout the cloud layer. Also in the case of boundary layer clouds, the main source of buoyancy generation of turbulent kinetic energy often arises from the flux of sensible and latent heat from the earth's surface. Therefore, the entrainment rate at the top of boundary layer clouds is often simply related to the input heat fluxes at the base of the cloud layer. This is not the case for elevated cloud layers, where it is necessary to find a relationship between the bottom and top heat fluxes. In cirrus clouds, radiative heating is the greatest source of destabilization of the cloud layer and results in the production of turbulent kinetic energy in the cloud layer; Lilly made the dubious assumption that latent heating effects can be ignored. As seen in Starr and Cox's model calculations, vapor-deposition growth of ice crystals and their sublimation are sources and sinks of latent heating. Nonetheless, Lilly estimated vertical velocity and time scales for radiatively induced cirrus turbulence, which suggested that they can support active turbulence comparable to boundary layer clouds.

As Lilly noted, however, can a mixed layer develop before the ice water that drives radiative heating has precipitated out? Starr and Cox found that precipitation from cirrus is a major contributor to cirrus heat and moisture budgets. Lilly pointed out that, in cirrus clouds generated by outflow from convective cells, large concentrations of cloud condensation and ice nuclei would be activated in the rapidly rising convective updrafts. As a result, a more colloidally stable cloud is produced that can lead to sufficiently high ice-water contents to drive a radiative–convective mixed layer. In slowly ascending air in the warm conveyor belt of extratropical cyclones, however, few CCN and IN are activated and the depletion of the ice-water content by settling ice crystals is favored. Thus, a radiative–convective equilibrium may not be established in many cirrus clouds.

11.4.6 Parameterization of Middle- and High-Level Clouds in Large-Scale Models

Because of energy transformations associated with changes of phase of water and also because of their enormous effect on infrared and shortwave radiation, middle- and high-level clouds must be considered in large-scale numerical models. Large scale refers to models with horizontal resolutions of about 100 km or greater, including general circulation models (GCMs) used for climate simulations, global models used for forecasting, and limited-area models used in forecasting or research. While historically a distinction has been made between models used for forecasting and climate simulations, the problems in climate simulation and medium- and long-range forecasting are related. For example, a major portion of the forecast error beyond a few days is contributed by systematic errors caused by the model's drift toward its own climate rather than by the real climate. Thus, there is increasing recognition that parameterization of physical processes, such as cloud and radiation effects, is important for forecast models as well as for climate models. Even on time scales as short as a day, evidence exists that clouds, through their effect on the surface-energy budget, affect circulations (Benjamin and Carlson, 1986).

11.4.6.a Effect of Clouds on Large-Scale Net Radiation Balance

Clouds play a dual role in the heat budget of the earth and atmosphere. On the one hand, their high albedo reflects incoming solar radiation in the visible wavelengths, thus cooling the earth. On the other hand, clouds are opaque to infrared radiation and their presence reduces the loss of IR radiation to space, thus warming the earth. Consequently, it is not possible to make general statements about the impact of changes in average cloud cover on the earth's climate. Instead, the effect of changes in cloud cover on climate depends critically on the level (temperature), season, and latitude at which the changes occur and on the optical properties of the clouds. The optical properties of clouds depend on the microphysical characteristics of the cloud, including the concentrations and size of embedded aerosol particles and the concentration, size, and water phase of the hydrometeors

(Chapter 5). For example, an increase of cirrus clouds at high altitudes in the winter would produce a warming because the additional solar radiation reflected would be minimal while the reduction in the net IR cooling would be substantial. In contrast, an increase of stratocumulus at low latitudes would result in a cooling, since the reflection of solar radiation would increase markedly while the reduction of IR loss to space would be small.

In addition to the importance of latitude and elevation of clouds in determining their effect on the net radiation at the earth's surface, variations in the optical properties of clouds can determine whether they warm or cool the surface. Cox (1971) indicates that cirrus clouds may cool or warm the surface, depending upon their infrared emittance. Measurements show that the emittances of tropical cirrus are high enough (0.6–0.8) to produce a net warming, while the emittances of midlatitude cirrus can be low enough (less than 0.6) to produce a net cooling.

Schneider (1972) used a one-dimensional radiation model to study the opposing effects of clouds on climate. Although this model could not address changes resulting from dynamical feedbacks of the general circulation with the changes in net radiation due to variations in cloud cover, the study is instructive in illustrating the properties of clouds that affect the radiation budget.

Figure 11.37 shows the effect of fraction of coverage of a single cloud layer, with tops at various elevations, on the infrared radiation lost to space (F_{IR}) and on the absorbed solar energy (Q_{ABS}), where Q_{ABS} is the globally averaged incident radiation (one-fourth times the solar constant Q_{SC}) modified by the reflection from clouds

$$Q_{ABS} = (Q_{SC}/4)(1-\alpha), \tag{11.21}$$

where α is the cloud albedo. The absorption Q_{ABS} is not a function of cloud height and shows a linear decrease from a value of about 300 W m^{-2} with zero cloud cover to 174 W m^{-2} for total cloud cover.

The reduction of IR loss to space does depend on cloud height, with clouds at higher altitudes and colder temperatures being more effective in reducing F_{IR}. Thus, increasing the coverage of a cloud layer at 3.5 km from 0 to 100% causes a reduction in IR loss from about 280 to about 240 W m^{-2}, while the same increase in a cloud layer at 8.5 km results in a reduction to about 160 W m^{-2}.

Because the slope of Q_{ABS} is greater than that of F_{IR} for all cloud tops, Fig. 11.37 indicates that an increase of global cloud cover would produce a decrease of globally averaged surface temperature, provided that the horizontal and vertical distribution of clouds and their optical properties remained unchanged.

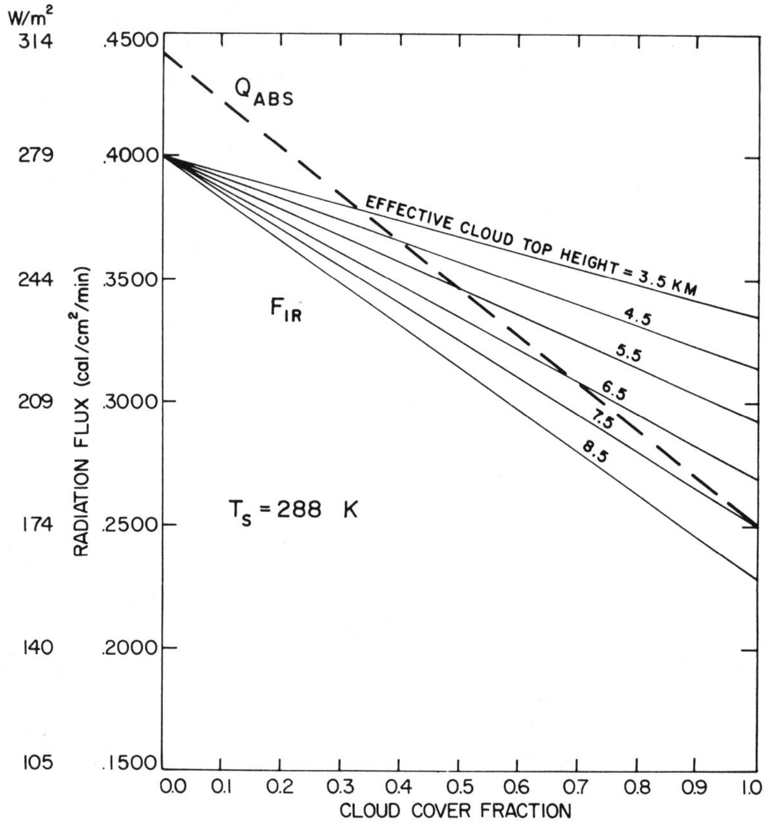

Fig. 11.37. Infrared flux to space F_{IR} emitted from earth-atmosphere system and absorbed solar energy Q_{ABS} as a function of amount of cloud cover and for several values of effective cloud-top height. [From Schneider (1972).]

Figure 11.37 may also be used to estimate the effect of changing the average cloud heights without changing the amount of cloud cover and the cloud albedo. For a constant cloud cover, increasing the cloud top reduces the IR loss to space and, therefore, results in a warmer climate.

The opposing effects of increases in cloud-cover amount and cloud height are illustrated in Fig. 11.38, which shows three possible equilibrium states—cloud heights, cloud amount, and surface temperature. Each curve represents the locus of possible cloud heights and amounts for which radiative equilibrium is maintained at the specified surface temperature. Thus, increasing cloud cover, if accompanied by the appropriate increase in cloud height, can result in zero surface temperature change. The arrows in the

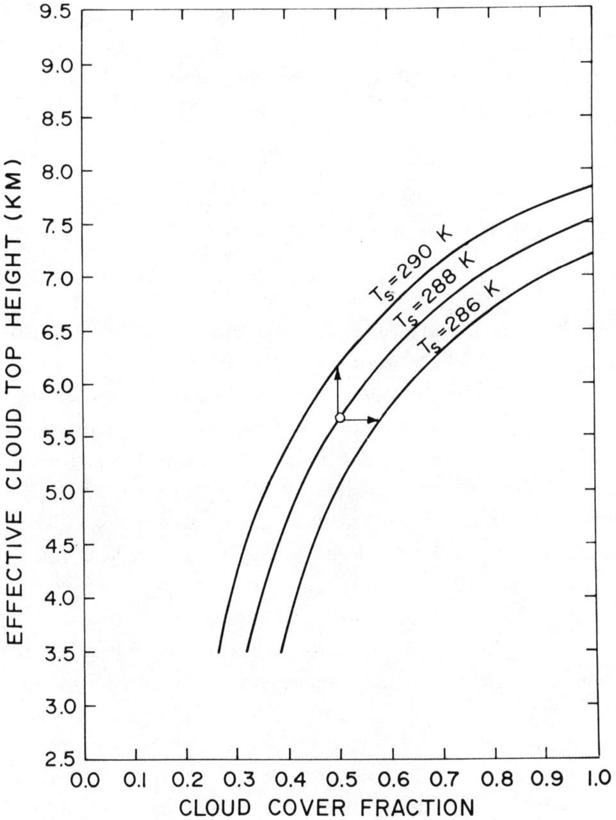

Fig. 11.38. Cross plot of intersections on Fig. 11.37 of the Q_{ABS} curve with various F_{IR} curves for three values of surface temperature. Each curve represents a locus of possible equilibrium values of cloud-top heights and cloud-cover amounts that are consistent with a constant value of surface temperature. [From Schneider (1972).]

figure illustrate the effect of changes in cloud cover or cloud height alone on surface temperature. In the example, increasing the cloud height from 5.5 to 6.1 km with no change in cloud cover raises the surface temperature by 2 K. Alternatively, an increase of cloud cover from 50 to 58% decreases the surface temperature by 2 K. All of these calculations assume that the cloud properties remain unchanged and that convective and radiative processes together maintain the same lapse rate.

The above discussion pertains to the global-average situation. The effect of changes in cloud cover on the net radiation budget has a large latitudinal and seasonal dependence, however, because the shortwave radiation varies strongly with these parameters. The seasonal and latitudinal effects can be illustrated qualitatively by considering the changes in the difference between Q_{ABS} and F_{IR} for various latitudes and seasons when the fraction of cloud cover increases from 0 to 100%. This difference, δ_i, where i pertains to latitude, is given by

$$\delta_i = \Delta Q^i_{ABS}/\Delta A_c - \Delta F^i_{IR}/\Delta A_c. \tag{11.22}$$

In this example, $\Delta A_c = 1.0$ and the Q^i_{ABS} is

$$Q^i_{ABS} = Q^i_s[1 - (A_c \alpha_{c_i} + (1 - A_c)\alpha_{s_i})], \tag{11.23}$$

where Q^i_s is the zonal average value of incoming solar radiation, α_{c_i} is the zonal-average albedo of the clouds in the ith latitude zone, and α_{s_i} is the zonal-average albedo in the zone under cloud-free conditions.

A plot of δ as a function of latitude for January and July is shown in Fig. 11.39. In July, when Q_{ABS} is greatest, an increase of cloud cover results in a decrease of net radiation at latitudes south of about 70°N. In January, the greatly reduced Q_{ABS}, especially at high latitudes, results in a decrease of net radiation south of about 40°N, with an increase poleward of this latitude.

In summary, the effect of clouds on the net radiation balance depends on season, geographic location, and elevation of clouds as well as cloud optical properties. In general, increases in cloud cover produce decreases in surface net radiation at low and middle latitudes in summer, with greater decreases associated with lower tropospheric clouds. Conversely, increases in cloud cover at high latitudes in the winter produce increases in net radiation, with the greatest increases occurring with high clouds.

11.4.6.b Parameterizations in GCMs

Even with horizontal resolutions as high as 100 km (high by historical standards in global models), it is not always possible to calculate explicitly the cloud and precipitation effects and the interactions with radiation as in models like that of Starr and Cox (1985a). Thus, it is necessary to model the major cloud effects as a function of the parameters predicted by the large-scale model. Until recently, very simple parameterizations of nonprecipitating and precipitating middle- and high-level clouds were used in global models. In most GCMs, it is assumed that nonconvective, stratiform middle, and high clouds occur when the resolvable-scale relative humidity exceeds some critical value RH_c. For the ECMWF, GFDL, GLAS, UCLA, and GISS models, this value is 100%. For the NCAR model, RH_c is assumed to be 80% because of the possibility of having a fraction of the model's minimum resolvable scale covered by clouds when the average relative humidity is below 100%.

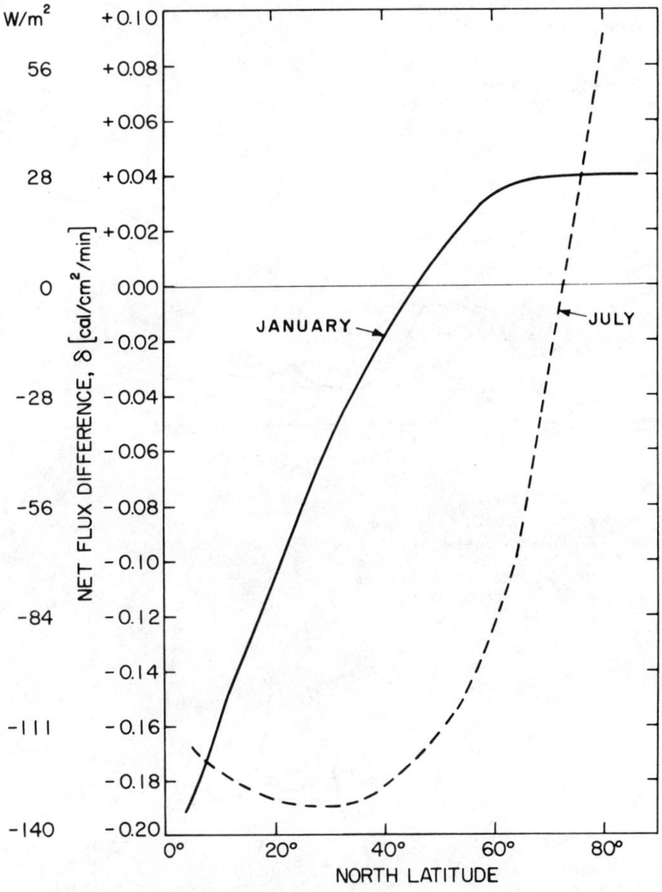

Fig. 11.39. Net flux difference $\delta\ [=(\partial Q_{\mathrm{ABS}}/\partial A_{\mathrm{c}})-(\partial F_{\mathrm{IR}}/\partial A_{\mathrm{c}})]$ computed as a function of north latitude. Although the hemispheric average value of δ is negative, δ is positive in high latitudes. [From Schneider (1972).]

Once clouds are diagnosed by the criterion $\mathrm{RH} > \mathrm{RH_c}$, the models treat differently the precipitation process and the interaction with radiation. In the GFDL and NCAR models, it is assumed that the water vapor excess over $\mathrm{RH_c}$ falls to the ground immediately as precipitation (no evaporation), whereas in the ECMWF, GLAS, UCLA, and GISS models, evaporation is allowed in the unsaturated lower layers. Only when the humidity of all lower layers equals $\mathrm{RH_c}$ does precipitation reach the ground.

Stephens (1984) reviewed the radiation schemes of several GCMs (NCAR, GFDL, GISS, and GLAS). In early GCMs (e.g., Holloway and

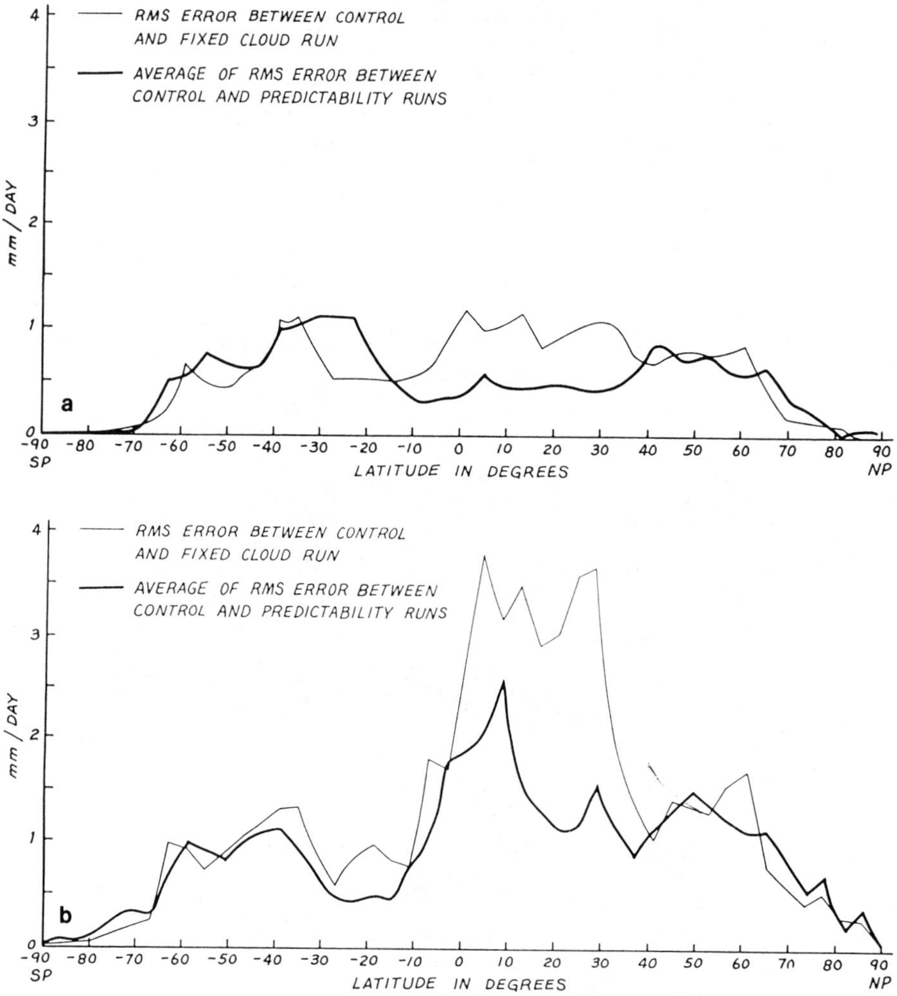

Fig. 11.40. (a) Zonally averaged root-mean-square (rms) difference of monthly mean evaporation between control and fixed cloud runs, and their averages for control and each predictability run. (b) Zonally averaged rms difference of monthly mean precipitation between control and fixed cloud runs, and their averages for control and each predictability run. [From Shukla and Sud (1981).]

Manabe, 1971) the water vapor and clouds predicted or diagnosed by the model were not allowed to interact with the radiation parameterization. Instead, water vapor and cloud effects on radiation were computed by assumed temporally constant fields of water vapor and clouds specified according to climatological fields. In recent GCMs, this assumption is relaxed by utilizing the model-predicted cloud and water-vapor fields in the calculation of radiation effects. Shukla and Sud (1981) compared two climate simulations, one with prescribed clouds and another with interactive clouds. They found significant differences in the two climates; the zonally asymmetric radiative heating associated with the fixed-cloud simulation produced greater generation of eddy available energy, more variance in the cyclone-scale transient waves, and large differences in the hydrologic cycle over the oceans. Figure 11.40 shows the root-mean-square differences between the zonally averaged monthly mean evaporation and precipitation between the two simulations. These and other differences indicate the importance of the cloud-radiation feedback processes.

Hansen *et al.* (1983) tested the sensitivity of the GISS GCM to variations in the treatment of clouds and their interaction with radiation processes. The sensitivity experiments are summarized in Table 11.5. In the control simulation, designated I-1, the fraction of the grid occupied by nonconvective clouds is computed under the assumption that the mixing ratio is constant in the grid volume but that temperature perturbations with a prescribed variance exist over the grid. The fraction of cloud cover is that fraction of the grid which is supersaturated owing to the temperature perturbations.

In the first sensitivity experiment, I-35, the subgrid-scale temperature perturbation is eliminated in the calculation of the fraction of nonconvective cloud cover, i.e., grid-scale saturation is required for the presence of clouds and so the fraction of cloud cover is either 0 or 100%. This simulation was

Table 11.5

Sensitivity of GISS GCM to Variations in Treatment of Clouds[a]

Experiment	Control
I-1	Control
I-35	No subgrid-scale temperature variance assumed in calculating fraction of grid covered by clouds
I-36	Large-scale rainfall calculated every 5 h rather than 1 h
I-37	Fixed annually averaged clouds from control simulation
I-38	Fixed annually and longitudinally averaged clouds from control simulation
I-39	Local temperature 0°C criterion for saturation over water or ice
I-40	Local temperature −40°C criterion for saturation over water or ice
I-41	Local temperature −65°C criterion for saturation over water or ice
I-42	Optical thickness of cirrus clouds reduced to $\tau = \frac{1}{3}$
I-43	Optical thickness of other nonconvective clouds reformulated

[a] Adapted from Hansen *et al.* (1983).

judged to be inferior to the control simulation in that an unrealistically large, low-level cloud cover was calculated over the oceans, particularly in the subtropical summer hemisphere. This result indicates that a parameterization that forces the extreme conditions of either no cloud or total cloud cover over grid scales of order 4° latitude by 5° longitude is unrealistic.

In experiment I-36, the frequency for calculating large-scale precipitation was decreased from every hour to every 5 h. This allowed the relative humidity on the resolvable scale to build up to greater values before being reduced by precipitation and resulted in an increase in global cloud cover from a value of 40% to a more realistic value of 45%. Although this increase resulted in a climate closer to the observed, the result is somewhat disturbing because of the arbitrary nature of the frequency at which precipitation is computed.

In sensitivity experiment I-37, the annual mean cloud cover produced by the control simulation was specified rather than allowing a time-dependent cloud cover to evolve. This experiment is different from the one studied by Shukla and Sud (1981), who used observed cloud cover rather than mean cloud cover generated by their model. Even with a model-generated temporally averaged cloud cover that varied with latitude, longitude, and height, the climate of I-37 differed substantially from the control. For example, in the Aleutian region, which has more clouds in winter than in summer, the use of annual mean clouds produced a statistically significant weakening of the Aleutian low in winter. The average cloud cover for January–March decreased by 6.2%, while the mean sea level pressure increased by 8.9 mbar in experiment I-37 compared to the control.

In experiment I-38, the cloud cover was specified not only as an annual average from the control simulation but also as the longitudinal average. Specification of a zonally uniform cloud cover reduced longitudinal heating differences and led to a large (25%) decrease in eddy kinetic energy in the low-latitude troposphere. These results are consistent with those of Shukla and Sud (1981).

Experiments I-39, I-40, and I-41 test the sensitivity of the model to the temperature below which sublimation of vapor to ice is assumed rather than condensation from vapor to water. In the control simulation, sublimation is assumed at all layers if the lowest layer is below 0°C, while condensation is assumed if the lowest layer is above 0°C. Experiments I-39, I-40, and I-41 use the *local* layer temperature to determine whether condensation or sublimation occurs, with threshold temperatures of 0°C, −40°C, and −65°C, respectively.

Because of the lower saturation vapor pressure over ice compared to water, the model more readily predicts saturation with respect to ice, and experiment I-39 shows an increase of high (300 mbar) clouds compared to

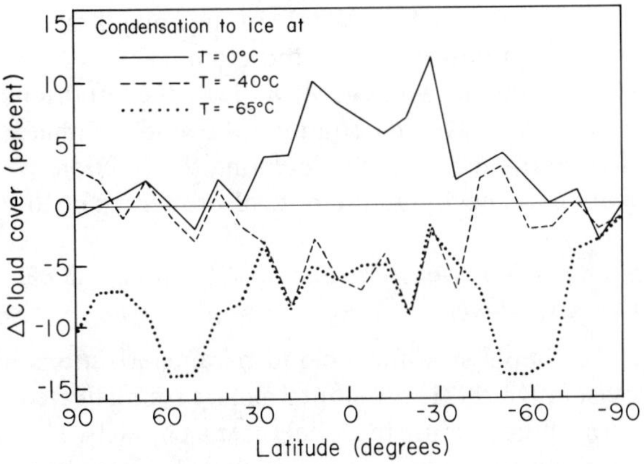

Fig. 11.41. Change of January cloud cover at 300 mbar if condensation to ice occurs at indicated local temperature. Run I-1 is control. [From Hansen *et al.* (1983).]

the control simulation (Fig. 11.41). Decreasing the threshold temperature from 0 to −40 or −65°C leads to large regions of air, supersaturated with respect to ice but undersaturated with respect to water, and a reduction of high clouds (Fig. 11.41).

In the final two sensitivity experiments involving the treatment of clouds, the optical thickness τ of the clouds was varied. In the control simulation, convective clouds were assigned a value of τ equal to 8 per 100 mbar of thickness. For nonconvective clouds, τ is given by

$$\tau = 100\left(\frac{p-100}{900}\right)\left[\frac{\min(\Delta p, 100)}{1000}\right], \tag{11.24}$$

where Δp is the thickness of the model layer in millibars and no clouds are allowed above 100 mbar. According to Eq. (11.24), a low-level cloud 100 mbar thick has $\tau \approx 10$, while a high-level cloud 100 mbar thick has $\tau \approx 1/3$.

In experiment I-42, Eq. (11.24) was replaced by the assumption that all nonconvective clouds with temperatures less than −15°C used $\tau = 1/3$. Experiment I-43 used the following simpler specification for τ for all nonconvective clouds:

$$\begin{aligned} \tau &= 1/3, && T < -15°\text{C}; \\ \tau &= 0.0133(p-100), && T > -15°\text{C}. \end{aligned} \tag{11.25}$$

The changes in the formulation of τ had a relatively minor effect on the simulations, compared to the other sensitivity experiments. The main difference was a reduction by 5–10 W m^{-2} of the annual mean net radiation gain at the top of the atmosphere and the ground.

In summary, experiments with GCMs indicate the importance of interactive, time-dependent clouds in determining the model climate. The models are also sensitive to the method of calculating the fraction of cloud cover, but not as sensitive to specifying the optical properties of the clouds.

11.4.6.c Sundqvist's Parameterization of Clouds in a Global Forecast Model

In contrast to the simple stratiform cloud parameterizations used in most GCMs, Sundqvist (1978, 1981) developed a more sophisticated parameterization of stratiform clouds for use in global forecast models. This parameterization includes a prognostic equation for cloud water and allows for subgrid-scale condensation, i.e., condensation when the grid-scale relative humidity is less than 100%. When a threshold humidity RH_c (~80%) is reached, a fraction a of the grid is assumed to be saturated and to contain cloud water. Part of the cloud water is assumed to be evaporating in the cloud-free portion of the grid. The net heating rate Q (J kg^{-1} s^{-1}) resulting from condensation and evaporation is

$$Q = aQ_c - (1-a)(E_c + E_r), \tag{11.26}$$

where Q_c is the release of latent heat in the cloudy portion and E_c and E_r are the cooling rates due to evaporation of cloud and rain water, respectively.

Sundqvist's parameterization is summarized as follows. The stratiform cloud parameterization scheme is activated for a convectively stable column in which the relative humidity exceeds RH_c at one or more levels. The sequence of the parameterization scheme is as follows:

(1) Evaluate the supply of water vapor per unit time available for condensation and moistening a grid volume as well as the supply of internal energy to a grid volume.

(2) Diagnose the fraction of the grid volume that is saturated, the rate of conversion of cloud water to precipitation water, and the evaporation rate of rain in the subsaturated layers.

(3) Evaluate the time rate of change of relative humidity.

(4) Evaluate the diabatic heating rate based on the supply of moisture and the time rate of change of relative humidity.

(5) Evaluate the local rate of change of potential temperature and water vapor mixing ratio.

(6) Predict changes in cloud water due to vertical and horizontal advection, settling of droplets, net heating due to condensation and evaporation, and conversion to precipitation.

Sundqvist (1981) tested the above scheme for parameterizing stratiform clouds in a version of the ECMWF global numerical model. The model was initialized at 0000 GMT, 16 January 1979, in a conventional way for the usual meteorological fields and cloud water set to zero everywhere. Figure 11.42 shows the global total cloud water (kg m^{-2}) in the model atmosphere as a function of time, together with the total rate of precipitation from stratiform clouds, convective clouds, and the evaporation at the earth's surface. Because of the lack of cloud water initially, as well as other imbalances between the model physics and initial fields, the model required about 24 h to reach a quasisteady state in the precipitation rate. The average evaporation rate of about 1 mm day^{-1}, as well as the total precipitation rate (also about 1 mm day^{-1}) in the model forecast after 120 h, was only about 40% of observed values, a discrepancy Sundqvist hypothesized was caused by not using properly tuned parameters.

A vertical cross section of cloud water content and fraction of cloud cover at forecast Day 5 along longitude 157.5°W is shown in Fig. 11.43. The section crosses two frontal systems, one around 45°N and the other between 55 and 60°N. Deep clouds are associated with these fronts, with maximum cloud water content and fraction of area covered about 0.63 g kg^{-1} and 90%, respectively. The parameterization indicates a cirrus shield extending northward of the northern front at elevations of 6–12 km.

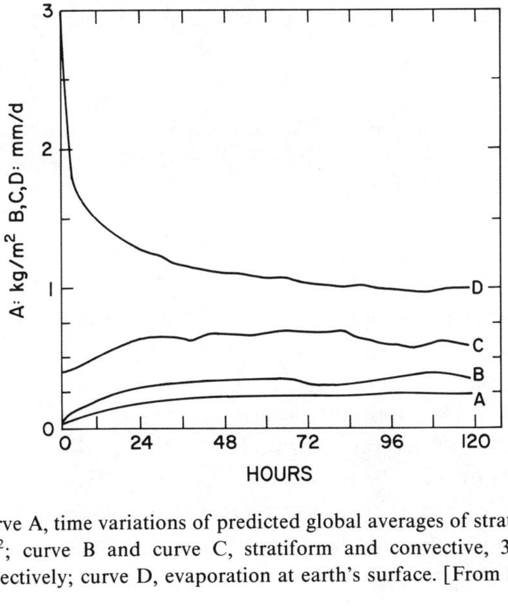

Fig. 11.42. Curve A, time variations of predicted global averages of stratiform cloud water content in kg m^{-2}; curve B and curve C, stratiform and convective, 3-h mean rates of precipitation, respectively; curve D, evaporation at earth's surface. [From Sundqvist (1981).]

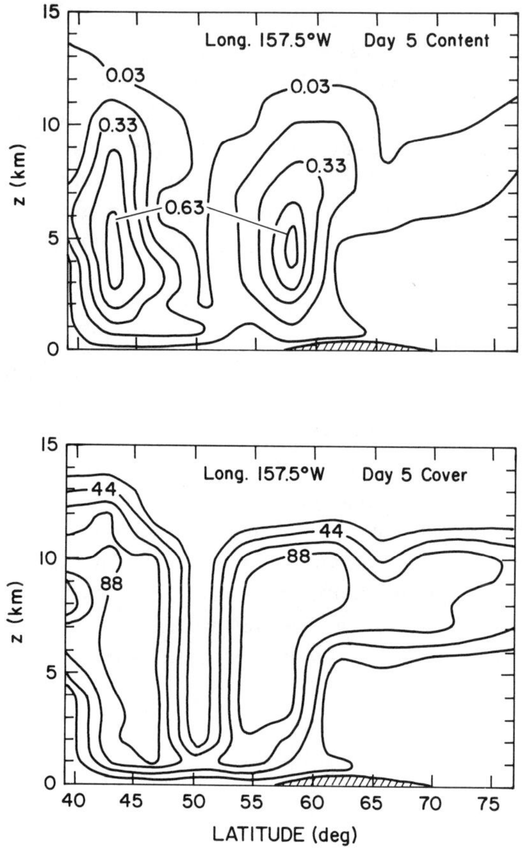

Fig. 11.43. Height–latitude cross section at forecast Day 5 along longitude 157.5°W of cloud water content in g kg^{-1} (top) and cloud cover in percent (bottom). Model land topography is hatched. [From Sundqvist (1981).]

Sundqvist's (1981) hemispheric maps of diagnosed low-, middle-, and high-level cloud cover from the above simulation show a considerable resemblance to satellite photographs. These results suggest that improved predictions and climate simulations are possible with properly tuned cloud parameterizations of the type described here. In addition to selecting values for the constants in the scheme, a proper tuning must consider the interactions of the stratiform cloud parameterization with the other physical parameterizations in the model, including the cumulus convection, radiation, and PBL processes.

References

Atkinson, B. W. (1981). "Meso-scale Atmospheric Circulations." Academic Press, New York.

Austin, P. M. (1960). Microstructure of storms as described by quantitative radar data. *Geophys. Monogr.* **5**, 86–92.

Austin, P. M., and R. A. Houze, Jr. (1972). Analysis of the structure of precipitation patterns in New England. *J. Appl. Meteorol.* **11**, 926–935.

Baldwin, D., E.-Y. Hsie, and R. A. Anthes (1984). Diagnostic studies of a two-dimensional simulation of frontogenesis in a moist atmosphere. *J. Atmos. Sci.* **41**, 2686–2700.

Benjamin, S. G., and T. N. Carlson (1986). Some effects of surface heating and topography on the regional severe storm environment. Part I: Three-dimensional simulations. *Mon. Weather Rev.* **114**, 307–329.

Bennetts, D. A., and B. J. Hoskins (1979). Conditional symmetric instability—A possible explanation for frontal rainbands. *Q. J. R. Meteorol. Soc.* **105**, 945–962.

Bennetts, D. A., and P. Ryder (1984). A study of mesoscale convective bands behind cold fronts. Part I: Mesoscale organization. *Q. J. R. Meteorol. Soc.* **110**, 121–145.

Blackadar, A. K. (1957). Boundary layer wind maxima and their significance for the growth of nocturnal inversions. *Bull. Am. Meteorol. Soc.* **38**, 283–290.

Boatman, J. F., and R. F. Reinking (1984). Synoptic and mesoscale circulations and precipitation mechanisms in shallow upslope storms over the western high plains. *Mon. Weather Rev.* **112**, 1725–1744.

Bonner, W. D. (1966). Case study of thunderstorm activity in relation to the low-level jet. *Mon. Weather Rev.* **94**, 167–178.

Bonner, W. D. (1968). Climatology of the low level jet. *Mon. Weather Rev.* **96**, 833–850.

Braham, R. R., Jr. (1967). Cirrus cloud feeding as a trigger for storm development. *J. Atmos. Sci.* **24**, 311–312.

Braham, R. R., Jr. (1983). The midwest snow storm of 8–11 December 1977. *Mon. Weather Rev.* **111**, 253–272.

Braham, R. R., Jr., and R. D. Kelly (1982). Lake-effect snow storms on Lake Michigan, U.S.A. *In* "Cloud Dynamics," (E. M. Agee and T. Asai, eds.), pp. 87–101. Reidel, Dordrecht, Netherlands.

Braham, R. R., Jr., and P. Spyers-Duran (1967). Survival of cirrus crystals in clear air. *J. Appl. Meteorol.* **6**, 1053–1061.

Browning, K. A. (1971). Radar measurements of air motion near fronts. Part II: *Weather* **26**, 320–340.

Browning, K. A., and T. W. Harrold (1969). Air motion and precipitation growth in a wave depression. *Q. J. R. Meteorol. Soc.* **95**, 288–309.

Businger, S. (1985). The synoptic climatology of polar low outbreaks. *Tellus* **37A**, 419–432.

Carbone, R. E. (1982). A severe frontal rainband. Part I: Stormwide hydrodynamic structure. *J. Atmos. Sci.* **39**, 258–279.

Carlson, T. N. (1980). Airflow through midlatitude cyclones and the common cloud pattern. *Mon. Weather Rev.* **108**, 1498–1509.

Churchill, D. D., and R. A. Houze (1984). Development and structure of winter monsoon cloud clusters on 10 December 1978. *J. Atmos. Sci.* **41**, 933–960.

Cox, S. K. (1971). Cirrus clouds and climate. *J. Atmos. Sci.* **28**, 1513–1515.

Cunningham, R. M. (1951). Some observations of natural precipitation processes. *Bull. Am. Meteorol. Soc.* **32**, 334–343.

Davis, L. G., R. L. Lavoie, J. I. Kelley, and C. L. Hosler (1968). Lake-effect studies. Final Rep. to Environ. Sci. Serv. Adm., Contract No. E22-80-67(N) Dep. Meteorol., Pennsylvania State Univ.

Ellenton, G. E., and M. B. Danard (1979). Inclusion of sensible heating in convective parameterization applied to lake-effect snow. *Mon. Weather Rev.* **107**, 551–565.

Emanuel, K. A. (1979). Inertial instability and mesoscale convective systems. Part I: Linear theory of inertial instability in rotating viscous fluids. *J. Atmos. Sci.* **36**, 2425–2449.

Emanuel, K. A. (1982). Inertial instability and mesoscale convective systems. Part II: Symmetric CISK in a baroclinic flow. *J. Atmos. Sci.* **39**, 1080–1097.

Emanuel, K. A. (1983a). On assessing local conditional symmetric instability from atmospheric soundings. *Mon. Weather Rev.* **111**, 2016–2033.

Emanuel, K. A. (1983b). The Lagrangian parcel dynamics of moist symmetric instability. *J. Atmos. Sci.* **40**, 2369–2376.

Emanuel, K. A. (1983c). Symmetric instability. *In* "Mesoscale Meteorology—Theories, Observations and Models" (D. K. Lilly and T. Gal-Chen, eds.), pp. 217–229. Reidel, Dordrecht, Netherlands.

Emanuel, K. A. (1983d). Conditional symmetric instability: A theory for rainbands within extratropical cyclones. *In* "Mesoscale Meteorology—Theories, Observations and Models" (D. K. Lilly and T. Gal-Chen, eds.), pp. 231–245. Reidel, Dordrecht, Netherlands.

E.S.S.A. (1966). "Storm Data," Monthly publication, December. U.S. Gov. Print. Off., Washington, D.C.

Forbes, G. S., and J. H. Merritt (1984). Mesoscale vortices over the Great Lakes in wintertime. *Mon. Weather Rev.* **112**, 377–381.

Hall, W. D., and H. R. Pruppacher (1976). The survival of ice particles falling from cirrus clouds in subsaturated air. *J. Atmos. Sci.* **33**, 1995–2006.

Hansen, J., G. Russell, D. Rind, P. Stone, A. Lacis, S. Lebedeff, R. Ruedy, and L. Travis (1983). Efficient three-dimensional global models for climate studies: Models I and II. *Mon. Weather Rev.* **111**, 609–662.

Harrold, T. W. (1973). Mechanisms influencing the distribution of precipitation within baroclinic disturbances. *Q. J. R. Meteorol. Soc.* **99**, 232–251.

Herzegh, P. H., and P. V. Hobbs (1980). The mesoscale and microscale structure and organization of clouds and precipitation in midlatitude cyclones. II. Warm-frontal clouds. *J. Atmos. Sci.* **37**, 597–611.

Heymsfield, A. J. (1975a). Cirrus uncinus generating cells and the evolution of cirriform clouds. Part I: Aircraft observations of the growth of the ice phase. *J. Atmos. Sci.* **32**, 799–808.

Heymsfield, A. J. (1975b). Cirrus uncinus generating cells and the evolution of cirriform clouds. Part II: The structure and circulations of the cirrus uncinus generating head. *J. Atmos. Sci.* **32**, 809–819.

Heymsfield, A. J. (1977). Precipitation development in stratiform ice clouds: A microphysical and dynamical study. *J. Atmos. Sci.* **34**, 367–381.

Heymsfield, A. J., and R. G. Knollenberg (1972). Properties of cirrus generating cells. *J. Atmos. Sci.* **29**, 1358–1366.

Hjelmfelt, M. (1988). Numerical sensitivity study of the morphology of lake-effect snow storms over Lake Michigan. *Prepr., Int. Cloud Phys. Conf., 10th, Int. Assoc. Meteorol. Atmos. Phys., Bad Homburg, F.R.G.*, pp. 413–415.

Hjelmfelt, M. R., and R. R. Braham, Jr. (1983). Numerical simulation of the airflow over Lake Michigan for a major lake-effect snow event. *Mon. Weather Rev.* **111**, 205–219.

Hobbs, P. V. (1978). Organization and structure of clouds and precipitation on the mesoscale and microscale in cyclonic storms. *Rev. Geophys. Space Phys.* **16**, 741–755.

Hobbs, P. V., T. J. Matejka, P. H. Herzegh, P. H. Locatelli, and R. A. Houze, Jr. (1980). The mesoscale and microscale structure and organization of clouds and precipitation in midlatitude cyclones. I. A case study of a cold front. *J. Atmos. Sci.* **37**, 568–596.

Holloway, J. L., Jr., and S. Manabe (1971). Simulation of climate by a global general circulation model. *Mon. Weather Rev.* **79**, 335–370.

Holton, J. R. (1979). "An Introduction to Dynamic Meteorology," 2nd Ed. Academic Press, New York.

Hoskins, B. J., and F. P. Bretherton (1972). Atmospheric frontogenesis models: Mathematic formulation and solution. *J. Atmos. Sci.* **29**, 11–37.

Houze, R. A., Jr., and P. V. Hobbs (1982). Organization and structure of precipitating cloud systems. *Adv. Geophys.* **24**, 225–315.

Houze, R. A., Jr., S. A. Rutledge, T. J. Matejka, and P. V. Hobbs (1981). The mesoscale and microscale structure and organization of clouds and precipitation in midlatitude cyclones. III. Air motion and precipitation growth in a warm-frontal rainband. *J. Atmos. Sci.* **38**, 639–649.

Hsie, E.-Y., R. A. Anthes, and D. Keyser (1984). Numerical simulation of frontogenesis in a moist atmosphere. *J. Atmos. Sci.* **41**, 2581–2594.

Hsu, H.-M. (1987). Mesoscale lake-effect snowstorms in the vicinity of Lake Michigan: Linear theory and numerical simulations. *J. Atmos. Sci.* **44**, 1019–1040.

Johnson, D. R., C. H. Wash, and R. A. Petersen (1976). The mass and absolute angular momentum budgets of the Alberta cyclone of 30 March—2 April, 1971. *Prepr., Conf. Weather Forecasting Anal., 6th, Albany, N.Y.*, pp. 350–356. Am. Meteorol. Soc., Boston, Massachusetts.

Juisto, J. E. (1967). Nucleation factors in the development of clouds. Ph.D. Thesis, Pennsylvania State Univ.

Keyser, D., and R. A. Anthes (1982). The influence of planetary boundary layer physics on frontal structure in the Hoskins–Bretherton horizontal shear model. *J. Atmos. Sci.* **39**, 1783–1802.

Keyser, D., and M. J. Pecnick (1985a). A two-dimensional primitive equation model of frontogenesis forced by confluence and horizontal shear. *J. Atmos. Sci.* **42**, 1259–1282.

Keyser, D., and M. J. Pecnick (1985b). Diagnosis of ageostrophic circulations in a two-dimensional primitive equation model of frontogenesis. *J. Atmos. Sci.* **42**, 1283–1305.

Keyser, D., and M. A. Shapiro (1986). A review of the structure and dynamics of upper-level frontal zones. *Mon. Weather Rev.* **114**, 452–499.

Lavoie, R. L. (1972). A mesoscale numerical model of lake-effect storms. *J. Atmos. Sci.* **29**, 1025–1040.

Lilly, D. K. (1988). Cirrus outflow dynamics. *J. Atmos. Sci.* **45**, 1594–1605.

Locatelli, J. D., P. V. Hobbs, and J. A. Werth (1982). Mesoscale structures of vortices in polar air streams. *Mon. Weather Rev.* **110**, 1417–1433.

Matejka, T. J., P. V. Hobbs, and R. A. Houze (1978). Microphysical and dynamical structure of mesoscale cloud features in extratropical cyclones. *Prepr., Conf. Cloud Phys. Atmos. Electr., Issaquah, Wash.* pp. 292–299. Am. Meteorol. Soc., Boston, Massachusetts.

Matveev, L. T. (1984). "Cloud Dynamics." Reidel, Dordrecht, Netherlands.

McVehil, G. E., and R. L. Peace, Jr. (1965). Some studies of lake effect snowfall from Lake Erie. *Proc. Conf. Great Lakes Res., 8th, Great Lakes Res. Div., Univ. Michigan* Publ. No. 13, pp. 262–372.

Mullen, S. L. (1979). An investigation of small synoptic scale cyclones in polar air streams. *Mon. Weather Rev.* **107**, 1636–1647.

Nuss, W. A., and R. A. Anthes (1987). A numerical investigation of low-level processes in rapid cyclogenesis. *Mon. Weather Rev.* **115**, 2728–2743.

Ogura, Y., H.-M. Juang, K.-S. Zhang, and S.-T. Soong (1982). Possible triggering mechanisms for severe storms in SESAME-AVE IV (9–10 May 1979). *Bull. Am. Meteorol. Soc.* **63**, 503–515.

Parsons, D. B., and P. V. Hobbs (1983). The mesoscale and microscale structure and organization of clouds and precipitation in midlatitude cyclones. VII: Formation, development, interaction and dissipation of rainbands. *J. Atmos. Sci.* **40**, 559–579.

Passarelli, R. E., Jr., and R. R. Braham, Jr. (1981). The role of the winter land breeze in the formation of Great Lake snow storms. *Bull. Am. Meteorol. Soc.* **62**, 482-491.

Peace, R. L., Jr., and R. B. Sykes, Jr. (1966). Mesoscale study of a lake effect snow storm. *Mon. Weather Rev.* **94**, 495-507.

Pease, S. R., W. A. Lyons, C. S. Keen, and M. Hjelmfelt (1988). Mesoscale spiral vortex embedded within a Lake Michigan snow squall band: High resolution satellite observations and numerical model simulations. *Mon. Weather Rev.* **116**, 1374-1380.

Rasmussen, E. (1979). The polar low as an extratropical CISK-disturbance. *Q. J. R. Meteorol. Soc.* **105**, 531-549.

Rasmussen, E. (1981). An investigation of a polar low with a spiral cloud structure. *J. Atmos. Sci.* **38**, 1785-1792.

Rasmussen, E. (1985). A case study of a polar low development. *Tellus* **37A**, 407-418.

Raymond, D. J. (1983). Wave-CISK mass flux form. *J. Atmos. Sci.* **40**, 2561-2572.

Reed, R. J. (1979). Cyclogenesis in polar air streams. *Mon. Weather Rev.* **107**, 38-52.

Roebber, P. J. (1984). Statistical analysis and updated climatology of explosive cyclones. *Mon. Weather Rev.* **112**, 1577-1589.

Rogers, E., and L. F. Bosart (1986). An investigation of explosively deepening oceanic cyclones. *Mon. Weather Rev.* **114**, 702-718.

Rosinski, J., C. T. Nagamoto, G. Langer, and E. P. Parungo (1970). Cirrus clouds as collectors of aerosol particles. *J. Geophys. Res.* **75**, 2961-2973.

Rothrock, H. J. (1969). An aid in forecasting significant lake snows. Tech. Memo. WBTM CR-30, U.S. Dep. Commer., E.S.S.A.

Rutledge, S. A., and P. V. Hobbs (1983). The mesoscale and microscale structure and organization of clouds and precipitation in midlatitude cyclones. VIII: A model for the "Seeder-Feeder" process in warm frontal rainbands. *J. Atmos. Sci.* **40**, 1185-1206.

Ryan, R. T., H. H. Blan, Jr., P. C. van Thuna, and M. L. Cohen (1972). Cloud microstructure as determined by an optical cloud particle spectrometer. *J. Appl. Meteorol.* **11**, 149-156.

Sanders, F., and J. R. Gyakum (1980). Synoptic-dynamic climatology of the "bomb." *Mon. Weather Rev.* **108**, 1589-1606.

Sardie, J. M., and T. T. Warner (1985). A numerical study of the development mechanisms of polar lows. *Tellus* **37A**, 460-477.

Schneider, S. H. (1972). Cloudiness as a global climatic feedback mechanism: The effects on the radiation balance and surface temperature of variation in cloudiness. *J. Atmos. Sci.* **29**, 1413-1422.

Sekhon, R. S., and R. C. Srivastava (1970). Snow size spectra and radar reflectivity. *J. Atmos. Sci.* **27**, 299-307.

Shapiro, M. A., T. Hample, and D. W. van de Kamp (1984). Radar wind profiles observations of fronts and jet streams. *Mon. Weather Rev.* **112**, 1263-1266.

Sheridan, L. W. (1941). The influence of Lake Erie on local snows in western New York. *Bull. Am. Meteorol. Soc.* **22**, 393-395.

Shukla, J., and Y. Sud (1981). Effect of cloud-radiation feedback on the climate of a GCM. *J. Atmos. Sci.* **38**, 2337-2353.

Staley, D. O., and R. L. Gall (1977). On the wavelength of maximum baroclinic instability. *J. Atmos. Sci.* **34**, 1679-1688.

Starr, D. O'C., and S. K. Cox (1980). Characteristics of middle and upper tropospheric clouds as deduced from rawinsonde data. Atmos. Sci. Pap. No. 327, Colorado State Univ. (US ISSN 0 067-0340).

Starr, D. O'C., and S. K. Cox (1985a). Cirrus clouds. Part I: A cirrus cloud model. *J. Atmos. Sci.* **42**, 2663-2681.

Starr, D. O'C., and S. K. Cox (1985b). Cirrus clouds. Part II: Numerical experiment on the formation and maintenance of cirrus. *J. Atmos. Sci.* **42**, 2682–2694.

Stephens, G. L. (1984). The parameterization of radiation for numerical weather prediction and climate models. *Mon. Weather Rev.* **112**, 826–867.

Sundqvist, H. (1978). A parameterization scheme for nonconvective condensation including prediction of cloud water content. *Q. J. R. Meteorol. Soc.* **104**, 667–690.

Sundqvist, H. (1981). Prediction of stratiform clouds: Results from a 5-day forecast with a global model. *Tellus* **33**, 242–253.

Sykes, R. B. (1966). The blizzard of 1966 in Central New York—Legend in its time. *Weatherwise* **19**, 241–247.

Wiggin, B. L. (1950). Great snows of the Great Lakes. *Weatherwise* **3**, 123–126.

Chapter 12 The Influence of Mountains on Airflow, Clouds, and Precipitation

12.1 Introduction

The emphasis of this chapter is on wintertime clouds and cloud systems that are forced at least in part by orography, but excluding deep convective clouds. We include, however, convective clouds that can be triggered by the release of potential instability during the passage of extratropical cyclones. Our focus is on air motions over mountainous terrain that are conducive to the formation and spatial distribution of precipitation. We also consider the interaction between flow over mountains and larger scale precipitating weather systems such as extratropical cyclones.

We briefly review the theories and models of orographic flows, and then turn our attention to the formation and distribution of precipitation over mountainous terrain.

12.2 Theory of Flow over Hills and Mountains

The classical theories of flow over small-amplitude hills are based on the linearized equations of motion, the thermodynamic energy equation, and mass-continuity equation. The basic approach is similar to the scale-analysis procedures employed in Chapter 2. For a more extensive review of this topic, we recommend the excellent review by Smith (1979).

As in Chapter 2, we decompose each variable into a base-state value with a subscript 0 and a perturbation from the reference state. Thus, for purely two-dimensional flow,

$$
\begin{aligned}
u(x, z) &= u_0(z) + u'(x, z),\\
w(x, z) &= w'(x, z),\\
\rho(x, z) &= \rho_0(z) + \rho'(x, z),\\
p(x, z) &= p_0(z) + p'(x, z),\\
T(x, z) &= T_0(z) + T'(x, z),
\end{aligned}
\tag{12.1}
$$

where vertical motion in the unperturbed atmosphere is zero $[w_0(z) = 0]$. Substituting Eq. (12.1) into the governing equations (Chapter 2), assuming the flow field is steady state, and linearizing, we obtain

$$\rho_0[u_0(\partial u'/\partial x) + w'(\partial u_0/\partial z)] = -\partial p'/\partial x, \tag{12.2}$$

$$\rho_0[u_0(\partial w'/\partial x)] = -\partial p'/\partial z - \rho' g, \tag{12.3}$$

for the equations of motion. The linearized mass-continuity equation is

$$u_0(\partial \rho'/\partial x) + w'(\partial \rho_0/\partial z) = -\rho_0(\partial u'/\partial x + \partial w'/\partial z). \tag{12.4}$$

If we now differentiate the equation of state for a dry atmosphere with respect to time,

$$dp/dt = RT(d\rho/dt) + \rho R(dT/dt), \tag{12.5}$$

and substitute the thermodynamic energy equation for adiabatic motion into Eq. (12.5), we obtain after some manipulation

$$dp/dt = (c_p/c_v)RT(d\rho/dt). \tag{12.6}$$

The quantity,

$$(c_p/c_v)RT = \gamma RT = c^2, \tag{12.7}$$

where c^2 is the speed of sound for dry adiabatic motion. Equation (12.6) thus becomes,

$$dp/dt = c^2(d\rho/dt). \tag{12.8}$$

Expansion of the total derivatives in Eq. (12.8) into local and advective contributions and linearization gives

$$u_0(\partial p'/\partial x) + w'(dp_0/dz) = c_0^2[u_0(\partial \rho'/\partial x) + w'(d\rho_0/dz)]. \tag{12.9}$$

Equation (12.9) can be rearranged into the form

$$\underset{\text{(a)}}{\frac{u_0}{\rho_0}\frac{\partial \rho'}{\partial x}} = w'\left(\underset{\text{(b)}}{-\frac{1}{\rho_0}\frac{d\rho_0}{dz}} + \underset{\text{(c)}}{\frac{g}{c_0^2}}\right) - \underset{\text{(d)}}{\frac{u_0}{\rho_0 c_0^2}\frac{\partial p'}{\partial x}}. \tag{12.10}$$

Equation (12.10) describes the formation of density anomalies which give rise to accelerations in vertical velocity. Term (a) represents the change in density experienced by an observer moving horizontally downstream at a speed $u_0(z)$. Term (b) represents the change in density due to vertical displacement of the base-state air mass. Term (c) represents the effects of adiabatic expansion or compression of the air parcel during vertical displacement. Term (d) accounts for the pressure deviation during vertical displacement, which is normally small for slow vertical displacements and is only important in the generation of fast acoustic waves.

Employing the definition of potential temperature (Chapter 2) and assuming a hydrostatic base state, we find that terms (b) and (c) can be combined to yield

$$-(1/\rho_0)(d\rho_0/dz)+(g/c_0^2)=(1/\theta_0)(d\theta_0/dz)=\beta. \tag{12.11}$$

Thus, terms (b) and (c) are a measure of static stability. Ignoring term (d), Eq. (12.10) can be written as

$$(u_0/\rho_0)(\partial\rho'/\partial x)=w'\beta=w'N^2/g, \tag{12.12}$$

where $N^2=g/\theta(\partial\theta/\partial z)$, and N is the Brunt–Väisälä frequency. Substitution of Eqs. (12.11) and (12.10) into Eq. (12.4) results in the simplified continuity equation,

$$(\partial u'/\partial x)+(\partial w'/\partial z)=(g/c_0^2)w'. \tag{12.13}$$

Following a straightforward, though lengthy, process of elimination of variables (i.e., pressure in particular), the above equations can be combined to form a single equation for vertical velocity,

$$\frac{\partial^2 w'}{\partial x^2}+\frac{\partial^2 w'}{\partial z^2}-S_0\frac{\partial w'}{\partial z}+\left(\frac{\beta g}{u_0^2}+\frac{S_0}{u_0}\frac{\partial u_0}{\partial z}-\frac{1}{u_0}\frac{\partial^2 u_0}{\partial z^2}\right)w'=0, \tag{12.14}$$

where

$$S_0=(d/dz)\ln\rho_0(z). \tag{12.15}$$

If we now introduce the new dependent variable,

$$\tilde{w}=[\rho_0(z)/\rho_0(0)]^{1/2}w', \tag{12.16}$$

then (12.13) becomes

$$(\partial^2\tilde{w}/\partial x^2)+(\partial^2\tilde{w}/\partial z^2)+l^2(z)\tilde{w}=0, \tag{12.17}$$

where

$$l^2(z)=\frac{\beta g}{u_0^2}-\frac{1}{u_0}\frac{\partial^2 u_0}{\partial z^2}+\frac{S_0}{u_0}\frac{\partial u_0}{\partial z}-\frac{1}{4}S_0^2+\frac{1}{2}\frac{\partial S_0}{\partial z}. \tag{12.18}$$

Equation (12.17) represents the foundation for two-dimensional, steady, linear mountain-wave theory. The parameter $l(z)$, often referred to as the Scorer parameter, is dominated by the stability and shear terms [first two terms in Eq. (12.18)]. Neglect of the latter terms involving S_0 is equivalent to making the Boussinesq approximation (Chapter 2) in which density variations are ignored except where multiplied by gravity.

In a Fourier decomposition of the horizontal structure of Eq. (12.17), the vertical velocity $\tilde{w}$ must satisfy the wave solution of the form

$$\tilde{w} = \hat{w}(z)\, e^{ikx}. \tag{12.19}$$

The amplitude of the kth component $\hat{w}(z)$ is then

$$[\partial^2 \hat{w}(z)/\partial z^2] + [l^2(z) - k^2]\hat{w}(z) = 0. \tag{12.20}$$

Solutions to Eq. (12.20) can be obtained for specified lower and upper boundary conditions. For an upper boundary condition, a radiation condition (e.g., Sommerfeld, 1912, 1948) at $z = \infty$ is imposed to prevent the reflection of upward-propagating waves that could interfere with the waves of interest in the middle and upper troposphere.

A free-slip lower boundary condition is applied such that

$$w(0, x) = u(0, x)[\partial h(x)/\partial x], \tag{12.21}$$

where $h(x)$ is the height of the topography. A variety of orographic shapes have been imposed as lower boundaries. Lyra (1943) used a rectangular orographic profile—which a number of investigators have criticized because it could force unrealistic gravity wave modes.

Queney (1947) obtained solutions to the mountain-wave problem for infinite sinusoidal orography. More general isolated orographic shapes can be obtained from the Fourier transform of the mountain shape,

$$\tilde{h}(k) = \frac{1}{\pi}\int_{-\infty}^{\infty} h(x)\, e^{-ikx}\, dx. \tag{12.22}$$

Queney (1947, 1948) used the bell-shaped, so-called "witch of Agnesi" profile,

$$h(x) = z_{\mathrm{m}} a^2/(x^2 + a^2), \tag{12.23}$$

which has the simple Fourier transform,

$$\tilde{h}(k) = z_{\mathrm{m}} a\, e^{-ka}, \tag{12.24}$$

where z_{m} is the height of the mountain and a is the mountain half-width.

With the top and bottom boundary conditions specified and with suitable definitions of $l^2(z)$, analytic solutions can be obtained to Eq. (12.20) for small-amplitude hills (i.e., on the order of 10 m or so). The problem is then to find solutions $\hat{w}(z)$ to Eq. (12.20) for all values of k and then evaluate the Fourier integral of the corresponding component solutions yielding the total disturbance in the flow field over a finite mountain barrier. The solutions to Eq. (12.20) are frequently extended to larger amplitude hills by multiplying the results by the relative scaling of the actual mountain. This presumes, of course, that the linear solutions are valid for the larger amplitude mountains.

12.2.1 Queney Models

Both Lyra and Queney obtained analytic solutions to the linearized mountain-wave equations for the case of constant stability and wind speed, which implies that $l(z)$ is independent of height [Eq. (12.18)]. Queney obtained several special case solutions to Eq. (12.20). For steady mountain waves to exist, they must have a phase velocity C^{p} relative to the fluid, which is equal and opposite to the mean flow u_0, and a group velocity directed upward away from the mountain. In this way, the waves produced by the flow against the mountain can remain steady against the prevailing flow. As air parcels flow over a mountain of wavelength L, they will experience an orbital frequency $2\pi u_0/L$. The highest natural frequency for gravity waves in the atmosphere is the Brunt–Väisälä frequency N. The condition

$$2\pi u_0/L = N \tag{12.25}$$

identifies an important cutoff frequency for pure gravity waves. Thus, for any atmosphere we can identify a cutoff wave number k_{s}, where

$$k_{\mathrm{s}} = N/u_0. \tag{12.26}$$

When the earth's rotation is considered, free oscillations of the geostrophic wind occurs. The natural frequency for such oscillations is the Coriolis parameter f, with a corresponding cutoff frequency

$$2\pi u_0/L = f \tag{12.27}$$

and a cutoff wave number

$$k_{\mathrm{f}} = f/u_0. \tag{12.28}$$

Inertial-gravity waves, therefore, have frequencies between k_{f} and k_{s}.

Queney obtained solutions to the linearized wave equation for the three cases in which a finite mountain half-width is (1) $a \ll 1/k_{\mathrm{s}}$, (2) $a \sim 1/k_{\mathrm{s}}$, and (3) $a \sim 1/k_{\mathrm{f}}$. He also obtained several other special solutions (not discussed here). The first two cases can be described in terms of a Froude number Fr (Clark and Peltier, 1977), defined as

$$\mathrm{Fr} = \frac{2\pi/N}{2a/u_0},$$

the ratio of the Brunt–Väisälä period to the mountain-forcing period of the waves. Therefore, case (1) corresponds to $\mathrm{Fr} \gg 0(1)$ and case (2) corresponds to $\mathrm{Fr} = 0(1)$.

12.2.1.a *Case $a \sim 1/k_s$; $Fr = 0(1)$*

Case $a \sim 1/k_s$ should apply to moderate-sized mountain ranges (10–20 km wide). Figure 12.1 illustrates the perturbation streamlines for this case. The solution is periodic in the vertical such that the flow exhibits an inverted ridge shape at $z = \pi/L$ which reverts to an upright ridge shape at $z = 2\pi/L$. The flow can be described as a field of nondispersive, vertically propagating gravity waves. At low levels, the streamlines exhibit a pronounced crest immediately upwind from the mountain crest and a weak trough on the lee side of the mountain. The upwind crest becomes weaker with height and the lee trough becomes more intense. Both features exhibit an upstream tilt of the phase lines associated with downward momentum transport. The wind speed is lowest on the windward slope of the ridge and fastest on the lee slope. There is a corresponding pressure difference across the ridge with high pressure upwind and lower pressure downwind.

At least qualitatively, the solutions exhibit features similar to the observed warm and dry foehn or chinook winds in the lee of major mountain ranges such as the Alps, Sierras, and Rockies. There is no evidence of lee-wave solutions for this case.

12.2.1.b *Case $a \sim 1/k_f$*

Queney also obtained solutions (Fig. 12.2) for the case of a broad mountain for which the mountain half-width $a \sim 1/k_f$. The actual dimension of a is 100 km. The waves that form are long, having a wavelength of approximately 600 km. Because the waves are located exclusively on the lee side of the mountain range, they give the appearance of trapped lee waves. These waves are hydrostatic and vertical accelerations are small. Trapped lee waves have a different character. The waves in Fig. 12.2, more properly called inertial-gravity waves, exhibit a horizontal displacement as the motion attempts to adjust to a geostrophic balance.

12.2.1.c *Case $1/a \gg k_s$; $Fr \gg 0(1)$*

Queney obtained solutions for a small mountain for which $a \sim 1$ km and $a \ll 1/k_s$ (Fig. 12.3). The effects of the earth's rotation are negligible for this case. At low levels, the flow resembles Fig. 12.1; at higher levels and particularly to the lee of the mountain, the flow resembles lee waves. Smith (1979) referred to these as a "dispersive tail" of nonhydrostatic waves with k less than, but not much less than, $l(z)$. As noted by Holmboe and Klieforth (1957), the structure of these lee waves differs markedly from observed lee waves, which have maximum amplitude at low levels and small, if any, tilt in the phase lines with height. Moreover, the amplitude of the observed lee waves is maintained for appreciable distances downwind, while those shown in Fig. 12.3 decay in amplitude rapidly downstream.

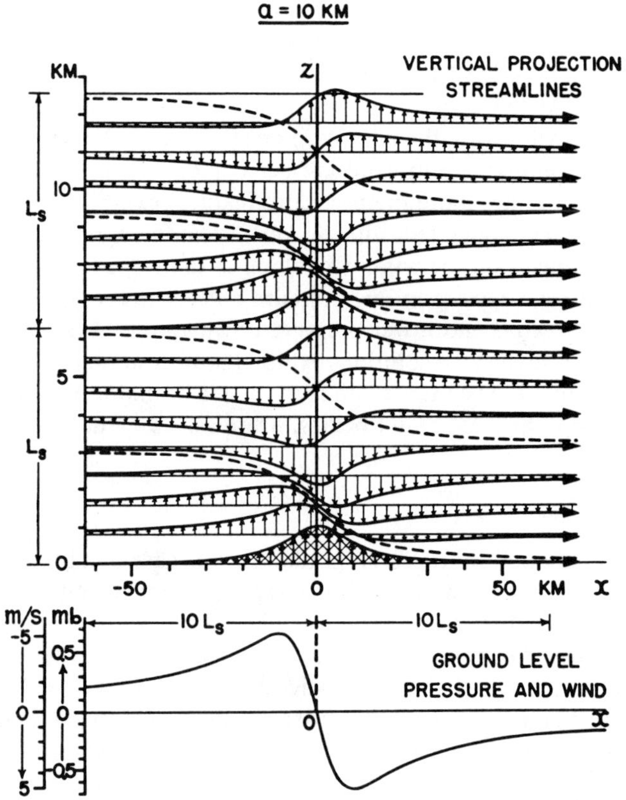

Fig. 12.1. Perturbation by a medium-size typical mountain range ($a = L_s/2\pi = 10$ km) in an unlimited uniform stratified current ($u = 10$ m s^{-1}). Upper part: vertical projection of the streamlines; displacement indicated by small arrows. Lower part: perturbation of pressure and longitudinal wind velocity at ground level. The horizontal displacement is negligible. [From Queney (1948).]

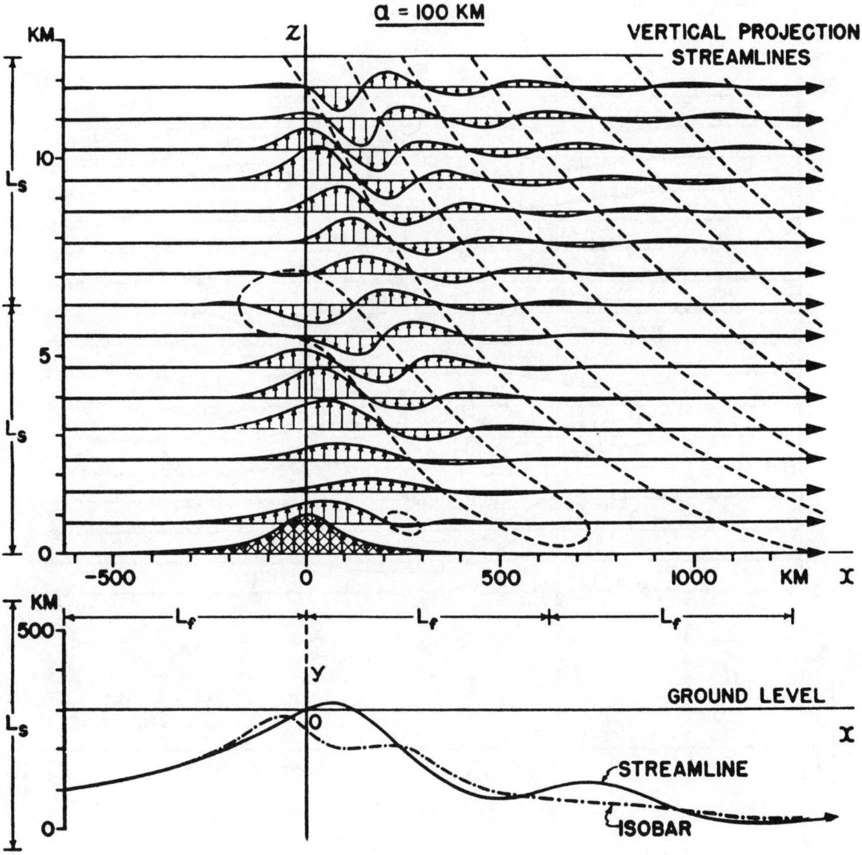

Fig. 12.2. Perturbation by a broad typical mountain range ($a = L_f/2\pi = 100$ km) in an unlimited uniform stratified current ($u = 10$ m s^{-1}). Upper part: vertical projection of the streamlines; displacement indicated by small arrows. Lower part: horizontal projection of a streamline and of the asymptotic isobar at ground level. [From Queney (1948).]

12.2.2 Linear Theory of Trapped Lee Waves

Queney's and Lyra's solutions were obtained for the case in which l^2 in Eq. (12.18) was a constant. Scorer (1949, 1953, 1954) divided the atmosphere into two and three layers, each having a different constant value of l^2. By applying the wave equation to each separate layer, he found probably the most important dynamic requisite for periodic lee-wave development, namely, that l^2 must be less in an upper layer than in a lower layer of sufficient depth. This can be accomplished either by the wind speed

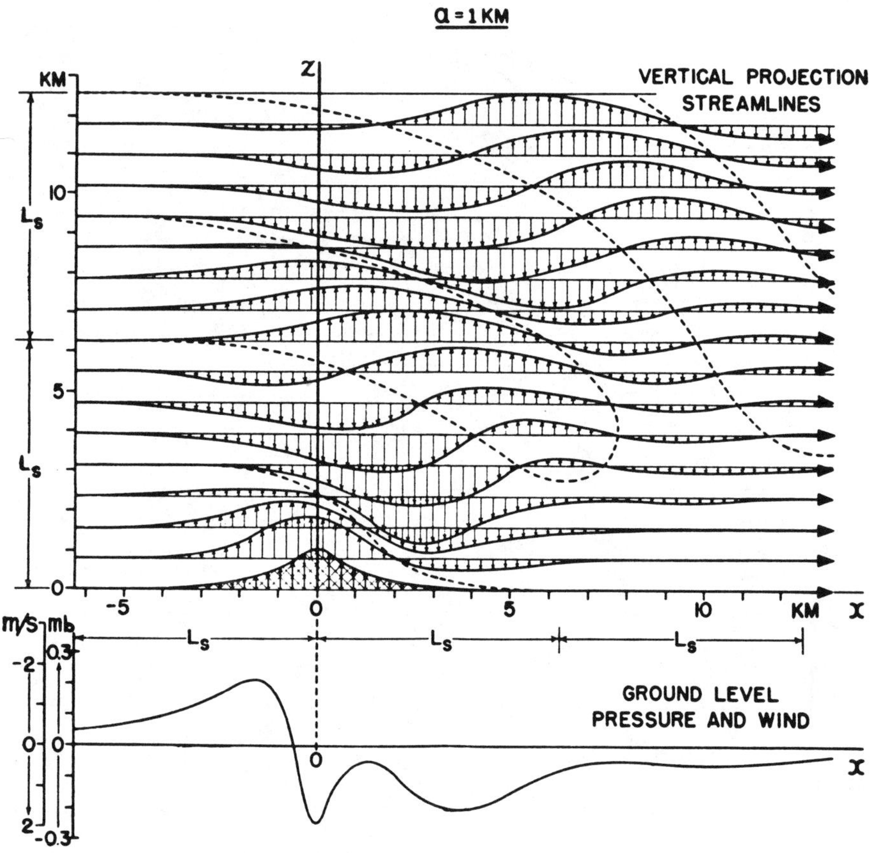

Fig. 12.3. Perturbation by a narrow typical mountain range ($a = L_s/2\pi = 1$ km) in an unlimited uniform stratified current ($u = 10 \text{ m s}^{-1}$). Upper part: vertical projection of the streamlines; displacement indicated by small arrows. Lower part: perturbation of pressure and longitudinal wind velocity at ground level. The horizontal displacement is negligible. [From Queney (1948).]

Fig. 12.4. Displacement of streamlines computed for an idealized airstream with a layer of high l^2 values near the ground. The left-hand section of the diagram shows the assumed potential temperature (θ), wind velocity (U), and Scorer's parameter (l^2) as function of height. The right-hand section of the diagram shows the computed displacements of the streamlines as functions of horizontal distance x at 1-km intervals in the vertical. The vertical displacement is plotted for each level on the same scale as the mountain profile at the bottom of the figure (mountain half-width, $b = 2$ km). [From Sawyer (1960).]

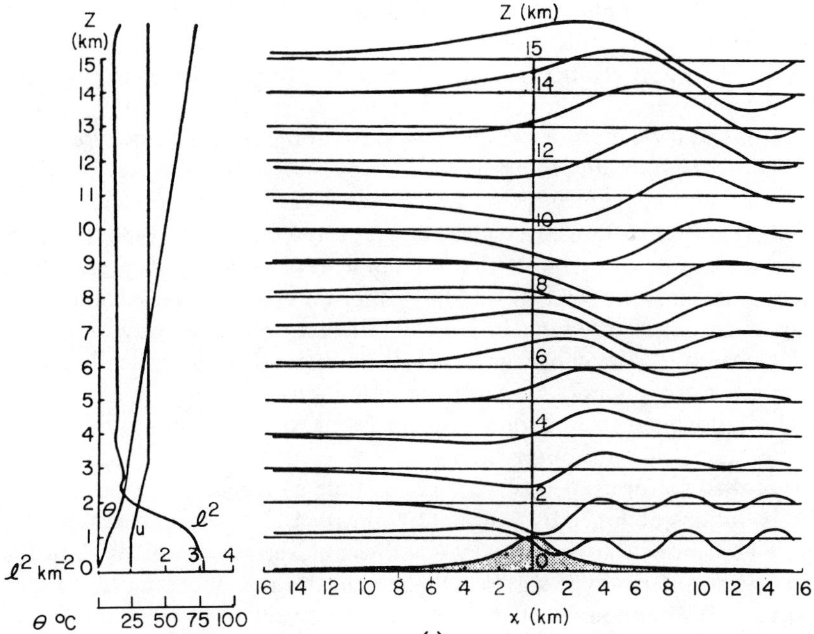
Z (km)
θ
u
ℓ^2
ℓ^2 km^{-2} 0 2 3 4
θ °C 25 50 75 100
u m/s 0 10 20 30 40
Z (km)
16 14 12 10 8 6 4 2 0 2 4 6 8 10 12 14 16
x (km)

(a)

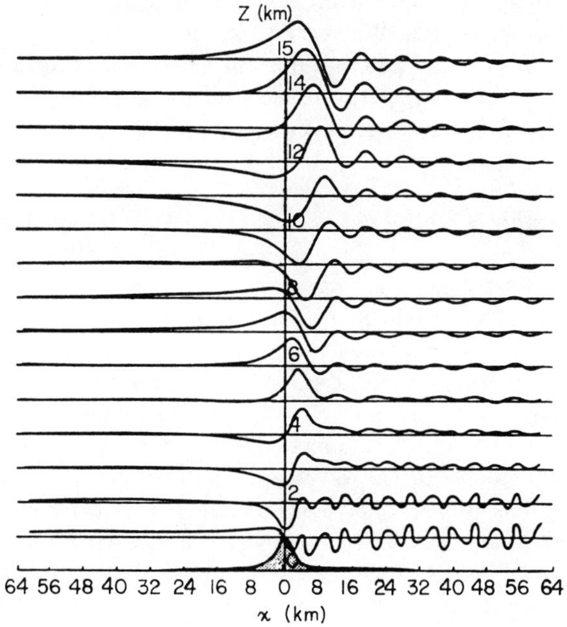
Z (km)
64 56 48 40 32 24 16 8 0 8 16 24 32 40 48 56 64
x (km)

(b)

increasing with height, the presence of a very stable layer in the lower atmosphere, or a combination of the two.

A disparity between the Queney and Scorer solutions has led to considerable debate over the years. The discrepancies arise mainly from different upper boundary conditions for long waves. Both Queney and Lyra chose solutions corresponding to upstream sloping phase lines. Queney obtained a unique solution by introducing a weak viscosity similar to Rayleigh (1883) while Lyra (1943) followed a technique pioneered by Kelvin (1886) in which disturbances far upstream of the mountain were removed. Scorer, on the other hand, chose the correct boundary conditions for the waves that traveled downstream, thus allowing him to obtain a trapped lee-wave solution. Unfortunately, Scorer chose the incorrect conditions for the vertically propagating waves: his main wave over the mountain crest exhibits phase lines which tilt downstream, a property inconsistent with a stationary mountain-wave pattern and with observations.

That the two forms of waves are compatible with each other can be seen in the solutions obtained by Sawyer (1960) with a 17-layer numerical model. Figure 12.4 shows trapped lee waves at lower elevations as well as vertically propagating Queney-type waves at higher elevations. Conceptually, trapped lee waves can be thought of in terms of the movement of wave packets in the atmosphere in which l^2 decreases with height. In the stable lower atmosphere, waves with $k^2 < l^2$ propagate upward and downwind of the mountain barrier. When the waves encounter a level where $k^2 > l^2$, the wave cannot propagate so that the wave energy is reflected downward. The wave energy then repeatedly reflects off the ground (or an adiabatic layer near the surface) and the layer where $k^2 > l^2$. The wave energy is said to be trapped, and leads to a standing wave pattern in which the phase lines have no tilt.

Smith (1979) noted several other ways in which trapped lee waves may occur. For example, in an abrupt change in stability at the tropopause level, a vertically propagating wave will be partially reflected downward. The reflected wave will then rebound off the surface of the earth, where it can partially reflect off the tropopause again. Each time, the wave loses a fraction of its energy, because the reflection is only partial; the wave decays downstream.

Scorer's discovery spawned nearly two decades of analytical and observational studies of lee waves and lee-wave clouds. Models were extended to include a number of layers (Wurtele, 1953; Corby and Sawyer, 1958; Sawyer, 1960; Danielsen and Bleck, 1970) and even an infinite number of layers (Palm, 1955). Small-amplitude linear theory has been applied to three-dimensional simulations of trapped lee waves (Scorer and Wilkinson, 1956; Palm, 1958; Sawyer, 1962; Crapper, 1962; Gjevik and Marthinsen, 1977).

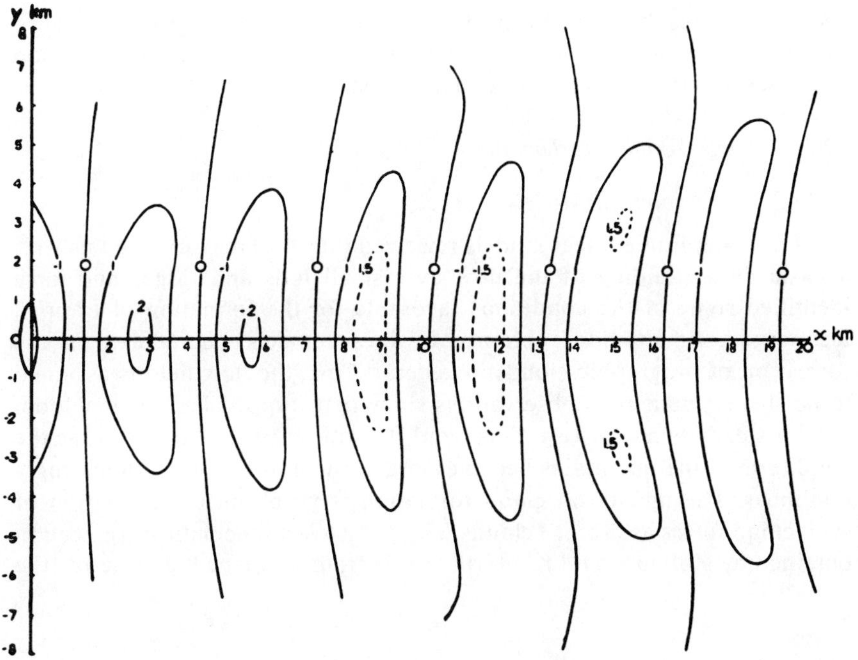

Fig. 12.5. Lee-wave component of vertical velocity at 1.5 km for two-layer airstream. Velocities in m s^{-1} for a 100-m hill. [From Sawyer (1962).]

The three-dimensional wave solutions obtained by Sawyer (1962), shown in Fig. 12.5 for a small isolated hill, resemble surface waves behind a moving ship. The wave pattern is wedge shaped with its apex at the mountain crest.

Clark and Gall (1982) have also investigated some three-dimensional properties of trapped lee waves with a three-dimensional, nonhydrostatic, numerical model. They simulated the flow about Elk Mountain, an isolated peak in southern Wyoming. Although from the ground the mountain appears as an isolated peak, a high ridge starts ~15 km south of the mountain and extends southward. One expects that under prevailing westerly flows or even northeasterly flow, the ridge, being largely downstream, would have little influence on the flow near Elk Mountain. Clark and Gall found that, when the atmosphere supported lee waves, the lee waves formed by the ridges south of Elk Mountain interacted with lee waves generated by Elk Mountain. The resultant constructive and destructive interference among the lee-wave families caused significant changes in the wind field east of Elk Mountain. When the atmosphere supported only freely propagating waves, Elk Mountain remained dynamically isolated from the neighboring ridges. These results show that properly formulated and constructed numerical models are useful in the investigation of flow fields and precipitation processes about finite-amplitude mountains.

12.2.3 Large-Amplitude Theories

12.2.3.a Analytical Models

The finite-amplitude linear models reviewed thus far have provided us with a basic understanding of the flow over small hills and ridges and have identified some of the conditions favorable for the formation of trapped lee waves. The solutions obtained by these models cannot explain the formation of orographic clouds in general. First, the flow field is assumed to be steady; numerous observations show that trapped lee waves seldom exhibit steady behavior (e.g., Starr and Browning, 1972). Moreover, as the amplitude of the mountains becomes larger, the flow becomes increasingly nonlinear. The impact on cloud formation is visible in the formation of such cloud forms as "rotor" clouds. Figure 12.6 is a schematic cross section obtained by Holmboe and Klieforth (1957) from sailplane flights across the

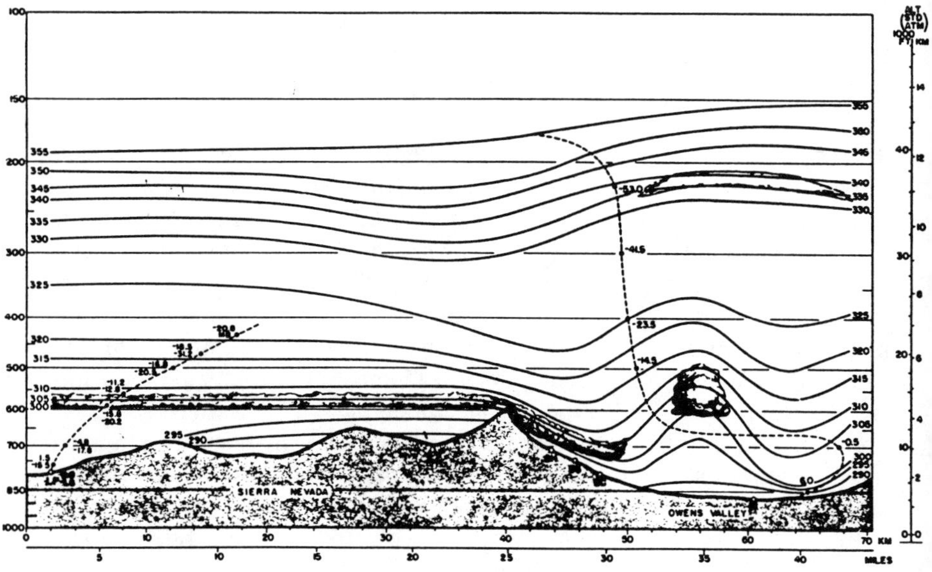

Fig. 12.6. West to east vertical cross section of potential temperature across the Sierra Nevada. Dashed line represents sailplane soundings. Observed Chinook arch or Foehn wall cloud is illustrated over barrier crest, as well as rotor cloud at low levels to the east and lenticular cloud at higher levels. [From Holmboe and Klieforth (1957).]

Sierra Nevada. Noteworthy are the foehn, or wall cloud, that extends some distance upwind and downwind of the mountain crest, the high-level lenticular cloud, and the low-level rotor cloud to the lee of the barrier. Many of these features, especially the rotor cloud, cannot be adequately explained by small-amplitude linear theory.

Long (1953a, b, 1954, 1955, 1972) developed a series of laboratory and nonlinear, steady-state, hydraulic flow models. The flow features produced by the laboratory hydraulic flow models and the analytic models resemble qualitatively the flow over the Sierra Nevada, particularly the formation of rotor clouds. Kuettner (1958) hypothesized that the flow creating rotor clouds is like a hydraulic jump. Long's models represent special mathematical and physical cases of the governing equations for which the theory becomes exactly linear. Variations in wind field and stability result in different special-case solutions for which analytic solutions are not available. In spite of the apparent qualitative success of Long's laboratory and analytic models, their applicability to atmospheric flow remains questionable

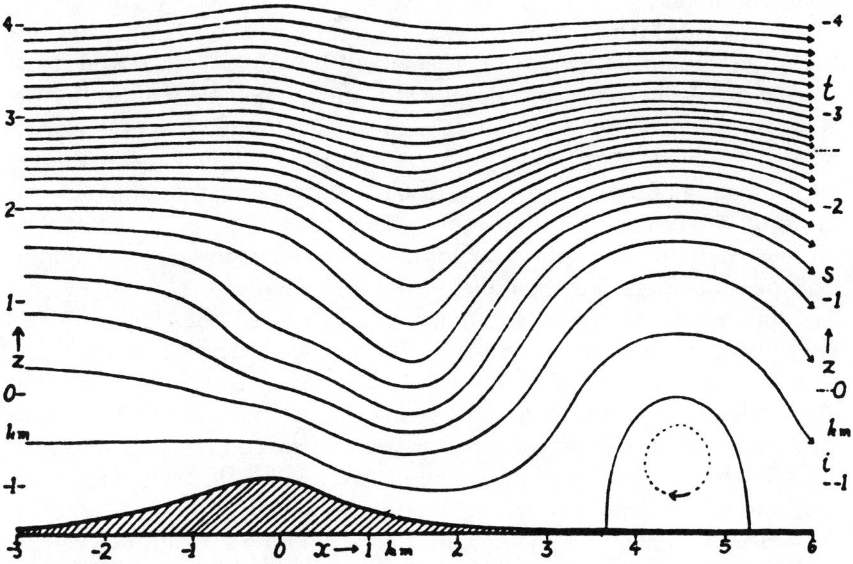

Fig. 12.7. Streamlines for flow over a two-dimensional ridge. Theoretically, only the first of an infinite train of waves is shown. However, the rotor is necessarily turbulent so that the second lee wave would probably be of smaller amplitude. This flow pattern, though particular, could occur in a restricted class or airsteam, but qualitatively it is what might occur in a much wider class. The chief restriction is that the airstream have periodic lee waves of large amplitude. [From Scorer and Klieforth (1959).]

because both laboratory fluid flow and analytic models contain a reflective free surface as an upper boundary. Because the energy reflected off the upper boundary can constructively or destructively interfere with wave energy emitted by the lower topographic surface, the character of the flow may not resemble atmospheric flow. We examine more extensively the application of hydraulic models to orographic flow in our discussion of downslope windstorms and associated cloud formation.

Scorer and Klieforth (1959) obtained solutions for special cases of large-amplitude flow over mountains. They showed that the presence or absence of rotors did not depend on the nonlinearity of the flow. An example of rotorlike flow obtained with linearized equations is shown in Fig. 12.7. The calculated lee waves and rotors, however, actually extend indefinitely downstream of the mountain.

12.2.3.b Numerical Models

As the mountain height increases, nonlinear effects on the flow are increasingly likely. Beginning in the mid-1960s, numerical models were applied to the simulation of flow over large-amplitude mountains (Eliassen and Palm, 1960; Krishnamurti, 1964; Hovermale, 1965; Foldvik and Wurtele, 1967; Eliassen, 1968; Mahrer and Pielke, 1975; Deaven, 1976; Clark and Peltier, 1977; Anthes and Warner, 1978; Klemp and Lilly, 1978; Nickerson, 1979; Durran and Klemp, 1982b; Tripoli and Cotton, 1982; Cotton *et al.*, 1986; Nickerson *et al.*, 1986). These models differ in the formulation of numerical operators, vertical coordinate systems, top and lateral boundary conditions, the evaluation of pressure (i.e., hydrostatic versus nonhydrostatic), and the inclusion of elasticity (incompressible versus elastic).

The details of the numerical formulation are important in obtaining realistic simulations. For example, the numerical operators should be nondamping, energy-conservative forms that allow a realistic propagation of wave energy through the atmosphere. The top and lateral boundary conditions should be nonreflective to prevent wave energy from reflecting off the top and lateral boundaries and interfering with the wave energy emitted by the mountain. The selection of a vertical coordinate can affect the solutions. To simulate flow over a variety of complex orographic shapes and to reduce truncation error near the surface, a terrain-following coordinate system is desirable. A number of investigators have found an isentropic vertical coordinate to be a natural coordinate for dry isentropic motions in the free atmosphere. Unless a hybrid vertical coordinate is used, however, isentropic coordinates prevent the use of unstable soundings near the earth's surface. If gravity waves break in the free atmosphere, the resultant turbulent mixing can generate locally unstable lapse rates which cannot occur in an isentropic coordinate system.

The technique for evaluating pressure has a substantial bearing on the simulated flow field. As the width of a mountain increases, the direction of the group velocity relative to the mountain becomes increasingly vertical. The hydrostatic limit is thus approached where the group velocity for standing waves is exactly vertical. One consequence of making the hydrostatic assumption is that trapped lee waves are completely filtered out of the solution. In some hydrostatic models (e.g., Hovermale, 1965), a form of lee wave is simulated, but the properties of such a lee-wave solution are probably a computational artifact or a result of the earth's rotation. Hydrostatic models are more suitable for simulating the response of the atmosphere to large mountains of the order of 100 km in width. They can realistically simulate the propagation of inertial-gravity waves emitted by such large mountains (e.g., Eliassen and Rekustad, 1971).

12.2.3.c Severe Downslope Windstorms

The consequences of nonlinearity in the simulation of flow over mountains and in orographic cloud formation are not well understood. For example, let us consider the controversy that has arisen regarding the factors involved in generating severe downslope windstorms in the lee of the Rocky Mountains near Boulder, Colorado. The focus of the controversy is on the explanation of the flow structure and the severe downslope winds that occurred in a well-documented case described by Lilly and Zipser (1972). Figure 12.8 illustrates a vertical cross section of potential temperature, clouds, and aircraft profiles obtained on 11 January 1972. Over the mountain crest is a rather typical flow pattern over a broad mountain. A stationary orographic cloud exists over the highest peaks. Directly to the lee of the higher peaks, the flow descends abruptly to the plains elevation. Evidence of trapped lee waves, including a lenticular cloud, can be seen over the plains. At higher levels is a deep trough in which air originating near stratospheric heights descends to below 500 mbar. This very high-amplitude wave is believed to be instrumental in causing surface winds in excess of 50 m s^{-1}. A vertical cross section of horizontal wind speeds for the same case (Fig. 12.9) shows that the maximum wind speeds occur along the lee slope of the mountain barrier at low levels.

Klemp and Lilly (1975) first explained the severe downslope wind phenomena with a two-dimensional, linearized, hydrostatic model in isentropic coordinates. Based on these linear results, they concluded that the mechanism leading to strong amplification of the wave is associated with the partial reflection of upward-propagating wave energy by variations in thermal stability. They argued that a strong wave response occurs whenever the mean vertical wavelength is such that an integral number of half-wavelengths can be confined between the ground and the tropopause.

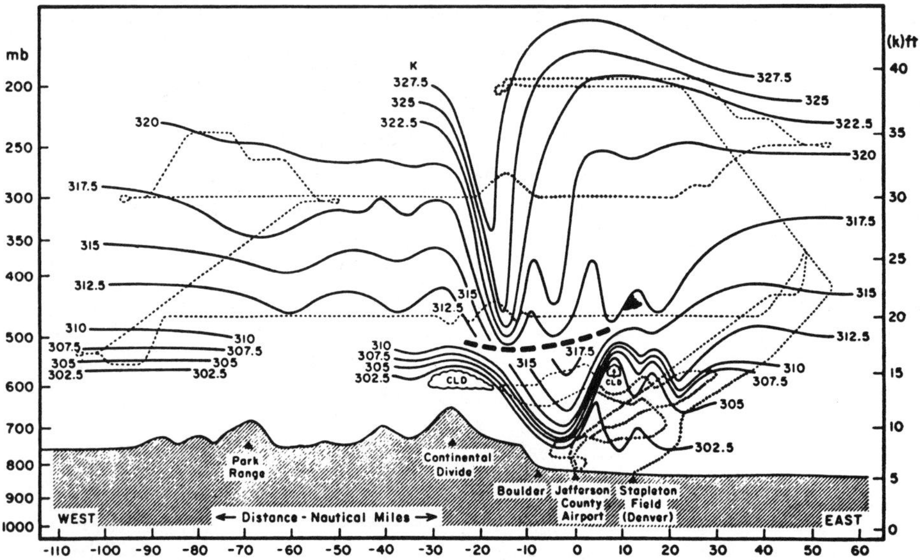

Fig. 12.8. Cross section of the potential temperature field in degrees Kelvin over the mountains and foothills as obtained from analysis of NCAR Queen Air and Sabreliner data on 11 January 1972. Data above 500 mbar are exclusively Sabreliner from 1700–2000 MST. Data below 500 mbar are primarily Queen Air from 1330–1500 MST. Flight tracks are indicated by the dashed lines, except by crosses in turbulent portions. It is not possible to determine if apparent westward displacement with height of the major features is real or related to the time difference between the two flights. Windstorm conditions on the ground extend eastward to where the isentropes rise sharply, a few miles east of the origin at the Jefferson County airport. [From Lilly and Zipser (1972).]

Constructive interference can then result between wave energy reflected off the surface and wave energy partially reflected from the tropopause temperature inversion.

Subsequently, Klemp and Lilly (1978) refined this concept with a two-dimensional, time-dependent, nonlinear, hydrostatic model also cast in isentropic coordinates. Figure 12.10 illustrates that the simulated flow field exhibits many features in common with the observed flow field (Figs. 12.8 and 12.9). Noteworthy is the large-amplitude standing wave over the lee of the mountain. Strong winds are also predicted just above the lee slope. Klemp and Lilly interpreted their linear model results as being generally consistent with the observations. Some evidence of the influence of non-linearity was noted—upstream blocking of the low-level flow by the mountain barrier. One effect of blocking was to alter the vertical wavelength of the gravity waves generated by the mountain, resulting in a linear model's

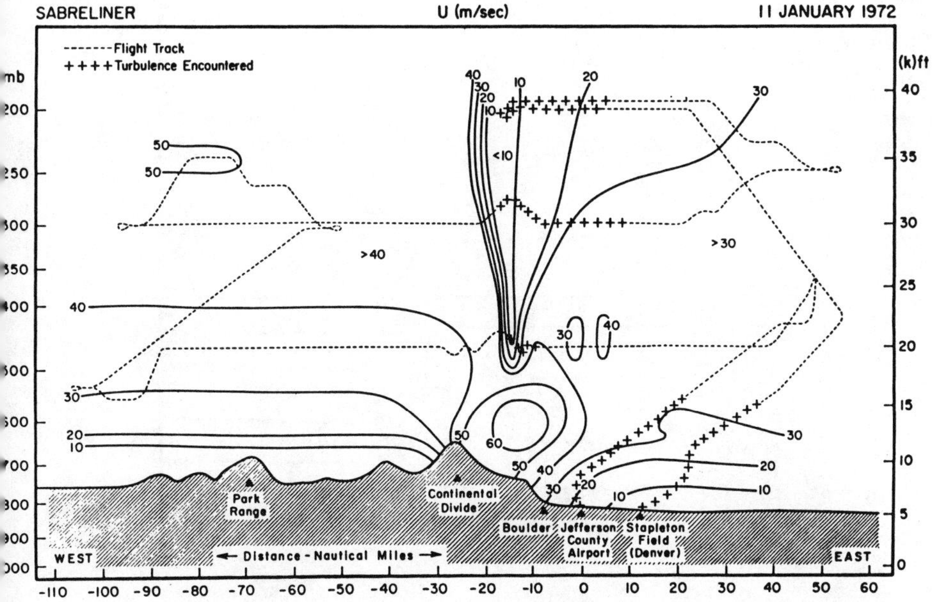

Fig. 12.9. Contours of horizontal velocity (m s^{-1}) along an east–west line through Boulder, Colorado, as derived from the NCAR Sabreliner data on 11 January 1972. The analysis below 500 mbar was partially obtained from vertical integration of the continuity equation, assuming two-dimensional steady-state flow. [From Klemp and Lilly (1975).]

incorrectly predicting the conditions when wave amplification can occur between the tropopause and the mountain. Klemp and Lilly noted other nonlinear effects, especially in the vicinity of critical layers (i.e., layers of zero wind or flow reversal). Linear theory predicts that total absorption of upwelling wave energy will take place in critical layers without reflection (Booker and Bretherton, 1967). Klemp and Lilly's nonlinear model, on the other hand, predicted partial reflections in the vicinity of critical layers. We shall see that the amount of reflection in the vicinity of critical layers is limited by the isentropic vertical coordinate. As the isentropes become vertically oriented, Klemp and Lilly had to activate eddy viscosity to prevent the isentropes from folding over. This prevented the formation of turbulent mixed layers that are very reflective to wave energy.

Using a two-dimensional, time-dependent, nonhydrostatic, numerical prediction model, Clark and Peltier (1977) and Peltier and Clark (1979, 1980) examined the consequences of nonlinear behavior in mountain waves and downslope windstorms. Like Klemp and Lilly (1978), the nonlinear model was first tested for its ability to reproduce linear solutions

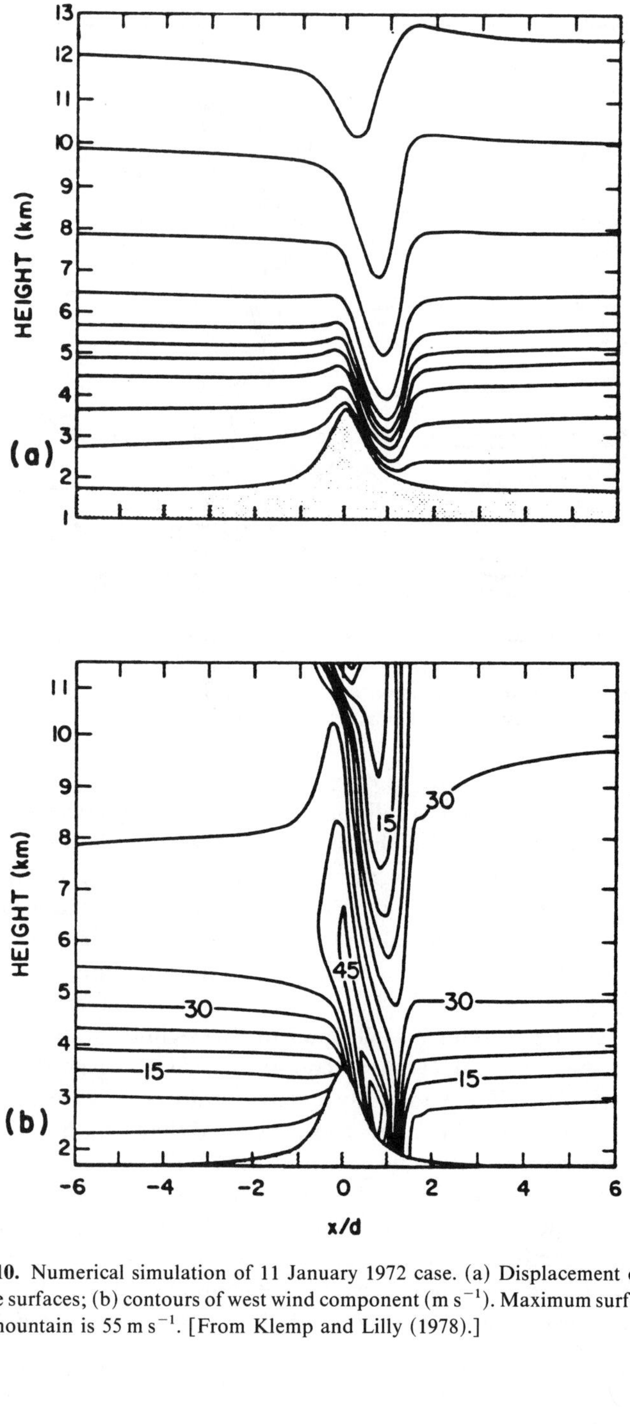

Fig. 12.10. Numerical simulation of 11 January 1972 case. (a) Displacement of potential temperature surfaces; (b) contours of west wind component ($m\ s^{-1}$). Maximum surface velocity lee of the mountain is $55\ m\ s^{-1}$. [From Klemp and Lilly (1978).]

when the small-amplitude conditions are applicable. This served as a foundation of credibility before the models are extended into the unknown territory of nonlinear wave behavior. They obtained several integrations of the nonlinear system, including a simulation of the 11 January 1972 windstorm. Figure 12.11 and Fig. 12.12 illustrate the time evolution of the simulated isentropic fields and winds for the 11 January 1972 case. Note the evolution of the high-amplitude standing wave pattern to the lee of the mountain crest. The amplitude of this wave is quite similar to the observed. Peltier and Clark also obtained trapped lee waves in the lower troposphere, similar to that in Fig. 12.8. Trapped lee waves were filtered out of Klemp and Lilly's solutions by the hydrostatic approximation. Peltier and Clark noted that the wavelength of the lee wave predicted with the nonlinear model was consistent with the linear theory predictions. The amplitude of the wave was much greater in the nonlinear solutions, having a maximum vertical velocity of $\sim 8\ \mathrm{m\ s^{-1}}$ compared to $\sim 1\ \mathrm{m\ s^{-1}}$ for the linear solutions.

The peak wind speeds near the surface along the lee slope shown in Fig. 12.12 reach $58\ \mathrm{m\ s^{-1}}$, within a few percent of the observed maximum. Although Peltier and Clark's results are qualitatively similar to those obtained by Klemp and Lilly (except the lee wave), interpretation of the sequence of events leading to the amplification of the large amplitude standing wave differs substantially. Like Klemp and Lilly, Peltier and Clark found a steepening of the isentropes to near-vertical orientation in the lower stratosphere when an integral number of half-wavelengths could be contained between the tropopause and the ground. In contrast to Klemp and Lilly, however, Peltier and Clark found that the wave actually broke, leading to a local wind reversal and a layer of constant potential temperature. As a consequence, wave energy reflected from the earth's surface became trapped between the resultant critical layer and the ground. This reflection cavity produced the large-amplitude streamline deflections that resulted in the strong surface winds. When the stratospheric wind profile was modified to prevent wave breaking, the final phase of wave amplification did not occur, and the results were then similar to linear theory. The surface drag caused by the strong lee-slope winds was over 300% of the value predicted by linear theory, even though the aspect ratio of the mountain was only slightly in excess of the critical value predicted by linear theory.

Clark and Peltier (1984) and Clark and Farley (1984) have elaborated on the role of breaking internal waves in the formation of severe downslope windstorms. Clark and Peltier (1984) showed that whenever a flow reversal or critical level occurred at a height $\frac{3}{4}\lambda_z$ above the level of forcing, where λ_z is the vertical hydrostatic wavelength of the internal waves, the direct and reflected waves interfered constructively and intense resonant growth of the low-level wave occurred. Clark and Farley (1984) extended the

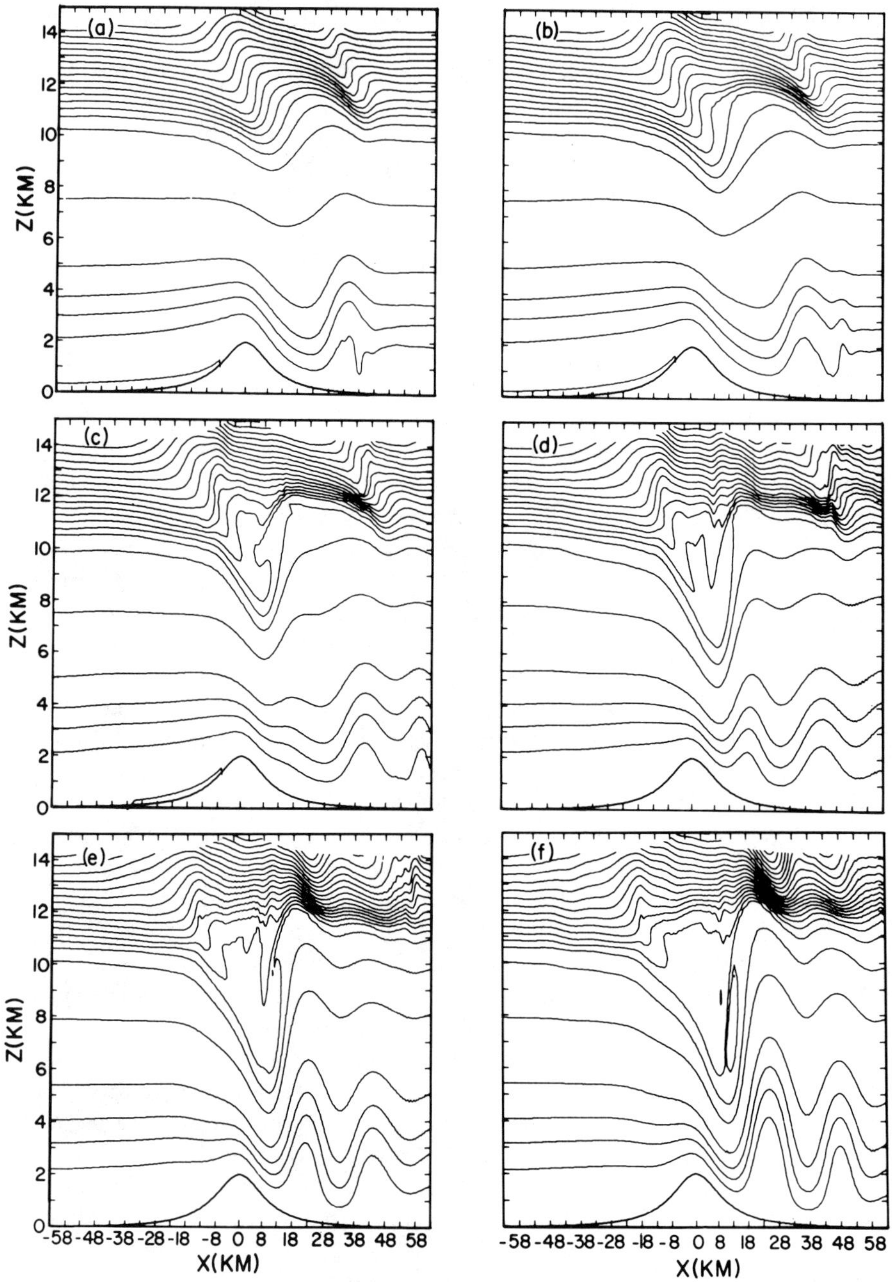

Fig. 12.11. Potential temperature for the second Boulder, Colorado, windstorm simulation in which the mixing coefficients are determined through first-order closure. Times are (a) 3200, (b) 4160, (c) 5120, (d) 6020, (e) 7040, and (f) 8000 s. Note the extreme deflection of the tropopause in the last frame. [From Peltier and Clark (1979).]

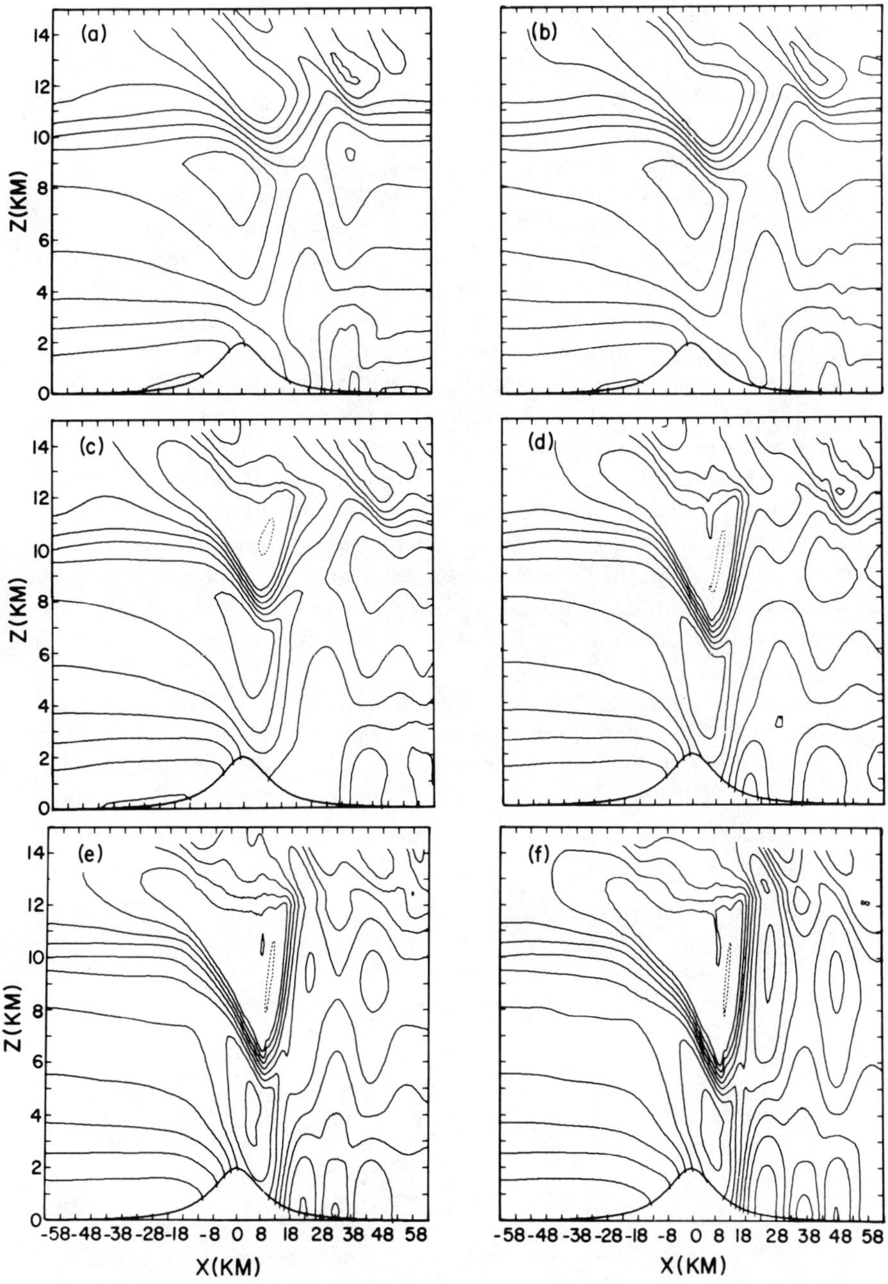

Fig. 12.12. Total horizontal velocity field for the Boulder, Colorado, windstorm simulation. Times are (a) 3200, (b) 4160, (c) 5120, (d) 6020, (e) 7040, and (f) 8000 s. Contour interval is 8 m s^{-1}. In (f) the horizontal wind maximum in the lee of the peak is in excess of 60 m s^{-1}. [From Peltier and Clark (1979).]

downslope windstorm simulations to three dimensions. The three-dimensional simulations exhibited strong surface gustiness which they attributed to the development of turbulent eddies in the convectively unstable region of the wave. Downdrafts at the leading edge of the convectively unstable region then transported the turbulence to the surface.

Smith (1985) and Durran (1986) presented further evidence of the transition from strong smooth stratified flow, dominated by a large-amplitude wave, to a deep turbulent, high surface-wind state, characterized by the development of highly nonlinear waves. Smith (1985) obtained analytic solutions to Long's equation for flow beneath a breaking wave by assuming that breaking of the high-amplitude wave produces a layer of neutral stability. Figure 12.13 is a schematic of the hydraulic flow corresponding to a layer beneath a mixed layer induced by a breaking wave. The height H_0 represents the dividing streamline height; above H_0, only weak waves are assumed to be present. At H_0, the flow splits with the lower branch descending in smooth, steady, nondissipative, hydrostatic flow. Between the split streamlines on the lee slope, the air is turbulent, well-mixed, and has little mean motion. As in Long (1955), an equation calculating the vertical displacement of a streamline following air moving over and to the lee of the mountain below H_0 is obtained.

Smith's theoretical predictions agree qualitatively with Clark and Peltier's numerical results. Some quantitative discrepancy exists but, in general, the hydraulic model is a reasonable representation of severe downslope windstorms in spite of the fact that it does not explicitly contain wave reflection from a critical level. Smith noted, however, "free boundary" resonances

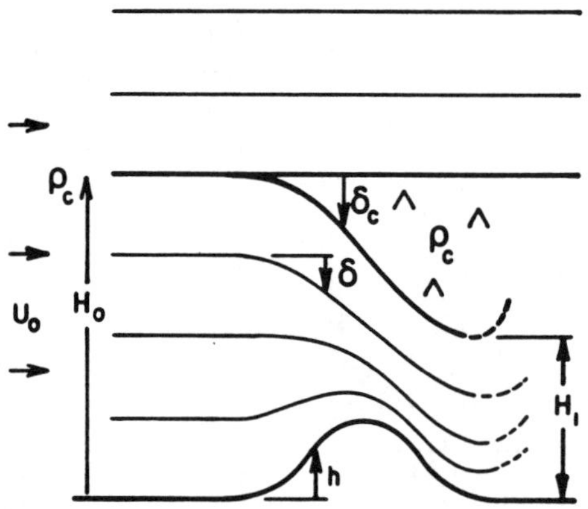

Fig. 12.13. Schematic of the idealized high-drag flow configuration, derived from aircraft observations and numerical simulations. A certain critical streamline divides and encompasses a region of uniform density. The disturbance aloft is small compared to that below. [From Smith (1985).]

appeared in the linear theory as special points in the finite-amplitude hydraulic theory. In the hydraulic theory, the free boundary condition is not applied at a fixed level giving a coherent reflection, but along a strongly deflected streamline of flow. Smith speculated that the linear resonant conditions predicted by linear models such as Klemp and Lilly's may be useful in predicting the conditions suitable for the onset of severe downslope windstorms even though they are not suitable for predicting the final severe-wind state. That is, the linear models may be useful in predicting the level where wave breaking and a midlevel mixed layer may form. This midlevel mixed layer is essential to the application of the hydraulic theory to severe downslope windstorm simulation. It is also consistent with the formation of a resonant cavity as hypothesized by Clark and Peltier. Durran (1986), however, does not believe that linear models are useful in diagnosing conditions where wave breaking will occur.

Smith speculated that a middle-level inversion can trigger a hydraulic jump and the formation of a severe windstorm if the Froude number, based on the height and strength of the inversion $\mathrm{Fr} = u_0/(g'H)^{0.5}$, is less than one, but not too small. Using a nonhydrostatic model, Durran (1986) showed that initial amplification of the surface wind and pressure drag occurred when a low-level inversion was displaced downward along the lee slope producing "supercritical flow." The concept of supercritical flow in hydraulic theory can be explained as follows. As a parcel of air ascends a windward slope of a mountain, it slows and converts kinetic energy to potential energy. After passing the crest, the parcel accelerates as potential energy is converted to kinetic energy. In supercritical flow, nonlinear advection dominates the pressure gradient force and the resultant acceleration is in the same sense as the gravitational force. As a result, kinetic energy is no longer returned to potential energy and the air continues to accelerate as it descends the leeward slope. Durran showed that removal of the low-level inversion prevented the initial amplification of the wave and no windstorm developed. The importance of elevated inversions to the generation of severe downslope windstorms has also been suggested in climatological studies of downslope windstorm phenomena reported by Colson (1954) and Brinkmann (1974).

Durran concludes that Clark and Peltier's critical-layer reflection mechanism and the concept of supercritical flow in hydraulic theory may be alternative frameworks for describing the severe downslope windstorm phenomena. The breaking of high-amplitude waves is analogous to the transition to supercritical flow in hydraulic theory. Durran's numerical experiments suggest, however, that wave breaking is not necessary and sufficient for the transition to supercritical flow and the formation of at least moderately strong windstorms.

This series of papers presents a clear perspective of the limitations of linear theory and hydrostatic flow models to the simulation of flow over mountains and cloud formation. Up to this point, we have not examined the influence of cloud processes on orographic flow.

12.2.4 Effect of Clouds on Orographic Flow

Cloud processes can influence mountain-wave flow and thereby feed back on the formation of precipitation in orographic clouds. In deriving the linearized wave equation, Eq. (12.20), we introduced the Scorer parameter $l(z)$. The leading-order term affecting $l(z)$ in Eq. (12.18) is $(\beta g)/(u_0^2)$, which can be equivalently expressed in terms of the Brunt–Väisälä frequency N, giving

$$l^2(z) = N^2/u_0^2. \tag{12.29}$$

We have defined N with respect to dry atmospheric motions. A number of investigators have shown that, when condensational heating and evaporational cooling are present, N should be modified accordingly.

Fraser *et al.* (1973) defined a moist Brunt–Väisälä frequency N_m as

$$N_m^2 = (g/T)(dT/dz + \gamma_m), \tag{12.30}$$

where γ_m represents the wet adiabatic lapse rate. They introduced wet processes in a Scorer-type linear model by simply replacing N by N_m in the wave equation. The inclusion of the effects of clouds in the flow model resulted in a less stable flow because N_m is less than N. Because the buoyancy-restoring force is decreased, the amplitude of the mountain wave under certain conditions can be significantly weakened. As expected from Eq. (12.29), the importance of the effect of clouds on the simulated flow fields depends on the wind profile as well.

Using a two-dimensional hydrostatic model, Barcilon *et al.* (1979) demonstrated that cloud processes can significantly modify momentum drag created by flow over mountains. The effect of cloud processes on lee waves could not be examined because they used the hydrostatic approximation. The effect is quite substantial.

If virtual temperature corrections, including condensate, are considered in the buoyancy term of the vertical equation of motion, Lalas and Einaudi (1974) showed that an additional term should be present in N_m,

$$N_m^2 = (g/T)(dT/dz + \gamma_m)(1 + Lr_s/RT) - g/(1 + r_w)(dr_w/dz), \tag{12.31}$$

where r_s is the saturation mixing ratio, and $r_w = r_s + r_l$, where r_l is the liquid-water mixing ratio. Durran and Klemp (1982a) defined Eq. (12.31) in terms of the wet conservative variable θ_q (Paluch, 1979; see also

Chapter 8). The saturated Brunt–Väisälä frequency is then

$$N_{\mathrm{m}}^{2}=\frac{g}{(1+r_{\mathrm{w}})}\left[\frac{\gamma_{\mathrm{m}}}{\gamma_{\mathrm{d}}}\left(\frac{d\ln\theta_{\mathrm{q}}}{dz}\right)-\frac{dr_{\mathrm{w}}}{dz}\right]. \qquad (12.32)$$

Durran and Klemp (1983) applied a two-dimensional version of the Klemp and Wilhelmson (1978a or b) convective cloud model to stable orographic flow. They applied the model to the simulation of the 11 January 1972 severe downslope windstorm over Boulder, Colorado. The addition of moisture decreased the downslope wind speed from 45 to 25 m s^{-1}, a result of a weakened mountain-wave amplitude. Durran and Klemp also noted that lee-side warming, referred to as the Chinook or Alpine foehn, is often attributed to the release of latent heat on the windward side of the barrier in precipitating clouds and to dry adiabatic descent on the lee side. In a precipitating cloud simulation, they noted that the lee-side temperatures were several degrees cooler than those in nonprecipitating flow. This suggests that the most important factor influencing Chinook or foehn wind lee-side temperatures is the amplitude of the mountain wave, which is larger in the dry case.

An even more dramatic illustration of the effects of moist processes on orographic flow was presented by Durran and Klemp (1982b) in their simulation of trapped lee waves. In one case, the Scorer parameter exhibited a sharp decrease with height (Fig. 12.14), which is capable of supporting trapped lee waves in a dry atmosphere. They then added moisture to the atmosphere until the Scorer parameter no longer exhibited a sharp drop-off with height for a saturated atmosphere (Fig. 12.14b). The results of their simulations are shown in Fig. 12.15. In the dry atmosphere, a distinct trapped lee wave is evident in their solutions. The addition of a layer with 90% relative humidity results in the formation of clouds over the mountain crest and in the regions of upward motion of the trapped lee waves. The wave structure is modified somewhat. As a result of adding a 100% saturated layer, the flow is modified so that the wavelength of the partially trapped waves is increased significantly. Finally, by adding 0.2 g kg^{-1} of liquid water to the saturated layer, a cloud could be maintained in the wave troughs as well as the wave crests. This so altered the resultant vertical profile of the Scorer parameter that the lee waves became untrapped.

Durran and Klemp also investigated a case in which the Scorer parameter decreased with height, thus supporting dry lee waves. With moisture added, however, the layer became conditionally unstable. In the previous case, latent heat release could not overcome environmental stability. Figure 12.16 illustrates the transformation from an atmosphere supporting steady trapped lee waves to transient convection as moisture is added. The lee-wave structure is eventually destroyed by the convection. They also examined the

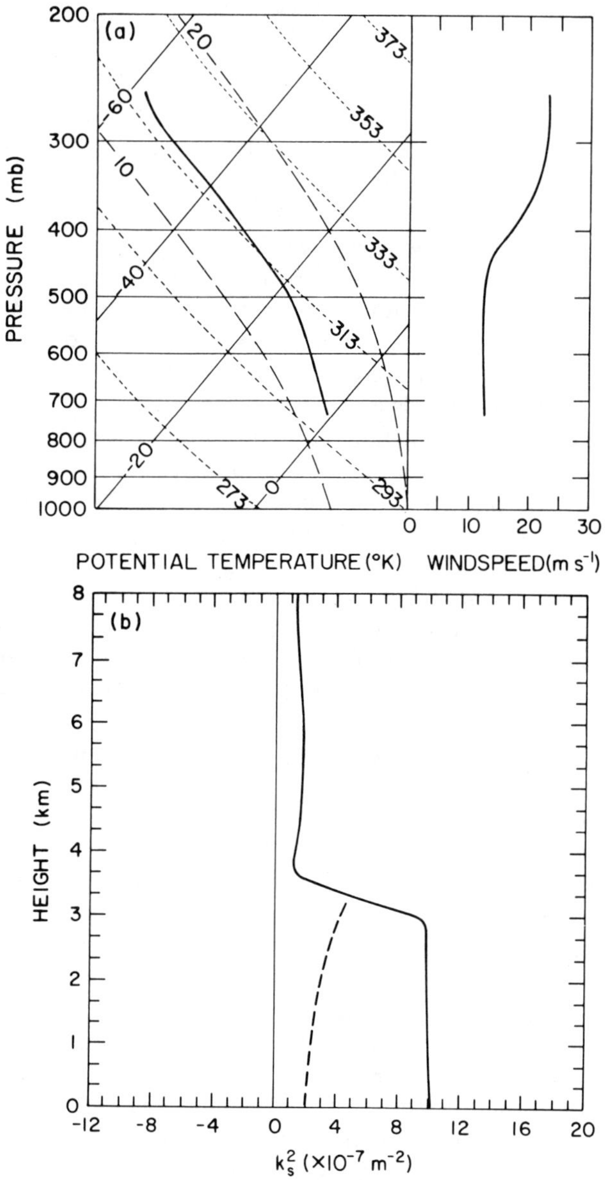

Fig. 12.14. Absolutely stable atmosphere favorable for the development of dry lee waves. (a) Temperature and wind speed profiles; dry adiabats are marked with a short-dash line; moist pseudoadiabats are marked with a long-dash line. (b) Scorer parameter (l^2) profiles; the dry l^2 is marked with a solid line, the equivalent saturated l^2 is a dashed line. [From Durran and Klemp (1982b).]

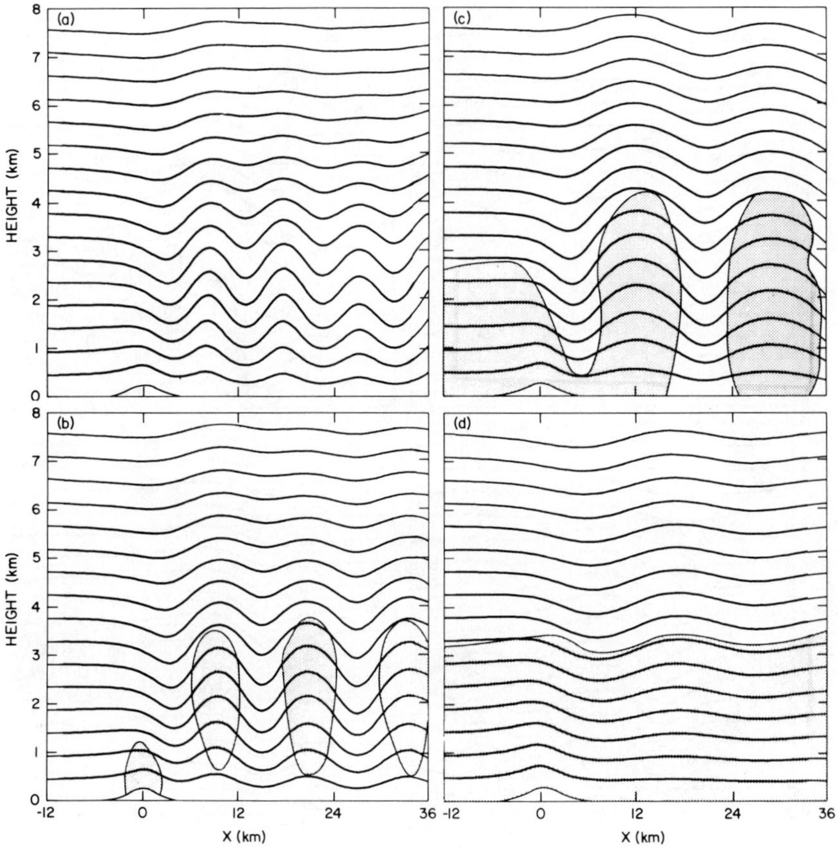

Fig. 12.15. Streamlines produced by a 300-m-high mountain in the flow for relative humidity (RH): (a) RH = 0%, (b) RH = 90%, (c) RH = 100%, and (d) RH = 100% with 0.2 g kg^{-1} of cloud, in the lowest layer upstream. Cloudy regions are shaded. [From Durran and Klemp (1982b).]

influence of moist layers in the middle troposphere on "detuning" trapped lee waves.

Thus far, we have focused on the influence of nonprecipitating clouds on stable orographic flow. Lilly and Durran (1983) extended the Durran–Klemp calculations to precipitating clouds as well. Using a simple Kessler-type warm rain parameterization (Chapter 4), they investigated the effects of cloud processes, including precipitation, on vertical momentum fluxes over orographic barriers. Figure 12.17 illustrates the calculated vertical momentum fluxes for (a) a case having low clouds and (b) a case saturated everywhere. The fluxes are normalized to fluxes expected for linear mountain

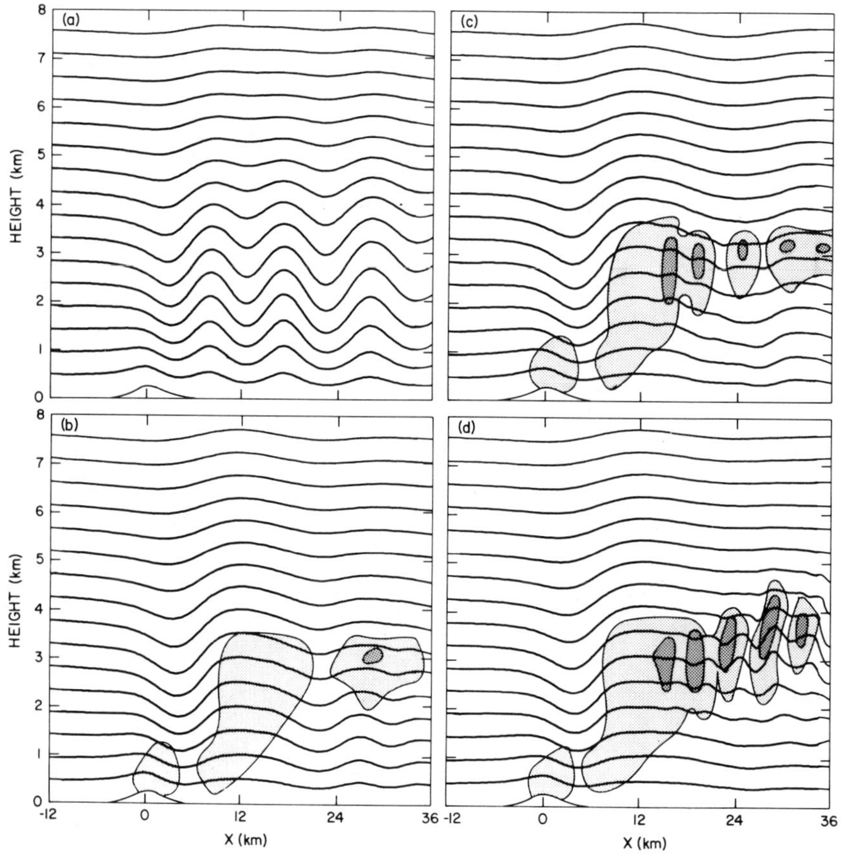

Fig. 12.16. Streamlines produced by a 300-m-high mountain in the flow. (a) Steady solution for RH = 0%. Time-dependent flow for RH = 90% in the lowest upstream layer at (b) $t = 8000$ s, (c) $t = 12{,}000$ s, and (d) $t = 16{,}000$ s. Cloud regions are shaded; dark shading indicates cloud densities exceeding 0.3 g kg^{-1}. [From Durran and Klemp (1982b).]

waves (M_{LC}). In the dry case, linear theory predicts that the vertical momentum flux remains constant through a vertically homogeneous layer (Eliassen and Palm, 1960). The model-predicted momentum flux profile is relatively constant below 11 km. Above 11 km, a viscous layer is imposed to absorb upward-propagating wave energy and prevent its reflection into the domain of interest. The momentum fluxes are greater than predicted by linear theory (i.e., $M/M_{LC} \sim 1.4$) because the calculations were performed for a 1000-m mountain. Miles and Huppert (1969) and Smith (1977) predicted that nonlinearity forced by large-amplitude mountains should create an amplification of momentum fluxes. In nonprecipitating saturated flow, the momentum flux is reduced by a factor of three or more relative to dry flow,

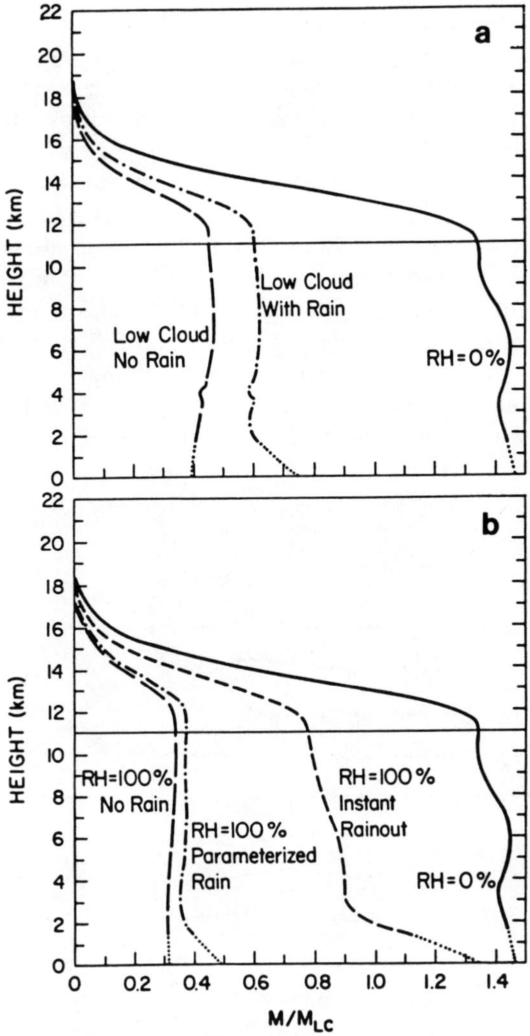

Fig. 12.17. The effects of rain on the vertical profiles of momentum flux produced by upstream moisture profiles in which (a) there are low clouds between the heights of 667 and 3000 m, and (b) RH = 100% everywhere. The fluxes are normalized by M_{LC}, the flux associated with linear mountain waves. [From Lilly and Durran (1983).]

a result of the effective reduction in stability of the atmosphere by condensational heating and evaporational cooling. Precipitation increases the wave momentum flux amplitudes relative to that of the nonprecipitating saturated flow. The momentum fluxes remain well below the dry case. Apparently, precipitation reduces the total condensate on the windward side of the barrier so that the descending flow warms adiabatically sooner than in the nonprecipitating case. The instant rainout case removes all condensed water, leaving no water available for evaporation on the lee slope. The results for this case exhibit a stronger wave response, which is closer to the dry case.

Smith and Lin (1982) examined the influence of asymmetric heating profiles across a mountain barrier induced by precipitation. Without precipitation, air following a streamline experiences heating as it rises over the windward slope and evaporative cooling as it descends along the lee slope. Smith and Lin imposed various heating functions at different locations across the mountain barrier. The amplitude of the heating function was inferred from observed surface precipitation rates. Figure 12.18 illustrates the simulated flow field for (a) the case with heating above the windward slope and cooling above the lee slope, (b) the precipitating case with heating only over the windward slope, and (c) the precipitating case with heating well upwind of the mountain crest. Smith and Lin claim that case (c) corresponds to the observation that the region of maximum condensation

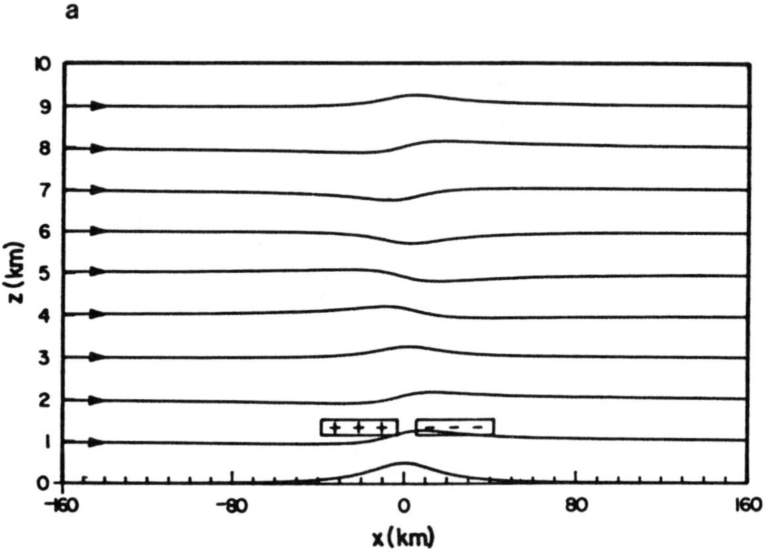

Fig. 12.18. (a) Hydrostatic flow with combined thermal and orographic forcing. Heating occurs over the windward slope while cooling is specified over the leeward slope. This flow is calculated with a heating rate $Q = 1107\ \mathrm{W\ m\ kg^{-1}}$, a wind speed $\bar{U} = 10\ \mathrm{m\ s^{-1}}$, a Brunt-Väisälä frequency $N = 0.01\ \mathrm{s^{-1}}$ and a mountain with half-width of 20 km and maximum height of 0.5 km. (b) Same as (a), except heating is applied only over windward slopes. (c) Same as (b), but with the isolated heating centered farther upstream ($c = 40$ km) to correspond to the observation that the region of maximum condensation rate often occurs well upstream of the mountain. [From Smith and Lin (1982).] (*Figure continues.*)

b

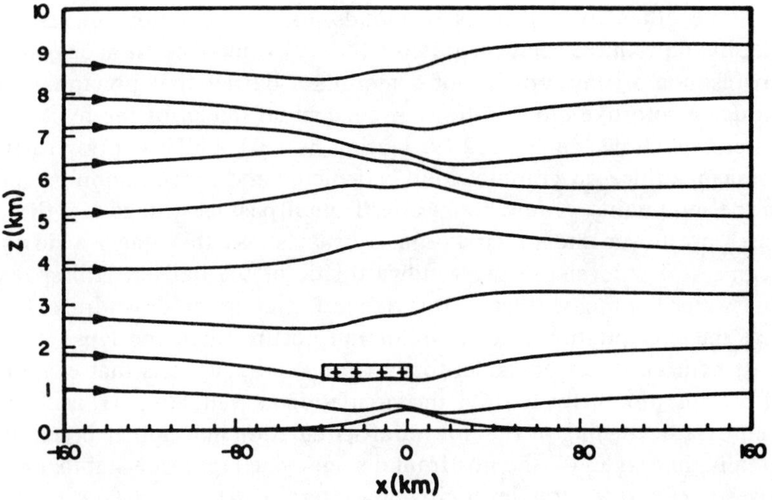

c

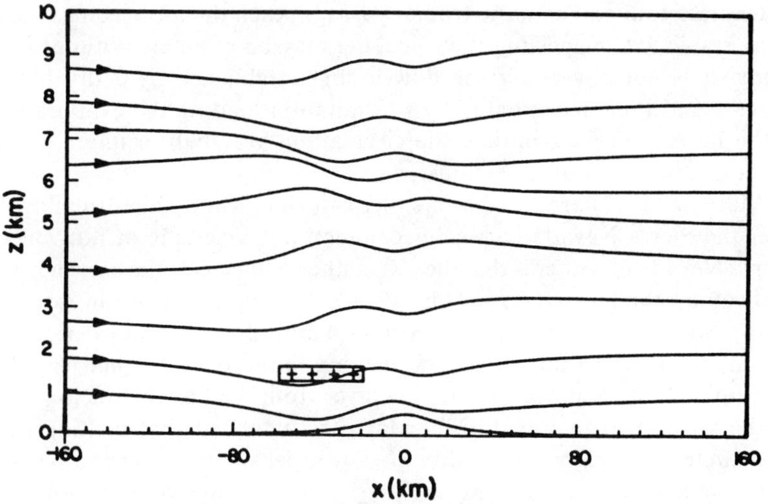

Fig. 12.18 (*continued*)

rate often occurs well upstream of the mountain. These examples show that precipitation-induced asymmetric heating can make significant changes in the flow fields. The magnitudes of the imposed heating rates correspond to small precipitation rates. Precipitation rates of 5–10 mm h^{-1} would produce a disturbance so large that the assumptions of linear theory would be violated.

Some other consequences of clouds and precipitation on stable orographic flows have been suggested. Hill (1978) inferred from special serial rawinsonde ascents upwind of a mountain barrier that precipitation can induce a rotorlike blocked flow. A conceptual model of the hypothesized process is shown in Fig. 12.19. During Stage I and just prior to frontal passage, a thick orographic cloud is depicted and precipitation is confined to the mountains. A few hours after frontal passage (Stage II), the cloud bases are lower and precipitation is heaviest. At this stage, a downdraft begins at low levels on the windward side of the barrier. Hill speculates that water loading is the major factor in initiating the downdraft. The fact that the precipitation-forced circulation occurs when the lapse rate is at least neutral, if not unstable, for wet processes, suggests that evaporation of precipitation helps to drive the circulation as well. Stage III is characterized by a deepening of the downdraft circulation and lighter precipitation. Finally, in Stage IV—the postfrontal stage—the lapse rate stabilizes, cloud bases rise, and the low-level circulation terminates.

The flow in Stage III is similar to the flow when blocking takes place, as dry stable air rises over a mountain barrier. The flow reversal and downdraft coincide with the frontal passage when the air becomes increasingly stabilized, suggesting that blocking may be a factor. Without further study, it is not obvious if the flow is first established by a dry blocking process and then accentuated by precipitation loading the evaporation or if the latter processes initiate the circulation. We shall examine blocking processes in subsequent sections.

Marwitz (1983) argued that melting can significantly alter the flow field over the Sierra Nevada barrier. A common characteristic of flow over the Sierra Nevada in winter is that the 0°C isotherm intersects the barrier roughly midway up the barrier slope (Fig. 12.20). Melting produces an isothermal layer; Atlas *et al.* (1969) estimated that 4 mm of melted snow produces a 700-m deep isothermal layer. A melting-induced isothermal layer can decouple the flow above the melting layer from the flow below it—evident in the wind field in Fig. 12.20b. Marwitz calculated that melting should contribute to a 3.3-mbar positive pressure perturbation on the barrier at levels sandwiched below the melting level and the barrier slope. This enhances the blocking effect of stable air being lifted over the barrier, and, as a result, winds below 2 km are slowed appreciably.

Marwitz *et al.* (1985) applied a two-dimensional version of the Anthes and Warner (1978) hydrostatic model to the simulation of the effects of melting on the flow over the Sierra Nevada. In one simulation, melting decreased the westerly component of flow upwind of the middle of the

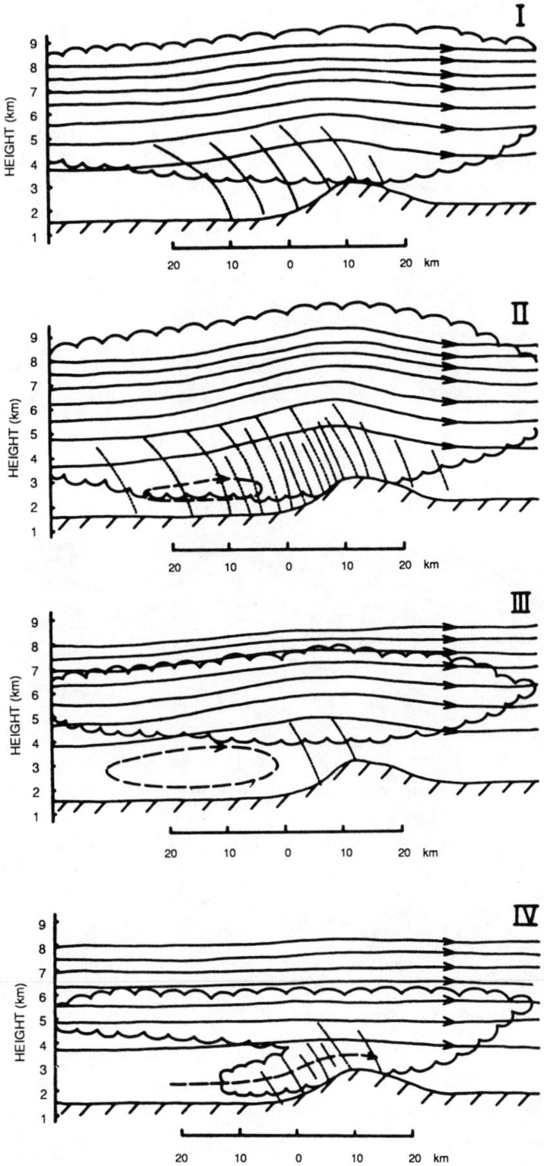

Fig. 12.19. Conceptual model of the development and dissipation of a precipitation-forced circulation during the course of a winter orographic storm. Airflow is indicated by streamlines with direction arrows also shown; cloud outlines are depicted by wavy outlines and precipitation is depicted by short dashed curves. Four stages are (I) just prior to or at time of frontal passage, (II) period of heavy precipitation and development of low-level precipitation forced circulation, (III) time of strong precipitation forced circulation and sharply reduced precipitation, and (IV) return of low-level orographic flow; increased precipitation, but with drier conditions aloft compared to stage II. [From Hill (1978).]

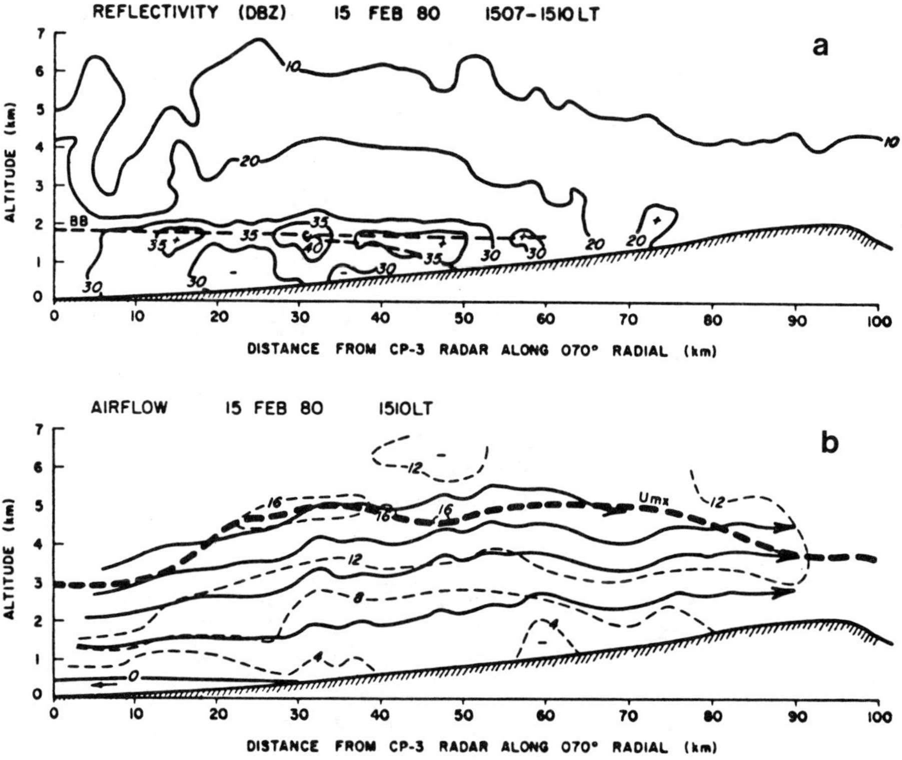

Fig. 12.20. Cross sections of (a) reflectivity (dBZ), and (b) airflow along 070° radial at 1507–1510 LST. The dashed line in (a) is the bright band. The wide dashed line in (b) is the altitude of the maximum wind (U_{mx}); the solid lines are streamlines and the thin dashed lines are isotachs in m s^{-1}. [From Marwitz (1983).]

barrier by 7.5 m s^{-1} and increased the southerly component of flow by 10 m s^{-1}. This southerly component of flow is referred to as the "barrier jet" (Parish, 1982) and is common along the western slopes of the Sierra Nevada.

Similar barrier-parallel flows are observed along the Brooks Range in Alaska (Schwerdtfeger, 1974) and Antarctica (Schwerdtfeger, 1975). The barrier-parallel jet is produced as nearly geostrophic flow impinges against a mountain barrier. As stably stratified air is lifted up the mountain slope, it becomes cooler than the undisturbed air at similar levels, resulting in a positive pressure perturbation along the windward slopes. Schwerdtfeger (1979) estimated pressure perturbations on the order of 4–8 mbar in the boundary layer away from the barrier, depending on initial winds and stratification. The resultant adverse-pressure gradient slows the wind component directed perpendicular to the barrier. This causes an imbalance in the previously geostrophic flow, which turns the wind toward low pressure or, in the case of the Sierra barrier, into a southerly component of flow. If the flow persists for several hours or more, a new southerly, terrain-locked, steady flow is established between the local mountain blocking-induced pressure gradient and the Coriolis turning of the winds.

Marwitz *et al.* (1985) calculated that melting can result in a doubling in intensity of the barrier jet if a stable orographic storm is sufficiently persistent. They concluded that for melting to produce such dramatic and consistent effects, two conditions must be met. First, the component of flow normal to the barrier should be less than 10 m s^{-1} within the melting layer. This is necessary to prevent the adiabatically cooled air from being driven over the barrier crest and not significantly affecting the flow on the windward side. Second, the barrier-normal component of flow must be 15–20 m s^{-1} immediately above the 0°C isotherm to produce an orographic ascent rate large enough to yield precipitation rates of 3–4 mm h^{-1}. The higher precipitation rates, in turn, increase the cooling by melting. Marwitz *et al.* noted that the effects of melting on the airflow are maximized when the maximum vertical shear of the horizontal wind is centered near 0°C.

12.3 Orogenic Precipitation

We turn our attention to the processes involved in the formation of orogenic precipitation—precipitation caused by orography. The distribution of orogenic precipitation is influenced by forcing by both orography and microphysical responses. Simple orographic flow models can be useful predictors of orogenic precipitation (e.g., Elliott and Shaffer, 1962; Rhea, 1978). Such models are strictly forced by direct barrier lifting and are not at all affected by gravity waves. Their apparent success is a result of two factors. First, the heaviest precipitation comes from large orographic barriers. While trapped lee waves may form to the lee of the larger mountain barriers, the air flows through the lee wave clouds so quickly that there is not enough time for significant precipitation to form. Second, in a stably stratified air mass, the highest moisture-mixing ratios are confined to the lowest levels. As a result, liquid-water production is largest at the lower levels of the air mass lifted over the barrier. A model forced simply by continuity of air flowing over a barrier can capture the dominant lifting of the moisture-rich air capable of producing precipitation. Whether or not the moisture condensed by lifting on the windward side of the barrier is realized as precipitation depends on air motions and precipitation processes occurring at higher levels and upstream of the mountains. That is, the actual precipitation efficiency of the cloud system, which may vary from 0 to 100%, is controlled to a large degree by the more detailed and complicated air motions and cloud microphysical processes, especially those at higher elevations upstream of the main mountain barrier.

12.3.1 Seeder–Feeder Process and Distribution of Precipitation

Relatively small hills, only 50 m or so above the general terrain height, can produce an increase in precipitation on the order of 25–50% (Douglas and Glasspoole, 1947; Bergeron, 1949, 1960, 1967–73; Holgate, 1973; Browning *et al.*, 1975). These observations pertain to relatively warm low-level clouds in which the dominant precipitation process is collision and coalescence. Because the time scale for air flowing through such small clouds is so short, there is insufficient time for precipitation to be initiated by collision and coalescence in the low-level orographic cloud. Bergeron (1949, 1965) hypothesized that a significant fraction of the water produced in the orographic cloud may be washed out by precipitation settling from higher level clouds that form by large-scale ascent. Precipitation in the higher level clouds may originate by collision and coalescence or by ice-phase precipitation processes. A conceptual model of the process developed by Bergeron (1965) (Fig. 12.21) depicts a moist, saturated low-level flow impinging on a hill. The forced ascent of the flow over the hill causes the formation of an orographic "feeder" cloud. Above the feeder cloud is a "seeder" cloud, which forms as a result of large-scale ascent. Precipitation from the seeder cloud (P_0) is steady stratiform rainfall of light to moderate intensity. In the absence of the orographic cloud, it results in a surface rainfall (P_1). If the seeder cloud rainfall encounters the water-rich environment of the feeder cloud, the precipitation elements collect cloud droplets, thereby enhancing

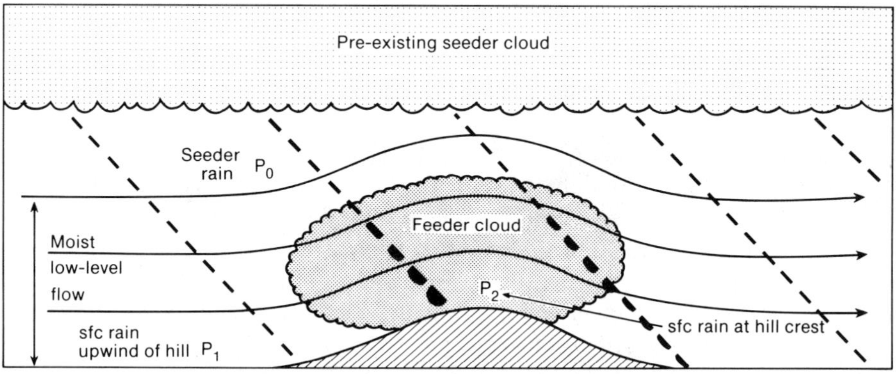

Fig. 12.21. Conceptual model illustrating the orographic enhancement of rain. [From Browning's (1979) adaptation of Bergeron's (1965) figure.]

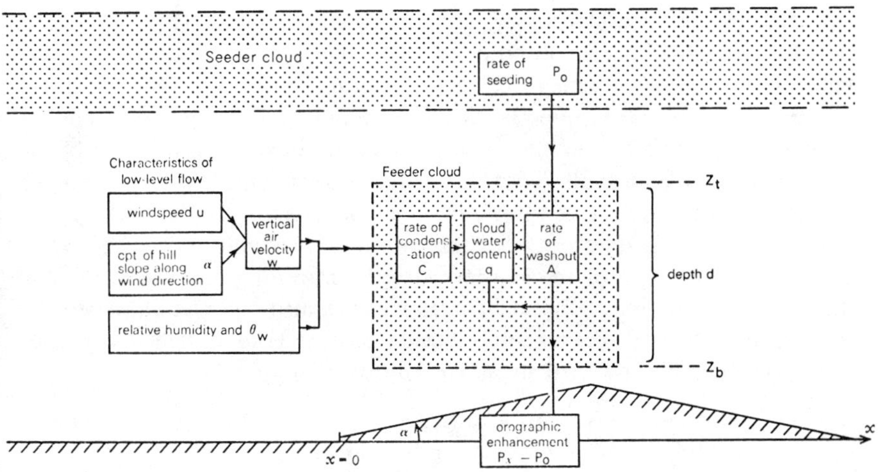

Fig. 12.22. Schematic of dependence of orographic enhancement on the characteristics of the low-level flow and the rate of seeding from above. [From Browning (1979).]

the rainfall to an amount P_2. The precipitation is enhanced by an amount $P_2 - P_0$, although Browning has noted that what is normally measured is $P_2 - P_1$. A number of investigators have developed quantitative models of the process (Hobbs *et al.*, 1973; Storebo, 1976; Bader and Roach, 1977; Gocho, 1978; Carruthers and Choularton, 1983; Choularton and Perry, 1986). Figure 12.22 illustrates some of the processes affecting the orographic enhancement of precipitation. Factors important to the orographic enhancement of precipitation are (1) relative humidity and θ_e or θ_w of low-level air, (2) slope of the hill perpendicular to the wind direction, (3) strength of the wind component normal to the mountain, (4) depth of the feeder cloud, (5) precipitation rate from the feeder cloud, (6) rate of production of condensate in the feeder cloud, (7) cloud water content, and (8) rate of accretion or washout by the precipitation emanating from the seeder cloud. The relative humidity of the layer of air between the seeder and feeder clouds is also key in determining whether or not the seed particles survive descent through the layer.

The seeder–feeder mechanism is most effective in the warm sector of an extratropical cyclone. In this region, the warm, moist conveyor belt often characterized by a low-level jet results in nearly saturated air impinging on small hills and mountains (Chapter 11). This is common in the winter in western Europe. While the magnitude of θ_w or θ_e seems to be a factor in determining the liquid-water production of the feeder cloud, Browning (1979) noted that the highest frequency of seeder–feeder rainfall in the

British Isles occurs in December, when θ_w is relatively low. This only points out the overwhelming importance of the passage of cyclonic storms that reach a maximum frequency from November to February. The warm sector of a cyclonic storm is typified by low-level convergence which moistens the air mass impinging on the hills. The evaporation of stratiform rainfall from middle and high clouds also contributes to moistening the low-level air mass.

For saturated flow crossing a given orographic barrier, Bader and Roach's model calculations demonstrated that the rate of condensation is proportional to the strength of the low-level winds. In noting that the horizontal drift of precipitation at higher wind speeds is also significant, especially for smaller hills, Carruthers and Choularton (1983) identified two factors of importance to wind drift of precipitation:

(1) The horizontal drift of a raindrop of radius r falling at a speed V_r through a cloud of depth Z_c by a wind of speed u_0,

$$L_d \approx Z_c u_0 / V_r .$$

(2) The scale of horizontal variation of liquid-water content (r_c),

$$L_{r_c} \approx r_c / (dr_c / dx) \simeq L,$$

where L is the half-width of the hill.

If $L_{r_c} \leq L_d$, such as over a short hill, wind drift will decrease the magnitude of maximum orographic enhancement because some of the seeder drops entering the feeder cloud summit would fall downwind in the region where both the feeder cloud droplets and seeder drops will evaporate. For hills having $L < 10$ km, reduction of rainfall as a result of wind drift can be appreciable.

Bader and Roach and Carruthers and Choularton concluded that orographic enhancement of rainfall increased with the seeder cloud precipitation rate P_0. Browning (1979) noted that not only is P_0 of consequence, but so also is the size distribution of the seeder cloud precipitation elements. Rainfall rate is proportional to the mean drop volume, whereas the net washout rate is proportional to the net cross-sectional area of the droplets. For a given P_0, relatively smaller drops would collect feeder cloud drops more efficiently than large drops. If the seeder drops are too small, the collection efficiencies diminish and the smaller drops become more susceptible to wind drift. Thus, an optimum combination of wind speed, hill size, precipitation rate, and precipitation particle size determines the efficiency of washout.

Figure 12.21 illustrates a separation of the seeder and feeder cloud. This is not necessarily the case or important to the seeder–feeder concept. In some intense cyclonic storms, the atmosphere may be saturated through a great depth. In that case, the feeder cloud may be just a low-level maximum in liquid-water content on the windward side of the barrier.

In the modeling studies of the seeder–feeder process discussed thus far, the airflow is first specified or calculated; then a low-level feeder cloud is calculated on the basis of the given low-level airflow and moisture content of the air mass; finally, an upper-level seeder cloud precipitation rate is specified. Richard *et al.* (1987) simulated the seeder–feeder process in a two-dimensional hydrostatic cloud model. In their model, variations in moisture content of the low-level airflow, the flow speeds, and the resultant cloud microphysical processes could feed back into the airflow dynamics. Only the seeder-cloud precipitation rates and particle sizes were specified. As a result of the nonlinear interactions among the flow dynamics, thermodynamics, and microphysical processes, factors such as the cloud droplet spectrum, the size of seeding raindrops, and the seeder cloud precipitation rate exhibited only a small influence on the orographic enhancement of rainfall. However, orographic rainfall increased dramatically with low-level wind speed because the wind speed enhanced the maximum vertical velocity upstream of the barrier crest. Moisture also had a large effect because of its influence on mountain-wave dynamics. As noted previously, moisture reduces the Brunt–Väisälä frequency which, in turn, reduces the mountain-wave amplitude and, consequently, reduces orographic enhancement of precipitation. The modeling investigation by Richard *et al.* (1987) illustrates that models in which cloud microphysics and cloud dynamics are not allowed to interact yield an exaggerated importance to cloud microphysical processes.

Orographic enhancement of precipitation by the seeder–feeder concept increases slowly with the precipitation rate P_0 (Carruthers and Choularton, 1983). In the simple models such as Bader and Roach's, P_0 is a specified value based on estimates of stratiform precipitation rates well upstream of hills or mountains. Operationally, one could determine P_0 well upstream of the orographic features by radar and then apply Bader and Roach's simple feeder model to estimate orogenic rainfall. This application presumes that orographic influences do not extend upward to the level of the seeder cloud. In the case of larger orographic barriers, blocking of the upstream flow can result in upward motion extending to the seeder cloud levels. Moreover, such orographically induced vertical motion can trigger the release of potential instability which can create or enhance P_0. Elliott and Shaffer (1962) and Elliott and Hovind (1964) noted that heavy orographic rain is frequently associated with potential instability. Potential instability can also be released by large-scale vertical motion. Blocking of the low-level flow can induce differential thermal advection, causing convective instability and seeder cloud precipitation. Because hills and mountains are rarely isolated, upward motion induced by upstream hills can trigger seeder clouds (Rauber, 1981; Cotton *et al.*, 1983, 1986).

The Seeder–Feeder Process in Ice-Phase Clouds

The seeder–feeder process is also relevant to ice-phase precipitation processes. Choularton and Perry (1986) considered precipitation formation by ice-crystal vapor deposition growth and by riming of cloud droplets. They ignored aggregation processes. Riming growth of ice particles is similar to accretion of cloud droplets by seeder cloud precipitation elements in an all-water cloud. The major differences are due to the variety of cross-sectional areas, fall velocities, and collection efficiencies presented by the various forms of ice particles. Owing to the lower saturation vapor pressure with respect to ice compared to water, significant vapor-deposition growth of ice particles can occur in a water-saturated cloud having no significant liquid-water content. Depending on atmospheric pressure, the largest vapor-deposition growth rates occur between −14 and −16°C. Ice-crystal habit also modulates the vapor-deposition growth rates. Dendritic crystal habits favored in a water-supersaturated environment at temperatures between −12 and −16°C yield the highest vapor-deposition growth rates. A secondary maximum in vapor deposition growth occurs near −6°C, where needle growth prevails. The secondary maximum occurs in spite of the fact that, in a water-saturated cloud, the supersaturation with respect to ice is relatively small. Choularton and Perry, however, ignored the effects of ice-particle habit on crystal growth rates. A seeder–feeder process can operate quite effectively in an ice-phase cloud even though the feeder cloud may contain very small amounts of liquid water. This is true if the feeder cloud has a top in the −12 to −16°C temperature range, the range of dendrite crystal growth. Nonetheless, the maximum orographic enhancement of precipitation occurs when the feeder cloud has sufficient liquid water production to support riming growth. Choularton and Perry found that the orographic enhancement of precipitation by snowfall exceeded that of rainfall. In the case of a small hill, they calculated the maximum enhancement of precipitation of the ice-phase cloud to be $3.2\ \mathrm{mm\ h^{-1}}$, while under similar conditions the enhancement for rainfall was $1\ \mathrm{mm\ h^{-1}}$. The effects of wind drift are greater in an ice-phase system. As a result, increasing wind speed increases the maximum enhancement of snowfall over a long hill, whereas it has a reverse effect over a short hill.

Wind-drift effects and the trajectories of precipitation particles are complicated in ice clouds by the variety of crystal habits and particle forms. Variation in ice-crystal concentrations has impact on particle trajectories; if concentrations are low, individual precipitation elements are larger and fall faster. They also grow by riming of cloud droplets at faster rates, which, in turn, increases their settling rates. Using the orographic flow model developed by Fraser *et al.* (1973), Hobbs *et al.* (1973) computed the trajectories of ice particles growing by vapor deposition and riming. Figure 12.23 illustrates the computed trajectories of precipitation elements starting at points A and B. For this case, the particles are carried farther downwind with increasing particle concentration, and the amount of precipitation reaching the ground increases. In some cases, ice particles settling in the dry sinking air on the lee side of a mountain may sublimate, resulting in a reduction of precipitation with increasing particle concentration. Aggregation did not substantially alter the particle trajectories relative to riming growth. However, if particle concentrations were high and if aggregation occurred over a thick layer, aggregation growth could be more rapid than growth by riming.

Rauber (1981) also determined air motions and liquid-water production using a linear, orographic flow model and calculated trajectories of ice particles. Windward of the orographic barrier, he found that the particle trajectories can sometimes be nearly horizontal. That is, the increase in

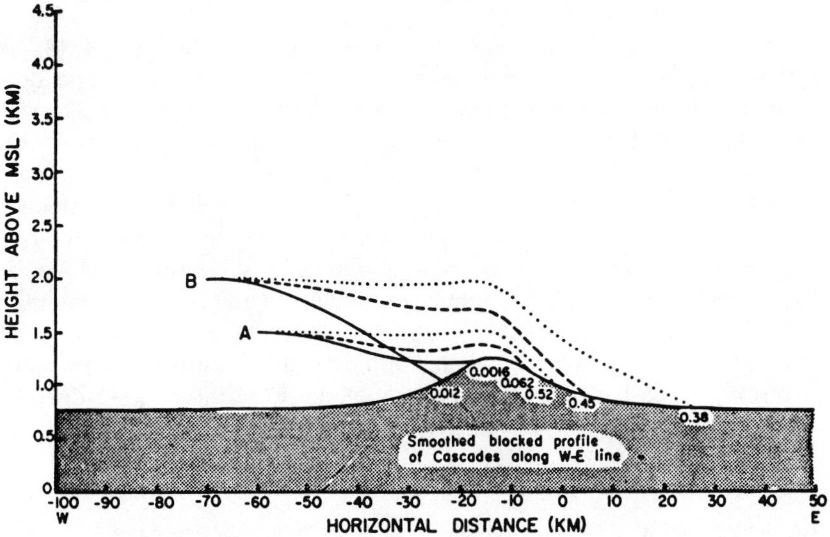

Fig. 12.23. Calculated trajectories for precipitation particles originating at A and B and growing by deposition riming over the Cascade Mountains in a westerly airstream with simulated blocking for the following specified concentrations (liter^{-1}) of ice particles: 1 (solid line), 25 (dashed line), 100 (dotted line). The number at the end point of each trajectory is the total mass (milligrams) of precipitation that reaches the ground at that point originating in a volume of 1 liter at the starting point of the trajectory. [From Hobbs *et al.* (1973).]

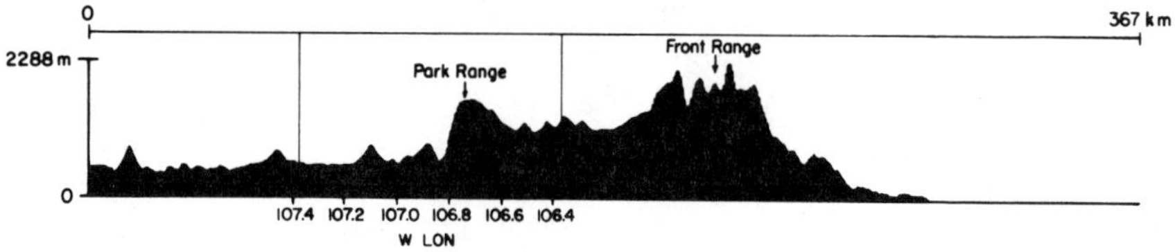

Fig. 12.24. The 30-sec longitude resolution topography at 40°30′N latitude. [From Cotton *et al.* (1986).]

particle fall speed as the crystal grows to larger dimensions is largely offset by the increase in updraft speed as the air flows over the barrier. Rauber calculated that, in order for upper-level clouds to contribute to surface precipitation over the mountain slopes, the precipitation would have to settle out of those clouds as much as 100 km upwind of the mountain. This is because the precipitation falling from upper-level wintertime seeder clouds is generally in the form of pristine crystals with low terminal velocities. As in the clouds studied by Browning and colleagues over the British Isles, such distant seeder clouds would be formed by large-scale lifting processes.

One deficiency of the modeling approach used by Rauber (1981) and Hobbs *et al.* (1973) is that liquid-water production is calculated independent of ice-particle depletion of liquid water. Ice-particle growth is calculated following each particle as they settle through the atmosphere and drift downstream. Thus, the growing ice particles cannot compete with each other for the available liquid water. An alternate approach—to calculate microphysical processes, including precipitation formation in an Eulerian framework—has the advantage that continuity of water substance can be maintained. In addition, the thermodynamic consequences of precipitation processes such as phase changes (e.g., melting, freezing, evaporation) can be considered in the thermodynamic and dynamic equations. A deficiency of this approach is that it is not computationally feasible to keep detailed track of the history of ice-particle growth processes. One cannot, for

---

Fig. 12.25. Microphysical structure of simulated orographic cloud at 5 h of simulation time (2 h after microphysics activated) for the control experiment (experiment 1) over the Park Range of Colorado. (a) Contours of cloud water mixing ratio at intervals of 0.01 g kg^{-1}, (b) ice-crystal mixing ratio at intervals of 0.05 g kg^{-1}, and (c) the aggregated contour that represents the boundary between regions of condensate and clear air. The values of x are relative to the domain extending from $x = -183.5$ to $x = +183.5$ km. The observation site is at $x = -50$ km. The vertical structure is displayed up to 10 km, although the actual domain extends to 16 km. Height coordinates are relative to a base height of 1.4 km MSL, corresponding to the lowest topographic height simulated. [From Cotton *et al.* (1986).]

a

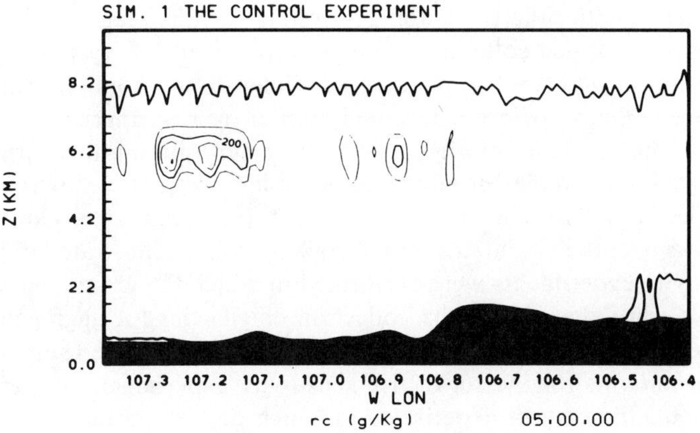
SIM. 1 THE CONTROL EXPERIMENT
Z(KM)
8.2
6.2
4.2
2.2
0.0
200
107.3 107.2 107.1 107.0 106.9 106.8 106.7 106.6 106.5 106.4
W LON
rc (g/Kg)
05:00:00

b

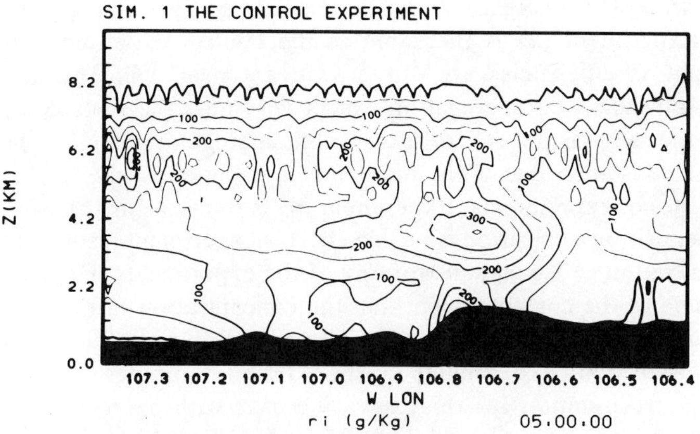
SIM. 1 THE CONTROL EXPERIMENT
Z(KM)
8.2
6.2
4.2
2.2
0.0
100
200
300
107.3 107.2 107.1 107.0 106.9 106.8 106.7 106.6 106.5 106.4
W LON
ri (g/Kg)
05:00:00

c

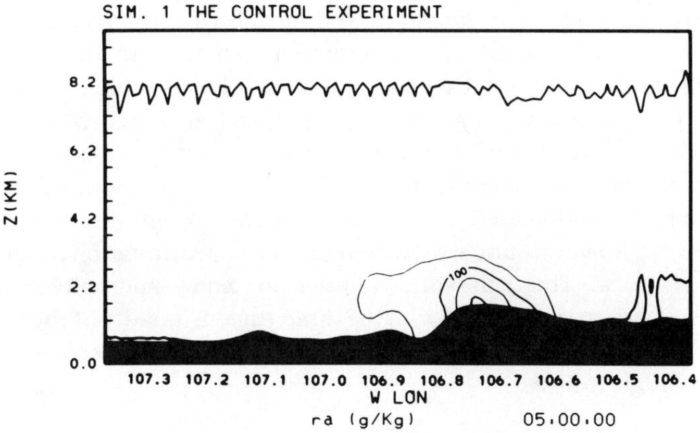
SIM. 1 THE CONTROL EXPERIMENT
Z(KM)
8.2
6.2
4.2
2.2
0.0
100
107.3 107.2 107.1 107.0 106.9 106.8 106.7 106.6 106.5 106.4
W LON
ra (g/Kg)
05:00:00

example, identify the transition of crystals such as plates to dendrites, or columns to capped columns, or unrimed to rimed ice crystals.

Cotton *et al.* (1986) applied a two-dimensional, nonhydrostatic model to the simulation of orogenic precipitation over northern Colorado (Fig. 12.24). The processes of ice-particle vapor-deposition growth, riming, and aggregation, as well as primary and secondary ice-particle nucleation, were included. Condensed water substance fields produced as the stably stratified air flows over the mountains for the control case are illustrated in Fig. 12.25. Sensitivity experiments were performed in which the ice-crystal nucleation models were altered and the collection efficiencies for aggregation of ice crystals were changed. Figure 12.26 illustrates the simulated surface precipitation amounts for the set of five sensitivity experiments. Experiment (1) represents the control experiment in which deposition nucleation, contact nucleation, and rime-splinter secondary ice-particle production are modeled along with the investigators' best estimate of aggregation collection efficiencies. Experiment (2) is the same as the control experiment except that aggregation efficiencies are only significant near 0°C. This modification essentially turns off aggregation except very near the surface. Experiment (3) is the same as the control except that aggregation efficiencies are set to a large constant value. This maintains aggregation at the largest rates for this series of experiments. Experiment (4) is the same as the control except that deposition nucleation is neglected; this experiment results in the lowest concentration of ice crystals for any of the experiments. Experiment (5) is the same as the control except that the concentration of deposition nuclei is assumed to be small in the "clean" upper troposphere.

The sensitivity experiments illustrated the importance of aggregation to surface precipitation for this case. The case with aggregation essentially turned off (experiment 2) exhibited about a 25% reduction in precipitation over the higher peaks. Turning off deposition nucleation caused the largest downwind precipitation transport and, as a result, the greatest amount of precipitation downwind of the mountain barriers. In this case, a 50% enhancement of precipitation was predicted relative to the control case over North Park, a valley lying between the Park Range and the Front Range (Fig. 12.24). These sensitivity experiments clearly show that uncertainties in cloud microphysical processes contribute to uncertainties of 25–50% in surface precipitation amounts in stably stratified orographic flow situations. The model, however, underpredicted observed precipitation amounts by an order of 50%. Thus, the uncertainties in cloud microphysics or other factors—perhaps associated with the three-dimensionality of the flow field—contributed to errors in the absolute prediction of orogenic precipitation on the order of 50%.

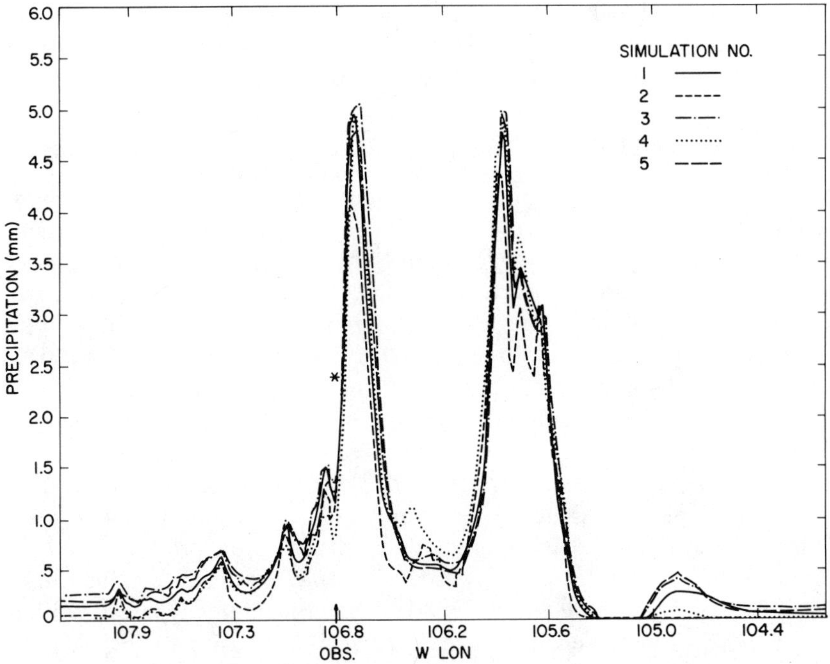

Fig. 12.26. Liquid equivalent precipitation accumulated at the ground across the domain for experiments (1) to (5). The asterisk represents the observed precipitation amount. OBS indicates the observation point. [From Cotton *et al.* (1986).]

12.4 Orographic Modification of Extratropical Cyclones and Precipitation

The heaviest precipitation events in mountainous terrain in the winter are associated with extratropical cyclones. In the case of low-amplitude hills or mountains or deep intense storm systems passing over larger mountains, the organization of vertical motion and precipitation associated with the passage of a cyclonic storm is similar to that in the model developed by the Norwegian school (Bjerknes, 1919; Bjerknes and Solberg, 1921). Many cyclones are modified significantly as they pass over the larger mountain barriers. In some cases, the storm is so strongly disturbed that certain characteristics of the Norwegian model, such as the warm front, are absent or not discernible in the sounding data (Hobbs, 1973). Many large mountain barriers are responsible for the high frequency of cyclogenesis located to the lee of the mountains (Petterssen, 1956; Radinovic and Lalic, 1959; Radinovic, 1965; Reitan, 1974; Chung *et al.*, 1976). Here, we concentrate on the effects of blocking by large mountains on the evolution of cyclonic storms and precipitation. We also examine the development of upslope storms on the eastern slopes of the Rocky Mountains.

12.4.1 Differential Thermal Advection Caused by Orographic Blocking

As stably stratified flow impinges on a mountain barrier, adiabatic cooling of the low-level lifted air results in the formation of a positive pressure perturbation along the windward slopes of the mountain. This produces a pressure gradient force which is directed upstream of the mountain barrier. The incoming airflow at low levels is thereby slowed down and in some cases it actually reverses in direction. When a baroclinic zone or cold front advances toward a mountain barrier, at low levels the cold air slows as it encounters the mountain-induced pressure gradient directed upstream of the barrier. At higher levels, the adverse mountain-induced pressure gradient is reduced in magnitude or even reversed in direction. The cold air advances at a greater speed at higher levels than it does below the mountain crest. This mountain-induced differential thermal advection process results in cooler air being advected over lower-level warmer air, leading to the formation of deep convection or "embedded" stratocumulus-type convection. Smith (1982) estimates changes in the horizontal wind using Queney's (1947) model for flow over a bell-shaped mountain. He assumed that the motion of the front was primarily due to horizontal advection, rather than wavelike propagation. If vertical motion is neglected, the horizontal position of the front as a function of altitude $x_{f(z,t)}$ is altered by advection as follows:

$$dx_f/dt = u_f(x, z). \qquad (12.33)$$

The advecting wind field (u_f) determined by Queney's (1947) model is

$$u_f(x, z) = u_0\left[1 + (hbl)\left|\frac{-b \sin lz + x \cos lz}{b^2 + x^2}\right|\right], \qquad (12.34)$$

where l is given by Eq. (12.29). The parameters h and b are the height and half-width of the mountain. Figure 12.27 illustrates the model's position of a cold front with an unperturbed slope of 1:50. Near the ground, the flow is slowed and the movement of the front is retarded. Because the flow is actually increased in speed aloft, the front moves more rapidly there. Approaching the mountain crest, the frontal surface becomes distorted and eventually turns over. Because the cold air has overridden the warmer low-level air, the atmosphere has been transformed to convectively unstable.

Convective instability is common as a frontal system approaches a mountain barrier (Hobbs *et al.*, 1975; Marwitz, 1980; Rauber *et al.*, 1986; Reynolds and Dennis, 1986). In some cases, convective instability may be

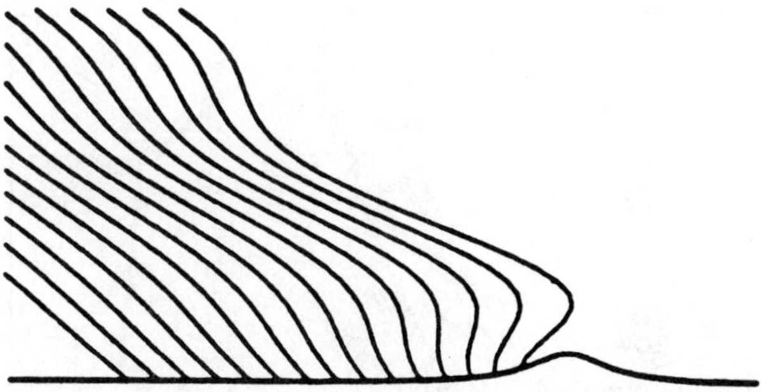

Fig. 12.27. A schematic depiction of the position of a cold front, at 2-h intervals, as it approaches and is influenced by a mountain range. The distortion of the frontal surface is from slowing of the low-level flow by the mountain and the acceleration aloft. This differential advection causes the cold air behind the front to override the warm air, producing an unstable air column. The resulting small-scale convection enhances precipitation upstream of the mountain and on its windward slopes. This diagram is constructed for $u_0 = 10\ \mathrm{m\ s^{-1}}$, $N = 0.01\ \mathrm{s^{-1}}$, $b = 20$ km, $h = 800$ m, $x_0 = -100$ km, and $a = 1/50$. The vertical exaggeration is 12:1. [From Smith (1982).]

a result of local orographic lifting; in others it may be a property of the unperturbed thermodynamic structure of an extratropical cyclone as it passes over mountainous terrain. Heggli and Reynolds (1985) and Reynolds and Dennis (1986), for example, applied Browning and Monk's (1982) split-front model to the description of the evolution of precipitation and liquid water during frontal passages in that region. The split-front model adapted to mountainous terrain is shown schematically in Fig. 12.28. Browning (1985) noted that split fronts are common in the United Kingdom. Whether or not local orography or the perturbing influence of the British Isles affects their formation in that region has not been determined. Browning and Monk's criterion for a split front is that the upper- and lower-level fronts be separated by more than 100 km; many are separated by more than 500 km. D.W. Reynolds (personal communication) noted that cyclonic storms often exhibit split-frontal characteristics in satellite imagery several hundred kilometers off the west coast of the United States. It is unknown if the continental/orographic massif of the North American continent plays a role in generating the split-front characteristics by inducing blocking well west of the continent. Although the horizontal separation of upper- and lower-level fronts is often less over the Sierras, the progression of cloud and precipitation events is consistent with the observations reported by Heggli and Reynolds and Reynolds and Dennis. As the cold front encounters low-level blocking, low θ_w air aloft advances ahead of the surface cold front, creating the upper-level front identified as UU in Fig. 12.28a.

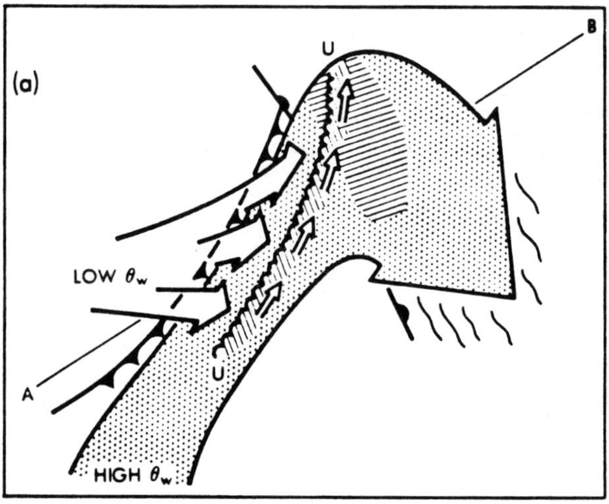

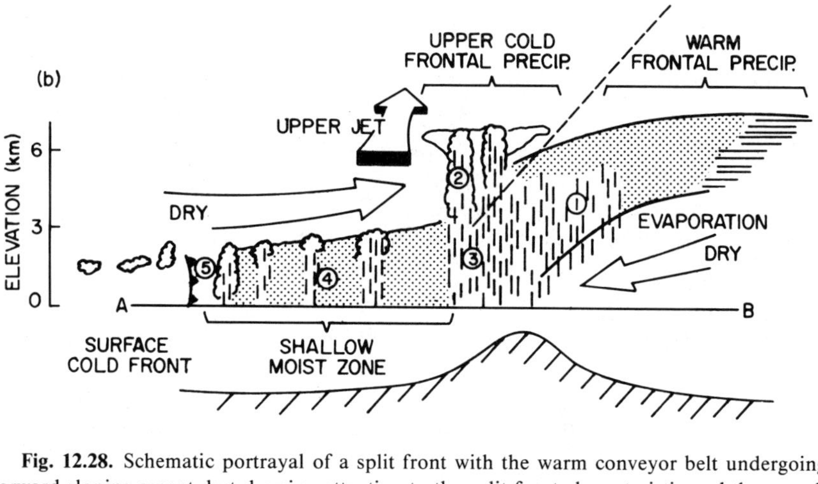

Fig. 12.28. Schematic portrayal of a split front with the warm conveyor belt undergoing forward-sloping ascent, but drawing attention to the split-front characteristic and the overall precipitation distribution: (a) plan view, (b) vertical section along AB in (a). In (a) UU represents the upper cold front. The hatched shading along UU and ahead of the warm front represents precipitation associated with the upper cold front and warm front, respectively. Numbers in (b) represent precipitation type as follows: (1) warm-frontal precipitation; (2) convective precipitation-generating cells associated with the upper cold front; (3) precipitation for the upper cold-frontal convection descending through an area of warm advection; (4) shallow moist zone between the upper and surface cold fronts characterized by warm advection and scattered outbreaks of mainly light rain and drizzle; (5) shallow precipitation at the surface cold front itself. [Adapted from Browning and Monk (1982), Browning (1985), and Reynolds and Dennis (1986). Reproduced with the permission of the Controller of Her Britannic Majesty's Stationery Office.]

Frequently, air ahead of the upper cold front is warm and moist because of the slantwise slow ascent of the warm conveyor belt of air. Precipitation in this region is typically steady and stratiform in character. Associated with the passage of the upper cold front, convective instability ensues and convective precipitation-generating cells prevail throughout much of the depth of the troposphere (Fig. 12.28b). The seeder–feeder mechanism operates very efficiently in this region. A shallow moist zone prevails behind or west of the upper cold front. Orographic lifting of this air leads to the formation of shallow orographic clouds and stratocumulus clouds. Because of the dry air mass aloft, seeder clouds and the seeder–feeder process are absent.

In some cases, low-level orographic blocking cuts off the warm moist conveyor belt on the windward side of the barrier. Hobbs *et al.* (1975) found that a shallow tongue of high θ_w air became trapped along the windward slopes of the Cascade Mountains. As a result, cloudiness decreased and precipitation was delayed until the front passed over the higher terrain. Leeward of the barrier, orographic sinking motion combined with the trapping of the high θ_w air on the windward side led to a complete disappearance of any frontal precipitation by the time the front was 37 km east of the Cascade crest.

Not surprisingly, the mesoscale features of extratropical cyclones such as prefrontal and postfrontal rainbands can be altered to a greater or lesser extent, depending on the size of the orographic features, the airflow in the various sectors of the storm relative to the topography, and the type of rainband. Parsons and Hobbs (1983) described a cyclone in which the warm-sector and postfrontal rainbands appeared to be triggered by orographic lifting of potentially unstable air along the windward slopes of a mountain barrier. In contrast, preexisting, postfrontal rainbands completely dissipated to the lee of the Olympic Mountains, Washington. Parsons and Hobbs noted that smaller hills did not have any substantial impact on precipitation from warm-sector rainbands. Both narrow and wide cold-frontal rainbands appeared to be affected less by the larger mountain barriers. The orientation and magnitude of the precipitation associated with these bands were altered somewhat, but the bands remained intact as they passed over the large orographic features.

12.4.2 Lee Upslope Storms

Orogenic precipitation events that occur on the lee side of a barrier with respect to the prevailing middle- to upper-level large-scale flow are frequent on the eastern slopes of the Rocky Mountains extending from Alberta, Canada, to northern New Mexico in the United States. Lilly (1981) suggests that a similar phenomenon occurs on the north side of the Alps, along the east coast of southern Mexico, and in the so-called "backdoor" cold fronts and subsequent high-pressure ridges that develop along the southern Appalachians. In an extensive review of upslope precipitation events, Reinking and Boatman (1986) distinguished between anticyclonic systems, which are shallow, and cyclonic systems, which are deeper and produce the major precipitation events. As in most meteorological classification schemes, these two classifications represent the extremes of a continuum of events; nonetheless, we adhere to this classification.

12.4.2.a Anticyclonic Upslope Storms

In the case of lee upslope events along the eastern slopes of the Rocky Mountains, cold Pacific and Arctic air masses are often not deep enough to cross the high mountain ridges of the Continental Divide. The barrier is hardly uniform in vertical extent, however; instead, the barrier is "leaky," allowing cold air to spill over the lower-level ridges and drain southeastward. One of the more leaky portions of the Continental Divide is the region just north of the Colorado border in southern Wyoming (Fig. 12.29). Here, cold Pacific air masses can leak through the lower-lying terrain and, as they flow southeastward, the turning of the flow to the right of its direction of motion

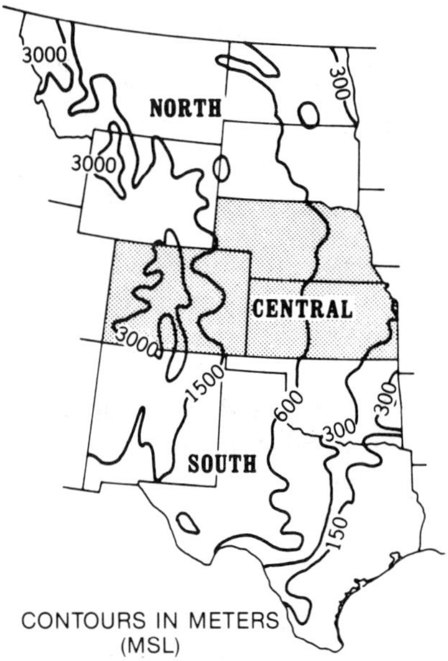

Fig. 12.29. Elevation high-plains topography doubles with each contour. [From Reinking and Boatman (1986), after Whiteman (1973).]

by the earth's rotation (Coriolis turning) causes the cold air to become trapped along the eastern slopes of the Rocky Mountains. A similar phenomenon occurs with the deeper Arctic air masses, except the topographic trapping of the air mass may not take place locally but occurs all along the north–south extent of the Rocky Mountain barrier from Alberta southward into New Mexico. The speed of southward movement of the topographically trapped cold air mass has been likened to a density current (Charba, 1974; Shapiro *et al.*, 1985; Mass and Albright, 1986) and to a topographically trapped Kelvin wave, such as ascribed to coastal lows off southern Africa (Gill, 1977). The southward surge of the cold air mass occasionally resembles the behavior of the "Southerly Buster" that moves northward along the southeast coast of Australia (Baines, 1980; Colquhoun *et al.*, 1985). The rapid southward progression of the cold air mass along the eastern slopes of the Rocky Mountains often gives rise to blizzardlike conditions.

Some evidence suggests that shallow cold fronts move faster during night than during day over the High Plains of the United States (Wiesmueller, 1982; Toth, 1987). A shallow cold front will be stalled during the day by low-lying ridges such as the Cheyenne Ridge extending along the Colorado–Wyoming border. As darkness ensues, the shallow cold air mass commences its southward movement again. Thus shallow upslope events occur more frequently during nighttime.

Toth (1987) showed by numerical experiment that, because heating is distributed over a deeper layer on the warm side of a cold front during the day, frontolysis occurs at the surface, leading to a slowing of the frontal movement. Furthermore, heating at the western higher terrain side of the cold air mass inhibits the development of horizontal pressure gradients favorable to northerly flow. Westerly winds over the higher terrain are heated during the day and flow downslope, mixing and eroding the cold air at the surface.

Figure 12.30 illustrates the southward progression of a topographically trapped cold front which moved along the eastern slopes of the Colorado Rockies. In anticyclonic events, the cloud mass forms as the air lifts upward along the eastern slopes of the barrier. It is shallow, being about 2 km in depth, and the moist air mass is absolutely stable. Using single-Doppler radar data, Lilly and Durran (1983) estimated average ascent rates of the cold air at 0.02–0.03 m s^{-1} over the gently sloping plains and 0.05–0.1 m s^{-1} over the foothills to the west. Lilly and Durran examined what happens to the shallow rising air mass as it reaches its maximum vertical extent. No easterly or southerly return flow is evident below the capping inversion. Lilly and Durran speculated that the rising cold air is entrained into the westerlies above the inversion. The cold air entrained through the inversion is then carried away to the east by the overriding westerlies. They provided some Doppler radar evidence that turbulent processes take place near the inversion, although details of the entrainment process were lacking. They argued that much of the entrainment took place over the foothills where

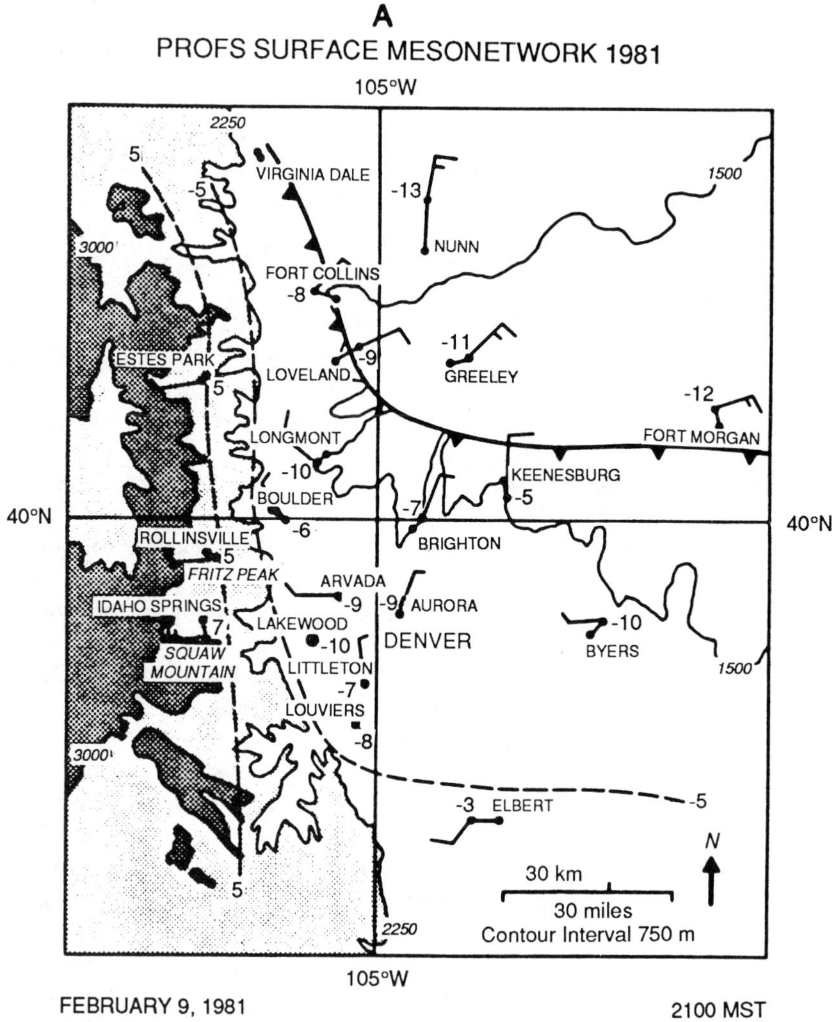

Fig. 12.30. Surface analysis at (A) 2100, (B) 2300, and (C) 0100 local time, 9–10 February (04–08Z), from PROFS surface net in north-central Colorado. Altitude contours are in meters, temperatures in °C, reduced dry adiabatically to the altitude in Boulder, Colorado. A full wind barb is 10 kt. The frontal position is based mainly on wind speed. The station locations and the nearest towns are denoted by solid dots. [From Lilly and Durran (1983).] (*Figure continues.*)

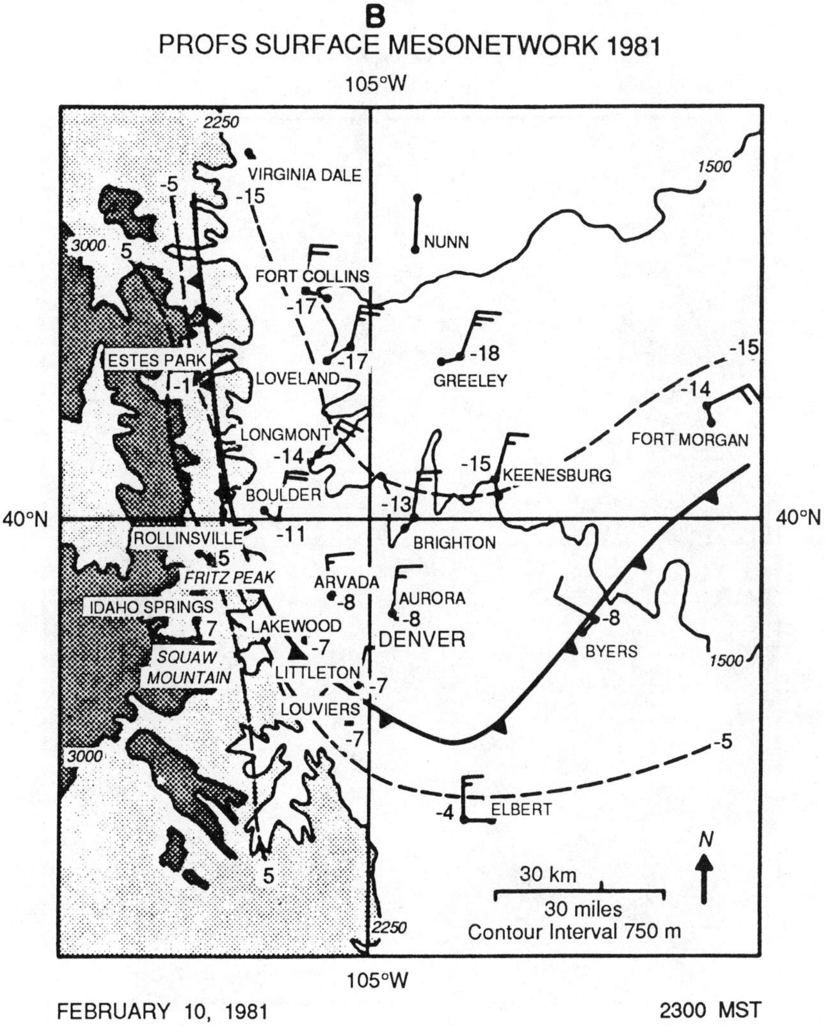

Fig. 12.30 *(continued)*

ground clutter prevented adequate radar coverage. The foothills are a preferred location for convective instability induced by differential thermal advection. The resultant convection would efficiently vent the cold air mass and mix it with the overlying westerly airstream.

For the period of steady upslope analyzed, Lilly and Durran noted that the weak upslope must be balanced by downward heat flux, through turbulent entrainment, in a self-regulating way. If a greater flux of cold air

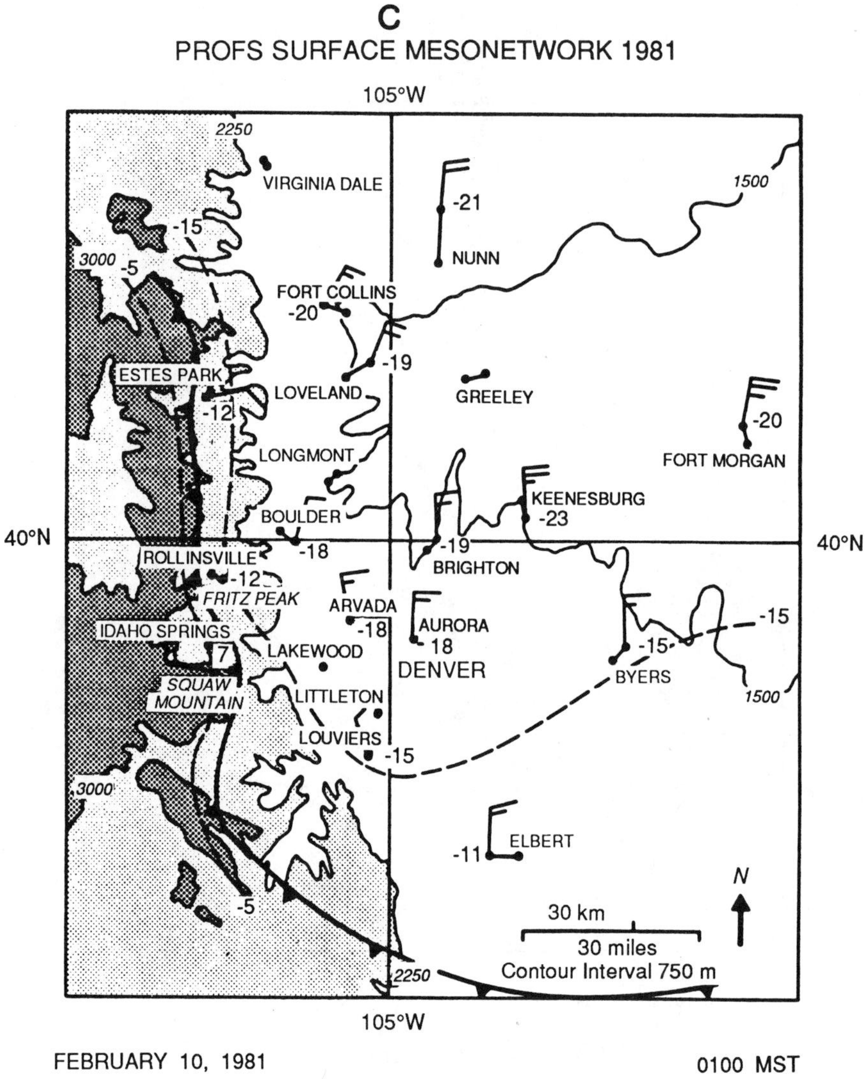

Fig. 12.30 (*continued*)

was moved up the mountain than could be entrained away by the upper westerlies, an adverse pressure gradient would be established that would inhibit the upslope flow. Conversely, excessive entrainment would result in a shallowing of the inversion and allow more cold air to be drawn up the slope.

Precipitation from shallow anticyclonic upslope storms can range from none to light to moderate. Reinking and Boatman argued that the seeder–feeder process is important to controlling the occurrence and amounts of precipitation from these shallow storms. Both aircraft observations (Walsh, 1977) and radar data (Lilly and Durran, 1983) suggest that precipitation settling from the overlying westerly airstream into the water-rich, upslope feeder cloud can serve as seeder elements. The seeder–feeder process is fundamental to precipitation formation in the shallow anticyclonic upslope events in which cloud tops are likely to be relatively warm (i.e., warmer than −10°C). For such warm-topped clouds in which liquid-water contents are relatively low (i.e., typically less than 0.2 g m^{-3}), primary and secondary production of ice crystals is quite low in the upslope feeder cloud. Ice crystals formed over the higher terrain at cold temperatures and advected eastward by the overlying westerlies can be effective seeder crystals.

Weickmann (1981) noted that small, heavily rimed graupel pellets are common in shallow upslope events with cloud tops in the −6 to −10°C range. Referring to the laboratory experiments of Fukuta *et al.* (1982, 1984), Reinking and Boatman (1986) suggested that this temperature range is a region of suppressed vapor-deposition growth of ice crystals. The preferred ice-crystal shapes are rather isometrically shaped columnar forms of crystals. Fukuta *et al.*'s experiments suggest that the faster fall speeds of these crystals enhance the ice particles' ability to switch over to the heavily rimed graupel mode of growth. This more rapid switchover to the graupel mode of growth compensates for some of the otherwise suppressed precipitation growth by vapor deposition.

12.4.2.b Cyclonic Systems

By far, the heaviest upslope precipitation events along the eastern slopes of the Colorado Rockies are associated with lee cyclogenesis that often commences in the Four Corners region (intersection of the states of Colorado, Utah, Arizona, and New Mexico). Figure 12.31 illustrates the 850-mbar analysis at the time a closed low had moved into southeastern Colorado. In the more intense storm systems, such as the 1982 Christmas Eve blizzard illustrated here, a closed low may extend upward to 500 or even 300 mbar. This results in vigorous northeasterly upslope flow through a large depth of the atmosphere. The circulation of the surface cyclone also draws polar air southward where it becomes topographically trapped along the eastern slopes of the Rocky Mountains. The advection of cold air up the eastern slopes of the Rocky Mountains can be particularly strong when the cold conveyor belt (see Fig. 11.4) injects a jet of cold air against the eastward-facing slopes. Warm moist air drawn from the Gulf of Mexico is

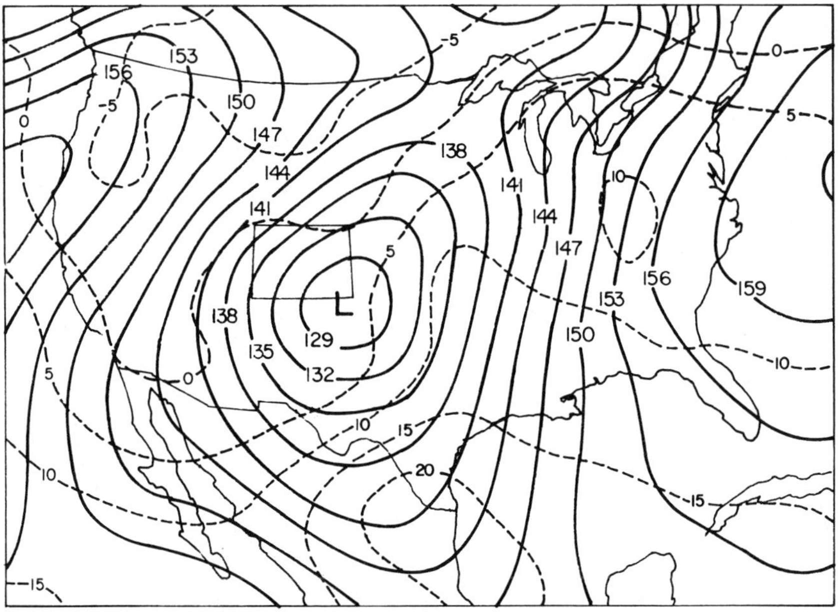

Fig. 12.31. The 850-mbar analysis for 0500 MST, 24 December 1982. [From Abbs and Pielke (1986).]

undercut by cold polar air, and the interaction of these two air masses can produce complex and heavy snowfall patterns. The distribution, intensity, and maximum amounts of snowfall depend upon the location of formation of the cyclone, its track and moisture supply, speed of movement, and local terrain effects.

Condensate production in these storms is a result of large-scale lifting by the cyclonic storm as well as terrain lifting. Auer and White (1982) extensively analyzed the Thanksgiving Day storm of 20–21 November 1979 along the Front Range of the Rocky Mountains. Reported snowfall amounts were 36 cm at Casper, Wyoming, 25 cm at Laramie, Wyoming, 66 cm in Cheyenne, Wyoming, 46 cm in Fort Collins, Colorado, and 43 cm in Denver, Colorado. To show local terrain influences, one author (WRC) measured 111 cm of snowfall at his house in the foothills at ~2400 m MSL, 24 km west of Fort Collins, Colorado. Vertical velocities forced by the large-scale dynamics of the storm were estimated at 0.02–0.06 m s^{-1} in the northeast quadrant of the storm. Low-level, terrain-induced vertical velocities due to small-amplitude ridges over the plains were estimated at 0.02 m s^{-1} near Denver and 0.04–0.07 m s^{-1} near Cheyenne. Higher terrain-induced vertical velocities were obviously present over the foothills.

Abbs and Pielke (1986) used a three-dimensional hydrostatic flow model to illustrate terrain influences on vertical velocity for the 1985 Christmas blizzard over eastern Colorado. They showed that the terrain effects are strongly dependent on the orientation of the prevailing flow with respect to the terrain features (Fig. 12.32). With northeasterly flow, the southern flanks of the Cheyenne Ridge and the Palmer Divide are under the influence

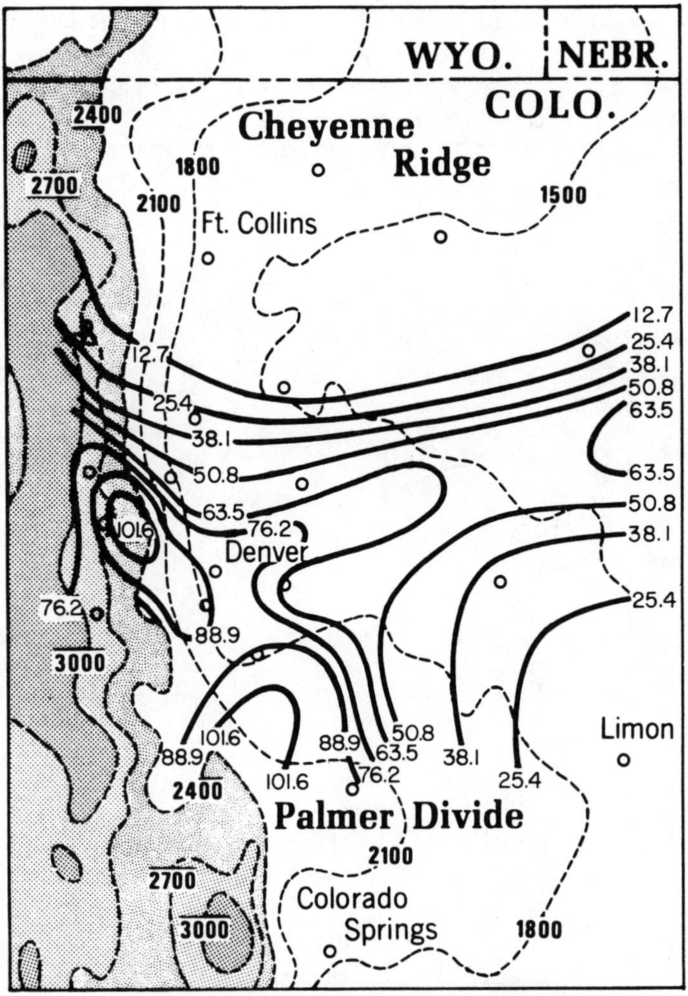

Fig. 12.32. Snow depth in centimeters (solid contours) and terrain height in meters (dashed contours) from the 1982 Christmas Eve blizzard measured within the PROFS mesonetwork of northeast Colorado. [Adapted from Reinking and Boatman (1986), after Schlatter *et al.* (1983).]

of downslope flow, but the north-facing slopes of the ridges are exposed to upslope flow. The vertical velocity field predicted by Abbs and Pielke (Fig. 12.33) depicts magnitudes in excess of 0.3 m s^{-1} along the northern slopes of the Palmer Divide and the eastern slope of the Continental Divide. The predicted regions of topographically induced ascent correspond to the regions of observed heaviest snowfall. In some locations along the windward slopes of the Palmer Divide, snowfall amounts in excess of 112 cm were reported. Similar terrain influences on cyclonic storm events over the

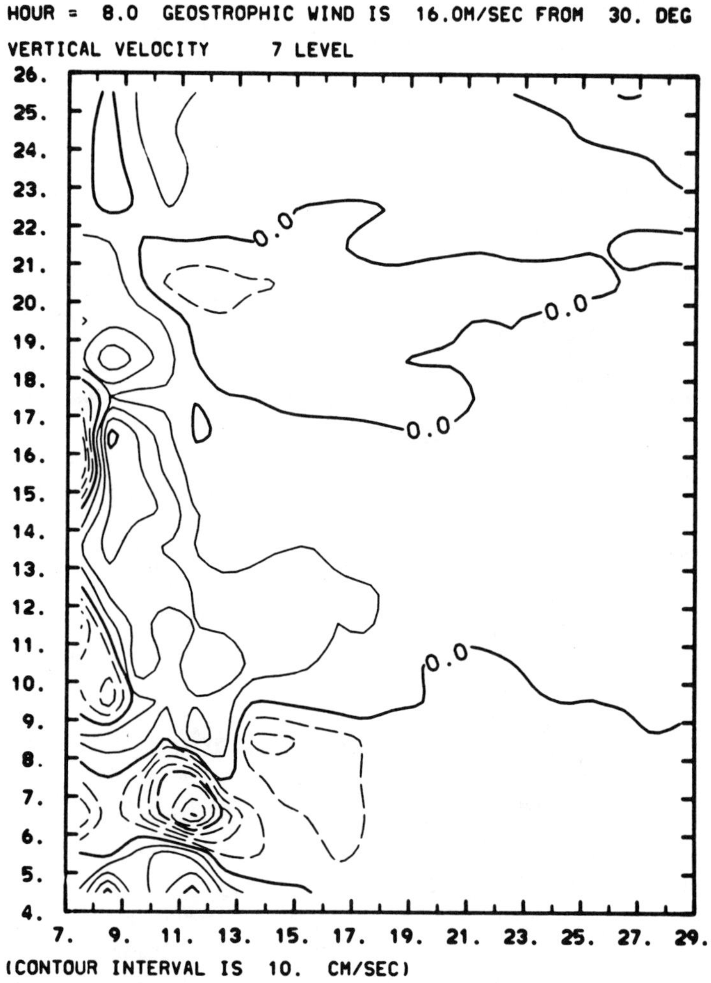

Fig. 12.33. Vertical velocity field at 250 m for run 2 after 8 h of integration. [From Abbs and Pielke (1986).]

Canadian western plains were estimated by Raddatz and Khandekar (1979) using a single-layer, two-dimensional flow model. In the Colorado-Wyoming Thanksgiving Day storm of 1979, Auer and White (1982) noted that the importance of terrain-induced vertical motion increased as the large-scale dynamics of the storm system decreased.

While computing the large-scale divergence field and vertical motion for the 1979 Thanksgiving Day storm, Auer and White noted that the level of nondivergence, and hence the level of maximum vertical velocity, corresponded to the temperature range of −13 to −17°C. This roughly corresponds to the dendritic mode of ice-crystal growth, in which the maximum rates of vapor depletion occurs. Moreover, the presence of dendritic crystals favors the operation of an aggregation process through a deep layer. Auer and White (1982) and Reinking and Boatman (1986) note that aggregates are the dominant precipitation form in the heaviest upslope precipitation events. Auer and White conclude that whenever the maximum vertical velocity of a storm system corresponds to roughly the −13 to −17°C level, the maximum in water production rate coincides with the maximum in ice-crystal growth rates (other factors remaining equal). Auer and White (1982) provided climatological evidence that the maximum snowfall events across the United States, correspond to the level of maximum large-scale vertical motion in the temperature range −13 to −17°C. They also provide evidence that the heavier orographic snowfalls are characterized by the occurrence of the maximum orographic lifting being coincident with the dendritic ice-crystal growth habit.

Although no major scientific studies have dealt with the role of convection in either shallow or deep upslope storm events, there is circumstantial evidence that embedded convection occasionally occurs. Reinking and Boatman noted that variations in precipitation rates suggest the presence of a banded storm structure similar to that frequently found in extratropical cyclones. Moreover, it is common to observe large heavily rimed graupel particles during the early stages of major upslope events. Such large graupel particles are generally favored in convective cells with high liquid water content. One would expect blocking-induced differential thermal advection to be the most pronounced during the early stages of an upslope storm event. The instability so caused would favor the formation of embedded convective elements.

12.5 Distribution of Supercooled Liquid Water in Orographic Clouds

The magnitude and spatial distribution of supercooled liquid water in wintertime orographic clouds are important to a number of applications.

The presence of supercooled water is a major factor in determining the amount and type of precipitation. The presence of supercooled water is also a hazard to aircraft operations. Thus an assessment of the amount and distribution of supercooled water is important to predicting aircraft icing conditions. The radiative properties of a cloud are also affected by the amount and spatial distribution of liquid water. The acidity of precipitation is affected by the degree of ice particle riming versus vapor deposition growth (Borys *et al.*, 1983). Finally, the opportunity for precipitation enhancement by cloud seeding is dependent upon the amount and spatial distribution of supercooled liquid water. It is this last application which has motivated a number of observational and diagnostic studies of the amount and distribution of supercooled liquid water in the western United States.

The liquid-water content at any location in an orographic cloud is the integrated consequence of several liquid-water production processes (FP) and liquid-water depletion processes (FD). Liquid-water production is determined by the cooling rate of the air as it rises over a mountain. The cooling rate, in turn, is primarily due to adiabatic ascent. Net radiative cooling is also a factor, at least in weaker wind situations. The major liquid-water depletion processes are due to vapor deposition and riming growth of ice crystals, coalescence among cloud droplets, and removal as precipitation and by entrainment of dry air. The depletion rates are therefore a function of the concentration of ice particles, the temperature and habits of the ice particles, the cloud supersaturation and liquid-water content, concentrations of cloud condensation nuclei, and the characteristics of cloud-mixing processes.

Whenever FP is greater than FD, liquid water will accumulate in the cloud to levels that can be detected by direct or remote-probing systems. If FP is less than or equal to FD, liquid water will be depleted so rapidly that only solid precipitation elements will be detectable. As a simple example of the competition between FP and FD, consider the diagram in Fig. 12.34 (Chappell, 1970). Estimated water-production rates for airflow over the Climax, Colorado, mountain barrier are a function of the 500-mbar temperature in that region. Chappell used the 500-mbar temperature as a crude index of the cloud-top temperature in that region. As the cloud-top temperatures become colder, cloud bases do too, because the bases are limited by terrain height. Saturation mixing ratios diminish and thereby decrease FP. The long-dashed curve in Fig. 12.34 illustrates Chappell's estimate of FD, assuming that the major depletion process was vapor-deposition growth of ice particles. He also assumed that the concentration of ice crystals could be determined by the so-called Fletcher ice nuclei spectra (Chapter 4). This figure illustrates that, at temperatures warmer than −20°C, FP is much

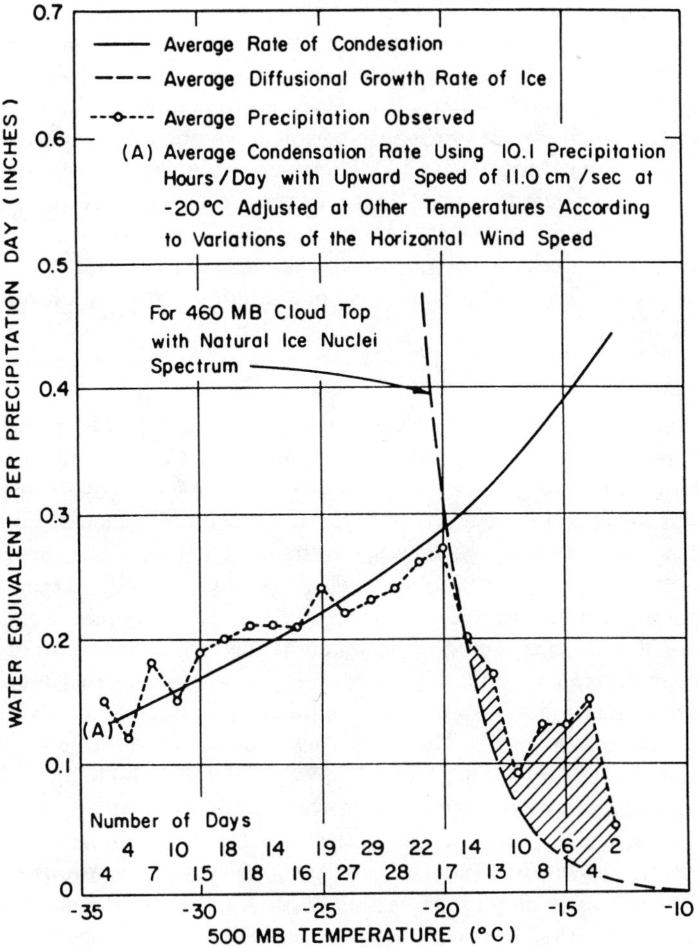

Fig. 12.34. Distribution of nonseeded precipitation at High Altitude Observatory, Climax, Colorado as a function of 500-mbar temperature compared to a theoretical distribution computed using the mean diffusional model. Precipitation data are from Climax I sample (251) and values are a running mean over a 2°C temperature interval. [From Chappell (1970).]

greater than FD; thus, the opportunities for significant liquid-water production are large. At temperatures colder than −20°C, the more numerous ice crystals deplete liquid water at a rate faster than it is produced. Also shown in Fig. 12.34 are the observed surface precipitation rates. At temperatures colder than −20°C, the observed precipitation is comparable in magnitude to the calculated FP. This simply shows that, at colder temperatures, FD is limited by FP and little excess liquid water is produced. At temperatures warmer than −20°C, the observed precipitation rate is considerably less than Chappell's estimate of FP, showing an opportunity for production of considerable liquid water. The observed precipitation rate exceeds Chappell's estimate of FD probably because Chappell assumed that only vapor deposition contributed to FD. However, riming growth of ice crystals could have been significant at these warmer temperatures, where high liquid-water contents are likely. Also noted in Chapter 4, the Fletcher ice nuclei curve is frequently a poor estimator of observed ice-crystal concentrations at warmer temperatures because of the effects of ice multiplication.

12.5.1 Stable Orographic Clouds

The system that allows easiest interpretation of supercooled water is the "pure" orographic cloud system which occurs over mountainous regions when large-scale frontal activity is absent. It occurs when a strong cross-barrier pressure gradient drives strong winds containing moist air across the barrier. These cloud systems are well-simulated by linear orographic flow models. As found by Hobbs (1973) over the Cascade Mountains of Washington and by Rauber (1985) over the Park Range of northern Colorado, the supercooled liquid-water content is largest in regions of strong orographic forcing on the windward side of steep rises in topography. Rauber also noted that in some cases liquid water varied uniformly through the depth of the orographic cloud layer without any discernible changes in the airflow properties of the cloud. The only detectable change in cloud structure was a variation in precipitation rate, probably a result of variations in ice-crystal concentrations. The causes of such concentration variations are not well known, but P.J. DeMott (personal communication) suggests that enhanced nucleation of ice particles occurs in localized regions of high supersaturation associated with upward vertical motions near cloud top.

12.5.2 Supercooled Liquid Water in Orographically Modified Cyclonic Storms

Variations in the distribution of liquid water in orographic clouds are frequently associated with variations in thermodynamic stratification and vertical motion as cyclonic storm systems pass over a mountain. Based on aircraft observations of liquid water and microphysical structure of orographic clouds over the San Juan Mountains of southern Colorado, Marwitz (1980) and Cooper and Marwitz (1980) identified four stages in thermodynamic stratification of a passing storm system. The four stages shown at the top of Fig. 12.35 can roughly be identified with the passage of a cyclonic

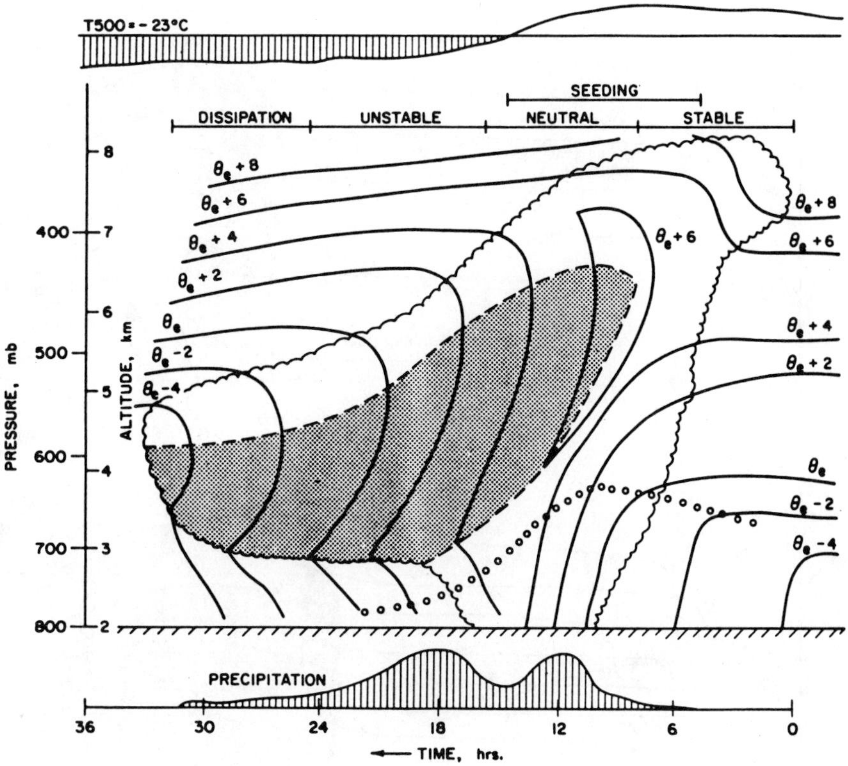

Fig. 12.35. Schematic height-time cross section for a typical storm over the San Juans. The solid lines are lines of constant equivalent potential temperature (θ_e), and the shaded area indicates convective instability. The typical sequence of 500-mbar temperature is shown at the top. The reference baseline is −23°C. The level below which the wind directions are <160° is indicated by circles and is defined as blocked flow. The four stages of the storm and a typical seeding period are indicated. [From Cooper and Marwitz (1980).]

storm. Figure 12.36 illustrates a conceptual model of the liquid-water distribution for three of the four stages. During the "stable stage," which corresponds to the prefrontal stage of a cyclonic storm, moisture supplied by the rising warm conveyor belt contributes to a deep stable cloud layer with cold tops. The cloud system is nearly completely glaciated owing to the high ice-crystal concentrations associated with the cold cloud tops. Liquid water occurs only in short-lived patches where gravity waves create transient regions of rapid ascent. Any liquid water produced is small—less than $0.1\ g\ m^{-3}$. Substantial blocking of the low-level flow occurs during this period (Fig. 12.36a). As a cyclonic storm advances across the mountain barrier, the atmosphere transforms to near-neutral stability throughout much

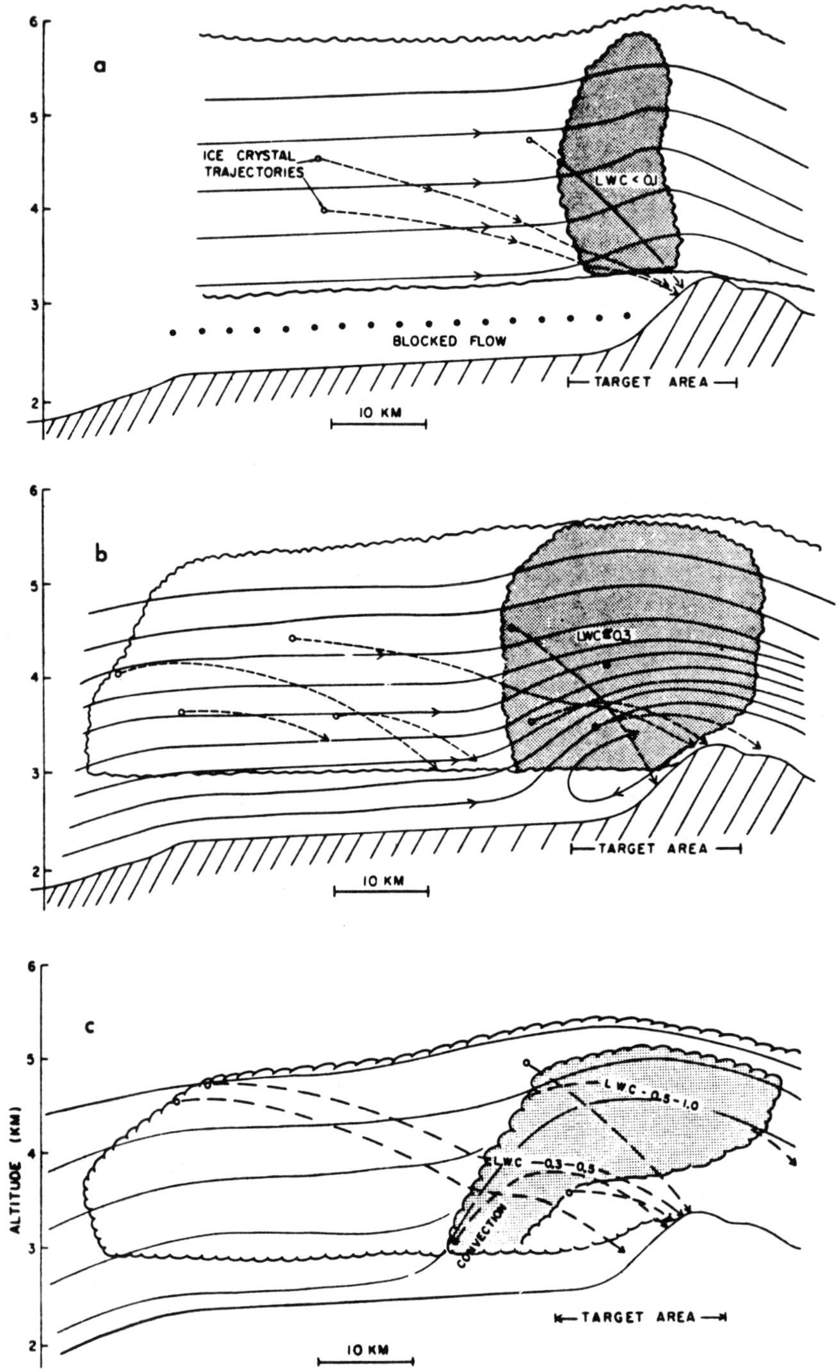
a
ICE CRYSTAL TRAJECTORIES
LWC < 0.1
BLOCKED FLOW
TARGET AREA
10 KM
b
LWC ≤ 0.3
TARGET AREA
10 KM
c
ALTITUDE (KM)
LWC - 0.5 - 1.0
LWC - 0.3 - 0.5
CONVECTION
TARGET AREA
10 KM

of the depth of the troposphere. Liquid-water contents rise to values on the order of 0.3 g m^{-3} (Fig. 12.36b).

Following the passage of the 500-mbar trough and in some cases a discernible front, the atmosphere becomes convectively unstable at lower levels upwind of the mountains and above a surface zone of convergence. Cooper and Marwitz reported that liquid-water contents were highest during this stage of the storm system, approaching 1.0 g m^{-3} in the upper cloud levels. Hobbs (1975) and Heggli *et al.* (1983) also found the highest liquid-water contents in postfrontal convective clouds observed over the Cascade Mountains and the California Sierras, respectively. Like Cooper and Marwitz, their observations were obtained primarily by aircraft penetrations. One deficiency of aircraft observations is that, for safety reasons, aircraft are limited to flying 600 to 1000 m above mountain tops. Also, for convective cells embedded in stratiform clouds, time continuity of the convective cells cannot be obtained. Remote-probing systems, such as microwave radiometers (Hogg *et al.*, 1983) and polarization-diversity lidar (Sassen, 1984), provide an opportunity to sense the presence of liquid water in regions previously inaccessible by aircraft. The data from these new observational systems are substantially altering our view of the distribution of liquid water in orographic clouds.

Using a combination of aircraft and passive microwave radar observations, Rauber (1985), Rauber *et al.* (1986), and Rauber and Grant (1986) described the distribution of supercooled liquid water observed in the Park Range of northern Colorado during passage of extratropical cyclones. In the region just ahead of and during frontal passage, clouds were characterized by wide areas of stratiform precipitation and clouds with frequently superimposed deeper clouds and heavier precipitation. Many of the heavier precipitating regions were convective, but some occurred in a stably stratified or neutral atmosphere. Liquid-water production in the more stratiform region of the prefrontal environment was strongly influenced by the orientation of the approaching front relative to the orientation of the mountain barrier. Over the Park Range, midlevel flow with a strong westerly component generated orographic lifting and was generally of higher moisture content. Convection, sometimes organized in bands was often observed in the prefrontal environment. Liquid water was found in the convective clouds during their initial stages of development, but it was rapidly depleted by

---

Fig. 12.36. Typical appearance of San Juan storms during the (a) stable, (b) neutral, and (c) unstable storm stages. The airflow is from the left at ~15 m s^{-1}. The shaded regions are regions where liquid water is generally found, and some typical liquid-water contents (g m^{-3}) are indicated. Solid lines with arrows show air trajectories, and the irregular lines indicate cloud boundaries. Some typical ice crystal trajectories are also shown as dashed lines. The farthest downwind trajectory shows the fall of a hydrometeor with a terminal velocity of 1 m s^{-1}. [From Cooper and Marwitz (1980).]

efficient precipitation processes. In general, liquid-water amounts reached a minimum within 15–20 min after the onset of convection.

The postfrontal clouds over the Park Range were also characterized by wide areas of stratiform precipitation. Orographic forcing was particularly strong because the westerly to northwesterly midlevel winds are normal to the barrier orientation. The highest liquid-water amounts were 10–15 km upwind of the barrier crest and reached a maximum over the windward slopes where the condensation rates were a maximum. Two regions of convective instability were observed in the postfrontal region; the first occurred just after frontal passage, and the second occurred during the passage of the 500-mbar trough as the system dissipated. The distribution of liquid water in the postfrontal zone was complicated by the fact that superimposed on steady orographic production of liquid water was transient convective activity. Again, the convection produced liquid water early in its life cycle, but later produced substantial numbers of ice crystals which depleted liquid water in the convective regions as well as in the lower-level feeder regions.

Figure 12.37 is a schematic of the observed distribution of supercooled liquid water in (A) shallow orographic clouds, (B) a deep stratiform cloud system, and (C) a convective region embedded in a shallow stratiform cloud system. In all three systems, liquid water was observed in the feeder zone directly upwind and over the barrier crest. A second region of liquid water was inferred between cloud base and $-10°C$. In this region, planar crystals falling into it did not grow rapidly by vapor-deposition growth. The more rapidly growing needles and columnar form of crystals were characteristically absent. The liquid-water depletion mechanisms (FD) were at a minimum, yet water production (FP) at these warmer temperatures was quite high. A third region near the cloud tops in Fig. 12.37 was also a region of frequent liquid water. Rauber and Grant (1986) suggested that this zone of liquid water results from an imbalance between liquid-water production by adiabatic cooling and depletion by vapor deposition. Large ice crystals, capable of extracting large quantities of vapor as they grow (and hence depleting liquid water by the Bergeron–Feindeisen process), rapidly settled out of the cloud-top layer. Remaining were numerous small ice crystals incapable of extracting water substance at the rates being supplied by adiabatic ascent. Another possible contribution to the observed liquid water near cloud top was that liquid-water production could be intensified by radiative cooling. Radiative cooling rates estimated by Chen and Cotton (1987) for stratocumulus clouds having relatively warm cloud-top temperatures can approach $200°C\ day^{-1}$ in a layer of 5-mbar thickness near cloud top. For wet adiabatic ascent, the cooling rate (CR),

$$CR = 665w \quad (°C\ day^{-1}), \tag{12.35}$$

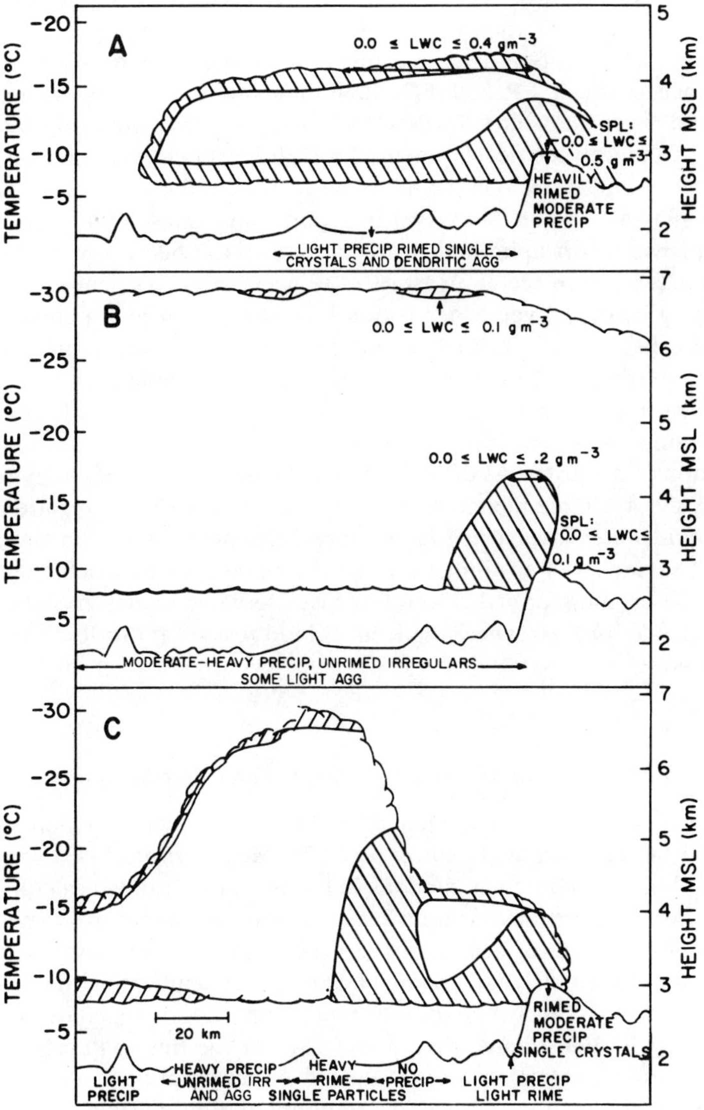

Fig. 12.37. Conceptual models of supercooled liquid-water distribution in Park Range cloud systems for (A) a shallow stratiform cloud system with a warm cloud-top temperature (CTT> −22°C); (B) a deep stratiform system with a cold cloud-top temperature; and (C) a deep convective region embedded in a shallow stratiform system. The characteristic precipitation type is listed for each case along the bottom of the figure. Magnitudes of liquid-water contents measured with aircraft or at SPL are shown on the figure. [From Rauber and Grant (1986).]

where w is the vertical motion in meters per second. Thus, an updraft of $0.3\ \mathrm{m\ s^{-1}}$ is required to produce the same rate of cooling as cloud-top radiation. Cloud-top radiative cooling, therefore, can contribute to an imbalance between FP and FD. Preliminary numerical experiments by Mulvilhill over the Colorado Park Range suggest that cloud-top radiative cooling enhanced the development of a high-level blanket-type cloud with enhanced liquid water relative to a case without radiative effects.

One may ask if a shallow layer of liquid water near cloud top is of any consequence to precipitation formation, especially because it occurs at a level inaccessible to accretion growth and because the residence time of ice particles growing by vapor deposition is so short. Consider a cloud whose top is in the −12 to −16°C temperature range. This temperature range is suitable for the growth of dendrites, provided that the cloud is supersaturated with respect to water. Rauber (1985) showed that, under suitable cloud-top temperature conditions, this cloud-top layer of liquid water can favor the formation of dendrites. As the dendrites settle into lower levels, they rapidly deplete the available moisture and can readily grow by aggregation. This thin liquid-water layer can trigger more efficient precipitation processes, and if it extends well upwind of the main mountain barrier in a blanket-type cloud, ice crystals formed in the layer can serve as seeder crystals in the seeder–feeder process where the feeder cloud resides at low levels over the barrier crest.

12.5.3 Observations of Supercooled Liquid Water over the Sierra Nevada

Aircraft observations of supercooled liquid water in winter cloud systems over the Sierra Nevada (Lamb *et al.*, 1976; Heggli *et al.*, 1983) suggested that the largest amounts of supercooled liquid water resided in convective towers. More recent observations using passive microwave radiometers as well as observations of heavily rimed ice crystals from shallow orographic clouds led Reynolds and Dennis (1986) to conclude that the shallow nonconvective clouds contained substantial amounts of supercooled liquid water. Heggli (1986) noted that, about 75% of the time, saturated clouds possibly containing liquid water existed below minimum aircraft flight altitudes and that the vertical distribution of liquid-water content was bimodal. A low-level maximum was observed at temperatures between −2 and −4°C and a secondary maximum similar to that observed by Rauber and Grant (1986) occurred between −10 and −12°C. Heggli also noted that supercooled liquid water occurred in multiple layers separated by 100 m or more of unsaturated air. The mechanisms responsible for such layering of the liquid-water zones have not been identified.

Heggli and Reynolds (1985) and Reynolds and Dennis (1986) interpreted the behavior of liquid-water variations during the passage of cyclonic storms over the Sierra Nevada in terms of Browning and Monk's (1982) split-front model (Fig. 12.28). Following the passage of the frontal rainband where the heaviest precipitation was observed, a shallow orographic cloud, often a stratocumulus cloud, prevails. The highest liquid-water contents are observed in this shallow cloud. Reynolds (1986) attributed this increase in supercooled liquid water to the disappearance of ice crystals settling from higher clouds. Deep clouds prevailing during the passage of the elevated frontal band yield an environment favorable for the seeder–feeder process. High-level seeder crystals readily deplete liquid water being generated by the lower-level, embedded feeder clouds. Following the passage of the frontal rainband, only shallow cloud layers prevail and the seeder–feeder process ends.

Heggli and Rauber (1988) described the results of a comprehensive analysis of the occurrence of supercooled liquid water in Sierra Nevada orographic clouds using a dual-channel microwave radiometer. The supercooled liquid-water content varied depending on the strength, evolution, and trajectory of motion of extratropical cyclonic storms. They examined the presence of supercooled liquid water relative to two basic types of cyclonic storm types—those having a predominant zonal flow component and those having a predominant meridional flow component. Storms with a zonal component of flow contained supercooled liquid water in postfrontal shallow stratocumulus and cumulus clouds. Storms with a meridional component of flow exhibited supercooled liquid water within prefrontal bands in the warm conveyor belt region. Little supercooled liquid water was found in the narrow or broad cold-frontal bands, presumably because the seeder–feeder process operates efficiently in that region.

In summary, supercooled liquid water in winter orographic clouds varies with the passage of cyclonic storm systems, the depth of the cloud system, the organization and type of cyclonic storm system, and proximity to mountain slopes. Liquid water is also found near the tops of the cloud system and at temperatures between −10 and −12°C, where liquid-water depletion processes are not efficient. There is also evidence of multiple layering of liquid water in some orographic clouds.

12.6 Efficiency of Orographic Precipitation and Diurnal Variability

12.6.1 Precipitation Efficiency of Orographic Clouds

Knowledge of the precipitation efficiency of orographic clouds and the factors causing its variability is important to the development of simple physical models for diagnosing or predicting orographic precipitation and in evaluating the efficacy of complex models that treat explicitly cloud microphysical processes. The effectiveness of an orographic cloud in scavenging pollutants and contributing to acid deposition is also controlled largely by the precipitation efficiency of the cloud system. The opportunity for artificially enhancing precipitation from orographic clouds is determined by the natural precipitation efficiency of the cloud system. If the precipitation efficiency is low, then an opportunity exists to artificially enhance precipitation, perhaps by cloud-seeding techniques.

Browning *et al.* (1975) examined the efficiency of orographic clouds from low hills over the British Isles and determined the efficiency to be

$$E_1 E_2 = W/C, \tag{12.36}$$

where E_1 is the efficiency influenced by microphysical processes and E_2 is a factor that takes into account any subsaturation in the initial flow upwind of the hills. The depth W of orographic water reaching the ground per unit time per unit width for a section across the mountain of length x_0 is given by

$$W = \int_{-x_0}^{x=0} R\, dx, \tag{12.37}$$

where R is the orographic component of rainfall. The parameter C represents the total depth of water condensed over a layer of 3-km depth. Browning *et al.* estimated rainfall efficiencies ranging from 0.1 to 0.7 and interpreted their results in terms of the efficiency of the seeder–feeder process and whether or not the low-level airstream was saturated. Cases of high efficiency were characterized by efficient natural seeding and an initially saturated flow due to large-scale ascent.

Elliott and Hovind (1964) calculated precipitation efficiencies over the San Gabriel and Santa Ynez ranges in California. They estimated the condensation rate by applying the Myers (1962) hydraulic airflow model to those ranges. Precipitation efficiencies ranged from 17 to 25% for cases in which the environment was stably stratified and from 26 to 28% for unstable cases. The deeper vertical penetration of the convective elements compensated for water loss by entrainment, yielding greater condensate and precipitation production.

Chappell (1970) estimating condensate supply rates by assuming values of orographic vertical motion, and dr_s/dt through the estimated mean cloud depth for Wolf Creek Pass and Climax, Colorado. Figure 12.34, showing the observed precipitation rates and inferred supply rates, suggests that precipitation efficiencies were near 100% at temperatures colder than −20°C and were about 25 to 35% at the warmer temperatures.

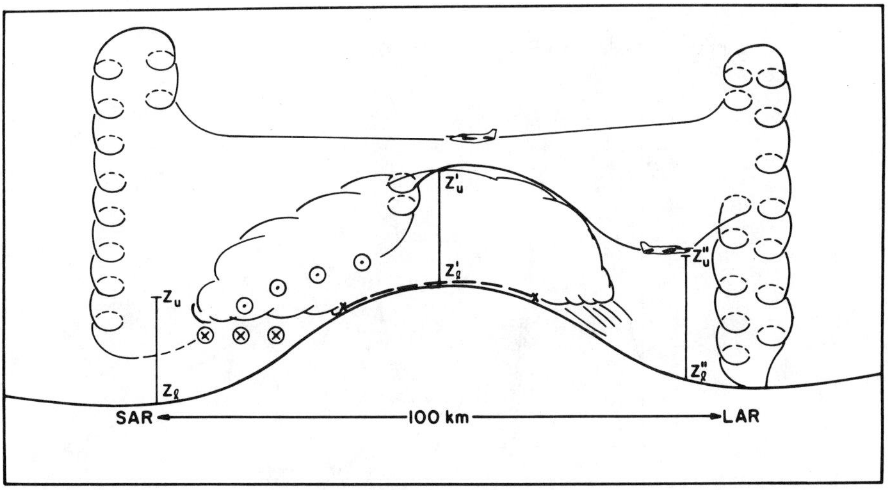

Fig. 12.38. Schematic of experimental design showing aircraft operations with respect to the orographic cloud. Airflow is from left to right. The columns represent the cloud-producing layer upstream ($z_l - z_u$), at maximum vertical lifting ($z'_l - z'_u$), and the downstream ($z''_l - z''_u$) along a vertical stream surface. The path, *x----x*, represents the region in which ground teams operate. The *x* in a circle represents cloud base flights normal to the airflow and the dots in a circle represents cloud penetrations normal to the airflow. [From Dirks (1973).]

Rather than use a model for estimated condensate production, Dirks (1973) employed aircraft data to measure directly precipitation efficiencies over a relatively isolated mountain in Wyoming. Aircraft soundings (Fig. 12.38) were used to calculate condensate production and precipitation rates. In all observed cases, the soundings were slightly unstable. Precipitation efficiencies ranged from 25 to 80%. The lowest efficiencies were at very cold temperatures and strong winds, and the highest efficiencies were under moderately cold cloud top temperatures and moderate wind speeds.

Normally, precipitation measurements for precipitation efficiency calculations are based on surface-precipitation gauge measurements or aircraft-precipitation samples. Evidence suggests that a significant contributor to the water content of the snowpack arises from rime deposits on trees and other objects located on mountain peaks; this deposit is then shed to the snow surface. Hindman *et al.* (1982), for example, estimated that between 4 and 11% of the water content of snowpack near the peak of the Park Range, Colorado, was the result of rime deposits.

12.6.2 Diurnal Variations in Wintertime Orographic Precipitation

Some limited observations suggest a significant diurnal variation in orographic precipitation. Grant *et al.* (1969) analyzed the diurnal variations in

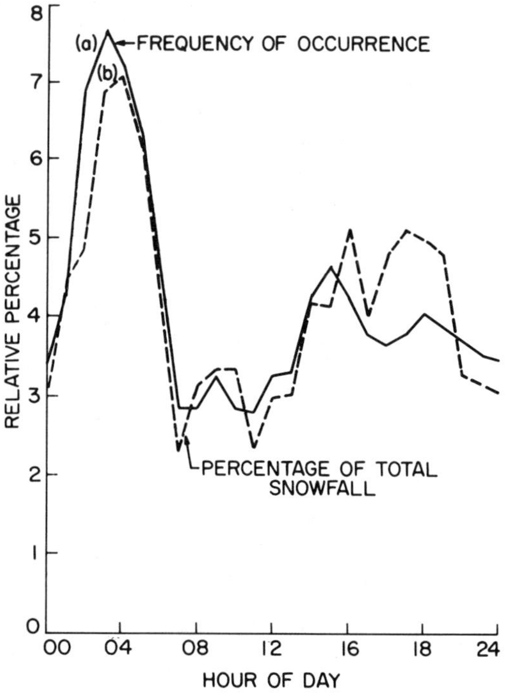

Fig. 12.39. Distribution of snowfall at Climax, Colorado, as a function of the hour of day, November through May for 1964 to 1967. [From Grant *et al.* (1969).]

the percentage of snowfall amounts and the frequency of snowfall events for a number of mountain precipitation stations in Colorado. Figure 12.39 depicts the percentage of total snowfall and frequency for Climax, Colorado, a central Colorado mountain site. A pronounced early-morning maximum at 0300 local time is evident in the frequency and amount of precipitation measured. A minimum occurs shortly after sunrise, and a secondary maximum is evident in the afternoon and early evening. An early morning maximum was also found at several other stations in the central and northern Colorado mountains. The secondary maxima in the afternoon were generally less pronounced in the more northern Colorado stations. A different behavior in the diurnal variation of precipitation applies to the southern Colorado San Juan Mountains. Stations on the lee slopes of the San Juans exhibited an early morning maximum and a secondary maximum in the afternoon in frequency and amount. Stations upwind of the mountain crest and over the mountain peaks, however, did not show an early morning maximum in precipitation frequency. Instead, a pronounced afternoon maximum in precipitation frequency was evident (Fig. 12.40). Unfortu-

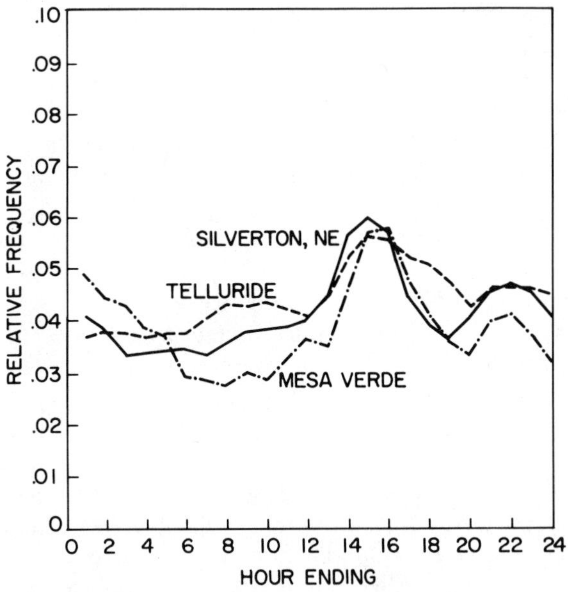

Fig. 12.40. Diurnal frequency of snowfall at Silverton, Telluride, and Mesa Verde, Colorado, November through April for 1948 to 1968. [From Grant *et al.* (1969).]

nately, the relative frequency of daily amounts of snowfall was not measured at the higher-elevation stations. The differences in behavior between the northern and southern Colorado precipitation stations are related to the climatological differences in airflow between the two regions. Southern Colorado experiences a generally moister higher value θ_e airflow than does northern Colorado. As a result, afternoon heating is more likely to lead to convective destabilization over the windward side of the San Juans than in the more northern Colorado mountains.

The pronounced early-morning maximum in precipitation frequency and amounts in the more northern Colorado stations are more difficult to explain; one can only speculate on possible mechanisms. It is likely that the early-morning maximum is the result of longwave radiation cooling at the earth's surface or at the top of a moist layer and/or orographic cloud layer. We noted earlier that orographic clouds often exhibit a large blanket stratiform cloud. One expects that, after sunset, when solar heating in the orographic cloud layer ceases, the longwave radiative cooling leads to a slow buildup of liquid water in the upper levels of the cloud deck. Eventually, the liquid water can build to levels where precipitation can commence or be enhanced. The actual presence of a blanket cloud deck may not be necessary for a radiative enhancement of liquid-water production. One auther (WRC) has frequently observed a clear afternoon sky at mountain locations slowly transform to a gradually thickening cloud deck after sunset which becomes deep enough to precipitate lightly in the early morning hours. The cloud deck then rapidly dissipates after sunrise. This suggests that radiative cooling at the top of a moist layer may trigger the onset of a stratiform cloud which, through cloud-top cooling, transforms to a precipitating cloud. The role of orographic lifting here is to lift the air to near water saturation so that only modest rates of radiative cooling will transform the layer into a saturated layer.

An even more complex dynamic linkage to radiative cooling may be considered. Tripoli (1986) found that enhanced cloud-top cooling at night can lead to local destabilization of the cloud top or moist layer and a reduction of the Brunt–Väisälä frequency (Chapter 10). As a result, gravity-wave energy dispersed from the lifting of air over the mountain barrier becomes trapped by the cloud-top unstable layer resulting from radiative cooling. This leads to higher amplitude gravity waves between the trapping cloud-top zone and the surface and results in enhanced orographic lifting, thus favoring deeper, wetter clouds. Radiative effects, and perhaps a nocturnal maximum in precipitation, may be less in higher wind situations. In a simulation of a Sierra Nevada orographic cloud in a high-wind situation, M. Meyers (personal communication) found that radiative cooling had little influence on the simulated cloud microstructure.

Another likely importance of nocturnal radiative cooling is that surface cooling will lead to enhancement of the positive pressure perturbation on the windward side of the barrier, thus strengthening orographic blocking. As suggested by Marwitz (1980), this will increase the effective orographic lifting of the mountain. In most cases, more localized lifting near the windward side of the barrier during the afternoon will be transformed by a larger cold pool on the windward side of the barrier into smoother, larger scale ascent commencing well upwind of the mountain. This favors the formation of a blanket-type cloud, which can respond to cloud-top radiative cooling. Moreover, initiation of cloud and precipitation farther upwind of a barrier favors operation of the seeder–feeder process near the barrier crest.

Observations of an early-morning maximum in orographic precipitation are not limited to the Colorado Mountains. Lee (1986) analyzed the climatology of precipitation events over the Sierra Nevada in the Sierra Cooperative Pilot Project (SCPP). A well-defined early morning maximum of heavy precipitation events was observed during December. No consistent pattern in heavy precipitation amounts was found in the other winter months, however. Light precipitation amounts exhibited a midday maximum, probably reflecting the greater tendency to develop convective instability in the warmer, moister air masses typical of the Sierra Mountains. Lee also determined the diurnal variations in precipitation echo types (Reynolds and Dennis, 1986). Surprisingly, major bands or well-organized radar-echo features, which exhibit elongated areas of reflectivity, occur about 35% of the time between 0300 to 0500 local time from January through March. The characteristic orographic cloud radar-echo type near mountain top occurs between 0430 and 1100 PST 25% of the time. Not surprisingly, the more convective echo types exhibit a midday maximum. Preference for an early morning maximum in the banded precipitation echo type is not well understood. It may be linked to changes in blocking of fronts or other factors noted earlier that favor the faster movement of fronts at night.

References

Abbs, D. J., and R. A. Pielke (1986). Numerical simulation of orographic effects on NE Colorado snowstorms. *Meteorol. Atmos. Phys.* **37**, 1–10.

Anthes, R. A., and T. T. Warner (1978). Development of hydrodynamic models suitable for air pollution and other mesometeorological studies. *Mon. Weather Rev.* **196**, 1045–1078.

Atlas, D., R. C. Tetehira, R. C. Srivastava, W. Marker, and R. Carbone (1969). Precipitation-induced mesoscale wind perturbations in the melting layer. *Q. J. R. Meteorol. Soc.* **95**, 544–560.

Auer, A. H., Jr., and J. M. White (1982). The combined role of kinematics, thermodynamics and cloud physics associated with heavy snowfall episodes. *J. Meteorol. Soc. Jpn.* **60**, 500–507.

Bader, M. J., and W. T. Roach (1977). Orographic rainfall in warm sectors of depressions. *Q. J. R. Meteorol. Soc.* **103**, 269–280.

Baines, P. B. (1980). The dynamics of the southerly buster. *Aust. Meteorol. Mag.* **28**, 175–200.

Barcilon, A., J. C. Jusem, and P. G. Drazin (1979). On the two-dimensional hydrostatic flow of a stream of moist air over a mountain ridge. *Geophys. Astrophys. Fluid Dyn.* **13**, 125–140.

Bergeron, T. (1949). The problem of artificial control of rainfall on the globe. II. The coastal orographic maxima of precipitation in autumn and winter. *Tellus* **1**, 15–32.

Bergeron, T. (1960). Operation and results of "Project Pluvius." *In* "Physics of Precipitation," Geophys. Monogr. No. 5, pp. 152–157. Am. Geophys. Union, Washington, D.C.

Bergeron, T. (1965). On the low-level redistribution of atmospheric water caused by orography. *Suppl., Proc. Int. Conf. Cloud Phys. Tokyo* pp. 96–100.

Bergeron, T. (1967–1973). "Mesometeorological Studies of Precipitation," Vols. I–V. Meteorol. Inst., Uppsala.

Bjerknes, J. (1919). On the structure of moving cyclones. *Geofys. Publ.* **1**, 1–8.

Bjerknes, J., and H. Solberg (1921). Meteorological conditions for the formation of rain. *Geofys. Publ.* **2**, 1–60.

Booker, J. R., and Bretherton, F. P. (1967). The critical layer for internal gravity waves in shear flow. *Q. J. R. Meteorol. Soc.* **27**, 513–539.

Borys, R. D., P. J. DeMott, E. E. Hindman, and D. Feng (1983). The significance of snow crystal and mountain-surface riming to the removal of atmospheric trace constituents from cold clouds. *In* "Precipitation Scavenging, Dry Deposition, and Resuspension, Vol. 1, Precipitation Scavenging" (H. R. Pruppacher, R. G. Semonin, and W. G. N. Slinn, coords.), pp. 181–189. Elsevier, New York.

Brinkmann, W. A. R. (1974). Strong downslope winds at Boulder, Colorado. *Mon. Weather Rev.* **102**, 592-602.

Browning, K. A. (1979). Structure, mechanism and prediction of orographically enhanced rain in Britian. Global Atmos. Res. Programme, Ser. No. 23, pp. 88-114. World Meteorol. Organ.

Browning, K. A. (1985). Conceptual models of precipitation systems. *Meteorol. Mag.* **114**, 293-318.

Browning, K. A., and G. A. Monk (1982). A simple model for the synoptic analysis of cold fronts. *Q. J. R. Meteorol. Soc.* **108**, 435-452.

Browning, K. A., C. W. Pardoe, and F. F. Hill (1975). The nature of orographic rain at wintertime cold fronts. *Q. J. R. Meteorol. Soc.* **101**, 333-352.

Carruthers, D. J., and W. T. Choularton (1983). A model of the feeder-seeder mechanism of orographic rain including stratification and wind-drift effects. *Q. J. R. Meteorol. Soc.* **109**, 575-588.

Chappell, C. F. (1970). Modification of cold orographic clouds. Ph.D. Thesis, Atmos. Sci. Pap. No. 173, Dep. Atmos. Sci., Colorado State Univ.

Charba, J. (1974). Application of gravity current model to analysis of squall line gust front. *Mon. Weather Rev.* **102**, 140-156.

Chen, C., and W. R. Cotton (1987). The physics of the marine stratocumulus-capped mixed layer. *J. Atmos. Sci.* **44**, 2951-2977.

Choularton, T. W., and S. J. Perry (1986). A model of the orographic enhancement of snowfall by the seeder-feeder mechanism. *Q. J. R. Meteorol. Soc.* **112**, 335-345.

Chung, Y. S., K. D. Hage, and E. R. Reinelt (1976). On lee cyclogenesis and airflow in the Canadian Rocky Mountains and the east Asian mountains. *Mon. Weather Rev.* **104**, 879-891.

Clark, T. L., and R. D. Farley (1984). Severe downslope windstorm calculations in two and three spatial dimensions using anelastic interactive grid nesting: A possible mechanism for gustiness. *J. Atmos. Sci.* **41**, 329-350.

Clark, T. L., and R. Gall (1982). Three-dimensional numerical model simulations of airflow over mountainous-terrain: A comparison with observations. *Mon. Weather Rev.* **110**, 766-791.

Clark, T. L., and W. R. Peltier (1977). On the evolution and stability of finite-amplitude mountain waves. *J. Atmos. Sci.* **34**, 1715-1730.

Clark, T. L., and W. R. Peltier (1984). Critical level reflection and the resonant growth of nonlinear mountain waves. *J. Atmos. Sci.* **41**, 3122-3134.

Colquhoun, J. R., D. J. Shepherd, C. E. Coulman, R. K. Smith, and K. McInnes (1985). The southerly buster of South Eastern Australia: An orographically forced cold front. *Mon. Weather Rev.* **113**, 2090-2107.

Colson, DeVer (1954). Meteorological problems in forecasting mountain lee waves. *Bull. Am. Meteorol. Soc.* **35**, 363-371.

Cooper, W. A., and J. D. Marwitz (1980). Winter storms over the San Juan Mountains. Part III: Seeding potential. *J. Appl. Meteorol.* **19**, 942-949.

Corby, G. A., and J. S. Sawyer (1958). The air flow over a ridge—The effects of the upper boundary and high-level conditions. *Q. J. R. Meteorol. Soc.* **84**, 25-37.

Cotton, W. R., R. L. George, P. J. Wetzel, and R. L. McAnelly (1983). A long-lived mesoscale convective complex. Part I: The mountain-generated component. *Mon. Weather Rev.* **111**, 1893-1918.

Cotton, W. R., G. J. Tripoli, R. M. Rauber, and E. A. Mulvihill (1986). Numerical simulation of the effects of varying ice crystal nucleation rates and aggregation processes on orographic snowfall. *J. Climate Appl. Meteorol.* **25**, 1658-1680.

Crapper, G. D. (1962). Waves in the lee of a mountain with elliptical contours. *Philos. Trans. R. Soc. London, Ser. A* **254**, 601–624.

Danielsen, E. F., and R. Bleck (1970). Tropospheric and stratospheric ducting of stationary mountain lee waves. *J. Atmos. Sci.* **27**, 758–772.

Deaven, D. G. (1976). A solution for boundary problems in isentropic coordinate models. *J. Atmos. Sci.* **33**, 1702–1713.

Dirks, R. A. (1973). The precipitation efficiency of orographic clouds. *J. Rech. Atmos.* **7**, 177–184.

Douglas, C. K. M., and J. Glasspoole (1947). Meteorological conditions in heavy orographic rainfall. *Q. J. R. Meteorol. Soc.* **73**, 11–38.

Durran, D. R. (1986). Another look at downslope windstorms. Part I: The development of analogs to supercritical flow in an infinitely deep, continuously stratified fluid. *J. Atmos. Sci.* **43**, 2527–2543.

Durran, D. R., and J. B. Klemp (1982a). On the effects of moisture on the Brunt–Väisälä frequency. *J. Atmos. Sci.* **39**, 2152–2158.

Durran, D. R., and J. B. Klemp (1982b). The effects of moisture on trapped mountain lee waves. *J. Atmos. Sci.* **39**, 2490–2506.

Durran, D. R., and J. B. Klemp (1983). A compressible model for the simulation of moist mountain waves. *Mon. Weather Rev.* **111**, 2341–2361.

Eliassen, A. (1968). On meso-scale mountain waves on the rotating earth. *Geophys. Norv.* **27**, 1–15.

Eliassen, A., and E. Palm (1960). On the transfer of energy in stationary mountain waves. *Geophys. Norv.* **22**, 1–23.

Eliassen, A., and J.-E. Rekustad (1971). A numerical study of meso-scale mountain waves. *Geophys. Norv.* **28**, 1–13.

Elliott, R. D., and E. L. Hovind (1964). On convection bands within Pacific Coast storms and their relation to storm structure. *J. Appl. Meteorol.* **3**, 143–154.

Elliott, R. D., and R. W. Shaffer (1962). The development of quantitative relationships between orographic precipitation and airmass parameters for use in forecasting and cloud seeding evaluation. *J. Appl. Meteorol.* **1**, 218–228.

Foldvik, A., and M. G. Wurtele (1967). The computation of the transient gravity wave. *Geophys. J. R. Astron. Soc.* **13**, 167–185.

Fraser, A. B., R. C. Easter, and P. V. Hobbs (1973). A theoretical study of the flow of air and fallout of solid precipitation over mountainous terrain: Part I. Air flow model. *J. Atmos. Sci.* **30**, 801–812.

Fukuta, N., M. Kowa, and N.-H. Gong (1982). Determination of ice crystal growth parameters in a new supercooled cloud tunnel. *Prepr., Conf. Cloud Phys., Chicago, Ill.* pp. 325–328. Am. Meteorol. Soc., Boston, Massachusetts.

Fukuta, N., H.-H. Gong, and A.-S. Wang (1984). A microphysical origin of graupel and hail. *Proc., Int. Conf. Cloud Phys., 9th, Tallinn, USSR* pp. 257–260.

Gill, A. E. (1977). Coastally trapped waves in the atmosphere. *Q. J. R. Meteorol. Soc.* **103**, 431–440.

Gjevik, B., and T. Marthinsen (1977). Three-dimensional lee-wave pattern. *Q. J. R. Meteorol. Soc.* **104**, 947–957.

Gocho, Y. (1978). Numerical experiment of orographic heavy rainfall due to a stratiform cloud. *J. Meteorol. Soc. Jpn.* **56**, 405–422.

Grant, L. O., C. F. Chappell, L. W. Crow, P. W. Mielke, Jr., J. L. Rasmussen, W. E. Shobe, H. Stockwell, and R. A. Wykstra (1969). An operational adaptation program of weather modification for the Colorado River Basin. Interim Rep. to Bur. Reclam., July 1968–June 1969, Contract No. 14-06-D-6467.

Heggli, M. F. (1986). A ground based approach used to determine cloud seeding opportunity. *Proc. Conf. Weather Modif. 10th, Arlington, Va.* pp. 64–67. Am. Meteorol. Soc., Boston, Massachusetts.

Heggli, M. F., and R. M. Rauber (1988). The characteristics and evolution of supercooled water in wintertime storms over the Sierra Nevada: A summary of radiometric measurements taken during the Sierra Cooperative Pilot Project. *J. Appl. Meteorol.* **27**, 989–1015.

Heggli, M. F., and D. W. Reynolds (1985). Radiometric observations of supercooled liquid water within a split front over the Sierra Nevada. *J. Climate Appl. Meteorol.* **24**, 1258–1261.

Heggli, M. F., L. Vardiman, R. E. Stewart, and A. Huggins (1983). Supercooled liquid water and ice crystal distributions within Sierra Nevada winter storms. *J. Climate Appl. Meteorol.* **22**, 1875–1886.

Hill, G. E. (1978). Observations of precipitation-forced circulations in winter orographic storms. *J. Atmos. Sci.* **35**, 1463–1472.

Hindman, E. E., R. D. Borys, and P. J. DeMott (1982). Hydrometeorological significance of rime ice deposits on trees in the Colorado Rockies. *Prepr., Int. Symp. Hydrometeorol., Am. Water Resour. Assoc.* pp. 95–99.

Hobbs, P. V. (1973). Anomalously high ice particle concentrations in clouds. *Invited Rev. Pap., Int. Conf. Nucleation, 8th, Leningrad.*

Hobbs, P. V. (1975). The nature of winter clouds and precipitation in the Cascade Mountains and their modification by artificial seeding. *J. Appl. Meteorol.* **14**, 783–858.

Hobbs, P. V., R. C. Easter, and A. B. Fraser (1973). A theoretical study of the flow of air and fallout of solid precipitation over mountainous terrain: Part II. Microphysics. *J. Atmos. Sci.* **30**, 813–823.

Hobbs, P. V., R. A. Houze, Jr., and T. J. Matejka (1975). The dynamical and microphysical structure of an occluded frontal system and its modification by orography. *J. Atmos. Sci.* **32**, 1542–1562.

Hogg, D. C., F. O. Guiraud, J. B. Snider, M. T. Decker, and E. R. Westwater (1983). A steerable dual-channel microwave radiometer for measurement of water vapor and liquid in the troposphere. *J. Appl. Meteorol.* **22**, 789–806.

Holgate, H. T. D. (1973). Rainfall forecasting for river authorities. *Meteorol. Mag.* **102**, 33–38.

Holmboe, J., and H. Klieforth (1957). Investigation of mountain lee waves and the air flow over the Sierra Nevada. Final Rep. to Geophys. Res. Dir., Air Force Cambridge Res. Cent. Contract No. AF 19(604)-728, March.

Hovermale, J. B. (1965). A non-linear treatment of the problem of airflow over mountains. Ph.D. Thesis, Dep. Meteorol. Pennsylvania State Univ.

Kelvin, Lord (1886). On stationary waves in flowing water. *Philos. Mag.* **5**, 353–357, 445–552, 517–530.

Klemp, J. B., and D. K. Lilly (1975). The dynamics of wave-induced downslope winds. *J. Atmos. Sci.* **32**, 320–339.

Klemp, J. B., and D. K. Lilly (1978). Numerical simulation of hydrostatic mountain waves. *J. Atmos. Sci.* **35**, 78–107.

Klemp, J. B., and R. B. Wilhelmson (1978a). The simulation of theee-dimensional convective storm dynamics. *J. Atmos. Sci.* **35**, 1070–1096.

Klemp, J. B., and R. B. Wilhelmson (1978b). Simulations of right- and left-moving storms produced through storm splitting. *J. Atmos. Sci.* **35**, 1097–1110.

Krishnamurti, T. N. (1964). The finite amplitude mountain wave problem with entropy as a vertical coordinate. *Mon. Weather. Rev.* **92**, 147–160.

Kuettner, J. (1958). Moazagotl und Foehnwell. *Contrib. Atmos. Phys.* **25**, 79–114.

Lalas, D. P., and F. Einaudi (1974). On the correct use of the wet adiabatic lapse rate in stability criteria of a saturated atmosphere. *J. Appl. Meteorol.* **13**, 318–324.

Lamb, D., K. W. Nielsen, H. E. Klieforth, and J. Hallett (1976). Measurements of liquid water content in winter cloud systems over the Sierra Nevada. *J. Appl. Meteorol.* **15**, 763–775.

Lee, T. F. (1986). Seasonal and interannual trends of Sierra Nevada Clouds and precipitation. *J. Climate Appl. Meteorol.* **26**, 1270–1276.

Lilly, D. K. (1981). Doppler radar observations of upslope snowstorms. *Proc. Conf. Radar Meteorol., 20th*, pp. 638–645. Am. Meteorol. Soc., Boston, Massachusetts.

Lilly, D. K., and D. R. Durran (1983). Stably stratified moist airflow over mountainous terrain. *Proc. Sino-Am. Workshop Mountain Meteorol., 1st, 1982, Beijing* (E. R. Reiter, Z. Baozhen, and Q. Yongfu, eds.), pp. 569–608. Science Press, Beijing and Am. Meteorol. Soc., Boston, Massachusetts.

Lilly, D. K., and E. J. Zipser (1972). The Front Range windstorm of 11 January 1972. A meteorological narrative. *Weatherwise* **25**, 56–63.

Long, R. R. (1953a). Some aspects of the flow of stratified fluids. I. A theoretical investigation. *Tellus* **5**, 42–58.

Long, R. R. (1953b). A laboratory model resembling the "Bishop-Wave" phenomenon. *Bull. Am. Meteorol. Soc.* **34**, 205–211.

Long, R. R. (1954). Some aspects of the flow of stratified fluids. II. Experiments with a two-fluid system. *Tellus* **6**, 97–115.

Long, R. R. (1955). Some aspects of the flow of stratified fluids. III. Continuous density gradients. *Tellus* **7**, 341–357.

Long, R. R. (1972). Finite amplitude disturbances in the flow of inviscid rotating and stratified fluids over obstacles. *Annu. Rev. Fluid Mech.* **4**, 69–92.

Lyra, G. (1943). Theorie der stationaren Leewellenstromung in freier Atmosphare. *Z. Angew. Math. Mech.* **23**, 1–28.

Mahrer, Y., and R. A. Pielke (1975). A numerical study of the air flow over mountains using the two-dimensional version of the University of Virginia mesoscale model. *J. Atmos. Sci.* **32**, 2144–2155.

Marwitz, J. D. (1980). Winter storms over the San Juan Mountains. Part I: Dynamical processes. *J. Appl. Meteorol.* **19**, 913–926.

Marwitz, J. D. (1983). The kinematics of orographic airflow during Sierra storms. *J. Atmos. Sci.* **40**, 1218–1227.

Marwitz, J., K. Waight, B. Martner, and G. Gordon (1985). Cloud physics studies in SCPP during 1984–85. Rep. to Div. Atmos. Resour. Res., Bur. Reclam., U.S. Dep. Inter. Contract No. 2-07-81-V0256, September.

Mass, C. F., and M. D. Albright (1986). Coastal southerlies and alongshore surges of the West Coast of North America: Evidence of mesoscale topographically trapped response to synoptic forcing. *Mon. Weather Rev.* **115**, 1707–1738.

Miles, J. W., and H. E. Huppert (1969). Lee waves in a stratified flow. Part 4. Perturbation approximations. *J. Fluid Mech.* **35**, 497–525.

Myers, V. A. (1962). Airflow of the windward side of a large ridge. *J. Geophys. Res.* **67**, 4267–4291.

Nickerson, E. C. (1979). On the numerical simulation of airflow and clouds over mountainous terrain. *Contrib. Atmos. Phys.* **52**, 161–175.

Nickerson, E. C., E. Richard, R. Rosset, and D. R. Smith (1986). The numerical simulation of clouds, rain, and airflow over the Vosges and Black Forest Mountains: A meso-β model with parameterized microphysics. *Mon. Weather Rev.* **114**, 398–414.

Palm, E. (1955). Multiple-layer mountain wave models with constant static stability and shear. Sci. Rep. No. 3, Contract No. AF 19(604)-728, Air Force Cambridge Res. Cent., Cambridge, Massachusetts.

Palm, E. (1958). Two-dimensional and three-dimensional mountain waves. *Geophys. Norv*, **20**.

Paluch, I. R. (1979). The entrainment mechanism in Colorado cumuli. *J. Atmos. Sci.* **36**, 2467–2478.

Parish, T. (1982). Barrier winds along the Sierra Nevada Mountains. *J. Appl. Meteorol.* **21**, 925–930.

Parsons, D. B., and P. V. Hobbs (1983). The mesoscale and microscale structure and organization of clouds and precipitation in midlatitude cyclones. IX: Some effects of orography on rainbands. *J. Atmos. Sci.* **40**, 1930–1949.

Peltier, W. R., and T. L. Clark (1979). The evolution and stability of finite-amplitude mountain waves. Part II: Surface wave drag and severe downslope windstorms. *J. Atmos. Sci.* **36**, 1498–1529.

Peltier, W. R., and T. L. Clark (1980). Reply. *J. Atmos. Sci.* **37**, 2122–2125.

Petterssen, S. (1956). "Weather Analysis and Forecasting," 2nd Ed. McGraw-Hill, New York.

Queney, P. (1947). "Theory of Perturbations in Stratified Currents with Applications to Air Flow over Mountain Barriers," Misc. Rep. No. 23. Univ. of Chicago Press, Chicago, Illinois.

Queney, P. (1948). The problem of air flow over mountains: A summary of theoretical studies. *Bull. Am. Meteorol. Soc.* **29**, 16–26.

Raddatz, R. L., and M. L. Khandekar (1979). Upslope enhanced extreme rainfall over the Canadian Western Plains: A mesoscale numerical simulation. *Mon. Weather Rev.* **107**, 650–661.

Radinovic, D. (1965). Cyclonic activity in Yugoslavia and surrounding areas. *Arch. Meteorol. Geophys. Bioklimatol.* **A14**, 391–408.

Radinovic, D., and D. Lalic (1959). Ciklonska aktivnost a Zapadnom Sredozemliju. *Rasprave Stud.—Mem.* **7**, 1–57.

Rauber, R. M. (1981). Microphysical processes in two stably stratified orographic cloud systems. M.S. Thesis, Atmos. Sci. Pap. No. 337, Dep. Atmos. Sci., Colorado State Univ.

Rauber, R. M. (1985). Physical structure of northern Colorado river basin cloud systems. Ph.D. Thesis, Atmos. Sci. Pap. No. 390, Dep. Atmos. Sci., Colorado State Univ.

Rauber, R. M., and L. O. Grant (1986). The characteristics and distribution of cloud water over the mountains of northern Colorado during wintertime storms. Part II: Spatial distribution and microphysical characteristics. *J. Climate Appl. Meteorol.* **25**, 489–504.

Rauber, R. M., D. Feng, L. O. Grant, and J. B. Snider (1986). The characteristics and distribution of cloud water over the mountains of northern Colorado during wintertime storms. Part I: Temporal variations. *J. Climate Appl. Meteorol.* **25**, 468–480.

Rayleigh, R. J. (1883). The form of standing waves on the surface of running water. *Proc. London Math. Soc.* **15**, 69–78.

Reinking, R. F., and J. F. Boatman (1986). Upslope precipitation events. *In* "Mesoscale Meteorology and Forecasting" (P. S. Ray, ed.), pp. 437–471. Am. Meteorol. Soc., Boston, Massachusetts.

Reitan, C. H. (1974). Frequencies of cyclones and cyclogenesis for North America, 1951–1970. *Mon. Weather Rev.* **102**, 861–868.

Reynolds, D. W. (1986). A randomized exploratory seeding experiment on widespread shallow orographic clouds: Forecasting suitable cloud conditions. *Proc. Conf. Weather Modif., 10th, Arlington, Va.* pp. 7–12. Am. Meteorol. Soc., Boston, Massachusetts.

Reynolds, D. W., and A. S. Dennis (1986). A review of the Sierra Cooperative Pilot Project. *Bull. Am. Meteorol. Soc.* **67**, 513–523.

Rhea, J. O. (1978). Orographic precipitation model for hydrometeorological use. Atmos. Sci. Pap. No. 287, Dep. Atmos. Sci., Colorado State Univ.

Richard, E., N. Chaumerliac, J. F. Mahfouf, and E. C. Nickerson (1987). Numerical simulation of orographic enhancement of rain with a mesoscale model. *J. Climate Appl. Meteorol.* **26**, 661-669.

Sassen, K. (1984). Deep orographic cloud structure and composition derived from comprehensive remote sensing measurements. *J. Climate Appl. Meteorol.* **23**, 568-583.

Sawyer, J. S. (1960). Numerical calculation of the displacements of a stratified airstream crossing a ridge of small height. *Q. J. R. Meteorol. Soc.* **86**, 326-345.

Sawyer, J. S. (1962). Gravity waves in the atmosphere as a three-dimensional problem. *Q. J. R. Meteorol. Soc.* **88**, 412-425.

Schlatter, T. W., D. V. Baker, and J. F. Henz (1983). Profiling Colorado's Christmas Eve blizzard. *Weatherwise* **36**, 60-66.

Schwerdtfeger, W. (1974). Mountain barrier effect on the flow of stable air north of the Brooks Range. *Proc. Alaskan Sci. Conf., 24th, Geophys. Inst., Univ. Alaska, Fairbanks* pp. 204-208.

Schwerdtfeger, W. (1975). The effect of the Antarctic Peninsula on the temperature regime of the Weddell Sea. *Mon. Weather Rev.* **103**, 45-51.

Schwerdtfeger, W. (1979). Meteorological aspects of the drift of ice from the Weddell Sea toward the mid-latitude westerlies. *J. Geophys. Res.* **84**, 6321-6327.

Scorer, R. S. (1949). Theory of waves in the lee of mountains. *Q. J. R. Meteorol. Soc.* **75**, 41-56.

Scorer, R. S. (1953). Theory of airflow over mountains: II—The flow over a ridge. *Q. J. R. Meteorol. Soc.* **79**, 70-83.

Scorer, R. S. (1954). Theory of airflow over mountains: III—Airstream characteristics. *Q. J. R. Meteorol. Soc.* **80**, 417-428.

Scorer, R. S., and H. Klieforth (1959). Theory of mountain waves of large amplitude. *Q. J. R. Meteorol. Soc.* **85**, 131-143.

Scorer, R. S., and M. Wilkinson (1956). Waves in the lee on an isolated hill. *Q. J. R. Meteorol. Soc.* **82**, 419-427.

Shapiro, M. A., T. Hampel, D. Rötzoll, and F. Mosher (1985). The frontal hydraulic head: A micro-α-scale ($\sim$1 km) triggering mechanism for mesoconvective weather systems. *Mon. Weather Rev.* **113**, 1166-1183.

Smith, R. B. (1977). The steepening of hydrostatic mountain waves. *J. Atmos. Sci.* **34**, 1634-1654.

Smith, R. B. (1979). The influence of mountains on the atmosphere. *Adv. Geophys.* **21**, 87-230.

Smith, R. B. (1982). A differential advection model of orographic rain. *Mon. Weather Rev.* **110**, 306-309.

Smith, R. B. (1985). On severe downslope winds. *J. Atmos. Sci.* **42**, 2597-2603.

Smith, R. B., and Y.-H. Lin (1982). The addition of heat to a stratified airstream with application to the dynamics of orographic rain. *Q. J. R. Meteorol. Soc.* **108**, 353-378.

Sommerfeld, A. (1912). Die greensche funktion der schwingungsgleichung. *Jahresber. Dtsch. Math.-Ver.* **21**, 309-353.

Sommerfeld, A. (1948). "Vorlesungen uber theoretische Physik," 2nd Rev. Ed., Vol. VI. Akad. Verlagsges., Leipzig.

Starr, J. R., and K. A. Browning (1972). Observations of lee waves by high-power radar. *Q. J. R. Meteorol. Soc.* **98**, 73-85.

Storebo, P. B. (1976). Small scale topographical influences on precipitation. *Tellus* **28**, 45-59.

Toth, J. J. (1987). Interaction of shallow cold surges with topography on scales of 100-1000 km. Ph.D. Thesis, Dep. Atmos. Sci., Colorado State Univ.

Tripoli, G. J. (1986). A numerical investigation of an orogenic mesoscale convective system. Ph.D. Thesis, Atmos. Sci. Pap. No. 401, Dep. Atmos. Sci., Colorado State Univ.

Tripoli, G. J., and W. R. Cotton (1982). The Colorado State University three-dimensional cloud/mesoscale model—1982. Part I: General theoretical framework and sensitivity experiments. *J. Rech. Atmos.* **16**, 185-220.

Walsh, P. A. (1977). Cloud droplet measurements in wintertime clouds. M.S. Thesis, Dep. Atmos. Sci., Univ. of Wyoming.

Weickmann, H. (1981). Mechanism of shallow winter-type stratiform cloud systems. NOAA Environ. Res. Lab. U.S. Gov. Print. Off. 1982-576-001/1220.

Whiteman, C. D. (1973). Some climatological characteristics of seedable upslope cloud systems in the high plains. NOAA Tech. Rep. ERL 268-APCL-27, U.S.D.C., Boulder, Colorado. U.S. Gov. Print. Off., 1972–784214/1150 Region No. 8, Washington, D.C.

Wiesmueller, J. L. (1982). The effect of diurnal heating on the movement of cold fronts through Eastern Colorado. NWS Tech. Memo., NWS-CR-66.

Wurtele, M. (1953). On lee wave in the interface separating two barotropic layers. Final Rep., Sierra Wave Proj., Contract No. AF 19(122)-263, Air Force Cambridge Res. Cent., Cambridge, Massachusetts.

Epilogue

The focus of this book is on the macroscopic aspects of clouds and storms and, in particular, the dynamics of precipitating mesoscale systems. We have attempted to show, moreover, that the air motions, thermodynamics, and precipitation processes associated with clouds do not operate in isolation. Instead they are dependent upon, and interact with, a broad range of scales of motion spanning from the scale of the global atmosphere, to scales of motion having dimensions of continents, down to scales of turbulent eddies of a few tens of meters. Moreover, the physical processes responsible for the formation of precipitation, and the thermodynamics of clouds commence with the nucleation of submicrometer-sized aerosol particles, creating cloud particles of a few tens of micrometers, followed by the growth of precipitation particles having dimensions of the order of 1000 μm. Eventually the resulting release of latent heat and radiative effects of cloud particles may affect the global atmospheric behavior. In spite of what may seem to be overwhelming obstacles, we have made great progress in furthering our understanding of the dynamics and physics of clouds and cloud systems.

Nonetheless, there remains a great deal to be learned about the behavior of clouds, storms, and mesoscale weather systems. Many of the scientific problems that remain must be resolved before major improvements in severe and short-range weather forecasting can be expected. Furthermore, many of the scientific unknowns about clouds and storms serve as roadblocks to further progress across a broad spectrum of the atmospheric sciences and related disciplines, including atmospheric chemistry on regional and global scales, the earth's general circulation, climate, and climate variability.

Fortunately the prospects for improving our understanding of the behavior of clouds, cloud processes, and precipitating mesoscale systems are great. New observing systems, capable of sampling the atmosphere at unprecedented temporal and spatial resolution, are being developed and implemented as regular weather-observing systems. No longer will we be constrained to twice-daily soundings of the atmosphere. A prototype network of wind-profiling systems will be in operation in the central United States by the early 1990s. This network will produce wind soundings several times an hour with high vertical resolution and with horizontal spacing capable of resolving the meso-β-scale features of weather systems. Under development are ground-based remote-sensing systems capable of producing temperature and moisture soundings having similar temporal and spatial resolution to the wind-profiling systems. A network of Doppler radars called NEXRAD will also be in place by the early 1990s. These radars will be capable of scanning the entire volume of the troposphere within its domain, producing an average wind profile, mappings of horizontal divergence, reflectivity from precipitation elements, and even local wind shears such as produced by tornadic thunderstorms and thunderstorm downdrafts. Similarly, new satellites soon to be launched will have greatly improved microwave and infrared sensors to measure temperature and water vapor over the entire earth with mesoscale horizontal and temporal resolution. A network of automated, telemetered surface network stations will be placed in the United States operational network in the next few years. As only one example of the many scientific uses of these data, the data obtained from satellites and a network of radars and surface stations can be combined to produce a four-dimensional budget of water and its various phases that can be used to understand global and regional hydrologic cycles and to infer latent heating of cloud systems for initialization of forecast models.

Taking advantage of these new observational systems and networks, a Stormscale Operational and Research Meteorology (STORM) program is to be implemented in the United States in the 1990s as a sequence of field experiments and data-assimilation/mesoscale forecasting studies. Supplemented by research radars, surface networks, soundings systems, and instrumented aircraft, the prototype operational network will allow observation of storms and cloud systems over a broader range of scales than has been possible in the past. For the first time it will be possible to observe storm systems spanning the entire mesoscale regime from the meso-γ-scale to the meso-α-scale, and these observations will be used to develop and verify new theories and models of clouds, storms, and other mesoscale atmospheric systems.

Index

M

N

O

T

U

International Geophysics Series

EDITED BY

RENATA DMOWSKA
Division of Applied Science
Harvard University

JAMES R. HOLTON
Department of Atmospheric Sciences
University of Washington
Seattle, Washington

*Out of print.

Volume 13 E. Palmén and C. W. Newton. Atmospheric Circulation Systems: Their Structure and Physical Interpretation. 1969

Volume 14 Henry Rishbeth and Owen K. Garriott. Introduction to Ionospheric Physics. 1969*

Volume 15 C. S. Ramage. Monsoon Meteorology. 1971*

Volume 16 James R. Holton. An Introduction to Dynamic Meteorology. 1972*

Volume 17 K. C. Yeh and C. H. Liu. Theory of Ionospheric Waves. 1972*

Volume 18 M. I. Budyko. Climate and Life. 1974*

Volume 19 Melvin E. Stern. Ocean Circulation Physics. 1975

Volume 20 J. A. Jacobs. The Earth's Core. 1975*

Volume 21 David H. Miller. Water at the Surface of the Earth: An Introduction to Ecosystem Hydrodynamics. 1977

Volume 22 Joseph W. Chamberlain. Theory of Planetary Atmospheres: An Introduction to Their Physics and Chemistry. 1978*

Volume 23 James R. Holton. An Introduction to Dynamic Meteorology, Second Edition. 1979*

Volume 24 Arnett S. Dennis. Weather Modification by Cloud Seeding. 1980

Volume 25 Robert G. Fleagle and Joost A. Businger. An Introduction to Atmospheric Physics, Second Edition. 1980

Volume 26 Kuo-Nan Liou. An Introduction to Atmospheric Radiation. 1980

Volume 27 David H. Miller. Energy at the Surface of the Earth: An Introduction to the Energetics of Ecosystems. 1981

Volume 28 Helmut E. Landsberg. The Urban Climate. 1981

Volume 29 M. I. Budyko. The Earth's Climate: Past and Future. 1982

Volume 30 Adrian E. Gill. Atmosphere to Ocean Dynamics. 1982

Volume 31 Paolo Lanzano. Deformations of an Elastic Earth. 1982*

Volume 32 Ronald T. Merrill and Michael W. McElhinny. The Earth's Magnetic Field: Its History, Origin, and Planetary Perspective. 1983

Volume 33 John S. Lewis and Ronald G. Prinn. Planets and Their Atmospheres: Origin and Evolution. 1983

Volume 34 Rolf Meissner. The Continental Crust: A Geophysical Approach. 1986